Biology Today

SANDRA S. GOTTFRIED

Assistant Professor of Biology and Education
Departments of Biology and Educational Studies
University of Missouri—St. Louis

with ***777*** *illustrations*

Mosby

St. Louis Baltimore Boston Chicago London Philadelphia Sydney Toronto

Mosby
Dedicated to Publishing Excellence

Editor-in-Chief: **James M. Smith**
Editor: **Robert J. Callanan**
Developmental Editor: **Kathleen Scogna**
Editorial Assistant: **Jennifer Collins**
Photo Researchers: **Holly Leicht, Jennifer Collins**
Project Manager: **Carol Sullivan Wiseman**
Senior Production Editor: **Linda McKinley**
Book Designer: **Susan Lane**
Art Coordinator: **E. Rhone Rudder**
Computer Art: **Pagecrafters, Inc.**
Artists: **Molly Babich, Scott Bodell, Raychel Ciemma, Barbara Cousins, Carlyn Iverson, Christy Krames, Bill Ober, Lili Robins, Kevin Somerville, Kate Sweeney, Nadine B. Sokol**
Cover Photographs: Landscape © Animals Animals/Richard Kolar
Plongee diving © Animals Animals/Henry Ausloos
Grizzlies © Animals Animals/Johnny Johnson
Peacock © Animals Animals/Miriam Austerman

Printed in the United States of America

Mosby–Year Book, Inc.
11830 Westline Industrial Drive
St. Louis, MO 63146

Library of Congress Cataloging in Publication Data

Gottfried, Sandra S.
Biology today / Sandra S. Gottfired.
p. cm.
Includes index.
ISBN 0-8016-7464-6
1. Biology. 2. Human biology. I. Title.
QH308.2.G68 1993
574—dc20 92-30357
CIP

93 94 95 96 97 GW/CD/VHP 9 8 7 6 5 4 3 2 1

Preface

Is human papilloma virus #16 a threat to your health? The answer is on p. 486. Why are anabolic steroids a health risk? Find out on p. 315. How is the AIDS epidemic affecting college students today? If you want to know, see pp. 344-345. If these questions intrigue you, then reading *Biology Today* may be more than a "textbook" experience. These topics and others, such as genetic engineering, the global population crisis, rain forest destruction, and the ozone hole, link the concepts of biology today to the students of today in a context of becoming an informed citizen of the world. Providing students with information that is vital to their health, well-being, and general education, *Biology Today* explores traditional biological content within societal, technological, and personal contexts.

I wrote *Biology Today* for the introductory biology student at a community college or university. Of my 20-plus year career in biology education, I spent 5 years teaching at a community college and have spent the last 4 years teaching introductory students at a university. I feel I am well-aware of my audience and their needs. My training in scientific research, biology, and education and experience as a science writer and teacher enabled me to write this book in a way that truly embodies the philosophy of educating introductory students to become better-informed citizens at personal, local, national, and international levels. To accomplish this task, I first assumed that my audience of students had no prior knowledge of biology. Therefore I used a writing style that is easy and direct, developing concepts within a context of rich description to provide a scaffold for learning and to make concepts more concrete. Today's introductory students need explanation, not simply a list of vocabulary terms.

In addition, *Biology Today* places traditional biology in a context both relevant and important to the student. This is done in two ways. First, everyday situations or concerns that students have encountered in the media are used as examples wherever possible. Second, the text has a strong human focus, with the human body serving as the focal point for the physiology section (Part Four, Human Biology: The Structure and Function of the Body). Students are interested in how their bodies function. They also need accurate information about their own physiology to make informed choices about activities that will affect their bodies, such as making decisions regarding birth control and choosing what foods to eat.

In summary, *Biology Today* portrays biology as a process rather than a product, helping students better understand how scientists do their work and develop theories about the natural world. Introductory students need a broad understanding and an appreciation of how the biological world affects their lives. *Biology Today* attempts to fill this need.

Organization

Biology Today is organized into eight parts. Chapters were designed to stand alone so that the professor may use the text in the order in which it was written or in an alternative order. The chapter opening pedagogy provides "Highlights" for each chapter that serve as a "menu" to the chapter content. These Highlights will help professors determine the chapter order that best suits the organization of their course. Page references to concepts previously defined and explained are sprinkled throughout the chapters.

In general, the book is organized in a traditional fashion. Basic biological concepts are presented in the first chapter with chemistry, cell biology, metabolism, physiology, and genetics composing the first five parts of the text. The last three parts present evolution, diversity, and ecology.

Part One
An introduction to biology and chemistry

This part begins with Chapter 1 (The Themes of Biology Today), which provides an overview of the scientific method and an introduction to biological themes such as unity and diversity that are emphasized throughout the book. In addition, other introductory concepts, such as the kingdoms of life, are introduced and explained. Chapter 2 (The Chemistry of Life) presents basic chemistry, with an emphasis on biologically relevant chemical principles.

Part Two
Cells: The basic unit of life

Part Two contains information about all aspects of cells: their basic structure and function (Chapter 3, Cell Structure and Function), the cell membrane and its function (Chapter 4, Cell Membranes), and the processes of mitosis and meiosis (Chapter 5, Cell Division).

Part Three
How living things transform energy

Cellular metabolism is the focus of this unit. A chapter on chemical reactions introduces this unit (Chapter 6, The Flow of Energy within Organisms) and prepares the students to study cellular respiration (Chapter 7, Cellular Respiration: How Cells Release Stored Energy from Food Molecules) and photosynthesis (Chapter 8, Photosynthesis: How Plants Capture and Store Energy from the Sun).

Part Four
Human biology: The structure and function of the body

This unit covers aspects of the structure and function of the human body. The chapters focus on homeostasis and how structure is related to function. The numerous boxed essays in this unit were designed to interest students. Boxes include such relevant issues "How to Avoid Heart Disease" in Chapter 12 (Circulation), "AIDS on the College Campus" in Chapter 13 (Defense Against Disease), and "Solving the Mystery of Alzheimer's Disease" in Chapter 16 (The Nervous System).

Part Five
How humans reproduce and pass on biological information

This unit covers human reproduction and genetics, beginning with Chapter 20 (Sex and Reproduction) and moving on to Chapter 21 (Human Development Before Birth). Mendelian genetics (Chapter 22, Patterns of Inheritance), human genetics (Chapter 23, Human Genetics), and molecular genetics (Chapter 24, The Molecular Basis of Inheritance) follow. Again, emphasis is placed on student-oriented information. In this unit, students learn about how they inherited their particular blood type, how color-blindness is passed along in the genes, and how genetic engineering is helping those with diabetes.

Part Six
Evolution: How living things change over time

Current information about evolution is presented in this unit. Chapter 25 (The Scientific Evidence for Evolution) presents the evidence contained in fossils, the geological record, and the molecular record that support the theory of evolution. Chapter 26 (The Evolution of the Five Kingdoms of Life) follows scientists' thinking as they attempt to construct scenarios of how life began. A separate, complete chapter on human evolution, including vertebrate evolution (Chapter 27), is a hallmark of this section.

Part Seven
The diversity and unity of living things

Structure, function, and evolutionary position of the five kingdoms of life are covered in this unit. The first two chapters describe the Kingdoms Monera, Protista, and Fungi (Chapter 28, viruses, bacteria, and genetic engineering and Chapter 29, protists and fungi). The reproductive patterns of plants are described and followed by their patterns of structure and function (Chapter 30, Plants: Reproductive Patterns and Diversity and Chapter 31, Plants: Patterns of Structure and Function). Chapters on invertebrates (Chapter 32, Invertebrate Animals: Patterns of Structure, Function and Reproduction) and vertebrates (Chapter 33, Vertebrate Animals: Patterns of Structure, Function, and Reproduction) describe patterns of body symmetry, body cavity structure, and embryological development, linking these concepts to the evolutionary history of these animal groups. Characteristics of animal phyla and vertebrate classes are discussed with a comparative approach.

Part Eight
How living things interact with each other and with their environment

These seven chapters discuss the interaction of organisms with the environment. The unit begins with two chapters on animal behavior: Chapter 34 (Innate Behavior and Learning in Animals) and Chapter 35 (Social Behavior in Animals). Chapter 36 (Population Ecology), Chapter 37 (Interactions within Communities of Organisms), and Chapter 38 (Ecosystems) present an overview of ecology. Chapter 39 (Biomes and Life Zones of the World) describes the large climatic areas of the world along with their distinctive plant and animal populations. This chapter includes a discussion of fresh water and saltwater environments. The unit ends with the ecological problems facing the world today in Chapter 40 (The Biosphere: Today and Tomorrow).

Features

Biology Today boasts several pedagogical features that make the book easy to use. Students should find these features a great help when reading their assignments for class and reviewing for exams and quizzes.

Each chapter opens with a short "vignette" that is designed to spark student interest. Sometimes these vignettes explain something that may already be familiar to the student in a context that relates to the chapter material. For example, Chapter 19 (Hormones) opens with a vignette about the dangers of anabolic steriods. Other vignettes focus on unusual phenomena that nevertheless have a biological explanation. Chapter 14 (Excretion) opens with a discussion about why some species of turtles "cry" and how this crying is related to salt and water balance.

In addition to the vignettes, the chapter openers contain a list of chapter "Highlights" that detail in short statements the chapter content. These highlights will help students condense the chapter material and provide a quick review. An "outline" of the chapter accompanies the highlights and also assists students with organizing and segmenting the chapter concepts.

Within each chapter are "Concept summaries" that provide spot summaries at key points in the chapter. They are short (usually no more than three sentences long) and effectively review the preceding material. Students will find them invaluable when reviewing for examinations.

"Boxed essays" are found throughout each chapter. The four themes of the boxed essays were specially designed with our students in mind. The "Biology and You" boxes link biology to students' everyday lives, highlighting the personal relevance of many biological concepts. "Biology, Technology, and Society" boxes highlight the links among biology, the fast-paced technology of today's world, and societal issues. "Biology in Focus" describes the work of the scientist in detail, often answering the question, "How do we know that?" And lastly, "Biology and Evolution" boxes emphasize the change of organisms over time and their evolutionary relationships to one another.

Each chapter closes with a "Summary" that lists all the key concepts in the chapter, a list of "Key terms" with the page number on which the term appears, a list of "Review questions" that tests student comprehension of the chapter content, and a few "Thought questions" that are designed to initiate class discussion and debate. Each chapter also has a short list of "Further Readings," chosen with the introductory student in mind. In addition to the usual *Scientific American* and *Science* articles, there are articles from *Smithsonian, National Wildlife,* and *National Geographic.*

Two appendixes provide additional support for the student. Appendix A provides answers to the review questions. Appendix B is a table of classification of the five kingdoms of life. Because of the human focus of this book, animals are classified through class, and the vertebrates are classified through order.

The glossary provides the pronunciation, definition, and derivation for each key term in the book and also gives the page reference for each term. This glossary is complete and easy to use and will help students immeasurably as they make their way through the text.

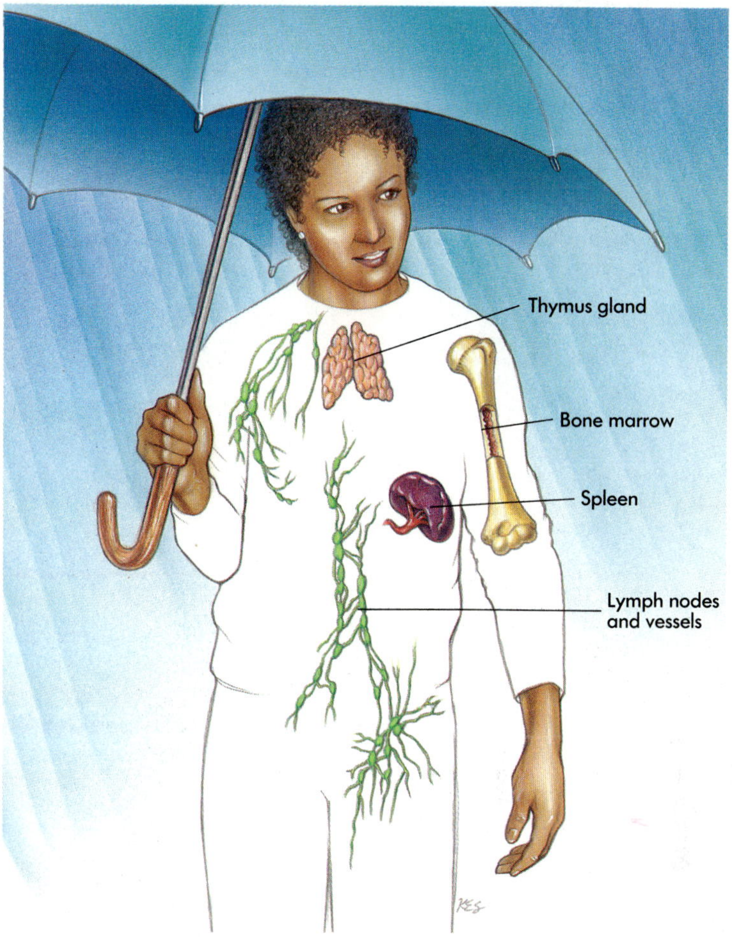

The art program

No biology text today can be a pedagogical success without a top-notch art program. Given the visual orientation of today's students—the computer graphics that they are bombarded with on television and the special effects that enliven today's feature films—it makes sense that a biology textbook use visual impact to teach concepts. *Biology Today* accomplishes this task by applying some basic principles to the art program:

- The illustrations and photographs should make abstract concepts more concrete to enhance student understanding.
- The content of text and art should match—there should not be concepts or structures presented in the art that are not discussed in the narrative.
- There should be an emphasis on simple, clear illustrations that teach process using layout and sometimes, for very complex processes, numbered steps.
- Legends should be a helpful guide to the illustration and should not introduce material that is not in the text.
- Art should be visually appealing so that students are prompted to "take a look."

We have followed these principles to the letter in *Biology Today,* with the result that the art program represents another important focus in the pedagogical program and is an element in itself.

Ancillaries

Biology Today has a full range of ancillary materials that meets the needs of both students and instructors. Carefully designed and executed, the ancillaries supplement and support *Biology Today* and enhance its student-oriented approach.

Instructor's resource manual

For the instructor, *Teaching Biology Today* is an invaluable teaching resource. Written by Dr. Ann Lumsden of Florida State University, each chapter contains a complete synopsis, suggested topics for class discussion, lecture outlines, teaching ideas, and a listing of applicable audiovisual resources. A text bank of over 2000 questions is also included in the manual. In addition, transparency masters of key illustrations in *Biology Today* are included in *Teaching Biology Today* to enhance visual learning.

Computerized test bank

The test questions included in *Teaching Biology Today* are also available as *Computest,* a computerized test generation system. *Computest* is available in MacIntosh and IBM formats. It has many features that make it easy for the instructor to design tests and quizzes. The instructor can browse and select questions for inclusion on an examination using several different criteria, including question type and level of difficulty.

Transparency acetates

Also available for the instructor are 150 full-color transparency acetates that reproduce the important illustrations in *Biology Today*. Labeling is clear, bold, and large enough for even students seated in the back of a large room to see.

Videodisc

Mechanisms of Life: Stability and Change Videodisc is a state-of-the art instructional medium that combines versatility, compactness, and ease of use. This videodisc contains thousands of still photographs, selected artwork, and film clips on biological processes to provide visual reinforcement in the classroom. Other features include the ability to search and display images or animated sequences and extensive use of full motion along with still images.

Laboratory manual

Exploring Biology Today is a unique laboratory manual written by Ann Wilke, Director of Undergraduate Laboratories at the University of Missouri—St. Louis. This manual responds to the needs of introductory students by guiding both formal and concrete thinkers through a laboratory experience that teaches them biology and how to become more critical thinkers. Based on the learning cycle, each of the 30 2-hour laboratory exercises in *Exploring Biology Today* leads students through their investigation of the biological world using three steps: exploration, in which students develop new understandings by means of hands-on experiences; concept introduction, in which the concepts students explored are articulated and labeled; and concept application, in which the students apply their knowledge and understanding to new situations. *Exploring Biology Today* is accompanied by an instructor's manual that offers support for both the laboratory set-up and the learning cycle approach.

Study guide

The student study guide, *Learning Biology Today,* by Dr. David Cotter of Georgia College, enhances and explores the concepts in the text by building from fact-based knowledge to the use of higher-order thinking skills. In each chapter, students are first tested for their mastery of basic concepts. Subsequent sections then test for deeper understanding of the chapter concepts, with the last section stressing application of these concepts to new situations. This "building block" approach facilitates the critical thinking necessary to solve problems using knowledge acquired from the text, without short-changing the straightforward learning of new concepts and ideas.

Acknowledgments

The highly professional and skilled staff at Mosby–Year Book, Inc. have helped make *Biology Today* an outstanding learning tool. Linda McKinley, production editor, made an extremely valuable contribution with her flawless copy editing and attention to detail. I also thank her for her talented page make-up. Susan Lane, book designer, provided a lovely design that is both simple and dynamic. I extend many thanks to Jennifer Collins and Holly Leicht for their persistent efforts and excellent work in acquiring the outstanding photographs used throughout the book. Very special thanks goes to the developmental editor, Kathleen Scogna, whose excellent editing and management skills played vital roles in *Biology Today*. I extend a sincere personal thanks to Kathleen for her exemplary work. My deepest gratitude goes to Bob Callanan, acquisitions editor, for his constant focus on excellence and unceasing support of innovation to make the entire *Biology Today* program a pedagogically sound and appropriate learning tool for today's students.

The reviewers of *Biology Today* provided countless suggestions that enhanced the quality of this textbook and ensured its accuracy. Their names are listed and to each one I extend a personal "Thank you." I would like to give special thanks to my colleagues at the University of Missouri—St. Louis who were always available to review manuscript pages and provide instant feedback: Ed Joern, Randy Nolan, Carl Thurman (now at the University of Northern Iowa), and Ann Wilke.

William Barnes, *Clarion University*
Kathy Burt-Utley, *University of New Orleans*
James Conkey, *Truckee Meadows Community College*
Neil Crenshaw, *Indian River Community College*
Don Emmeluth, *Fulton-Montgomery Community College*
Robert Grammar, *Belmont College*
Ronald Hoham, *Colgate University*
Ed Joern, *University of Missouri—St. Louis*
Alan Karpoff, *University of Louisville*
Miriam Kitrell, *Kingsborough Community College*
John Knesel, *Northeast Louisiana University*
Kenneth Mace, *Stephen F. Austin State University*
Virginia Michelich, *Dekalb College*
Randy Nolan, *University of Missouri—St. Louis*
Michael Postula, *Parkland College*
June Ramsey, *Pensacola Junior College*
David Rayle, *San Diego State University*
Michael Smiles, *State University of New York—Farmingdale*
Carl Thurman, *University of Northern Iowa*
Ann Wilke, *University of Missouri—St. Louis*
Nancy Yurko, *Prince Georges Community College*

Personal thanks go to George Johnson of Washington University in St. Louis for contributing boxed essays to *Biology Today*. Also, a very special personal thank you goes to Arthur Tarrow of Melbourne, Florida, for publishing my first book, which was a step to further publications and, of course, to *Biology Today*.

This book is dedicated to my parents, Evo and Florence Sebastianelli of Farmington, Connecticut, and to my son, Marc Gottfried of St. Louis, Missouri.

SANDRA S. GOTTFRIED

Contents in Brief

PART ONE AN INTRODUCTION TO BIOLOGY AND CHEMISTRY

1 **The themes of biology today** 2

2 **The chemistry of life** 16

PART TWO CELLS THE BASIC UNIT OF LIFE

3 **Cell structure and function** 44

4 **Cell membranes** 67

5 **Cell division** 80

PART THREE HOW LIVING THINGS TRANSFORM ENERGY

6 **The flow of energy within organisms** 100

7 **Cellular respiration** 114
How cells release stored energy from food molecules

8 **Photosynthesis** 130
How plants capture and store energy from the sun

PART FOUR HUMAN BIOLOGY
THE STRUCTURE AND FUNCTION OF THE BODY

9 **Levels of organization in the human body** 146

10 **Digestion** 165

11 **Respiration** 184

12 **Circulation** 198

13 **Defense against disease** 216

14 **Excretion** 232

15 **Nerve cells and how they transmit information** 249

16 **The nervous system** 262

17 **The senses** 280

18 **Protection, support, and movement** 294

19 **Hormones** 314

PART FIVE HOW HUMANS REPRODUCE AND PASS ON BIOLOGICAL INFORMATION

20 **Sex and reproduction** 334

21 **Human development before birth** 354

22 **Patterns of inheritance** 372

23 **Human genetics** 388

24 **The molecular basis of inheritance** 404

PART SIX EVOLUTION HOW LIVING THINGS CHANGE OVER TIME

25 **The scientific evidence for evolution** 426

26 **The evolution of the five kingdoms of life** 444

27 **Human evolution** 463

PART SEVEN THE DIVERSITY AND UNITY OF LIVING THINGS

28 **Viruses, bacteria, and genetic engineering** 486

29 **Protists and fungi** 502

30 **Plants** 526
Reproductive patterns and diversity

31 **Plants** 544
Patterns of structure and function

32 **Invertebrate animals** 560
Patterns of structure, function, and reproduction

33 **Vertebrate animals** 580
Patterns of structure, function, and reproduction

PART EIGHT HOW LIVING THINGS INTERACT WITH EACH OTHER AND WITH THEIR ENVIRONMENT

34 **Innate behavior and learning in animals** 598

35 **Social behavior in animals** 610

36 **Population ecology** 624

37 **Interactions within communities of organisms** 638

38 **Ecosystems** 656

39 **Biomes and life zones of the world** 671

40 **The biosphere** 694
Today and tomorrow

Contents

PART ONE

AN INTRODUCTION TO BIOLOGY AND CHEMISTRY

1 The themes of biology today 2

Scientific process: The unifying theme among all sciences 3
  The pathway of thinking in the scientific method 3
How biologists do their work 6
  Summarizing the steps in the scientific method 7
  Theory building 7
The unifying themes of biology today 7
  Living things display both diversity and unity 7
  Living things are composed of cells and are hierarchically organized 7
  Living things interact with each other and with their environments 8
  Living things transform energy and maintain a steady internal environment 10
  Living things exhibit forms that fit their functions 11
  Living things reproduce and pass on biological information to their offspring 11
  Living things change over time, or evolve 11
Classification: A reflection of evolutionary history 12
Biology today and you 14
Summary 14

2 The chemistry of life 16

Atoms 17
Molecules 18
  Nature of the chemical bond 19
  Ions and ionic bonds 19
  Covalent bonds 20
The cradle of life: Water 22
  Water is a powerful solvent 24
  Water organizes nonpolar molecules 24
  Water ionizes 25
Inorganic versus organic chemistry 26
The chemical building blocks of life 28
  Carbohydrates 30
  Fats and lipids 32
  Proteins 34
  Nucleic acids 39
The use of isotopes in medicine 40
Summary 40

PART TWO

CELLS
THE BASIC UNIT OF LIFE

3 Cell structure and function 44

The cell theory 45
Why aren't cells larger? 46
Two kinds of cells 47
Eukaryotic cells: An overview 49
  The plasma membrane 49
  The cytoplasm and cytoskeleton 50
  Membranous organelles 53
How cells crawl 54
  Bacteria-like organelles 58
  Cilia and flagella 60
  Cell walls 61
Prokaryotic cells 61
  Cell walls 62
  Simple interior organization 63
A comparison of prokaryotic and eukaryotic cells 65
Summary 65

4 Cell membranes 67

The foundation of membranes 68
Membrane channels and cystic fibrosis 70
Membrane proteins 70
How substances move across cell membranes 72
  Movement that does not require energy 72
  Movement that does require energy 75
Summary 78

5 Cell division 80

Chromosomes 81
Cell division, growth, and reproduction 83
Cancer and mitosis 84
Mitosis 86
  Prophase 86
  Metaphase 89
  Anaphase 90
  Telophase 90
Cytokinesis 91
  Cytokinesis in animal cells 91
  Cytokinesis in plant cells 91
Meiosis 92
  The stages of meiosis 93
The importance of meiotic recombination 97
Summary 97

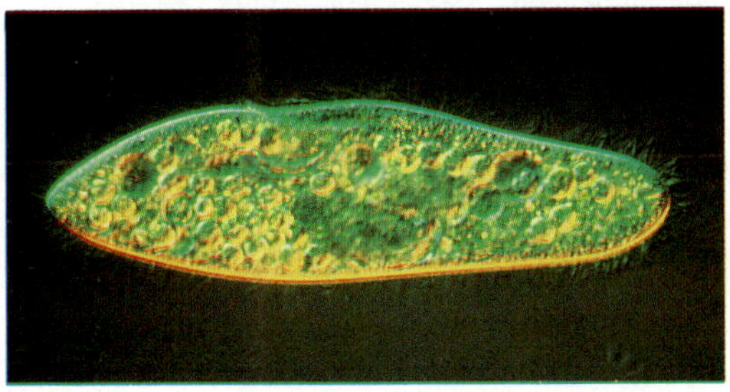

PART THREE

HOW LIVING THINGS TRANSFORM ENERGY

6 The flow of energy within organisms 100

Starting chemical reactions 101
Energy flow and change in living systems 103
Regulating chemical reactions 105
  Enzymes: Biological catalysts 105
Nonfattening sweets 107
Storing and transferring energy 110
  ATP: The energy currency of living things 110
  Carrying and transferring energy using ATP 111
Summary 112

7 Cellular respiration 114
How cells release stored energy from food molecules

Using chemical energy to drive metabolism 115
How cells make ATP: Variations on a theme 116
How oxygen-using organisms release ATP from food molecules 116
Carbohydrates and winter depression 118
  Oxidation-reduction 119
  Electron carriers 119
  A more detailed look at cellular respiration 120
How organisms release ATP from food molecules without using oxygen 127
Summary 128

8 Photosynthesis 130
How plants capture and store energy from the sun

Producers and consumers of food 131
The energy in sunlight 132
Capturing light energy in chemical bonds 132
An overview of photosynthesis 134
C_4 and CAM photosynthesis 136
Light-dependent reactions: Making ATP and NADPH 136
  How photosystems capture light 136
Light-independent reactions: Making glucose 141
Relationships: Photosynthesis and cellular respiration 143
Summary 143

PART FOUR

HUMAN BIOLOGY
THE STRUCTURE AND FUNCTION OF THE BODY

9 **Levels of organization in the human body** 146

How the human body is organized 147
Tissues 149
  Epithelial tissue 149
  Connective tissue 152
  Muscle tissue 158
  Nervous tissue 159
Organs 160
Organ systems 160
The organism: Coordinating it all 162
Summary 163

10 **Digestion** 165

The nutrition-digestion connection 166
Where it all begins: The mouth 170
The journey of food to the stomach 172
Preliminary digestion: The stomach 174
Terminal digestion and absorption: The small intestine 175
  Accessory organs that help digestion 175
Ulcers: What they are and how to prevent them 176
  Digestion 176
  Absorption 178
Concentration of solids: The large intestine 179
Diet and nutrition 180
Summary 182

11 **Respiration** 184

Respiration 185
The pathway of air into and out of the body 186
  Nasal cavities 186
  The pharynx 186
  The larynx 186
  The trachea 186
  The bronchi and its branches 188
  The alveoli: Where gas exchange takes place 188
How to "catch" cancer 189
The mechanics of breathing 190
  Inspiration 190
  Expiration 191
  Deep breathing 191
  Lung volumes 192
Gas transport and exchange 192
  External respiration 192
  Internal respiration 194
Choking: A common respiratory emergency 195
Chronic obstructive pulmonary disease 196
Summary 196

12 **Circulation** 198

Functions of the circulatory system 199
  Nutrient and waste transport 199
  Oxygen and carbon dioxide transport 199
  Temperature maintenance 199
  Hormone circulation 201
The heart and blood vessels 201
  Arteries and arterioles 201
  Capillaries 202
  Veins and venules 203
  The heart: A double pump 204
How to avoid heart disease 206
The blood 208
  Blood plasma 208
  Types of blood cells 208
The lymphatic system 211
Diseases of the heart and blood vessels 213
Summary 214

13 **Defense against disease** 216

Nonspecific versus specific resistance to infection 217
Nonspecific resistance 217
Specific resistance 218
Discovery of specific resistance: The immune response 218
The cells of the immune system 221
The immune response: How it works 221
  Sounding the alarm 221
  The two branches of the immune response 222
Helping the immune system fight cancer 225
How do immune receptors recognize antigens? 226
Immunization: Protection against infection 227
Defeat of the immune system: AIDS 227
Allergy 229
Summary 230

14 Excretion 232

What substances does the body excrete? 233
The organs of excretion 234
An overview of how the kidney works 234
A closer look at how the kidney works 235
The anatomy of the kidney 235
The workhorse of the kidney: The nephron 235
Testing urine for drugs 242
The urinary system 244
The kidney and homeostasis 244
Problems with kidney function 245
Kidney stones 245
Renal failure 246
Treatments for renal failure 247
Summary 247

15 Nerve cells and how they transmit information 249

The communication systems of your body 250
The nerve cell, or neuron 250
The nerve impulse 252
The neuron at rest: The resting potential 253
Conducting an impulse: The action potential 254
Transmission of the nerve impulse: A propagation of the action potential 255
Speedy neurons: Saltatory conduction 256
Transmitting information between cells 257
Neuron-to-muscle cell connections 259
Neuron-to-neuron connections 259
Summary 260

16 The nervous system 262

The organization of the nervous system 263
The central nervous system 265
The brain 265
Solving the mystery of Alzheimer's disease 268
The spinal cord 271
Memories 273
The peripheral nervous system 273
Sensory pathways 273
Integration 275
Motor pathways 275
Summary 278

17 The senses 280

The nature of sensory communication 281
Sensing the body's internal environment 282
Sensing the body'sposition in space 282
Sensing the external environment 283
The general senses 283
The special senses 284
Summary 292

18 Protection, support, and movement 294

How skin, bones, and muscles work together 295
Skin 295
Bones 296
The skeletal system 298
Muscles 305
The muscular system 305
Can special shoes make you run faster? 308
Summary 312

19 Hormones 314

Endocrine glands and their hormones 315
The pituitary gland 318
Living with diabetes 321
The thyroid gland 322
The parathyroid glands 325
The adrenal glands 325
The pancreas 327
The pineal gland 328
The thymus gland 328
The ovaries and testes 328
Nonendocrine hormones 328
Summary 330

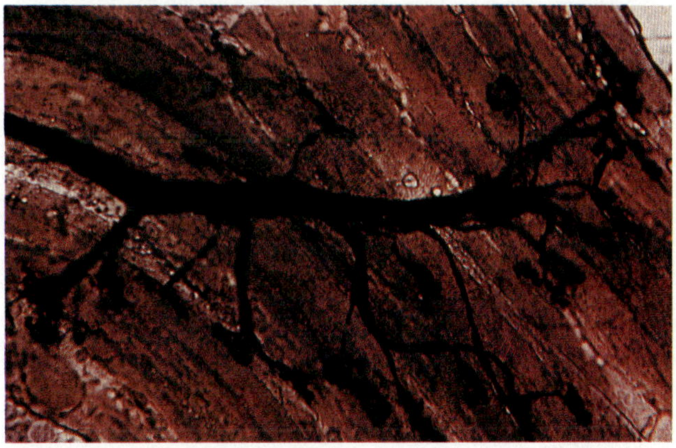

PART FIVE

HOW HUMANS REPRODUCE AND PASS ON BIOLOGICAL INFORMATION

20 Sex and reproduction 334

Reproduction 335
The male reproductive system 335
  The production of sperm: The testes 335
  Maturation and storage: The epididymis and vas deferens 338
  Nourishment of the sperm: The accessory glands 338
  The penis 338
The female reproductive system 339
  The production of eggs: The ovaries 339
  Passage to the uterus: The uterine tubes 340
  The site of prenatal development: The uterus 341
  The reproductive cycle 341
  The vagina 343
  The external genitals 343
AIDS on the college campus 344-345
  The mammary glands 346
The sexual response 346
Contraception and birth control 347
  Abstinence 347
  Sperm blockage 350
  Sperm destruction 350
  Prevention of egg maturation 350
  Surgical intervention 351
Summary 352

21 Human development before birth 354

Development 355
Fertilization 355
The first and second weeks of development: The pre-e .ıbryo 356
Early development of the extraembryonic membranes 360
Development from the third to eighth week: The embryo 361
  The third week 361
Fetal alcohol syndrome 362
  The fourth week 365
  The fifth through eighth week 366
Development from the ninth to thirty-eighth week: The fetus 366
  The third month 366
  The fourth through sixth month 367
  The seventh through ninth month 368
Birth 368
Physiological adjustments of the newborn 369
Summary 370

22 Patterns of inheritance 372

Inheritance and variation within species 373
  Historical views of inheritance 373
  The birth of the study of inheritance 374
Gregor Mendel's experiments to determine inheritance patterns 374
  Conclusions Mendel drew from his experiments 377
  Analyzing Mendel's experiments 378
  How Mendel tested his conclusions 379
  Further questions Mendel asked 380
  Analyzing the results of Mendel's dihybrid crosses 381
The connection between Mendel's factors and chromosomes 383
Sex linkage 383
Solving the mysteries of inheritance 385
Summary 385

23 Human genetics 388

Studying inheritance patterns in humans using karyotypes 389
  The inheritance of abnormal numbers of autosomes 389
  The inheritance of abnormal numbers of sex chromosomes 391
  Changes in chromosome structure 392
Curing genetic disorders by gene transfer 393
  Gene mutations 394
Studying inheritance patterns in humans using pedigrees 395
Dominant and recessive genetic disorders 398
Incomplete dominance and codominance 399
Multiple alleles 400
Genetic counseling 400
Using blood typing to test for paternity 401
Summary 402

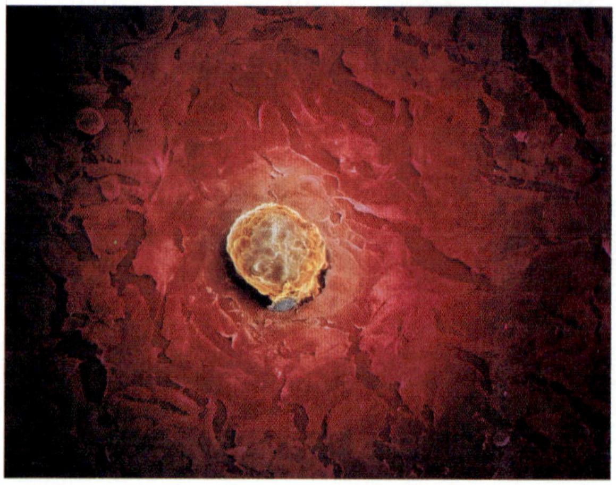

24 **The molecular basis of inheritance** 404

The location of the hereditary material 405
The chemical nature of the hereditary material: Nucleic acids 406
The nucleic acids: DNA and RNA 406
A major scientific breakthrough: The structure of DNA 408
How DNA replicates 410
The human genome project 411
Genes: The units of hereditary information 412
Gene expression: How DNA directs the synthesis of polypeptides 413
An overview 413
Transcribing the DNA message to RNA 413
The Hershey-Chase experiment 414
Translating the transcribed DNA message into a polypeptide 416
Differences in prokaryotic and eukaryotic genes 419
Differences in prokaryotic and eukaryotic gene expression 420
Prokaryotic gene expression 420
Eukaryotic gene expression 422
Gene therapy: Changing genes to cure disease 422
Summary 423

PART SIX

EVOLUTION
HOW LIVING THINGS CHANGE OVER TIME

25 **The scientific evidence for evolution** 426

The development of Darwin's theory of evolution 427
Darwin's observations 427
Factors that influenced Darwin's thinking 429
Natural selection: A mechanism of evolution 432
An example of natural selection at work: Darwin's finches 432
The publication of Darwin's theory 433
Testing the theory 434
The fossil record 434
The age of the Earth: Rock and fossil dating 436
Comparative anatomy 438
Comparative embryology 440
Molecular biology 440
Summary 442

26 **The evolution of the five kingdoms of life** 444

Theories about the origin of organic molecules 445
Theories about the origin of cells 447
The history of life 448
The Archean Era: Oxygen-producing cells appear 448
The Proterozoic Era: The first eukaryotes appear 452
The Paleozoic Era: The occupation of the land 452
Mass extinctions 455
The Mesozoic Era: The age of reptiles 455
Was Protoavis the first bird? 457
Changes in the Earth: The impact on evolution 458
The Cenozoic Era: The age of mammals 458
Summary 461

27 **Human evolution** 463

Evolution of the vertebrates 464
Evolution of the fishes 465
Evolution of the amphibians, reptiles, and birds 466
Evolution of the mammals 468
All about Eve, or how scientists are attempting to construct the human family tree 469
Evolution of the primates 471
The prosimians 473
The anthropoids 474
Evolution of the anthropoids 476
Evolution of the hominids 477
The first hominids: *Australopithecus* 477
The first humans: *Homo habilis* 480
Human evolution continues: *Homo erectus* 480
Modern humans: *Homo sapiens* 480
Summary 482

PART SEVEN

THE DIVERSITY AND UNITY OF LIVING THINGS

28 Viruses, bacteria, and genetic engineering 486

Viruses 487
The discovery of viruses 487
Viral replication 489
Viruses and cancer 490
The classification of viruses 490
Bacteria 491
Bacterial reproduction 491
Bacterial diversity and classification 492
Bacteria as disease producers 493
Genetic engineering 494
Natural gene transfer among bacteria 494
The new agriculture 496
Human-engineered gene transfer among bacteria 496
Applications of recombinant DNA technology 498
Medically important products 498
Gene therapy 499
Agriculture 499
Summary 500

29 Protists and fungi 502

Protists 503
Animal-like protists: protozoans 504
Malaria today 508
Plant-like protists: Algae 509
Fungus-like protists 514
Fungi 516
Zygote-forming fungi 518
Sac fungi 519
Club fungi 520
Imperfect fungi 522
Lichens 523
Summary 524

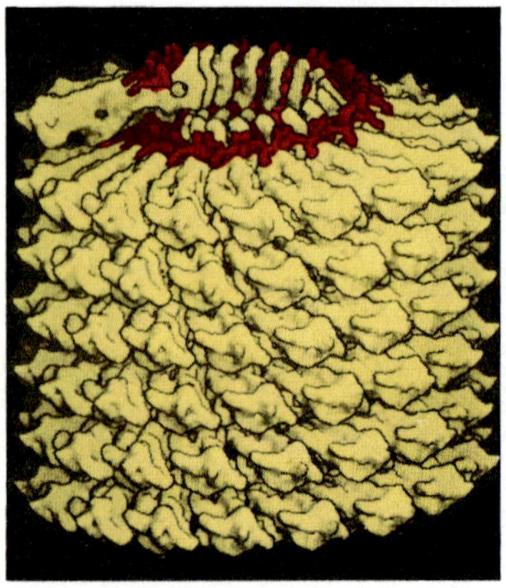

30 Plants 526
Reproductive patterns and diversity

Characteristics of plants 527
The general pattern of reproduction in plants 527
Patterns of reproduction in nonvascular plants 529
The bryophytes 529
Patterns of reproduction in vascular plants 531
Seedless vascular plants 531
Vascular plants with naked seeds 533
Vascular plants with protected seeds 534
Mechanisms of pollination 535
Fruits and their significance in sexual reproduction 536
Finding new food plants 537
Seed formation and germination 538
Types of vegetative propagation in plants 540
Regulating plant growth: Plant hormones 541
Summary 542

31 **Plants** 544
Patterns of structure and function
The organization of vascular plants 545
Tissues of vascular plants 545
Fluid movement: Vascular tissue 546
Food storage: Ground tissue 547
Protection: Dermal tissue 547
Growth: Meristematic Tissue 548
Organs of vascular plants 548
Roots 548
Shoots 551
Movement of water and dissolved substances in vascular plants 555
The organization of nonvascular plants 557
Summary 558

32 **Invertebrate animals** 560
Patterns of structure, function, and reproduction
Characteristics of animals 561
Patterns in symmetry 562
Patterns in body cavity structure 562
Patterns in embryological development 564
Invertebrates: Diversity in symmetry and coelom 565
Asymmetry: Sponges 565
Radial symmetry: Hydra and jellyfish 566
Bilateral symmetry: Variation on a theme in many phyla 568
Parasitic flatworms 572
Summary 578

33 **Vertebrate animals** 580
Patterns of structure, function, and reproduction
Chordates and vertebrates: Unity in symmetry and coelom 581
Tunicates 581
Lancelets 583
Vertebrates 583
Summary 595

PART EIGHT

HOW LIVING THINGS INTERACT WITH EACH OTHER AND WITH THEIR ENVIRONMENT

34 **Innate behavior and learning in animals** 598
Ethology: The biology of behavior 599
The link between genetics and behavior 600
Innate (instinctive) behaviors 600
Kineses 600
Taxes 600
Reflexes 602
Fixed action patterns 602
Learning 603
Imprinting 603
Habituation 605
Trial-and-error learning (operant conditioning) 605
Classical conditioning 606
Insight 606
Yawn! 608
Summary 608

35 **Social behavior in animals** 610
Types of social behavior 611
Competitive behaviors 611
Elephant talk 614
Reproductive behaviors 615
Parenting behaviors 618
Group behaviors 618
Human behavior 621
Summary 622

36 **Population ecology** 624
An introduction to ecology 625
Population growth 625
Exponential growth 625
Carrying capacity 627
Population size 627
Population density and dispersion 628
Regulation of population size 628
Density-independent limiting factors 628
Density-dependent limiting factors 628
The urban explosion 629
Mortality and survivorship 630
Demography 631
The human population explosion 632
Summary 636

37 **Interactions within communities of organisms** 638
Ecosystems 639
Communities 639
Types of interactions within communities 640
Competition 640
Predation 642
Competition and killer bees 646
Symbiosis 649
Changes in communities over time: Succession 651
Summary 654

38 **Ecosystems** 656
Populations, communities, and ecosystems 657
The flow of energy through ecosystems 658
Food chains and webs 659
Food pyramids 661
Hubbard Brook and the cycling of nutrients in ecosystems 663
The cycling of chemicals within ecosystems 664
The water cycle 664
The carbon cycle 665
The nitrogen cycle 666
The phosphorus cycle 667
Summary 669

39 **Biomes and life zones of the world** 671
Biomes and climate 672
The sun and its effects on climate 672
Atmospheric circulation and its effects on climate 672
Life on land: The biomes of the world 675
Tropical rain forests 676
The El Niño southern oscillation 678
Savannas 678
Deserts 680
Temperate grasslands 681
Temperate deciduous forests 682
Taiga 682
Tundra 684
Life in fresh water 684
Estuaries: Life between rivers and oceans 685
Life in the oceans 687
The intertidal zone 687
The neritic zone 688
The open-sea zone 690
Summary 692

40 **The biosphere** 694
Today and tomorrow
The biosphere 695
The land 696
Diminishing natural resources 696
Species extinction 700
Ten things you can do to change the world 702
Solid waste 702
The water 703
Surface water pollution 703
Ground water pollution 705
Acid rain 705
The atmosphere 707
Air pollution 707
Ozone depletion 708
Overpopulation and environmental problems 708
Summary 709

APPENDIXES

A **Answers to review questions** A-1
B **Classification of organisms** B-1

Glossary G-1

Credits C-1

PART ONE

AN INTRODUCTION TO BIOLOGY AND CHEMISTRY

1 The Themes of Biology Today

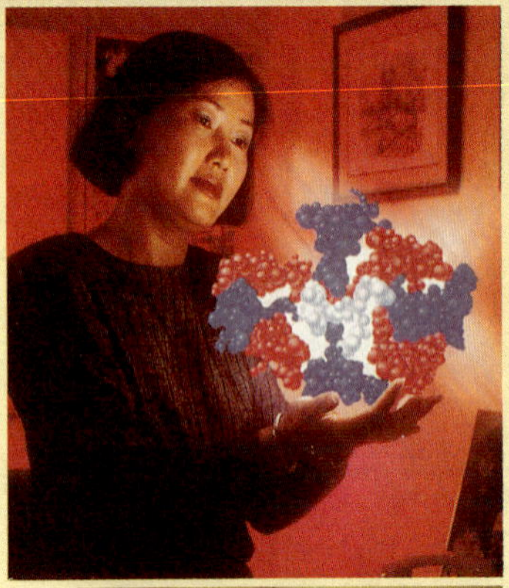

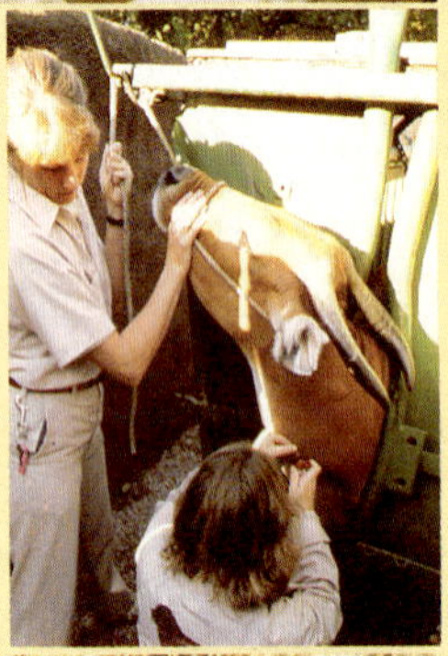

Biology today has many faces. Molecular biologists study life at the chemical level, probing the workings of the hereditary material and the molecular "chain of command" within cells. Cell biologists study individual cells or groups of cells, often by growing cells outside of organisms in cultures, and ask questions regarding cell-to-cell interactions and the effects of the environment on cells. Working at the organism level, biologists study animals, plants, and other multicellular organisms. For example, in one of the photos, zoologists are shown collecting blood from a zoo animal to study the changes in its hormone levels during its reproductive cycle. Other biologists study populations—individuals of the same species that occur together at one place and at one time—and are interested in interactions among them. Population biologists study topics such as the changes in population sizes and the causes of these fluctuations. Some biologists work with a global orientation and are interested in questions having worldwide impact. These biologists study the effects of burning billions of acres of tropical rain forest on global weather patterns and trends.

HIGHLIGHTS

▼

All scientists study the natural world by using a process called the *scientific method* in which they ask questions, make predictions, and then test their predictions by means of experimentation.

▼

Scientific explanations, developed by means of the scientific method and supported by overwhelming evidence over time, are called *theories*—powerful concepts that help scientists make dependable predictions about the world.

▼

Rising out of this abundance of tested predictions are many themes, or accepted explanations, regarding living things.

▼

Modern taxonomy, the classification of the diverse array of living things, categorizes organisms based on their common ancestry while taking into account their existing similarities and differences.

OUTLINE

Scientific process: The unifying theme among all sciences

The pathway of thinking in the scientific method

Summarizing the steps in the scientific method

Theory building

The unifying themes of biology today

Living things display both diversity and unity

Living things are composed of cells and are hierarchically organized

Living things interact with each other and with their environments

Living things transform energy and maintain a steady internal environment

Living things exhibit forms that fit their functions

Living things reproduce and pass on biological information to their offspring

Living things change over time, or evolve

Classification: A reflection of evolutionary history

Biology today and you

Today, the study of biology includes these focuses and a myriad of others. Biologists ask questions that probe the intricacies of life from the perspectives of both distinct subdisciplines and blended subdisciplines of biology, calling on the knowledge and techniques of related fields. They answer questions that add to our knowledge base, such as "How much of our protective ozone layer has been lost?" and "How do our defensive white blood cells recognize, move toward, and destroy invading cells?" Biologists probe into areas such as the transfer of hereditary material from organism to organism, the relationship between diet and disease, and the development of new food plants. This probing advances knowledge in applied fields such as medicine, agriculture, and industry and produces products that enhance and lengthen lives. As you read *Biology Today,* you will find out more about these and other topics that are a part of biology today. As the ancient Greeks put it, you will have a *bios logos*—a discourse on life.

Scientific process: The unifying theme among all sciences

Science often seems a mysterious process to the nonscientist. White-coated men working in sterile laboratories and engaging in difficult-to-understand processes embody the stereotype of the typical scientist and *his* work. However, as you can see in the opener photos, scientists are men *and* women, and are a culturally and ethnically diverse group. In addition, not all scientists work in laboratories. In fact, science takes place just about anywhere!

Although scientists and "their" science may differ in their focuses, all scientists—biologists, chemists, physicists, and so forth—study the natural world and go about their work in a similar way. This similar way is referred to as the **scientific method.** The scientific method is a process, or series of steps, scientists use to answer the questions they ask. In fact, asking questions is part of the method! In addition, all scientists ask questions that can be investigated by means of experimentation. In other words, they ask questions that can be tested, producing results that are open to verification by others in their discipline. Scientists do not ask philosophical or religious questions such as "What is the meaning of life?" or "Does life exist after death?" However, in a film released in the early 1990s, a group of medical students, planning experiments to answer the latter question, almost died while engaged in a fictional interpretation of scientific inquiry!

The scientific method is a set of procedures used to answer questions that is common among the various scientific disciplines.

The pathway of thinking in the scientific method

The scientific method usually starts with observations that lead to questions. In fact, one of the most important processes of science is observation, but observation alone does not constitute science. The predecessors of scientists—the ancient Greeks and Romans such as Plato and Aristotle—were skilled observers but were not scientists by today's definition. Rather, their science was called **natural history** and they were called **naturalists.** Until about the mid-1800s, biology continued to have a natural history perspective, focusing on descriptions and comparisons of organisms and providing explanations to questions based on observation alone. Although the scientific method had already taken hold in the physical sciences, it was not widely used in biology because many people felt that the living world transcended physical laws—experimentation could not answer questions regarding living things. A critical shift occurred during the nineteenth century, the shift that made biology a "true" science. A preoccupation with form moved to an investigation of function and process. Scientists went beyond observation to put forth **hypotheses,** or plausible answers to their questions, *testing them by experimentation.*

Hypothesis formation (often characterized as an educated guess that tentatively answers a question) and experimentation are key pieces to the scientific method. Before scientists formulate hypotheses, however, they gather information regarding their questions. Only after searching the scientific literature and synthesizing all the information they can find regarding their questions do scientists generate hypotheses. The hypothesis, therefore, is not just a guess but a well-thought-out prediction based on available knowledge and generalizations made from observations. This pattern of thought—developing generalizations from specific instances—is called **inductive reasoning.**

The scientific method usually starts with observations that lead to questions. Scientists then generate hypotheses, or plausible answers to their questions, by making generalizations based on their observations and the knowledge available to them. Scientists test their hypotheses by experimentation.

For example, you may observe that many of your friends grow houseplants that look healthy and robust. Since you struggle to keep yours alive, you might ask them their "secret." Many may answer that one key to growing healthy houseplants is giving them regular doses of Brand X fertilizer. You may have also read advertisements and other literature stating that Brand X fertilizer causes houseplants to grow 30% taller and produce healthier, greener plants. From these observations and testimonials, you might make the inductive generalization that Brand X fertilizer helps houseplants grow well. This generalization becomes a hypothesis when stated in an "If . . . then . . . " format. Your hypothesis might be the following: If houseplants are given Brand X fertilizer, then they will grow at least 30% taller and be greener than plants with no fertilizer. As you can see, a hypothesis is a *prediction.*

In analyzing the simplified hypothesis in our example (hypotheses can be much more complex), notice that the factor you are manipulating (the fertilizer) is stated after the "if" part of the hypothesis. This factor is called the **independent variable,** or manipulated variable. The word *variable* refers to a factor that changes, or varies. And, in fact, to test your hypothesis, you will need to add different

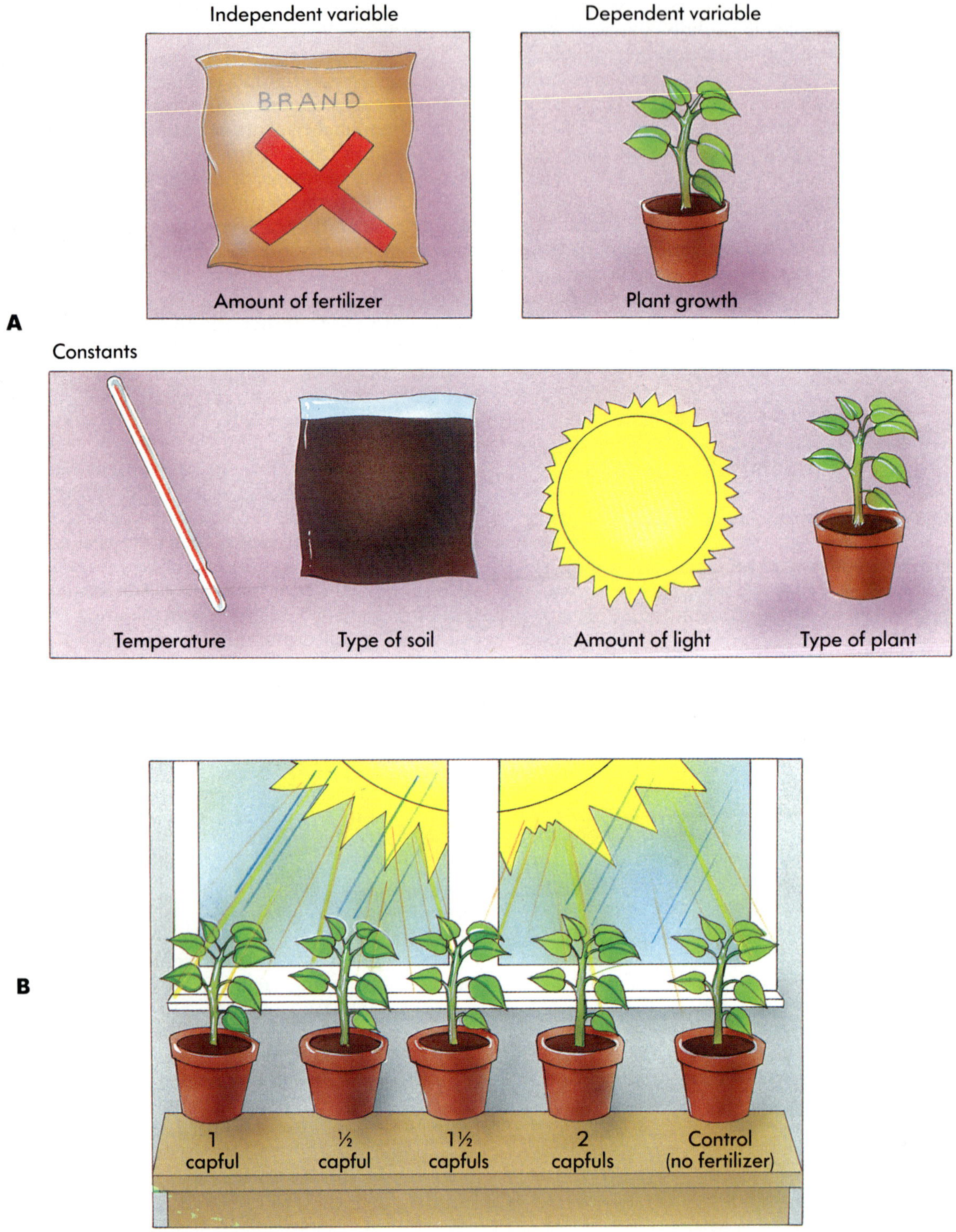

FIGURE 1-1 An experiment to determine how Brand X fertilizer influences plant growth.
A The variables in experiments are independent (manipulated) and dependent (responding). Other potential variables are held constant. In the plant experiment, the independent variable is the amount of fertilizer, the dependent variable is plant growth, and the constants are the type of soil, temperature, amount of light, and type of plant.
B The experimental set-up for the fertilizer-plant growth experiment.

amounts of Brand X fertilizer to houseplants to determine the effects on plant growth and color. Also notice that the factor stated after the "then" part of the hypothesis depends on the manipulated (independent) variable. This factor is called the **dependent variable,** or responding variable. It varies in response to changes in the independent variable.

Scientists test hypotheses by means of **controlled experiments.** An experiment tests a hypothesis by producing data that either support or refute the prediction made in the hypothesis. This "testing" process involves the use of deduction, or **deductive reasoning.** Deduction begins with a general statement (the prediction about Brand X fertilizer and its effects on houseplants) and proceeds to a specific statement (the effects of Brand X fertilizer on *your* houseplants).

In a controlled experiment the independent variable is manipulated and changes in the dependent variable are observed while all other possible variables are kept constant; that is, they are controlled. The variables to be kept constant include any factors in the experimental set-up that can change. In this case, they include the type of soil, the amount of time the plants are exposed to the light, the temperature in which the plants are grown, and the type of plant tested. Can you think of any other possible variables that may need to be controlled?

As you design a controlled experiment to test your hypothesis, you should first outline the way in which you would manipulate the independent variable. In this case, you may decide to give one plant the "dose" of fertilizer suggested on its container and give some plants slightly more and others slightly less. If the suggested dose is one capful of liquid fertilizer, you may decide to give a second plant one-half capful, a third plant one and one-half capfuls, and a fourth plant two capfuls. Each plant should be the same type of plant such as all philodendrons of approximately the same size. You should pot them all in the same soil in the same type and size of pot, place them near one another on the same windowsill, and water them all with the same amount of water at the same time. These plants will receive the **treatment,** in this case, the fertilizer. They are the experimental plants. A fifth plant should be the same as all the rest in every respect *except* that it will not receive the treatment (fertilizer). This plant serves as the **control** (Figure 1-1). The control is the standard against which the treatment plants can be compared. Any changes in the control are the result of factors other than the treatment.

Over a preselected time, you should observe your plants and **collect data** regarding leaf color and plant height or in the case of philodendrons (vine-like plants), vine length. The data regarding the vine length of your philodendrons are called **quantitative data** because these data are based on numerical measurements (centimeters, for example). The data regarding leaf color are called **qualitative data** because these data are descriptive and not based on numerical measurements.

At the beginning of your experiment, you should measure the total length of the vines of each plant and record these data. At the end of your predetermined growing period, 6 weeks for example, you measure the vines again, recording these data. Your data table may look like this:

TOTAL VINE LENGTH IN CENTIMETERS (cm)

DOSE OF FERTILIZER	0 cf	½ cf	1 cf	1½ cf	2 cf
0 weeks	145	150	148	151	149
6 weeks	166	180	181	179	184
Increase in vine length	21	30	33	28	35

KEY: cf = capfuls of Brand X fertilizer

Regarding the collection of qualitative data, your data table may look like this:

DESCRIPTION OF LEAF COLOR ON SELECTED VINES

DOSE OF FERTILIZER	0 cf	½ cf	1 cf	1½ cf	2 cf
0 weeks	mg	mg	mg	mg	mg
6 weeks	lg	mg	dg	dg	dg
Change in vine color	neg	nc	pos	pos	pos

KEY: y = yellow mg = medium green
lg = light green dg = dark green
neg = turned lighter (negative change)
pos = turned darker (positive change)
nc = no change

Do your data support your hypothesis: If I add Brand X fertilizer to the soil of my houseplants, they will grow at least 30% taller and be greener than plants with no fertilizer? Looking at the qualitative data, the plants with fertilizer were greener than the control with no fertilizer. One capful to two capfuls, in this experiment, produced the greenest plants.

Looking at the quantitative data, can you determine if the vines were at least 30% longer than the control? First, let's compare the plant treated with one-half capful fertilizer to the control plant. The control plant grew 21 centimeters and the test plant grew 30 centimeters—9 centimeters more than the control. Is this difference at least 30% more (30% greater than the control growth)? This question is really asking "What percent of 21 is 9?" To find out, you divide 9 by 21, which yields 0.43. Multiplying by 100 to change this number into a percentage yields 43%. In other words, the vines of the plant receiving one-half capful of Brand X fertilizer grew 42% longer than the control! Using the same reasoning (1) the vines of the plant receiving one capful of Brand X fertilizer grew 57% longer, (2) the vines of the plant receiving one and one-half capfuls grew 33% longer, and (3) the vines of the plant receiving two capfuls grew 67% longer. In summary, these data do support your hypothesis; in fact, your test data suggest that Brand X fertilizer may help philodendrons grow greener and longer than the manufacturer suggests!

Why do you think the manufacturer of Brand X fertilizer is not making greater claims regarding this product? Your experiment certainly shows much greater than a 30% increase in growth. The answer lies in the limitations of your experiment. First, you experimented with one type of houseplant: philodendrons. Brand X fertilizer is probably sold for use on a variety of houseplants. Not all plants may grow as vigorously in response to the fertilizer as your philoden-

BIOLOGY IN FOCUS

How biologists do their work

The Consent

Late in November, on a single night
Not even near to freezing, the gingko trees
That stand along the walk drop all their leaves.
In one consent, and neither to rain nor to wind
But as though to time alone: the golden and green
Leaves litter the lawn today, that yesterday
Had spread aloft their fluttering fans of light.
What signal from the stars? What senses took it in?
What in those wooden motives so decided
To strike their leaves, to down their leaves,
Rebellion or surrender? and if this
Can happen thus, what race shall be exempt?
What use to learn the lessons taught by time
If a star at any time may tell us: Now.

HOWARD NEMEROV

What is bothering the poet Howard Nemerov is that life is influenced by forces he cannot control or even identify. It is the job of biologists to solve puzzles such as the one he poses—to identify and try to understand those things that influence life.

Nemerov asks, "Why do gingko trees drop all their leaves at once?" To find an answer to questions such as this, biologists and other scientists pose *possible* answers, or alternative hypotheses, and test them to see which answers are upheld by evidence. Tests of alternative hypotheses are experiments. To learn why the gingko trees drop all their leaves simultaneously, a scientist might formulate several alternative possible hypotheses:

Hypothesis 1: Gingko trees possess an internal clock that times the release of leaves to match the season. On the day Nemerov describes, this clock sends a "drop" signal (perhaps a chemical) to all the leaves at the same time.

Hypothesis 2: The individual leaves of gingko trees are each able to sense day length, and when the days in the fall attain a specific ratio of day to night, each leaf responds independently by falling.

Hypothesis 3: A strong wind arose the night before Nemerov made his observation, blowing all the leaves off the gingko trees.

Next the scientist attempts to eliminate one or more of the hypotheses by conducting an experiment and predicting an expected outcome to this experiment. To test hypothesis 2, the scientist might cover some of the leaves so that these leaves cannot use light to sense day length. The scientist predicts that the covered leaves should not fall when the others do, since they did not receive the same information. Suppose, however, that despite the covering of some of the leaves, all the leaves fall together. The data do not uphold hypothesis 2. Either of the other hypotheses and many others remain as possible explanations.

This simple experiment with gingkos serves to point out the essence of scientific progress: science does not prove that certain explanations are true but can only provide data to uphold or refute them. Hypotheses that are not consistent with experimental results are rejected. Hypotheses that are not proven false by an experiment or series of experiments are provisionally accepted—but they may be rejected in the future when more information becomes available if they are not consistent with the new information.

Generally speaking, a hypothesis that stands the test of time, often tested and never rejected, is called a theory. In biology the hypothesis that tiny organelles (mitochondria) within your cells are descendents of bacteria is a theory. This idea cannot be proven, but the overwhelming weight of evidence supports the hypothesis and most biologists accept this concept. There is no absolute truth in science, however, only varying degrees of uncertainty, and the possibility always remains that future evidence will cause a theory to be revised or discarded. The scientific community's acceptance of a theory is always provisional.

Some theories are so strongly supported that the likelihood of their being rejected in the future is negligible. The theory of evolution, for example, is so broadly supported by different lines of inquiry that most biologists accept it with as much certainty as they do the theory of gravity. The theory of evolution provides the conceptual framework that unifies biology as a science.

As you proceed through this text, you will encounter a great deal of information, often coupled to explanations. These explanations are hypotheses that have stood the test of repeated experiment action. Many will continue to do so; others will be revised. Biology, like all sciences, is in a constant state of flux, with new ideas gaining support and replacing old ones.

drons. Second, you experimented on five plants. Scientists perform experiments over and over again, or **repeated trials,** before drawing conclusions from their data. The Brand X fertilizer manufacturing company probably performs hundreds of experiments on a variety of houseplants before drawing conclusions from their data and generalizing to the population of houseplants grown throughout the United States (or possibly throughout the world).

Summarizing the steps in the scientific method

As you can see from this example, the scientific method is a procedure for problem solving that is appropriate for everyday use as well as for use by professional scientists. In fact, the scientific method is often referred to as *organized common sense*. In summary, the steps of the scientific method are:

- Posing a question
- Forming a hypothesis
- Testing the hypothesis by means of a controlled experiment
- Recording and analyzing the data
- Drawing a conclusion as to whether the data support the hypothesis
- Repeating the work

Evidence that contradicts the prediction of a hypothesis shows it to be false; that is, it disproves the hypothesis. However, evidence that is in keeping with the prediction a hypothesis makes does *not* prove that further testing will also produce supporting evidence. Factors may affect the outcome of a future experiment that may be unforeseen at the time. In your experiment, for example, the data collected using your plants supported your hypothesis. But data collected using other plants in someone else's home may not support your hypothesis. In summary, then, scientists cannot actually *prove* hypotheses; they can only *disprove* hypotheses or *support* hypotheses with evidence.

Theory building

As scientists repeat, or replicate, each other's work, their data may uphold a hypothesis again and again. As the explanations of such consistently supported and related hypotheses are woven together, "grand explanations" that account for existing data and that consistently predict new data are developed. Interwoven hypotheses upheld by overwhelming evidence over time are called **theories.** The word *theory* in a scientific context has a much stronger meaning than the everyday use of the term. A theory is a synthesis of hypotheses that have withstood the test of time and is therefore a powerful concept that helps scientists make dependable predictions about the world. Theories are supported by such an overwhelming weight of evidence that they are accepted as scientifically valid statements. The possibility always remains, however, that future evidence will cause a theory to be revised, since scientific knowledge grows and changes as new data are collected, analyzed, and then synthesized with information that came before.

The unifying themes of biology today

Biology is often viewed by the nonscientist as an accumulation of facts, but it is much more than that! As the beginning of this chapter shows, biology (as well as all sciences) is a way, or process, of understanding the world. As a result of observing living things, their environments, and the interactions between them, scientists have posed questions, have put forth hypotheses based on those observations, and have tested their predictions of the living world. Rising out of this abundance of tested predictions are the themes, or accepted explanations, that permeate the science of biology. Although the specific details of these themes may be updated or changed as biologists modify their hypotheses, the themes themselves transcend time.

Living things display both diversity and unity

The diversity, or variety, of living things is astounding. Biologists estimate that from 5 million to 30 million *different* species exist on Earth. A **species** is a group of related organisms that share common characteristics and are able to interbreed, producing offspring that can reproduce. Only a sampling of the array of the species humans have seen and categorized is shown in Figure 1-2. Although these organisms are so very different from one another, each displays characteristics that are common among all species. The remaining themes describe ways in which organisms are diverse in their types but unified in their patterns, and the reason this is so.

Living things are composed of cells and are hierarchically organized

A cell is a microscopic mass of protoplasm, a chemically active mixture of complex substances suspended in water that is bounded by a membrane and contains hereditary material. Until the invention of the microscope around 1600, naturalists could not probe this invisible level of organization of living things. And it was not until the mid-1800s when advances in the technology of the microscope allowed botanist Matthias Schleiden and zoologist Theodor Schwann to determine (through repeated lines of inquiry) that the unit of structure of all living things is the cell. Shortly thereafter, the German medical microscopist Rudolf Virchow argued that all cells can arise only from preexisting cells. The cells, he wrote, are "the last constant link in the great chain of mutually subordinated formations that form tissues, organs, systems, the individual. Below them is nothing but change."

Virchow's statement refers to the hierarchy, or levels, of organization seen in all living things. Living things, or **organisms,** are either multicellular (composed of many cells) or unicellular (composed of a single cell). Some organisms called **colonial organisms** are single-celled organisms that work and live together as a team. But whether unicellular, multicellular, or colonial, every organism has the cell as its

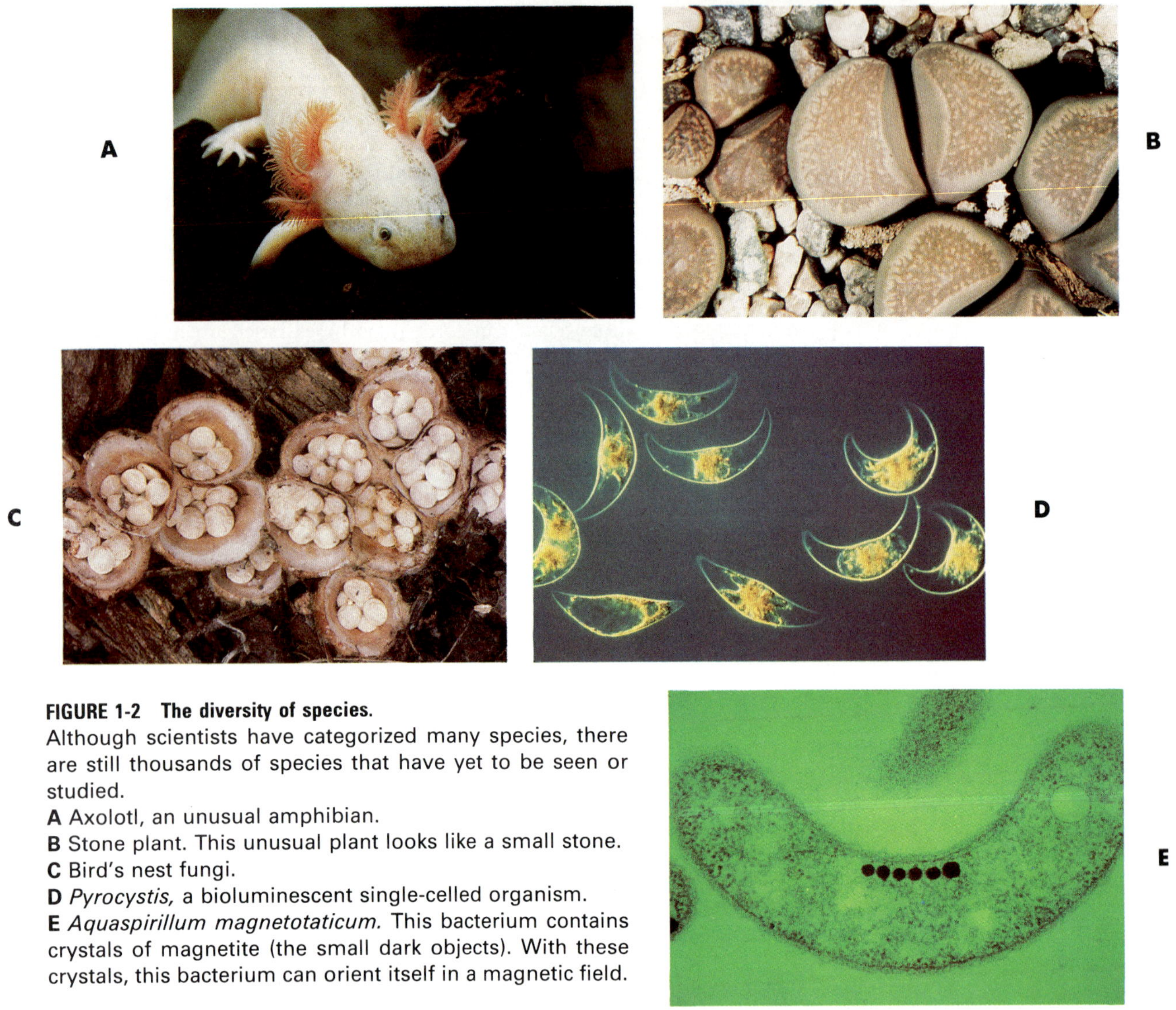

FIGURE 1-2 The diversity of species.
Although scientists have categorized many species, there are still thousands of species that have yet to be seen or studied.
A Axolotl, an unusual amphibian.
B Stone plant. This unusual plant looks like a small stone.
C Bird's nest fungi.
D *Pyrocystis,* a bioluminescent single-celled organism.
E *Aquaspirillum magnetotaticum.* This bacterium contains crystals of magnetite (the small dark objects). With these crystals, this bacterium can orient itself in a magnetic field.

"lowest level" of structure and function. Smaller units, such as atoms and molecules, make up cells but are not *living* units.

As shown in Figure 1-3, cells in multicellular organisms are organized to form the structures and perform the functions of the organism. The next "highest," or more inclusive, level of organization is the tissue. **Tissues** are groups of similar cells that work together to perform a function. Grouped together, tissues form a structural and functional unit called an **organ.** An **organ system** is a group of organs that function together to carry out the principal activities of the organism—its highest level of organization.

Interactions take place within an organism among its levels of organization. For example, when a fox sees and then chases a mouse, a complex series of events occur within the fox. First, the nerve *cells* embedded in the back wall of its eyes *(organs)* conduct impulses to its brain *(an organ)*. The vision center within the brain (*nervous tissue*) interprets these impulses, resulting in the fox seeing the mouse. Impulses speed to other brain centers *(nervous tissues)*, which integrate this visual message with messages from olfactory *cells* in its nose. The brain coordinates these impulses, then sends out a response in the form of nerve impulses to the muscles *(organs)*. Skeletal muscles contract, allowing the animal to chase after its prey.

Living things interact with each other and with their environments

Interactions occur not only among the levels of organization within organisms but also between organisms and their external environments. Biologists usually classify the interactions between organisms and their environments into the following hierarchy of levels of organization: populations, communities, ecosystems, and the biosphere. These levels of organization build on the levels of organization of individual organisms and are depicted in Figure 1-3.

A population consists of the individuals of a given species that occur together at one place and at one time. To continue the example, the foxes living in a small forest make up a population. The foxes interact with one another in a variety of ways. Sometimes they compete for the same limited resources and for mates. Conversely, individuals within

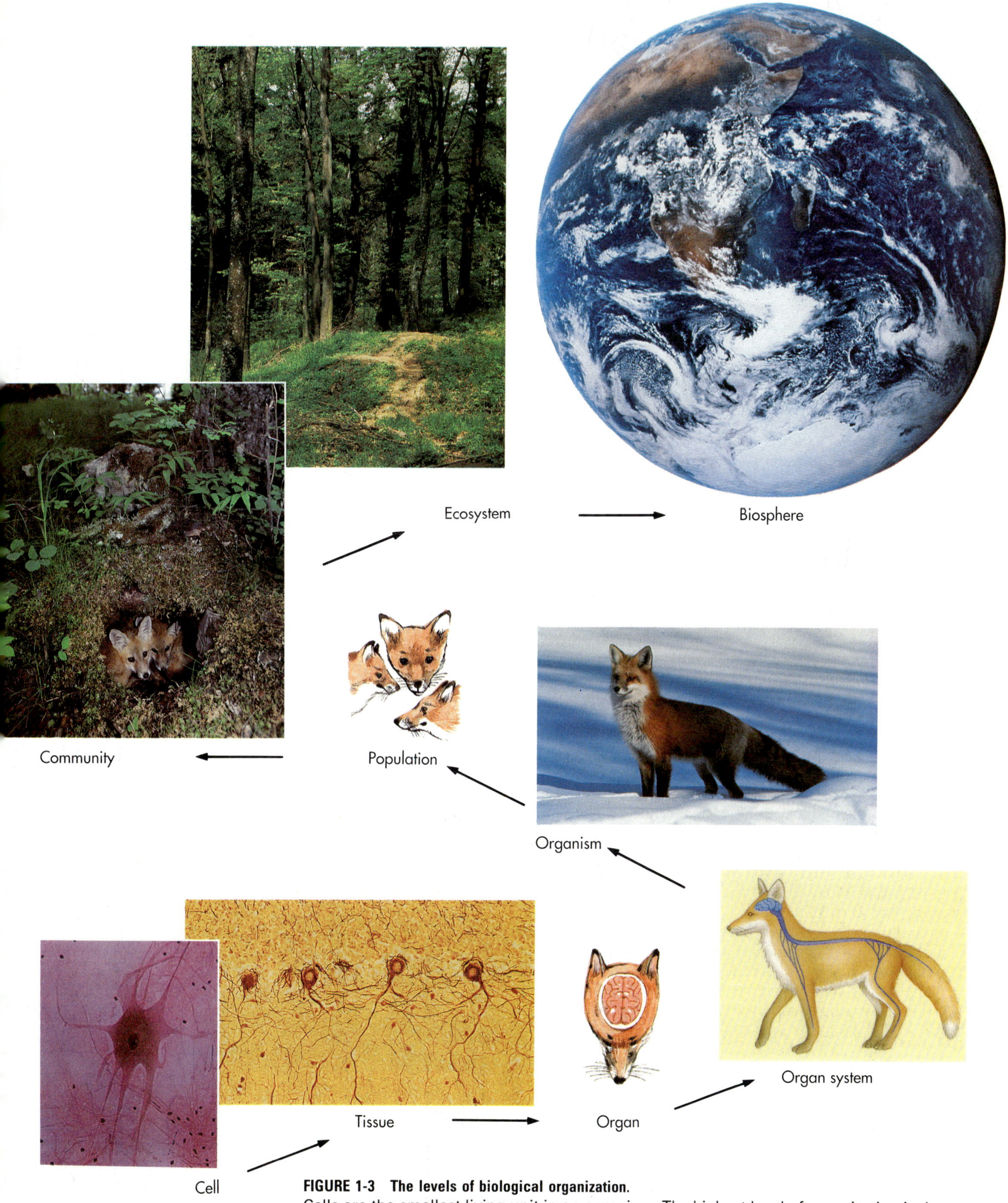

FIGURE 1-3 The levels of biological organization.
Cells are the smallest living unit in an organism. The highest level of organization is the biosphere, which is the area on Earth where biological activity takes place.

animal populations may also work cooperatively for common purposes. Foxes often hunt together, for example, working with one another to overtake and trap prey.

Populations of different species that interact with one another make up a community of organisms. A forest community may be made up of populations of bacteria, fungi, earthworms, plant-eating and animal-eating insects, mice, deer, salamanders, foxes, snakes, hawks, trees, and grasses. These populations within the forest community compete with one another for resources as do organisms within populations. The foxes, snakes, and hawks, for example, compete with one another to capture the mice for food. Other types of interactions may also exist within the community, such as mutualistic relationships in which two different species live in a close association for the benefit of both. Fungi and plants are often interdependent on one another. The fungi live near the roots of many plants, such as trees in a forest. The fungi envelop the roots and send billions of minute cell extensions into the soil. These microscopic "fingers" of fungus absorb water and nutrients better than the roots could alone, and they pass the substances to the plant. In turn, the fungus uses certain products that the plant makes by photosynthesis.

An ecosystem is a community of plants, animals, and microorganisms that interacts with one another and their environments, and that are interdependent on each other for their survival. The forest ecosystem includes all the organisms previously discussed as well as nonliving components of the environment such as air and water, which contribute substances needed for the ecosystem to function.

The biosphere is the part of the Earth where biological activity exists. Most people refer to the biosphere as simply "the environment." Within this global environment, living things interact with each other and with nonliving resources in a myriad of ways. Organisms other than humans use only renewable, or replaceable, resources. When they die, decomposers return the nutrients held within the dead organisms to the soil and air. Humans, however, use many nonrenewable resources and fill the land with wastes. Scientists are testing and refining alternative renewable energy sources such as solar, wind, and water energy to help curb the use of nonrenewable energy sources such as coal, oil, and natural gas. The most severe crisis of natural resource destruction is occurring today in the tropical rain forests. Species are dying out as their rain forest habitats, or homes, are lost. In addition, humans are overpopulating the land and polluting the water and the air. The destiny of future generations of all living things depends on our solving the problems humans have created and our learning to live on this planet without harming it and everyone's future existence.

Living things transform energy and maintain a steady internal environment

Organisms need energy to do the "work" of living. This work involves many processes, such as movement, cell repair, reproduction, and growth. It also involves maintaining a stable internal environment in spite of a differing external environment. Energy "drives" the chemical reactions that underlie all of these activities.

What is the source of the energy that fuels life processes and helps organisms maintain an inner equilibrium? Ultimately, it comes from the sun. For example, certain organisms within our forest ecosystem such as the trees and grasses make their own food by capturing energy from the sun in a process called **photosynthesis.** These organisms are called **producers.** During photosynthesis, producers (primarily green plants) convert the energy in sunlight into chemical energy by locking it within the bonds of the food molecules they synthesize. Organisms that cannot make their own food, such as the insects, mice, deer, salamanders, foxes, snakes, and hawks in the forest, are called **consumers.** They feed on the producers and on each other, passing energy along that was once captured from the sun (Figure 1-4). Both the producers and the consumers release the stored energy in food by breaking down its molecules bit by bit. As food molecules are broken down, much of the energy that is released is used to do work, but some of it is lost as heat and is therefore unusable. Consequently, organisms need a continual input of energy to fuel the chains of chemical reactions that move, store, and free energy needed to perform the activities of life. Decomposers such as the bacteria and fungi break down the organic molecules of dead organisms, serving as the last link in the flow of energy through an ecosystem and contributing to the recycling of nutrients within the environment.

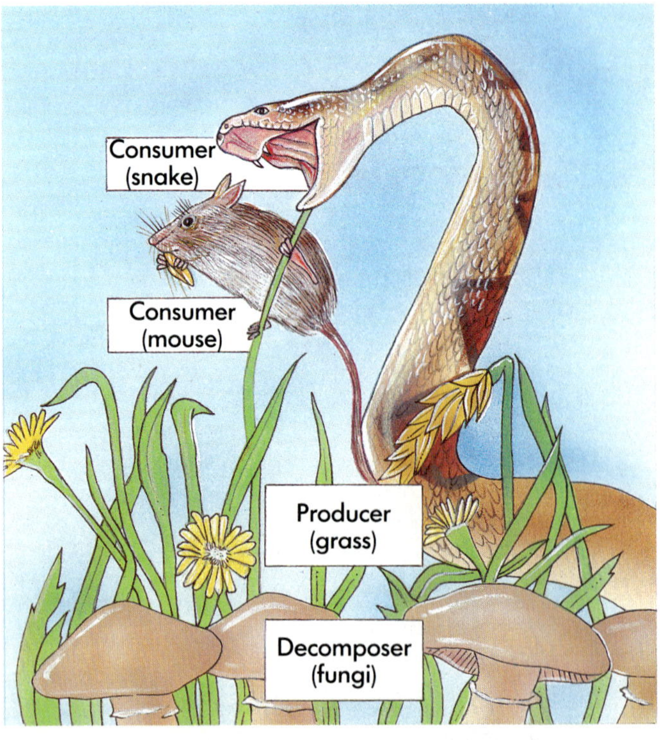

FIGURE 1-4 An energy pyramid in a forest ecosystem. Producers make their own food by means of photosynthesis. Consumers such as mice eat producers. Some consumers such as snakes eat other consumers. Decomposers, such as the fungi, decompose dead organisms and recycle their nutrients within the environment.

Living things exhibit forms that fit their functions

Would it make sense to try to turn a screw with a hammer or eat soup with a knife? Of course not—tools and kitchen utensils are structured in specific ways to do specific jobs. Just as the shape and structure of a spoon or a screwdriver fits its function, the structures of living things fit their functions. Biologists sum up this idea with the phrase "form fits function." By analyzing form, inferences can be made regarding function. Conversely, knowing function gives insights into form.

For example, look at the variety of bird feathers shown in Figure 1-5. By analyzing their forms, determine which is used to insulate the bird against the cold. First, determine the characteristics of a good insulator from your everyday experience. Are you warmer in cold weather when you wear one thick sweater, or are you warmer when you wear many thinner layers that together may have the same thickness as the sweater? The answer is that many layers provide better insulation against the cold because they trap air between the layers. This trapped air is then warmed by the heat radiating from your body. In essence, you create a "blanket" of warm air around your body. Looking at the bird feathers, which feather might trap air within its structure?

A

B

C

FIGURE 1-5 Different types of feathers: form fits function. By examining the feathers shown here, can you determine which feather is the best insulator? The feather in **A** is used for flight. The feather in **B** is a display feather and is used by the male peacock to attract females. The feather in **C** is a down feather and used for insulation. Arranged in layers, these feathers trap air so that it can be warmed by the bird's body.

Living things reproduce and pass on biological information to their offspring

All organisms give rise to organisms of their same kind—they reproduce. They reproduce both sexually and asexually. In sexual reproduction, two parents give rise to offspring; in asexual reproduction, only one parent gives rise to offspring. In either case, genes—the units of heredity—are passed from one generation to the next.

Genes are made up of molecules of DNA, or deoxyribonucleic acid. In these molecules of DNA lies the "code of life," instructions that are translated into a working organism. Interestingly, these instructions take the form of molecular subunits of the DNA molecule called **nucleotides.** Using only four different nucleotides, DNA codes for all the structural and functional components of an organism.

The secret to DNA's ability to carry information regarding the variety of structural and functional components lies in its code. The four nucleotides of DNA are used like code letters to produce code words, each composed of three letters. These code words are then sequenced to produce code sentences, which translated in a living organism results in the formation of a specific molecule designed to do a specific job. Thus the diversity of life is produced by the same code letters that produce code words and code sentences unique to each of the many species of organisms of the living world.

Living things change over time, or evolve

All organisms have common characteristics because they are related to one another. Just as you have a history and a family tree, so does the Earth's family of organisms. Yet all organisms have differences because as they changed over time, they diverged from one another. The Earth itself has changed from its beginnings some 4.6 billion years ago.

Scientists know very little about what "incubator Earth" was like nearly 4 billion years ago, but they do agree that it was a harsh environment different from today's environment. Under these conditions, many scientists hypothesize that the elements and simple compounds of the primitive atmosphere reacted with one another, forming complex molecules. In a way that scientists can only hypothesize, biochemical change took place over time and resulted in the appearance of single-celled organisms approximately 3.5 billion years ago. The remains of these early cells (and of any organisms) preserved in rocks are called **fossils.** Fossils provide scientists with a record of the history of life and document the changes in living things that have taken place over billions of years. By 1.4 to 1.2 billion years ago, the fossil record documents the existence of cells more complex than the first cells, and by 500 million years ago, an abundance of multicellular organisms—with members of groups similar to those that exist today—had appeared.

The fossil record is only one piece of evidence suggesting that living things have changed over time. Using this and other types of evidence, Charles Darwin, a nineteenth century English naturalist, developed what was then a hypothesis of organismal change over time or, as he put it, "descent with modification." He termed his hypothesis **evolution.** Darwin also proposed a mechanism by which evolution took place. Since Darwin's time, his hypothesis has been consistently supported by an overwhelming amount of scientific data and has therefore become a well-accepted theory. This theory embodies the ideas that organisms alive today are descendants of organisms that lived long ago and that organisms have changed and diverged from one another over billions of years. Scientists still ponder, examine, and develop hypotheses regarding details of the mechanisms of evolution but agree that evolution has and is taking place.

Classification: A reflection of evolutionary history

As in any family tree, some organisms are more closely related than others. The species of the world are no exception. In the family trees of species, organisms having a common ancestor in the not-too-distant past are said to be *closely related.* Organisms having a common ancestor further down the family tree are said to be *distantly related.* Modern **taxonomy,** the classification of the diverse array of species, categorizes organisms based on their common ancestry (their evolutionary history) while taking into account their existing similarities and differences. Therefore organisms that are close relatives and resemble one another in a variety of ways are placed in common groupings that reflect their similarities and closeness; organisms that are not closely related are placed in groupings separate from one another that reflect their differences.

Taxonomists group all living things into one of five broad categories, called kingdoms: Monera, Protista, Plantae, Fungi, and Animalia (Figure 1-6). Kingdom Monera includes the bacteria. This kingdom differs from the other four kingdoms in that its cells lack membrane-bounded intracellular structures called **organelles.** In addition, the hereditary material of the cell is not bounded by a membrane. Such cells are called **prokaryotes.** Finding these organisms

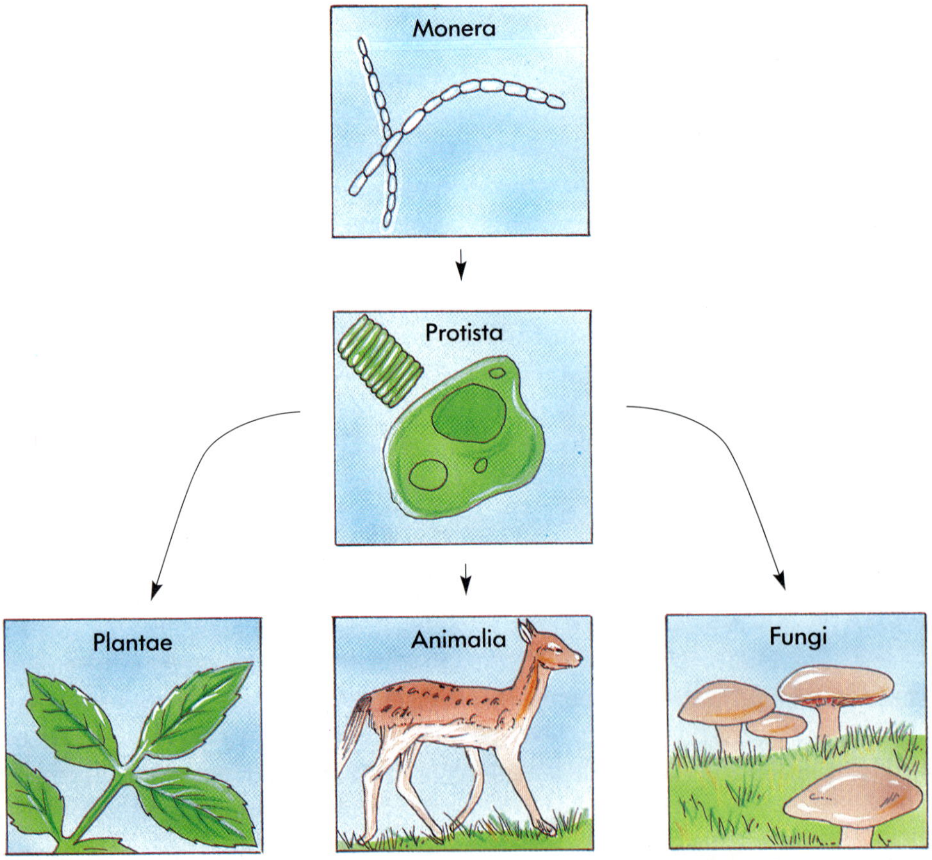

FIGURE 1-6 The five kingdoms of life.
All life can be categorized into one of these five broad groups. The *arrows* indicate the direction of evolutionary development.

is an easy task—bacteria live almost everywhere. If you took a moist cotton swab and drew it across any surface (except one that had been sterilized) you will collect an array of bacteria that is visible under a microscope. The organisms of the other four kingdoms are made up of cells that do have organelles and a membrane-bounded nucleus. Such cells are called **eukaryotes.**

Ancestors of the monerans gave rise to organisms in the kingdom Protista. The protist kingdom consists of primarily single-celled eukaryotes, whose cells are much more complex and much larger than the bacteria. By collecting a small amount of pond water and looking at it under a microscope, you can see some of the organisms in this kingdom. Some of the differences between the single-celled protists and the bacteria can be easily seen in Figure 1-6. This kingdom also includes the algae, which include both multicellular and unicellular organisms that seem to be more closely related to the protists than to the plants, animals, or fungi.

The remaining three kingdoms originated from the protists. The plants are multicellular organisms that live on land and make their own food by using carbon dioxide from the air and light energy from the sun. Examples of plants are mosses, ferns, pine trees, and flowering plants. Animals are multicellular organisms that cannot make their own food and that obtain food by eating and digesting other organisms. Fungi are decomposers (as are the bacteria) and survive by breaking down substances and absorbing the breakdown products. Many of these multicelluar eukaryotes live off organisms that are no longer alive, such as fallen trees and leaves and dead animals. Some fungi, however, attack living organisms, causing human diseases or conditions such as athlete's foot and ringworm or plant diseases such as potato blight.

The kingdoms of organisms are further subdivided into groupings that reflect an increasing closeness in evolutionary history, as well as organisms' closeness in their existing behavioral (in the case of animals) and biochemical characteristics. Kingdoms Monera, Plantae, and Fungi are each subdivided into **divisions.** Kingdoms Protista and Animalia are each subdivided into **phyla** (singular, phylum). The animal kingdom, for example, has approximately 19 phyla depending on the particular classification system that is referenced. Figure 1-7 shows some representatives of the animal kingdom. One of the phyla within this kingdom is the phylum Chordata: animals with a backbone. Notice that the starfish shown as a member of the kingdom Animalia is not included in phylum Chordata, yet all the other animals originally shown are included. These animals are more closely related to one another than to the starfish.

The next subcategory is **class.** The class Mammalia is a subgroup of phylum Chordata. The snake is no longer included in this group because it is not a mammal. Mammals are characterized by having skin with hair and by nourishing their young with milk secreted by mammary glands. The next subcategory is **order.** One order of class Mammalia is the order Rodentia, which includes most gnawing mammals. Humans are not gnawing animals and are not included in this subgroup. The last three subgroupings, in order of increasing relatedness, are **family, genus,** and **species.** Of all the organisms shown, which one is most closely related to squirrels?

FIGURE 1-7 Classification scheme for a squirrel. The classification of organisms reflects their evolutionary, behavioral, and biochemical closeness. Organisms of the same genus are the most closely related.

The last two categories in the hierarchy of classification, genus and species, provide the scientific name of an organism. This system is known as **binomial nomenclature**—literally, "two-name naming." These names are usually Latin names that describe the organism. In this way, scientists worldwide have a common language and a reference for organisms that have been described and named. For example, humans are *Homo sapiens,* meaning "wise man." House cats are *Felis domesticus,* meaning "domesticated cat." Notice that the genus name is capitalized and the species name begins with a lower case letter. Both are italicized. What is the scientific name for the squirrel in our example?

Biology today and you

Learning about biology today is more than learning the scientific names for the organisms around you or learning biological "facts." It also involves learning about problem solving, how scientists go about their work, and worlds you may never have known existed, such as the world of microbes, the world of overpopulation, and the world of modern gene technology. Learning about biology today is learning how to protect the planet and how to take care of your body. But most of all, learning about biology today is learning about yourself—your evolutionary roots and your connectedness to all things living on this Earth.

Summary

1. All scientists study the natural world by using the scientific method. The scientific method is a process, or series of steps, scientists use to answer the questions they ask. Scientists ask questions that can be investigated by means of experimentation, producing results that are open to verification by others in their discipline.
2. To answer their questions, scientists develop hypotheses, or plausible answers to their questions, by generalizing from their observations of the world and information learned from scientific literature. Scientists then test their hypotheses by means of controlled experiments in which they keep all variables constant except the one being tested. An experiment tests a hypothesis by producing data that either support or refute the prediction made in the hypothesis.
3. As they weave together explanations of consistently supported and related hypotheses, scientists develop grand explanations that account for existing data and that consistently predict new data. Theories are interwoven hypotheses upheld by overwhelming evidence over time. Theories are powerful concepts that help scientists make dependable predictions about the world.
4. Although scientific knowledge grows and changes, certain themes, or accepted explanations, rise out of the abundance of tested predictions and permeate the science of biology. Although the specific details of these themes may be updated or changed as biologists modify their hypotheses, the themes themselves transcend time.
5. Seven themes of life are described in this textbook:
   a. Living things display both diversity and unity.
   b. Living things are composed of cells and are hierarchically organized.
   c. Living things interact with each other and their environments.
   d. Living things transform energy and maintain a steady internal environment.
   e. Living things exhibit forms that fit their functions.
   f. Living things reproduce and pass on biological information to their offspring.
   g. Living things change over time, or evolve.
6. Modern taxonomy, the classification of the diverse array of species, categorizes organisms based on their common ancestry (their evolutionary history) while considering their existing similarities and differences. Therefore organisms that are close relatives and resemble one another in a variety of ways are placed in common groupings that reflect their similarities and closeness. Organisms that are not closely related are placed in groupings separate from one another that reflect their differences.
7. Taxonomists group all living things into one of five broad categories, or kingdoms: Monera, Protista, Plantae, Fungi, and Animalia. Ancestors of the monerans (the bacteria) gave rise to organisms in the kingdom Protista (single-celled eukaryotes and the algae). Ancestors to the protists gave rise to the remaining three kingdoms of organisms: plants, animals, and fungi. The kingdoms of organisms are further subdivided into groupings that reflect an increasing closeness in evolutionary history as well as other characteristics.

REVIEW QUESTIONS

1. Describe three different topics that are all part of the science of biology.
2. Distinguish between inductive and deductive reasoning. Give an example of each from your daily life.
3. You've been hearing about how regular exercise (say, walking for 1 hour 4 days a week) could help people lose weight. You decide to do a series of controlled experiments to test this idea and persuade 50 students at your school to participate. State your hypothesis and identify the independent and dependent variables.
4. What is a control? How would you set up the control in the preceding experiment?
5. What is the scientific method? Summarize its steps.
6. The text states that scientists cannot actually prove hypotheses to be true. Explain this statement.
7. What is a theory? Can scientists state with certainty that a theory is true? Explain your answer.
8. Summarize the seven unifying themes of biology.
9. Distinguish among the four levels of organization found in the body of a living multicellular organism.
10. Distinguish among the four levels of the hierarchy that describes how organisms interact with their environments.
11. How do human beings' interactions with the biosphere differ from those of all other organisms?
12. From what ultimate source do you obtain the energy that keeps you alive? Explain how you obtain energy from that source.
13. What is DNA? Explain its role in living things.
14. Briefly state the theory of evolution. What are fossils, and how do they relate to this theory?
15. Name the five kingdoms of living things, and give an example of an organism from each. To which kingdom do you belong?
16. Place the following terms in their correct sequence, starting with the term that refers to the group containing the largest number of organisms: phylum, species, genus, order, family, class, kindom.
17. Suppose that two organisms are members of the same order. Are the following statements true or false? If they're false, explain why:
    a. You can assume the organisms belong to the same class.
    b. You can assume the organisms belong to the same genus.
    c. You can assume the organisms will have the same "first name" in their binomial nomenclature.

KEY TERMS

binomial nomenclature 14
controlled experiments 5
deductive reasoning 5
dependent variable 5
evolution 12
hypotheses 3
independent variable 3
inductive reasoning 3
natural history 3
naturalists 3
organ 8
organ system 8
organism 7
qualitative data 5
quantitative data 5
scientific method 3
species 13
taxonomy 12
theories 7
tissues 8

THOUGHT QUESTIONS

1. If all living things are composed of one or more cells, it follows that a computer-driven machine cannot be alive. And yet such a machine can transform energy, do work, make copies of itself, and evolve over time to better suit the challenges of its environment. In a recent series of movies, such machines even attempt to replace humans as the world's dominant "life form." What is your opinion of whether machines can be alive? If you feel they can be, what *is* life?
2. Though all life on the Earth's surface gets its energy from the sun, deep-sea vents (spots on the ocean floor where heated water forces its way through the crust) create ecosystems on the ocean bottom below where light can ever penetrate. Where does the energy that powers these ecosystems come from?

FOR FURTHER READING

Gould, S. (1987, January). Darwinism defined: The difference between fact and theory. *Discover,* pp. 64-70.
This is a clear account of what biologists do and do not mean when they refer to the theory of evolution.

Gould, S. (1989). *Wonderful life: The Burgess Shale and the nature of history.* New York: W.W. Norton & Company, Inc.
This marvelous book is about the early evolution of animals and about evolution in general.

2 The chemistry of life

Water . . . without it you would die. In fact, life on Earth could not go on if deprived of this amazing liquid, and evolution could never have taken place without it. What is it about water that makes it so important to life?

The clues needed to answer this question lie in the chemical structure of water. Put another way, the importance of water is due to its characteristics, which in turn depend on its "parts" and how they are put together. In the photo, you can easily observe some of these characteristics. Notice how the water "sticks together" to form droplets in the air. Notice, too, that its surface behaves like a thin film, caving in and bulging out in tiny waves as the airborne droplets fall on it. The story of how the parts of water relate to its characteristics, and the answer to why water is important to you, will soon be told—as you begin your study of the chemistry of life.

HIGHLIGHTS

▼ The chemical interactions of atoms, the smallest units of elements, are largely determined by the distribution of their electrons.

▼ Atoms may combine by means of chemical bonds to form molecules.

▼ Water has properties that are extremely important to life.

▼ The main groups of biological molecules are carbohydrates, lipids, proteins, and nucleic acids.

OUTLINE

Atoms
Molecules
- Nature of the chemical bond
- Ions and ionic bonds
- Covalent bonds

The cradle of life: Water
- Water is a powerful solvent
- Water organizes nonpolar molecules
- Water ionizes

Inorganic versus organic chemistry
The chemical building blocks of life
- Carbohydrates
- Fats and lipids
- Proteins
- Nucleic acids

Atoms

Chemistry is the science of matter, the physical material that makes up everything in the universe. Matter is anything that takes up space and has a measurable amount of substance, or mass. All matter (including water) is made up of tiny parts called **atoms**. Although scientists know a great deal about atoms, a simplified explanation provides a good starting point in understanding their complex structures. One explanation was proposed in 1913 by the Danish physicist Niels Bohr. Bohr proposed that every atom is made up of particles tinier than the atom itself. These subatomic particles are of three types: **protons, neutrons,** and **electrons.** Protons and neutrons are found at the core, or nucleus, of the atom. Electrons orbit the nucleus.

Protons have mass and carry a positive (+) charge. Neutrons, although similar to protons in mass, are neutral and carry no charge. Electrons have very little mass and carry a negative (−) charge. They have so little mass, in fact, that they contribute only about one eyelash's worth of mass to your entire body! For this reason, the *atomic mass* of an atom is defined as the combined mass of all its protons and neutrons without regard to its electrons.

The number of protons in an atom, called the *atomic number*, is the same as the number of electrons in that atom. Atoms are therefore electrically neutral; the positive charges of the protons are balanced by the negative charges of the electrons. The number of neutrons in an atom, however, may not equal the number of protons. Atoms that have the same number of protons but different numbers of neutrons are called **isotopes** (Figure 2-1). Isotopes of an atom differ in atomic mass but have similar chemical properties because they have the same number of electrons. The properties of an atom are determined by its electrons because atoms interact with one another by means of their electrons and not their protons or neutrons.

An atom is a core (nucleus) of protons and neutrons surrounded by orbiting electrons. The electrons largely determine the chemical properties of an atom.

The key to the chemical behavior of atoms lies in the arrangement of their electrons. Although scientists cannot precisely locate the position of any individual electron at a particular time, they can predict where an electron is most likely to be. This volume of space around a nucleus where an electron is most likely to be found is called an **orbital** of that electron.

Atoms can have many electron orbitals. Some are simple spheres enclosing the nucleus like a wrapper, whereas others resemble dumbbells and other complex shapes. Each orbital of an atom contains one or two electrons. Atoms having more than two electrons have more than one orbital. These orbitals can be the same or a different distance from the

A

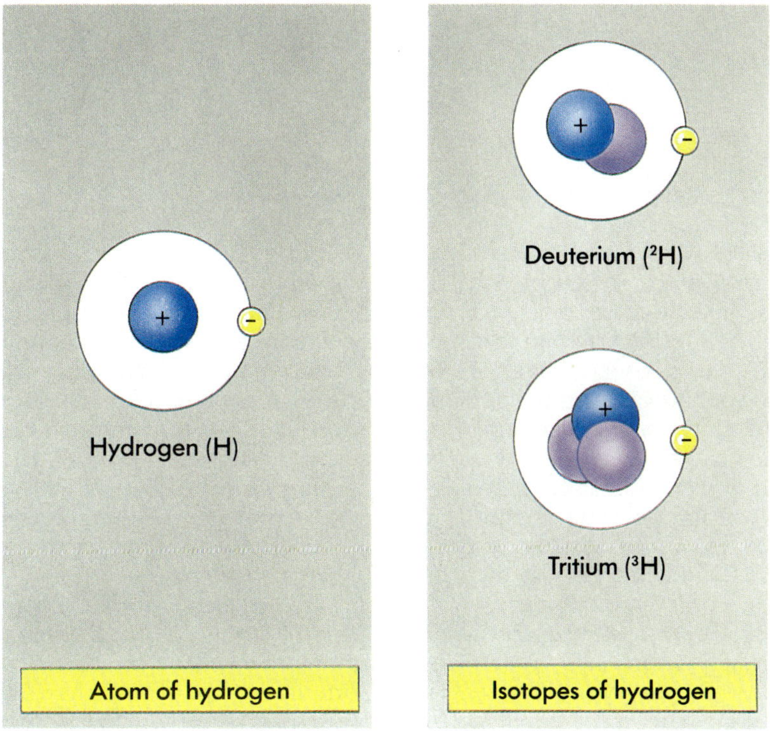

B

FIGURE 2-1 Isotopes.
Isotopes of an atom differ in atomic mass but have the same chemical properties.
A The smallest atom is hydrogen. Its nucleus consists of a single proton.
B Two naturally occurring isotopes of hydrogen also exist. Both isotopes have a single proton in their nucleus (like hydrogen) but have different numbers of neutrons. Deuterium has one neutron, and tritium has two.

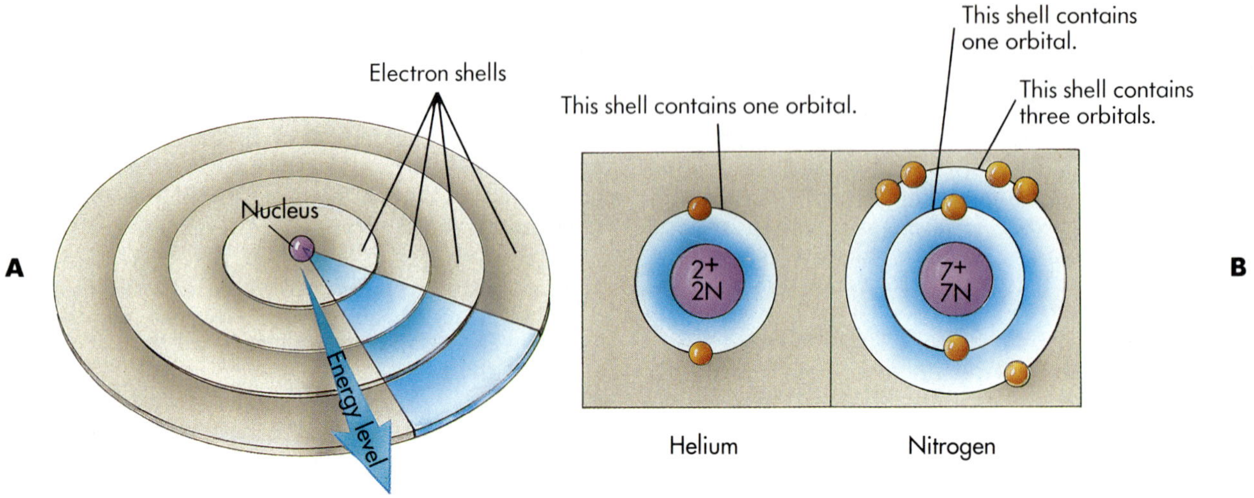

FIGURE 2-2 How atoms are organized.
A Electron shells, or energy levels, in an atom. Each concentric circle represents a different distance from the nucleus and thus a different electron energy level.
B Electron shells for helium and nitrogen. The electrons are indicated by orange dots. Within each shell, pairs of electrons occupy orbitals, which are not shown in this diagram because of their complex shapes. For example, the innermost shell of nitrogen has one orbital containing two electrons. The outermost shell has three orbitals—two of these orbitals have two electrons each, and one orbital has only one electron, for a total of five electrons in the outermost shell.

nucleus. Orbitals at the same distance from the nucleus occupy the same shell.

In Figure 2-2, *A* the nucleus is shown as a small circle surrounded by concentric rings. These rings represent electron shells. In Figure 2-2, *B* the electrons are clustered to signify that they occupy different orbitals within that shell. The actual three-dimensional shapes of the orbitals are not shown. Notice that the atom of nitrogen pictured in Figure 2-2, *B* has electrons occupying two shells. The innermost shell has one orbital containing two electrons. The second shell has three orbitals; two of these orbitals have two electrons each and one orbital contains only one electron. This outer shell of nitrogen, then, has a total of five electrons.

Because the energy of electrons increases as their distance from the attractive force of the nucleus increases, the various electron shells of atoms are also called **energy levels**. Electrons occupying increasingly distant shells from the nucleus have a step-wise increase in their levels of energy. The energy of electrons at the various energy levels is also influenced by the number of protons in the nucleus, since the attractive force of the nucleus increases as its number of protons increases.

The farther an electron is from the nucleus, the more energy it has.

Molecules

The identity of an atom is determined by its number of protons and electrons. For example, an atom containing two protons and two electrons is helium, a gas you have probably seen used to blow up balloons. An atom possessing seven protons and seven electrons is nitrogen (Figure 2-2, *B*).

Elements are pure substances that are made up of a single kind of atom and that cannot be separated into different substances by ordinary chemical methods. The atoms of most elements interact with one another to form **molecules,** combinations of tightly bound atoms. Molecules can be made up of atoms of the same element or atoms of different elements.

The oxygen in the air consists of molecules made up of pairs of atoms of the same element—oxygen. These pairs are represented by the **chemical formula** O_2. A chemical formula is a type of "shorthand" used to describe a molecule. The atoms are represented by symbols, such as *O* for oxygen. Other chemical symbols are shown in Figure 2-2. A subscript shows the number of these atoms present in the molecule. Molecules made up of the atoms of two or more elements are called **compounds.** An example of a compound is water, which is H_2O. In this chemical formula the symbol *H* stands for the element hydrogen.

Three different types of chemical forces influence

whether an atom will interact with other atoms to form molecules. In addition, these forces influence the type of interactions that are likely to take place. These forces are (1) the tendency of electrons to occur in pairs, (2) the tendency of atoms to balance positive and negative charges, and (3) the tendency of the outer shell, or energy level, of electrons to be full. This third chemical force is often called the **octet rule.**

The word *octet* means "eight objects" and refers to the fact that the outer electron shell of most atoms contains a maximum of eight electrons. The first energy level is an exception to this rule because it contains a maximum of two electrons. The octet rule states that an atom with an unfilled outer shell has a tendency to interact with another atom or atoms in ways that will complete this outer shell. Although the octet rule does not apply to all atoms, it does apply to all biologically important ones—those involved in the structure, energy needs, and information systems of living things. These atoms and the molecules they make up are discussed later in this chapter.

The identity of an atom is defined by its number of protons and electrons. Its interaction with other atoms is determined by chemical forces that are influenced by the number and arrangement of its electrons.

Atoms of elements that have equal numbers of protons and electrons, no unpaired electrons, and full outer-electron energy levels are the only ones that exist as single atoms. Atoms with these characteristics are called **noble gases,** or inert gases, because they do not react readily with other elements. Many stories explain the reason they are called noble, but they all center around the concept of nobility—those who have everything (in this case a full outer shell), need nothing and interact little with others. Most of the noble gases—helium, neon, argon, krypton, xenon, and radon—are rare. In addition, they are relatively unreactive and unimportant in living systems. Helium and radon are probably the most well known and the least rare noble gases. Radon, in fact, has gained much notice in recent years. Formed in rock or soil particles from the radioactive decay of radium, radon gas can seep through cracks in basement walls and remain trapped in homes that are not well ventilated. Prolonged exposure to radioactive radon in levels greater than those normally found in the atmosphere is thought to lead to lung cancer.

Nature of the chemical bond

An atom having an incomplete outer shell can satisfy the octet rule in one of three ways:

1. It can gain electrons from another atom.
2. It can lose electrons to another atom.
3. It can share one or more electron pairs with another atom.

Such interactions among atoms result in **chemical bonds,** forces that hold atoms together. If the force is caused by the attraction of oppositely charged particles formed by the gain or loss of electrons, the bond is called **ionic.** If the force is caused by the electrical attraction created by atoms sharing electrons, the bond is called **covalent.** Other, weaker kinds of bonds also occur.

Ions and ionic bonds

Electrons stay in their orbits because they are attracted to the positive charge of the nucleus. However, electrons far from the nucleus are not held as tightly as electrons closer to the nucleus. The electrons more distant from the nucleus are said to be less stable than the electrons closest to the nucleus, and they interact with other atoms more easily than close, stable, tightly held electrons. Therefore atoms typically interact with other atoms by means of the electrons in their outermost (highest) energy levels, or shells. The types of interactions that occur tend to result in atoms with completed outer shells, thus satisfying the octet rule.

An atom tends to lose electrons if it has an outer shell needing many electrons to be complete. An atom tends to gain electrons if it has a nearly completed outer shell. Interestingly, one of these types of atoms tends to interact with the other type, resulting in a completed outer shell for each. The atom with a nearly completed outer shell tends to "take" enough electrons to complete its outer shell from an atom having only one or two atoms in an outer shell that needs eight for completion. Once these outer electrons are gone from the atom giving up electrons, the next shell in becomes its new, complete outer shell. As a result of this interaction, neither atom is electrically neutral; the number of protons no longer equals the number of electrons in either atom. The atom taking on electrons acquires a negative charge. The atom giving up electrons acquires a positive charge. They are not called atoms because they are no longer electrically neutral. Such charged particles are called **ions.**

Figure 2-3 illustrates the formation of ions with a specific example of electron "give and take." Sodium (Na) is an element with 11 protons and 11 electrons. It is a soft, silver-white metal that occurs in nature as a part of compounds. One familiar compound is sodium chloride, or table salt. Of sodium's 11 electrons, 2 are in its innermost energy level (full with 2 electrons), 8 are at the next level, and 1 is at the outer energy level. Because of this distribution of electrons, the outer electron is unpaired ("free") and has a strong tendency to form a pair. In addition, its outer energy level is not full, and therefore the octet rule is not satisfied.

Chlorine (Cl) is an element with 17 protons and 17 electrons. It is a greenish-yellow gas that is poisonous and irritating to the nose and throat. In the compound sodium chloride, however, it does not have these characteristics. Of chlorine's 17 electrons, 2 are at its innermost energy level, 8 at the next energy level, and 7 at the outer energy level. Chlorine, like sodium, has an unpaired electron and an outer energy level that is not full. Sodium and chlorine atoms can interact with one another in a way that results in both having paired electrons and full outer energy levels.

When placed together, the metal sodium and the gas chlorine react explosively. The single, unpaired electrons in the outer energy levels of the sodium atoms fly off, lost to the sodium. These electrons pair with the unpaired electrons

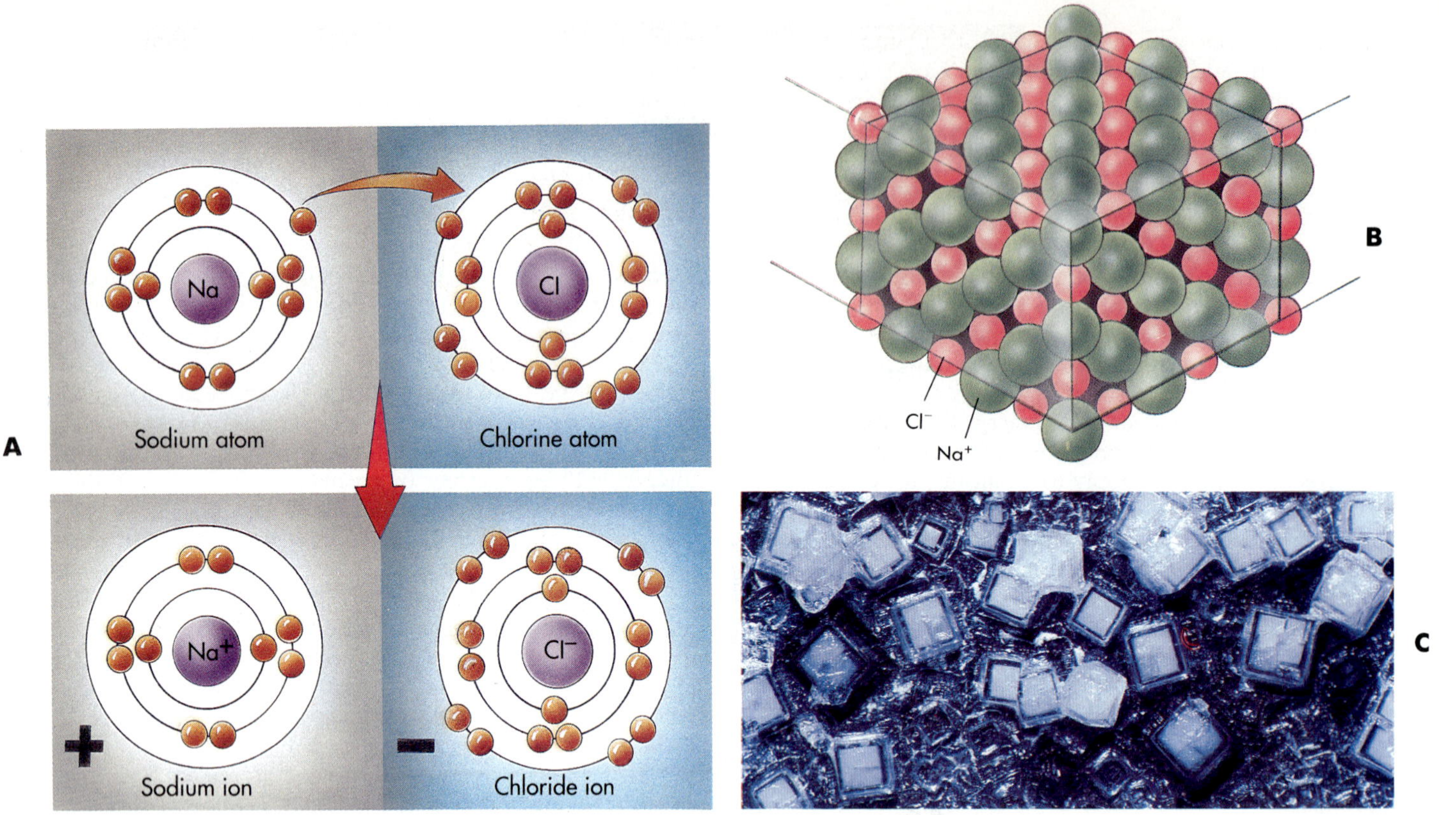

FIGURE 2-3 The formation of an ionic bond.
A When a sodium atom donates an electron to a chlorine atom, the sodium atom, lacking that electron, becomes a positively charged sodium ion. The chlorine atom, having gained an extra electron, becomes a negatively charged chloride ion.
B and **C** Positive and negative ions cluster so that each charge is surrounded by ions of the opposite charge. This is what happens when Na^+ and Cl^- ions come together to form crystals of salt.

in the outer shells of the chlorine atoms. The result is the production of Na^+ and Cl^- ions. Ions, however, are unbalanced with respect to their positive and negative charges. This balance is achieved when these ions come together as their opposite charges attract one another. This type of attraction is called *electrostatic attraction* and results in ionic bonding of the sodium and chloride ions. As these ions are drawn to one another, they form geometrically perfect crystals of salt (Fig. 2-3, *B* and *C*).

An ionic bond is an attraction between ions of opposite charge.

The transfer of electrons between atoms is an important chemical event—one type of chemical reaction. In fact, this type of chemical interaction has a special vocabulary. When an atom loses an electron, it is said to be **oxidized.** The reaction is called an *oxidation,* a term that comes from the original meaning (an alternate definition still valid today), which is "to combine with oxygen." For example, when iron combines with oxygen in the presence of moisture, it becomes oxidized. The product of this oxidation is commonly known as *rust*. Conversely, when an atom gains an electron (or loses oxygen), it is said to be **reduced.** The reaction is called a *reduction*. Oxidations and reductions always occur together because an electron lost from an atom cannot exist by itself—it will be transferred to another atom. Therefore the shortened term **redox reaction** is often used to describe this paired transfer.

The loss of an electron by an atom is called an *oxidation.* The gain of an electron is called a *reduction.* Together, these paired electron transfers between atoms are called *redox reactions.*

Covalent bonds

Covalent bonds form when two atoms share electrons. Hydrogen is a simple example of an atom that usually shares electrons with other atoms. As you can see in Figure 2-4, a hydrogen atom has a single, unpaired electron and an unfilled outer electron shell. A filled outer shell at this energy level requires only two electrons. When hydrogen atoms are close enough to one another, an interesting thing happens.

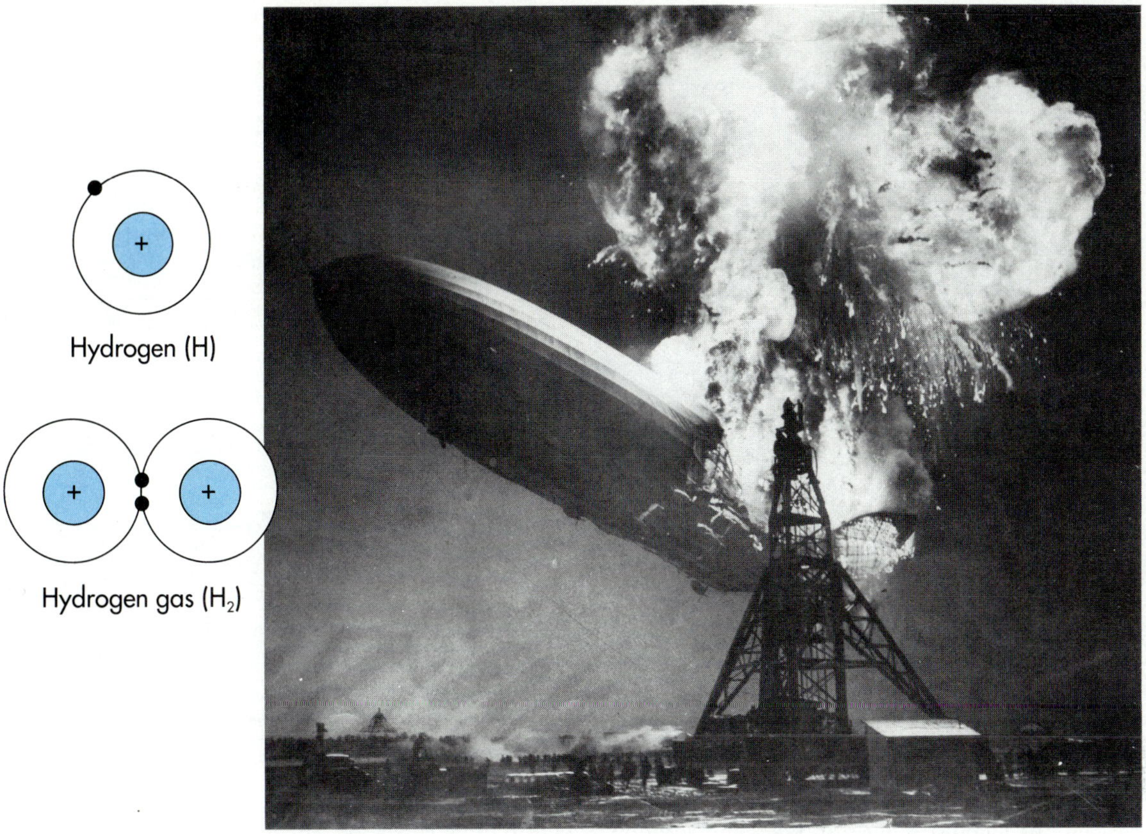

FIGURE 2-4 Covalent bonds.
Hydrogen gas is a diatomic molecule composed of two hydrogen atoms. Each hydrogen atom shares its electrons with another hydrogen atom. A more stable molecule is formed when two hydrogen atoms share their electrons with an oxygen atom, forming water. The flash fire that consumed the *Hindenburg* occurred when the hydrogen gas used to inflate the airship combined explosively with oxygen gas in the air to form water.

They form pairs, with each of their single electrons moving around the two nuclei. These paired atoms of hydrogen are called *diatomic molecules* and are represented by the chemical formula H_2.

As a result of this sharing of electrons, the diatomic hydrogen gas molecule is electrically balanced because it now contains two protons and two electrons. In addition, each hydrogen atom has two orbiting electrons in its outer shell, completing this shell. This relationship also results in the pairing of two free electrons. Thus, by sharing their electrons, the two hydrogen atoms form a stable molecule.

A covalent bond is formed by the sharing of one or more pairs of electrons.

Covalent bonds can be very strong, that is, difficult to break. **Double bonds,** those bonds in which two pairs of electrons are shared, are stronger than **single bonds,** covalent bonds sharing only one pair of electrons. As you might expect, **triple bonds,** those bonds in which three pairs of electrons are shared, are the strongest of these three types of covalent bonds. In chemical formulas that show the structure of covalently bonded molecules, single bonds are represented by a single line between two bonded atoms, double bonds by two lines, and triple bonds by three lines. For example, the structural formula of hydrogen gas is H—H, oxygen gas is O═O, and nitrogen gas is N≡N.

An atom can also form covalent bonds with more than one other atom. Carbon (C), for example, contains six electrons: two at the inner level and the other four in the outer shell. To satisfy the octet rule, it must gain four additional electrons by sharing its four outer-shell electrons with another atom or atoms, forming four covalent bonds. Because there are many ways that four covalent bonds may form, carbon atoms are able to participate in many different kinds of molecules in living systems, such as proteins and carbohydrates.

The "strength" of a covalent bond refers to the amount of energy needed to make or break that bond. The energy that goes into making the bond is held within the bond and is released when the bond is broken. Therefore covalent bonds are actually a storage place for energy as well as a type of chemical "glue" that holds molecules together. Living things store and use energy by means of making and breaking covalent bonds, thereby using molecules as a type of "energy currency."

FIGURE 2-5 Water takes many forms.
As a liquid, it fills our rivers and sometimes falls in great cascades. The iceberg on which the penguins are congregating was formed in Antarctica from huge blocks of ice breaking away into the ocean water. When water cools below 0° C, it forms beautiful crystals. However, water is not always plentiful. In a dry creek bed, there is no hint of water except for the broken patterns of dry mud.

FIGURE 2-6 Water is the cradle of life.
Life most likely originated in the warm waters of ancient Earth. Many kinds of organisms, like these small frogs seen through the transparent walls of their eggs, begin life in water.

The cradle of life: Water

One covalently bonded molecule that plays a major role in living systems is water—H_2O. In fact, water is the most abundant molecule in your body, making up about two thirds of your body weight! Although it seems to be a simple molecule, water has many surprising properties. For example, of all the common molecules on earth, only water exists as a liquid at the earth's surface (Figure 2-5). When life on earth was beginning, this liquid provided a medium in which other molecules could move around and interact. Life evolved as a result of these interactions. And life, as it evolved, maintained these ties to water (Figure 2-6). Three fourths of the earth's surface is covered by water. Where water is plentiful, such as in the tropical rain forests, the land abounds with life. Where water is scarce, such as in the desert, the land seems almost lifeless except after a rainstorm. No plant or animal can grow and reproduce without some amount of water.

The chemistry of life, then, is water chemistry. You might think that such an important molecule with such interesting properties would be complicated and hard to understand. Fortunately, this is not the case. Water has a

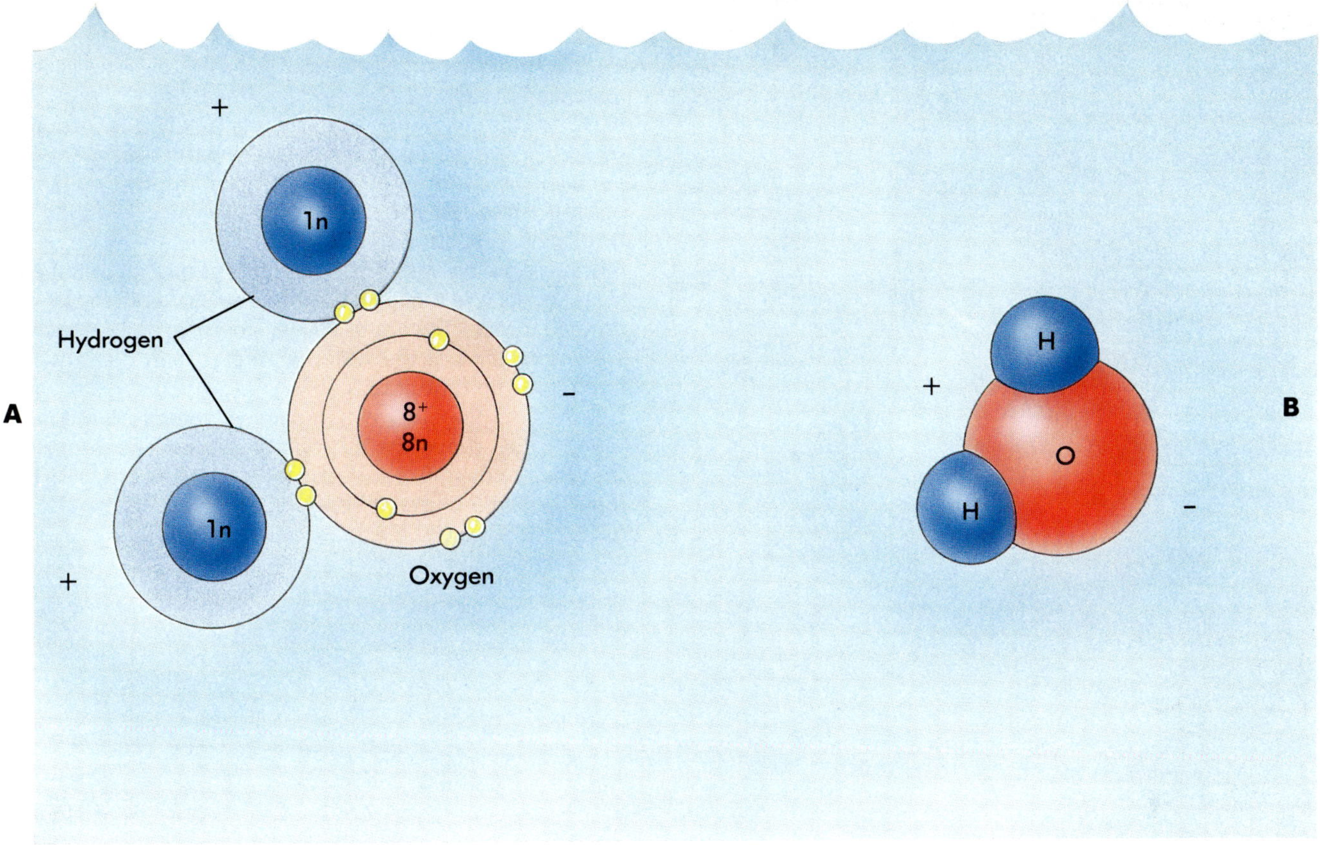

FIGURE 2-7 The structure of water.
A Each water molecule is composed of one oxygen atom and two hydrogen atoms. The oxygen atom shares a pair of electrons with each participating hydrogen atom.
B In this model the + and − represent the partial charges at opposite sides of the water molecule.

simple atomic structure: one oxygen atom bonded by single covalent bonds to two hydrogen atoms. The resulting molecule is stable. It satisfies the octet rule and has no unpaired electrons, and its positive and negative charges are balanced.

This "simple" water molecule has one outstanding chemical property: it attracts other molecules and forms a special type of chemical bond with them. These bonds, called **hydrogen bonds,** have approximately 5% to 10% the strength of covalent bonds. The ability of water to form weak bonds with other molecules is the reason for much of the organization and chemistry of living things.

The story of water's ability to attract other molecules lies in its structure and in the movement of its electrons. The oxygen atom in a water molecule, like all oxygen atoms, contains eight protons in its core, or nucleus. Each hydrogen atom contains only one proton. As a result, the electron pair shared in each of the oxygen-hydrogen covalent bonds of a water molecule is more strongly attracted to the oxygen nucleus than to either of the hydrogen nuclei. Although the electron orbitals surround both the oxygen and hydrogen nuclei, the negatively charged electrons are far more likely to be found near the oxygen nucleus at a given moment than near one of the hydrogen nuclei. Because of this situation, the "oxygen end" of the water molecule acquires a partial negative charge. The "hydrogen end" acquires a partial positive charge (Figure 2-7). Molecules such as water that have opposite partial charges at different ends of the molecule are called **polar molecules.** Water is one of the most polar molecules known.

Polar molecules interact with one another. Hydrogen bonds form when the partial negative charge at one end of the molecule is attracted to the partial positive charge of another polar molecule. In fact, water forms bonds with other water molecules in just this way. Although hydrogen bonds are weak, these bonds are constantly made and broken. (Each lasts only 1/100,000,000,000 of a second!) The cumulative effect of very large numbers of hydrogen bonds is responsible for the many important physical properties of water (Figure 2-8) and is the answer to how water "sticks together" as in the chapter opener photo.

Much of the biologically important behavior of water results because its oxygen atom attracts electrons more strongly than its hydrogen atoms do. As a result, water molecules each have electron-rich (−) and electron-poor (+) regions, giving them positive and negative poles.

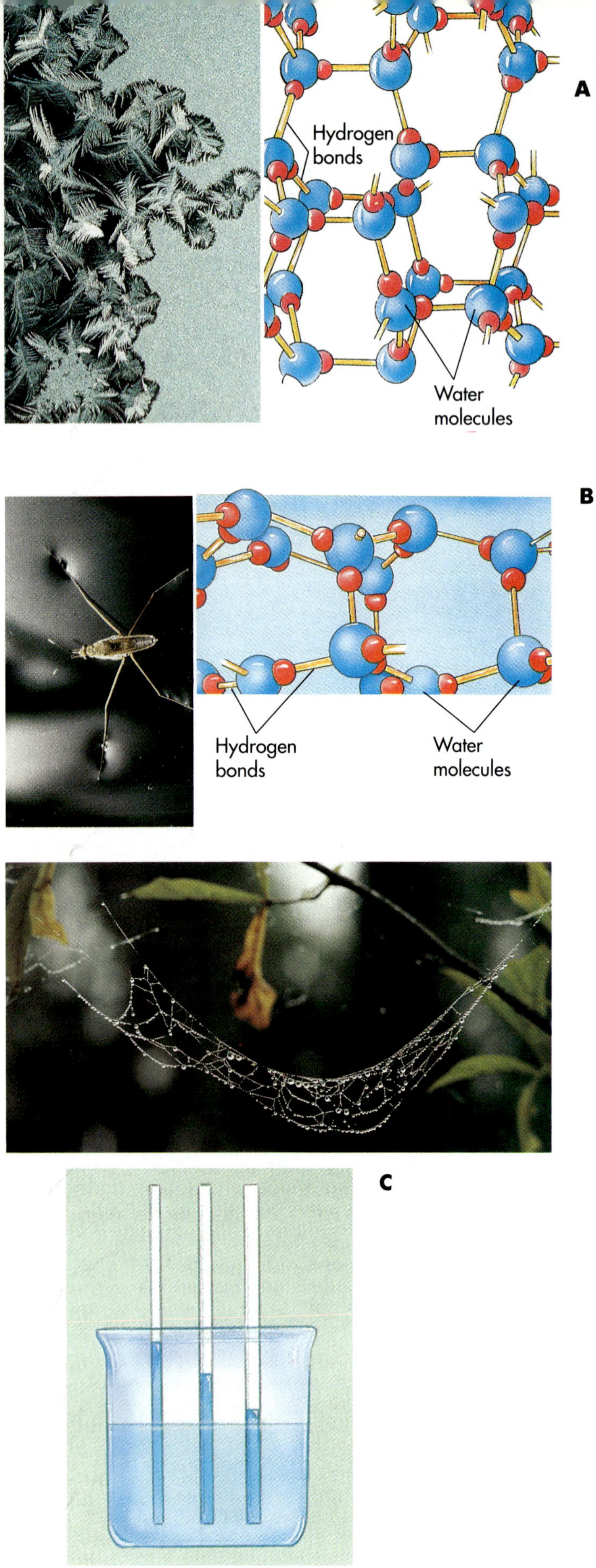

FIGURE 2-8 Many of the physical properties of water depend on hydrogen bonding.
A Ice formation. When water cools below 0° C, it forms a regular crystal structure in which the partial charges of each atom in the water molecule interact with opposite charges of atoms in other water molecules, forming H bonds.
B Surface tension. Some insects, such as the water strider, literally walk on water. In this photograph you can see the dimpling its feet make on the water as its weight bears down on the surface. The water strider does not sink because the surface tension of the water is greater than the downward force that the foot produces.
C Adhesion. Capillary action causes water within a narrow tube to rise above the surrounding fluid. The adhesion of the water to the glass surface, drawing it upward, is stronger than the force of gravity drawing the water down.

Water is a powerful solvent

Water molecules gather closely around any molecule that exhibits an electrical charge, such as ions and polar molecules. For example, sodium chloride (table salt) is made up of the positively charged sodium (Na^+) and negatively charged chloride (Cl^-) ions. These ions are attracted to one another and cluster in a regular pattern, forming crystals. When you put salt in a pot of boiling water, some ions break away from the crystals. Water molecules instantly form hydrogen bonds with these ions, surrounding each of them. This **hydration shell** prevents each ion from associating with other ions. The salt is said to be *dissolved* (Figure 2-9).

Similarly, hydration shells form around all polar molecules and ions. Molecules that dissolve in water this way are said to be **soluble** in water. Chemical interactions readily take place in water because so many kinds of molecules are water soluble and therefore move among water molecules as separate molecules or ions.

Water organizes nonpolar molecules

Remember the old saying that "oil and water don't mix"? This statement is true because oil is a nonpolar molecule and cannot form hydrogen bonds with water. Instead, the water molecules form hydrogen bonds with each other, causing the water to exclude the nonpolar molecules. It is almost as if nonpolar molecules move away from contact with the water. For this reason, nonpolar molecules are referred to as being **hydrophobic.** The word *hydrophobic* comes from the Greek meaning "water" (hydros) "fearing" (phobos).

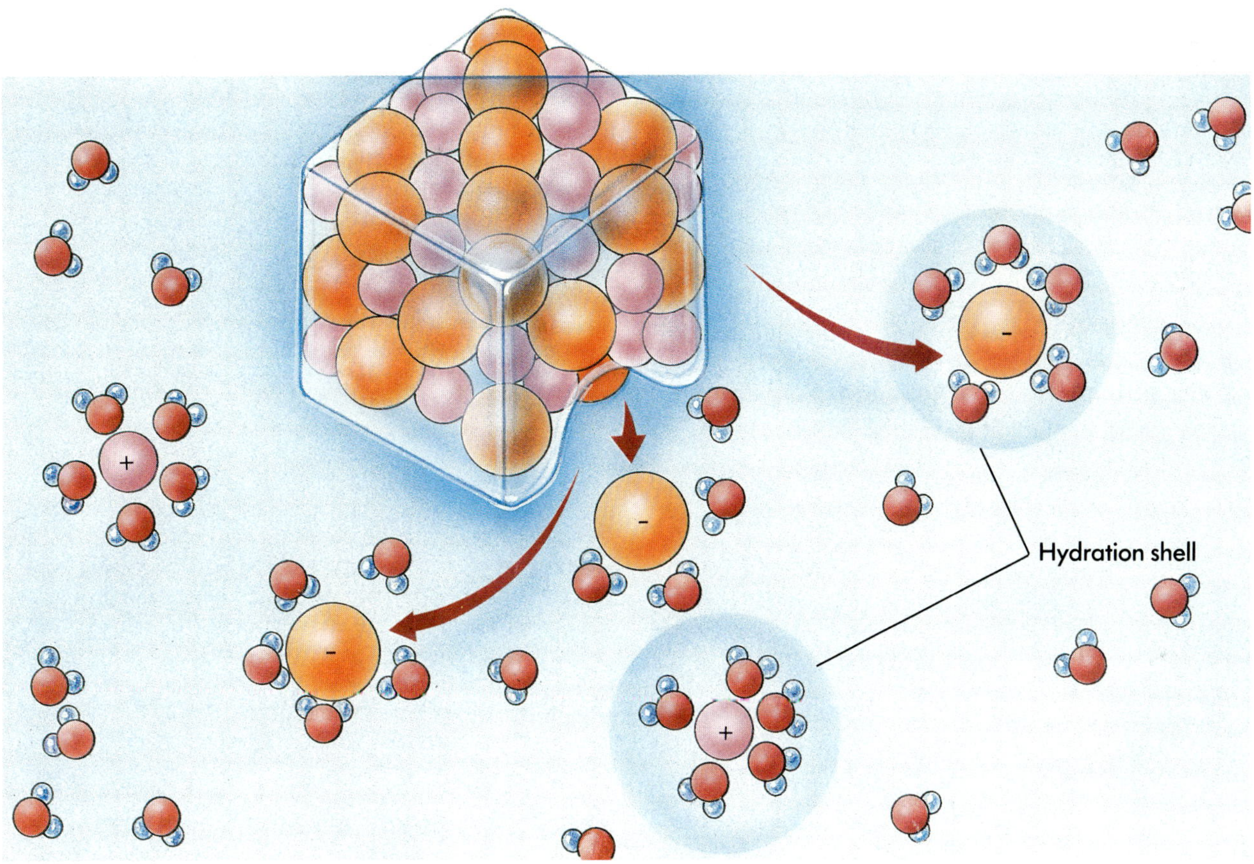

FIGURE 2-9 How salt dissolves in water.
The hydration shells formed by hydrogen bonds prevent individual Na^+ and Cl^- ions from associating. The salt is said to be *dissolved.*

This tendency for nonpolar molecules to band together in a water solution is called **hydrophobic bonding.** Hydrophobic forces determine the three-dimensional shapes of many biological molecules, which are usually surrounded by water within organisms.

Water ionizes

The covalent bonds of water molecules sometimes break spontaneously. When this happens, one hydrogen atom nucleus (a proton) dissociates from the rest of the water molecule, leaving behind its electron. Because its positive charge is no longer balanced by an electron, it is a positively charged hydrogen ion, H^+. The remaining part of the water molecule now has an "extra" electron. It is therefore a negatively charged hydroxyl ion, OH^-. This process of spontaneous ion formation is called **ionization:**

$$H_2O \rightarrow OH^- + H^+$$

Only very few water molecules are ionized at a single instant in time. Scientists calculate that the fraction of water molecules that dissociate (ionize) in pure water is 0.0000001. This tiny number can be written another way by using exponential notation. This is done by counting the number of places to the right of the decimal point. Because there are seven places, this number is written as 10^{-7}. The minus sign means that the number is less than 1.

To indicate the concentration of H^+ ions in a solution, scientists have devised a scale based on the slight degree of spontaneous ionization of water. This scale is called the **pH scale.** The pH values on this scale are expressed as positive numbers, 1 to 14. The pH of a solution is determined by taking the negative value of the exponent of its hydrogen ion concentration. For example, pure water has a hydrogen ion concentration of 10^{-7} and therefore a pH of 7. When water ionizes, hydroxyl ions (OH^-) are produced in a concentration equal to the concentration of hydrogen ions. In pure water the concentrations of H^+ and OH^- ions are equal at 10^{-7}. Because these ions join spontaneously, water, at pH 7, is neutral.

Any substance that dissociates to form H^+ ions when it is dissolved in water is called an **acid.** The more hydrogen ions an acid produces, the stronger an acid it is. Although an acid produces a higher concentration of H^+ ions than pure water (0.00001 as opposed to 0.0000001, for example), its pH is lower. Using the above numbers to illustrate: the first number (0.00001 or 10^{-5}) represents the hydrogen ion concentration of an acid. The negative value of its exponent results in a pH of 5. The second number (0.0000001 or 10^{-7}) represents the hydrogen ion concentration of water. The negative value of its exponent results in a pH of 7. Because pH is a logarithmic scale, a change of 1 reflects a tenfold change in pH. For example, an acid of pH 6 is ten times more acidic than neutral. An acid with pH 5 is 100 times more acidic than neutral.

Figure 2-10 shows many common acids and their pH values. The pH of champagne, for example, is about 4. This low pH is due to the dissolved carbonic acid that causes champagne to bubble. Some bodies of water are acidic, such as peat bogs (pH 4 to 5). The hydrochloric acid (HCl) of your stomach ionizes completely to H^+ and Cl^-. It forms a strong acid with a pH of 2 to 3. Stronger acids are rarely found in living systems.

pH refers to the relative concentration of H^+ ions in a solution. Low pH values indicate high concentrations of H^+ ions (acids), and high pH values indicate low concentrations.

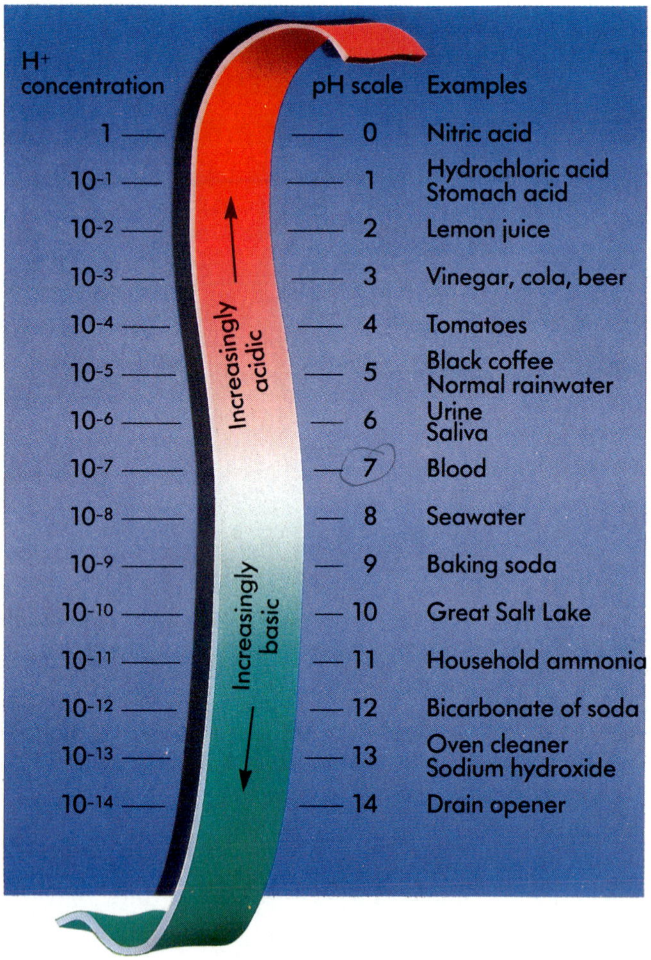

FIGURE 2-10 The pH scale.
A fluid is assigned a value according to the number of hydrogen ions present in a liter of that fluid. The scale is logarithmic, so a change of only one means a tenfold change in the concentration of hydrogen ions. Lemon juice is 100 times more acidic than tomato juice, and seawater is 10 times more basic than pure water.

Any substance that combines with H^+ ions, as OH^- ions do, is said to be a **base.** Any increase in the concentration of a base lowers the H^+ ion concentration. Bases therefore have pH values higher than water's neutral value of 7. For example, the environment of your small intestine is kept at a basic pH of between 7.5 and 8.5. Strong bases such as sodium hydroxide (NaOH) have pH values of 12 or more. As with acids, a change of 1 in the pH value of a base reflects a tenfold change in pH.

Inorganic versus organic chemistry

With the exception of water, most molecules that are formed by living organisms and that make up their structures contain the element carbon. This fascinating element lends itself to being the basis of living material because of its ability to form four bonds with other atoms and therefore interact with other molecules in a myriad of ways. It is also an extremely abundant element, more so than silicon, for example, an atom with similar bonding properties. Silicon is used by some organisms, however, in building "outer skeletons" or shells. An example of a group of such organisms is the diatoms, one of the most abundant protists on this planet (see Figure 29-14).

The carbon-containing molecules that make up living things are called **organic molecules.** Organic molecules are often very large and usually interact with each other by means of covalent bonding. The study of organic molecules and their interactions is called *organic chemistry*. The study of molecules and substances not containing carbon is called *inorganic chemistry*. Inorganic molecules are often quite small and usually interact with one another by means of ionic bonding.

Although carbon is the underlying theme of biological molecules, 10 other elements are common in living organisms. These elements are listed in Table 2-1 and are compared according to their frequency in the earth's crust. Notice that the great majority of atoms in living things (96.3% in fact) are either nitrogen, oxygen, carbon, or hydrogen. Another representation of these data is shown in Figure 2-11, the periodic table of the elements. Elements in the periodic table are arranged according to their atomic number (see p. 28).

TABLE 2-1

The most common elements on Earth and their distribution in the human body

ELEMENT	SYMBOL	ATOMIC NUMBER	APPROXIMATE PERCENT OF EARTH'S CRUST BY WEIGHT	PERCENT OF HUMAN BODY BY WEIGHT	IMPORTANCE OR FUNCTION
Oxygen	O	8	46.6	65.0	Necessary for cellular respiration, component of water
Silicon	Si	14	27.7	Trace	—
Aluminum	Al	13	6.5	Trace	—
Iron	Fe	26	5.0	Trace	Critical component of hemoglobin in the blood
Calcium	Ca	20	3.6	1.5	Component of bones and teeth, trigger for muscle contraction
Sodium	Na	11	2.8	0.2	Principal positive ion bathing cells, important in nerve function
Potassium	K	19	2.6	0.4	Principal positive ion in cells, important in nerve function
Magnesium	Mg	12	2.1	0.1	Critical component of many energy-transferring enzymes
Hydrogen	H	1	0.14	9.5	Electron carrier, component of water and most organic molecules
Manganese	Mn	25	0.1	Trace	—
Fluorine	F	9	0.07	Trace	—
Phosphorus	P	15	0.07	1.0	Backbone of nucleic acids, important in energy transfer
Carbon	C	6	0.03	18.5	Backbone of organic molecules
Sulfur	S	16	0.03	0.3	Component of most proteins
Chlorine	Cl	17	0.01	0.2	Principal negative ion bathing cells
Vanadium	V	23	0.01	Trace	—
Chromium	Cr	24	0.01	Trace	—
Copper	Cu	29	0.01	Trace	Key component of many enzymes
Nitrogen	N	7	Trace	3.3	Component of all proteins and nucleic acids
Boron	B	5	Trace	Trace	—
Cobalt	Co	27	Trace	Trace	—
Zinc	Zn	30	Trace	Trace	Key component of some enzymes
Selenium	Se	34	Trace	Trace	—
Molybdenum	Mo	42	Trace	Trace	Key component of many enzymes
Tin	Sn	50	Trace	Trace	—
Iodine	I	53	Trace	Trace	Component of thyroid hormone

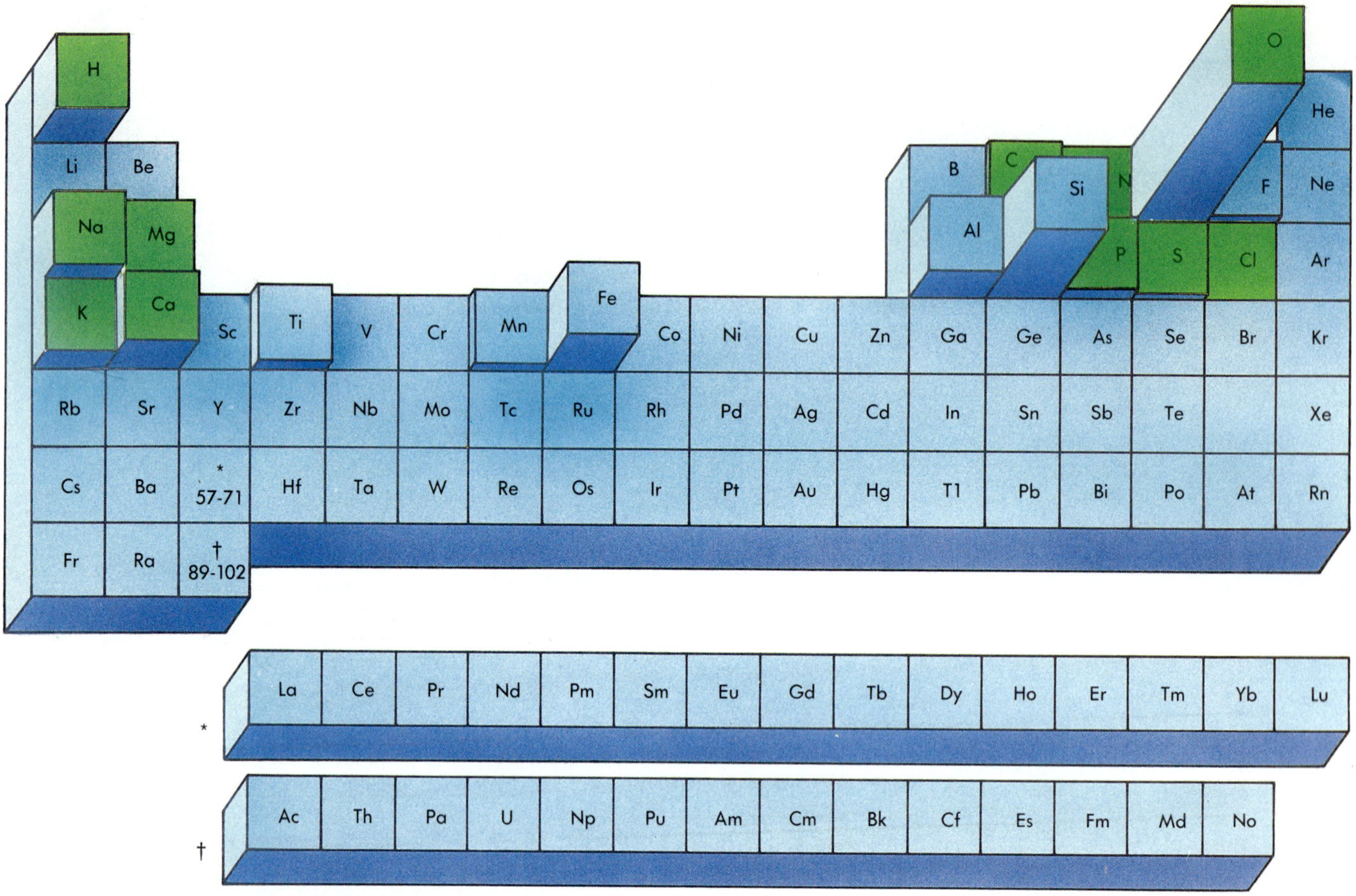

FIGURE 2-11 Periodic table of the elements.
In this representation the frequency of elements that occur in the Earth's crust in more than trace amounts is indicated in the vertical dimension. Thus the most frequent element, oxygen, rises the farthest above the plane of the page, whereas iron, not as common, does not rise as far. Elements found in significant amounts in living organisms are shaded in green.

The chemical building blocks of life

It is often helpful to think of an organic molecule as a carbon-based core with other special parts attached. Each of these special parts is really a group of atoms called a **functional group** and has definite chemical properties. A hydroxyl group (—OH), for example, is a functional group. The most important functional groups are illustrated in Figure 2-12. These groups are important because most chemical reactions that occur within organisms involve the transfer of a functional group from one molecule to another. Other frequent chemical reactions involve the breaking of carbon-carbon bonds.

Some of the molecules of living things are simple organic molecules, often having only one functional group. Other molecules are far larger with many functional groups and are called **macromolecules.** Most of these giant molecules are **polymers,** macromolecules that are built by forming covalent bonds between similar building blocks, or **monomers,** to form long chains. Four major groups of biologically important macromolecules are polymers: complex carbohydrates, lipids, proteins, and nucleic acids. Lipids are different from the rest, however, because they are made up of two types of monomers rather than just one. For this reason, lipids are also termed **composite molecules.** The four major classes of macromolecules are presented in Table 2-2.

Although the four major classes of macromolecules are each composed of different building blocks, the process by which these building blocks are put together is the same. **Dehydration synthesis** is the process by which monomers are put together to form polymers. During dehydration synthesis, one molecule of water is removed (dehydration) from each two monomers that are joined (synthesis). One monomer loses its hydroxyl group (—OH), and the other loses an atom of hydrogen (H). Having lost electrons they were sharing in covalent bonding, both monomers, bond covalently with one another (see Figure 2-22). The process of dehydration synthesis uses energy, which is stored in the bond that is made, and takes place with the help of special molecules called *enzymes* (see Chapter 6).

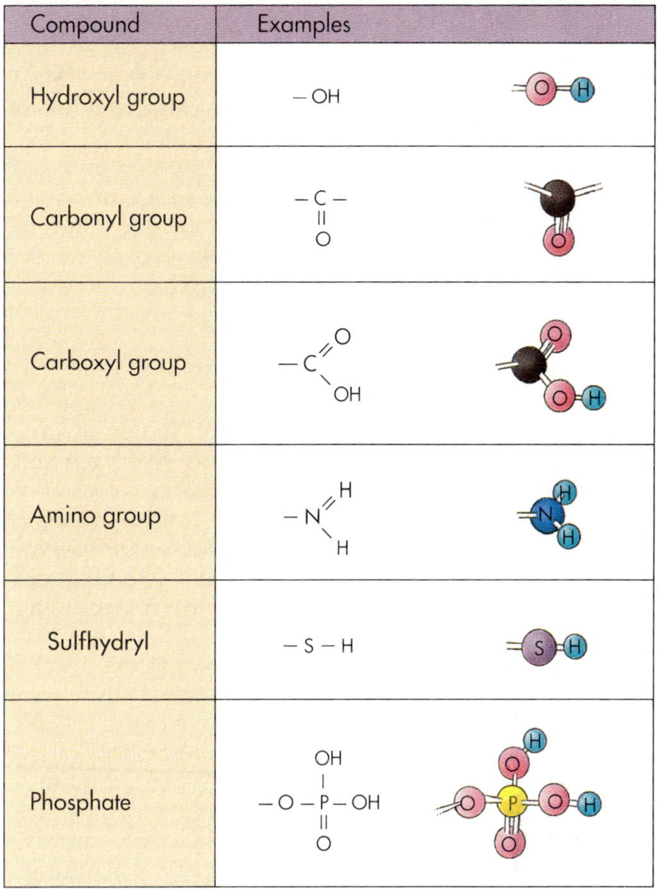

FIGURE 2-12 The principal functional chemical groups. These groups tend to act as units during chemical reactions. They also confer specific chemical properties on the molecules that possess them. Hydroxyl groups make a molecule more basic, for example, because this group forms OH^- ions when the molecule is dissolved in water. Carboxyl groups make a molecule more acidic because this group forms H^+ ions.

TABLE 2-2

Macromolecules

MACROMOLECULE	SUBUNIT	FUNCTION	EXAMPLE
CARBOHYDRATES			
Starch, glycogen	Glucose	Stores energy	Potatoes
Cellulose	Glucose	Makes up cell walls in plants	Paper
Chitin	Modified glucose	Makes up the exterior skeleton in some animals	Crab shells
LIPIDS			
Fats	Glycerol + 3 fatty acids	Store energy	Butter
Phospholipids	Glycerol + 2 fatty acids + phosphate	Make up cell membranes	All membranes
Steroids	4 carbon rings	Act as chemical messengers	Cholesterol, estrogen
PROTEINS			
Globular	Amino acids	Help chemical reactions take place	Hemoglobin
Structural	Amino acids	Make up tissues that support body structures and provide movement	Muscle
NUCLEIC ACIDS			
DNA	Nucleotides	Helps code hereditary information	Chromosomes
RNA	Nucleotides	Helps decode hereditary information	Nucleolus of cells

Polymers are disassembled in an opposite process called **hydrolysis** (see Figure 2-22). During hydrolysis the bonds are broken between monomers with the addition of water (and in the presence of enzymes). In fact, the term *hydrolysis* literally means "to break apart" (lysis) "by means of water" (hydro). The hydroxyl group of a water molecule bonds to one monomer, and the hydrogen atom bonds to its neighbor. The energy held in the bond is released.

Carbohydrates

Carbohydrates are molecules that contain carbon, hydrogen, and oxygen, with the concentration of hydrogen and oxygen atoms in a 2:1 ratio. Abundant energy is locked in their many carbon-hydrogen bonds. Plants, algae, and some bacteria produce carbohydrates by the process of photosynthesis (see Chapter 8). Most organisms use carbohydrates as an important fuel, breaking these bonds and releasing energy to sustain life.

Among the least complex of the carbohydrates are the simple sugars or **monosaccharides.** This word comes from two Greek words meaning "single" (monos) and "sweet" (saccharon) and reflects the fact that monosaccharides are individual sugar molecules. Some of these sweet-tasting sugars have as few as three carbon atoms. The monosaccharides that play a central role in energy storage, however, have six. The primary energy-storage molecule used by living things is glucose ($C_6H_{12}O_6$), a six-carbon sugar with seven energy-storing carbon-hydrogen bonds. Notice in Figure 2-13 that glucose, like other sugars, exists as a straight chain or as a ring of atoms.

Sugars are among the most important energy-storage molecules in living things.

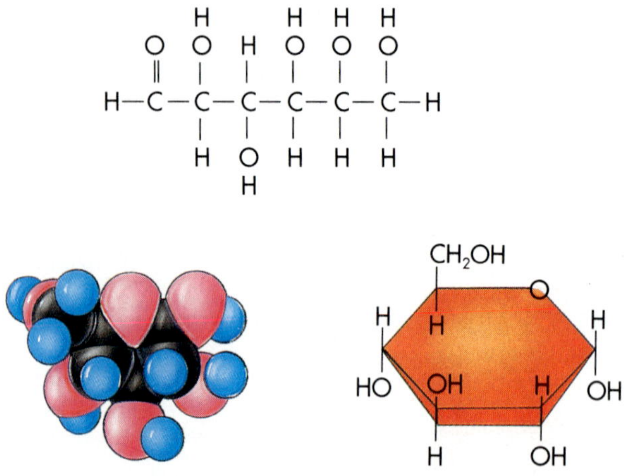

FIGURE 2-13 Structure of a glucose molecule.
Glucose exists as a linear six-carbon molecule or as a ring of atoms. In the ring structure on the right, carbon atoms are located at each point in the hexagon except where the oxygen atom is shown.

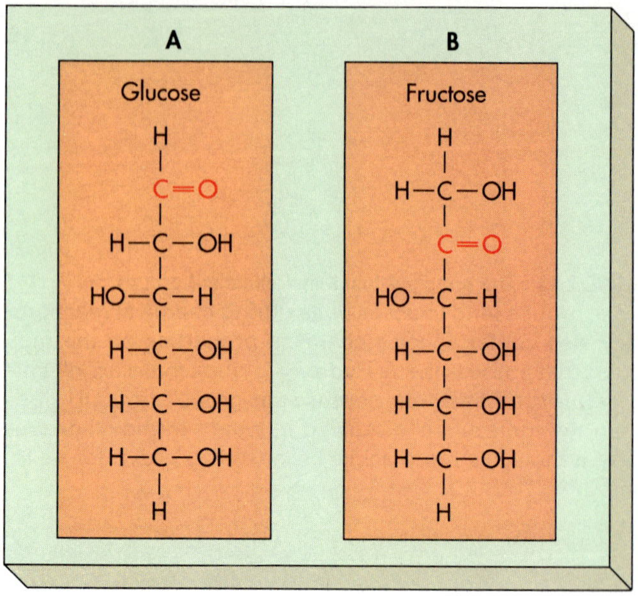

FIGURE 2-14 Isomers.
A structural isomer of glucose **(A)**, such as fructose **(B)**, has identical chemical groups bonded to different carbon atoms.

Glucose is not the only sugar with the formula $C_6H_{12}O_6$. Other monosaccharides having this same formula are fructose and galactose. Because these molecules have the same molecular formula as glucose but are put together slightly differently, they are called **isomers,** or alternative forms, of glucose (Figure 2-14). Your taste buds can tell the difference: fructose is much sweeter than glucose.

In living systems, two of these three sugars are often found covalently bonded to one another. Two monosaccharides linked together is a **disaccharide.** Many organisms, such as plants, link monosaccharides together to form disaccharides that are less readily broken down while being transported within the organism (Figure 2-15). Sucrose (table sugar) is a disaccharide formed by linking a molecule of glucose to a molecule of fructose. It is the common transport form of sugar in plants. Lactose, or milk sugar (glucose + galactose), is a disaccharide produced by many mammals to feed their young.

Not only do organisms unlock, use, and transport the energy within carbohydrate molecules, they can store this energy. To do this, however, organisms must convert soluble sugars such as glucose to an insoluble form to be stored. Sugars are made insoluble by joining them together into long polymers called **polysaccharides.** To store energy, plants use glucose to form polysaccharides called **starches.** The starch amylose, for example, is made up of hundreds of glucose molecules linked together in long, unbranched chains. Most plant starch is a branched version of amylose called *amylopectin.* Animals store glucose as highly branched chains called **glycogen** (Figure 2-16).

Starch and glycogen, both types of polysaccharides, are storage forms of glucose. Plants store sugar as starch, whereas animals store sugar as glycogen.

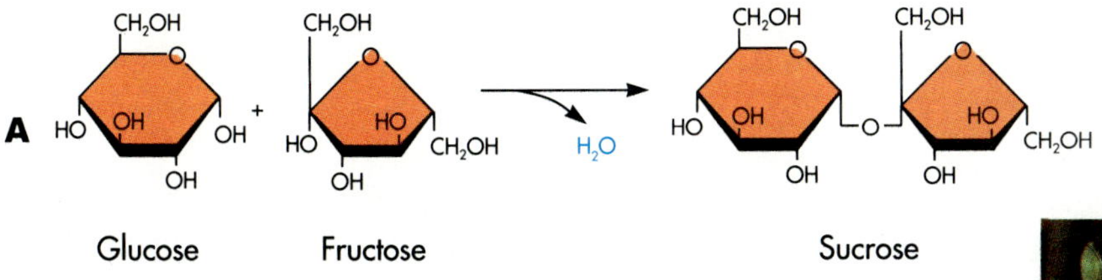

B

FIGURE 2-15 Disaccharides.
A The disaccharide shown here, sucrose, is formed by linking the two monosaccharides glucose and fructose. **B** Sucrose is the most common transport form of sugar in plants. This hungry butterfly has its uncoiled mouthparts extended down into a sugar solution, which it is sucking from a flower.

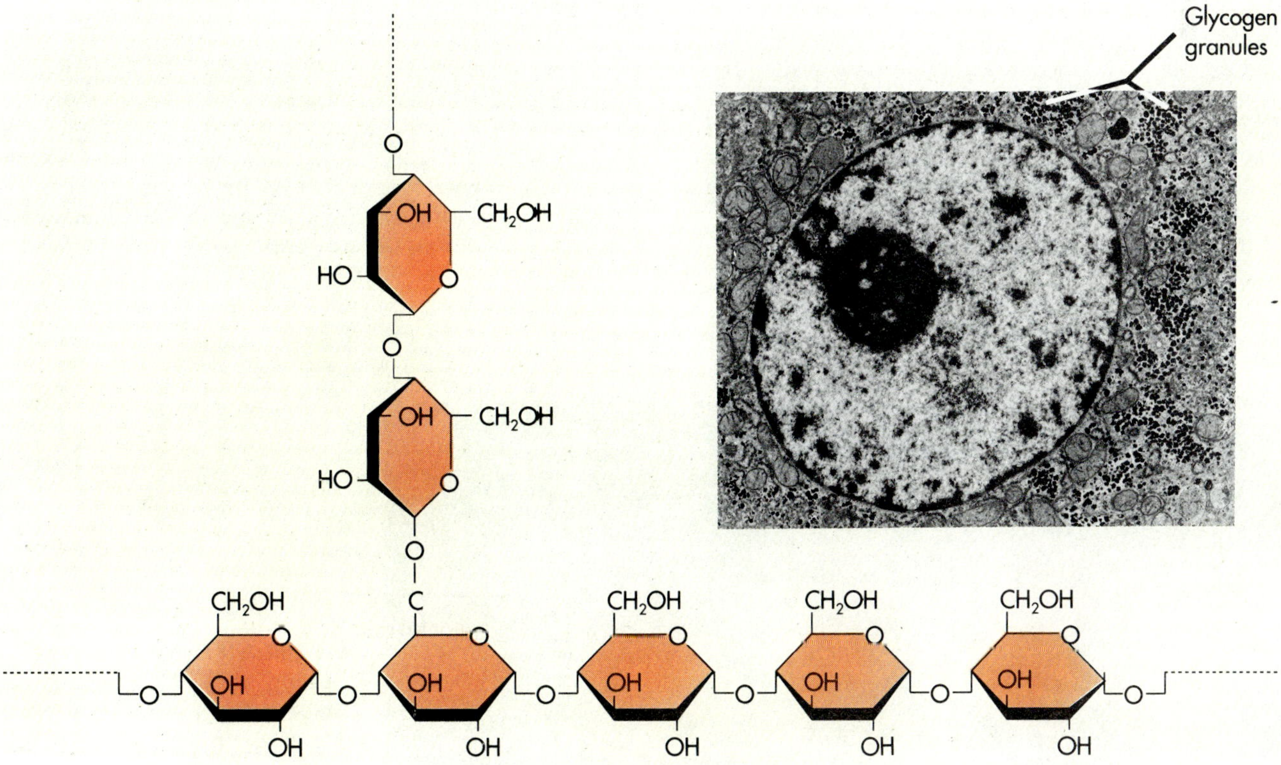

FIGURE 2-16 Glycogen, a starch.
The nuclear membrane of the liver cell in the micrograph is surrounded by glycogen, a highly branched starch. Plants store glucose as starch, whereas animals store it as glycogen.

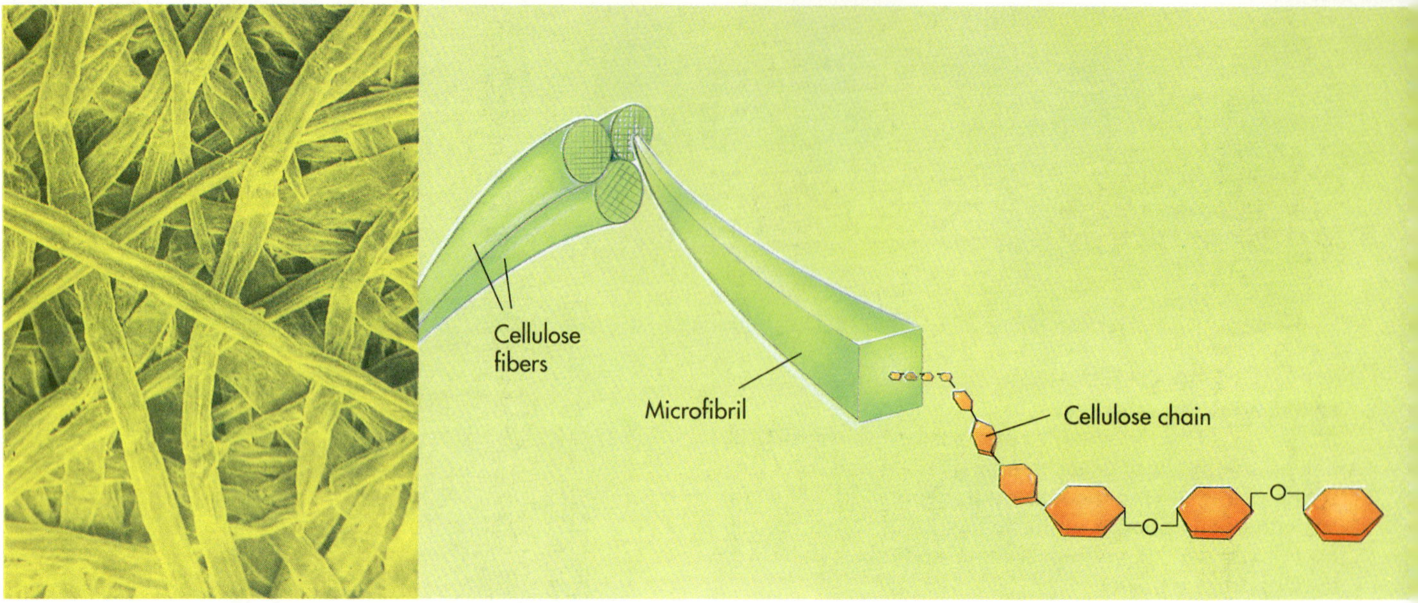

FIGURE 2-17 A journey into wood.
The jumble of cellulose fibers *(left)* is from a ponderosa pine. Each fiber is composed of microfibrils, which are bundles of cellulose chains. Cellulose fibers can be very strong, which is one reason wood is such a good building material.

The chief component of plant cell walls is a polysaccharide called **cellulose.** Cellulose is chemically similar to amylose but is bonded together in a way that most organisms cannot digest (Figure 2-17). For this reason, cellulose works well as a biological structural material and occurs widely in this role in plants. The structural material in insects, many fungi, and certain other organisms is a modified form of cellulose called **chitin** (Figure 2-18). Chitin is a tough, resistant surface material that is also relatively indigestible.

Fats and lipids

When organisms store glucose molecules for long periods, they usually store them as **fats** rather than as carbohydrates. Fats are large molecules made up of carbon, hydrogen, and oxygen as are the carbohydrates, but their hydrogen to oxygen ratio is higher than 2:1. For this reason, fats contain more energy-storing carbon-hydrogen bonds than carbohydrates. In addition, fats are nonpolar, insoluble molecules, so they work well as storage molecules.

FIGURE 2-18 Chitin.
Chitin is the principal structural element in the external skeletons of many invertebrates, such as this lobster.

Fats are only one kind of **lipid**. Lipids include a wide variety of molecules, all of which are soluble in oil but insoluble in water. This insolubility is due to the fact that almost all the bonds in lipids are nonpolar carbon-carbon or carbon-hydrogen bonds. Three important categories of lipids are (1) oils, fats, and waxes; (2) phospholipids; and (3) steroids.

Lipids are composite molecules: that is, they are made up of more than one component. Oils and fats are built from two different kinds of subunits:

1. Glycerol: Glycerol is a three-carbon molecule with each carbon bearing a hydroxyl (—OH) group. The three carbons form the backbone of the fat molecule.
2. Fatty acids: Fatty acids have long **hydrocarbon** chains (chains consisting only of carbon and hydrogen atoms) ending in a carboxyl (—COOH) group. Three fatty acids are attached to each glycerol backbone (Figure 2-19). Because there are three fatty acids, the resulting fat molecule is called a **triglyceride:**

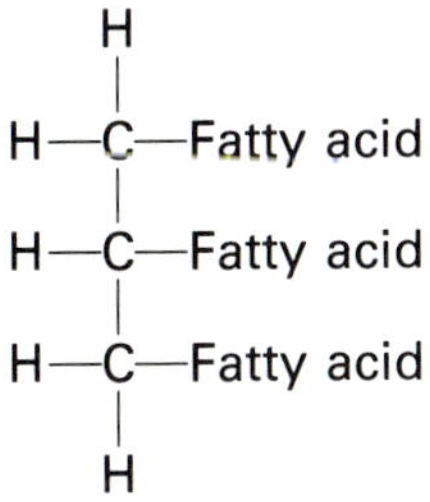

The difference between fats and oils has to do with the number of double bonds in their fatty acids. As Figure 2-20 shows, a fatty acid with only single bonds between its carbon atoms can hold more hydrogen atoms than a fatty acid with double bonds between its carbon atoms. A fatty acid that carries as many hydrogen atoms as possible, such as the fatty acid in Figure 2-20, *A* is saturated. Fats composed of fatty acids with double bonds are unsaturated because the double bonds replace some of the hydrogen atoms. If a fat has more than one double bond, it is polyunsaturated. Polyunsaturated fats have low melting points and are therefore liquid fats, or oils. The fatty acids of most plant triglycerides such as vegetable oils are unsaturated. (Exceptions are the tropical oils.) Animal fats, in contrast, are often saturated and occur as hard fats. Human diets with large amounts of saturated fats appear to upset the normal balance of fatty acids in the body, a situation that can lead to diseases of the circulatory system.

Fats are important energy-storing molecules. They are made up of three fatty acid chains attached to a glycerol backbone.

Waxes, which are used by land plants and some animals as a waterproofing material, differ from fats and oils by having a chemical backbone slightly different than glycerol. Phospholipids are also similar to oils except that one of their fatty acids is replaced by a phosphate group attached to a

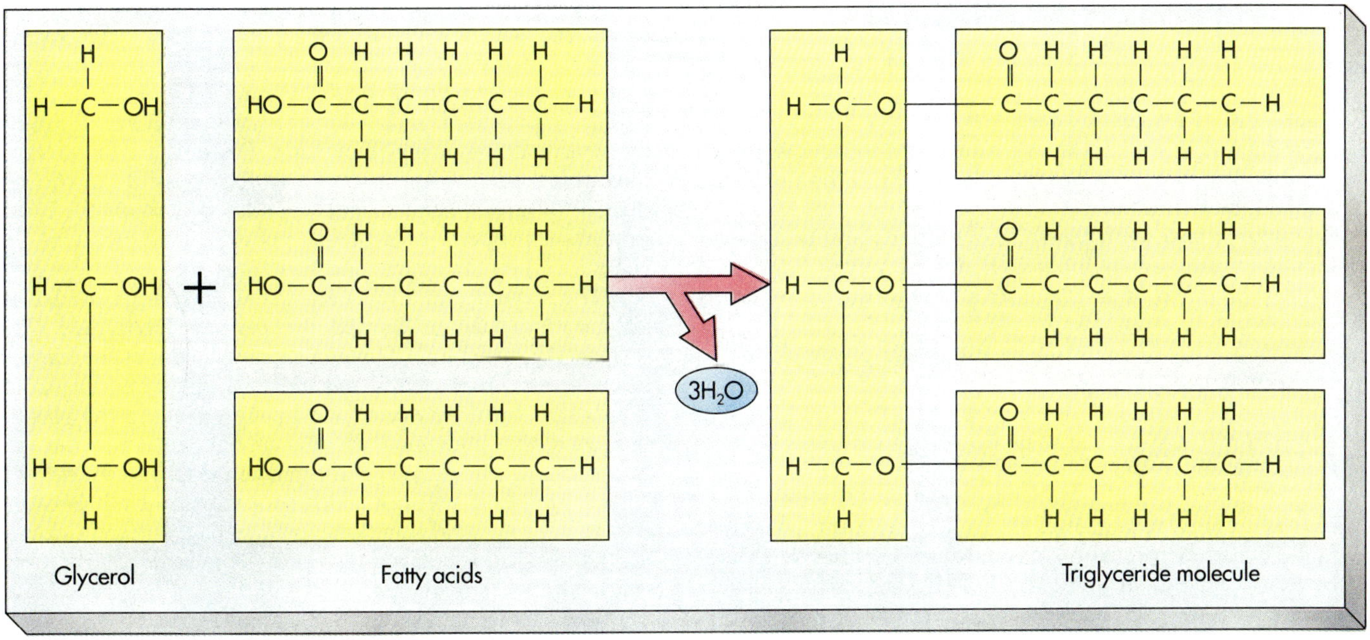

FIGURE 2-19 Structure of a triglyceride.
Triglycerides are composite molecules, made up of three fatty acid molecules bonded to a single glycerol molecule.

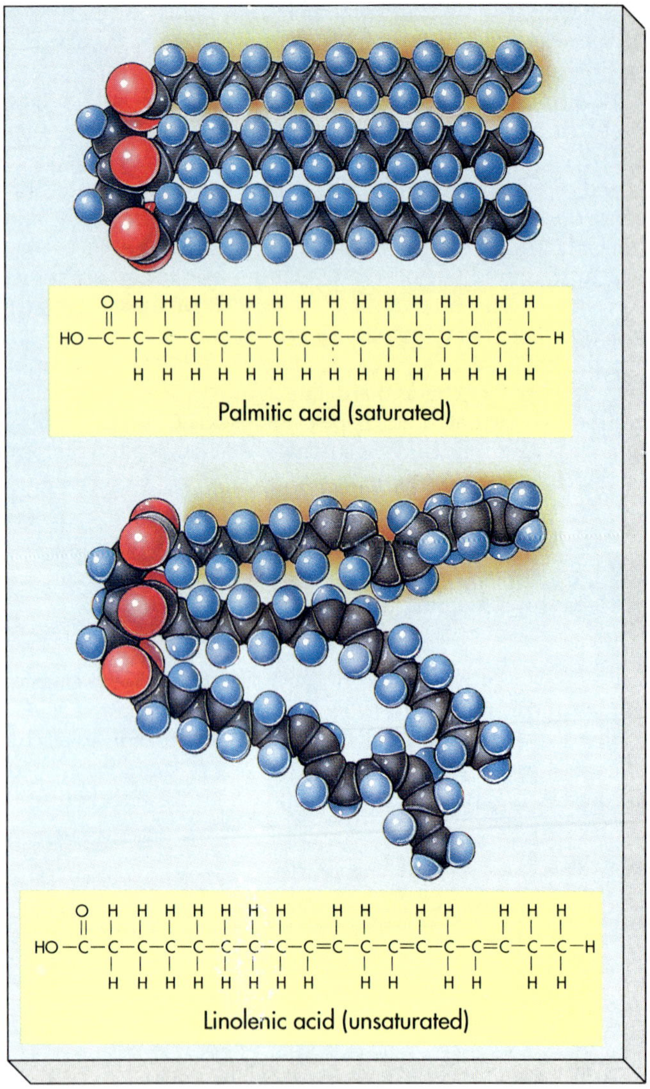

FIGURE 2-20 Saturated and unsaturated fats.

A Palmitic acid, with no double bonds and thus a maximum number of hydrogen atoms bonded to the carbon chain, is a saturated fatty acid.

B Linolenic acid, with three double bonds and thus fewer than the maximum number of hydrogen atoms bonded to the carbon chain, is an unsaturated fatty acid.

nitrogen-containing group. Phospholipids play a key role in the structure of cell membranes (see Chapter 4). Membranes often contain steroids, a lipid having a structure very different from oils. Steroids are composed of four carbon rings. Most of your cell membranes contain the steroid cholesterol. Male and female sex hormones (discussed in Chapters 19 and 20) are also steroids.

Proteins

Proteins are the third major group of macromolecules that make up the bodies of organisms. Proteins play diverse roles in living things. Perhaps the most important proteins are **enzymes,** proteins capable of speeding up specific chemical reactions. Other short proteins called **peptides** are used as chemical messengers within your brain and throughout your body. Collagen, a structural protein, is an important part of bones, cartilage, and tendons. Despite their varied functions, all proteins have the same basic structure: a long chain of amino acids linked end to end.

Amino acids are small molecules containing an amino group (—NH_2), a carboxyl group (—COOH), a hydrogen atom, a carbon atom, and a functional group that differs among amino acids. In a generalized formula for an amino acid, the functional group is shown as *R*. The identity and unique chemical properties of each amino acid are determined by the nature of the functional (R) group.

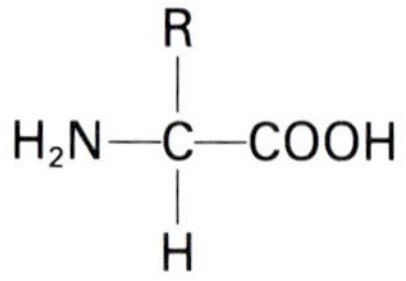

Only 20 different amino acids make up the diverse array of proteins found in living things. Each protein differs according to the amount, type, and arrangement of amino acids that make up its structure. These 20 "common" amino acids are illustrated in Figure 2-21. They are grouped according to the chemical nature of their functional groups.

When ionized, each amino acid has a positive (amino, or —NH_3^+) group at one end and a negative (carboxyl, or —COO^-) group at the other end. During dehydration synthesis, each of these groups on separate amino acids loses a molecule of water between them, forming a covalent bond that links the two amino acids (Figure 2-22, *A*). This bond is called a **peptide bond.** A long chain of amino acids linked by peptide bonds is a **polypeptide.** Proteins are long, complex polypeptides. The great variability possible in the sequence of amino acids in polypeptides is perhaps the most important property of proteins, permitting tremendous diversity in their structures and functions.

The sequence of amino acids that make up a particular polypeptide chain is termed the **primary structure** of a protein (Figure 2-23). This sequence determines the further levels of structure of the protein molecule resulting from bonds that form between these groups. Having the proper sequence of amino acids, then, is crucial to the functioning of a protein. Put simply, if the protein does not assume its correct shape, it will not work properly or at all. Because different amino acid functional groups have different chemical properties, the shape of a protein may be altered by a single amino acid change.

The functional groups of the amino acids in a polypeptide chain interact with their neighbors, forming hydrogen bonds. In addition, portions of a protein chain with many nonpolar functional groups (see Figure 2-21) tend to be

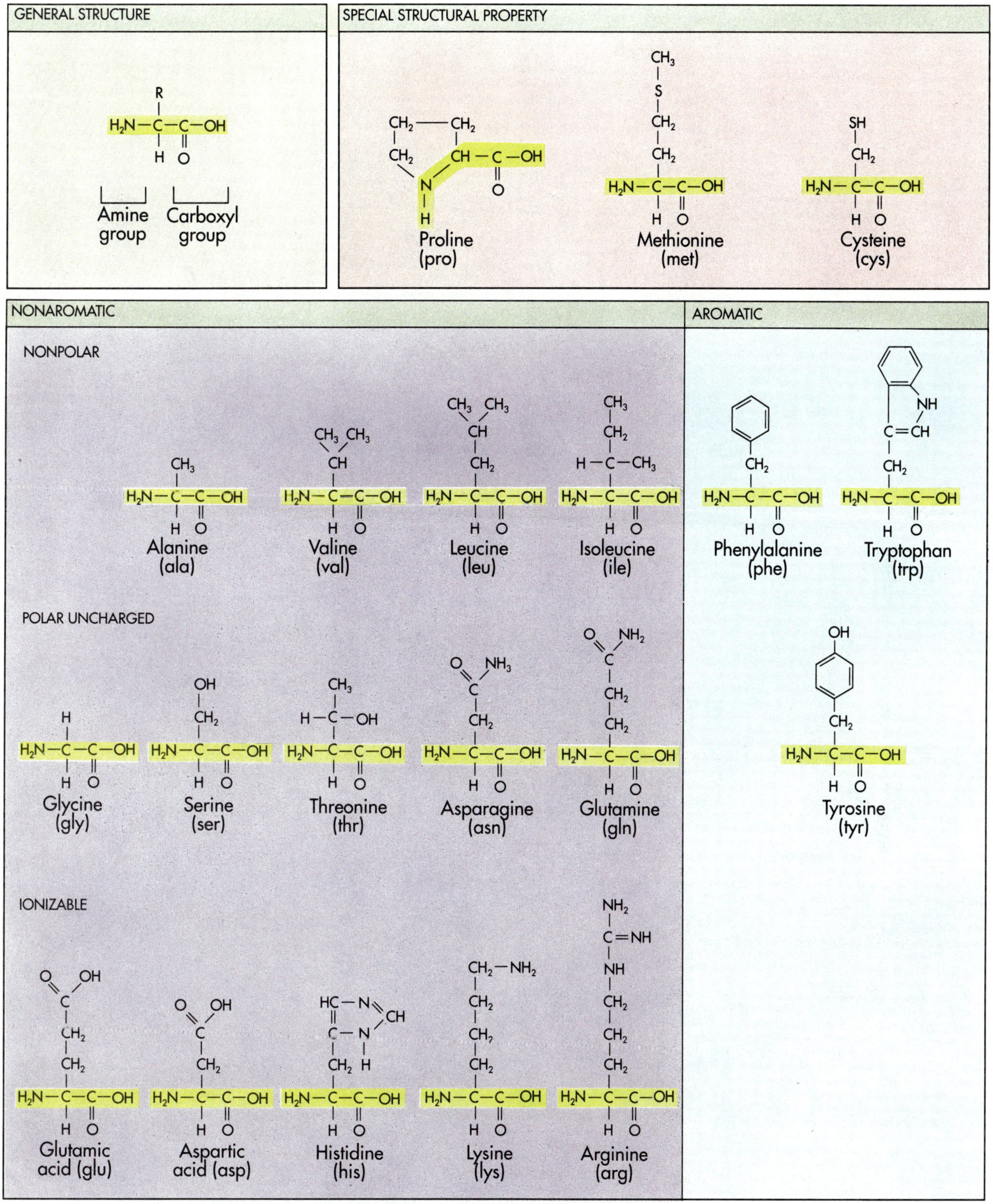

FIGURE 2-21 The 20 common amino acids.
Each amino acid has the same chemical backbone but differs from the others in the side, or R, group. Six of the amino acid R groups are nonpolar. The ones containing ring structures are called the *aromatic amino acids.* Another six are polar but uncharged; these differ from one another in their polarity. Five more are polar and capable of ionizing to a charged form; under typical conditions some of these five are acids, others bases. The remaining three have special chemical (structural) properties that play important roles in forming links between protein chains or forming kinks in their shapes.

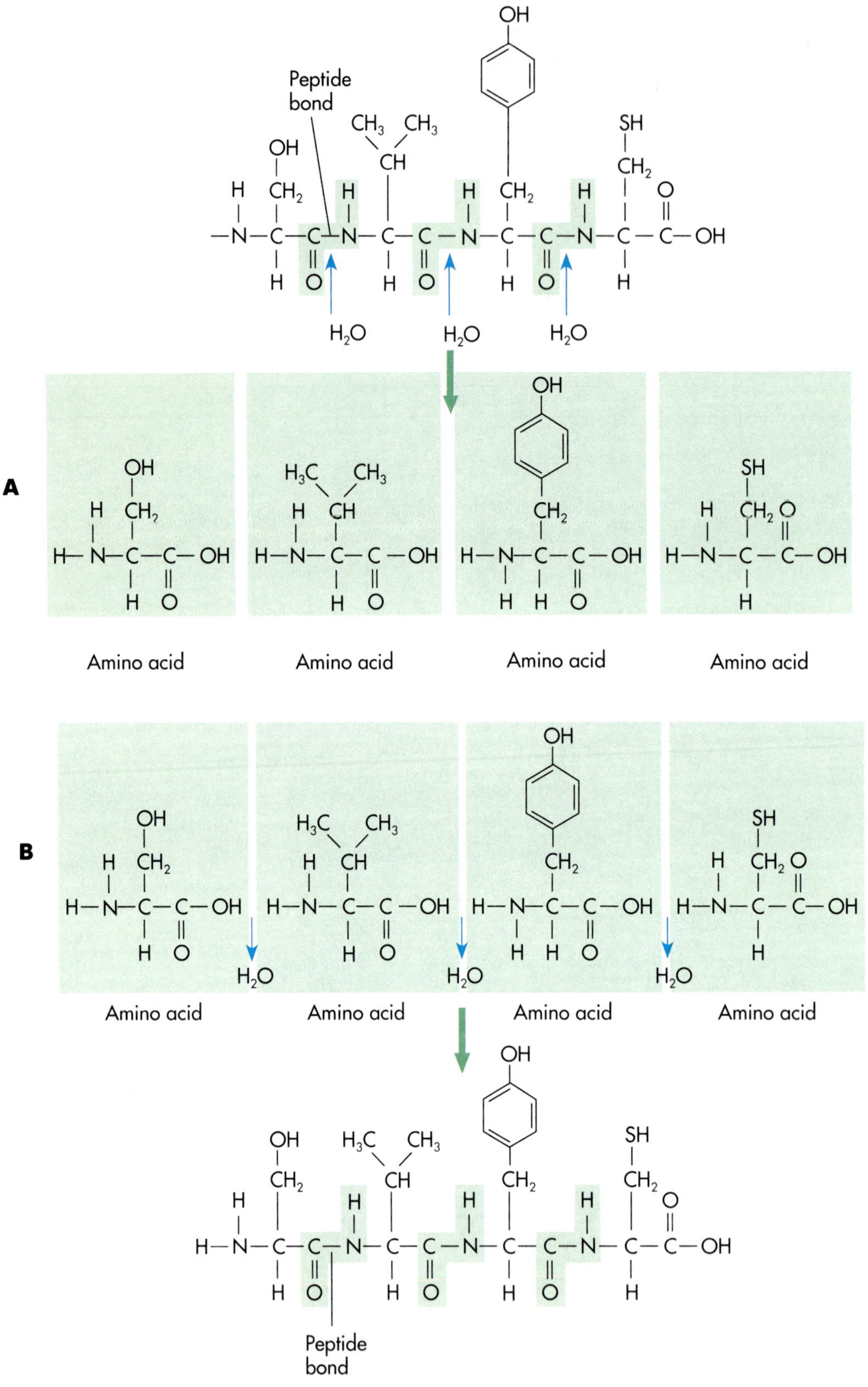

FIGURE 2-22 How a polypeptide chain is formed and broken.
A During hydrolysis, a molecule of water is added to each peptide bond that links adjacent amino acids, breaking the bonds between them.
B During dehydration synthesis a peptide bond is formed between two amino acids. A chain of amino acids is called a *polypeptide.*

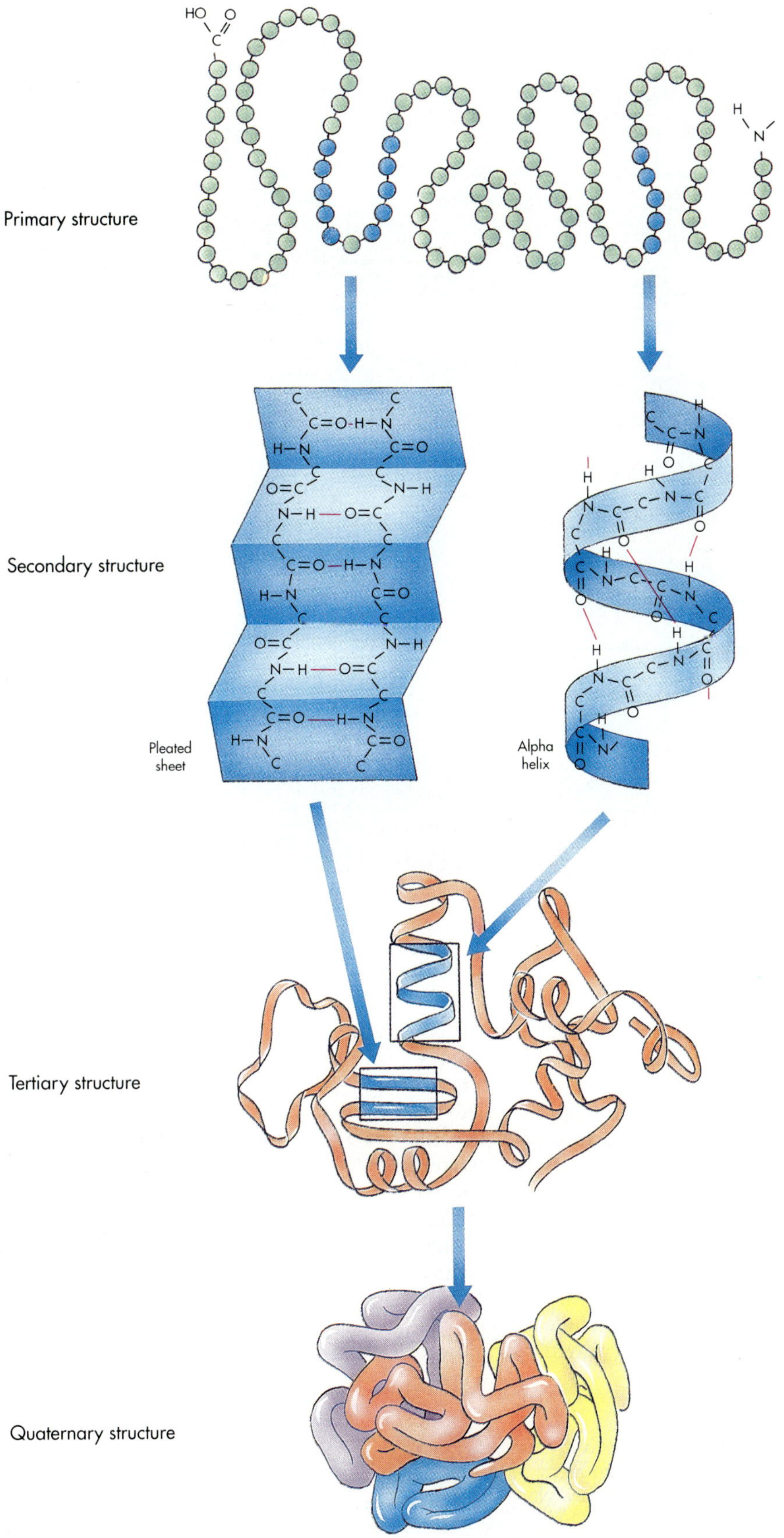

FIGURE 2-23 How primary structure determines a protein's shape.
The amino acid sequence of the enzyme protein lysozyme, called its *primary structure,* encourages the formation of hydrogen bonds between nearby amino acids, producing coils and foldbacks called the *secondary structure.* The lysozyme protein assumes a three-dimensional shape like a cupped hand; this is called its *tertiary structure.* Many proteins (not lysozyme) aggregate in clusters called the *quaternary structure* of the protein.

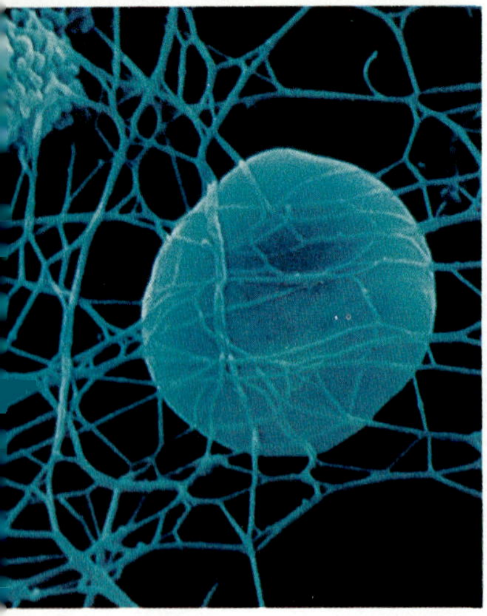

A

B

C

D

E

shoved into the interior of the protein because of their hydrophobic properties. Because of these interactions, polypeptide chains tend to fold spontaneously into sheets or wrap into coils. This folded or coiled shape is called its **secondary structure.** The primary structure of a protein also determines its three-dimensional shape, or **tertiary structure.** Proteins made up largely of sheets often form fibers (Figure 2-24). Proteins that have regions forming coils frequently fold into globular shapes. When two polypeptide chains associate to form a functional unit, the chains are termed *subunits,* and the functional unit is the protein. For proteins that consist of subunits, the way these subunits are assembled into a whole is called the **quaternary structure.**

Approximately 20 different amino acids are used in various sequences and combinations to make up a wide variety of proteins. These amino acids have different properties from one another. These differences and the sequence in which the amino acids are arranged determine the shape of a protein, which is crucial to its proper functioning.

FIGURE 2-24 Many proteins play structural roles.
A Fibrin—electron micrograph of a blood clot.
B Collagen—strings of a tennis racket.
C Keratin—a peacock feather.
D Silk—a spider's web.
E Hair—a woman's hair.

Nucleic acids

Organisms store information about the structures of their proteins in macromolecules called **nucleic acids.** Nucleic acids are long polymers of repeating subunits called **nucleotides.** Each nucleotide is made up of three smaller building blocks (Figure 2-25):

1. A five-carbon sugar
2. A phosphate group ($—PO_4^{\equiv}$)
3. An organic nitrogen-containing molecule called a *base*

To form the nucleic acid chain, the sugars in nucleotides are linked together in a line by means of the phosphate and hydroxyl groups. A nitrogenous base protrudes from each sugar (Figure 2-26). The order in which the nucleotides are linked together forms a code that ultimately specifies the order of amino acids in a particular protein.

> **Organisms store and use hereditary information by coding the sequence of the amino acids of each of their proteins as a sequence of nucleotides in nucleic acids.**

Organisms have two forms of nucleic acid. One form, deoxyribonucleic acid (DNA), stores the information for making proteins. The other form, ribonucleic acid (RNA), directs the production of proteins. Details of the structure of DNA and the ways it interacts with RNA are presented in Chapter 24.

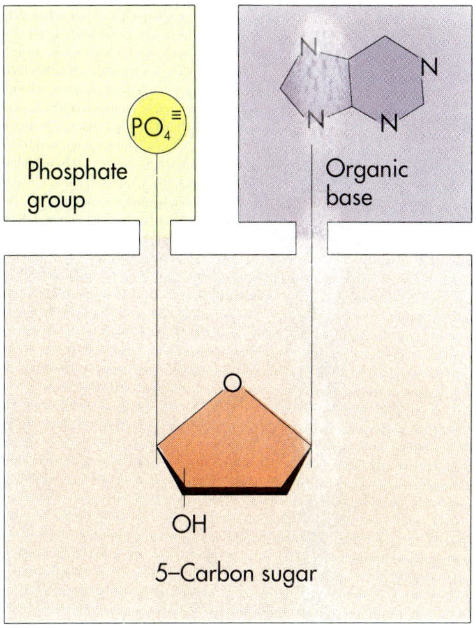

FIGURE 2-25 The structure of a nucleotide.
A nucleotide is made up of three elements: a five-carbon sugar, a phosphate group, and an organic nitrogen-containing molecule called a *base.*

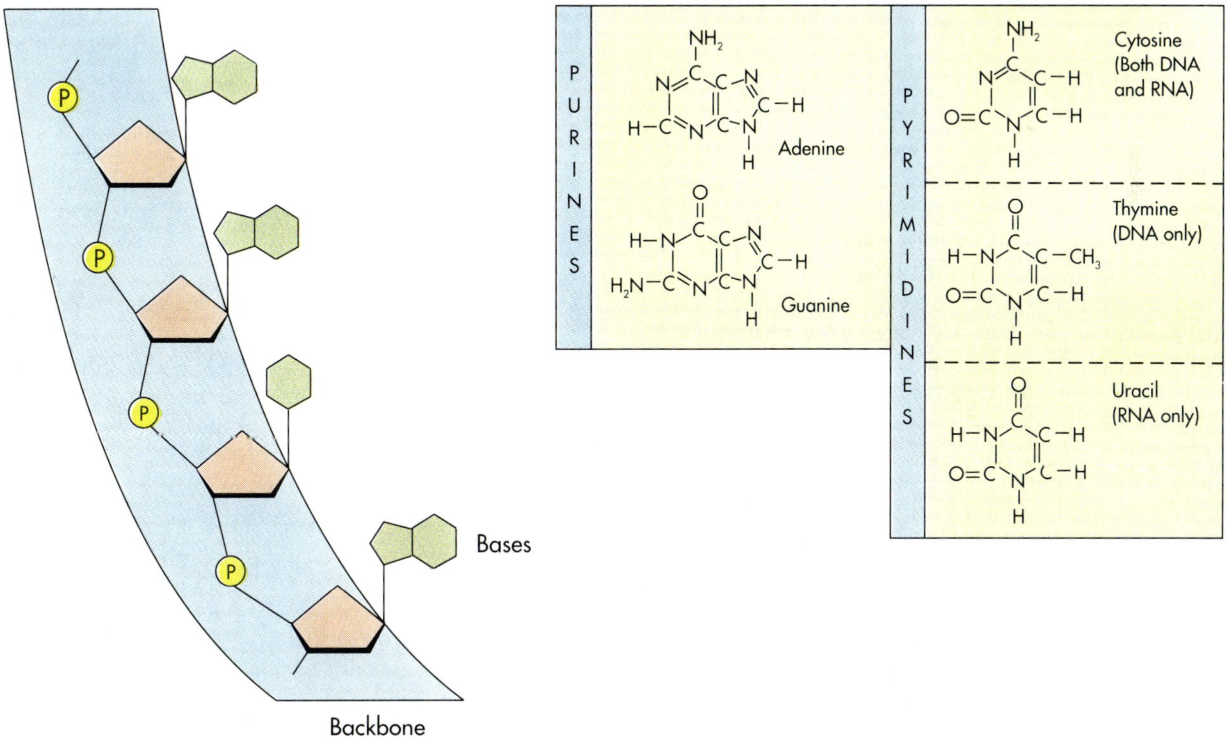

FIGURE 2-26 The five nitrogenous bases of nucleic acids.
Only four of these bases occur in DNA and four in RNA. In DNA the base thymine replaces the base uracil found in RNA.

BIOLOGY & you

The use of isotopes in medicine

Isotopes are elements that have the same number of protons and electrons but have different numbers of neutrons. For example, hydrogen (one proton and no neutrons) has two isotopes: deuterium (one proton and one neutron) and tritium (one proton and two neutrons) (see Figure 2-1, *B*). Because only the number of neutrons is different, all isotopes of an element have similar chemical properties. In fact, tritium can "substitute" for hydrogen in chemical reactions!

Some isotopes are unstable and can emit subatomic particles: neutrons, protons, or electrons. These unstable isotopes are termed **radioactive.** When a radioactive isotope emits subatomic particles, a new, more stable element is formed. This process is called **decay.**

Because of their radioactivity and the fact that isotopes are nearly indistinguishable to organisms, doctors use isotopes in many ways to detect and treat disease.

When used in diagnosis, isotopes are called **tracers** or **labels** because they can be used to follow the fate of certain substances in the body. For example, suppose your doctor wants to investigate the rate at which your thyroid gland is using iodine, a substance in your diet that your body uses to make thyroid hormone. To determine this rate, your doctor injects you with a tracer—a solution of radioactive iodine, which your body will use in the same way it uses the "normal" iodine you acquire in your food. Equipment that counts the emission of particles from the nuclei of the radioactive iodine is placed over your thyroid gland and measures how fast the tracer iodine is entering your thyroid. The doctor can then see *exactly* how fast your thyroid is using iodine by simply reading the information on the detector (Figure 2-A).

Another way isotopes are used is in the treatment of cancer, especially those cancers (such as skin cancers) that are present on the surface of the body. A patch containing powerful radioactive isotopes is taped over the cancerous tumor. The particles emitted from the isotope bombard the tumor and destroy the cancerous tissue. Treatment with isotopes does have some drawbacks, however. Because the isotopes are radioactive, the patient receiving this type of treatment must be kept under strict quarantine. Isotope therapy also causes nausea and breakdown of the "normal" tissue surrounding the tumor. Unfortunately, treatment with isotopes is not a cure for cancer—the isotopes can only destroy localized tumors. Cancer cells that have migrated to other parts of the body are unaffected by isotope therapy. Still, treatment with isotopes offers hope to those in the early stages of cancer.

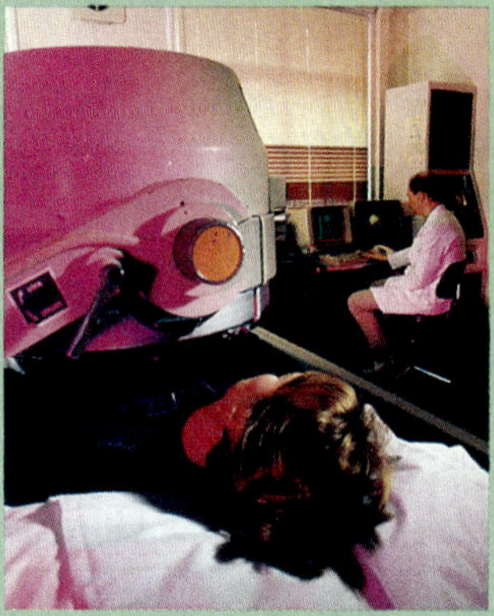

FIGURE 2-A Isotopes are used to diagnose and treat illness.

Summary

1. Atoms make up all matter and are composed of three types of subatomic particles: protons, neutrons, and electrons. Protons and neutrons are found at the core of the atom, and electrons orbit this core.
2. The chemical behavior of an atom is largely determined by the distribution of its electrons, particularly the number of electrons in its outermost energy level. Atoms are most stable when their outer energy levels are filled. Electrons are lost, gained, or shared until this condition is reached.
3. The atoms of most elements interact with one another to form molecules, combinations of tightly bound atoms. Compounds are molecules made up of the atoms of two or more elements.
4. Water is an extremely important molecule to living things. It has one outstanding chemical property: it attracts other charged particles to its positive and negative ends. For this reason, water is a polar molecule.
5. Water dissolves polar substances and excludes nonpolar substances. Chemical interactions readily take place in water because so many kinds of molecules are water soluble and therefore move among water molecules as separate molecules or ions.
6. The chemistry of life is concerned with the interactions among biological molecules, or organic molecules. Organic molecules have a carbon "core" with various functional groups attached to the carbon atoms.
7. Most organisms use carbohydrates as an important fuel. The most important of the energy-storing carbohydrates is glucose, a six-carbon sugar.
8. Certain lipids—the fats—are important in the long-term storage of energy. Phospholipids are a major component of cell membranes.
9. Proteins are linear polymers of amino acids. Because the 20 amino acids that occur in proteins have functional groups with very different chemical properties, the shape and therefore the functioning of a protein are critically affected by its particular sequence of amino acids.
10. Hereditary information is stored as a sequence of nucleotides in the nucleotide polymer DNA. A second form of nucleic acid, RNA, directs the production of proteins.

REVIEW QUESTIONS

1. a. Draw a diagram of an atom with an atomic number of 1. Label the nucleus and the subatomic particles, and show the electrical charge of each particle.
   b. Draw an isotope of the same atom.
2. Fill in the blanks: The number of __________ in an atom determines the element to which it belongs. The atom's chemical properties are largely determined by its __________ .
3. Distinguish among atoms, molecules, and compounds. Give an example of each.
4. Explain the three types of chemical forces that influence how an atom interacts with other atoms. What is the significance of the octet rule?
5. Distinguish between ionic and covalent bonds, and give an example of each.
6. What characteristics of water make it so unusual? How do these traits make life on earth possible? Relate these properties to water's molecular structure.
7. After an oil spill in the ocean, why is it necessary to clean up the water?
8. One substance has a pH of 4.2; another has a pH of 8.6. Explain what pH means and how it is measured. What is each substance called?
9. Distinguish between organic and inorganic molecules. Which would you study if you wanted to learn more about the human body?
10. What do polymers and composite molecules have in common? How do they differ? Name a biologically important example of each.
11. Distinguish among monosaccharides, disaccharides, and polysaccharides. To what group do they belong, and why are they important?
12. Identify and describe the function of the following in living systems:
    a. $C_6H_{12}O_6$
    b.

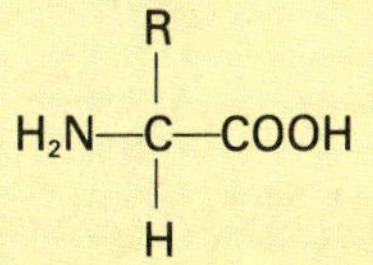

13. Discuss the three major types of lipids. What are their principal functions?
14. While doing your grocery shopping, you find this label on a product:

Nutritional Information Per Serving

% of calories from fat	100%
Polyunsaturated fat	2 grams
Saturated fat	2 grams
Protein	0 grams
Carbohydrates	0 grams

Explain the difference between polyunsaturated and saturated fats. What type(s) of macromolecules does this product contain?

15. What are proteins, and why are they important?
16. Distinguish among the primary, secondary, tertiary, and quaternary structures of a protein.
17. Name the two forms of nucleic acids, and explain their significance.

KEY TERMS

acid 26
atoms 17
base 26
carbohydrates 30
cellulose 32
chemical bonds 19
chemical formula 18
compounds 18
covalent bond 19
dehydration synthesis 28
disaccharides 30
electrons 17
elements 18
energy levels (shell) 18
fats 32
functional group 28
glycogen 30
hydration shell 24
hydrolysis 30
ionic bond 19
ionization 25
ions 20
isotopes 17
lipid 33
macromolecules 28
molecules 18
monosaccharides 30
neutrons 17
nucleic acid 39
nucleotide 39
octet rule 19
orbital 17
organic molecules 26
peptide bond 34
pH scale 25
polar molecules 23
polypeptide 34
polysaccharides 30
protein 34
protons 17
starches 30

THOUGHT QUESTIONS

1. What is the maximum number of covalent bonds that any atom can take part in?
2. A jar of old-fashioned peanut butter, if left on the shelf for a long time, will separate into two layers, a layer of peanut oil forming over the butter. Modern polyunsaturated peanut butter does not separate in this way on the shelf. Why not?

FOR FURTHER READING

Cox, T. (1990, February 3). The origin of the chemical elements. *New Scientist*, pp. 1-4.
The heavier elements making up most of the Earth—and you—were created through the birth and death of the stars.

Fitzgerald, J., & Taylor, G. (1989, May). Carbon fibres stretch the limits. *New Scientist*, pp. 48-53.
This is a fascinating discussion of the ways in which the principles discussed in this chapter are used to produce modern, synthetic materials.

Sutton, C. (1989, February). Subatomic forces [supplement]. *New Scientist*, pp. 1-4.
This article is a contemporary review of the interactions between particles within the atom.

PART TWO

CELLS
THE BASIC UNIT OF LIFE

3 Cell structure and function

HIGHLIGHTS

▾
All organisms on Earth are cells or are made up of cells.
▾
Most cells are microscopically small, providing sufficient surface area with respect to volume for cell survival.
▾
All cells can be grouped into two broad categories: prokaryotic cells and eukaryotic cells.
▾
Prokaryotic cells, which are characterized by the bacteria, are simpler in structure than eukaryotic cells, which include all other types of cells.

OUTLINE

The cell theory
Why aren't cells larger?
Two kinds of cells
Eukaryotic cells: An overview
- The plasma membrane
- The cytoplasm and cytoskeleton
- Membranous organelles
  - The endoplasmic reticulum
  - Golgi bodies
  - Lysosomes
  - Vacuoles
  - The nucleus
- Bacteria-like organelles
  - Mitochondria
  - Chloroplasts
- Cilia and flagella

Prokaryotic cells
- Cell walls
- Simple interior organization

A comparison of prokaryotic and eukaryotic cells

The illustration may not look like any highway you've ever seen, but that's exactly what it is. This highway, and billions more like it, makes up a transportation network within your body. The threads, called *microtubules*, shuttle particles within your body's cells. Amazingly, they can move particles in two directions at the same time!

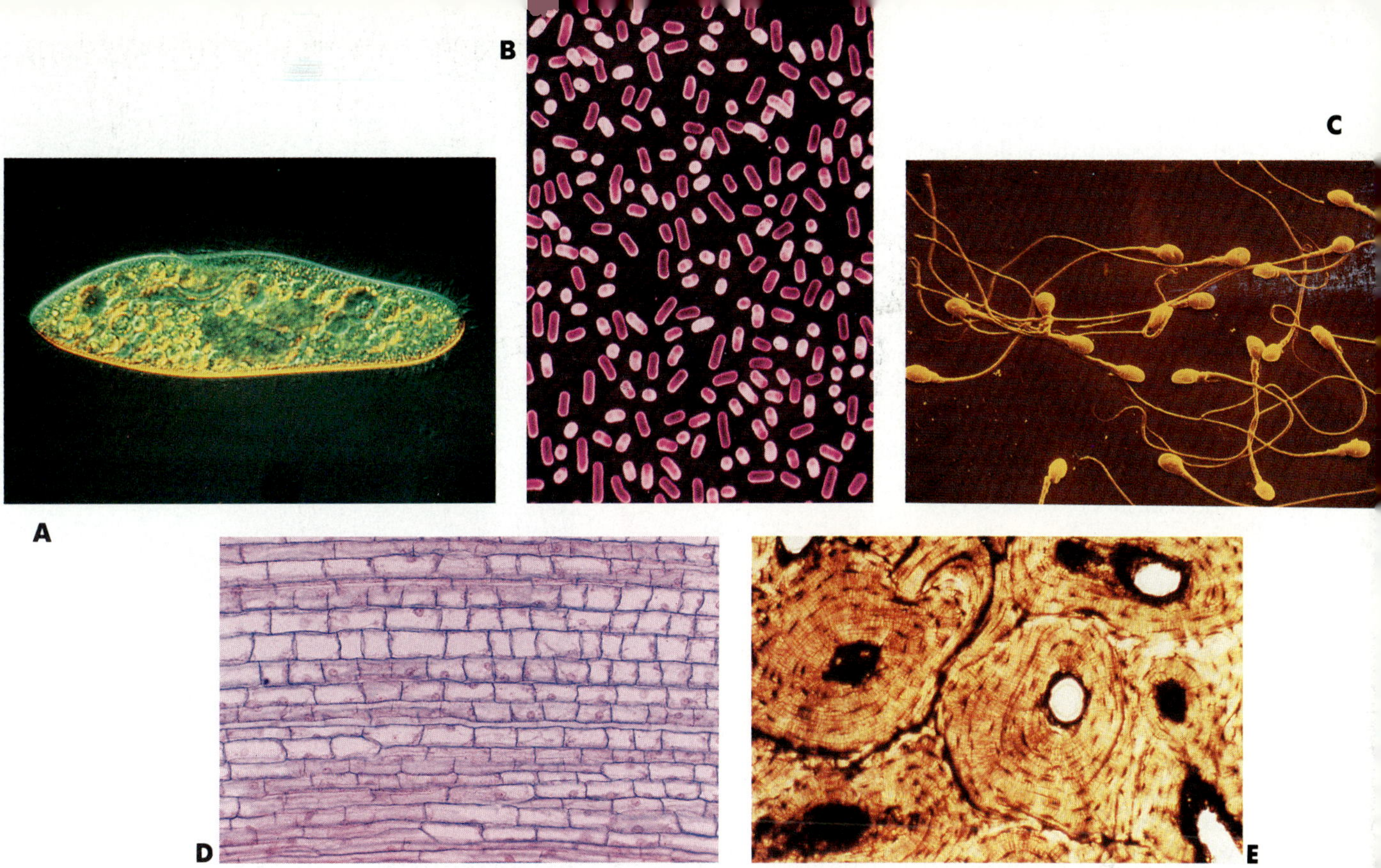

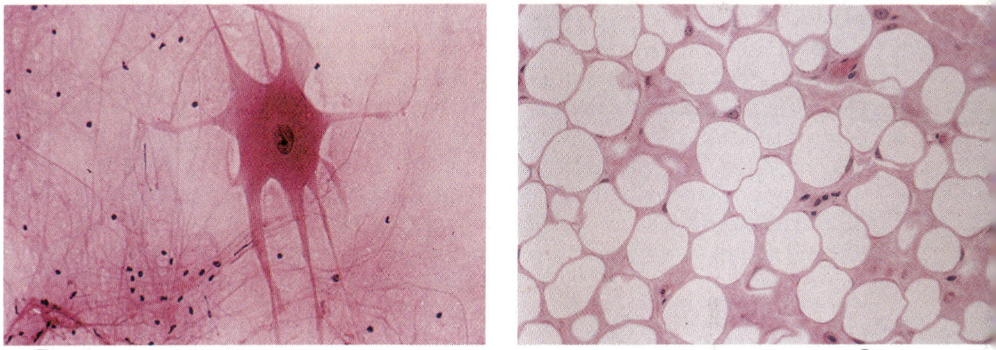

FIGURE 3-1 Cells—wondrously diverse in shape, structure, and function.
A Paramecium.
B Bacteria.
C Sperm cells.
D Plant cells.
E Bone cells.
F Nerve cell.
G Fat cells.

Microtubules are only one type in a myriad of structures found within cells. Many other subcellular structures do jobs such as providing energy, manufacturing molecules, and directing the activities of cells. But what are these complex structures called "cells"?

Cells are much more than the empty "boxes" seen by the British microscopist Robert Hooke in the mid-1600s. Scientists today know that cells are the smallest unit of life that can exist independently. Cells can take in nutrients, break them down to release energy, and get rid of wastes. They can reproduce, react to stimuli, and maintain an internal environment different from their surroundings. In multicellular organisms, cells work together to maintain life. In single-celled organisms, each cell survives independently. This chapter will help you become familiar with the structure of cells and how they work. By studying their structure and function, you will also see their tremendous diversity and complexity (Figure 3-1).

The cell theory

In 1839, botanist Matthias Schleiden and zoologist Theodor Schwann formulated the theory that all living things are made up of cells. In other words, they realized that cells make up the structure of such things as houseplants and people—but not of rocks or soil. It took another 50 years and the work of Rudolf Virchow for scientists to understand another basic concept about cells. Living cells can only be produced by other living cells. As Virchow put it, "all cells from cells." Together, these concepts are called **the cell theory.**

The cell theory is a profound statement regarding the nature of living things. It includes three basic principles:

1. All living things are made up of one or more cells.
2. The smallest *living* unit of structure and function of all organisms is the cell.
3. All cells arise from preexisting cells.

When these statements were formulated in the mid-1800s, scientists discarded the idea of **spontaneous generation:** that living things could arise from the nonliving. This theory suggested that frogs could be born of the mud in a pond and that rotting meat could spawn the larvae of flies. After the development of the cell theory, scientists recognized that life arose directly from the growth and division of single cells. Today, scientists recognize that life on earth represents a continuous line of descent from the first cells that evolved on Earth.

All organisms on Earth are cells or are made up of groups of cells. In addition, all organisms are descendants of the first cells.

Why aren't cells larger?

Most animal cells are extremely small, ranging in diameter from about 10 to 30 microns. To help you understand this size, there are 1000 microns in 1 millimeter (mm)—the width of a paper clip's wire. As you might expect, most cells are invisible to the naked eye without the aid of a microscope. Your red blood cells, for example, are so small that it would take a row of about 2500 of them to span the diameter of a dime. Only a few kinds of cells are large. Individual cells of the marine alga *Acetabularia,* for example, are up to 5 centimeters long (Figure 3-2). If you had eggs for breakfast, you were eating single cells! Other kinds of cells are long and thin, such as nerve cells that run from your spinal cord to your toes or fingers. But not very many cells are as large as a hen's egg or as long as a nerve cell.

To understand why cells are so small, you must first realize that most cells are constantly working, doing such jobs as breaking down molecules for energy, producing substances that cells need, and getting rid of wastes. Each cell is a hotbed of activity and must move substances in and out across its boundary—the cell membrane—quickly enough to meet its needs. Therefore the amount of membranous surface area a cell has in relationship to the volume it encloses is crucial to its survival.

To illustrate, suppose that a cell is a cube like the one pictured in Figure 3-3. Each side of the "cell" is 1 centimeter

FIGURE 3-2 The marine green alga *Acetabularia.*
The marine green alga *Acetabularia* is a large, single-celled organism with clearly differentiated parts, such as the stalks and elaborate "hats" visible here. Each cell is a different individual several centimeters tall.

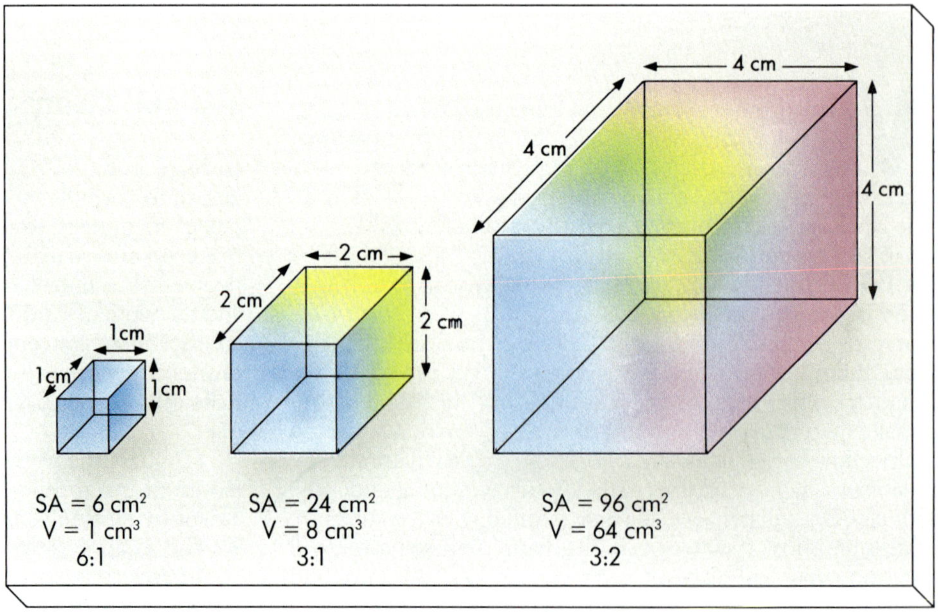

FIGURE 3-3 Cells maintain a large surface-to-volume ratio.
As the cell increases in volume, the ratio of its surface area to volume gets smaller. Multicellular organisms are made up of many microscopic cells. This organization increases the total surface area of the cell membranes within living things.

(cm) in length. The surface area of the membrane of each "side" of the cell is 1 cm × 1 cm, or 1 cm^2. Therefore the total area of the six surfaces or sides is 6 cm^2. The volume of this cell is calculated by multiplying length × width × height. Its volume is therefore 1 cm × 1 cm × 1 cm, or 1 cm^3. The ratio, or relationship between the surface area and volume of this cell, can be expressed as 6:1.

What happens as the cell grows? As Figure 3-3 points out, as the cell grows to 2 centimeters on a side, the total surface area is 24 cm^2. The volume increases to 8 cm^3 (2 cm × 2 cm × 2 cm). This surface-to-volume ratio is 24: 8, or 3:1. As the cell grew, it went from having 6 cm^2 of surface area for each 1 cm^3 of volume, to only 3 cm^2 for each 1 cm^3. The diagram also illustrates that this ratio only gets smaller as the cell continues to grow to 4 centimeters on a side.

Therefore one reason that cells are microscopic is that they must maintain a large surface area-to-volume ratio or they cannot move substances in and out (across their membranes) fast enough to meet their needs. Another reason that cells generally do not grow large has to do with their controlling centers, or **nuclei.** Scientists think that a nucleus could not control all the activities of an active cell if it grew too large. In fact, some large complex cells, such as the unicellular paramecia, have two nuclei. Large cells with only one nucleus, such as unfertilized egg cells, are usually inactive cells.

Most cells are microscopically small so that substances can move in and out of them, across their cell membranes, quickly enough to meet their needs.

Two kinds of cells

All cells can be grouped into two broad categories: **prokaryotic cells** and **eukaryotic cells.** Cells are placed into one of these categories based on their types of structure. Prokaryotic cells, or prokaryotes, have a simpler structure than eukaryotes and were the first type of cell to exist as life arose on Earth billions of years ago. Eukaryotic cells evolved from these simpler cells.

You know the prokaryotes as *bacteria*. The cells of organisms within the other four kingdoms of life—plants, animals, protists, and fungi—are eukaryotes. Although the structure of cells in either group may vary among species, prokaryotes and eukaryotes each have distinctive features common to each group. In addition, because members of both groups are living cells, they also have some of the same features. Almost all cells have the following four characteristics:

1. A surrounding membrane
2. A thick fluid enclosed by this membrane that, along with the other cell contents, is called *protoplasm*
3. Organelles, or "little organs," located within the protoplasm (of eukaryotes) that carry out certain cellular functions
4. A control center called a *nucleus* (in eukaryotes) or *nucleoid* (in prokaryotes) containing the hereditary material DNA

Look at Figures 3-4, 3-5, and 3-6 to identify these characteristics. Also, refer often to Table 3-2 located at the end of this chapter to help you clarify and classify the differences among three cell types: plant cells and animal cells (both eukaryotes) and bacterial cells (prokaryotes).

FIGURE 3-4 The structure of a bacterial cell. Bacterial cells are prokaryotic and have a simpler structure than eukaryotic cells. Prokaryotic cells have no membrane-bound organelles and fewer compartmentalized functions. The cells of all forms of life, however, have a plasma membrane, DNA, and ribosomes.

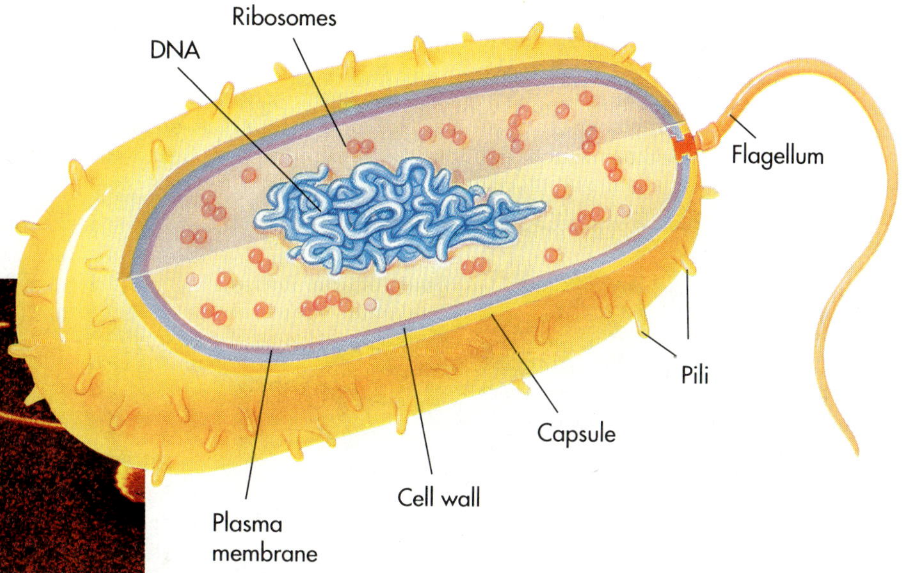

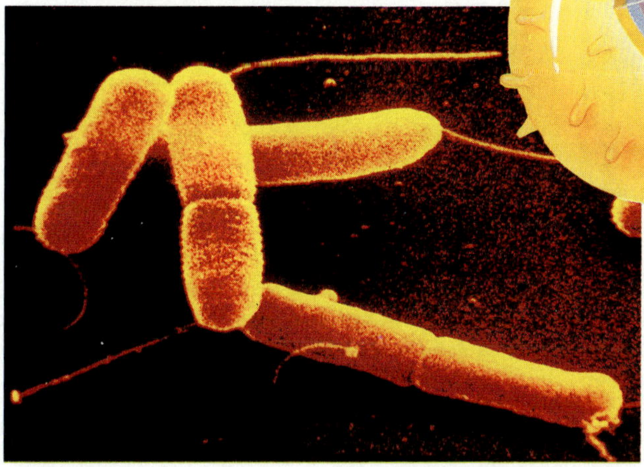

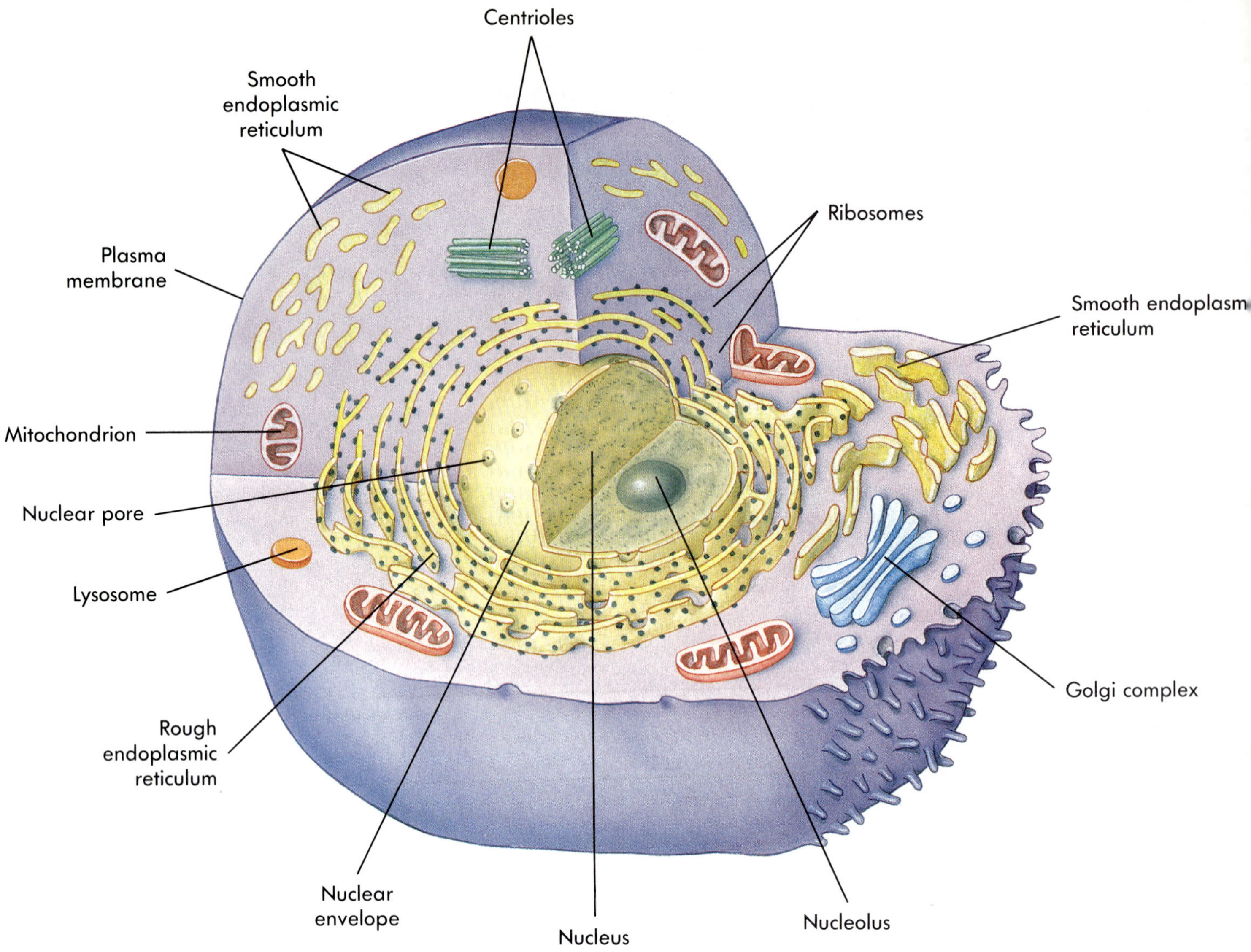

FIGURE 3-5 An animal cell.
This diagram shows the major eukaryotic organelles. Membranes and membranous organelles create anatomical and functional compartments, including a complex system of transport channels. Organelles found in eukaryotic cells include the true nucleus, mitochondria, endoplasmic reticulum, and Golgi bodies.

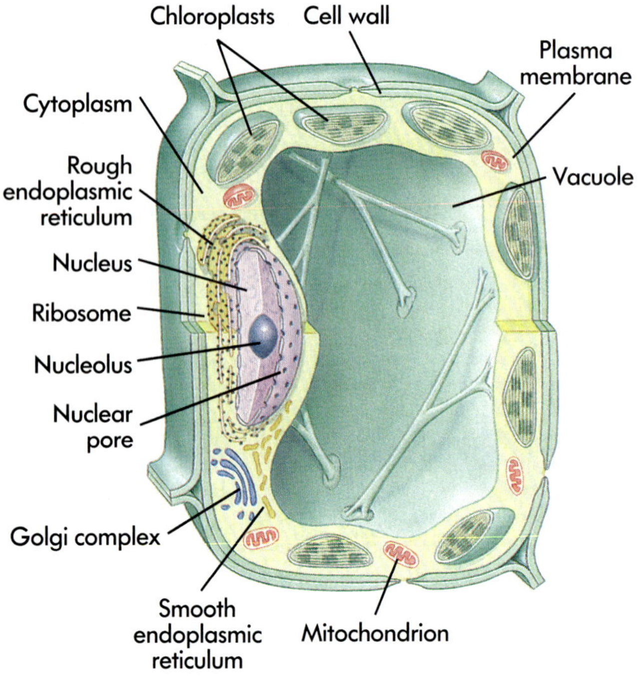

FIGURE 3-6 A plant cell.
Many plant cell organelles are similar to those in animal cells, since both are eukaryotic. But the typical plant cell has chloroplasts that carry out photosynthesis, has a rigid cellulose cell wall, and frequently has large water-filled vacuoles occupying a major part of the cell volume.

Eukaryotic cells: An overview

Although eukaryotic cells are quite a diverse group, they share a basic architecture. They are all bounded by a membrane called the **plasma membrane.** This membrane encloses a semifluid material called the *cytosol,* which is crisscrossed with a supporting framework of protein called a **cytoskeleton.** All eukaryotic cells also possess many organelles (Table 3-1). These organelles are of two general kinds: (1) membranes or organelles derived from membranes and (2) bacteria-like organelles.

Most biologists agree that the bacteria-like organelles in eukaryotes were derived from ancient **symbiotic bacteria.** The word **symbiosis** means that two or more organisms live together in a close association. An organism that is symbiotic within another is called an **endosymbiont.** The major endosymbionts that occur in eukaryotic cells are mitochondria, which occur in all but a very few eukaryotic organisms, and chloroplasts, which occur in algae and plants.

The plasma membrane

All eukaryotic cells are bounded by a membrane called the ***plasma membrane.*** It gets this name because it encloses the protoplasm, the semifluid cell contents, and cell organelles. Although the plasma membrane is a thin, nonrigid structure, in many cells it is molded into a specific shape. Human red blood cells, for example, look like a round pillow that has been punched in the center (Figure 3-7, *A*). The muscle cells of the stomach have tapered ends and bulging middles. In cells that have no other supporting structure (such as a rigid cell wall), cell shapes are determined by a protein network lying beneath the plasma membrane. Were it not for this protein network, or cytoskeleton, these cells would look like shapeless blobs.

The plasma membrane has two main components: phospholipids and proteins. Lipids, as you may recall from Chapter 2, are made up of glycerol and fatty acids. The phospholipids of plasma membranes are composed of two fatty acids bonded to a glycerol backbone. A polar, nonlipid mol-

TABLE 3-1

Eukaryotic cell structures and their functions

STRUCTURE	DESCRIPTION	FUNCTION
STRUCTURAL ELEMENTS		
Cell wall	Outer layer of cellulose or chitin, or absent	Protection, support
Plasma membrane	Lipid bilayer in which proteins are embedded	Regulation of what passes in and out of cell, cell-to-cell recognition
Cytoskeleton	Network of protein filaments	Structural support, cell movement
Flagella (cilia)	Cellular extensions with 9 + 2 arrangement of pairs of microtubules	Motility or moving fluids over surfaces
ORGANELLES		
Endoplasmic reticulum	Network of internal membranes	Formation of compartments and vesicles
Ribosomes	Small, complex assemblies of protein and RNA, often bound to ER	Sites of protein synthesis
Nucleus	Spherical structure bounded by double membrane, site of chromosomes	Control center of cell, directing of protein synthesis and cell reproduction
Chromosomes	Long threads of DNA associated with protein	Sites of hereditary information
Nucleolus	Site on chromosome of rRNA synthesis	Assembly of ribosomes
Golgi complex	Stacks of flattened vesicles	Packaging of proteins for export from cell
Lysosomes	Membranous sacs containing digestive enzymes	Digestion of worn-out mitochondria and cell debris, role in cell death
Mitochondria	Bacteria-like elements with inner membrane highly folded	Power plant of the cell
Chloroplasts	Bacteria-like elements with vesicles containing chlorophyll found in plant cells and algae	Site of photosynthesis in plant cells and algae

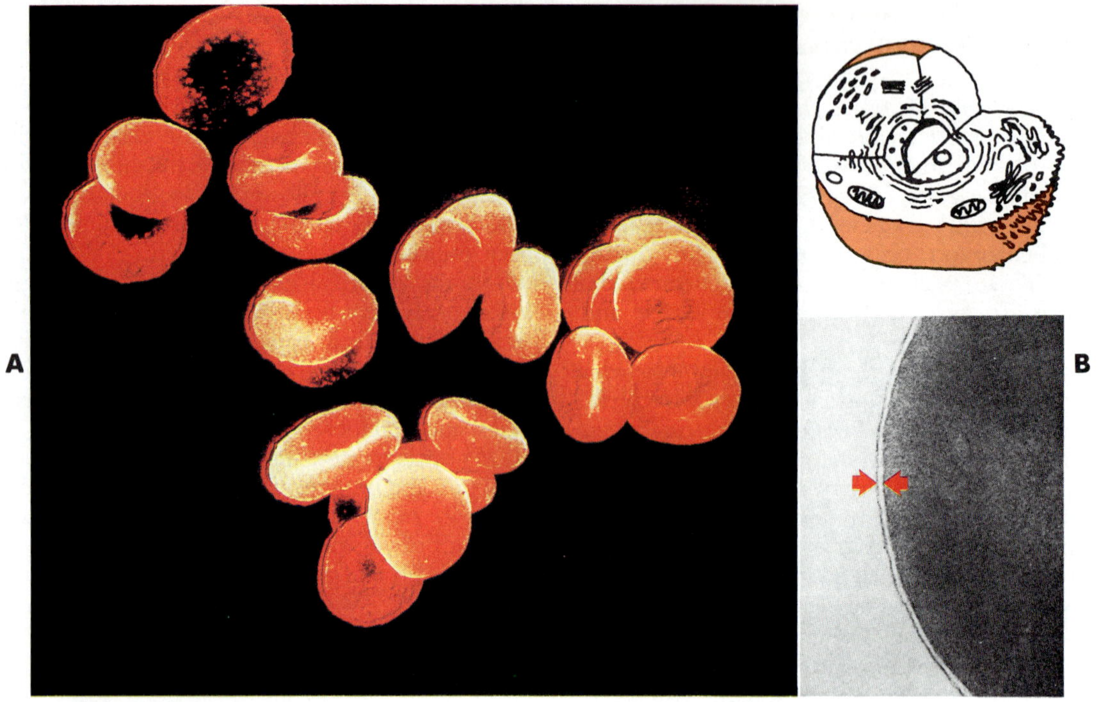

FIGURE 3-7 Red blood cells.
A The human red blood cell is a disk with a concave depression on both sides.
B This electron micrograph of a thin section of a red blood cell ($\times$200,000) clearly shows the double-layered plasma membrane, which is indicated by *arrows.*

ecule is also attached to the glycerol backbone by means of a phosphate group. In the plasma membrane, these molecules form a double layer (Figure 3-7, *B*).

Nothing can enter or leave the cell without passing across the plasma membrane. Interestingly, very little does cross the lipid layer itself. Instead, a variety of proteins penetrates the lipid layer and controls the interactions of the cell with its environment. Some of these proteins, called *channels,* act as doors that let specific molecules into and out of the cell. Other proteins, called *receptors,* recognize certain chemicals such as hormones that signal cells to respond in particular ways (see Chapter 19). Receptor proteins cause changes within the cell when they come in contact with such chemical "messengers" or signals. A third type of protein works to identify a cell as being of a particular type. These cell surface markers help cells identify one another, an ability that helps cells in a multicellular organism (such as your body) function correctly. (Chapter 4 presents a more detailed look at cell membranes.)

The plasma membrane is a thin, nonrigid structure that encloses the cell and regulates interactions between the cell and its environment.

The cytoplasm and cytoskeleton

The **cytoplasm** of the cell is a viscous fluid containing all cell organelles except the nucleus. The word *cytoplasm* literally means "living gel" (plasm) "of the cell" (cyto). The major components of the cytoplasm are (1) a gel-like fluid, (2) storage substances, (3) a network of interconnected filaments and fibers, and (4) cell organelles. The fluid part of the cytoplasm is made up of approximately 75% water and 25% proteins. The proteins, mostly enzymes and structural proteins, make the cytoplasm viscous—much like thickening gelatin. Nonprotein molecules involved in the various chemical reactions of the cell, such as ions and ATP (an important energy-storing molecule), are also dissolved in this fluid.

Storage substances vary from one type of cell to another. For example, the cytoplasm of liver cells contains large molecules of glycogen, a storage form of glucose. Fat cells contain a large lipid droplet.

The cytoplasm also contains a cytoskeleton—a network of filaments and fibers that do many jobs. The cytoskeleton is made up of three different types of fibers: **microfilaments, microtubules,** and **intermediate fibers.** The microfilaments are the thinnest—a twisted double chain of protein—and are invisible with an ordinary light microscope. The microtubules, a chain of proteins wrapped in a spiral to form a tube, are the thickest members of the cytoskeleton. As the name suggests, the intermediate fibers are an in-between size. These are thread-like protein molecules that wrap around one another to form "ropes" of protein (Figure 3-8).

The microfilaments and intermediate fibers provide the protein network that lies beneath the plasma membrane to help support and shape the cell (Figure 3-9). This network of proteins extends into the cytoplasm and looks something like the web of proteins shown in Figure 3-10. Microtubules

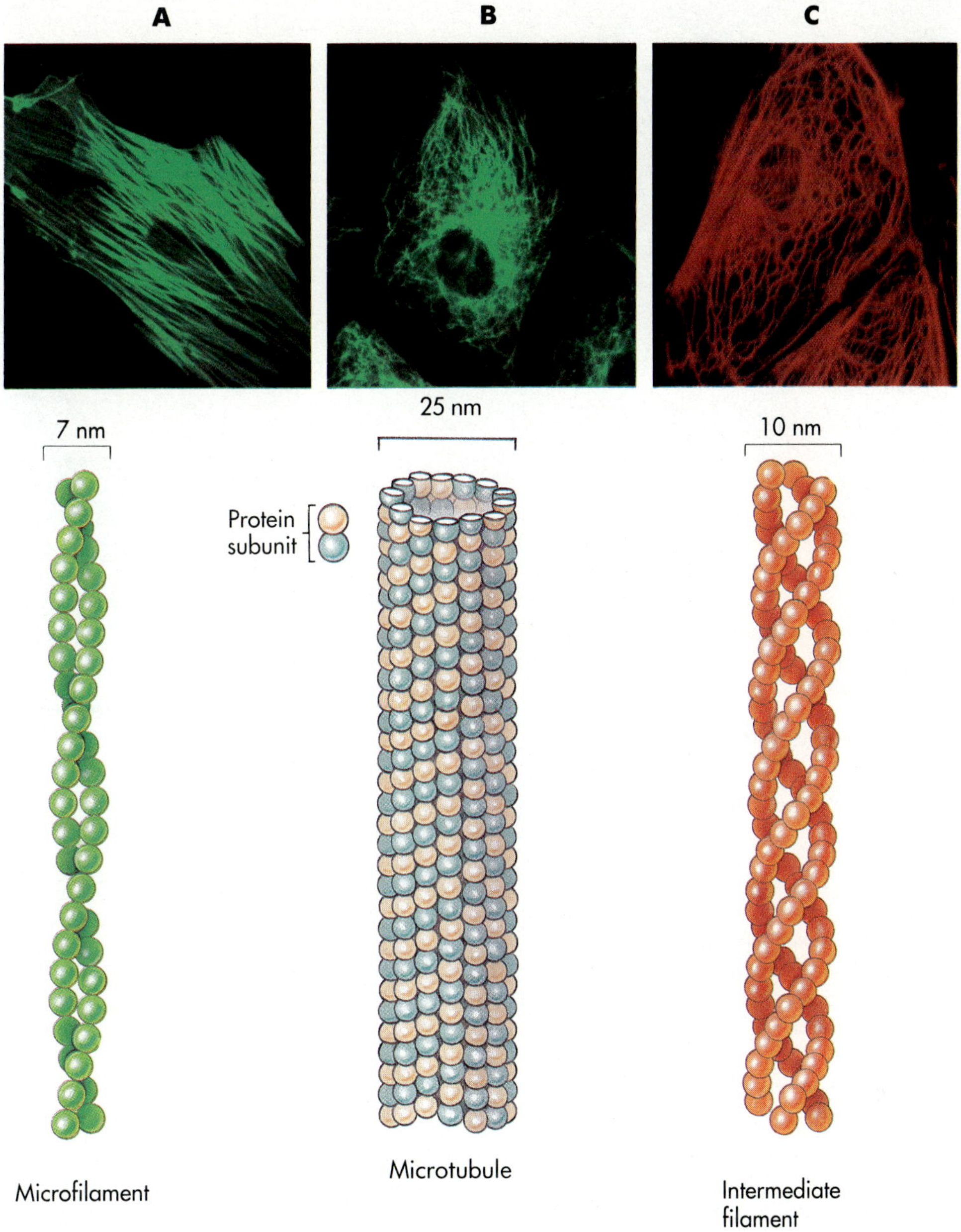

FIGURE 3-8 Molecules that make up the cytoskeleton.
A Microfilaments run parallel to the cell membrane in bundles.
B Microtubules are composed of a spiral array of protein subunits. The microtubules in the micrograph radiate from an area near the nucleus (the dark region in the center). Microtubules provide intracellular support in the nondividing cell.
C Intermediate filaments are probably made up of three subunits that are wound together in a coil, interrupted by uncoiled regions. In a skin cell, such as the one shown, intermediate filaments form thick, wavy bundles that probably provide structural support.

are also part of this "skeleton" of the cytoplasm. Together, these three types of protein fibers provide the cell with mechanical support and help anchor the organelles. They also help move substances from one part of the cell to another. The filaments forming the highway in the chapter opener photo are microtubules found within nerve cells.

The cytoskelton is a network of filaments and fibers within the cytoplasm that helps maintain the shape of the cell, move substances within cells, and anchor various structures in place.

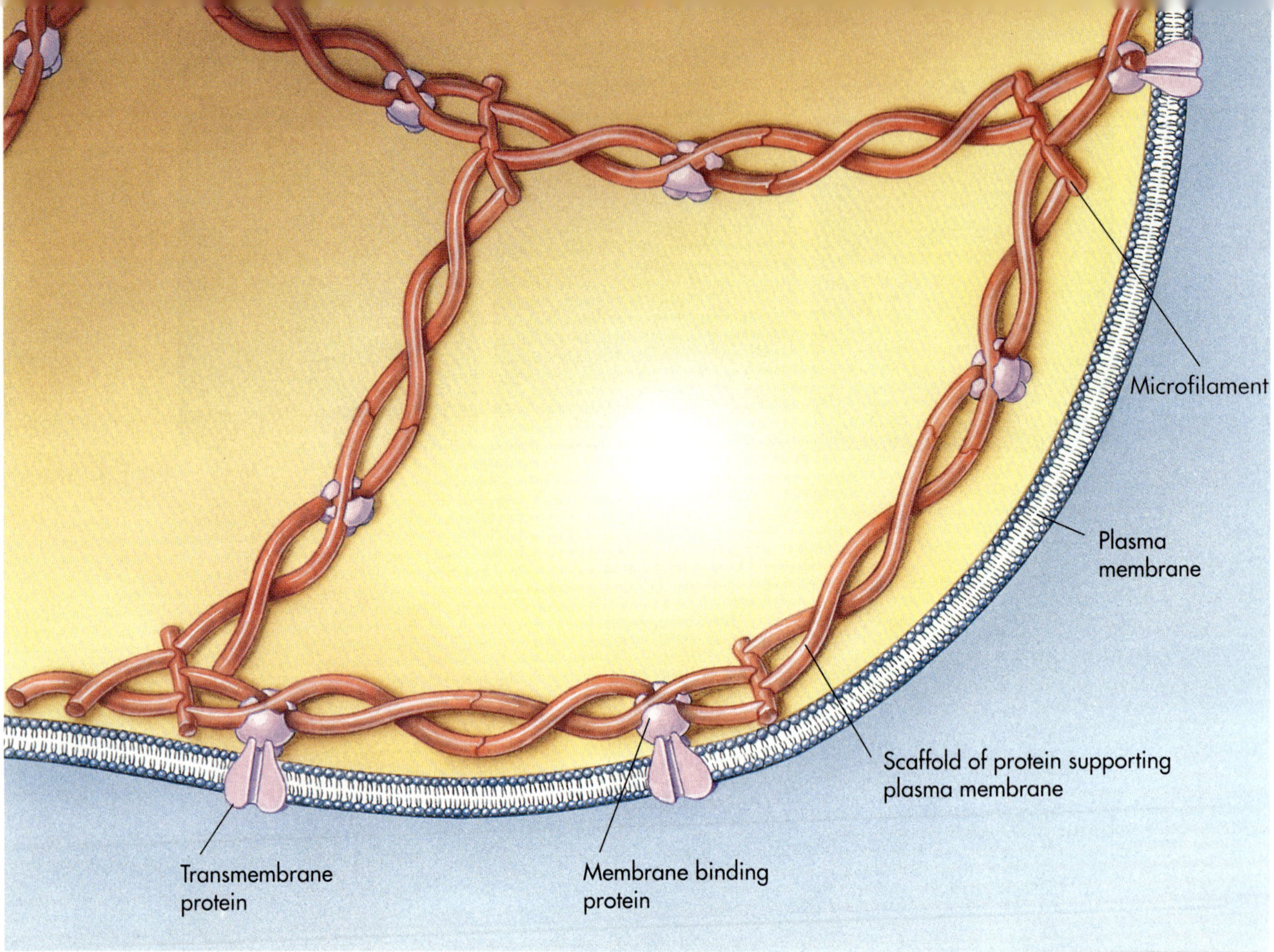

FIGURE 3-9 Membrane support.
Cell membranes are reinforced by a network of protein fibers. Here you can see a portion of the network, attached to the inside of the plasma membrane by proteins inserted into the lipid bilayer.

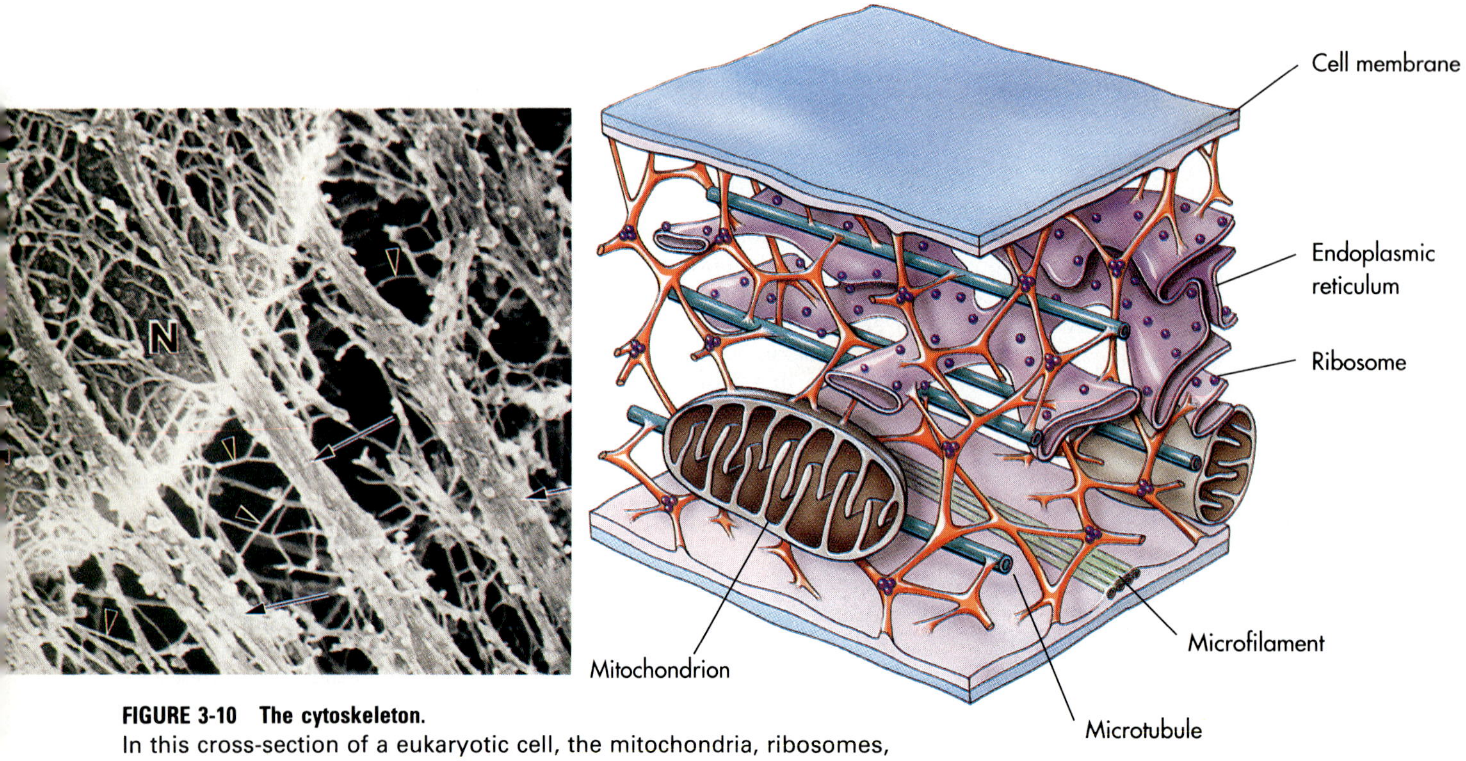

FIGURE 3-10 The cytoskeleton.
In this cross-section of a eukaryotic cell, the mitochondria, ribosomes, and endoplasmic reticulum are all supported by a fine network of filaments. Microtubules run through the filaments, linking various portions of the cell. In the micrograph, the *arrows* point to the tiny filaments that anchor the nucleus to the cytoplasm.

Membranous organelles

Enmeshed within the cytoskeletal fibers, the cell organelles constantly work, each contributing in a special way to the life and well-being of the cell. Most organelles of eukaryotic cells are bounded by membranes. Many different enzymes—molecules that help specific chemical reactions take place—are attached to these membranes. In this way, these subcellular structures form organized compartments within the cytoplasm and make the cell an efficiently running living machine.

The endoplasmic reticulum

The **endoplasmic reticulum,** or ER, is an extensive system of interconnected membranes that forms flattened channels and tube-like canals within the cytoplasm, almost like a cellular subway system. The channels are used to help move substances from one part of the cell to another. The name *endoplasmic reticulum* may sound very strange. It is, however, descriptive of its location and appearance. The word *endoplasmic* means "within the cytoplasm." The word *reticulum* comes from a Latin word meaning "a little net." And that is exactly what the ER looks like—a net within the cytoplasm.

There are two types of endoplasmic reticula in cells: **rough ER** and **smooth ER.** The *rough* refers to the minute, spherical structures called **ribosomes** covering their surfaces. With ribosomes dotting their surfaces, these ER membranes look much like long sheets of sandpaper (Figure 3-11).

Ribosomes are the places where proteins are manufactured. Some ribosomes are attached to the rough ER, but others float in the cytoplasm bound to cytoskeletal fibers. The cytoplasmic ribosomes help produce proteins for use in the cell, such as those making up the structure of the cytoskeletal fibers or certain organelles. The proteins made at the ribosomes of the rough ER are most often destined to leave the cell. Cells specialized for secreting proteins, such as the pancreatic cells that manufacture the hormone insulin, contain large amounts of rough ER. After the proteins are manufactured at the ribosomes on the surface of the ER, they enter the inner space of the ER. Within this channel the proteins may be changed by enzymes bound to the inner surface of the ER membrane. Carbohydrate molecules are often added to them.

When the proteins reach the end of their journey in the ER, they are encased in tiny membrane-bounded sacs called **vesicles.** These vesicles are formed by sections of smooth ER continuous with the rough ER. "Buds" of the smooth ER pinch off as the newly formed proteins reach them. These vesicles eventually fuse with the membranes of another organelle called the **Golgi body.**

Smooth ER has no ribosomes attached to its surfaces. Therefore it does not have the grainy appearance of rough ER and does not manufacture proteins. Instead, smooth ER has enzymes bound to its inner surfaces that help build carbohydrates and lipids. Cells specialized for the synthesis of these molecules, such as animal cells that produce male or female sex hormones, have abundant smooth ER.

The endoplasmic reticulum (ER) is an extensive system of membranes that divides the interior of eukaryotic cells into compartments and channels. Rough ER makes and transports proteins destined to leave the cell. Smooth ER helps build carbohydrates and lipids.

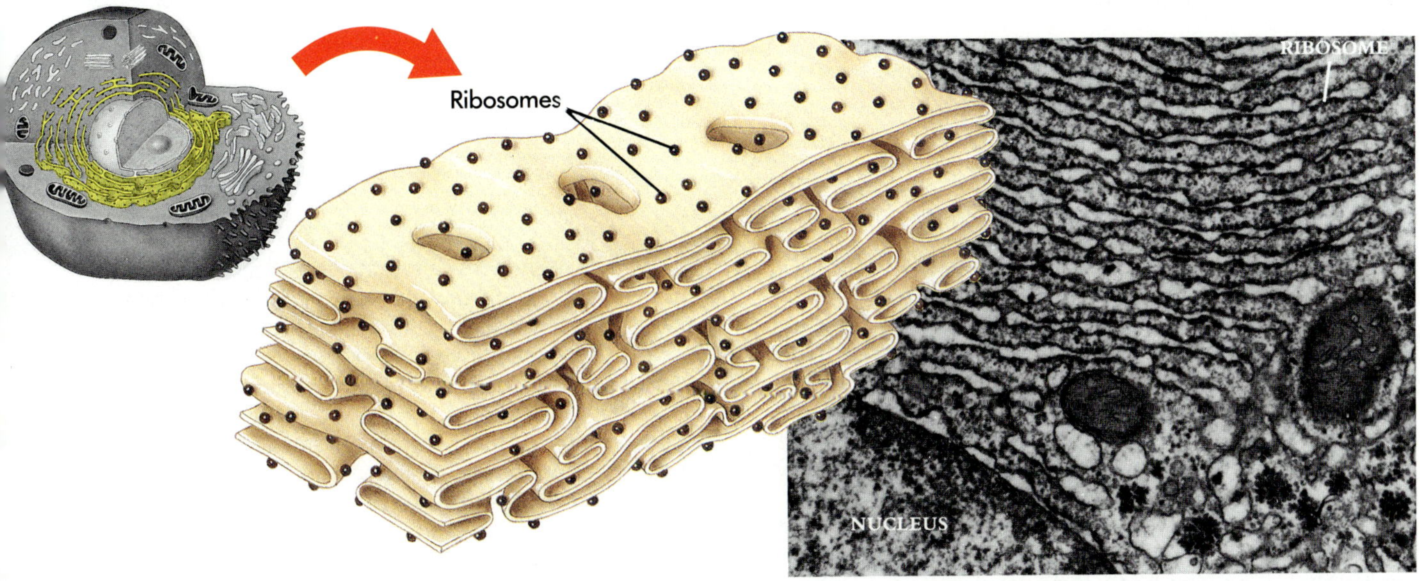

FIGURE 3-11 Rough endoplasmic reticulum (ER).
The electron micrograph is of a rat liver cell, rich in ribosomes attached to ER. As the drawing indicates more clearly, you can see that the ribosomes are associated with only one side of the rough ER. Proteins are synthesized at the ribosomes along the rough ER, secreted into the ER channel, and may be either transported to other parts of the cell or secreted outside the cell.

BIOLOGY IN FOCUS

How cells crawl

You usually think of your body cells as not moving, stuck together in solid masses of tissue—each liver cell a part of your liver, each skin cell a part of your skin. The cells in your blood are usually thought of as being driven through the circulation system by the pumping of the heart. However, many cells in your body move about actively, going from one place to another. In your blood, for example, is a kind of white blood cell called a *phagocyte* that "hunts down" and destroys any bacteria that invade your body (Figure 3-A). Reacting to molecules released by the bacteria, the phagocyte follows this chemical trail to their quarry. After they find it, they engulf and digest their prey. Your body contains more than 100 billion of these phagocytes. If you add up all the distance they creep each day, the total distance is greater than 60,000 miles!

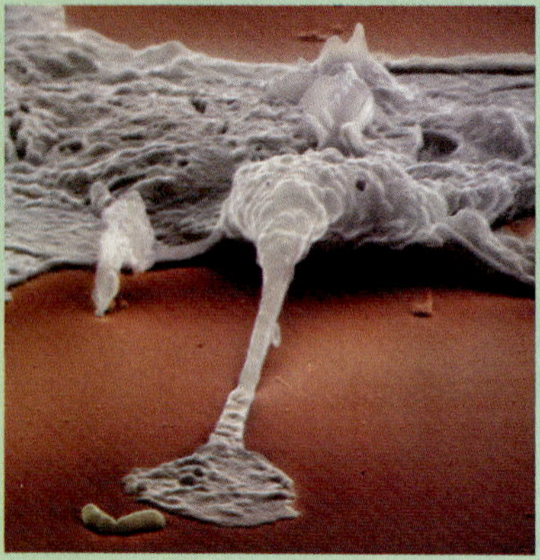

FIGURE 3-A In this electron micrograph, a phagocyte is chasing the rod-shaped bacterium in the foreground. Notice the pseudopod, or false foot, that inches toward the invader.

How does a phagocyte "chase" a bacterium? Most research now favors what has been called the *toothpaste-tube hypothesis.* The cytoplasm lying just within the cell membrane of each phagocyte is rich in actin, a protein that can aggregate to become firm and gel-like. This stiff layer of actin beneath the membrane gives the cell form and shape.

When a molecule released by a bacterium collides with the phagocyte's surface and binds to receptor proteins, the actin gel breaks down in the zone immediately below the membrane. This breakdown creates a deep pool of unattached actin proteins. What happens next is dramatic. Each of the unattached actin molecules is, in chemical terms, a solute. Breaking down the gel into its component actin molecules creates a high concentration of solutes just below the membrane surface. Water always moves from a region of low solute concentration to one of high concentration, a process called *osmosis.* The pool of concentrated actin causes water to be drawn outward from the cell interior. The rush of water into the region beneath the membrane creates a strong osmotic pressure that literally pushes the cell membrane out at that point, creating an outward extension of the cell called a *pseudopod,* which means "false foot."

Now you can understand how a phagocyte "hunts" bacteria so successfully. Remember that it was a chemical released from a nearby bacterium that set off this chain of events. It is at the point of the chemical encounter that the pseudopod shoots out, *in the direction the bacterial molecule came from*—just as if the pseudopod had been deliberately aimed at the bacterium.

Golgi bodies

First described by the physician Camillo Golgi in the last half of the nineteenth century, Golgi bodies look like microscopic stacks of pancakes in the cytoplasm. Animal cells each contain 10 to 20 sets of these flattened membranes. Plant cells may contain several hundred because the Golgi bodies are involved in the synthesis and maintenance of plant cell walls (a structure animal cells do not have). Collectively, the Golgi bodies are referred to as the **Golgi complex.**

Molecules come to a Golgi body in vesicles pinched off from the ER. The membranes of the vesicles fuse with the membranes of a Golgi body. Once inside the space formed by the Golgi membranes, the molecules may be modified by the formation of new chemical bonds or by the addition of carbohydrates. For example, mucin, which is a protein with attached carbohydrates that forms a major part of the mucous secretions of the body, is put together in its final form in the Golgi. This "refining" of molecules occurs in stages in the Golgi, with different parts of the stack of membranes containing enzymes that do specific jobs. When molecular products are ready for transport, they are sorted and pinched off in separate vesicles (Figure 3-12). Each vesicle travels to its destination and fuses with another membrane. The vesicles containing those molecules that are to be secreted from the cell fuse with the plasma membrane. In this way the contents of the vesicle are liberated from the cell (Figure 3-13). Other vesicles fuse with the membranes of organelles such as lysosomes, delivering the new molecules to their interiors.

The Golgi complex is the delivery system of the eukaryotic cell. It collects, modifies, packages, and distributes molecules that are made at one location within the cell and used at another.

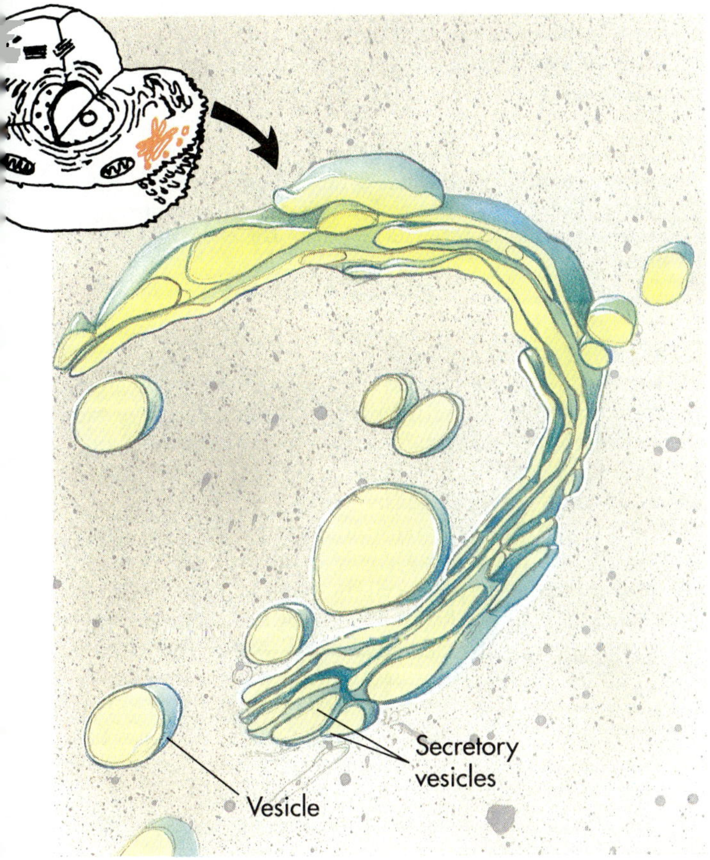

FIGURE 3-12 A Golgi complex.
The Golgi complex is a structure that collects, modifies, packages, and distributes molecules, particularly proteins made at the rough ER and lipids made at the smooth ER. These molecules accumulate in the Golgi vesicles, where they may be modified by chemicals into glycoproteins or glycolipids. When ready, the molecules are sorted and pinched off in separate secretory vesicles.

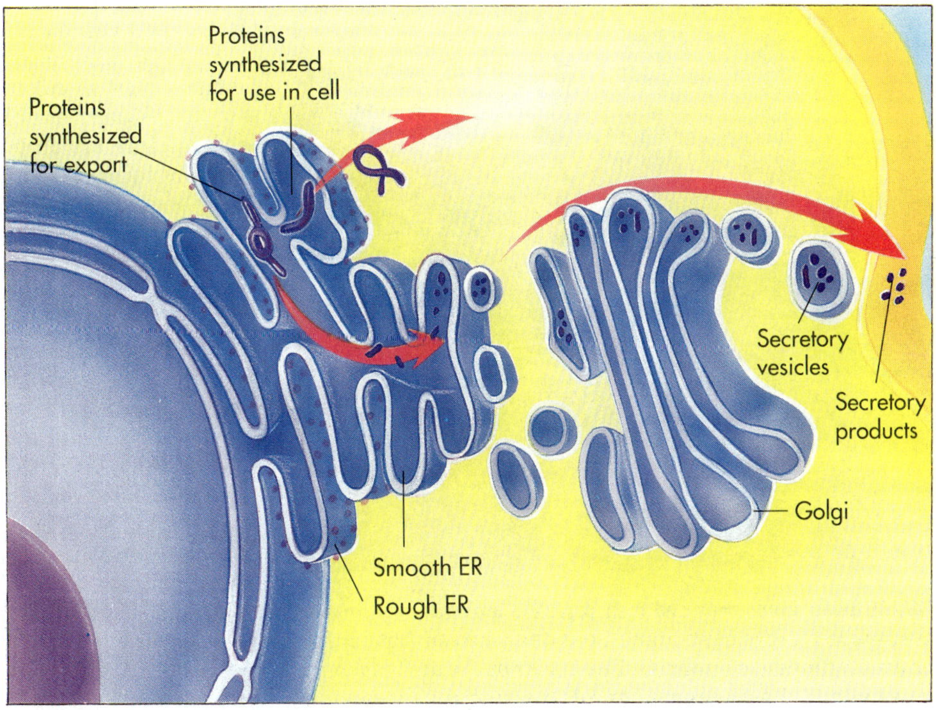

FIGURE 3-13 How a molecule is packaged for transport in a Golgi body.
Molecules, in this case proteins, are synthesized at the ribosomes attached to the ER. A protein can either move out into the cytoplasm for use by the cell, or it can be packaged for export by the Golgi body. The proteins destined for export move through the Golgi body, are modified by other chemicals, and are finally encased in secretory vesicles. These vesicles then fuse with the cell's plasma membrane, liberating the proteins from the cell.

Lysosomes

The "new" molecules delivered to **lysosomes** are digestive enzymes—molecules that help break large molecules into smaller molecules. Lysosomes are, in fact, membrane-bounded "bags" of many different digestive enzymes (Figure 3-14). Several hundred of these organelles may be present in one cell alone!

Lysosomes and their digestive enzymes are extremely important to the health of a cell. They help cells function by aiding in cell renewal, constantly breaking down old cell parts as they are replaced with new cell parts. During development, lysosomes help remodel tissues, such as the reabsorption of the tadpole tail as the tadpole develops into a frog. In some cells, lysosomes also break down substances brought into the cell from the environment. For example, one job of certain white blood cells is to get rid of bacteria invading the body. The cytoplasm of these cells flows around their "prey," engulfing them in a membrane-bounded sac called a **vacuole.** A lysosome then fuses with this vacuole, and digestion of the invader begins. In the past, lysosomes were thought to be active during the aging process, breaking open and destroying whole cells. Scientists now know this idea to be false, since lysosome enzymes work only in the acid environment of the lysosome and become inactive at the relatively neutral pH of the cytoplasm.

What happens to the digestion products of lysosomes? Substances such as parts of bacteria are packaged in a vesicle, transported to the plasma membrane, and exit the cell. Other molecules, such as the breakdown products of old cell parts, may simply be released into the cytoplasm. These cellular "building blocks" can be recycled—used once again to build new cell parts.

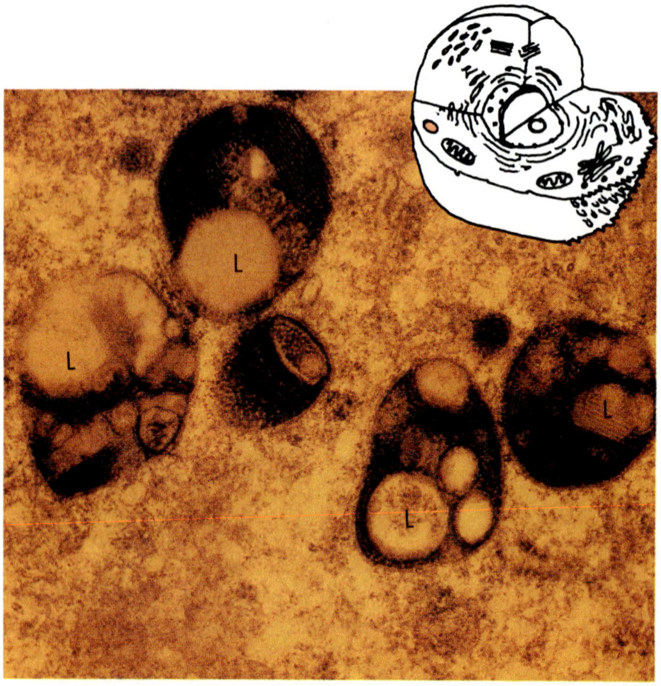

FIGURE 3-14 Lysosomes and their digestive enzymes. Lysosomes (*L*) are membrane-bounded vesicles that contain powerful digestive enzymes. The enzymes are the blackened areas within the lysosomes in this electron micrograph.

Lysosomes are vesicles that contain digestive enzymes. These enzymes break down old cell parts or materials brought into the cell from the environment.

Vacuoles

The word *vacuole* comes from a Latin word meaning "empty." However, these membrane-bounded storage sacs only look empty. Within them can be found such substances as water, food, and wastes. Their number, kind, and size vary in different kinds of cells.

Vacuoles are most often found in plant cells. These giant water-filled sacs play a major role in helping plant tissues stay rigid. In a mature plant cell the vacuole is often so large that it takes up most of the interior of the cell. Certain single-celled eukaryotes (protists) and some yeasts also contain vacuoles. Some fresh water protists contain contractile vacuoles. These organelles take up "extra" water that tends to flow into these organisms. Periodically, these vacuoles fuse with the plasma membrane and expel their contents, keeping the cell from swelling and bursting. Many protists have food vacuoles, places where food is stored and eventually digested by fusion with lysosomes.

The nucleus

The **nucleus,** or control center of the cell, is made up of an outer, double membrane that encloses the chromosomes and one or more nucleoli. In fact, the word *eukaryote* means "true nucleus" and refers to the fact that the nucleus is a closed compartment bounded by a membrane. Other than large fluid-filled vacuoles in plants, this compartment is the largest in eukaryotic cells.

The outer, double membrane of the nucleus is called the **nuclear envelope.** The inner of the two membranes actually forms the boundary of the nucleus, and the outer membrane is continous with the endoplasmic reticulum. At various spots on its surface, the double membrane fuses to form openings called **nuclear pores.** Figure 3-15 shows the surface of the nuclear membrane of a cell that was frozen and then cracked during a special type of preparation for the electron microscope. The pores look like pock marks on the surface of the membrane. These ring-like holes are lined with proteins and serve as passageways for molecules entering and leaving the nucleus.

Within the nucleus is the hereditary material deoxyribonucleic acid, or **DNA.** DNA determines whether your hair is blonde or brown or whether a plant flowers in pink or white. It controls all activities of the cell. To accomplish these amazing feats, DNA performs one job: it directs the synthesis of ribonucleic acid, or RNA, which in turn directs the synthesis of proteins. (The structure, interactions, and roles of both DNA and RNA are discussed further in Chapter 24.)

A

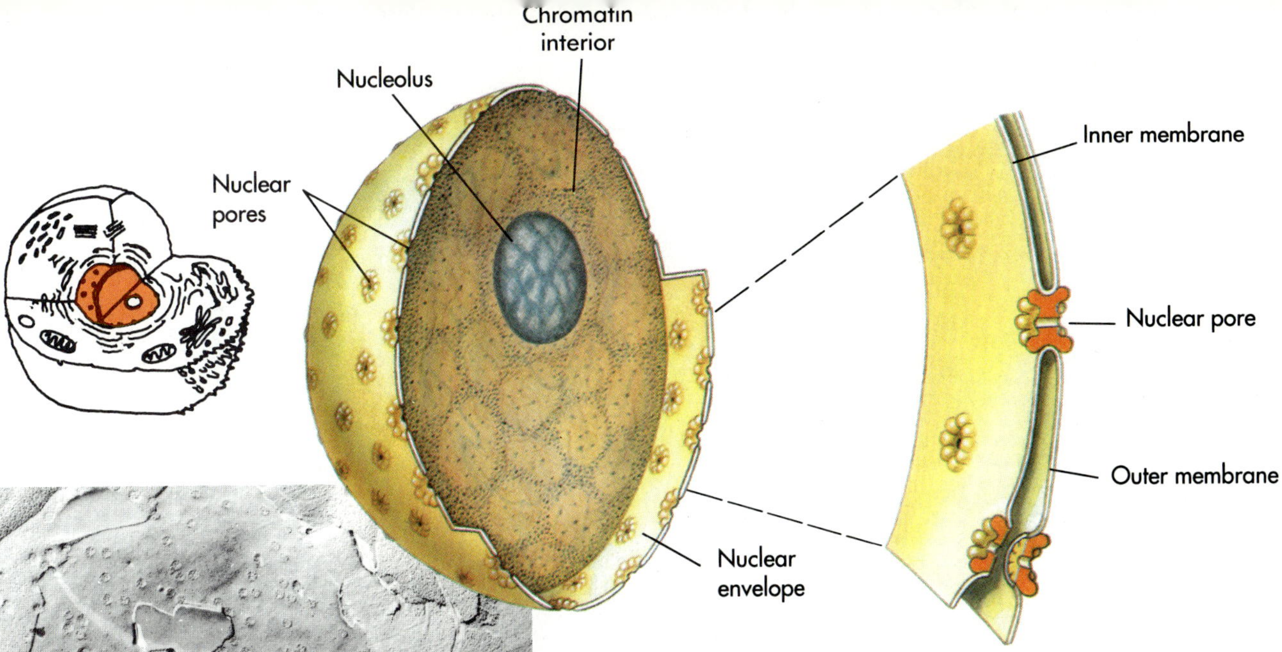

B

C

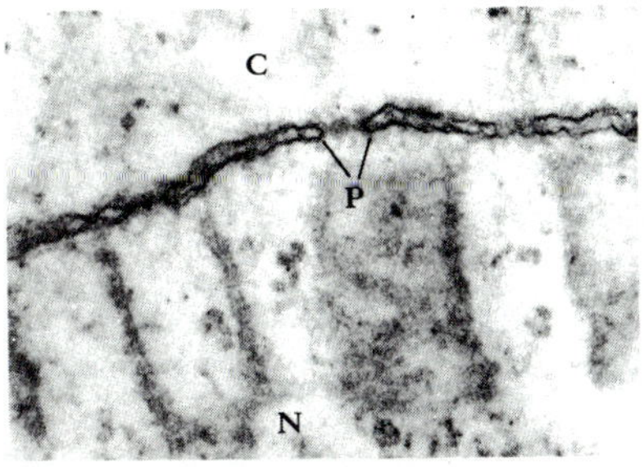

FIGURE 3-15 The nucleus.
The nucleus is surrounded by a double membrane called the nuclear envelope **(A)**. The membrane is covered with pores, visible as depressions in the electron micrograph **(B)**. These pores are visible in **C**, designated by *P*. *C* is the cytoplasm.

DNA is bound to proteins in the nucleus, forming a complex called *chromatin*. In a cell that is not dividing, the chromatin is "strung out," looking like strands of microscopic pearls. But as a cell begins to divide, the DNA coils more tightly around the proteins, condensing to form shortened, thickened structures called **chromosomes** (Figure 3-16).

The nucleus of a eukaryotic cell contains the hereditary material, or DNA. When the cell divides, the DNA condenses into compact structures called *chromosomes*.

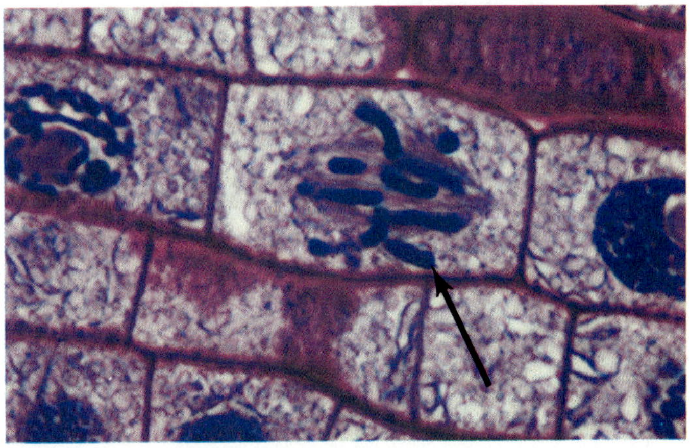

FIGURE 3-16 Eukaryotic chromosomes within an onion root-tip cell.
As a cell begins to divide (a process called *mitosis*), long strands of DNA and protein thicken, shorten, and condense into chromosomes *(arrow)*.

The **nucleolus** (plural nucleoli) is a darkly staining region within the nucleus (Figure 3-17). Most cells have two or more nucleoli. Making up the bulk of the nucleolar material is a "special" area of DNA that directs the synthesis of ribosomal ribonucleic acid, or **rRNA.** As rRNA is made at this DNA, it forms clumps of molecules that are structural components of ribosomes. Proteins, another component of ribosomes, can also be found in the nucleolus. They are

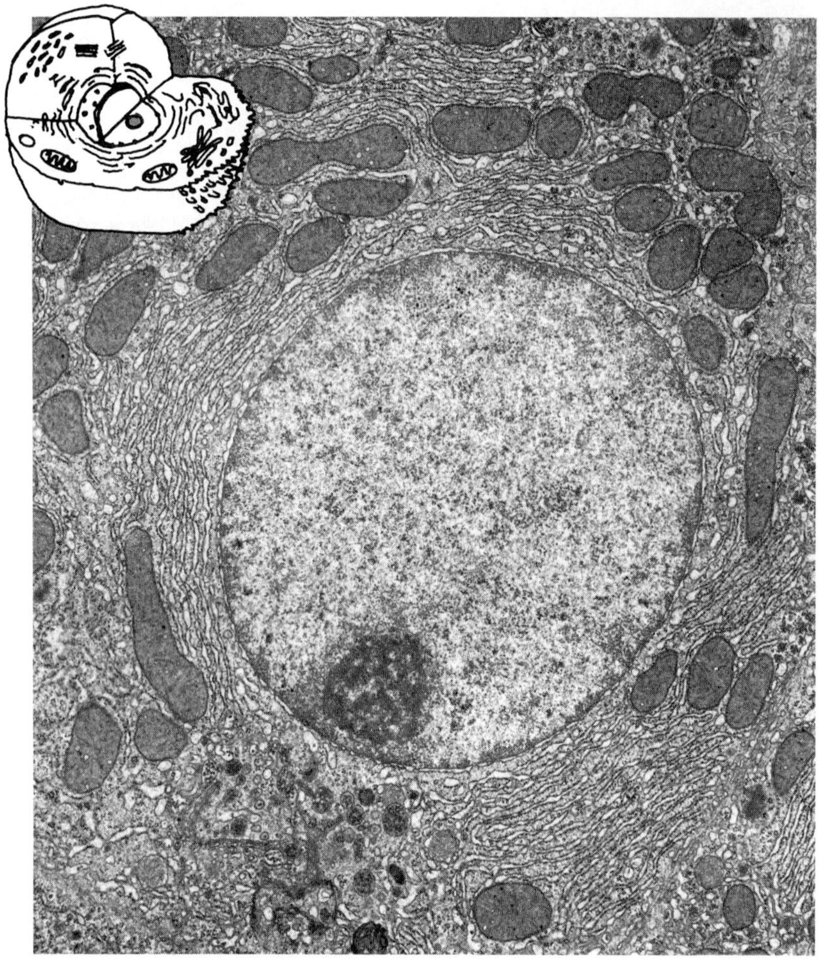

FIGURE 3-17 The nucleolus.
Appearing as a darkly staining region within the nucle[...] the nucleolus is the site of ribosomal RNA synthesis [...] consists of ribosomal RNA plus some ribosomal protei[...]

brought into the nucleus from the cytoplasm. After the ribosomes are manufactured (or partly manufactured) in the nucleolus, they move through the pores in the nuclear membrane and pass into the cytoplasm.

Bacteria-like organelles

The idea of the symbiotic origin of mitochondria and chloroplasts has had a controversial history, and a few biologists still do not accept it. The endosymbiotic hypothesis is, however, the most widely agreed-on model at this time.

Mitochondria

The **mitochondria** that occur in most eukaryotic cells are thought to have originated as symbiotic bacteria. According to this hypothesis the bacteria that became mitochondria were engulfed by eukaryotic cells early in their evolutionary history. Before they had acquired these bacteria, the host cells were unable to carry out chemical reactions necessary for living in an atmosphere that had increasing amounts of oxygen. The engulfed bacteria were able to carry out these reactions and are considered to be the precursors to mitochondria.

Mitochondria are oval, sausage-shaped, or thread-like organelles about the size of bacteria. They are bounded by a double membrane. The outer of the two membranes is smooth and defines the shape of the organelle. The inner membrane, however, has many folds called **cristae** that dip into the interior of the mitochondrion. These cristae resemble the folded membranes that occur in various groups of bacteria. Notice in Figure 3-18 that this arrangement of membranes forms two mitochondrial compartments.

The job of the mitochondria is to break down fuel molecules, releasing energy for cell work. The two most important fuels of cells are glucose and fatty acids. Some organisms, such as plants, make their fuel (glucose) using the raw materials of carbon dioxide, water, and sunlight. Other organisms, such as animals, eat food and digest it to produce glucose and fatty acids, which can then be transported to the cells. In both plants and animals, these fuels are broken down by cells, releasing energy by means of a series of oxygen-requiring reactions called **cellular respiration.** The energy released during these reactions is stored for later use

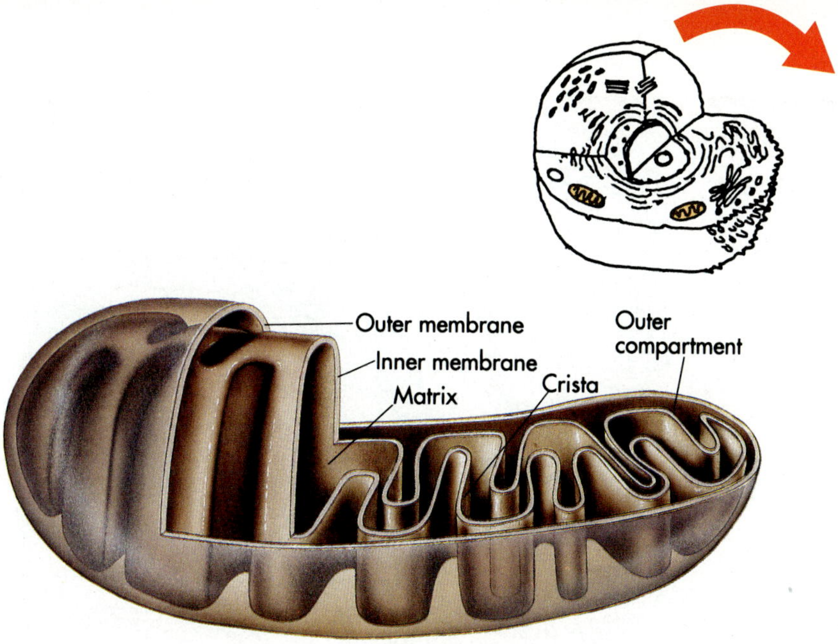

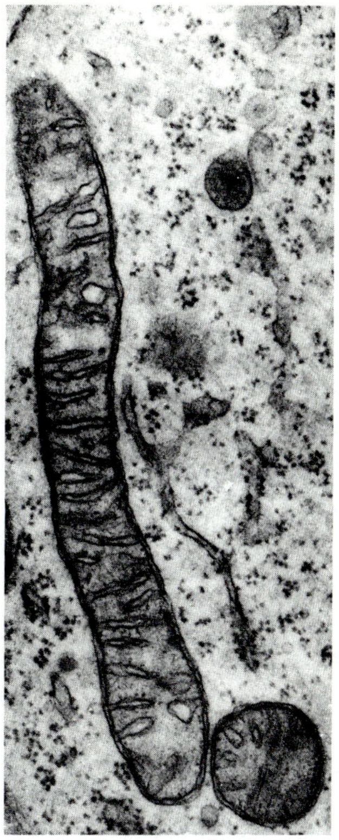

FIGURE 3-18 The mitochondrion.
Mitochondria have a smooth outer membrane and an extensively folded inner membrane. The drawing reveals the intricate structure of the cristae and the inner and outer compartments. The similarity of the electron micrograph to a rod-shaped bacterium (such as in Figure 3-25) is striking.

in special molecules called ATP. (See Chapter 6 for a detailed description of cellular respiration and ATP.) Using chemical symbols, the reactions of cellular respiration can be summarized as follows:

$C_6H_{12}O_6$	+	$6O_2$	→	$6CO_2$	+	$6H_2O$	+	energy
Glucose		Oxygen		Carbon dioxide		Water		

Cellular respiration begins in the cytoplasm, but most of the energy from the breakdown of glucose and fatty acids is generated in the mitochondria. The enzymes that are used in this breakdown are bound to the membranes of the mitochondrion. In this way the complex series of reactions needed to break apart glucose and to capture the liberated energy and store it are accomplished in an orderly and efficient way separated from the rest of the cell within this closed compartment.

The mitochondria apparently originated as endosymbiotic bacteria. In a complex series of reactions using oxygen, cell fuel is broken down to release energy within mitochondria.

Chloroplasts

Symbiotic events similar to those postulated for the origin of mitochondria also seem to have been involved in the origin of **chloroplasts.** They are thought to be derived from symbiotic photosynthetic bacteria. These energy-producing organelles are found in the cells of plants and algae. In these organelles the energy in sunlight is used to power the reactions that make the "cellular fuel"—glucose—by using molecules of carbon dioxide from the air. Together, the complex series of chemical reactions that perform these tasks is known as **photosynthesis.** Using chemical symbols, this series of reactions can be summarized as follows:

$6CO_2$	+	$6H_2O$	+	Sun's energy	→	$C_6H_{12}O_6$	+	$6O_2$
Carbon dioxide		Water				Glucose		Oxygen

The glucose can then be broken down in the mitochondria to release its energy to ATP for immediate use or short-term storage, or the glucose can be stored as complex carbohydrates for later use.

Chloroplasts apparently originated as endosymbiotic photosynthetic bacteria. Within chloroplasts, carbon dioxide, water, and light energy are used to produce the cell fuel glucose during a series of reactions called *photosynthesis*.

Chloroplasts have a structure similar to the mitochondria. Like mitochondria, chloroplasts are bounded by a double membrane. In addition, the inner of these membranes forms folds, invading the interior of the organelle. In chlo-

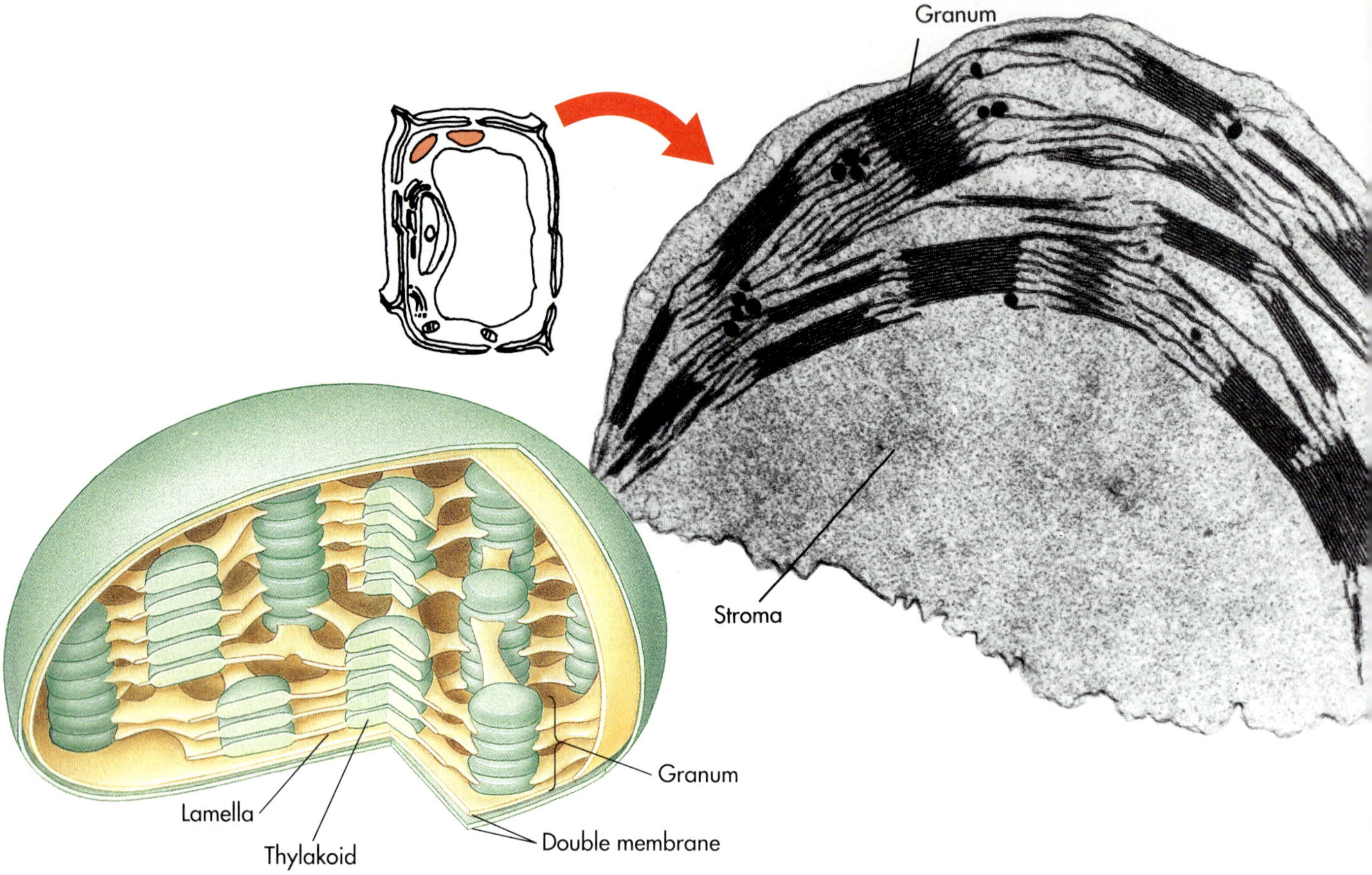

FIGURE 3-19 The chloroplast.
Chlorophasts have a double outer membrane and a system of interior membranes that form sacs called *thylakoids.* A stack of thylakoids is called a *granum.* Chlorophyll and the other light-gathering pigments for photosynthesis are embedded in the numerous grana.

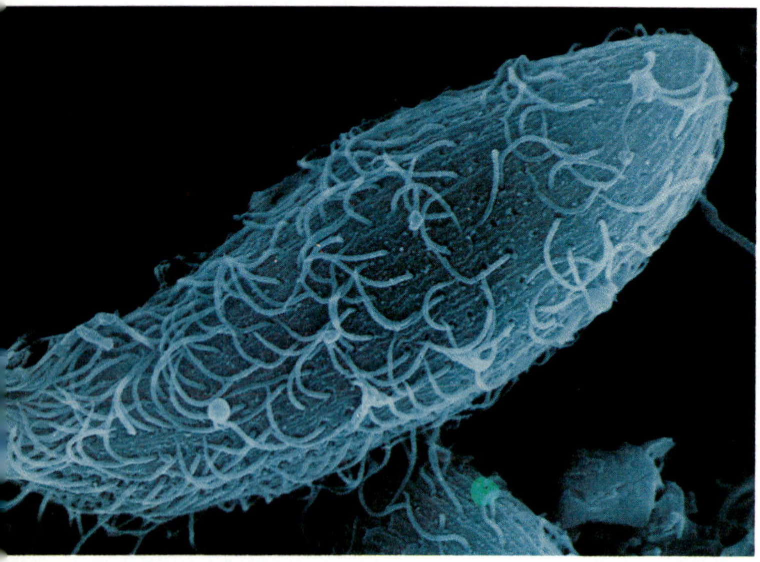

FIGURE 3-20 Cilia.
The surface of this single-celled protist is virtually covered with hair-like cilia, which wave in an undulating motion to propel the cell.

roplasts, however, these membranes are extensive and form sacs called **thylakoids.** These sacs, stacked one on another, are called **grana** (Figure 3-19). Chlorophyll, a chemical that can absorb light energy from the sun and that allows photosynthesis to take place, is found within the grana. As with the mitochondria, chloroplasts provide an orderly, closed compartment within the cell in which a series of reactions can occur.

Cilia and flagella

Single-celled eukaryotic cells are often motile—able to move within their environments. One way in which these cells move is by means of cell extensions that look somewhat like hairs. These structures, called **cilia,** are short and often cover a cell (Figure 3-20). In some cells, cilia are found only on certain sections of the cell and are used to move substances across these surfaces. Some cells have whip-like extensions called **flagella.** Used strictly for movement, these structures are longer than cilia, but fewer are usually present on a cell.

Although cilia and flagella differ in length, they have the same structure: they are bundles of microtubules covered with the plasma membrane of the cell. As you can see in Figure 3-21, nine pairs of microtubules surround a single,

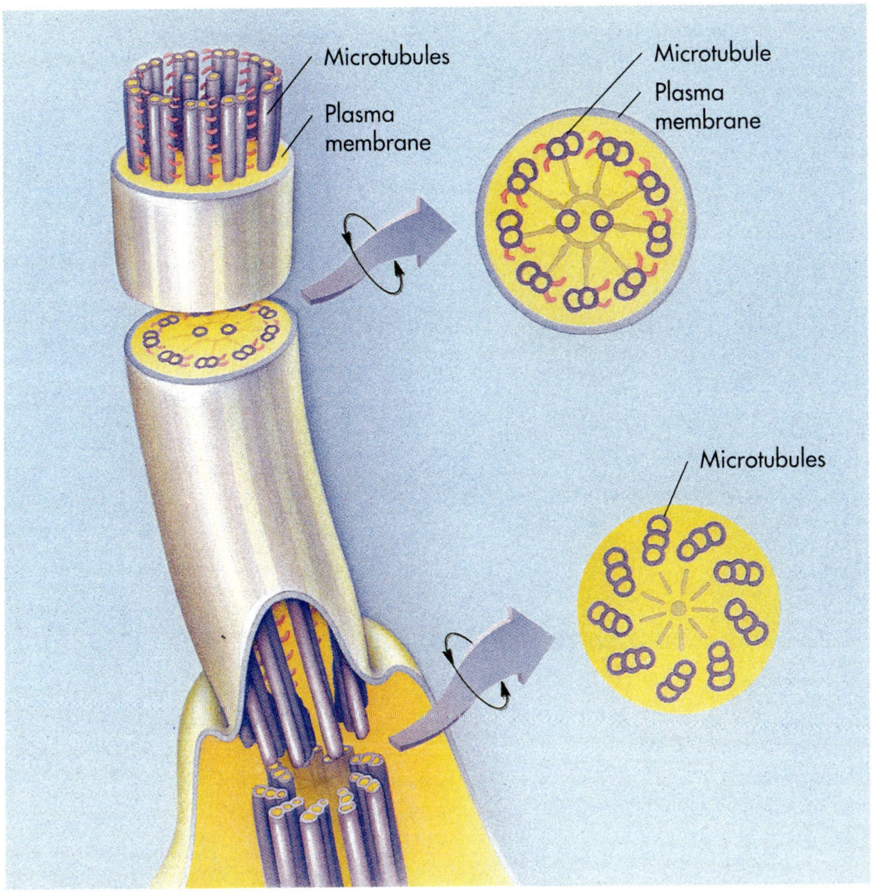

FIGURE 3-21 Structure of a flagellum. A eukaryotic flagellum is composed of a ring of nine pairs of microtubules, with two microtubules in its core. Anchoring the flagellum to the cell is a basal body of nine microtubule triplets. Cilia have the same pattern of microtubules.

central pair. As these microtubules dip into the cell beneath the level of the plasma membrane, they connect with another structure called the *basal body*. Also composed of microtubules, a basal body serves to anchor a cilium or flagellum to the cell.

Cilia and flagella are whip-like organelles of motility that protrude from some cells.

Cell walls

Many plants, algae, and fungi have a rigid structure called a **cell wall** that surrounds the plasma membrane. Almost all plant cell walls are made up of cellulose, large molecules formed by the linking of glucose units. Many single-celled eukaryotes have a similar type of cell wall. However, other single-celled eukaryotes, such as *Paramecium* and *Euglena* species, have an outer structure called a *pellicle* that is made up of protein. Fungi have cell walls that contain chitin, the same substance that is found in the shells of organisms such as grasshoppers and lobsters.

Cell walls perform many jobs for cells. In plants and fungi, they help impart a stiffness to the tissues. They also provide some protection from a drying environment. In single-celled organisms, cell walls give shape to the organisms and help protect them. Pellicles are even more specialized in their function, often serving as anchors for defense mechanisms.

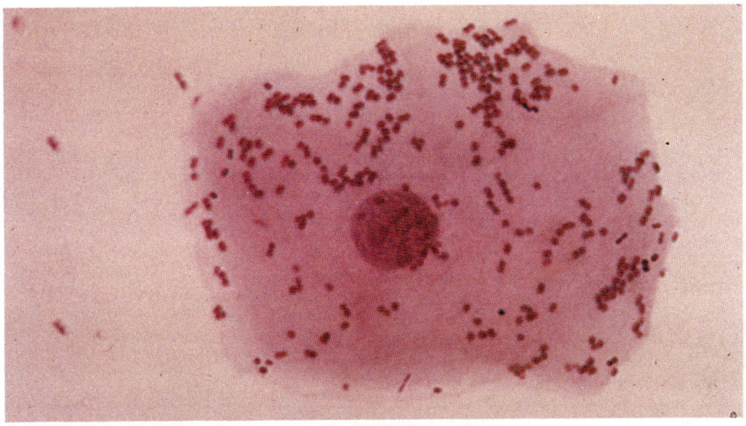

FIGURE 3-22 Bacteria are smaller than eukaryotic cells. The small dots that you see in this photo are bacteria. The large, somewhat rectangular shape in the background is a human cheek cell. As you can see, the eukaryotic cell dwarfs the tiny bacteria that are on it.

Prokaryotic cells

Prokaryotic cells include the bacteria and the cyanobacteria (blue-green bacteria). Smaller than eukaryotic cells, the tiniest bacteria have a diameter about 100 times smaller than one of the cells in your body! The largest bacteria are about two and one-half times smaller. Figure 3-22 shows how

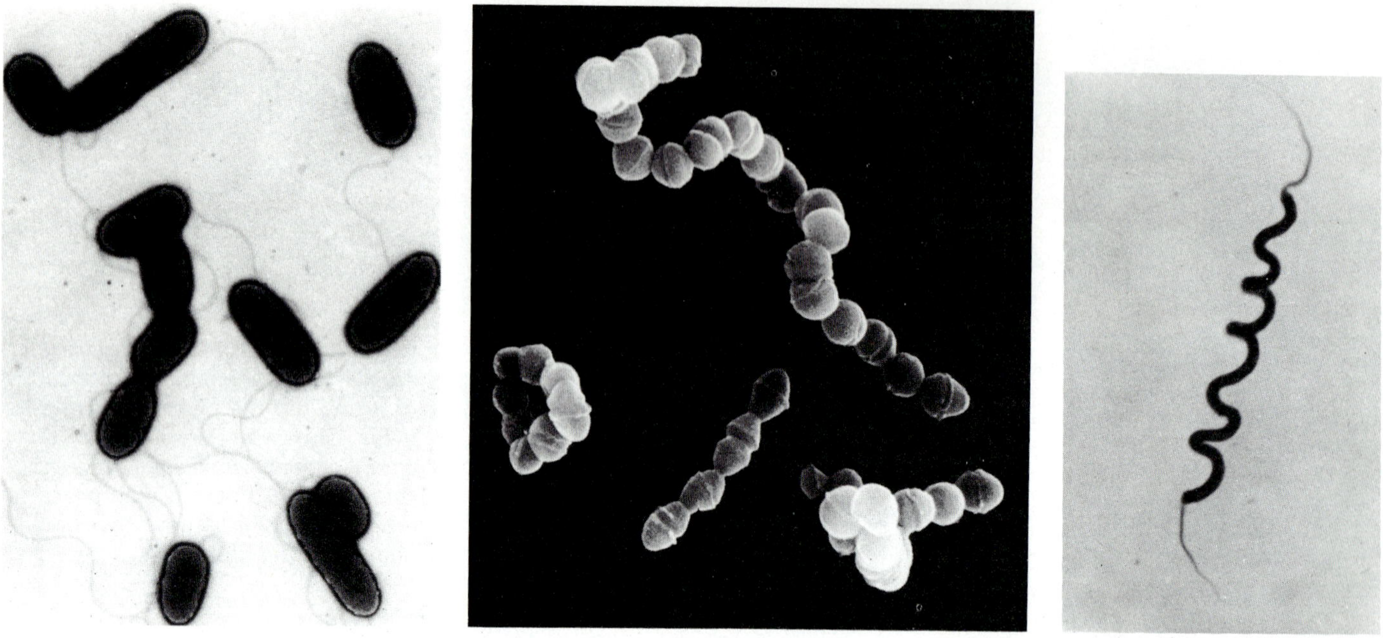
A B C

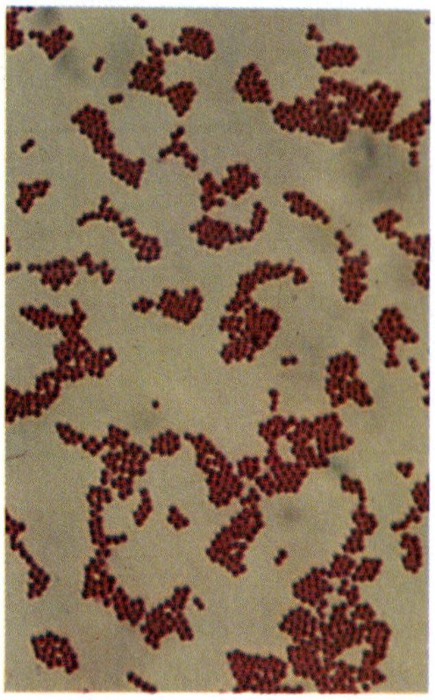
D

FIGURE 3-23 Bacterial cells have several different shapes.
A A rod-shaped bacterium, *Psuedomonas* species, a type associated with many plant diseases.
B *Streptococcus* species, a more or less spherical bacterium in which the individuals adhere in chains.
C *Spirillum* species, a spiral bacterium. This large bacterium has a tuft of flagella at each end.
D *Staphylococcus* species, spherical bacteria that group in grape-like bunches.

bacteria compare in size to a eukaryotic cell. Along with being smaller than eukaryotic cells, bacteria also have a much simpler internal organization containing no organelles other than ribosomes.

Although bacteria are single-celled organisms, you may be fooled if you observed them under a microscope. Notice the bacteria in Figure 3-23. Some types form long chains of spherical cells, whereas other bacteria characteristically form arrangements that look like bunches of grapes. However, each member of a group of bacteria is an individual organism. Bacterial shapes vary too. Some bacteria are rod shaped; others look like corkscrews. Still others are minute spheres.

Cell walls

With the exception of only one group, all bacteria are enclosed by a rigid cell wall. But this cell wall is unlike that of plants, fungi, or protists. The cell wall of bacteria is made up of a network of large molecules called *peptidoglycan*. This network is shown in Figure 3-24. Notice that two types of molecules make up the network: sugars and amino acids. This network of molecules performs many jobs for the cell, such as keeping out many harmful substances and keeping the cell from bursting. Certain antibiotics such as penicillin kill bacteria (and not your body cells) because they interfere with the synthesis of the bacterial cell wall.

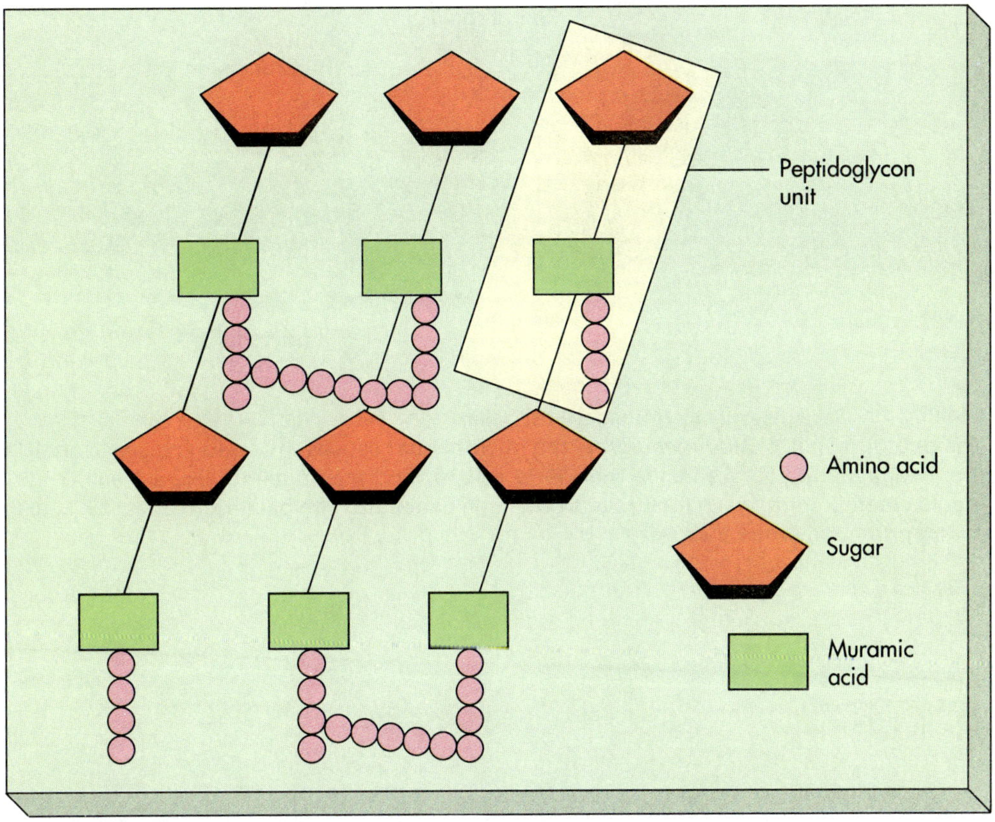

FIGURE 3-24 Peptidoglycan gives structural strength to the bacterial cell wall.
The structural strength of the bacterial cell wall is due to a network of sugar derivatives and unusual amino acids called *peptidoglycan.* The backbone of the wall is alternating molecules of an acidic sugar and an amino sugar, and extending down as pillars off of each acidic sugar is a chain of four or five amino acids. The backbones are united and stabilized against osmotic pressure by crosslinking bridges of amino acids.

Simple interior organization

Beneath the bacterial cell wall lies the plasma membrane. Similar in structure to that of eukaryotic cells, this thin, fragile membrane surrounds the cytoplasm of the cell. The cytoplasm of prokaryotic cells is made up of about 80% water, an assortment of molecules involved in the chemical reactions of the cell, DNA, ribosomes, and several kinds of reserve materials known as *inclusions*. Types of inclusions vary from one kind of bacterium to another.

The cytoplasm of prokaryotes does not contain the varied organelles of eukaryotes. Ribosomes, however, are packed into the cytoplasm of bacteria in great numbers, giving it a "grainy" appearance. These ribosomes are smaller than eukaryotic ribosomes. The single, circular strand of DNA found in bacteria is not surrounded with a membrane as in eukaryotes and is not a "true nucleus." In fact, the word *prokaryote* means "before the nucleus." Therefore this area of the cell is termed a **nucleoid,** rather than a nucleus. In addition, bacteria often contain small "rings" of DNA called **plasmids.**

Although motile bacteria have flagella like some eukaryotic cells, flagellar structure is different. Bacterial flagella (and bacterial cytoplasm) do not contain microtubules. Instead, bacterial flagella are composed of several chains of protein that wrap in corkscrew fashion around a hollow core. These proteins are attached to others that anchor the flagellum to the cell wall and membrane, and are covered by a flexible tube. Bacteria swim by rotating their flagella rather than by "whipping" them as eukaryotes do (Figure 3-25).

Most bacteria contain infoldings of the plasma membrane known as *mesosomes*. Mesosomes are connected to the plasma membrane and play a role in reproduction. In addition, enzymes needed for certain chemical reactions of the cell to take place are often bound to the mesosome and plasma membrane.

Prokaryotic cells, the bacteria and the blue-green bacteria, have a simple interior organization with a single, circular strand of hereditary material that is not enclosed with a membrane.

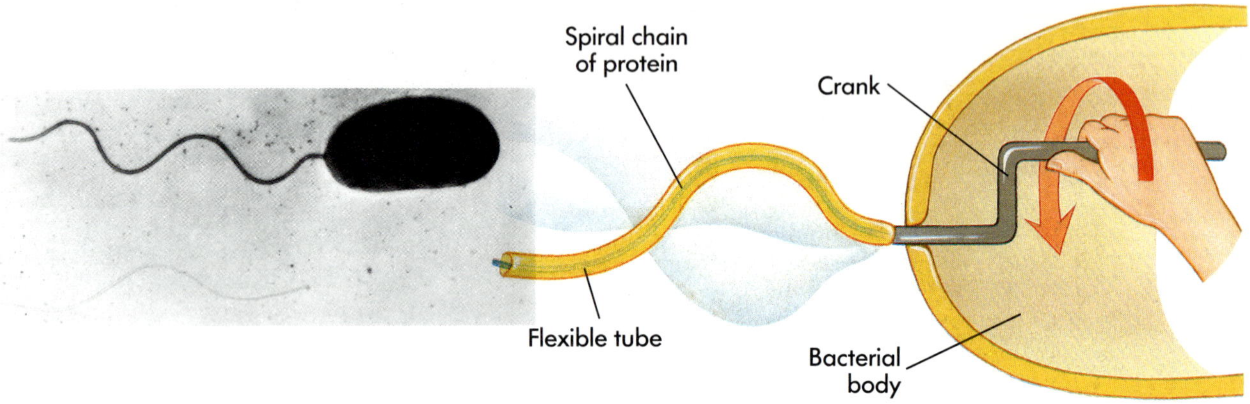

FIGURE 3-25 Bacteria swim by rotating their flagella.
The photograph is of *Vibrio cholorae,* the microbe that causes the serious disease cholera. Imagine that you are inside the *Vibrio* cell, turning the flagellum like a crank. You would create a spiral wave that travels down the flagellum. The bacterium uses this kind of rotary motion when it swims.

TABLE 3-2

A comparison of bacterial, animal, and plant cells

	BACTERIUM	ANIMAL	PLANT
EXTERIOR STRUCTURE			
Cell wall	Present (protein-polysaccharide)	Absent	Present (cellulose)
Cell membrane	Present	Present	Present
Flagella	Sometimes present (one strand)	Sometimes present	Absent except in sperm of a few species
INTERIOR STRUCTURE			
ER	Absent	Usually present	Usually present
Microtubules	Absent	Present	Present
Centrioles	Absent	Present	Absent
Golgi bodies	Absent	Present	Present
ORGANELLES			
Nucleus	Absent	Present	Present
Mitochondria	Absent	Present	Present
Chloroplasts	Absent	Absent	Present
Chromosomes	A single circle of naked DNA	Multiple units, DNA associated with protein	Multiple units, DNA associated with protein
Ribosomes	Present	Present	Present
Lysosomes	Absent	Usually present	Present
Vacuoles	Absent	Absent or small	Usually a large single vacuole in mature cell

A comparison of prokaryotic and eukaryotic cells

As you can see, prokaryotes and eukaryotes differ greatly in their structure and complexity (Table 3-2). The structure of eukaryotic cells is much more complicated and diverse than the structure of the prokaryotic cells, even though they share common ancestors. The most distinctive difference between these two cell types is the extensive subdivision of the interior of eukaryotic cells by membranes. The most visible of these membrane-bounded compartments, the nucleus, gives eukaryotes their name. There are no equivalent membrane-bounded compartments within prokaryotic cells. The membranes of some photosynthetic bacteria are extensively folded inward (Figure 3-26), but they do not isolate any portion of the cell from any other portion. Although these two types of cells are different, both have many unifying features and bustle with activity.

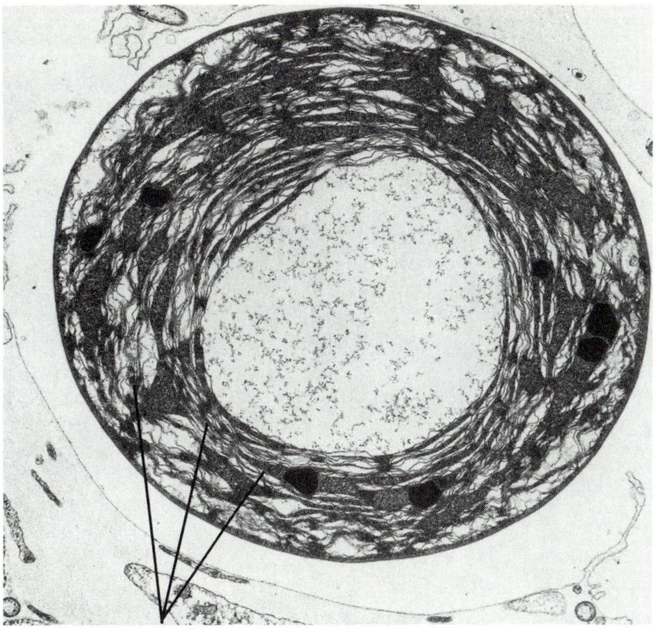

FIGURE 3-26 Photosynthetic bacteria have extensive membranes, but not membranous organelles.
This purple photosynthetic bacterium shows numerous photosynthetic lamellae (membranes), but they form flat sheets in the cell, not the compartments of a eukaryotic cell.

Summary

1. A cell is a membrane-bounded unit containing hereditary material and cytoplasm. It can release energy from fuel and use that energy to grow and reproduce.
2. The cell theory was first formulated in the mid-1800s. It states that all living things are made up of cells, the smallest unit of life. In addition, all cells arise from the growth and division of other cells.
3. Most cells divide before they grow too large, maintaining a large surface area-to-volume ratio. In this way, cells are able to move substances across their cell membranes quickly enough to meet their needs.
4. All cells can be classified as either prokaryotic or eukaryotic. Prokaryotic cells are the bacteria and have a relatively simple structure. Eukaryotic cells are more complex than prokaryotes and evolved from ancestors of today's prokaryotes. Eukaryotic cells make up the bodies of plants, animals, protists, and fungi.
5. The eukaryotic cells represent a highly diverse group. However, all eukaryotic cells are bounded by a plasma membrane and contain membrane-bounded organelles: the endoplasmic reticulum, Golgi bodies, lysosomes, vacuoles, and nucleus. In addition, two types of eukaryotic organelles, chloroplasts and mitochondria, are derived from "live in" ancient bacteria.
6. The cytoplasm of eukaryotic cells contains a network of filaments and fibers that help support and shape the cell. Prokaryotic cells lack this network.
7. The most distinctive difference between prokaryotic and eukaryotic cells is the extensive subdivision of the cytoplasm of eukaryotic cells by membranes. These membranes form channels that compartmentalize the workings of the cell. In addition, they serve as a "subway system" for moving molecules from one place to another within the cytoplasm. Substances can move unimpeded within prokaryotic cells because this level of organization is not present.
8. Both prokaryotic and eukaryotic cells contain ribosomes, the sites of protein synthesis. Prokaryotic ribosomes are smaller than eukaryotic ribosomes.
9. Cilia and flagella are composed of microtubules and help propel eukaryotic and prokaryotic cells. In some eukaryotic cells, cilia are used to move substances across surfaces. Some prokaryotic cells have flagella, but their structure is different from eukaryotic flagella.

REVIEW QUESTIONS

1. What are cells? What function(s) can they perform?
2. What are the basic principles that make up the cell theory?
3. Why must cells maintain a large surface area-to-volume ratio?
4. Some of the skeletal muscle cells that allow you to move your legs can be as long as 30 to 40 centimeters. These cells have many nuclei. Explain why.
5. What four characteristics are shared by almost all cells?
6. Identify the type(s) of cells (prokaryotic or eukaryotic) that you would expect to find in each of the following:
   a. A fungus on a piece of bread
   b. A venus flytrap (an insect-eating plant)
   c. *Escherichia coli* bacteria in the human intestine
   d. Your dog
7. Draw a picture of a "typical" eukaryotic cell. Label the plasma membrane, cytoplasm, nucleus, etc.
8. Fill in the blanks: All eukaryotic cells are bounded by a thin, nonrigid __________ __________.
9. Distinguish between the cytoplasm and the cytoskeleton. What are their functions?
10. Make a table that briefly describes each of the following and lists its function(s): plasma membrane, flagella (cilia), cytoskeleton, endoplasmic reticulum, ribosomes, nucleus, chromosomes, nucleolus, Golgi complex, lysosomes, mitochondria, and chloroplasts.
11. Distinguish between rough ER and smooth ER. What are their functions?
12. What would happen inside your cells if the lysosomes stopped working?
13. What are DNA and RNA?
14. Fill in the blank: When a eukaryotic cell begins to divide, its DNA condenses to form __________.
15. What do mitochondria and chloroplasts have in common? How are they different?
16. Fill in the blanks: Sperm cells are the only cells in the human body that move by means of whip-like extensions called __________.
17. Name three ways in which prokaryotic cells differ from eukaryotic cells.
18. You're examining three cells. You know that one is a bacterium, one is from a plant, and the third is from an animal. From their structures, how can you tell which is which?

KEY TERMS

cell theory 45
cell wall 61
chloroplasts 59
chromosomes 57
cilia 60
cristae 58
cytoplasm 50
cytoskeleton 48
DNA 56
endoplasmic reticulum 53
endosymbiont 48
eukaryotic cells 47
flagella 60
Golgi body 54
Golgi complex 54
grana 60
intermediate fibers 50
lysosomes 56
microfilaments 50
microtubules 50
mitochondria 58
nuclear envelope 56
nuclear pores 56
nucleoid 63
nucleolus 57
nucleus 56
plasma membrane 48
plasmid 63
prokaryotic cells 47
ribosomes 53
rough ER 53
rRNA 57
smooth ER 53
symbiosis 48
symbiotic bacteria 48
thylakoids 60
vacuole 56
vesicles 53

THOUGHT QUESTIONS

1. Most of the 10 trillion cells in your body are small, in large part because small cells have a large surface for a given amount of volume. Calculate the total cell surface area of your body, assuming the average cell is a sphere with a diameter of 20 micrometers.
2. Lysosomes are cell organelles that recycle worn-out mitochondria and other cellular debris. Each lysosome contains a high concentration of enzymes that attack and digest membranes. Can you think of any reasons why a lysosome does not digest itself?

FOR FURTHER READING

De Duve, C. (1986). *A guided tour of the living cell.* Vols. 1 and 2. New York: Scientific American Books.
This review has excellent illustrations and presents a wide variety of the techniques used to study cells.

McDermott, J. (1989, August). A biologist whose heresy redraws Earth's tree of life. *Smithsonian,* pp. 71-81.
This is an engaging account of the scientific approach taken by Lynn Margulis, leading contemporary advocate of the endosymbiotic theory of the origin of some organelles.

Symmons, M., Prescott, A., & Warn, R. (1989, February). The shifting scaffolds of the cell. *New Scientist,* pp. 44-47.
New findings show that the microtubules of which cells are built are constantly appearing and disappearing.

Cell membranes

4

The doors of Ireland are beautifully unique. Some are larger than others, some sturdier than others. Some have fancy pillars marking their opening, whereas others have a simple casing. Yet in spite of their great diversity, they share a common function: all serve as passageways into and out of the buildings of Ireland.

OUTLINE

The foundation of membranes

Membrane proteins

How substances move across cell membranes

- Movement that does not require energy
  - Diffusion
  - Osmosis
  - Facilitated diffusion
- Movement that does require energy
  - Active transport
  - Endocytosis
  - Exocytosis

HIGHLIGHTS

- The basic foundation of the cell membrane is a lipid bilayer.
- A variety of proteins floats on or within the membrane lipid bilayer, providing the means by which the cell interacts with its environment.
- Most substances move into and out of the cell by means of protein transport molecules located within the cell membrane.
- The cell, by means of its transport proteins, can regulate the amount and types of molecules and ions that enter and leave.

The cells within your body have doors, too. And like the doors of Ireland, they are beautifully diverse and unique. Your cells' "doors" are proteins. These large molecules are embedded in the membranes of your cells. Without these proteins, your membranes would act like impermeable plastic bags enclosing your cell contents. With their protein doors, however, your cells can interact with the environment as substances move across their membranes, an activity vital to their—and your—survival.

The foundation of membranes

Amazingly, the membranes that surround all living cells are only a few molecules thick. In fact, it would take more than 10,000 membranes piled on one another just to equal the thickness of this page! But although membranes are thin, they are not simple in their construction. If we take a membrane "apart," however, its structure can be easy to understand.

Cell membranes have two main components: **phospholipids** and proteins. Phospholipids are the molecules that make up the "foundation" of membranes. Like the fat and oil molecules you studied in Chapter 2, a phospholipid has a backbone of a three-carbon molecule called *glycerol*. Attached to this backbone are two long chains of carbon atoms, with hydrogen atoms attached to the carbons. These hydrocarbon chains are called *fatty acids*. A molecule of fat or oil has three such chains. A phospholipid molecule, in contrast, has only two. The third position is occupied instead by one of a variety of phosphate functional groups (Figure 4-1). The word *phosphate* is used because these functional groups are compounds containing the element phosphorus. The term *phospholipid*, then, is really descriptive of a molecule composed of a phosphate functional group attached to a portion of a fat, or lipid, molecule.

As you may recall from Chapter 2, functional groups have definite chemical properties. In the case of phospholipids, the phosphate functional groups are polar molecules. In other words, these groups form hydrogen bonds with

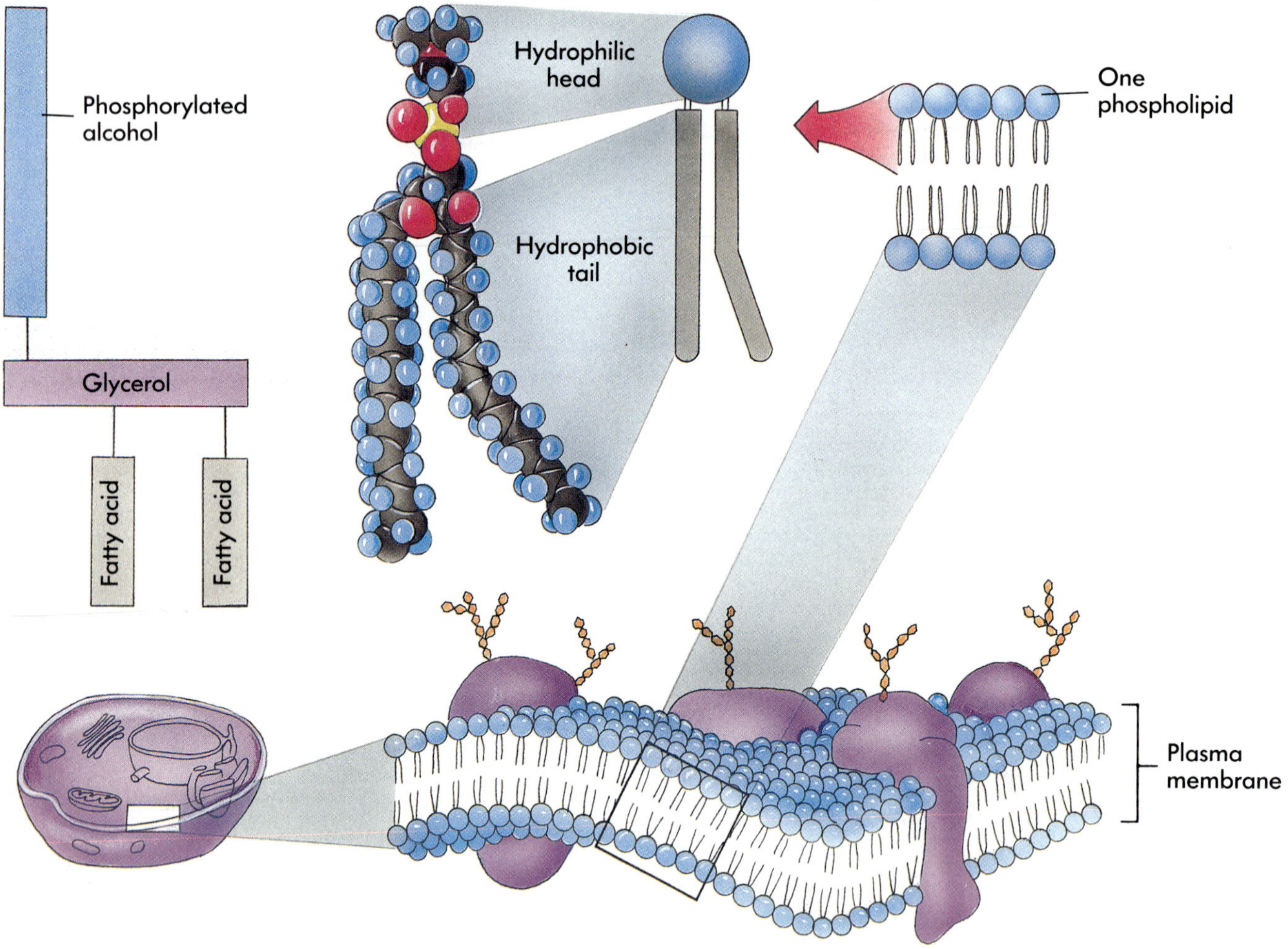

FIGURE 4-1 Structure of the plasma membrane.
The plasma membrane of a cell is composed of two layers of phospholipids. Each phospholipid is composed of a glycerol molecule attached to two chains of fatty acids. The phosphorylated alcohol is the phosphate functional group for this particular phospholipid. The head portion of the phospholipid is hydrophilic (water loving) and points outward. The tail portion is hydrophobic (water fearing) and points inward.

water. They are therefore termed **hydrophilic,** or "water loving." Consequently, they are soluble, or dissolve in, water. The two fatty acid chains are nonpolar molecules and do not form hydrogen bonds with water. They are termed **hydrophobic,** or "water fearing." Unlike the rest of the molecule, these fatty acid chains are not soluble in water.

Imagine what happens when a collection of phospholipid molecules is placed in water. The long nonpolar "tails" of the phospholipid molecules are pushed away by the water molecules that surround them. They are shouldered aside as the water molecules move toward the molecules that do form hydrogen bonds with them—the phosphate functional groups. As a result of these movements, the phospholipids form *two* layers of molecules, with their hydrophobic tails oriented inward (away from the water) and their hydrophilic "heads" oriented outward (toward the water). Put another way, these molecules form two layers with their tails facing each other. This structure is called a **lipid bilayer** and is the foundation of the membranes of all living things (Figure 4-1).

The basic foundation of biological membranes is a lipid bilayer. Lipid bilayers form spontaneously, with the nonpolar tails of the phospholipid molecules pointed inward, and the polar heads pointed outward.

Did you ever blow soap bubbles when you were a child? Take a moment to think about how the bubbles looked. Can you remember how the soap film forming the bubble seemed to swirl and move as the bubble floated in the air? Lipid bilayers are "fluid" much like the "shell" of the soap bubble you remember. The fluid nature of the bilayer is caused by movement of the phospholipid molecules. Although water constantly forces the phospholipid molecules to form a stable bilayer, it is indifferent to where individual phospholipid molecules are located within the bilayer. In fact, the hydrogen bonds between the water molecules and the hydrophilic heads of the phospholipids are constantly breaking and re-forming. Therefore individual molecules move about somewhat freely.

A lipid bilayer forms a fluid, flexible covering for a cell and keeps the watery contents of the cell on one side of the membrane and the watery environment on the other.

If cells were encased with pure lipid bilayers, however, they would be unable to survive because the interior of the lipid bilayer is completely nonpolar. It repels most water-soluble molecules that attempt to pass through it. In fact, as Figure 4-2 shows, the lipid bilayer allows only a few uncharged polar molecules, such as individual water (H_2O) and ammonia (NH_3) molecules, to pass through. (Ammonia is a common cellular waste product formed from the breakdown of proteins.) In addition, substances that are soluble in oil, such as vitamins A, D, and E, can also move across the lipid bilayer. It prevents, however, any water-soluble substances such as sugars, amino acids, and proteins and charged particles (important ions) such as calcium (Ca^{2+}) and potassium (K^+) from entering or leaving the cell. Such molecules pass through the cell membrane by means of proteins that are embedded in the lipid bilayer. This model of the cell membrane is widely accepted today. It is called the **fluid mosaic model,** a name that describes the fluid nature of a lipid bilayer studded with a mosaic of proteins.

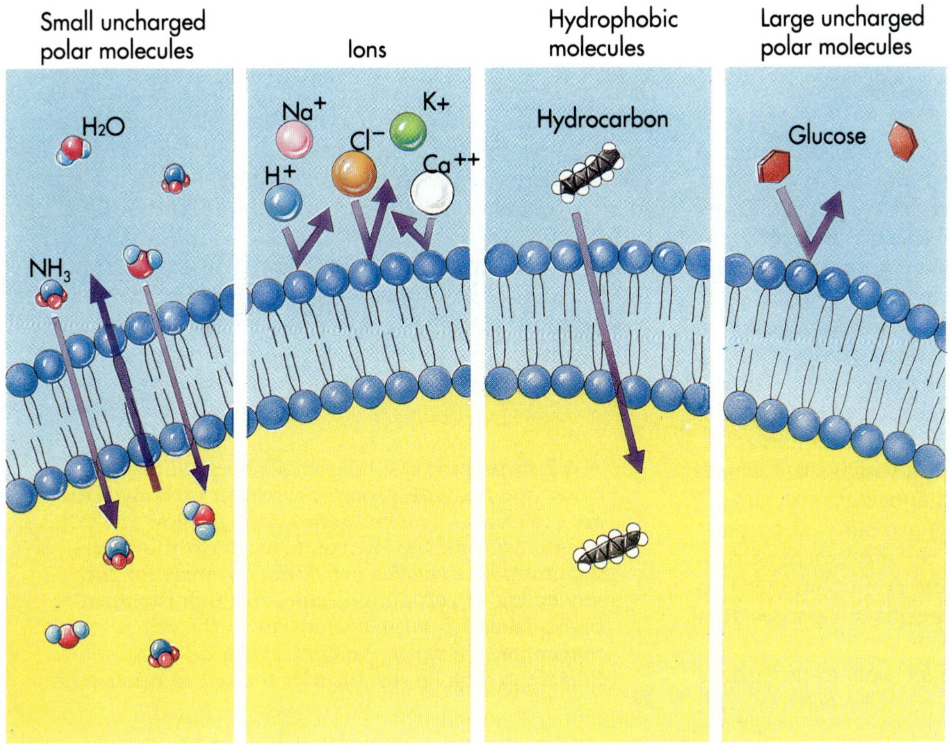

FIGURE 4-2 Some molecules pass across bilayer membranes, others do not.
Lipid bilayer membranes are permeable to oxygen, lipids, and small uncharged molecules if they are polar like water. Membranes are not permeable to large molecules if they are polar or to anything that is charged, such as ions or proteins.

BIOLOGY & you

Membrane channels and cystic fibrosis

Cystic fibrosis is a fatal disease of human beings in which affected individuals secrete a thick mucus that clogs the airways of the lungs. These same secretions block the ducts of the pancreas and liver so that the few patients who do not die of lung disease die of liver or pancreas failure. Cystic fibrosis is usually thought of as a children's disease because in the past years, few affected individuals lived long enough to become adults. With modern therapy, the average affected individual now lives 23 years (Figure 4-A). There is no known cure.

Cystic fibrosis is a hereditary disorder passed down from parent to child. It is the most common fatal hereditary disease of Caucasians, affecting about 40,000 people in the United States. Difficult to study, cystic fibrosis affects many organs. Until recently, it was impossible to identify the nature of the hereditary material—the defective gene—responsible for the disease. In 1985 the first clear clue was obtained. An investigator, Paul Quinton, seized on a commonly observed characteristic of cystic fibrosis patients, that their sweat is abnormally salty, and performed the following experiment. He isolated a sweat duct from a small piece of skin and placed it in a solution of salt (NaCl) that was three times as concentrated as the NaCl inside the duct. He then monitored the movement of ions. Both the Na^+ and the Cl^- ions diffuse into the duct because of the higher outer ion concentrations. In skin isolated from normal individuals, Na^+ ions indeed entered the duct, transported by the sodium-potassium pump; Cl^- ions followed, passing through a passive channel. Both ions crossed the membrane easily. In skin isolated from individuals with cystic fibrosis, the sodium potassium pump transported Na^+ ions into the ducts, but no Cl^- entered. The passive chloride channels were not functioning in these individuals.

Researchers now know that cystic fibrosis results from a defect in a protein within plasma membranes. In normal individuals, this protein acts as a channel for the passage of Cl^- ions across the membrane. The defective gene for this critical transport protein was isolated in 1987, and its position on a particular human chromosome was pinpointed in 1989. Now scientists are searching for ways to transfer working copies of the gene into cystic fibrosis patients. For the first time, there is hope for a cure. In January 1992, researchers inserted the gene into a common cold virus and tested the use of inhalers to get the virus (and the needed gene) into the lung tissue of patients.

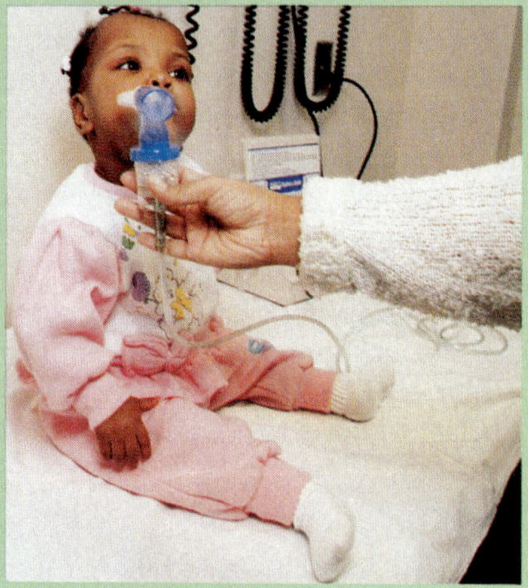

FIGURE 4-A A child with cystic fibrosis using an inhaler, which disperses medicine into a mist that can then be inhaled.

Membrane proteins

A collection of proteins float within or on the phospholipid bilayer of a membrane. These protein doors are not open to every molecule. In fact, the wide variety of protein doors of the cell membrane have specific jobs to do. And different cells having different needs have different doors! The variety of doors shown in Figure 4-3 is only a hint of the wide variety of protein doors that exists in cell membranes.

Some proteins span the distance from one side of the membrane to the other. Some are anchored firmly in the membrane, whereas others move continually. Held to the interior of the membrane by their hydrophobic amino acids, these proteins serve as channels through which small water-soluble substances can move. These channels are specialized, however, and only certain molecules are allowed to pass. Other proteins take an active part in the movement of substances through their channels. All work in some way to regulate the kind and amount of specific substances that move into and out of the cell.

Other membrane proteins are bound only to the surface of the bilayer (see Figure 4-3). These proteins perform such jobs as identifying messages sent to the cell and forming junctions with other cells. Consequently, the varied proteins of the phospholipid bilayer enable a cell to interact with its environment in numerous ways. Figure 4-4 shows how cell identity markers (labeled *carbohydrate chains*) are scattered over the membrane surface. These molecules are short chains of sugars (carbohydrates) attached to some of the phospholipids and proteins. These molecules appear to act like cell "name tags," helping cells to recognize one another and, unfortunately, for viruses to recognize certain cells.

A cell membrane is a bilayer of phospholipid molecules studded with proteins to which carbohydrates are attached. The membrane proteins allow the cell to interact with the environment, performing various functions such as providing channels for the movement of certain molecules through the membrane, identifying molecules sent to the cell as messengers, and forming junctions with other cells. Surface carbohydrates identify the cell to other cells.

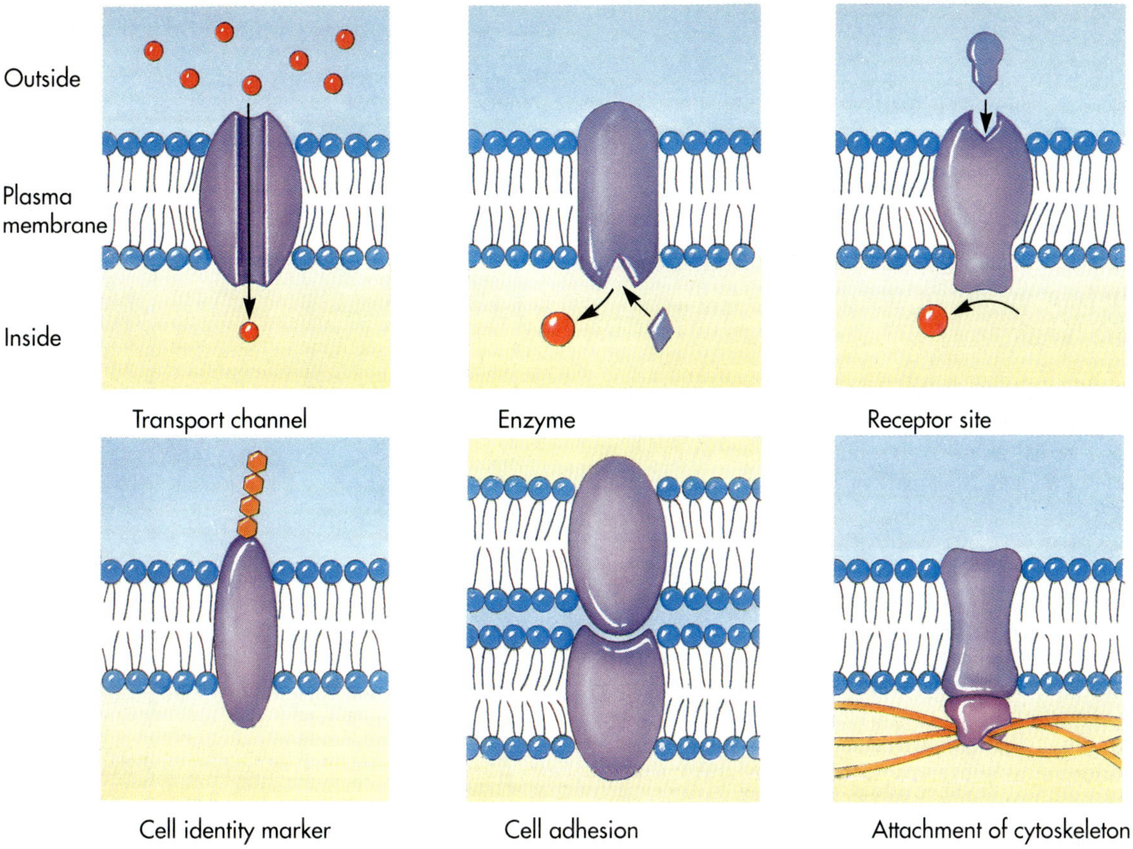

FIGURE 4-3 Functions of plasma membrane proteins.
The different proteins shown *(purple shapes)* perform many important functions. Some proteins act as *channels* through which certain substances enter and leave cells. Enzymes help speed up chemical reactions (see Chapter 6). Some proteins act as *receptor sites,* which bind specific chemicals. This binding prompts the cell to begin a specific function, such as the synthesis of a hormone. *Cell identity markers* are proteins that let other cells in the body know that they are not foreign invaders. Still other proteins perform structural roles: some proteins serve as *attachment sites* for other cells so that cells stick together, and other proteins anchor a cell's cytoskeleton in place.

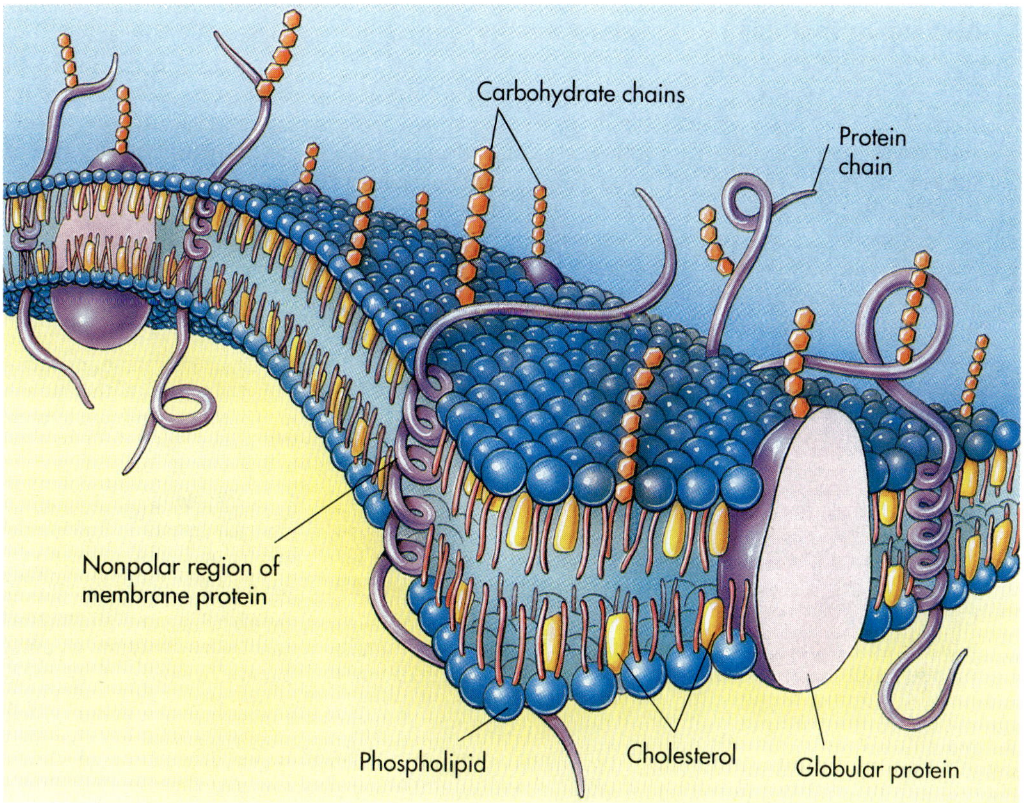

FIGURE 4-4 Cell identity markers are found on the surface of the membrane.
The purple tail-like structures shown in the diagram are proteins that enable the cell to interact with its environment.

Eukaryotic cells contain many membranes, such as the endoplasmic reticulum, the nuclear membrane, and the membranes of the Golgi complex. These membranes are not all identical. They all share, however, the basic architecture of the cell membrane.

How substances move across cell membranes

Cells usually live in an environment where they are bathed in water. If you realize that bacteria live on your body, or that protists live in the soil, this fact may not seem to be true. However, all cells must move substances across their membranes to survive, and water is the liquid in which their molecules are dissolved. In addition, water provides a fluid environment within which molecules can move. Therefore water must surround cells, even in microscopic amounts that may not be readily apparent.

How do molecules get where they are going? To answer this question, you must first understand that molecules cannot move *purposefully* from one spot to another. Molecules move randomly. All molecules and small particles have a constant, inherent "jiggling" motion called *Brownian movement*. And because molecules are always jiggling, they tend to bump into things, such as other molecules. A bump may push a molecule in a particular direction until it bumps into another molecule and gets pushed again. This random motion in all directions often results in a net movement of molecules in a particular direction in response to differences in concentration, pressure, or electrical charge. These differences are referred to as a **gradient.** The term *net movement* means that although individual members of a group of molecules are moving in different directions, the resulting movement of the group is in one direction. This movement results in a uniform distribution of molecules.

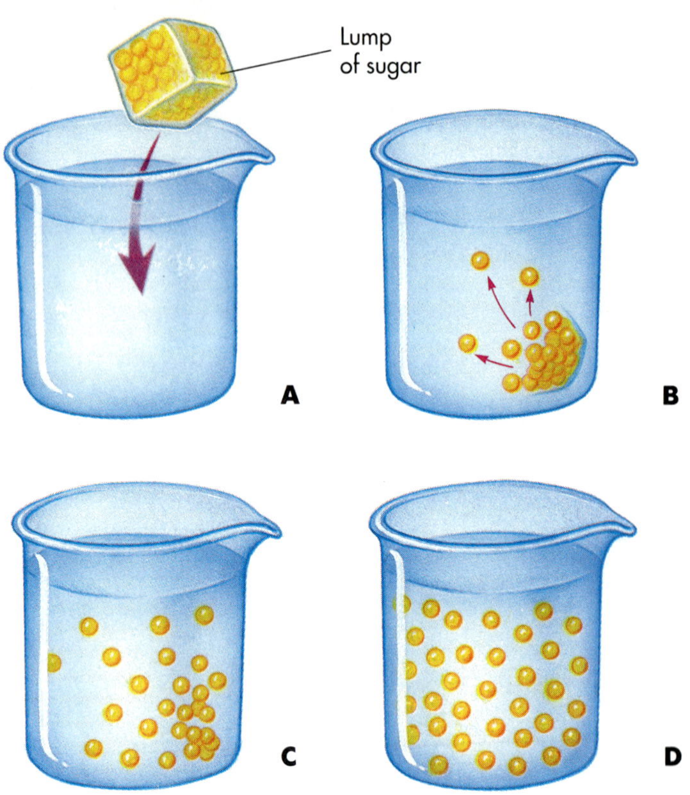

FIGURE 4-5 Diffusion.
If a lump of sugar is dropped into a beaker of water **(A),** its molecules dissolve and diffuse (**B** and **C**). Eventually, diffusion results in an even distribution of sugar molecules throughout the water **(D).**

Movement that does not require energy

The net movement of molecules often requires no input of energy. In this type of movement, molecules move from regions of high concentration to low concentration or from regions of high pressure to regions of low pressure. Ions, because they each carry an electrical charge, move toward unlike charges or away from like charges.

As molecules or ions move from regions of high concentration to regions of low concentration, they are referred to as going *down* a concentration, pressure, or electrical gradient. Some molecules move into and out of cells as they move down gradients. Molecular movement down a gradient but across a cell membrane is called **passive transport.** There are three types of passive transport: diffusion, osmosis, and facilitated diffusion.

Diffusion

Everyone has had experience with molecules in the air moving down a concentration gradient. Did you ever wake up to the smell of fresh brewed coffee? Did you ever spill a bottle of a liquid that had a strong odor, such as ammonia or perfume? The molecules that reached the smell receptors in your nose moved from an area of high concentration (the spill) to an area of low concentration (your nose). They moved down the concentration gradient, eventually becoming evenly spread.

You can perform a simple experiment to demonstrate diffusion. Take a small jar, fill it with ink to the brim, and cap it. Place the jar at the bottom of a full bucket of water, and gently remove the cap. The ink molecules will slowly diffuse out of the jar until there is a uniform concentration of ink in the bucket and in the jar. This process is illustrated in Figure 4-5 with a lump of sugar.

Diffusion is the net movement of molecules from a region of higher concentration to a region of lower concentration, eventually resulting in a uniform distribution of the molecules. This movement is the result of random, spontaneous molecular motions.

Osmosis

Osmosis is a special form of diffusion in which water molecules move from an area of higher concentration to an area of lower concentration across a differentially permeable membrane. A **differentially permeable membrane** allows only certain types of molecules to pass through it, or permeate it, freely. Cell (plasma) membranes are differentially permeable membranes (and are also called *semipermeable* or *selectively permeable membranes*). In fact, most types of molecules that occur in cells cannot pass freely across the plasma membrane. Small organic molecules such as sugars and amino acids, for example, can dissolve in water but not in lipids, so they are unable to move among the lipid molecules of the bilayer to traverse it. Water molecules, however, are small molecules that are able to pass through slight imperfections in the lipid bilayer and diffuse across the membrane.

The cytoplasm of a cell consists of many different types of molecules and ions dissolved in water. The mixture of these molecules and water is called a **solution.** Water, the most common of the molecules in the mixture, is called the **solvent.** The other kinds of molecules dissolved in the water are called **solutes.**

Because of diffusion, both solvent and solute molecules in a cell move from regions where the concentration of each is greater to regions where the concentration of each is less. When two regions are separated by a membrane, however, what happens depends on whether the molecule can pass freely through that membrane.

Osmosis is the diffusion of water molecules across a differentially permeable membrane.

The cytoplasm of living cells contains approximately 1% dissolved solutes. If a cell is immersed in pure water, interesting things begin to happen. Because the water has a lower concentration of solutes than the cell, water molecules begin to move into the cell. The reason is simple: water moves down the concentration gradient, or from a region of higher concentration of water molecules to a region of lower concentration of water molecules. The pure water is said to be **hypotonic** (Greek *hypo*, "under") with respect to the cytoplasm of the cell because its concentration of solutes is less than the concentration of solutes in the cell.

As you can see in Figure 4-6, a cell in a hypotonic solution begins to blow up like a balloon. Likewise, if a cell is placed in a **hypertonic** (Greek *hyper*, "over") solution, one with a solute concentration higher than the cytoplasm of the cell, water will move out of the cell, and the cell will shrivel. If a cell is placed in a solution with the same concentration of solutes as its cytoplasm, an **isotonic** (Greek *isos*, "equal") solution, water will move into and out of the cell, but no net movement will take place.

A solution with a solute concentration lower than that of another fluid is said to be *hypotonic* to that solution. A solution with a solute concentration higher than that of another fluid is said to be *hypertonic* to that solution. Solutions having equal solute concentrations are isotonic to one another.

Intuitively you may think that as "new" water molecules diffuse into a cell placed in a hypotonic solution, the pressure of the cytoplasm pushing against the cell membrane builds. This is indeed what happens. At the same time, however, the water molecules that continue to diffuse into the cell also exert a pressure—**osmotic pressure.** Because the pressure of the cytoplasm within the cell opposes osmotic pressure in this case, diffusion of water molecules into the cell will not continue indefinitely. The cell will eventually reach an equilibrium—a point at which the osmotic force driving water inward is counterbalanced exactly by the pressure out-

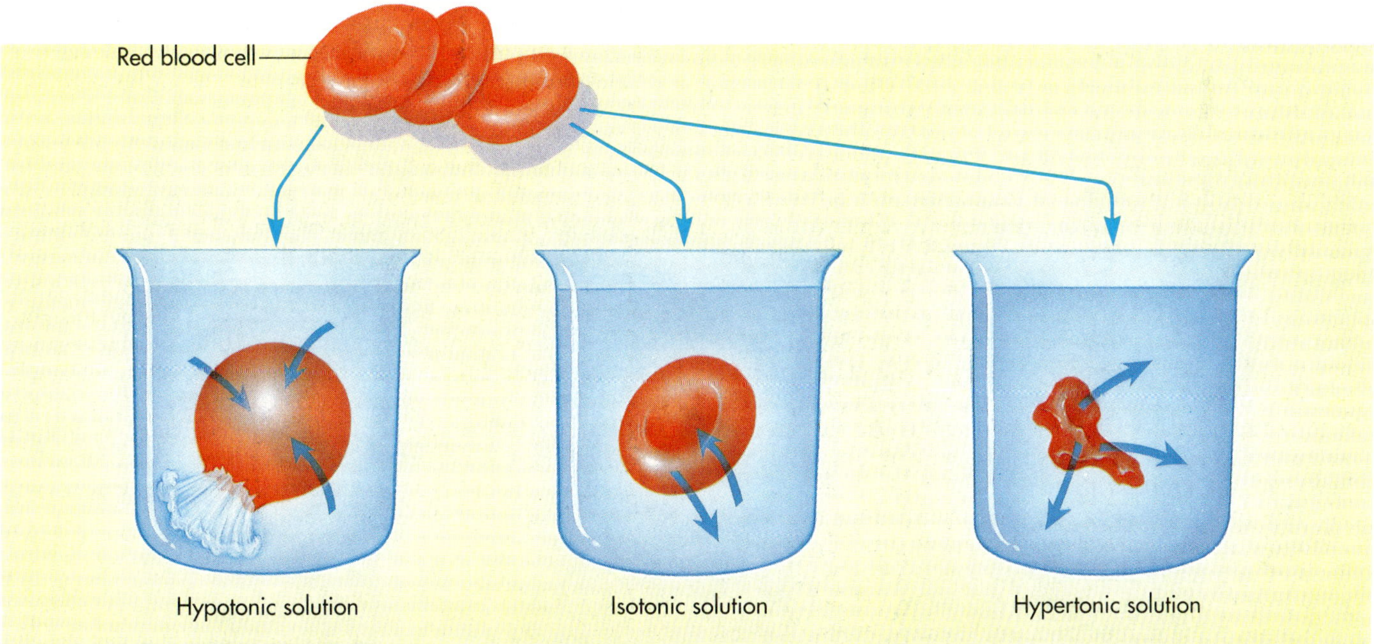

FIGURE 4-6 Osmosis in a red blood cell.
When the outer solution is hypotonic, water moves into the cell and causes the cell to eventually burst, or lyse. When the outer solution is hypertonic, water moves out of the cell, and the cell shrinks. *Arrows* show the direction of net movement of the water.

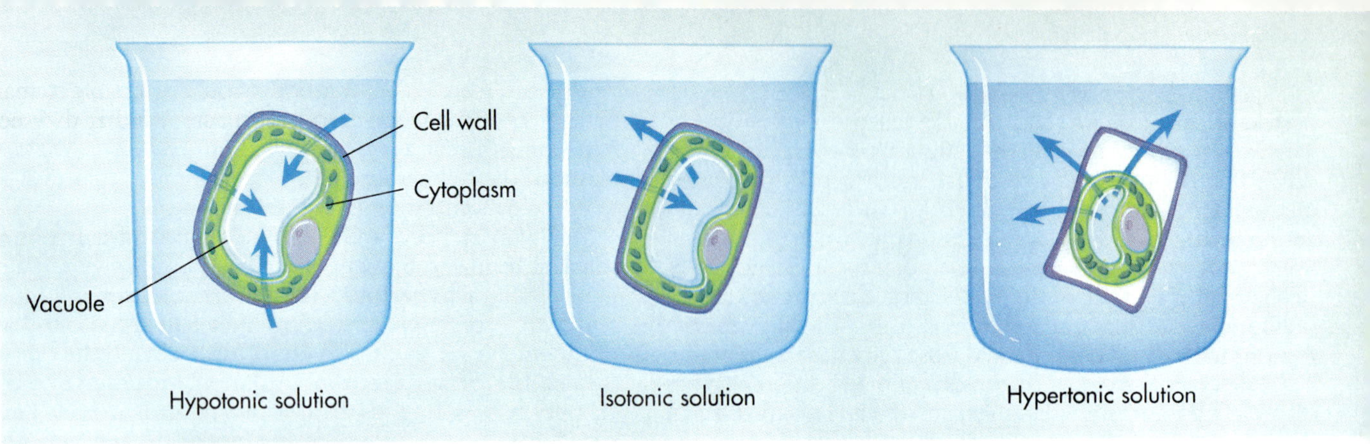

FIGURE 4-7 Osmosis in a plant cell.
In plant cells, the large central vacuole contains a high concentration of solutes, so water tends to diffuse inward, causing the cells to swell outward against their rigid cell walls. The cell continues to take on water until the osmotic pressure equals the turgor pressure. If a plant cell is immersed in a high-solute (hyptertonic) solution, water leaves the cell, causing the cytoplasm to shrink and pull in from the cell wall (plasmolysis).

ward of the cytoplasm. However, the pressure within the cell may become so great that the cell bursts like an over-inflated balloon. The cell is said to **lyse.**

Water moves into cells placed in hypotonic solutions and out of cells placed in hypertonic solutions.

Single-celled and multicellular organisms have various mechanisms that work to keep their cells from swelling and bursting or shriveling like prunes. Single-celled organisms that live in fresh water, for example, battle a constant influx of water. Some of these organisms have one or more organelles that collect water from the cell's interior and transport it to the cell surface. Many multicellular animals such as humans circulate a fluid through their bodies that bathes cells in isotonic liquid. By controlling the composition of its circulating body fluids, a multicellular organism can control the solute concentration of the fluid bathing its cells, adjusting it to match that of the cells' interiors. The blood in your body, for example, contains a high concentration of a protein called *albumin,* which serves to elevate the solute concentration of the blood to match that of your tissues so that the movement of water molecules is regulated.

Plant cells have cell walls that support their membranes, so they behave differently in hypertonic and hypotonic solutions than cells with no walls. To demonstrate what happens, you can perform an experiment. Take two stalks of celery, cut a slice off the bottom of each, and place one stalk in a glass of pure water and one in a glass of very salty water. After a few hours, observe the two stalks. The stalk in pure water should be firm and crisp. The stalk in salty water will most likely be limp. What happened?

The pure water is hypotonic to the cells of the celery. Therefore water moved into the celery cells by osmosis. Pressure began to build within each cell as its water vacuole grew larger and pressed against the cell wall. In plants, this pressure is called **turgor pressure.** Because each cell membrane of a plant is supported by a cell wall, the swollen cells did not burst. Instead, water moved into the cells until the osmotic pressure equaled the turgor pressure (Figure 4-7). These plump, stiff cells made the entire stalk of celery rigid.

The salt water, conversely, is hypertonic to the cells of the celery. Therefore water moved out of these cells, causing **plasmolysis,** a shrinking of the cytoplasm. As you can see in Figure 4-7, during plasmolysis the cell membrane pulls away from the wall of the cell, and the cell becomes limp—as does the entire stalk of celery.

Facilitated diffusion

One of the most important properties of any cell is its ability to move substances necessary for survival into its interior and to get rid of unnecessary or harmful substances. Your cells, for example, move glucose into the cytoplasm from the bloodstream to be used for fuel. Without the ability to move glucose into your cells, you would die.

Cells can perform these feats because they have differentially permeable membranes. The cell membrane can select which molecules are to enter and leave and in which concentrations because most molecules must enter or leave the cell through protein doors, or channels. These doors are not open to every molecule that presents itself; only a particular molecule can pass through a given kind of door. Some of the channels in the cell membrane help certain molecules and ions enter or leave the cell and speed their movement by providing them with a passageway. These passageways are most likely transport proteins that extend from one side of the membrane to the other. After the transport protein binds with a solute molecule, its shape changes, and the molecule moves across the membrane. This type of transport process, in which molecules move down the concentration gradient without an input of energy by the cell, is called **facilitated diffusion** (Figure 4-8). Facilitated diffusion helps rid the cell of certain molecules present in high concentrations and moves molecules into the cell that are present on the outside in high concentrations.

Facilitated diffusion is the movement of selected molecules across the cell membrane by specific transport proteins along the concentration gradient and without an expenditure of energy.

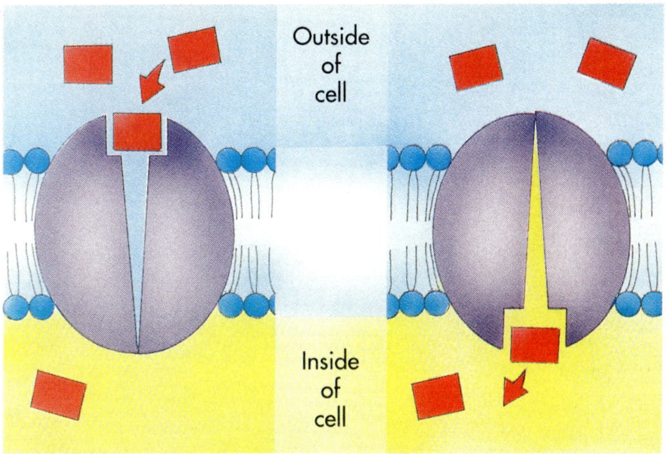

FIGURE 4-8 Facilitated diffusion.
Facilitated diffusion moves molecules across the cell membrane by means of protein channels *(purple oval shapes).* It is a passive process because the molecules move in the direction of lower concentration.

Facilitated diffusion has two essential characteristics: (1) it is *specific*, with only certain molecules being able to traverse a given channel, and (2) it is *passive*, the direction of net movement being determined by the relative concentrations of the transported molecule inside and outside the membrane.

Movement that does require energy

Cells often move substances into or out of the cell *against* the gradients of concentration, pressure, and electrical charge. A cell uses energy to move molecules against a gradient much like you might use energy to move something against gravity. For example, if you are driving downhill, you can put your car into neutral and coast (although that is not the safest way to drive downhill). The car will continue to move without a "push" from the engine. As soon as you come to a hill, however, you must put the car in gear and press on the accelerator or the car will soon come to a stop. Cells, too, need to use energy to move molecules "uphill," or against a gradient. Cells also need to expend energy to move large molecules or particles into the cell that cannot move across the cell membrane.

Active transport

There are many molecules that a cell takes up or eliminates against a concentration gradient. Some such as sugars and amino acids are molecules the cell needs to use for energy or to build new cell parts. Others are ions such as sodium and potassium that play a critical role in functions such as the conduction of nerve impulses. These many kinds of molecules enter and leave cells by way of a variety of selectively permeable transport channels. In all these cases, a cell must expend energy to transport these molecules against the concentration gradient and maintain the concentration difference. This type of transport is called **active transport.** Active transport is one of the most important functions of any cell. Without it, the cells of your body would be unable to maintain the proper concentrations of substances that they need for survival.

Active transport is the movement of a solute across a membrane against the concentration gradient with the expenditure of chemical energy. This process requires the use of a transport protein specific to the molecule(s) being transported.

More than one third of all the energy expended by a cell that is not dividing is used to actively transport sodium (Na^+) and potassium (K^+) ions. The type of channel by which *both* ions are transported across the cell membrane *in opposite directions* is a **coupled channel,** which has binding sites for both molecules on one membrane transport protein. This remarkable coupled channel is called the **sodium-potassium pump,** and it uses energy to move these ions across the cell membrane. Sodium ions are moved out of the cell to maintain a low internal concentration relative to the concentration outside of the cell. Conversely, potassium ions are moved into the cell to maintain a high internal concentration relative to the concentration outside the cell (Figure 4-9). (Three sodium ions are moved out for every two potassium ions that are moved in by the channel.) This transport mechanism is important in most cells of the human body but is particularly important for the proper functioning of muscle and nerve cells.

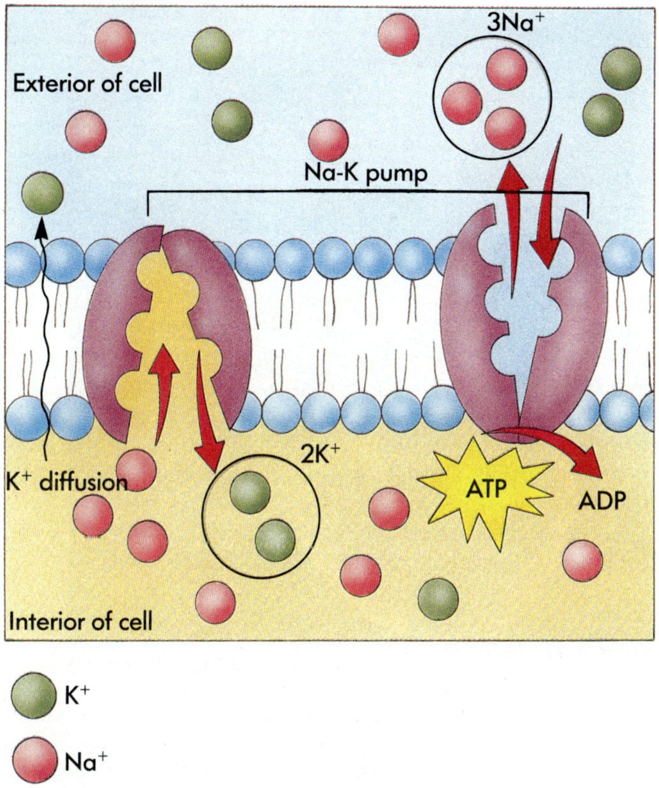

FIGURE 4-9 The sodium-potassium pump.
The pump is represented by the *purple shapes,* which show the two states of one protein. The molecules move in the direction of higher concentration.

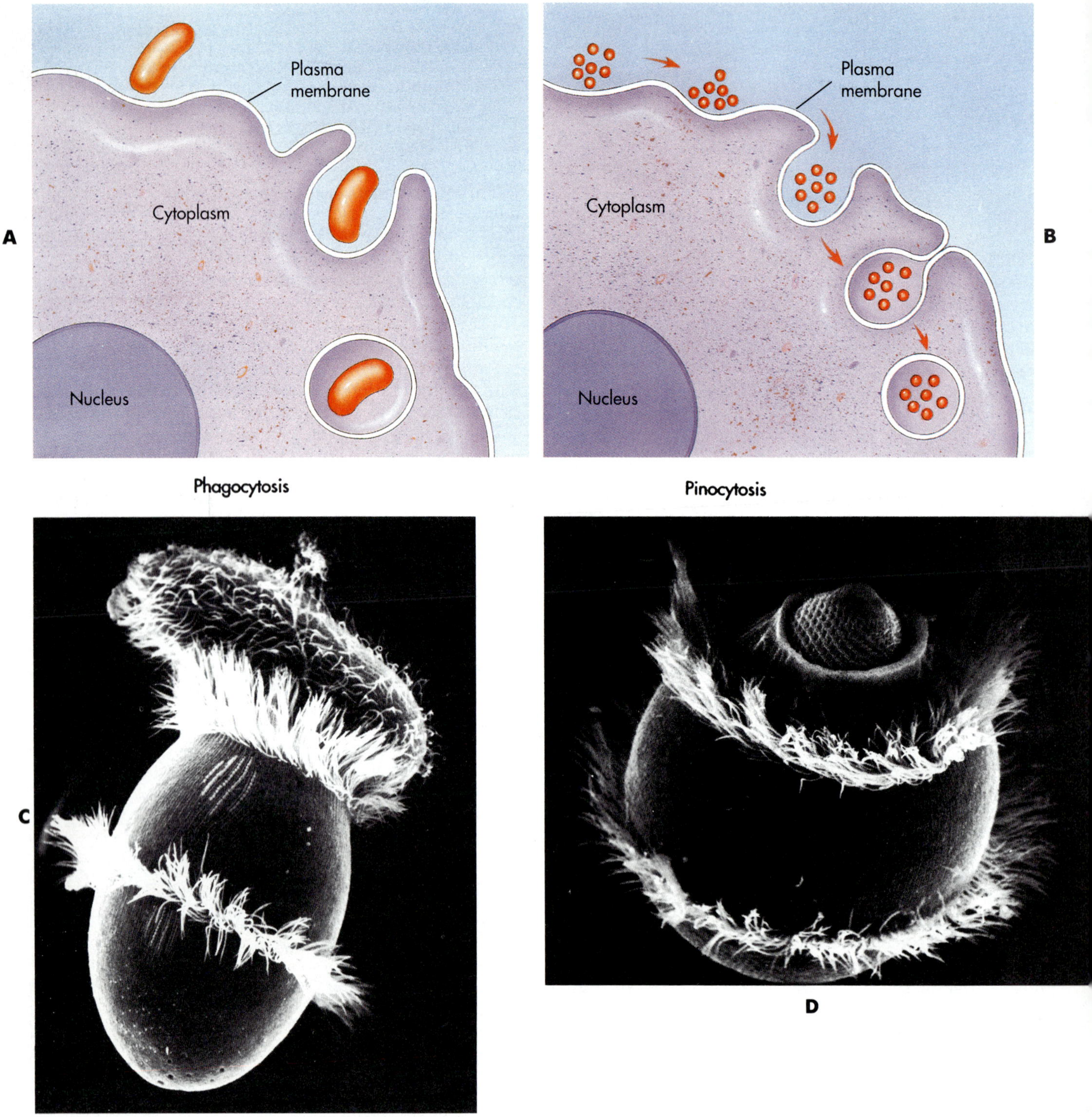

FIGURE 4-10 A cell ingests particles or organisms by endocytosis.
A Phagocytosis is the ingestion of an organism or particulate matter.
B Pinocytosis is the ingestion of dissolved materials (liquids).
C This large protist *(Didinium nasutum)* has begun to phagocytize a smaller protist (*Paramecium* species).
D The paramecium is almost completely ingested.

Endocytosis

Certain types of cells transport particles, small organisms, or large molecules such as proteins into their cells. In humans, for example, white blood cells police the body fluids and ingest substances as large as invading bacteria. In nature, some single-celled organisms often eat other single-celled organisms whole! How can cells move such large substances into their interiors?

Cells such as these ingest particles or molecules that are too large to move across the membrane by a process called **endocytosis,** which literally means "into the cell." Even if the substance is not being transported against a concentration gradient, the cell must use energy for this process to occur. During endocytosis a cell envelops the particle with finger-like extensions of the membrane-covered cytoplasm (Figure 4-10). The edges of the membrane eventually meet on the other side of the particle. Because of the fluid nature of the lipid bilayer, the membranes fuse together, forming a vesicle around the particle.

Endocytosis is a process in which cells engulf large molecules or particles and bring these substances into the cell packaged within vesicles.

If the material that is brought into the cell contains an organism (Figure 4-10, *A*) or some other fragment of organic matter, the endocytosis is called **phagocytosis** (Greek *phagein*, "to eat," and *cytos*, "cell"). If the material brought into the cell is liquid—contains dissolved molecules—the endocytosis is referred to as **pinocytosis** (Greek *pinein*, "to drink"). Pinocytosis is common among the cells of multicellular animals (Figure 4-10, *B*). Human egg cells, for example, are "nursed" by surrounding cells that secrete nutrients that the maturing egg cell takes up by pinocytosis.

Virtually all animal cells are constantly carrying out endocytosis, trapping extracellular fluid in vesicles and ingesting it. Within the cell, these vesicles fuse with lysosomes, tiny cellular "bags" of digestive enzymes, to break down these large particles into molecules usable to the cell. Rates of endocytosis vary from one cell type to another but can be surprisingly large. Some types of white blood cells, for example, ingest 25% of their cell volume each hour!

Exocytosis

The reverse of endocytosis is **exocytosis.** As with endocytosis, a cell must expend energy for exocytosis to occur. During exocytosis, a cell discharges material by packaging it in a vesicle and moving the vesicle to the cell surface. The membrane of the vesicle fuses with the cell membrane, and the contents are expelled (Figure 4-11). In plants, exocytosis is the main way that cells move the materials from the Golgi body and out of the cytoplasm to construct the cell wall. In animals, many cells are specialized for secretion using the mechanism of exocytosis, such as cells that produce and secrete digestive enzymes or hormones.

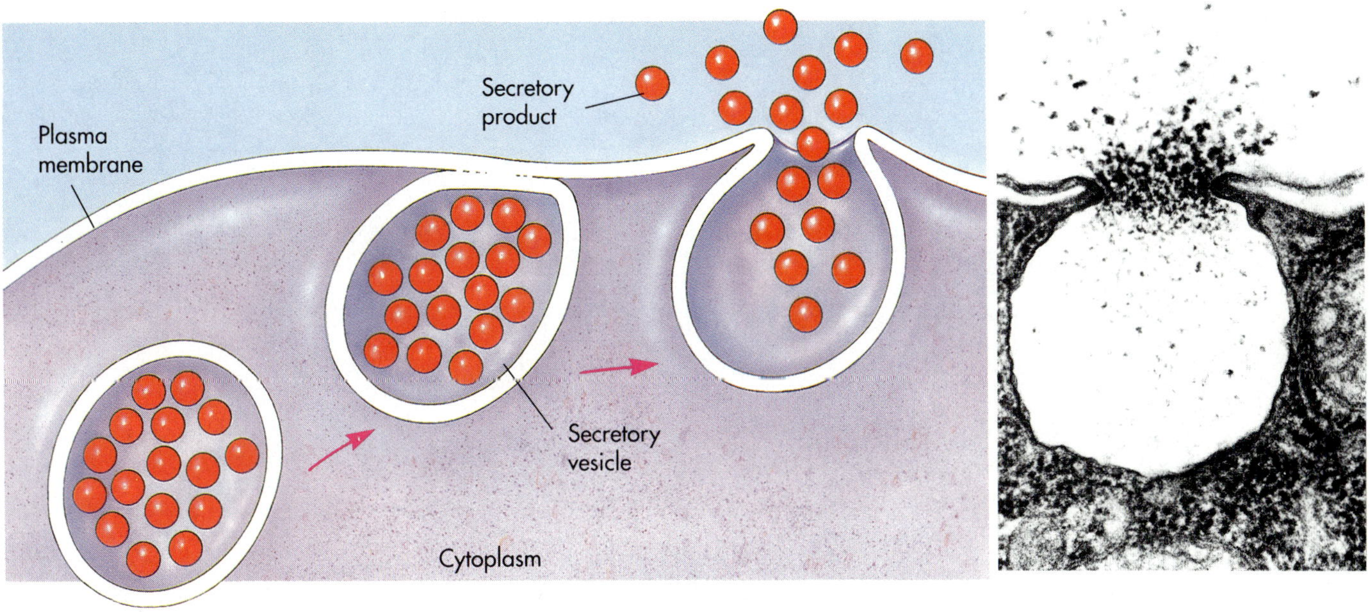

FIGURE 4-11 Exocytosis.
Proteins and other molecules are secreted from cells in small pockets called *vesicles,* whose membranes fuse with the cell membrane. When this fusion occurs, the contents of the vesicles are released to the cell surface.

Summary

1. Phospholipids are molecules made up of a portion of a fat molecule with a phosphate functional group attached. Because these molecules have polar and nonpolar parts, they form a double layer of molecules (a bilayer) when in a watery environment. A phospholipid bilayer is the foundation of cell membranes.
2. Lipid bilayers allow only certain molecules to pass across them. Membrane proteins facilitate the movement of many types of molecules into and out of the cell.
3. Proteins float within or on the phospholipid bilayer of the membrane and perform a variety of jobs in addition to the transport of molecules. Some of these jobs are identifying messages sent to the cell and forming junctions with other cells. Carbohydrates attached to the proteins and lipids play a role in cell recognition.
4. Molecules move randomly with a constant jiggling motion. However, molecules and ions undergo net movement in response to differences, or gradients, in concentration, pressure, and electrical charge, resulting in an equal distribution of these molecules and ions. Movement "down" a gradient requires no added energy to take place.
5. Molecular movement down a gradient but across a cell membrane is called *passive transport*. Three types of passive transport are diffusion, osmosis, and facilitated diffusion.
6. When cells move substances against a gradient, energy is required. This type of movement is called *active transport*.
7. The transport of particles or molecules into the cell that are too large to move across the membrane takes place by a process called *endocytosis*. The reverse of endocytosis is exocytosis. These processes require energy to take place.

REVIEW QUESTIONS

1. Describe the structure of a lipid bilayer, and explain its significance for a cell.
2. How do molecules move into and out of cells? Why is this movement important?
3. What is the role of the surface carbohydrates in a cell membrane?
4. What is a gradient? How does it affect molecules?
5. Distinguish between passive transport and active transport.
6. Identify the type(s) of molecular movement involved in each of the following. Which are active and which are passive?
   a. Glucose molecules leave your bloodstream (where they are usually in higher concentrations) and enter your cells by attaching to specialized carrier proteins.
   b. You lounge in a bathtub full of warm tap water until your skin looks swollen and "puckered."
   c. You dissolve a spoonful of instant coffee crystals in a mug of hot water.
   d. Your cells "import" amino acids by transporting them against the concentration gradient.
7. In question 6c, identify the solute, solvent, and solution.
8. Fill in the blanks: If a cell is placed in a(n) ________ solution, water will move into the cell. If the solution is ________, water will leave the cell. No net water movement will occur if the solution is ________.
9. Explain the difference between diffu sion and facilitated diffusion.
10. What is the sodium-potassium pump, and why is it important? What type of transport is involved?
11. Draw a diagram of a cell engaging in endocytosis. Give an example of this process in the human body.
12. Draw a diagram of a cell performing exocytosis. Give an example of how organisms use this process.

KEY TERMS

active transport 75
coupled channel 75
diffusion 72
endocytosis 77
exocytosis 77
facilitated diffusion 74
fluid mosaic model 69
gradient 72
hydrophilic 69
hydrophobic 69
hypertonic 73
hypotonic 73
isotonic 73
lipid bilayer 69
lyse 74
osmosis 72
osmotic pressure 73
passive transport 72
phagocytosis 77
phospholipids 68
pinocytosis 77
plasmolysis 74
sodium-potassium pump 75
solutes 73
solution 73
solvent 73
turgor pressure 74

THOUGHT QUESTIONS

1. The interior of a membrane's lipid bilayer is a hydrophobic (nonpolar) zone through which polar molecules cannot freely pass. Yet many cells manufacture highly polar proteins for use outside the cell. How do these polar macromolecules manage to cross the cell membrane?
2. The lipid bilayer that forms the foundation of a cell membrane is driven to assume its structure by hydrophobic interactions with water. What happens to the outermost skin cells of your fingers when you dip your hand into a nonwater environment, say in vegetable oil? Why?
3. A typical cell of your body—for example, one lining your intestine—expends one third of all its energy running the sodium-potassium pump. Why? What is it doing pumping all these sodium and potassium ions that is so important?
4. A typical person drinks liquids in rather large volumes a few times each 24 hours, rather than sipping continuously all through the day and night. How does the blood stay isotonic at all times if it is diluted by water so irregularly?

FOR FURTHER READING

Bretscher, M. S. (1985, October). The molecules of the cell membrane. *Scientific American*, pp. 100-108.
This contains a good description of the structure of the cell membrane and explains the way transmembrane proteins are anchored within the lipid bilayer.

Hakomori, S. (1986, May). Glycosphingolipids. (1986, May). *Scientific American*, pp. 44-53.
Carbohydrate chains attached to lipid molecules play an important role in cell-to-cell recognition, an area of intensive present-day research.

Unwin, N, & Henderson, R. (1984, February). The structure of proteins in biological membranes. *Scientific American*, pp. 78-94.
This is a lucid account of how proteins are anchored within membranes by means of nonpolar segments.

5 Cell division

HIGHLIGHTS

The hereditary material of cells condenses into discreet bodies called *chromosomes* before cell division; in this form, the hereditary material can be distributed in an orderly fashion during cell division.

Cell division includes division of the cytoplasm and division of the nuclear material.

Mitotic cell division produces new cells during the growth and repair of multicellular organisms and produces sex cells in plants; mitosis is also a mechanism of reproduction in some simpler organisms.

Meiotic cell division produces sex cells in animals and spores in plants.

OUTLINE

Chromosomes
Cell division, growth, and reproduction
Mitosis
- Prophase
- Metaphase
- Anaphase
- Telophase

Cytokinesis
- Cytokinesis in animal cells
- Cytokinesis in plant cells

Meiosis
- The stages of meiosis
  - Meiosis I
  - Meiosis II

The importance of meiotic recombination

When you were born, did you come with instructions? You may not know it, but you did. The brightly stained bodies in the photo are human "instructions" called **chromosomes,** but they can't be read in the usual way. Instead, using a special code that can be "understood" by other special molecules, they direct all the chemical reactions that take place in your body. Instructions in the chromosomes determine such things as whether your hair is red or brown and whether you are male or female. But, most importantly, they hold the key to the mystery of what makes "you."

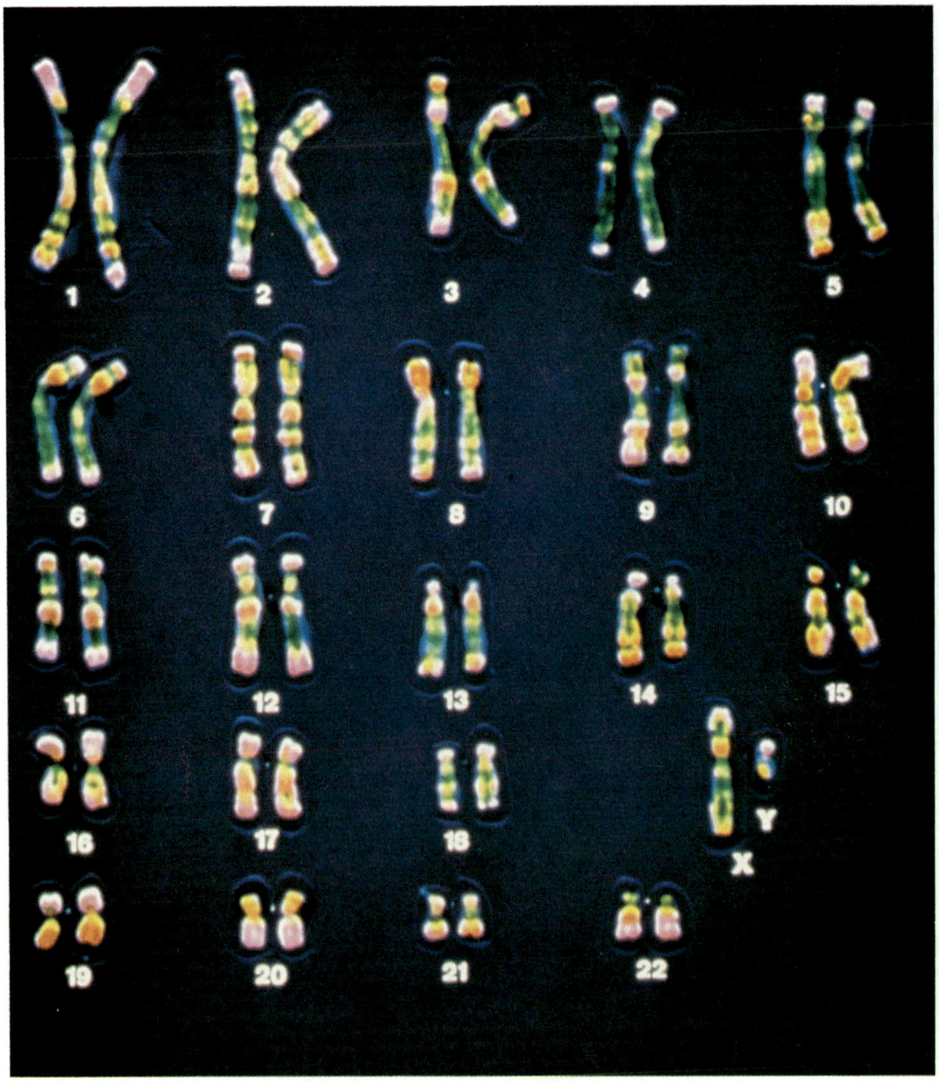

Chromosomes

You might title the chapter opening photo, "The Human Body: Instructions for Growth, Development, and Maintainance." Scientists would call it a **karyotype.** A karyotype is the particular array of chromosomes that belong to an individual. Scientists take this type of picture by specially treating a cell just as it is about to divide. The cell is then photographed, and the chromosomes are cut out of the photograph and arranged on a piece of paper. Each chromosome has a mate, so they are paired up and arranged in order of decreasing size.

Notice the chromosomes labeled *X* and *Y*. These are called **sex chromosomes.** They carry the **genes,** or pieces of hereditary information, that determine the gender of an individual as well as certain other characteristics. The presence of the Y chromosome in this karyotype means that these chromosomes belong to a male. A female does not have the smaller Y chromosome but has another X chromosome instead. Humans have 22 pairs of chromosomes in addition to this one pair of sex chromosomes. These other 22 pairs are called **autosomes.** They carry the majority of the genetic information but have no genes that determine gender. Species differ in their number of chromosomes. For example, humans have 46, goldfish have 92, and fruitflies have 8.

In the century since the discovery of chromosomes, scientists have learned a great deal about their structure and function. The chromosomes of eukaryotes are made up of a complex of deoxyribonucleic acid (DNA) and protein. (The chemical structure of DNA is discussed in Chapter 24, and that of proteins is discussed in Chapter 2.) This complex is called **chromatin.** The DNA is the part of the chromatin that contains your instructions, commonly called the *code of life*. In fact, the amount of information contained in the DNA of one chromosome alone would fill about 2000 books of 1000 pages each!

Proteins and the hereditary material, DNA, make up the chromatin material of the cell.

The DNA of each chromosome exists as one very long, double-stranded molecule that extends unbroken through the chromosome's entire length. If the strand of DNA from a single chromosome were laid out in a straight line, it would stretch about 5 centimeters, or 2 inches (Figure 5-1). That may not sound long, but 23 pairs of these strands fit into the nucleus of each of the microscopic cells of your body.

To fit into cells, these long strands of DNA are coiled, much like you might coil up your garden hose. The coiling of DNA within cells, however, is much more complicated. Figure 5-2 shows how this coiling is accomplished to provide organization and orderliness. First, the double strand of DNA winds like thread on a spool around groups of tiny proteins called **histones.** In this form the DNA looks similar to a string of pearls (Figure 5-3). This string of pearls then wraps up into larger coils called **supercoils.** These supercoils are looped and "packaged" with other proteins to form chromatin. And as Figure 5-2 shows, the chromatin can be condensed by further coiling to form the chromosomes. This condensation occurs just before and during cell division. At this time, each chromosome is composed of two identical structures that are called **sister chromatids** because each DNA molecule makes a copy of itself, or replicates, before the condensing of the chromatin. Chromatids are held together at a constricted region called the **centromere.**

Just before and during cell division, chromatin material coils and condenses to form discreet bodies called chromosomes.

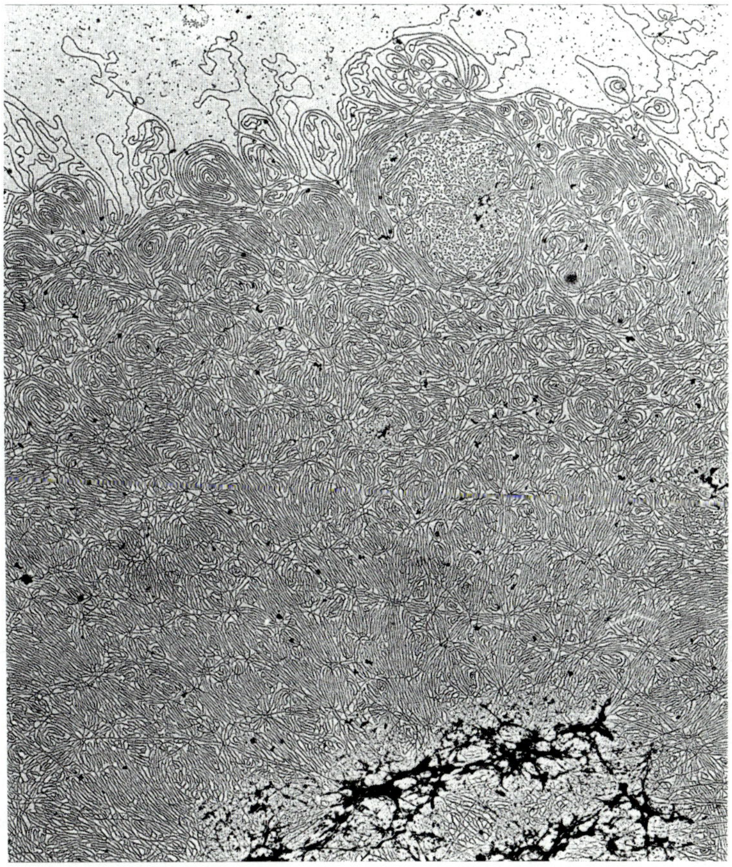

FIGURE 5-1 A human chromosome contains an enormous amount of DNA.
The dark element at the bottom of the photograph is the protein component of a single chromosome. All of the surrounding material is the DNA of that chromosome.

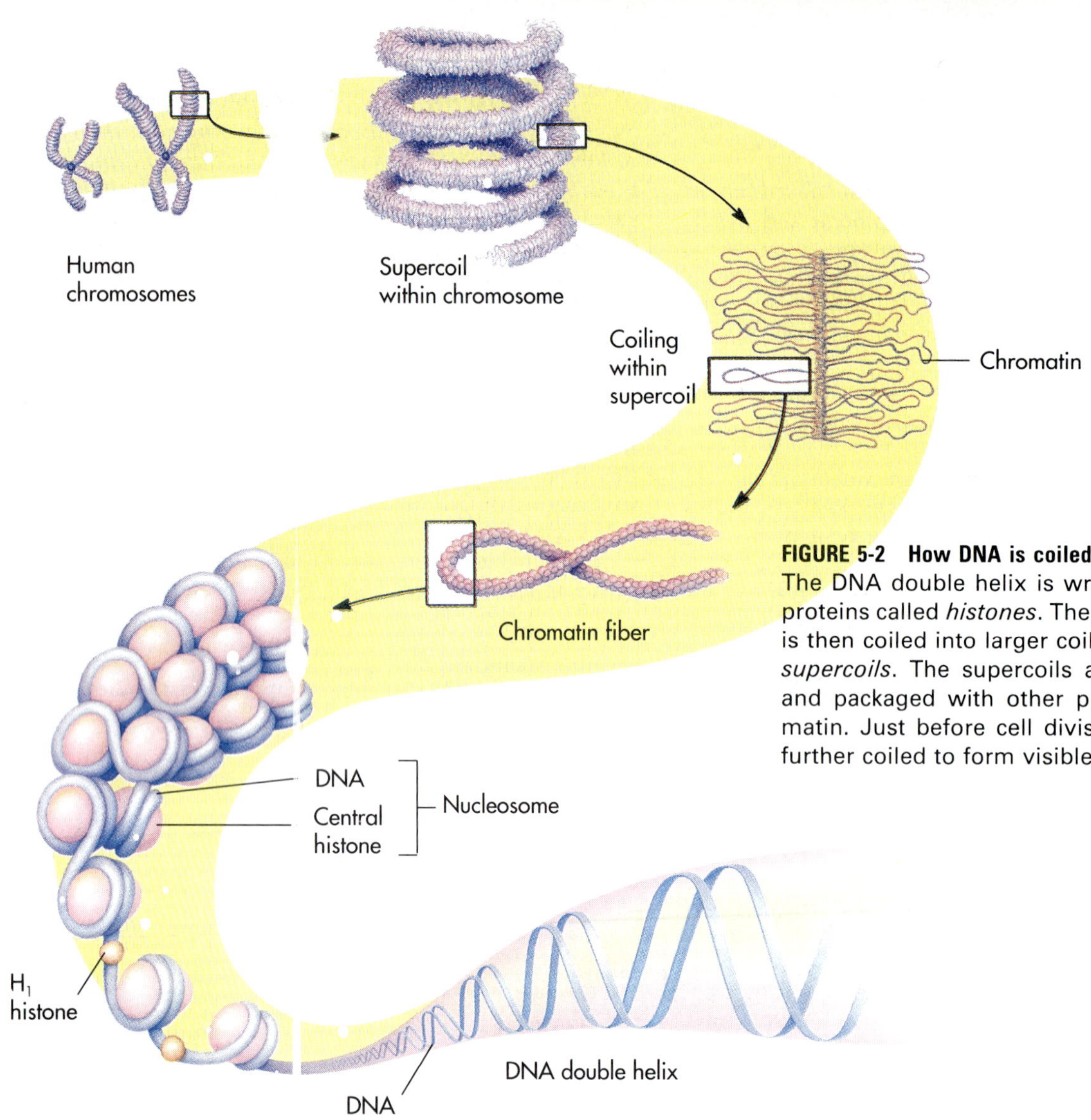

FIGURE 5-2 How DNA is coiled within chromosomes.
The DNA double helix is wrapped around special proteins called *histones*. The DNA-histone package is then coiled into larger coils appropriately called *supercoils*. The supercoils are themselves coiled and packaged with other proteins to form chromatin. Just before cell division, the chromatin is further coiled to form visible chromosomes.

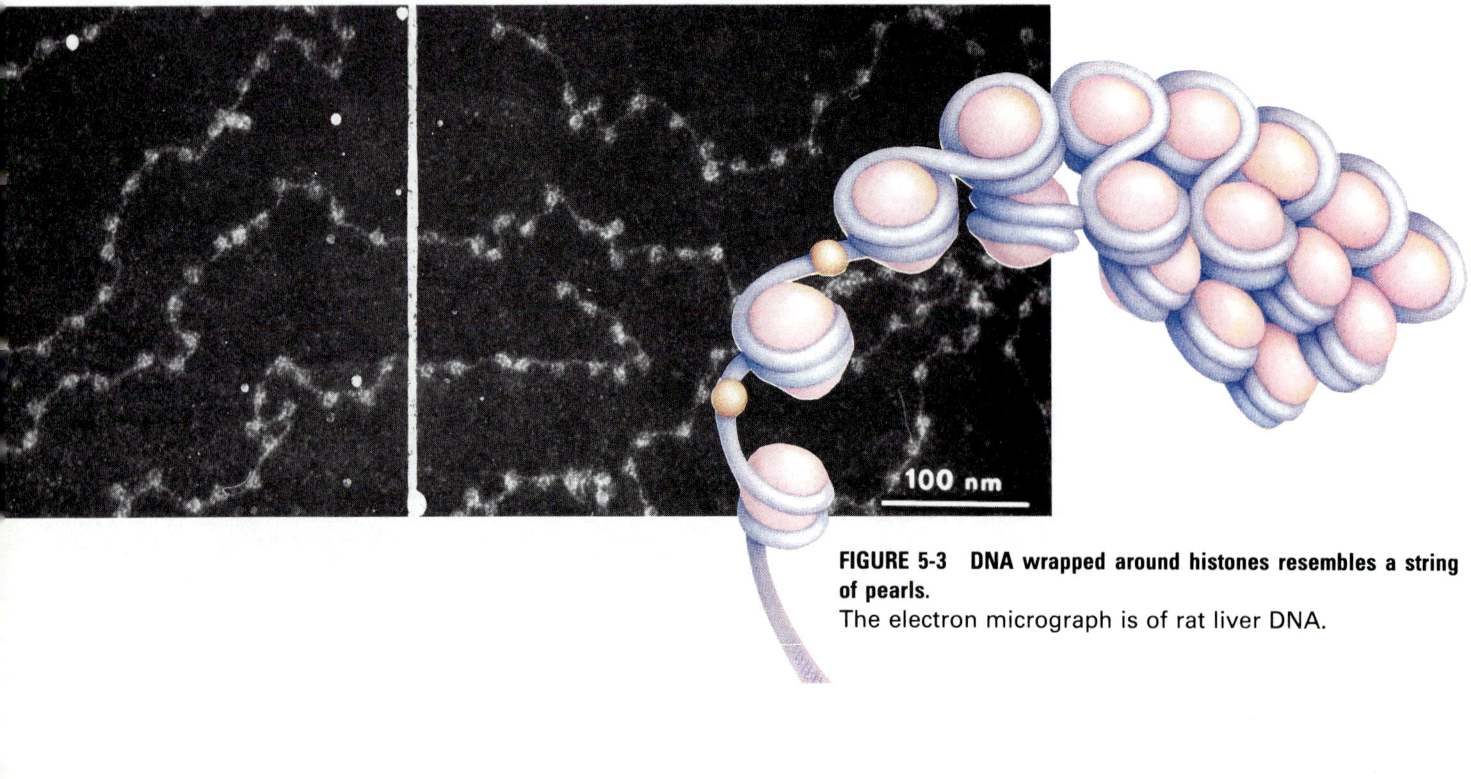

FIGURE 5-3 DNA wrapped around histones resembles a string of pearls.
The electron micrograph is of rat liver DNA.

Cell division, growth, and reproduction

The fact that DNA is packaged neatly in chromosomes not only allows a tremendous amount of information to reside in each cell of your body but also allows your cells to separate this genetic material in an organized way during cell division. If the coiling and supercoiling of DNA were not orderly, the hereditary material of dividing cells would become hopelessly tangled.

Cell division occurs in eukaryotic cells as a means of cellular growth, repair, and reproduction. To illustrate, look at the human life cycle diagrammmed in Figure 5-4. A **life cycle** is the progression of stages an organism passes through from its conception until it conceives another similar organism. The human life cycle is representative of the life cycle of all animals. The baby in the diagram represents that part of the life cycle during which a new individual has been produced by the fusion of **gametes,** or sex cells, from a male and a female of the same species. The female gamete is the egg, and the male gamete is the sperm.

After a person (or other animal) grows to sexual maturity, the sex organs within its body begin to produce gametes by a type of cell division called **meiosis.** During meiosis, one parent cell produces four sex cells, but these cells are *not* identical to the parent cell. Each sex cell is **haploid;** that is, it contains *half* the amount of hereditary material of the original parent cell. It is a single set of the genetic information. Because of this reduction in chromosome number, one sex cell from each of two parent organisms can join together in a process called **fertilization** to form the first cell of a new individual that has a full complement of hereditary material. This "new" cell is **diploid.** That is, it contains double the haploid amount—a double set of the genetic information. This type of reproduction, which involves the fusion of gametes to produce the first cell of a new individual, is called **sexual reproduction.**

The first cell of any multicellular organism divides, and those cells divide, and so on, eventually producing a complete, properly developed individual. This type of cell division is called **mitosis** and produces two "daughter" cells from one parent cell. These daughter cells have genetic information that is identical with one another and with the original parent cell. As the growth and development of the individual continues, certain cells are triggered by other developmental processes to begin to differentiate, or become different, from one another while continuing to divide. The

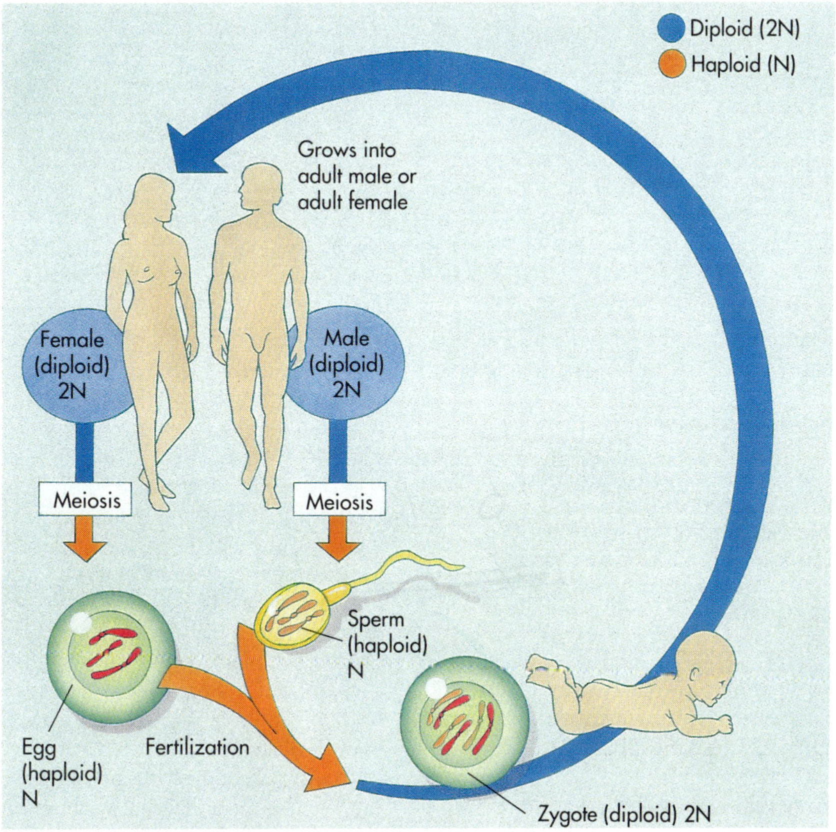

FIGURE 5-4 The human life cycle.
Mitosis, which produces two identical diploid daughter cells, is responsible for the growth of a human being into an adult. Meiosis, on the other hand, produces four haploid gametes (sex cells) that are *not* identical—each gamete contains half the hereditary material of the original cell. Meiosis in human females produces one functional gamete and three nonfunctional gametes. During fertilization, one gamete from each of the two parents comes together to form a new individual.

BIOLOGY & you

Cancer and mitosis

Cancer is perhaps one of the most dreaded human diseases. The National Cancer Society estimates that 520,000 people died of cancer in 1992 in the United States. Everyone has had family or friends affected by the disease. Many of us will die of cancer. Not surprisingly, a great deal of effort has been expended to learn the cause of this disease. Using new techniques of molecular biology, much progress has been made in understanding the molecular basis of cancer—how normal cells are transformed into cancerous ones.

Cancer is a growth disorder of cells. Normally, cells are triggered to begin mitosis by special proteins that are growth factors. Growth factors trigger cell division by binding to receptor sites embedded in the plasma membrane. When this binding occurs, the cell is signaled to begin cell division. In cancer, an apparently normal cell begins to divide in an uncontrolled and invasive way. The result is a mass of cells, or tumor, that constantly grows in size as the cells within it undergo rapid mitosis. Cancer is initiated by an abnormality either in the growth factors or in the receptors that prompt the cell to begin dividing.

Using the technique of transfection, scientists have been able to determine the exact nature of this abnormality. In transfection, a researcher tests a tumor cell's DNA for genes that direct the chemical composition of the growth factors and the shape of receptor sites (genes are the small segments of DNA that direct all the cell's activities). By using transfection techniques, researchers have found that a single gene isolated from a cancer cell can transform normally dividing cells into cancerous ones. The researchers called such a cancer-inducing gene an *onc gene,* from the Greek word *oncos,* meaning "cancer."

Onc genes are "normal" genes that have mutated, or changed. From their transfection experiments, scientists have determined that there are three types of *onc* genes that cause uncontrollable cell division. One type of *onc* gene produces a protein that mimics the growth factor necessary for cell division. This type of *onc* gene constantly makes this protein, which binds to receptor sites and results in rapid mitosis. The second type of *onc* gene produces a protein that mimics the receptor site. When additional receptor sites are available for binding, cells divide more often. The third type produces proteins that lower the amount of growth factor necessary to initiate cell division. As a result, the cell with the *onc* gene divides at a faster rate than a cell with the normal gene.

What causes an otherwise normal gene to turn into a deadly *onc* gene? Heredity is one factor, as are the mutations that sometimes occur when DNA is replicated in preparation for mitosis. Other causes are environmental. Chemicals in cigarette smoke (see the boxed essay in Chapter 11), asbestos, and other substances have been implicated in cancer. These chemicals, or carcinogens, are capable of changing normal genes into *onc* genes.

Scientists are currently looking into ways to control cancer by controlling cell division. Their research has revealed interesting information about the mechanisms that control cell division, but for the time being, they have not yet uncovered a cure for cancer.

animal life cycle begins again when the growth and development of the individual is complete and sexual maturity is reached.

Mitosis is a process of cell division that produces two identical cells from an original parent cell. Meiosis is a process of cell division that produces four cells from one parent cell. Each of these four cells has one set of genetic information rather than two sets like the parent cell.

Plants and animals are similar in that they both use the process of mitosis to grow. Also, the life cycles of both plants and animals have two phases: a haploid phase and a diploid phase. In animals, however, the diploid phase of the life cycle predominates. The haploid phase consists only of single-celled gametes. In most organisms, these gametes live only a day or two after meiosis is complete unless fertilization takes place. During fertilization, two haploid gametes join to form the first cell in the diploid phase of the life cycle. Plant life cycles differ from animal life cycles because most plants have distinct *multicelled* haploid as well as diploid phases. One phase usually dominates over the other, but both phases are always present.

In plant life cycles, the haploid phase is the **gametophyte** (gamete-plant) generation and the diploid phase is the **sporophyte** (spore-plant) generation. Compare the generalized plant life cycle in Figure 5-5 with the animal life cycle in Figure 5-4. Although haploid plants are common, haploid animals are rare. This results in some striking differences in the life cycles of animals and plants. In animals, gametes are usually produced in diploid individuals by the process of meiosis. In plants, gametes are produced in haploid individuals by the process of mitosis. The product of fertilization in plants is the first cell of a diploid spore-producing plant, not a diploid gamete-producing organism as in animals. (Spores are haploid, single-celled reproductive bodies capable of growing into the haploid gametophyte plant.) Because sporophytes are diploid, they produce spores by meiosis, reducing the number of chromosomes by half during this process.

In plants, unlike animals, mitosis occurs during both growth and reproduction. Haploid plants produce gametes by mitosis, whereas diploid plants produce spores by meiosis. The life cycle of a plant consists of the alternation of the growth and reproduction of each of these plant types.

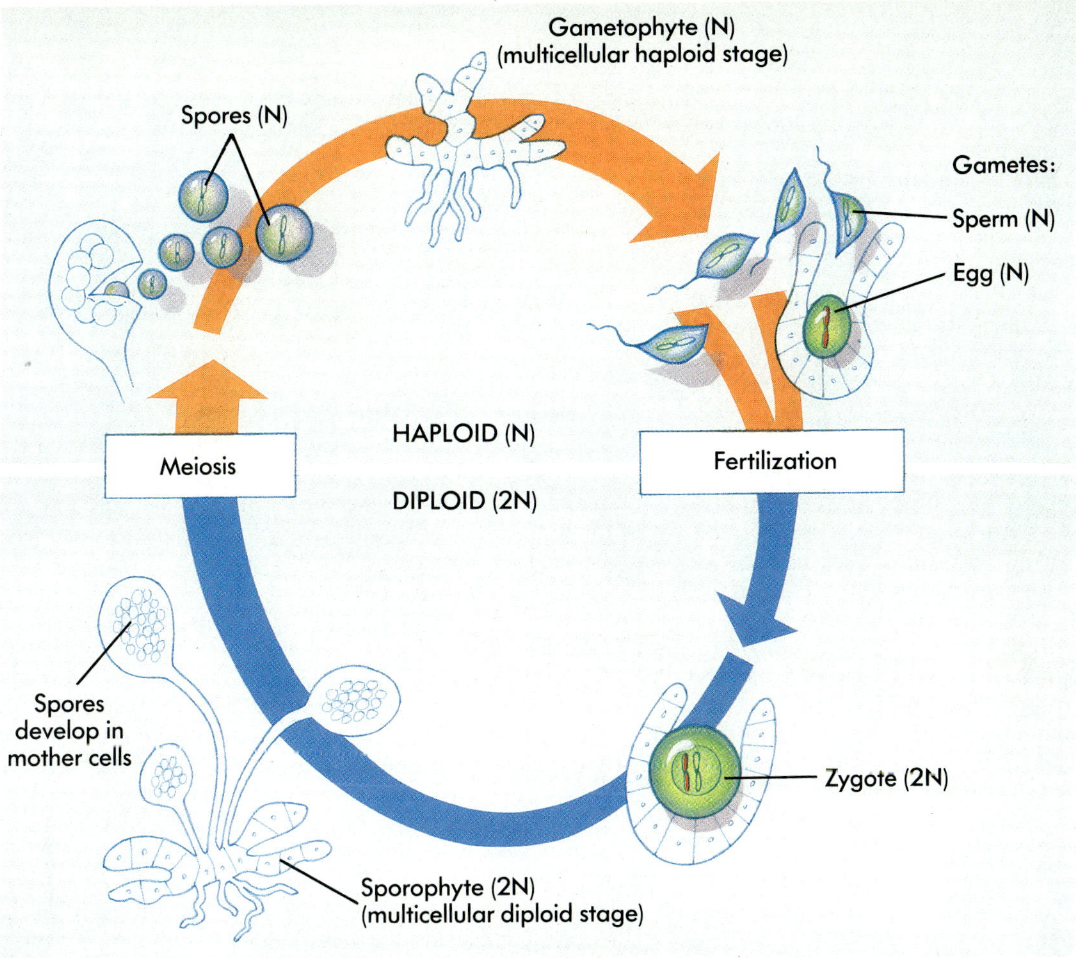

FIGURE 5-5 The life cycle of plants.
Notice how the sporophyte phase of a plant, which is diploid, alternates with the gametophyte phase, which is haploid.

Single-celled eukaryotes (such as many of the protists) generally reproduce by mitosis; this process plays no role in the growth of single-celled organisms but plays a role in the growth of populations of single-celled organisms. During reproduction, a parent organism divides by mitosis, producing two identical organisms. This type of reproduction is called **asexual reproduction.** In addition to dividing by mitosis, single-celled organisms sometimes exchange genetic information with compatible members of their species. Although this exchange of hereditary material is important because it increases the genetic diversity within the population, it does not result in the production of new individuals. Therefore chromosome exchange in single-celled organisms is not in itself considered a reproductive process. Some simple animals such as the hydra pictured in Figure 5-6 and many plants are also capable of reproducing asexually by mitosis.

Many of the single-celled eukaryotes reproduce asexually by mitosis.

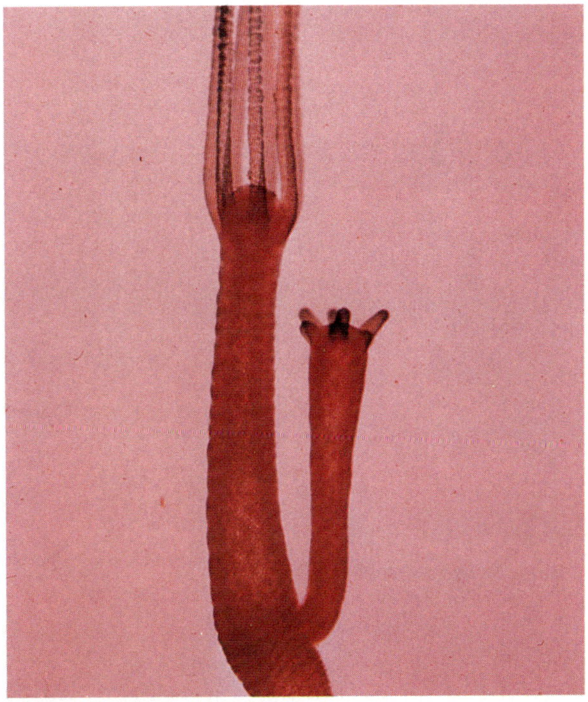

FIGURE 5-6 Asexual reproduction.
Many organisms, such as this hydra, are capable of reproducing asexually by mitosis.

Mitosis

Notice in Figure 5-7 that only a small portion of a cell's "life," called the **cell cycle,** is involved with mitosis. The activities of **interphase** occupy the majority of the cell cycle.

In the past, interphase was nicknamed the "resting stage." Scientists now know that a cell in interphase is far from resting! In fact, it doubles its size and replicates its DNA as well as carrying out its normal life functions. During the G_1 stage of interphase (the "gap" of time between the last mitosis and the start of DNA synthesis), the cell is growing. For many organisms, this growth period occupies the major portion of the cell's life span—possibly days, months, or years. During the S (synthesis) stage of interphase, a complete replica of the cell's DNA is synthesized. The cell now contains two complete, identical copies of hereditary information. These two copies of hereditary information can be seen as sister chromatids when the chromosomes later become visible.

During the G_2 of interphase (the time gap between the end of DNA synthesis and the beginning of mitosis), the supercoils of DNA, normally somewhat diffuse or "strung out" within the cell nucleus, begin the long process of **condensation.** During this process the complex of DNA and proteins coil into more tightly compacted bodies that will become visible as chromosomes during mitosis. In addition, during the G_2 phase the cell begins to assemble the "machinery" (microtubules that make up the spindle) that it will later use to move and divide the chromosomes. In animal cells, for example, the **centrioles** replicate. These organelles are surrounded by microtubule-organizing centers that play a key role in the mitotic process; most plants and fungi lack centrioles.

Interphase is the portion of the cell cycle in which the cell grows and carries out normal life functions. During this time the cell also produces an exact copy of the hereditary material, DNA, as it prepares for cell division.

Mitosis is a continuous sequence of events that occurs just after interphase, and that results in the division of the chromosomes duplicated during its S phase. To more easily understand this process, scientists divide its events into the following phases: prophase, metaphase, anaphase, and telophase (Figure 5-8).

Prophase

The first stage of mitosis, **prophase,** begins when the chromosomes have condensed to the point where they become visible under a light microscope. As prophase continues, the chromosomes continue to shorten and thicken, looking much bulkier at the end of prophase than at its beginning. The nucleolus, which was previously conspicuous, disappears. This disappearance is due to the nucleolus being unable to make ribosomal RNA (rRNA) when the part of the chromosome bearing the rRNA genes is condensed. And it is rRNA that makes up most of the substance of the nucleolus (see Chapter 3).

Prophase is the stage of mitosis characterized by the appearance of visible chromosomes.

While the chromosomes are condensing, another series of equally important events is also occurring: special microtubules (thin, tube-like, protein structures [see Chapter 3]) called the **spindle fibers** are being assembled. In animal cells, these spindle fibers extend from a pair of related microtubular structures called *centrioles* (Figure 5-8). Although the centrioles were once thought to play a role in forming the spindle, recent evidence suggests that this is not the case. Instead, the microtubules of the spindle appear to form from granules surrounding the centrioles, called the *microtubule organizing center*, or MTOC. Interestingly, plant cells do not

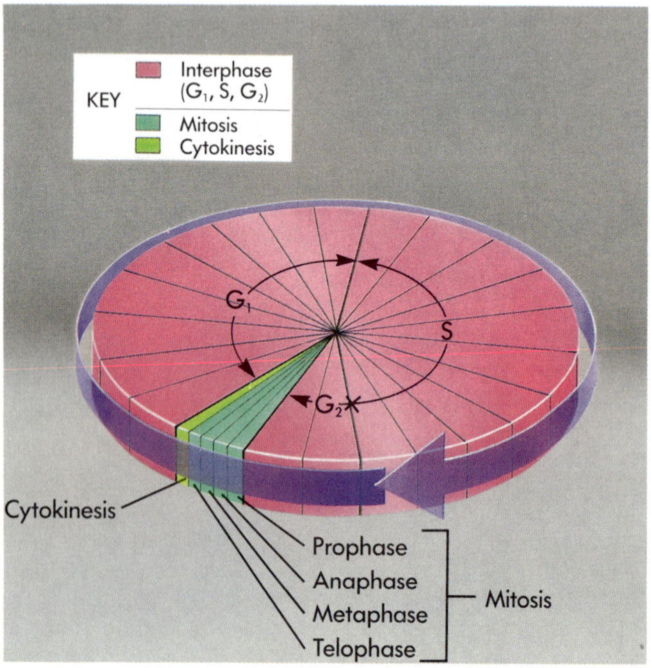

FIGURE 5-7 The cell cycle
In this diagram, each wedge represents 1 hour of the 22-hour division cycle of human liver cells. Actual cell division makes up a very small part of the cycle.

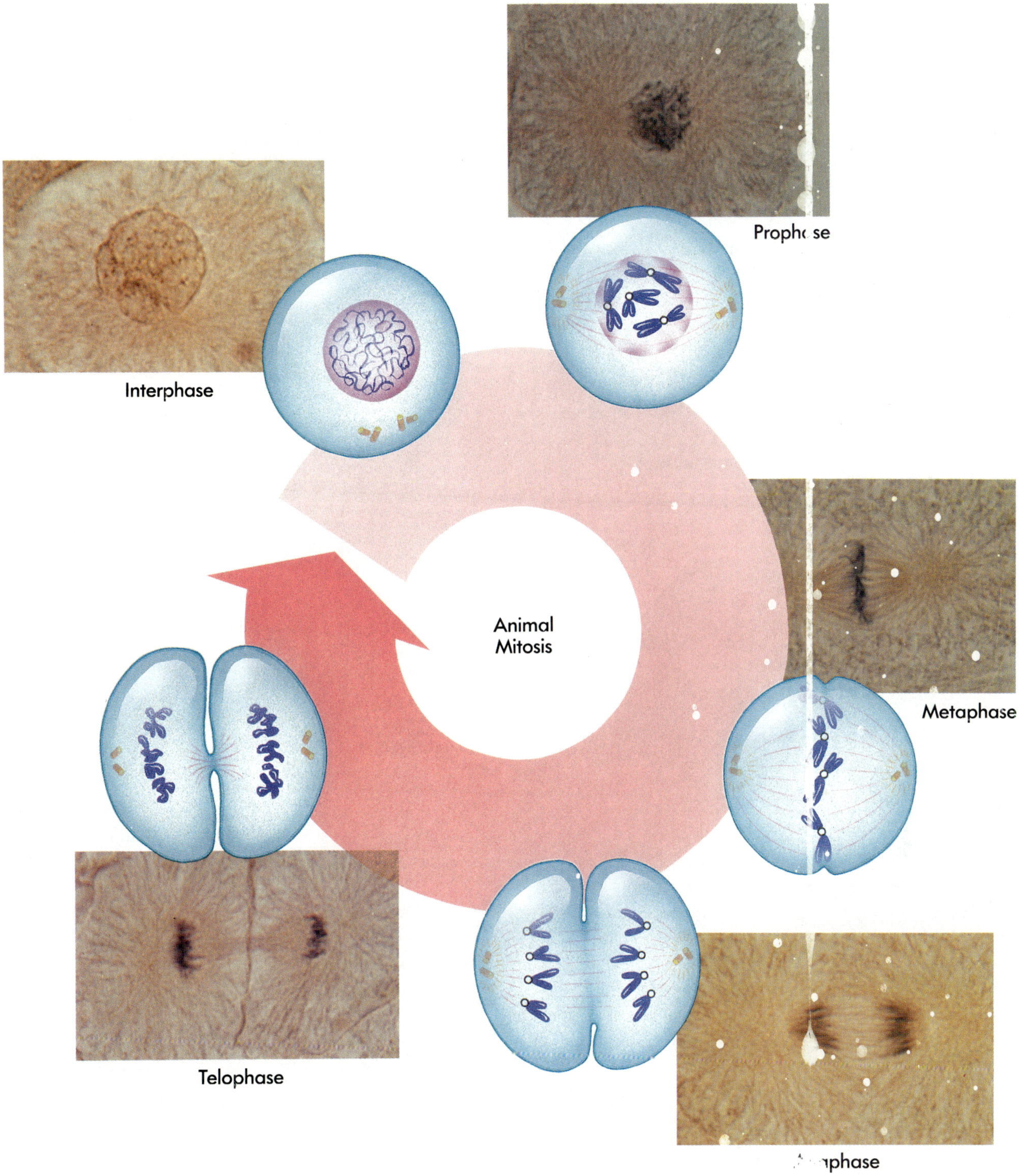

FIGURE 5-8 The stages of mitosis in an animal cell.
The stages of mitosis are shown here in photographs and drawings of a d   white-fish cell. Although they are barely noticeable in the photographs, centrio   a key role in animal cell mitosis. Although interphase is shown here to complete   cycle, it is not a stage of mitosis.

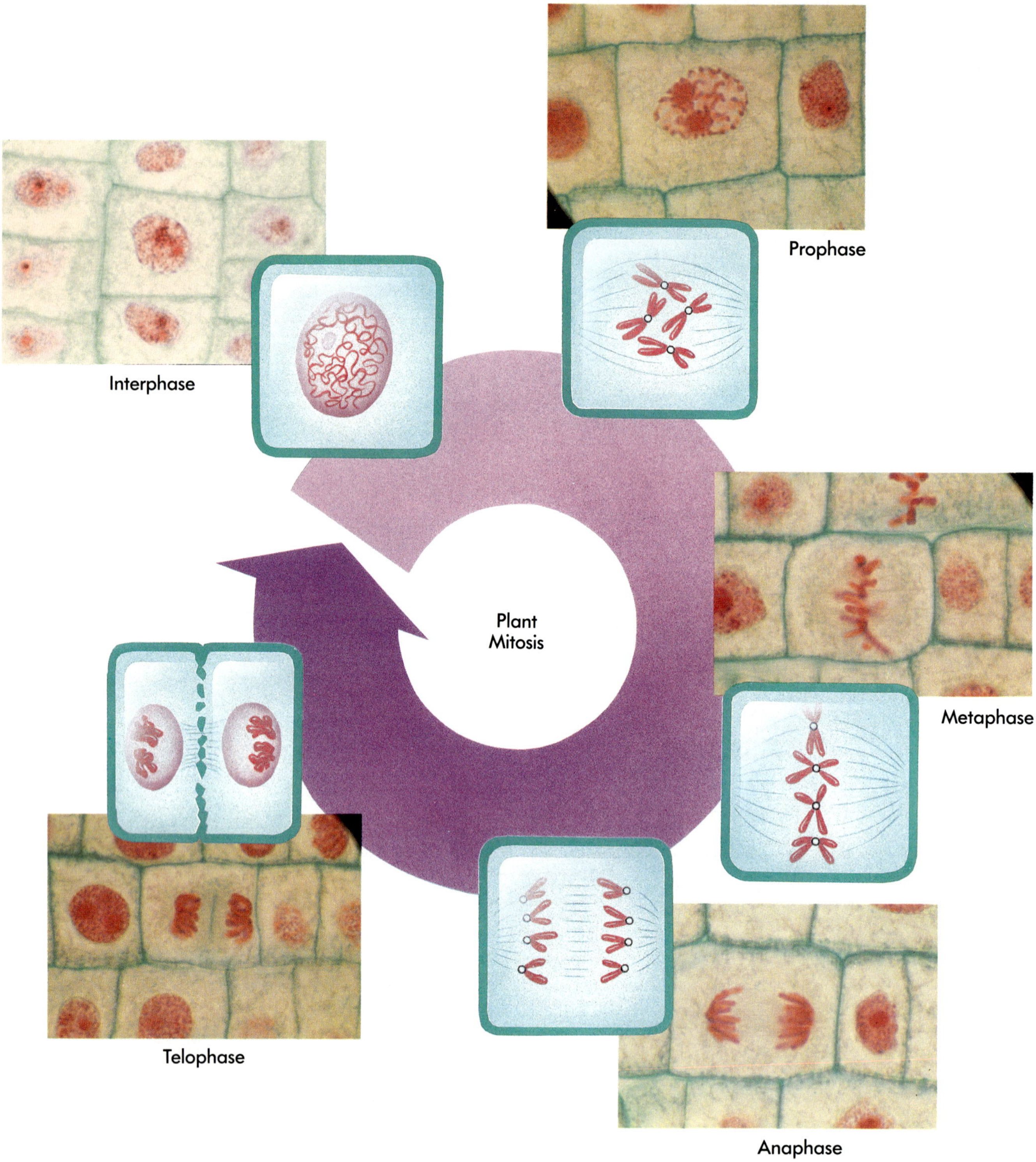

FIGURE 5-9 The stages of mitosis in a plant cell.
The stages of mitosis are shown here in photos and drawings of a dividing allium root cell. A major difference between animal and plant mitosis is that plant cells lack centrioles, although they do contain spindle fibers. Although interphase is shown here to complete the cell cycle, it is not a stage of mitosis.

contain centrioles, but spindle fibers form in plant cells (Figure 5-9). At this time researchers do not know what structure acts as a centriole or an MTOC in plants.

In early prophase the centrioles of animal cells begin to move away from one another. By the end of prophase, each member of the pair has moved to an opposite end, or pole, of the cell. As the spindle fibers form, the nuclear envelope breaks down, forming small vesicles that disperse in the cytosol. Without the nucleus in the way, the spindle fibers form a bridge between the centrioles, spanning the distance from one pole to the other. Some spindle fibers extending from each pole attach to "their" side of the centromere of each sister chromatid. By this time, each chromatid has developed a **kinetochore** at its centromere. A kinetochore is the structure to which several pole-to-centromere microtubules will attach. The effect is to attach one sister chromatid to one pole and the other sister chromatid to the other pole. During later stages of mitosis the sister chromatids separate and move to opposite poles of the cell, pulled by the spindle fibers. The proper attachment of each spindle fiber to the kinetochore is therefore critical to the process of mitosis; a mistake is disastrous. The attachment of two sister chromatids to the same pole, for example, results in their not separating, so they end up in the same daughter cell. Such a situation can result in a cell with an extra chromosome. Depending on the type of cell, it may be abnormal or it may die.

When the centrioles reach the poles of the cell in animal mitosis, they radiate an array of microtubules outward (toward the cell membrane) as well as inward (toward the chromosomes). This arrangement of microtubules is called an **aster.** The function of the aster is not well understood. Evidence suggests that the aster probably acts as a support during the movement of the sister chromatids.

By the end of prophase, spindle fibers that radiate from opposite poles of the cell attach to each kinetochore at the centromere.

Metaphase

The second phase of mitosis, **metaphase,** begins when the chromatid pairs line up in the center of the cell. In reality, the chromosomes are not in a line but form a circle. They only appear to form a line when viewed two dimensionally with a light microscope as in Figures 5-8, *C* and 5-9, *C*. Figure 5-10 is a diagram of a three-dimensional view of an animal cell at metaphase. As you can see, the chromosomes form a circle perpendicular to the direction of the spindle fibers. Positioned by the microtubules attached at their centromeres, all the chromosomes are equidistantly arranged between the two poles at the "equator" of the cell. The region of this circular arrangement, called the **metaphase plate,** is not a physical structure but indicates approximately where the future axis of cell division will be.

Metaphase is the stage of mitosis characterized by the alignment of the chromosomes in a ring, equidistant from the two poles of the cell.

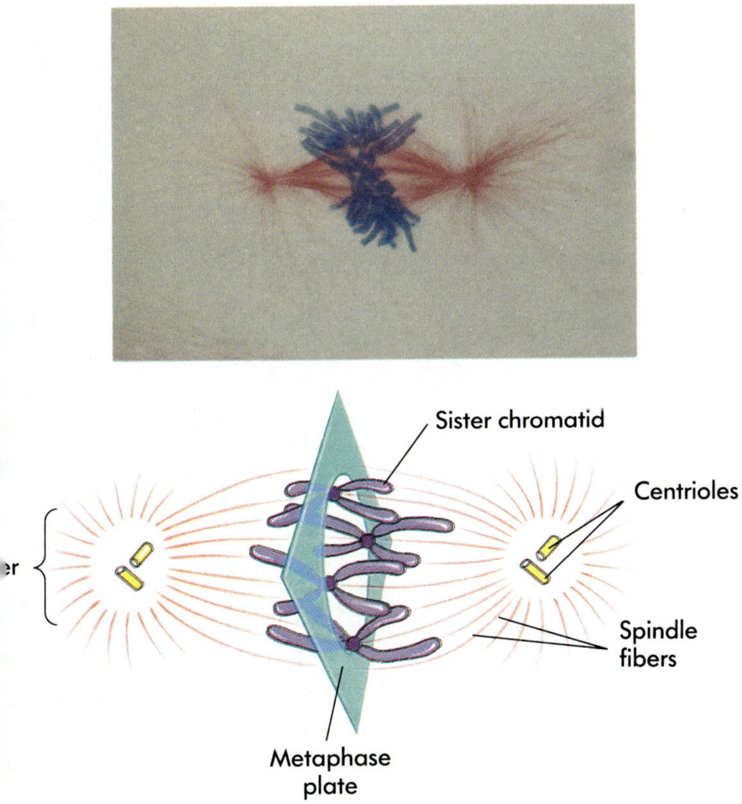

FIGURE 5-10 The metaphase plate.
In metaphase, the chromosomes array themselves around the spindle midpoint.

Anaphase

Of all the stages of mitosis, anaphase is the most beautiful to watch but is also the shortest. At the beginning of anaphase the sister chromatids separate at the centromere, freeing them from their attachment to one another. Before this split, the chromatids are tugged in two directions at once by opposing microtubules—somewhat like a cellular tug-of-war. With the separation of the chromatids (now called chromosomes), they move rapidly toward opposite poles of the cell, each pulled at its kinetochore by attached, shortening microtubules.

Anaphase is the stage of mitosis characterized by the physical separation of sister chromatids and their movement to opposite poles of the cell.

Telophase

The separation of the sister chromatids in anaphase equally divides the hereditary material that replicated just before mitosis. Therefore each of the two new cells that are forming receive a complete, identical copy of the chromosomes. This partitioning of the genetic material is the essence of the process of mitosis.

The events of this last phase of mitosis, **telophase,** ready the cell for **cytokinesis,** or division of the cytoplasm. The spindle fibers are chemically disassembled and therefore "disappear." The nuclear envelope re-forms around each set of what were sister chromatids, now chromosomes in their own right. These chromosomes begin to uncoil, returning to their normally strung out, more diffuse state. The DNA of the nucleolus begins making ribosomal RNA once again; this rRNA is visible, resulting in the "reappearance" of the nucleolus. In summary, the events of telophase are very much like the reverse order of the events of prophase. (Figure 5-11 compares mitosis in an animal cell and a plant cell.)

Telophase is the stage of mitosis during which the mitotic apparatus assembled during prophase is disassembled, the nuclear envelope is reestablished, and the normal use of the genes present in the chromosomes is reinitiated.

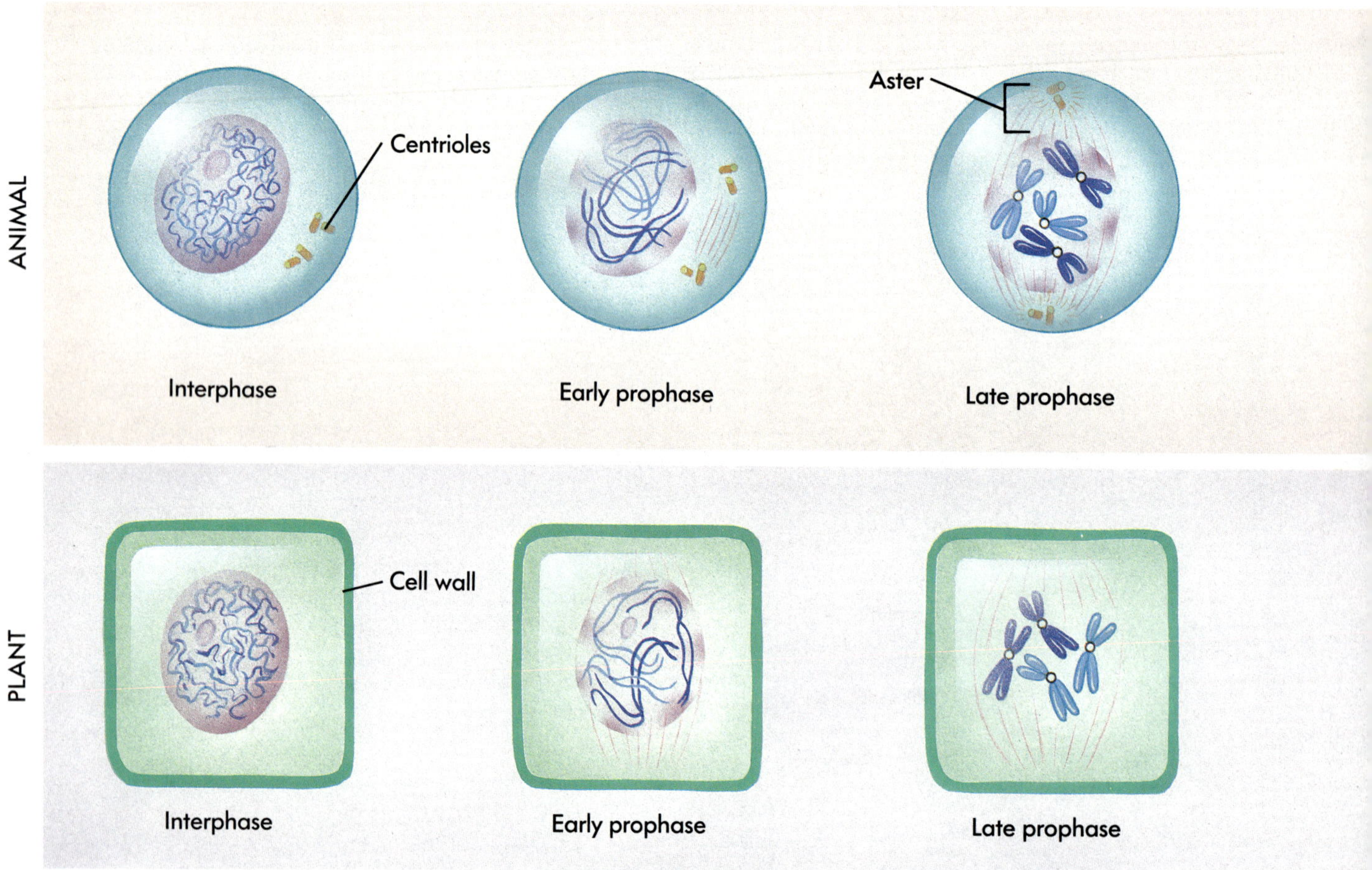

FIGURE 5-11 A comparison of interphase and the stages of mitosis in animal and plant cells. Differences include the presence of centrioles in animal cells and the formation of the cell plate in plant cells.

Cytokinesis

At the end of telophase, mitosis is complete. The eukaryotic cell has divided its duplicated hereditary material into two nuclei that are positioned at opposite ends of the cell. While this process has been going on, the cytoplasmic organelles, such as the mitochondria and (if present) the chloroplasts, have been somewhat equally distributed to each "side" of the cytoplasm. These organelles replicate throughout interphase.

At this point the process of cell division is still not complete. The division of the cytoplasm—that portion of the cell outside the nucleus—has not yet begun. The stage of the cell cycle at which cell division actually occurs is called *cytokinesis*. Cytokinesis generally involves the division of the cell into approximately equal halves.

Cytokinesis is the physical division of the cytoplasm of a eukaryotic cell into two daughter cells.

Cytokinesis in animal cells

In animal cells and in the cells of other eukaryotes that lack cell walls, cytokinesis occurs by a "pinching" of the cell in two. This pinching is accomplished by a belt of microfilaments that encircles the cell at the metaphase plate. These microfilaments contract, forming a **cleavage furrow** around the circumference of the cell (Figure 5-12). As contraction proceeds, the furrow deepens until the opposing edges of the membrane make contact with one another. Then the membranes fuse, separating the one cell into two new cells.

Cytokinesis in plant cells

In plants and some algae, a rigid wall surrounds the cell. Therefore cytokinesis involves the laying down of a new cell wall between the two daughter cells rather than a pinching in of the cytoplasm. Plants manufacture new sections of membrane and wall from tiny vesicles most likely derived from the Golgi complex. These membrane "parts" accu-

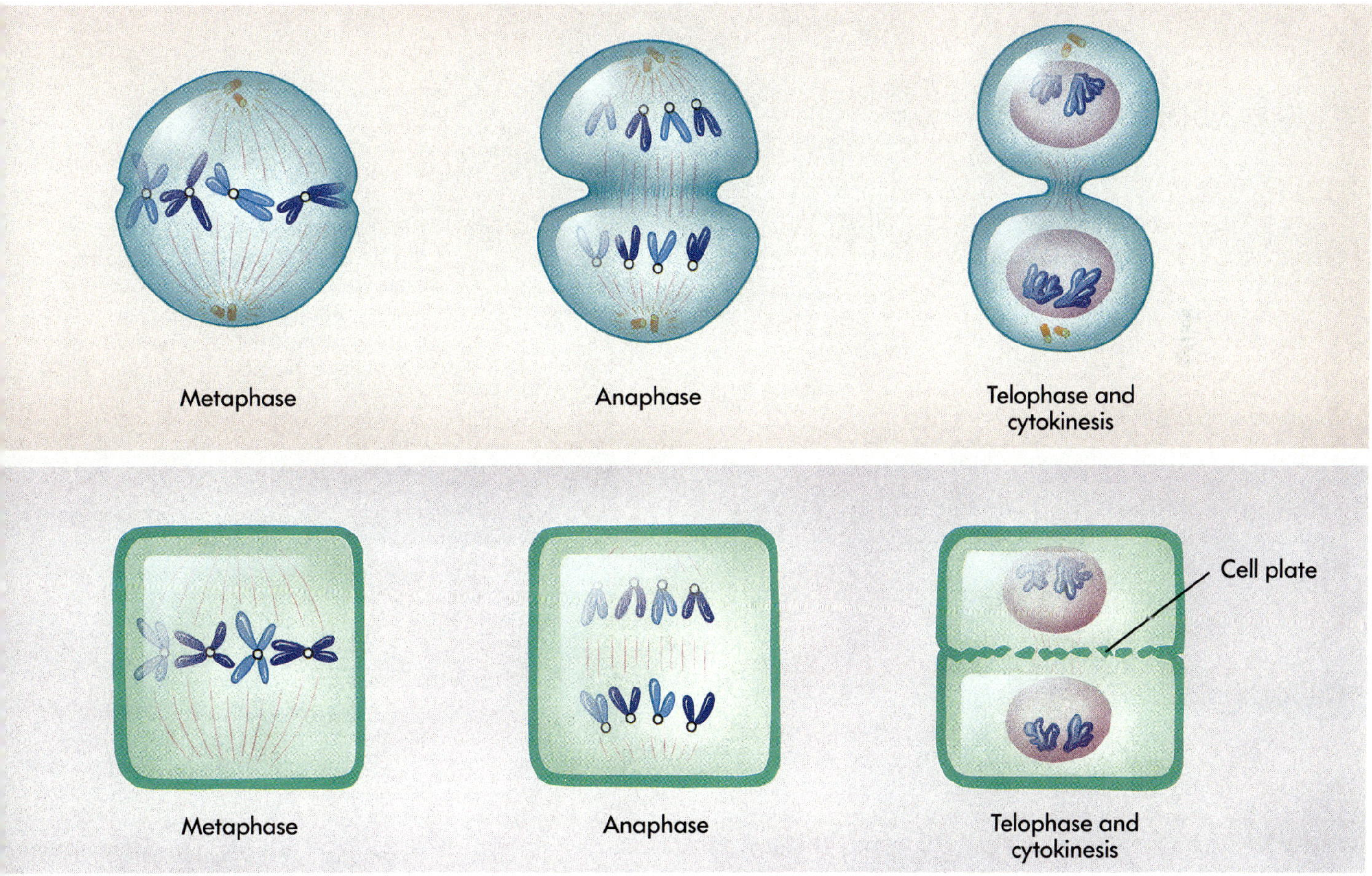

FIGURE 5-11—cont'd

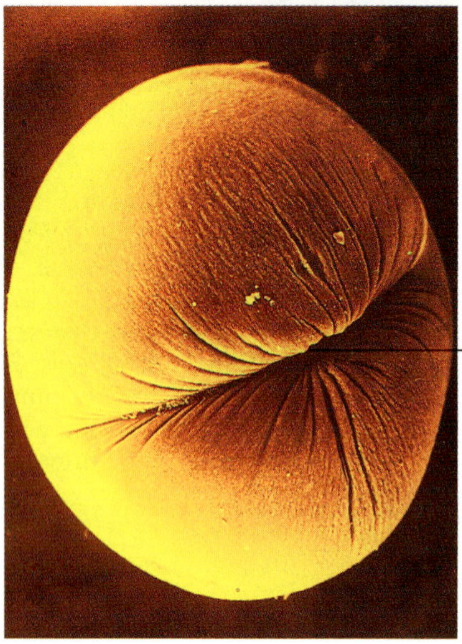

FIGURE 5-12 Cytokinesis in an animal cell.
A cleavage furrow is forming around this dividing sea urchin egg.

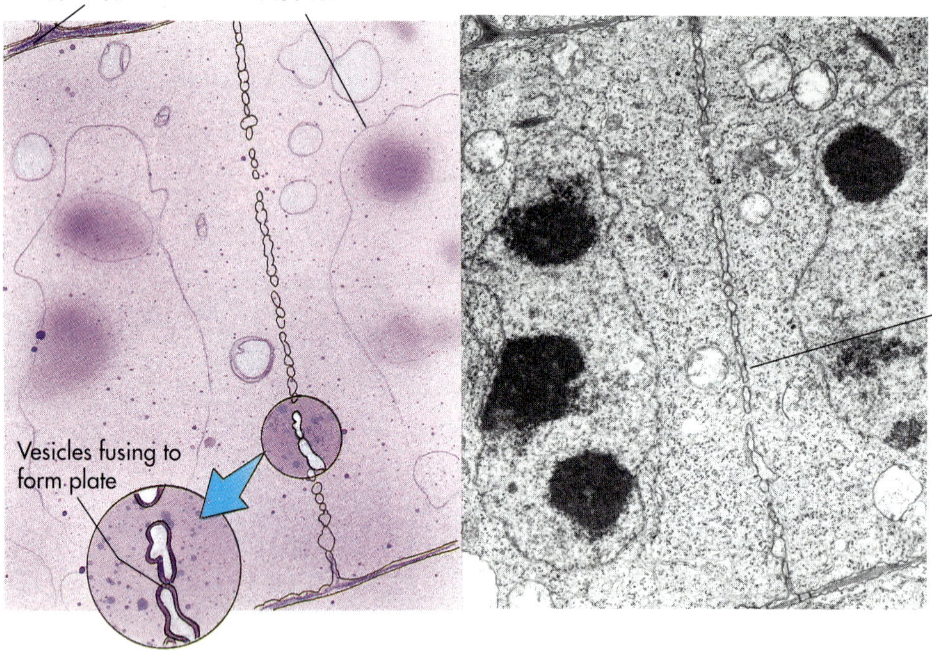

FIGURE 5-13 Cytokinesis in plant cells In this photograph and companion drawing, a cell plate is forming between daughter nuclei. The cell plate is composed of pieces of cell membrane (vesicles) that fuse together.

mulate at the metaphase plate and fuse, beginning the formation of a partition called the **cell plate.** A new cell wall forms between the two membranes of the cell plate. Figure 5-13 shows cytokinesis in a plant cell and the formation of a cell plate.

Meiosis

Most animals, plants, algae, fungi, and certain protists reproduce sexually. In sexual reproduction, gametes of opposite sexes or mating types (sometimes just termed + *cells* and − *cells*) unite in the process of fertilization, producing the first cells of new individuals (see Figure 5-4).

Humans have 46 chromosomes in all of their body cells—the diploid amount. If the two cells that joined in fertilization each contained 46 chromosomes, however, the first cell of the future offspring would have 92 chromosomes. An individual born after 10 generations would have more than 47,000 chromosomes! Such a continuing addition of chromosomes to each new individual is obviously an unworkable situation. Even early investigators realized that there must be some mechanism during the course of gamete formation to reduce the number of chromosomes. They reasoned that if sex cells were formed with half the number of chromosomes characteristic of the cells of that species, then the fusion of these cells during fertilization would produce cells of new individuals with the proper number of chromosomes. Investigators soon oberved this special type of cell division and named it *meiosis,* from a Greek word meaning "less."

During meiosis the diploid number of chromosomes is reduced by half, forming haploid cells. For this reason, meiosis is often called **reduction-division**. Four daughter cells result from two divisions of one original parent cell, although in human females, only one of the four cells is functional (see Chapter 20). In animals, meiosis occurs in the cells that produce gametes.

> **Meiosis is a process of nuclear division in which the number of chromosomes in cells is halved, forming gametes in most animals and spores in plants.**

The stages of meiosis

Although meiosis is a continuous process, scientists describe it by dividing it into stages, just as is done in describing the process of mitosis. The two forms of cell division have much in common. They are both special forms of nuclear division, although these processes are often referred to as forms of cell division for convenience. Meiosis consists of two "rounds" of divisions. These two rounds, or stages, of meiosis are called **meiosis I** and **meiosis II.** Each stage is further subdivided into prophase, metaphase, anaphase, and telophase, just as in mitosis. In meiosis, however, prophase I is much more complicated.

Meiosis is preceded by an interphase that is similar to the interphase of mitosis. During interphase the chromosomes are replicated, resulting in each chromosome consisting of two genetically identical sister chromatids held together at the constricted centromere. The centrioles also replicate.

Meiosis I

In prophase I, the individual chromosomes condense as their DNA coils more tightly, thus becoming visible under a light microscope. Because this DNA replicated before meiosis, each chromosome consists of two sister chromatids joined at the centromere.

Chromosomes, as you can see in the chapter opener photo, come in pairs. Each pair is formed during fertilization as an egg and sperm fuse. Each of these gametes contributes one chromosome to each pair of chromosomes. Thus, you received half your chromosomes (one set of genetic information) from your father, and half (a second set of genetic information) from your mother. Likewise, all organisms produced by sexual reproduction receive half their chromosomes from one parent organism and half from the other. These pairs of chromosomes are called **homologous chromosomes** or **homologues** (Figure 5-14).

Homologous chromosomes each contain genes that code for the same inherited traits, such as eye color or hair color. During prophase I, homologous chromosomes line up side-by-side in a process called **synapsis.** Because each chromosome is made up of two sister chromatids, the paired homologous chromosomes together have four chromatids. These synapsed homologues are therefore called **bivalents** (meaning "two strong"), or **tetrads** (meaning "groups of four").

> **Early in prophase I of meiosis, homologous chromosomes pair up in a process called *synapsis.***

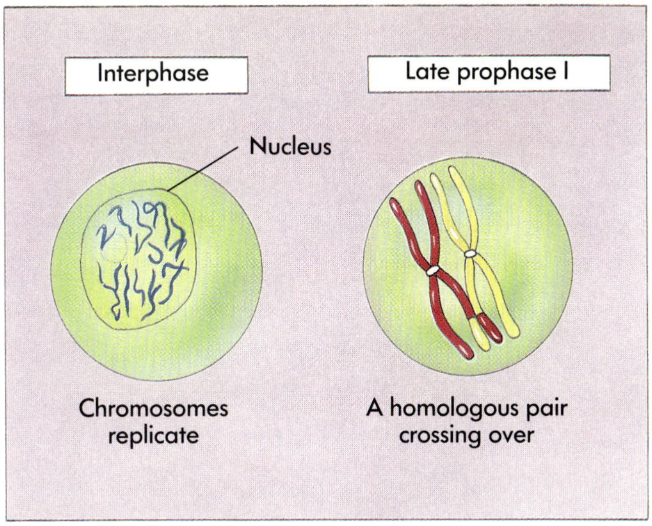

FIGURE 5-14 Interphase and late prophase in meiosis I. Chromosomes are replicated during interphase. Each chromosome is paired (homologous chromosomes). During prophase I, they line up side by side (called *synapsis*) and exchange genetic material (called *crossing over*).

The process of synapsis begins a complex series of events called **crossing-over.** During crossing-over, homologous *nonsister* chromatids actually cross over one another. These "crossed-over" pieces break away from the chromatids to which they are attached and reattach to the non-sister chromatid—literally exchanging parts (Figure 5-15). Once crossing-over is complete, the nuclear envelope dissolves and the homologues begin to move apart. These four chromatids cannot separate from one another, however, because (1) the sister chromatids are held together at their centromeres and (2) the paired homologues are held together at the points where crossing-over occurred. However, as the chromosomes move apart somewhat, the points of crossing-over can be seen (under a light microscope) as X-shaped structures called **chiasmata** (Figure 5-16).

> **Synapsis initiates the process of crossing-over. During this process, homologous chromosomes exchange genetic material.**

Crossing-over is a significant event in meiosis because it produces new combinations of genes. This process provides one way in which offspring have a different genetic makeup than either of its parents.

In metaphase I of meiosis I, the microtubules have formed a spindle, just as in mitosis. A crucial difference exists, however, between this metaphase and that of mitosis: the chromosomes line up with their homologues "double-file" in meiosis, not "single-file" as in mitosis. For each pair of homologues, however, the orientation on the metaphase plate is random; which homologue is oriented toward which pole is a matter of chance.

> **The two similar chromosomes of homologous pairs orient in tetrads randomly on the metaphase plate.**

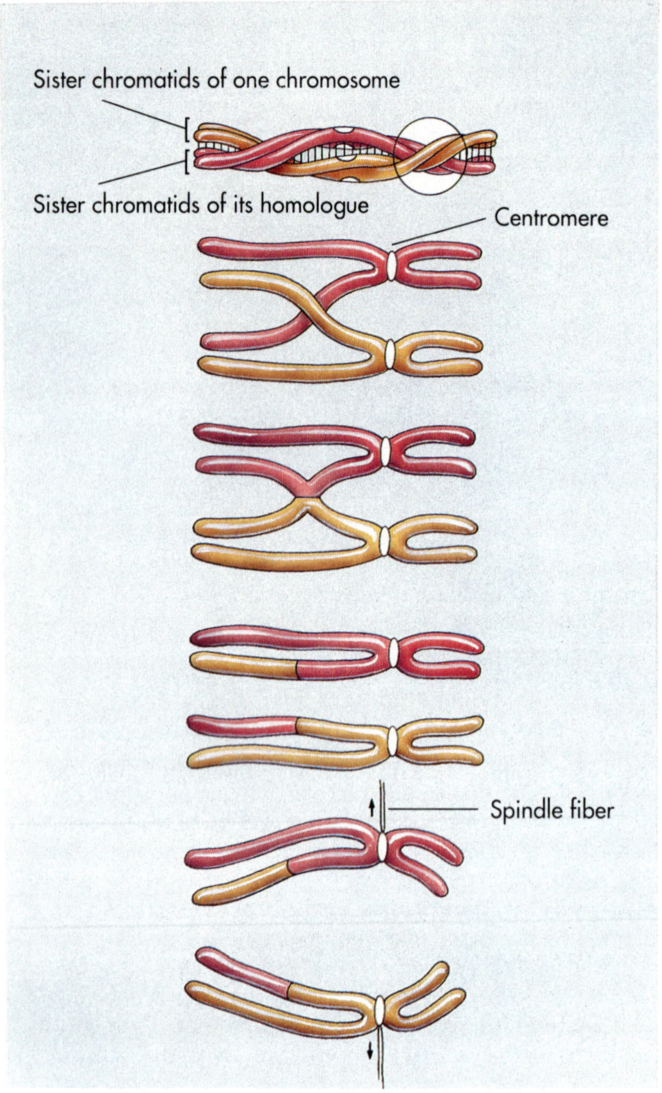

FIGURE 5-15 Crossing over.
The *circle* highlighting the upper chromosome indicates one region where homologous chromosomes are being held together tightly. Crossing over occurs when the paired chromosomes exchange genetic material at such locations.

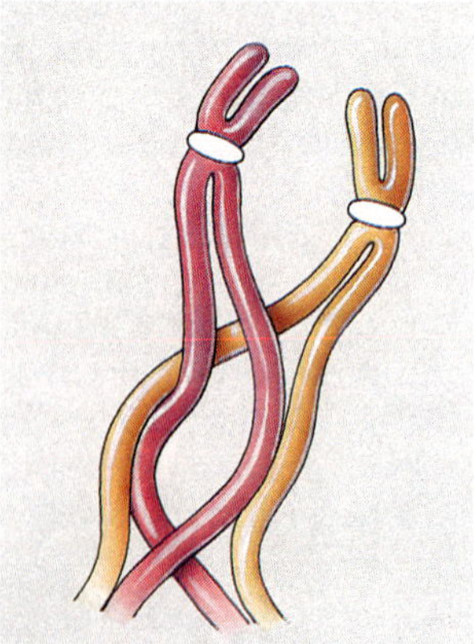
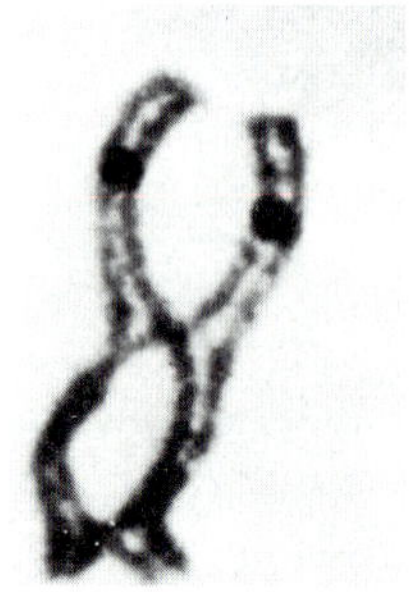

FIGURE 5-16 Chiasmata.
Chiasmata are the points of crossing over, where genetic material has been exchanged.

After spindle attachment to the kinetochore of each homologous pair is complete, the homologues begin to move toward opposite poles of the cell. As this movement occurs, the homologues pull apart at their cross-over points. By the end of anaphase I, each pole has one member of each chromosome pair. Each chromosome still has two sister chromatids. These chromatids are no longer perfectly identical, however, due to crossing-over. During the last stage of meiosis I—telophase I—the two groups of chromosomes gather together at their respective poles, forming two chromosome clusters. A nuclear membrane forms around each group of chromosomes.

An interphase-like period of variable length often occurs between meiosis I and meiosis II. During this time, *there is no replication of DNA*. In some cells, cytokinesis occurs and the cells separate completely. In cells in which cytokinesis does not occur, telophase I simply merges with prophase II.

The first meiotic division is traditionally divided into four stages as well as interphase:

1. **Prophase I—Homologous chromosomes pair and exchange pieces of genetic material.**
2. **Metaphase I—Homologous chromosomes align on a central plane.**
3. **Anaphase I—Homologous chromosomes move toward opposite poles. Chromatids do *not* separate.**
4. **Telophase I—Individual chromosomes gather together at the two poles and the nuclear membranes re-form.**
5. **Interphase—The haploid cells separate completely.**

Meiosis II

At the end of anaphase I, each pole of the original cell has a haploid complement of chromosomes. That is, each pole has *half* the normal amount of chromosomes—only one member of each homologous pair. Each chromosome, however, still has two sister chromatids. Meiosis II separates these sister chromatids. Because of crossing-over in the first phase of meiosis, these sister chromatids are *no longer identical* to one another.

The two haploid cells formed during meiosis I divide during meiosis II. The nuclear envelopes disappear and the spindle fibers form. The chromosomes line up, their sister chromatids separate, and they move to the opposite poles of each cell. At this point the nucleoli reorganize and nuclear envelopes form around each set of chromosomes (Figure 5-17).

Meiosis II results in the production of four daughter cells, each with a haploid number of chromosomes. The cells that contain these haploid nuclei function as gametes for sexual reproduction in most animals and as spores in plants.

During meiosis II, four haploid daughter cells are produced from the two haploid daughter cells of meiosis I. These two haploid cells were produced during meiosis I from one original diploid parent cell.

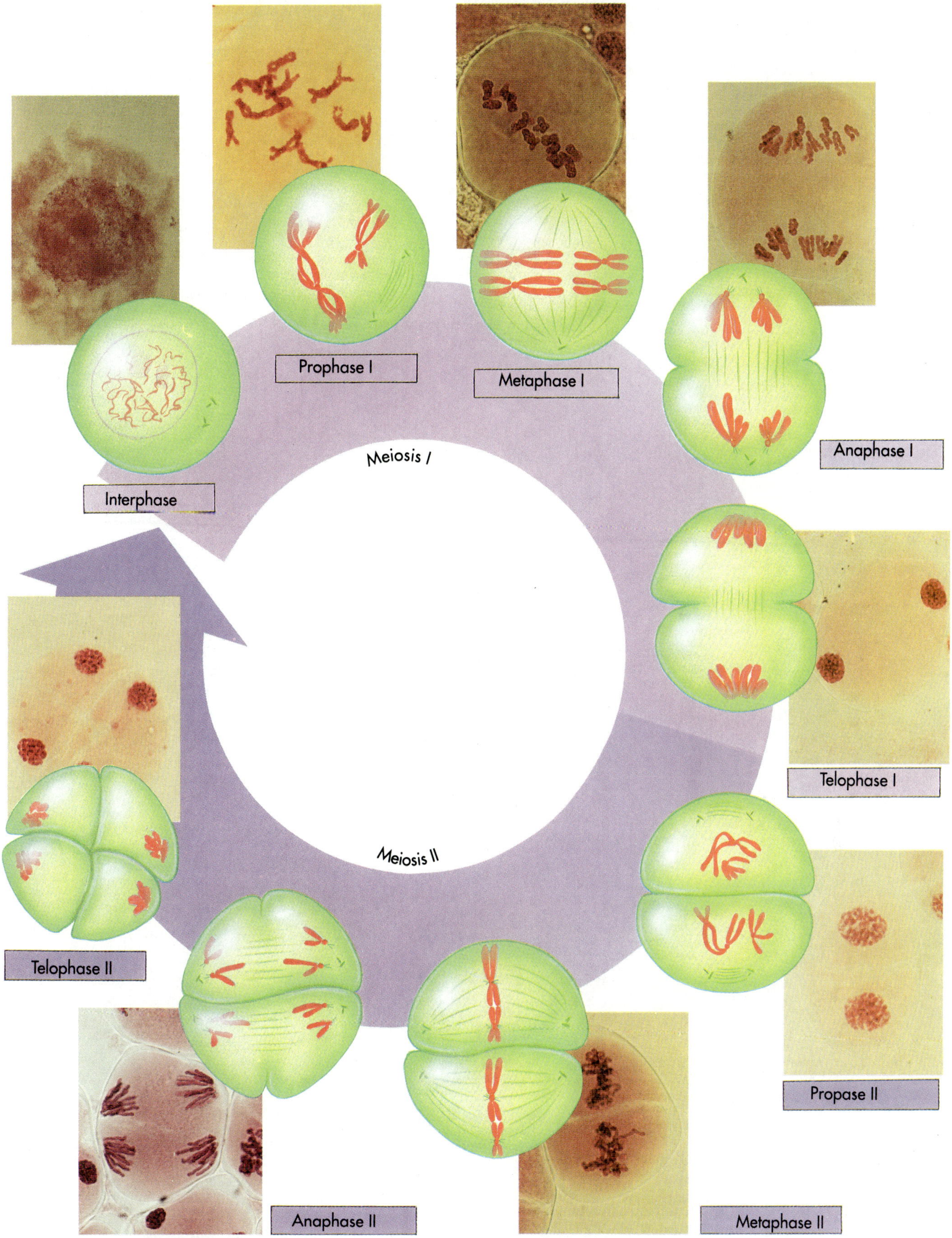

FIGURE 5-17 Meiosis.
Meiosis ensures the reassortment of genetic material. An interphase-like period of variable length often occurs between meiosis I and meiosis II. During this time there is no replication of DNA. Although interphase is shown here before prophase I, it is not a part of meiosis.

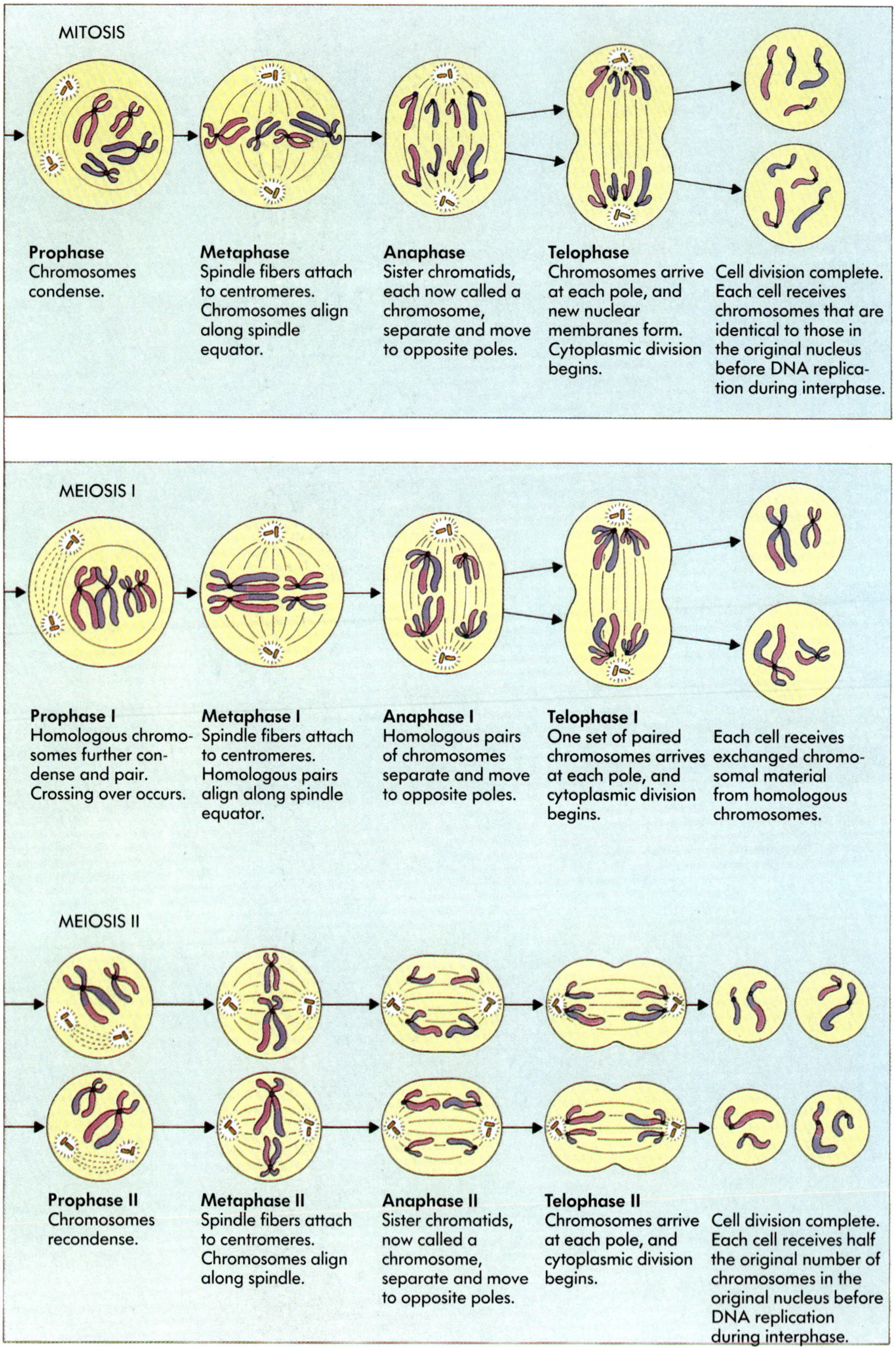

FIGURE 5-18 A comparison between mitosis and meiosis.
Meiosis involves two nuclear divisions with no DNA replication between them. Meiosis therefore produces four daughter cells, each with half the original amount of DNA. Mitosis produces two identical daughter cells.

The importance of meiotic recombination

The reassortment of genetic material that occurs during meiosis (Figure 5-18) is the principal factor that has made the rapid evolution of eukaryotic organisms over the past 1.5 billion years possible. Meiosis represents an enormous advance in the ability of organisms to generate variability in their hereditary material. To understand why this is true, remember that most organisms have more than one chromosome. Human beings, for example, have 23 different pairs of homologous chromosomes, one of each pair from the father and one of each pair from the mother. Each human gamete receives one of the two copies of each of the 23 different chromosomes, but which copy of a particular chromosome it receives is random. For example, the copy of chromosome 14 that a particular human gamete receives has no influence on which copy of chromosome 5 that it will receive. Each of the 23 pairs of chromosomes goes through meiosis independently of all the others, so there are 2^{23} (more than 8 million) different possibilities for the kinds of gametes that can be produced, and no two of them are alike. In addition, crossing-over adds even more variability to the random assortment of chromosomes. The subsequent union of two gametes thus creates a unique individual, a new combination of the 23 chromosomes that probably has never occurred before and probably will never occur again.

Summary

1. The hereditary material is deoxyribonucleic acid, or DNA. In eukaryotes, DNA is combined with protein to form a complex called *chromatin*. In preparation for cell division, the chromatin coils and supercoils, forming structures called *chromosomes* that are visible under a light microscope.
2. The "packaging" of DNA into chromosomes allows a tremendous amount of information to reside in each cell of a eukaryote and also allows this hereditary material to be separated in an organized way during cell division.
3. Eukaryotic cells divide during the growth, repair, and reproduction of organisms.
4. Mitosis is the process of nuclear division that distributes a complete, identical set of chromosomes to each of two daughter cells. This type of nuclear division is used in the growth and repair of multicellular organisms and in the asexual reproduction of single-celled eukaryotes and some simple animals. It produces cells that are identical to the parent cell.
5. Meiosis is the process of nuclear division that distributes half the complement of chromosomes to each of four daughter cells. This type of nuclear division is used to produce gametes for sexual reproduction in most animals and spores in plants.
6. Crossing-over is an essential element of meiosis. This process produces sister chromatids that are not identical with each other. Crossing-over and the reassortment of genetic material that occurs during meiosis provides for new combinations of hereditary material during sexual reproduction.
7. Genetic reassortment by meiosis is the principal factor that has made the rapid evolution of eukaryotic organisms possible.

REVIEW QUESTIONS

1. What is chromatin and of what does it consist? Why is it important?
2. Distinguish among karyotype, sex chromosome, autosome, and gene.
3. Fill in the blanks: Within cells, a double strand of DNA coils around groups of tiny proteins called ________ and then wraps up into larger coils called ________. Each chromosome consists of two identical ________ ________ joined together at the ________.
4. Why do eukaryotic cells undergo mitosis? Why do they undergo meiosis?
5. Compare the cells that result from mitosis with those that result from meiosis. How are they different?
6. What is mitotic interphase? What activities are going on during this phase?
7. Place these steps in the correct sequence and label each stage. What process is being described?
   a. The chromosomes line up in a ring in the center of the cell.
   b. The nuclear envelope forms and the chromosomes uncoil.
   c. The chromosomes grow shorter and thicker, and spindle fibers form.
   d. The sister chromatids separate and move to opposite poles of the cell.
8. Place these steps in the correct sequence and label each stage. What process is being described?
   a. Homologous chromosomes move toward opposite poles of the cell; chromatids do not separate.
   b. Chromosomes gather together at the two poles of the cell and the nuclear membranes re-form.
   c. Homologous chromosomes pair and exchange segments.
   d. Homologous chromosomes align on a central plane.
   e. The haploid cells separate completely.
9. What is cytokinesis? Compare cytokinesis in the cells of a cat and the cells of a pine tree. What are the differences, if any?
10. Fill in the blanks: Cells that contain the full complement of chromosomes are called ________. Cells with half the characteristic number of chromosomes are ________.
11. Fill in the blanks: ________ chromosomes each contain genes that code for the same characteristics. During prophase I of meiosis, these chromosomes line up side-by-side in a process called ________.
12. Compare the cells resulting from meiosis to their mother cell. What functions do these cells serve?
13. Explain the importance of the genetic recombination that occurs during meiosis.

KEY TERMS

anaphase 90
asexual reproduction 85
autosomes 81
cell cycle 86
cell plate 92
centrioles 86
centromere 81
chiasmata 93
chromatin 81
chromosomes 81
cleavage furrow 91
crossing over 93
cytokinesis 90
diploid 83
gametes (sex cells) 83
gene 81
haploid 83
homologous chromosomes (homologues) 93
interphase 86
karyotype 81
kinetochore 89
meiosis 83
metaphase 89
metaphase plate 89
mitosis 83
prophase 86
reduction-division 92
sex chromosomes 81
sexual reproduction 83
sister chromatids 81
spindle fibers 87
synapsis 93
telophase 90
tetrads 93

THOUGHT QUESTIONS

1. Your body, which contains some 10 trillion cells, started its life as a single fertilized egg. Every cell in your body today (except your gametes) has been produced by a series of mitotic divisions from that first cell. What is the minimum number of cell generations required to do this?
2. One of the first steps in cell division is the condensation of the DNA into compact chromosomes. What drives this process—where does the energy required to wrap up so much DNA come from?
3. Nondisjunction occurs when both spindle fibers attached to a centromere attach to the same pole of the cell. What stops this from happening 50% of the time? How does a particular spindle fiber "know" where to attach?
4. Do you think crossing-over happens between sister chromatids? Why do you think this?

FOR FURTHER READING

The cell cycle. (1989). *Science, 246*(4930), 545-640.
This is a comprehensive account of all aspects of modern research into the cell cycle.

Kornberg, R. D., & Klug, A. (1981, February). The nucleosome. *Scientific American*, pp. 52-64.
This article describes how the DNA helix is wound on a series of protein spools.

Murray, A., & Kirschner, M. (1991, March). What controls the cell cycle? *Scientific American*, pp. 56-63.
This article is a modern look at how cell division is controlled.

Sloboda, R. D. (1980). The role of microtubules in cell structure and cell division. *American Scientist, 68*, 290-298.
This article contains a good description of how the spindle apparatus works.

PART THREE

HOW LIVING THINGS TRANSFORM ENERGY

6 The flow of energy within organisms

Did you know that your body takes constant "rollercoaster rides" that you can't even feel? All the chemical reactions that take place in your body use and change energy just like a rollercoaster. At the beginning of a rollercoaster ride, the cars are pulled up a high hill by a cable. For the cable to pull the cars, it must be run by an engine that supplies energy for the cable to do its work. Energy is, in fact, the ability to do work. In the case of the rollercoaster, energy is needed to pull the cars up the hill, otherwise they would go nowhere. After the cars reach the top and are freed of the cable, they swoop down the hill, releasing this input of energy and "coasting" around the rails at breakneck speed.

HIGHLIGHTS

- Energy, the capacity to do work, continually flows throughout living systems to maintain life.
- Energy can never be destroyed, but as it flows throughout living systems, much of it is changed to heat, becoming unusable for performing work.
- Metabolic pathways, which are regulated by enzymes, are the means by which energy is moved, stored, carried, and freed in living systems.
- ATP is a molecule used universally by living cells to capture and release energy in chemical reactions.

OUTLINE

Starting chemical reactions

Energy flow and change in living systems

Regulating chemical reactions

Enzymes: Biological catalysts

How enzymes work

Factors that affect enzyme activity

Cofactors and coenzymes

Storing and transferring energy

ATP: The energy currency of living things

Carrying and transferring energy using ATP

All the chemical reactions within your body—your metabolism—take place in a similar way. Chemical reactions are changes in molecules in which one substance is changed into another. However, for chemical reactions to take place, they need an input of energy to get started. This input of energy helps molecules "climb" an energy "hill" and interact with one another.

Starting chemical reactions

Figure 6-1 is a graph of the varying energy states of the molecules in a chemical reaction. Doesn't this graph look like the beginning hill of a rollercoaster? The **substrates** are the molecules entering into the chemical reaction—the rollercoaster cars. The energy needed to "pull" the substrates up the energy hill is called the **free energy of activation**. This energy input destabilizes the bonds of the substrates. Molecules with unstable bonds interact with one another more easily than molecules with stable bonds do. In this way, the free energy of activation allows the chemical reaction to "go." The substrates undergo a chemical change resulting in new bonding arrangements between the molecules. This chemical change may involve the breaking of bonds and the making of new bonds. The changed substrates are called **products**.

Free energy of activation is needed to start a chemical reaction.

Sometimes substrates are chemically broken down to yield products. When a substrate is broken down, energy is released from the chemical bonds that were holding it together. This type of reaction is called **exergonic**, meaning

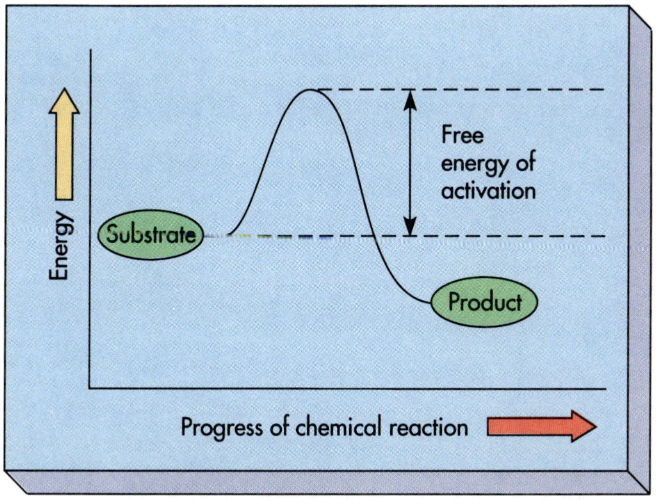

FIGURE 6-1 A chemical reaction.
Before a chemical reaction occurs, energy must be supplied to the substrates to destabilize existing chemical bonds. This energy is called the *free energy of activation*.

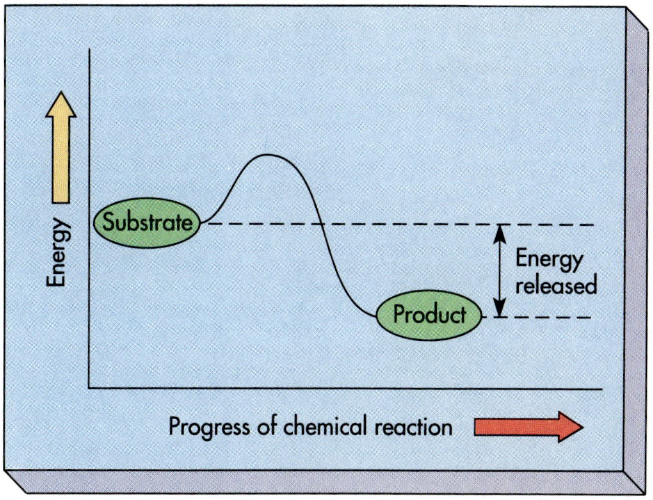

FIGURE 6-2 An exergonic reaction.
In an exergonic reaction the products contain less energy than the reactants. The excess energy is released.

"energy out." In an exergonic reaction the products contain less energy than the substrate. The excess energy is released (Figure 6-2). Exergonic reactions happen *spontaneously;* that is, they absorb free energy from their surroundings and occur with no additional input of energy from another chemical reaction.

You may have seen an example of an exergonic reaction if you have ever used baking soda and water to remove the light-blue substance that can build up on the terminals of a car's battery. This substance is formed when battery acid seeps from inside the battery and reacts with the lead on the battery terminals. As you pour a baking soda solution onto the battery, it fizzes and bubbles and the material that has built up disappears. What happened? The baking soda and acid reacted to form two new substances: carbon dioxide gas and water. In addition, energy was released as heat (Figure 6-3). Exergonic reactions take place in your body too. In fact, a series of exergonic reactions break down the glucose in your cells to supply your body with energy. The heat released from these reactions helps keep your body warm. Reactions such as this, which release energy by breaking down complex molecules into simpler molecules, are called **catabolic reactions.**

In some chemical reactions, substrates are chemically joined to yield a product. When substrates are joined, energy is used to build the chemical bonds holding the product together. This type of reaction is called **endergonic**, meaning "energy in." In an endergonic reaction, the product contains more energy than the substrates (Figure 6-4). Endergonic reactions *do not occur spontaneously*; that is, they require more free energy of activation to "drive" the reaction than the substrates are able to absorb from their surroundings. Endergonic reactions play important roles in your body, such as putting molecules together to build muscles and bones. Reactions such as this, which use energy to build complex molecules from simpler molecules, are called **anabolic reactions.**

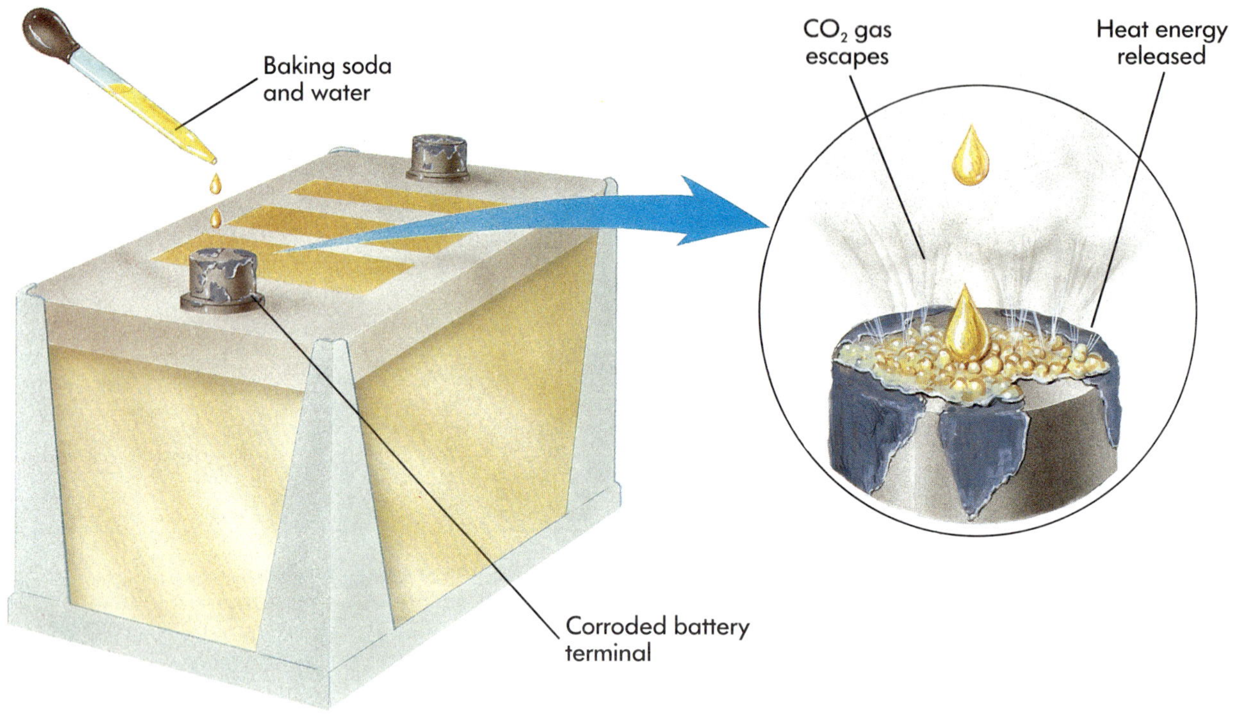

FIGURE 6-3 An example of an exergonic reaction.
When you clean your car battery with baking soda and water, these two substances react with the acid on the battery terminals, releasing carbon dioxide gas and water. Energy in the form of heat is also released.

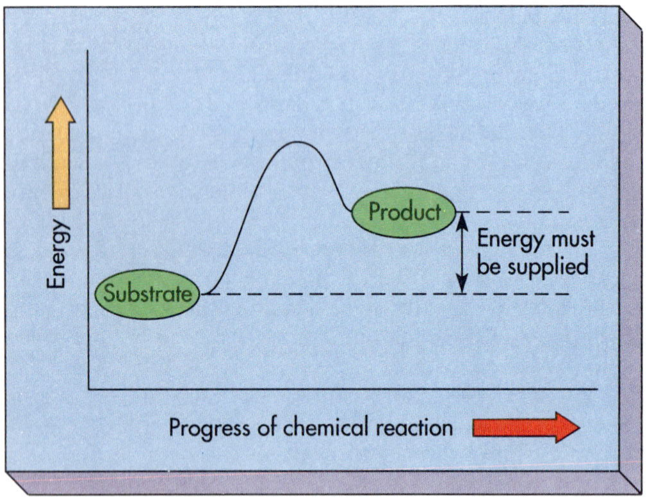

FIGURE 6-4 An endergonic reaction.
In an endergonic reaction the products of the reaction contain more energy than the reactants, so the extra energy must be supplied for the reaction to proceed.

In living things, exergonic and endergonic reactions are **coupled reactions,** meaning that they occur in conjunction with one another. In this way the energy released when molecules are split (exergonic reactions) is used to power the combining of molecules (endergonic reactions) (Figure 6-5).

Reactions that break apart substrates release energy. These reactions are exergonic. Reactions that bond substrates together store energy. These reactions are endergonic. In living systems, exergonic reactions are coupled with endergonic reactions, supplying the energy for endergonic reactions to take place.

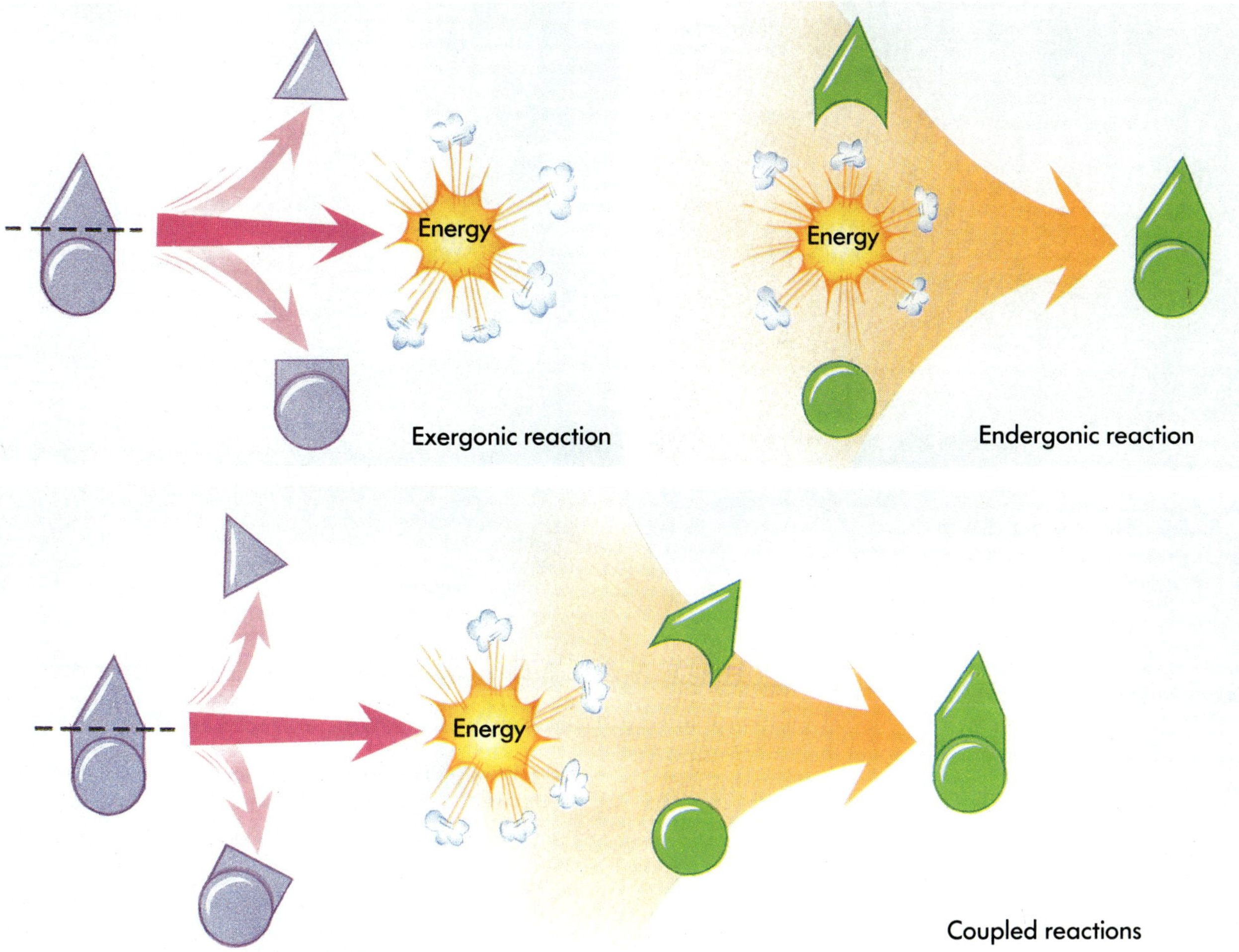

FIGURE 6-5 A coupled reaction.
In a coupled reaction the energy released in an exergonic reaction is used to drive an endergonic reaction.

Energy flow and change in living systems

Life can be viewed as a constant flow of energy that is channeled by organisms to do the work of living. As you learned at the beginning of this chapter, energy is the ability to do work, and work is many things, such as the pull of a cable on a rollercoaster car and the swift dash of a horse. It is also heat, such as the blast from an explosion or a warming fire. Energy can exist in many forms, including mechanical force, heat, sound, electricity, light, radioactivity, and magnetism. All these forms of energy are able to create change—to do work.

Energy exists in two states. Energy not actively doing work but having the capacity to do so is called **potential energy.** The rollercoaster cars perched atop that first hill, for example, possess this type of "stored" energy. They have the capacity to roll downhill, having stored the energy from their pull up the hill. Then, as the cars roll down the hill, they are actively engaged in doing work. This form of energy is called **kinetic energy,** or the energy of motion. Much of the work performed by living organisms involves changing potential energy to kinetic energy (Figure 6-6). The exergonic reactions that supply your body with energy, for example, change the potential energy stored in the food you eat to kinetic energy for cellular work.

Energy is the capacity to do work, or bring about change. It can exist in many forms such as mechanical force or heat. Energy actively doing work is called *kinetic energy;* stored energy is called *potential energy.*

All of the changes in energy that take place in the universe, from nuclear explosions to the buzzing of bees, are governed by two laws called the **laws of thermodynamics.**

FIGURE 6-6 The energy of motion.
Transforming potential energy into kinetic energy, this cheetah has just sprung into motion.

The **first law of thermodynamics** states that energy can change from one form to another and from one state to another, but it can never be lost. Nor can any new energy be made. The total amount of energy in the universe remains constant.

The first law of thermodynamics states that energy cannot be created or destroyed; it can only be changed from one form or state to another.

The **second law of thermodynamics** states that all objects in the universe tend to become more disordered and that the disorder in the universe is continually increasing. Stated simply, disorder is more likely than order. You can relate to this concept as you try to keep your personal environment in order. Your bedroom becomes continually messy, for example, unless you make an effort to keep it clean. Interestingly, molecules also become increasingly "disordered."

Disorder with respect to molecules refers to their random motion. As you may recall from Chapter 4, this random motion, or disorder, is an inherent jiggling that is characteristic of all molecules and small particles. Their disorder increases as energy is added to molecules in the form of heat. This heat is the energy that is "lost" as energy transfers are made between molecules during chemical reactions. This lost heat energy is often called the *kinetic energy of molecular motion.* Although the total amount of energy in the universe does not change, the amount of useful energy available to do work decreases as progressively more energy is degraded to heat. With the addition of this heat, molecules move more quickly, becoming more disordered.

The second law of thermodynamics states that disorder in the universe constantly increases. Energy spontaneously converts to less organized forms.

FIGURE 6-7 All the energy that powers life is captured from sunlight.
These trees in a Michigan forest absorb sunlight and use it in the process of photosynthesis.

The energy lost to disorder is referred to as **entropy.** In fact, entropy is a measure of the disorder of a system. Sometimes the second law of thermodynamics is simply stated as "entropy increases." So although energy cannot be destroyed, the universe is constantly moving toward increasing entropy. Eventually, the universe will have "wound down" like an abandoned clock. Scientists speculate this will occur approximately 100 billion years from now.

Although energy cannot come into or go out of the universe, the Earth is constantly receiving "new" energy from the sun. Much of this energy heats up the oceans and continents. Some of it is captured by photosynthetic organisms such as green plants (Figure 6-7). In photosynthesis, the energy from sunlight is changed to chemical energy, combining small molecules into more complex molecules by means of endergonic reactions. This stored energy can be shifted to other molecules by forming different chemical bonds. In addition, this stored energy can be changed into kinetic energy: motion, light, electricity—and heat. Thus energy continuously flows into and through the biological world, with new energy from the sun constantly flowing in to replace the energy that is lost as heat.

Photosynthetic organisms such as green plants change energy from the sun to other forms of energy that drive life processes. This energy is never destroyed, but as it is exchanged in chemical reactions, much of it is changed to heat, a form of energy that is not useful for performing work.

Regulating chemical reactions

The flow of energy within an organism such as you consists of a long series of coupled reactions. Energy is "moved" from one molecule to the next by means of exergonic and endergonic reactions. Some of this energy is stored in the bonds of the molecules that make up the structure of your body. Some is freed to do cellular work. And some is lost as heat. These chains of reactions that move, store, and free energy are called **metabolic pathways.** Metabolic pathways accomplish jobs such as obtaining energy from the food you eat and repairing tissues that are worn out or damaged. By means of such pathways, your body and the bodies of all living things work to maintain order and avoid increasing entropy. Put simply, all the metabolic pathways in your body work to help you survive. But for these chemical reactions to occur, they must be "pushed" over the hill of activation energy. In addition they must be controlled so that they occur "on schedule."

Enzymes: Biological catalysts

As you read earlier in this chapter, chemical reactions require activation energy to get started. This energy is needed for various reasons. In some chemical reactions, it is used to help break old bonds so that new bonds can be formed. In others, it is used to excite electrons—to help them achieve a higher energy level or orbit—so that they will pair up in covalent bonds. And in others, this energy helps molecules overcome the mutual repulsion of their many electrons so that they can get close enough to react. In any chemical reaction the free energy of activation performs one or more of these jobs.

If molecules are moving very quickly and bump into one another forcefully, the kinetic energy of the bump can provide enough energy for activation. This situation rarely occurs in living organisms, however, because the chemical reactions of living things take place in the moderate temperatures of living cells. But living systems contain proteins called **enzymes** that lower (or lessen) the free energy of activation, allowing chemical reactions to take place.

Enzymes are proteins that act as **catalysts.** A catalyst is a substance that increases the rate of a chemical reaction but is not chemically changed by the reaction. It lowers the barrier of the free energy of activation—reduces the amount of energy needed for the reaction to go—by bringing substrates together so they can react with one another or by placing stress on the bonds of a single substrate, making it more reactive (Figure 6-8). Enzymes are biological catalysts and control all the chemical reactions making up the metabolic pathways in living things. Life is therefore a process regulated by enzymes.

Enzymes are biological catalysts that reduce the amount of free energy of activation needed for a chemical reaction to take place, speeding up the reaction.

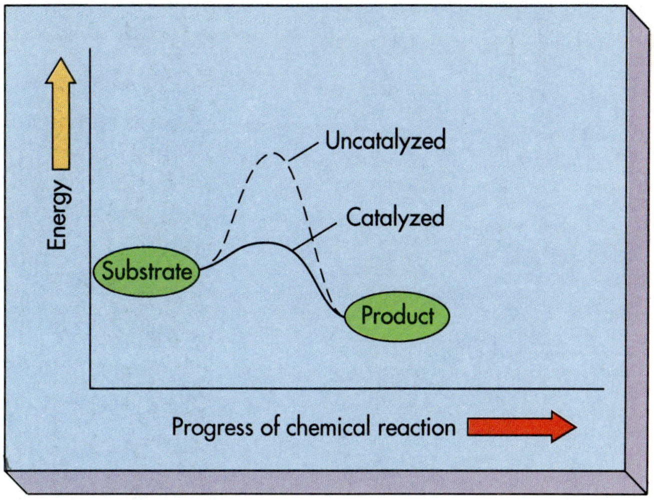

FIGURE 6-8 The function of enzymes.
Enzymes are able to catalyze particular reactions because they lower the amount of activation energy required to initiate the reaction.

How enzymes work

Enzymes are proteins that have one or a few grooves or furrows on their surfaces, much like the deep creases in a prune (Figure 6-9). These surface depressions are called **active sites** and are determined by the side groups of the amino acids making up the protein. Nonpolar side groups of amino acids, for example, tend to be shoved into the interior of the protein due to their hydrophobic (water-fearing) interactions, thereby imparting a particular shape to that portion of the protein (see Chapter 2).

The active sites are the locations on the enzyme where a reaction is catalyzed—where **catalysis** takes place. For catalysis to occur, a substrate must fit into the surface depression of an enzyme so that many of its atoms nudge up against atoms of the enzyme. The fit between a substrate and an enzyme is much like putting your foot into a tight-fitting shoe. However, the binding of a substrate in some cases causes the enzyme to adjust its shape slightly, allowing a better fit called an *induced fit*. Just as your feet will only fit into certain shoes, substrates can only fit into, or bind with, the active site of certain enzymes. Therefore enzymes typically catalyze only one or a very few different chemical reactions. Because of this specificity, each cell in your body contains from 1000 to 4000 different types of enzymes.

Enzymes catalyze both endergonic and exergonic reactions. Figure 6-10 shows how an enzyme catalyzes an exergonic reaction. In this example, a molecule of sucrose binds to the active site of an enzyme. After binding takes place, certain atoms within the active site of the enzyme chemically interact with the sucrose. This interaction causes a slight change in the shape of both the enzyme and the sucrose molecule. This change in shape places stress on the bonds joining the glucose and fructose subunits, reducing the amount of free energy of activation needed to be ab-

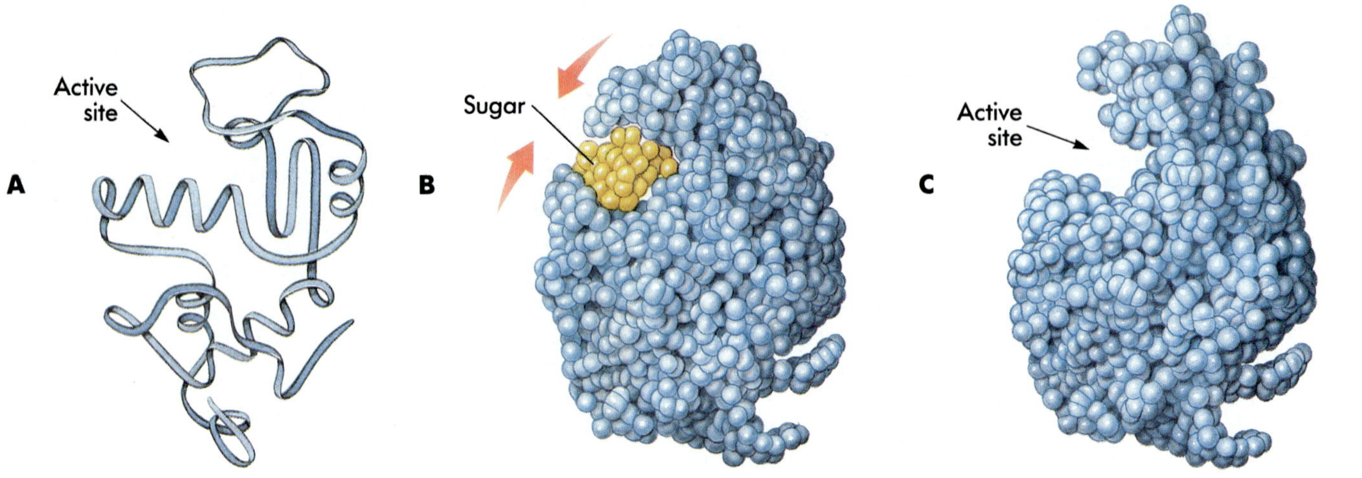

FIGURE 6-9 How the active sites in the enzyme lysozyme work to break apart bacterial cell walls. The tertiary structure of the enzyme lysozyme, shown as a ribbon in **A** and a three-dimensional model in **B** and **C**, forms a groove called the *active site* through the middle of the protein. This groove fits the shape of a substrate such as chains of sugars within bacterial cell walls. A chain of sugars, indicated in *yellow* in **C**, slides into the groove, its entry inducing the protein to alter its shape slightly and embracing the substrate more intimately. This *induced* fit positions a glutamic acid in the protein right next to the bond between two adjacent sugars in the substrate, causing stress to be put on the bond. The bond between the two sugars eventually breaks, releasing the individual sugars. In this way, the enzyme lysozyme breaks apart bacterial cell walls, killing the bacteria.

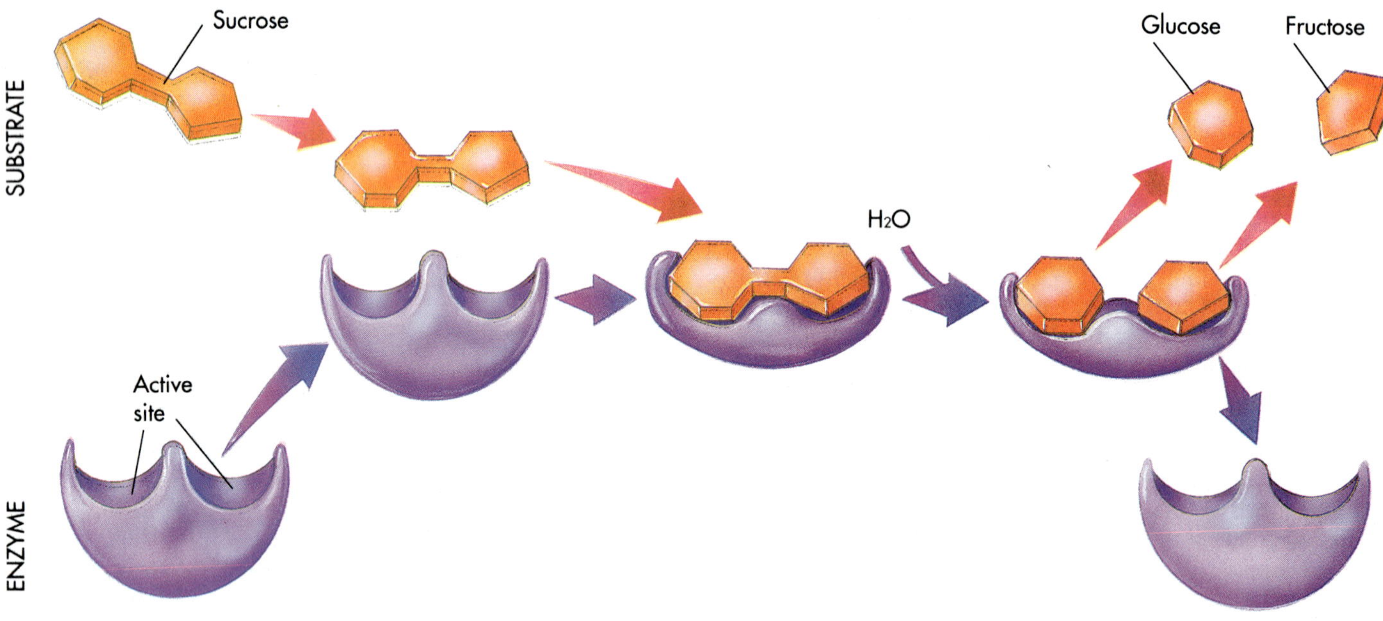

FIGURE 6-10 How an enzyme catalyzes an exergonic reaction.
In this reaction, sucrose binds to the active site of an enzyme. A reaction takes place between the active site atoms and the sucrose atoms, causing a change in shape of the enzyme and the sucrose molecule. As a result of this change in shape, stress is placed on the bonds holding the glucose and fructose subunits together in the sucrose molecule. This stress lowers the amount of energy needed to break the bonds. When the bond breaks, the glucose and fructose are released from the enzyme.

BIOLOGY TECHNOLOGY & society

Nonfattening sweets

Americans crave sweets and a slender figure. In an attempt to satisfy both cravings, many avoid high-calorie sweets, substituting instead low-calorie foods that have been artificially "sweetened" with chemicals. Americans today can choose between three additives that make low-calorie food taste sweet: saccharine (300 times as sweet as sucrose), aspartame (180 times as sweet, marketed as "Nutrasweet"), and acesulfame (200 times as sweet, marketed as "Sunette"). Three others now undergoing government testing will probably soon join the list: cyclamate (30 times as sweet as sucrose, now cleared of charges that it leads to cancer), alitame (a mixture of amino acids 2000 times sweeter than sucrose), and sucralose (a chlorinated version of sucrose 600 times sweeter). The problem with all these additives is that sugar does more than just sweeten food. It adds bulk and texture, color and aroma (Figure 6-A). The perfect sweetener would be sugar itself . . . if only eating too much of it did not make you fat and lead to tooth decay.

Imagine a kind of table sugar that looks, tastes, and cooks like the real thing but has no calories or harmful side effects. You could eat mountains of candy made from such a sweetener without gaining weight. The stuff a dieter's dreams are made of, nonfattening sugars are very much like the fattening kind you are familiar with as sucrose (table sugar) but with one very important difference—they are left handed!

As Louis Pasteur discovered in the late 1800s, most sugars are "right-handed" molecules, the hydroxyl group that binds a key carbon atom being located on the right side. However, sugars can be readily made in the laboratory that are left-handed—the critical hydroxyl ion is on the left side, producing a mirror-image chemical twin of the natural form. The two twin forms are made of the same chemical bonds and behave the same way chemically, so the artificial left handed sugar tastes just as sweet as the natural kind.

But the enzymes that break down sugars in the human digestive system can tell the difference. To digest a sugar molecule, an enzyme must first grasp it, much like a shoe fitting onto a foot, and all the body's enzymes are right-handed! A left-handed sugar just doesn't fit, any more than a right-handed shoe fits onto a left foot.

FIGURE 6-A Forbidden sweets such as ice cream may become nonfattening in the future if they are sweetened with "left-handed" sugars.

These new left-handed sugars are called *l*-sugars. They do not occur in nature except as traces in snail eggs and seaweed. Laboratory-manufactured sweeteners made with the *l*-sugars are still under development and have not yet made it to market. Because they pass through the body without being used, *l*-sugars can let diet-conscious sweet lovers have their cake and eat it too. Nor will they contribute to tooth decay—bacteria cannot metabolize them either.

sorbed by the molecules for them to react. The bond then breaks, and the products are released from the enzyme.

Enzymes also catalyze reactions that bind two substrates together (Figure 6-11). Both substrates bind at the active site on the enzyme. The binding of the substrates to the enzyme distorts and weakens their chemical bonds. In addition, their interaction with the enzyme orients the substrates so that their reactive sites are in contact with one another and react. Put simply, the enzyme is like a dating service. It gets the substrates together so that their meeting does not simply occur by chance. Interestingly, the Chinese characterize enzymes in a similar way, calling them *tsoo mei*, or marriage brokers. Moreover, enzymes do not get used up after one interaction takes place. They are only intermediaries in chemical reactions and are then released, available to catalyze yet another reaction.

Enzymes work by bringing substrates together so that they react more easily and by placing stress on bonds, which lowers the amount of free energy of activation that must be absorbed by the substrates for them to react.

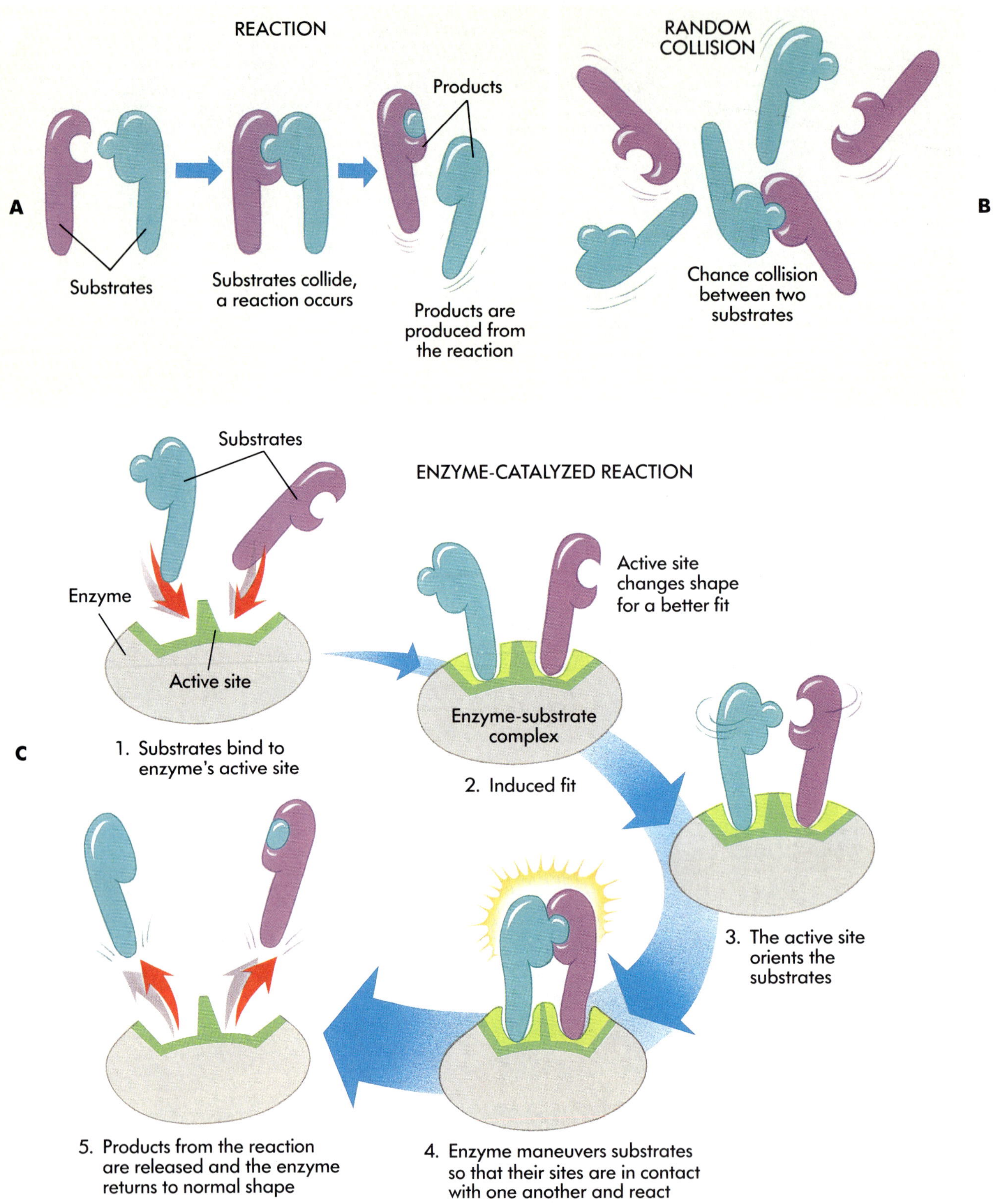

FIGURE 6-11 Enzymes also catalyze reactions between two substrates.
A A collision between two substrates causes a reaction to occur, resulting in products.
B Without an enzyme, substrates only collide by chance.
C With an enzyme, a reaction is more likely to occur. The binding of the substrates to the active site of the enzyme weakens their chemical bonds so that less energy is needed for a reaction to occur. The enzyme also orients the substrates in the best possible position for a productive reaction.

Factors that affect enzyme activity

The activity of an enzyme is affected by anything that changes its three-dimensional shape. If it loses its shape, or is denatured, an enzyme cannot bind with a substrate. Three environmental conditions that can affect enzyme activity in this way are temperature, pH, and the binding of specific chemicals. You can see that the protein in an egg become denatured when you cook it. The heat of the cooking process permanently changes the shape of the egg proteins, which results in the egg becoming firmer and more opaque.

Most human enzymes function best between 36° C and 38° C, close to body temperature. Likewise, the enzymes of other living things work best at temperature ranges specific to the organism. Bacteria that live in hot springs, for example, have enzymes that function at temperatures between 74° C and 76° C (Figure 6-12, *A*). At temperatures colder or warmer than the enzyme's optimum range, the bonds between the amino acids that are responsible for the enzyme's three-dimensional shape become weak or rigid, changing the shape of the active site and causing chemical reactions to stop.

Enzymes also work best within a particular range of pH values. As previously stated, pH is a measure of the acidity or alkalinity of a solution (see Chapter 2). Most enzymes work best within the range of pH 6 to 8. Some enzymes, however, function in very acidic environments. For example, the enzyme pepsin digests proteins in your stomach at pH 2, a very acidic level (Figure 6-12, *B*).

The activity of an enzyme is sensitive not only to temperature and pH, but also to the presence of specific chemicals that bind to the enzyme and cause changes in its shape. By means of these specific chemicals, a cell is able to turn enzymes "on" and "off." When the binding of a chemical changes the shape of the protein and the active site so that it can catalyze a chemical reaction, the chemical is called an *activator*. When the binding causes a change in the active site that shuts off enzyme activity, the chemical is called an *inhibitor*. Enzymes usually have special binding sites for the activator and inhibitor molecules that affect them, and these binding sites are different from their active sites.

Enzyme action is often regulated by inhibitors in a process called **negative feedback.** In this process the enzyme catalyzing the first step in a series of chemical reactions has an inhibitor binding site to which the end product of the pathway binds. As the concentration of the end product builds up in the cell, it begins to bind to the first enzyme in the metabolic pathway, shutting off that enzyme. In this way, the end product is feeding information back to the first enzyme in the pathway, shutting the pathway down when additional end product is not needed (Figure 6-13). Such **end-product inhibition** is a good example of the way many enzyme-catalyzed processes within cells are self-regulated. For example, the enzyme threonine deaminase catalyzes the first reaction in a series that converts the amino acid threonine to another amino acid isoleucine. The cells of the body

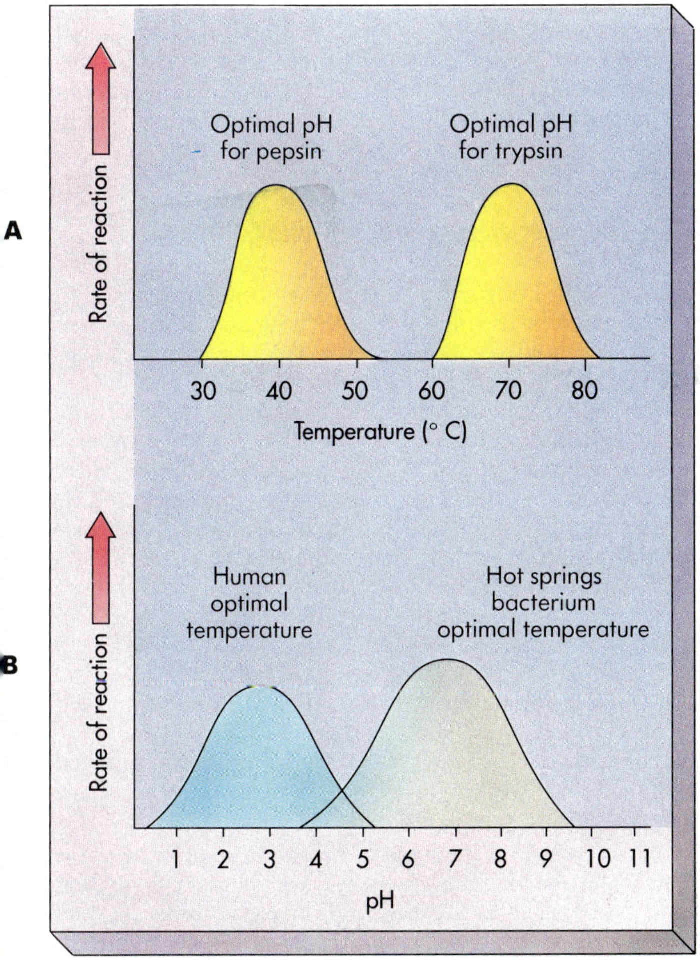

FIGURE 6-12 Enzymes are sensitive to the environment.
The activity of an enzyme is influenced both by temperature **(A)** and pH **(B)**. Human enzymes tend to work best at temperatures of about 37° C and within a pH range of 6 to 8.

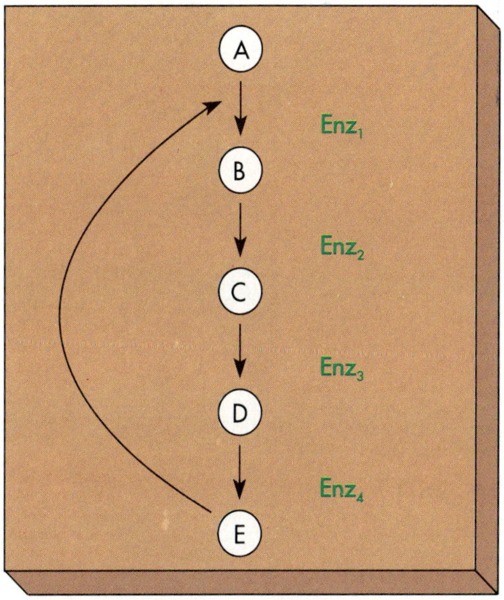

FIGURE 6-13 End-product inhibition.
Product *E* controls the rate of its own synthesis by acting on Enz_1, the enzyme that catalyzes the first reaction in the pathway.

need both amino acids in proper concentrations to make proteins. The presence of an adequate amount of isoleucine for protein manufacture "shuts off" the enzyme threonine deaminase and any further conversion of threonine to isoleucine.

The activity of enzymes is regulated by changes in enzyme shape; these changes result when activator and inhibitor molecules bind to specific enzymes.

Cofactors and coenzymes

Enzymes often have additional parts to their structures that are made up of molecules other than proteins (Figure 6-14). These additional chemical parts are called **cofactors**. Cofactors help enzymes catalyze chemical reactions. For example, many enzymes have metal ions such as zinc, iron, or copper locked into their active sites. These ions help draw electrons from substrate molecules. One of your digestive enzymes, carboxypeptidase, breaks down proteins in foods by using a zinc ion to draw electrons away from the bonds being broken in the food. Many trace elements necessary for your health such as manganese help enzymes in this way. When the cofactor is a nonprotein organic (carbon-containing) molecule, it is called a **coenzyme.** Many of the vitamins that your body requires, such as members of the B-vitamin complex, are used to synthesize coenzymes to maintain health. These coenzymes perform many jobs in the body, playing key roles in the reactions of cellular respiration, amino acid synthesis, and protein metabolism.

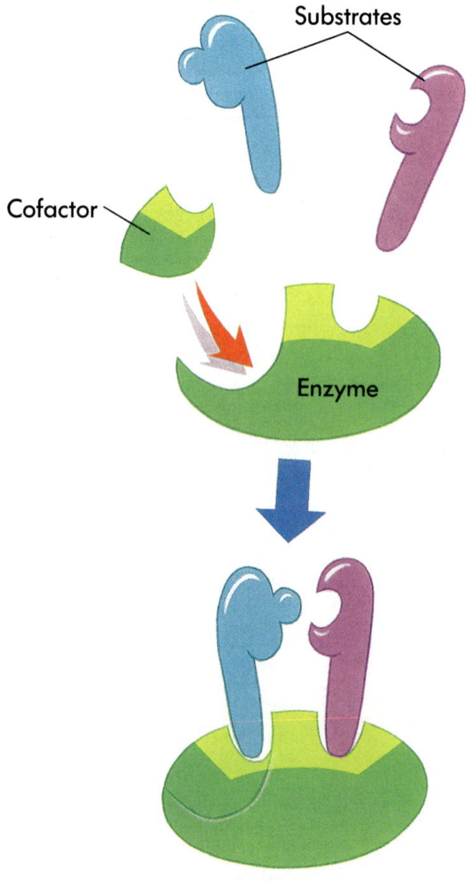

FIGURE 6-14 Cofactors.
Cofactors assist enzymes in catalyzing chemical reactions. An example of a cofactor is zinc, which helps a digestive enzyme break down food by drawing electrons away from the bonds being broken in the food.

Special nonprotein molecules called *cofactors* help enzymes catalyze chemical reactions. Cofactors that are nonprotein organic molecules are called *coenzymes*.

Storing and transferring energy

Many metabolic pathways in your body break down complex substances into simpler substances, releasing energy in the process. What happens to the energy that is released in these exergonic reactions? If your body could only use this energy when it was released, you might run out of energy a few hours after lunch, leaving you without enough energy to eat another meal. Obviously, living things could not survive under such circumstances. They have evolved ways to capture energy when it is released so that it can be used to do cell work.

ATP: The energy currency of living things

One way your body stores energy is by converting it to fat. In fact, many of us think we have too much of this stored energy. Fat, however, cannot serve our cells' immediate energy needs. It is used for long-term storage.

Glycogen is another molecule used by the body to store energy. Although glycogen can be more readily converted into energy than body fat can, it is still unable to meet immediate energy demands. The primary molecule used by cells to capture energy and supply it at a moment's notice is a molecule called **adenosine triphosphate,** or **ATP.**

Each ATP molecule is made up of three subunits (Figure 6-15):

1. A five-carbon sugar called *ribose*
2. A double-ringed molecule called *adenine*
3. Three phosphate groups (PO_4) linked in a chain called a *triphosphate group*

Together, the ribose sugar and the adenine rings are called *adenosine*. The "working end" of the molecule, however, is the triphosphate group.

Notice in the diagram that the bonds linking the phosphate groups in ATP are shown as wavy lines. These symbols stand for special bonds called **high-energy bonds.** As their name suggests, high-energy bonds require a lot of energy to form. These bonds are also special in another way: they are fairly unstable and require very little free energy of activation to be broken.

ATP is often referred to as the energy currency of the cell because it is used much like money. When cells break

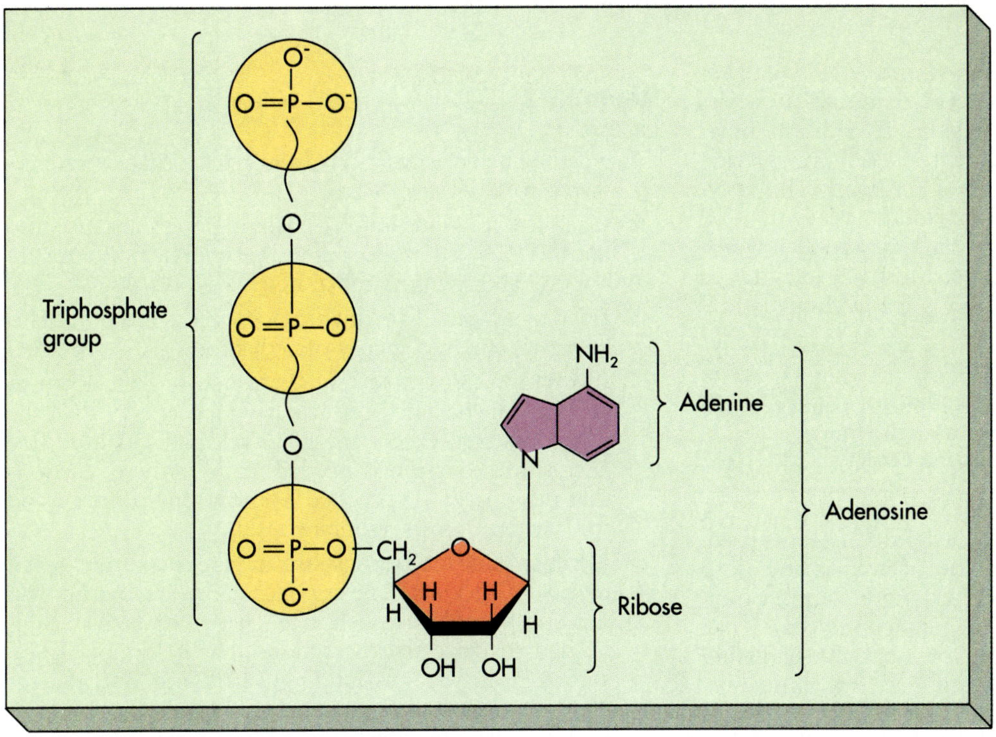

FIGURE 6-15 ATP is the primary energy currency of the cell.
ATP has three subunits: a five-carbon sugar called *ribose*, a double-ringed structure called *adenine*, and three phosphate groups called a *triphosphate group*. The phosphate groups are linked together by high-energy bonds.

down molecules in exergonic reactions, the energy that is released can be captured in molecules of ATP and carried there until it is needed. Likewise, when cells need energy to drive endergonic reactions, the cell can "spend" ATP to provide this energy.

ATP, the universal energy currency of all cells, can capture energy in its high-energy bonds and later release this energy.

Carrying and transferring energy using ATP

When ATP is used to drive an endergonic reaction, the bond that links the last phosphate group to the rest of the ATP molecule is broken, releasing needed energy (Figure 6-16). The molecule that remains is called **adenosine diphosphate,** or **ADP.** In addition, the phosphate group (often symbolized as P_i, or inorganic phosphate) exists on its own. Thus ATP → ADP + P_i + energy.

Cells always contain a pool of ATP, ADP, and phos-

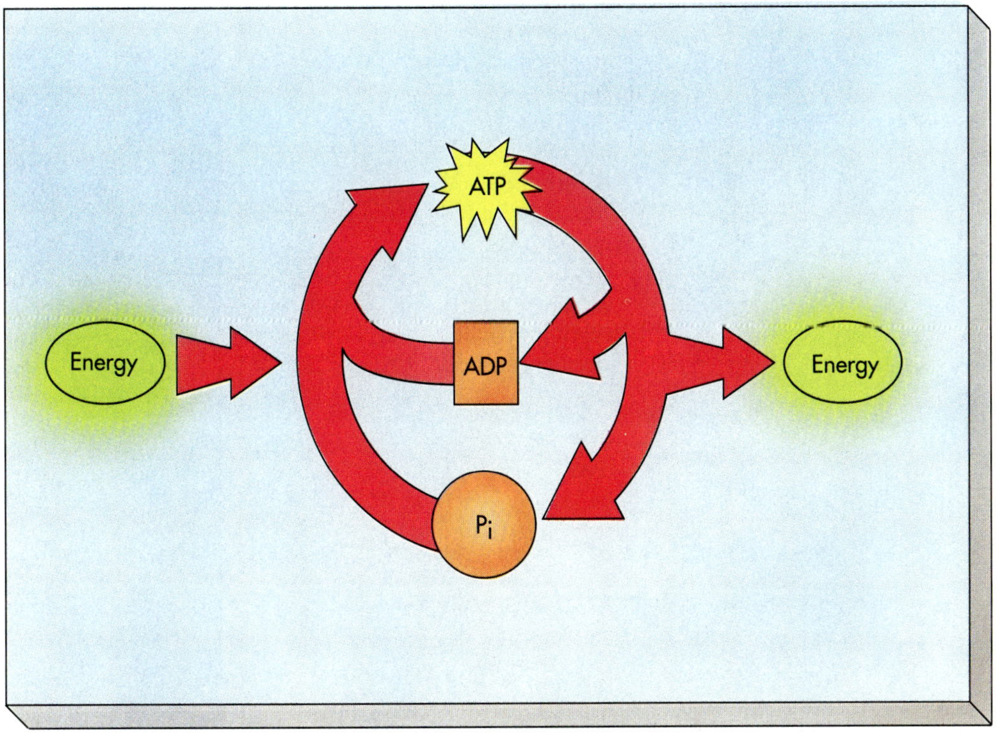

FIGURE 6-16 The ATP-ADP cycle.
When ATP is used to drive the energy-requiring activities of living things, the high-energy bond that links the last phosphate group to the ATP molecule is broken, releasing energy. Cells always contain a pool of ATP, ADP, and phosphate (symbolized by P_i), so ATP is continually being made from ADP, phosphate, and the energy released from exergonic reactions.

phate. ATP is constantly being cleaved into ADP and phosphate to drive the endergonic, energy-requiring processes of the cell by means of coupled reactions. In addition, however, ATP is continually being made from ADP, phosphate, and energy during coupled exergonic reactions. This recycling happens quickly. In fact, if you could mark every ATP molecule in your body at one instant in time and then watch them, they would be gone in a flash. Most cells maintain a particular molecule of ATP for only a few seconds before using it.

ATP, ADP, and phosphate are continually being recycled within living cells, performing the function of capturing, carrying, and releasing energy.

ATP is used to fuel a variety of cell processes. It is used, for example, in the light production of fireflies and deep water fishes. It is used to build larger, more complex molecules like proteins from smaller, simpler molecules like amino acids. It also provides energy for you to move, fueling the reactions that take place in your muscles that cause muscle fibers to contract. Cells also use ATP to help move substances against a gradient—it is the fuel of active transport. Last, the energy in ATP is changed into electrical energy primarily in nerves.

Where does the energy come from that is captured in molecules of ATP? The ultimate source of energy, of course, is the energy of the sun that is captured by plants during the process of photosynthesis. Plants and the animals that eat them serve as food sources for other animals. Your body breaks down the food you eat into molecules such as glucose that are usable for fuel by your cells. Your cells break down these fuel molecules, releasing energy that is then captured in molecules of ATP.

Summary

1. All the chemical reactions in a living organism are referred to as its *metabolism.*
2. Chemical reactions involve changing one substance into another. The substances entering the reaction are substrates. The changed substrates are products.
3. Energy is needed to initiate a chemical reaction. This energy is the free energy of activation.
4. In exergonic reactions, substrates are broken down to yield products. The products have less energy than the substrates; the excess energy is released. So little free energy of activation is needed to drive these reactions that this energy is absorbed from the environment and the reaction occurs spontaneously.
5. In endergonic reactions, substrates are chemically joined to yield a product. The product has more energy than the substrates. So much free energy of activation is needed to drive these reactions that it cannot be absorbed from the surroundings. In living systems, endergonic reactions are coupled with exergonic reactions to provide the energy needed to drive the endergonic process.
6. Energy exists in two states. Stored energy, or energy not actively doing work, is potential energy. Energy of motion, or energy actively doing work, is kinetic energy. Energy can be changed from one form to another during chemical reactions.
7. Energy cannot be created or destroyed; it can only be changed from one form or state to another. However, energy is continually being "lost" as heat during chemical reactions. Heat is a form of energy that is not useful for performing work.
8. Chains of chemical reactions that move, store, carry, and release energy in living systems are called *metabolic pathways.* Special proteins called *enzymes* regulate these series of reactions.
9. Enzymes lower the free energy of activation and speed up reactions by bringing substrates together so that they can react or by stressing chemical bonds that must break before a reaction can take place. Enzyme activity is affected by temperature, pH, and the binding of certain chemicals.
10. Adenosine triphosphate, or ATP, is the energy currency of the cell in that it can "capture" energy from an exergonic reaction, "carry" this energy, and then "spend" it when needed.

REVIEW QUESTIONS

1. Fill in the blanks: __________ is the energy needed to start a chemical reaction. The __________ are the molecules that enter into the reaction; as the reaction occurs, these molecules change to become __________.
2. Distinguish among exergonic, endergonic, and coupled reactions.
3. Fill in the blanks: Energy actively doing work is called __________, __________; stored energy is __________ __________.
4. What are the first and second laws of thermodynamics?
5. When asked about his messy room, a friend who's studied science simply replies, "Entropy increases." Explain what this means.
6. Explain the role played by photosynthetic organisms in the biological flow of energy. Why is it important?
7. What are metabolic pathways? Explain their significance.
8. What are enzymes, and why are they important?
9. Explain how enzymes catalyze reactions.
10. If you are ill with a high fever, how does this affect the enzymes in your body?
11. What are cofactors and coenzymes? Name a biologically important example of each.
12. Name three ways in which your body stores energy for later use. Why is it important to store energy?
13. ATP has been called the "energy currency of the cell." Explain what this means.
14. Explain the relationship among ATP, ADP, and phosphate. Why are these substances important?
15. Industrial pollution can change the pH of a pond or river to make the water more acidic. How can this affect the metabolic pathways of the plants that live in the water?

KEY TERMS

active sites 105
adenosine diphosphate (ADP) 111
adenosine triphosphate (ATP) 110
anabolic reactions 101
catabolic reactions 101
catalysis 105
catalysts 105
coenzyme 110
cofactors 110
coupled reactions 102
endergonic reaction 101
end-product inhibition 109
entropy 105
enzymes 105
exergonic reaction 101
first law of thermodynamics 104
free energy of activation 101
kinetic energy 103
metabolic pathways 105
negative feedback 109
potential energy 103
product 101
second law of thermodynamics 104
substrate 101

THOUGHT QUESTIONS

1. A typical person at rest (that is, sitting down or lying in bed) consumes 1400 kilocalories per day. Each ATP molecule releases 7.3 kilocalories of energy when it is cleaved to ADP + P_i. Your body contains about 35 grams of ATP. Estimate how long an individual ATP molecule lasts before being converted to ATP.
2. Which is the more efficient at storing energy (useful calories stored per pound), fat or carbohydrate? Why?
3. Many of a cell's most important reactions are endergonic, including the reactions involved in producing proteins and DNA. Endergonic reactions never take place spontaneously because their products require more energy than the starting materials contain. How then does the cell induce these reactions to occur?

FOR FURTHER READING

Hinkle, P. C., & MacCarty, R. E. (1978, March). How cells make ATP. *Scientific American*, pp. 104-123.
This article describes how cells use electrons stripped from foodstuffs to make ATP.

Lehninger, A. L. (1961, September). How cells transform energy. *Scientific American*, pp. 62-73.
Written more than 25 years ago, this article still provides one of the clearest explanations of how cells channel energy through metabolism.

7 Cellular respiration

How cells release stored energy from food molecules

Did you know that you and the car in the photo have a lot in common? Both you and the car are complex machines and you both burn fuel for energy. But you and a car burn fuel in ways that have important differences.

In the car's engine, gasoline is burned in the presence of oxygen. During exergonic reactions, molecules of gasoline are broken down to carbon dioxide and water, releasing energy as heat. Explosive bursts of heat produce pressure that moves the pistons in the engine, propelling the car. When the driver presses down on the accelerator and "gives it the gas," the rate of these exergonic reactions increases. Because the car has no way to store the energy liberated from these reactions, this energy is used immediately, and the car speeds ahead.

OUTLINE

Using chemical energy to drive metabolism

How cells make ATP: Variations on a theme

How oxygen-using organisms release ATP from food molecules

- An overview of cellular respiration
- Oxidation-reduction
- Electron carriers
- A more detailed look at cellular respiration
  - Glycolysis
  - The Krebs cycle
  - The electron transport chain

How organisms release ATP from food molecules without using oxygen

HIGHLIGHTS

All the activities that organisms perform require energy to power the endergonic chemical reactions that underlie their processes.

Most living things release the stored energy in food molecules needed to drive endergonic reactions by an oxygen-dependent process called *cellular respiration.*

Cellular respiration captures energy in molecules of ATP, an energy-carrying molecule, during three complex series of reactions: glycolysis, the Krebs cycle, and the electron transport chain.

Organisms that do not use oxygen, such as certain microorganisms, make ATP by a process known as *fermentation.*

In the engines of each of your cells—the mitochondria—the products of glucose breakdown are burned in the presence of oxygen. During a series of exergonic reactions, molecules of glucose are broken down to carbon dioxide and water, releasing energy. Most of this energy is released as heat and helps maintain your body temperature. But about 38% of it is captured in molecules of ATP—the energy currency of your cells. In fact, you do a better job of converting the energy in food to useful energy for your body—the car can convert only about 25% of the energy in gasoline into useful energy. And by using ATP, your body can capture, carry, and "spend" energy when it is needed, an ability that even the car of the future does not have.

Using chemical energy to drive metabolism

All the activities that organisms perform—the swimming of bacteria, the purring of a cat, your reading these words—use energy. These activities require energy because endergonic chemical reactions underlie all life activities. For example, energy-requiring chemical reactions cause muscles to contract and are involved in the active transport of molecules across nerve cell membranes, maintaining the crucial balance of various ions necessary for their functioning. In addition, living things constantly use energy to build larger molecules from smaller ones. Chains of reactions move, store, and free the energy needed to perform these activities of life. As you may recall from Chapter 6, such chains of reactions are called *metabolic pathways*.

In metabolic pathways, exergonic reactions—those that release energy—occur spontaneously with the help of enzymes, absorbing enough free energy as heat from their surroundings to "go." Endergonic reactions—those that require an input of energy—do not occur spontaneously because they are unable to absorb enough free energy from their environment to "drive" the reaction. So where does the energy come from to drive endergonic reactions? All living things get this energy in the same way: they *couple* endergonic reactions with exergonic reactions—those that release energy from molecules such as ATP. In this way, the energy of ATP is spent to enable an endergonic reaction to go (Figure 7-1).

Chemical energy powers metabolism by driving endergonic reactions. Chemical energy from exergonic reactions is captured in molecules of ATP, and the splitting of ATP is coupled to endergonic reactions, providing the energy necessary to drive them.

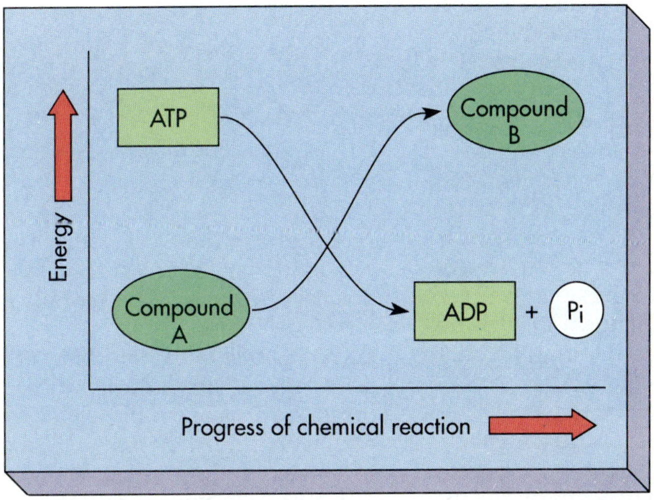

FIGURE 7-1 A coupled reaction.
The energy that is released from the conversion of ATP to ADP + P_i (an exergonic reaction) is used to drive the conversion of compound A to compound B (an endergonic reaction).

FIGURE 7-2 Photosynthesis.
These vigorously growing corn plants capture energy from the sun, producing glucose from carbon dioxide and water.

How cells make ATP: Variations on a theme

The energy used to make ATP is actually energy from the sun, captured by plants, algae, and certain bacteria in a process called **photosynthesis.** During this process, the energy from the sun powers the synthesis of glucose molecules, using the raw materials of carbon dioxide and water (Figure 7-2). Glucose is a sugar that plants use as a source of energy and as "building blocks" to construct larger molecules such as the cellulose found in plant cell walls. Plants also manufacture starch from glucose, a food that can be stored in plant cells and then converted to glucose when energy is needed. By eating plants and by eating animals that eat plants, humans and other animals can harvest the energy from the sun that was originally captured and stored by plants (see Chapter 8).

Living things, including photosynthetic organisms, make ATP by breaking down nutrient molecules such as glucose to release the energy stored in its bonds. The complex series of chemical reactions during which ATP is made from nutrient molecules in the presence of oxygen is known as **cellular respiration.** The term *cellular* refers to this process taking place in the cells of living things. The term *respiration* describes the process: the breakdown of fuel molecules in the presence of oxygen, with a resulting release of energy. Cellular respiration is called an **aerobic** process because this series of reactions is oxygen dependent. Do not confuse this process with the breathing of oxygen gas that your body carries out, which is also called respiration. However, the process of cellular respiration depends on the oxygen you breathe. *All* organisms that take in oxygen—including the plants, which produce oxygen as a by-product of photosynthesis—break down nutrient molecules by means of cellular respiration.

Organisms that do not "breathe" oxygen, such as certain microorganisms, make ATP by a process known as **fermentation.** This process consists of glycolysis (an initial series of reactions of cellular respiration) and one or two additional reactions that take place **anaerobically**—literally, "without" (an) "oxygen" (aerobically). Baker's yeast, for example, can generate ATP by fermentation. During this process the yeast produces carbon dioxide gas, an end product that causes bread dough to rise. Certain yeasts produce alcohol during fermentation and are used in the production of beers, wines, and other alcoholic beverages.

Interestingly, *all* organisms use either cellular respiration or fermentative pathways to generate ATP, suggesting a strong evolutionary relationship among all living things. The only organisms that use neither of these two ATP-generating processes are single-celled parasites called *chlamydia.* One species of chlamydia causes certain diseases of the eyes and also one of the most prevalent sexually transmitted diseases in the United States: nongonococcal urethritis. Another species causes diseases in animals. These organisms, among the smallest of the bacteria, can grow only within other cells and obtain their ATP from the cells they infect. For this reason, scientists call chlamydia "energy parasites." Chlamydia are thought to have lost the ability to generate ATP during their evolution.

All cells, with few exceptions, make ATP by the process of cellular respiration or by the process of fermentation.

How oxygen-using organisms release ATP from food molecules

An overview of cellular respiration

The process of cellular respiration is described chemically in the following formula:

$$C_6H_{12}O_6 + 6O_2 \xrightarrow{36\,ADP + 36P_i \;\rightarrow\; 36\,ATP} 6CO_2 + 6H_2O$$

Glucose + Oxygen → Carbon dioxide + Water

Although the above formula is stated as if glucose were broken down to carbon dioxide and water in a single step, this is not the case. The formula is a summary of a complex series of reactions, much like you might summarize the details of a book in a single paragraph. This simplified formula shows that as cellular respiration occurs, the net effect is the breakdown of one molecule of the substrate glucose in the presence of six molecules of oxygen to yield end products of six molecules of carbon dioxide and six molecules of water. During this process, enough energy is liberated from the glucose as it is being cleaved to power 36 endergonic reactions: 36 molecules of ADP are bonded to 36 atoms of inorganic phosphate . . . an energy yield of 36 ATPs! Therefore the breakdown of glucose during cellular respiration

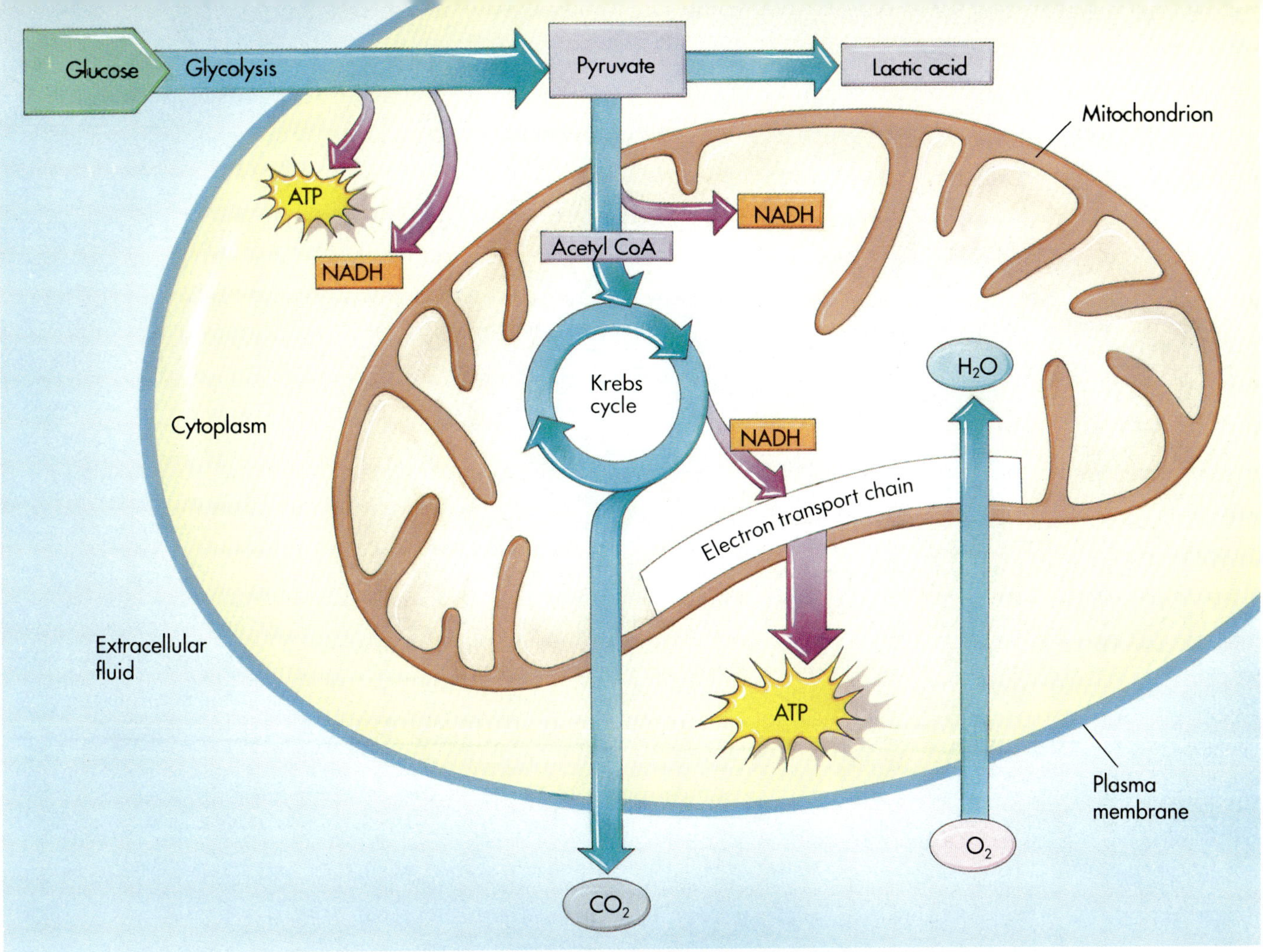

FIGURE 7-3 An overview of cellular respiration.
Living things, including plants, use cellular respiration to make ATP.

releases energy (is exergonic), which helps to produce molecules of ATP, a type of endergonic reaction.

> **Cellular respiration is a series of chemical reactions that captures energy in molecules of ATP. During cellular respiration, glucose is broken down in the presence of oxygen, yielding carbon dioxide and water and capturing energy in molecules of ATP.**

The complex series of reactions of cellular respiration can be divided into three parts: **glycolysis,** the **Krebs cycle,** and the **electron transport chain.** The word *glycolysis* comes from Greek words meaning "to break apart" (lysis) a "sugar" (glyco). Glycolysis takes place in the cytosol of the cell and is the first stage of extracting energy from glucose. No oxygen is needed for it to take place. It is a metabolic pathway in which ATP is generated, but the total yield of ATP molecules is small—only two ATPs for each original glucose molecule. When glycolysis is completed, the six-carbon glucose has been cleaved in half, yielding two three-carbon molecules called **pyruvate,** or pyruvic acid. The two pyruvate molecules still contain most of the energy that was present in the one original glucose molecule.

The Krebs cycle takes place in the mitochondria of the cell and is the second stage of extracting energy from glucose. Named after the biochemist Sir Hans Krebs, this cycle of reactions begins with a two-carbon molecule that is produced by the removal of a molecule of carbon dioxide (CO_2) from each pyruvate of glycolysis. Each of these two-carbon molecules combines with a four-carbon molecule to form a six-carbon molecule called *citric acid.* For this reason, the Krebs cycle is also often called the **citric acid cycle.** For every two pyruvate molecules entering the Krebs cycle as the result of the glycolytic breakdown of one glucose molecule, two more ATP molecules are made and a large number of electrons are removed from the substrates in the cycle. These electrons are passed to special molecules, also located in the mitochondria, called *electron carriers.* These electron carriers are part of the electron transport chain. By means of chemical reactions and processes that take place along the chain, 32 ATPs are produced. The process of cellular respiration is summarized in Figure 7-3.

> **Three series of reactions make up the process of cellular respiration: glycolysis, the Krebs cycle, and the electron transport chain.**

BIOLOGY & *you*

Carbohydrates and winter depression

Every winter as the days grow shorter, some people suffer from what has come to be known as *seasonal affective disorder (SAD)* or, more simply, winter depression (Figure 7-A). The symptoms include depression, lethargy, an inability to concentrate, and often a profound craving for carbohydrates. This craving often leads to overeating and excessive weight gain. What does eating have to do with this kind of depression, and why is the problem seasonal?

In the last few years a wealth of information has emerged that casts light on this unusual disorder. It now appears that winter depression reflects biochemical disturbances in two distinct biological systems. One system involves the hormone melatonin, which affects mood and subjective energy levels; the other involves the chemical serotonin that has a strong influence on appetite for carbohydrates.

Melatonin is a little-understood hormone produced by the pineal gland at the base of the brain. Its overproduction appears to be the underlying cause of winter depression. For reasons not understood, rates of melatonin production are affected by day length—the shorter the day, the more hormone produced. That is the reason why melatonin-induced depression is seasonal; in the winter the days are shorter. That is also the reason why people in the north are more affected by winter depression—the incidence is much greater in Minnesota where winter days are short (more than 100 people per 1000) than in Florida where winter days are hours longer (fewer than 6 people per 1000). In the southern hemisphere the seasons are reversed, and peak winter depression occurs in June and July when day length is shortest there.

So what does this have to do with craving carbohydrates? Melatonin seems to inhibit the release of serotonin from serotonin-releasing nerve cells, leading to a decrease in serotonin levels. Serotonin mediates a feedback mechanism in the brain affecting the amount of carbohydrate an individual chooses to eat: eating carbohydrates increases the brain's production of serotonin, which acts on the brain's appetite centers to reduce carbohydrate craving. When this feedback mechanism is disturbed, as happens when rising melatonin levels in winter depression cause serotonin levels to fall, the brain fails to respond when carbohydrates are eaten, and the desire for them persists longer than it should.

Both winter depression and the carbohydrate craving associated with it can be treated successfully by artificially increasing daylength. Patients with SAD treated with light as bright as daylight for 2 hours in the early morning often recover completely after only a few days of treatment. Melatonin is secreted in the dark, and this treatment shortens the nightly period of melatonin secretion by several hours.

FIGURE 7-A *"Yes, I'm somewhat depressed, but seasonally adjusted I'm probably happy enough."*

Oxidation-reduction

As chemical reactions take place in metabolic pathways such as cellular respiration, energy stored in chemical bonds is transferred to new chemical bonds, with the electrons shifting from one energy level to another. In some (but not all) chemical reactions, electrons actually pass from one atom or molecule to another. This class of chemical reaction is called an **oxidation-reduction reaction.** Oxidation-reduction reactions are critically important to the flow of energy through living systems and essential to the flow of energy in cellular respiration.

When an atom or molecule gives up an electron, it is **oxidized.** The process by which this occurs is called **oxidation.** The name reflects that in biological systems, oxygen, which strongly attracts electrons, is the most frequent electron acceptor.

When an atom or molecule gains an electron, it becomes **reduced.** The process is called **reduction.** This name reflects that the addition of an electron *reduces* the charge by one. For example, if a molecule had a charge of +2, the addition of an electron (-1) would reduce the molecule's charge to +1.

Oxidation and reduction always take place together because every electron that is lost by one atom (oxidation) is gained by some other atom (reduction) (Figure 7-4). Together they are therefore called **redox reactions.** In a redox reaction the charge of the oxidized atom is increased, and the charge of the reduced atom is lowered.

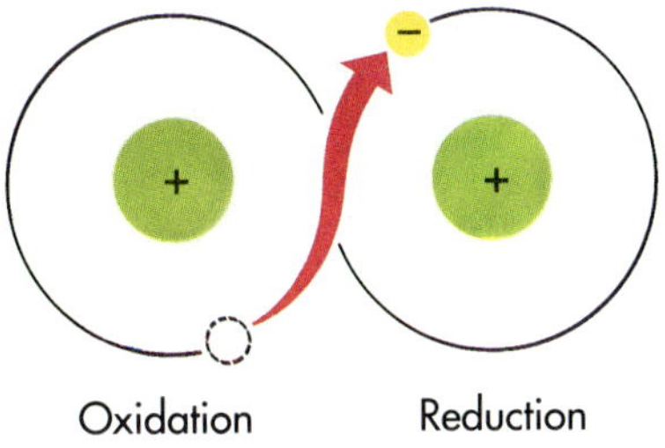

FIGURE 7-4 Oxidation-reduction.
Oxidation is a loss of an electron; reduction is the gain of one.

Oxidation is the loss of an electron by an atom. Reduction is the gain of an electron by an atom.

Electron carriers

In biological systems, electrons often do not travel alone from one atom to another, but instead take along a proton. As you may recall from Chapter 2, a proton and an electron together make up a hydrogen atom. In living things, special electron carriers transfer electrons as hydrogen atoms from one molecule to another. One such carrier is the molecule nicotinamide adenine dinucleotide, or NAD^+, a molecule synthesized from the vitamin niacin. The + sign indicates that NAD is the oxidized form of the molecule. It can accept two electrons and one proton (H^+) to become reduced to NADH. Notice that this molecule no longer bears a positive charge because it has been reduced. In addition, the extra H^+ and electron (hydrogen atom) is shown by the addition of a H to NAD.

Flavin adenine dinucleotide, or FAD, is another special electron carrier. It can accept *two* electrons and *two* protons to be reduced to $FADH_2$. The subscript *2* indicates the number of hydrogen atoms that FAD has accepted.

Both NADH and $FADH_2$ can pass on the electrons and protons they accept to other carriers. They are key players in the electron transport chain, one of the three series of reactions in cellular respiration. By passing electrons and protons in a series of redox reactions, the molecules in this chain create a flow of energy that works in special ways to produce most of the ATP of cellular respiration (Figure 7-5).

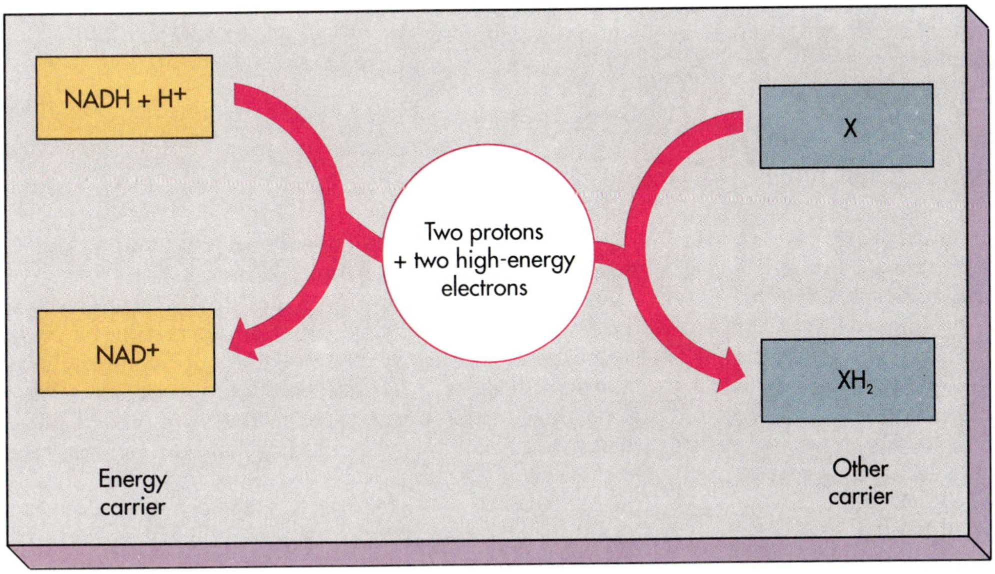

FIGURE 7-5 $NADH_2$, an electron carrier.
The high-energy electrons carried by $NADH_2$ are transferred to proton pumps or other energy-requiring components (indicated by *X*).

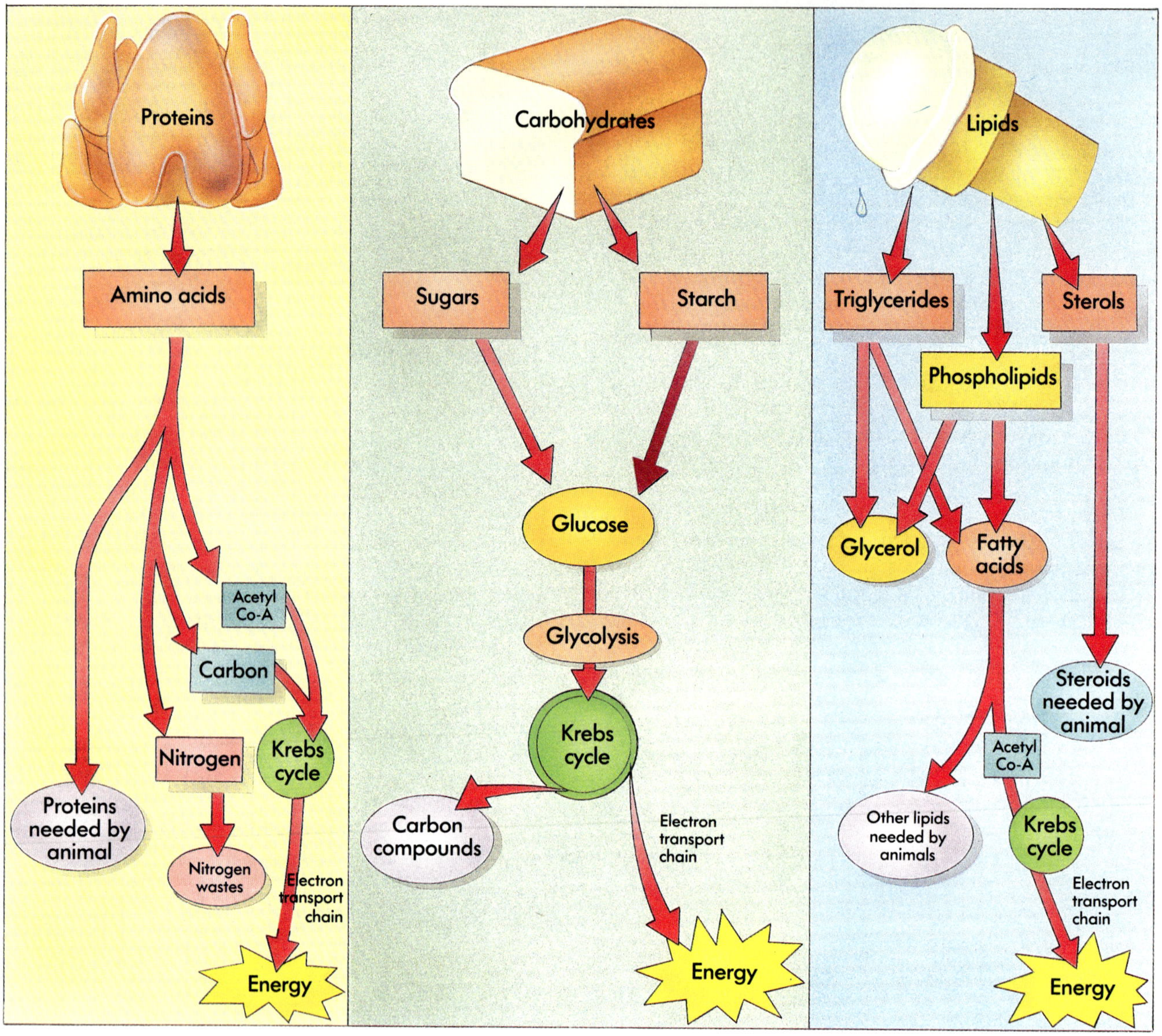

FIGURE 7-6 The complex molecules in food must be broken down into simpler molecules before they can be used by cells as fuel.
Proteins are broken down into amino acids. Carbohydrates are broken down into simpler sugars. Lipids are broken down into fatty acids and glycerol. Note that proteins and lipids, once broken down into their simpler subunits, enter the Krebs cycle without undergoing glycolysis.

A more detailed look at cellular respiration

Food is a complex mixture of sugars, lipids, proteins, and other molecules. The first thing that happens in its journey toward ATP production is that digestive system enzymes break down complex molecules to simple ones. Complex sugars like those found in vegetables and pasta, for example, are split into simple sugars such as glucose or into sugars that are changed to glucose. Proteins, which are found in foods such as meats and nuts, are broken down to amino acids. Complex lipids like those found in oil-based salad dressings are split into fatty acids and glycerol. These steps taken by the digestive system yield no usable energy, but they change a diverse array of complex molecules into a small number of simpler molecules that can be used by your cells as fuel.

The primary metabolic pathways in your body that break apart these simple molecules to release their stored energy and capture it in molecules of ATP are glycolysis, the Krebs cycle, and the electron transport chain. Once they are changed to glucose, carbohydrates are completely metabolized this way. After the digestion of proteins to amino acids, their carbon portions are chemically modified and are then metabolized by the Krebs cycle and the electron transport chain, "skipping over" glycolysis. Likewise, some of the breakdown products of the fats in your diet are converted to substances that can also be metabolized by these two series of reactions (Figure 7-6).

Glycolysis

Scientists think that glycolysis was one of the first metabolic processes to evolve. One reason is that this process uses no molecular oxygen and therefore occurs readily in an environment devoid of oxygen—a characteristic of the atmosphere of the primitive earth. In addition, all of the reactions of glycolysis occur free in the cytoplasm; none are associated with any organelle or membrane structure. "Early" cells most certainly had no specialized structures such as organelles, so metabolic processes that evolved independently of membrane systems are probably evolutionarily older than those associated with membrane systems.

All living things (except the chlamydia) possess glycolytic enzymes. However, most present-day organisms are able to extract considerably more energy from glucose molecules than glycolysis does. For example, of the 36 ATP molecules you obtain from each glucose molecule that you metabolize, only 2 of these are obtained by glycolysis. Why, then, is glycolysis still maintained even though its energy yield is comparatively meager?

This simple question has an important answer: evolution is a step-by-step process. Change occurs during evolution by improving on past successes. In catabolic metabolism, glycolysis was an improvement. A **catabolic process** is one in which complex molecules are broken down into simpler ones. Cells that could not carry out the catabolic reactions of glycolysis were at a competitive disadvantage; cells that were capable of glycolysis survived the early competition of life. Later improvements in catabolic metabolism built on this success. Glycolysis was not discarded during the course of evolution but rather was used as the starting point for the further extraction of chemical energy.

The catabolic pathway of glycolysis is shown in Figure 7-7. Its reactions can be divided into four stages:

1. Stage A—**Glucose mobilization:** During reactions 1 and 2, glucose is changed into a compound that can be split into two molecules. During two of these reactions, ATP molecules have their outermost phosphate groups enzymatically removed. The bonds holding these phosphate groups release a high amount of free energy when broken. The energy in this case is used to attach the phosphate group to the glucose molecule. Before energy can be released from glucose, your body must spend two ATP to "prime" the glucose "pump."
2. Stage B—**Cleavage:** During reaction 3, the six-carbon product of glucose mobilization is split into two three-carbon molecules.
3. Stage C—**Oxidation:** During reaction 4, an H^+ and two electrons are removed from each three-carbon molecule and are donated to NAD^+. NAD^+ acts as an electron carrier, forming two molecules of NADH for each original glucose molecule. As this chemical change takes place, an inorganic phosphate molecule is bonded to each resulting three-carbon molecule by means of high-energy bonds.
4. Stage D—**ATP generation:** During reactions 5 through 7, the two high-energy phosphate groups on each three-carbon molecule are removed by enzymes and bonded to two ADP molecules, producing four ATP molecules for each original glucose molecule. In addition, a molecule of water is removed from each molecule. The end product of these reactions is two three-carbon molecules called pyruvate.

Because each glucose molecule is split into *two* three-carbon molecules, the overall net reaction sequence yields two ATP molecules as well as two molecules of pyruvate:

−2ATP	Stage A
+ 2(2ATP)	Stage D
+2 ATP	

Although this is not a great amount of energy (only 2% of the energy in the glucose molecule), glycolysis does generate ATP. During the first anaerobic, or "airless," stages of life on Earth, this reaction sequence was the only way for living things to extract energy from food molecules.

The sequence of reactions of glycolysis generates a small amount of ATP by "reshuffling" the bonds of glucose molecules. Glycolysis is a very inefficient process, capturing only about 2% of the available chemical energy of glucose.

Three changes take place during glycolysis:

1. Glucose is converted to pyruvate.
2. ADP + P_i is converted to ATP.
3. NAD^+ is converted to NADH.

These three products can continually be formed as long as the substrates that are used to produce them are available. The glucose is supplied by the food you eat. ADP and P_i continually become available as ATP is broken down to do cellular work. But cells contain only a small amount of NAD^+, and the supply is quickly depleted unless NADH passes along its hydrogen atom to another electron carrier. In this way NADH is oxidized to form NAD^+ once again.

Cells "recycle" NADH in one of two ways:

1. *Aerobic metabolism:* In the presence of oxygen gas, NADH passes its hydrogen atom to another electron carrier in the electron transport chain. The oxygen, an excellent electron acceptor, forms the last "link" in the chain. Each atom of oxygen accepts two H atoms to form one molecule of water. In addition, three molecules of ATP are formed during various chemical processes that take place (see p. 127).
2. *Anaerobic metabolism:* When oxygen is not available, another organic molecule must accept the hydrogen atom instead. This process occurs when organisms such as bacteria or yeast grow without oxygen and is called *fermentation* (Figure 7-8). Fermentation is described in more detail later in this chapter.

In all aerobic organisms—like you—the oxidation of glucose continues with the further oxidation of the product of glycolysis, which is pyruvate. In eukaryotic organisms, aerobic metabolism takes place in the mitochondria. Glycolysis, however, takes place in the cytoplasm outside the mitochondria. Therefore the cell must spend two ATPs (for each original molecule of glucose) to move the two NADH molecules produced during glycolysis into the mitochondrion. So, although the entire process of aerobic cellular

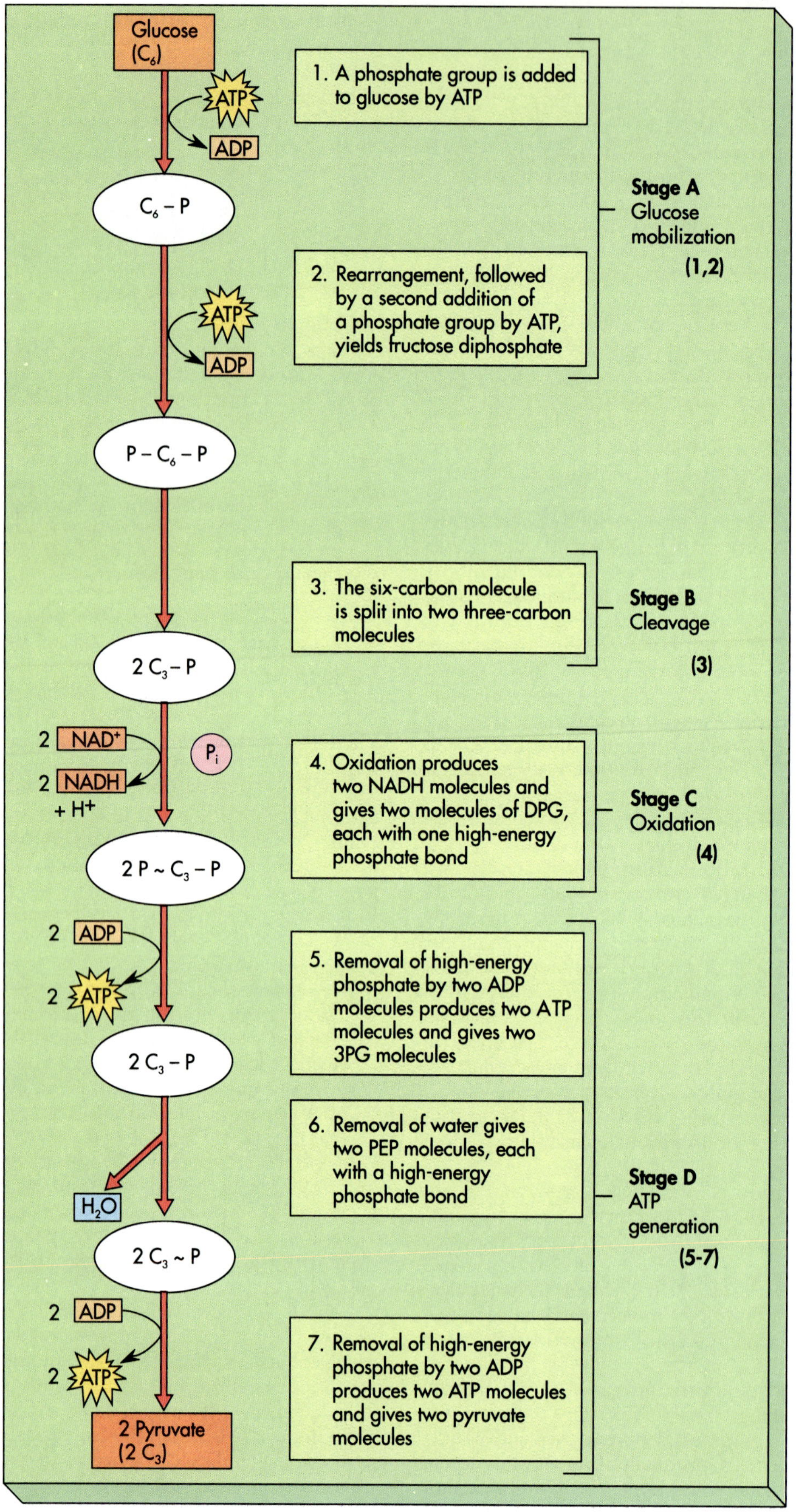

FIGURE 7-7 Glycolysis.
Glycolysis produces a net yield of two ATP molecules for each original glucose molecule, as well as two molecules of NADH and two molecules of pyruvate.

FIGURE 7-8 Fermentation.
In the absence of oxygen, yeast metabolizes pyruvate to ethyl alcohol. This process takes place naturally in grapes left to ferment on vines and artificially in fermentation vats of crushed grapes. As the concentration of ethyl alcohol in the vats rises to about 10%, further growth of the yeast is inhibited. What is left is wine.

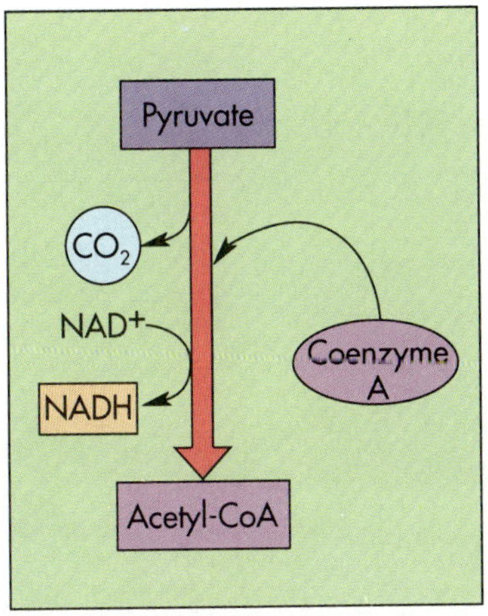

FIGURE 7-9 How pyruvate enters the Krebs cycle.
First, a carbon atom from each pyruvic acid molecule is split off as carbon dioxide gas, and a molecule of NADH is formed. Coenzyme A, a carrier molecule, is added to what is left of the pyruvic acid molecule, forming acetyl-CoA. Acetyl-CoA is the starting compound for the Krebs cycle.

respiration produces 38 ATPs, the net result is 36 ATPs for each molecule of glucose. Pyruvate, which is in high concentration in the cytoplasm, simply diffuses into the mitochondrion.

Before entering the reactions of the Krebs cycle, each pyruvate molecule is cleaved into a two-carbon molecule called an **acetyl group.** The left-over carbon atom from each pyruvate is split off as carbon dioxide gas (CO_2). In addition, an H^+ and two electrons reduce NAD^+ to NADH. In the course of these reactions the two-carbon acetyl fragment removed from pyruvate is added to a carrier molecule called *coenzyme A,* or CoA, forming a compound called **acetyl-CoA.** This process is summarized in Figure 7-9.

These reactions produce a molecule of NADH, which is later used to produce ATP molecules. Of greater importance, however, is the acetyl-CoA. Acetyl-CoA is important because it is the first substrate in the Krebs cycle, which generates many more molecules of NADH. It is produced not only during the breakdown of glucose but also during the metabolic breakdown of proteins, fats, and other lipids. This molecule is the point at which many of the catabolic processes of eukaryotic cells converge (see Figure 7-6).

The Krebs cycle

The Krebs cycle, which oxidizes acetyl-CoA, consists of nine enzyme-mediated reactions. These reactions are shown in Figure 7-10. The cycle has two stages:

1. Stage a—**Preparation reactions:** These three reactions set the scene. In the first reaction, acetyl-CoA joins a four-carbon molecule from the "end" of the cycle to form the six-carbon molecule citric acid. In the next two reactions, chemical groups are rearranged.
2. Stage b—**Energy extraction:** Four of the remaining six reactions are oxidations in which H ions and electrons are removed from the intermediate compounds in the cycle to form three NADH molecules for each acetyl-CoA, six for each original glucose molecule. In addition, one molecule of $FADH_2$ is formed, two for each original glucose. During the cycle a molecule called *GTP* is produced, which is quickly used to produce an ATP; two ATP are generated for each original molecule of glucose.

Together, the nine reactions make up a cycle that begins and ends with the same four-carbon molecule. At every turn of the cycle, acetyl-CoA enters and is oxidized to CO_2 and H_2O, and the H ions and electrons are donated to electron carriers.

In the process of aerobic respiration the glucose molecule has been consumed entirely. Its six carbons were first split into three-carbon units during glycolysis. One of the carbons of each three carbon unit was then lost as CO_2 in the conversion of pyruvate to acetyl-CoA, and the other two were lost during the oxidations of the Krebs cycle. Part of the glucose molecule's energy and its electrons, which are preserved in four ATP molecules and the reduced state of 12 electron carriers, are all that is left.

The breakdown of the two pyruvate molecules produced from one glucose molecule during glycolysis generates two ATP molecules, 10 molecules of NADH, and 2 molecules of $FADH_2$. NADH and $FADH_2$ can be used to generate ATP.

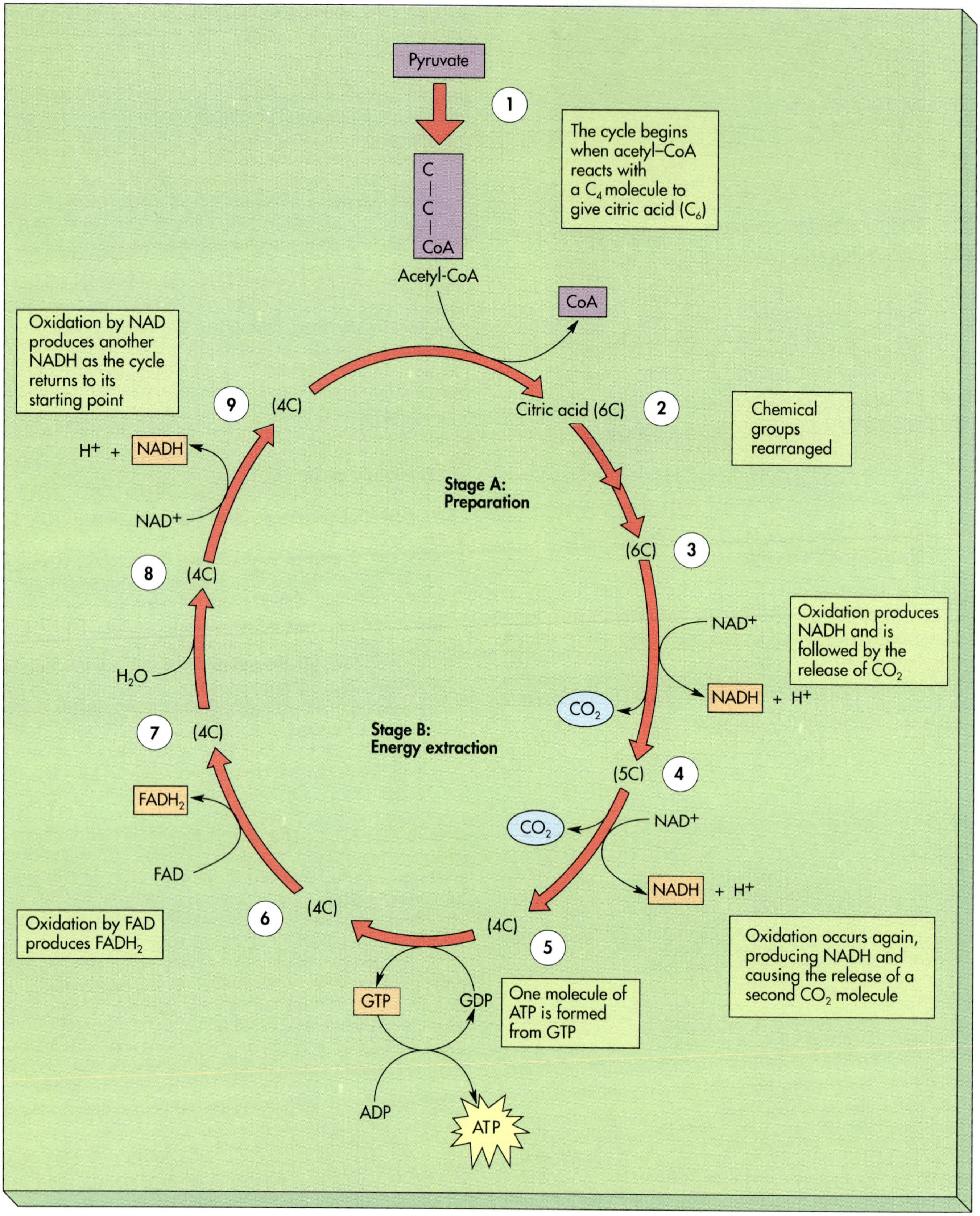

FIGURE 7-10 The Krebs cycle.
The Krebs cycle yields three molecules of NADH, one molecule of $FADH_2$, and one molecule of ATP for each acetyl-CoA molecule that enters the cycle.

The electron transport chain

Embedded within the inner mitochondrial membrane is a series of electron carrier proteins known as the **electron transport chain** (Figure 7-11). Here, electrons are passed from one carrier protein to another. The NADH molecules formed during glycolysis and the subsequent oxidation of pyruvate carry their electrons to this membrane. As you may recall, each NADH contains a pair of electrons and a proton gained when NADH was formed from NAD^+. The $FADH_2$ molecules are already attached to this membrane, each containing two protons and two electrons gained when $FADH_2$ was formed from FAD.

Three of the electron carrier proteins in the electron transport chain act as proton pumps. A proton pump is an active transport mechanism in cell membranes that forces hydrogen ions (protons) out of a cell. As each of these carriers accepts electrons in turn, the electrons fall to lower energy levels, releasing energy in the process. In addition, these carriers change shape as they accept the electrons, forming a channel from the inner compartment, or matrix, of the mitochondrion to the outer compartment—the space between the mitochondrial membranes. The energy released from the electrons being passed along the chain pumps protons through these channels and against the concentration gradient into the outer compartment of the mitochondrion. As a high concentration of protons builds up in the outer compartment, they begin to diffuse back across the membrane. But the only way they can get back in is through special channels called **respiratory assemblies,** visible in electron micrographs as projections on the inner surface of the membrane (Figure 7-12). As the protons travel inward through these channels, proteins associated with the channels couple the energy from the movement of the protons to the reaction of bonding P_i to ADP, forming ATP (Figure 7-13). Oxygen is the final acceptor for both the electrons and protons. Water is therefore an end product of cellular respiration, with one molecule of water produced for each NADH or $FADH_2$. Because the electron transport chain uses oxygen as its terminal electron acceptor, this series of reactions is also termed the *respiratory chain.*

The electrons harvested from glucose and transported to the mitochondrial membrane by NADH drive protons out across the inner membrane. The return of the protons by diffusion generates ATP.

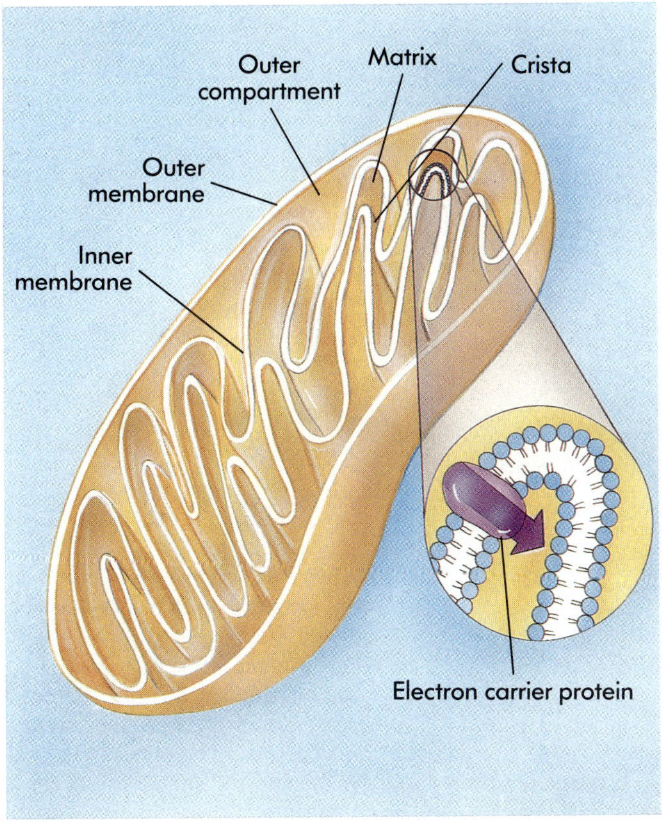

FIGURE 7-11 Site of the electron transport chain.
Located deep within the inner mitochondrial membrane, electron carrier proteins accept electrons from the NADH and $FADH_2$ formed during glycolysis and the Krebs cycle.

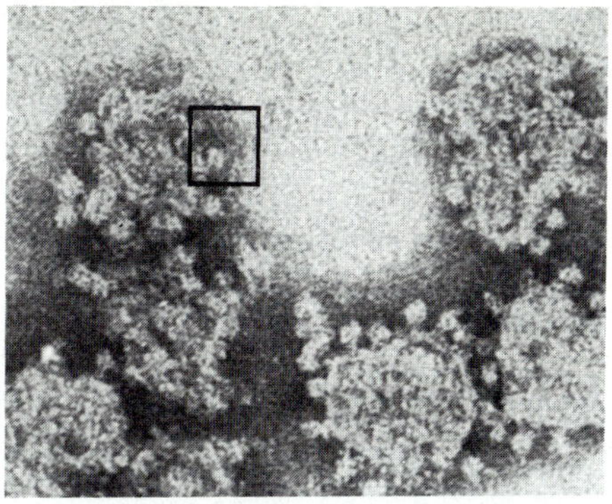

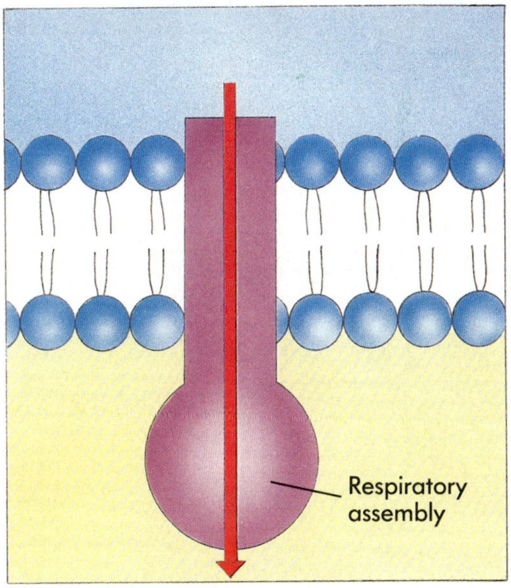

FIGURE 7-12 The respiratory assembly.
Protons in the outer compartment of a mitochondrion pass back into the inner compartment through special channels called *respiratory assemblies*. Each passage of a proton back into the inner compartment through a respiratory assembly is coupled to the synthesis of an ATP molecule.

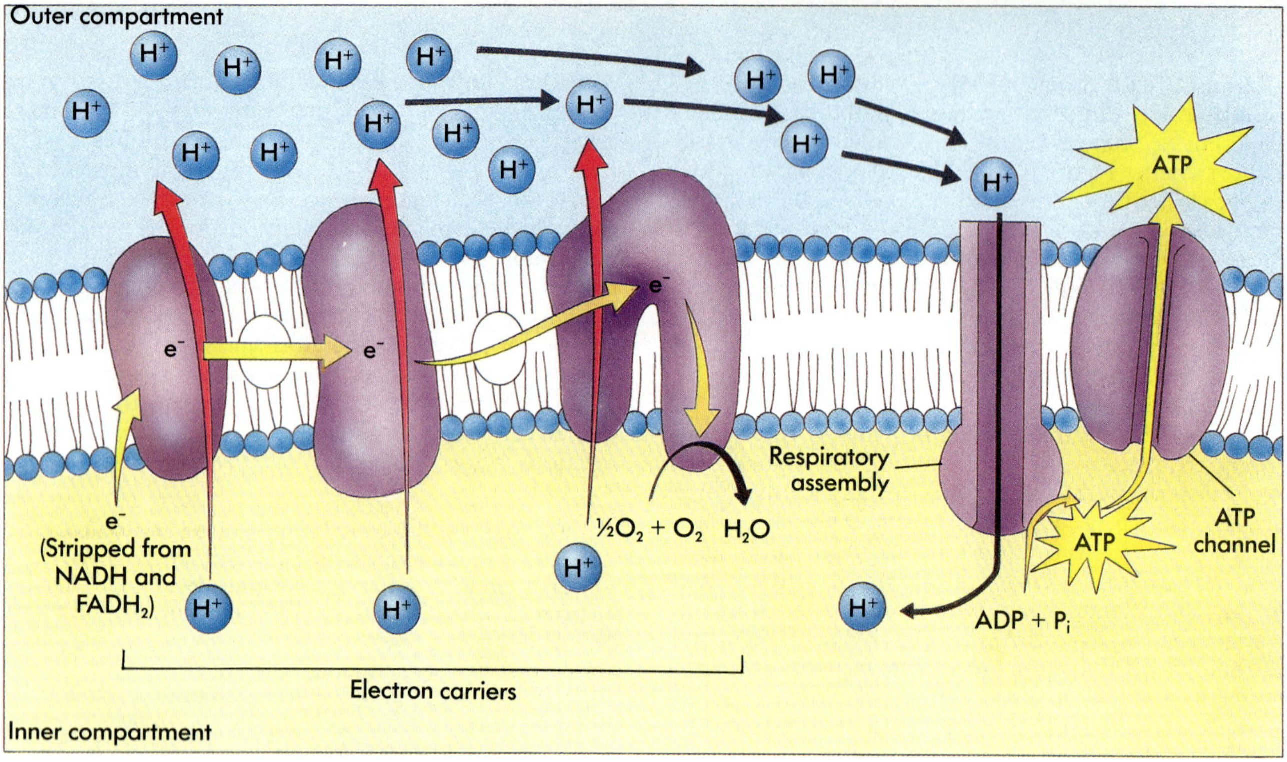

FIGURE 7-13 The electron transport chain.
The NADH and $FADH_2$ formed during the Krebs cycle pass their electrons to electron acceptors located in the inner membrane of the mitochondrion.

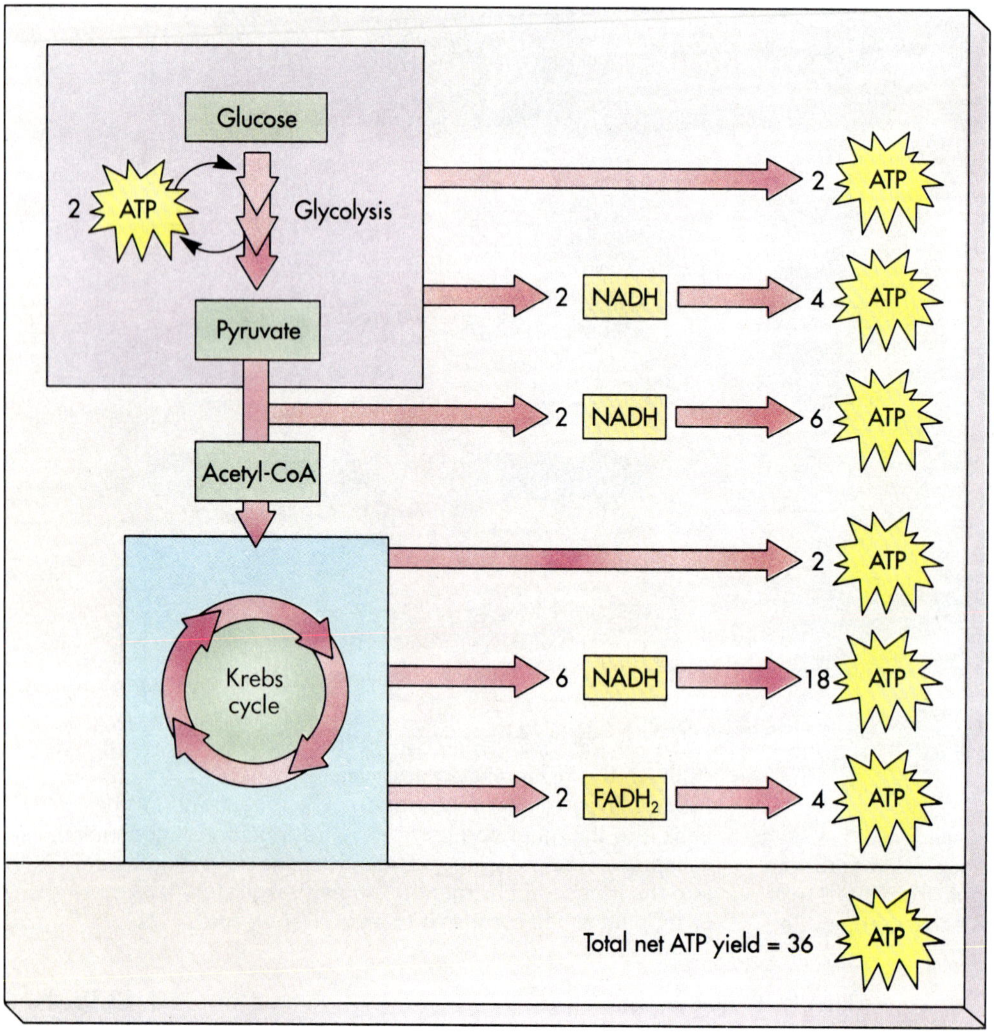

FIGURE 7-14 An overview of the energy extracted from the oxidation of glucose.
During glycolysis, each NADH actually produces three ATP molecules. However, because eukaryotes carry out glycolysis in the cell cytoplasm and the Krebs cycle in their mitochondria, it takes energy to transport their electron-carrying NADH molecules into the mitochondrion—one ATP molecule per NADH, a total of two ATPs per molecule of glucose. Therefore each NADH molecule produced during glycolysis yields only two molecules of ATP instead of three.

Each NADH molecule produced during cellular respiration ultimately causes the production of three ATP molecules, since its electrons activate three pumps. Each $FADH_2$, which activates two of the pumps, leads to the production of two ATP molecules. However, eukaryotes carry out glycolysis in the cytoplasm and the Krebs cycle within the mitochondria. This separation of the two processes within the cell requires transporting the electrons of the NADH created during glycolysis across the mitochondrial membrane, using one ATP molecule per NADH. Thus each glycolytic NADH produces only two ATP molecules instead of three in the final total. Figure 7-14 describes the total energy yield from the aerobic metabolism of one molecule of glucose.

How organisms release ATP from food molecules without using oxygen

In the absence of oxygen, aerobic metabolism cannot take place because of the lack of a final electron acceptor in the electron transport chain. Not only will the reactions of the electron transfer chain come to a halt, the Krebs cycle reactions will not take place because NADH molecules will not be "recycled" to NAD^+ molecules, needed electron acceptors in the cycle. In such a situation, cells must rely on glycolysis to produce ATPs. However, they must regenerate NAD^+ from NADH for use in the glycolytic pathway. During aerobic respiration, cells accomplish this task by means of the reactions of the electron transport chain. During anaerobic respiration, bacteria and yeasts produce an organic molecule from pyruvate that will accept the hydrogen atom from NADH and thus re-form NAD^+. These reactions, including the glycolytic pathway, are termed **fermentation.** The end products of fermentation depend on the organic molecule that is produced from pyruvate.

Bacteria and yeasts carry out many different sorts of fermentations. In fact, more than a dozen fermentative processes have evolved among bacteria, each process using a different organic molecule as the hydrogen acceptor. Often the resulting reduced compound is an acid. In some organisms such as yeasts, an organic molecule called *acetaldehyde* is formed from pyruvate and then reduced, producing an alcohol (Figure 7-15, *A*). This particular type of fermentation is of great interest to people because it is the source of the ethyl alcohol in wine and beer. However, ethyl alcohol is an undesirable end product for yeast because it becomes toxic to the yeast when it reaches high levels. That is why natural wine contains only about 12% alcohol—12% is the amount it takes to kill the yeast fermenting the sugars.

Muscle cells undergo a similar fermentative process when they produce ATP without sufficient oxygen (Figure 7-15, *B*). During strenuous exercise, muscle cells break down large amounts of glucose to produce ATP. However, because muscle cells are packed with mitochondria, the pace of glucose breakdown far outstrips the blood's ability to deliver oxygen for aerobic respiration. Therefore the muscle cells switch from aerobic respiration to fermentation. Muscle cells do not change pyruvate to another organic molecule to be reduced as the bacteria and yeasts do but instead directly reduce pyruvate to lactic acid. This acid builds up

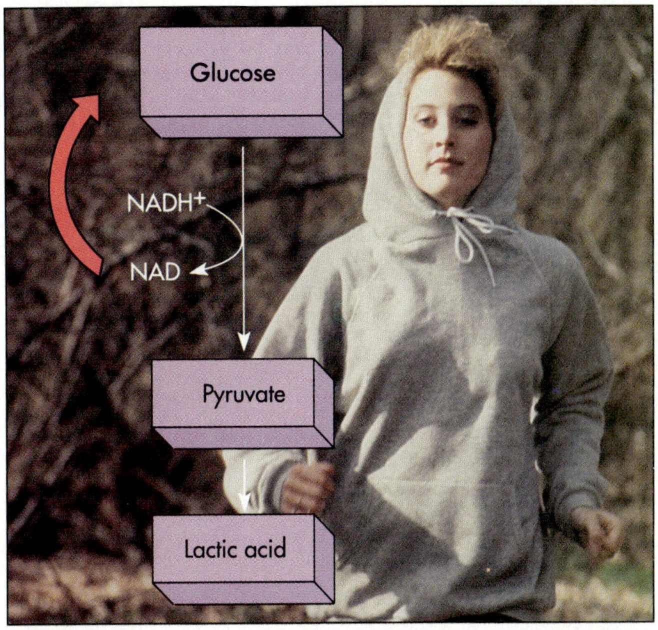

A

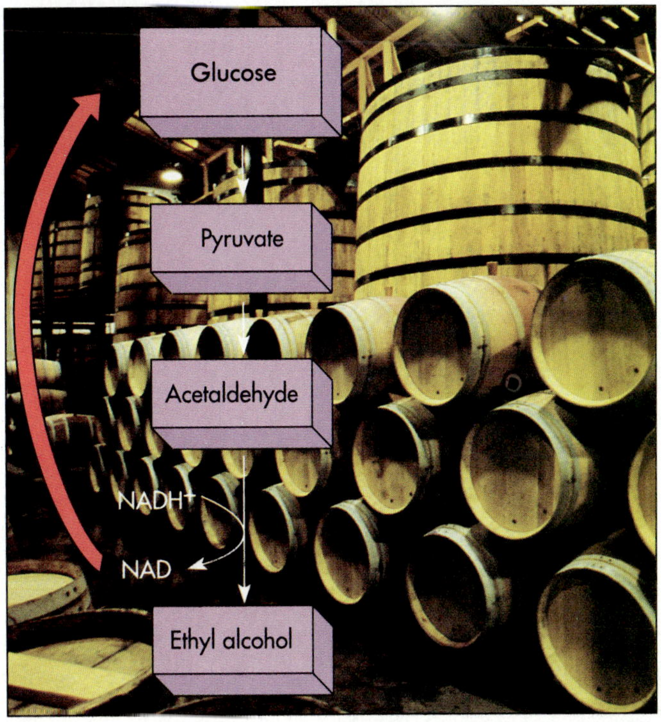

B

FIGURE 7-15 Anaerobic metabolism.
A Yeasts use acetaldehyde as the hydrogen acceptor. The result is the ethyl alcohol of wine and beer.
B When animal cells do not have sufficient oxygen, pyruvate accepts the hydrogen atom from NADH to reform the NAD^+ needed to continue glycolysis. This converts pyruvate into lactic acid, which builds up in the muscles until oxygen becomes available. This is what happens when you exercise strenuously. The ache in your muscles as you run is the result of a buildup of lactic acid.

in the muscle and tends to produce a sensation of muscle fatigue. Gradually, however, the lactic acid is carried by the blood to the liver, where it is broken down aerobically. The oxygen needed to break down this lactic acid produces what is termed an **oxygen debt** and is the reason a person continues to pant after completing strenuous exercise.

In fermentations, which are anaerobic processes, pyruvate molecules or organic compounds produced from pyruvate are reduced, accepting the electrons of NADH generated in the glycolytic breakdown of glucose. These last reactions of the fermentative process produce the NAD^+ needed for the anaerobic breakdown of glucose to continue.

Summary

1. All the activities performed by living things involve chains of reactions that use, move, carry, store, and free energy. These chains of reactions are called *metabolic pathways*. In metabolic pathways, organisms couple exergonic reactions that release energy from molecules such as ATP with endergonic reactions to enable them to take place.
2. Almost all living things, including photosynthetic organisms, make ATP by a process called *cellular respiration*. Certain organisms make ATP by an anaerobic process called *fermentation*.
3. Cellular respiration is composed of three series of chemical reactions: glycolysis, the Krebs cycle, and the electron transport chain. During these reactions, glucose is broken down in the presence of oxygen, capturing energy in molecules of ATP. The by-products of this process are carbon dioxide and water.
4. The energy stored in chemical bonds can be transferred from one atom or molecule to another. In some chemical reactions, electrons pass from one atom or molecule to another. This class of chemical reaction is called an *oxidation-reduction reaction*.
5. NAD^+ and FAD are important electron carriers in living systems. Both become reduced by accepting electrons and protons and can then pass on the electrons and protons they accept to other carriers. Such carriers play important roles in the production of ATP.
6. In the human body, most carbohydrates are changed to glucose and completely metabolized by cellular respiration. Parts of protein and lipids molecules are also metabolized by part of this metabolic pathway.
7. Cellular respiration yields 36 ATP molecules from glycolysis, the Krebs cycle, and the electron transport chain.
8. Organisms that metabolize nutrient molecules anaerobically by the process of fermentation use pyruvate or a molecule derived from pyruvate to accept the electrons produced during the glycolytic breakdown of glucose. This process produces end products that are frequently acids or alcohols.

REVIEW QUESTIONS

1. What is ATP, and why is it important to life?
2. Distinguish between cellular respiration and fermentation. What purpose do they serve?
3. Explain what this formula means. What process does it describe?

$$C_6H_{12}O_6 + 6O_2 \rightarrow 6CO_2 + 6H_2O$$

4. Fill in the blanks: The three series of reactions that make up cellular respiration are ________, ________, and ________.
5. Fill in the blanks: When an atom or molecule gives up an electron, it is ________ and its electrical charge ________. When it gains an electron, it is ________ and its charge ________.
6. Identify the chemical process described in question 5 and explain its significance.
7. You stop at a fast-food restaurant to eat a burger and fries. Explain how your body converts these foods into usable energy.
8. Summarize the changes that occur during glycolysis.
9. The chapter says that glycolysis is inefficient; explain why. If it is inefficient, why do our cells still carry out the process?
10. Summarize what happens during the Krebs cycle.
11. Create a table that summarizes the output of aerobic metabolism for glycolysis, the oxidation of pyruvate, and the Krebs cycle.
12. What is the electron transport chain? Explain its significance.
13. Draw a diagram that summarizes the stages, products, and by-products of cellular respiration. Show which events occur outside the mitochondrion and which take place inside the organelle. Include the number of ATP molecules produced at each stage (see below).

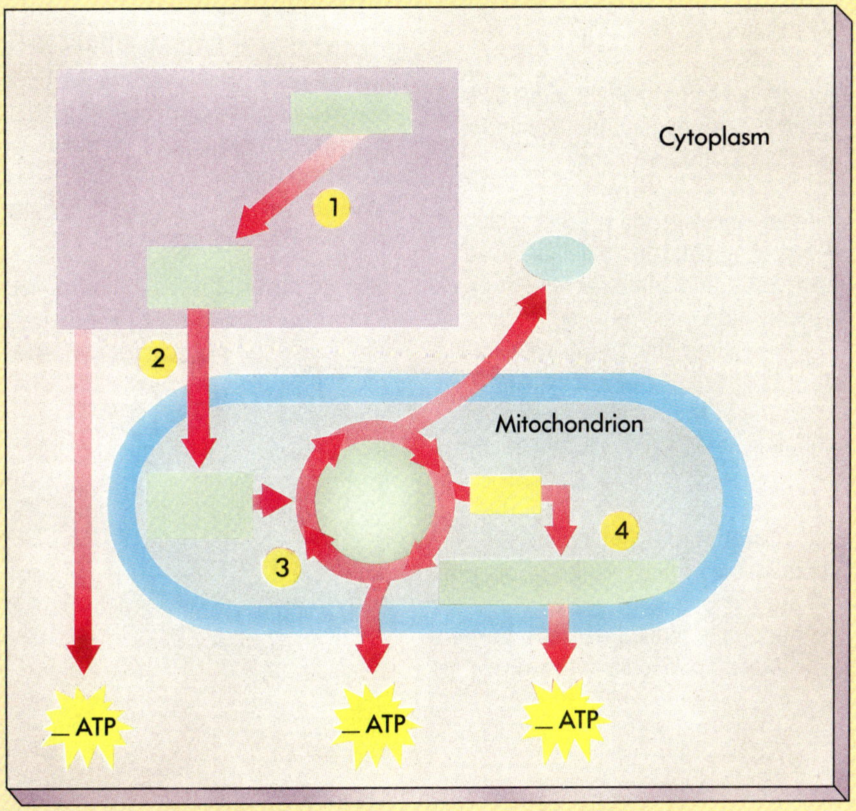

KEY TERMS

acetyl-CoA 123
aerobic 116
anaerobic 116
catabolic process 121
cellular respiration 116
electron carrier 119
electron transport chain 117
fermentation 116
glycolysis 117
Krebs cycle 117
oxidation 119
oxidation-reduction reaction 119
oxygen debt 128
pyruvate 117
reduction 119
respiratory assembly 125

THOUGHT QUESTIONS

1. When a person falls into water and stays submerged for long lengths of time without breathing air, the person usually drowns. However, if the water is very cold, a submerged individual can sometimes survive as long as half an hour under water. Why do you think the temperature of the water plays such a critical role?
2. If you were able to reach within a cell with a microscopic probe and punch a hole in a mitochondrion, would that mitochondrion still "work"—that is, be capable of generating ATP by oxidative respiration? Explain.

FOR FURTHER READING

Dickerson, R. (1980, March). Cytochrome *c* and the evolution of energy metabolism. *Scientific American*, pp. 136-154.
This is a superb description of how the metabolism of modern organisms evolved.

Hinkle, P. & McCarty, R. How cells make ATP. *Scientific American*, pp. 104-125.
This is a good summary of oxidative respiration, with a clear account of the events that happen at the mitochondrial membrane.

Levine, M., Muirhead, H., Stammers, D., & Stuart, D. (1978). Structure of pyruvate kinase and similarities with other enzymes: possible implications for protein taxonomy and evolution. *Nature, 271*, 626-630.
This advanced article recounts how the enzymes of oxidative metabolism may have evolved and is well worth the effort of reading it.

8 Photosynthesis

How plants capture and store energy from the sun

HIGHLIGHTS

▼ Virtually all living things on the surface of the Earth depend on the sun for their existence.

▼ Photosynthesis is the process whereby plants, algae, certain single-celled eukaryotes, and some bacteria produce organic molecules from carbon dioxide by harvesting energy from sunlight.

▼ Some of the complex events of photosynthesis require light energy and some do not.

▼ The summary equations of photosynthesis and cellular respiration are the reverse of one another, although the individual steps do not make up reverse pathways.

OUTLINE

Producers and consumers of food

The energy in sunlight

Capturing light energy in chemical bonds

An overview of photosynthesis

Light-dependent reactions: Making ATP and NADPH

How photosystems capture light

Noncyclic electron flow

Cyclic electron flow

Light-independent reactions: Making glucose

Relationships: Photosynthesis and cellular respiration

An eclipse of the sun is an awesome phenomenon to observe. But did you ever think about what might happen if the light of the sun were blocked from the Earth forever? You might predict that the temperature of the Earth's atmosphere would drop, and you would be correct. Obviously, you might also think, the Earth would become shrouded in darkness. And that's true too. But more devastating consequences would occur—more devastating than the death and destruction that occurred when the atomic bomb was dropped on Hiroshima during World War II. Life on the surface of the Earth—as we know it today—would totally and completely end.

The living organisms of Earth depend on the sun and have for billions of years. Your life depends on a process called *photosynthesis,* a process powered by the sun. By means of photosynthesis, energy from the sun is captured and stored in molecules of food. This food ultimately builds and fuels the bodies of all living things.

FIGURE 8-1 Autotrophs and heterotrophs.
A Photosynthesizer (autotroph). The grass growing under these palms grows actively during the hot, rainy season, and after capturing the energy of the sun, the grass converts the energy to molecules of glucose that are stored as starch.
B Herbivore (heterotroph). These large antelopes, known as *wildebeests*, consume the grass and incorporate some of its stored energy into their own bodies.
C A carnivore (heterotroph). This lion feeds on wildebeests and other animals, incorporating part of the prey's stored energy into their own bodies.

Producers and consumers of food

Plants, algae, certain single-celled eukaryotes, and some bacteria are the members of the living world able to carry out photosynthesis. Photosynthetic organisms harvest light energy from the sun and change it into stored chemical energy within food. Organisms that produce their own food by photosynthesis, along with a few others that use chemical energy in a similar way, are called **autotrophs.** The word *autotroph* literally means "self-feeder." All organisms live on the food produced by autotrophs, including the autotrophs themselves.

Organisms that cannot produce their own food are called **heterotrophs,** literally "other-feeders." Heterotrophs consume other organisms and, ultimately, the food produced by autotrophs. At least 95% of the species of organisms on Earth—all animals, all fungi, and most protists and bacteria—are heterotrophs. They live by feeding on the chemical energy that is fixed (incorporated into organic molecules) by photosynthesis. Therefore all living things—plants, bacteria, and you—share the same ultimate dependency on the sun (Figure 8-1).

Most autotrophs manufacture their own food by converting the energy in sunlight into chemical energy by means of the process of photosynthesis. Heterotrophs cannot make their own food and ultimately depend on autotrophs—and the sun—for survival.

The energy in sunlight

How do photosynthetic organisms "harvest" light energy from the sun? And how do they use this energy to create chemical bonds in food molecules? First, it is important to understand what light energy is.

As you may recall from Chapter 6, energy can exist in many forms, such as mechanical force, heat, sound, electricity, light, radioactivity, and magnetism. All these forms of energy are able to create change—to do work. Sunlight is a form of energy known as **electromagnetic energy,** also called **radiation.**

Electromagnetic energy travels as waves. The waves can be envisioned as ripples on the surface of a still pond after a pebble has been thrown into the water. Electromagnetic "ripples," however, are disturbances in electric and magnetic fields rather than in water.

The array of electromagnetic waves coming from the sun varies in length. Their lengths are measured from the crest of one wave to the crest of the next and are expressed in meters or in billionths of meters (nanometers, or nm). The shortest wavelengths are gamma rays; the longest are radio waves.

The full range of electromagnetic radiation in the universe is called the **electromagnetic spectrum** (Figure 8-2). Visible light is only a small part of this spectrum. But it too is made up of many different wavelengths. Remember the last rainbow you saw? Its array of colors was caused by the separation of the various wavelengths of visible light as they passed through tiny droplets of water in the air.

Although electromagnetic energy travels as waves, it also behaves like individual particles—discrete "packets" of energy—called **photons.** The energy content of each photon is inversely proportional to the wavelength of the radiation. Put simply, short-wavelength radiation (such as gamma rays and x-rays) contains photons of a high energy level; long-wavelength radiation (such as microwaves and radio waves) contains photons of a low energy level. Within the range of visible light, then, the shorter wavelengths of violet light carry a greater "energy punch" than the longer wavelengths of red light.

Sunlight can be described both as waves, or disturbances in electric and magnetic fields, and as discrete packets of energy called *photons.*

Capturing light energy in chemical bonds

A molecule "captures" the energy of sunlight when a photon of energy boosts an electron of the molecule to a higher energy level than it already occupies. This electron is then said to be "excited." (See Chapter 2 for a description of the energy levels of electrons.) Boosting an electron to a higher energy level requires just the right amount of energy—no

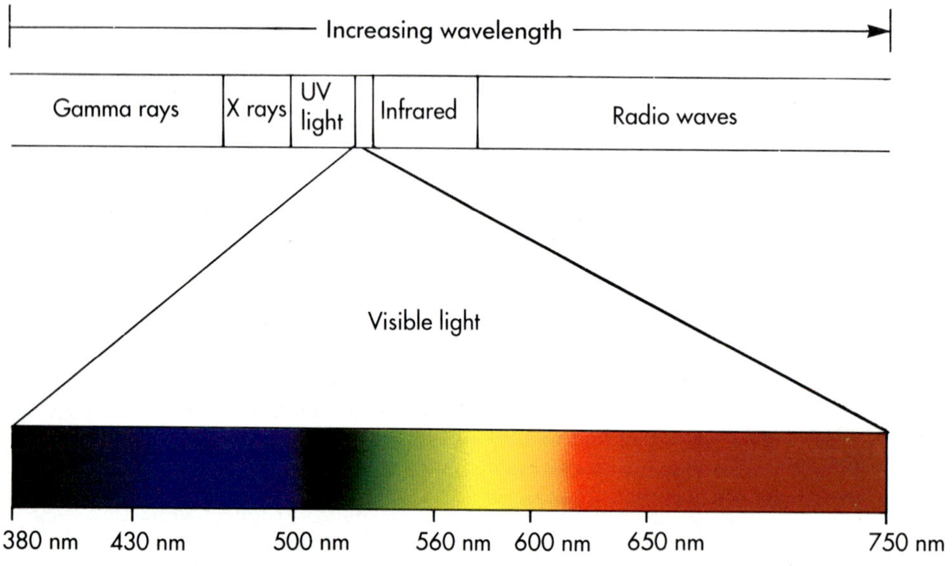

FIGURE 8-2 The electromagnetic spectrum.
Light is a form of electromagnetic energy and is conveniently thought of as a wave. The shorter the wavelength of light, the greater the energy. Visible light represents only a small part of the electromagnetic spectrum, that between 380 and 750 nanometers.

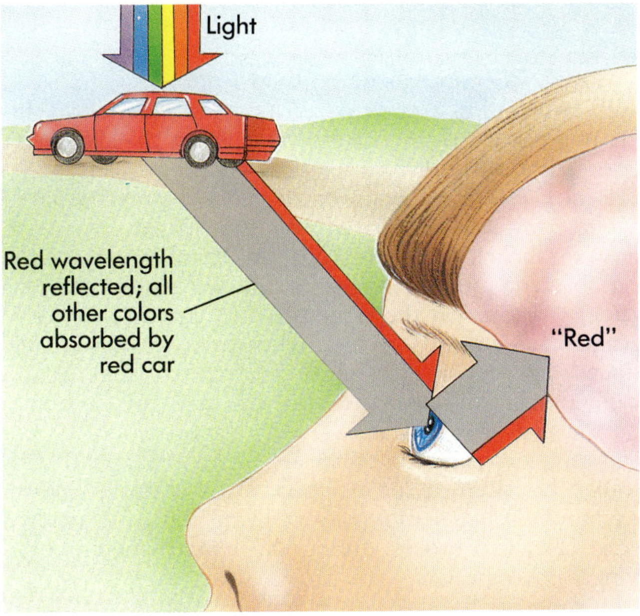

FIGURE 8-3 How you see color.
A red car is painted with a pigment that absorbs all the wavelengths of light *except* red. Red wavelengths bounce off the pigment and are reflected back to your eyes, where special cells in your eyes absorb the red light. The brain then interprets impulses sent by these cells as red.

more and no less. For example, when you climb a ladder, you must raise your foot just so far to climb a rung, not 1 centimeter more or less. Likewise, specific atoms can absorb only certain photons of light—those that correspond to available energy levels. Therefore a given atom or molecule has a characteristic range, or **absorption spectrum,** of photons it is capable of absorbing depending on the electron energy levels that are available in it.

Molecules that absorb some visible wavelengths of light and reflect others are called **pigments.** Pigments are responsible for all the colors you see. For example, a car looks red because it is painted with a pigment that absorbs all the wavelengths of light *but* red. The red wavelengths of light bounce off the pigment and are reflected back to your eyes. Pigments located in special cells within your eyes then absorb the photons of this red light. The resulting electron excitations ultimately generate nerve impulses that are sent to the brain. The brain interprets these impulses as red (Figure 8-3). Likewise, white objects absorb no light and reflect all the wavelengths of visible light to your eye. A black object absorbs all wavelengths, reflecting none.

A *pigment* is a molecule that absorbs some wavelengths of light and reflects others. The wavelengths absorbed by a particular pigment depend on the energy levels available in the molecule to which light-excited electrons can be boosted.

Organisms have evolved a variety of pigments. Two groups of pigments that are important to the process of photosynthesis are the **carotenoids** and the **chlorophylls.** Carotenoids absorb photons of green, blue, and violet wavelengths and reflect red, yellow, and orange. Chlorophylls absorb photons of violet-blue and red wavelengths and reflect green and yellow (Figure 8-4). Chlorophyll is used as the primary light gatherer in all plants and algae and in

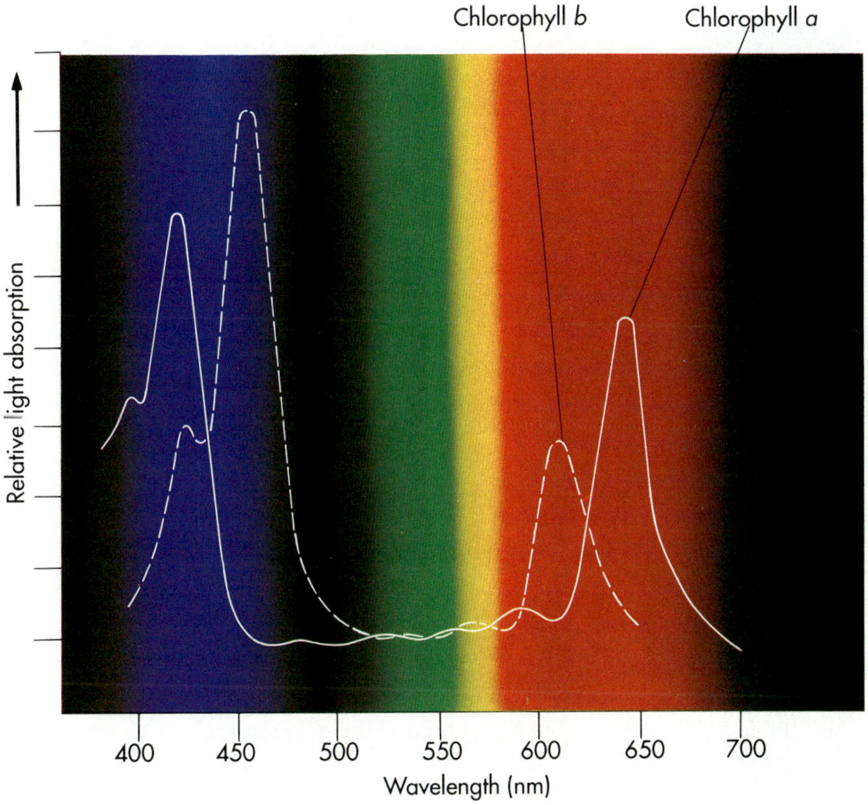

FIGURE 8-4 Absorption spectra for chlorophylls.
Chlorophylls, such as chlorophyll *a* and chlorophyll *b*, absorb predominantly violet-blue and red light in two narrow bands of the spectrum.

certain photosynthetic bacteria—the cyanobacteria. The carotenoids are important, however, because they absorb wavelengths of light chlorophyll cannot and then pass the energy to chlorophyll, thereby increasing the spectrum of light that can be absorbed. Chlorophyll masks the presence of the carotenoids, which only become visible when chlorophyll breaks down. This breakdown occurs in autumn (in the areas of the world that have seasons) as the days shorten and the amount of sunlight decreases. As a result, the carotenoids become visible, providing the magnificent yellows and oranges of fall foliage. The bright reds are caused by the unmasking of other non–light-gathering pigments.

An overview of photosynthesis

The process of photosynthesis is described chemically in the following equation:

$$6CO_2 + 6H_2O + \text{Light energy} \rightarrow C_6H_{12}O_6 + 6O_2$$

Carbon dioxide + Water + Light energy → Glucose + Oxygen

This chemical equation represents the beginning and end points in a series of chemical reactions and merely summarizes complex events. It shows that the net effect of photosynthesis is the production of one molecule of glucose and six molecules of oxygen from six molecules of carbon dioxide and six molecules of water in the presence of sunlight.

A more accurate summary of the chemical reactions of photosynthesis is represented by the following equation:

$$6CO_2 + 12H_2O + \text{Light energy} \rightarrow C_6H_{12}O_6 + 6H_2O + 6O_2$$

Here, water appears on both sides of the equation—as a substrate and as a product. Writing the equation in this way points out that the water molecules on the left side of the equation and those on the right are *not* the same water molecules. The six molecules of oxygen (12 atoms) produced by photosynthesis are derived from those in the water molecules on the left side of the equation. Plants, algae, and the cyanobacteria produce oxygen in this way. In addition, the water molecules on the right side of the equation derive their oxygen atoms from the molecules of carbon dioxide. Glucose (an organic molecule) is constructed from the carbon and oxygen in carbon dioxide (an inorganic molecule) and the hydrogen in water. During the process of photosynthesis, inorganic molecules are broken apart and their atoms are reshuffled and put back together again to form organic molecules—molecules of living things. Photosynthesis, then, is a process during which the living and nonliving worlds meet.

Photosynthesis, an energy-requiring series of reactions, produces glucose from carbon dioxide and water. In addition, plants, algae, and cyanobacteria release oxygen as a by-product of photosynthesis.

The complex events of photosynthesis involve three different kinds of chemical processes (Figure 8-5). During the first process, ATP and NADPH molecules are made using energy captured from sunlight. As you learned in Chapter 6, ATP is the primary molecule used by cells to capture and carry energy and to supply energy at a moment's notice. NADPH is an electron carrier. It behaves much like the NADH of cellular respiration (see Chapter 7). Its oxidized form, $NADP^+$, can accept two electrons and one proton, or H^+, to become reduced to NADPH. The electrons $NADP^+$ accepts contain a great deal of energy, thus producing molecules that carry energy in the form of energized

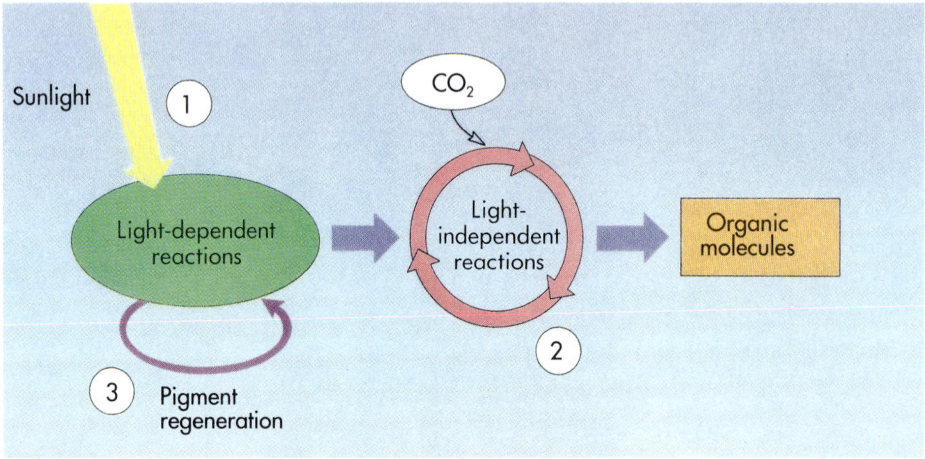

FIGURE 8-5 Overview of the three stages of photosynthesis.
The light-dependent reactions produce NADPH and ATP. The light-independent reactions use this NADPH and ATP to drive the formation of glucose from carbon dioxide. Pigment regeneration replenishes the pigment that absorbed the light so that it can begin another light-dependent reaction.

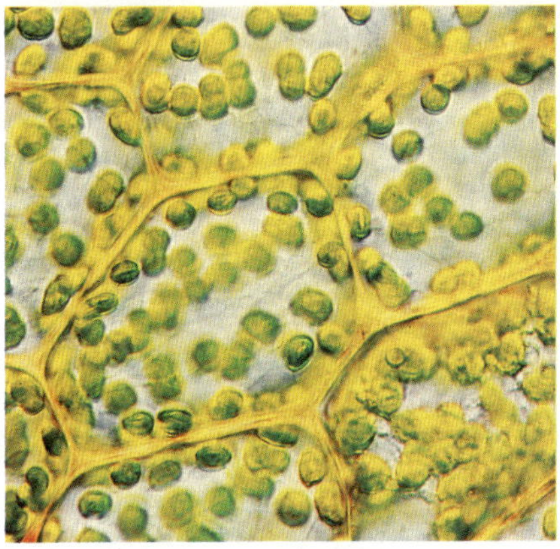

FIGURE 8-6 All the reactions of photosynthesis take place in chloroplasts.
These moss cells are densely packed with bright green chloroplasts.

electrons. The reactions of photosynthesis that produce ATP and NADPH are called the **light-dependent reactions** because they take place only in the presence of light.

Second, a series of reactions uses this newly made ATP and NADPH to drive the formation of glucose from carbon dioxide. These reactions are called the **light-independent reactions** because they do not need light to take place as long as ATP and NADPH are available. Third, the pigment that absorbed the light in the first place is readied to begin another light reaction.

The reactions of photosynthesis take place on **photosynthetic membranes.** In photosynthetic bacteria these membranes are infoldings of the cell membrane. In plants, algae, and the photosynthetic single-celled protists called *euglenoids*, all the reactions of photosynthesis are carried out within cell organelles called **chloroplasts** (Figure 8-6). Chloroplasts, which were probably derived evolutionarily from photosynthetic bacteria, have a system of internal membranes. These membranes are organized into flattened sacs called **thylakoids.** Stacks of thylakoids are referred to as **grana** (Figure 8-7). Each thylakoid is a closed compartment, a fact that plays an important role in the formation of ATP. Surrounding the thylakoids is a fluid called the **stroma.** It contains the enzymes of the light-independent reactions.

Photosynthesis takes place in the chloroplasts of photosynthetic eukaryotic cells. It involves three processes: the synthesis of ATP and NADPH, the use of ATP and NADPH to produce glucose, and the recycling of the photosynthetic pigment.

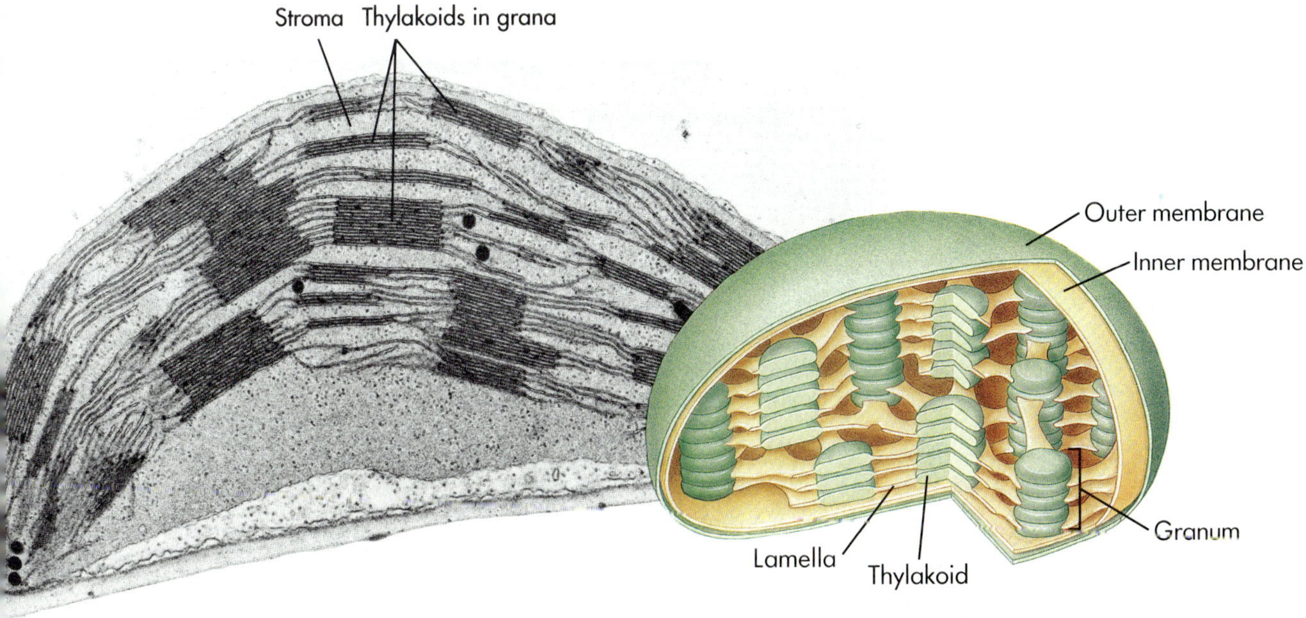

FIGURE 8-7 Structure of a chloroplast.
Within a chloroplast the photosynthetic membranes are organized into closed sacs called *thylakoids.* These thylakoids are stacked into structures called *grana* (singular, granum). (It should be noted, however, that certain groups of algae lack grana.) The fluid surrounding the grana is called the *stroma.* The stroma contains the enzymes necessary for the light-independent reactions. Lamellae (singular, lamella) are the membranes connecting the grana.

BIOLOGY IN FOCUS

C_4 and CAM photosynthesis

The method of photosynthesis presented in this chapter is C_3 photosynthesis, since the first product produced in this method is a three-carbon compound. There are two additional types of photosynthesis, and they are each used by two very different types of plants.

The C_4 photosynthetic pathway is used by plants that favor high light intensity, high temperature, and low soil moisture—plants and weeds such as corn, sugarcane, and pigweed that evolved in the dry tropics (Figure 8-A). In C_4 photosynthesis, CO_2 enters the plant through the stomata, tiny openings on the surface of the leaves that open and close to admit water and CO_2. When CO_2 is absorbed on the wet surface of the internal cells of the leaf, it forms carbonic acid, which immediately dissociates (separates) into H^+ and HCO_3^- (bicarbonate). The bicarbonate is then captured by the enzyme PEP carboxylase. PEP carboxylase connects the carbon in the HCO_3^- to phosphoenolpyruvate (a three carbon molecule, or PEP for short), forming a four-carbon molecule. It is the formation of this four-carbon molecule that lends C_4 photosynthesis its name. The four-carbon molecule is then decarboxylated (carbon is dropped off) to form a three-carbon molecule, which then enters into the normal Calvin cycle pathway. Although C_4 photosynthesis seems like a cumbersome way to generate three carbon molecules, it is actually a very efficient system that concentrates carbon under conditions of low CO_2 in the atmosphere, conditions that are found in the dry tropics.

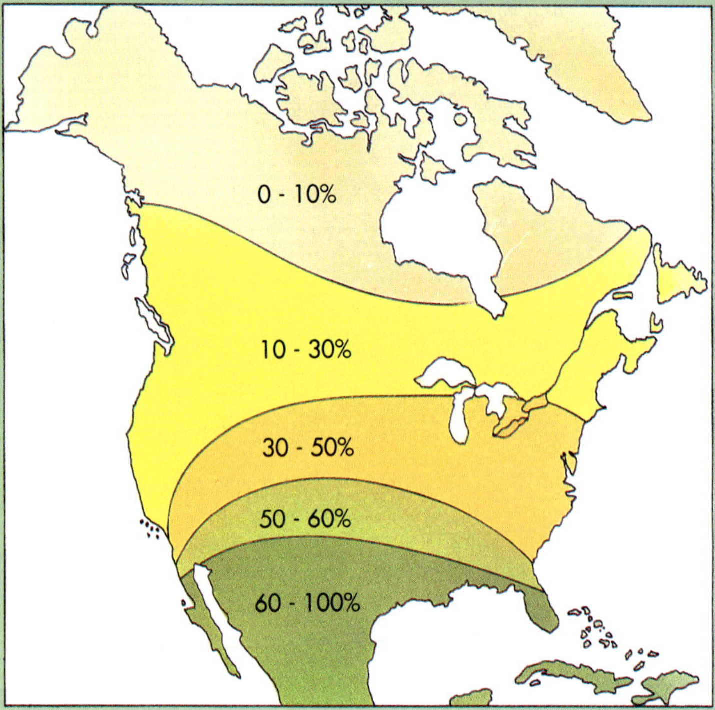

FIGURE 8-A Distribution of C_4 grasses in North America. Many more C_4 grasses occur toward the south, where the average temperatures during the growing season are higher.

CAM photosynthesis, which is named in part for a variety of desert plants called the *Crassulaceae,* uses many of the same biochemical aspects as C_4 photosynthesis. PEP carboxylase is used to transform carbon into a four-carbon compound. Instead of being immediately decarboxylated and then entering into the Calvin cycle, the four-carbon compound is stored until the next day, when it enters into the Calvin cycle. Why should a plant absorb CO_2, transform it into a four-carbon compound, and then feed it into the Calvin cycle the next day? The answer lies in the fact that CAM plants open their stomata only at night, in contrast to C_3 and C_4 plants, which open their stomata during the day and night. When sunlight reaches a CAM plant, the stomata close and CO_2 cannot enter the plant. CAM photosynthesis allows CAM plants to survive in extremely arid regions.

Light-dependent reactions: Making ATP and NADPH

The light-dependent reactions of photosynthesis perform the important job of changing the energy in sunlight into usable chemical energy: ATP and NADPH. These molecules are used in the light-independent reactions to make glucose from carbon dioxide. The first event leading to ATP and NADPH production is the capturing of a photon of light by a pigment.

How photosystems capture light

In all but the most primitive bacteria, light is captured by individual networks of chlorophylls, carotenoids, and other pigment molecules. These networks, located in the thylakoid membranes of the chloroplasts, are called **photocenters.** The pigment molecules within the photocenters are arranged in such a way that they act like sensitive "antennae." These antennae capture and funnel photon energy to special molecules of chlorophyll called **chlorophyll *a*.** The variety of

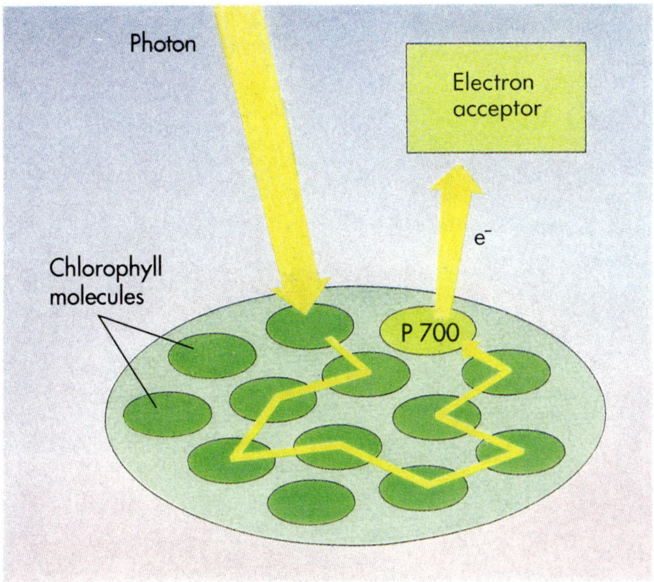

FIGURE 8-8 How photocenters work.
When light of the proper wavelength strikes any chlorophyll molecule within a photocenter, the photon's energy is absorbed by it. The energy passes from one chlorophyll molecule to another until it encounters P700, which channels the energy out of the photocenter to the first electron acceptor in the electron transport chain.

pigments in the photocenters allows a plant to capture energy from a range of wavelengths of light and use it to power photosynthesis.

When light of the proper wavelength strikes any pigment molecule of the photocenter, one of its electrons is boosted to a higher energy level. This pigment molecule passes the energy, or the high-energy electron, to a neighboring molecule within the photocenter. The energy is passed from one pigment molecule to the next, until it reaches the molecule of chlorophyll *a* that will participate in photosynthesis. This molecule is called **P700.** The *P* stands for pigment. The *700* refers to the fact that this molecule of chlorophyll *a* absorbs wavelengths in the 700 nanometer range, which is in the far-red portion of the light spectrum (Figure 8-8).

A simple analogy to the form of energy transfer in the photocenter is the initial break in a game of pool. If the cue ball squarely hits the point of the triangular array of 15 pool balls, the two balls at the far corners of the triangle fly off. None of the central balls move at all. The energy is transferred through the central balls to the most distant ones. In a similar way, the pigment molecules of the photocenter channel excitation energy to a molecule of chlorophyll *a* in the form of excited electrons.

A photocenter is an array of pigment molecules within the thylakoid membranes of a chloroplast. It acts like a light antenna, capturing and directing photon energy toward a single molecule of chlorophyll *a* that will participate in photosynthesis.

Plants and algae have two different types of photosystems that play a role in photosynthesis. These photosystems are named after the order in which they evolved in photosynthetic organisms: **Photosystem I** and **Photosystem II.** In Photosystem I, energy is transferred to a molecule of chlorophyll *a*, P700 (see Figure 8-6). In Photosystem II, energy is transferred to a molecule of chlorophyll *a* called **P680,** whose name refers to its absorption peak of 680 nanometers. These two kinds of chlorophyll *a* act as reaction centers in each type of photosystem.

Noncyclic electron flow

Both photosystems absorb light at the same time. However, to understand the events of the light-dependent reactions, it is easiest to first describe what happens in Photosystem II. When two photons of light are absorbed by pigment molecules in Photosystem II, the energy is funneled to P680. This energy causes two electrons to be ejected from the P680 molecule. The electrons are captured by an electron acceptor located within the thylakoid membrane. However, an electron "hole" is now left in the P680 molecule. The P680 pulls two electrons from a molecule of water (H_2O), normally a hard task to accomplish. The P680 strongly attracts these electrons, however, causing the molecule of water to be split into two hydrogen ions and an oxygen atom. Oxygen atoms from two split molecules of water quickly join to form a molecule of oxygen (O_2)—a by-product of photosynthesis and the gas that helps keep us alive (Figure 8-9).

Meanwhile, the electron acceptor passes the two energized electrons from the P680 to the next in a series of electron acceptors. This series of electron acceptors or carrier

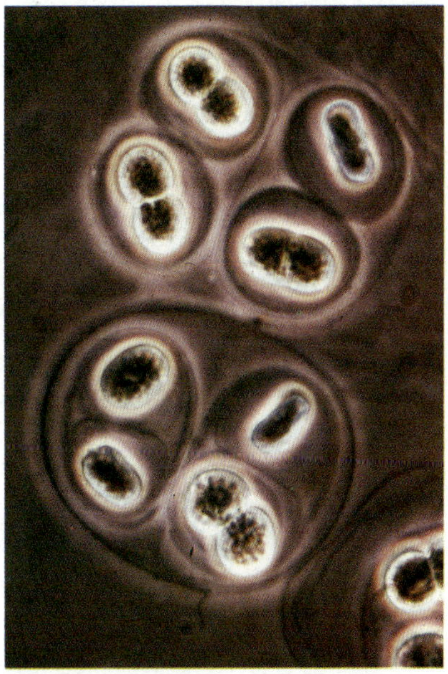

FIGURE 8-9 All of the oxygen you breathe, indeed all the oxygen in the Earth's atmosphere, has been generated by photosynthesis. In the roughly 3 billion years since the advent of the cyanobacteria, enough oxygen has accumulated to account for 21% of the atmosphere.

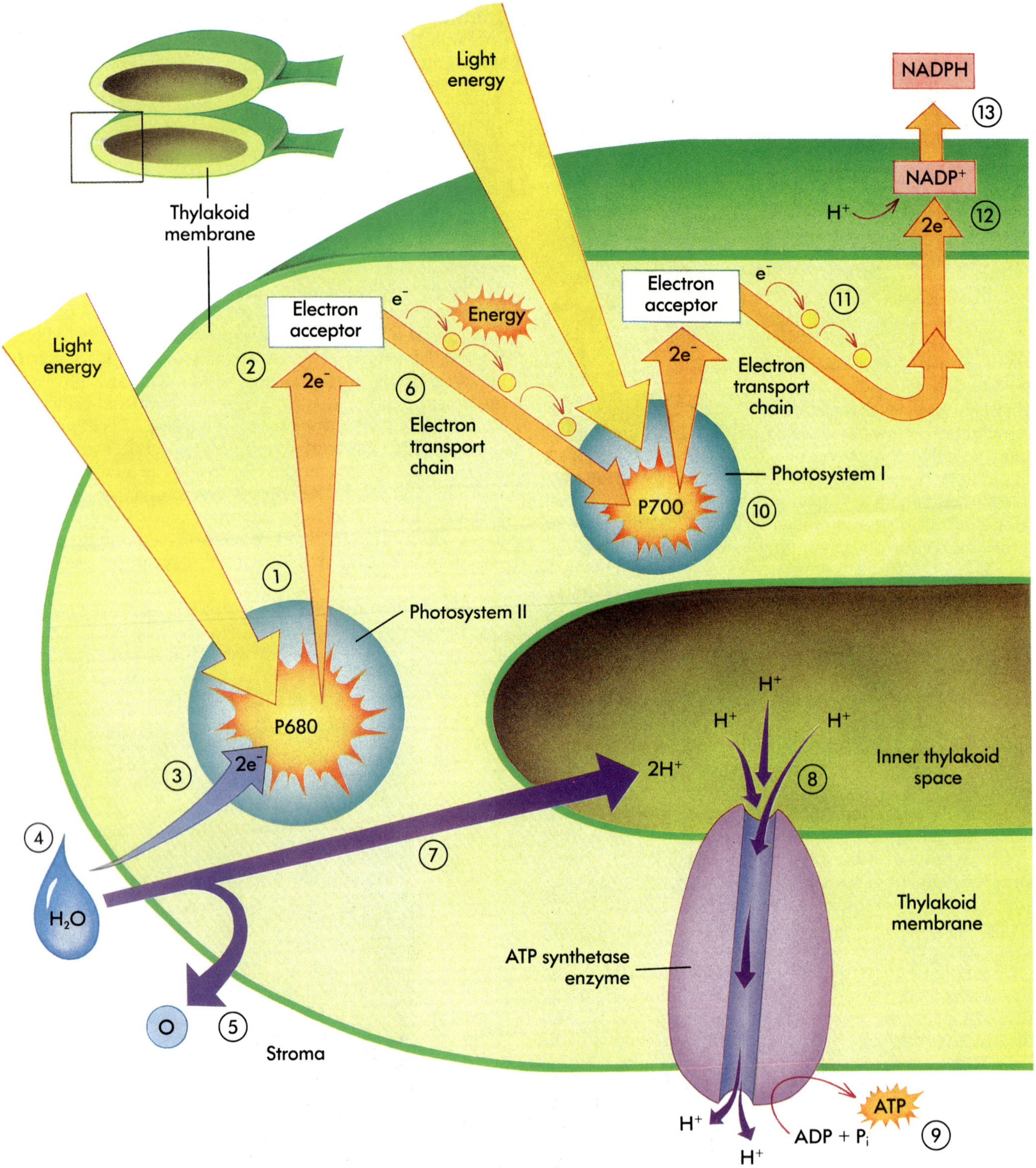

FIGURE 8-10 Light-dependent reactions: noncyclic electron flow.
Noncyclic electron flow harvests energy in a process that involves Photosystems I and II working together to produce NADPH and ATP. These two products are used as the energy source in the light-independent reactions to power the generation of glucose from carbon dioxide.

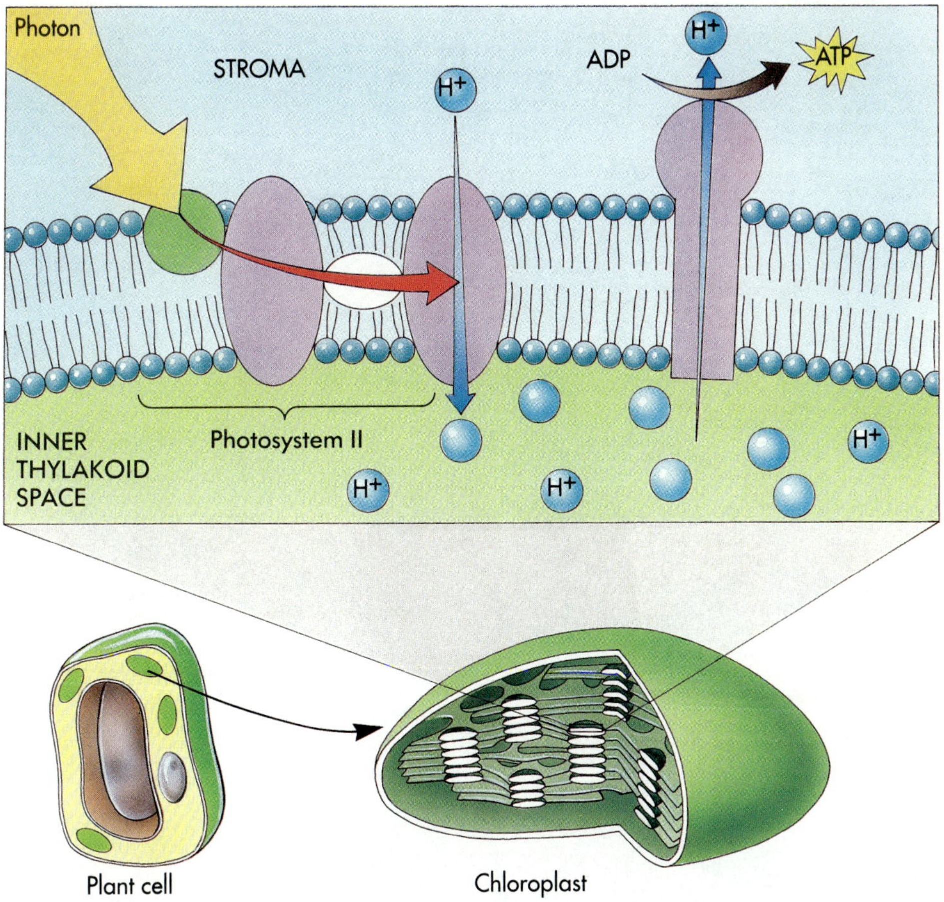

FIGURE 8-11 How ATP is made in a chloroplast.
As a high concentration of protons builds up in the thylakoid space, protons are pumped out of the space through special channels. Enzymes couple the force of this proton movement to the synthesis of ATP from ADP and P_i.

proteins is called an *electron transport chain* as in cellular respiration (see Chapter 7). Here, electrons are passed from one carrier protein to another. As each of these carriers accepts electrons in turn, the electrons fall to lower energy levels, releasing energy in the process. The energy released from the electrons being passed along the chain pumps the protons (H^+) left over from the splitting of water molecules through the thylakoid membrane and into the interior of the thylakoid compartment (Figure 8-10). As a high concentration of protons builds up within this thylakoid space, they begin to diffuse back through the membrane. But the only way they can get back out is through special channels, visible in electron micrographs as knobs on the "outside" surface of the thylakoid membrane. As the protons travel outward through these channels, enzymes couple the energy from the movement of the protons to the reaction of bonding P_i to ADP, forming ATP (Figure 8-11).

When the electrons finish their journey through the electron transport chain, they are then accepted by the chlorophyll *a* molecule of Photosystem I—P700. This chlorophyll molecule, as you recall, has its electrons energized by photons of light at the same time as P680 is energized. So it, too, develops electron holes as energized electrons are ejected from their orbits. The electrons of P680, now devoid of their "boost" of energy, fill these electron holes. The energized electrons of P700 have meanwhile been passed down the very short electron transport chain of Photosystem I. ATP molecules are not generated in this transport chain. Instead, the last acceptor is a molecule of $NADP^+$, which picks up a proton in addition to the two electrons to become reduced to NADPH. Because the electrons it accepts are still highly energized, NADPH serves as a storage depot for this energy. This molecule is considered a source of "reducing power" for the light-independent reactions. The processes that occur in Photosystems I and II make up the light-dependent reactions of eukaryotic photocenters.

Plants and algae use a two-stage photosystem to carry out the light-dependent reactions of photosynthesis. During noncyclic electron flow, energized electrons ejected from Photosystem II are passed down an electron transport chain, triggering events of ATP production. The electron hole in Photosystem II is filled by electrons from the breakdown of water, a process that releases oxygen as a by-product. Energized electrons ejected from Photosystem I, along with hydrogen ions, reduce $NADP^+$ to NADPH.

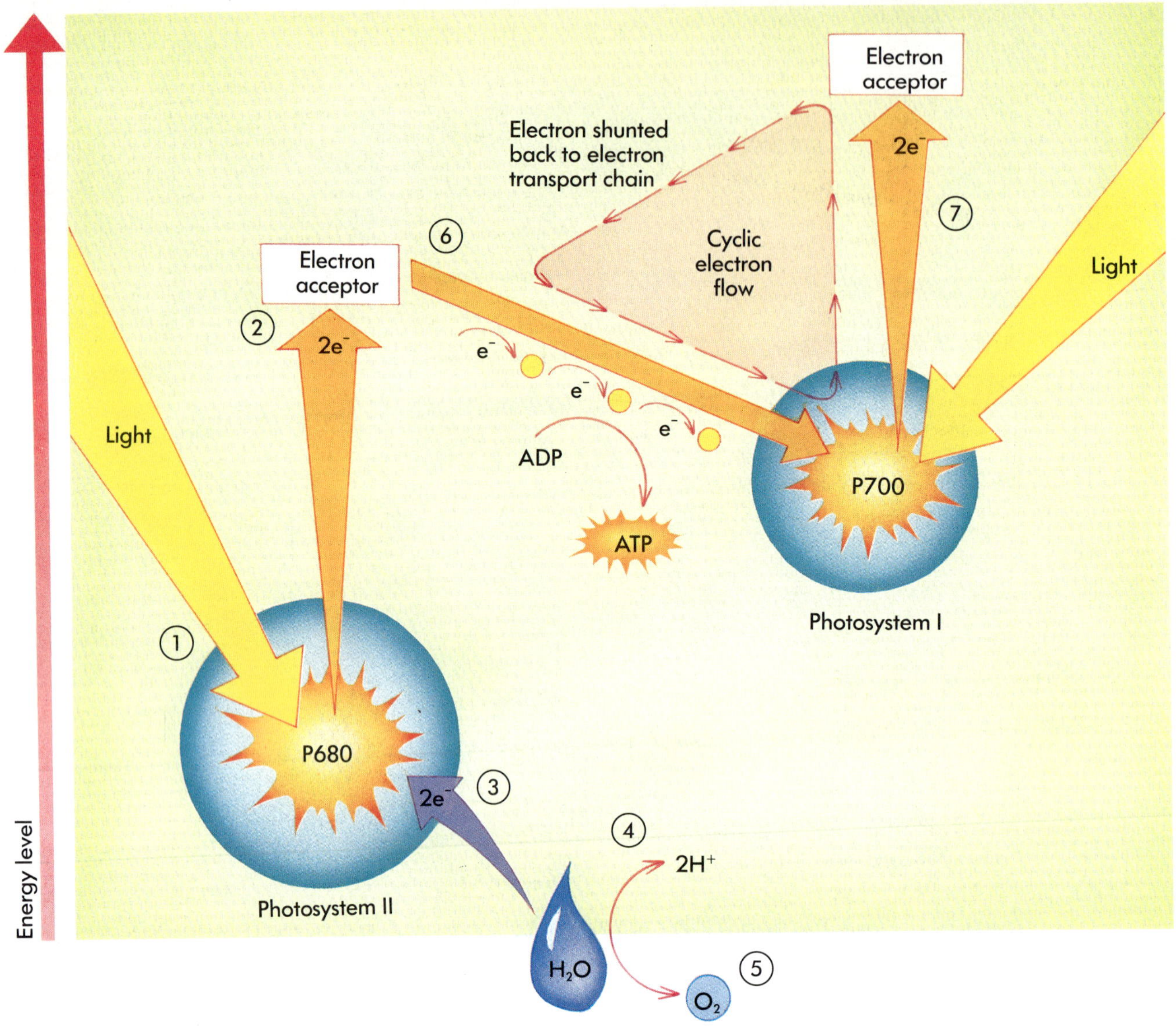

FIGURE 8-12 Light-dependent reactions: cyclic electron flow.
Cyclic electron flow harvests energy in a process that involves Photosystem I working independently of Photosystem II. This method of generating energy produces ATP, but it does not directly produce NADPH as noncyclic electron flow does.

Cyclic electron flow

Sometimes Photosystem I works on its own in a modified way, independent of Photosystem II (Figure 8-12). In this situation, the excited electrons ejected from the P700 molecule are not accepted by $NADP^+$. Instead, they are shunted to the electron transport chain that connects Photosystems I and II. Here, the electrons are passed along the chain, triggering the events of ATP production. The electrons, no longer in an excited state, are then accepted back by P700 to complete a cycle.

Cyclic electron flow provides eukaryotic organisms with ATP when there is no need for additional NADPH. Many photosynthetic bacteria (groups that are different from the cyanobacteria mentioned previously) have only a single photosystem (Photosystem I) and always produce ATP by means of cyclic electron flow. They do not, however, split molecules of water but use other inorganic molecules such as hydrogen sulfide (H_2S) instead. Simple bacteria were probably the first organisms to harness the energy of sunlight to produce ATP with the evolution of Photosystem I. However, even though these bacteria could produce ATP, they did not have a ready means of generating NADPH. They had to make NADPH in a roundabout way and expend a lot of ATP to do so.

About 3 billion years ago, a great evolutionary advance took place when the cyanobacteria originated. These bacteria evolved Photosystem II and solved the NADPH problem. Today, plants and algae, unlike most photosynthetic bacteria, use a two-stage photosystem.

Photosystem I, the first photosynthetic pathway to evolve, is used by most bacteria to produce ATP using energy from the sun. Photosynthetic organisms having a two-stage photosystem can use Photosystem I in a modified way to produce ATP without the production of NADPH.

Light-independent reactions: Making glucose

The light-dependent reactions of photosynthesis described in the preceding section use light energy to produce metabolic energy in the form of ATP and to produce reducing power in the form of NADPH. But this is only half the story. Photosynthetic organisms use the ATP and NADPH produced by the light-dependent reactions to build organic molecules from atmospheric carbon dioxide, a process called **carbon fixation.** This phase of photosynthesis does not need light to drive its reactions. It is therefore referred to as the *light-independent reactions*. In addition, it is called the **Calvin cycle** after the researcher who discovered its steps. In plants, algae, and the euglenoids, these reactions are carried out by a series of enzymes located in the stroma of the chloroplast.

The six-step Calvin cycle has three stages, which produce one three-carbon sugar from three molecules of carbon dioxide (Figure 8-13). The three-carbon sugar that is made during the Calvin cycle is called **glyceraldehyde phosphate.** Two molecules of this sugar are used to manufacture one

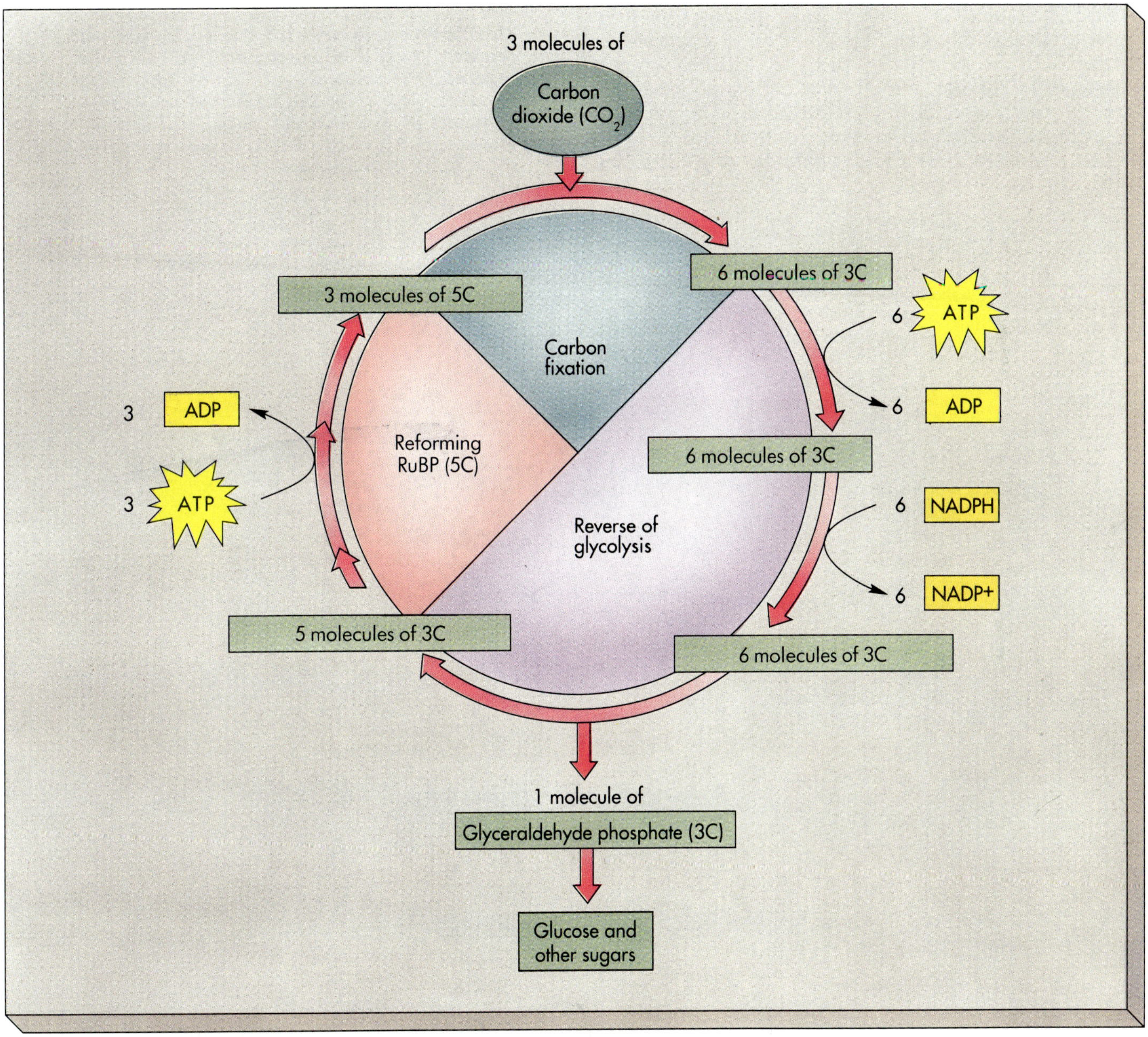

FIGURE 8-13 The Calvin cycle.
For every three molecules of carbon dioxide that enter the cycle, one molecule of the three-carbon compound glyceraldehyde is produced. Notice that the process requires energy, supplied as ATP and NADPH. ATP and NADPH are generated by the light-dependent reactions.

glucose molecule and other sugars. Referring back to the equation of photosynthesis, it takes six molecules of carbon dioxide to produce one molecule of glucose. For one molecule of glyceraldehyde phosphate to be produced, the cycle must take place three times, and it must take place six times to produce one molecule of glucose.

In the first stage of the Calvin cycle, carbon dioxide joins a five-carbon molecule called **ribulose bisphosphate (RuBP)** from the "end" of the cycle. This process is called *carbon fixation*. That is, the inorganic carbon molecule carbon dioxide is attached to a biological molecule, making it available for use to synthesize glucose and other complex carbohydrates. For three turns of the cycle, three carbon dioxide molecules join with three RuBP molecules and form three six-carbon molecules. However, these molecules are unstable. In the second stage of the cycle, these unstable molecules split quickly into six three-carbon molecules that are "chemically reshuffled" to produce glyceraldehyde phosphate molecules. Only one of these six glyceraldehyde phosphate molecules is used in the manufacture of glucose (and other sugars). The other five are used to resynthesize three molecules of RuBP (the third stage of the cycle). These molecules of RuBP are needed to keep the cycle going. In this way, RuBP is used as a "handle" to fix carbon dioxide, and the three-carbon molecules produced are used to build sugars for energy and storage and to reconstruct or recycle molecules of RuBP.

Interestingly, the three-carbon molecules formed during the first step of the Calvin cycle are also found in the process of glycolysis (see Chapter 7). Using the ATP and NADPH from the light reactions, these molecules are bonded together and are chemically reduced by steps that are actually a reversal of the steps in glycolysis.

Three main events take place during the light-independent reactions of photosynthesis: (1) carbon is fixed with the attachment of carbon dioxide to an organic molecule, (2) glucose is produced from molecules of carbon dioxide, and (3) the organic molecule that hooks to carbon dioxide is re-formed to begin the cycle again.

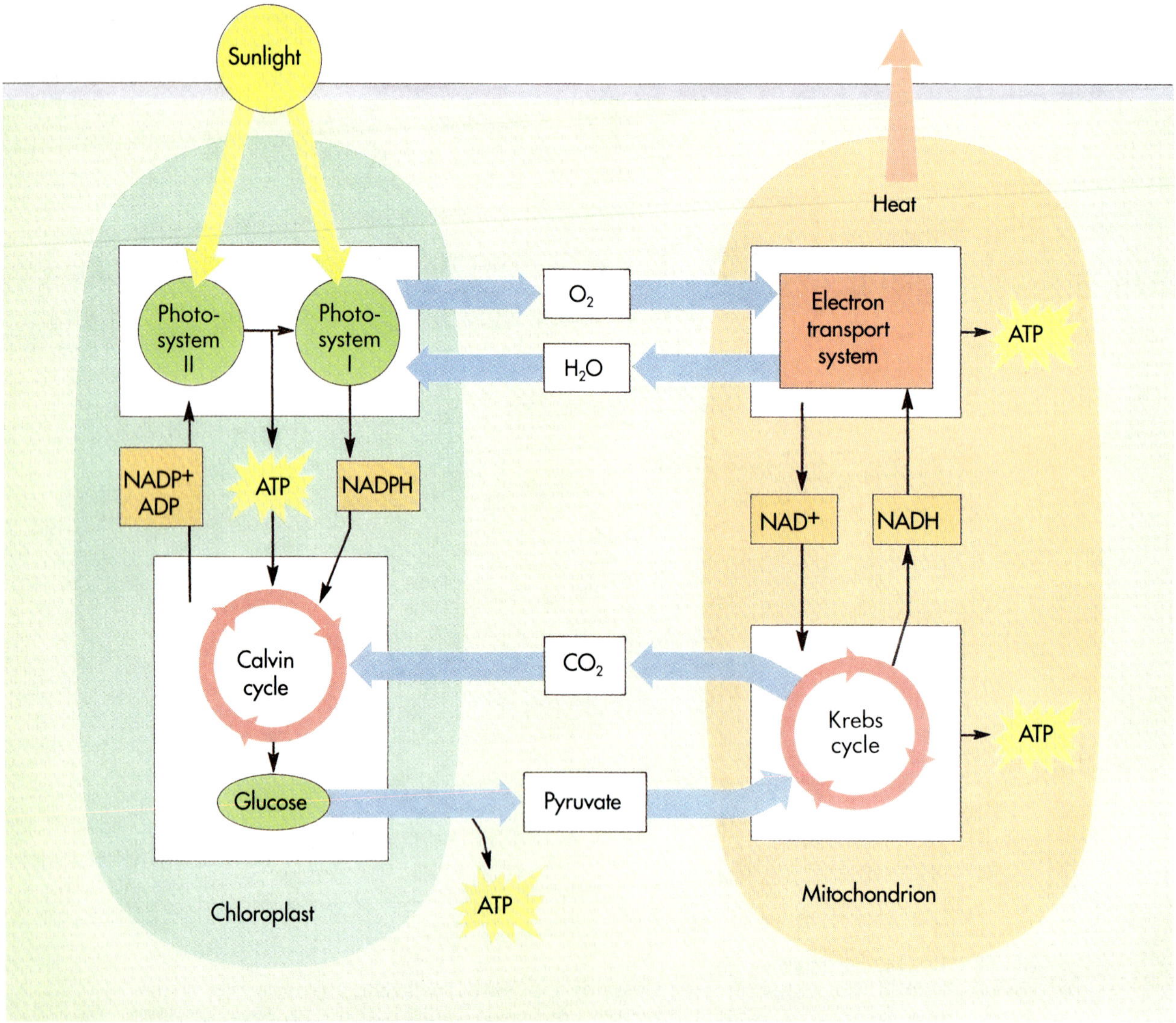

FIGURE 8-14 The metabolic machine.
This diagram shows how the processes of photosynthesis and respiration are connected.

Relationships: Photosynthesis and cellular respiration

Chapter 7 told the story of cellular respiration: the breakdown of glucose and the capture of its energy in molecules of ATP. Interestingly, the by-products of these series of reactions are carbon dioxide and water—the substrates of photosynthesis. Conversely, oxygen gas, which is the by-product of eukaryotic photosynthesis, is an essential substrate of cellular respiration. The photosynthetic product glucose is the initial substrate for cellular respiration.

Clearly, the metabolic pathways of photosynthesis and cellular respiration are connected. In fact, they can together be thought of as one great metabolic "machine" (Figure 8-14). Within the chloroplast, sunlight drives the production of glucose, using up carbon dioxide and water while generating oxygen. This oxygen is converted back to water in the mitochondrion by accepting electrons harvested from glucose molecules after the electrons have been used by the electron transport system to drive the synthesis of ATP. Water and oxygen gas thus cycle between chloroplasts and mitochondria within a plant cell, as do glucose and carbon dioxide. Cells without chloroplasts, such as your own cells, require an outside source of glucose and oxygen, and generate carbon dioxide and water.

Summary

1. Photosynthesis is a process whereby energy from the sun is captured by living organisms and used to produce molecules of food. Organisms capable of making their own food—plants, algae, euglenoid protists, and some bacteria—are called *autotrophs*. Organisms not capable of making their own food, called *heterotrophs,* depend on the food produced by autotrophs.
2. The energy from the sun is electromagnetic radiation and behaves as both waves and as discrete packets of energy called *photons*. Visible light is only a part of the full range of the electromagnetic spectrum.
3. Special molecules called *pigments* absorb some wavelengths of light and reflect others. When a molecule of a pigment absorbs this energy, one of its electrons is boosted to a higher energy level than it already occupies. Two important types of pigments that play a role in photosynthesis are the chlorophylls and the carotenoids. These pigments are organized into photocenters and serve to capture light energy and change it into chemical energy useful in photosynthesis.
4. During photosynthesis, atmospheric carbon dioxide is used to produce molecules of sugar. The events of this synthesis involve three different types of chemical processes. During the first process, ATP and NADPH molecules are produced and oxygen is given off as a by-product. During the second process the ATP and NADPH are used to power the synthesis of glucose. During the third process, the chlorophyll pigments are readied to absorb light once again.
5. The light-dependent reactions of photosynthesis are those that produce ATP and NADPH. These molecules are used during the light-independent reactions to produce glucose.
6. Cellular respiration and photosynthesis are intimately connected biochemical pathways. The by-products of cellular respiration—carbon dioxide and water—are the substrates of photosynthesis. Conversely, the by-product of photosynthesis—oxygen—and its product glucose are the substrates of cellular respiration. These major metabolic pathways are the energy crossroads of cells.

REVIEW QUESTIONS

1. Compare autotrophs and heterotrophs. Which are you?
2. You hold a prism in front of a window and the incoming sunlight forms a rainbow on the floor. Explain what happened.
3. What is a pigment? Explain its role in how you see colors.
4. Fill in the blanks: Two groups of pigments, the ________ and the ________ are important in photosynthesis.
5. Explain what this equation means. What process is being described, and why is it important?

$$6CO_2 + 6H_2O + \text{Light energy} \rightarrow C_6H_{12}O_6 + 6O_2$$

6. Summarize the three processes involved in photosynthesis. What is meant by the terms *light-dependent reactions* and *light-independent reactions*?
7. Define "photocenter" and "chlorophyll *a*" and explain their significance.
8. Distinguish between Photosystem I and Photosystem II. Which organisms use which?
9. Describe the two-stage photosystem used by plants and algae to carry out light reactions.
10. Summarize the main events that take place during the light-independent reactions of photosynthesis.
11. Large areas of tropical rain forest are cut down every year to open up the land for human use. Explain how this could affect the relationship between plant and animal life in those areas.
12. Fill in the blank: The reactions of photosynthesis take place within organelles called ________.
13. Some scientists feel that a large nuclear explosion could blanket the earth with a thick haze of atmospheric dust that would last for years. Explain how this would affect life on earth.
14. Fill in the blanks: Sunlight is a form of energy known as ________ ________, also called ________. Sunlight travels as ________ and as discrete packets of energy called ________.

KEY TERMS

absorption spectrum 133
autotroph 131
Calvin cycle 141
carbon fixation 141
carotenoids 133
chlorophyll *a* 136
chlorophylls 133
chloroplast 135
cyclic electron flow 140
electromagnetic energy 132
electromagnetic spectrum 132
grana 135
heterotrophs 131
light-dependent reactions 135
light-independent reactions 136
P680 137
P700 137
photocenter 136
photons 132
photosynthetic membranes 135
Photosystem I 137
Photosystem II 137
pigment 133
radiation 132
stroma 135
thylakoid 135

THOUGHT QUESTIONS

1. Imagine you place a potted plant in your classroom window after weighing it carefully. Pot and all, it weighs just 3 pounds. Watering it properly, you let the plant grow for a year and then reweigh it. Pot and all, it now weighs 4½ pounds. Where precisely did the 1½ pounds of extra body mass come from?
2. The roots of plants are not exposed to sunlight—they are under the ground. How do they manufacture ATP?
3. If ATP is made inside chloroplasts, how does it get out?
4. Too much sun is dangerous to people because it increases the likelihood of skin cancer. Plant leaves, on the other hand, are constantly exposed to sunlight, and yet plants don't get "sunburn" or epidermal cancer. Why not?

FOR FURTHER READING

Barber, J. (1989, September). A quantum leap forward in understanding photosynthesis. *Plants Today,* pp. 165-169.
This is an engaging account of the discovery of Photosystem II, for which the Nobel prize was given in 1988.

Govindjee, R., & Coleman, W. (1990, February). How plants make oxygen. *Scientific American,* pp. 50-58.
This article explains how plants and some bacteria exploit solar energy to split water molecules into oxygen gas.

Govindjee, P., & Govindjee, R. (1974, December). The absorption of light in photosynthesis. *Scientific American,* pp. 69-87.
This article gives a good account of how a photocenter works.

PART FOUR

HUMAN BIOLOGY

THE STRUCTURE AND FUNCTION OF THE BODY

9 Levels of organization in the human body

OUTLINE

How the human body is organized
Tissues
- Epithelial tissue
- Connective tissue
  - Cells that defend
  - Cells and tissues that shape and bind
    - Fibroblasts
    - Cartilage cells
    - Bone cells
  - Cells and tissues that isolate
- Muscle tissue
  - Smooth muscle
  - Skeletal muscle
  - Cardiac muscle
- Nervous tissue

Organs
Organ systems
The organism: Coordinating it all

HIGHLIGHTS

▼

There are several levels of organization in the human body: atoms and molecules are organized into cells, cells are organized into tissues, tissues are organized into organs, and organs are organized into organ systems.

▼

Four basic types of tissues make up the organs: epithelial, connective, muscle, and nervous.

▼

The human body can be divided into 11 different organ systems: digestive, respiratory, circulatory, immune, urinary, nervous, skeletal, integumentary, muscular, endocrine, and reproductive.

▼

The organ systems interact with one another to maintain a stable internal environment, or "steady state," called *homeostasis.*

Did you ever blow soap bubbles through a straw when you were a child, or see a bubble-blowing machine at a science museum? The bubbles form one on top of the other, looking much like the pattern in the photo. But although the "bubbles" in the photo are filled with air, they are not bounded by soap. They are enclosed by cells and are filled with the air in your lungs. In fact, these bubbles are not bubbles at all. They are microscopic air sacs that help deliver the oxygen in the air you breathe to your bloodstream.

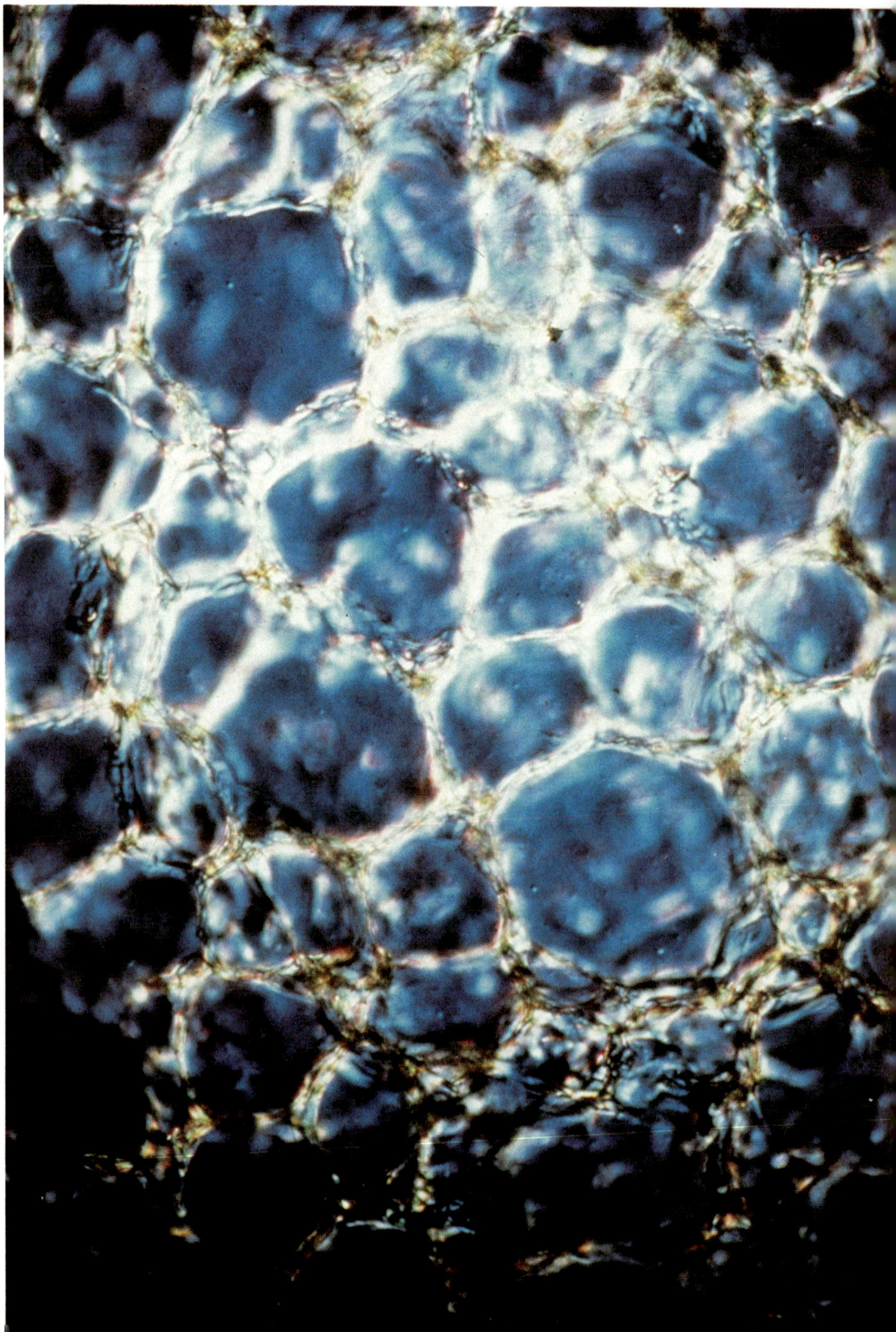

Although several hundred million air sacs can be found in each of your lungs, your lungs are not made up solely of air sacs. They include many other structures, such as blood vessels and air passageways. Together, these structures and others help the lungs do their job—transporting oxygen to the bloodstream and ridding the blood of the waste gas carbon dioxide. A complex structure such as the lung is called an *organ,* and forms only one of the levels of organization of the human body.

How the human body is organized

The human body, like the bodies of all other multicellular animals, is made up of many different types of cells. In fact, the human body contains over one hundred different kinds of cells! These cells are not distributed randomly but are organized to form the structures and perform the functions of the human body, just like workers in a factory may be organized to manufacture a product.

The cells of the body are organized into **tissues.** Tissues are groups of similar cells that work together to perform a function. Traditionally, tissues are divided into four basic types based on their function: **epithelial, connective, muscle,** and **nervous** (Figure 9-1). For example, the cells making up the walls of the air sacs of your lungs are a type of epithelial tissue—tissue that covers body surfaces and lines its cavities. All four tissue types are described later on in this chapter.

Two or more tissues grouped together to form a structural and functional unit are called **organs.** Your heart is an organ. It contains cardiac muscle tissue wrapped in con-

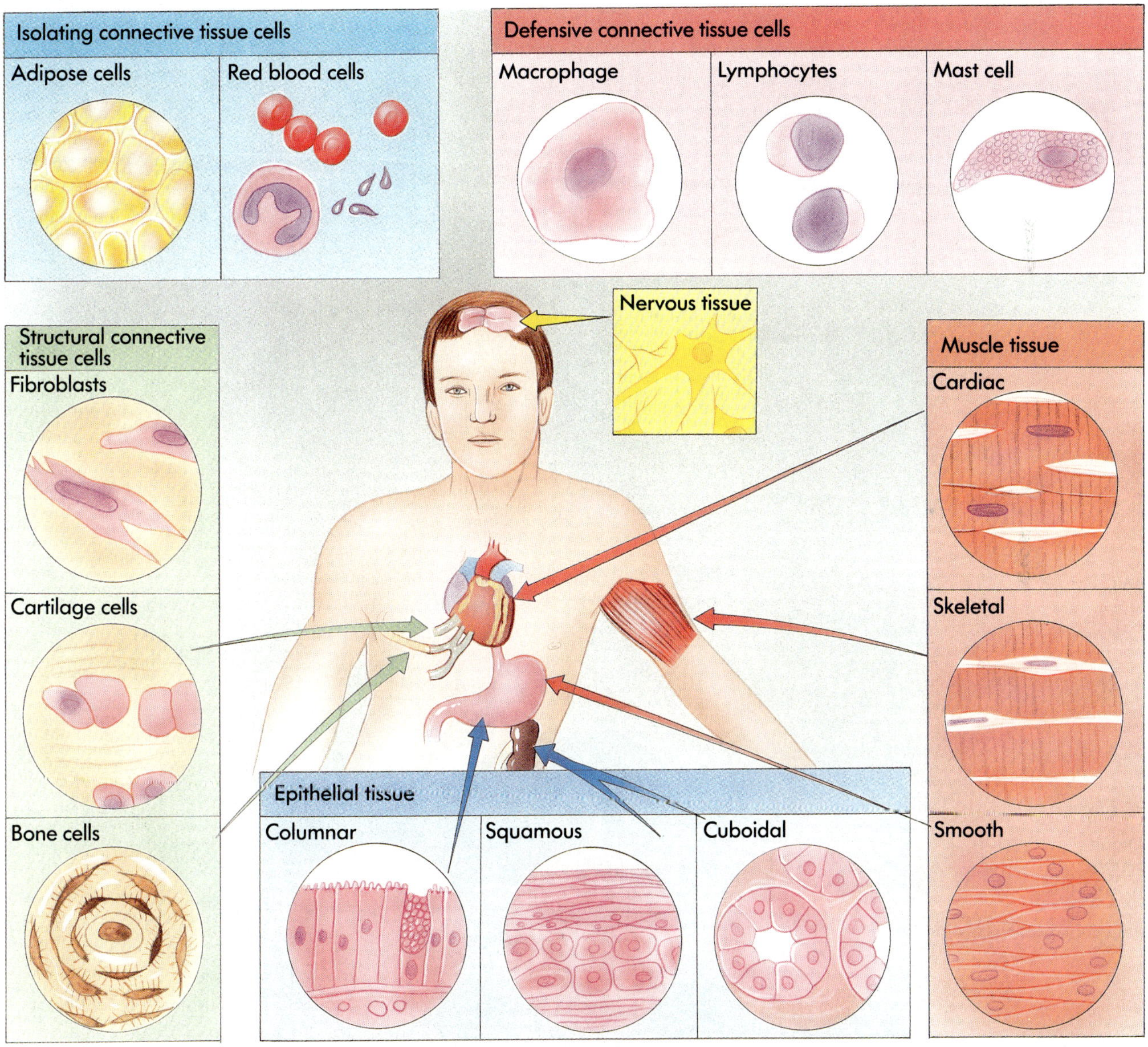

FIGURE 9-1 The basic tissue types and their cells.
The tissues in the body are divided into four basic types based on their function: epithelial, connective, muscle, and nervous.

TABLE 9-1

The major human organ systems

SYSTEM	FUNCTIONS	COMPONENTS	CHAPTER
Digestive	Breaks down food and absorbs breakdown products	Mouth, esophagus, stomach, intestines, liver, and pancreas	10
Respiratory	Supplies blood with oxygen and rids it of carbon dioxide	Trachea, lungs, and other air passageways	11
Circulatory	Brings nutrients and oxygen to cells and removes waste products	Heart, blood vessels, blood, lymph, and lymph structures	12
Immune	Helps defend body against infection and disease	Lymphocytes, macrophages, and antibodies	13
Urinary	Removes wastes from bloodstream	Kidney, bladder, and associated ducts	14
Nervous	Receives and helps body respond to stimuli	Nerves, sense organs, brain, and spinal cord	15, 16, 17
Skeletal	Protects the body and provides support for locomotion and movement	Bones, cartilage, and ligaments	18
Muscular	Produces body movement	Skeletal, cardiac, and smooth muscle	18
Integumentary	Covers and protects the body	Skin, hair, nails, and sweat glands	18
Endocrine	Coordinates and regulates body processes and functions	Pituitary, adrenal, thyroid, and other ductless glands	19
Reproductive	Produces sex cells and carries out other reproductive functions	Testes, ovaries, and other associated reproductive structures	20

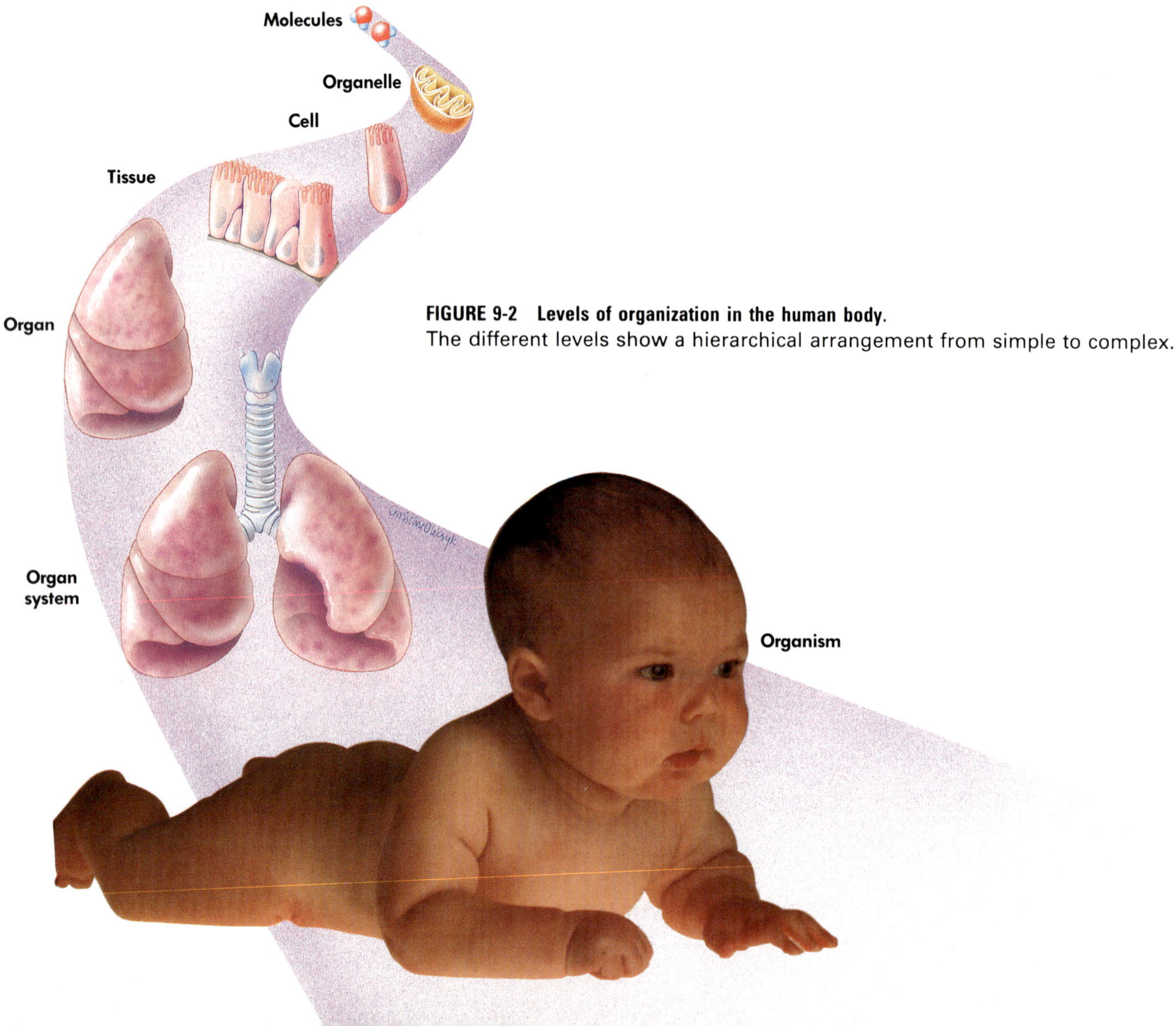

FIGURE 9-2 Levels of organization in the human body. The different levels show a hierarchical arrangement from simple to complex.

nective tissue and "wired" with nerves. All of these tissues work together to pump blood through your body. Other examples of organs are the stomach, skin, liver, and eyes.

An **organ system** is a group of organs that function together to carry out the principal activities of the body. For example, the digestive system is composed of individual organs concerned with the breaking up of food (teeth), the passage of food to the stomach (esophagus), the storage and partial digestion of food (stomach), the digestion and absorption of food and the absorption of water (intestine), and the expulsion of solid waste (rectum). The human body contains 11 principal organ systems (Table 9-1), which are discussed in later chapters.

The organ systems are all encased in the human body (Figure 9-2), which is an incredible living "machine." The human body has the same general body architecture that all animals with a backbone have (Figure 9-3). It includes a long tube that travels from one end of the body to the other, from mouth to anus. This tube is suspended within an internal body cavity called the **coelom.** In human beings the coelom is divided into two main parts: (1) the thoracic cavity, which contains the heart and lungs, and (2) the abdominal cavity, which contains organs such as the stomach, intestines, and liver. The lower portion of the abdominal cavity is often referred to as the *pelvic cavity*. The body is supported by an internal scaffold, or skeleton, made up of bones that grow as the body grows. A bony skull surrounds the brain, which is located in a cavity separate from the coelom, called the *cranial cavity*. In addition, a column of bones called the *vertebral column* forms the backbone, or spine, of the human body. These bones surround a nerve cord, the spinal cord, that relays messages between the brain and other parts of the body.

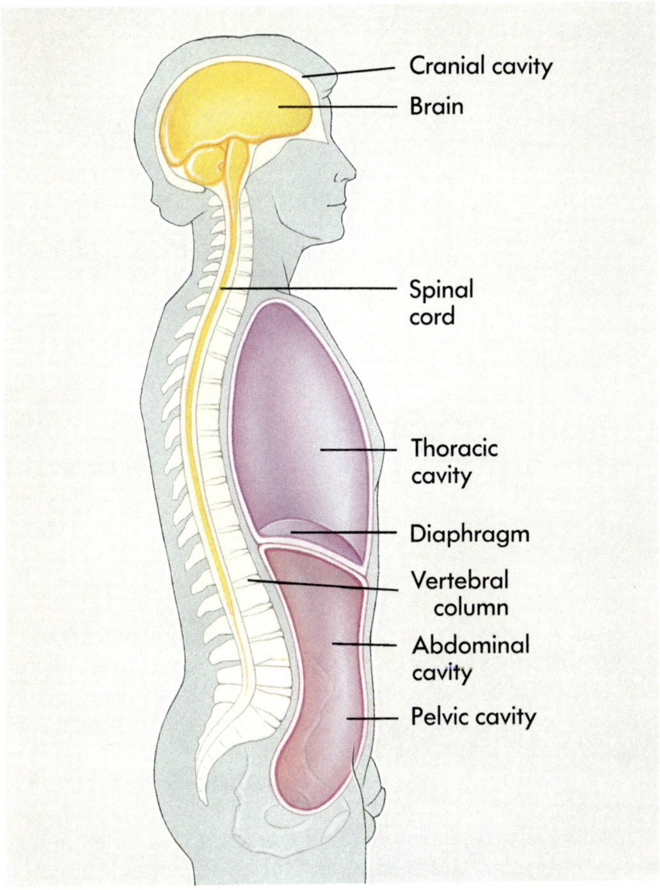

FIGURE 9-3 Architecture of the human body.
Humans, like all vertebrates, have a spinal cord and brain enclosed in the vertebral column and skull. In mammals a muscular diaphragm divides the coelom into the thoracic cavity and the abdominal cavity.

Tissues

Epithelial tissue

Epithelial cells guard and protect your body; these cells are really the "doors" to the inner you. Put simply, epithelial cells cover and line the surfaces of the body, both internal and external. Remember, you can think of your body as a tube, with your skin forming the outside of the tube and your digestive system forming the inside of the tube. Therefore the surfaces of your inner tube as well as your skin are made up of epithelial cells. And, because epithelial cells cover the surfaces of the body, they determine which substances enter and which do not.

All epithelial cells, collectively called the **epithelium,** are broadly similar in form and function. The epithelial layers of the body function in six different ways:

1. **Protection:** They protect the tissues beneath them from drying out, mechanical injury, and invasion by microorganisms.
2. **Absorption:** They provide a barrier that can help or hinder the movement of materials into the tissue beneath. Because epithelium encases all of the body's surfaces, every substance that enters or leaves the body must cross an epithelial layer.
3. **Sensation:** Many sensory nerves end in epithelial cell layers. The epithelium therefore provides a sensory surface.
4. **Secretion:** Certain epithelial cells are specialized to produce and discharge substances; they are called *glands*. Glands may be single cells, such as the mucus-secreting cells lining the intestine. Most glands, however, are multicellular structures, such as the thyroid gland located in your neck or the pituitary gland that hangs like a tiny pea from the underside of your brain. Even your sweat glands are made up of many cells. Because many glands lie deep within your body and do not cover or line a surface, they may seem inappropriately described as being composed of epithelial tissue. However, glands form during embryological development from infoldings of epithelial cell layers and do fit in this category.
5. **Excretion:** Specialized epithelial cells in the kidney excrete waste products during the formation of urine. Also, the epithelium forming the air sacs of the lungs excrete the waste gas carbon dioxide.
6. **Surface transport:** Some epithelial cells have hair-like projections called *cilia*. These cilia beat in unison, causing a wave-like movement in the thin film of mucus that bathes the cells' surfaces, sweeping particles along in the

TABLE 9-2

Epithelial tissue

TISSUE	LOCATION	FUNCTIONS	DESCRIPTION
Simple squamous epithelium 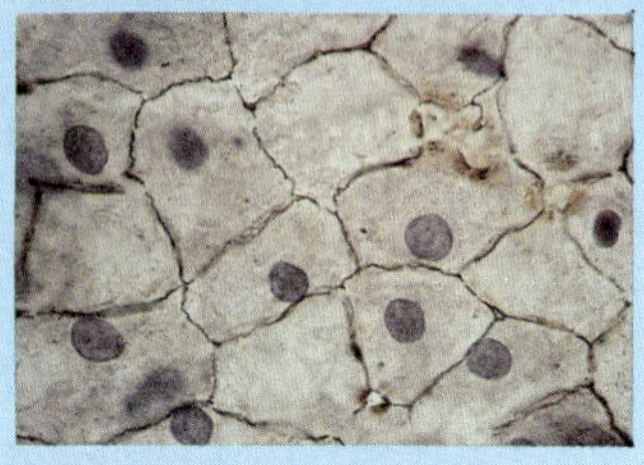	Lining of blood vessels, air sacs of lungs, kidney tubules, and lining of body cavities	Diffusion, filtration, and passage of materials where little protection is needed	The simplest of all epithelial tissues, simple epithelium consists of a single layer of thin, flat cells.
Cuboidal epithelium 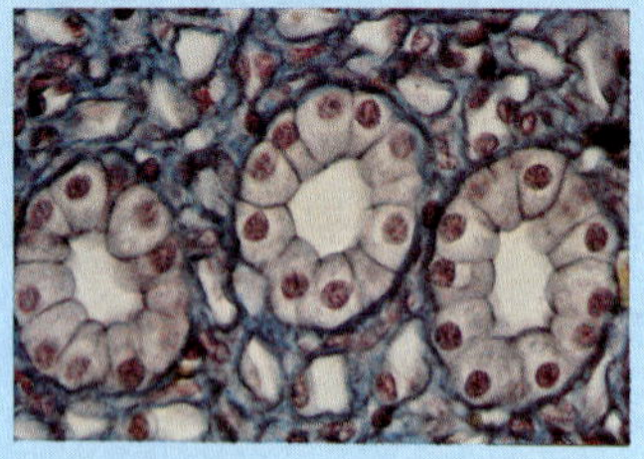	Kidney tubules, glands and their ducts, terminal bronchioles of lungs, and surface of ovaries and retina	Secretion, absorption, or movement of substances	The structure is a single layer of cube-shaped cells. Some have microscopic extensions called *microvilli.* Some have cilia that protrude from their surfaces.
Columnar epithelium 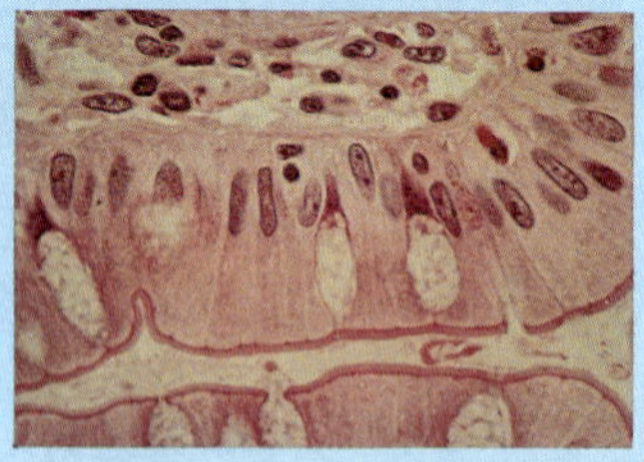	Lining of the digestive and upper part of respiratory tracts, auditory and uterine tubes	Secretion of mucus	The structure is a single layer of tall, narrow cells. Some have microvilli or cilia.

process. The cells lining parts of your respiratory passageways, for example, use this technique to effectively keep many foreign particles from entering the lungs. Cigarette smoking damages these cilia, resulting in accumulations of mucus in the throat and respiratory passageways that leads to "smoker's cough" and, over time, to serious upper respiratory problems.

Structurally, epithelial tissues share some common characteristics. Epithelial tissues are usually only one or a few cells thick (with the exception of the skin), are packed together and stacked very tightly, and have very few blood vessels running through them. The circulation of nutrients, gases, and wastes in epithelial tissue occurs by diffusion from the capillaries of neighboring tissue. In addition, although the chemical reactions of many types of epithelial cells take place at a very slow rate, many have amazing regenerative powers. Your skin cells, for example, are continually being replaced throughout your lifetime. Your liver, a gland having epithelial tissue on its absorptive and secretory surfaces, can readily regenerate substantial portions of tissue if parts are surgically removed or are damaged by certain diseases.

There are three main shapes of epithelial cells: **squamous, cuboidal,** and **columnar** (Table 9-2). Squamous cells are thin and flat. They are found in places such as the air sacs of the lungs, the lining of blood vessels, and the skin. Cuboidal cells have complex shapes, but look like cubes when the tissue is cut at right angles to the surface. These cells are found lining tubules in the kidney and the ducts of glands. Columnar cells (as their name suggests) look like

TISSUE	LOCATION	FUNCTIONS	DESCRIPTION
Stratified squamous epithelium 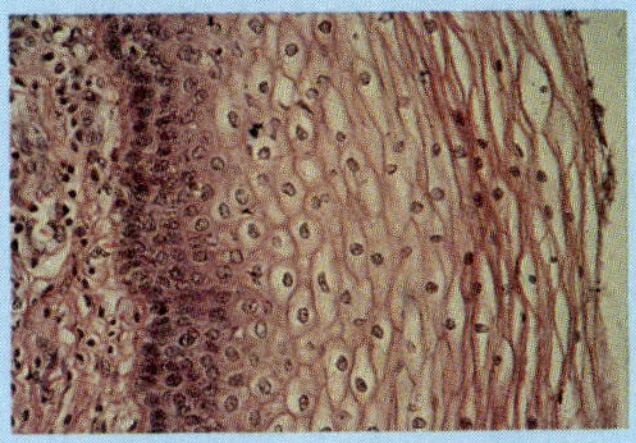	Skin, mouth, and throat lining; vaginal lining; anal lining; and cornea	Protection, hard outer layer continuously removed by friction and replaced from below	The structure has several layers of cells. The lower layers are columnar and active. The upper layers are flattened at the surface.
Pseudostratified epithelium 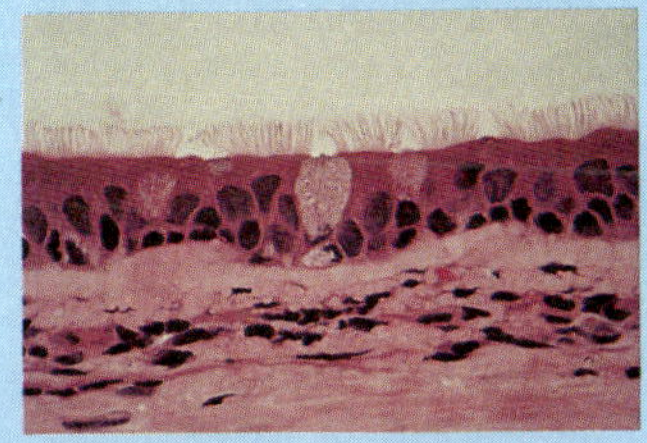	Nasal cavity and sinuses, ducts of some glands, and some ducts of the male reproductive system	Protection and secretion of mucus	The structure is a single layer of cells. It is similar to columnar epithelium, but all cells are not the same height. Some reach the surface and others do not. Some have cilia, and some may have microvilli.
Transitional epithelium 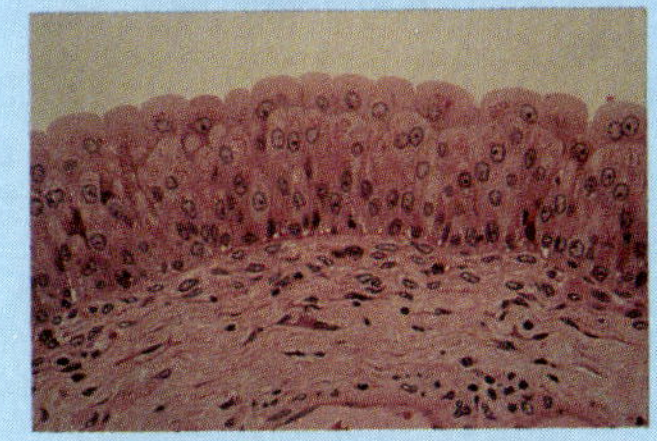	Lining of urinary bladder and ureters	Stretches easily to accomodate fluid fluctuations in an organ or tube	There are several layers of columnar cells beneath layers of surface cells. Cells are flattened when tissue is stretched.

tiny columns when viewed from the side. Much of the digestive tract is lined with columnar epithelium.

The various types of epithelial cells may also be arranged in various ways. Epithelial tissue that is only one cell thick is referred to as **simple epithelium.** It is usually found in areas where substances diffuse through the tissue and where substances are secreted, excreted, or absorbed. The cells lining your blood vessels are simple squamous epithelium, for example, whereas much of your digestive tract is lined with simple columnar epithelium. **Stratified epithelium** is made up of two or more layers. These layers usually serve a protective function. The surface of your skin is stratified squamous epithelium. **Pseudostratified epithelium** only looks as though it is layered. In reality, this tissue is made up of only one layer of cells—some tall, some short. The sides of the top portions of the tall cells rest on top of the shorter cells, looking like two or more layers of cells. Some of your airways are lined with pseudostratified epithelium. **Transitional epithelium** is tissue that can stretch. The cells of transitional epithelium are cube-like; when stretched, the cells appear thin and flat. Transitional epithelium is found in the bladder and ureters of the urinary tract, an organ and tubes that accomodate fluid fluctuations.

Epithelial cells cover and line the internal and external surfaces of the body and compose the glands. These cells are of three types and are arranged in ways that best suit their functions.

Connective tissue

The cells of connective tissue provide the body with structural building blocks and potent defenses. In addition, connective tissue joins the other tissues of the body. Connective tissue performs an assortment of important jobs for the body. The varied composition of the different types of connective tissue reflect this diversity. However, connective tissue is generally made up of cells that are usually spaced well apart from one another and are embedded in a nonliving substance called a **matrix.** In fact, connective tissue is made up of a great deal more matrix than cells! It varies in consistency among the different types of connective tissue—from a fluid to a gel to crystals.

The different types of connective tissues and cells are categorized in many different ways. One way to group them is by function: **defensive, structural,** and **isolating connective tissue.** Defensive cells, those that protect the body from attack, float in a matrix of blood plasma. They roam the circulatory system, "hunting" invading bacteria and foreign substances. An example of this type of cell is the lymphocyte, a special type of white blood cell (Figure 9-4). Structural connective tissue cells, such as bone and cartilage cells, stay in one place, secreting proteins into the empty spaces between them. These proteins provide structural connective tissue with a fibrous matrix, giving it the strength it needs to support the body and provide connections among tissues. Isolating connective tissue cells act as storehouses, accumulating specific substances such as fat, the skin pigment melanin, and hemoglobin.

Connective tissue and its cells provide a framework for the body, join its tissues, help defend it from foreign invaders, and act as storage sites for specific substances.

Cells that defend

The three principal defensive cell types are **lymphocytes** (see Figure 9-4), **macrophages,** and **mast cells.** All of these cells are dispersed throughout the body either in the blood or among other tissues.

Lymphocytes are a type of white blood cell that circulates in the blood or resides in the organs, vessels, and nodes of the lymphatic system. Both these cells and the lymphatic system itself play complex, key roles in the body's defense against infection. Your body has an amazing trillion or so lymphocytes ready to attack foreign cells or viruses that enter the body or produce specific antibodies that can act against specific substances. (How lymphocytes and antibodies function in the body's defense against disease is described in Chapter 13.)

Macrophages are abundant in the bloodstream and also in the fibrous mesh of many tissues, such as the lungs, spleen, and lymph nodes. They develop, or differentiate from, white blood cells called *monocytes.* Usually macrophages freely move about, but sometimes they stay in one place, attached to fibers. These cells may be thought of as the janitors of the body, "cleaning up" cellular debris and invading bacteria by a process known as *phagocytosis*—an engulfing and digesting of particles.

Mast cells produce substances that are involved in the body's inflammatory response to physical injury or trauma. One important substance produced by mast cells is histamine, a chemical that causes blood vessels to dilate, or widen. As more blood then flows through the vessels, it brings added oxygen and nutrients and dilutes any toxins, or poisons. The increased blood flow also aids the movement of defensive leukocytes coming to the area. Mast cells, although important in the inflammation response, also play a role in allergic reactions (see Chapter 13).

Protective connective tissue cells defend the body against foreign invaders. Lymphocytes attack foreign cells or viruses that enter the body or produce specific antibodies that can act against specific substances, macrophages then engulf and digest the invader. Mast cells enlarge the blood vessels in response to trauma, speeding the healing process.

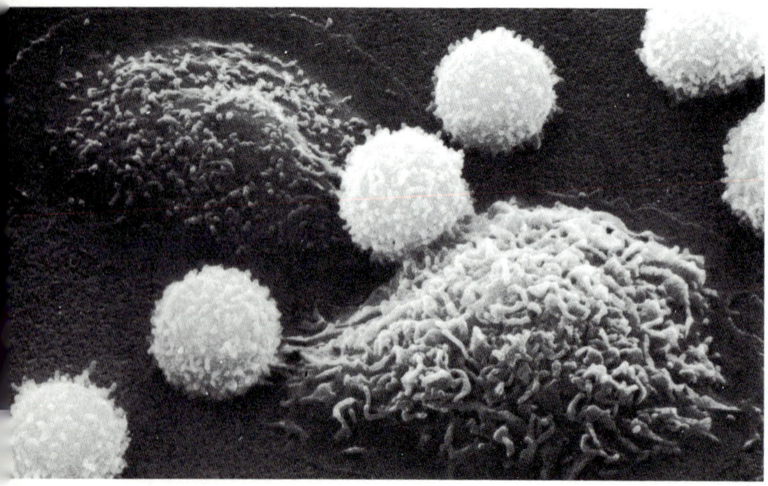

FIGURE 9-4 Lymphocytes and macrophages.
The lymphocytes are small and spherical; the macrophages are larger and more irregular in form. Both types of cells play key roles in the body's defense against infection.

Cells and tissues that shape and bind

The three principal types of cells found in structural connective tissue are **fibroblasts,** cartilage cells **(chondrocytes),** and bone cells **(osteocytes).** These cells produce substances that cause the tissues of which they are a part to have distinctive characteristics.

Fibroblasts. Of all the connective tissue cells, fibroblasts are the most numerous. They are flat, irregular, branching cells that secrete fibers into the matrix between them (Figure 9-5). These fibers are of three basic types: collagen, reticulin, and elastic.

Both the collagen fibers and the reticulin fibers are made up of the protein collagen, the most abundant protein in the human body (Figure 9-6). Collagen fibers are strong and wavy. These properties allow the connective tissues that they make up to be somewhat flexible and stretchy, without the fibers themselves stretching. **Dense fibrous connective tissue** is primarily made up of collagen fibers. In Table 9-3 you can see bundles of collagen fibers with widely spaced rows of fibroblasts. This type of connective tissue is very strong and is found as tendons connecting muscles to bones, making up the lower layer of the skin, and making strong attachments between organs.

Reticulin is a fine-branching fiber that forms the framework of many glands such as the spleen and the lymph nodes. It also makes up the junctions between many tissues. The tissue formed by fibroblasts and reticulin alone is called **reticular connective tissue** (Table 9-3).

Elastic fibers, as the name suggests, act much like rubber bands. They are not made of collagen but of a protein called *elastin,* a "stretchy" protein. **Elastic connective tissue** is made up of branching elastic fibers with fibroblasts interspersed throughout (see Table 9-3). This type of tissue is found in structures that expand and then must return to their original shape, such as the lungs and large arteries.

Loose connective tissue contains various connective tissue cells and fibers within a semi-fluid matrix (see Table 9-3). Fibroblasts and macrophages are the most common cells in loose connective tissue. It also contains loosely packed elastic and collagen fibers, so it is therefore a somewhat strong but very flexible tissue. Loose connective tissue is distributed widely throughout the body and is found wrapping nerves, blood vessels, and tissues; filling spaces between body parts; and attaching the skin to the layers beneath it. If you've ever skinned chicken before cooking, for example, you've seen the loose connective tissue that binds the skin to the muscle beneath.

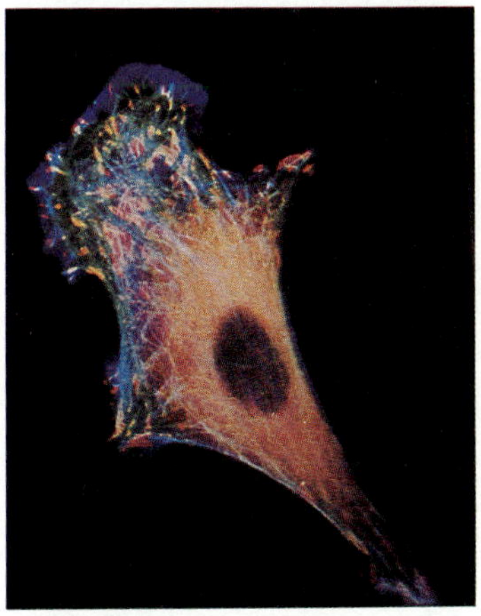

FIGURE 9-5 A fibroblast. Note the flat, irregular, branching shape. Fibroblasts secrete fibers of protein, constructing a web that supports other fibroblasts.

FIGURE 9-6 Collagen fibers. Each fiber is composed of many individual collagen strands and can be very strong.

Cartilage cells. Chondrocytes are the cells that produce cartilage, a specialized connective tissue that is hard and strong. It is found in many places such as the ends of long bones, the airways of the respiratory system, and the spaces between the vertebrae. Cartilage is made up of cells that secrete a matrix consisting of a semisolid gel and fibers. The fibers are laid down along the lines of stress in long parallel arrays (groups or arrangements). The result of this process is a firm and flexible tissue that does not stretch.

As the cartilage cells secrete the matrix, they wall themselves off from it and eventually come to lie in tiny chambers called **lacunae.** In Table 9-3, you can see that although there are three different types of cartilage—**hyaline cartilage, elastic cartilage,** and **fibrocartilage**—they all have cartilage cells within lacunae. Their differences lie in the matrix.

Hyaline cartilage has very fine collagen fibers in its matrix that are almost impossible to see under the light microscope. During your development before birth, most of your skeleton was composed of hyaline cartilage. As an adult, you have hyaline cartilage on the ends of your long bones, cushioning the places where these bones meet. Hyaline cartilage also rings the windpipe, keeping this airway "propped" open, and makes up parts of your ribs and nose. Elastic cartilage, as the name suggests, has elastic fibers embedded in its matrix. It is found where support with flexibility is needed, such as in the external ear. Fibrocartilage has collagen fibers embedded in its matrix. It is therefore a very tough substance and is used in places of the body where "shock absorbers" are needed. It is found, for example, as discs between the vertebrae and in the knee joint.

TABLE 9-3

Connective tissue

TISSUE	LOCATION	FUNCTIONS	DESCRIPTION
Dense fibrous connective tissue 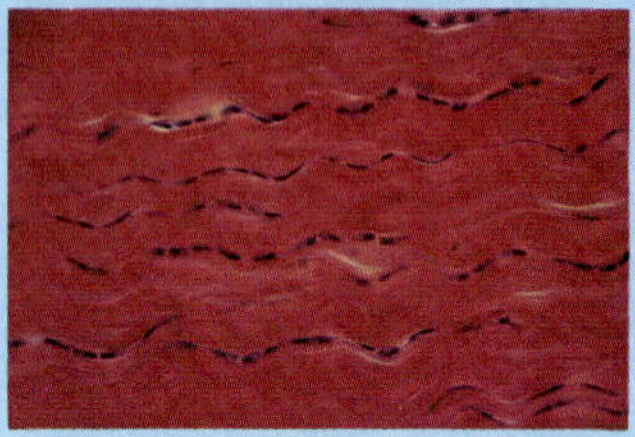	Tendons (attach muscle to bone) and ligaments (attach bones to bones), attachments between organs and dermis of the skin	Support, ability to withstand great pulling forces in the direction in which the fibers are oriented	Structure consists of mostly collagen fibers with occasional rows of collagen-producing cells.
Reticular connective tissue 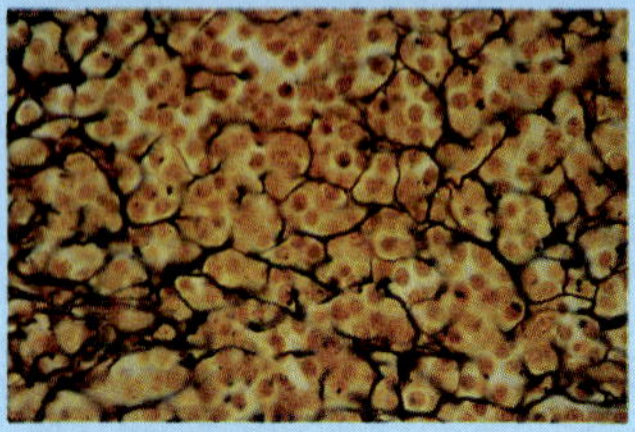	Liver, lymph nodes, spleen, and bone marrow	Support	The structure is intertwined reticular fibers.
Elastic connective tissue 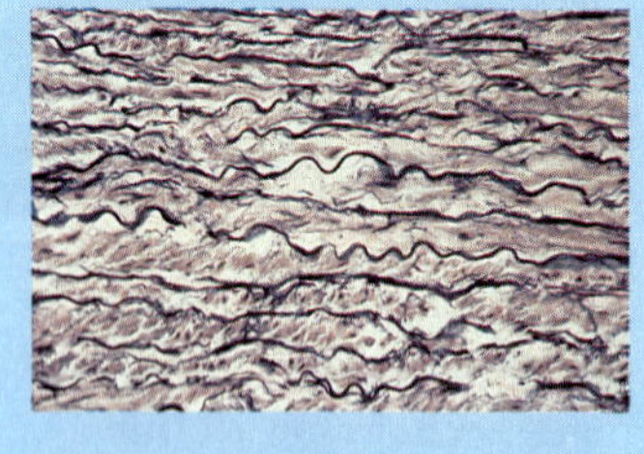	Lung tissue, arteries	Strength with stretching and recoil	The structure consists of elastic fibers dotted with cells.
Loose connective tissue 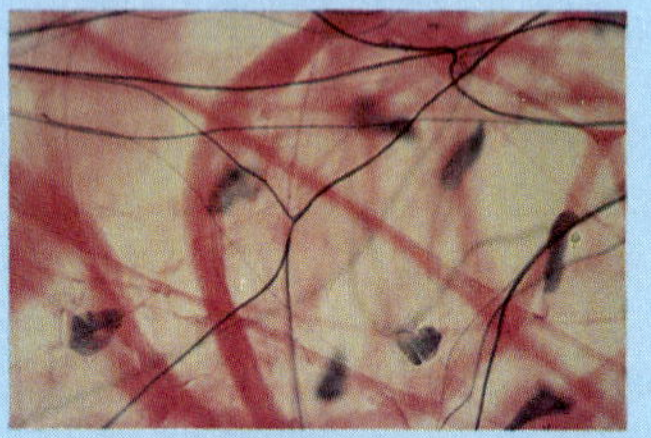	Widely distributed throughout body; packing between glands, muscles, and nerves; attachments between the skin and underlying tissues	Support, loose packing	The structure consists of cells within a fine network of fibers (mostly collagen), which they produce. The cells and fibers are separated from each other by fluid-filled spaces.
Hyaline cartilage 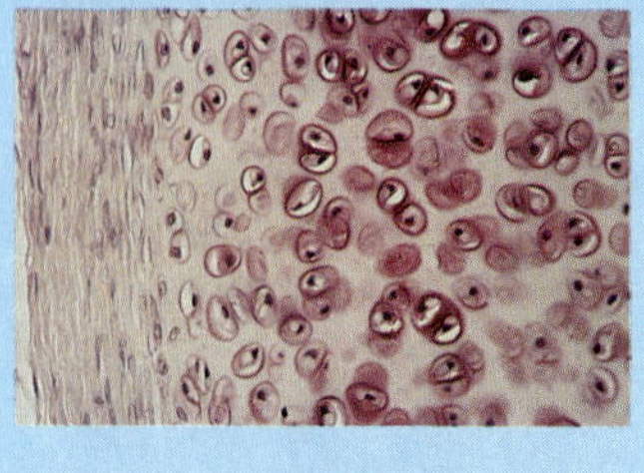	Ends of long bones, joints, respiratory tubes, costal cartilage of ribs, nasal cartilage, and embryonic skeleton	Flexible support, reduction of friction between movable bones	Cartilage cells are found in lacunae within a rigid, transparent matrix. Collagen fibers are small and not visible.

TISSUE	LOCATION	FUNCTIONS	DESCRIPTION
Elastic cartilage 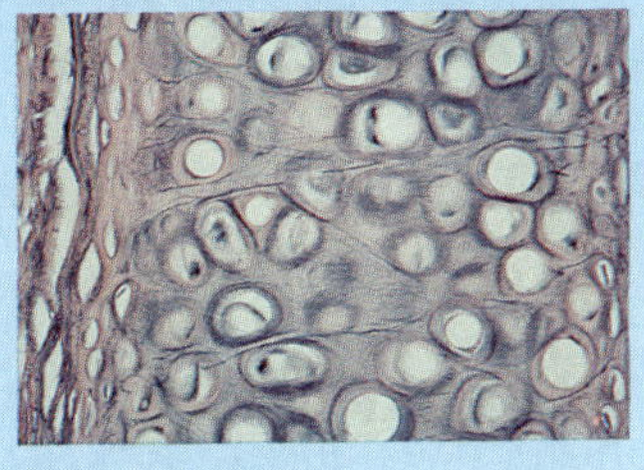	Auditory tube, external ear, epiglottis	Rigidity with flexibility, returning to original shape after being stretched	The structure resembles hyaline cartilage but has elastic fibers.
Fibrocartilage 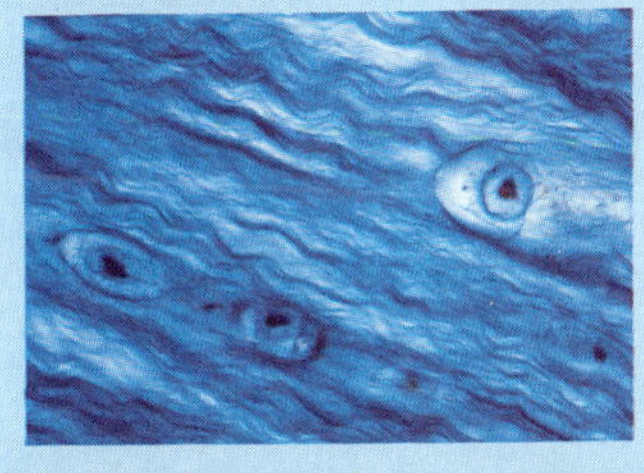	Connection between pubic bones, intervertebral disks	Support, connection, shock absorption, ability to withstand considerable pressure	The structure resembles hyaline cartilage but has thick bundles of collagen fibers.
Bone 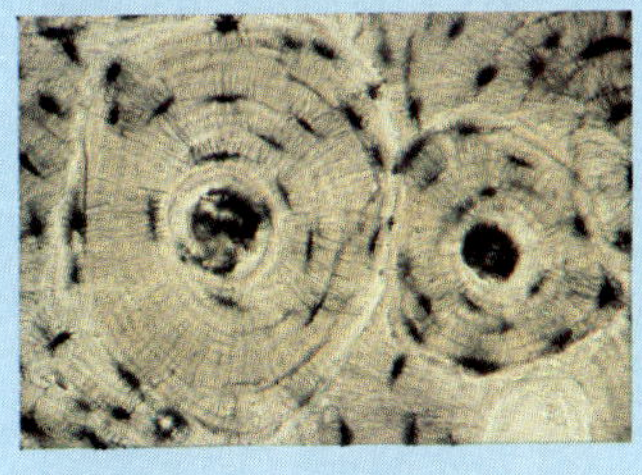	Bones of skeleton	Strength, support, and protection of internal organs; storage of calcium; attachment for muscles	Bone-making cells (osteocytes) are found in lacunae. In compact bone (as shown) the lacunae are arranged in circles around the Haversian canals, which contain blood vessels and nerves. The structure is a hard, mineralized matrix.
Adipose connective tissue 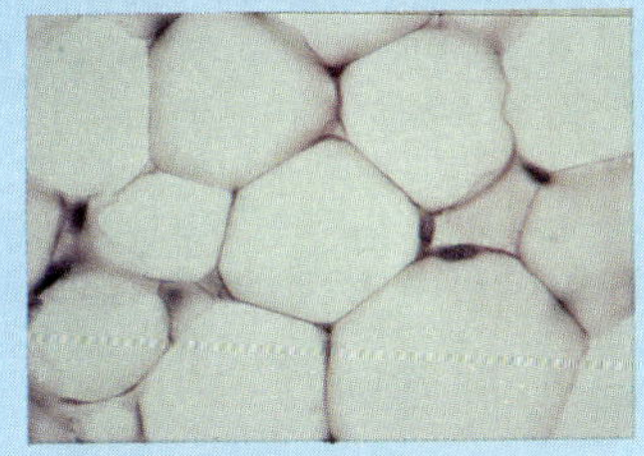	Found under the skin, insulation of organs such as the heart, kidneys, and breasts	Storage, insulation, energy, support of organs	The structure consists of lipid-filled, ring-shaped cells packed together.
Blood 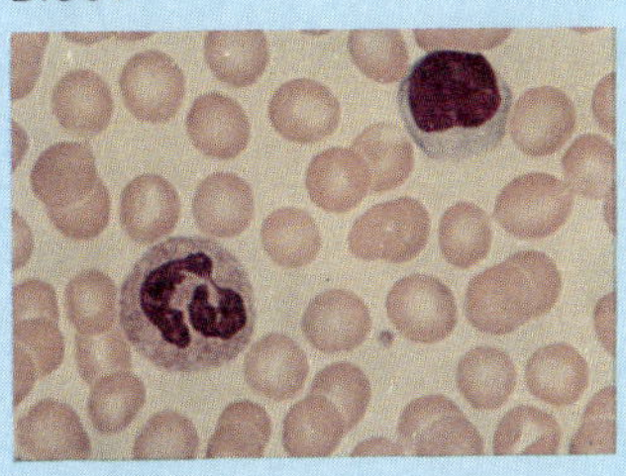	Blood vessels, heart	Protection of body from infections; transportation of oxygen, nutrients, wastes, and other materials; regulation of body temperature	The structure consists of blood cells in a fluid matrix.

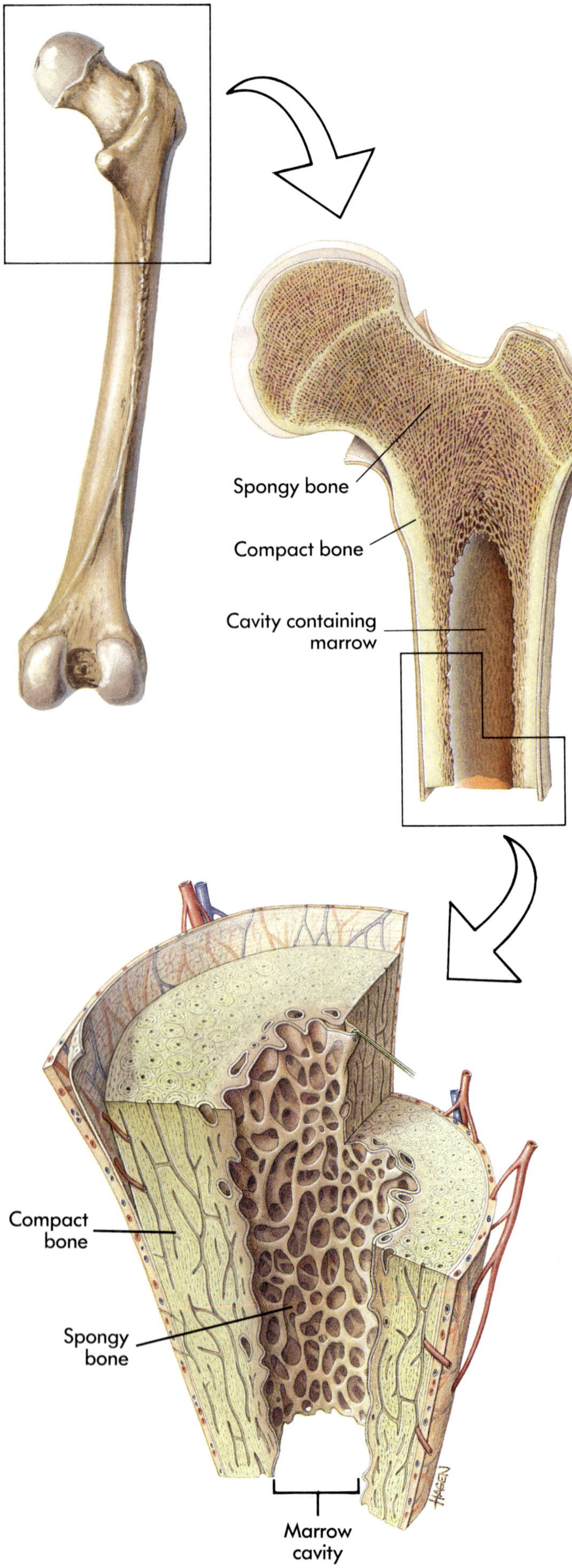

Bone cells. Osteocytes are the cells that produce bone. As in cartilage, these cells are isolated in lacunae. They lay down a matrix of collagen fibers that become coated with small, needle-shaped crystals of calcium. The calcium imparts a rigidity to the bone, while the fibers keep the bone from being brittle. Bone makes up the adult skeleton, which serves to support the body and protect many of the organs. It is also a storehouse for calcium.

The bones of your skeleton have two types of internal structure: **spongy bone** and **compact bone.** Spongy bone makes up the ends of long bones; and the interiors of long bones, flat bones, and irregular bones (Figure 9-7). It is composed of an open lattice of bone that supports the bone just as beams support a building (Figure 9-8). It also helps keep bones somewhat lightweight. The spaces within the latticework of bone are filled with red bone marrow, the substance that produces most of the body's blood cells.

Compact bone is more dense than spongy bone and gives the bone the strength to withstand mechanical stress. In compact bone, the cells lay down matrix in thin concentric rings, forming tubes of bone around narrow channels or canals (see Table 9-3). These canals run parallel to the length of the bone, are interconnected, and contain blood vessels and nerves. The blood vessels provide a lifeline to the bone-forming cells, and the nerves control the diameter of the blood vessels and thus the flow through them.

The three different types of structural connective tissue cells each produce substances that cause the tissues of which they are a part to have distinctive characteristics. Fibroblasts produce fibers of different types that are found in dense and loose connective tissue, reticular connective tissue, and elastic connective tissue. Cartilage cells lay down a matrix of gel and fibers. Bone cells lay down a matrix of fibers that become coated with calcium.

Cells and tissues that isolate

The third general class of connective tissue is composed of cells that specialize in accumulating and transporting particular molecules. Isolating connective tissues include the fat cells of adipose tissue as well as pigment-containing cells.

As you can see in Table 9-3, groups of fat cells as seen under a microscope bear a striking resemblance to chicken wire. The "wire" is really the cytoplasm and nucleus of each cell pushed against the cell membrane by a large fat droplet.

FIGURE 9-7 Structure of a long bone.
A long bone, such as the bone located in your upper leg, is composed of spongy bone and compact bone. Spongy bone adds support and (along with the hollowed-out shaft) keeps the bone somewhat lightweight. Compact bone adds strength.

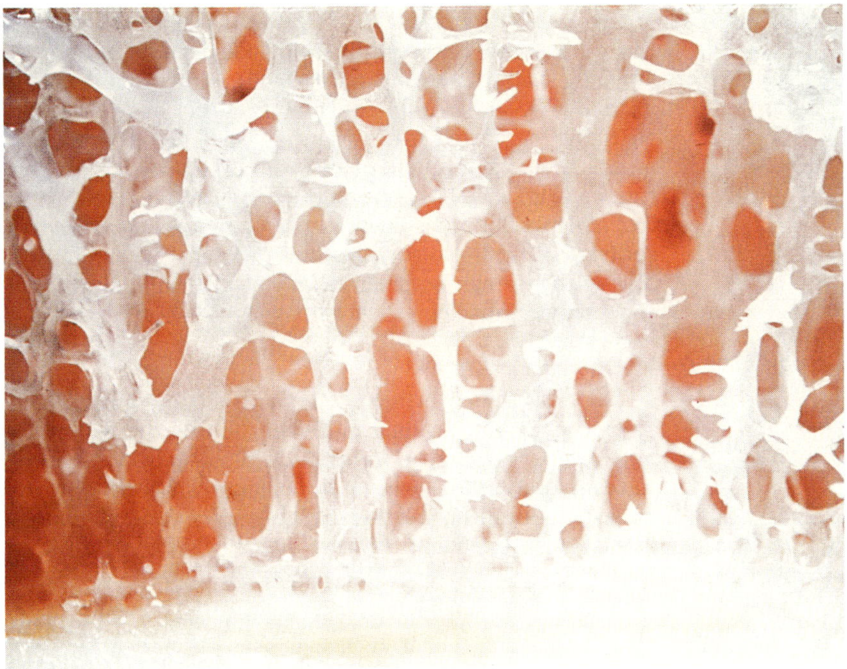

FIGURE 9-8 The interior of spongy bone.
While you may think of bone as hard and solid, the interior of spongy bone is composed of a delicate latticework. Bone, like most tissues present in your body, is a dynamic structure, constantly renewing itself.

The fat takes up much of the cell and is released when the body needs it for fuel. Although many people feel that they have more fat cells than they need for "fuel emergencies," it serves other purposes too. Adipose tissue helps shape and pad the body and insulates against heat loss.

Possibly the most important "isolating" cells are red blood cells, one of the solids that float in the fluid connective tissue called blood. Blood cells are classified according to their appearance (Figure 9-9), either as **erythrocytes** (red blood cells) or **leukocytes** (white blood cells). Some types of white blood cells were described earlier in this chapter: they include the macrophages and lymphocytes that defend the body. The role of red blood cells is very different. They act as mobile transport units, picking up and delivering gases. Cell fragments called *platelets* are also present in the blood. These cell "pieces" play an important role in the clotting of blood.

Red blood cells are the most common of the blood cells. There are about 5 billion in every milliliter of blood. During their maturation in the red bone marrow, they lose their nuclei and mitochondria and their endoplasmic reticula dissolve. As a result of these processes, red blood cells are relatively inactive metabolically, but they still perform the life-sustaining job of carrying oxygen to your tissues. This oxygen is carried by the iron-containing pigment hemoglobin. This pigment imparts the color to red blood cells. Hemoglobin is produced within the red bone marrow as the red blood cells are formed. An amazing 300 million molecules per cell become isolated within each red blood cell.

Blood cells float in a fluid intercellular matrix, or plasma. This fluid is both the banquet table and the refuse heap of your body, since practically every substance used and discarded by cells is found in the plasma. These substances include the sugars, lipids, and amino acids that are the fuel of the body, as well as the products of metabolism such as the waste gas carbon dioxide. The plasma also contains minerals such as calcium used to form bone; fibrinogen, which helps the blood to clot; albumin, which gives the blood its viscosity; and antibody proteins produced by lymphocytes. Every substance secreted or discarded by cells is also present in the plasma.

Many connective tissues are specialized for accumulating particular classes of molecules, such as fats and pigments. Red blood cells accumulate the oxygen-carrying pigment hemoglobin.

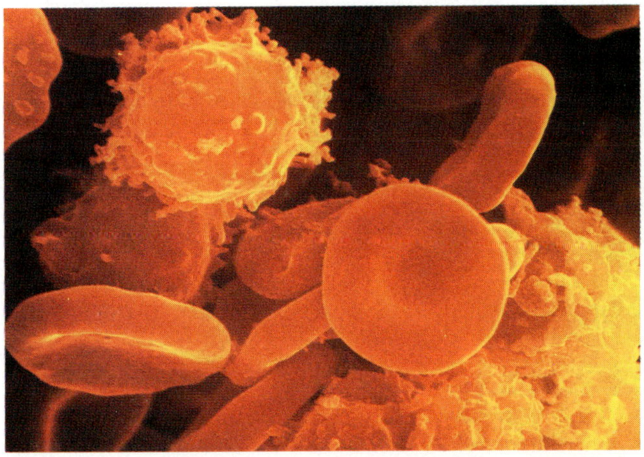

FIGURE 9-9 Blood cells.
White blood cells, or leukocytes, are roughly spherical and have irregular surfaces. Red blood cells, or erythrocytes, are disks with depressed centers.

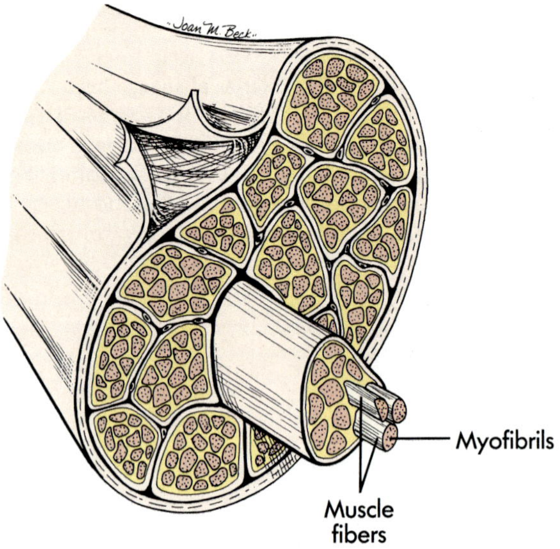

FIGURE 9-10 A skeletal muscle fiber.
Muscle fibers (cells) are composed of many myofibrils. Myofibrils, in turn, are made up of microfilaments, which are responsible for muscle contraction.

Muscle tissue

Muscle cells are the workhorses of your body. The distinguishing characteristic of muscle cells is the abundance of special thick and thin microfilaments. These microfilaments are highly organized to form strands called **myofibrils.** Each muscle cell is packed with many thousands of these myofibril strands (Figure 9-10). The myofibrils shorten when the microfilaments slide past each other, causing the muscle to contract. Table 9-4 shows the three different kinds of muscle cells of the human body: **smooth muscle, skeletal muscle,** and **cardiac muscle.**

Smooth muscle

Smooth muscle cells are long, with bulging middles and tapered ends and a single nucleus. The cells are organized into sheets, forming smooth muscle tissue. This tissue contracts involuntarily—you cannot consciously control it. Because it is found in the organs, or viscera, smooth muscle tissue is also often called *visceral muscle tissue.*

TABLE 9-4 Muscle tissue

TISSUE	LOCATION	FUNCTIONS	DESCRIPTION
Smooth muscle 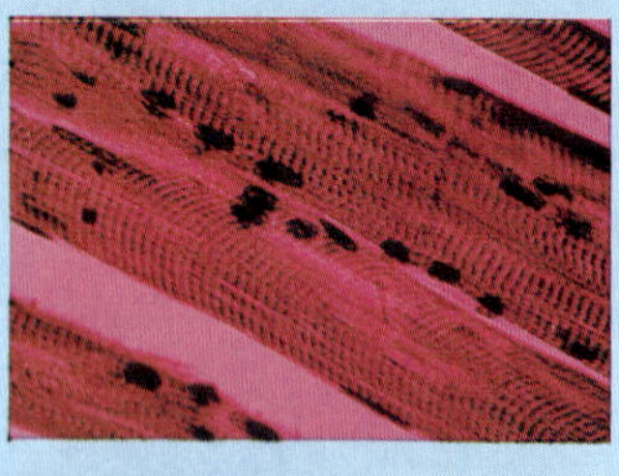	Walls of hollow organs, pupil of eye, skin (attached to hair), and glands	Regulation of size of organs, forcing of fluid through tubes, control of amount of light entering eye, and production of "gooseflesh" in the skin; functioning under involuntary control	The tissue is not striated. The spindle-shaped cells have a single, centrally located nucleus.
Skeletal muscle 	Attachment to bone	Movement of the body, functioning under voluntary control	The tissue is striated. Cells are large, long, and cylindrical with several nuclei.
Cardiac muscle 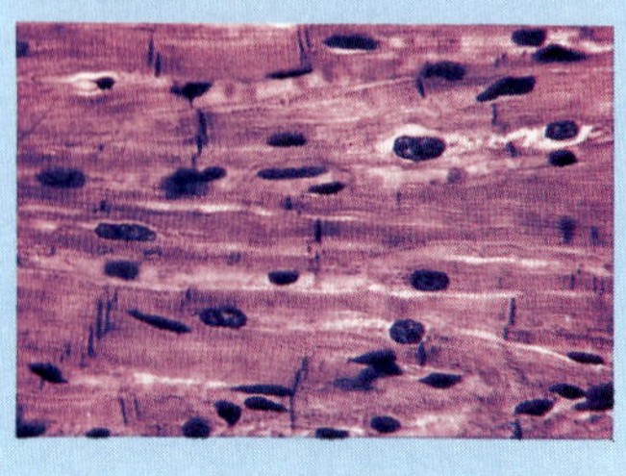	Heart	Pumping of blood, functioning under involuntary control	The tissue is striated. Cells are cylindrical and branching with a single, centrally located nucleus.

Some smooth muscle contracts when it is stimulated by a nerve or hormone. Examples of smooth muscle that contract in this way are the muscles found lining your blood vessels and those that make up the iris of your eye. But nerves do not reach each muscle cell. In many cases, impulses may be able to pass directly from one smooth muscle cell to another so that a wave of contraction can pass through a layer of some kinds of smooth muscle all by itself. In other smooth muscle tissue such as that found in the wall of the intestines, individual cells may contract spontaneously when they are stretched, leading to a slow, steady "squeeze" of the tissue.

Skeletal muscle

Skeletal muscles are attached to your bones and allow you to move your body. These muscles are called *voluntary muscles* because you have conscious control over their action. They are also called striated muscles because the tissue has "stripes"—microscopically visible bands, or striations (see Table 9-4). These striations are due to the organization of thick and thin microfilaments within the myofibrils and the alignment of the myofibrils with one another. The myofibrils are organized so that all the myofibrils within a single cell contract at the same time when the muscle cell is stimulated by a nerve.

Striated muscle cells are extremely long. A single muscle cell, or **fiber,** may run the entire length of a muscle! Each fiber has many nuclei that are pushed to the edge of the cell and lie just under the cell membrane. Bundles of these muscle cells are wrapped with connective tissue and joined with other bundles to form the muscle itself.

Cardiac muscle

The heart is composed of striated muscle fibers, but these fibers are arranged differently from their arrangements in skeletal muscle. Instead of very long cells running the length of the muscle, heart muscle is composed of chains of single cells. These chains of cells are organized into fibers that branch and interconnect, forming a latticework (see Table 9-4). This lattice structure is critical to how heart muscle functions, and it allows an entire portion of the heart to contract at one time.

Muscle cells contain microfilaments that are capable of contraction. Smooth muscle contracts involuntarily and is located in the walls of certain internal structures such as blood vessels and the stomach. Skeletal muscle is connected to bones and allows you to move your body. Cardiac muscle makes up the heart, acting as a pump for the circulatory system.

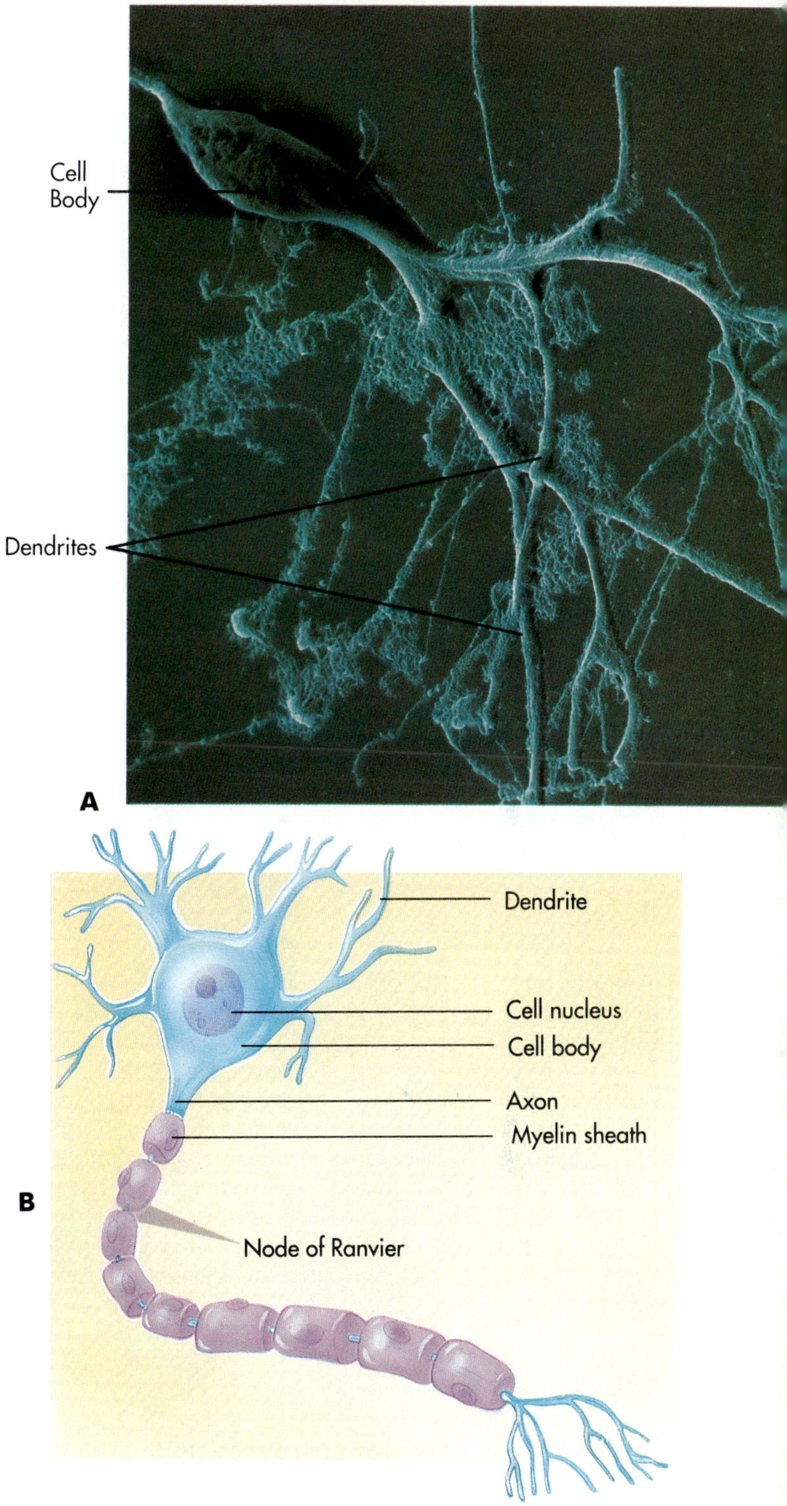

FIGURE 9-11 A human neuron.
A The cell body is at the upper left, with the axon extending up out of view. The branching network of fibers extending down from the cell body is made up of dendrites, which carry signals to the cell body.
B This diagram illustrates the generalized structure of a vertebrate neuron.

Nervous tissue

The fourth major class of tissue in humans is nervous tissue. It is made up of two kinds of cells: (1) **neurons,** which transmit nerve impulses; and (2) **supporting cells,** which nourish and protect the neurons.

Neurons are cells specialized to conduct an electric current (Figure 9-11). The cell body of a neuron contains the nucleus of the cell. Two different types of cell projections extend from the cell body. One set of projections, the **dendrites,** act as antennae for the reception of nerve impulses and conduct these impulses toward the cell body. A single projection called an **axon** conducts impulses away from the cell body. When axons or dendrites are long, they are referred to as **nerve fibers.** Some nerve fibers are so long, in fact, that they can extend from your spinal cord all the way to your fingers or toes! Single neurons over a meter in length are common.

The **nerves,** which appear as fine white threads when they are viewed with the naked eye, are actually composed of clusters of axons and dendrites. Like a telephone trunk cable, they include large numbers of independent communication channels—bundles of hundreds of axons and dendrites, each connecting a different nerve cell with a different muscle fiber or sensory receptor. In addition, the nerve contains numerous supporting cells bunched around the nerve fibers. In the brain and spinal cord, which together make up the central nervous system, these supporting cells are called **glial cells**. The supporting cells associated with nerve fibers of all other nerve cells, which make up the peripheral nervous system, are called **Schwann cells.**

Neurons are cells that are specialized to conduct electrical signals. Nerve tissue is made up of these cells and supporting cells.

Organs

The four major classes of tissues that we have discussed in this chapter (see Figure 9-1) are the building blocks of the human body. These tissues form the organs of the body. Each organ contains several different types of tissue coordinated to form the structure of the organ and to perform its function. A muscle, for example, is composed of muscle cells that together make up the muscle tissue. Bundles of this tissue are wrapped in connective tissue and wired with nervous tissue. Muscles can help you walk, pump your blood, and help you digest your food. Different combinations of tissues are found in different organs that perform different functions.

Organ systems

An organ system is a group of organs that function together to carry out the principal activities of the body. Figure 9-12 shows the eleven major organ systems of the human body.

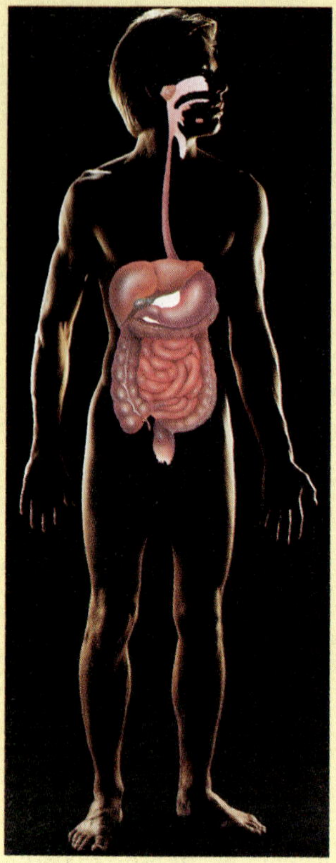

Digestive system

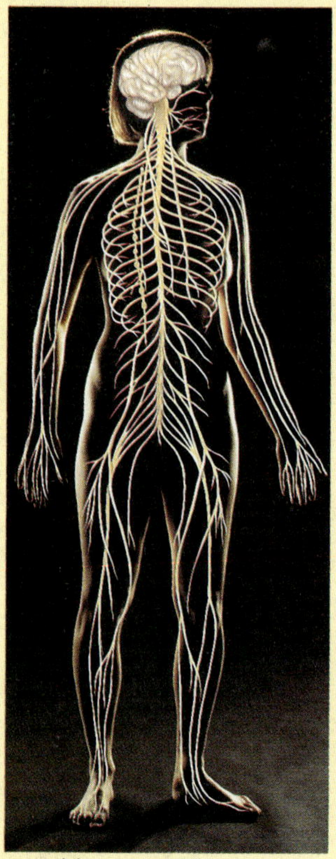

Nervous system

FIGURE 9-12 The major organ systems of vertebrates.
The body is composed of combinations of the four types of tissue, assembled in various ways. The many organs working together to carry out the principal activities of the body are traditionally grouped together as organ systems.

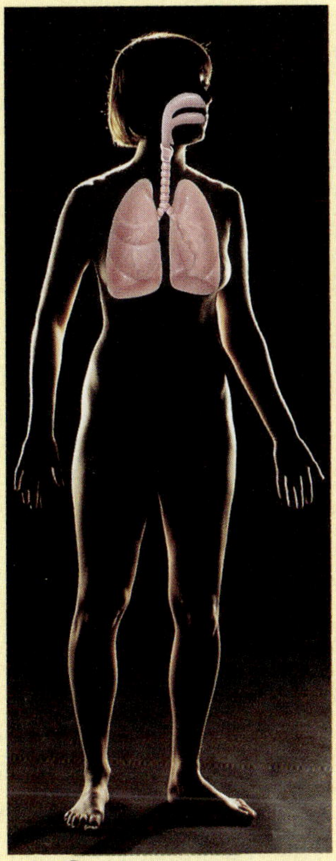

Respiratory system

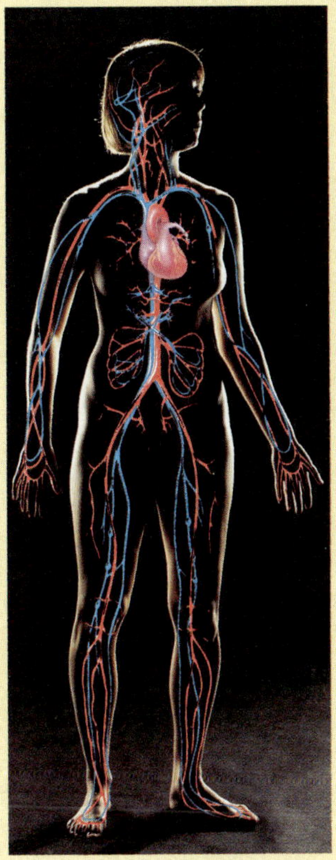

Circulatory system

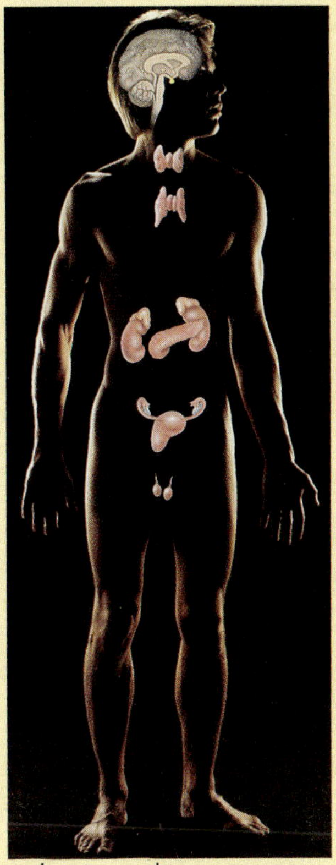

Endocrine and immune systems

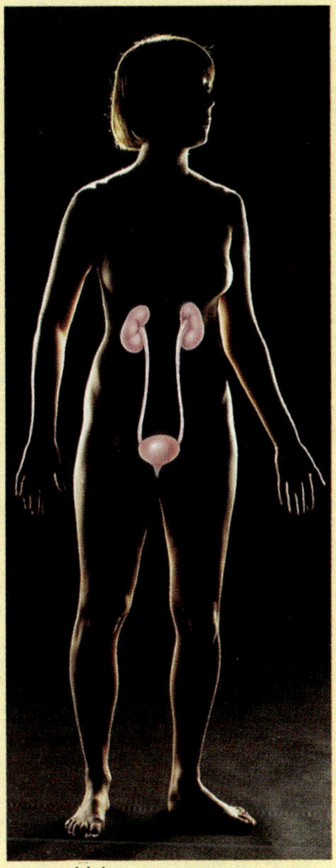

Urinary system

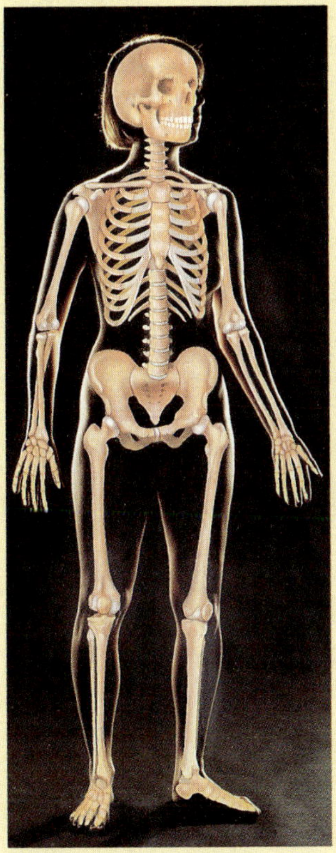

Skeletal and integumentary systems

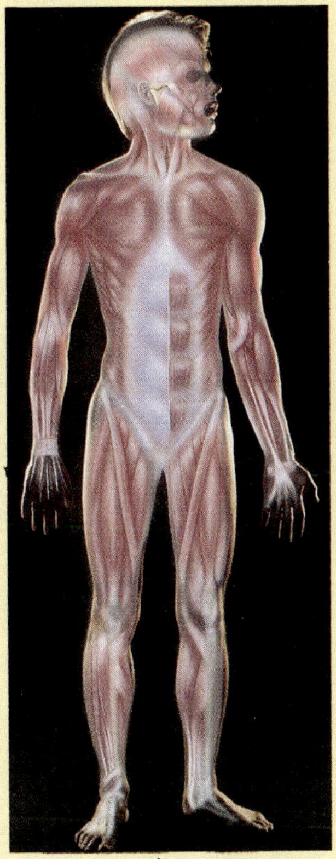

Muscular system

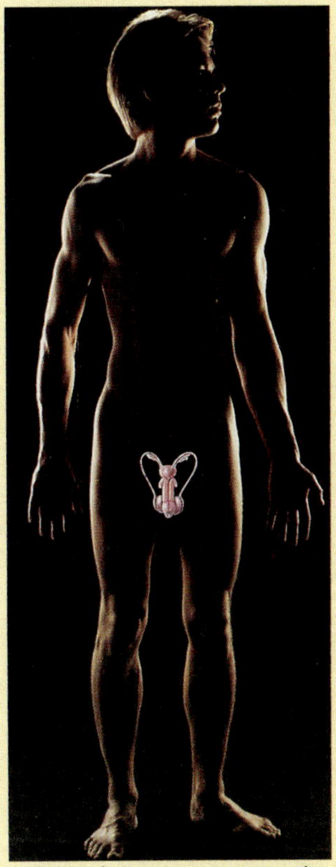

Reproductive system—male

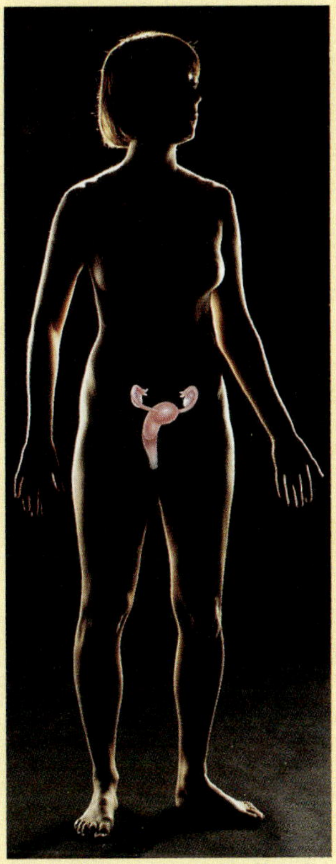

Reproductive system—female

The skeletal system supports and protects your body. It is moved by the large, voluntary muscles of the muscular system. Other muscles in this system help move internal fluids throughout your body. The nervous system regulates most of the organ systems. It can sense conditions in both your internal and external environment and help your body respond to this environmental information. The organs of your endocrine system secrete chemicals called *hormones* that also regulate body processes and functions. The circulatory system is the transportation system of the body. It brings nutrients and oxygen to your cells and removes the waste products of metabolism. Along with the immune and integumentary (skin) systems, it also helps defend the body against infection and disease. The respiratory system works hand in hand with the circulatory system, supplying the blood with oxygen and ridding it of the waste gas carbon dioxide. The food you eat is broken down by the digestive system and is absorbed through the intestinal walls into the bloodsteam. Solid wastes are also eliminated from the body by this organ system. Liquid wastes are eliminated by the urinary system after it collects waste materials and excess water from the bloodstream. And to ensure the continuance of the human race, the reproductive system produces gametes, or sex cells, that can join in the process of fertilization to produce the first cell of a new individual.

The organism: Coordinating it all

As you can see from their descriptions, the organ systems interact with one another to keep the organism—you—alive and well. This state of "wellness" is called **homeostasis.** Put another way, homeostasis is the maintainance of a stable internal environment in spite of what may be a very different external environment. To maintain this internal equilibrium, all your molecules, cells, tissues, organs, and organ systems must work together.

Your body maintains a "steady state" by means of feedback systems or **feedback loops.** Feedback loops are mechanisms by which information regarding the status of a physiological situation or system is fed back to the system so that appropriate adjustments can be made. The thermostat in your house or apartment works by means of a feedback loop and is a good analogy to use to understand feedback loops in your body. If you set the temperature in your house at 68° F, for example, the furnace will go on until that temperature is reached. A sensor in the thermostat monitors the temperature in the room, and this information triggers the electrical stimulation of the furnace. As soon as the temperature near the thermostat reaches 68° F, the thermostat sends an electrical signal to the furnace to turn off. In this way the temperature fluctuates within a small range. Such a feedback loop is called a **negative feedback loop** because the change that takes place is negative—the furnace turns off—to counter the rise in temperature. In other words, the response of the regulating mechanism (the thermostat turning off the furnace) is negative with respect to the output (the heat of the furnace). Figure 9-13 shows a negative feedback loop.

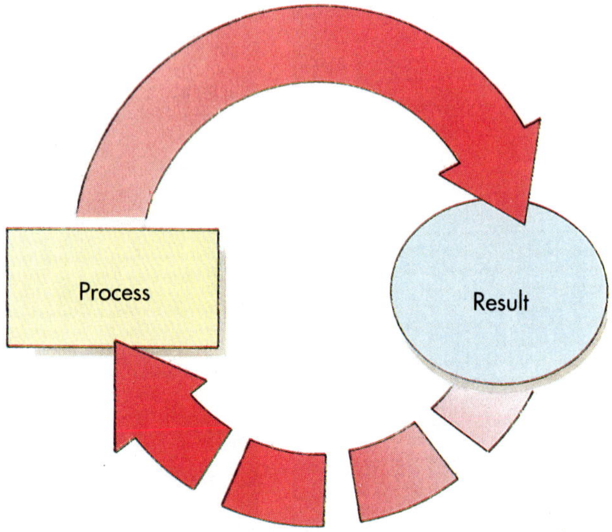

FIGURE 9-13 A negative feedback loop.
A body process produces a result such as a change in temperature or an increase in levels of a hormone. In a negative feedback loop, an increase in the magnitude of this result decreases the rate of the process. Negative feedback loops are instrumental in maintaining homeostasis within the body.

Most regulatory mechanisms of your body work by means of negative feedback loops. The secretion of hormones is regulated in this way (see Chapter 19). Sufficient levels of specific hormones in the blood trigger mechanisms that result in the shutdown of their secretion. Likewise, your body temperature is maintained by means of negative feedback, much like the temperature of your home is maintained. When your body temperature rises too high, your thermostat (a special portion of the brain called the *hypothalamus*) senses this rise and counters it. Messages race along your nerve fibers to your sweat glands, triggering the release of sweat, which evaporates and cools the body. Blood vessels near the surface of the skin dilate, or widen, bringing blood near the surface, dissipating its heat. Likewise, when your body temperature falls, the hypothalamus triggers tiny muscles under the skin to contract, causing a shiver that generates heat. Blood vessels near the surface of the skin constrict, lessening the flow of blood there and the amount of heat that radiates from the body.

Very few body mechanisms are regulated by means of **positive feedback loops.** In positive feedback loops the response of the regulating mechanism is positive with respect to the output. You can see the problems inherent with this type of "regulatory" mechanism—a situation continues to intensify rather than shutting down. For example, in a positive feedback loop, your furnace would be stimulated to stay on when the temperature in the house reached a particular setting. Positive feedback loops most often disrupt the steady state and can even lead to death. In circulatory shock, for example, a severe loss of blood can result in such a decreased blood volume that the body cannot compensate—homeostasis cannot be maintained. The blood pressure drops and the blood flow is reduced so much that the heart and brain do not receive enough oxygen. As the heart weakens, the blood flow decreases more, weakening the heart even further. As the body becomes damaged from the loss of oxygen and nutrients, the shock worsens and can result in death. There are, however, some examples of and good uses for positive feedback in the body, including oxytocin release during labor.

The molecules, cells, tissues, organs, and organ systems that make up an organism must all work together to maintain homeostasis—a stable internal environment. Homeostasis is maintained by means of negative feedback loops, regulatory mechanisms that slow down or shut down output systems when they reach certain levels. For example, a high internal body temperature triggers mechanisms that lower the temperature.

Summary

1. The human body is organized hierarchically; that is, small units of structure and function together form larger units of structure and function, eventually forming the whole organism. Specifically, the cells of the body are organized into tissues. Several different tissues grouped together in a structural and functional unit make up an organ. Organs that work together to carry out particular body activities are called *organ systems*. The human body is an organism made up of 11 major organ systems.
2. The four basic types of tissue are epithelial tissue, connective tissue, muscle tissue, and nerve tissue.
3. Epithelium covers the surfaces of the body. The lining of the major body cavities is composed of simple epithelium, whereas the exterior skin is composed of stratified epithelium. The major gland systems are also derived from epithelium.
4. The defensive connective tissue cells are macrophages, which engulf foreign bacteria and antibody-coated cells or particles; lymphocytes, certain of which produce antibodies; and mast cells, which release inflammation-producing chemicals at sites of trauma.
5. The structural connective tissue cells include cartilage cells (chondrocytes), bone cells (osteocytes), and fibroblasts.
6. The isolating connective tissue cells include pigment-producing cells, such as red blood cells, and fat cells.
7. Muscle contraction provides the force for mechanical movement of the body. There are three kinds of muscle tissue: smooth muscle, which is found in internal organs; skeletal muscle, which moves the body parts; and cardiac muscle, which makes up the heart—the pump of the circulatory system.
8. Nerve cells provide the body with a means of rapid communication. There are two kinds of nerve cells: neurons and supporting cells. Neurons are specialized for the conduction of electrical impulses.
9. Organs are body structures composed of several different tissues grouped into a structural and functional unit. An organ system is a group of organs that function together to carry out the principal activities of the body.
10. The human body is organized so that its parts form an integrated whole. These parts, work together at each level of organization to maintain homeostasis—a stable internal environment within the human organism.

KEY TERMS

cardiac muscle 158
chondrocytes 153
coelom 149
columnar 150
compact bone 156
connective tissue 147
cuboidal 150
defensive connective tissue 152
dense fibrous connective tissue 153
elastic cartilage 153
elastic connective tissue 153
epithelial cell 149
epithelial tissue 147
erythrocytes 157
fibroblast 153
fibrocartilage 153
homeostasis 162
hyaline cartilage 153
isolating connective tissue 152
lacunae 153
leukocytes 157
loose connective tissue 153
lymphocyte 152
macrophage 152
mast cell 152
matrix 152
muscle tissue 147
negative feedback loop 162
nervous tissue 147
neuron 159
organ 147
organ system 149
osteocytes 153
positive feedback loop 163
pseudostratified epithelium 151
reticular connective tissue 153
simple epithelium 151
skeletal muscle 158
smooth muscle 158
spongy bone 156
squamous 150
stratified epithelium 151
structural connective tissue 152
supporting cell 159
tissue 147
transitional epithelium 151

REVIEW QUESTIONS

1. Distinguish among cells, tissues, organs, and organ systems. Give an example of each.
2. What are the four basic types of tissue in the human body? Give an example of each.
3. What are epithelial cells? Summarize their functions.
4. Over time, regular cigarette smoking paralyzes the cilia that line the respiratory passageways. How could this affect the smoker's health?
5. Draw and label the three shapes of epithelial cells. Give an example of where each is found in your body.
6. What is connective tissue, and what are its functions?
7. What do lymphocytes, erythrocytes, and fibroblasts have in common? How do they differ?
8. Your cat scratches your finger, and the skin surrounding the scratch turns reddish and feels warm. Explain why.
9. Fill in the blanks. The three principal types of cells in structural connective tissue are ________, ________, and ________.
10. Fill in the blanks: ________ produce cartilage, and ________ produce bone. ________bone is composed of an open lattice of bone; it is less dense than ________ bone.
11. Match each term with the most appropriate phrase:
    Platelets
    Plasma
    Erythrocytes
    Leukocytes
    Hemoglobin
    a. Red blood cells
    b. White blood cells
    c. A pigment containing iron
    d. The fluid intercellular matrix of blood
    e. Cell fragments that are important in the clotting of blood.
12. Everybody hates fat cells. Do they serve a purpose, and if so, what?
13. Identify the three kinds of muscle cells in your body, and give an example of each.
14. Fill in the blanks: Nervous tissue consists of two kinds of cells: ________, which ________, and ________, which ________.
15. Create a table that lists the major function(s) and at least two components of each of the principal organ systems in your body.
16. What is homeostasis? How is it achieved?

THOUGHT QUESTIONS

1. Certain tissues in your body do not repair themselves if damaged (nervous tissue is a good example), whereas other tissues like skin rapidly repair themselves. Why do you think nerve tissue does not repair itself as readily as skin?
2. Which tissue types have the greatest numbers of cells in your body? The least? How is this related to their function?
3. Some cell types actively move around. Which ones can you name?

FOR FURTHER READING

Caplan, A. (1984, October). Cartilage. *Scientific American*, pp. 84-97.
This is an interesting account of the many roles played by cartilage in the vertebrate body.

Houk, J. C. (1988, February). Control strategies in physiological systems. *FASEB Journal*, p. 97.
This general discussion of control systems is written for a general audience.

National Geographic Society: *The incredible machine*, National Geographic Society, Washington, D.C., 1986.
This series of outstanding articles on the human body focuses on its major organ systems. It is beautifully illustrated and fun to read.

Digestion

10

What's wrong with this picture? Take a critical look at the photo and solve this nutrition "puzzle."

Nothing looks awry? Fish is certainly a healthy choice, you might be thinking, and you are correct. In fact, evidence suggests that a group of substances in fish oil protects the human body against heart attack. In addition, fish is generally lower in calories than beef or poultry and has other "healthy" qualities. The problem is that the typical American dinner usually contains double, triple, or sometimes quadruple the amount of protein needed for the entire day! The half-pound of fish here is too hefty a portion of protein.

Vegetables and a potato are also healthy food choices. However, you may not be able to see the butter that is saturating these foods. This fat, plus the fat in the salad dressing and ice cream, dooms this meal to having at least half its calories come from fat—far too many when you realize that one tablespoon of vegetable oil satisfies your daily requirement.

The sugary soda and dessert probably caught your eye. As you may know, refined sugar, or sucrose, is a relatively simple carbohydrate that provides only calories—something most Americans already consume in excess. Sugar is therefore often called a source of "empty" calories, and it also promotes tooth decay.

HIGHLIGHTS

- Carbohydrates, fats, and proteins supply the body with energy and the raw materials to build the substances it needs.
- Vitamins and minerals help the chemical reactions of the body take place; water provides the medium for these reactions.
- Digestion breaks down the "energy nutrients"—carbohydrates, fats, and proteins—into smaller molecules that can be used to produce ATP by means of cellular respiration.
- Parts of the digestive system are specialized to perform both the mechanical and chemical processes of digestion.

OUTLINE

The nutrition-digestion connection

Where it all begins: The mouth

The journey of food to the stomach

Preliminary digestion: The stomach

Terminal digestion and absorption: The small intestine

Accessory organs that help digestion

Digestion

Absorption

Concentration of solids: The large intestine

Diet and nutrition

How would you change the composition of this meal to make it healthier? Do you know which types of foods to add and which to cut down on or omit? To do this task, you may need to know the answers to a few questions. What are nutrients? Why do we need them to survive? What amounts and types of nutrients do we need? And how do our bodies get these nutrients from the food we eat? The answers to some of these questions begins the story of digestion. By the time you finish reading this chapter, you should be able to answer all the questions.

The nutrition-digestion connection

As you may recall from Chapter 8, humans are heterotrophs, organisms that cannot produce their own food. Heterotrophs must ingest, or take in, food. At least 95% of the species of organisms on Earth—all animals, all fungi, and most protists and bacteria—are heterotrophs. Food provides these "other-feeders" with two things: (1) energy and (2) the raw materials to build the substances they need. The energy in food is described in units called **calories,** or more properly kilocalories (thousands of calories, or kcal). The raw materials are called **nutrients.**

There are six classes of nutrients: **carbohydrates, fats** (lipids), **proteins, vitamins, minerals,** and **water.** These are the substances, in fact, that make up most of your body. Assuming that you are a proper weight for your height, your body is made up of about 60% water and about 20% fat. The other 20% is mostly protein, carbohydrate, combinations of these two substances, and two major minerals found in your bones: calcium and phosphorus. Other minerals and vitamins make up less than 1% of you.

As you may recall from Chapter 2, carbohydrates, lipids, and proteins are organic (carbon-containing) compounds and are used by your body as a source of energy, or kilocalories. They are therefore often referred to as the "energy" nutrients. (Proteins, however, are not a "preferred" energy source.) In addition, these three organic compounds are used as building blocks for growth and repair as well as to produce other substances your body may need.

Your body does not obtain energy from vitamins, minerals, or water. Vitamins are organic molecules that perform functions such as helping your body form red blood cells and "unlock" the energy in carbohydrates, fats, and proteins. Although required only in small amounts, each of the 13 different vitamins plays a vital role in your body (Table 10-1). Many, such as the B vitamins, are used for the synthesis of coenzymes, small molecules that help enzymes work. For example, niacin and riboflavin, two of the B vitamins, are used to make the coenzymes NAD and FAD used in cellular respiration (see Chapter 7). The roles and interactions of some other vitamins, such as vitamin C, are not as well understood.

Minerals are inorganic substances and are transported around the body as ions dissolved in the blood and other body fluids. Your body uses a variety of minerals, and they perform a variety of functions (see Table 10-1). Calcium, for example, does many jobs, including making up a part of the structure of your bones and teeth and helping your blood to clot. Sodium plays a key role in regulating the fluid balance within your body. Magnesium is an important player in the process of releasing energy from carbohydrates, fats, and proteins.

Water provides the medium in which all the body's reactions take place. Along with dissolved substances and (sometimes) suspended solids, it bathes your cells, makes up much of the cytosol within your cells, and courses through your arteries and veins. It even lubricates your joints and cushions organs such as the brain and spinal cord. The atoms that make up water molecules are also held within the bonds of the energy molecules. These atoms are key players when bonds are broken (by hydrolysis) or formed (by dehydration synthesis) in these molecules. (See Chapter 2 for a description of the processes of hydrolysis and dehydration synthesis.)

Your body obtains the energy and raw materials it needs to survive from six classes of nutrients. Carbohydrates, fats, and proteins provide energy and building blocks. Vitamins, minerals, and water help body processes take place. Some minerals are also incorporated into body structures.

To obtain these nutrients and energy from the food you ingest, your body must first digest the food. **Digestion** is a process in which food particles are broken down into small molecules that can be absorbed by the body. Digestion is carried out mechanically in the mouth as food is crushed by the teeth and in the stomach as food is churned by the stomach's muscular walls. Digestion is carried out chemically in three ways: (1) by hydrochloric acid (HCl), which denatures, or unfolds, protein molecules and disrupts the protein "glue" that holds cells together; (2) by bile salts, which emulsify, or separate, large lipid droplets into much smaller lipid droplets; and (3) by a variety of highly specific enzymes that help cleave certain chemical bonds (Table 10-2). As you read in Chapter 6, enzymes are proteins that speed up the rate of chemical reactions in living things. In fact, without enzymes, you would die before needed chemical reactions took place!

TABLE

10-1

Major vitamins and minerals

	FUNCTION	DIETARY SOURCE	RECOMMENDED DAILY ALLOWANCE	DEFICIENCY SYMPTOMS
VITAMINS				
Vitamin A (retinol)	Used in making visual pigments, maintenance of epithelial tissues	Green vegetables, milk products, liver	1 mg	Night blindness, flaky skin
B-complex vitamins				
B_1	Coenzyme in CO_2 removal during cellular respiration	Meat, grains, legumes	1.5 mg	Beriberi, weakening of heart, edema
B_2 (riboflavin)	Part of coenzymes FAD and FMN, which play metabolic roles	Many different kinds of foods	1.8 mg	Inflammation and breakdown of skin, eye irritation
B_3 (niacin)	Part of coenzymes NAD^+ and $NADP^+$	Liver, lean meats, grains	20 mg	Pellagra, inflammation of nerves, mental disorders
B_5 (pantothenic acid)	Part of coenzyme-A, a key connection between carbohydrate and fat metabolism	Many different kinds of foods	5 to 10 mg	Rare disorder, fatigue, loss of coordination
B_6 (pyridoxine)	Coenzyme in many phases of amino acid metabolism	Cereals, vegetables, meats	2 mg	Anemia, convulsions, irritability
B_{12} (cyanocobalamin)	Coenzyme in the production of nucleic acids	Red meat, dairy products	0.003 mg	Pernicious anemia
Biotin	Coenzyme in fat synthesis and amino acid metabolism	Meat, vegetables	Minute	Rare disorder, depression, nausea
Folic acid	Coenzyme in amino acid and nucleic acid metabolism	Green vegetables	0.4 mg	Anemia, diarrhea
Vitamin C	Important in forming collagen, cement of bone, teeth, connective tissue of blood vessels; possible help in maintaining resistance to infection	Fruit, green leafy vegetables	45 mg	Scurvy, breakdown of skin, blood vessels
Vitamin D (calciferol)	Increase of absorption of calcium and promotion of bone formation	Dairy products, cod liver oil	0.01 mg	Rickets, bone deformities
Vitamin E (tocopherol)	Protection of fatty acids and cell membranes from oxidation	Margarine, seeds, green leafy vegetables	15 mg	Rare disorder
Vitamin K	Essential to blood clotting	Green leafy vegetables	0.03 mg	Severe bleeding

Continued.

TABLE

10-1

Major vitamins and minerals—cont'd

	FUNCTION	DIETARY SOURCE	RECOMMENDED DAILY ALLOWANCE	DEFICIENCY SYMPTOMS
MINERALS				
Calcium (Ca)	Constituent of bones and teeth; participant in blood clotting, nerve transmission, muscle action, cell membrane permeability, enzyme activity	Milk, milk products, green leafy vegetables, whole grains, egg yolk, legumes, nuts	800 mg	Rickets, osteoporosis, tetany (decrease in serum Ca level)
Chlorine (Cl)	Water balance, acid-base balance, regulation of HCl synthesis in digestion	Salt	1.7-5.1 mg	Acid-base imbalance
Chromium (Cr)	Association with enzymes in glucose metabolism	Cereals, whole grains, brewer's yeast, animal protein	0.05-0.2 mg	Unknown
Cobalt (Co)	Component of vitamin B_{12}, red blood cell production	Same as vitamin B_{12}	Unknown, evidently minute	Anemia
Copper (Cu)	Hemoglobin and melanin production, electron transport system	Liver, meat, seafood, whole grains, legumes, nuts	2-3 mg	Anemia and loss of energy
Fluoride (F)	Extra strength in teeth, prevention of dental caries	Fish, tea, drinking water	1.5-4 mg	No deficiency symptoms
Iodine (I)	Thyroid hormone production, maintenance of normal metabolic rate	Iodized salt, seafood	150 μg	Decrease of normal metabolism
Iron (Fe)	Hemoglobin synthesis, oxygen transport	Liver, meat, eggs, whole grains, dark green vegetables, legumes, nuts, enriched breads and cereals	10-18 mg	Anemia, decreased oxygen transport, energy loss

The enzymes that help digest the energy nutrients—proteins, carbohydrates, and lipids—are of three basic types:

1. **Proteases,** which break down proteins to smaller polypeptides and polypeptides to amino acids.
2. **Amylases,** which break down starches and glycogen to sugars. (Both starch and glycogen are storage forms of polysaccharides.)
3. **Lipases,** which break down the triglycerides in lipids to fatty acids and glycerol.

(Proteins, carbohydrates, and lipids, along with their breakdown products, are described in Chapter 2.)

During digestion, proteins are unfolded by hydrochloric acid and are then cleaved into peptides and individual amino acids by protease enzymes. Starch and glycogen are digested to sugars by amylases. Triglycerides are digested to fatty acids and glycerol by lipases.

	FUNCTION	DIETARY SOURCE	RECOMMENDED DAILY ALLOWANCE	DEFICIENCY SYMPTOMS
MINERALS—cont'd				
Magnesium (Mg)	Bone formation, muscle and nerve function, coenzyme constituent	Milk, milk products, meat, seafood, whole grains, legumes, nuts	300-350 mg	Increased nervous system irritability, dilation of blood vessels
Manganese (Mn)	Hemoglobin synthesis, growth, activation of several enzymes	Cereals, whole grains, legumes, soybeans, leafy vegetables	2.5-5 mg	Tremors and convulsions
Phosphorus (P)	Bone and teeth formation, important in energy transfer (ATP), component of nucleic acids	Milk, milk products, meat, egg yolk, whole grains, legumes, nuts	800 mg	Loss of energy and cellular function
Potassium (K)	Muscle and nerve function	Fruits, vegetables, legumes, nuts, whole grains, meat	1.8-5.6 mg	Muscle weakness, abnormal electrocardiogram
Sodium (Na)	Osmotic pressure regulation, nerve and muscle function	Table salt, milk, milk products, meat, eggs, carrots, beets, spinach, celery	1.1-3.3 mg	Nausea, vomiting, exhaustion and dizziness
Sulfur (S)	Component of hormones, several vitamins, and proteins	Meat, eggs, milk, milk products, legumes, nuts	Unknown	Unknown
Zinc (Zn)	Component of several enzymes, carbon dioxide transport and metabolism, necessary for protein metabolism	Seafood, liver, meat, milk, milk products, eggs, whole grains	15 mg	Deficient carbon dioxide transport and deficient protein metabolism

TABLE 10-2

Digestive enzymes

SOURCE	ENZYME	SUBSTRATE	DIGESTION PRODUCT
Salivary gland	Amylase	Starch, glycogen	Disaccharides
Stomach	Pepsin	Proteins	Short polypeptides
Small intestine	Peptidases	Short peptides	Amino acids
	Disaccharidases: Lactase, Maltase, Sucrase	Disaccharides	Glucose, monosaccharides
Pancreas	Pancreatic amylase	Starch, glycogen	Disaccharides
	Trypsin	Proteins	Polypeptides
	Chymotrypsin		
	Carboxypeptidase		

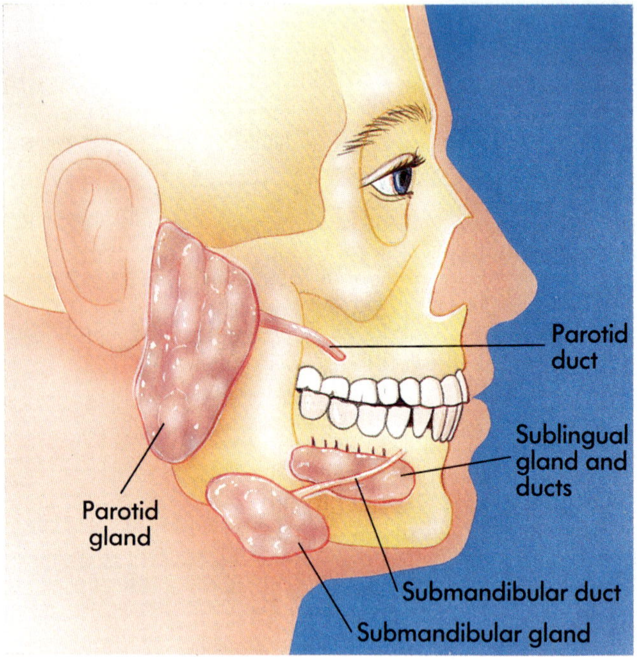

FIGURE 10-1 The salivary glands.
Saliva, the solution that these glands secrete, contains salivary amylase, an enzyme that breaks down starch into maltose.

Where it all begins: The mouth

Does your mouth ever "water" when you think about your favorite food? Does the smell of some of your favorite foods evoke this reaction too? Do you know why this reaction occurs?

The water in your mouth is really a secretion from a set of glands called **salivary glands.** The locations and names of each of these glands are shown in Figure 10-1. Their secretion, **saliva,** is a solution that consists primarily of water, mucus, and the digestive enzyme **salivary amylase.** Other substances can be found in smaller amounts, such as antibodies and a bacteria-killing enzyme.

Salivary amylase breaks down starch into molecules of the disaccharide maltose. During this reaction, the starch is broken down on the addition of water—a process called **hydrolysis.**

$$\text{Starch} + \text{Water} \xrightarrow{\text{Salivary amylase}} \text{Maltose}$$

Enzymes that break down substances by hydrolysis are called *hydrolyzing enzymes*. Along with playing this role in digestion, the saliva also moistens and lubricates food so that it is swallowed easily and does not scratch the tissues of the throat.

The secretions of the salivary glands are controlled by the nervous system. Although the nervous system works to maintain a constant secretion of saliva in the mouth, it "speeds up" the secretion when it is stimulated by the presence (or sometimes the sight or smell) of food. When food is in the mouth, nerve endings called *chemoreceptors,* which are sensitive to the presence of certain chemicals, send a signal to the brain, which responds by stimulating the salivary glands. Did you ever suck on a slice of lemon? Then you know that the most potent stimuli are acid solutions: lemon juice can increase the rate of salivation eightfold.

Mechanical digestion begins when the teeth tear food apart into tiny pieces. In this way the surface area of the food is increased, allowing the digestive enzymes to mix with the food and break it down more quickly and completely. The teeth of humans as well as those of other organisms are specialized in different ways. This specialization depends on the type of food an organism eats and how it obtains its food. Human beings are **omnivores.** An omnivore eats both plant and animal foods. As a result, human teeth are structurally intermediate between the pointed, cutting teeth characteristic of **carnivores,** or meat eaters, and the flat, grinding teeth characteristic of **herbivores,** or plant eaters. In fact, the teeth of humans are like carnivores in the front and herbivores in the back.

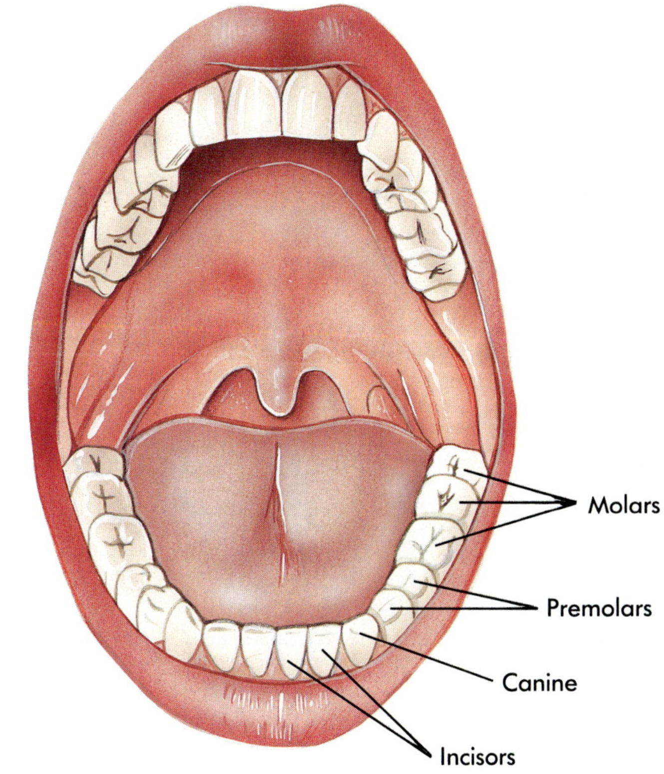

A

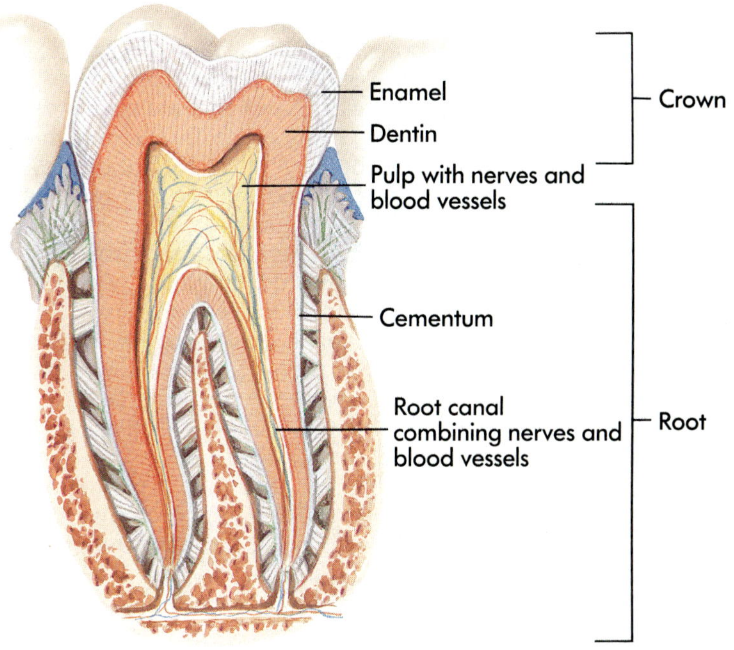

B

FIGURE 10-2 Teeth in humans.

A The four front teeth are incisors, used for biting. Canines are used for tearing food. The premolars and molars are used for grinding and crushing.

B Although you may forget until you are in the dentist's chair, each tooth in your mouth is alive and contains nerves and blood vessels.

The four front teeth in the upper and lower jaws of humans are **incisors** (Figure 10-2, *A*). These teeth are sharp and chisel shaped and are used for biting. On each side of the incisors are pointed teeth called **canines,** which are used in tearing food. Behind each canine, on each side of the mouth and along both top and bottom jaws, are two **premolars** and three **molars,** all of which have flattened ridged surfaces for grinding and crushing food. In early childhood, however, humans do not have these 32 adult teeth but only 20 "baby teeth." These first teeth are lost during childhood and are replaced by the 32 adult teeth.

Each tooth is alive and is rooted in the bones of the upper and lower jaw. The gums cover this bone; the portion of the tooth protruding above the gumline is called the **crown** (Figure 10-2, *B*). The portion of the tooth that extends into the bone is called the **root.** Inside lies a central, nourishing **pulp** containing nerves, blood vessels, and connective tissue. The nerves and blood vessels pass out of the tooth through holes at the root. The actual chewing surface of the tooth is made up of enamel—the hardest substance in the body. It is layered over the softer dentin that forms the body of the tooth.

Teeth shred and grind the plant and animal material that humans ingest as food. Saliva secreted into the mouth moistens the food, which aids its journey into the rest of the digestive system and begins its enzymatic digestion.

The journey of food to the stomach

As food is chewed and moistened, the tongue forms it into a ball-like mass called a **bolus** and pushes it into the **pharynx.** The pharynx is the upper part of the throat (Figure 10-3). As this happens, the soft palate raises up, sealing off the nasal cavity and preventing any food from entering this chamber. The soft palate is the tissue at the back of the roof of the mouth. The pressure of the food in the pharynx stimulates nerves in its walls that begin the swallowing reflex, an involuntary action. As part of this reflex action, the voice box, or **larynx,** raises up to meet the **epiglottis,** a flap of tissue that folds back over the opening to the larynx. With this action the epiglottis acts much like a trapdoor, closing over the **glottis,** the opening to the larynx and trachea (your windpipe) so that food will not go down "the wrong way." If you place your hand over your larynx (Adam's apple), you can feel it move up when you swallow.

After passing into the pharynx and then bypassing the windpipe, the food enters the **esophagus,** a food tube that connects the pharynx to the stomach (Figure 10-4). The esophagus, which is about 25 centimeters long (a bit less than a foot), pierces the diaphragm before it connects to the stomach. The diaphragm is a sheet-like muscle that forms the floor of the chest cavity and is a muscle of breathing (see Chapter 11). The opening in the diaphragm that allows the esophagus to pass is called the *esophageal hiatus* (opening). Sometimes the stomach, intestine, or other abdominal organ pushes through this opening up into the chest, resulting in occasional heartburn. Such a condition, called a *hiatal hernia,* is quite common, especially in people over 50 years of age, but it can be treated with a bland diet, antacids, and sleeping with the head elevated.

The esophagus ends at the "door" to the stomach. This door is called the **cardiac opening** and is ringed by a circular muscle called the **lower esophageal sphincter.** When the ring of muscle contracts, or tightens, it closes the cardiac opening. When it relaxes, or loosens, it opens this stomach door. Although it is in the abdominal cavity, the cardiac opening is located very close to the heart, which lies just above it in the thoracic cavity. Have you ever had heartburn? This distress is caused when the acid contents of the stomach splash back into the esophagus at the cardiac opening. The acid burns the esophagus—but it feels like your heart is on fire!

The esophagus does not take part in digestion but instead acts like an escalator, moving food down toward the stomach. The sets of longitudinal and circular muscles in its walls work in tandem to produce successive waves of contractions called **peristalsis** (Figure 10-5). The movement of food to the stomach is therefore not dependent on gravity, so even astronauts in zero gravity—or you standing on your head—can swallow without difficulty.

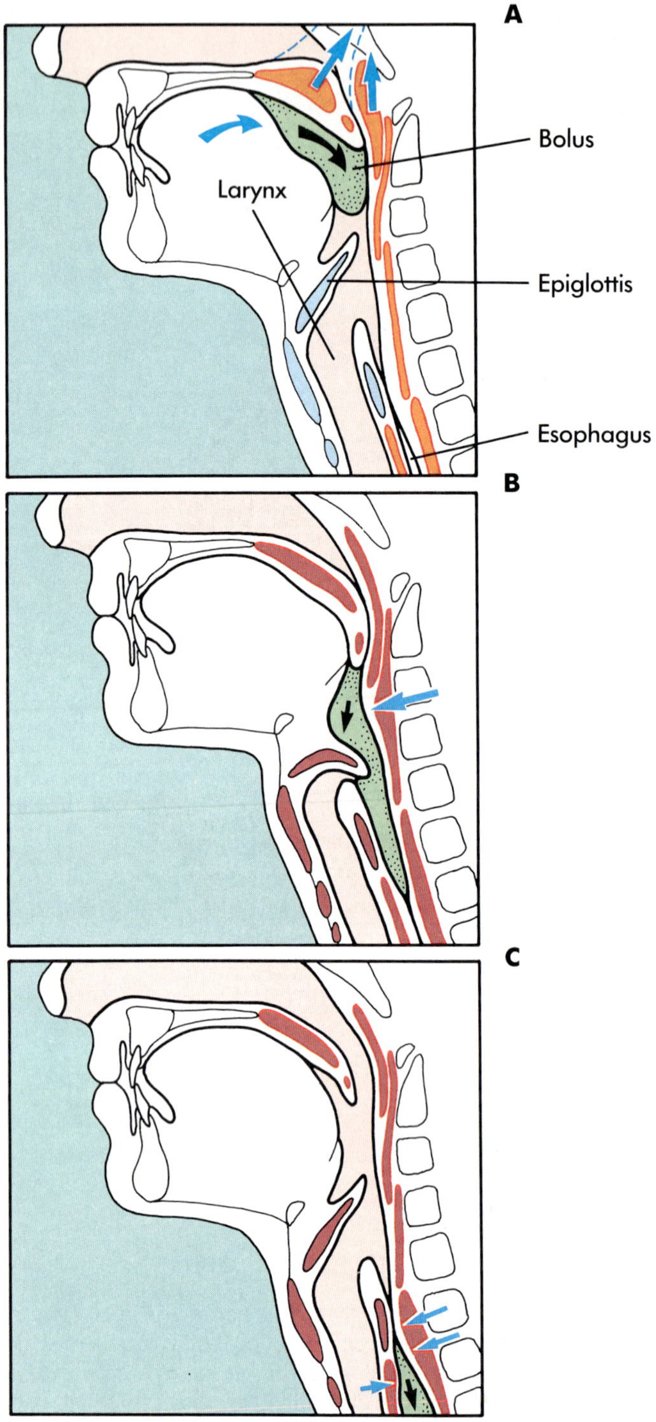

FIGURE 10-3 How humans swallow.
As food passes past the rear of the mouth **(A)**, muscles raise the soft palate against the back wall of the pharynx, sealing off the nasal passage. In addition, a flap of tissue called the *epiglottis* folds down **(B)**, sealing the respiratory passage. After the food enters the esophagus the soft palate lowers and the epiglottis is raised **(C)**, opening the respiratory passage between the nasal cavity and the trachea.

When the tongue pushes food into the pharynx, nerves stimulate the swallowing reflex. The food then enters the esophagus, or food tube, and is moved to the stomach by rhythmic muscular contractions called *peristalsis.*

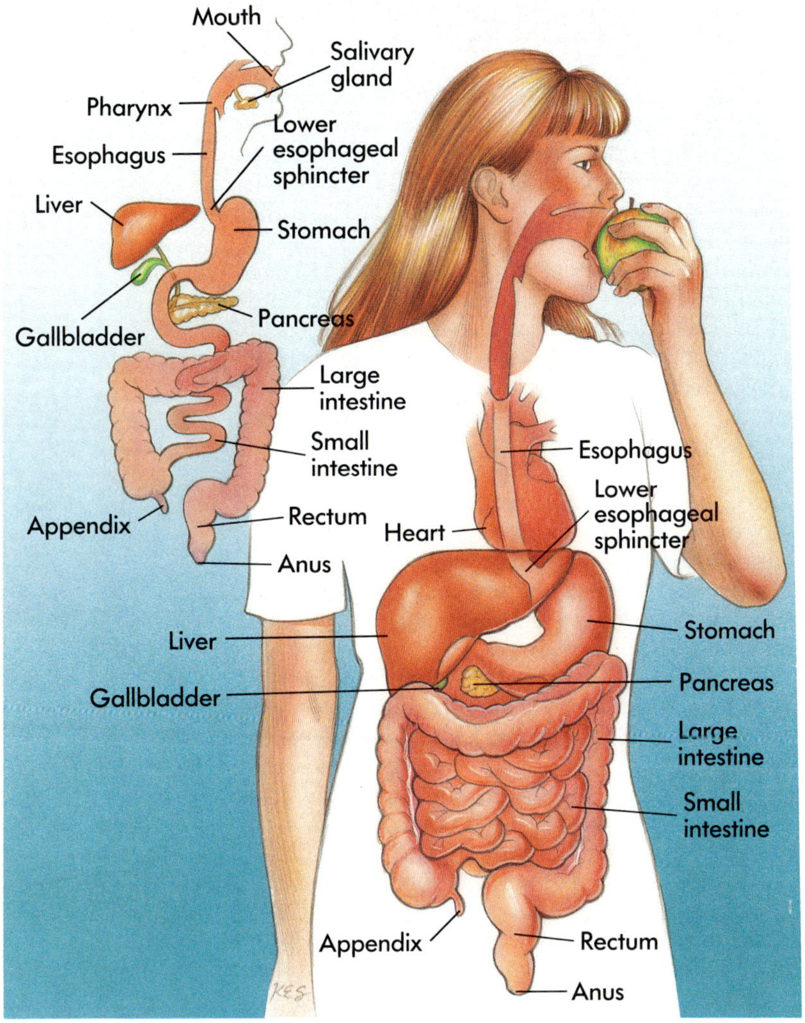

FIGURE 10-4 The human digestive system. Note the location of the cardiac opening, which acts as a "door" to the stomach. The pain of heartburn is caused when acid in the stomach splashes through this opening, burning the esophagus.

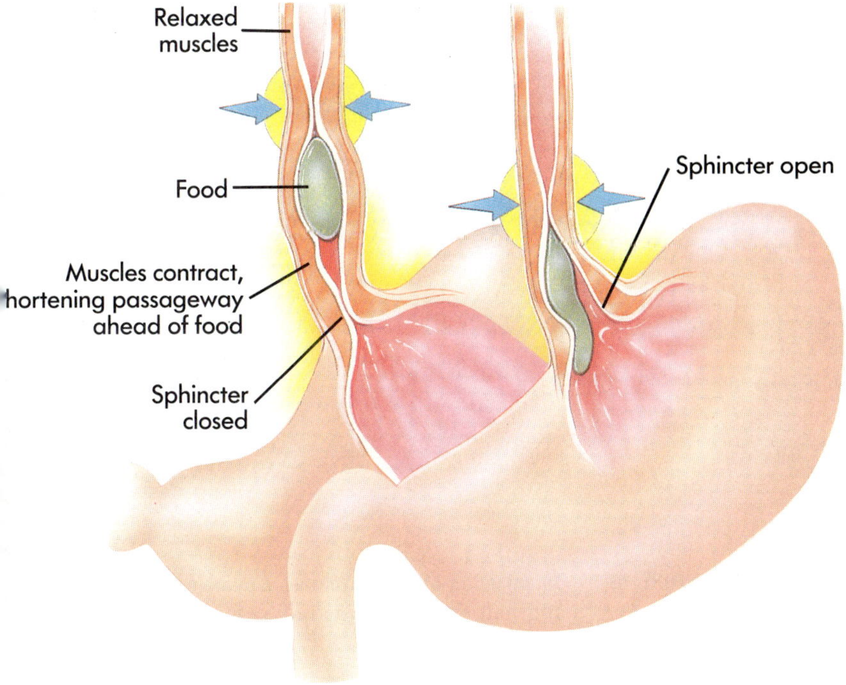

FIGURE 10-5 Peristalsis. Successive waves of contraction move food down the esophagus to the stomach.

Preliminary digestion: The stomach

The lower esophageal sphincter relaxes when food reaches it, allowing food to enter the **stomach.** The stomach is a muscular sac in which the food is collected and partially digested by hydrochloric acid and proteases (Figure 10-6). The stomach then "feeds" this food, little by little, to the primary organ of digestion, the small intestine.

The stomach and intestinal tract have the same basic structural plan (Figure 10-7). Their interiors are lined with a layer of tissue called the **mucosa.** The mucosa consists of epithelial cells, blood and lymph vessels, and a thin layer of muscle. It covers a deeper, thicker layer of connective tissue, the **submucosa,** which is rich in blood vessels and nerves. Surrounding the connective tissue are layers of smooth muscle tissue—three in the stomach and two in the intestines. An envelope of tough connective tissue called **serosa** serves as the outer covering of the digestive tract (as well as the other abdominal organs). Thin sheets of connective tissue called **mesentery** are attached to the serosa along most of the intestinal tract, holding it "in place" and serving as a highway for blood vessels and nerves.

The inner surface of the stomach is dotted with **gastric glands** that extend from the epithelium but dip deeply into the mucosa. Two different kinds of cells in these glands secrete a gastric juice made up of hydrochloric acid and the protein pepsinogen. After the secretion of both, the acid chemically interacts with the pepsinogen, converting it to the protein-hydrolyzing enzyme **pepsin.** The hydrochloric acid also softens the connective tissue in foods; denatures, or unfolds, large protein molecules; and kills most bacteria that may have been ingested with the food. The pepsin digests only proteins, breaking them down into short polypeptides. Other epithelial cells are specialized for the secretion of mucus. This mucus, produced in large quantities, lubricates the stomach wall and protects the stomach from digesting itself.

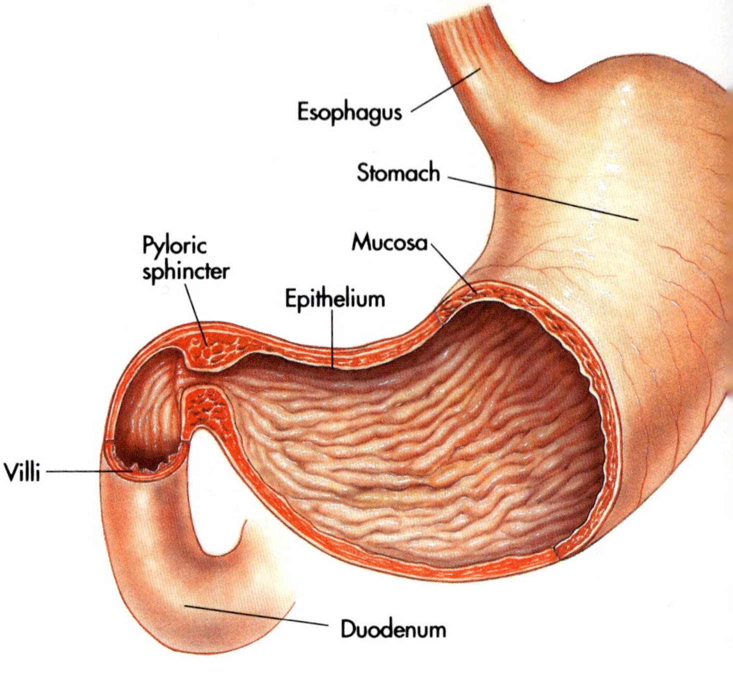

FIGURE 10-6 The stomach.
The stomach stores food, partially digests it with hydrochloric acid and proteases, and moves the food to the small intestine.

The stomach controls the production of gastric juice by means of a digestive hormone called **gastrin** (Table 10-3). Hormones are regulating chemicals that are made at one place in the body and work in another. Gastrin is produced by endocrine (hormone-secreting) cells that are scattered throughout the epithelium of the stomach. Some stomachs greatly overproduce gastrin, however, which results in excessive acid production. The causes of this overproduction include such factors as stress, heredity, diet, and smoking. This extra acid may attack the walls of the first portion of

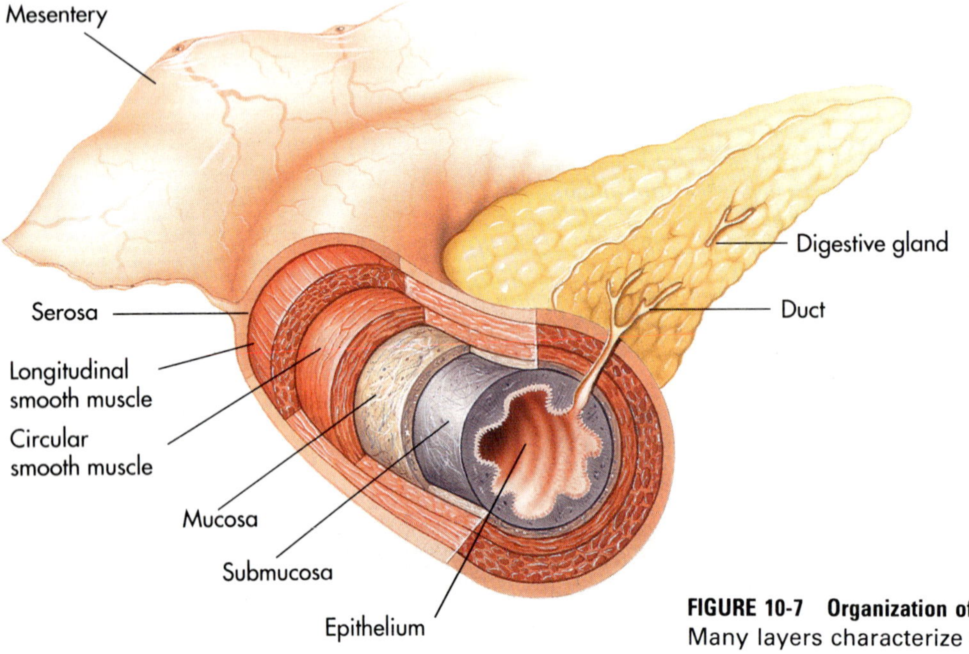

FIGURE 10-7 Organization of the vertebrate digestive tract.
Many layers characterize the digestive tract.

TABLE 10-3

Hormones of digestion

HORMONE	SOURCE	STIMULUS	ACTION	NOTE
Gastrin	Stomach	Entry of food into stomach	Regulates secretion of HCl	Unusual in that it acts on same organ that secretes it
Cholecystokinin (CCK)	Duodenum	Arrival of food in small intestine	Stimulates gall bladder contraction and therefore the release of bile salts into intestine, stimulates secretion of digestive enzymes by pancreas	CCK bears a striking structural resemblance to gastrin
Secretin	Duodenum	HCl in duodenum	Stimulates pancreas to secrete bicarbonate, which neutralizes stomach acid	The first hormone to be discovered (1902)

the small intestine, or duodenum, burning holes through the wall. These holes are called **duodenal ulcers.** Because the contents of the small intestine are not normally acidic, this organ is much less able to withstand the disruptive actions of stomach acid than the wall of the stomach is. For this reason, over 90% of all ulcers are duodenal, although other ulcers sometimes occur in the stomach.

Food stays in the stomach for approximately 2 to 6 hours. During this time the contractions of the muscular wall of the stomach churn the food, mixing it with the gastric juice and mucus. By the time it is ready to leave the stomach, the food has the consistency of pea soup—a substance called **chyme.** The gate to the small intestine, the **pyloric sphincter,** opens to allow just a bit of the chyme to pass through. When the acid in this chyme is neutralized and the food is digested, the pyloric sphincter is signaled by the nervous system to open again, allowing the next bit of chyme to pass.

In the stomach, concentrated acid breaks up connective tissue and unfolds proteins. These proteins are digested by pepsin into short polypeptides. Starches and lipids are not digested in the stomach.

Terminal digestion and absorption: The small intestine

Most of the digestion of food takes place in the small intestine. Within this organ, starches and glycogen are broken down to sugars, proteins to amino acids, and lipids to fatty acids and glycerol. These products of digestion then pass through the cells of the intestinal mucosa and diffuse into the blood in underlying blood vessels. The triglycerides move into the lymph in neighboring lymphatic vessels.

All of this activity takes place in the **small intestine,** the tube-like portion of the digestive tract that begins at the pyloric sphincter and ends at its T-shaped junction with the large intestine. Although the small intestine is approximately 6 meters long—long enough to stretch from the ground to the top of a two-story building—only the first 25 centimeters (8 inches) is actively involved in digestion. This initial portion of the small intestine is called the **duodenum.** The other portions of the small intestine, the **jejunum** and the **ileum,** are highly specialized (along with the duodenum) to aid in the absorption of the products of digestion by the blood and lymph.

Accessory organs that help digestion

Some of the enzymes necessary for digestion are secreted by the salivary glands, epithelial cells of the stomach, and epithelial cells of the duodenum. The others are secreted by the **pancreas,** a long gland that lies beneath the stomach and is surrounded on one side by the curve of the duodenum (see Figure 10-4). A tiny duct runs from the pancreas to the small intestine and serves as the passageway for the pancreatic juice. As you can see from Table 10-2, this secretion of the pancreas includes a number of digestive enzymes.

The **liver** is another organ that works with the duodenum to digest food. This organ, which weighs over 3 pounds, lies just under the diaphragm (see Figure 10-4). It is one of the most complex organs of the body and performs over 500 functions! Although the liver produces no digestive enzymes, it does help in the digestion of lipids by secreting

BIOLOGY & you

Ulcers: What they are and how to prevent them

Everyone has seen the portrayal of the hardhitting executive who barks orders, has no time for relaxation, and is the proud owner of a "bleeding" ulcer. Unfortunately, this portrayal is no longer confined to business people who thrive on stress. Many people now have ulcers, and sufferers are from all walks of life, including children and college students. In addition, the incidence of ulcers among women is on the rise.

What is an ulcer? Basically, an ulcer is a hole that develops in the lining of the digestive tract (Figure 10-A). These holes or sores develop because of an excess of hydrochloric acid secretion in the digestive tract that literally burns a hole through the protective mucus in the tract. Ulcers are divided into two types: peptic and duodenal. Peptic ulcers occur in the lining of the stomach itself. Duodenal ulcers are more common and, as their name implies, are located in the duodenum, the first segment of the small intestine that leads from the stomach. The duodenum is especially sensitive to acid, since it is essentially a basic environment. Acid mixed with the stomach contents, if it is not neutralized before it leaves the stomach and enters the small intestine, can irritate the duodenum and cause ulcers. Duodenal ulcers are more likely than peptic ulcers to perforate (punch through) the lining of the tract, causing bleeding and potential infection of the surrounding organs by the contents of the stomach.

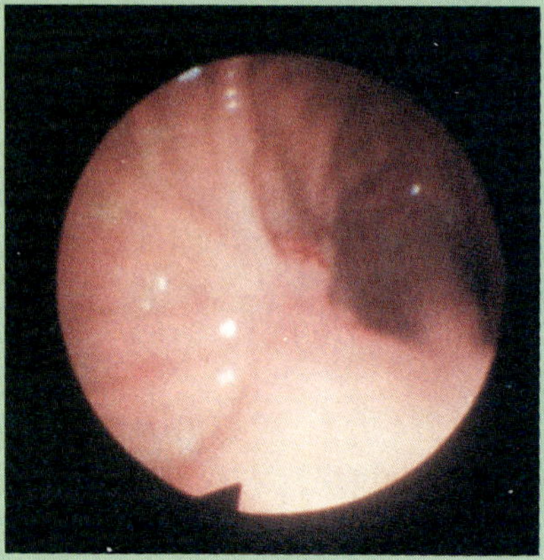

FIGURE 10-A View of an ulcer from the end of an endoscope.

The most common symptom of ulcers is pain. The classic symptom of an ulcer is a boring pain in the stomach that occurs 1 to 2 hours after eating. The pain is relieved when the affected individual eats again. The pain is caused by excess acid irritating the ulcer; when food is introduced into the stomach again, the acid is neutralized and the pain subsides.

The factors contributing to ulcers are numerous. Smoking seems to be linked to ulcers, since cigarette smoking stimulates acid production in the stomach. In fact, experts think that the increase in smoking among women is a prime factor in their increase in ulcers. Prolonged aspirin intake has also been implicated in the incidence of ulcers. Of course, stress and emotional tension have always been linked to ulcers. Recently, researchers have discovered that a bacterium called *Heliobacter pylori* might cause ulcers. This bacterium is extremely common and enters the body through human contact or dirty water. Researchers estimate that as many as 50% of all Americans harbor this bacteria in their stomachs. The good news is that once diagnosed, the patient can be treated with antibiotics, which destroy the bacteria and cure the ulcers.

You can do a lot to prevent ulcers. First, don't smoke. Watch your intake of aspirin, and if it is a problem for you, learn to deal more effectively with stress. By protecting your digestive tract with common sense, you can avoid the pain of ulcers.

a collection of molecules called **bile.** One of the many components of bile is bile pigments, breakdown products of the hemoglobin from old, worn-out red blood cells. These pigments give bile its greenish color. The bile also contains bile salts, which are substances that act much like detergents, breaking lipids up into minute droplets of triglycerides, similar to droplets of cream suspended in milk. The liver manufactures bile salts from cholesterol. Excess bile is stored and concentrated in the **gallbladder** on the underside of the liver. A bile duct brings bile from the liver and gallbladder to the small intestine.

The liver and pancreas, although not organs of the digestive system, help digestion take place. The pancreas secretes a number of digestive enzymes. The liver produces bile as one of its numerous and diverse functions. One component of bile, bile salts, aids in lipid digestion.

Digestion

As the food enters the small intestine, some of it has already been partially digested. Salivary amylase has broken some of the bonds in the starches and glycogen, producing the disaccharide maltose. However, much starch and glycogen remains undigested. In the small intestine, **pancreatic amylase** breaks down these carbohydrates to maltose. Maltose, sucrose (table sugar), and lactose (milk sugar) are digested to the monosaccharides glucose, fructose, and galactose by enzymes called **disaccharidases** that are produced by specialized epithelial cells of the small intestine. Some people do not produce the enzyme to digest lactose, however, and are therefore unable to digest milk. These persons cannot drink milk without experiencing cramps and, in some cases, diarrhea.

Some of the proteins in the food entering the small intestine have also been partially digested. The hydrochloric acid of the stomach has unfolded these proteins, and pepsin has cleaved some of them to shorter polypeptides. Three

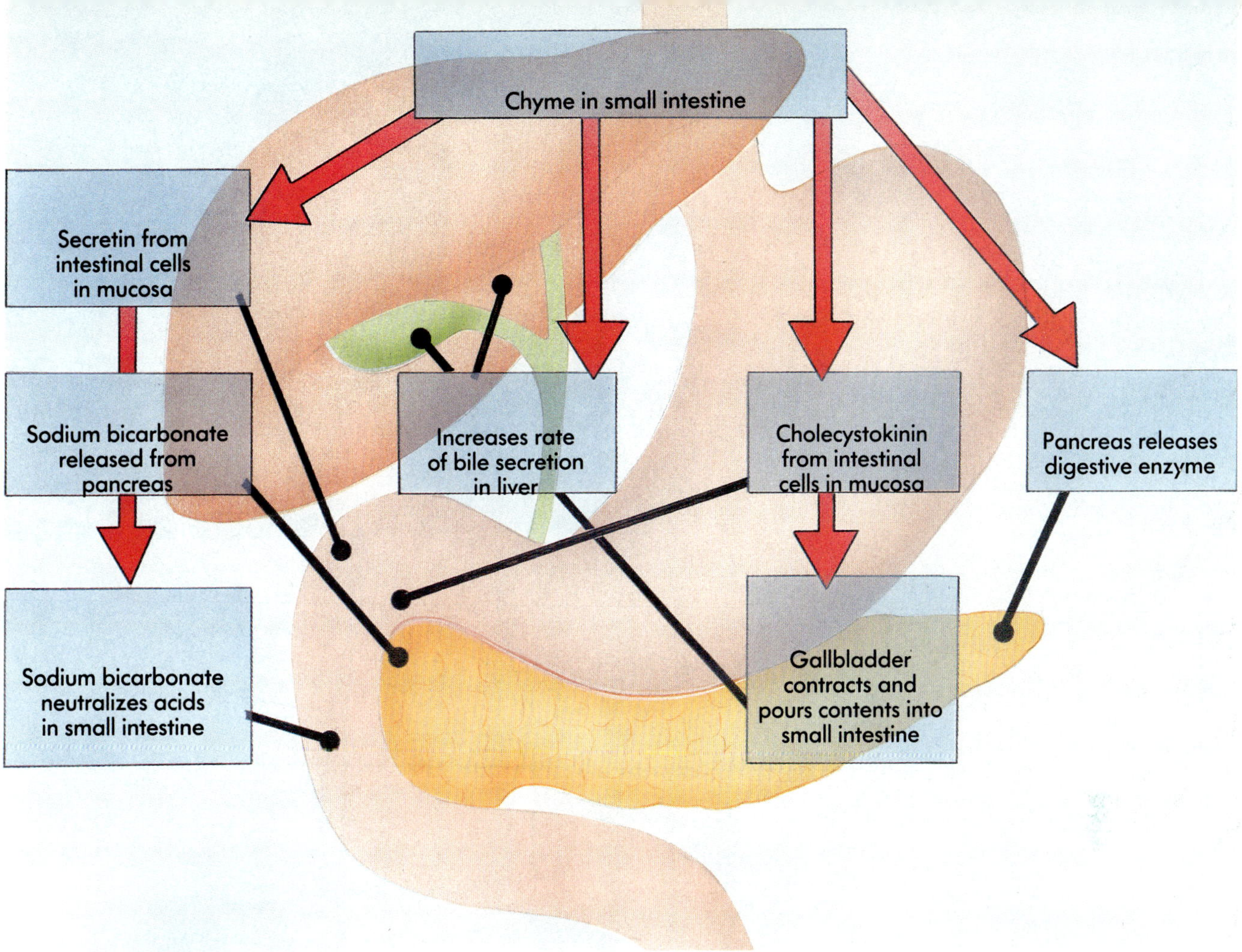

FIGURE 10-8 Digestive hormones manage many digestive enzymes. Secretin and cholecystokinin control and integrate digestive processes in the small intestine. The boxes describe events that can take place during digestion. A line from each box shows the approximate location of the event. *Arrows* indicate how one event triggers another in the digestive process.

other enzymes produced by the pancreas complete the digestion of proteins: **trypsin, chymotrypsin,** and **carboxypeptidase.** These enzymes work as a team with **peptidases** produced by cells in the intestinal epithelium, breaking down polypeptides into shorter chains and then to amino acids. These enzymes are secreted in an inactive form and become active in the presence of a particular enzyme secreted by cells in the intestinal epithelium. This way, they do not digest their way down the pancreatic duct to the small intestine.

The lipids are not digested until they reach the small intestine. Because they are insoluble in water, they tend to enter the small intestine as globules. Before these lipids can be digested, they are emulsified, or made soluble, by bile salts. When the lipid globules are in the form of triglycerides, **pancreatic lipase** cleaves them into fatty acids and glycerol.

In the first part of the small intestine, the duodenum, digestion is completed. Undigested starch and glycogen are digested to disaccharides and then to monosaccharides. Undigested proteins and other polypeptides are digested to shorter peptides and then to amino acids. Triglycerides are made soluble and are then digested to fatty acids and glycerol.

As you can see, the digestion of food involves so many players that the "digestive team" could use a manager! In fact, the key players in digestion—the liver, gallbladder, pancreas, stomach and small intestine—have more than one manager. These managers of digestion are hormones (see Table 10-3). Earlier in this chapter you read how the hormone gastrin controls the release of hydrochloric acid in the stomach. Two other hormones, **secretin** and **cholecystokinin** (CCK), control digestion in the small intestine (Figure 10-8).

When chyme enters the small intestine, the acid in it stimulates cells in the intestinal mucosa to produce secretin. This hormone does two things: first, it stimulates the release of an alkaline fluid called *sodium bicarbonate* from the pancreas. This solution neutralizes the acid in the chyme so that it will not damage the wall of the small intestine and produces the proper pH in which the pancreatic and intestinal enzymes will work. In addition, secretin increases the rate of bile secretion in the liver.

The presence of fatty acids and partially digested proteins in the chyme stimulates the mucosa to produce CKK. This enzyme signals the gallbladder to contract and pour its contents into the small intestine. It is also a stimulus to the pancreas to release its digestive enzymes.

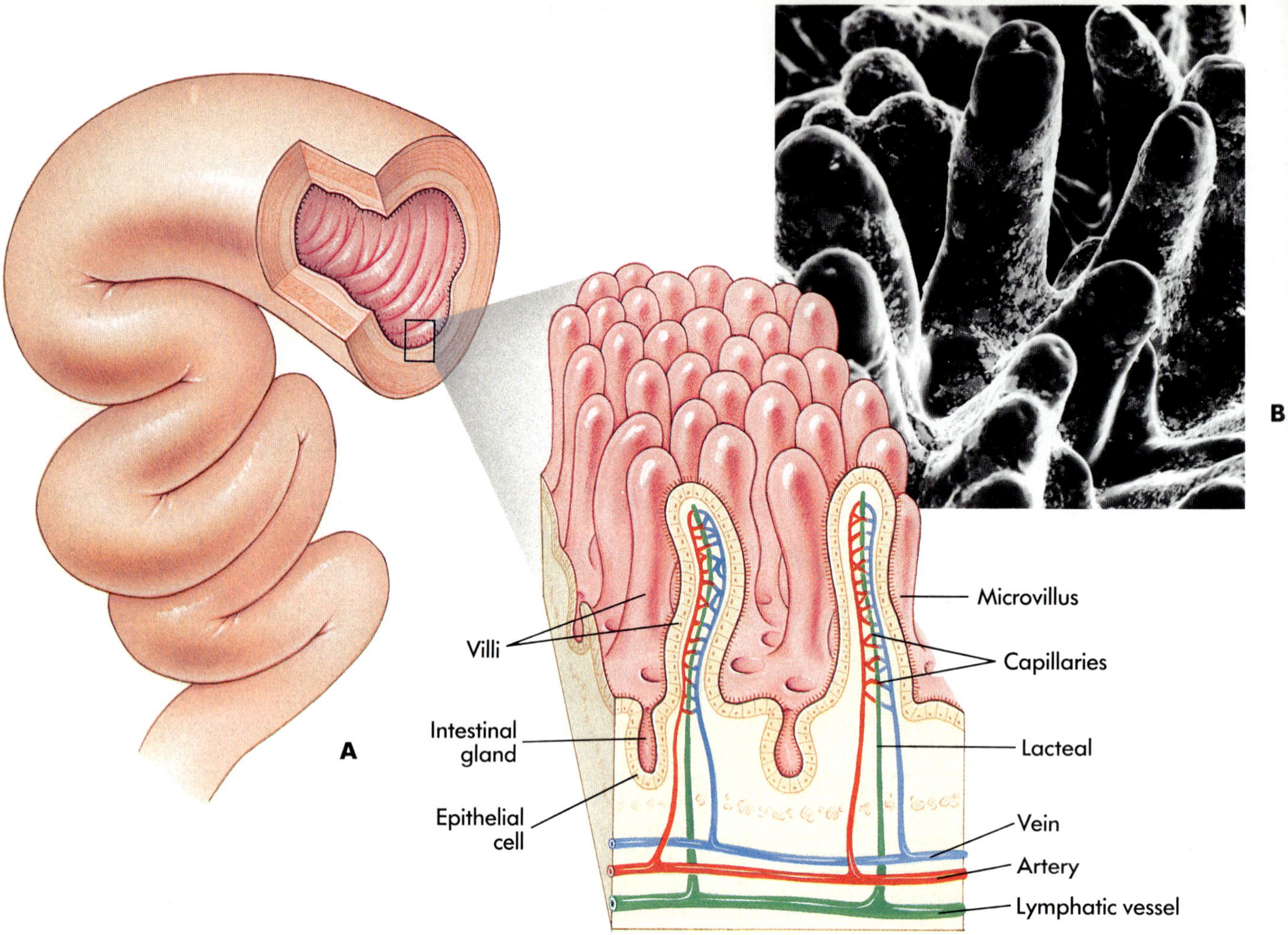

FIGURE 10-9 Structure of villi.
A Villi are finger-like projections of tissue located on the inner surface of the small intestine. The villi, and the microvilli on the surface of the villi, greatly increase the surface area of the small intestine, providing more room for absorption.
B Electron micrograph of intestinal villi.

Absorption

The amount of material passing through the small intestine is startlingly large. An average person consumes about 800 grams of solid food and 1200 milliliters of fluid each day, for a total volume of about 2 liters, or a little more than 2 quarts. To this amount is added about 7 liters of fluid secreted by the salivary glands, stomach, pancreas, liver, and the small intestine itself. The total is a remarkable 9 liters, or almost 2½ gallons! Of the 800 grams of solid food and 9 liters of liquid that enter the digestive tract, only about 50 grams of solid and 100 milliliters of liquid leave the intestinal tract as waste, or **feces.** The small intestine absorbs the 750 grams of nutrients per day and most of the water; the large intestine absorbs the remaining water—approximately 750 milliliters.

To aid in the absorption of nutrients, the internal surface area of the small intestine increases dramatically in three ways. First, the mucosa and submucosa of the small intestine are thrown into folds; they do not have a smooth inner surface like a garden hose. This folded surface, in turn, is covered by fine fingerlike projections of the epithelium. These projections are called **villi** (sing. villus) and are so small that it takes a microscope to see them (Figure 10-9). In addition, the epithelial cells of the villi are covered on their exposed surfaces by cytoplasmic projections called **microvilli.** The infoldings, the villi, and the microvilli provide the small intestine with a surface area of about 300 square meters, or 2700 square feet, an area greater than the floor space in many homes!

Within each villus is a network of capillaries and a lymphatic vessel called a **lacteal.** Although the nutrients from digested food do not have far to travel—through a single layer of epithelial cells and a single layer of cells forming the wall of the capillary or lacteal—a great deal of work must be done. Each monosaccharide and amino acid must "catch a ride" on a carrier molecule to get into an epithelial cell. Energy is needed to ferry many of these molecules across cell membranes—a process called *active transport* (see Chapter 4). Others are taken up by facilitated diffusion. Once the monosaccharides and amino acids are in the epithelial cells, however, they accumulate and eventually diffuse through the base of the cell and into the blood. When in the bloodstream, they are quickly swept away to the liver for processing and storage. When blood levels of glucose are sufficient to supply all your cells with this fuel of cellular respiration (see Chapter 7), the liver stores glucose as glycogen. When more glucose is needed, such as between meals, the liver readily converts the glycogen back to glucose. In this way, the liver is your "metabolic bank," making deposits and withdrawals in the "currency" of glucose molecules.

The internal surface area of the small intestine is increased by the presence of inner folds and projections, which results in a more efficient absorption of nutrients and water. Monosaccharides and amino acids are absorbed into the intestinal epithelium by means of facilitated diffusion and active transport and then diffuse into the bloodstream. Epithelial cells re-form fatty acids and glycerol to triglycerides and shuttle them to the blood by means of the lymphatic system. Water is absorbed by osmosis.

The absorption of lipids takes place somewhat differently. After lipids are broken down into fatty acids and glycerol, they become surrounded by bile salts. Packaged in this way, they move to the cell membranes of the villi. As you may recall from Chapter 4, one of the main ingredients in cell membranes is lipid. Therefore when the fatty acids and glycerol from digestion come into contact with these cell membranes, they discard their "shell" of bile salts and easily move across the membrane and into the cell. Short-chained fatty acids are absorbed directly into the bloodstream. Longer-chained fatty acids are reassembled into triglycerides by the endoplasmic reticulum and are then encased in protein. After this processing, they pass out of the epithelial cells and into the lacteal. These protein-coated triglycerides are then transported in the lymphatic fluid through a system of vessels that drains the lymph into the blood at the left subclavian vein, a major blood vessel at the base of the neck.

Concentration of solids: The large intestine

The large intestine, or **colon,** is much shorter than the small intestine—only about a meter and a half, or 5 feet, long. It is wide, however, having a diameter slightly less than the width of your hand. In contrast, the small intestine has a diameter only slightly larger than the width of two of your fingers. The small intestine joins the large intestine about 7 centimeters up from its end, creating a blind pouch out of the beginning of the large intestine (Figure 10-10). Hanging from this pouch is the **appendix,** a structure that serves no essential purpose in humans. An infection of the appendix is called *appendicitis* and can be quite serious and painful. In

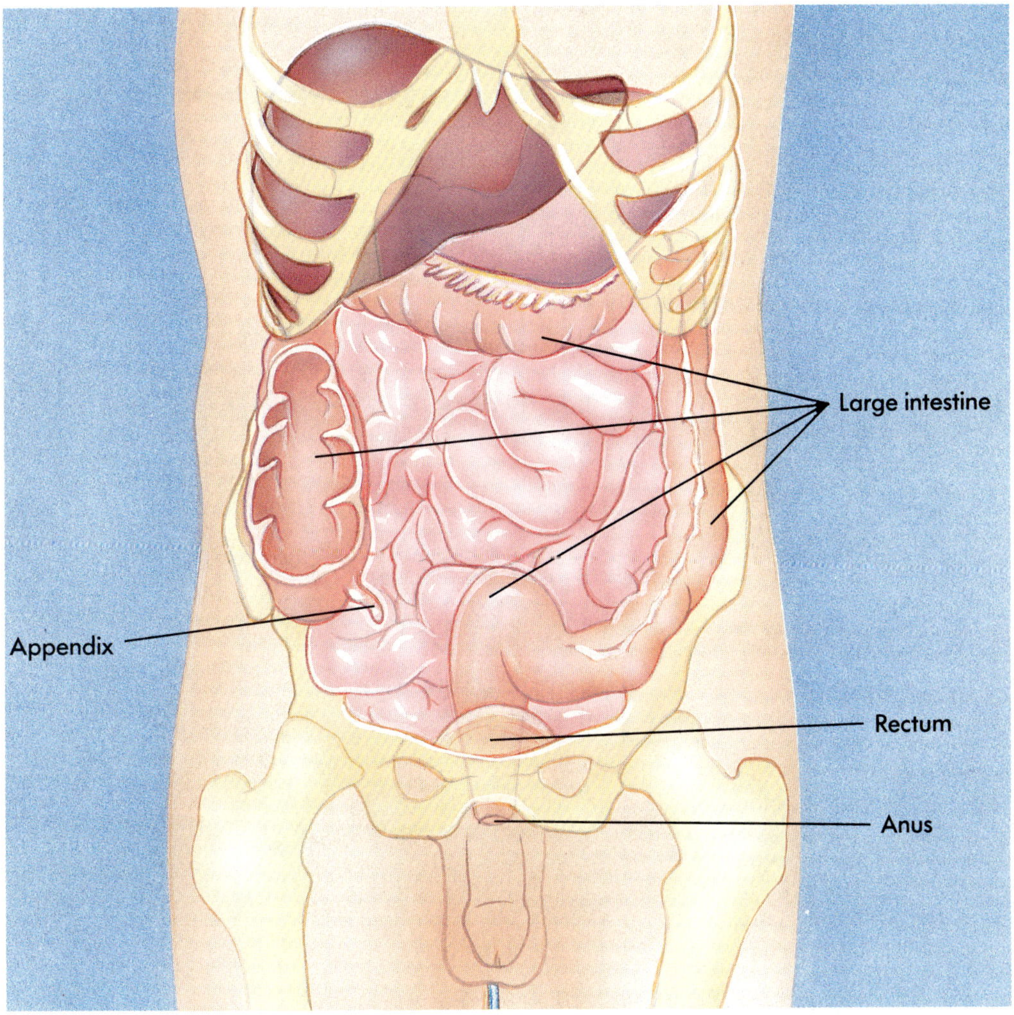

FIGURE 10-10 Location of the large intestine.
The large intestine, or colon, terminates at the rectum. The appendix hangs from a blind (dead-end) pouch where the small intestine meets the colon. The appendix serves no function in humans.

certain animals the blind pouch is more fully developed and is called the *cecum.* It serves as a place where the cellulose of plant cell walls is digested by the activity of intestinal bacteria and other microorganisms. Humans cannot digest cellulose, so it becomes a digestive waste. This waste, however, is important to the regular movement of the feces through the large intestine. Also called **dietary fiber,** undigested plant material provides bulk against which the muscles of the large intestine can push.

The junction of the small and large intestines is in the lower right side of the abdomen. From there the large intestine goes up the right side of the abdomen to the liver. It then turns a left, crossing the abdominal cavity just under the diaphragm. On the left side of the abdominal cavity, it turns downward, ending at a short portion of the colon called the **rectum.** The rectum terminates at the **anus,** the opening for the elimination of the feces.

No digestion takes place within the large intestine. Its job is to absorb sodium and water, to eliminate wastes, and to provide a home for friendly bacteria. These bacteria help keep out disease-causing microbes and produce certain vitamins, especially vitamin K.

Waste materials move slowly along the smooth interior of the large intestine as water and sodium are slowly reabsorbed. As they move along, the wastes become more compacted. If the wastes move too slowly through the colon, too much water may be reabsorbed, leading to a difficulty in elimination called constipation. Conversely, if the wastes move too quickly, as happens with certain intestinal illnesses, not enough water may be removed, resulting in diarrhea. Eventually, the solids within the colon pass into the rectum as a result of the peristaltic contractions of the muscles encasing the large intestine. From the rectum the solid material passes out the anus through two anal sphincters. The first of these is composed of smooth muscle. It opens involuntarily in response to a pressure-generated nerve signal from the rectum. The second sphincter, in contrast, is composed of skeletal muscle. It is subject to voluntary control from the brain, thus permitting a conscious decision to delay **defecation,** or the elimination of waste.

The large intestine serves primarily to reabsorb water and sodium from digestive wastes and eliminate the remainder, or feces.

Diet and nutrition

The digestion of food yields no usable energy but changes a diverse array of complex molecules into a small number of simpler molecules that can be used by your cells as fuel for cellular respiration. During cellular respiration, ATP molecules are produced—the energy currency of your body. As you read in Chapter 7, glucose is completely broken down by the processes of glycolysis, the Krebs cycle, and the electron transport chain to yield 36 usable molecules of ATP. In addition, the carbon portions of amino acids and the glycerol backbones of triglycerides are converted to substances that can also be metabolized by two of these pathways. Fatty acids are metabolized by another metabolic pathway to yield ATP molecules.

Any intake of food in excess of that required to maintain the blood sugar (glucose) level and the glycogen reserve in the liver results in one of two consequences. Either the excess glucose is metabolized by the muscles and other cells of the body or it is converted to fat and stored within fat cells. Only when all the body's energy needs have been met—the energy needed to run chemical reactions, move muscles, and digest food—will kilocalories be stored as fat. Think of your body as a giant scorecard, keeping track of kilocalories eaten and kilocalories used. If you eat more than you use, you will gain weight. Unfortunately, 10% to 25% of all teenagers and 25% to 50% of all adults in the United States are obese. In other words, they weigh at least 20% more than the average weight for their height (Figure 10-11).

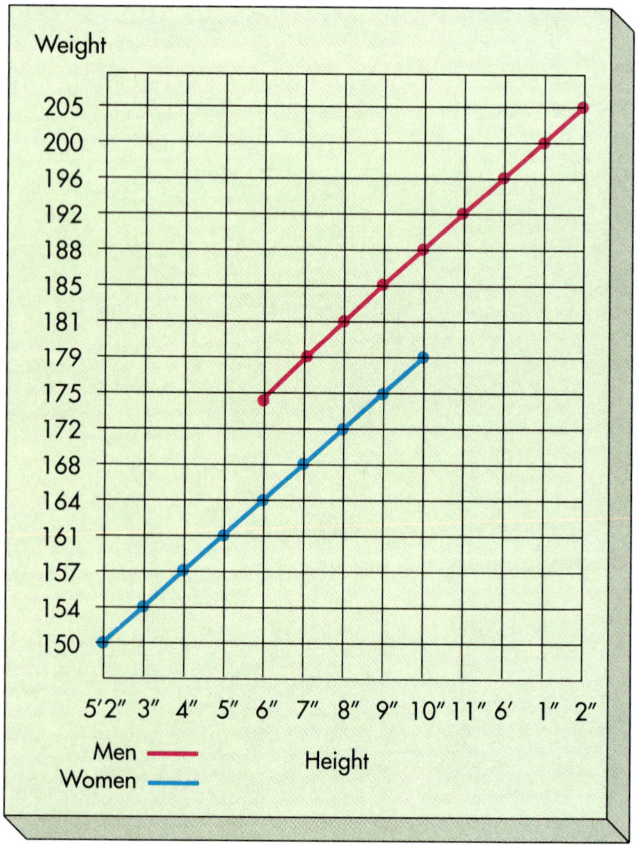

FIGURE 10-11 Obesity ranges for men and women.
Obesity is usually characterized as the state of being more than 20% heavier than the average person of the same sex and height. The graph charts height versus weight, showing the lowest weights at which a person of an indicated height is considered obese by most physicians.

As the digestive process breaks down food into molecules usable in cellular respiration, it also "unlocks" the vitamins and minerals from these foods. Vitamins fall into two general categories: water soluble and fat soluble (see Table 10-1). The water-soluble vitamins enter the cells of the intestinal mucosa and move into the bloodstream. The fat-soluble vitamins are carried across the membranes of the intestinal cells associated with the fatty acids and glycerols. They too are transported to the bloodstream by means of the lymphatic system. Minerals are absorbed into the bloodstream as ions. (The recommended daily intake of the various vitamins and minerals is listed in Table 10-1.) Most people who eat a sufficient amount and variety of foods get the vitamins they need in the food they eat.

Proteins, lipids, and carbohydrates provide more than energy for the diet—they provide the raw materials to build the substances the body needs. In fact, proteins are used primarily as building blocks and not as a source of energy. Of the 20 different amino acids that make up proteins, humans can manufacture only 12. Therefore 8 of the amino acids, called **essential amino acids,** must be obtained by humans from proteins in the food they eat (Figure 10-12). Protein foods that contain the essential amino acids in amounts proportional to the body's need for them are called high-quality, or complete proteins. Unfortunately, many high-quality protein foods such as meat, cheese, and eggs are high in animal fat and cholesterol as well. A high percent of animal fat and cholesterol in the diet has been shown to result in weight gain and increase the risk of heart attack because of the clogging of blood vessels with cholesterol deposits. To reduce fat intake yet get the amino acids you need, you should combine any legume (dried peas, beans, peanuts, or soy-based food) with any grain, nut, or seed or you should combine any grain, legume, nut, or seed with small amounts of milk, cheese, yogurt, eggs, red meat, fish, or poultry.

FIGURE 10-12 The protein content of a variety of common foods. Eight of the essential amino acids humans need must be obtained from proteins in the diet.

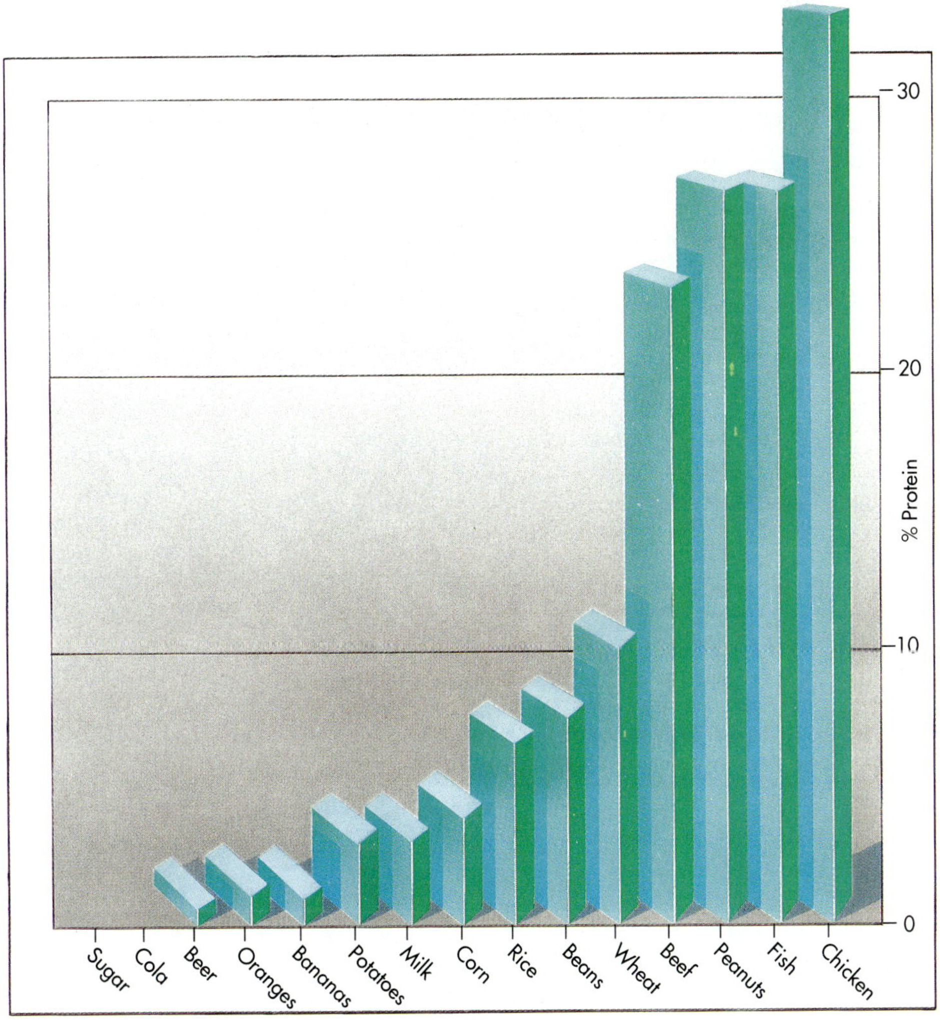

So how much protein, fat, and carbohydrate *should* you have in your diet? The Senate Committee on Dietary Goals for the United States recommends that 12% of your daily intake of kilocalories come from proteins, 30% *or less* come from fats, and 58% from carbohydrates (Figure 10-13). You may want to approximate the number of kilocalories in the meal shown at the beginning of the chapter. How many of its kilocalories come from proteins, fats, and carbohydrates? Although you may not be analyzing all of its nutritional aspects, what recommendation can you make for it to be a healthier meal?

FIGURE 10-13 What proportion of the energy nutrients should you eat?
Experts agree that 12% of your total daily intake of kilocalories should be from proteins, 30% or less from fats, and 50% from carbohydrates. Most Americans, however, eat too much protein and fat and not enough carbohydrates. How does your diet compare to the experts' recommendations?

Summary

1. Nutrients are substances in food that are used to help the body grow and repair and sustain itself. There are six classes of nutrients: carbohydrates, fats (lipids), proteins, vitamins, minerals, and water.
2. Digestion is the process whereby food is broken down into a form that is usable by the body's cells. Carbohydrates are digested to monosaccharides, lipids to fatty acids and glycerol, and proteins to amino acids. Amino acids are primarily used as building blocks. Monosaccharides and glycerol are primarily used as the fuel of cellular respiration and other metabolic pathways that produce ATP molecules, the energy currency of the body.
3. Vitamins and minerals are not broken down during digestion but are released from the food of which they are a part. These nutrients help various chemical reactions and bodily processes take place.
4. The digestive tract leads from the mouth and pharynx, through the esophagus, to the stomach. The digestion of carbohydrates begins in the mouth with the action of salivary amylase. The stomach juices contain a concentrated acid in which the protein-digesting enzyme pepsin is active.
5. Food passes from the stomach to the small intestine where the pH is neutralized. A variety of enzymes, many synthesized in the pancreas, acts to complete digestion. Most digestion occurs in the first 25 centimeters of the small intestine, a portion called the *duodenum*.
6. The products of digestion are absorbed across the walls of the small intestine, which possess numerous villi and achieve a large surface area. Amino acids and monosaccharides pass into the bloodstream. Fatty acids and glycerols pass into the lymphatic system and, coated with proteins, are transported to the bloodstream.
7. Glucose and other metabolic products of digestion do not enter the general circulation directly but instead flow to the liver. The liver removes and stores any excess metabolic products and maintains blood glucose levels within narrow bounds.
8. The large intestine has little digestive or absorptive activity; it functions principally to absorb water from the waste that is left over and eliminate it from the body.
9. The Senate Committee on Dietary Goals for the United States recommends that 12% of the daily intake of kilocalories come from proteins, 30% or less from fats, and 58% from carbohydrates.

KEY TERMS

amylases 168
anus 180
appendix 179
bile 176
bolus 172
calories 166
carbohydrates 166
carboxypeptidase 177
carnivores 170
cholecystokinin 177
chyme 175
chymotrypsin 177
colon 179
defecation 180
digestion 166
duodenum 175
epiglottis 172
esophagus 172
essential amino acids 181
fats 166
feces 178
gallbladder 175
gastrin 174
glottis 172
herbivores 170
hydrolysis 170
ileum 175
jejunum 175
lacteal 178
larynx 172
lipases 168
liver 175
minerals 166
nutrients 166
omnivores 170
pancreas 175
pepsin 174
peptidase 177
peristalsis 172
pharynx 172
proteases 168
proteins 166
pyloric sphincter 175
rectum 180
salivary amylase 170
salivary glands 170
secretin 177
small intestine 175
stomach 177
trypsin 177
villi 178
vitamins 166

REVIEW QUESTIONS

1. Name the six classes of nutrients, and explain why you need each one in your diet.
2. Describe the digestive fates of proteins, lipids, and carbohydrates. What three types of enzymes are involved?
3. What is saliva, and what is its function?
4. Fill in the blanks: A(n) ________ eats only plants, a(n) ________ eats only meat, and a(n) ________ eats both. Which are you?
5. You bite off a mouthful from a crunchy apple and chew it thoroughly. Describe the role(s) played by each type of tooth in your mouth.
6. Fill in the blanks: When the tongue pushes food into the ________ (the upper part of the throat), nerves stimulate the ________. The ________ acts as an escalator that moves the food to the stomach by rhythmic contractions called ________.
7. Create a table that lists the location, substrate, and digestion product for each of the following: salivary amylase, pepsin, peptidases, pancreatic amylase, lactase, maltase, sucrase, lipase, trypsin, and chymotrypsin.
8. Fill in the blank: Most of the digestion of food takes place in the ________.
9. Identify two accessory organs that help digest foods. What role(s) does each play in the process?
10. What is the duodenum? Summarize the digestive activities that take place there.
11. Identify and give a function for each of the following: gastrin, secretin, and cholecystokinin. What do they have in common?
12. For years, baking soda (sodium bicarbonate) has been a popular home remedy for indigestion and heartburn. Explain why.
13. Discuss the features of the small intestine that increase its internal surface area. Why is this increase important?
14. Place the following in the correct digestive sequence and summarize the function(s) of each: anus, large intestine, stomach, and small intestine.
15. Summarize how your body produces ATP from glucose, amino acids, triglycerides, and fatty acids.
16. What does the term *essential amino acids* mean?
17. Keep a record of everything you eat and drink for 3 days. Then (this could be painful!) analyze your diet. Compare it to the recommendations in Figure 10-13. How could you improve your nutritional intake?

THOUGHT QUESTIONS

1. Prepare a personal energy budget: *OUT:* For the last 48 hours, write down what proportion of the time you spent at rest (sleeping or sitting), what proportion walking, and what proportion running or doing hard physical work. Now estimate the number of kilocalories you have expended, using the 1-2-3 rule (1 kcal/min sitting; 2 kcal/min walking; 3 kcal/min running). *IN:* Write down everything you have eaten and drunk in the last 48 hours. Now estimate the number of calories you have consumed, adding up those from carbohydrate (4 kcal/gram), from fat (9 kcal/gram), and from protein (3 kcal/gram). Compare out with in. Explain.

FOR FURTHER READING

Harris, A. R. (1988). Why do more women get ulcers? *Chatelaine, 61*, 32.
This article discusses the link between ulcers in women and risk factors such as smoking, alcohol, and stress.

Moog, F. (1981, November). The lining of the small intestine. *Scientific American*, pp. 154-176.
This is a clear description of the most important absorptive surface in the human body.

Winick, M. (1988). *Controls of appetite, current concepts in nutrition.* New York: John Wiley & Sons, Inc.
This book describes appetite regulation by control centers in the brain and receptors in the gastrointestinal tract and discusses that the CNS pathways involved in satiety involve endogenous opiates and mediate responses to stress and reward.

11 Respiration

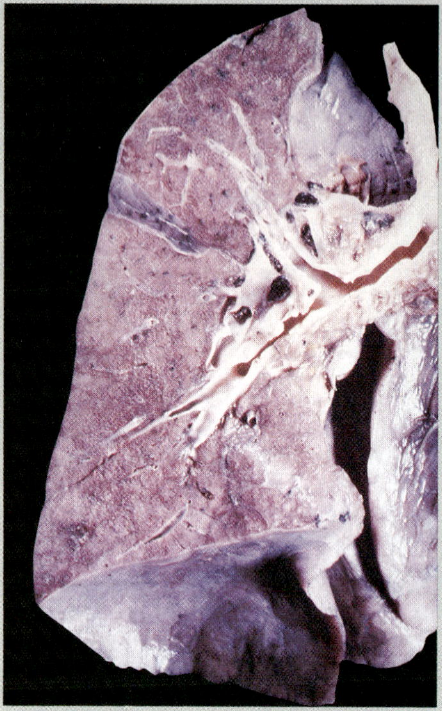

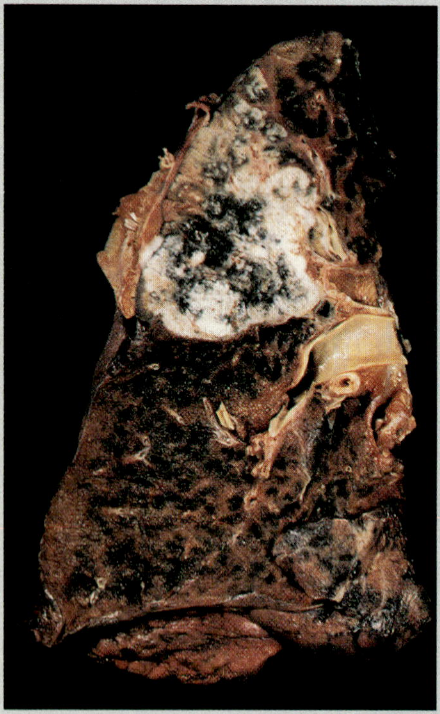

Which lung would you choose? Probably the healthy, red lung in the top photo. Unbelievably, however, over 50 million Americans choose to have lungs similar to that in the bottom photo. Why? Because these people choose to smoke cigarettes.

HIGHLIGHTS

- In its journey to the lungs, air passes through the nostrils, nasal cavities, pharynx, larynx, trachea, bronchi, and bronchioles.
- In the process of breathing, a negative pressure within the thoracic cavity causes air to be pulled into the lungs; a positive pressure causes air to be pushed out of the lungs.
- Oxygen diffuses into the blood and carbon dioxide diffuses out of the blood at the lungs; the reverse exchange of gases occurs at the body cells.
- Cigarette smoking damages respiratory structures and can lead to such disorders as lung cancer, chronic bronchitis, and emphysema.

OUTLINE

Respiration

The pathway of air into and out of the body

- Nasal cavities
- The pharynx
- The larynx
- The trachea
- The bronchi and its branches
- The alveoli: Where gas exchange takes place

The mechanics of breathing

- Inspiration
- Expiration
- Deep breathing
- Lung volumes

Gas transport and exchange

- External respiration
- Internal respiration

Choking: A common respiratory emergency

Chronic obstructive pulmonary disease

When people smoke cigarettes, they inhale many particles and gases. Carbon monoxide is one of the colorless, odorless gases present in cigarette smoke (and in your car's exhaust). It binds to the oxygen-carrying molecules of hemoglobin in your blood, reducing their oxygen-carrying capability by as much as 15%! Tobacco also contains nicotine, a drug that is physically addicting. In addition, nicotine stimulates the nervous system in a way that causes the heart to race and the blood pressure to rise. The blood vessels also narrow—including the arteries that supply the heart with blood. Both the nicotine and carbon monoxide work in different ways to decrease the amount of oxygen that can get to the heart. If the decrease is severe, a part of the heart muscle can die, an experience commonly referred to as a *heart attack*.

Tar is another substance found in cigarette smoke. It is the residue of smoke—what is left over after the nicotine and moisture have been removed. In the bottom photo, you can easily see how tar blackens lung tissue. The whitish areas are no longer functional—they are cancerous tumors caused by the tar. This person did not survive the addiction to cigarettes.

Respiration

Your lungs play an important role in the process of **respiration** (Figure 11-1). The term *respiration* is used in many ways when speaking about the processes of the human body, so its definition can be confusing. Used alone, respiration refers to gas exchange—the uptake of oxygen (O_2) and the release of the waste gas carbon dioxide (CO_2) by the whole body. The phrase *cellular respiration* uses the term respiration in a more specific way: it is the chemical process by which cells break down fuel molecules to release energy, using oxygen and producing carbon dioxide. Cellular respiration is a process that uses the oxygen you breathe in and produces the carbon dioxide you breathe out. (This process is described in detail in Chapter 7).

At the millions of tiny air sacs, or **alveoli,** that make up most of your lungs, oxygen enters the blood and carbon dioxide leaves. The exchange of these gases between the blood and the alveoli is known as **external respiration.** The movement of air into and out of the lungs is **breathing.**

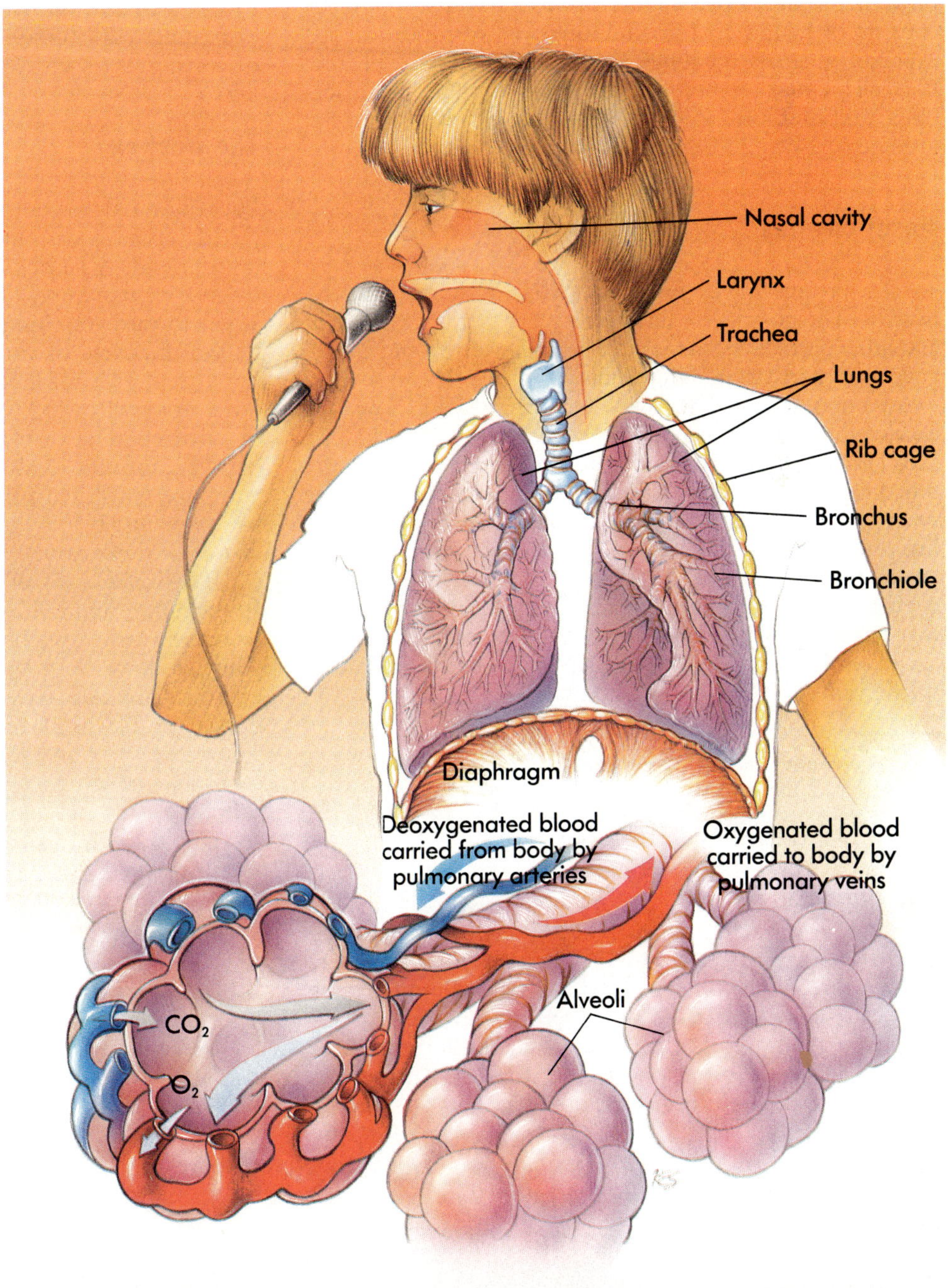

FIGURE 11-1 The human respiratory system. A section of the lung is enlarged to show the alveoli.

After the blood "picks up" oxygen at the lungs and "gets rid of" carbon dioxide, it travels to the heart and gets a push out to the rest of the body. The cells of the body receive the oxygen they need and get rid of the waste gas carbon dioxide as the blood moves past the cells within microscopic blood vessels. At the cells, carbon dioxide moves by diffusion from the tissue fluid into the blood. Tissue fluid (also called *interstitial fluid* or *intercellular fluid*) is a water-like fluid derived from the blood that bathes all the cells of the body. Oxygen moves in the opposite direction—from the blood into the tissue fluid. This process is called **internal respiration,** the exchange of oxygen and carbon dioxide between the blood and the tissue fluid.

Respiration is the uptake of oxygen and the release of carbon dioxide by the body. The processes of cellular respiration, internal respiration, and external respiration are all part of the general process of respiration.

The pathway of air into and out of the body

How many times per minute do you breathe? A breath consists of taking air into the lungs, or **inspiration,** and expelling air from the lungs, or **expiration.** Try to breathe normally, count the number of breaths you take in 15 seconds, and multiply by four. Is your breath rate within the average of 14 to 20 breaths per minute?

Nasal cavities

As you breathe in, air first enters your body through the nostrils. The nostrils are lined with hairs that filter out dust and other particles from the air. The air is warmed and moistened as it swirls around in the **nasal cavities.** These cavities, located above your oral cavity and behind your nose, are bordered by projections of bone covered with epithelial tissue. This tissue stays moist with mucus secreted by its many mucous glands. This sticky fluid helps trap dirt and dust that you breathe in. The epithelium is also covered with tiny hair-like projections called **cilia** (see Figure 11-3). The word *cilia* comes from a Latin word meaning "eyelashes." These "cell eyelashes" beat in unison, creating a current in the mucus that carries the trapped particles toward the back of the nasal cavity. From here, the mucus drips into the throat and is swallowed—at a rate of over a pint per day!

As air moves about within the nasal cavities, it is warmed and moistened.

The pharynx

After passing through the nasal cavities, air enters the **pharynx,** or throat. The pharynx extends from behind the nasal cavities to the openings of the esophagus and larynx. The esophagus, as you may recall from Chapter 10, is the "food tube," a passageway for food to the stomach. The **larynx,** or voice box, lies at the beginning of the trachea. The **trachea** is the air passageway that runs down the neck in front of the esophagus and that brings air to the lungs.

The larynx

The larynx is a cartilaginous box shaped somewhat like a triangle. Stretched across the upper end of the larynx are the **vocal cords** (Figure 11-2). The vocal cords are two pieces of elastic tissue covered with a mucous membrane. Muscles within the larynx pull on the cartilage, which in turn pulls or relaxes the tension on the vocal cords. When the vocal cords are stretched tightly, the space between them—the **glottis**—is closed. In this way the vocal cords provide a "backup" for the epiglottis that flaps over the glottis during swallowing. Both structures work to prevent food and drink from "going down the wrong way." The vocal cords also produce the sounds you make as air rushes by and causes them to vibrate. You can illustrate this principle by stretching a rubber band between your fingers. Have someone repeatedly pluck the band. At the same time increase and then decrease the stretch on the band. What happens? Your vocal cords work in much the same way to produce a variety of pitches of sound. But you also have a mouth with lips and a tongue to form the sounds into words. Your lungs add a power supply and volume control to your personal musical instrument—your voice.

The trachea

Put your hand on your larynx, or "Adam's apple," and then picture about 4 or 5 inches of garden hose attached to its bottom end. A garden hose is about the diameter of your trachea, or windpipe, which extends downward from your larynx toward your lungs. The trachea has thin walls, similar to the thickness of those in the hose. Garden hoses are reinforced with materials such as rubber or vinyl to keep them from collapsing; your trachea is reinforced with rings of cartilage. The cartilage wraps around the trachea only partway, forming C shapes that begin and end where the windpipe lies next to the esophagus. Press gently on your windpipe just below your Adam's apple and you can feel a couple of these cartilaginous rings.

The inner walls of the trachea are lined with ciliated epithelium (Figure 11-3). Certain cells in the epithelium secrete mucus. Together, the cilia and the mucus provide your windpipe with an "up escalator" for any particles or microbes you may have inhaled. This escalator brings substances up to the pharynx where they are swallowed and eliminated through the digestive tract. In the trachea of a cigarette smoker, however, action of the cilia is impaired, causing mucus to build up in the airway. The result is that the tars in cigarettes are not caught and expelled with the action of cilia. They move easily into the lungs and settle there. A chronic cough, often called *smoker's cough,* is triggered by accumulations of mucus below the larynx.

Air passes from the nasal cavities and into the throat, or pharynx. From the pharynx, it passes over the vocal cords that are stretched across the larynx. From this voice box, air moves into the trachea, or windpipe, on its way to the lungs.

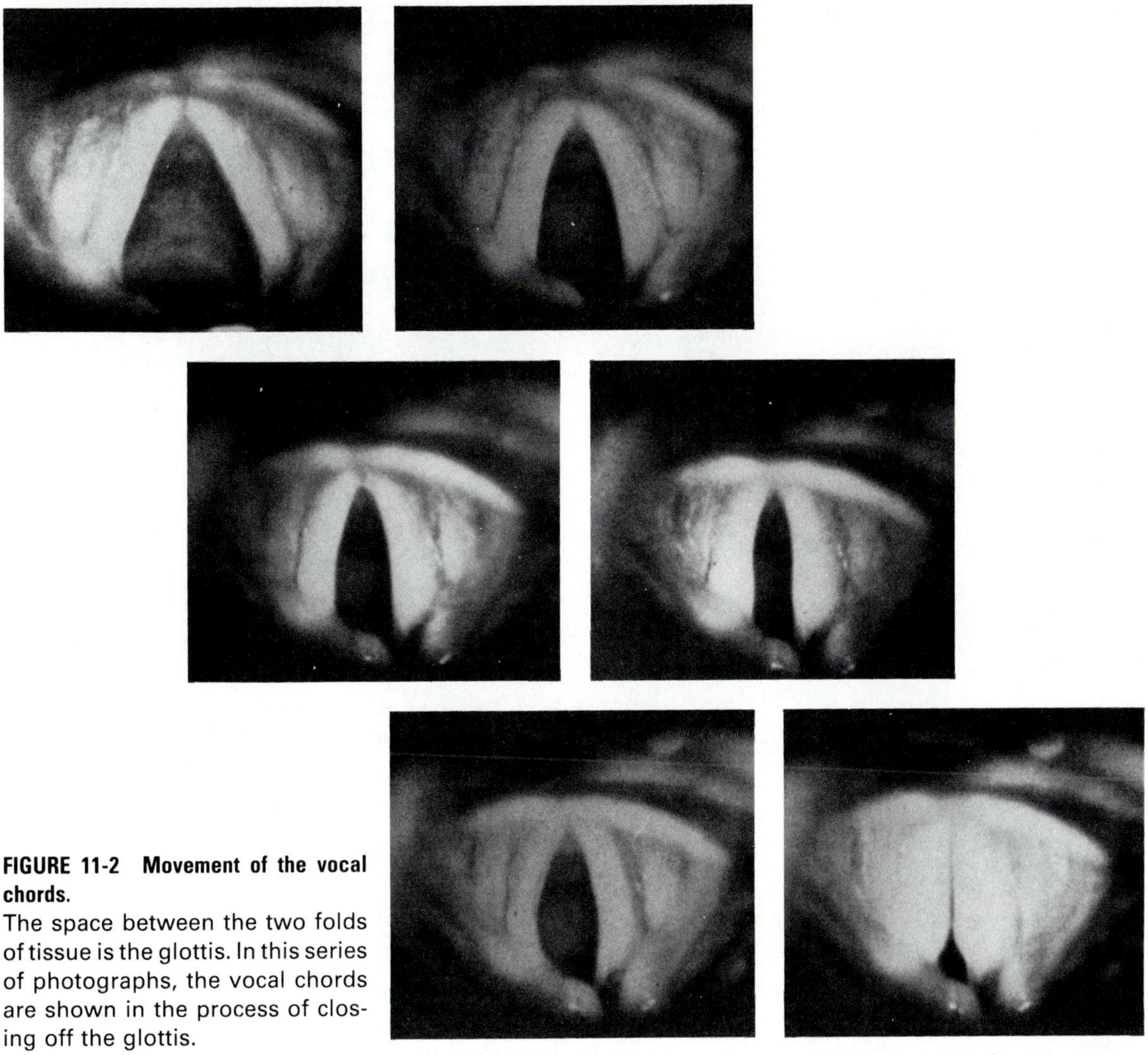

FIGURE 11-2 Movement of the vocal chords.
The space between the two folds of tissue is the glottis. In this series of photographs, the vocal chords are shown in the process of closing off the glottis.

FIGURE 11-3 Cilia such as these line the trachea.
Cilia sweep trapped particles out of the respiratory tract.

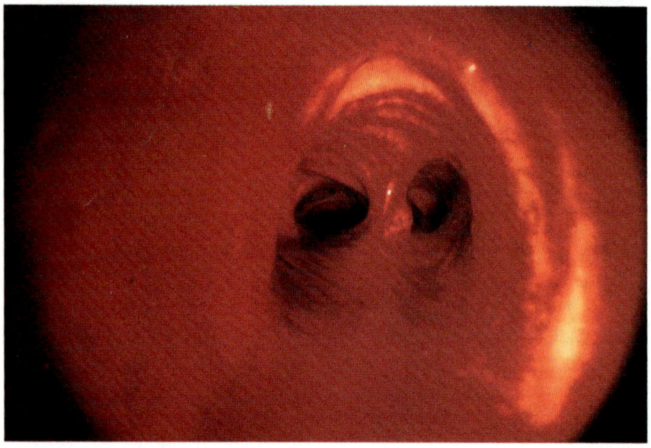

FIGURE 11-4 The trachea.
The two holes that you see are the primary bronchi. One bronchus leads to each lung.

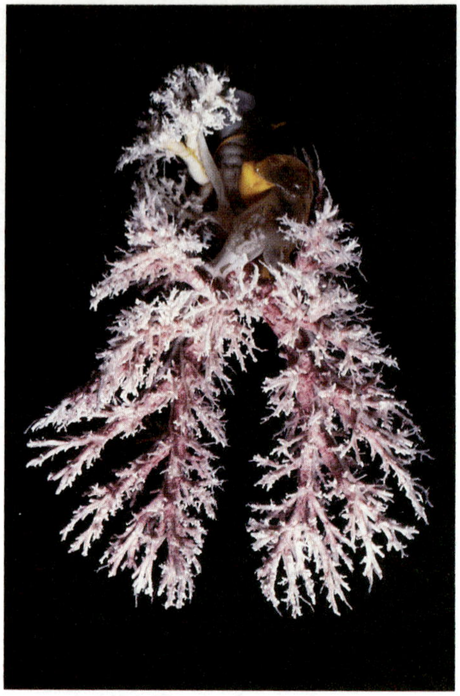

FIGURE 11-5 Secondary bronchioles.
Secondary bronchi divide into smaller and smaller branches, resembling an upside-down tree or a system of roots.

The bronchi and its branches

If you could look down your trachea as in Figure 11-4, the two "black holes" you would see are your **primary bronchi.** These airways are structured much like the trachea but are smaller in diameter. One goes to each lung, branching into three right and two left **secondary bronchi** serving the three right and two left lobes of the lungs. The heart is nestled into the left side of the lungs, taking up some of the space that a third lobe might occupy.

The secondary bronchi divide into smaller and smaller branches, looking much like an upside-down tree (Figure 11-5), until they end in thousands of passageways called **respiratory bronchioles.** These airways have a diameter less than that of a pencil lead. Their walls have clusters of tiny pouches that, along with the respiratory bronchioles, are the sites of gas exchange. These pouches, or air sacs, are called **alveoli.**

> **Air moves from the trachea into the two bronchi that supply each lung. The bronchi divide into smaller and smaller branches ending in respiratory bronchioles having outpouchings called *alveoli*, the sites of gas exchange.**

The alveoli: Where gas exchange takes place

The alveoli provide a perfect place for carbon dioxide and oxygen to diffuse between the air in the lungs and the blood. These clusters of microscopic air sacs are bounded by membranes made up of a single layer of epithelial cells. A network of capillaries tightly clasps each alveolar sac. The capillary walls are also only one cell thick and press against the alveolar epithelium. These two adjacent membranes provide the thinnest possible barrier between the blood in the capillaries and the air in the alveoli.

The alveoli also provide another important component of efficient gas exchange: a large surface area. In fact, if the epithelial membrane of all your alveoli was lain flat, it would cover a tennis court! The capillaries cover this enormous surface, creating patterns much like tightly woven spider webs, providing nearly a continuous "sheet" of blood over the alveolar surface.

Large white blood cells called *macrophages* are also found at the alveoli. These cells work to remove any particles or microbes that have escaped the other defenses of the airways. The macrophages transport the invaders to the bronchioles or to the lymphatic system. Sometimes the job is too big, however, and particles remain in the lungs. In smokers the action of the macrophages is impaired, making the lungs more susceptible to disease and injury.

> **The alveoli provide a thin, enormous surface area over which gas exchange can take place.**

How to "catch" cancer

Of the 2.2 million people who died in the United States last year, one-fourth died of cancer and 25% of these—106,000 people—died of lung cancer. About 150,000 cases of lung cancer were diagnosed each year from 1980 to 1992, and 90% of these persons died or will die within 3 years. Of those who die, 96% will be cigarette smokers.

Smoking is a popular pastime in the United States. Almost one third of the U.S. population smokes. American smokers consumed over 550 billion cigarettes in 1991. These cigarettes emit smoke tars and other chemicals in their tobacco that are potent mutagens, and these cancer-causing chemicals are introduced to your lungs when you inhale cigarette smoke. Absorbed by the cells lining your lungs, these chemicals lead to cancer as surely as if you had smoked powerful radioisotopes. Cigarette smoke also contains nicotine, a chemical mimic of the neurotransmitter acetylcholine that is powerfully addictive. That is why it is so difficult to stop smoking once you have started—your body becomes addicted. Smoking is drug abuse with a very dangerous drug.

Among cigarette manufacturers, it has been popular to argue that the causal connection between smoking and cancer has not been proven, that somehow the relationship is coincidental. Look carefully at the data presented in Figure 11-A and see if you agree. The upper graph presents data collected on the incidence of smoking and lung cancer for American males from the turn of the century until now. Note that as late as 1920, lung cancer was a rare disease. With a lag of some 20 years behind the increase in smoking, it became progressively more common.

Now look at the lower graph, which presents data on American females. Significant numbers of American females did not smoke until after World War II, when many social conventions changed. As late as 1963, when lung cancer among males was near current levels, this disease was still rare in females. In the United States that year, only 6588 females died of lung cancer. But as their smoking increased, so did their incidence of lung cancer, with the same inexorable lag of about 20 years. American females today have achieved equality with their male counterparts inthe numbers of cigarettes that they smoke—and their lung cancer death rates are now approaching those of males. In 1991, more than 49,000 females died of lung cancer in the United States.

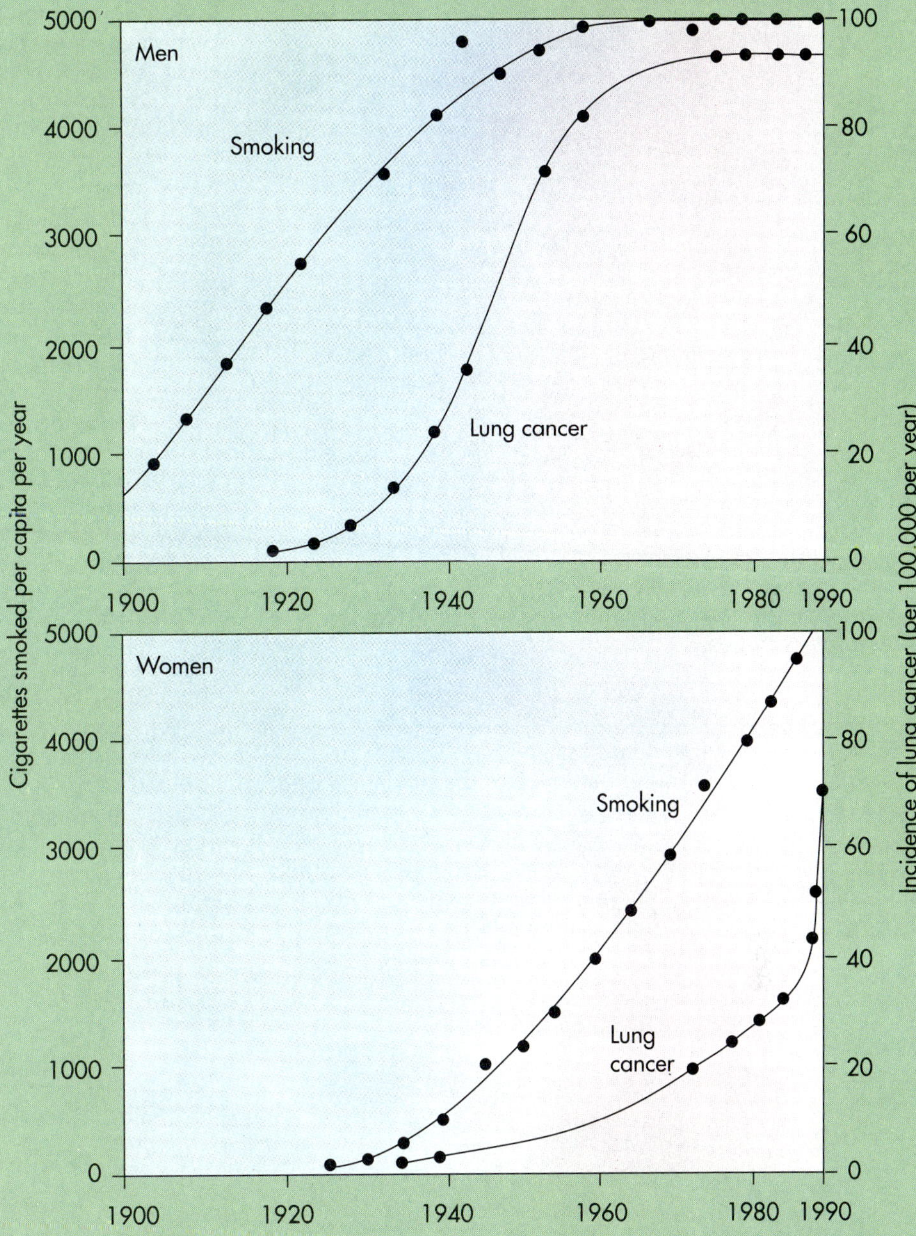

Among smokers, the current rate of death from lung cancer is 180 per 100,000, or about 2 of every 1000 smokers each year. Life insurance companies have computed that, on a statistical basis, smoking a single cigarette lowers your life expectancy 10.7 minutes. (That is more time than it takes to smoke the cigarette!) Every pack of cigarettes bears an unwritten label: *The price of smoking this pack of cigarettes is 3½ hours of your life.*

If you don't smoke, don't start. If you do smoke, quit now. Your body gradually replaces its tissues, and your chance of getting lung cancer will fall until it eventually is no greater than that of nonsmokers.

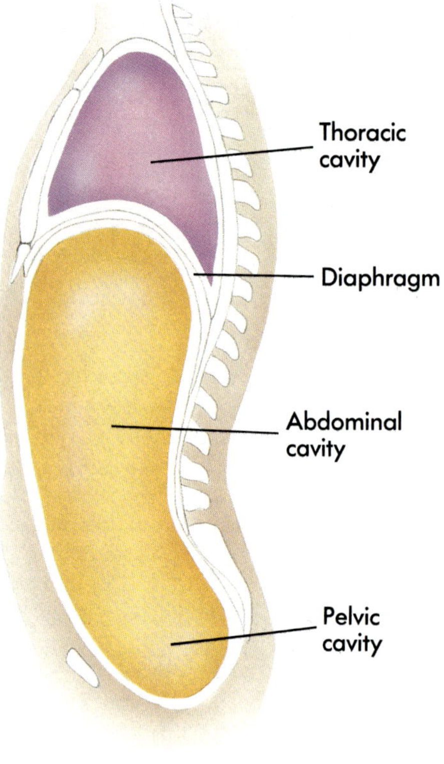

FIGURE 11-6 Location of the diaphragm.
The diaphragm forms a partition between the thoracic and abdominal cavities.

FIGURE 11-7 How a human breathes.
Inhalation: The diaphragm flattens and the ribs are raised, increasing the volume of the chest cavity. As a result of the larger volume, air is pulled in through the trachea.
Exhalation: The diaphragm resumes its dome shape and the ribs fall, reducing the volume of the chest cavity and forcing air outward through the trachea.

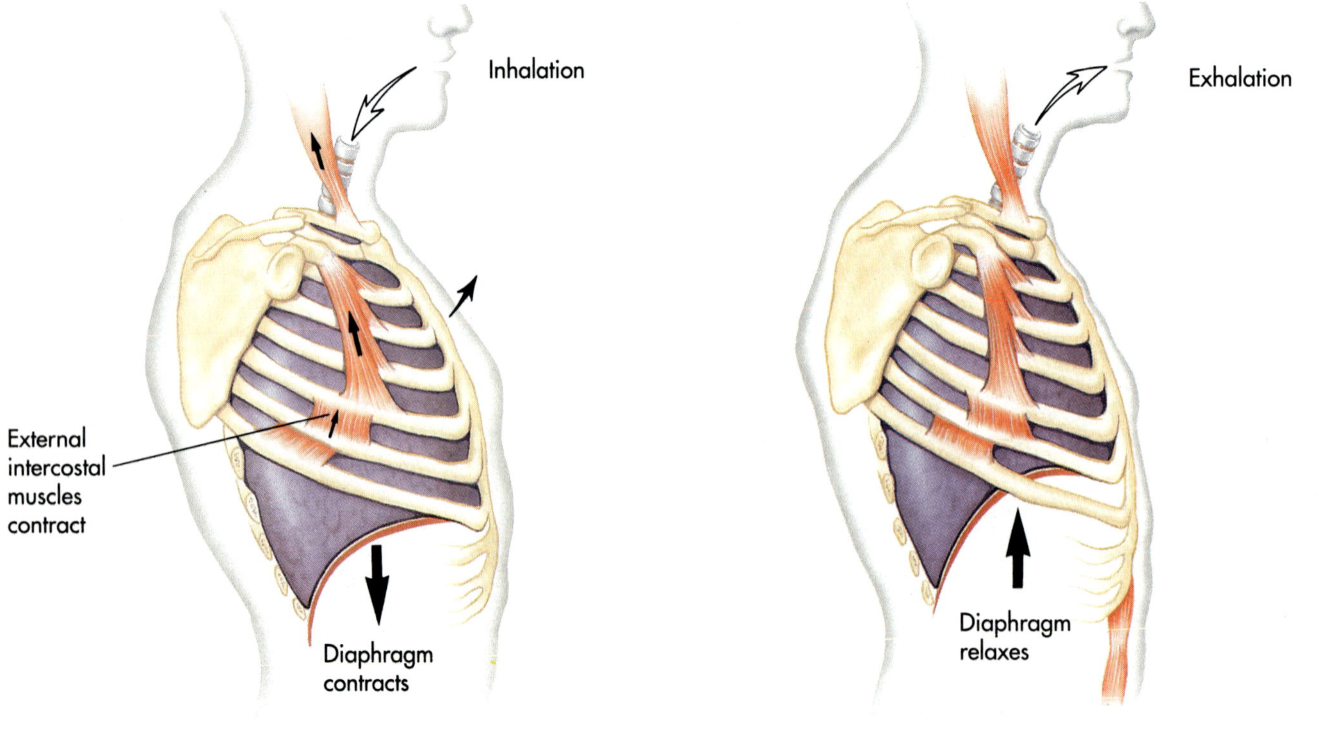

The mechanics of breathing

Air moves into and out of your lungs as the volume of your thoracic cavity is made larger and smaller by the action of certain muscles. The **thoracic cavity,** or chest cavity, is within the trunk of your body above your diaphragm and below your neck. The **diaphragm** is a sheet of muscle that forms the horizontal partition between the thoracic cavity and the abdominal cavity. Various blood vessels and the esophagus puncture it as they traverse these two body cavities. The position of the diaphragm is shown in Figure 11-6.

The diaphragm is assisted by other muscles of breathing. These muscles extend from rib to rib—from the lower border of each rib to the upper border of the rib below—and are called **intercostal muscles.** The word *intercostal* literally means "between" (inter) "the ribs" (costal). You have two sets of intercostals: the **internal intercostals** and the **external intercostals.** The internal intercostals are those that lie closer to the interior of the body (as their name suggests) and that have fibers extending obliquely downward and backward (from front to back). The **external intercostals** extend from back to front, having fibers that are directed downward and forward.

Inspiration

When you are breathing quietly—not exerting yourself physically—your diaphragm and external intercostals alone are responsible for the change in the size of your thoracic cavity. This change in size results in the movement of air into and out of your lungs. During **inspiration** the dome-shaped diaphragm contracts, flattening somewhat and thereby lowering the floor of the thoracic cavity. The external intercostals contract, raising the rib cage. Notice in Figure 11-7 how these two actions increase the size of the thoracic cavity.

The interior walls of the thoracic cavity are lined with a thin, delicate, sheet-like membrane called the **pleura.** The pleura folds back on itself to cover each lung. These two parts of the membrane are close to one another, separated only by a thin film of fluid. As the volume of the thoracic cavity increases during inspiration, the lungs also expand, held to the wall of the thoracic cavity by cohesion of the water molecules between the two membranes. Put simply, the lungs stick to the thoracic wall and move with it.

As the lungs expand in volume, the air pressure within the lungs decreases because there are fewer air molecules per unit of volume. As a result, air from the environment outside the body is pulled into the lungs, equalizing the pressure inside and outside the thoracic cavity. By means of this process, you breathe in 13,638 liters (more than 3000 gallons) of air every day!

Expiration

The lungs contain special nerves called *stretch receptors*. When the lungs are stretched to their normal inspiratory capacity, these receptors send a message to a respiratory center in the brain. This respiratory center is located in parts of the brainstem called the *medulla* and the *pons*. (see Chapter 16). The respiratory center stops sending "contract" messages to the muscles of breathing, which causes them to relax—a passive process in contrast to the active process of inspiration. As the diaphragm relaxes, it assumes its dome-like shape, reducing the volume of the thoracic cavity. Likewise, as the external intercostals relax, the rib cage drops, reducing the volume of the thoracic cavity further. The volume of the lungs, in turn, decreases, aided by the recoil action of the lungs' elastic tissue. The reduced volume of the lungs results in an increase in the air pressure within them. Air is forced out of the lungs, equalizing the pressure outside and inside the thoracic cavity. The cycle of inspiration and expiration is shown in Figure 11-8.

The brainstem also contains areas sensitive to changes in the levels of carbon dioxide in the blood that control the rate of the cycle of inspiration and expiration. Hydrogen ions (H^+) that are produced from a series of chemical reactions involving carbon dioxide also play an extremely important role in influencing the activity of the respiratory center (see Internal respiration, p. 194). So when you have a high level of carbon dioxide in your blood, the level of hydrogen ions in your blood increases (blood pH decreases), and your breathing rate is "stepped up."

Deep breathing

When you breathe deeply, such as during physical exercise, the internal intercostals as well as other muscles in the chest and abdomen help out the muscles of respiration. During inspiration, certain muscles attached to the breastbone—the bone in the center of your chest to which your ribs are attached—pull up on it. In addition, other muscles pull up

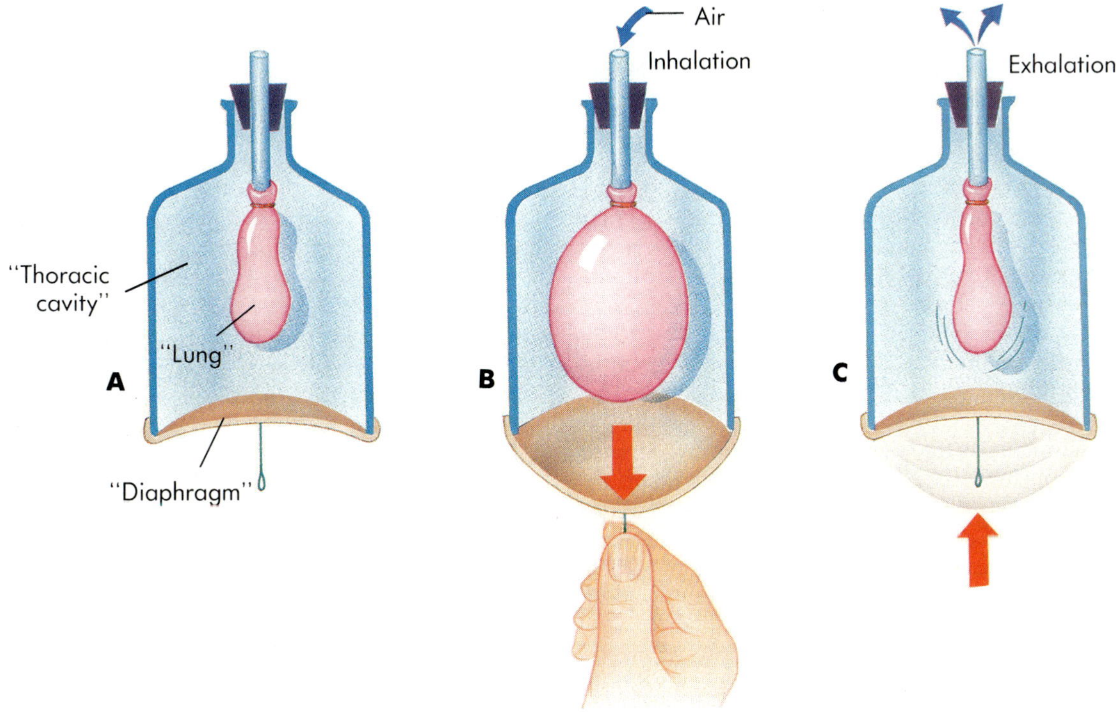

FIGURE 11-8 A simple experiment that shows how you breathe.
In the jar is a balloon **(A)**. When the diaphragm is pulled down, as shown in **B**, the balloon expands. When it is relaxed **(C)**, the balloon contracts. In the same way, air is taken into your lungs when your diaphragm contracts and flattens, expanding the volume of your chest cavity. When your diaphragm relaxes and resumes its dome shape, the volume of your chest cavity decreases and air is expelled.

on the first two ribs. This action increases the volume of the chest cavity more than during quiet breathing, so more air flows into the lungs. By deep breathing and increasing the rate of breathing, world champion runners have been shown to increase their air intake fifteenfold.

Expiration, a passive process during quiet breathing, becomes an active process during deep breathing. The internal intercostals pull down on the rib cage. Abdominal muscles also contract, pulling down on the lower ribs and compressing the abdominal organs, which results in a push up on the diaphragm. These actions cause an additional decrease in the volume of the thoracic cavity and cause more air to be expelled than during quiet breathing.

Inspiration occurs when the volume of the thoracic cavity is increased and the resulting negative pressure causes air to be pulled into the lungs. Expiration occurs when the volume of the thoracic cavity is decreased and the resulting positive pressure forces air out of the lungs.

Lung volumes

How much air do you move into and out of your lungs during inspiration and expiration? You can find out by performing this simple procedure: Fill a large jar with water and invert it in a pan of water. Be careful not to let any of the water seep out of the jar while you are turning it upside down. Mark the level of the water in the jar with tape or a wax pencil. Take about a foot-long piece of rubber tubing or garden hose and put one end up into the jar. Put the other end into your mouth and exhale normally. The air you breathe out will displace the water. Mark the new level of the water. Remove the jar from the water and fill it to this line with water. Now measure the volume of the water between the two lines with a measuring cup or a graduated cylinder. Its volume equals the volume of air you breathed out. The average adult male breathes out 500 milliliters of air during quiet breathing, or slightly more than one pint. This volume of air—the amount inspired or expired with each breath—is called the **tidal volume.**

Of the 500 milliliters of air you normally breathe in, only about 350 milliliters reaches the alveoli. The other 150 milliliters is either on its way into or out of the lungs, occupying space in the nose, pharynx, larynx, trachea, and bronchial tree. This space is called **dead air space** because it serves no useful purpose in gas exchange. Some of this air, in fact, will never reach the lungs. And some air, called **residual air,** remains in the lungs—even during deep breathing.

Gas transport and exchange

The transport of gases throughout the body is assisted by the circulatory system. Without the help of a "highway" of blood, scientists estimate that it would take a molecule of oxygen 3 years to diffuse from your lung to your toe! Obviously, humans could not survive if gas transport were this slow.

Although the pathway of blood throughout the body will be described in detail in the next chapter, it is helpful to understand the basic routing of blood when discussing gas transport. Notice in Figure 11-9 that the upper right chamber of the heart collects incoming blood from the upper and lower body. This blood has given up its supply of oxygen to the tissues, so it is oxygen poor. Along its route, however, it "collected" the waste product of cellular respiration, which is carbon dioxide. Deoxygenated blood such as this (more carbon dioxide than oxygen) is shown as blue in the diagram. The deoxygenated blood passes from the upper to the lower right chamber of the heart, where it is pumped to the lungs. Some of this blood goes to the right lung, and some to the left.

At the lungs, carbon dioxide within the blood of the capillaries surrounding the alveoli and the oxygen in the air of the alveoli are exchanged. External respiration, the exchange of gases at the lungs, works by the process of diffusion. It converts deoxygenated blood to oxygenated blood (more oxygen than carbon dioxide). This blood is shown as red in Figure 11-9. The oxygenated blood flows from the lungs to the left side of the heart, where the lower left chamber pumps it out to other parts of the body.

The circulatory system aids in respiration by transporting gases throughout the body.

External respiration

Earlier in this chapter, the way molecules of air exert pressure is discussed. Air is made up of many different kinds of molecules, such as oxygen, nitrogen, and carbon dioxide. Each of these gases exerts a pressure that depends on the number of molecules of the specific gas present per unit of volume. Molecules of a gas in a liquid (such as blood) also exert a pressure. Differences in pressures of a particular gas in two adjoining locations within a living system produce a *pressure gradient.* Molecules of a gas such as oxygen or carbon dioxide tend to move from an area of higher pressure to an area of lower pressure, or "down" the pressure gradient. Each gas moves according to its own pressure gradient, unaffected by the pressure gradients of other gases with which it might be mixed. This movement is a type of diffusion (see Chapter 4).

As deoxygenated blood (blue in Figure 11-9) arrives at the lungs, the pressure of the carbon dioxide in the blood is greater than the pressure of carbon dioxide in the air within the alveoli. Therefore carbon dioxide diffuses out of the blood and into the alveoli. Likewise, the pressure of the oxygen in the air within the alveoli is greater than the pressure of the small amount of oxygen in the blood. Therefore oxygen diffuses out of the alveoli and into the blood.

Most of the carbon dioxide (70%) in the blood travels around "hitched" to water molecules in the fluid portion of the blood. The carbon dioxide and water combine chemically to form molecules of carbonic acid (H_2CO_3) but quickly break apart, or dissociate, into bicarbonate ions (HCO_3^-) and hydrogen ions (H^+). About one fourth (23%) of the carbon dioxide is carried in the red blood cells,

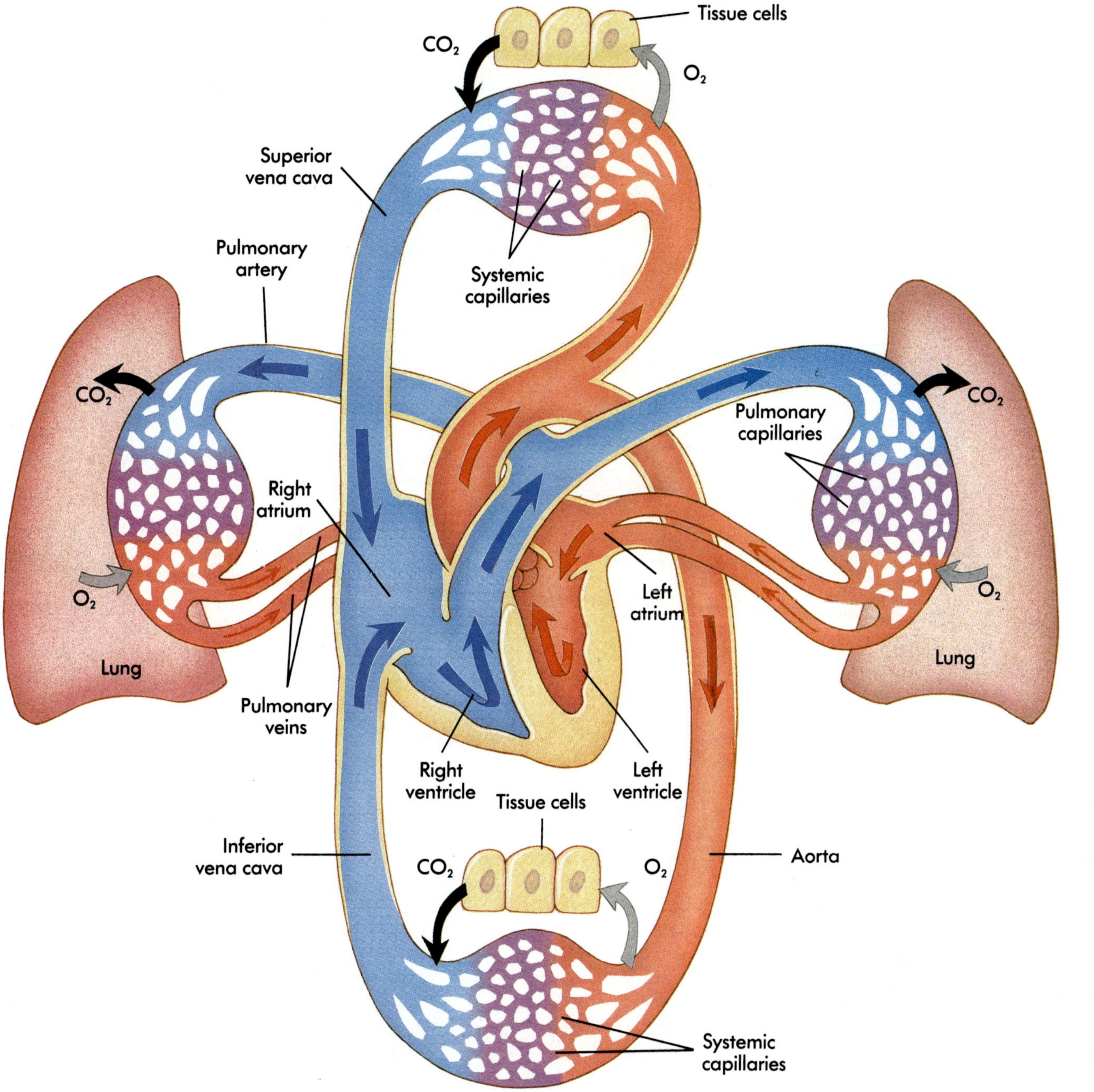

FIGURE 11-9 The route of blood during gas transport.
As blood moves through the body, it gives up its supply of oxygen (O_2) to the tissues and collects carbon dioxide (CO_2). This deoxygenated blood is then pumped through the heart to the lungs, where the CO_2 in the blood and O_2 in the alveoli are exchanged. The reoxygenated blood flows to the heart and is pumped to the rest of the body tissues.

bound to the oxygen-carrying molecule **hemoglobin** (Hb). It is carried, however, by a different portion of the molecule than oxygen is. A small amount (7%) of the carbon dioxide is simply dissolved in the blood. At the alveoli, the dissolved carbon dioxide first moves out of the blood. This decrease in its concentration triggers a reversal of the chemical reactions described above:

$$H^+ + HCO_3^- \rightarrow H_2CO_3 \rightarrow H_2O + CO_2$$

Carbon dioxide also dissociates from hemoglobin, and the now freed carbon dioxide molecules diffuse into the alveoli.

As oxygen diffuses from the alveoli, very little of it dissolves in the fluid portion of the blood. Instead, it combines with hemoglobin within the red blood cells. When the pressure gradient is high, as in the alveoli, hemoglobin binds with large amounts of oxygen. When oxygenated, hemoglobin turns bright red, which makes blood look red. You will notice, however, that the blood vessels on the underside of your wrists look blue. These blood vessels are

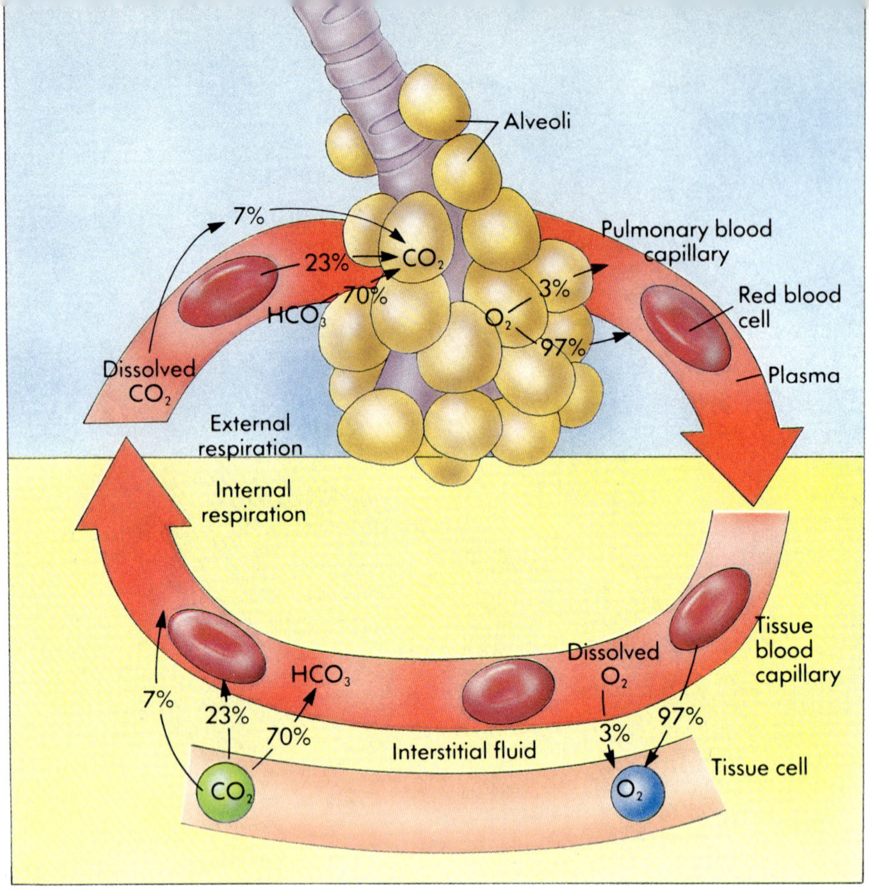

FIGURE 11-10 How gases are transported during external and internal respiration. External respiration takes place in the lungs, as red blood cells exchange CO_2 for O_2. During internal respiration, oxygenated blood travels to the tissues, distributing O_2 and collecting CO_2.

veins carrying deoxygenated blood. Deoxygenated blood is dark red but appears blue through layers of skin.

At high altitudes, such as in mountainous regions, the air is thinner and the pressure of the oxygen molecules within the air is lower than at sea level. Therefore less of a pressure gradient is created at the alveoli. As a consequence, less oxygen diffuses into the blood, causing shortness of breath, nausea, and dizziness. This condition is referred to as **high altitude sickness.** Mountain climbers, athletes working out or playing at high altitudes (such as Mile High Stadium in Denver, Colorado), and tourists visiting an area of high altitude (such as Mexico City, Mexico) need to slowly work up to their normal levels of activity to give their bodies time to adjust to the lower oxygen pressure.

During external respiration, carbon dioxide moves down a pressure gradient, diffusing from the blood in capillaries surrounding the alveoli to air in the alveoli. Likewise, oxygen moves down a pressure gradient, diffusing from the alveolar air to the blood in the capillaries surrounding the alveoli. Carbon dioxide is carried in the fluid portion of the blood primarily as bicarbonate ions; oxygen is carried within the red blood cells by the oxygen-carrying molecule hemoglobin.

Internal respiration

Oxygenated blood is pumped out to the body by the left ventricle, or lower left chamber, of the heart. It travels through the aorta to other large arteries that soon branch into smaller and smaller arteries. Eventually the blood vessels become so small that only one red blood cell at a time can pass. These vessels are the capillaries and are the sites of internal respiration.

Internal respiration is the exchange of oxygen and carbon dioxide between the blood within capillaries at the tissues and the tissue fluid that bathes the cells. As this exchange takes place, oxygenated blood becomes progressively deoxygenated. In essence, the processes of internal respiration are the reverse of the processes of external respiration. As the blood begins its journey from the lungs around the body, it is oxygen rich. The oxygen molecules in the blood exert a higher pressure than those in the cells because oxygen is continuously being used in the cells in the process of cellular respiration. Consequently, oxygen molecules begin to dissociate from the hemoglobin, diffuse into the tissue fluid, and from there diffuse into the cells. Conversely, the levels of carbon dioxide are higher in the cells than in the blood because carbon dioxide is continuously being produced during cellular respiration. Consequently, carbon dioxide diffuses from the cells into the tissue fluid and then into the blood.

As gas exchange at the capillaries continues, an interesting thing happens. The oxygen supply of the blood decreases, thereby decreasing the pressure of oxygen in the blood. Diffusion of oxygen into the tissue fluid does not slow down, however, because the pH at the capillaries lowers as carbon dioxide diffuses into the blood. Remember, carbon dioxide is carried in the blood primarily as bicarbonate ions. As these ions are formed, hydrogen ions are also produced. The build-up of these hydrogen ions makes the blood increasingly acidic as more carbon dioxide diffuses in. (See Chapter 2 for a discussion of the relationship between hydrogen ions and pH.) This acid environment helps split more oxygen from hemoglobin, thereby enhancing oxygen's diffusion into the tissue fluid. (Figure 11-10 shows the processes of internal and external respiration.)

Temperature also helps split oxygen from hemoglobin. During exercise or any type of exertion, your body needs more oxygen delivered to its cells. However, active cells also produce more carbon dioxide and heat, thereby helping to split oxygen from hemoglobin and meet their own needs. This is one example of how your body works to maintain homeostasis, a state of internal equilibrium.

During internal respiration, carbon dioxide moves down a pressure gradient, diffusing from the tissue fluid surrounding the body cells to the blood. Likewise, oxygen moves down a pressure gradient, diffusing from the blood within capillaries to the tissue fluid. As carbon dioxide diffuses into the blood, the resultant increase in hydrogen ions increases the dissociation of oxygen from hemoglobin.

Choking: A common respiratory emergency

Have you ever been eating with someone who started to choke? Did you know what to do? Choking is caused when food or a foreign object becomes lodged in the windpipe. When you are with someone who is choking, first notice whether the person can talk, breathe, or cough. If so, stay with the person until the airway is cleared by coughing. Do not try to slap the person on the back. The slapping may only cause the food to become more deeply lodged in the windpipe.

If a person cannot talk or cough and appears not to be breathing, administer several short, quick abdominal thrusts. This technique is called the *Heimlich maneuver,* after Dr. Henry Heimlich who developed the procedure. First, stand behind the choking victim. Put your arms around the person, placing your fists just below the breastbone. The proper placement is shown in Figure 11-11. Then give a series of quick, sharp, upward and inward thrusts. These thrusts push in on the diaphragm and the thoracic cavity, suddenly decreasing its volume. This sudden decrease creates a surge in air pressure below the obstruction, which usually projects it forcefully from the windpipe.

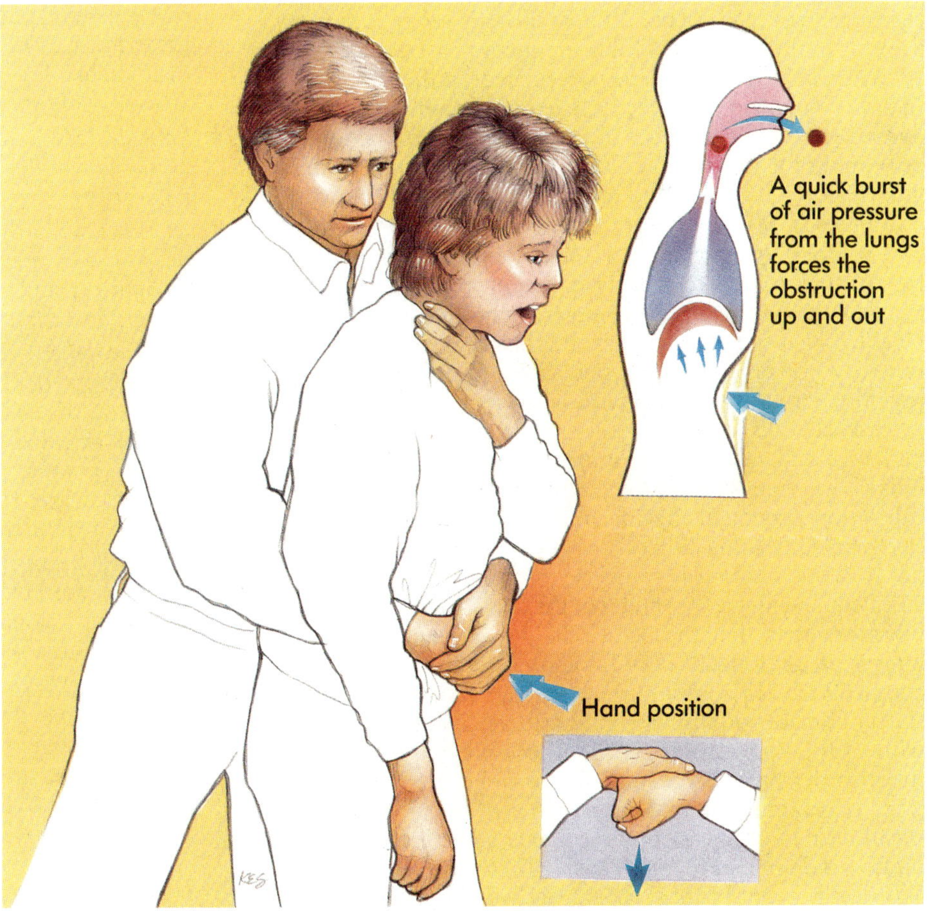

FIGURE 11-11
The Heimlich maneuver.
Learning the Heimlich maneuver could save someone you know from choking.

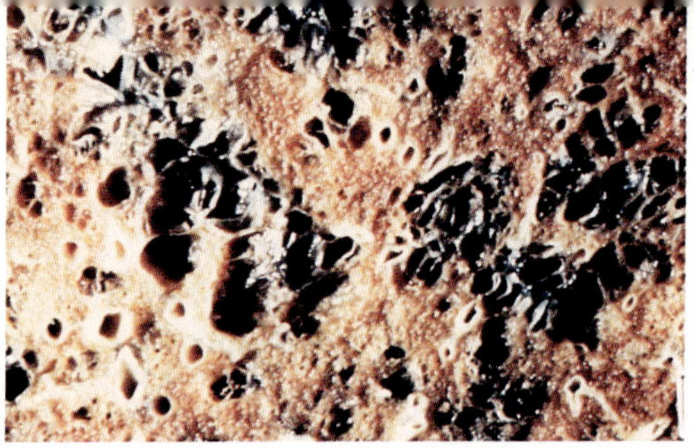

FIGURE 11-12 Alveoli ruptured as a result of emphysema. Emphysema permanently damages lungs and, along with cancer, is one of the dangerous risks of smoking.

Chronic obstructive pulmonary disease

The term **chronic obstructive pulmonary disease (COPD)** is used to refer to disorders that "block up" the airways and impair breathing. COPD affects 25 million people in the United States alone and is responsible for at least 50,000 deaths per year. Most doctors agree that it is one of the fastest growing health problems in this country. Two disorders commonly included in COPD are **chronic bronchitis** and **emphysema.**

The term *chronic* refers to something that occurs over an extended time. Chronic bronchitis, then, does not refer to an isolated, one-time infection or inflammation. It refers to an inflammation of the bronchi and bronchioles that lasts for at least 3 months each year for 2 consecutive years with no accompanying disease as a cause. The primary cause is cigarette smoking. Air pollution and occupational exposure to industrial dust are much less frequent causes.

Cigarette smoking paralyzes ciliated epithelial cells so that they can no longer effectively remove incoming particles and microbes. It also causes increased mucus production by cells lining the trachea. A continued buildup of mucus provides food for bacteria, and infection can result. The mucus also plugs up the respiratory "plumbing."

Normally, bronchioles widen, or dilate, during inspiration; they narrow, or constrict, during expiration. If mucus is plugging various bronchioles, some air may therefore be able to get to the alveoli beyond the plugged bronchioles but may not be able to get out. Coughing spells produce pressure within these continuously inflated alveoli that ruptures their walls, decreasing the surface area over which gas exchange can take place (Figure 11-12). In addition, the lungs lose their elasticity and the ability to recoil during exhalation, staying filled with air. This disorder is aptly called *emphysema,* meaning "full of air." A person with emphysema has to work voluntarily to exhale.

People with COPD find that they have more respiratory infections than they did before the disorder and that these infections last longer. They have a morning cough or may cough all day. In addition, they may tire easily and become short of breath with minimal physical exertion. Some people with COPD feel as though they cannot breathe at times. As the disorder progresses, some find it difficult to do a day's work or accomplish the daily activities of living. Periods of breathlessness increase. Some persons suffer bouts of respiratory failure and must be hospitalized. COPD is serious and deadly—but in most cases, it is avoidable.

Summary

1. Respiration is the uptake of oxygen and the release of carbon dioxide by the body. It includes the processes of cellular respiration, internal respiration, external respiration, and breathing.
2. Breathing consists of taking air into the lungs, or inspiration, and expelling air from the lungs, or expiration. The average rate of breathing is 14 to 20 breaths per minute.
3. Air passes through many respiratory structures on its way to the lungs. The nasal cavities, located behind the nostrils, warm and moisten incoming air. The air then passes down the throat, or pharynx, and into the larynx. The larynx, or voice box, sits on top of the windpipe, or trachea. The trachea divides into two bronchi, which further divide into smaller and smaller bronchioles, ending in microscopic air sacs called *alveoli.*
4. The epithelium of the trachea, the bronchi, and some of the bronchioles produce mucus that helps trap foreign particles. In addition, cilia beat to produce a current in the mucus that brings particles up to the throat to be swallowed.
5. During inspiration, the diaphragm, a sheet-like muscle that forms the floor of the thoracic cavity, flattens somewhat as it contracts, increasing the volume of the thoracic cavity. Muscles between the ribs pull up on the rib cage, also increasing this volume. The negative pressure that is created pulls air into the lungs. During expiration, these muscles relax, decreasing the volume of the thoracic cavity. The positive pressure that is created pushes air out of the lungs.
6. Oxygen in the air and carbon dioxide in the blood are exchanged at the alveoli. This exchange is called *external respiration.*
7. Deoxygenated blood from the body is pumped by the heart to the lungs. Here it is oxygenated during external respiration and is then returned to the heart to be pumped to the rest of the body.
8. At the capillaries in the body tissues, oxygen in the blood diffuses out of the blood and into the tissue fluid surrounding the cells and then into the cells for use during cellular respiration. The waste of cellular respiration, carbon dioxide, diffuses out of the cells, into the tissue fluid, and then into the blood. This process is called *internal respiration.*
9. To assist a person who is choking and cannot talk or breathe, administer several short, quick, abdominal thrusts with your arms around the victim and your fists just below the breastbone.
10. *Chronic obstructive pulmonary disease (COPD)* refers to disorders that impair movement of the air in the respiratory system. Chronic bronchitis and emphysema are two common COPD disorders that are most frequently caused by cigarette smoking.

REVIEW QUESTIONS

1. Distinguish among respiration, cellular respiration, internal respiration, and external respiration.
2. Fill in the blanks: A breath consists of taking air into the lungs (__________) and expelling it from the lungs (__________).
3. Your biology instructor poses a problem in class, and you suggest a solution. Explain how you produced the necessary sounds.
4. Fill in the blanks: After passing into the throat and then the larynx, air moves from the __________ into the two __________ that supply each lung. These passageways branch repeatedly until they end in __________. Their walls contain tiny pouches called __________ that are the sites of gas exchange.
5. Fill in the blanks: Inspiration occurs when the volume of the thoracic cavity __________. Expiration occurs when the volume of the thoracic cavity __________.
6. Explain how differences in air pressure help you to breathe.
7. Fill in the blanks: The amount of air inspired or expired with each breath is called the __________ __________. Some __________ __________, however, always remains in the lungs.
8. Distinguish between oxygenated and deoxygenated blood.
9. How does the circulatory system assist the process of respiration? Why is this help important?
10. Explain what happens during gas exchange at the alveoli. What gases are exchanged, and why does the exchange take place?
11. While on vacation in the mountains, you find that you feel light headed and short of breath. Explain why.
12. How does the pH of blood change within the capillaries? Why is this important?
13. Explain the process of gas exchange at the capillaries. What gases are exchanged, and why does an exchange of gases occur?
14. Lung cancer is an often fatal disease that is becoming more common in the United States. Over 90% of lung cancer patients smoke cigarettes. Explain how cigarette smoking affects the respiratory passageways and lungs and how this could lead to cancer.

KEY TERMS

alveoli 185
breathing 185
cilia 186
chronic bronchitis 196
chronic obstructive pulmonary disease 196
dead air space 192
diaphragm 190
emphysema 196
expiration 192
external respiration 185
glottis 186
hemoglobin 193
inspiration 190
intercostal muscles 190
internal respiration 186
larynx 186
nasal cavities 186
pharynx 186
pleura 191
primary bronchi 188
residual air 192
respiration 185
respiratory bronchioles 188
secondary bronchi 188
thoracic cavity 190
tidal volume 192
trachea 186
vocal cords 186

THOUGHT QUESTIONS

1. The US Surgeon General reports that heart disease killed over 800,000 Americans in 1990, the single largest cause of death. Of these deaths, one-third are directly attributable to smoking cigarettes. Why do you think smoking increases the risk of heart disease?
2. In the James Bond spy novel *Goldfinger,* a woman is murdered by painting her body with gold paint so that all of her skin is covered. Agent 007 states knowingly that if even a small square of skin had been left uncovered at the base of her skull, she would have lived but that painting over *all* of her skin caused her to suffocate. What do you think of this possibility?

FOR FURTHER READING

Feder, M., & Burggren, W. (1985, November). Skin breathing in vertebrates. *Scientific American*, pp. 126-142.
This article discusses how many vertebrates do a significant portion of their breathing through their skin.

Perutz, M. F. (1978, December). Hemoglobin structure and respiratory transport. *Scientific American*, pp. 92-125.
This is an account of how hemoglobin changes its shape to facilitate oxygen binding and unloading by the man who won a Nobel Prize for unraveling the structure of hemoglobin.

Raloff, J. (1987, August 8). New clues to smog's effect on lungs. *Science News, 132,*86.
This article describes how the lung's defense mechanisms can be damaged by environmental contamination.

12 Circulation

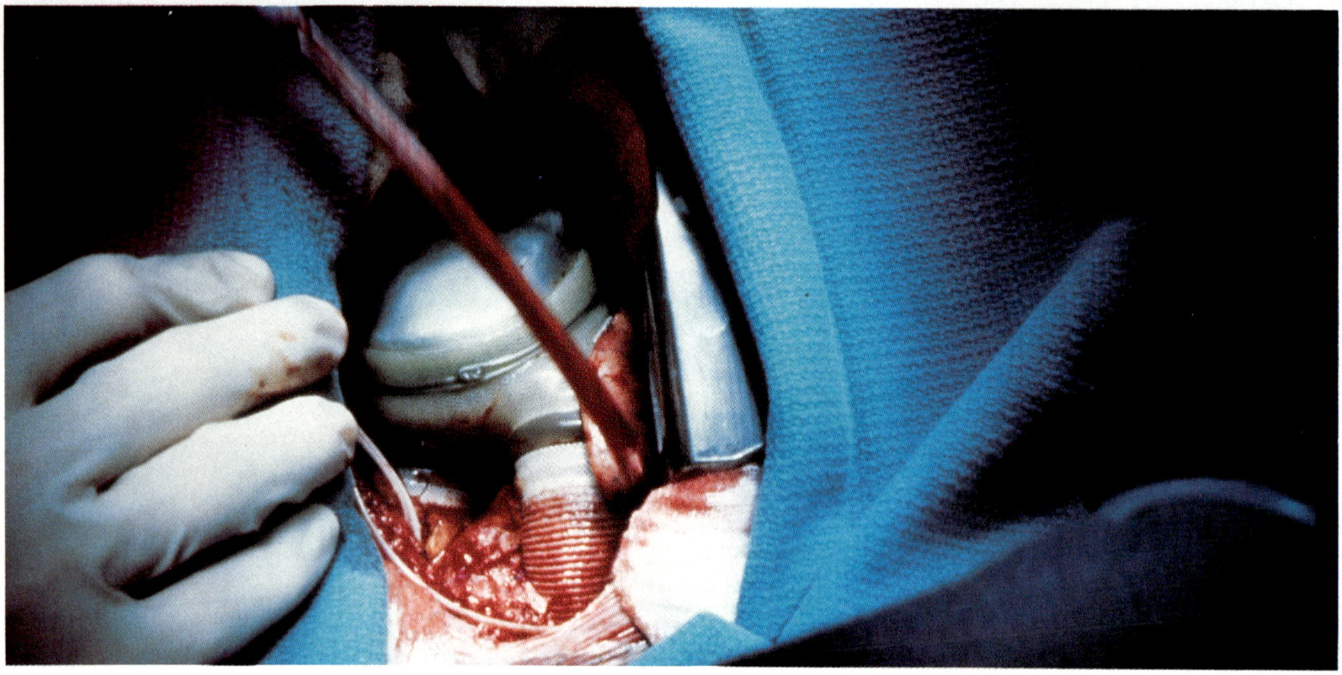

OUTLINE

Functions of the circulatory system
- Nutrient and waste transport
- Oxygen and carbon dioxide transport
- Temperature maintenance
- Hormone circulation

The heart and blood vessels
- Arteries and arterioles
- Capillaries
- Veins and venules
- The heart: A double pump
  - Circulatory pathways
  - How the heart contracts
  - Monitoring the heart's performance

The blood
- Blood plasma
- Types of blood cells
  - Erythrocytes
  - Leukocytes
  - Platelets

The lymphatic system

Diseases of the heart and blood vessels

History was made in 1982 when a 61-year-old dentist from Seattle, Washington, had his diseased heart replaced with a total artificial heart (TAH) in an operation similar to the one in the photo. This bridge to life—the Jarvik-7—was named for the doctor who designed it. A product of over three decades of research on mechanical replacements for the heart, the Jarvik-7 was constructed primarily of polyurethane and a Dacron polyester mesh and was actually held together by Velcro!

HIGHLIGHTS

▼
The circulatory system is made up of the heart, the blood vessels, and the blood; the heart, a muscular pump, circulates the blood throughout a system of vessels that permeates the body.

▼
The circulatory system is the transportation system of the body, carrying nutrients, oxygen, and hormones to the cells and removing metabolic wastes.

▼
The lymphatic system helps the circulatory system by collecting, filtering, and returning the fluid that moves out of the blood and into the tissues and that does not directly return to the blood by itself.

▼
Cardiovascular diseases, or diseases of the heart and blood vessels, are the leading causes of death in the United States.

Since 1982, the Jarvik-7 has been implanted in approximately 180 patients, but it lost its approval for use from the Food and Drug Administration. Air tubes that supplied strong pulses of air to the artificial heart, "pushing" the blood from its chambers, caused problems; the place where they entered the body became infected easily. Also, the artificial heart often produced blood clots that traveled to the brain, resulting in strokes. Now, however, history is being made again with a new Jarvik heart—the Jarvik 2000.

Called a *left ventricular assist device,* this new design helps a person's diseased heart; it does not replace it. The small egg-shaped pump sits in the left lower chamber of the patient's heart. It connects to a battery pack worn outside the body and is regulated by microprocessors (computer chips). As amazing as these "high tech" replacements and assist devices are, scientists still have not been able to duplicate the body's own remarkable pump—the heart.

Functions of the circulatory system

Your circulatory system is made up of three components: (1) the **heart,** a muscular pump; (2) the blood vessels, a network of living tube-like vessels that permeate the body; and (3) the blood, which circulates within these vessels (see Figure 12-3). Together, the heart and blood vessels—the "plumbing" of the circulatory system—are known as the **cardiovascular system.** The prefix *cardio* refers to the heart, and the suffix *vascular* refers to blood vessels. However, the terms *cardiovascular system* and *circulatory system* are often used interchangeably.

Your circulatory system is like a roadway that connects the various muscles and organs of your body with one another. It serves four principal functions:

1. Nutrient and waste transport
2. Oxygen and carbon dioxide transport
3. Temperature maintenance
4. Hormone circulation

In addition, proteins and ions in the fluid portion of the blood regulate the movement of water between the blood and the tissues. Specialized cells defend the body against invading microorganisms and other foreign substances. (These two functions are discussed later in the chapter.) Many biologists argue that the functions of the circulatory system are truly functions of the blood alone; however, the blood cannot perform these functions without circulating throughout the body within vessels, pushed by a powerful pump.

Nutrient and waste transport

The nutrient molecules that fuel cell metabolism are transported to the cells of the body by the circulatory system. Sugars and amino acids pass into the bloodstream at the small intestine, diffusing into a fine net of blood vessels below the mucosa. Most fatty acids are absorbed by the epithelial cells lining the small intestine. After being resynthesized into fats and "packaged" for transport, these fats pass into lymphatic vessels and are transported to the blood by the lymphatic system. The blood takes all these digestive products to the liver for processing. There, some of these molecules are converted to glucose, which is released into the bloodstream. Others, such as essential amino acids and vitamins, pass through the liver unchanged. Still others, such as excess "energy" molecules and excess amino acids, are used for the synthesis of glycogen and body fat and are stored for later use.

From the liver, the blood carries glucose and other energy molecules to all the body's cells. In addition, the blood also brings molecules such as amino acids to the cells, which are used as building blocks to produce other substances. The cells, in turn, release the waste products of metabolism into the bloodstream. The blood carries most of these wastes to a cleansing organ, the kidney, which captures and concentrates them for excretion in the urine (see Chapter 14). The cleansed blood then passes back to the heart.

Oxygen and carbon dioxide transport

The cells of the body carry out cellular respiration and need oxygen for this series of reactions to take place. Oxygen is transported to the cells of the body by the circulatory system. Within the lungs, oxygen molecules diffuse into the circulating blood through the walls of capillaries, which are very fine blood vessels (see Chapter 11). This oxygen passes into red blood cells suspended in the liquid portion of the blood. From the lungs the blood carries its cargo of oxygen to all cells of the body. At the same time, it picks up a waste product of cellular respiration, carbon dioxide. The blood then returns to the lungs where the carbon dioxide is released and a fresh supply of oxygen is captured.

Temperature maintenance

As you read in Chapter 6, energy is continuously being "lost" as heat during the chemical reactions that take place in the cells of your body. The blood distributes this heat as it circulates, helping to maintain your body temperature. As the blood circulates, it passes through delicate, microscopic networks of blood vessels that lie under your skin. As the blood passes through these vessels, it "gives up" heat because the environment is usually cooler than the body's temperature of 98.6° F. The body works to balance the amount of heat produced and the amount of heat lost to maintain a stable internal temperature.

To maintain this balance, a regulatory center in the brain, the hypothalamus, acts like your own personal thermostat, constantly monitoring body temperature and stimulating regulatory processes. If your body temperature drops, for example, signals from this center cause surface blood vessels to narrow, or constrict. Constriction of these blood vessels limits blood flow to the surface of the skin and lessens heat loss. Conversely, if your body temperature rises, signals from this center cause surface blood vessels to widen, or dilate. Dilation increases blood flow to the surface of the skin and increases heat loss (Figure 12-1).

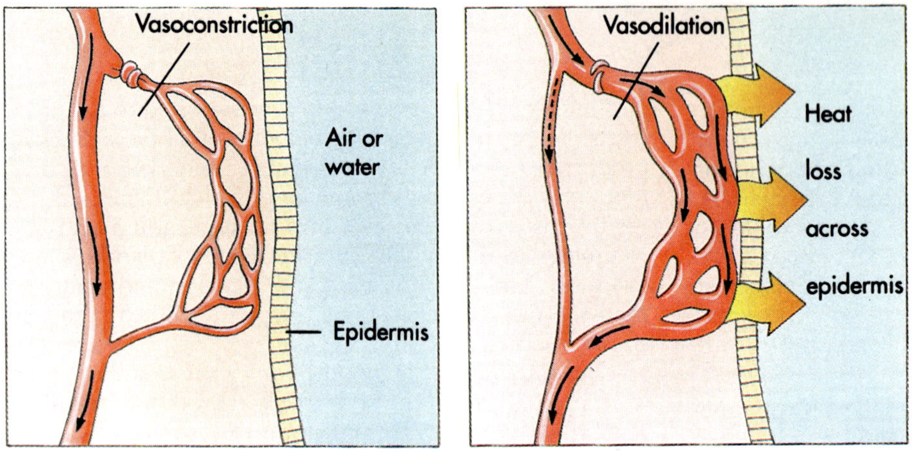

FIGURE 12-1 Regulation of heat loss. The amount of heat lost at the body's surface can be regulated by controlling the flow of blood to the surface. Constriction of the surface blood vessels limits flow and lessens heat loss. Dilation of the surface blood vessels increases blood flow and thus increases heat loss.

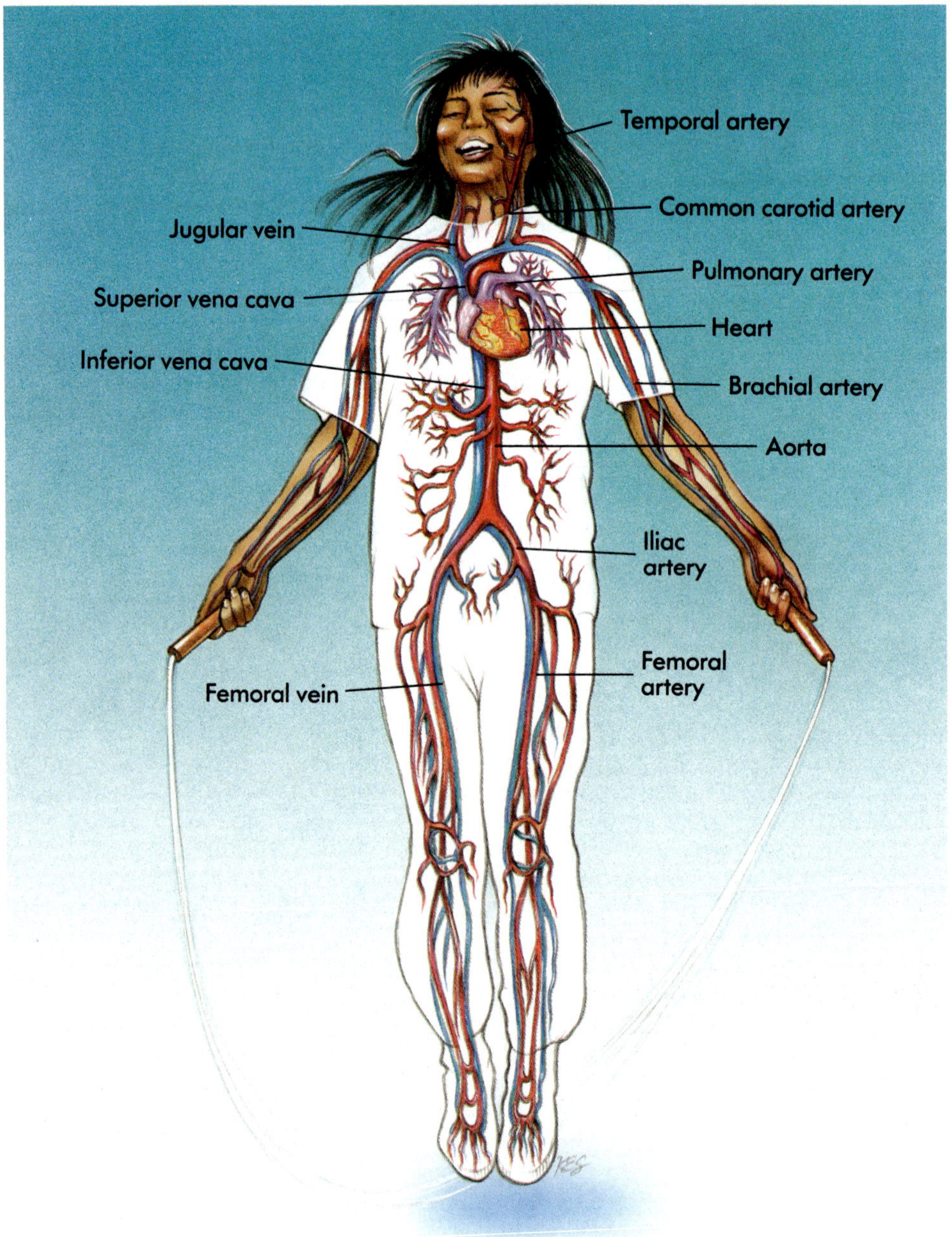

FIGURE 12-2 The human circulatory system. The circulatory system is composed of the heart, blood vessels, and blood.

Hormone circulation

The chemical reactions and other activities of the body are coordinated by nerve signals and hormones. Hormones are the body's chemical "messengers"; they are produced in one place and produce an effect in another. The circulatory system is the "highway" within which hormones travel throughout the body from their site of production and, in this way, reach the target tissues that are capable of responding to them.

The circulatory system transports oxygen and nutrients to cells, transports carbon dioxide and metabolic wastes away from cells, helps maintain a stable internal temperature, and carries chemical messengers called *hormones* throughout the body. Substances within the blood regulate the movement of water between the blood and the tissues and defend the body against disease.

The heart and blood vessels

As the blood is pushed along by the heart to begin its journey throughout the body, it leaves the heart within vessels known as **arteries** (Figure 12-2). From the arteries, the blood passes into a network of **arterioles,** or small arteries. From these, it eventually is forced through the **capillaries,** a fine latticework of very narrow tubes, which get their name from the Latin word *capillus,* meaning "a hair." As blood passes through these capillaries, gases are exchanged between the blood and the tissues, nutrients are "delivered" to the tissues, and wastes are "picked up" from the tissues. After its journey through the capillaries, the blood passes into a third kind of vessel: the **venules,** or small **veins.** A network of venules and larger veins collects the circulated blood and carries it back to the heart.

Arteries and arterioles

Arteries are vessels that carry blood away from the heart. The adult human heart pumps about 70 milliliters of blood—a little over 2 ounces—into the arteries with each beat. On average, your heart beats about 75 times per minute, pumping over 5 liters, or 5½ quarts, of blood into your arteries each minute of your life. The vessels leading out from the heart expand slightly in diameter and then recoil before the next heartbeat in response to the pressure of blood surging into them.

The walls of the arteries are made up of three layers of tissue (Figure 12-3, *A*) and have a hollow core called the *lumen* through which blood flows. *Endothelial cells* (the "inner" epithelium of blood vessels and the heart) line arteries and are in contact with the blood. Surrounding these cells is a thick layer of smooth muscle and elastic fibers. The elastic tissue allows the artery to expand and recoil in response to the pulses of blood. The steady contraction of the muscle layer strengthens the wall of the vessel against overexpansion. This muscle and elastic layer is encased within an envelope of protective connective tissue.

Arterioles differ from arteries in that they are smaller in diameter. The walls of the largest arterioles are constructed very much like those of arteries. However, as arterioles get smaller in diameter as they near the capillaries, they consist of nothing but a layer of endothelium wrapped with a few scattered smooth muscle cells.

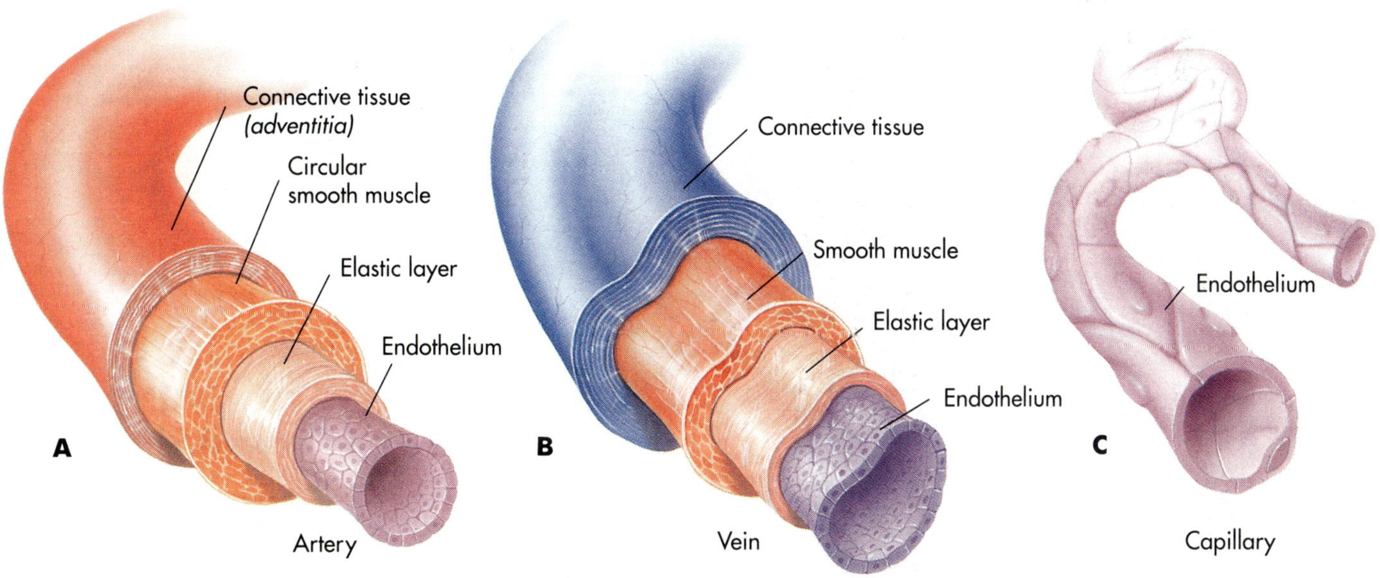

FIGURE 12-3 The structure of some important blood vessels.
Blood leaves the heart through arteries **(A)** and returns to the heart through veins **(B)**. The capillaries **(C)** are the tubes through which the blood is forced in the tissues, exchanging gases, delivering nutrients, and collecting wastes.

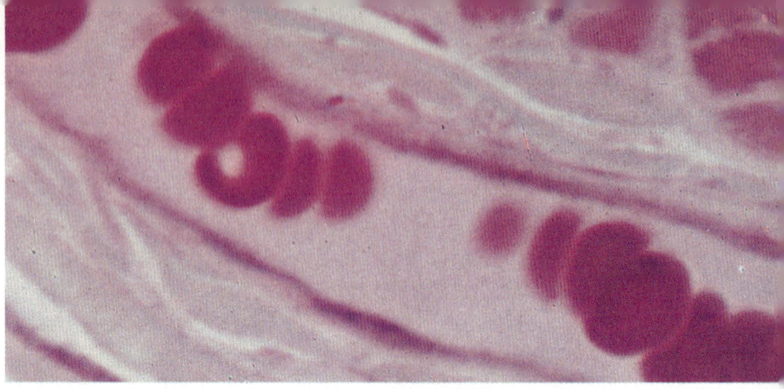

FIGURE 12-4 Red blood cells in a capillary passing along in single file.
Many capillaries are even smaller than those shown here from the bladder of a monkey. Red blood cells will even pass through capillaries narrower than their own diameter, pushed along by the pressure generated by a pumping heart.

The muscle cells within the walls of arterioles tighten or relax in response to messages from nerves and hormones. When these muscles relax, the arteriole dilates and the blood flow through it increases. This response is exactly what is happening when you blush. When you are embarrassed (or when you become overheated), signals from the nerve fibers connected to muscles surrounding the arterioles are inhibited, which relaxes the smooth muscle and causes the arterioles in the skin to dilate. The increased blood flow brings heat to the surface for escape and causes you to have a red face! Conversely, when you are scared or cold, the muscles in the walls of the arterioles in your skin contract. When the muscles contract, the arteriole constricts and the blood flow through it decreases, conserving your body heat. When you are scared, the body constricts these vessels to route more blood to other tissues such as your skeletal muscles.

Capillaries

Capillaries are microscopic blood vessels that connect arterioles with venules. They have a simple structure (Figure 12-3, *C*) and are little more than tubes one cell thick with a length that would barely stretch across the head of a pin. The internal diameter of the capillaries is about the same as that of red blood cells, causing these cells to squeeze through the capillaries single file (Figure 12-4). The closeness between the walls of the capillaries and the membranes of the red blood cells facilitates the diffusion of gases, nutrients, and wastes between them—a "swap" of oxygen and nutrients for carbon dioxide and other metabolic waste products.

Your entire body is permeated with a fine mesh of capillaries, networks that amount to several thousand kilometers in overall length. In fact, if all the capillaries in your body were laid end to end, they would extend across the United States! These networks of capillaries are called **capillary beds.** In a capillary bed, some of the capillaries connect arterioles and venules directly (Figure 12-5) and are called **thoroughfare channels.** From these channels, loops of true capillaries—those not on the direct flow route from arterioles to venules—leave and return. Almost all exchange between the blood and the cells of the body occurs through these loops. The entry to each loop is guarded by a ring of muscle that, when contracted, blocks flow through the capillary. Restricting blood flow in capillary beds near the sur-

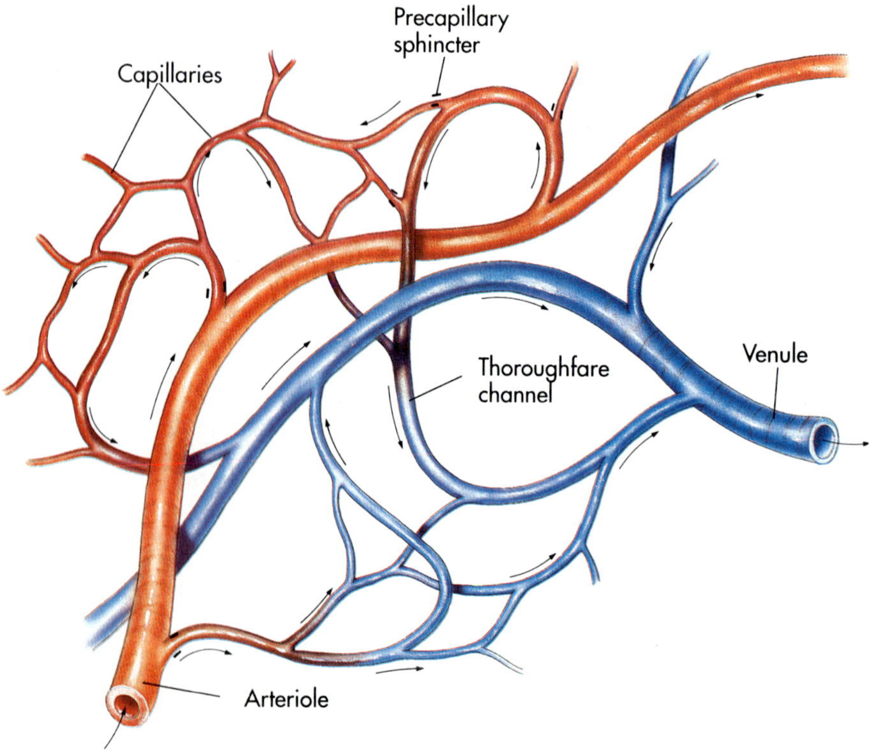

FIGURE 12-5 The capillary bed connects arteries with veins.
The most direct connection is via thoroughfare channels that connect arterioles directly to venules. Branching from these thoroughfare channels is a network of finer capillary channels. Most of the exchange of gases, nutrients, wastes, and ions between body and red blood cells occurs while the red blood cells are in this capillary network.

face of the skin is a powerful means by which the body can limit heat loss in addition to the constriction of surface arterioles. In addition, the body can cut down on the flow within a capillary bed when heavy flow is not needed—to your muscles, for example, when you are resting. Likewise, the body can increase the flow within a capillary bed when the need increases—to your small intestine, for example, after a meal. Interestingly, you do not have enough blood to fill all your capillary beds if they were all open at the same time. If such a situation occurred, you would faint because of lack of sufficient blood to the brain.

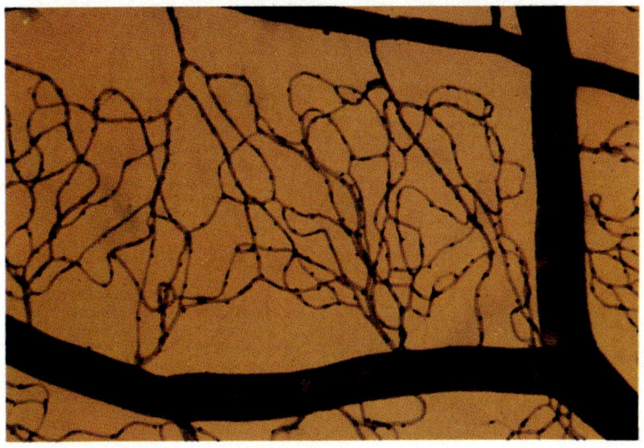

FIGURE 12-6 Venules.
Venules collect blood from the capillary bed and deliver it to the larger veins.

Veins and venules

Venules are small veins that collect blood from the capillary beds and bring it to larger veins (Figure 12-6) that carry it back to the heart. The force of the heartbeat is greatly diminished by the time the blood reaches the veins, so these vessels do not have to accommodate pulsing pressures as arteries do. Therefore the walls of veins, although similar in structure to those of the arteries, have much thinner layers of muscle and elastic fiber (Figures 12-3, *B* and 12-7). Lacking much of this supportive tissue, the walls of veins collapse when they are empty (although this situation would never occur in the living body). Empty arteries, on the other hand, stay open like tiny pipes. In addition, the lumen of veins is larger than that of arteries. This difference in size is related to the lower pressure of blood flowing within veins back toward the heart—large vessels present less resistance to the flow than smaller vessels do.

The pathway of blood back to the heart from much of the body is an uphill struggle. The pressure pushing the blood upward in the veins of your legs, for example, approximately equals the force of gravity pulling it down. As skeletal muscles contract, they press on veins and help move blood along. In addition, many veins have one-way valves that help blood move back toward the heart by preventing its backflow (Figure 12-8). If some of the valves in a vein are weak, however, gravity can force blood back through these valves, overloading a portion of a vein and pushing its walls outward. These "stretched out" veins, called *varicose veins,* are often seen in the legs at the surface of the skin.

Arteries and veins are the major vessels of the circulatory system and have walls composed of three layers: (1) an innermost, thin layer of endothelial tissue that is in contact with the blood; (2) a layer of elastic fibers and smooth muscle serviced by nerves; and (3) a layer of protective connective tissue. The walls of arteries have more elastic tissue and smooth muscle than veins do, which helps them accommodate pulses of blood pumped from the heart. Capillaries are microscopic blood vessels that have walls only one cell thick. The exchange of substances between the cells and the blood takes place at the capillaries.

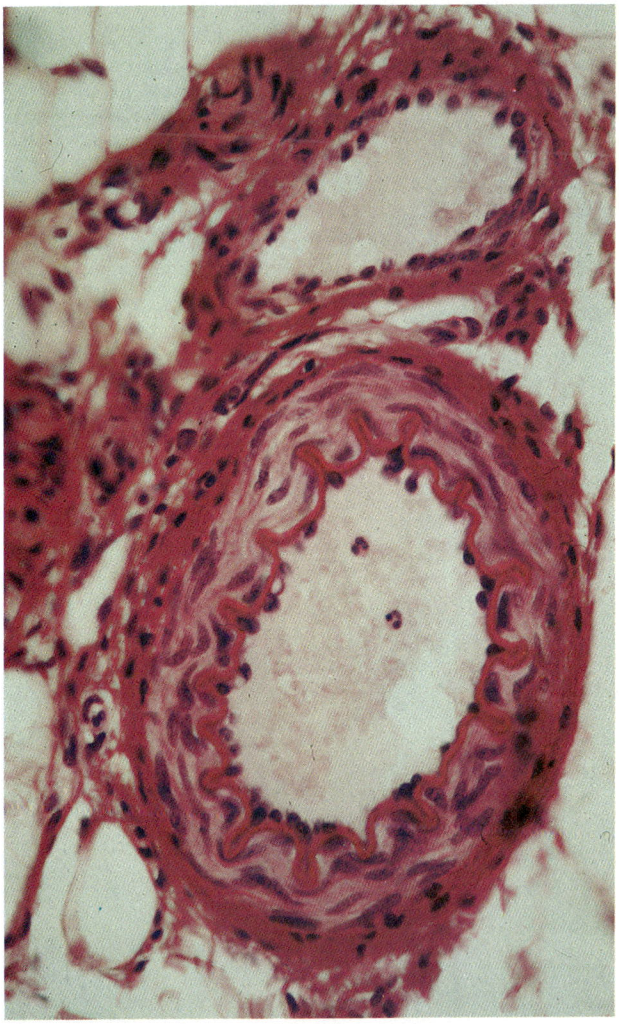

FIGURE 12-7 A closer look at blood vessels.
The vein *(top)* has the same general structure as an artery *(bottom)* but has much thinner layers of muscle and elastic fiber. An artery will retain its shape when empty, but a vein will collapse.

FIGURE 12-8 A one-way valve in a vein.
These valves assist the blood in its movement back toward the heart by preventing the blood from moving backwards.

The heart: A double pump

The human circulatory system is referred to as a *closed circulatory system* because blood flows throughout the body confined within blood vessels. The blood is therefore separated from the rest of the body's fluids and does not mix freely with them. The plumbing of a closed circulatory system requires not only a system of vessels through which the blood can continuously flow throughout the body but also a pump to push the blood along. The pump of the human circulatory system is the heart, but it is two pumps in one! Figure 12-9 is a diagram of a frontal section through the heart and shows the organization of this double pump. The left side (one pump) has two connected chambers as does the right side (the other pump). But the two sides, or pumps, of the heart are not directly connected with one another.

Circulatory pathways

The journey of blood around the body starts with the entry of oxygenated blood into the heart from the lungs. Oxygenated blood from the lungs enters the left side of the heart, emptying directly into the upper, left chamber of the heart, the **left atrium,** through large vessels called the **pulmonary veins.** These veins are unusual in that they carry oxygenated blood; other veins, since they carry blood back to the heart from the body tissues, carry deoxygenated blood. The word *pulmonary* refers to the lungs. The circulation of blood to and from the lungs is therefore called the **pulmonary circulation** (Figure 12-10).

From the left atrium of the heart, blood flows through a one-way valve, the **bicuspid valve,** into the lower, adjoining chamber, the **left ventricle.** Most of this flow (roughly 80%) occurs while the heart is relaxed. The atrium then contracts, pushing the remaining 20% of its blood into the ventricle. After a slight delay the ventricle contracts. The walls of the ventricle are far more muscular than those of the atrium, and as a result, this contraction is much stronger. It forces most of the blood out of the ventricle in a single strong pulse. The blood is prevented from going back into the atrium by the bicuspid valve, whose flaps are pushed shut as the ventricle contracts. Strong fibers that prevent the flaps from moving too far when closing are attached to their edges. If the flaps did move too far, they would project out into the atrium.

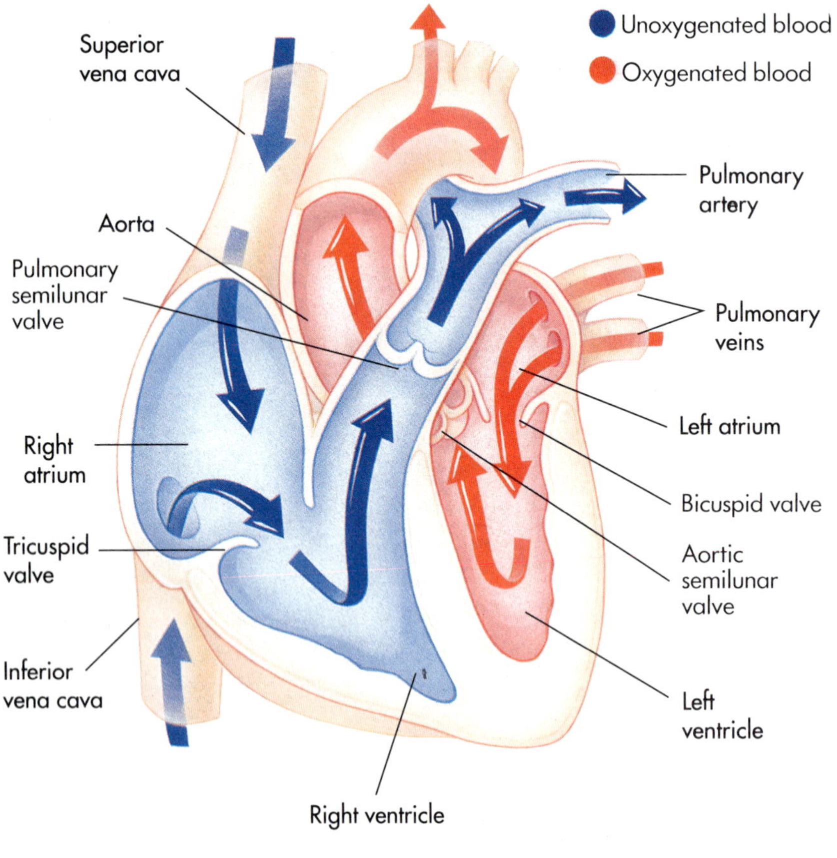

FIGURE 12-9 The path of blood through the human heart.
Oxygenated blood from the lungs enters the left atrium by way of the pulmonary veins and flows through the bicuspid valve to the left ventricle, from which it enters the general circulatory system by way of the aorta. Deoxygenated blood enters the right atrium by way of the superior and inferior vena cavae and flows through the tricuspid valve to the right ventricle. The blood then moves through the pulmonary semilunar valve to the pulmonary artery and lungs.

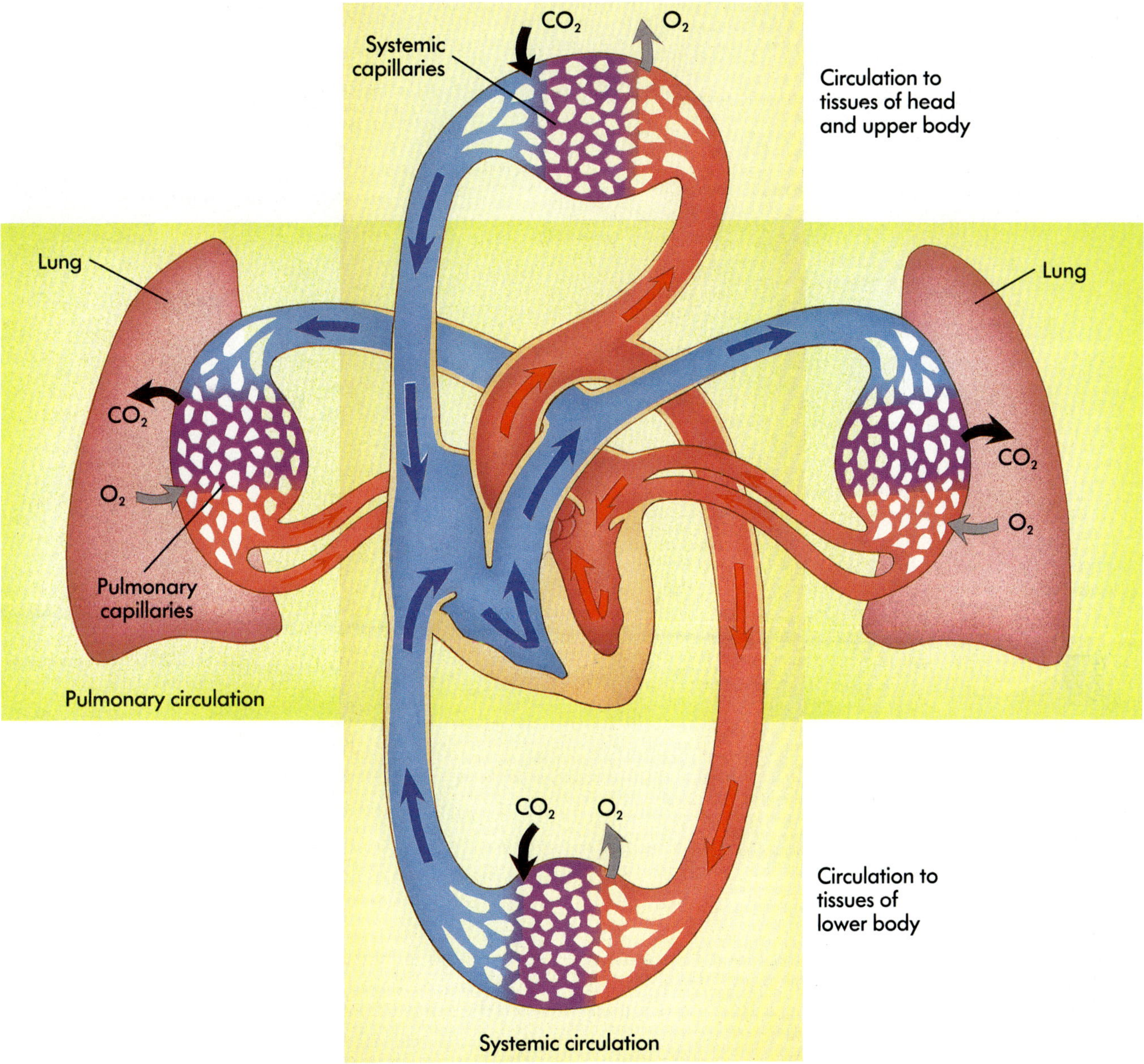

FIGURE 12-10 Pulmonary circulation versus systemic circulation.
The circulation of the blood to and from the lungs is called the *pulmonary circulation.* The circulation of the blood to the body regions and organs other than the lungs is called the *systemic circulation.*

Prevented from reentering the atrium, the blood takes the only other way out of the contracting left ventricle: an opening that leads into the largest artery in the body—the **aorta.** The aorta is closed off from the left venticle by a one-way valve, the **aortic semilunar valve.** It is oriented to permit the flow of the blood out of the ventricle. As the blood is pushed forcefully out of the left ventricle to make its trip around the body, it rushes into the aorta, causing the elastic, muscular walls of this artery to bulge slightly outward. Quickly, however, the aorta walls recoil as a stretched elastic band does when let go. This recoil action pushes on the blood; some of it is pushed backward against the valve. The valve is constructed in such a way that it snaps shut in response to this backflow. In fact, the other one-way valves found in the heart and blood vessels are constructed in a similar manner, preventing the backflow of blood.

Many arteries branch from the aorta, carrying oxygen-rich blood to all parts of the body. The pathway of blood vessels to the body regions and organs other than the lungs is called the **systemic circulation** (see Figure 12-10), with the aorta being the first and largest vessel in the circuit. The first arteries to branch off the aorta are the coronary arteries, which carry freshly oxygenated blood to the heart itself; the muscles of the heart do not obtain their supply of blood from directly within the heart. From the arch of the aorta,

BIOLOGY & *you*

How to avoid heart disease

Infectious diseases like AIDS are not the greatest killers of Americans. Every year over 80% of those who die in the United States (some 1.5 million people in 1991) die from one of five chronic diseases—what you often think of as the disorders of middle age. The top five causes of death in 1991 were the following:

1. Heart disease
2. Cancer
3. Stroke
4. Accidents
5. Chronic obstructive pulmonary disease

The most striking thing about this list is the preponderance of preventable illness. One fourth of the deaths are from cancer, and most of them are due to smoking. Even more striking, over 900,000—60% of all deaths—are due to heart disease and stroke. This carries a very important message to everyone because deaths from these chronic illnesses are influenced greatly by individual living habits. In fact, it is possible to predict in large measure how a person will die from how that person lives. As many as 70% of the deaths from these "top killers" could be prevented by altered living habits.

Heart disease is the greatest killer in the United States today. It is not, however, an inevitable result of being alive. A few simple changes in lifestyle can drastically lower an individual's risk of heart disease. Strokes, which are caused by a stoppage of blood flow to part of the brain, share many of the same risk factors as heart disease. The three principle risk factors are the following:

1. *Smoking:* About 29% of Americans smoke, leading to one third of heart disease deaths because smoking elevates cholesterol levels and thus blood vessel blockage, according to the U.S. Surgeon General. Smokers also run a 20% to 60% greater chance of stroke. It is not easy to quit smoking—nicotine has been declared an addictive drug—but the best way for smokers to avoid death by heart disease and stroke is to throw their cigarettes away.
2. *High cholesterol levels:* If your blood cholesterol level is high—over 200 for adults—it should be lowered by eating fewer saturated fats (eggs, meat, and tropical oils)(Figure 12-A). If the blood cholesterol level is over 240, many physicians recommend cholesterol-lowering drugs.
3. *High blood pressure:* Hypertension (high blood pressure) is a killer, causing up to one half of all strokes. If your blood pressure exceeds 140 over 90 mm, it should be lowered with a low-fat, low-salt diet.

FIGURE 12-A This unhealthy meal with its large doses of salt and fats and the cigarette beside it is a major contributor to heart attacks.

the carotid arteries branch off and bring blood to networks of vessels in the neck and head. The subclavian arteries bring blood to the shoulders and arms. The aorta then descends down the trunk of the body, with arteries branching off to supply various organs such as the kidneys, liver, and intestines. The aorta divides into two major vessels at the level of the lower back, one traveling to each leg.

The blood that flows into the arterial system eventually returns to the heart after delivering its supply of oxygen to the cells of the body and picking up the waste gas carbon dioxide. This exchange takes place at the capillaries. In returning, it passes through a series of veins, eventually entering the right side of the heart. Two large veins collect blood from the systemic circulation. The **superior vena cava** drains the upper body, and the **inferior vena cava** drains the lower body. These veins dump deoxygenated blood into the right atrium.

The right side of the heart is similar in organization to the left side. However, the muscular walls of the right ventricle are not as thick as those of the left. Blood passes from the right atrium into the right ventricle through a one-way valve, the **tricuspid valve.** It passes out of the contracting right venticle through a second valve, the **pulmonary semilunar valve,** into a single **pulmonary artery,** which subsequently branches into two pulmonary arteries that carry deoxygenated blood to the lungs. The blood then returns from the lungs to the left side of the heart, replenished with oxygen and cleared of much of its load of carbon dioxide. (The human circulatory system as a whole is outlined in Figure 12-2.)

The circulation of the blood to and from the lungs is called the *pulmonary circulation.* The circulation of the blood through the parts of the body other than the lungs is called the *systemic circulation.*

How the heart contracts

The contraction of the heart depends on a small cluster of specialized cardiac muscle cells that are embedded in the upper wall of the right atrium (Figure 12-11). This cluster of cells, called the **SA (sinoatrial) node,** automatically and rhythmically sends out impulses that initiate each heartbeat. The SA node is therefore nicknamed the "pacemaker of the heart."

The impulse initiated by the SA node causes both atria to contract simultaneously and also excites a bundle of cardiac muscle cells located at the base of the atria. These cells

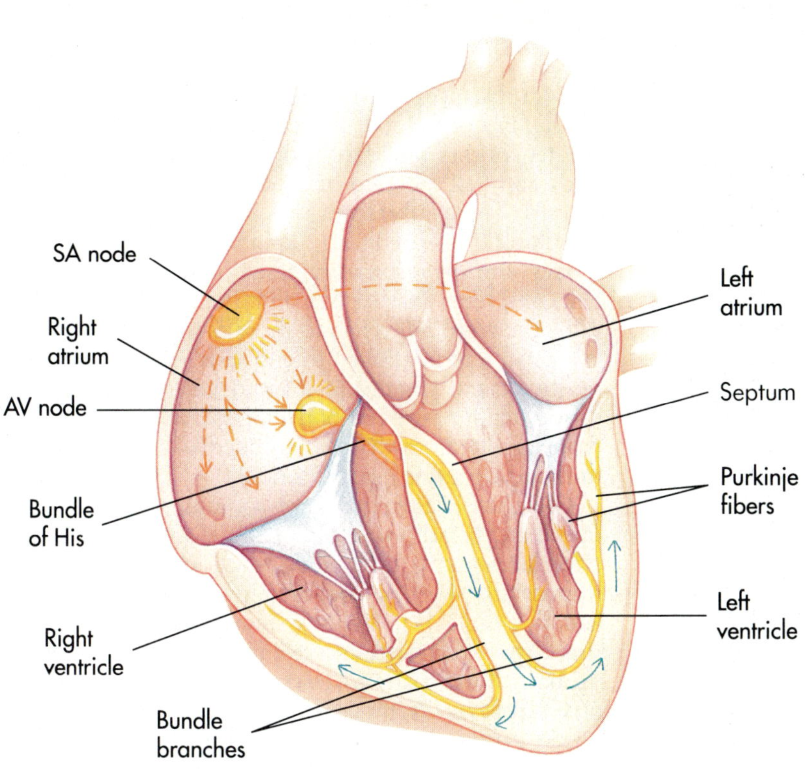

FIGURE 12-11 How the heart contracts. Contraction of the human heart is initiated by an impulse that begins at the SA node. After passing over the right and left atria and causing their contraction, the impulse reaches the AV node. From there it passes through the bundle of His, which has a bundle branch to each ventricle. From the tips of the ventricles, the depolarization is conducted rapidly over their surfaces by the branches of the Purkinje fibers.

are known as the **AV (atrioventricular) node.** The AV node conducts the impulse to a strand of specialized muscle in the **septum,** the tissue that separates the two sides of the heart. This strand of impulse-conducting muscle, known as the **bundle of His,** has branches that divide to the right and left. On reaching the apex (lower tip) of the heart, each branch further divides into conducting fibers called **Purkinje fibers,** which initiate the almost simultaneous contraction of all the cells of the right and left venticles. The conduction of the impulse from the atria to the ventricles takes approximately 0.2 seconds. During this "delay" the atria finish emptying their contents into the corresponding ventricles before the ventricles start to contract.

An excitory impulse is initiated in the SA node of the heart, which causes the atria to contract and the impulse to be passed along to other specialized groups of cardiac cells: the AV node, bundle of His, bundle branches, and Purkinje fibers. The passage of this excitory impulse causes the almost simultaneous contraction of all cells of the right and left ventricles.

Monitoring the heart's performance

The heartbeat is really a series of events that occur in a predictable order. A physician can gain information about the health of the heart and events occurring during the heartbeat by making several different kinds of observations. The simplest is to listen to the heart at work.

The first sound heard, a low-pitched lubb, is caused by the turbulence in blood flow created by the closing of the bicuspid and tricuspid valves at the start of ventricular contraction. A little later a higher-pitched dupp can be heard, signaling the closing of the pulmonary and aortic semilunar valves at the end of ventricular contraction. If the valves are not closing fully or if they open too narrowly, a slight backflow of blood occurs within the heart. This backflow can be heard as a "sloshing" sound; this condition is known as a *heart murmur.*

A second way to examine the events of the heartbeat is to monitor the blood pressure, the force exerted by the blood on blood vessel walls. During the first part of the heartbeat the atria are filling and contracting. At this time the pressure in the arteries leading from the left side of the heart out to the tissues of the body decreases slightly because blood is not being forced into them by the left ventricle. This period of relaxation (with respect to the ventricles) is referred to as the **diastolic period.** The force the blood exerts on the blood vessels at this time is the diastolic pressure. During the contraction of the ventricles, a pulse of blood is forced into the systemic arterial system by the left ventricle, immediately raising the blood pressure within these vessels. This period of contraction (with respect to the ventricles) ends with the closing of the aortic semilunar valve and is referred to as the **systolic period.** The force the blood exerts on the blood vessels at this time is the systolic pressure.

Blood pressure is measured in units called millimeters (mm) of mercury (Hg). These units refer to the height to which a column of mercury is raised in a tube by an equivalent pressure. Using this system of measurement, normal blood pressure values are 70 to 90 (mm of Hg) diastolic and 110 to 130 systolic. Blood pressure is expressed as the systolic pressure over the diastolic pressure, such as 110 over 70. When the inner walls of the arteries accumulate fats, as they do in the condition known as **atherosclerosis,** the diameters of the passageways are narrowed. Such narrowing is one cause of elevated systolic and diastolic blood pressures.

A third way to monitor the events of a heartbeat is to measure the electrical changes that take place as the heart's chambers both contract and relax. Because so much of the human body is made up of water, it conducts electrical currents rather well. Therefore as the impulses initiated at

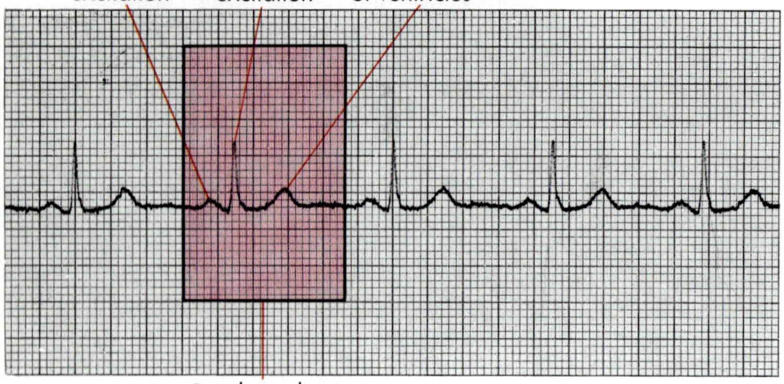

FIGURE 12-12 An electrocardiogram.
An electrocardiogram records the electrical impulses made throughout the body as the heart contracts.

the SA node pass throughout the heart as an electrical current, this current passes in a wave throughout the body. Although the magnitude of this electrical pulse is tiny, it can be detected with sensors placed on the skin. A recording made of these impulses (Figure 12-12) is called an **electrocardiogram,** or **ECG.**

Three successive electrical pulses are recorded in a normal heartbeat. The first pulse occurs when the atria contract. There is a much stronger pulse 2/10 of a second later, reflecting both the contraction of the ventricles and the relaxation of the atria. Finally, a third pulse occurs caused by the relaxation of the ventricles.

The heartbeat is a series of events that occurs in a predictable order. A physician can gather data about the health of the heart by monitoring the heartbeat in various ways: by listening to the sounds the heart makes, monitoring the blood pressure, and measuring the electrical changes that take place as it contracts and relaxes.

The blood

Your blood makes up about 8% of your body. It is a viscous, or thick, fluid made up of two parts: a liquid portion called **plasma** and a portion made up of **formed elements**—several different kinds of cells and cell parts suspended within the plasma.

Blood plasma

Blood plasma is a straw-colored liquid made up of water and dissolved substances. The dissolved substances can be grouped into three categories:

1. Nutrients, hormones, respiratory gases and wastes: These substances can be thought of as the "traffic" on the highway of blood. Dissolved within the plasma are substances that move from one place to another in the body and that are used or produced by the metabolism of cells. These substances include glucose, lipoproteins (a soluble form of lipid), amino acids, vitamins, hormones, and the respiratory gases.
2. Salts and ions: Plasma is a dilute salt solution. Chemically, the word *salt* refers to more than just table salt. It is a general term applied to any substance composed of positively and negatively charged ions. In water, salts dissociate into their component ions. An ion is a charged particle—an atom that has an unequal number of protons and electrons (see Chapter 2).

   The chief plasma ions are sodium (Na^+), chloride (Cl^-), and bicarbonate (HCO_3^-). In addition, there are trace amounts of other ions, such as calcium (Ca^{2+}), magnesium (Mg^{2+}), copper (Cu^{2+}), potassium (K^+), and zinc (Zn^{2+}). In living systems these ions are called *electrolytes*. Electrolytes serve three general functions in the body. First, many are essential minerals. Second, they play a role in the movement of water—osmosis—between various compartments within the body. Third, they help maintain the acid-base (pH) balance required for normal cellular activities.
3. Proteins: Blood plasma is approximately 90% water. As the plasma circulates past the cells of the body, water moves between the tissue fluid and the blood from an area of higher concentration of water to an area of lower concentration of water. Blood contains a concentration of proteins that balances that of the cells, thereby balancing the concentration of water between the cells and the blood. For this reason, water is not osmotically "sucked" out of the blood.

   The types of proteins in blood vary. Some of these proteins are antibodies and other proteins that are active in the immune system. Others are fibrinogen and prothrombin, key players in blood clotting. Taken together, however, these proteins make up less than half of the amount of protein that is necessary to balance the protein content of the other cells of the body. The rest consists of a protein called *serum albumin,* which circulates in the blood as an osmotic counterforce.

Blood plasma contains nutrients, hormones, respiratory gases, wastes, and a variety of ions and salts. It also contains high concentrations of the protein serum albumin, which functions to keep the blood plasma in osmotic equilibrium with the cells of the body.

Types of blood cells

Although blood is liquid, 45% of its volume is actually occupied by cells and pieces of cells, collectively called *formed elements.* There are three principal types of formed elements in the blood: erythrocytes, leukocytes, and platelets (Figure 12-13).

Blood cell		Life span in blood	Function
Erythrocyte	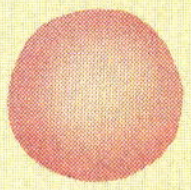	120 days	O_2 and CO_2 transport
Neutrophil	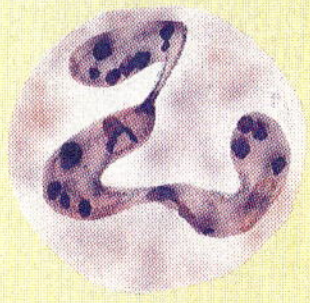	7 hours	Immune defenses
Eosinophil	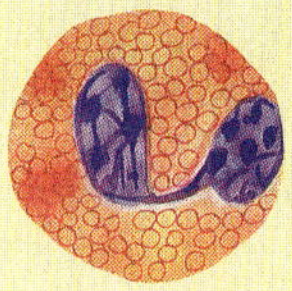	8-12 days	Defense against parasites
Basophil	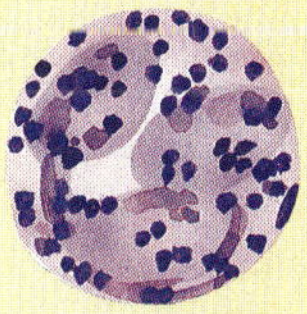	A few hours to a few days	Inflammatory response
Monocyte	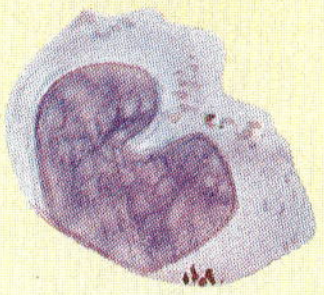	3 days	Immune surveillance (precursor of tissue macrophage)
B-lymphocyte	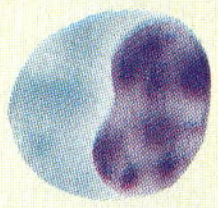	Memory cells may survive for years	Antibody production (precursor of plasma cells)
T-lymphocyte	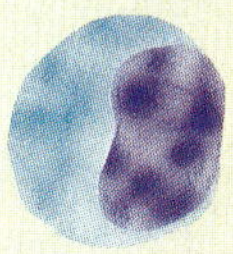	Memory cells may survive for years	Cellular immune response
Platelets	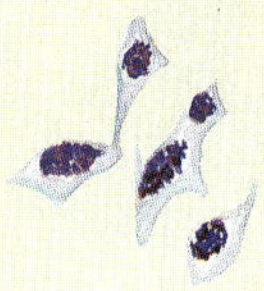	7-8 days	Blood clotting

FIGURE 12-13 Types of blood cells.
Different blood cells perform different functions and have varying lifespans, as this chart shows.

Erythrocytes

In only one teaspoonful of your blood, there are about 25 billion **erythrocytes,** or **red blood cells.** Each erythrocyte is a flat disk with a central depression (Figure 12-14), something like a doughnut with a hole that does not go all the way through. Almost the entire interior of each cell is packed with the oxygen-carrying molecule hemoglobin (Figure 12-15). One estimate, in fact, is that each erythrocyte can carry 280 million molecules of hemoglobin!

Mature erythrocytes do not have nuclei or the ability to manufacture proteins. Red blood cells are therefore unable to repair themselves and consequently have a rather short life span. Erythrocytes live only about 4 months. Old, worn-out erthrocytes are "processed" by macrophages in the spleen, liver, and bone marrow. New erythrocytes are constantly being synthesized and released into the blood by cells within the soft interior marrow of bones at the amazing rate of 2 million per second!

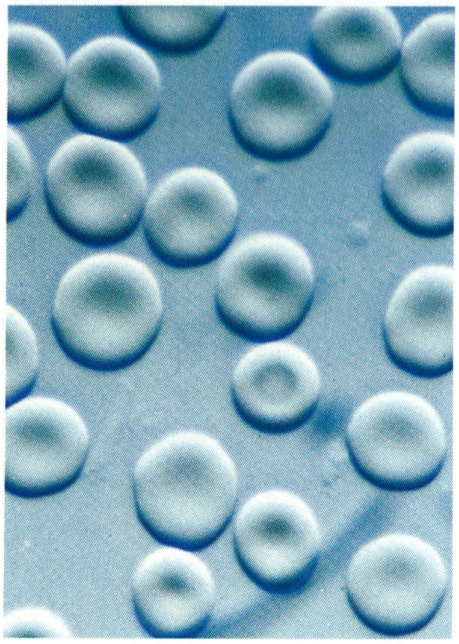

FIGURE 12-14 Human erythrocytes, magnified 1000 times.
Mature human erythrocytes have a characteristic collapsed appearance, like a pillow on which someone has sat.

Leukocytes

Less than 1% of the cells in human blood are **leukocytes,** or **white blood cells.** In fact, there are only about 1 or 2 leukocytes for every 1000 erythrocytes. Leukocytes are larger than red blood cells; they contain no hemoglobin, have nuclei, and are essentially colorless. There are several kinds of leukocytes and each has a different function. All functions, however, are related to the defense of the body against invading microorganisms and other foreign substances (see Chapter 13).

There are two major groups of leukocytes: **granulocytes** and **agranulocytes.** Granulocytes are circulating leukocytes and get their name from the tiny granules in their cytoplasm. In addition, they have lobed nuclei (see Figure 12-13). The granulocytes are classified into three groups by their staining properties. About 50% to 70% of them are **neutrophils,** cells that migrate to the site of an injury and stick to the interior walls of the blood vessels. They then form projections that enable them to push their way into the infected tissues, where they engulf, or *phagocytize,* microorganisms and other foreign particles. The term *phagocytosis* comes from a Latin word meaning "cell eating." **Basophils,** a second kind of granulocyte, contain granules that rupture and release chemicals that enhance the body's response to injury or infection. They play a role in causing allergic responses. The third kind of granulocyte, the **eosinophils,** are also believed to be involved in allergic reactions. In addition, they act against certain parasitic worms.

Agranulocytes have no cytoplasmic granules, nor are their nuclei lobed. One group of agranulocytes, the **monocytes,** circulate as the granulocytes do. Monocytes are attracted to the sites of injury or infection where they are converted into **macrophages**—enlarged, amoeba-like cells that entrap microorganisms and particles of foreign matter by phagocytosis. They usually arrive after the neutrophils and "clean up" any bacteria and dead cells. **Lymphocytes,** the other type of agranulocyte, recognize and react to substances that are foreign to the body, sometimes producing a protective immunity to disease. Occasionally, however, these cells produce an inflammation or an allergic response (see Chapter 13).

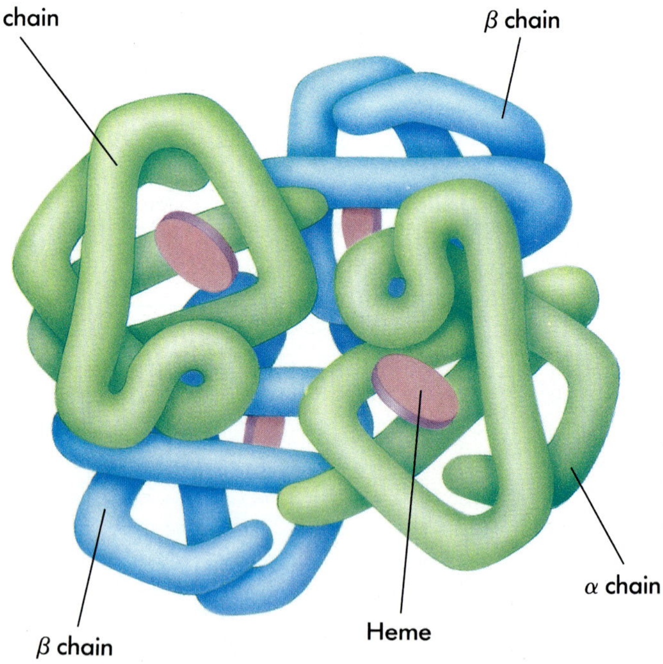

FIGURE 12-15 A hemoglobin molecule.
The interior of a red blood cell is packed with hemoglobin. Each hemoglobin molecule is composed of four polypeptide chains. At the center of each chain is a heme group that contains iron. Oxygen binds to the iron for transport.

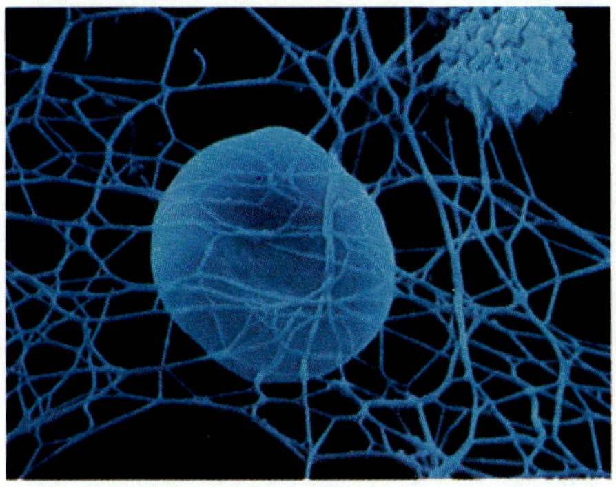

FIGURE 12-16 Fibrin threads have caught a red blood cell. Eventually, many red blood cells will become caught in this net, forming a clot.

Platelets

Certain large cells within the bone marrow called *megakaryocytes* regularly pinch off bits of their cytoplasm. These cell fragments, called **platelets,** enter the bloodstream and play an important role in controlling blood clotting, or coagulation. The clotting of blood is a complicated process initiated by damage to blood vessels or tissues.

When an injury occurs, platelets clump at the damaged area, temporarily blocking blood loss. The damaged tissues and platelets release a complex of substances called *thromboplastin.* The release of this complex begins a cascade of events, one dependent on the other. A simplified explanation is as follows: thromboplastin interacts with calcium ions, vitamin K, and other clotting factors to form an enzyme called *prothrombin activator.* Prothrombin activator brings about the conversion of prothrombin to thrombin. Thrombin converts fibrinogen, a soluble protein, to fibrin, an insoluble thread-like protein. Fibrin threads, along with trapped red blood cells (Figure 12-16), form the clot—a "plug" at the damaged area so that blood cannot escape.

The three principal types of formed elements in the blood are erythrocytes, leukocytes, and platelets. Erythrocytes, or red blood cells, carry oxygen; leukocytes, or white blood cells, defend the body against invading microorganisms and other foreign substances; and platelets play an important role in blood clotting.

The lymphatic system

Although the blood proteins and electrolytes help maintain an osmotic balance between the blood and the tissues, the blood loses more fluid to the tissues than it reabsorbs from them. Of the total volume of fluid that moves from the blood into and around the tissues, about 90% reenters the cardiovascular system. Where does the other 10% go? The answer is to the body's one-way, passive circulatory system, the **lymphatic system** (Figure 12-17). This system counteracts the effects of net fluid loss from the blood.

By osmosis and diffusion, blind-ended lymphatic capillaries (Figure 12-17) fill with tissue fluid (including small proteins that have diffused out of the blood). From these capillaries, the tissue fluid—now called **lymph**—flows through a series of progressively larger vessels to two large lymphatic vessels, which structurally resemble veins. These vessels drain into veins of the circulatory system near the base of the neck through one-way valves. Although the lymphatic system has no heart to pump lymph through its vessels, the fluid is driven through them when the vessels are squeezed by the movements of the body's muscles. The lymphatic vessels contain a series of one-way valves (Figure 12-18) that permit movement only in the direction of the neck.

Small, ovoid, "spongy" structures called **lymph nodes** are located in various places of the body along the route of the lymphatic vessels. They are clustered in areas such as the groin, armpits, and neck, filtering the lymph as it passes through. Some lymphocytes, the cells that activate the immune response, reside in the lymph nodes. (These cells are discussed in more detail in Chapter 13.) In addition, the lymphatic system has two organs: the **spleen** and the **thymus.** The spleen stores an emergency blood supply and also contains white blood cells. Specific types of white blood cells in the spleen serve to destroy old red blood cells, filter microorganisms out of the blood as it passes through, and initiate an immune response against the foreign microbes. The thymus plays an important role in the maturation of certain lymphocytes called *T cells,* which are an essential part of the immune system.

Approximately 10% of the fluid that moves out of the blood and into the cells and cell spaces at the capillaries does not return to the blood directly. It is collected and then returned to the blood by a system of one-way, blind-ended vessels called the *lymphatic system.*

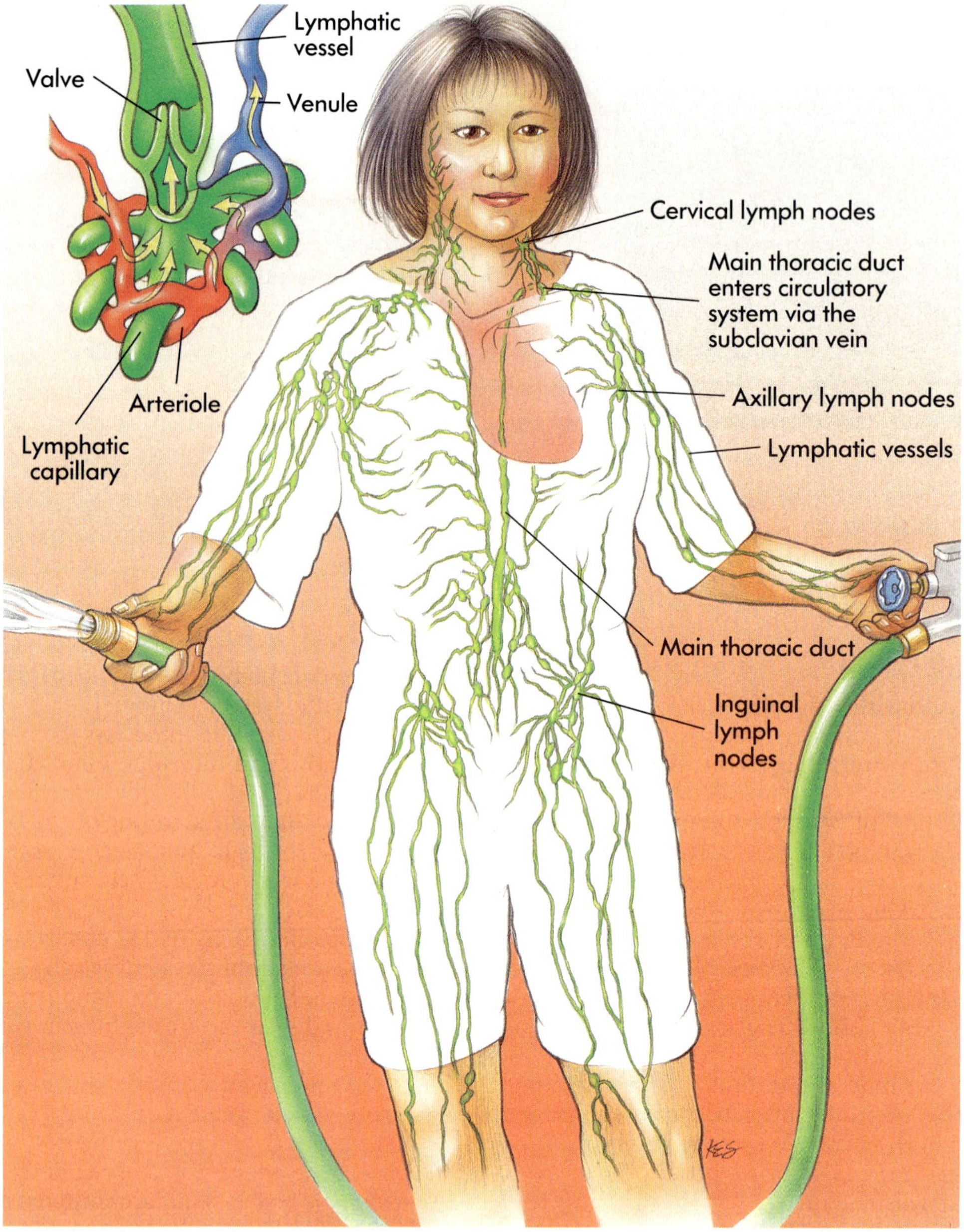

FIGURE 12-17 The human lymphatic vessels and nodes.
Other organs containing lymphatic tissue include the spleen, thymus, appendix, tonsils, and bone marrow. The enlarged diagram at the upper left shows the path of the excess fluid that leaves the arteriole end of a capillary bed, enters the adjacent tissue spaces, and is absorbed by lymphatic capillaries.

FIGURE 12-18 A lymphatic vessel magnified 25 times.
Flow from left to right is not impeded because this flow tends to force open the valve. Flow from right to left is prevented because "backward" flow tends to force the one-way valve closed.

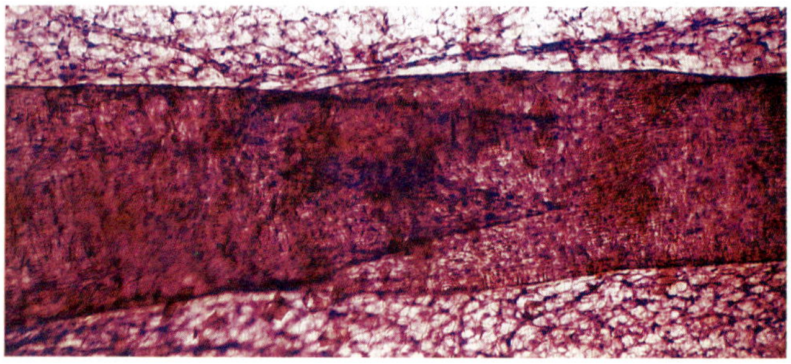

Diseases of the heart and blood vessels

Cardiovascular diseases are the leading cause of death in the United States. More than 42 million people in this country have some form of cardiovascular disease—about one person in five.

Heart attacks, the most common single cause of death in the United States, result from an insufficient supply of blood reaching an area of heart muscle. Heart attacks may be caused by a blood clot forming in one of the vessels that supplies the heart with blood, thereby blocking its blood supply. They may also result if a vessel is blocked sufficiently by fatty deposits, especially cholesterol and triglycerides. Recovery from a heart attack is possible if the segment of damaged heart tissue is small enough that other blood vessels in the heart can supply the damaged tissues. *Angina pectoris*, which literally means "chest pain," occurs for reasons similar to those that cause heart attacks, but it is not as severe. In this case, a reduced blood flow to the heart muscle weakens the cells but does not kill them. The pain may occur in the heart and often also in the left arm and shoulder.

The amount of heart damage associated with a small heart attack may be relatively slight and thus difficult to detect. It is important that such damage be detected, however, so that the overall condition of the heart can be evaluated properly. Electrocardiograms are very useful for this purpose because they reveal abnormalities in the timing of heart contractions, abnormalities that are associated with the presence of damaged heart tissue. Damage to the AV node, for example, may delay as well as reduce the second, ventricular pulse. Unusual conduction routes may lead to continuous disorganized contractions called *fibrillations*. In many fatal heart attacks, ventricular fibrillation is the immediate cause of death.

Many factors contribute to heart disease and heart attacks. One factor is heredity. You may inherit the predisposition to heart disease from your parents. Other factors involve eating too much saturated fat and cholesterol, not exercising, and being overweight. Cigarette smoking greatly increases a person's risk for heart attacks and strokes, although the physiological mechanisms involved are not totally understood at this time.

Strokes are caused by an interference with the blood supply to the brain and often occur when a blood vessel bursts in the brain. A stroke may be also caused by a blood clot, or *thrombus*, that forms in these vessels or by a blood clot that has traveled to the brain from another location. This type of clot is called an **embolus.** Blood clots may be caused by cancer or other diseases. The effects of strokes depend on the severity of the damage and the specific location of the stroke.

Arteriosclerosis is a thickening and hardening of the walls of the arteries. Blood flow through such arteries is restricted, and they lack the ability to dilate and so have difficulty accommodating the volume of blood pumped out by the heart. The narrowing of these vessels forces the heart to work harder. *Atherosclerosis* is a form of arteriosclerosis in which masses of cholesterol and other lipids build up within the walls of large and medium-sized arteries (Figure 12-19). These masses are referred to as **plaque.** The accumulation of plaque impairs the arteries' proper functioning. When this condition is severe, the arteries can no longer dilate and constrict properly and the blood moves through them with difficulty. The accumulation of cholesterol is thought to be the prime contributor to atherosclerosis, and diets low in cholesterol are now prescribed to help prevent this condition. Atherosclerosis contributes to both heart attacks and strokes.

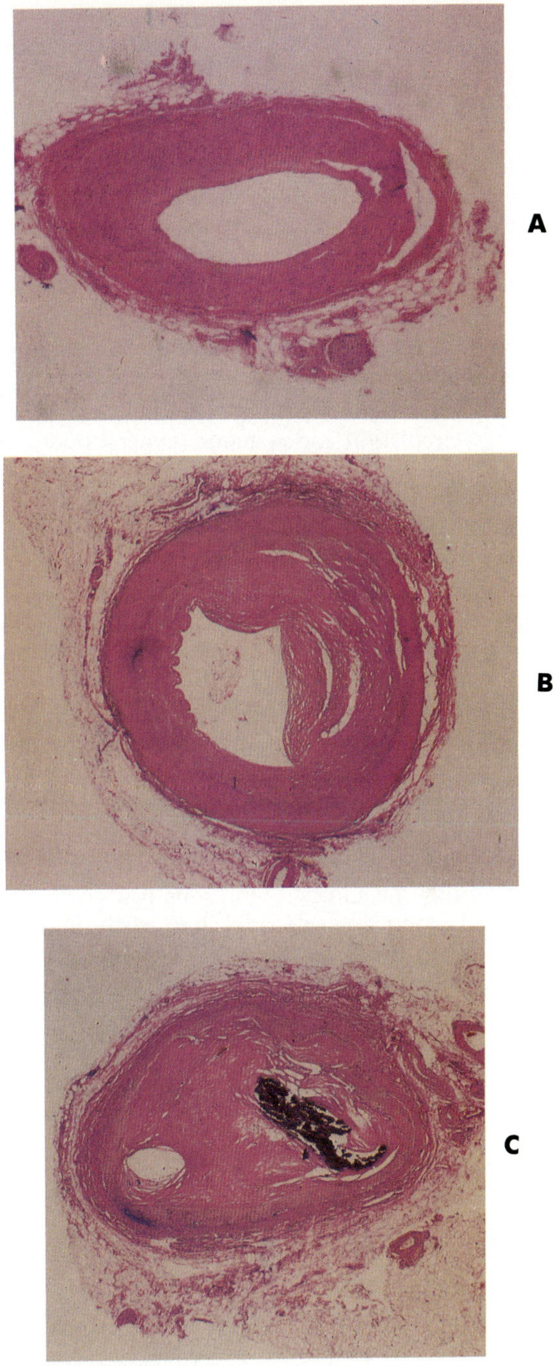

FIGURE 12-19 The path to a heart attack.
A The coronary artery shows only minor blockage.
B The artery exhibits severe atherosclerosis—much of the passage is blocked by buildup of cholesterol and other lipids on the interior walls of the artery.
C The coronary artery is almost completely blocked.

Summary

1. The *circulatory system* is made up of three components: the heart, the blood vessels, and the blood. Together, the heart and blood vessels—the "plumbing" of the circulatory system—is known as the *cardiovascular system*. These terms are often used interchangeably.
2. The circulatory system transports nutrients, wastes, respiratory gases, and hormones and plays an important role in temperature maintenance of the body.
3. The heart is a double pump, pushing both pulmonary (lung) circulation and systemic (general body) circulation. Because the two circulations are kept separate within the heart, the systemic circulation receives only fully oxygenated blood.
4. The general flow of blood circulation through the body is a circuit starting from the heart, which pumps blood out via muscled arteries to the capillary networks that interlace the tissues of the body; the blood returns to the heart from these capillaries via the veins.
5. The contraction of the heart is initiated at the SA node, or pacemaker, as a periodic spontaneous impulse. The impulse spreads across the surface of the two atrial chambers, causing all of these cells to contract.
6. The passage of the impulse to the ventricles is briefly delayed by tissue that insulates the two segments of the heart from one another. Only a narrow channel of specialized conducting cells is able to carry the impulse from the atria to the ventricles. The delay in the passage of the impulse permits the atria to empty completely into the ventricles before ventricular contraction occurs.
7. The heartbeat can be heard, or monitored, by tracking changes in blood pressure through the period of filling and contracting of the atria (the diastolic period) and the contraction of the left ventricle (the systolic period). The impulses can also be measured directly as an electrical current that passes in a wave throughout the body. A recording of this electrical wave is called an *electrocardiogram*.
8. The liquid portion, or plasma, of the circulating blood contains the proteins and ions that are necessary to maintain the blood's osmotic equilibrium with the surrounding tissues.
9. The formed elements of the blood are the red blood cells (erythrocytes), white blood cells (leukocytes), and platelets. The red blood cells transport oxygen, the white blood cells defend the body against disease, and the platelets are essential to the process of blood clotting.
10. The lymphatic system gathers fluid from the body that has been "lost" from the circulatory system by diffusion and returns it via a system of lymphatic capillaries, lymphatic vessels, and two large lymphatic ducts to veins in the lower part of the neck.
11. Cardiovascular diseases, diseases of the heart and blood vessels, are the leading cause of death in the United States. Heart attacks, strokes, and atherosclerosis are all serious cardiovascular diseases.

KEY TERMS

agranulocytes 210
aorta 205
aortic semilunar valve 205
arteriole 201
arteriosclerosis 213
artery 201
atherosclerosis 207
atrium 204
AV (atrioventricular) node 207
basophils 210
bicuspid valve 204
bundle of His 207
capillary 201
capillary bed 202
cardiovascular system 199
diastolic period 207
electrocardiogram (ECG) 208
eosinophils 210
erythrocytes (red blood cells) 210
formed elements 208
granulocytes 210
heart 199
leukocytes (white blood cells) 210
lymph 211
lymph nodes 211
lymphatic system 211
lymphocytes 210
macrophages 210
monocytes 210
neutrophils 210
plaque 213
plasma 208
platelets 211
pulmonary artery 206
pulmonary circulation 204
pulmonary semilunar valve 206
pulmonary veins 204
Purkinje fibers 207
SA (sinoatrial) node 206
spleen 211
systemic circulation 205
systolic period 207
thymus 211
tricuspid valve 206
vein 201
vena cava (superior and inferior) 206
ventricle 204
venule 201

REVIEW QUESTIONS

1. What are the principal functions of the circulatory system?
2. You've just finished a great lunch. Discuss what happens to the molecules of the food after they pass into your bloodstream.
3. It's a hot summer day, and as you ride your bike you start feeling very warm. Explain how your circulatory system helps to keep you from overheating.
4. Distinguish between the circulatory system and the cardiovascular system. What are the component(s) of each?
5. Fill in the blanks: Blood leaves the heart through vessels called __________ and passes into a network of __________. From there it's forced through the __________, where gas exchange occurs. The blood then passes into the __________ and ____ and is carried back to the heart.
6. Mystery novels sometimes describe people who "turn pale with fear." Describe the circulatory reason for this.
7. How can the body regulate blood flow within the capillary beds? Why is this important?
8. Compare the structure of arteries, capillaries, and veins. Relate any differences to their respective functions.
9. Draw a diagram of a frontal section through the human heart; show the relationship of the heart to the systemic and pulmonary circulations. Use arrows to show the direction of blood flow. Label the left and right ventricles, the left and right atria, the systemic circulation, and the pulmonary circulation.
10. Fill in the blanks: The circulation of blood to and from the lungs is the __________ __________. The pathway of blood vessels serving the other areas of the body is the __________ __________.
11. Fill in the blanks: If your inferior vena cava were blocked, this would affect blood flow from the __________ __________.
12. What are the SA node and the AV node? Explain their significance.
13. What is blood plasma, and what substance(s) does it contain?
14. Identify and give a function for each of the following. What do they have in common?
    a. Erythrocytes
    b. Leukocytes
    c. Platelets
15. Match each cell to its function. What do all four cells have in common?
    Neutrophils
    Basophils
    Macrophages
    Lymphocytes

    a. Contain granules that rupture and release chemicals enhancing the body's response to injury or infection
    b. Recognize and react to foreign substances, sometimes producing a protective immunity
    c. Migrate to the site of an injury and stick to the interior walls of the blood vessels
    d. Are enlarged, amoeba-like cells that entrap foreign particles by phagocytosis
16. What is the lymphatic system? Explain its significance.
17. Explain how a high-fat diet could place you at risk for cardiovascular disease.

THOUGHT QUESTIONS

1. How long do you imagine it takes a single red blood cell to tour your body, from the time it leaves the left ventricle of your heart to the time it returns there? How would you go about measuring this?
2. Why does atherosclerosis raise blood pressure? Won't a narrowing of an artery's diameter simply cause the blood to move faster through it, like closing down the nozzle on a garden hose?

FOR FURTHER READING

Goldstein G., & Belz, A. L. (1986, September). The blood-brain barrier. *Scientific American*, pp. 74-83.
Many chemicals will not pass from the bloodstream to the cells of the brain. This article explains that this "barrier" is the result of the structure of brain capillaries, which possess an unusual array of membrane channels.

Robinson T., Factor, S., & Sonneblink, E. (1986, June). The heart as a suction pump. *Scientific American*, pp. 84-91.
Between birth and death your heart beats millions of times. These authors argue that the heart is aided greatly in this Herculean task by a very clever trick: contraction compresses elastic elements within the heart muscles, which then bounces back to expand the ventricles.

13 Defense against disease

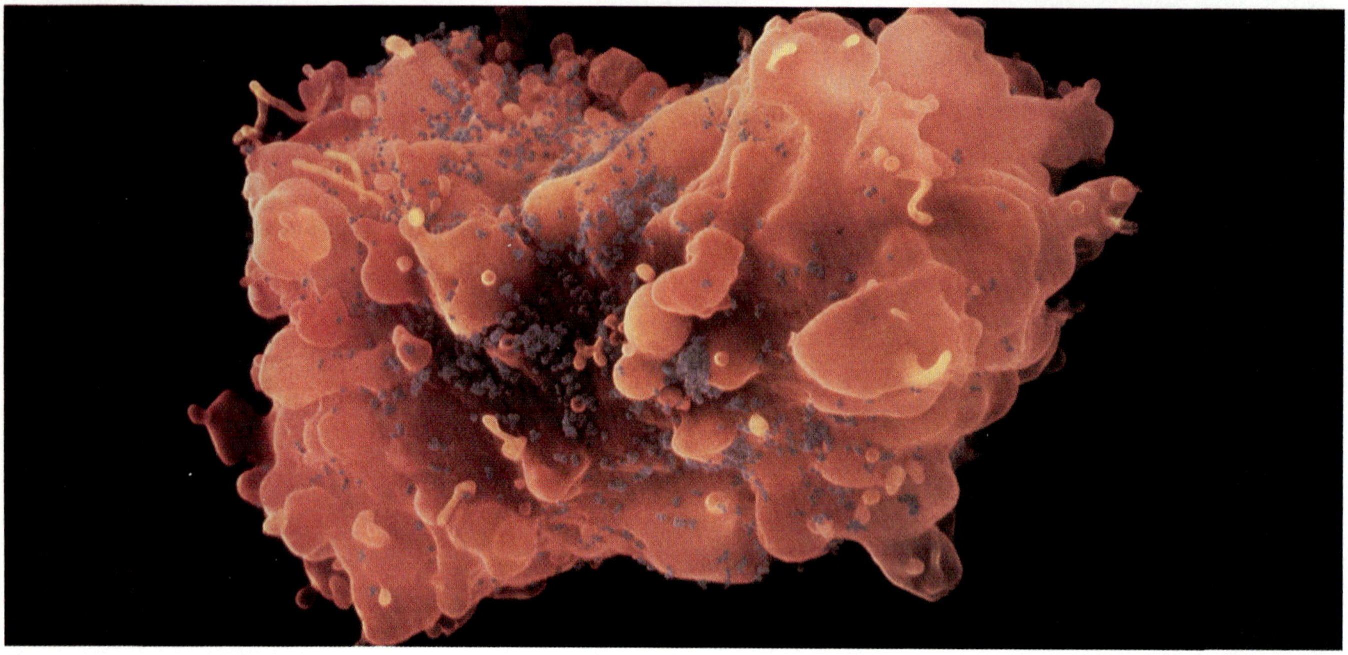

OUTLINE

Nonspecific versus specific resistance to infection

Nonspecific resistance

Specific resistance

Discovery of specific resistance: The immune response

The cells of the immune system

The immune response: How it works

- Sounding the alarm
- Two branches of the immune response
  - The cell-mediated immune response
  - The humoral immune response

How immune receptors recognize antigens

Immunization: Protection against infection

Defeat of the immune system: AIDS

Allergy

Every day you fight for your life. Some of your attackers swarm over your skin. Some enter your mouth, nose, eyes . . . any place where they may be able to gain a foothold in your tissues and spread. The battles between you and these invaders are silent and deadly. The fighting is one on one; chemical warfare is commonplace.

It is your immune system that fights and wins these thousands of battles, yet the war within you rages on continually. Each day thousands of people lose their inner wars, succumbing to invasions of the body by viruses, bacteria, fungi, or protists. People whose immune system is impaired, such as those infected with the deadly AIDS virus, have the hardest time winning the battle because the AIDS virus attacks the immune system itself, lowering its defenses.

HIGHLIGHTS

▼

The body has two defense systems: nonspecific defenses working to keep out any foreign invader and specific defenses working against each particular type of foreign cell or virus that may invade the body.

▼

The nonspecific defenses of the body are the skin and mucous membranes, chemicals that kill bacteria, and the inflammatory response.

▼

The specific defense of the body is the immune response, a two-branched response that works in a complex way to rid the body of a foreign invader and to develop a future resistance within the body to that particular invader.

▼

When the immune system is attacked, as during infection with the deadly AIDS virus, infections normally defeated by the immune system may overpower the body.

Looking at the photo, you can see the AIDS virus doing some of its "dirty" work, attacking one of the most important cells of the immune system—a helper T cell. Helper T cells identify foreign invaders and stimulate the production of other cells to fight an infection. Although thousands of times larger than the AIDS virus particles (colored blue for this photograph), the helper T cell is inactivated and sometimes killed by this virus. The body is then an easy mark for other invaders to enter—unnoticed—and win the war between life and death.

Nonspecific versus specific resistance to infection

Your body works in many ways to keep you healthy and free of infection. An **infection** results when microorganisms or viruses enter the tissues, multiply, and cause damage. Your body has one set of defenses against infection called **nonspecific defenses.** These defenses work to keep out *any* foreign invader, just like the walls and roof of your home protect you from any of "the elements" of the environment—rain, hail, wind, the sun's rays, insects, and the neighbor's dog, for example. But your body has another set of defenses too. This set of defenses are called **specific defenses** because they work against *each particular type* of microbe that may invade your body. A specific defense in your home, for example, may be a certain chemical you use to rid your kitchen of ants in the summer. Your nonspecific defenses consist of mechanical and chemical barriers and cells that attack invaders in general. Your specific defense is called **immunity,** and it consists of cellular and molecular responses to particular foreign invaders.

The defenses of the body are both nonspecific—acting against any foreign invaders — and specific—acting against particular invaders.

Nonspecific resistance

Unbroken skin is an effective mechanical barrier to any foreign substance entering the body. Most microorganisms cannot penetrate the skin of a healthy individual. However, if the skin is injured, many types of microorganisms will grow and multiply at the injured site, causing an inflammation or infection.

An **inflammation** occurs when cells are damaged by microbes, chemicals, or physical injuries. Characterized by redness, pain, heat, and swelling, an inflammation consists of a series of events that removes the cause of the irritation, repairs the damage that was done, and protects the body from further invasion and infection. During this nonspecific response, blood vessels dilate, or widen, bringing an increased supply of blood to the injured area. This increased flow brings defensive substances to the site of injury as fluid and phagocytic white blood cells pass out of the capillaries. Lymphatic drainage removes dissolved poisonous substances that may accumulate there. Blood clots wall off the area, preventing the spread of microbes or other injurious substances to other parts of the body. Phagocytes, both neutrophils and monocytes, migrate to the area and ingest microbes and other foreign substances. Nutrients stored in the body are also released to the area to support these defensive cells.

The mucous membranes of the body, such as those lining your mouth, nose, throat, eyelids, urinary tract, and genital tract, also defend the body against invaders. These membranes are "sticky" and trap microbes and foreign particles much like flypaper traps flies. Some of the mucous membranes of the body also have other physical or chemical aids for combating infection. For example, the membranes covering the eyes and eyelids are constantly washed in tears, a fluid that contains a chemical called lysozyme, which is deadly to most bacteria. The ciliated mucous membranes of the respiratory passageways have a comparatively thick coating of mucus that not only traps invaders but efficiently moves them from the respiratory passageways to be swallowed by the digestive system. In the digestive system, microorganisms are killed by the acid environment of the stomach. The environment of the urinary and genital tracts are somewhat acidic too—an unfavorable situation for many foreign bacteria. Foreign bacteria are also kept from invading the body by bacteria that are normal inhabitants of the body, colonizing such areas as the mouth, throat, colon, vagina, and skin. Figure 13-1 summarizes some of the nonspecific defense mechanisms of the body.

The nonspecific defenses of the body include the skin and mucous membranes, chemicals that kill bacteria, and the inflammatory process.

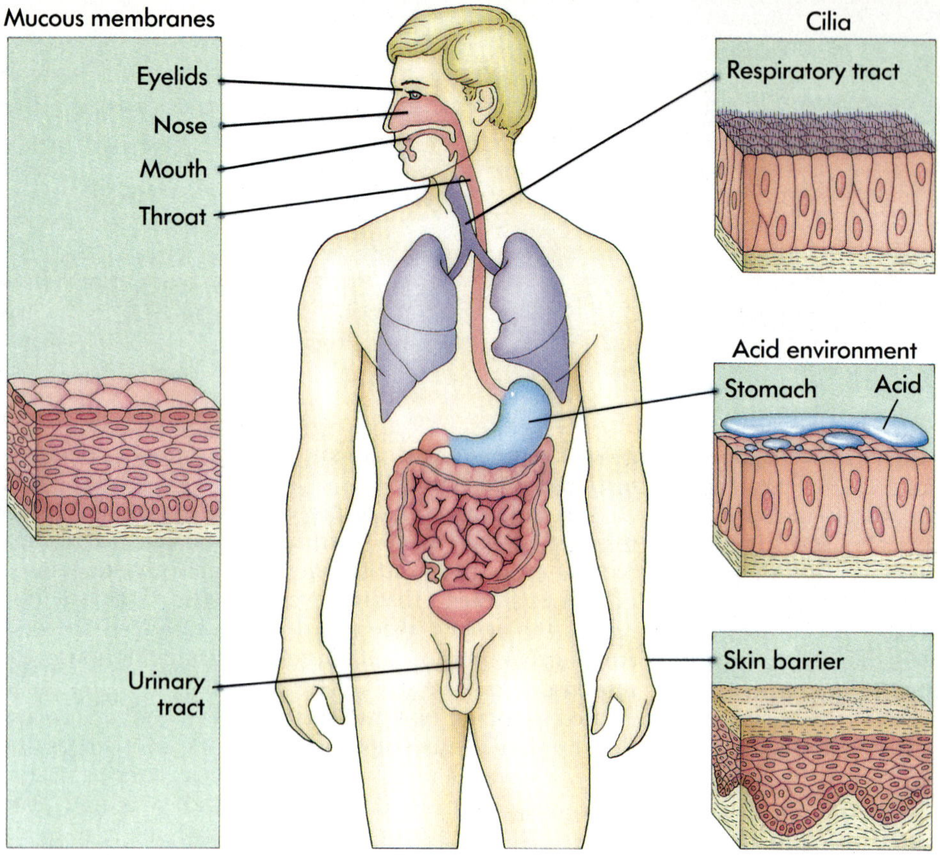

FIGURE 13-1 Nonspecific defenses in the human body.
These nonspecific defenses are the body's first line of defense against foreign invaders.

Specific resistance

Sometimes invaders "get by" your nonspecific defenses. This situation may occur if your body is invaded by many organisms at one time or if the invaders are extremely **virulent,** that is, if they are "good at" establishing an infection and damaging the body. In addition, the defenses of a person ill with certain diseases, such as diabetes mellitus or cancer, may not be as strong as persons who are free of disease.

Invaders that "get into" the blood, lymph, or tissues encounter a second line of defense—the **immune system,** your body's specific defense. Have you ever had measles, chickenpox, or mumps? All these infections are caused by specific viruses that enter the body through the respiratory system, invade cells, multiply, and spread. As you may know from experience, the body is usually effective in combating these diseases. In addition, having had one of these diseases confers an immunity, or resistance, to getting that same disease again. The mechanism that provides you with this resistance is the immune system—the backbone of your health—protecting you not only from measles, mumps, and chickenpox but also from a myriad of other diseases. And although the immune system is not always able to protect a person from infection, the immune system of a healthy person staves off most infections and usually overcomes those that take hold.

The specific defense of the body is the immune response.

Discovery of specific resistance: The immune response

In 1796, an English country doctor named Edward Jenner (Figure 13-2) carried out experiments that marked the beginning of immunology, the study of the responses of the body when it is challenged by molecules foreign to it. His experiments had to do with a disease called *smallpox,* a common and frequently deadly disease in Jenner's day. Although thousands died from smallpox, many survived. Jenner noted that these survivors were now immune from becoming reinfected. The only other group of people immune to smallpox were milkmaids who had caught another, much milder form of "the pox" called *cowpox.* Jenner hypothesized that cowpox somehow conferred protection against smallpox.

To test his hypothesis, Jenner deliberately infected healthy people with fluid he removed from the pox vesicles of sick milkmaids, causing the healthy people to develop cowpox. Many of the people Jenner infected with cowpox became immune to smallpox, just as he had predicted. Scientists now know that smallpox is caused by the variola virus and that cowpox is caused by a different but similar virus called the *vaccinia virus.* The word vaccinia is derived from the Latin vacca meaning "cow." Jenner's patients injected with fluid containing cowpox virus mounted a defense against the cowpox infection, a defense that was also effective against infection by the similar smallpox virus. Jenner's procedure of injecting a harmless microbe into a person or animal to confer resistance to a dangerous one is called **vaccination.**

FIGURE 13-2 The birth of immunology.
Edward Jenner is shown here inoculating patients with cowpox in the 1790s and thus protecting them from the disease smallpox. The underlying principles of vaccination were not understood until more than a century later.

It was a long time after Jenner's experiments before scientists understood how one microbe conferred resistance to another. Further important information was added about 85 years after Jenner by Louis Pasteur of France. Pasteur was studying fowl cholera, a serious bacterial disease of chickens. From diseased chickens, Pasteur isolated bacteria that brought on the disease when injected into healthy birds. These birds became seriously ill and died.

During a period of experimentation, Pasteur injected healthy birds with an "old" culture of fowl cholera—one that had sat on his benchtop for several weeks—because he had no fresh cultures. Interestingly, the injected birds became only slightly ill and then recovered. Pondering this outcome, Pasteur reasoned that the culture had become weakened as it sat for many days at room temperature. Trying to get his experiment to work, he then injected the recovered birds with massive doses of fresh cultures—active virulent fowl cholera bacteria. These birds did *not* develop fowl cholera! Chickens receiving the same injections but that had not previously recovered from a mild form of the illness all died. Pasteur began to realize that his old culture acted on the birds in a unique way. Clearly, these old, weakened bacteria possessed something that could confer resistance yet not cause the chickens to become seriously ill. Scientists now know what that "something" was: molecules protruding from the surface of the bacterial cells.

Every cell has a tangle of proteins, carbohydrates, and lipids on its surface (Figure 13-3). It was the presence of molecules on the surface of the cholera bacteria—molecules different from any of the birds' own—to which the chickens were responding. "Nonself" or "foreign" molecules such as these are called **antigens.** Chickens injected with weakened or killed fowl cholera bacteria are immune to later infection because the bacterial antigens cause the chickens to produce proteins called **antibodies.** These antibodies are able to recognize any future cholera invaders, weakened or normal, and prevent them from causing disease (Figure 13-4). The body's response to foreign molecules, such as the production of antibodies directed against a specific antigen, is called an **immune response.** The immune response is a result of the complex activities of recognition and defense carried out by the immune system.

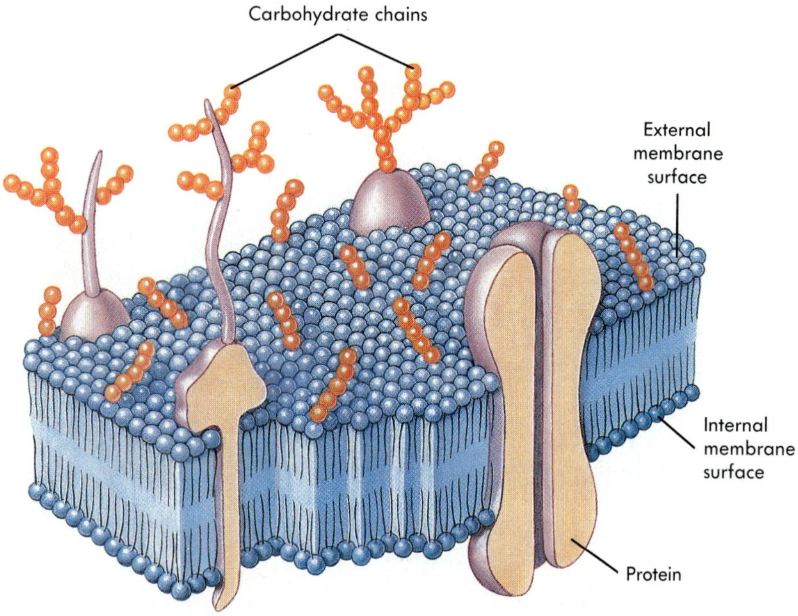

FIGURE 13-3 The outer surface of a cell is imbedded with a tangle of proteins, carbohydrates, and lipids.
The lipid molecules often serve as highly specific cell surface markers that identify specific cell types and also identify the cell as "self."

FIGURE 13-4 Immunity to smallpox.
The immunity to smallpox that Jenner's patients acquired was a direct result of their cowpox inoculation. The cowpox inoculation caused Jenner's patients to develop antibodies to the antigen. An exposure to smallpox later stimulated them to produce large amounts of the antibody in much larger amounts than before, resulting in immunity to smallpox.

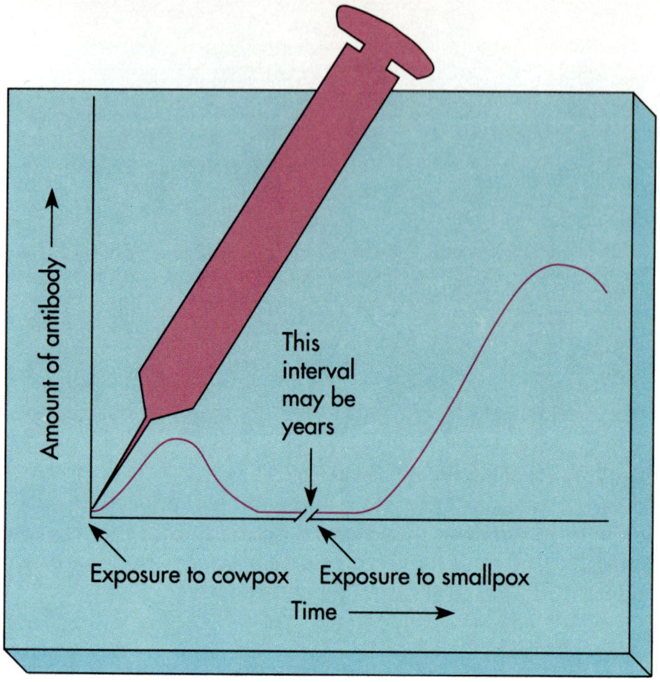

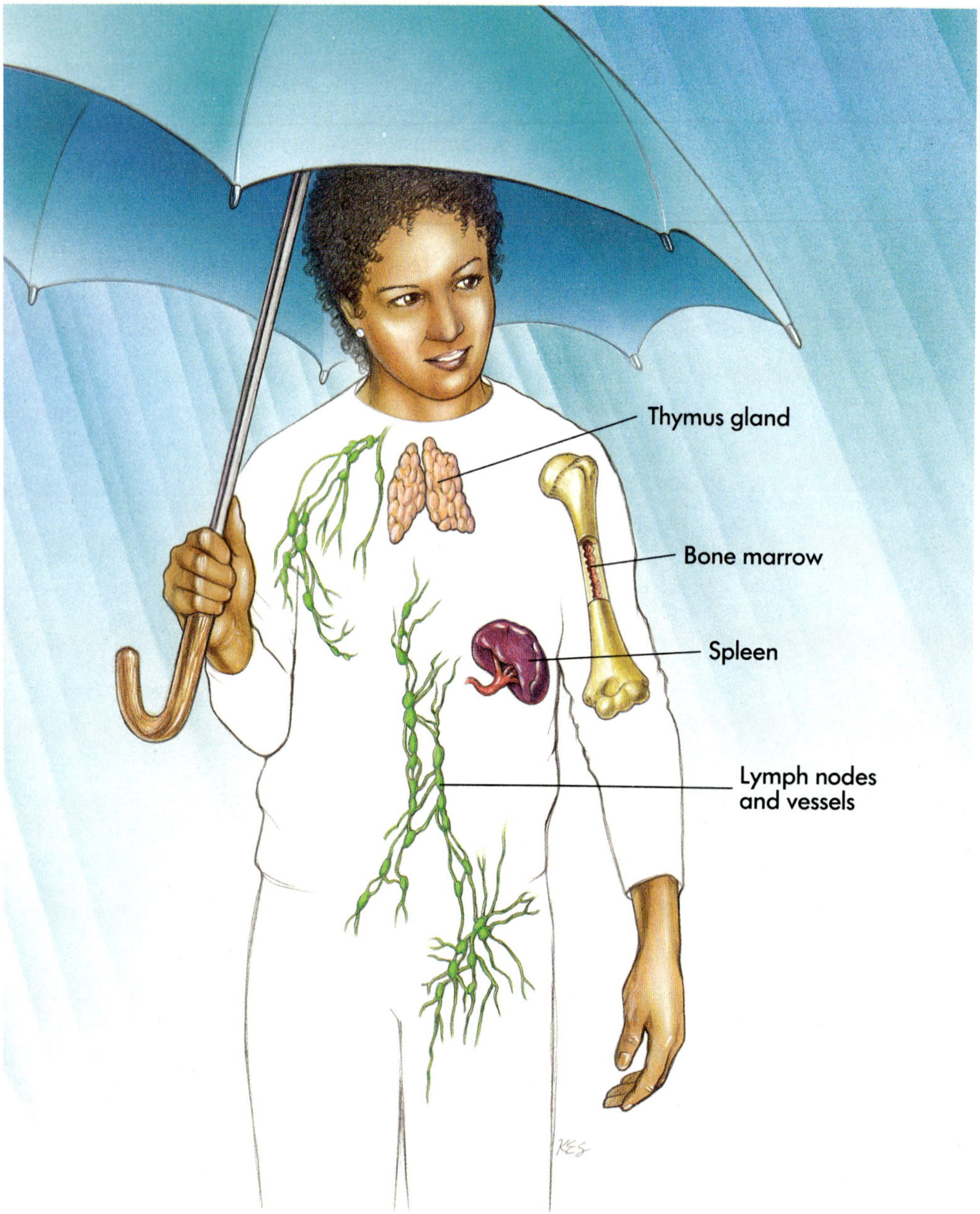

FIGURE 13-5 Organs in which immune cells reside.
The principal organs in which the cells of the immune system mature and reside include the thymus, bone marrow, spleen, lymph nodes, and lymph vessels.

The cells of the immune system

The immune system is different from the other body systems in that it cannot be identified by interconnected organs. In fact, the immune system is not a system of organs, nor does it have a single "controlling" organ. The immune system is, instead, made up of cells that are scattered throughout the body but have a common function: reacting to specific foreign molecules.

The cells of the immune system are white blood cells, or leukocytes (see Chapter 12). Two types of white blood cells are involved in the immune system: **phagocytes** and **lymphocytes.** Three classes of lymphocytes play roles in specific resistance: T cells, B cells, and **natural killer (NK) cells.** These cells arise in the bone marrow, circulate in the blood and lymph, and reside in the lymph nodes, spleen, liver, and thymus (Figure 13-5). Although not bound together, these white blood cells exchange information and act in concert as a functional, integrated system—your "army" of 200 billion defenders—that is called into action when antigens invade your body.

The immune system is composed of white blood cells, or leukocytes. Two types are involved: phagocytes and lymphocytes (T cells, B cells, and natural killer cells).

The immune response: How it works

The immune system responds in several ways to defend your body against foreign invaders. Some of the cells of the immune system react immediately to invasion by foreign antigens. Other cells work to protect your body against future attack. The immune response is a complex yet coordinated effort, and it involves several types of immune protection. An overview of the alarm sounding activation of the immune response is shown in Figure 13-6.

Sounding the alarm

Constantly patrolling your body are roving armies of white blood cells called *phagocytes*. A phagocyte is a cell that destroys other cells by engulfing and ingesting them. This process is called **phagocytosis.** Two types of white blood cells are phagocytes: neutrophils and macrophages. **Macrophages** are the phagocytes that play a key role in the body's immune response.

Large, irregularly shaped cells, the macrophages act as the body's scavengers. Macrophages phagocytize anything that is not "normal," including cell debris, dust particles in the lungs, and invading microbes. When the body is not under "attack," only a small number of macrophages circulate in the body's bloodstream and lymphatic system. In response

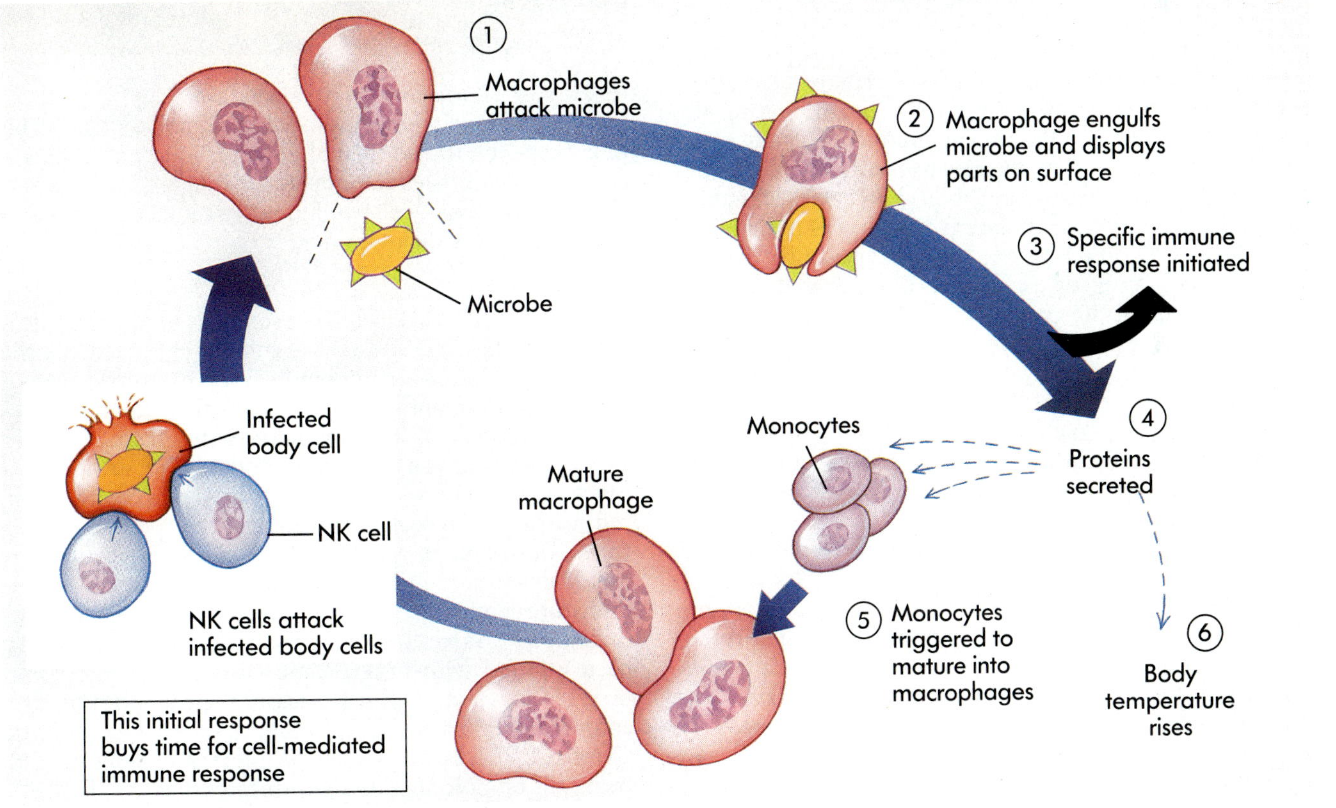

FIGURE 13-6 Activation of the immune response.
As macrophages encounter microbes *(1)*, they initiate a local inflammation. Macrophages engulf and display parts of microbes on their surfaces *(2)*, causing an activation of the specific immune response *(3)* and the secretion of proteins *(4)*. Some proteins stimulate the maturation of monocytes *(5)*; others cause fever *(6)*.

to infection, precursors of macrophages called **monocytes** develop into mature macrophages in large numbers.

When macrophages encounter foreign microbes in the body, they attack and engulf them and then "display" parts of these microbes on their surfaces. This display, as well as proteins secreted by the macrophages, is important in activating the immune response.

The macrophages secrete many different proteins. Some of these proteins trigger the maturation of monocytes into macrophages, thereby increasing their numbers. Another protein, **interleukin-1,** signals the brain to raise the body temperature, producing a **fever.** The higher temperature aids the immune response and inhibits the growth of invading microorganisms.

At the onset of the infection, immune system cells other than the monocytes also go to work. Reacting to cell surface changes that occur on cancer cells or virally-infected cells, *natural killer cells* attack them, breaking them apart. Rather than ingesting infected or diseased cells as macrophages do, natural killer cells secrete proteins that create holes in the membranes of the cells they attack. In addition, natural killer cells secrete toxins that poison these cells. The cells under attack burst open and die. This initial defense by macrophages and natural killer cells peaks within a day or two of the infection. This response "buys time" for the immune system to respond.

The two branches of the immune response

In addition to its other interactions mentioned previously, interleukin-1 activates helper T cells that have been "presented" with antigens by the macrophages, triggering the immune response. **T cells,** or **T lymphocytes,** develop in the bone marrow but then migrate to the thymus. The thymus, a small gland located in the neck, is the place where T (thymus) cells mature. In the thymus, T cells develop the ability to identify invading bacteria and viruses by the foreign molecules (antigens) on their surfaces. Tens of millions of different T cells are produced, each specializing in recognizing one particular foreign antigen. No invader can escape being recognized by at least a few T cells. There are four principal kinds of T cells:

1. **Helper T cells,** which initiate the immune response
2. **Cytotoxic T cells,** which break apart cells that have been infected by viruses and break apart foreign cells such as incompatible organ transplants
3. **Inducer T cells,** which oversee the development of T cells in the thymus
4. **Suppressor T cells,** which limit the immune response

When they are triggered by interleukin-1 and the presentation of antigens, the helper T cells can stimulate both branches of the immune response: the **cell-mediated immune response** and the **humoral immune response.**

The cell-mediated immune response

The activation of helper T cells by interleukin-1 and the binding of antigen to these activated helper T cells unleash a chain of events known as the *cell-mediated immune response.* The "main event" of this response is that cytotoxic T cells ("cell-poisoning" cells), also called *killer cells,* recognize and destroy infected body cells.

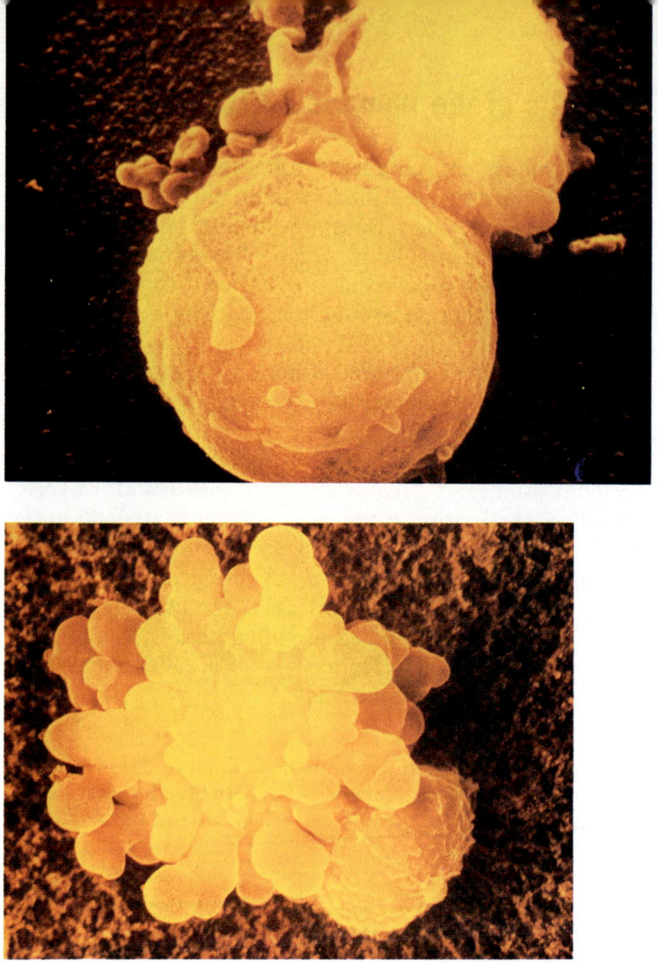

FIGURE 13-7 A cytotoxic T cell: a killer cell.
The cell-mediated response induces cytotoxic T cells to bind to antigens on the surface of an infected cell and secrete a chemical that breaks the infected cell apart. The top photo shows a cytotoxic T cell binding with a tumor cell; at bottom, the tumor cell is disintegrating.

When a helper T cell has been activated, it produces a variety of chemical substances collectively called **lymphokines.** One type of lymphokine attracts macrophages to the site of infection, and another inhibits their migration away from it. Another of the lymphokines stimulates T cells that are bound to foreign antigens to undergo cell division many times. This cell division produces enormous quantities of T cells capable of recognizing the antigens specific to the invader. Each type of activated T cell does a specific job.

The activated inducer T cells trigger the maturation of immature lymphocytes in the thymus into mature T cells. Activated cytotoxic T cells kill the body's own cells that have been infected with fungi, some viruses, and bacteria that produce slowly developing diseases such as tuberculosis. In addition, they act to kill cells that have become cancerous. Cytotoxic T cells bind to these infected or abnormal cells by means of molecules on their surfaces that specifically "fit" antigens, much like a key fits a lock. Because the entire cell binds to the abnormal cells (by means of specific cell-surface proteins), this response is called *cell-mediated.* The cytotoxic T cells then secrete a chemical that breaks apart the foreign cell (Figure 13-7). Unfortunately for organ transplant patients, cyctotoxic T cells also recognize foreign body cells in

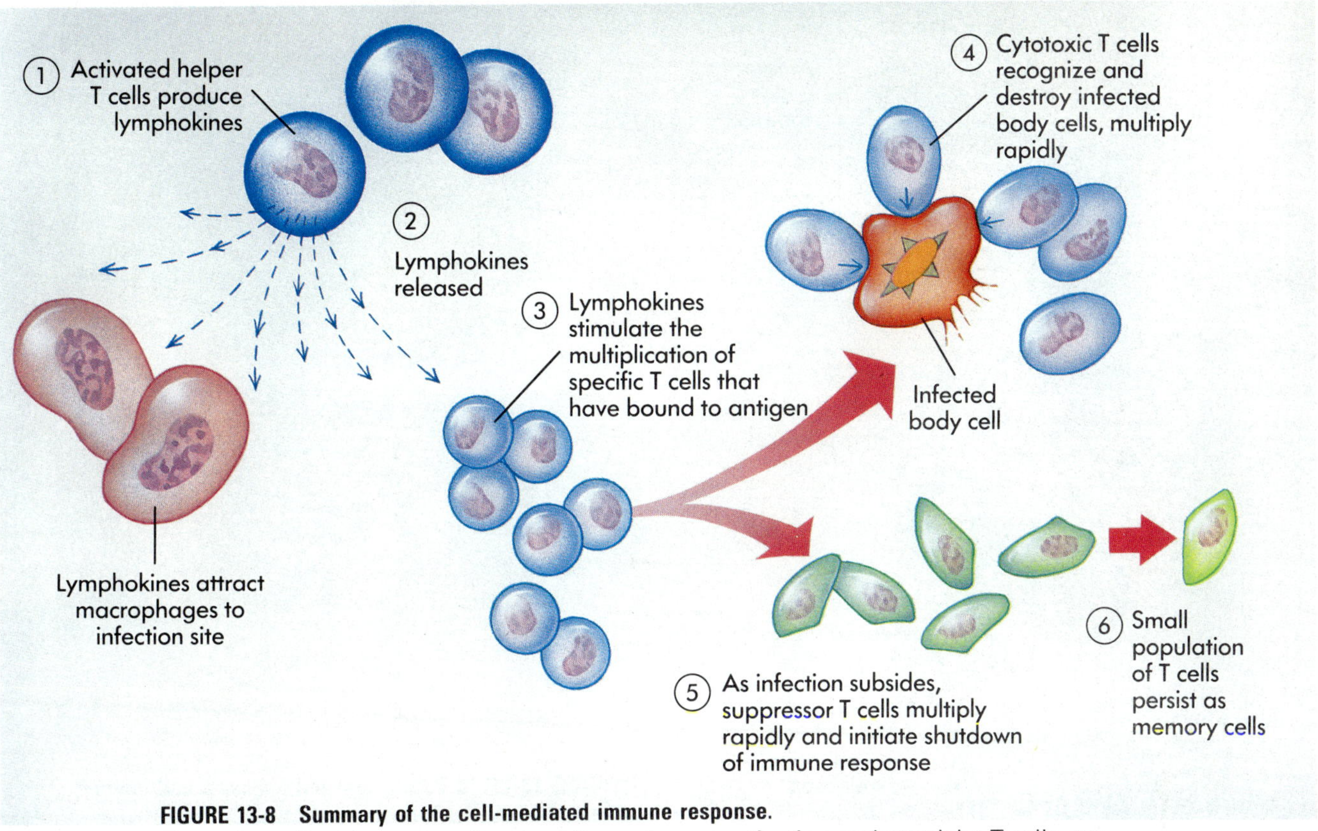

FIGURE 13-8 Summary of the cell-mediated immune response.
The cell-mediated response involves three stages: activation and attack by T cells, proliferation of suppressor cells that eventually shuts down the immune response, and the persistence of memory cells.

a similar way. Because of this, the immune system attacks transplanted tissue, leading to the rejection of transplanted organs.

After the cytotoxic cells and macrophages do their jobs, the cell-mediated immune response begins to shut down. The cells in charge of shut-down are the suppressor T cells. The number of suppressor T cells slowly begins to rise after activation by the helper T cells. However, it takes about 1 to 2 weeks for their numbers to increase to a point where they are able to suppress the cytotoxic T cell response.

After suppression, or shut-down, a population of T cells persists, probably for the life of the individual. Referred to as **memory T cells,** these helper and cytotoxic T cells provide an accelerated and larger response to any later encounter with the same antigens. A summary of the cell-mediated immune response is shown in Figure 13-8.

T cells carry out the cell-mediated immune response, during which cytotoxic T cells recognize and destroy body cells infected with certain bacteria, viruses, and fungi. In addition, they destroy transplanted cells and cancer cells. Helper T cells initiate the response, activating cytotoxic T cells, macrophages, inducer T cells, and, as the infection subsides, suppressor T cells.

The humoral immune response

When helper T cells are stimulated to respond to foreign antigens, they activate the cell-mediated immune response as described and activate a second, more long-range defense called the **humoral,** or antibody, **response.** Depending on the types of antigens present, the helper T cells may stimulate either or both of these branches of the immune response. The key players in humoral immunity are lymphocytes called **B cells** or **B lymphocytes.** The B cells are named after a digestive organ in birds called the *bursa of Fabricius* in which these lymphocytes were first discovered. However, B cells mature in the bone marrow of humans, which may be a convenient way for you to remember these cells. The term *humoral response* refers to the fact that B cells secrete antigen-specific chemicals into the bloodstream—one of the body fluids called "humours" long ago.

On their surfaces, B cells each have about 100,000 copies of a protein receptor that binds to antigens. Because different B cells bear different protein receptors, each recognizes a different, specific antigen. At the onset of a bacterial infection, for example, the receptors of one or more B cells bind to bacterial antigens. The B cells may bind to either free bacteria or bacterial antigens displayed by macrophages. These antigen-bound B cells are detected by helper T cells, which then bind to the antigen–B cell complex (Figure 13-9). After binding, the helper T cells release lymphokines that trigger cell division in the B cell.

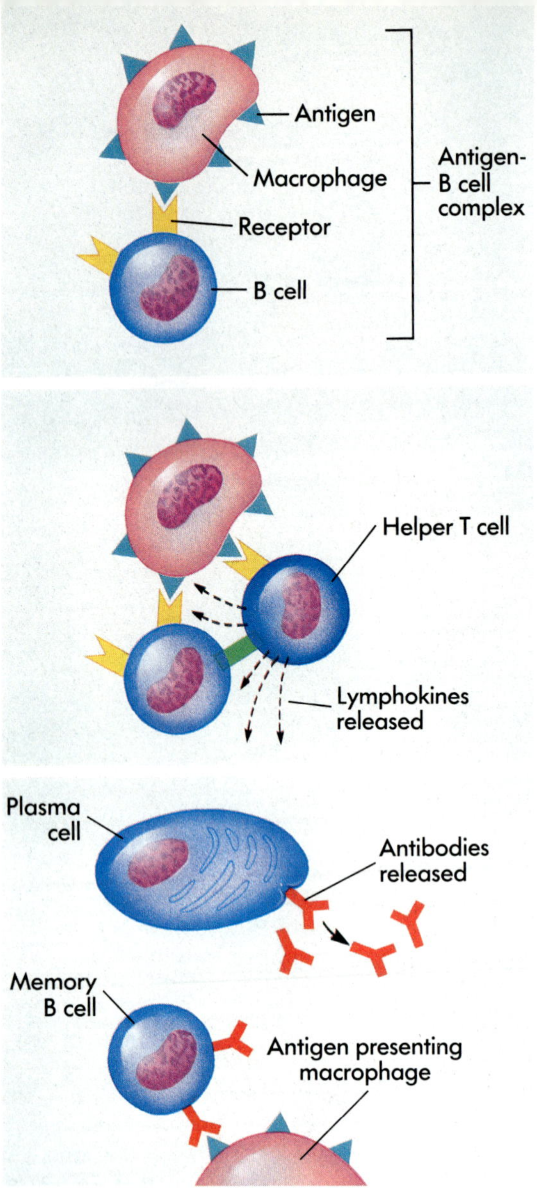

FIGURE 13-9 Summary of the humoral immune response. The humoral immune response involves the participation of three immune cell types: B cells, helper T cells, and antigen-presenting macrophages. The humoral response imparts a more long-range defense than the cell-mediated response through the production of antibodies.

After about 5 days and numerous cell divisions, a large clone of cells is produced from each B cell that was stimulated to divide. A **clone** is a group of identical cells that arise by repeated mitotic divisions from one original cell. (See Chapter 5 for a discussion of mitosis.) After these clones are formed, most of the B cells stop dividing but begin producing and secreting copies of the receptor protein that responded to the antigen. These receptor proteins are called *antibodies,* or **immunoglobulins.** The secreting B cells are called **plasma cells.** After B cells become plasma cells, they live for only a few days but secrete a great deal of antibody during that time (Figure 13-10). In fact, one plasma cell will typically secrete more than 2000 antibodies per second!

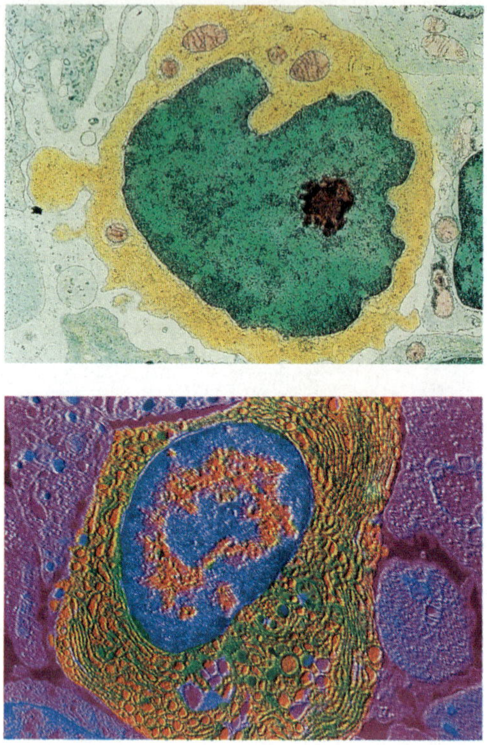

FIGURE 13-10 A B cell *(top)* and a plasma cell *(bottom)*. The plasma cell contains a large amount of endoplasmic reticulum, on which antibodies are made.

Antibodies do not destroy a virus or bacterium directly but rather mark it for destruction by one of two mechanisms (Figure 13-11):

1. *Complement:* **Complement** are proteins that kill foreign cells by creating a hole in their membranes. Water floods the cell through these holes, causing the cell to swell and burst.
2. *Macrophages:* The activity of these phagocytes is enhanced by the binding of both antibodies and complement to the antigen.

The B cell clones that did not become plasma cells live on as circulating lymphocytes called **memory B cells.** These cells provide an accelerated response to any later encounter with the stimulating antigen. This is why immune individuals are able to mount a prompt defense against infection. As in the case of the cell-mediated immune response, the antibody response is shut down after several weeks by suppressor T cells.

B cells carry out the humoral immune response in which B cells recognize foreign antigens and, if activated by helper T cells, proceed to produce large quantities of antibody molecules directed against the antigen. The antibodies bind to the antigens they encounter and "mark" them for destruction.

Helping the immune system fight cancer

BIOLOGY TECHNOLOGY & society

Malignant melanoma is a particularly lethal form of skin cancer. Tumors grow very rapidly, quickly subverting the body's defenses (Figure 13-A). On October 7, 1991, doctors injected genetically engineered cells into the thigh of a melanoma patient in an attempt to use gene therapy to help his immune system destroy the cancer. This is one of the first instances in which a human has been the direct target of genetic engineering.

The human cell that is the target of this pioneering effort is a special kind of white blood cell called a *tumor-infiltrating lymphocyte (TIL)*. This cell is part of the immune system's cancer surveillance system. It will normally seek out and attack a cancerous tumor but is not strong enough by itself to control the tumor. In previous years, researchers had tried removing TIL cells from patients with malignant melanoma, culturing billions of them in test tubes, and then returning the cells to the patient's bloodstreams to seek out the tumors from which they arose—but even in greatly increased numbers the cells were not strong enough to cure the cancer.

The key result, though, is that the TIL cells did seem to seek out the tumors from which they arose. This result, researchers reasoned, ought to make them ideal vehicles for delivering therapeutic agents to the tumors. In initial studies in the spring of 1991, researchers added a marker gene to TIL cells so that they could be sure that the engineered cells indeed home in on the malignant melanoma tumors. They did. After careful review of these results, researchers are now trying to use the TIL cells as "trucks" to deliver substances that attack cancer cells to the tumors.

Researchers remove TIL cells from the patient with malignant melanoma and insert into each cell's chromosomes a gene that will command the cell to produce a protein called *tumor necrosis factor (TNF)*. This protein kills tumor cells by preventing them from establishing a blood supply. The engineered TIL cells are then returned to the patient's bloodstream to seek out and invade the malignant melanoma tumors. As each genetically altered TIL cell finds and enters a tumor, it is able to attack using a much stronger weapon, TNF. The engineered TIL cell, in effect, becomes a factory that makes the tumor-killing protein inside the tumor itself.

This first attempt to help the human immune system fight cancer is experimental, and the degree to which it will be successful in treating malignant melanoma will not be known for a year or more. Because the approach shows great promise, it has medical researchers very excited. The chairman of the Recombinant DNA Advisory Committee that approved the therapy called the advance historic: "What we're doing today is adding gene therapy to vaccines, antibiotics and radiation in the medical arsenal. Medicine has been waiting thousands of years for this."

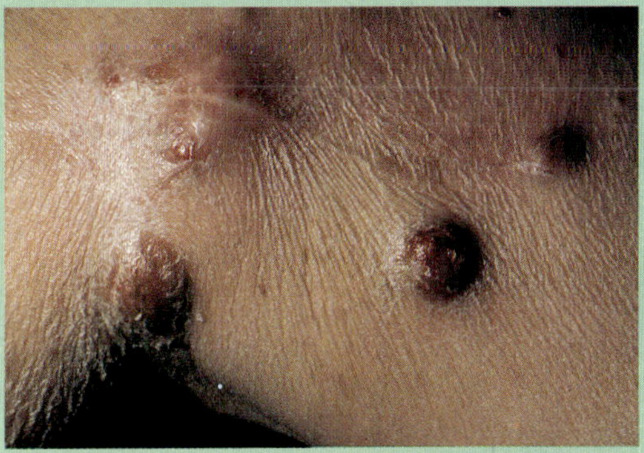

FIGURE 13-A Malignant melanoma. This cancer is one of the fastest-spreading cancers known and can be caused by excessive exposure to the sun.

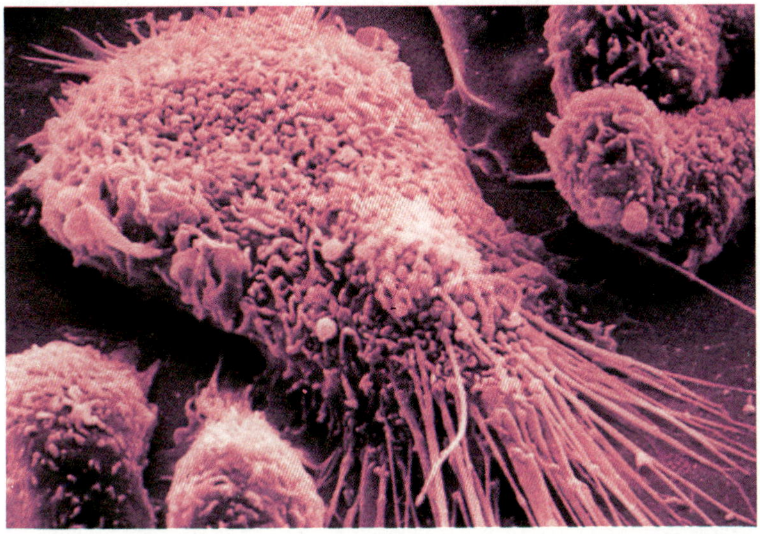

FIGURE 13-11 A macrophage ingesting cells. Macrophages engulf "marked" antigens by phagocytosis.

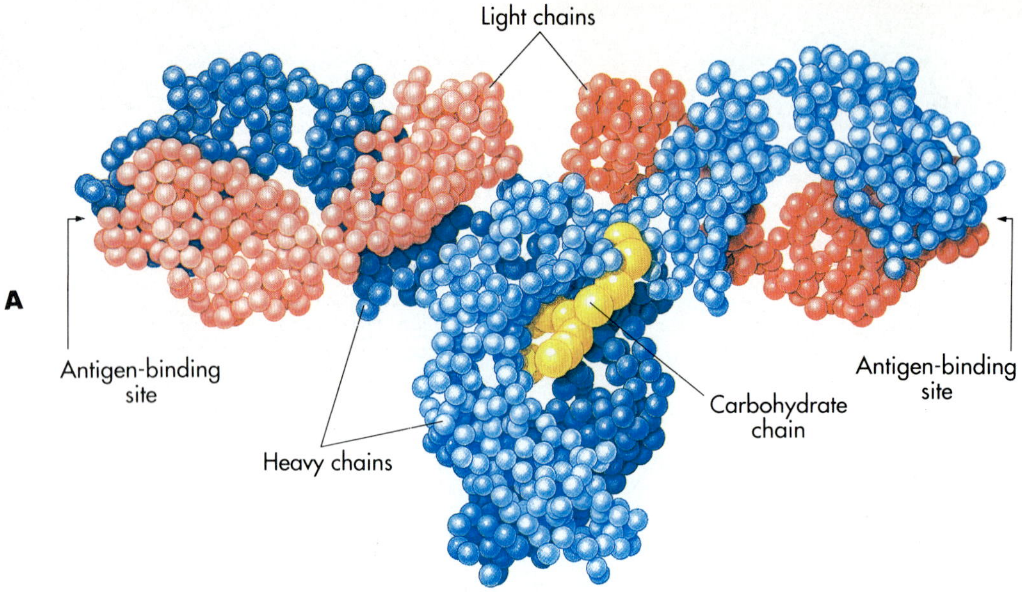

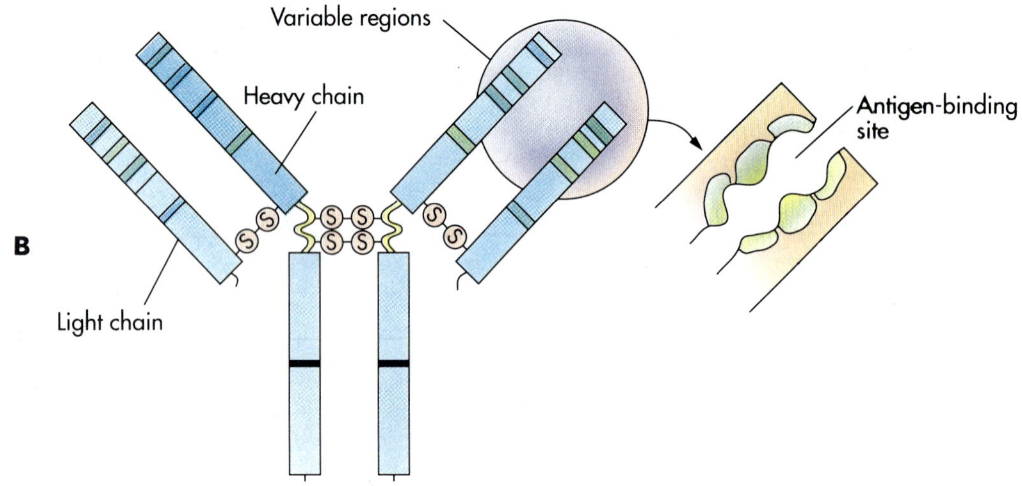

FIGURE 13-12 The structure of an antibody molecule.
A A molecular model of an antibody. Each amino acid is represented by a small sphere. The heavy chains are colored blue, the light chains, red. The four chains wind around one another to form a Y shape, with two identical antigen-binding sites at the arms of the Y.
B A schematic drawing of an antibody molecule. Each molecule is composed of two identical light chains and two identical heavy chains. Disulfide bonds hold the chains together.

How immune receptors recognize antigens

The cell surface receptors of lymphocytes (T cells and B cells) can recognize specific antigens with great precision. Even single amino acid differences between proteins can often be discriminated, with a receptor recognizing one form and not the other. This high degree of precision is a necessary property of the immune system, since without it the identification of foreign antigens would not be possible in many cases; the differences between self and foreign (nonself) molecules can be very subtle.

A typical antibody, or immunoglobulin molecule, consists of four polypeptide chains. A polypeptide is a chain of amino acids linked end-to-end by peptide bonds (see Chapter 2). There are two identical short strands, called **light chains,** and two identical long strands, called **heavy chains.** The four chains are held together by disulfide (-S-S-) bonds, forming a Y-shaped molecule (Figure 13-12, *A*).

The two "arms" of the Y determine which antigen will bind to the antibody. Antibodies "recognize," or "lock on," to antigens by means of special binding sites in their arms (Figure 13-12, *B*). These binding sites are made up of specific sequences of amino acids that determine the shape of the site. The specificity of the antibody molecule for an antigen depends on this shape. An antigen fits into the binding site on the antibody like a hand into a glove. Changes in the amino acid sequence of an antibody can alter the shape of this region and, by doing so, change the antigen that can bind to that antibody, just as changing the size of a glove will alter which hand it can fit.

An antibody molecule recognizes a specific antigen because it possesses binding sites into which an antigen can fit, much as a key fits into a lock. Changes in the amino acid sequence at the binding sites alter the antibody's shape and thus change the identity of the antigen that is able to fit into them.

The "stem" of the Y determines what role the antibody plays in the immune response. An antibody can have one of five different stems and therefore be a member of one of the five classes of antibodies. For example, the antibodies produced in the first week of an infection are primarily class M antibodies and are called *IgM* (immunoglobulin M). After the first week, IgG antibodies are primarily produced.

Immunization: Protection against infection

Although scientists have unraveled many of the mysteries of the immune response since Jenner's day, scientists still use the same basic idea of vaccination that Jenner used to help people become resistant to specific diseases. A vaccine is made up of disease-causing microbes or toxins (poisons) that have been killed or changed in some way so as not to produce the disease. Injection with these antigens causes an antibody (B cell) response, with the production of memory cells. A "booster shot" induces these memory cells to differentiate into antibody-producing cells and more memory cells. So, vaccination causes your body to build up antibodies against a particular disease without getting the disease. This type of immunity is called **active immunity.**

Scientists can now use the techniques of genetic engineering to produce vaccines (see Chapter 28). Instead of growing cultures of an agent of infection such as a bacterium or a virus, scientists can, in the laboratory, produce large quantities of antigenic proteins from disease-causing microbes. These antigens can then be used as vaccines.

Vaccination to produce active immunity works well for many diseases but is a somewhat slow process. After the injection of the antigen, it takes weeks for the body to develop sufficient antibodies and memory cells to combat the disease. However, another type of immunity, **passive immunity,** can be used when protection is needed quickly. One example is the use of **antitoxin** when a person has been bitten by a snake.

When a poisonous snake bites a person or other animal, it injects a toxin (poison) into the body. This toxin circulates in the bloodstream. Certain toxins damage the nerve cells they reach and may result in death. An antitoxin is a preparation of antibodies specific for a toxin. These antibodies are injected into the victim and they bind with the toxin, preventing the toxin from causing damage. Although effective, this type of "borrowed immunity" lasts for only a short time. The body soon uses or eliminates these antibodies, so they will not be available if another snake bite occurs.

Newborn babies have a type of passive immunity "borrowed" from their mothers. Antibodies that pass from mother to fetus through the placenta provide babies with a short-term immunity that subsides by the time the baby has produced its own system of antibodies. In addition, if a mother nurses her baby, antibodies in the mother's milk may provide the baby with some protection.

Defeat of the immune system: AIDS

Helper T cells are the key to the entire immune response because they initiate the proliferation of both T cells and B cells (Figure 13-13). Without helper T cells, the immune system is unable to mount a response to foreign antigens.

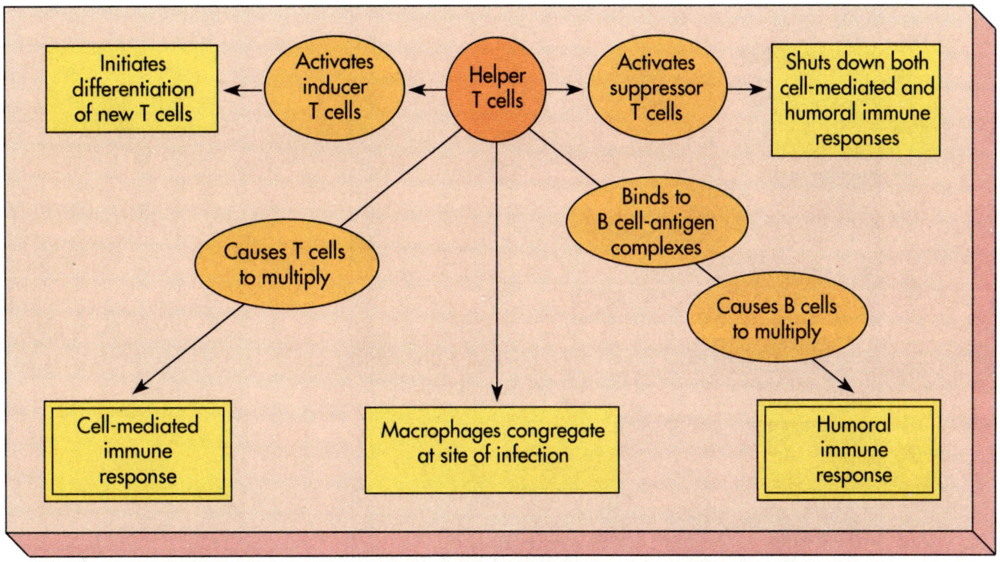

FIGURE 13-13 Helper T cells are the key to the immune response. This diagram outlines the many roles of the helper T cell.

The AIDS virus, called **human immunodeficiency virus, or HIV,** is deadly because it mounts a direct attack on T lymphocytes, mainly targeting the helper T cells. The HIV particles kill helper T cells by entering them, using the cells' own "machinery" to produce more virus particles, and then bursting open the cells to release the "new" viruses (see p. 229). These viruses infect other helper T cells until the entire population is destroyed. In a normal individual, helper T cells make up 60% to 80% of circulating T cells; in AIDS patients, helper T cells often become too rare to detect. Because the immune response cannot be initiated, any of a variety of infections proves fatal. This is the primary reason that AIDS is a particularly devastating disease (Figure 13-14).

The AIDS virus destroys the ability of the immune system to mount a defense against any infection because it attacks and destroys helper T cells.

Although the AIDS virus was discovered to be a cause of human disease only within the past decade, it is clear that it is a serious disease. The fatality rate of AIDS is virtually 100%; of the earliest identified AIDS patients, only a few are still alive. The disease is *not* highly infectious; that is, it is not transmitted from person to person by casual contact. It is only transmitted by the direct transfer of body fluids, typically in semen or vaginal fluid during sexual activity, and in blood during transfusions or by contaminated hypodermic needles. It also crosses the placenta to infect babies within the womb of an infected mother. Not all individuals exposed to AIDS (as judged by antibodies in their blood directed against the AIDS virus) have yet come down with the disease, but scientists expect between 95% and 100% of all HIV-positive people to develop AIDS.

Efforts to develop a vaccine against AIDS continue, both by splicing portions of the AIDS surface protein gene into a harmless virus (Figure 13-15) and by attempting to develop a harmless strain of AIDS. These approaches have not yet proved successful and are limited by the fact that different strains of AIDS virus seem to possess different surface antigens. Drugs that inhibit specific genes of the AIDS virus or that block the operation of enzymes critical to the synthesis of its genetic material are also being investigated, with a notable lack of success at this time.

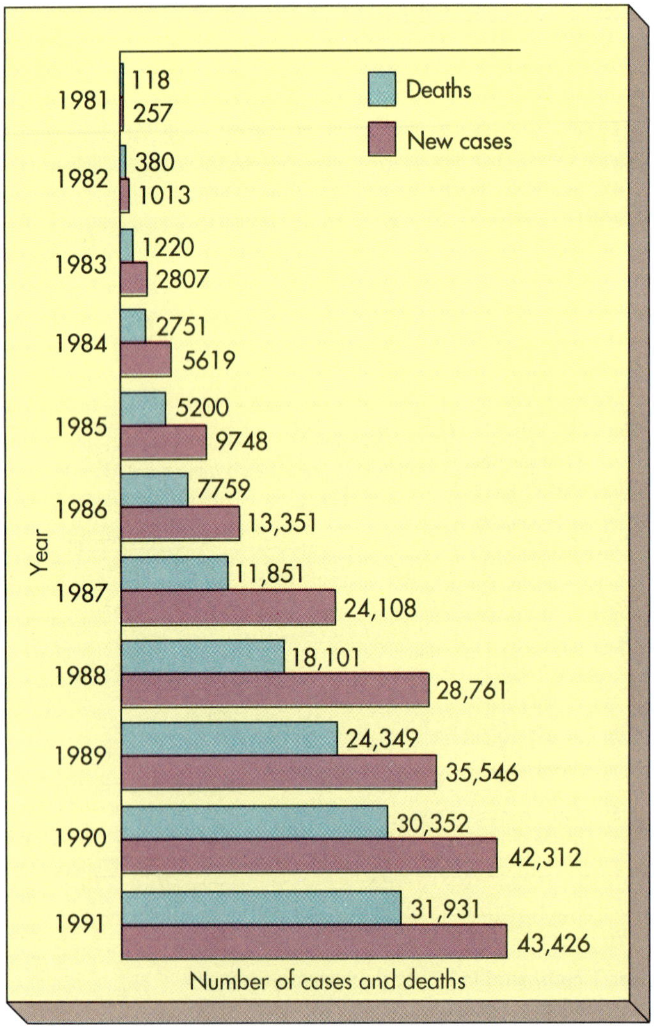

FIGURE 13-14 AIDS cases and deaths since 1981.
The U.S. Centers for Disease Control reports that by the end of 1991, there were 206,392 AIDS cases in the United States and 133,232 deaths. Over 1 million other individuals are thought to have been infected with HIV. By August 1989, 8 years into the epidemic, 100,000 AIDS cases had been reported; the next 100,000 cases took just 26 months.

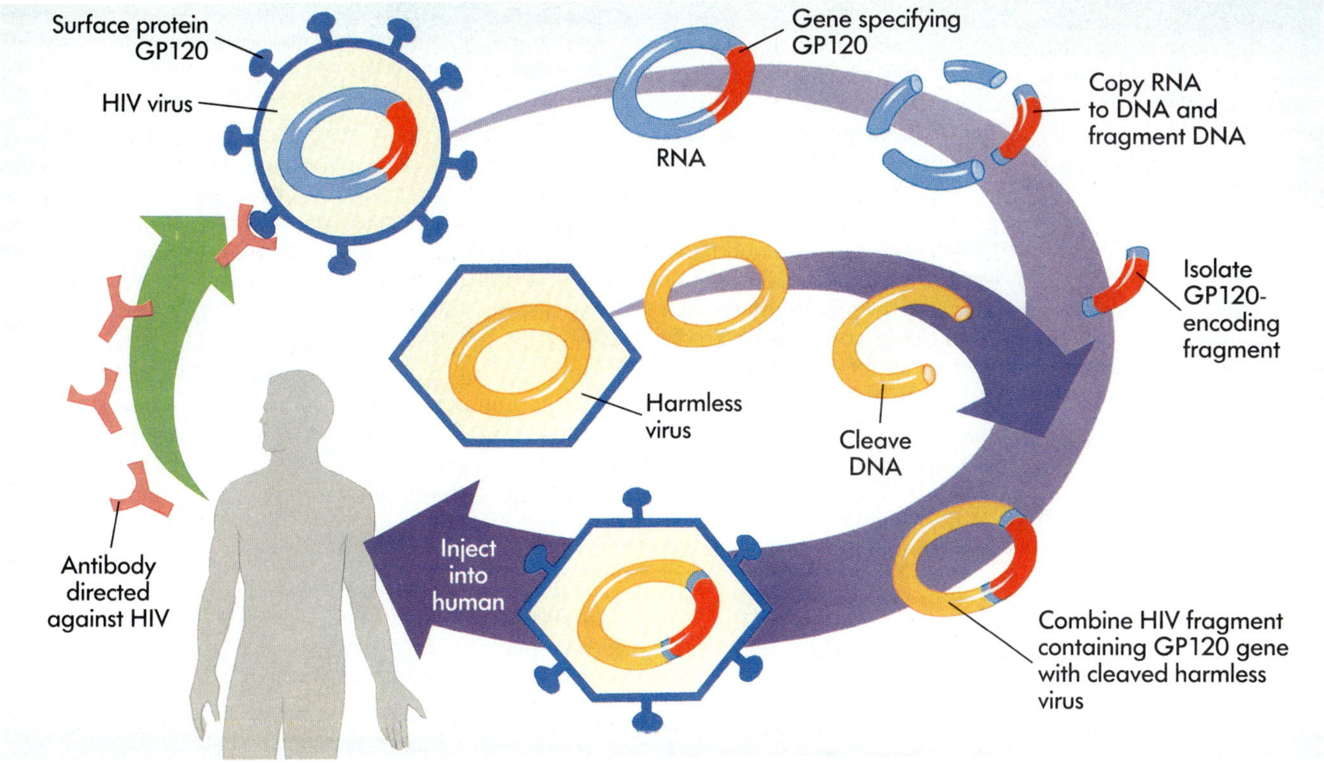

FIGURE 13-15 How scientists are attempting to construct a vaccine for AIDS.
Of the several genes of the human immunodeficiency virus (HIV), one is selected (GP120) that directs the synthesis of a certain surface feature on the virus. All of the other HIV genes are discarded. This one gene is not in itself harmful to humans. This one gene or fragment of it is inserted into the DNA of a harmless vaccinia cowpox virus, resulting in a harmless virus whose surface imitates the AIDS virus. Persons injected with the vaccinia virus do not become ill, but they do develop antibodies directed against the infecting virus surface. Because the surface contains HIV proteins, the new antibodies serve to protect the infected person against any subsequent exposure to the HIV virus.

Allergy

Although the human immune system provides very effective protection against viruses, bacteria, parasites, and other microorganisms, sometimes it does its job too well, mounting a major defense against a harmless antigen. Such immune responses are called **allergic reactions.** Hay fever, the sensitivity that many people exhibit to proteins released from plant pollen, is a familiar example of an allergy. Many other people are sensitive to proteins released from the feces of a minute house-dust mite (Figure 13-16). This microscopic insect lives in the house dust present on mattresses and pillows and eats the dead epithelial tissue that everyone sheds from their skin in large quantities daily. Many people sensitive to feather pillows are actually allergic to the feces of mites that are residents of the feathers.

What makes an allergic reaction uncomfortable and sometimes dangerous is involvement of class E (IgE) antibodies. The binding of antigens to these antibodies initiates an inflammatory response; powerful chemicals cause the dilation of blood vessels and a host of other physiological changes. Sneezing, runny noses, and fever often result. In some instances, allergic reactions can be far more dangerous, resulting in anaphylactic shock, a severe and life-threatening

FIGURE 13-16 The house-dust mite *Dermphagoides*.
Believe it or not, mites like this live in the dust in your house. Many people are allergic to the proteins released in these mites' feces.

response of the immune system. The chemicals that dilate blood vessels are released suddenly into the bloodstream, causing many vessels to widen suddenly. Their widening causes the blood pressure to fall. In addition, the muscles of the trachea, or windpipe, may contract, making breathing difficult.

Not all antigens are **allergens,** initiators of strong immune responses. Nettle pollen, for example, is as abundant in the air as ragweed pollen, but few people are allergic to it. Nor do all people develop allergies; the sensitivity seems to run in families. Allergies require both a particular kind of antigen and a high level of class E antibodies. The combination of an appropriate antigen on the one hand and inappropriately high levels of IgE antibodies on the other hand produces the allergic response.

Summary

1. Although immunity was discovered almost 200 years ago, it has only recently been understood that resistance to disease is achieved by populations of white blood cells, collectively called the *immune system.*
2. The human body has both nonspecific and specific defenses against disease. The nonspecific defenses, those acting against any foreign invader, include the skin and mucous membranes, chemicals that kill bacteria, and the inflammatory process. The specific defense of the body is the immune response, a complex series of processes carried out by the cells of the immune system.
3. There are two types of white blood cells involved in the immune system: phagocytes and lymphocytes. Three classes of lymphocytes play roles in specific resistance: T cells, B cells, and natural killer cells. These lymphocytes are capable of recognizing specific foreign substances (usually proteins) called *antigens.*
4. Lymphocytes recognize specific antigens by means of receptor proteins on the lymphocyte cell surface.
5. The immune response is initiated by groups of T cells called *helper T cells.* When stimulated by macrophages and antigens, helper T cells can activate the two branches of the immune response: the cell-mediated immune response (T cell response), and the humoral immune response (B cell response).
6. In the cell-mediated immune response, cytotoxic T cells attack infected body cells. In the humoral immune response, a longer-range defense than the T cell response, B cells are converted to plasma cells that secrete proteins called *antibodies.* These antibodies specifically bind circulating antigen and mark cells or viruses bearing antigens for destruction.
7. The exquisite specificity of an antibody for a particular antigen is caused by the three-dimensional shape of the ends of the arms, or antigen binding sites, of the molecule. Slight changes in the amino acid sequence alter the shape of these sites and thus the identity of molecules able to fit into it.
8. Vaccination, which is an injection with disease-causing microbes or toxins that have been killed or changed in some way so as to be harmless, causes the body to build up antibodies against a particular disease. Scientists can now use "laboratory-made" antigenic proteins as vaccines.
9. The AIDS virus is deadly because it primarily attacks helper T cells, the key to the entire immune response. Because helper T cells are no longer present to initiate the immune response, any of a variety of otherwise commonplace infections prove fatal.
10. An allergic reaction is an immune response against a harmless antigen. It is produced by the combination of a particular kind of antigen and a high level of a certain class of antibody.

REVIEW QUESTIONS

1. Distinguish between nonspecific and specific defenses. Which could be considered the first line of defense and why?
2. Name your body's nonspecific defenses, and summarize how each protects you against infection or injury.
3. What is the immune system? Explain its significance.
4. How is the immune system different from other body systems discussed so far?
5. Fill in the blanks: Nonself molecules on the surfaces of foreign cells are called ________.
6. Fill in the blanks: Four types of cells are involved in the immune system: ________, ________, ________, and ________. These are all ________.
7. Distinguish among monocytes, macrophages, and natural killer cells. How does each contribute to the body's immune response?
8. What are the four principal types of T cells? Summarize the function(s) of each.
9. Summarize the events of the cell-mediated immune response. What cells are involved?
10. You had measles as a child. Now your younger brother has measles, but you don't catch it from him. Explain why.
11. Summarize the events of the humoral immune response. What cells are involved?
12. Explain how an antibody molecule recognizes a specific antigen. Why is this important?
13. A nurse vaccinates a child against the tetanus bacterium. Later, the child receives a booster shot. Describe how these two vaccinations affect the child's immune response during subsequent exposures to tetanus.
14. Fill in the blanks: When a vaccination causes your body to build up antibodies against a particular disease, this is ________ immunity.
15. Which virus is responsible for AIDS? Explain why this virus is so deadly.
16. Many people—including you, perhaps—are allergic to something. Explain what an allergic reaction is and why it occurs. What causes its uncomfortable symptoms?

KEY TERMS

active immunity 227
allergens 230
allergic reaction 228
antibody 219
antigen 219
B cell (B lymphocyte) 223
cell-mediated immune response 222
clone 224
complement 224
cytotoxic T cell 222
fever 222
heavy chain 226
helper T cell 222
human immunodeficiency virus (HIV) 227
humoral immune response 222
immune response 219
immune system 221
immunity 217
immunoglobulin 224
inducer T cell 222
infection 217
inflammation 217
interleukin-1 222
light chain 226
lymphocytes 222
lymphokine 222
macrophage 221
memory B cell 224
monocytes 222
natural killer (NK) cells 221
nonspecific defense 217
passive immunity 227
plasma cell 224
phagocyte 221
suppressor T cell 222
specific defense 217
T cell (T lymphocyte) 222
vaccination 218

THOUGHT QUESTIONS

1. The flu epidemic of 1918 killed 22 million people in 18 months, many more than the number who died in the First World War that just preceded it. By 1920, however, the flu epidemic was over. What do you think stopped it?
2. When your body is fighting an infection, your immune system causes your body to raise its temperature. In severe infections the rise in temperature can be high enough to produce serious body damage and even death. Why do you think the body does not regulate the maximum rise in temperature during a fever it induces to fight an infection?

FOR FURTHER READING

Caldwell, M. (1992, April). The transplanted self. *Discover*, pp. 62-68.
Organ transplant donors in the future might donate their immune cells along with their organs to prevent organ rejection by the recipient.

Gallo, R. C., & Montagnier, L. (1987, October). AIDS in 1988. *Scientific American*, p. 40.
In their first collaboration the two investigators who established the cause of AIDS describe how HIV was isolated and linked to AIDS, the current status of AIDS research, and the prospects for an AIDS therapy.

Jaret, P. (1986, June). Our immune system: the wars within. *National Geographic*, pp. 702-736.
This is a very readable account of current progress in the study of the human immune system, with striking photographs by Lennart Nilsson.

Smith, K. A. (1990, March). Interleukin-2. *Scientific American*, pp. 50-57.
The first hormone of the immune system to be recognized, it helps the body to mount a defense against microorganisms by triggering the multiplication of only those cells that attack an invader.

14 Excretion

HIGHLIGHTS

▼

The major organs of excretion in the human body are the lungs and the kidneys.

▼

The formation of urine by the kidneys not only rids the body of metabolic wastes but controls the balance of water and salts, processes that are important to the maintenance of homeostasis.

▼

The kidneys work by removing most of the water and dissolved substances from the blood, then replacing what is needed by the body at that time.

▼

Other organs of the urinary system—the ureters, bladder, and urethra—temporarily store and then excrete the urine produced by the kidneys.

OUTLINE

What substances does the body excrete?

The organs of excretion

An overview of how the kidney works

A closer look at how the kidney works

- The anatomy of the kidney
- The workhorse of the kidney: The nephron
  - Filtration
  - Selective reabsorption
  - Tubular secretion
  - A summary of urine formation

The urinary system

The kidney and homeostasis

Problems with kidney function

- Kidney stones
- Renal failure
- Treatments for renal failure

The story of excretion is, to a large extent, a story about water and salt. And that story is the reason this sea turtle is crying.

Animals (like you and the sea turtle) not only live on land but live in salt water, in fresh water, in marshes . . . in a wide array of environments. In any environment, an animal needs water to live. Water bathes its cells and helps move substances within its body. Water is the medium in which substances are dissolved and in which chemical reactions take place. The amount of water in the body is crucial to survival—having too much water can cause as many problems as having too little.

In animal bodies, water is separated within compartments—any space bounded by the selectively permeable membranes of cells. Cells themselves are compartments, as are such structures as the interiors of the blood vessels and the heart. Water moves among the compartments within the body by osmosis, moving from areas of a high concentration of water to areas of a lower concentration of water. The concentration of salts—compounds that form ions in water—affects the concentration of water in the various body compartments and therefore water's movement. To regulate the water in the various body compartments, animals must regulate the amount of salt as well—a "balancing act" of sorts.

Animals have evolved various mechanisms that deal with their "water and salt problem." Salt water "tears" is only one mechanism. Because the turtle lives in a salty environment, water tends to leave its "less salty" body. To compensate, the turtle drinks a lot of seawater but then needs to get rid of the salt and keep the water. Interestingly, the sea turtle has evolved excretory organs located in the corners of its eyes that eliminate salt. So shedding salty tears is really one way of accomplishing its balancing act. Animals that live on land, such as humans, do not have the problem of water moving into or out of their bodies through the skin in response to the concentration of water in their environments. Therefore land animals can regulate the balance of water and salt in their bodies without shedding a sea of salty tears.

What substances does the body excrete?

Although the story of **excretion** is, to a large extent, a story about water and salt balance, this process also regulates the concentrations of other ions and molecules and removes metabolic wastes from the body. The metabolic wastes of excretion, however, are not the wastes of the digestive process removed during **elimination.** Unabsorbed digestive wastes are not involved in cellular metabolism—the chemical reactions of the body. Technically, they never pass *into* the body's cells or blood but remain "outside" as they pass down the "pipeline" of the digestive tract. Metabolic wastes, on the other hand, are substances produced within the body by various chemical processes such as cellular respiration.

The only "true" excretory products that are removed from the body during elimination are bile pigments and the salts of certain minerals. **Bile pigments** are produced by the liver from the breakdown of old, worn-out red blood cells. These pigment molecules are combined with other substances to form bile, which is carried by a duct from the liver, to the gallbladder (bile's storage pouch), and then to the small intestine. You may recall from Chapter 11 that bile plays a role in the emulsification of fat globules during the digestion of food. The bile pigments leave the body mixed with the digestive wastes and are the cause of the characteristic color of the feces. Due to its role in the excretion of the bile pigments, the liver is considered an organ of excretion. The large intestine is also considered an organ of excretion because cells lining its walls excrete the salts of some minerals such as calcium and iron (Figure 14-1).

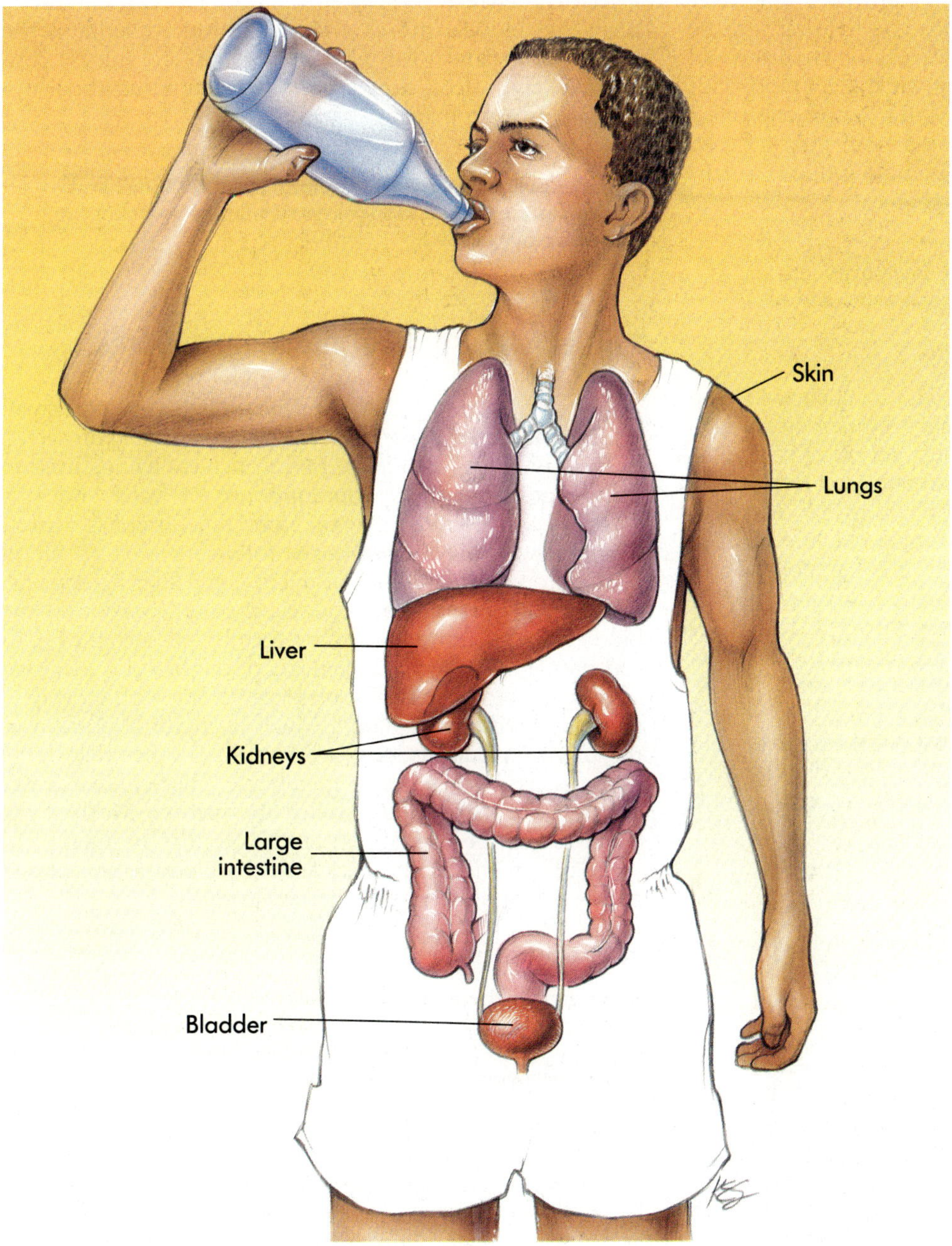

FIGURE 14-1 Organs that carry out excretion.
In addition to the kidneys, the skin, lungs, liver, and large intestine all play a role in the excretion of metabolic waste from the body.

Bile pigments excreted by the liver and the minerals excreted by the large intestine make up only a tiny fraction of the excretory products of the human body. The major products of excretion are water, salts (primarily NaCl—table salt), carbon dioxide, and nitrogen-containing molecules. Carbon dioxide (CO_2) is a gas produced during cellular respiration and is excreted primarily by the lungs (see Chapter 11). The lungs also excrete water in the form of water vapor—which is the reason you can "see" your breath on a cold day as the water vapor condenses to droplets. A small amount of salt and often a large amount of water also leave the body by means of glands in the skin (see Chapter 18).

Nitrogen-containing molecules, or **nitrogenous wastes,** are produced from the breakdown of proteins and nucleic acids. As discussed in Chapter 10, humans need to take in protein as a source of amino acids. These amino acids are used to construct body-building proteins. The liver breaks down any "extra" amino acids you may eat—those not needed for building new body parts or repairing old ones. In the liver, enzymes break down these amino acids by removing their amino groups ($—NH_2$), a process called **deamination.** The molecules that result can then be used to supply the body with energy. The amino group is a "leftover" that cannot be used.

The primary metabolic waste products of the human body are carbon dioxide, water, salts, and nitrogen-containing molecules.

In the liver, amino groups ($—NH_2$) are chemically converted to ammonia (NH_3). Ammonia is quite toxic, or poisonous, to all cells, and the body must get rid of it quickly, not allowing its concentration to increase. A bit of chemical "reshuffling" takes care of this problem, as the ammonia combines with carbon dioxide to form **urea,** the primary excretion product from the deamination of amino acids. In fact, approximately 90% of all nitrogenous wastes are eliminated from the human body in the form of urea. Two other nitrogenous wastes found in small amounts in the urine are uric acid and creatinine. **Uric acid** is formed from the breakdown of nucleic acids (DNA and RNA) found in the cells of the food you eat and from the metabolic turnover of your nucleic acids and ATP. **Creatinine** is derived primarily from a nitrogen-containing molecule called *creatine* found in muscle cells.

In humans, most nitrogenous (nitogen-containing) waste products are excreted as urea.

The organs of excretion

The skin, liver, and large intestine play a minor role in the excretion of metabolic wastes from the body. The primary organs of excretion are the lungs (see Chapter 11) and the **kidneys** (see Figure 14-1). As mentioned previously, the lungs excrete carbon dioxide and water vapor. The kidneys excrete the ions of salts, such as Na^+ (sodium), K^+ (potassium), Cl^- (chloride), Mg^{2+} (magnesium), and Ca^{2+} (calcium). In addition, they excrete the nitrogenous wastes urea, creatinine, and uric acid along with small amounts of other substances that may vary depending on diet and general health. These substances with water form the excretion product called **urine.**

The kidneys produce urine by first filtering out most of the molecules dissolved in the blood, then selectively reabsorbing useful components, and finally secreting a few other waste products into this remaining filtrate, which becomes urine. As the kidneys process the blood in this way, they regulate its chemical composition and water content and, in turn, the chemical and fluid environment of the body. For example, although almost no amino acids are removed from the blood by the kidneys, almost half its urea is removed and then excreted. In addition, the kidneys maintain the concentrations of all ions within narrow boundaries. This strict maintenance of specific levels of ion concentrations keeps the blood's pH at a constant value, maintains the proper ion balances for nerve conduction and muscle contraction, and affects the amount of water reabsorbed into the bloodstream from the filtrate.

The primary organs of excretion are the lungs and the kidneys. The kidney is a regulatory as well as an excretory organ.

An overview of how the kidney works

Your body has two kidneys, each about the size of a small fist, located in the lower back region and partially protected by the lower ribs (see Figure 14-1). Each is a living filtration plant that balances the concentrations of water and salts (ions) in the blood and, at the same time, excretes wastes. Put simply, kidneys work in the following way: As blood flows through the millions of microscopic filtering systems, or **nephrons,** within each kidney, the fluid portion of the blood is forced through capillary membranes into the nephron tubules. The formed elements of the blood (red blood cells, white blood cells, and platelets), along with most proteins, stay within the blood. Much of the fluid portion of the blood, including ions and molecules other than large proteins, passes through the membrane. The fluid that passes through the membrane is called the **filtrate.** Most of the water as well as selected ions and molecules are then reabsorbed from the filtrate back into the blood as the filtrate passes through the rest of the nephron.

The human kidney maintains the proper balance of water in the body, retains substances the body needs, and eliminates metabolic wastes by first filtering most substances out of the blood and then reabsorbing what is needed.

The fluid that passes through the membrane filters of each nephron contains many molecules that are of value to the body, such as glucose, amino acids, and various salts and ions. Humans (and all vertebrates) have evolved a means

FIGURE 14-2 Fresh water fishes produce large quantities of dilute urine.
Because these fish live in fresh water, they take on water by osmosis. To maintain a proper water balance, they drink very little and do not reabsorb water in the nephron.

of selectively reabsorbing these valuable molecules without absorbing the waste molecules that are also dissolved in the filtered waste fluid, or urine. Such selective reabsorption gives the vertebrates great flexibility, since the membranes of different groups of animals have evolved specific transport channels that reabsorb a variety of different molecules. This flexibility is a key factor underlying the ability of different vertebrates to function in many diverse environments. They can reabsorb small molecules that are especially valuable in their particular habitat and not absorb wastes. In all vertebrates the kidney carries out the processes of filtration, reabsorption, and excretion. It can function with modifications in fresh water, in the sea, and on land (Figures 14-2 and 14-3).

FIGURE 14-3 Kangaroo rats produce small quantities of concentrated urine.
In contrast to fresh water fishes, the kangaroo rat can concentrate urine to a high degree by reabsorbing water. As a result, it avoids losing any more moisture than necessary. Because kangaroo rats live in dry or desert habitats, this feature is extremely important to them.

A closer look at how the kidney works

The anatomy of the kidney

The kidneys look like two gigantic reddish-brown kidney beans. Substances enter and leave the kidneys through blood vessels that pierce the kidney near the center of its concave border. If you were to slice a kidney vertically, dividing it into front (ventral) and back (dorsal) portions, you would see an open area called the **renal pelvis** near the concave border. The urine produced by the nephrons within the kidneys is carried by **collecting ducts** that empty into this space. From here, urine exits the kidney via a tube called a **ureter.**

The kidney itself is made up of outer, reddish tissue called the **cortex** and more "inner" reddish-brown tissue called the **medulla.** Within the medulla are "lined" triangles of tissue called the **renal pyramids** (Figure 14-4). (The word *renal* means "kidney.") The cortex and pyramids are made up of approximately one million nephrons, their collecting ducts, and the blood vessels that surround them.

Figure 14-5 shows the plan of the major blood vessels within a kidney. Blood enters each kidney via the right and left **renal arteries.** Branches split off this major artery, travel up the "sides" of the pyramids, and meet at the interface of the cortex and medulla. Smaller branches extend into the cortex, giving rise to the arterioles that enter each individual nephron. The total volume of blood in your body passes through this network of blood vessels every 5 minutes!

The kidney is made up of two main types of tissue: an outer cortex and an inner medulla. These tissues are composed of microscopic filtering units called *nephrons,* their collecting ducts, and blood vessels.

The workhorse of the kidney: The nephron

Each nephron (Figure 14-6) is a tubule that is structured in a special way to accomplish three tasks:

1. **Filtration:** During filtration the blood is passed through membranes that separate blood cells and proteins from much of the water and molecules dissolved in the blood. Together the water and dissolved substances are referred to as the **filtrate.**
2. **Selective reabsorption:** During reabsorption, desirable ions and metabolites and most of the water from blood plasma are recaptured from the filtrate, leaving nitrogenous wastes, excess water, and excess salts behind for later elimination. The filtrate is now called *urine.*
3. **Tubular secretion:** During secretion the kidney *adds* materials such as potassium and hydrogen ions, ammonia, and potentially harmful drugs to the filtrate from the blood. These secretions rid the body of certain materials and control the blood pH.

Filtration

At the front end of each nephron tube is a filtration apparatus called a **Bowman's capsule.** The capsule is shaped

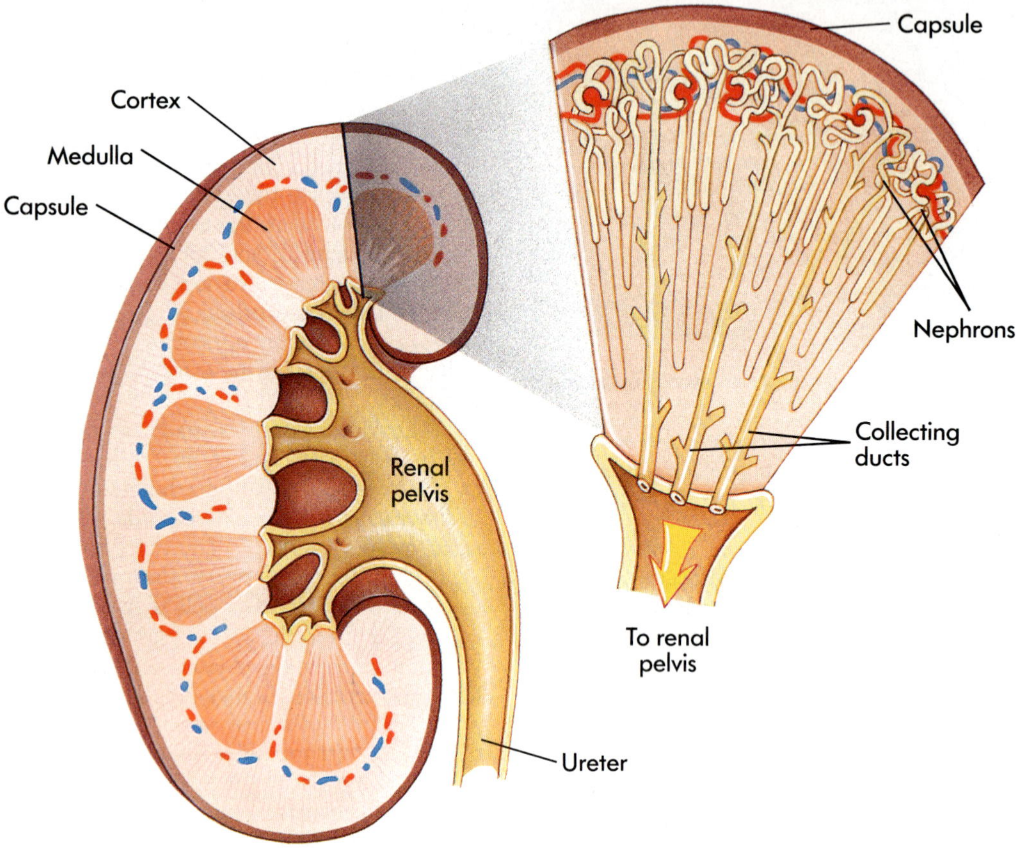

FIGURE 14-4 Structure of the kidney.
The enlarged portion of the diagram shows a detailed section of a renal pyramid and the cortex above. The renal pyramids are composed of nephrons, their collecting ducts, and surrounding blood vessels.

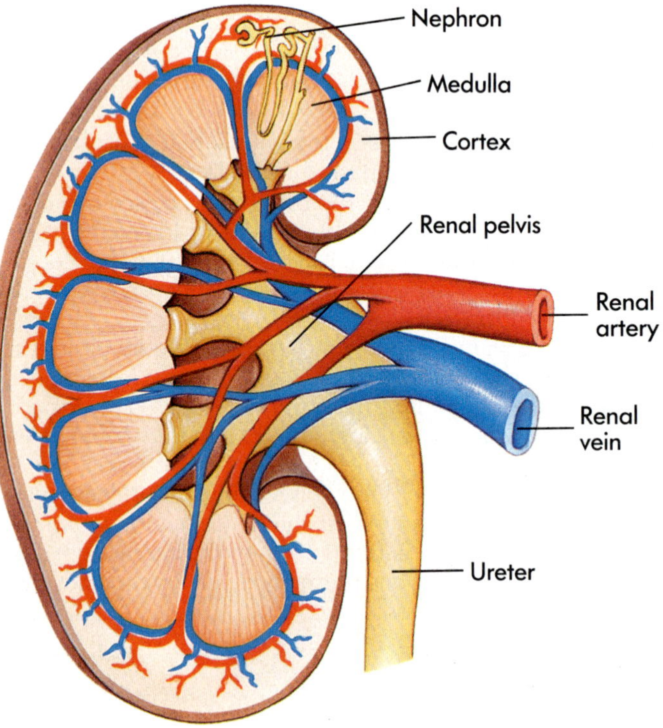

FIGURE 14-5 Plan of the major blood vessels in the kidney.
Blood enters the renal artery, which splits off into branches that travel into the cortex. Blood exits the kidney through the renal vein.

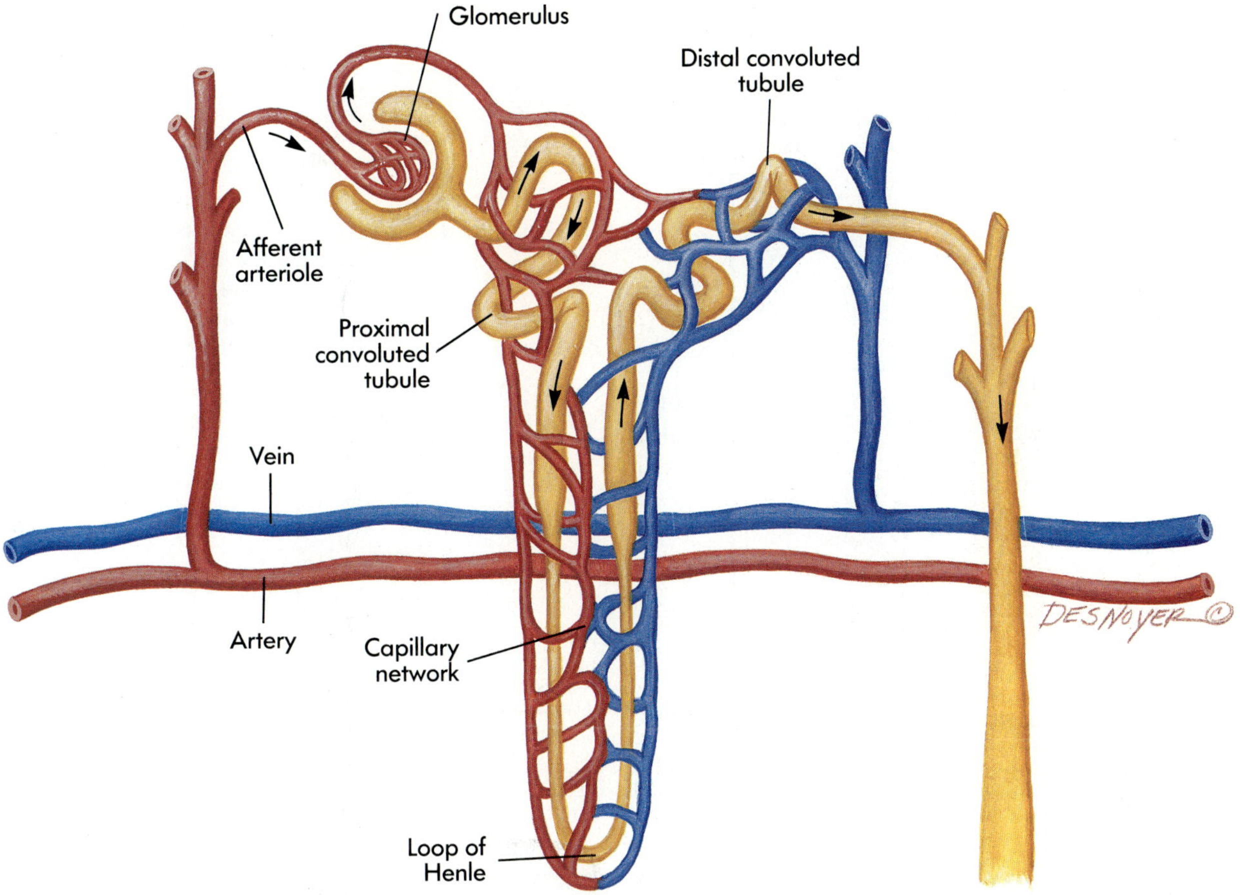

FIGURE 14-6 Structure of a nephron.
Part of the nephron is located in the cortex, and part is located in the medulla. Nephrons filter the blood, reabsorb substances to the blood from the blood filtrate, and secrete certain ions and potentially harmful substances into the filtrate. These processes take place along specific portions of the nephron tubule.

much like a caved-in tennis ball and surrounds a tuft of capillaries called the **glomerulus.** The glomerular capillaries branch off an entering arteriole termed the **afferent arteriole** (named after the Latin verb *affero* which means "going toward"). The walls of the glomerular capillaries, along with specialized cells of the capsule, act as filtration devices. The pressure of the incoming blood forces the fluid within the blood through the capillary walls and through spaces between capsular cells that surround the capillaries. Capillary walls are made up of a single layer of cells having differentially permeable membranes. These walls do not allow large molecules such as proteins and the formed elements of the blood to pass through. Water and smaller molecules such as glucose, ions, and nitrogenous wastes pass through easily.

This filtrate passes into the tubule of the nephron at Bowman's capsule. From there, it passes into a coiled portion of the nephron called the **proximal convoluted tubule.** This name describes the coiled (convoluted) tubule as that portion closest to (proximal to) Bowman's capsule. Both these structures lie in the cortex of kidney. As the filtrate passes through the proximal tubule, reabsorption begins.

Selective reabsorption

Reabsorption is carried out by the epithelial cells throughout the length of the tubule. During this extremely discriminating process, specific amounts of certain substances are reabsorbed depending on the body's needs at the time. Substances are reabsorbed by both active and passive transport mechanisms (see Chapter 4). In the proximal convoluted tubule, glucose and small proteins are put back into the blood. These substances move out of the tubule and into the blood within surrounding capillaries (Figure 14-7).

The filtrate moves from the proximal convoluted tubule to the descending arm of the loop of Henle. The loop of Henle is the part of the tubule that dips into the medulla of the kidney. The part that dips "down" from the cortex (where Bowman's capsule and the proximal convoluted tubule are) into the medulla is called the **descending arm of the loop of Henle.** The tubule then travels in the opposite direction, moving back up into the cortex. This portion of the nephron is called the **ascending arm of the loop of Henle.** The walls of the descending arm are impermeable to salt and urea. Put simply, the cells making up the walls

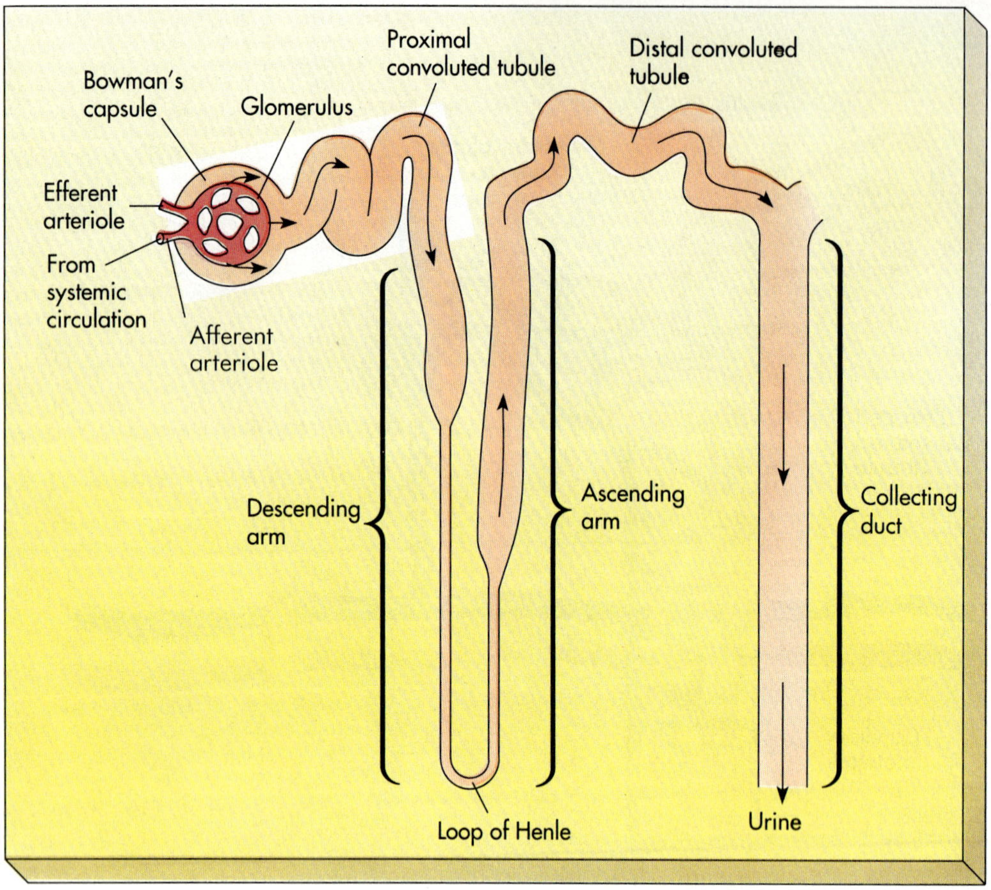

FIGURE 14-7 Path of the filtrate through the nephron.

In this figure and the five that follow, the filtrate will be traced through the nephron. In the first stage of filtration, the filtrate passes into the nephron at Bowman's capsule. The filtrate then enters the proximal convoluted tubule, where glucose and small proteins are reabsorbed by the blood.

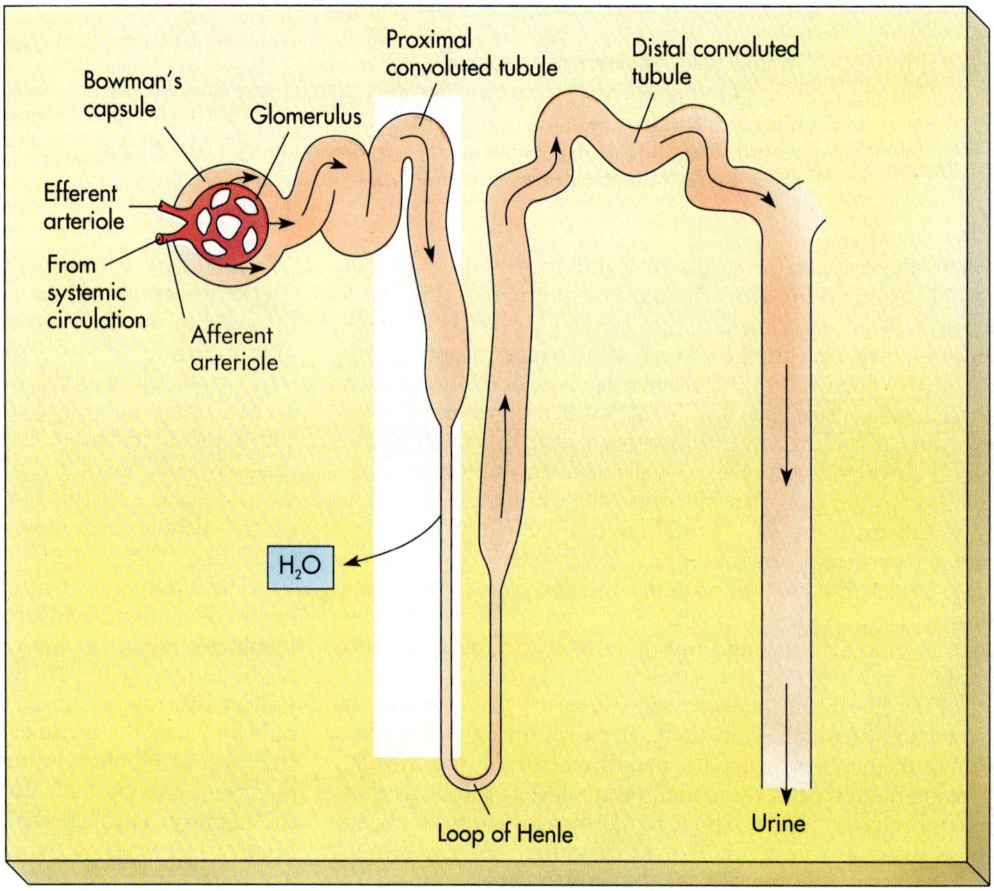

FIGURE 14-8 The descending loop of Henle.

In the descending loop of Henle, water moves out of the tubule into the surrounding tissue, leaving a more concentrated filtrate.

of the descending arm do not permit salt and urea to pass out of the tubule. However, the walls of the descending arm are freely permeable to water. Water moves out of the tubule by osmosis, passing into the surrounding tissue fluid and then into the blood within surrounding capillaries. As water passes out of the descending arm, it leaves behind a more concentrated filtrate (Figure 14-8).

At the turn of the loop, the walls of the tubule become permeable to salt but much less permeable to water (Figure 14-9). Therefore as the filtrate (which became quite concentrated as it moved down the descending arm) passes up the ascending arm, salt passes out into the surrounding tissue fluid by diffusion. This movement of salt produces a high concentration of salt in the tissue fluid surrounding the bottom of the loop.

Higher in the ascending arm, the walls of the tubule contain active transport channels that pump even more salt out of the filtrate within the kidney tubule (Figure 14-10). This active removal of salt from the ascending arm causes water to diffuse outward from the filtrate just above the ascending arm at the **distal convoluted tubule.** The walls of this portion of the kidney tubule are permeable to water, which is not the case in the ascending arm. Left behind in the filtrate are some water and the urea that initially passed through the glomerulus as nitrogenous waste; eventually the urea concentration becomes very high in the tubule.

Finally, the filtrate empties into a *collecting duct* that passes back into the medulla (Figure 14-11). The collecting ducts of all the nephrons bring the urine to the renal pelvis, an open area in which the urine collects before it flows out of the kidney. Unlike the other portions of the kidney tubule, the lower portion of the collecting duct is permeable to urea. During this final passage, some of the concentrated urea in the filtrate diffuses out into the surrounding tissue fluid, which has a lower urea concentration than the filtrate. A high urea concentration in the tissue fluid surounding the nearby loop of Henle results. This high concentration of urea produces the osmotic gradient that caused water to move out of the filtrate as it passed down the descending arm. Water also passes out of the filtrate as it moves down the collecting duct. This movement occurs as a result of the osmotic gradient created by both the movement of urea out

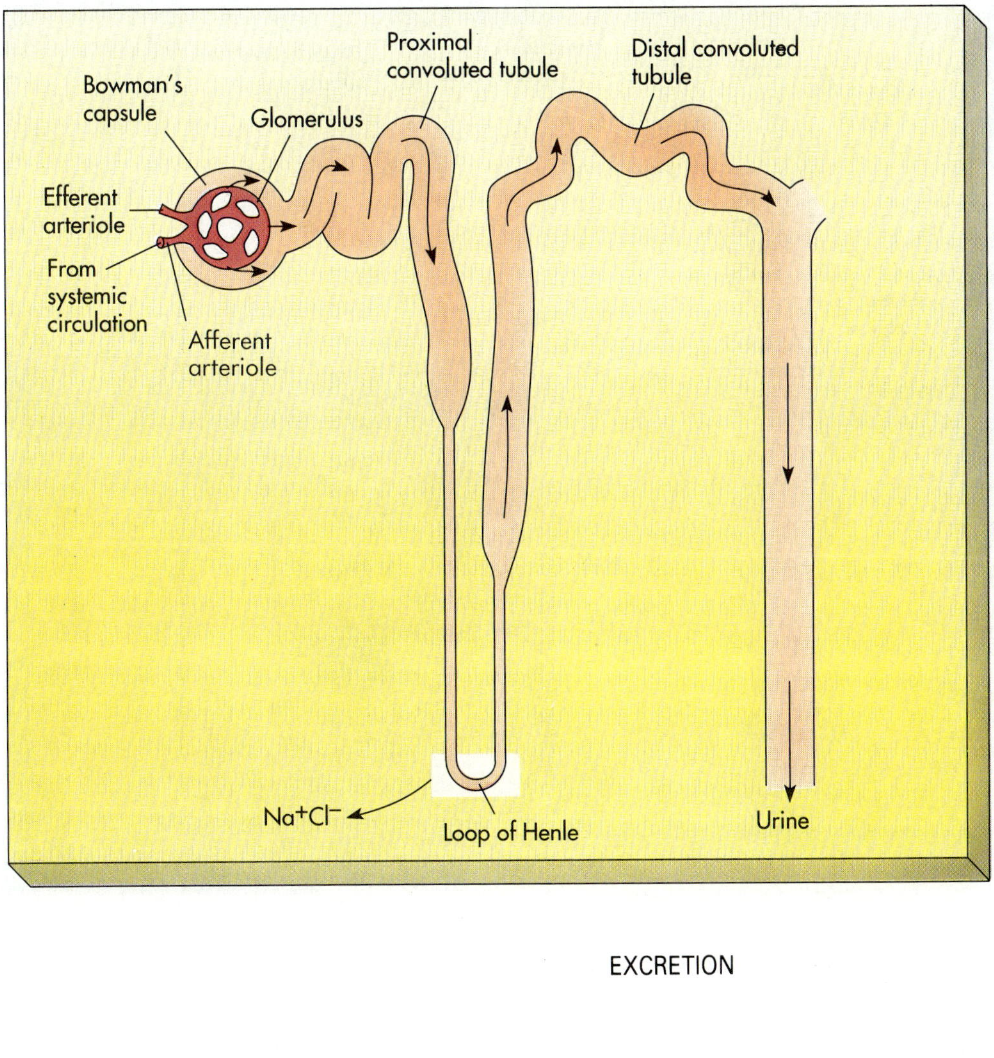

FIGURE 14-9 The turn of the loop of Henle.
At the turn of the loop of Henle, the walls of the tubule become more permeable to salt (Na^+ and Cl^- ions) but less permeable to water. As the concentrated filtrate moves through the turn of the loop, salt diffuses out of the tubule into the surrounding tissues. The area surrounding the bottom of the loop, then, acquires a high concentration of salt.

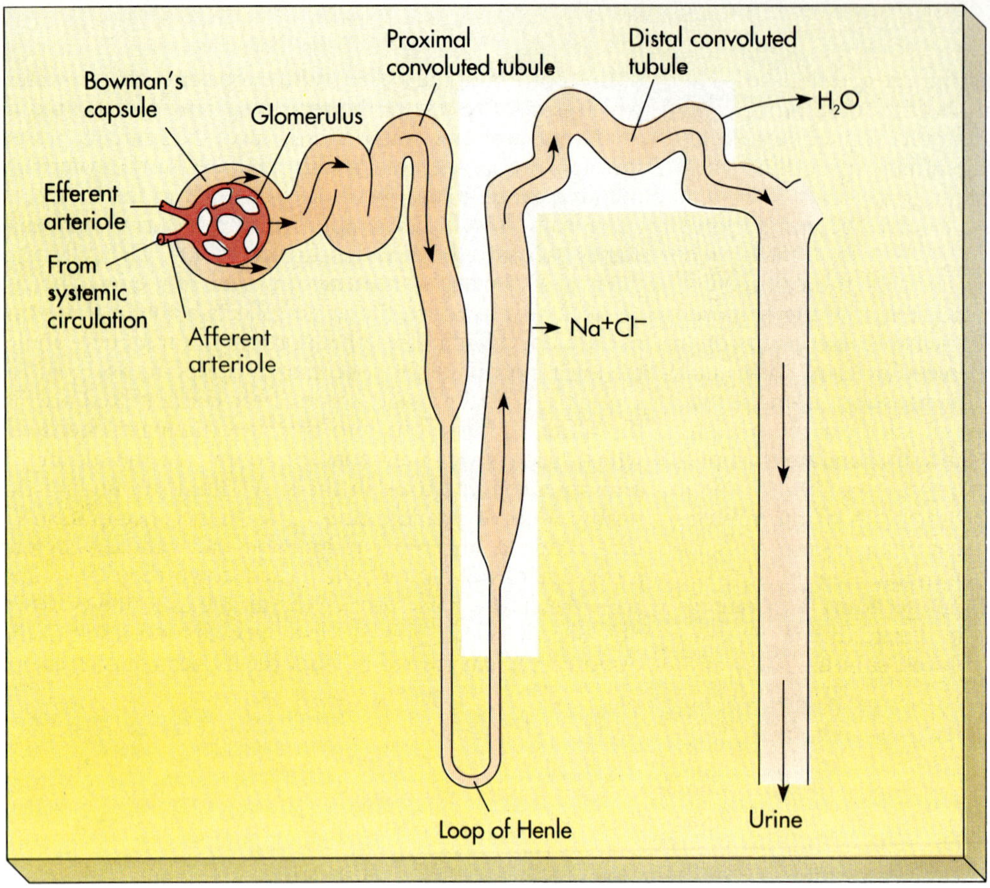

FIGURE 14-10 The ascending loop of Henle.

In the ascending loop, channels in the cell membranes of the tubule cells actively pump salt (Na^+ and Cl^-) out of the kidney tubule. At the distal convoluted tubule, water diffuses out of the filtrate. The filtrate now contains only some water and urea.

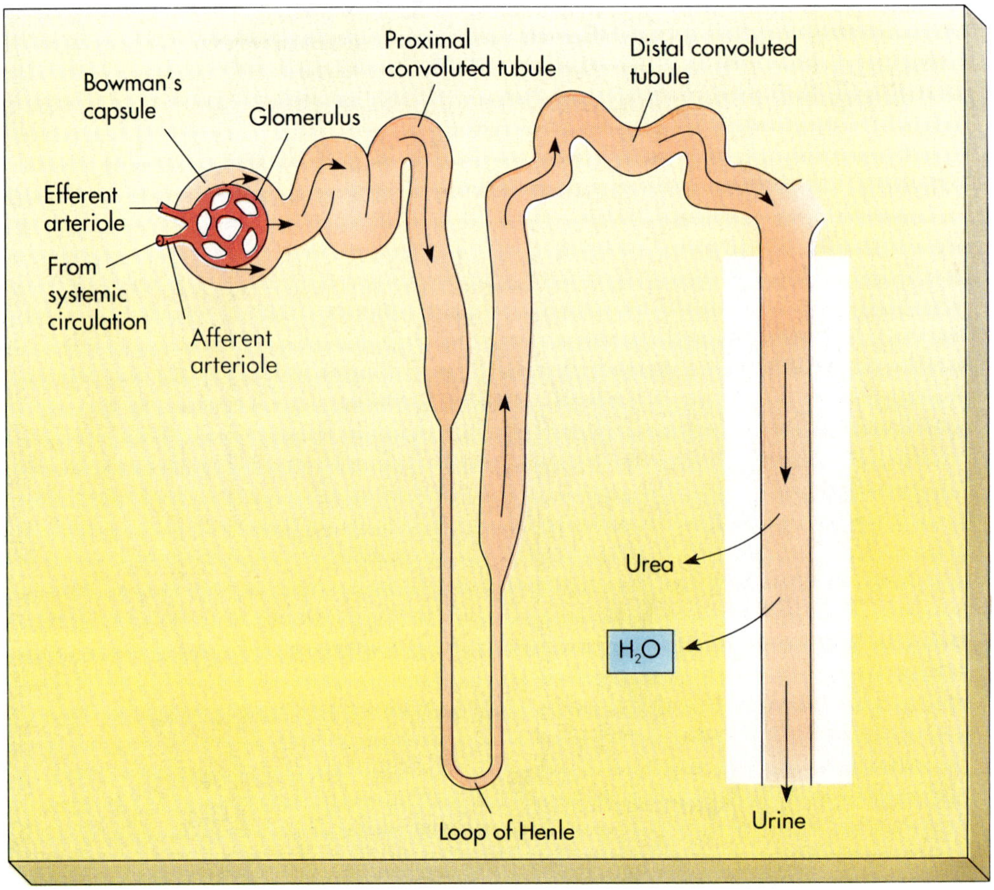

FIGURE 14-11 The collecting duct.

In the collecting duct, urea and water pass out of the filtrate. The filtrate, or urine, moves out of the collecting duct to the renal pelvis, in which urine collects before it flows from the kidney.

of the collecting duct and the movement of salt out of the ascending arm. The water is then collected by blood vessels in the kidney, which carry it into the systemic circulation.

In summary, the human kidney achieves a high degree of water reabsorption by using the salts and urea in the glomerular filtrate to increase the osmotic concentration of the tissue fluid surrounding the nephron. This high concentration of salts and urea creates an osmotic gradient that pulls water from the filtrate out into the surrounding tissue fluid. It is collected there by blood vessels impermeable to the high urea concentration but permeable to water.

Tubular secretion

A major function of the kidney is the elimination of a variety of potentially harmful substances that you may eat, drink, or inhale or that may be produced during metabolism. The human kidney has evolved the ability to detoxify our blood through the process of tubular secretion. During this process the cells making up the walls of the distal convoluted tubule take substances from the blood within surrounding capillaries and from within the surrounding tissue fluid and put them into the filtrate (urine) within the kidney tubule (Figure 14-12). Ammonia (NH_3) from the deamination of amino acids that has not been converted to urea is removed from the blood by tubular secretion. Certain prescription drugs such as penicillin are also removed by this process. In addition, the body rids itself of harmful drugs such as marijuana, cocaine, heroin, and morphine by tubular secretion. These drugs are processed by the liver and their breakdown products are secreted into the filtrate within the kidney tubules. Specific tests for the breakdown products of these drugs can be performed on urine samples to determine whether a person has particular drugs in the body. Testing the urine in this way, or **drug testing,** has become a controversial topic among groups of people such as athletes and employees involved in certain types of work such as airline pilots.

Tubular secretion can also be thought of as a fine-tuning mechanism. By removing specific amounts of hydrogen ions (H^+) from the blood and secreting them into the filtrate, the kidney can keep the pH of the blood at a constant level (7.35 to 7.45). (See Chapter 2 for an explanation of pH.) Likewise, the potassium ion (K^+) concentration of the blood is fine tuned by tubular secretion. The proper concentration of potassium ions is important to the proper functioning of muscles, including the heart.

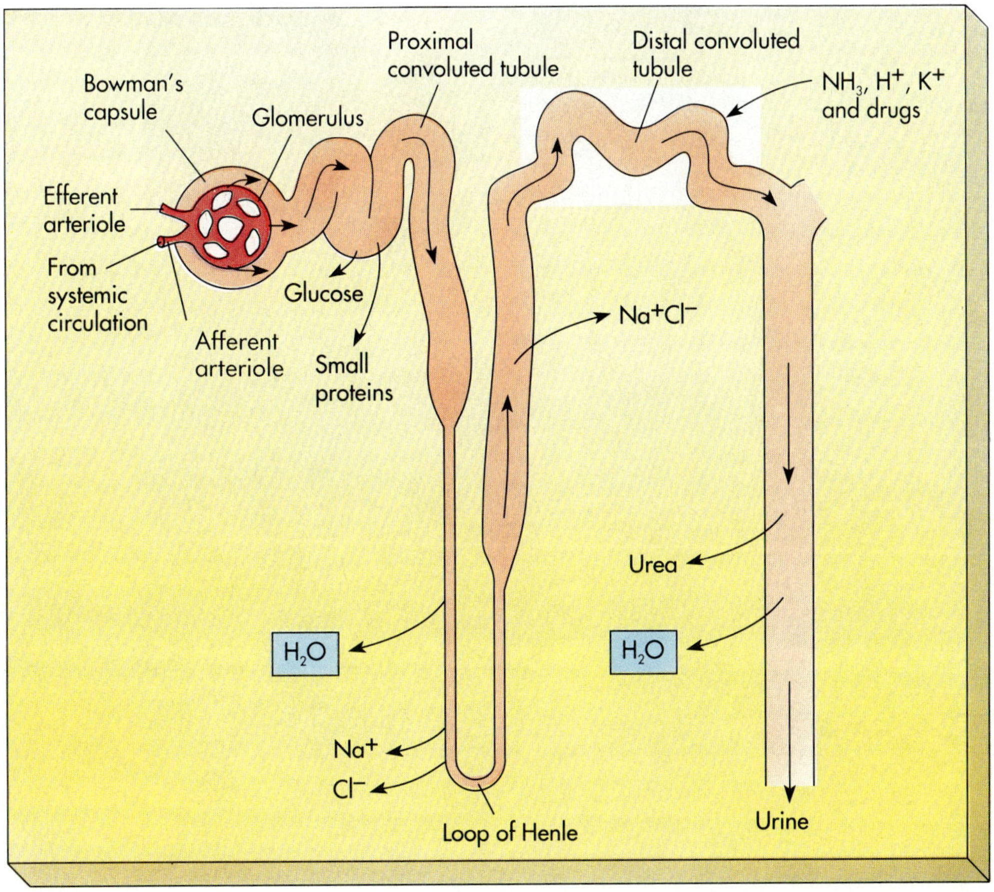

FIGURE 14-12 The distal convoluted tubule.
Another important function of the kidney is tubular secretion. The distal convoluted tubule secretes substances such as ammonia, drugs, and ions into the filtrate. This illustration also summarizes the process of selective reabsorption.

BIOLOGY TECHNOLOGY & society

Testing urine for drugs

In the late 1960s, the National Institute on Drug Abuse in Washington, DC, began to develop ways of monitoring heroin use among people at treatment centers to learn which therapies worked and which did not. They settled on analysis of urine, finding that telltale traces of heroin use could be detected in urine for days after drug use. Recall that substances such as drugs are removed from the blood during tubular secretion. These substances become part of the filtrate that eventually leaves the body as urine.

After 20 years, urine testing has become quite a sophisticated industry. In the last few years in particular, there has been a revolution in the industry because of the improved quality of test procedures.

The major improvements have come with the introduction of tests that use recombinant DNA techniques to amplify trace chemicals in urine. The Enzyme-Multiplied Immunoassay Technique (EMIT) has become the standard approach for screening large numbers of urine samples. EMIT can be used in a robot tester that employs a light sensor to read urine samples and prints out a value for each of five or six drugs present—it can churn out 18,000 results an hour. Traces of marijuana, cocaine, opiates, and amphetamines can be detected for days and even weeks after use using EMIT. The test is more than 98% accurate, with error biased toward nondetection.

FIGURE 14-A Cocaine paraphernalia.

Because EMIT responds to a broad range of opiate and amphetamine compounds, it sometimes kicks out a "positive" for harmless prescription drugs like ibuprofen and foods like poppy seeds. For this reason, samples that come up positive on EMIT are then retested by more cumbersome but 100% accurate procedures (gas chromatography-mass spectrometry). For example, eating poppy seeds may trigger a positive signal on EMIT, but the confirmation test tells that a key heroin breakdown product (*6-O*-acetylmorphine) is not present.

Enlistees in the armed services are routinely tested for evidence of drug use. The biggest use of urine testing, however, is in courts and prisons. There is strong evidence that heavy drug use is linked with criminal activity. In drug monitoring surveys carried out by the Department of Justice in ten major cities in 1988, for example, half of all arrestees, and in most cities three-quarters of them, tested positive for drugs.

Drug use by the drivers of public transportation (bus, train, subway, and airplane) has sometimes resulted in tragic accidents or near-accidents, and there has been considerable public pressure for mandatory urine testing of such key employees. Congress has not yet decided how broadly to impose this surveillance or whether the loss in privacy would be worth the gain in public safety. Thousands of private companies already test new job applicants—in 1990, over 40% of 1000 companies surveyed were using some sort of drug screening. In today's world, testing urine for evidence of drug use is becoming part of everyday life (Figure 14-A).

A summary of urine formation

The roughly 2 million nephrons that form the bulk of the two human kidneys receive a flow of approximately 2000 liters of blood per day—enough to fill about 20 bathtubs! The nephrons function by first filtering the blood—removing most of its water and all but its largest molecules and its cells. The nephron then selectively reabsorbs substances back into the blood. This process is "driven" by two factors: (1) the development of a high osmostic gradient surrounding the loop of Henle and (2) the varying permeability of the membranes of the cells lining the kidney tubule.

Because of the varying permeability of kidney tubule cells, substances move from the filtrate back into the blood by diffusion and active transport mechanisms at specific places along the length of the tubule. Each substance will only be reabsorbed to a particular threshold level, however, with the rest remaining in the filtrate. For example, glucose is reabsorbed at the proximal convoluted tubule to a threshold level of 150 milligrams per 100 milliliters of blood. Any glucose above this threshold will be excreted in the urine. The amount of glucose in the blood of a healthy person does not normally exceed this limit. However, persons with diabetes mellitus (sugar diabetes) excrete glucose because they fail to produce a hormone called *insulin* that promotes glucose uptake by the cells. Urea, on the other hand, has a very low threshold that is reached quickly. Therefore most urea stays in the filtrate (except the urea that diffuses from the collecting duct, adding to the osmotic gradient surrounding the loop of Henle).

The high osmotic gradient surrounding the loop of Henle causes most of the water filtered from the blood to be reabsorbed, or conserved, in the descending arm of the loop. Water conservation is, in fact, the job of this loop. Vertebrates that do not need to conserve water have no loop. For example, a freshwater fish drinks little and produces large amounts of urine. Because the body fluid of a freshwater fish is hyperosmotic (contains more dissolved substances and therefore has a lower concentration of water) to the water in which it lives, water is not reabsorbed in its nephrons. The excess water that enters its body passes instead through the relatively straight nephron tubes to be excreted as urine (Figure 14-13). In general, the longer the loop of Henle, the more water can be reabsorbed. Animals such as desert rodents that have highly concentrated urine have exceptionally long loops of Henle.

The nephron also controls blood pH, fine tunes the concentrations of certain ions and molecules, and removes potentially harmful drugs from the blood by the process of tubular secretion. In these ways the kidney contributes to the chemical and water balance of the body, thus functioning as a major organ of internal equilibrium, or homeostasis. The resulting fluid is urine, a waste that is excreted from the body.

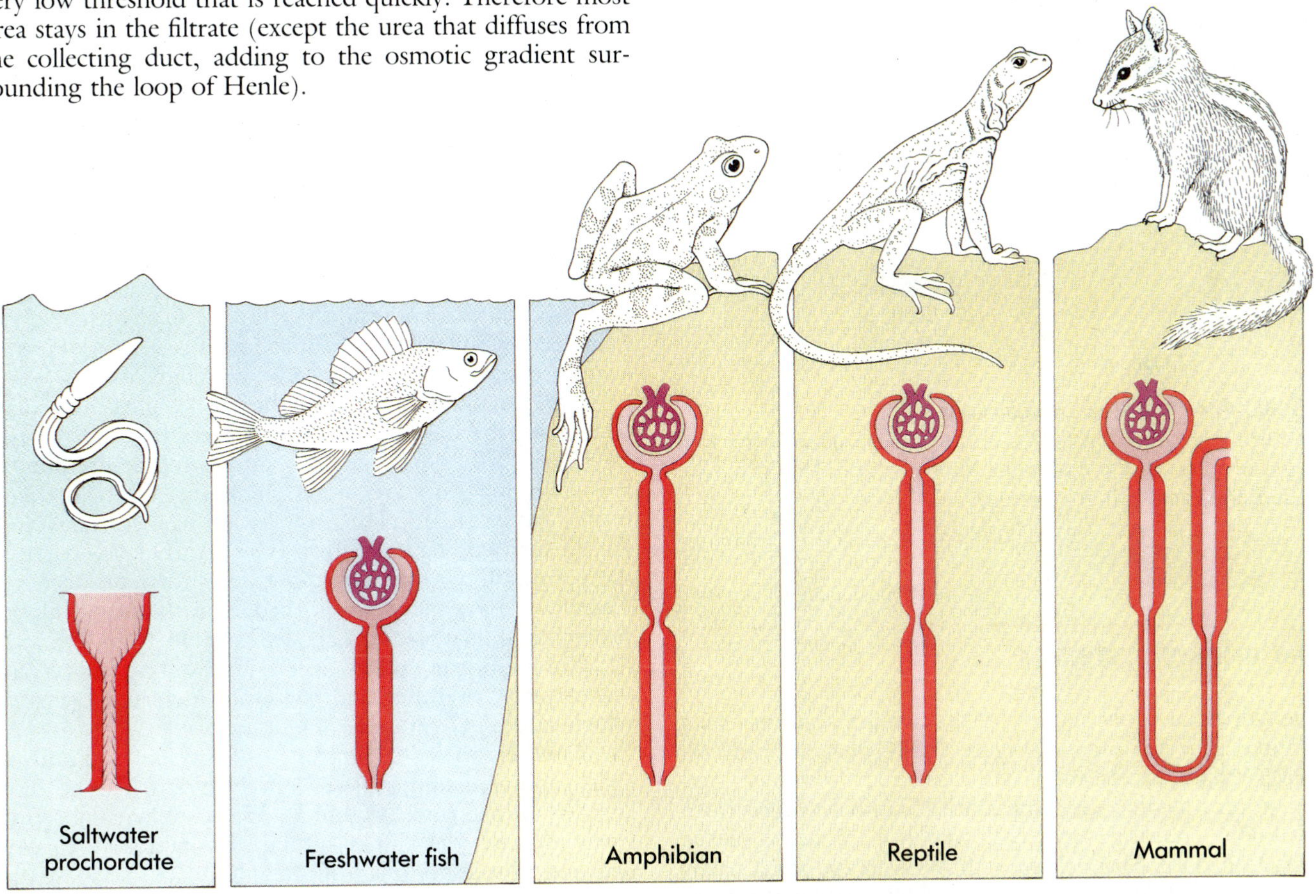

FIGURE 14-13 The function of the loop of Henle is the conservation of water. Animals that do not need to conserve water such as fresh water fishes do not have loops of Henle. Only mammals have a loop in the nephron tubule that allows the reabsorption of water.

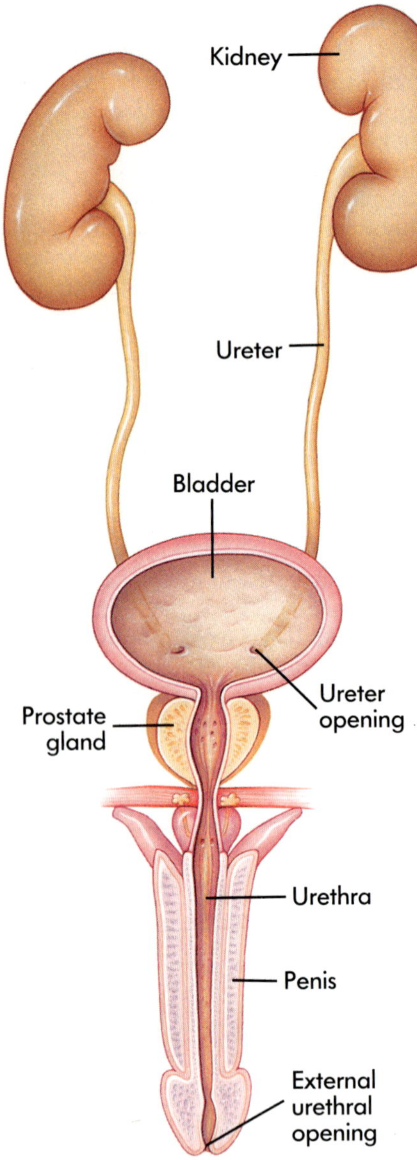

FIGURE 14-14 The human urinary system.
The urinary system in males and females includes the kidneys, ureters, urinary bladder, and urethra. This diagram shows the male urinary system.

The urinary system

The kidneys are only one part of the **urinary system**—a set of interconnected organs (Figure 14-14) that not only removes wastes, excess water, and excess ions from the blood but stores this fluid, or urine, until it can be expelled from the body. The urine exits each nephron by means of the collecting duct. From there, it flows into the renal pelvis (see Figure 14-4). This area narrows into a tube called the ureter that leaves the kidney on its concave border, near where the renal artery and vein enter and exit. By means of muscular contractions **(peristalsis),** these muscular tubes—one from each kidney—bring the urine to a storage bag called the **urinary bladder.**

The urinary bladder is a hollow muscular organ that, when empty, looks much like a deflated balloon. As the bladder fills with urine, it assumes a pear shape. On the average, the bladder can hold 700 to 800 milliliters of urine, or almost a quart. However, when less than half this amount is in the bladder, special nerve endings in the walls of the bladder, called *stretch receptors,* send a message to the brain that results in the desire to urinate.

The urinary bladder empties into a tube called the **urethra.** This tube leads from the underside of the bladder to the outside. The urethra in men plays a role in the reproductive system, carrying semen to the outside of the body during ejaculation. However, the urinary and reproductive systems have no connection in women. In addition, the urethra in women is much shorter than the urethra in men. For this reason, women contract infections more easily than men.

The kidney and homeostasis

The kidney is an excellent example of an organ whose principal function is homeostasis, the maintenance of constant physiological conditions within the body. It is concerned with both water balance and ion balance. To maintain homeostasis, however, the urinary system depends in part on the endocrine (hormone) system.

A **hormone** is a "chemical messenger" sent by a gland to other cells of the body. Hormones are produced in one part of the body and have an effect in another (see Chapter 19). The hormone that regulates the rate at which water is lost or retained by the body is **antidiuretic hormone,** or **ADH.** It is secreted by the pituitary gland at the base of the brain. Its effect on the excretion of water is easy to remember if you know that a **diuretic** is a substance that helps remove water from the body. Because ADH is an *anti*diuretic, it works to conserve, or retain, water.

ADH works by controlling the permeability of the distal convoluted tubules and collecting ducts to water (Figure 14-15, *A*). When ADH levels increase, the permeability of the collecting ducts to water increases. Water therefore moves out of the ducts by osmosis and back into the blood within surrounding capillaries. When ADH levels decrease, the permeability of the collecting ducts to water decreases. Therefore less water is reabsorbed from the urine (filtrate) and remains in the ducts and is excreted from the body. Alcohol inhibits the release of ADH, resulting in decreased reabsorption of water and therefore increased urination, which causes dehydration.

Another hormone regulates the level of sodium ions (Na^+) and potassium ions (K^+) in the blood. This hormone is called **aldosterone** (Figure 14-15, *B*). When aldosterone levels increase, the kidney tubule cells increase their reabsorption of sodium ions from the filtrate and decrease their reabsorption of potassium ions. Put simply, aldosterone promotes the retention of sodium and the excretion of potassium. In addition, because the concentration of sodium in the blood affects the reabsorption of water, the increase in sodium in the blood causes water to move by osmosis from the filtrate into the blood.

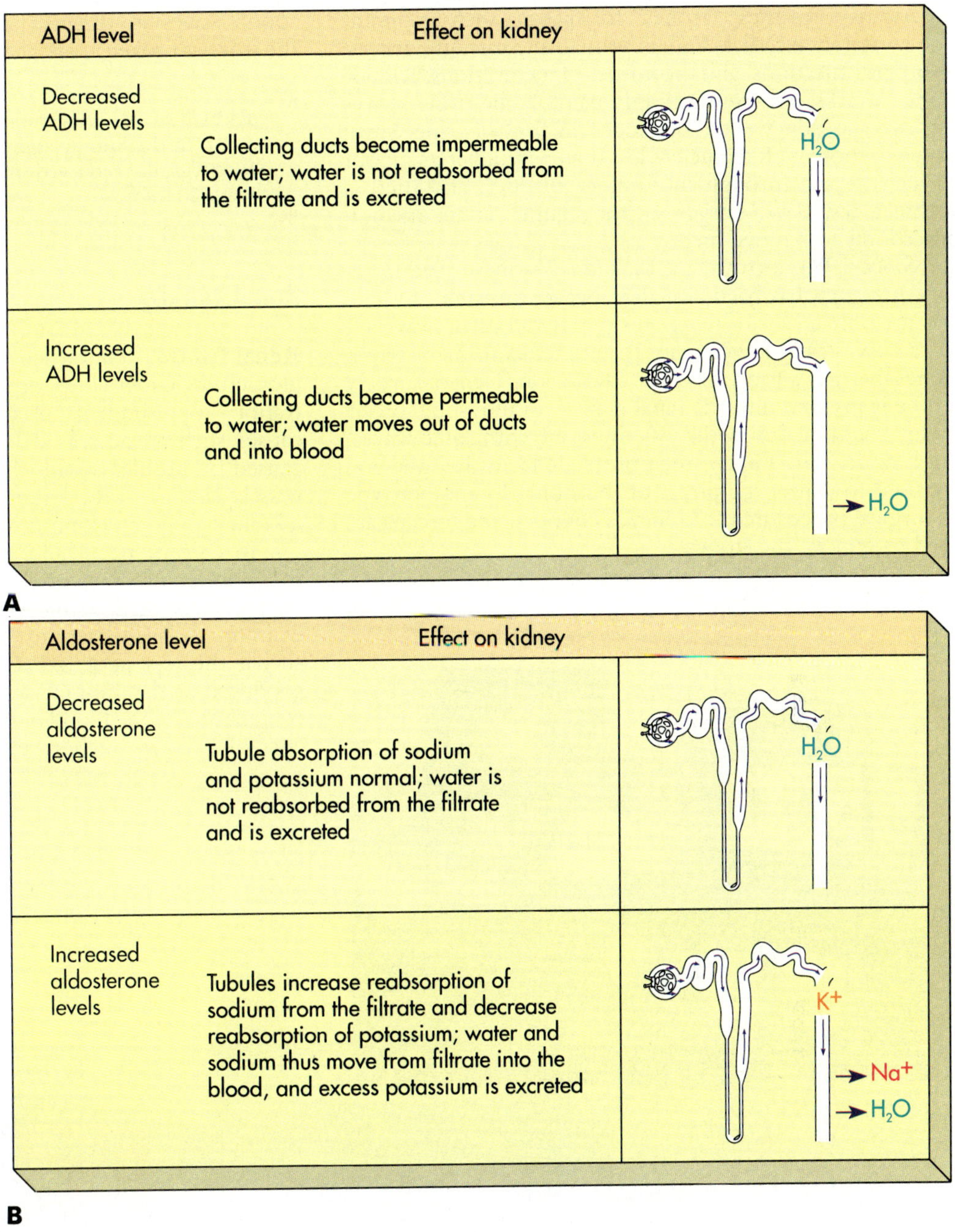

FIGURE 14-15 How hormones regulate kidney function.
A ADH is a hormone that helps the body conserve water by acting on the permeability of the distal convoluted tubules and collecting ducts to water. When ADH levels decrease, more water is excreted from the body. When ADH levels increase, more water is conserved.
B Aldosterone regulates the levels of sodium and potassium in the body by acting on the permeability of the distal convoluted and collecting tubule to these ions and water. When Na^+ levels are low and K^+ levels are high, aldosterone acts on the nephron tubules to correct the imbalance. Aldosterone causes Na^+ and water to be reabsorbed in the blood and excess K^+ to be excreted.

The principal function of the kidney is homeostasis—the maintenance of a constant internal environment. Two hormones help the kidney maintain homeostasis. The hormone ADH regulates the amount of water reabsorbed at the collecting duct. The hormone aldosterone promotes the conservation of sodium and therefore water and promotes the excretion of potassium.

Problems with kidney function

Kidney stones

During the formation of urine, salts and other wastes to be excreted from the body are normally dissolved in the water of the filtrate and pass with it out of the kidney as urine.

Sometimes, however, certain salts do not stay dissolved—most notably calcium salts or uric acid—but instead form crystals called **kidney stones.** Each year more than 300,000 Americans form these "stones" for reasons such as diminished water intake, diets high in protein and calcium, genetic disorders, infections, and the misuse of medications. Kidney stone "attacks" are excruciatingly painful; the stone moves through the kidney and then blocks the flow of urine. Small stones—those with diameters less than 5 millimeters (about ⅕ inch)—pass through the ureters, bladder, and urethra within a few days. Larger stones require certain medical procedures to remove them.

Since 1984, a treatment for kidney stones called *shock-wave lithotripsy* has been available. Before this time, surgery was the only means of relief. Surgery is still used, however, to remove stones larger than 2 centimeters in diameter or stones that are causing infection. During one form of surgery a tube is inserted into the renal pelvis. An ultrasonic probe within the tube first bombards the stone with sonic waves, crushing it. Tiny forceps then protrude from the tube and are used to collect the pieces of the stone. In another type of surgical procedure the kidney is cut open and forceps are used to remove the stones.

Shock-wave lithotripsy involves a procedure quite different from these two. The patient is lowered up to the neck into a tank of water. Guided by x-ray monitors, intense sound waves are directed at the stone, shattering it. Once the stone is broken apart, the pieces are excreted in the urine.

Kidney stones are crystals of certain salts that develop in the kidney and block urine flow. They may be surgically removed or broken apart with shock-wave therapy.

Renal failure

Renal failure occurs when the filtration of the blood at the glomerulus either slows or stops. In acute renal failure, filtration stops suddenly. Acute renal failure can have many causes, such as a decreased flow of blood through the kidneys caused by problems with the heart or blockage of a blood vessel, damage to the kidney by disease, or the presence of a kidney stone blocking urine flow.

In chronic renal failure, the filtration of the blood at the glomerulus slows gradually. Unfortunately, this condition is usually irreversible, since it is most commonly caused by injury to the glomerulus. These injuries have many causes, such as the deposit of toxins, bacterial cell walls, or molecules produced by the immune system within the glomerulus; the coagulation of the blood within the glomerulus; or the presence of a disease such as diabetes.

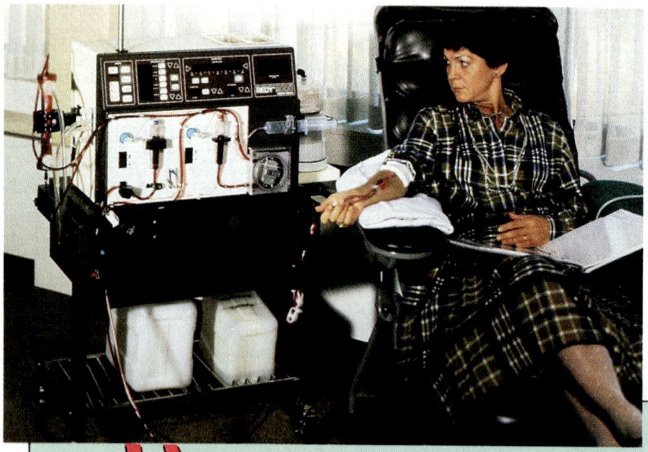

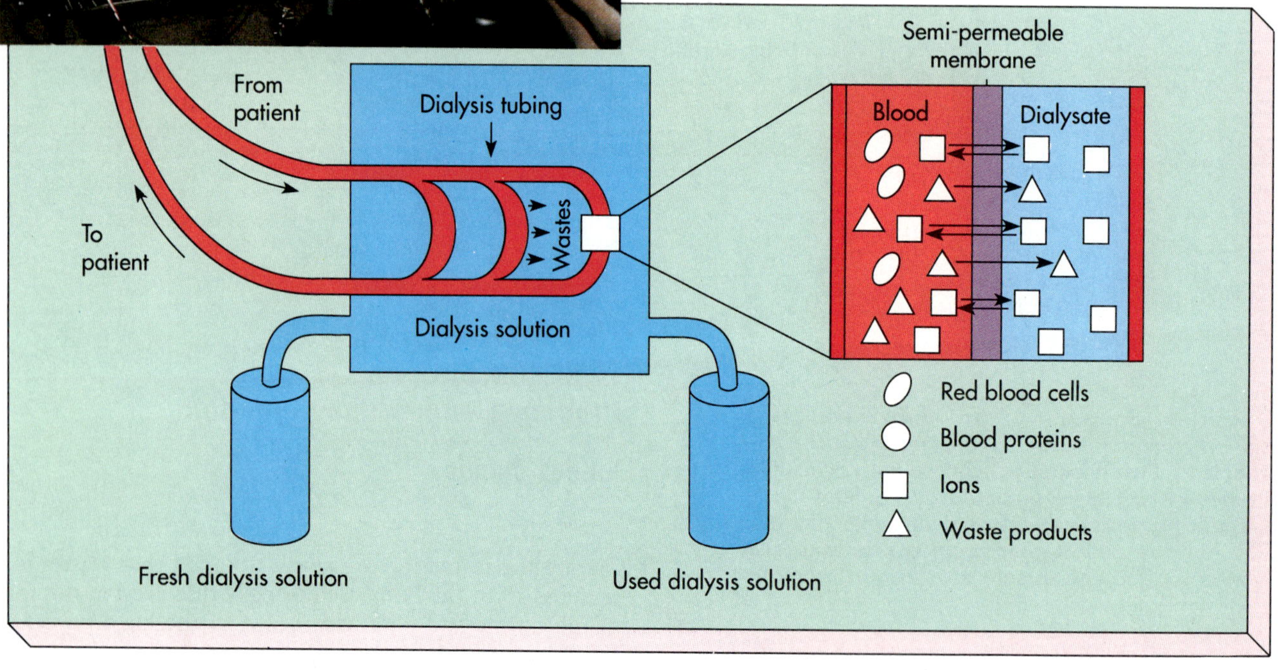

Treatments for renal failure

If the kidneys become unable to excrete nitrogenous wastes, regulate the pH of the blood, and regulate the ion concentration of the blood because of renal failure, the individual will die unless the blood is filtered. This job is accomplished by a machine called an **artificial kidney.** The process of filtering blood in this way is called **dialysis.** During dialysis, small amounts of the patient's blood are pumped through tubes to one side of a selectively permeable membrane (Figure 14-16). This membrane is a derivative of cellulose, very similar to material that makes up the cell walls of plants. On the other side of the membrane is a fluid called the *dialysate.* The dialysate contains the same concentration of ions that is normally found in the bloodstream. Because small molecules can pass across the membrane, any "extra" ions in the patient's blood move by diffusion into the dialysate until their concentrations on both sides of the membrane are equal. In addition, the dialysate contains no wastes, so the wastes in the patient's blood also diffuse into the dialysate. Unfortunately, dialysis is a slow process, usually taking from 6 to 8 hours. Recently, however, researchers have created a new synthetic membrane made of a plastic co-polymer said to be 25% more efficient than cellulose membranes. Other scientists are also currently developing tailor-made "artificial cells" to replace or help kidney cells. The artificial cells currently being developed are designed to change the toxic urea and ammonia that build up in patients with renal failure into useful amino acids.

In the United States, approximately 7000 renal failure patients receive kidney transplants per year. During a transplant operation a patient's nonfunctioning kidney is removed and a donor kidney is implanted. Although two thirds of these transplant patients suffer serious problems of organ rejection, many of their problems are alleviated with the use of immunosuppressive drugs. Recently, treating patients with monoclonal antibody therapy has been shown to reverse rejection in a significant number of patients.

Monoclonal antibodies are special proteins produced in the laboratory that are all descendants of a single cell. These proteins are therefore all alike. In addition, they are antibodies and will attack a specific antigen (see Chapter 13). The monoclonal antibodies developed for use with kidney transplant patients target the specific immune cells involved in the rejection of the donor kidney. The use of monoclonal antibodies and new drug therapies hold much promise for kidney transplant patients.

Renal failure is a reduction in the filtration rate of blood in the glomerulus. This condition has many possible causes and may be treated by dialysis or with a kidney transplant.

FIGURE 14-16 Dialysis.
The dialysis procedure uses a semipermeable membrane to imitate the nephron of the kidney. Blood from the patient on dialysis is separated from the dialysate by this membrane. Wastes and excess ions diffuse across the membrane into the dialysate. Although dialysis is a life-saving measure, it is no substitute for the selective reabsorption of the human kidney.

Summary

1. Excretion is a process whereby metabolic wastes, excess water, and excess salts are removed from the blood and passed out of the body. Elimination is a process that takes place as digestive wastes leave the body during defecation.
2. The excretory products of the body are bile pigments, nitrogen-containing molecules (nitrogenous wastes), carbon dioxide, water, and salts. The bile pigments are excreted by the liver and passed out of the body by means of the digestive system. Most carbon dioxide is excreted by the lungs. The kidneys excrete most of the nitrogenous wastes along with excess water and salts.
3. The kidneys excrete wastes as urine. Urine is formed within microscopic tubular units of the kidney called *nephrons.* During the formation of urine, most of the water and other small molecules are first filtered out of the blood. These substances are called the *filtrate.*
4. As the filtrate flows through the nephron tubule, substances are "selected" for reabsorption into the blood within surrounding capillaries. Various substances are reabsorbed along sections of the tubule because the permeability of its walls varies along its length. In addition, most of the water removed from the blood is reabsorbed from the filtrate because of a high osmotic gradient that surrounds certain sections of the tubule.
5. Different groups of vertebrates can function in a wide variety of environments because of adaptations in their nephrons allowing selective reabsorption of molecules valuable to their particular habitat.
6. By means of tubular secretion, the kidney also excretes the breakdown products of a variety of potentially harmful substances from the blood, such as marijuana, cocaine, heroin, morphine, and prescription drugs. The breakdown products of these substances can be detected in the urine.
7. The kidney is made up of an outer region called the *cortex* and an inner region called the *medulla.* Certain portions of the nephron lie in the cortex and another portion dips into and out of the medulla. Because of the high osmotic gradient created within the medulla, the reabsorption of water takes place there.
8. Urine leaves the kidneys by means of muscular tubes called *ureters.* It is conveyed to a storage pouch called the *urinary bladder.* A tube called the *urethra* brings urine to the outside. These interconnected organs make up the urinary system.
9. The principal function of the kidney is homeostasis. Two hormones help the kidneys control the balance of water and ions in the blood, thus helping maintain homeostasis.
10. Crystals of salts that sometimes form within the kidney are called *kidney stones.* If these stones do not pass out of the body on their own, they may need to be removed. Surgery and shock-wave lithotripsy are two ways in which kidney stones are removed.
11. When the filtration of the blood at the glomerulus is seriously impaired, the blood must be filtered by means of dialysis. In severe cases of renal failure, a patient may require a kidney transplant.

REVIEW QUESTIONS

1. What is excretion? Explain its importance to life.
2. What are the primary metabolic waste products that you excrete?
3. What do urea, urine, and uric acid have in common? How do they differ?
4. Fill in the blanks: In humans, the primary organs of excretion are the __________ and the __________.
5. Choose the correct answer: People eliminate most nitrogenous wastes in the form of (a) uric acid; (b) amino acids; (c) nitrogen; (d) urea.
6. Summarize the function(s) of your kidneys.
7. Draw a diagram of a human kidney showing the location of the renal pelvis, collecting ducts, ureter, cortex, medulla, renal pyramids, and renal arteries.
8. What is a nephron? Describe its function(s).
9. Summarize how your kidneys help your body to conserve water.
10. Explain why urine can reveal the use of certain drugs.
11. Do you think people's urine should ever be tested for drug use? Do you feel it would be justified in some cases more than others? Explain your answer.
12. Briefly summarize the process by which urine is formed.
13. What is the urinary system? Summarize its functions and its major components.
14. What is homeostasis? Explain how ADH and aldosterone are important in maintaining homeostasis.
15. Describe two problems that can occur with kidney function.

KEY TERMS

afferent arteriole 237
aldosterone 244
antidiuretic hormone (ADH) 244
artificial kidney 247
ascending arm of the loop of Henle 237
bile pigments 233
Bowman's capsule 235
collecting ducts 235
cortex 235
creatinine 234
deanimation 234
descending arm of the loop of Henle 237
dialysis 247
distal convoluted tubule 239
elimination 233
excretion 233
filtrate 234
filtration 235
glomerulus 237
kidney 234
medulla 235
nephron 234
nitrogenous wastes 234
proximal convoluted tubule 237
renal arteries 235
renal failure 246
renal pyramid 235
selected reabsorption 235
tubular secretion 235
urea 234
ureter 235
urethra 244
uric acid 234
urinary bladder 244
urinary system 244
urine 234

THOUGHT QUESTIONS

1. Imagine you are the sole survivor of a ship that sunk in the ocean, and you have managed to crawl onto a raft containing nothing but bottles of whiskey. After several days, you are very thirsty. Without water you may die. Should you drink the whiskey, the ocean water, or neither?
2. Among the many ways in which Americans can improve their diets is to decrease their intake of salt. What is wrong with a high-salt diet?

FOR FURTHER READING

Beeuwkes, R. (1982). Renal countercurrent mechanisms, or how to get something for (almost) nothing. In Taylor, C. R., et al, editors. *A companion to animal physiology.* New York: Cambridge University Press.
This is a clearly presented summary of current ideas about how the human kidney works.

Heatwole, H. (1978). Adaptations of marine snakes. *American Scientist, 66,* 594-604.
Several groups of snakes are able to live in the sea by clever adaptations that modify salt and water balance.

Nerve cells and how they transmit information

15

You won't find this machine among the Nautilus equipment at your local health club. It wasn't designed for physical fitness buffs but for paraplegics and quadriplegics—people who cannot move their legs. Impossible as it sounds, this machine can take over the job of nerves that "tell" leg muscles to move.

Nerves bring messages from the brain to other parts of the body by means of electrical impulses that sweep along their membranes. This stationary bicycle with its attached computer mimics this nervous activity. The computer generates pulses of electrical current and sends them to the patient's hamstrings, quadriceps, and buttocks by means of electrodes. These electrical pulses "fire" each leg muscle for an instant and in a specific order. The result is what you see in the photo—a pedaling motion that helps keep a patient's muscles, metabolism, and even mood in shape.

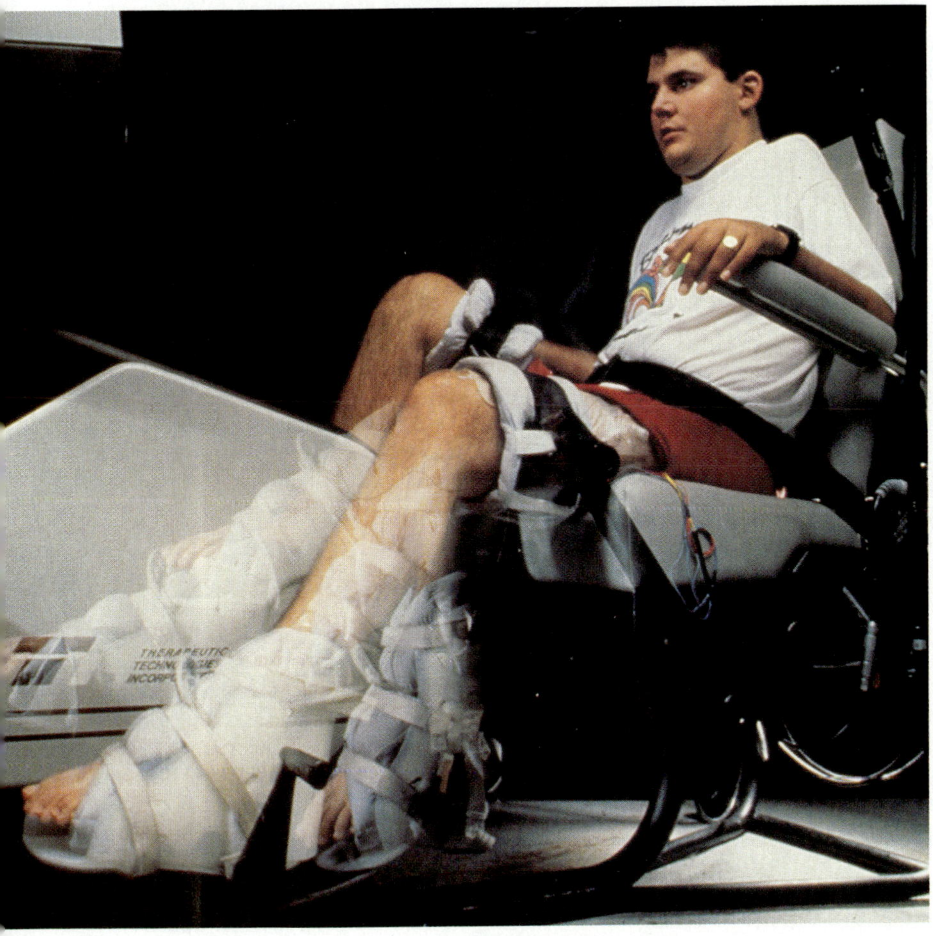

HIGHLIGHTS

- The tissues within your body communicate with one another by means of "slow messengers" called *hormones* and "fast messengers" called *nerve impulses.*
- Nerve impulses are electrical signals transmitted along the membranes of nerve cells, or neurons.
- The nerve impulse is initiated by a stimulus and is spread because of electrical and chemical forces at work along the length of the neuron.
- Nerve impulses are transmitted to other neurons, organs, or muscles by means of a variety of chemicals called *neurotransmitters.*

OUTLINE

The communication systems of your body

The nerve cell, or neuron

The nerve impulse

- The neuron at rest: The resting potential
- Conducting an impulse: The action potential
- Transmission of the nerve impulse: A propagation of the action potential
- Speedy neurons: Saltatory conduction

Transmitting information between cells

- Neuron-to-muscle cell connections
- Neuron-to-neuron connections

In the past decade, remarkable information has been learned about the nervous system. Scientists have discovered that cells in the peripheral nervous system—those outside the spinal cord and brain—are capable of limited regeneration. Some researchers are now experimenting with ways to enhance this process. For example, researchers in California have designed computer chips riddled with microscopic holes through which they are trying to get the severed ends of nerve cells to grow. They hope that the chips will not only help cut nerve cells reconnect but will relay messages along damaged nerve cells.

Scientists have had a more difficult time finding ways to encourage cells in the central nervous system (the brain and spinal cord) to repair themselves. Within the past decade, researchers have learned ways to stimulate nerve cell growth and regeneration in the brains of rats. Unfortunately, these techniques do not work in humans; the regeneration of human brain and spinal cord tissue is merely science fiction at this time. However, science and technology are entering arenas in which many paralysis victims, like the one in the photo, may be able to take a walk—or pedal—in ways they never before thought possible.

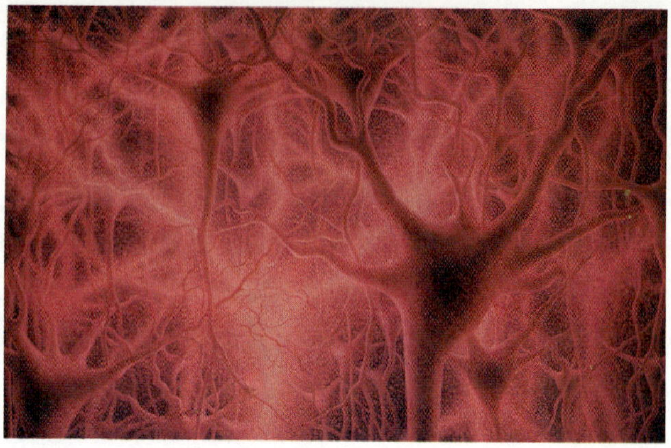

FIGURE 15-1 A network of neurons.
Nerve cells, or neurons, transmit nerve impulses throughout the body. Highly specialized, neurons are extremely sensitive to conditions within the body and to external conditions to which the body is subjected.

The communication systems of your body

The cells of your body communicate with one another in several ways. Your nervous system is only one means of communication. A simple way in which cells communicate with one another is by direct contact: open channels between adjacent cells allow ions and small molecules to pass freely from one cell to another. This method of communication allows cells to interact with their neighbors but is too slow and inefficient a method for interactions among cells that are far apart, like those in your toes and those in your fingers. It can be likened to running a large corporation with only face-to-face interactions between the people who sit next to each other! It would be better, in terms of distant communication, if managers of the various departments within a corporation sent memos to their staff members, instructing them what to do.

The varied tissues within the body communicate in a similar way by means of chemical "instructions" called **hormones.** The hormone "memos" are produced by one of several different **endocrine glands,** secreted into the bloodstream, and are carried around the body by the circulatory system. Each hormone, however, only interacts with certain "target tissues," just as a memo that is sent through interoffice mail is delivered only to those persons to whom it is addressed. (Hormones are discussed in more detail in Chapter 19.)

Although hormones are an important means of communication within your body, they do not serve all its communication needs. For example, if the message to be delivered to your leg muscles is "contract quickly, we are about to be hit by a car," hormones are too slow a message system. That is, a person in an emergency does not send a memo to the police but uses the telephone to get help quickly. That, in effect, is what your nervous system does.

The "quick" message system of your body is the **nervous system.** Your nervous system, as well as the nervous systems of all complex animals, is made up of nerve cells, or **neurons.** Neurons are specialized cells that transmit signals throughout the body (Figure 15-1). Bundled in groups, the long cell extensions of neurons make up nerves. Their message "signals" are called **nerve impulses.** Like the dots and dashes of Morse code, all nerve impulses are the same, differing only in their frequencies, their points of origin, and their destinations.

The command center of your nervous system is the **brain,** a precisely ordered but complicated maze of interconnected neurons—a large biological computer. The brain is connected by a network of neurons to both the hormone-producing glands and the individual muscles and other tissues. This dual channel of command permits great flexibility: the signals can be slow and persistent (hormones), fast and transient (nerve signals), or any combination of the two.

Two forms of communication integrate and coordinate body functions in humans, as well as in all other vertebrate animals and some invertebrates. Neurons are specialized cells that transmit rapid signals called *nerve impulses,* reporting information or initiating quick responses in specific tissues. Hormones are chemical messengers that trigger widespread prolonged responses, often in a variety of tissues.

The nerve cell, or neuron

Despite the fact that individual neurons vary widely in size and shape, all neurons have the same basic parts: a cell body and cell extensions called **dendrites** and **axons.** Some neurons have only a few cell extensions; others are bushy with these projections. Many of the neurons in your body, such as those extending from the base of your spinal cord to your toes, have extensions that are a meter or more long!

Figure 15-2 illustrates some of the structural differences that exist among neurons. These differences are intimately related to the varying functions of nerve cells. **Sensory neu-**

A
B
C

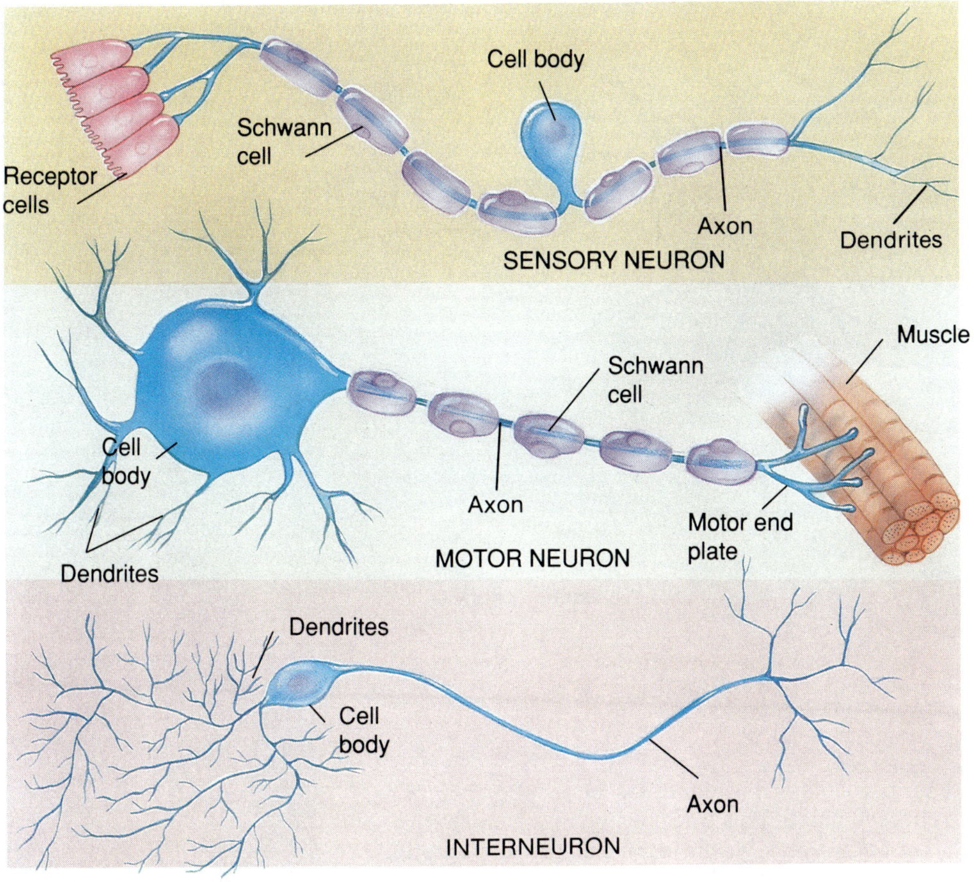

FIGURE 15-2 Types of vertebrate neurons.
A Sensory neurons carry signals from sense organs to the brain and spinal cord.
B The axons of motor neurons carry commands from the brain to muscles and glands.
C Neurons within the brain often have extensive, highly branched dendrites.

rons transmit information to the central nervous system (Figure 15-2, *A*). Each of these neurons has one long dendrite bringing messages from all parts of the body. The cell bodies of sensory neurons lie near the central nervous system. **Motor neurons** transmit commands away from the central nervous system (Figure 15-2, *B*). Each of these neurons has one long axon bringing messages to muscles and glands. The cell bodies of most motor neurons lie in or near the central nervous system. Axons carry messages from the cell body and are usually quite long. Many axons, for example, extend from the spinal cord to muscles or glands. As mentioned previously, the axons that control the muscular activity in your legs and feet can be more than a meter long (depending on your height). Even longer axons occur in larger mammals. In a giraffe, single axons extend from the toes in its front legs all the way to the neck and from the toes in its back legs all the way to the pelvis, each spanning a distance of several meters (Figure 15-3). **Interneurons,** those located within the brain or spinal cord, integrate incoming with outgoing messages (Figure 15-2, *C*). These neurons usually have a highly branched system of dendrites, which is able to receive input from many different neurons converging on a single interneuron.

FIGURE 15-3 Some nerve cells are quite long.
From the neck of this giraffe, axons of neurons extend all the way to the toes of the front legs—a distance of about 5 meters.

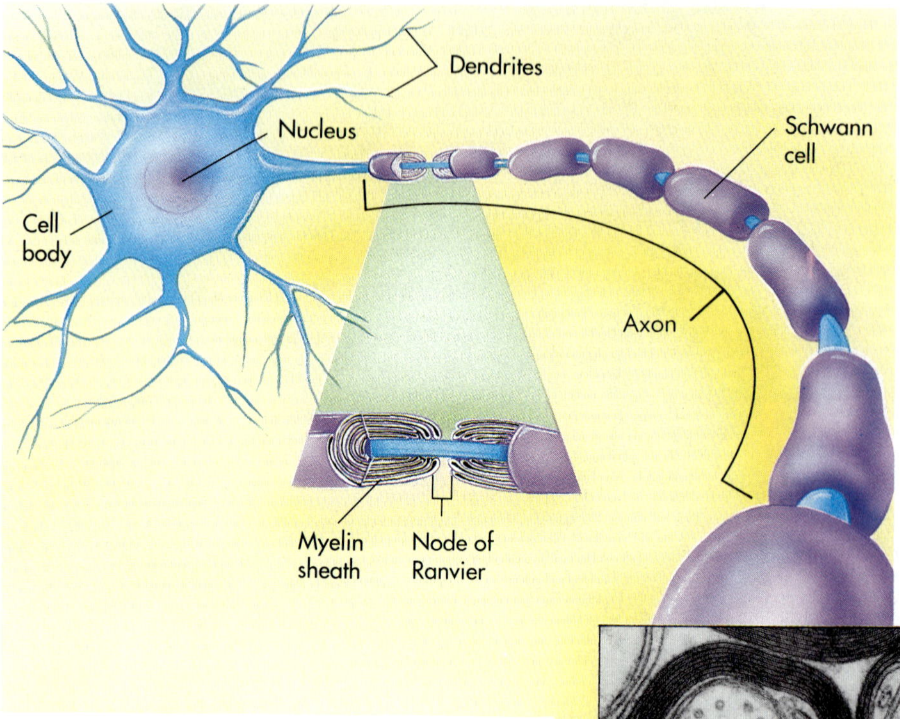

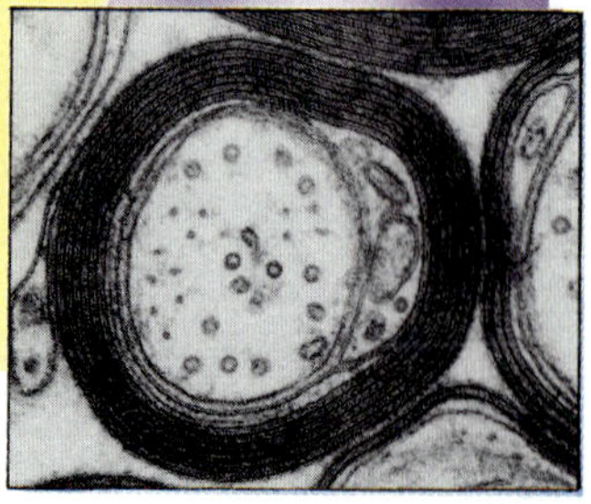

FIGURE 15-4 The insulation of neurons.
Long cell extensions of many neurons, particularly the dendrites of sensory neurons and the axons of motor neurons, are encased at intervals by Schwann cells, which form the myelin sheath. The myelin sheath has many layers, visible in the electron micrograph, that insulate the axon. The nerve impulse jumps from uninsulated node to uninsulated node and so moves very rapidly down the axon.

The **cell body** of a neuron has the structures typical of a eukaryotic cell, such as a membrane-bound nucleus. Surrounding the nucleus is cytoplasm, which contains the various cell organelles. The cell body is responsible for producing substances that are necessary for the nerve cell to live. The rough endoplasmic reticulum, for example, makes proteins that are used for the growth and regeneration of the nerve cell processes, or extensions. Substances such as these are able to leave the cell body and travel down the cell processes of the neuron, riding on "currents" within the cytoplasm or along "tracks" made by microtubules and microfilaments (see p. 44). In addition, some materials travel back to the cell body to be degraded or recycled.

A nerve cell, or neuron, is made up of a cell body that contains a nucleus and other cell organelles and two types of cellular projections: axons and dendrites. Dendrites bring messages to the nerve cell body; axons carry messages from the nerve cell body. Structural differences exist among neurons; these differences relate to functional differences among neurons.

Most neurons do not exist alone but have companion cells nearby. These companion cells are called **neuroglia** and provide nutritional support to neurons. Special neuroglial cells called **Schwann cells** are wrapped around many of the long cell processes of sensory and motor neurons of the peripheral nervous system. Such long cell extensions are nerve fibers. The fatty wrapping created by multiple layers of many Schwann cell membranes is called the **myelin sheath.** The myelin sheath insulates the axon; however, the Schwann cells are wrapped around the axon in such a way that uninsulated spots occur at regular intervals (Figure 15-4). These uncovered spots are called **nodes of Ranvier.** These nodes and the myelin sheath (discussed in more detail later in this chapter) create conditions that speed the nerve impulse as it is conducted along the surface of the axon.

The nerve impulse

How do nerve cells conduct nerve impulses? The answer is not obvious from examining the amazingly complex network of nerve cells that extends throughout the body. Looking at this network and trying to understand how the nervous system operates is a little like gazing at a telephone wire and trying to understand how it transmits your voice to a receiver far away—nothing can be seen moving along the wires. The key to understanding how your telephone (and your nerve cells) transmits information lies in an understanding of the abstract concept of electricity.

Electricity is a form of energy—an invisible form—and you can only see, hear, or feel its effects. Electricity is invisible because it is a flow of electrons, subatomic particles so minute that they cannot be visualized even with a high-powered microscope. In a telephone line a flow of electrons carries information from your telephone to the receiver of the person whom you called. The telephone changes the sound energy of your voice into electrical energy, transmits it over the telephone lines, and then changes it back to sound energy when it reaches its destination.

Neurons carry information in a similar way. The job of nerve cells is to transmit information from the environment to the spinal cord and brain, from one cell to another within the brain, and from the brain and spinal cord to other parts of the body. Nerve cells transmit this information in the form of electrical signals. The stimulation of specialized receptor cells, such as the rods and cones in your eyes, or specialized receptor endings of nerve cells, such as pressure receptors in your skin, causes electrical signals to be generated in these cells. Once an electrical signal is generated in a receptor, it can travel in the nervous system.

The neuron at rest: The resting potential

How do neurons conduct electricity? The story begins with the neuron "at rest"—a neuron not conducting an impulse. A resting neuron, interestingly, is electrically charged. This electrical charge has been measured in the laboratory (by means of microelectrodes) to be approximately −70 millivolts (mV) (Figure 15-5, *A*). The negative charge means that the inside of the cell (near the membrane where the microelectrode is placed) is negatively charged relative to extracellular fluid along the outside of the membrane (where a reference electrode is placed). A millivolt is one-thousandth of a volt, the unit measure of electrical potential. The term *electrical potential* refers to the amount of potential energy created by a separation of positive and negative charges (in this case, along the inside and outside of the cell membrane of the neuron). This potential energy comes from the work that is done to separate these charges. For example, work is necessary to roll a boulder up a hill. Some of the actively working energy (kinetic energy) used to roll the boulder uphill is stored in the boulder (potential energy) as it sits at the top of the hill. In the same way, positive and negative charges that are separated from one another possess potential energy. The energy it takes to separate these oppositely charged particles is released if they rejoin.

The charges that are separated from one another along the nerve cell membrane are carried as ions—atoms with unequal numbers of protons and electrons (see Chapter 2). The ions that play the principal role in the development of the electrical potential along the membrane of the neuron (and that are also important to many other body processes) are sodium ions (Na^+) and potassium ions (K^+).

Embedded within the cell membranes of neurons are enzymes called **sodium-potassium pumps** that actively transport these ions across the cell membrane (see Chapter 4 and Figure 15-6). These transmembrane proteins (enzymes) use the energy stored in molecules of ATP to move potassium ions into the neuron at the same time that they move sodium ions out of the neuron. However, the potassium ions can simply diffuse back out of the cell through channels or tunnels open to them. Put simply, potassium

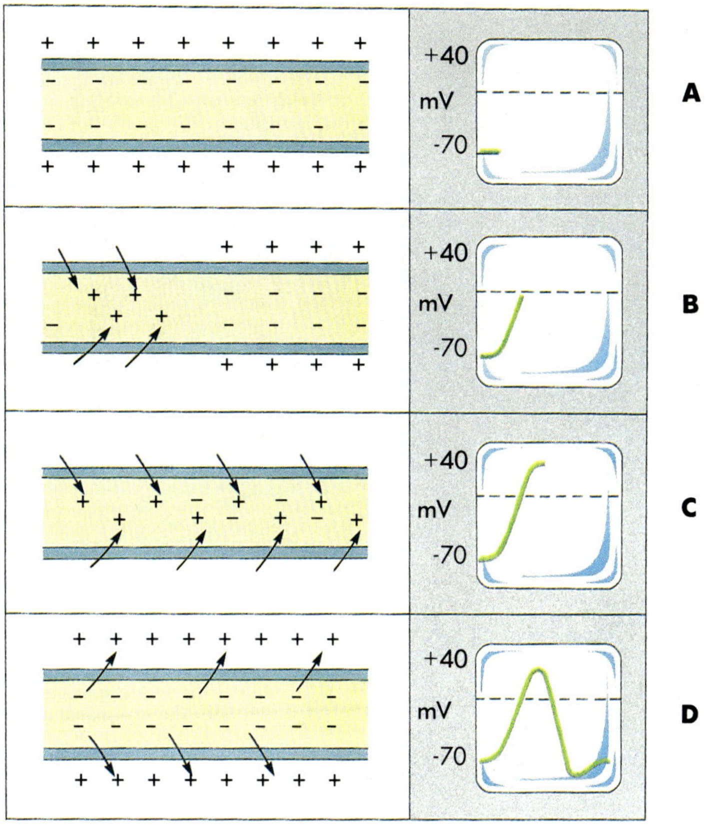

FIGURE 15-5 The stages of membrane depolarization.
A When a neuron is at rest, there is a net negative charge along the inside of the nerve cell membrane of −70 mV.
B When a neuron is sufficiently stimulated, Na^+ ions enter the cell, abolishing the voltage difference.
C Enough positive ions enter to establish a net interior positive charge.
D The exit of K^+ ions then returns the interior to a net negative charge.

ions actively enter the cell through one type of "door" but then exit by another. The situation with the sodium ions is different. These ions *cannot* move back into the cell even though they would normally do so by diffusion. There are no open inward channels in the neuron membrane through which the sodium ions can pass. In other words, they have no open door to get back into the cell. Sodium specific channels exist, but they are closed at −70 millivolts (the resting potential).

As a result of constantly pumping sodium ions out, the cell lowers the concentration of positively charged sodium ions within its interior. In this way a chemical gradient of sodium ions is established, with a higher concentration of sodium ions outside the cell than inside the cell. Negatively charged chloride ions (Cl^-) are attracted to the positively charged sodium but are unable to "follow" the sodium ions out of the cell because the neuron cell membrane is only slightly permeable to them. Therefore an electrical gradient is also established; the inside of a nerve cell along its membrane is more negatively charged than the outside of the cell

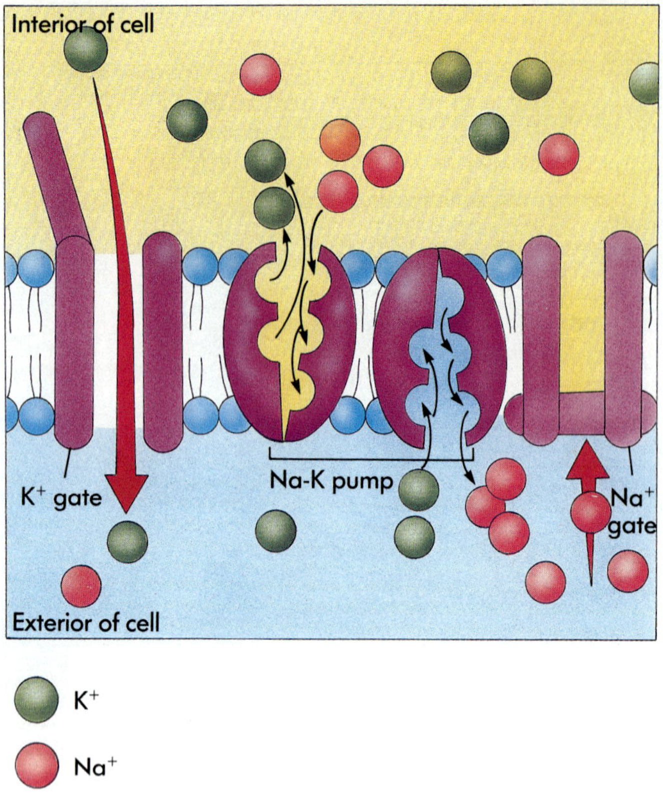

FIGURE 15-6 The sodium-potassium pump helps maintain the resting potential in a neuron.
The sodium-potassium pump consists of enzymes that are embedded in the cell membranes of neurons. The pump maintains the interior negative charge and exterior positive charge of the neuron as it pumps Na^+ ions outside the cell and K^+ ions inside the cell. However, K^+ ions are constantly diffusing back out of the cell and Cl^- ions remain within the cell, blocked from diffusing out because the nerve cell membrane is only slightly permeable to them. Due to these events, the interior of the cell becomes slightly negative. This separation of negative and positive charges across the cell membrane of a neuron is called *polarization.* Polarization characterizes the resting potential of the neuron.

along the membrane. The membrane of the neuron is said to be **polarized;** there is a difference in charge on the two sides of the membrane. Resting neurons—ones not conducting an impulse—have this difference in electrical charge on either side of the membrane. The difference in charge is called the **resting potential.** This electrical potential difference across the membrane is the basis for the transmission of signals by nerves.

The action of sodium-potassium transmembrane pumps and ion-specific membrane channels produces conditions that result in the separation of positive and negative ions along the inside and outside of the nerve cell membrane. This separation of charged particles creates an electrical potential difference, or electrical charge, along the membrane of the resting neuron. This electric potential difference is defined as the *resting potential.*

Conducting an impulse: The action potential

A neuron transmits a nerve impulse when it is excited by an internal or external environmental change called a **stimulus** (pl. stimuli). Examples of stimuli are pressure, chemical activity, sound, and light. Specialized **receptors** detect stimuli. For example, the rods and cones in the retina of your eye are sensitive to light, pressure receptors in your skin allow you to feel a hug, and certain cells in your nose allow you to smell your favorite dessert baking in the oven.

Stimuli cause a nerve impulse to be transmitted by initiating events that change the electrical potential difference of the receptor cell or nerve cell membrane. This change in electrical potential difference is called **depolarization.** The first event to occur when a neuron is sufficiently stimulated, depolarizing the membrane to a level called the **threshold potential** (Figure 15-5, *B*), is the opening of the sodium specific channels in the membrane. You can think of these channels as being operated by electricity. A change in voltage across the nerve cell membrane opens and closes them. What actually happens is that the sudden change in voltage changes the shape of the proteins forming the channels. As their shape changes, these channels become permeable to sodium ions and allow them to diffuse into the cell.

As the sodium specific channels open, a small amount of sodium ions moves rapidly into the cell, diffusing from where there are more sodium ions (outside the cell) to where there are less (inside the cell). This permeability for sodium ions across the cell membrane wipes out the local electrochemical gradient (the polarization) and further depolarizes the nerve cell membrane. Amazingly, although only about 1 in 10 million ions actually moves across the membrane, the interior of the cell develops a positive charge relative to the outside of approximately +40 millivolts (Figure 15-5, *C*), a 110 millivolt electrical difference from the −70 millivolts resting potential. This electrical difference occurs because a new electrical potential develops when the inwardly directed permeability for sodium ions develops across the cell membrane. This rapid change in the membrane's electrical potential is called the **action potential.**

Although it may seem hard to believe, the depolarization of the cell membrane (the action potential) lasts for only *a few thousandths of a second* (milliseconds) because the sodium channels close this quickly. They cannot reopen until after the resting potential is reestablished and another de-

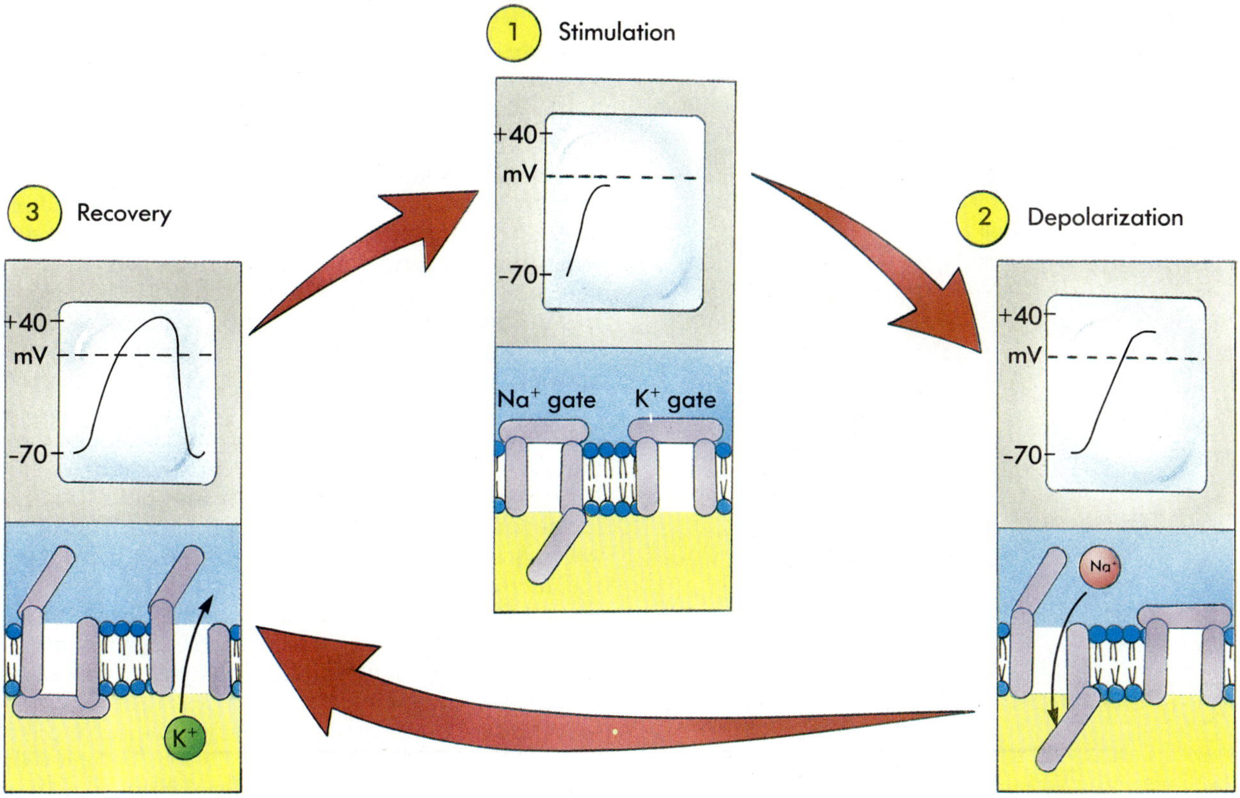

FIGURE 15-7 The action potential.
The stages of the action potential are *(1)* stimulation of a neuron, which causes Na^+ gates in the neuron to open; *(2)* depolarization, in which the Na^+ ions flood into the neuron, depolarizing the electrochemical gradient established during the resting potential; and *(3)* recovery, in which K^+ gates open, allowing K^+ ions to leave, thus repolarizing the membrane.

polarization occurs, triggering them again. During this inactive state of the sodium channels, a nerve impulse cannot be conducted. This period, which is only milliseconds long, is called the **refractory period.**

When the sodium channels close, potassium ions move outward as they usually do through potassium channels. In addition, many neurons contain voltage-sensitive potassium channels that open as the membrane depolarizes and the sodium channels shut down. It is this event—the movement of potassium ions out of the cell—that repolarizes the membrane (Figure 15-5, *D*). The sodium-potassium pump works to maintain the resting potential on a more long-term basis. The whole process of stimulation, depolarization, and recovery happens in the blink of an eye (Figure 15-7). In fact, 500 such cycles could occur, one after another, in the time it takes you to say the words "nerve impulse."

Transmission of the nerve impulse: A propagation of the action potential

An action potential at one point on the nerve cell membrane is a stimulus to neighboring regions of the cell membrane. The change in membrane potential causes sodium channels to open, depolarizing the adjacent section of membrane. Put simply, depolarization at one site produces depolarization at the next. In this way the initial depolarization passes outward over the membrane, spreading out in all directions from the site of stimulation (Figure 15-8). Like a burning fuse, the signal is usually initiated at one end and travels in one direction, but it would travel out from both directions if it were "lit" in the middle. This self-propagating "wave" of depolarization that travels along the nerve cell membrane is the nerve impulse.

A nerve impulse arises when a receptor is stimulated. This stimulation results in a rapid change in the electrical potential difference along the nerve cell membrane. This transient disturbance, or depolarization, has electrical consequences to which nearby transmembrane proteins respond, spreading the depolarization in wave-like fashion.

The nerve impulse is an **all-or-nothing response.** The amount of stimulation applied to the receptor or neuron must be sufficient to open enough sodium channels to generate an action potential. Otherwise, the cell membrane will simply return to the resting potential. For any one neuron the action potential is always the same. But any stimulus intense enough to open a sufficient number of sodium channels will depolarize the membrane to the threshold level and initiate an impulse. The neuron is said to have "fired." This wave of depolarization will have a constant amplitude (the height or strength of the wave). The speed at which impulses are conducted, however, varies among nerves.

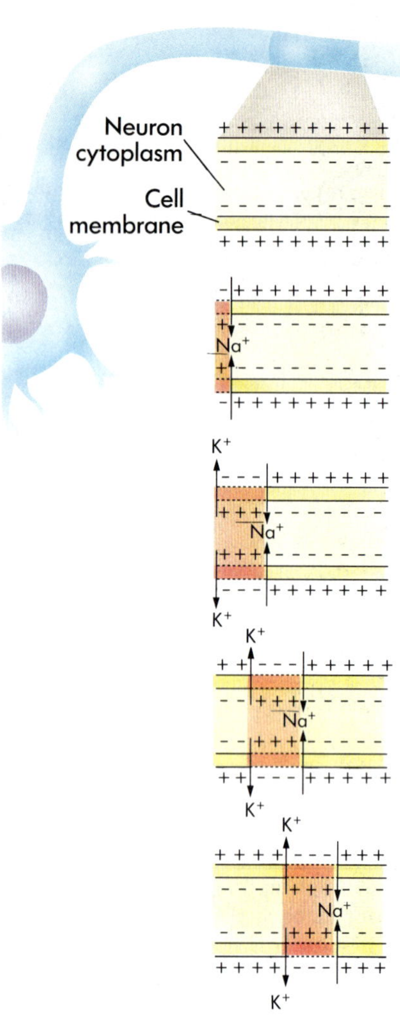

FIGURE 15-8 Transmission of a nerve impulse.
Depolarization moves laterally along a neuron in a self-propagating wave. The process of depolarization can be compared to the burning of a fuse because the signal usually starts at one end and travels in one direction. However, if the signal were able to start in the middle of the neuron, depolarization would travel in both directions outward.

Speedy neurons: Saltatory conduction

As mentioned earlier, many neurons have axons and axon-like dendrites that are covered by neuroglial Schwann cells. These cells envelop many of the long cell processes of sensory and motor neurons and certain interneurons as well (Figure 15-9), wrapping their cell membranes around them so many times that they form stacks of lipid-rich layers. In fact, some neurons have as many as 100 membrane layers surrounding them! This "cell wrapper" is the myelin sheath.

Schwann cells wrap around the length of an axon or axon-like dendrite, one after the other, with spaces separating one from the next (Figure 15-10). These spaces, the nodes of Ranvier, are critical to the propagation of the nerve impulse in myelinated cells. The myelin sheath is an insulator; it prevents the transport of ions across the neuron membrane beneath it. However, within the small gaps between Schwann cells, the surface of the axon is exposed to the intercellular fluid surrounding the nerve. An action potential can be generated only at these gaps. In fact, the pumps and channels that move ions across the neuron membrane are concentrated at the nodes of Ranvier, enhancing ion movements at these spots.

The action potential moves along the nerve cell membrane differently in myelinated cells than in unmyelinated cells. The wave of membrane depolarization that travels down the axon of unmyelinated neurons is impossible in myelinated axons due to the myelin sheath, which acts as an electrical insulator. Instead, the action potential "jumps" from one node to the next, causing a depolarization there. This depolarization opens the voltage-sensitive sodium channels at that node, resulting in the production of an action potential. This very fast form of nerve impulse conduction is known as **saltatory conduction,** from the Latin word *saltare,* meaning "to jump."

Impulses conducted by myelinated neurons travel much faster than impulses conducted by nerve fibers of the same diameter without this insulation. In fact, impulses conducted in large-diameter myelinated neurons travel up to 270 miles per hour (120 meters per second). These myelinated neurons can transmit a signal from your toes to your brain in less than half a second!

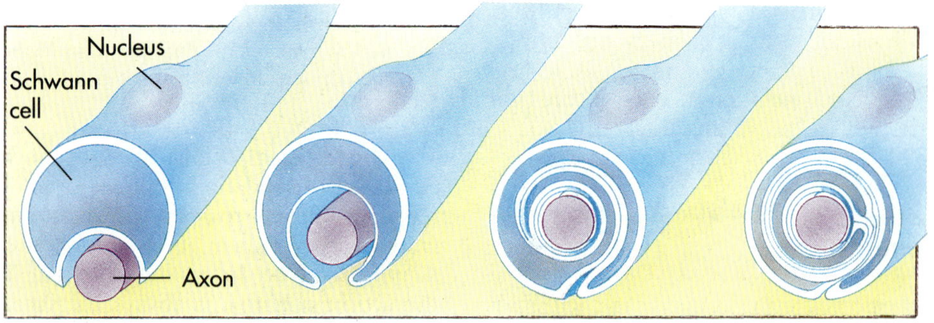

FIGURE 15-9 Schwann cells.
Schwann cells wrap around nerve cell processes forming stacks of lipid-rich layers. The progressive growth of the Schwann cell membrane around the process contributes the many membrane layers characteristic of myelin sheaths. These layers act as an excellent electrical insulator.

FIGURE 15-10 Saltatory conduction.
A In a fiber without Schwann cells, each portion of the membrane becomes depolarized in turn, like a row of falling dominoes.
B In neurons of the same diameter, the nerve impulse moves faster along a myelinated fiber because the wave of depolarization jumps from node to node without ever depolarizing the insulated membrane segments between nodes.

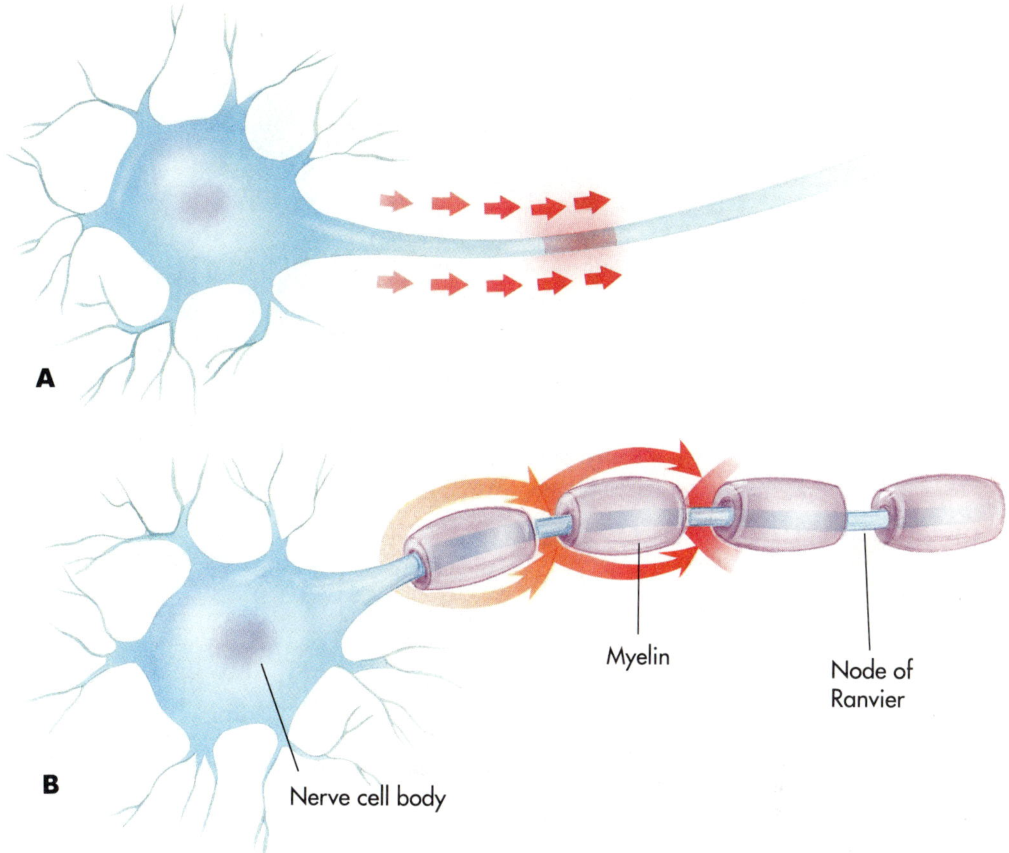

Multiple sclerosis (MS), one of the leading causes of serious neurological disease in adults, affects approximately one out of every 2000 people in the United States. This disease results in the destruction of large patches of the myelin sheath around neurons of the brain and spinal cord. Left behind are hardened scars (scleroses) that interfere with the transmission of nerve impulses. The resulting slowed transmission of signals in the nervous system results in a gradual loss of motor activity. Although the cause of MS is unknown, evidence strongly suggests that proteins within the myelin shealth are attacked by certain enzymes or by cells from the body's immune system.

Not all nerve impulses propagate as a continuous wave of depolarization spreading along the neuronal membrane. Along myelinated neurons, impulses travel by "jumping" along the membrane, leaping over insulated portions. Impulses travel much more quickly along these myelinated neurons than along unmyelinated ones.

Transmitting information between cells

When the nerve impulse reaches the end of an axon, it must be transmitted to another neuron or to muscle or glandular tissue. Muscles and glands, because they effect (or cause) responses when stimulated by nerves, are called **effectors.** This place—where a neuron communicates with another neuron or an effector cell—is called the **synapse.**

Most neurons do not actually touch other neurons or cells with which they communicate. Instead, there is a minute space (*billionths* of a meter across) separating these cells (Figure 15-11). This space or gap is called the **synaptic cleft.** The nerve impulse must cross this gap and does so by the direct passage of electrical current or by changing an electrical signal to a chemical signal. Chemical synapses are the prevalent type of synapse in humans (and all vertebrates).

The membrane on the axon side of the synaptic cleft is called the **presynaptic membrane.** In chemical synapses, when a wave of depolarization reaches the presynaptic membrane, it stimulates the release of organic molecules called **neurotransmitters** into the cleft. These molecules are stored in thousands of small membrane-bound sacs located at the tips of the axon. Each sac contains from 10,000 to 100,000 molecules of neurotransmitter. These chemicals diffuse to the other side of the gap. Once there, they combine with receptor molecules in the **postsynaptic membrane** (associated with either a dendrite or cell body) of the "target" cell. When they do, they cause ion channels to open.

A synapse is a junction between an axon tip and another cell, usually including a narrow gap separating the two cells. Passage of the impulse across the gap is by an electrical current or, more likely in vertebrates, a chemical signal from the axon.

Chemical junctions between neurons and other neurons or effector cells have a distinct advantage over direct electrical connections—flexibility. The chemical transmitters can be different in different junctions. Just as you might

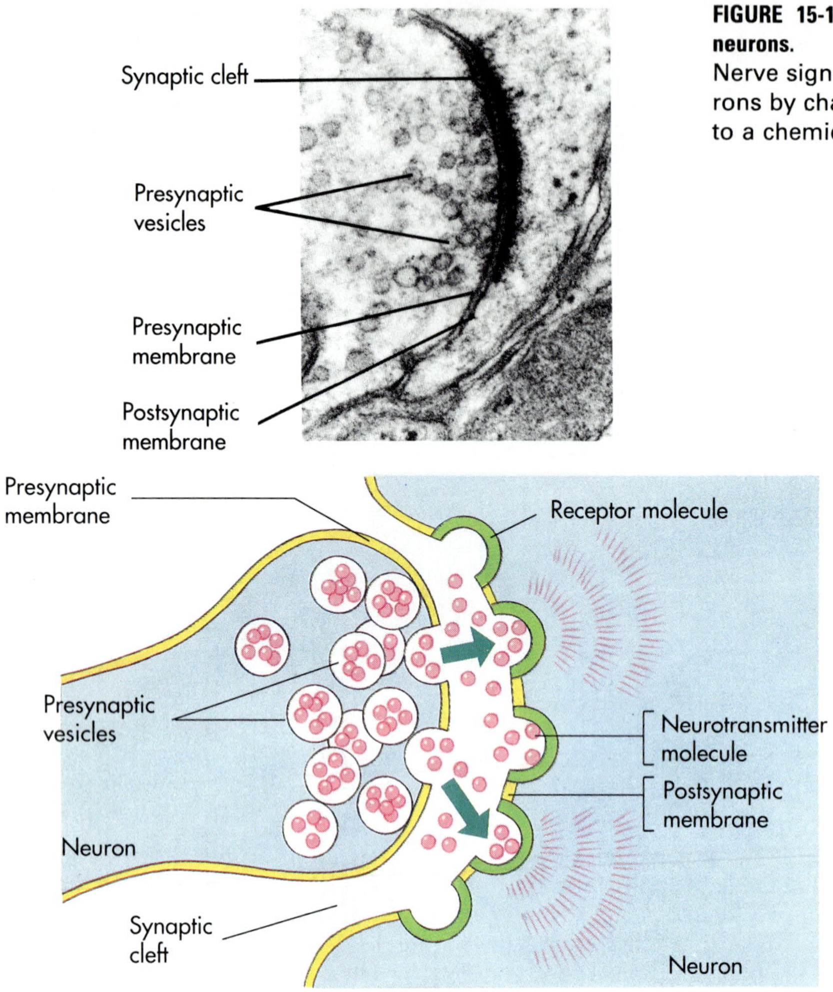

FIGURE 15-11 A synaptic cleft between two neurons.
Nerve signals cross the gap between neurons by changing from an electrical signal to a chemical signal.

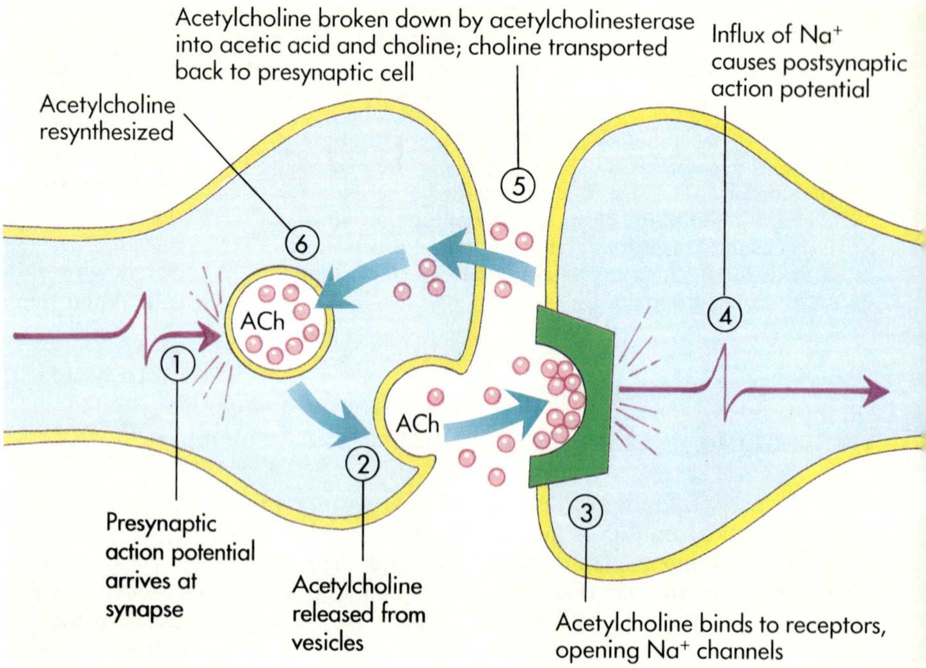

FIGURE 15-12 The sequence of events in synaptic transmission at a neuromuscular junction.
This diagram demonstrates how a nerve impulse causes muscle cells to contract. *ACH* is an abbreviation for acetylcholine.

take an aspirin to stop headache pain or cough syrup to subdue a cough, different neurotransmitters result in different kinds of responses. In fact, more than 60 different chemicals have been identified that act as neurotransmitters or that act to modify the activity of neurotransmitters.

Neuron-to-muscle cell connections

Synapses between neurons and skeletal muscle cells are called **neuromuscular junctions.** The neurotransmitter found at neuromuscular junctions is **acetylcholine.** Passing across the gap, the acetylcholine molecules bind to receptors in the postsynaptic (muscle cell) membrane, opening sodium channels. This influx of sodium ions depolarizes the muscle cell membrane, which initiates a wave of depolarization that passes down the muscle cell (Figure 15-12). This wave of depolarization releases calcium ions, which in turn trigger muscle contraction.

At a neuromuscular junction, acetylcholine released from an axon tip depolarizes the muscle cell membrane, releasing calcium ions that trigger muscle contraction.

After an impulse has been transmitted across the synaptic cleft, the neurotransmitter must be broken down or the postsynaptic membrane will remain depolarized. The breakdown products of the neurotransmitter then diffuse or are actively transported from the postsynaptic cell. In general, some or all of the breakdown products are transported back to the presynaptic cell to be reused. For example, the neurotransmitter acetylcholine is broken down to acetic acid and choline by an enzyme called **acetylcholinesterase**. Choline is transported back to the presynaptic cell, where it is used to make molecules of acetylcholine. Acetylcholinesterase is one of the fastest-acting enzymes in the body, breaking down one acetylcholine molecule every 40 microseconds. The fast work of acetylcholinesterase permits as many as 1000 impulses *per second* to be transmitted across the neuromuscular junction. Interestingly, nerve gases and the agricultural insecticide parathion work by blocking the action of acetylcholinesterase. These chemicals can cause death because they produce continuous neuromuscular transmission, which results in a continuous muscular contraction of vital muscles such as those involved in breathing and the circulation of blood. Many drugs work by affecting synapses also. For example, cocaine, local anaesthetics, and some tranquilizers work by destroying the control of neurotransmitter release.

Neuron-to-neuron connections

Impulses are transmitted from one neuron to another by a variety of neurotransmitters. Some neurotransmitters depolarize the postsynaptic membrane, which results in the continuation of the nerve impulse. This type of synapse is called an **excitatory synapse**. Other neurotransmitters have the reverse effect, reducing the ability of the postsynaptic membrane to depolarize. This type of synapse is called an **inhibitory synapse**. A single nerve cell can have *both* kinds of synaptic connections to other nerve cells. And, as you might expect, excitory signals cancel out inhibitory signals, modifying each other's effects. The postsynaptic neuron "keeps score" as the impulses reach its dendrites and cell body and responds accordingly. For this reason the postsynaptic neuron is called an **integrator** (Figure 15-13).

The dendrites and cell body of a postsynaptic neuron integrate the information they receive from presynaptic neurons. The summed effect of both excitory and inhibitory signals either facilitate depolarization or inhibit it.

Within your body, synapses are organized into functional units with definite patterns, like electrical circuits. Just like your house is wired in a definite pattern to provide electricity to various appliances and light fixtures, your body is "wired" in a specific manner. But the manner in which your body is wired is much more complex than that in your home, allowing you to maintain internal homeostasis as well as adapt to the outside environment.

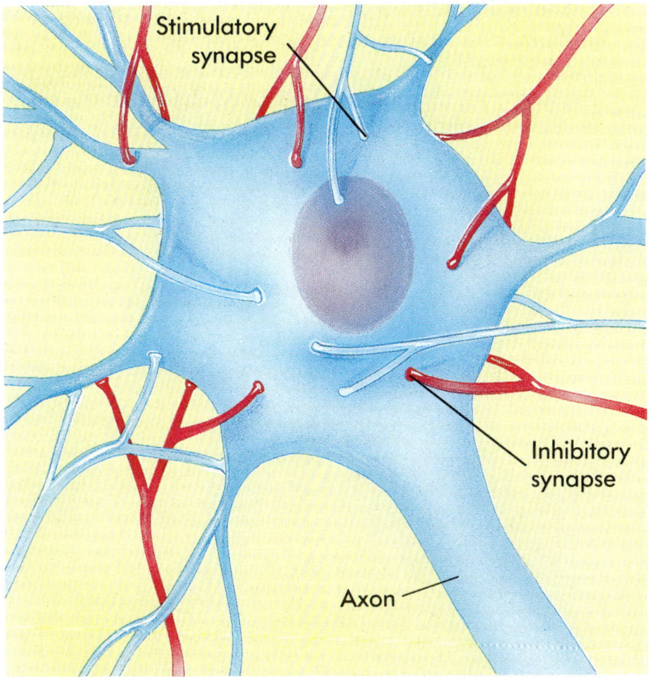

FIGURE 15-13 Integration of nerve impulses takes place on the neuron cell body.
The synapses made by some axons are inhibitory, tending to counteract depolarization of the postsynaptic membrane; these axons are indicated in *red.* The synapses made by other axons are stimulatory, tending to depolarize the postsynaptic membrane; these are axons indicated in *blue.* The summed influences of these inputs determine whether the postsynaptic membrane will be sufficiently depolarized to initiate a nerve impulse.

Summary

1. The human body has two primary means of internal communication: hormones and nerve impulses. Nerve impulses are quick and transient electrical signals. Hormones are slower, persistent chemical signals.
2. Nerve cells are called *neurons*. These cells have a cell body containing structures typical of a cell and cellular extensions called *axons* and *dendrites*. Dendrites receive incoming messages and bring them toward the cell body. Axons carry messages from the cell body.
3. Nerve cells transmit information in the form of electrical signals. A resting neuron (one not conducting an impulse) has an electrical potential difference across its cell membrane. This electrical potential, or resting potential, occurs because of separation of positively and negatively charged ions along the inside and outside of the nerve cell membrane.
4. A neuron transmits an impulse when it is excited by an environmental change, or stimulus. Specialized receptors detect stimuli and can initiate events that change the electrical potential difference along the nerve cell membrane. This change in electrical potential is called *depolarization*.
5. If a neuron is depolarized to a certain threshold, sodium channels in the membrane open. This permeability for sodium ions across the cell membrane wipes out the local electrochemical gradient and further depolarizes the nerve cell membrane. This rapid change in the membrane's electrical potential is called the *action potential*.
6. Within a few thousandths of a second, the sodium channels close. When these channels close, ion permeabilites are changed once again, and potassium tends to move out of the cell. This change in the electrochemical gradients at the cell membrane changes the electrical potential of the membrane once again, resulting in a return to the resting potential.
7. An action potential at one point on the nerve cell membrane causes depolarization of the adjacent section of membrane. A "wave" of depolarization continues along the nerve membrane. This self-propagating wave of depolarization is the nerve impulse.
8. The long cell processes of many neurons are wrapped in cells in which lipid-rich membranes act as insulation. Spaces between the cells making up the neuron "cell wrapper" help speed the nerve impulse as the action potential "jumps" from one node to the next. This very fast form of nerve impulse conduction is known as *saltatory conduction*.
9. When a nerve impulse reaches the axon tip of most vertebrate nerve cells, it causes the release of chemicals called *neurotransmitters*. These chemicals pass across a space called the *synaptic cleft* to the next cell. In neuron-to-muscle cell connections the neurotransmitter triggers muscle cell contraction. In neuron-to-neuron connections the neurotransmitter may cause either an excitory response or an inhibitory response.
10. The integration of nerve impulses occurs on the cell body membranes of individual neurons, which receive both excitory and inhibitory signals. These signals tend to cancel each other out, with the final amount of depolarization depending on the mix of the signals received.

REVIEW QUESTIONS

1. What do neurons and hormones have in common? How are they different?
2. Fill in the blanks: The "command center" of your nervous system is your __________. The nervous system is made up of nerve cells (__________) that transmit signals called __________ throughout your body.
3. What are dendrites and axons? What are their functions?
4. Fill in the blanks: Two ions that play an important role in the conduction of nerve impulses are __________ and __________.
5. Place these steps in the correct sequence. What process is being described?
   a. Sodium ions move into the cell.
   b. Potassium ions move out of the cell, and the cell membrane depolarizes.
   c. Sodium ion–specific channels in the cell membrane open.
   d. The membrane depolarizes to its threshold potential.
   e. An action potential develops.
6. Explain the term *resting potential.* What creates it, and why is it important?
7. A neuron transmits a nerve impulse when it is excited by an environmental change called a __________. Specialized __________ detect these changes, which leads to a change in electrical potential difference called __________.
8. Explain how a nerve impulse "travels" through the body.
9. What is saltatory conduction? Explain how it works and what advantages it offers.
10. Fill in the blanks: In neurons of the same diameter, nerve impulses travel more quickly along __________ neurons than along __________ neurons.
11. What is a synapse? How does a nerve impulse travel across a chemical synapse?
12. You decide to move your finger, and it moves. Explain how your nervous system communicated your decision to the muscles in your finger.
13. In question 12, explain why your finger muscles stopped contracting when you wanted them to stop.
14. Why is the postsynaptic neuron called an *integrator?* What does it integrate?

KEY TERMS

acetylcholine 259
acetylcholinesterase 259
action potential 254
all or nothing response 255
axon 250
brain 250
cell body 252
dendrite 250
depolarization 254
effector 257
excitatory synapse 259
inhibitory synapse 259
integrator 259
interneuron 251
motor neuron 251
myelin sheath 252
nerve impulse 250
nervous system 250
neuroglia 252
neuromuscular junction 259
neuron 250
neurotransmitter 257
node of Ranvier 252
polarization 254
postsynaptic membrane 257
presynaptic membrane 257
receptor 254
refractory period 255
resting potential 254
saltatory conduction 256
Schwann cell 252
sensory neuron 250
sodium-potassium pump 253
stimulus 254
synapse 257
synaptic cleft 257
threshold potential 254

THOUGHT QUESTIONS

1. In saltatory conduction an electrical current jumps from node to node along a myelinated nerve axon. What sort of membrane structure at the axon's nodes makes it possible for electricity to open sodium channels there?
2. If you take an addictive drug like heroin or cocaine, your nervous system soon becomes addicted to it. When the brains of addicts are examined closely, it is found that the brain neurons have greatly increased numbers of protein receptors for the addictive drug. What does the number of drug-specific receptors in neuron membranes have to do with addiction?

FOR FURTHER READING

Dunant, Y., & Israel, M. (1985, April). The release of acetylcholine. *Scientific American,* pp. 58-66.
This article challenges the accepted theory that acetylcholine is emitted by synaptic vesicles.

Kuffler, S. W., & Nicholls, J. G. (1984). *From neuron to brain: A cellular approach to the function of the nervous system.* ed. 2, Sunderland, Mass: Sinauer Associates, Inc.
This is a superb overview of the mechanisms of nerve excitation and transmission.

Morell, P., & Norton, W. T. (1980, May). Myelin. *Scientific American,* pp. 88-118.
This article still reflects current thought. It describes the composition of myelin, its synthesis, and how myelination increases the conduction velocity.

Reynolds, B. A., & Weiss, S. (1992, March 27). Generation of neurons and astrocytes from isolated cells of the adult mammalian central nervous system. *Science,* pp. 170-1710.
The development of mammalian neurons is believed to end just after birth. New research, however, shows that some neurons can be grown in vitro (in a test tube) with the help of epidermal growth factor.

16 The nervous system

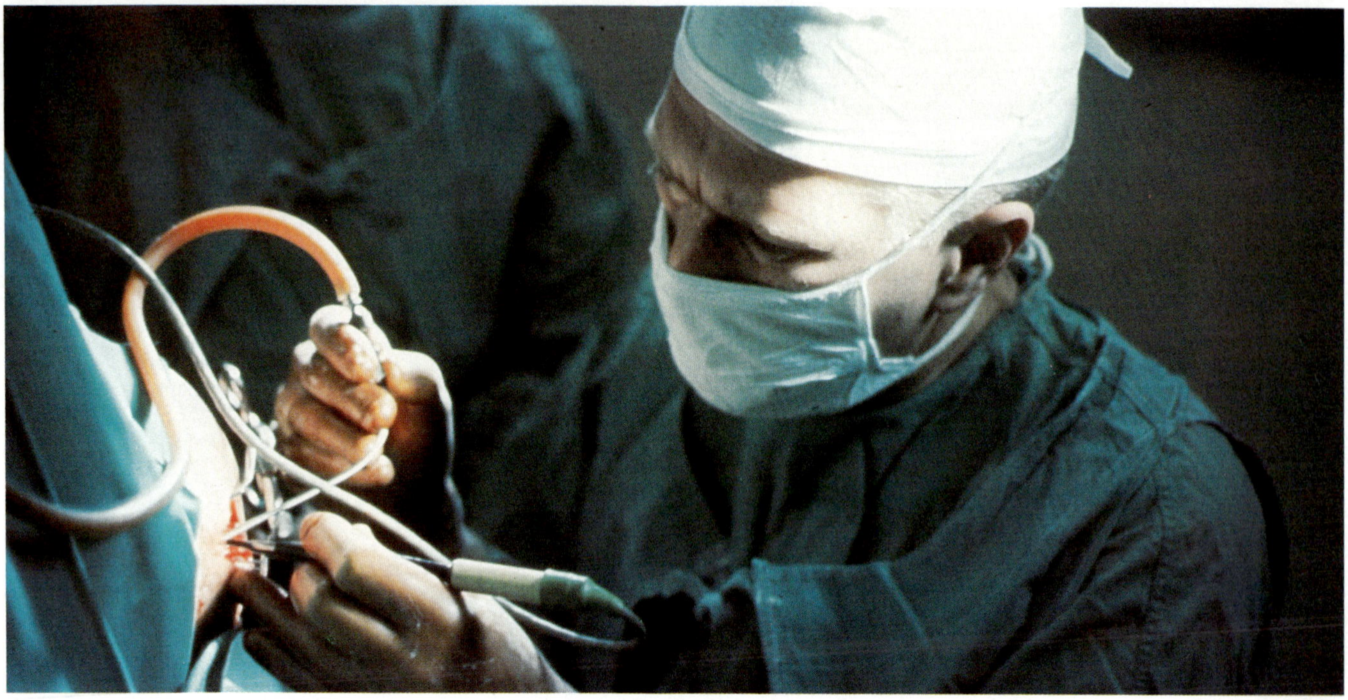

A brain transplant? Science fiction, of course. But scientists are performing types of experimental brain surgery today that would have seemed like science fiction only a decade ago. One of these intricate procedures is a tissue transplant called *brain grafting*.

OUTLINE

The organization of the nervous system

The central nervous system

- The brain
  - The cerebrum
  - The thalamus and hypothalamus
  - The cerebellum
  - The brainstem
- The spinal cord

The peripheral nervous system

- Sensory pathways
- Integration
- Motor pathways
  - The somatic nervous system
  - The autonomic nervous system

HIGHLIGHTS

▼

The nervous system is made up of the central nervous system, which consists of the brain and spinal cord, and the peripheral nervous system, which consists of nerves that extend from the brain and spinal cord to the muscles and glands.

▼

The cerebrum, the dominant part of the human brain, is covered with gray matter, or unmyelinated neurons, that perform "higher" cognitive functions such as perceiving, thinking, remembering, and feeling.

▼

Other parts of the brain and spinal cord perform jobs of integration such as controlling subconscious movements of the skeletal muscles and coordinating complex reflexes such as breathing.

▼

The sensory pathways of the peripheral nervous system bring messages from receptors to the brain and spinal cord; the motor pathways relay commands to the skeletal muscles by means of the somatic nervous system and to smooth and cardiac muscles and glands by means of two opposing branches of the autonomic nervous system.

Brain grafting involves taking tissue from another part of the body, like adrenal tissue found atop the kidneys, and surgically implanting it in part of the brain. This tissue takes over the function of brain tissue that is not working properly. One example of the use of this procedure is in patients with Parkinson's disease.

Parkinson's disease is a progressive disorder of the nervous system that affects approximately 50,000 Americans each year. It develops when one part of the brain does not produce enough of a neurotransmitter called *dopamine*. The disease first causes muscle tremors, resulting in uncontrolled movement such as shaking of the hands or arms. After awhile the muscles become rigid, and the victim can no longer write, eat, or perform other similar tasks. Medications help some persons but are not a cure.

In their search for alternative therapies, researchers transplanted minute bits of tissue from the adrenal glands into a part of the brain that helps coordinate movement. The adrenal glands, interestingly, manufacture dopamine as an intermediate in the production of other hormones. Early results of this "transplant" therapy have shown near-miraculous improvement in patients with Parkinson's. Scientists are now studying these first transplant patients to collect long-term data on the positive and negative effects of this procedure.

The organization of the nervous system

The brain is only one part of the complex network of neurons known as the **nervous system** (Figure 16-1). One job of the nervous system is to gather information about the body's internal and external environments. The nervous system then performs its other job—processing and responding to the information it has gathered. These nervous system responses are nerve impulses—commands sent out to the body's muscles and glands, directing them to react in an appropriate way.

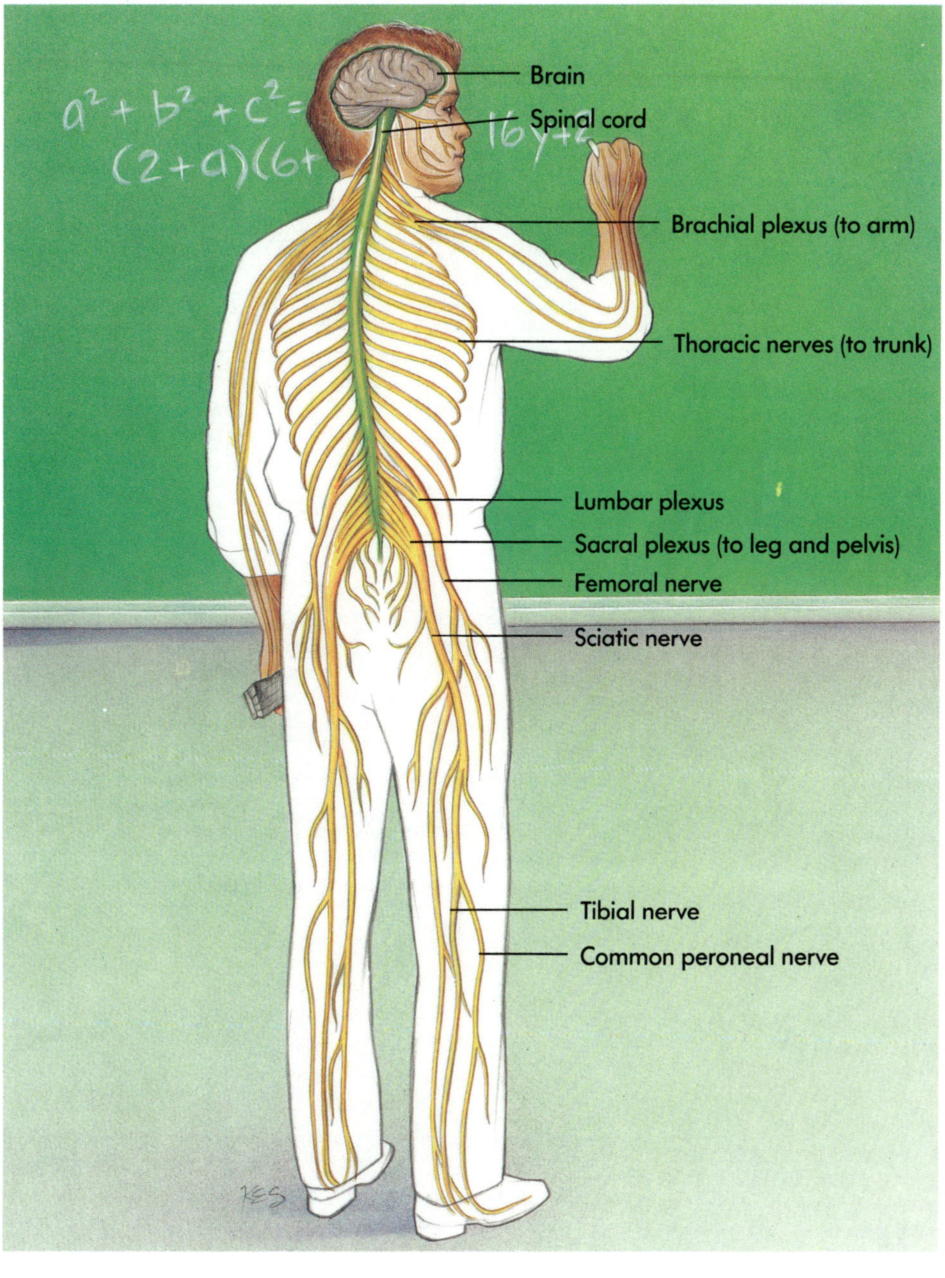

FIGURE 16-1 The human nervous system.
The central nervous system is shown in green and the peripheral nervous system in gold.

Structurally, the nervous system can be divided into two main parts (Figure 16-1): (1) the **central nervous system**—the site of information processing within the nervous system, which is made up of the **brain** and **spinal cord,** and (2) the **peripheral nervous system**—an information "highway," which is made up of nerves that bring messages to and from the brain and spinal cord. The nerves of the peripheral nervous system are made up of the long cell processes of nerve cells (nerve fibers), support cells (see Chapter 15), connective tissue, and blood vessels. The nerves of the peripheral nervous system contain the nerve fibers of two different types of neurons: **sensory neurons,** which transmit information to the central nervous system, and **motor neurons,** which transmit commands away from the central nervous system (Figure 16-2). Sensory neurons are also called **afferent neurons,** from the Latin prefix *affero* meaning "going toward." Motor neurons are also called **efferent neurons,** from the Latin prefix *effero* meaning "going away from."

The human nervous system is made up of the central nervous system, consisting of the brain and the spinal cord, and the peripheral nervous system. Within the peripheral nervous system, sensory pathways transmit information to the central nervous system, and motor pathways transmit commands from it.

One group of motor neurons controls **voluntary responses,** such as coordinating the movement of muscles in your legs so that you can cross the street. These responses are called voluntary because you consciously choose whether or not to do this activity. Motor neurons that control voluntary responses make up the **somatic nervous system.** The word *somatic* means "body" and refers to the fact that these neurons carry messages to your skeletal muscles—those that move the parts of your body. In addition, certain voluntary activities that may seem somewhat out of your control, such as blinking and breathing, are also directed by the somatic nervous system. Such activities are actually **reflexes,** automatic responses to stimuli that are mediated by the spinal cord or "lower" portions of the brain—those closest to the spinal cord.

Another group of motor neurons carries messages that control **involuntary responses.** These responses include such activities as mixing the food in your stomach with acid and enzymes after a meal, and "shooting" adrenaline into your bloodstream when you dodge an oncoming car. Motor neurons that carry messages about involuntary activities make up the **autonomic nervous system.** The word *autonomic* comes from Greek words meaning "self" *(auto)* "governing" *(nomos)*. This portion of the nervous sytem, then, literally takes care of you—by itself! In general, it works to promote homeostasis (a "steady state") within your body. The autonomic nervous system accomplishes this feat by

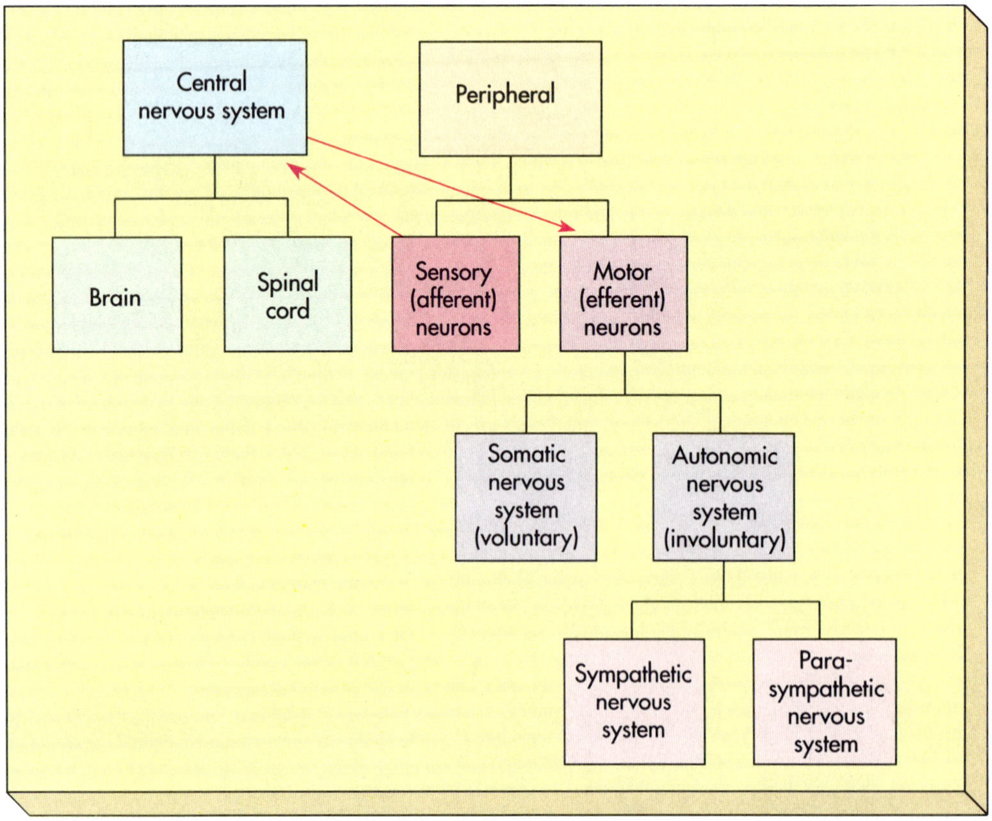

FIGURE 16-2 The organization of the human nervous system. The nervous system is divided into two parts: the central nervous system and the peripheral nervous system. The central nervous system is composed of the brain and spinal cord. The peripheral nervous system is composed of sensory pathways that transmit information to the central nervous system and motor pathways that transmit information away from the central nervous system.

carrying messages that speed up or slow down the activities of your glands, heart muscle, and smooth muscles such as those found in the digestive, excretory, and circulatory systems. In fact, these opposing messages are carried on separate neurons, dividing the autonomic nervous sytem functionally into two parts: the sympathetic and parasympathetic systems. (The effect of nerve impulses from each of these systems on various organs is shown in Figure 16-15.) An overview of the relationships of the various parts of the nervous system is shown in Figure 16-2.

One group of motor neurons makes up the somatic nervous system and controls voluntary responses. Another group of motor neurons makes up the autonomic nervous system and controls involuntary responses.

The central nervous system

In many ways, your central nervous system can be compared to the central processing unit (CPU) of your computer. Without your CPU, input (what you type on the keyboard, for example) is not processed by any of the programs in your computer. Therefore you will have no output either on your monitor or from your printer. In a similar but much more complex way, your brain and spinal cord make sense of incoming sensory information and then produce outgoing motor impulses. This function of the central nervous sytem is called **integration.** Each part of the central nervous system plays its own unique role in integration. Many "simple" nervous reactions, such as pulling your hand away from a hot stove, are integrated in the spinal cord. Other more complex nervous system reactions such as breathing are controlled in a lower portion of the brain called the *brainstem.* And the most complex or "highest" functions of the nervous system, such as thinking, remembering, and feeling, are all integrated in a portion of the brain called the *cerebrum* . . . a part of the brain that can do things a computer will *never* be able to do!

The brain

Your brain weighs about 3 pounds and contains an amazing 100 billion (100,000,000,000) neurons, a number that does not include neuroglial (supporting) cells. Both types of cells work together to form the intricate structure of the brain. Although a complicated whole, the human brain can be described as having four main parts: the **cerebrum,** the **cerebellum,** the **diencephalon** (thalamus and hypothalamus), and the **brainstem.** These four parts are shown in Figure 16-3.

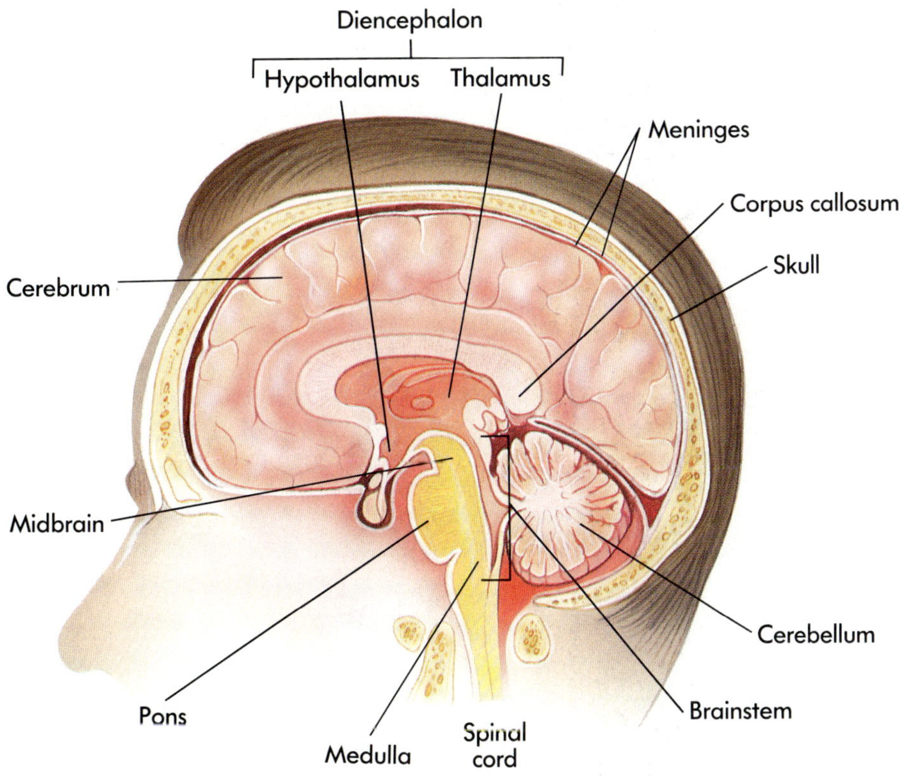

FIGURE 16-3 The human brain.
The brain has four main parts, each with different functions: the cerebrum, the cerebellum, the diencephalon, and the brainstem.

The cerebrum

The cerebrum, the dominant part of the human brain, is so large that it appears to wrap around and envelop the other three parts. In the brains of humans (and other primates) the cerebrum is split into two halves, or hemispheres, the right and left sides of the brain. These two sides of the cerebrum are connected by a single, thick bundle of nerve fibers called the **corpus callosum.** The corpus callosum is a communication bridge that allows information to pass from one side of the brain to the other, so each side "knows" what the other is thinking and doing.

Most of the activity of the cerebrum takes place within the **cerebral cortex,** a thin layer of tissue that forms its outer surface. The word *cortex* actually means "rind" or "bark" and helps give a sense of the relative thickness of this tissue to the underlying cerebral tissue. Your cortex "cap" is made up of unmyelinated neurons (see Chapter 15 for a discussion of myelin) and is densely packed with neuron cell bodies. It therefore appears gray and is referred to as **gray matter.**

Although the cerebral cortex is only a few millimeters thick, it has a tremendous surface area. It gains this surface area by lying in deep folds called **convolutions,** making the surface of the cerebrum look like a jumble of hills and valleys. These convolutions increase the surface area of the cerebral cortex 30 times. If laid out flat, it would cover an area of 5 square feet!

The cerebral cortex of each hemisphere of the brain is divided into four sections by deep grooves among its many convolutions. These four sections are called the **frontal, parietal, temporal,** and **occipital lobes** (Figure 16-4). By examining the effect of injuries to particular sites on the cerebrum, scientists have been able to determine the approximate location of the various activities of each lobe of the cerebral cortex.

There are three major types of activities that take place within the lobes of the cerebral cortex: motor, sensory, and association activity (see Figure 16-4). The **motor area,** the part of the brain that sends messages to move your skeletal muscles, straddles the rearmost portion of the frontal lobe. Each point on its surface is associated with the movement of a different part of the body (Figure 16-5). Right behind the motor area, on the leading edge of the parietal lobe, lies the **sensory area.** Each point on the surface of the sensory area represents sensory receptors from a different part of the body, such as the pressure sensors of the fingertips and the taste receptors of the tongue. Other sensory areas are located on other lobes. For example, the auditory area lies within the temporal lobe; different surface regions of this area correspond to different tonal patterns and rhythms. The visual area lies on the occipital lobe, with different sites corresponding to different positions on the retina.

The remaining areas within the cerebral cortex are referred to as **association areas.** These areas appear to be the site of higher cognitive activities, such as planning and contemplation. The associative cortex represents a far greater portion of the total cortex in primates than it does in any other mammals and reaches its greatest extent in human beings. In a mouse, for example, 95% of the surface of the cerebral cortex is occupied by motor and sensory areas. In humans, only 5% of the surface is devoted to motor and sensory functions; the remainder is associative cortex.

The cerebral cortex is the major site of higher cognitive processes such as sense perception, thinking, learning, and memory. It makes up a thin layer on the surface of the cerebrum, the largest and most dominant part of the brain.

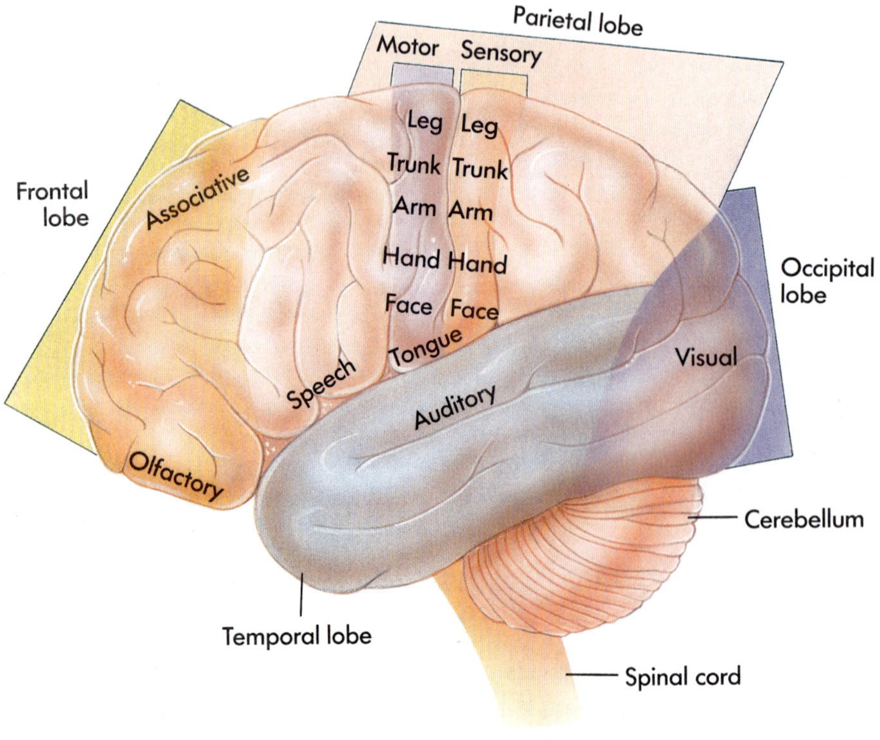

FIGURE 16-4 The cerebral cortex. The cerebral cortex has four sections: the frontal, parietal, temporal, and occipital lobes. Each of these lobes is responsible for various motor, sensory, and associative activities.

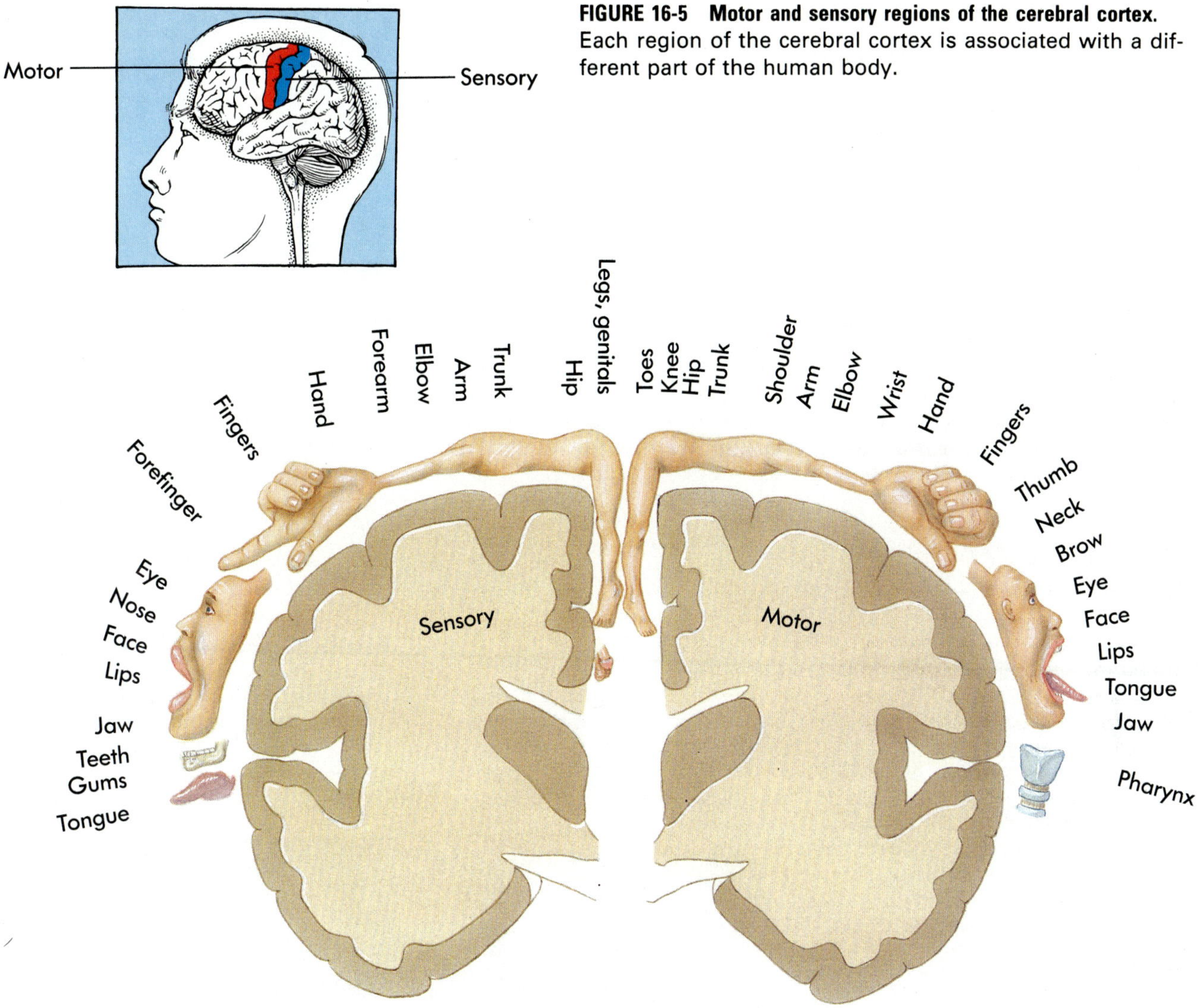

FIGURE 16-5 Motor and sensory regions of the cerebral cortex. Each region of the cerebral cortex is associated with a different part of the human body.

Although each hemisphere of the cerebrum contains motor, sensory, and association areas, the hemispheres are responsible for different associative activities. The right side of the cerebrum controls spatial relationships, musical and artistic ability, and expression of emotions. You might think of it as your "artistic" and "visual" side. The left side controls speech, writing, logical thought, and mathematical ability. It is your "logical" or "verbal" side. Injury to the left hemisphere of the cerebrum, for example, often results in the partial or total loss of speech, but a similar injury to the right side does not. Also, several speech centers control different aspects of speech. An injury to one speech center produces halting but correct speech; and injury to another speech center produces fluent, grammatical, but meaningless speech; and injury to a third center destroys speech altogether. Injuries to other sites on the surface of the brain's left hemisphere result in impairment of the ability to read, write, or do arithmetic. Comparable injuries to the right hemisphere have very different effects, resulting in impairment of three-dimensional vision, musical ability, and the ability to recognize patterns and solve inductive problems. The significance of this clustering of associative activities in different areas of the brain is not clear and remains a subject of much interest.

The right side of the cerebrum controls spatial relationships, musical and artistic ability, and expression of emotions. The left side controls speech, writing, logical thought, and mathematical ability.

In addition to the gray matter of the cerebral cortex, other masses of gray matter are located deep within the cerebrum. These islands of gray matter, shown in Figure 16-6, are often collectively referred to as the **basal ganglia.** Ganglia are groups of nerve cell bodies; these areas of gray matter are therefore described by their name as groups of nerve cell bodies located at the base, or lowest level, of the cerebrum. (Generally, however, groups of nerve cell bodies having a similar function are called **nuclei** when they are located within the central nervous system, and are called **ganglia** when they are located within the peripheral nervous

BIOLOGY & *you*

Solving the mystery of Alzheimer's disease

Alzheimer's disease was first described in 1901 by a Swiss psychiatrist, Dr Alois Alzheimer. One of Dr. Alzheimer's patients, aged 51, was having substantial memory difficulties. She continued to deteriorate, forgetting words and losing her ability to reason. She died 4 years later. At the autopsy, the woman's brain was found to be riddled with lesions, spaces where nerve cells had died.

The disease currently affects an estimated 2 million Americans. It is characterized by a gradual loss of memory and reasoning. Affected individuals cannot remember things that they heard or saw just a few minutes previously. They have trouble finding their way around and eventually forget how to talk, to feed themselves, even to swallow. There is no treatment.

Because Alzheimer's is most common among old people, doctors at first believed Alzheimer's to be a normal part of old age, as the brain simply wore out. Evidence soon mounted, however, that it was a genetic disorder—Alzheimer's showed up most often in particular families, for example. Studying how it is inherited, researchers soon focused on two human chromosomes, 19 and 21. Many if not all Alzheimer's cases could be linked to one of these two chromosomes.

In 1991, researchers identified the "Alzheimer's gene" on chromosome 21. They found that a single point mutation in this gene can lead to Alzheimer's disease. The same mutation was identified in two unrelated families who have Alzheimer's and does not occur in people who do not have the disorder. Although other genes (such as the one on chromosome 19) may also cause Alzheimer's, this gene offers the first clear look at the cause of the disorder.

What is the normal function of this gene? The gene at fault encodes a component of nerve cell membranes called *amyloid protein*. Amyloid is found in every nerve cell, but in Alzheimer's patients its synthesis goes awry, spitting out pieces of amyloid that pile up in sticky masses on the cell surface and eventually killing the nerve cell. The mutation appears to make the amyloid protein more fragile and liable to be chopped into fragments that accumulate in the brain.

As the population ages and people begin to live long enough for defective amyloid genes to begin their deadly work, Alzheimer's disease is becoming more common (Figure 16-A). It now affects 20% of the population that reaches age 85. Researchers are now focusing their efforts on finding ways to block the formation of amyloid fragments or their attachment to the surface of brain cells. For example, researchers are searching for ways to inhibit the enzyme that chops off the amyloid fragments. A cure will not be available tomorrow, but researchers finally seem to be on the road to success.

FIGURE 16-A The number of people over age 65 is steadily increasing. Therefore Alzheimer's disease will affect an increasing number of people unless a cure is found.

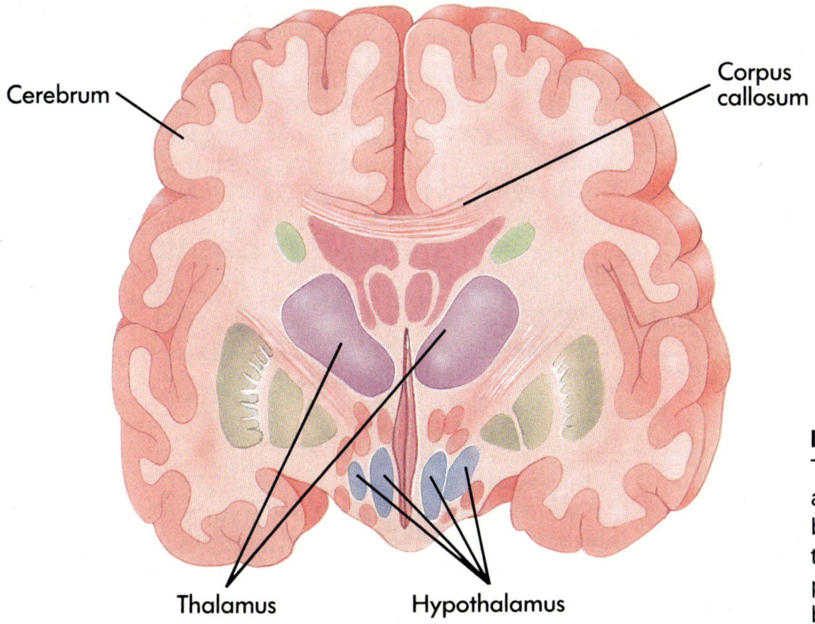

FIGURE 16-6 The basal ganglia.
The basal ganglia, shaded in bright green and olive green, are groups of nerve cell bodies. The ganglia are connected to the thalamus and hypothalamus and play a part in controlling many subconscious behaviors.

system.) The basal ganglia are connected to one another, to the cerebral cortex, and to parts of the brain not yet described—the thalamus and the hypothalamus—by bridges of nerve fibers. These ganglia play important roles in the control of large, subconscious movements of the skeletal muscles, such as swinging the arms while walking. Injury to the basal ganglia can result in an array of uncontrolled muscular activity, such as shaking, aimless movements, sudden jerking, and muscle rigidity. Parkinson's disease is associated with the degeneration of parts of the basal ganglia.

The cerebral nuclei, also called the *basal ganglia*, are groups of nerve cell bodies located at the base of the cerebrum that control large, subconscious movements of the skeletal muscles.

The cerebral cortex, or gray matter, covers an underlying solid white region of myelinated nerve fibers. These nerve fibers are the highways of the brain, bringing messages from one part of the brain to another. These highways, like the message highways in all parts of the nervous system, are composed of individual axons and long dendrites all bundled together like the strands of a telephone cable. Within the central nervous system, bundles of nerve fibers are called **tracts.** In the peripheral nervous system, they are called **nerves.** The tracts that make up the white matter of the cerebrum run in three principal directions: (1) from a place in one hemisphere of the brain to another place in the same hemisphere, (2) from a place in one hemisphere to a corresponding place in the other hemisphere, and (3) to and from the cerebrum to other parts of the brain and spinal cord.

The cerebral white matter, which lies beneath the cerebral cortex, consists of myelinated nerve fibers that send messages from one part of the central nervous system to another.

The thalamus and hypothalamus

At the base of the cerebrum but not part of it, lying close to the basal ganglia, are paired oval masses of gray matter called the **thalamus.** A view of these nuclei in Figure 16-6 shows them connected by a bridge of gray matter and indicates how the thalamus—meaning "inner chamber"—got its name. This structure acts as a relay station for most sensory information. This information comes to the thalamus from the spinal cord and certain parts of the brain. The thalamus then sends these sensory signals to appropriate areas of the cerebral cortex. In addition, the thalamus interprets certain sensory messages such as pain, temperature, and pressure.

The **hypothalamus,** located beneath the thalamus (*hypo* means "under"), controls the activities of various body organs. Its position is shown in Figure 16-6. The hypothalamus works to maintain homeostasis, a steady state within the body, by means of its various activities such as the control of body temperature, respiration, and the heartbeat. It also directs the hormone secretions of the pituitary, which is located at the base of the brain.

The hypothalamus is linked by a network of neurons to the cerebral cortex. This network, together with the hypothalamus, is called the **limbic system.** The term *limbic* is derived from a Latin word meaning "border." As shown in Figure 16-7, the neurons of the limbic system form a ringlike border around the top of the brainstem. The operations of the limbic system are responsible for many of the most deep-seated drives and emotions of vertebrates, including pain, anger, sexual drive, hunger, thirst, and pleasure.

The thalamus and hypothalamus are masses of gray matter that lie at the base of the cerebrum. The thalamus receives sensory stimuli, interprets some of these stimuli, and sends the remaining sensory messages to appropriate locations in the cerebrum. The hypothalamus contols the activities of various body organs and the secretion of certain hormones.

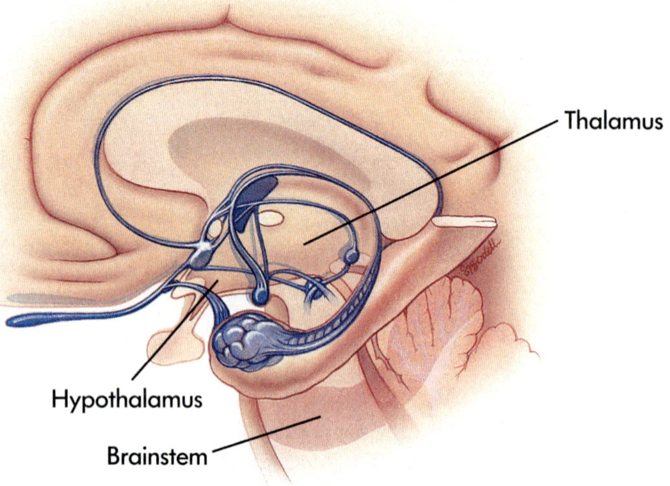

FIGURE 16-7 The limbic system.
Shaded in blue in this diagram, the limbic system is responsible for many of your most deep-seated drives and emotions.

The cerebellum

The cerebellum is a relatively large part of the brain, weighing slightly less than half a pound. It is located below the occipital lobes of the cerebrum, as shown in Figure 16-3. Although its name means "little cerebrum," the cerebellum does not perform cerebral functions. Instead, it coordinates subconscious movements of the skeletal muscles. Sensory nerves bring information to the cerebellum about the position of body parts relative to one another, the state of relaxation or contraction of the skeletal muscles, and the general position of the body in relation to the outside world. These data are gathered in the cerebellum and synthesized. The cerebellum then issues orders to motor neurons that result in smooth, well-coordinated muscular movements, contributing to overall muscle tone, posture, balance, and equilibrium.

The cerebellum coordinates subconscious movements of the skeletal muscles.

The brainstem

If you think of the brain as being shaped somewhat like a mushroom, the cerebrum, thalamus, and hypothalamus would be its cap. The brainstem would be the mushroom's stalk (see Figure 16-3). Its 3 inches of length is made up of three parts—the midbrain, pons, and medulla. Each part makes up about an equal length of the brainstem and contains tracts of nerve fibers that bring messages to and from the spinal cord. In addition, each portion of the brainstem contains nuclei that govern important reflex (automatic) activities of the body. Many of the cranial nerves, nerves that enter the brain rather than the spinal cord, enter at the brainstem. These cranial nerves bring messages to and from the regulatory centers of the brainstem or use the brainstem as a relay station. These nerves can be seen in Figure 16-8 and are discussed in more detail later in the chapter.

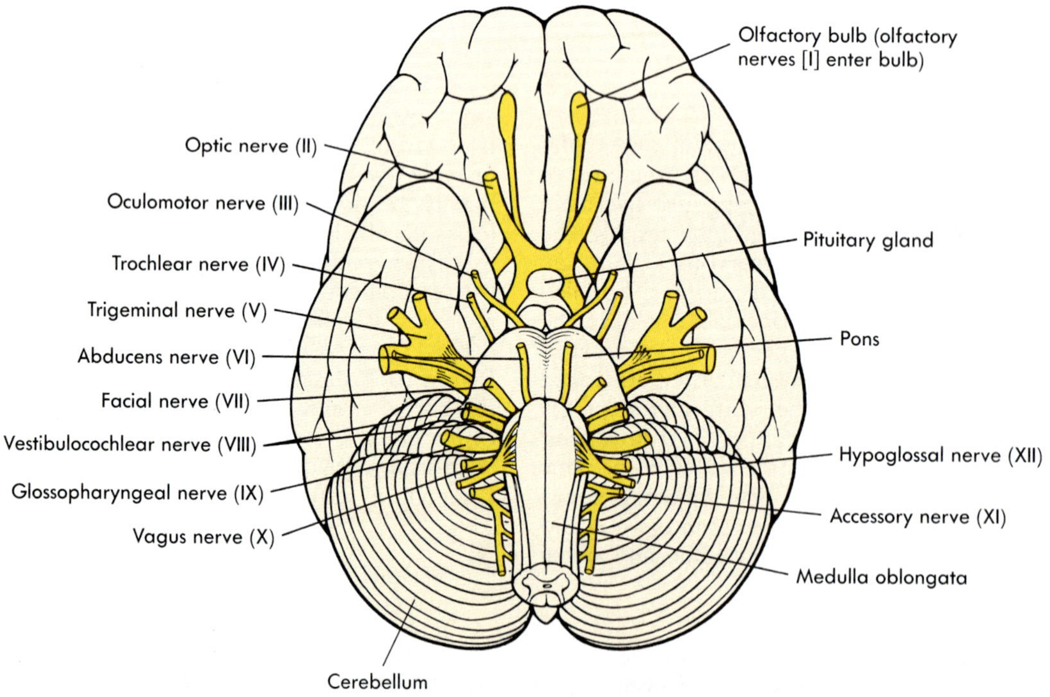

FIGURE 16-8 A view of the brain from its underside.
In this view, the origins of the cranial nerves can be clearly seen.

At the top of the brainstem sits the **midbrain,** extending down from the lower portion of the thalamus and hypothalamus about an inch—approximately the distance from the tip of your thumb to its first knuckle. If you were to cut open the midbrain, you would see both white and gray matter. The white matter consists of nerve tracts that connect the upper parts of the brain (cerebrum, thalamus, and hypothalamus) with lower parts of the brain (pons and medulla). In addition, the midbrain contains nuclei that act as reflex centers for movements of the eyeballs, head, and trunk in response to sights, sounds, and various other stimuli. For example, if a plate falls off the counter behind you, you probably turn around quickly and automatically. That is your midbrain at work.

The term **pons** means "bridge," and it is actually two bridges. One bridge consists of horizontal tracts that extend to the cerebellum, connecting this part of the brain to other parts and to the spinal cord. The other bridge consists of longitudinal tracts that connect the midbrain and structures above to the medulla and spinal cord below. In addition, the gray matter of the pons contains nuclei that work with certain nuclei in the medulla to help control respiration.

The **medulla,** the lowest portion of the brainstem, is continuous with the spinal cord below. Because of its location, a large portion of the medulla consists of tracts of neurons that bring messages up from the spinal cord and others that take messages down to the spinal cord. Interestingly, most of these tracts cross over one another within the medulla. Therefore sensory information from the right side of the body is perceived in the left side of the brain and vice versa. Likewise, the right side of the brain sends messages to the left side of the body; the left side of the brain controls the right side. In addition to these tracts, the medulla contains reflex centers that regulate heartbeat, control the diameter of blood vessels, and adjust the rhythm of breathing. Centers there also control less vital functions such as coughing, sneezing, and vomiting.

Throughout the entire length of the brainstem, but concentrated in the medulla, weaves a complex network of neurons called the **reticular formation.** All of the sensory systems have nerve fibers that feed into this system, which serves to "wiretap" all of the incoming and outgoing communications channels of the brain. In doing so, the reticular formation monitors information concerning the incoming stimuli and identifies important ones. The reticular formation also plays a role in consciousness, increasing the activity level of many parts of the brain when aroused and decreasing in activity during periods of sleep. Interestingly, this is the part of the brain that causes a knockout during boxing. When the head is twisted sharply and suddenly during a punch to the jaw, the brainstem (and reticular formation) is twisted, resulting in unconsciousness.

The brainstem, consisting of the midbrain, pons, and medulla, contains tracts of nerve fibers that bring messages to and from the spinal cord. In addition, nuclei located in various parts of the brainstem control important body reflexes such as the rhythm of the heartbeat and rate of breathing.

The spinal cord

The central nervous system also includes the brain's extension, the spinal cord. The spinal cord runs down the neck and back within an inner "tunnel" of the vertebral column, or spine. This tunnel is created by the stacking of vertebrae one on another, with their central spaces, or **foramina** (sing. **foramen**), in alignment with one another. This bony casing protects the spinal cord from injury, just as the bones of the skull protect the brain.

The spinal cord receives information from the body by means of **spinal nerves** (see Figure 16-1). It carries this information to the brain along organized tracts of myelinated nerve fibers (white matter) and similarly sends information from the brain out to the body. In addition, the gray matter of the spinal cord integrates responses to certain kinds of stimuli. These integrative pathways are called **reflex arcs** and are discussed in more detail later in the chapter.

The spinal cord, the part of the central nervous system that runs down the neck and back, receives information from the body, carries this information to the brain, and sends information from the brain to the body.

The white matter tracts of the spinal cord, unlike the white matter of the brain, are located on the exterior of the spinal cord with gray matter in the center (Figure 16-9). (The brain, as discussed earlier, is covered with gray matter, which surrounds the white matter hidden below.) In addition, the spinal cord has a tiny central canal that pierces

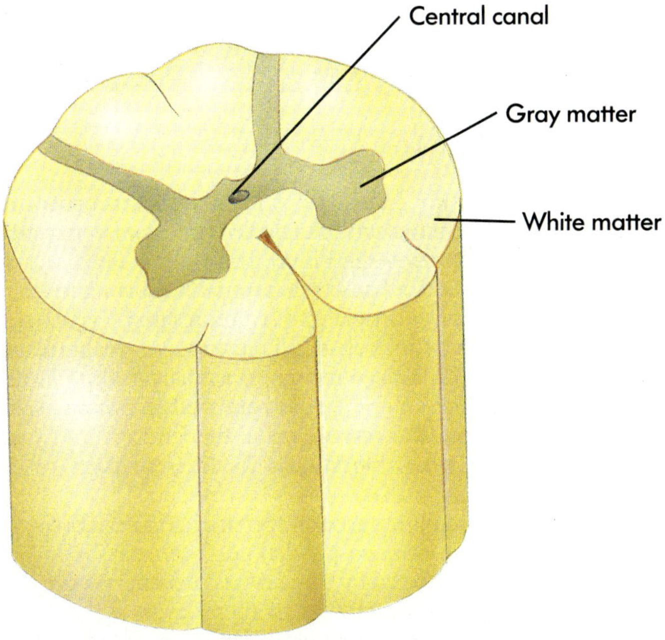

FIGURE 16-9 Section through the spinal cord.
The location of the gray matter, the white matter, and the central canal is shown here.

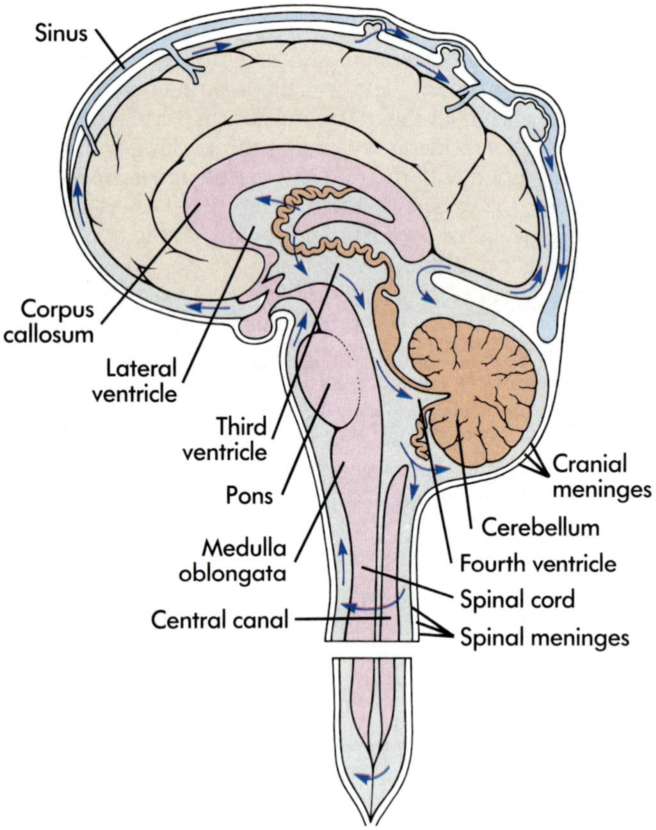

FIGURE 16-10 Flow of the cerebrospinal fluid. *Arrows* indicate the direction of flow. Notice that the fluid travels around the cerebrum and into the ventricles within the brain. The cerebrospinal fluid acts like a shock absorber for the brain and is also the medium in which gases such as CO_2 dissolve. The amount of gases in the cerebrospinal fluid triggers the cells in the brain to adjust such factors as the heartbeat and breathing rates.

its length. This tube-like space is filled with **cerebrospinal fluid** and is continuous with fluid-filled spaces, or **ventricles,** in the brain. (The arrows in Figure 16-10 show the direction of flow of this fluid.) The cerebrospinal fluid is a filtrate of the blood, formed by specialized structures in the ventricles. This fluid acts as a shock absorber, cushioning the brain and the spinal cord. In fact, your brain is actually *floating* in this fluid, although its volume would barely fill an average-sized glass. In addition, the cerebrospinal fluid brings nutrients, hormones, and white blood cells to different parts of the brain.

As well as being encased in the skull and vertebrae and cushioned by the cerebrospinal fluid, both the brain and spinal cord are protected by three layers of membranes called the **meninges.** The outermost of these layers is called the *dura mater,* from the medieval Latin meaning "tough mother." Surrounding the brain, the dura mater is made up of two parts: one that adheres to the inner side of the bones of the cranium and another that lies closer to the brain. Surrounding the spinal cord, however, the dura mater is only a single layer. In between it and the bones of the vertebrae is a space filled with fat, connective tissue, and blood vessels called the *epidural* (outside the dura) *space.* At the level of the lower back, the epidural space is where anaesthetics are injected to numb the lower body for certain operations or childbirth. Another space exists between the middle (arachnoid layer) and inner (pia mater) membranes of the meninges. The cerebrospinal fluid circulates within this space, called the *subarachnoid space.* The subarachnoid space is continuous with the central canal of the vertebral column and the ventricles of the brain (see Figure 16-10).

The brain and spinal cord are protected by a cushion of fluid called the *cerebrospinal fluid*, layers of membranous coverings called the *meninges*, and bones of the skull and vertebral column.

BIOLOGY IN FOCUS

Memories

On an ordinary day in 1989, a woman named Eileen Franklin-Lipsker glanced at her young daughter as she played in the kitchen. Something about the tilt of her daughter's face as she looked up at Eileen disturbed her. Suddenly, an avalanche of horrible images flooded Eileen—images of her father murdering a friend 20 years earlier as she looked on, a helpless witness. Troubled and confused by her memories, Eileen contacted the police and related her incredible story. Police arrested her father and charged him with the 1969 murder of Eileen's playmate, whose killer had never been found. In 1990, George Franklin was convicted of the murder. The most damaging evidence was Eileen's complete recall of the events of the murder, which she recounted to a stunned courtroom.

Many people have had the experience of a memory suddenly coming into consciousness, but the validity of such experiences gained credence when Vietnam veterans began to describe suddenly vivid memories of forgotten experiences they had had during the war. These long buried memories which had been repressed came out when a later emotional event triggered their release. In addition, the publicity given to the sexual abuse of children within the last decade has prompted many victims of this trauma to remember and relate their repressed experiences in "flashbacks" of memories of their abuse, which are instigated by an emotional or physical stimulus years later.

Although the exact physiological mechanism is unknown, researchers do have some idea of how a person can repress and suddenly recall memories. The middle portion of the brain, the hippocampus, acts as the brain's "file clerk." To store a memory, the hippocampus directs the parts of an experience, such as the sights, smells, and sounds, to different sectors of the brain through a network of neurons. Thus different parts of memory are stored in different places, and the neuron network, if it is fired often, allows access to the memory. If the neuron network is not fired, the memory can be forgotten—or simply repressed. If the proper stimulus, such as the glance from Eileen's daughter, can cause the network of neurons to fire, a long buried memory can come to the surface, often with fine details still intact. It is not unusual for people experiencing such memory flashbacks to be able to describe the time of day, the objects in the room, and the clothes they were wearing during the experience they are recalling.

It is because of these fine details that juries tend to believe the validity of flashbacks. Researchers continue to be fascinated with the brain's capacity to store and retrieve and are actively researching the brain's ability to yield "classified" information.

The peripheral nervous system

Just as your central nervous system can be compared to the CPU of your computer, your peripheral nervous sytem can be compared to the cables that connect the CPU to the other pieces of hardware—the "peripherals"—that make up your computer. In your computer, connector cables carry information to the monitor to be visualized on the screen or to the printer to be printed. So too the nerve "cables" of your peripheral nervous system bring information to your muscles, instructing them to contract and relax in patterns that result in the integrated and flowing movements of your body. For example, food in your stomach (the input) sends signals along your peripheral nervous system to the medulla to be interpreted, resulting in the release of gastric juice and the churning of the muscular stomach walls (the output). These are only two examples of how the peripheral nervous system works to connect the integrative portion of your nervous system with its **receptors** (structures that detect stimuli) and **effectors** (muscles and glands that respond to that stimuli).

Sensory pathways

Seeing a sunset, hearing a symphony play, feeling tortured with pain—all these stimuli travel along sensory neurons to the central nervous system and arrive there in the same form—as nerve impulses. Put simply, every nerve impulse is identical to every other one. How, then, does the brain distinguish between pleasure and pain or sight and sound? Interestingly, the information that the brain derives from sensory input is based solely on the source of the impulse and its frequency. Thus if the auditory nerve is artificially stimulated, the central nervous system perceives the stimulation as sound. If the optic nerve is artificially stimulated in exactly the same manner and degree, the stimulation is perceived as a flash of light. Increasing the intensity of the stimulation of either receptor will result (within limits) in an increase in the frequency of nerve impulses and therefore produce a perception of a brighter light or a louder noise.

Many kinds of receptors have evolved among vertebrates, with each receptor sensitive to a different aspect of the environment. Sensory receptors are able to change spe-

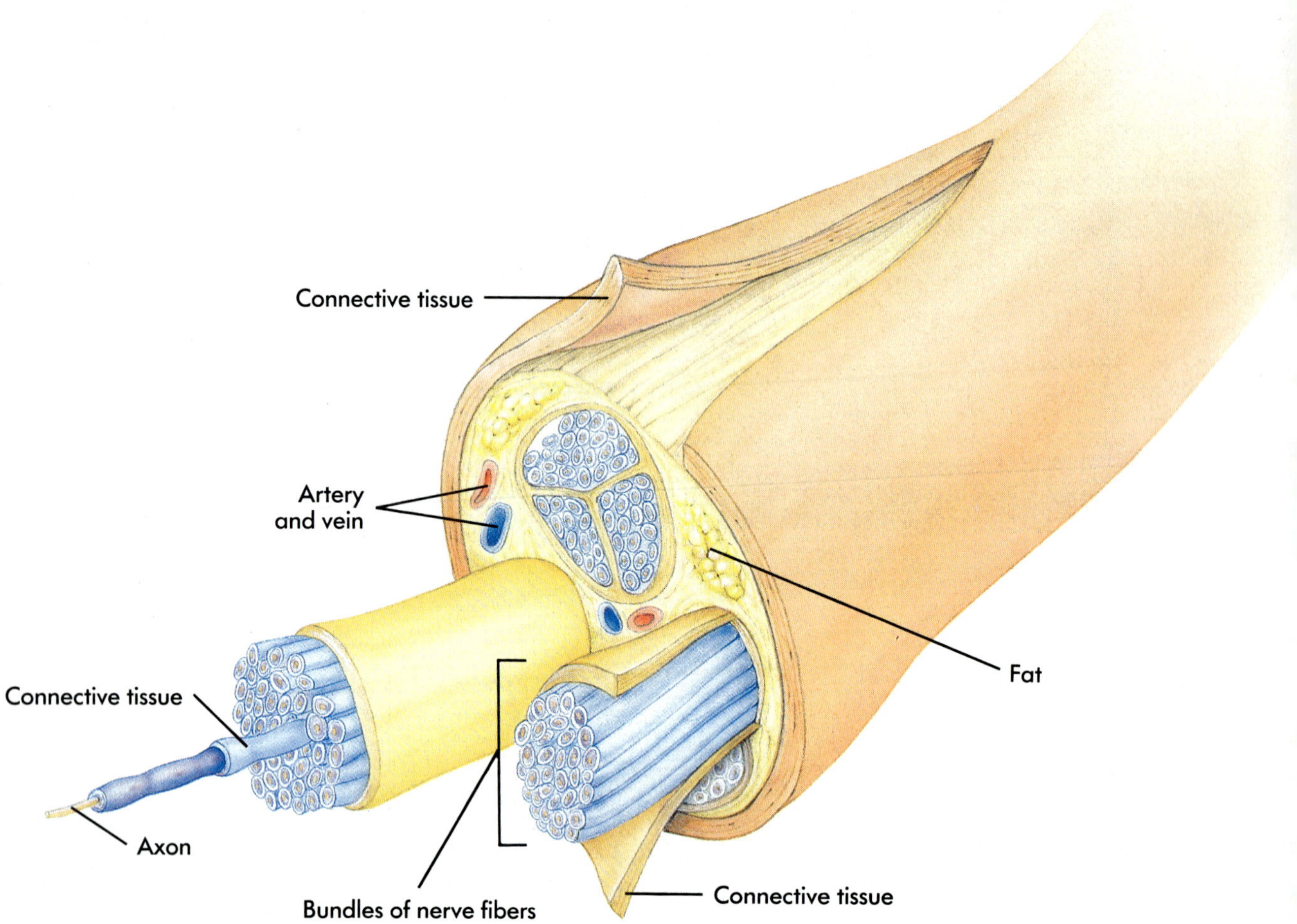

FIGURE 16-11 Structure of a nerve.
A nerve is composed of bundles of nerve fibers and various layers of connective tissue.

cific stimuli into nerve impulses by having low thresholds for specific types of stimuli and high thresholds for others. Receptors in the retina of your eye, for example, have a low threshold for light. Appropriate stimuli open ion channels within the membrane of the receptor cell, thus depolarizing it (see Chapter 15). This depolarization is called a **generator potential.** When the generator potential reaches the threshold level, it initiates an action potential in the sensory neurons with which the receptor synapses. Simple receptors, such as pain receptors in the skin, may be simply the dendrite endings of sensory neurons. The generator potential then initiates an action potential along that same neuron.

Sensory nerve fibers that enter the brain directly are axons, not dendrites as in the peripheral nervous system. The cell bodies and dendrites of these sensory neurons are located outside the brain in the sense organ. Their axons are bundled into groups with other sensory nerve fibers arising from the same area to form certain **cranial nerves.** The cranial nerves that conduct only sensory stimuli are those associated with sight, sound, smell, and equilibrium, such as cranial nerve I (see Figure 16-8); the olfactory nerve, which brings in "smell" messages from your nose; cranial nerve II, the optic nerve, which brings in "sight" messages from your eyes; and cranial nerve VIII, which brings impulses from your ears regarding hearing and balance. Other cranial nerves called **mixed nerves** are made up of both sensory (incoming or afferent) and motor (outgoing or efferent) nerve fibers, such as cranial nerve VII, which shuttles messages to and from your salivary and tear glands, and cranial nerve X (the vagus nerve), which regulates the function of your heart rate, respiration rate, and digestive activities. The motor nerve fibers of cranial nerves are also axons; the cell bodies of these motor neurons are located within the brain. Cranial nerve XII, which stimulates movement in your tongue, carries outgoing (motor) messages only.

Sensory and motor nerve fibers that enter the spinal cord directly make up the spinal nerves (Figure 16-11). Nerve fibers that serve the same general area of the body, whether of sensory or motor neurons, are bundled together to form these nerves (see Figure 16-1). Thus all spinal nerves are mixed nerves. Close to the spinal cord, however, motor and sensory fibers separate from one another. The sensory (afferent) fibers are myelinated axon-like dendrites that extend from the source of stimulation (your fingers or toes, for example) to a swelling near their entrance to the dorsal (back) side of the spinal cord. This swelling is a ganglion containing the nerve cell bodies of these sensory neurons. The short axons of these neurons extend from this ganglion to the gray matter of the spinal cord. Here, the axon ends of these sensory neurons synapse with neurons that play a role in integrating incoming messages with outgoing responses.

Integration

"Integrating" neurons, or **interneurons,** receive incoming messages and send appropriate outgoing messages in response. Interneurons extend from the spinal cord to the brain, make up part of the brain tissue itself, and in the case of certain reflex pathways, are located within the gray matter of the spinal cord. The pathway of a nerve impulse from stimulus to response may involve a single interneuron, as in the case of certain reflexes. In a few of the simplest reflexes, a sensory neuron may synapse directly with a motor neuron, with no interneuron(s) playing a role. Most pathways, however, involve many interneurons that, in sequence, receive the incoming message, direct it to other interneurons for interpretation (such as specific areas of the cerebrum), and bring the outgoing message to the appropriate motor neurons.

Motor pathways

The short dendrites of spinal motor neurons, located in anterior portions of the gray matter of the spinal cord, synapse with interneurons. These dendrites then conduct the impulses "picked up" from the interneurons to their cell bodies located within the same area of gray matter of the spinal cord. The nerve impulse sweeps along the membrane of the axon of each motor neuron to effectors: muscles or glands that produce a response. Motor pathways to skeletal muscles contain a single motor neuron; motor pathways to smooth and cardiac muscles and to glands are made up of a series of two motor neurons.

The bundles of axons leaving the spinal cord join with the dendrites of sensory neurons entering the same level of the cord, forming spinal nerves. A total of 31 pairs of spinal nerves brings messages to and from specific areas of the body. Figure 16-1 shows the approximate location of the spinal nerves and, in general, the areas of the body that they serve.

Sensory receptors change stimuli into nerve impulses. The nerve fibers of sensory neurons then bring this information to the brain or spinal cord where, except in the case of very simple reflexes, interneurons interpret them and send appropriate outgoing messages. The axons of motor neurons conduct these impulses to muscles and glands, the structures that effect a response.

The somatic nervous system

The central nervous system directs different types of responses in different ways. Movements of the skeletal muscles are controlled by messages from the brain and spinal cord via pathways containing a single motor neuron. These motor pathways are referred to as the *somatic nervous system.* Messages along these pathways coordinate your fingers when you grasp a pencil, spin your body when you dance, and outstretch your arms when you hug a friend. These muscular movements are primarily subject to conscious control by the associative cortex of the cerebrum. However, not all movements of the skeletal muscles are conscious. Blinking your eyes, putting one foot in front of the other when you walk, and breathing can be consciously controlled but most often take place without conscious thought.

The somatic branch of the peripheral nervous system consists of motor fibers that send messages to the skeletal muscles.

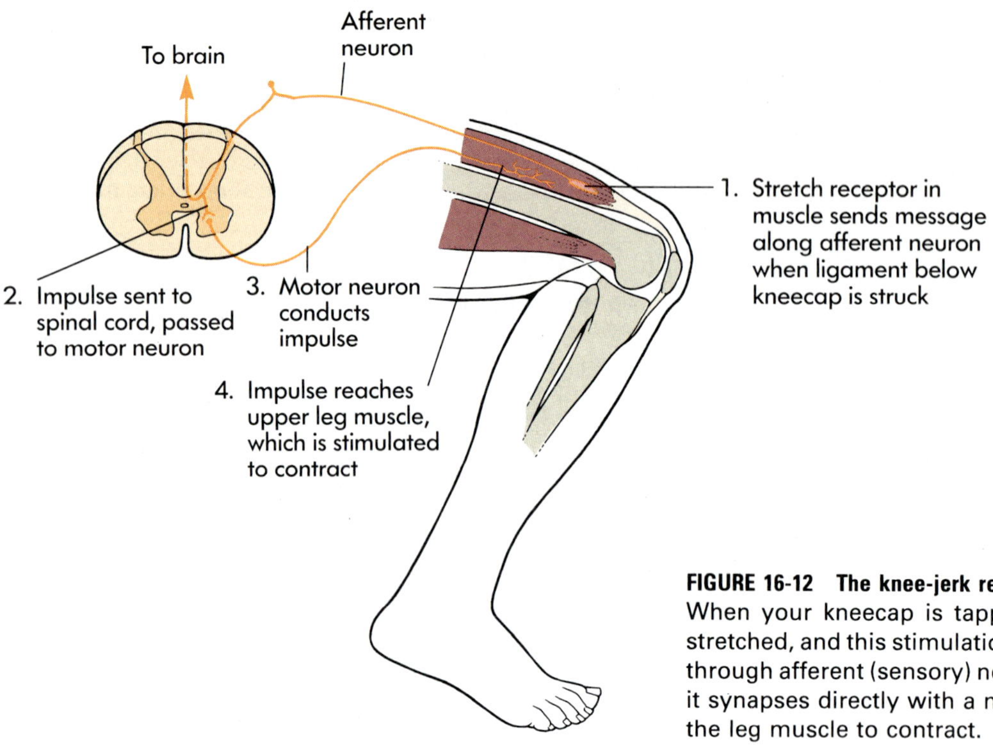

FIGURE 16-12 The knee-jerk reflex.
When your kneecap is tapped, the muscle in your leg is stretched, and this stimulation of the stretch receptor travels through afferent (sensory) neurons to the spinal cord, where it synapses directly with a motor neuron, which stimulates the leg muscle to contract.

A *reflex* is an automatic response to nerve stimulation. Very little, if any, integration (and certainly no thinking) takes place during reflex activity. The knee jerk is one of the simplest types of reflexes in the human body; a sensory neuron synapses directly with a motor neuron (Figure 16-12). This pathway—the pathway an impulse follows during reflex activity—is called a *reflex arc.* Simple reflex arcs such as the knee jerk are called **monosynaptic reflex arcs** (literally, "one synapse"). A monosynaptic reflex arc is not subject to control by the central nervous system. Most voluntary muscles within your body possess such monosynaptic reflex arcs, although usually in conjunction with other more complex reflex pathways. It is through these more complex paths that voluntary control is established. In the few cases in which the monosynaptic reflex arc is the only feedback loop present, its function can be clearly seen, such as in the knee jerk reaction. If the ligament just below the kneecap is struck lightly by the edge of your hand or by a doctor's rubber mallet, the resulting sudden pull stretches the muscles of the upper leg, which are attached to the ligament. Stretch receptors in these muscles immediately send an impulse along afferent nerve fibers to the spinal cord, where these fibers synapse directly with motor neurons that extend back to upper leg muscles, stimulating them to contract and the leg to jerk upward. Such reflexes play an important role in maintaining posture.

A reflex is an automatic response to nerve stimulation. The pathway of nervous activity in a reflex is called a *reflex arc.* Impulses travel along sensory neurons to the spinal cord, which sends a message out via motor neurons without involving the brain.

The autonomic nervous system

When you enter a dark movie theater, the pupils of your eyes dilate, or become larger. When you are frightened, adrenaline is pumped into your bloodstream and your heart beats faster. Both of these reactions are not governed by conscious thought—they occur automatically—and involve the action of smooth muscle, cardiac muscle, or glands. The motor pathways that control such involuntary and automatic responses of the glands and nonskeletal muscles of the body are referred to as the *autonomic nervous system.* The autonomic nervous system takes your temperature, monitors your blood pressure, and sees to it that your food is properly digested. The body's internal physiological condition is thus fine tuned within relatively narrow bounds, a regulatory process called *homeostasis.*

The autonomic nervous system is made up of two divisions, the **parasympathetic nervous system** and the **sympathetic nervous system,** that act in opposition to each other, speeding up or slowing down certain bodily processes (Figure 16-13). Each of these motor pathways consists of a series of two motor neurons. In the parasympathetic system the axons of the motor neurons leaving the spinal cord extend to ganglia located near the muscles or organs they affect. In the sympathetic system the axons leaving the spinal cord are much shorter and extend only to ganglia located near the vertebral column. The axons of second motor neurons in each system extend from these ganglia to their "targets."

The neurotransmitters used by each of these two branches of the autonomic nervous system differ at the axon ends of the second motor neuron—at the synapse with the target organ. The actions of these different neurotransmitters oppose each other. Thus because each gland (except the

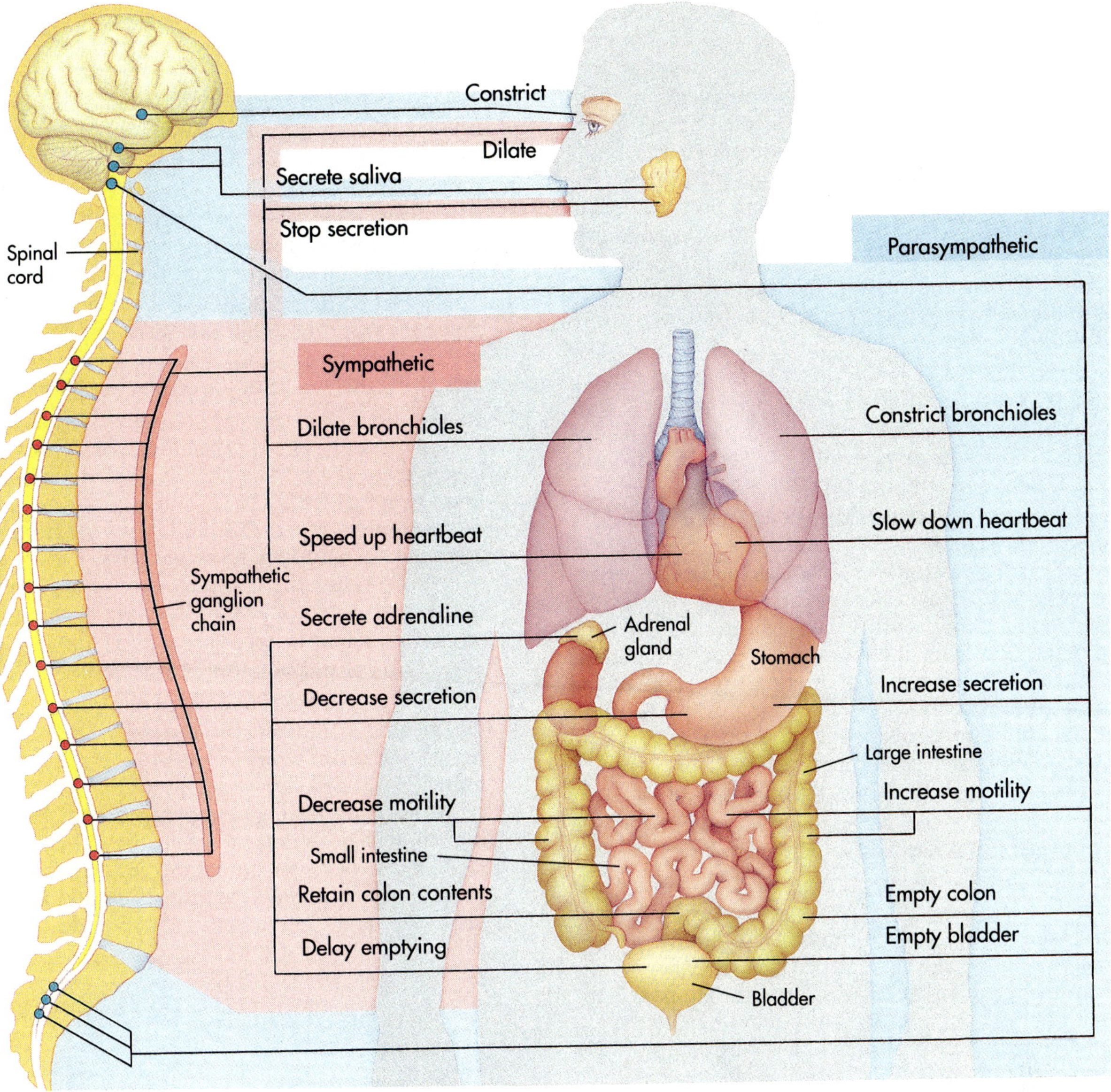

FIGURE 16-13 The sympathetic and parasympathetic nervous systems.
The ganglia of sympathetic nerves are located near the spine, and the ganglia of parasympathetic nerves are located far from the spine, near the organs they affect. A nerve runs from both of the systems to every organ indicated, except the adrenal gland.

inner portion of the adrenal gland), smooth muscle, and cardiac muscle is "wired" to *both* the systems, an arriving signal will either stimulate or inhibit the organ. For example, the sympathetic system speeds up the heart and slows down digestion, whereas the parasympathetic system slows down the heart and speeds up digestion. In general, the two opposing systems are organized so that the parasympathetic system stimulates the activity of normal body functions such as the churning of the stomach, the contractions of the intestine, and the secretions of the salivary glands. The sympathetic system, on the other hand, generally mobilizes the body for greater activity, as in increased respiration or a faster heartbeat. The "decision" to stimulate or inhibit a muscle, organ, or gland is "made" by the central nervous system.

The autonomic nervous system is a branch of the peripheral nervous system that consists of two antagonistic sets of motor fibers. Messages sent along these fibers control smooth and cardiac muscle as well as glands.

Summary

1. The nervous system consists of the peripheral nervous system and the central nervous system. The peripheral nervous system, made up of the nerves of the body, gathers information about the internal and external environments and brings it to the central nervous system via sensory nerve cells. The central nervous system, made up of the brain and spinal cord, then processes and responds to that information. These responses are sent out to the body via the motor nerve cells of the peripheral nervous system.
2. The cerebrum, the dominant part of the human brain, is split into two hemispheres connected by a nerve tract called the *corpus callosum*. Each hemisphere is divided further by deep grooves into four lobes: the frontal, parietal, temporal, and occipital lobes.
3. Various cognitive activities take place in the outer gray matter of the cerebrum, or cerebral cortex. Specific sensory areas are located in the various lobes of the cortex. One large sensory area lies on the leading edge of the parietal lobe. Other sensory areas, the auditory and visual areas, lie within the temporal and occipital lobes, respectively. The motor region is mostly a part of the frontal lobe. Association areas connect all parts of the cerebral cortex and govern such functions as memory, reasoning, intelligence, and personality.
4. The cerebral white matter, myelinated nerve fibers that lie beneath the cerebral cortex, send messages from one part of the central nervous system to another.
5. The basal ganglia, groups of nerve cell bodies located at the base of the cerebrum, control large, subconscious movements of the skeletal muscles. Additional groups of nerve cells bodies also located near the base of the cerebrum but not part of it are the thalamus and the hypothalamus. These parts of the brain receive and interpret certain sensory stimuli and control the activities of various body organs and glands, respectively.
6. The cerebellum is the part of the brain that coordinates unconscious movements of the skeletal muscles.
7. The brainstem consists of three parts: the midbrain, pons, and medulla. All three parts contain nerve fibers that bring messages to and from the spinal cord. In addition, the brainstem contains groups of nerve cell bodies that control important body reflexes.
8. The spinal cord brings messages to the brain and sends messages from the brain out to the body. It also integrates the incoming and outgoing information of reflex arcs, with the exception of certain simple reflexes.
9. In the peripheral nervous system, sensory receptors change stimuli into nerve impulses. These impulses travel along the nerve fibers of sensory neurons to the brain or spinal cord. There, "integrating" neurons called *interneurons* interpret this sensory information and send appropriate outgoing messages. The axons of motor neurons conduct these impulses to muscles and glands, the structures that effect a response.
10. No matter what the stimulus, a nerve impulse takes the same form: depolarization of a nerve fiber membrane.
11. The motor pathways are divided into somatic pathways, which relay commands to skeletal muscles, and autonomic pathways, which stimulate the glands and other muscles of the body. The autonomic pathways consist of nerve pairs having antagonistic neurotransmitters, one of which stimulates while the other inhibits. In general, the parasympathetic nerves stimulate the activity of normal internal body functions and inhibit alarm responses, and the sympathetic nerves do the reverse.

KEY TERMS

afferent neuron 264
association area 266
autonomic nervous system 264
basal ganglia 267
central nervous system 264
cerebellum 265
cerebral cortex 266
cerebrospinal fluid 272
cerebrum 265
corpus callosum 266
cranial nerve 275
diencephalon 265
effector 273
efferent neuron 264
foramen (foramina) 271
frontal lobe 266
ganglion (ganglia) 267
generator potential 275
gray matter 266
hypothalamus 269
integration 265
interneuron 275
limbic system 269
medulla 271
meninge 272
midbrain 271
mixed nerve 275
monosynaptic reflex arc 276
motor area 266
motor neuron 264
nerve 269
nervous system 263
nuclei 267
occipital lobe 266
parasympathetic nervous system 276
parietal lobe 266
peripheral nervous system 264
pons 271
receptor 273
reflex 264
reflex arc 271
reticular formation 271
sensory area 266
sensory neuron 264
somatic nervous system 264
spinal cord 264
spinal nerve 271
sympathetic nervous system 276
temporal lobe 266
thalamus 269
tract 269
ventricle 272

REVIEW QUESTIONS

1. Distinguish among the nervous system, central nervous system, and peripheral nervous system. What are the functions and components of each?
2. Fill in the blanks: Nerve cells that transmit information to the central nervous system are ________ ________ (also known as ________ ________); those that carry commands away from the central nervous system are ________ ________ (________ ________).
3. Distinguish between the somatic and autonomic nervous systems. What type of response does each control? Give an example of each type.
4. Draw a picture of the human brain and label its four main parts.
5. Fill in the blanks: The ________ ________ (also called the ________ ________) is the major site of "higher" cognitive processes. Most of it consists of ________ ________ where these higher activities take place. The ________ ________ sends message to move your skeletal muscles, whereas the ________ ________ receives sensory information from receptors throughout your body.
6. Distinguish between the thalamus and hypothalamus. What are their respective functions?
7. Identify and give a function for each of the following:
   a. Nerves
   b. Limbic system
   c. Cerebellum
   d. Cerebrospinal fluid
8. What is the brainstem? Name its three major components and summarize its functions.
9. Describe the spinal cord, and summarize its functions. Why is it important?
10. You're studying in a quiet library. Suddenly someone drops an armful of books; you hear the loud crash and turn your head toward the noise. Explain how your sensory and motor pathways allowed you to detect and respond to this stimulus.
11. In question 10, why did you perceive this stimulus as a loud noise (rather than, say, as a bright light)?
12. What is a reflex? Give an example.
13. Compare the effects of the parasympathetic and sympathetic nervous systems on your body. Why is this important?

THOUGHT QUESTION

1. The organs of your body are controlled by two branches of the autonomic nervous system, the sympathetic and parasympathetic. Each organ is reached by neurons from each system. Why *two* systems? Why use two neurons to communicate with an organ rather than one?

FOR FURTHER READING

Allport, S. (1986) *Explorers of the black box: The search for the cellular basis of memory.* New York: W.W. Norton & Co.
This is a vivid account of the pioneering studies of Eric Kandel and others in their efforts to demonstrate how we remember. Easy to read, this book shows scientists in action, gathering data and disputing among themselves about what the data mean.

Corina, D. P., Vaid, J., & Bellugi, V. (1992). The linguistic basis of left hemisphere specialization. *Science, 255,* 1258-1261.
The left hemisphere of the human brain seems to specialize in linguistics (spoken language) but not in sign language or symbolic (pantomine) language.

Kalaska, J. F., & Crammond, D. J. (1992, March 20). Cerebral cortical mechanisms of reaching mechanisms. *Science,* pp. 1517-1523.
This article reports on the role of the central nervous system in the planning and execution of a reaching movement.

Winter, P., & Miller, J. (1985, April). Anesthesiology. *Scientific American,* pp. 124-131.
This article is a clear description of how certain drugs act to block consciousness.

17 The senses

Can pleasure block out pain? These Pakistani men are flagellating themselves with knives at a Muslim festival. Flagellation is a religious disciplinary or devotional practice of beating oneself. This practice has been used among peoples of various religious sects and is documented as far back as ancient Rome. It serves various religious purposes, such as rites of penance, purification, and fertility. And, as you can see from the photo, those practicing this ritual do not appear to be experiencing tremendous pain.

OUTLINE

The nature of sensory communication

Sensing the body's internal environment

Sensing the body's position in space

Sensing the external environment

- The general senses
- The special senses
  - Smell
  - Taste
  - Sight
  - Hearing
  - Balance

HIGHLIGHTS

Sensory receptors are the body's "window on the world," detecting stimuli from both the external and internal environments.

Stimuli, or changes in the environment, are detected when they are of a sufficient magnitude to open ion channels within the membranes of receptor cells, depolarizing them and thus initiating action potentials along those neurons or adjacent neurons.

The body has a mosaic of receptors referred to as the *general senses,* which detect stimuli such as pain, pressure, touch, and temperature within the body's internal and external environments.

The body has a few specialized receptors referred to as the *special senses,* which detect smell, taste, sight, hearing, and balance.

Pain receptors are distributed widely throughout the body and are very important to your well-being; they are the body's way of knowing when it is in danger of being harmed. Simple in structure, pain receptors are the branching ends of dendrites of certain sensory neurons. These neurons bring the pain message to the spinal cord, where it is transferred to tracts of neurons that travel up the cord to the brain.

In 1965, Ronald Melzack and Patrick Wall described a means by which the body could short-circuit its own warning system. They called this revolutionary idea the *gate-control theory of pain*. Melzak and Wall hypothesized that signals coming down from the brain could put a stop to signals going up the spinal cord. In this way, the "gate" was shut to certain incoming sensory messages before they got to the brain. In the same way, they explained, people "tune out" background noise to focus on a task, or football players continue to play the game with a broken leg, not realizing their situation. Current physiological research supports the idea that various parts of the brain exert control over the perception of pain. In fact, psychologists who treat chronic pain sufferers use the brain's ability to focus on certain sensory information and screen out other data to help patients develop strategies for the control of pain.

The nature of sensory communication

Pain is not a type of environmental stimulus itself but can be caused by various types of stimuli such as heat, cold, and pressure. Sensory receptors can change environmental stimuli such as these into nerve impulses. Some receptors are composed of nervous tissue; others are not but are capable of initiating a nerve impulse in an adjacent neuron. Many kinds of receptors have evolved among vertebrates (organisms having a backbone), with each receptor sensitive to a different aspect of the environment (Figure 17-1).

For you to be aware of, or to sense, your internal or external environment, certain events must take place. First, a change in the environment (the stimulus) must be of sufficient magnitude to open ion channels within the membrane of the receptor cell, thus depolarizing it (see Chapter 15).

B

C

A

FIGURE 17-1 Types of sensory information processed by sensory receptors.
A A person who feels sick is relying on information about the body's internal environment. A headache, for example, causes pain because of the activation of pain receptors located within the body.
B This dancer maintains a proper stance by using information about his position in space.
C This person is reading, using her eyes to perceive patterns of light. In this way, the reader is sensing information about the external environment.

This depolarization is called a *generator potential*. When the generator potential reaches the threshold level, it initiates an action potential—a nerve impulse—in the sensory neurons with which the receptor synapses. In simple receptors (the dendrite endings of sensory neurons) the generator potential initiates an action potential along that same neuron. The impulse is conducted by nerve fibers to either the spinal cord or the brain.

Impulses conducted to specific sensory areas of the cerebral cortex produce conscious sensations (see Chapter 16). Only the cerebral cortex can "see" a flower, "hear" a symphony, or "feel" a paper cut. Although not a part of the cortex, the thalamus can sense pain, but it is unable to distinguish its source or intensity. Impulses that end at the spinal cord or brainstem do not produce conscious sensations. They may, however, result in reflex activity such as the rhythmic contraction of the muscles of breathing or the turning of your head toward a startling noise.

Receptors provide the body with three types of information:

1. Information about the body's internal environment
2. Information about the body's position in space
3. Information about the external environment

Receptors that sense the internal environment are located deep in the body within the walls of blood vessels and organs. They tell you when you are hungry, thirsty, sick, or tired. Receptors sensitive to stimuli ouside the body are located at or near the body surface. They allow you to see, hear, taste, and feel various stimuli in the environment. Receptors that provide information about body position and movement are located in the muscles, tendons, joints, and inner ear. They tell you whether you are lying down or standing up and where the various parts of your body are in relationship to one another.

Sensory receptors are cells that can change environmental stimuli into nerve impulses. Specific receptors detect certain stimuli that inform the body about both its external and internal environments and its position in space.

Sensing the body's internal environment

Many of the neurons that monitor body functions are simply nerve endings that depolarize in response to direct physical stimulation—to temperature, to chemicals like carbon dioxide diffusing into the nerve cell, or to a bending or stretching of the neuron cell membrane. Among the simplest of these neurons are those that report on changes in body temperature and blood chemistry.

Temperature-sensitive neurons in the hypothalamus, a tiny portion of the brain that lies above the brainstem, act as your body's thermostat. These neurons constantly "take your temperature" by monitoring the temperature of your blood. If the temperature of the blood rises, such as when you are sick or vigorously exercising, neurons in the hypothalamus trigger your body's heat loss mechanisms. Such mechanisms include dilating, or widening, the blood vessels closest to the skin so that excess heat can be lost to the environment. Likewise, if the temperature of the blood drops, such as when you are in a cold environment, neurons in the hypothalamus trigger your body's heat production mechanisms. Such mechanisms include shivering, a cycle of contraction and relaxation of your skeletal muscles producing body heat as a waste product of the cellular respiration that generates ATP for muscle contraction.

Other receptors are sensitive to the levels of carbon dioxide and oxygen in your blood, as well as its pH. These receptors are embedded within the walls of your arteries at several locations in the circulatory system. Bathed by the blood that flows through the arteries, these chemical receptors provide input to respiratory centers in the medulla and pons, which use this information to regulate the rate of respiration. When carbon dioxide and pH levels in the blood rise and the oxygen level falls, the respiratory centers respond by increasing the respiration rate.

A variety of other receptors that sense the environment within your body have membranes whose ion channels open in response to mechanical force. Put simply, twisting, bending, or stretching these nerve endings results in depolarization of their membranes, causing these nerves to "fire." These receptors differ from one another primarily in their locations and in the way that they are oriented with respect to the stimulus. Pain receptors, for example, are widely distributed throughout the body. They respond to many different types of stimuli when these stimuli reach a level that can endanger the body. Pain receptors deep within the body detect such internal environmental stresses as inadequate blood flow to an organ, excessive stretching of a structure, and spasms of muscle tissue.

Specialized mechanical receptors sensitive to pressure changes are attached to the walls of three major arteries of your body (the carotids and the aorta) and constantly monitor your blood pressure. These highly branched networks of nerve endings detect the stretching of the walls of these arteries caused by the "push" of the blood as it is pumped out of the heart. These neurons fire at a low rate at all times, sending a steady stream of impulses to the cardiac (heart) center in the medulla. A rise in blood pressure stretches the walls of these vessels, resulting in a higher frequency of nerve impulses, which inhibits the cardiac center. A drop in blood pressure lessens the stretch, resulting in a lower frequency of nerve impulses, which stimulates the cardiac center.

Receptors that sense the body's internal environment are located within the walls of blood vessels and organs.

Sensing the body's position in space

Your body uses receptors called **proprioceptors** to sense its position in space. The term *proprioceptor* comes from a Greek word meaning "self." Proprioceptors tell you how much your arms and legs are bent, where they are in relationship to your body, and where your head is relative to the ground . . . even if you cannot see! They also tell you whether your body is moving.

Proprioceptors buried deep within your muscles keep

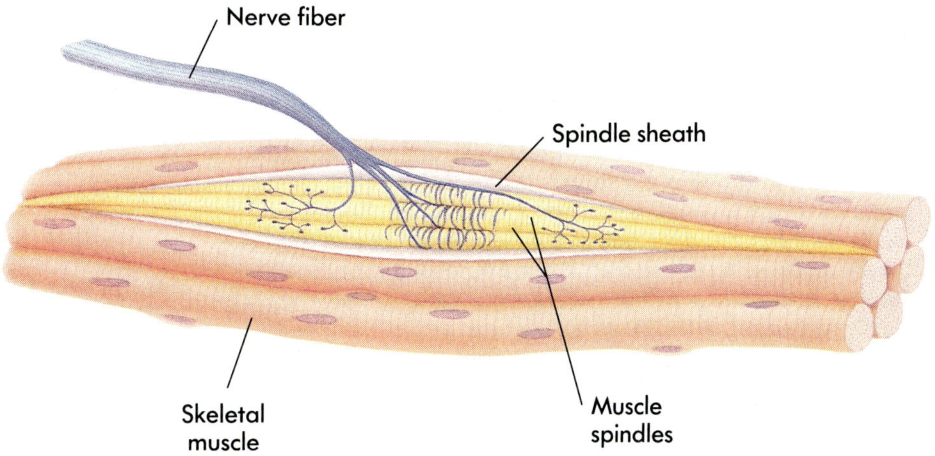

FIGURE 17-2 A stretch receptor embedded within skeletal muscle.
Stretching of the muscle elongates the spindle fibers, which deforms nerve endings, causing them to fire and send a nerve impulse out along the nerve fiber.

track of the degree to which your muscles are contracted (Figure 17-2). These receptors are actually specialized muscle cells called *muscle spindles*. Wrapped around each spindle is the end of a sensory neuron called a **stretch receptor.** When a muscle is stretched, the muscle spindle gets longer and stretches the nerve ending, repeatedly stimulating it to fire. Conversely, when the muscle contracts, the tension on the fiber lessens and the stretch receptor ceases to fire. The frequency of stretch receptor discharges along the nerve fiber "tells" the body the degree to which a muscle is contracted at any given moment. The central nervous system uses this information to control those movements that involve the combined action of several muscles, such as those involved in breathing and walking. Other proprioceptors are found in your tendons, the connective tissue that joins your muscles to your bones, and in the tissue surrounding your joints. These receptors are also stimulated when they are stretched and help protect your muscles, tendons, and joints from excessive tension and pulling.

Gravity and motion receptors are other types of proprioceptors. These receptors, both located in the inner ear, help you maintain your equilibrium, or balance. They are discussed later in the chapter.

Receptors called *proprioceptors,* located within the skeletal muscles, tendons, and inner ear, give the body information about the position of its parts relative to each other and to the pull of gravity.

Sensing the external environment

A variety of receptors is located at or near the body surface that senses changes in the external environment. Many types of simple, microscopic receptors are located within the skin. These receptors detect touch, pressure, pain, and temperature—the so-called general senses. Other, more structurally complex receptors are located in specific places at or near the body surface. These receptors detect smell, taste, sight, and hearing and help you keep your balance. They are your special senses.

The general senses

The body has a mosaic of mechanical receptors in the skin for touch, pressure, and pain. These receptors are the same as those found deeper in the body, but because they are located close to the surface of the skin, they provide information about the external rather than the internal environment. Figure 17-3 is a diagram of a cross section of human skin showing these specialized nerve endings.

Disk-shaped dendrite endings called *Merkel's disks* and egg-shaped receptors called *Meissner's corpuscles* are two types of touch receptors. (The word *corpuscle* means "little body.")

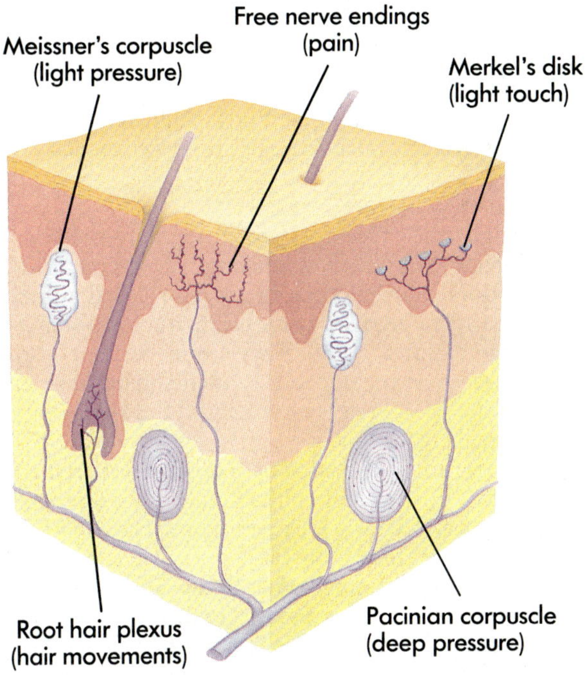

FIGURE 17-3 Receptors in the human skin.
These receptors, although they are found within the body, sense the external environment because they are located close to the surface of the skin.

Both are widely distributed in the skin but are most numerous in the hands, feet, eyelids, tip of the tongue, lips, nipples, clitoris, and tip of the penis. In addition, free nerve endings wrap around the roots of hairs and detect any stimulus, such as the wind or an insect, that moves body hair. Pressure receptors called *pacinian corpuscles* are located deeper within the skin. As shown in Figure 17-3, pacinian corpuscles look layered, much like an onion. The layers are made up of connective tissue with dendrites sandwiched between. These receptors are most numerous in the nipples and external genitals of both sexes. Pacinian corpuscles are also a part of the body's internal sensing system because of their locations around joints and tendons, in muscles, and in certain organs.

There are two populations of nerve endings in the skin that are sensitive to changes in temperature. One set is stimulated by a lowering of temperature (the cold receptors) and the other by a raising of temperature (the heat receptors). Scientists are unsure how these receptors work. However, evidence suggests that changes in the temperature cause changes in the shapes of the proteins that make up the sodium channels of their cell membranes, leading to the depolarization and "firing" of these neurons.

Various receptors in the skin sense touch, pressure, pain, and temperature stimuli from the external environment.

The special senses

The special senses—smell, taste, sight, hearing, and balance—have receptors that are much more specialized and complex than the receptors for the general senses. Smell and taste receptors sense chemicals, whereas hearing receptors sense sound waves. Other receptors in the ears detect motion and the effect of gravity on the body. Sight receptors sense wavelengths of electromagnetic energy known as *visible light*.

Smell

Although smell receptors are structurally the simplest of your special senses, they can distinguish several thousand different odors. Smell receptors are also called **olfactory receptors,** a name derived from Latin words meaning "to make" *(facio)* "a smell" *(oleo)*. They are *not* located (surprisingly) in your nostrils, but in a ½ inch square of tissue in the roof of the nasal cavity—just behind the bridge of your nose (Figure 17-4).

Olfactory receptors consist of neurons whose cell bodies are embedded in the nasal epithelium, the tissue that lines the nasal cavity. Dendrites extend from these cell bodies. Cilia, microscopic projections of the dendrites, poke out of the epithelium like minute tufts of hair and are bathed by the mucus that covers this tissue. Gases in the air dissolve in the mucus and come into contact with the cilia.

Scientific evidence suggests that the olfactory receptors detect different smells because of specific binding of airborne gases with receptor chemicals located within the cilia. This interaction opens ion channels within the membrane of the receptor so that a generator potential is developed, firing

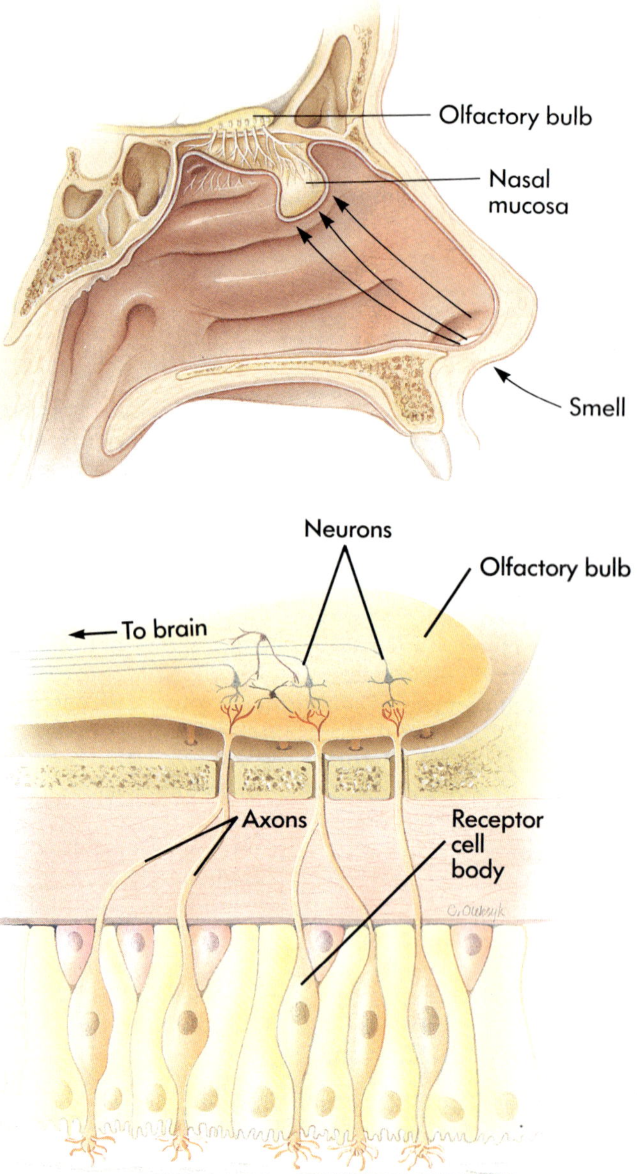

FIGURE 17-4 The sense of smell.
Human beings smell by means of olfactory receptor cells located in the lining of the nasal cavity. The receptor cells are neurons. Airborne gases combine with chemicals within the cilia on the neurons' dendrites, and the neuron is depolarized. Axons from these sensory neurons carry the impulse back through the olfactory nerve directly to the cerebral cortex in the brain.

the neuron when it reaches a particular threshold. From here, the nerve impulse travels to the olfactory area of the cerebral cortex to be interpreted. On its way, it travels through the brain's limbic system (see Chapter 16), the area of the brain responsible for many drives and emotions. Does the odor of baking cookies cause you pleasure? Does the smell of rotting garbage cause you to turn your head in disgust? That is your limbic system at work—in conjunction with your cerebrum and your sense of smell.

The smell, or olfactory, receptors are located in the nasal epithelium and detect airborne chemicals as they bind with receptor chemicals.

Taste

Taste receptors, or **taste buds,** detect chemicals in the foods you eat. Human beings have four kinds of taste buds located on the tongue that each respond to a different class of stimulus: salty, sweet, sour, and bitter. Each type of taste bud is concentrated on different areas of the tongue, with sweet and salty on the front, sour on the sides, and bitter at the back (Figure 17-5). Humans, of course, perceive a rich and diverse array of tastes that are much more complex than simply salty, sweet, sour, and bitter. Interestingly, these tastes are composed of different combinations of impulses from just these four types of chemoreceptors. In addition, the sense of taste interacts with the sense of smell to produce the taste sensation, as you may have noticed when you had a "stuffy nose" and could perceive only the four basic tastes—those not dependent on smell for their perception.

Nearly 9000 taste buds are packed within the short projections on your tongue known as *papillae*. Figure 17-5 shows a drawing of a typical papilla on the tongue and points out the location of taste buds deep within the papilla. Taste buds are microscopic structures shaped like tiny onions. Each is made up of 30 to 80 receptor cells bound together by support cells. Hair-like projections of the receptor cells poke through an opening in the taste bud called a *taste pore*. The receptor cells are stimulated by the various chemicals in food as these chemicals dissolve in the saliva and come into contact with the cellular "hairs." A certain threshold of stimulation produces generator potentials within the receptor cells, firing neighboring sensory nerve fibers. These nerve fibers travel from the tongue and throat to the cerebrum, medulla, and thalamus. Some of these nerve fibers also travel to the reticular formation (see Chapter 16), the part of the brain that monitors incoming stimuli, identifying "important" information.

The taste receptors, or taste buds, are microscopic chemoreceptors embedded within the papillae of the tongue. They work with the olfactory receptors to produce the taste "sensation."

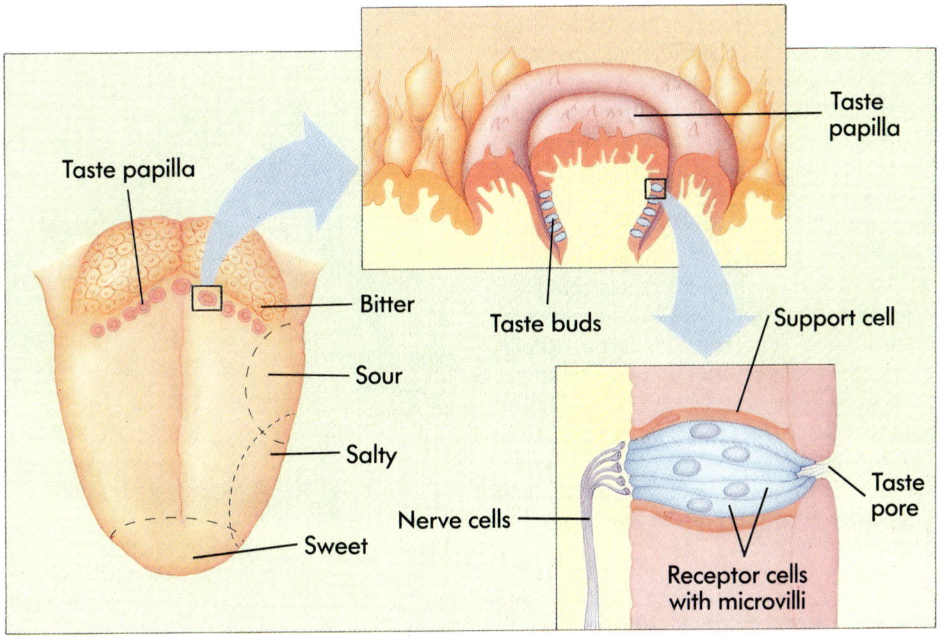

FIGURE 17-5 Taste.
Human beings have four kinds of taste buds (bitter, sour, salty, and sweet) located on different regions of the tongue. Groups of taste buds are typically organized in sensory projections called papillae. Individual taste buds are bulb-shaped collections of chemical receptor cells that open out into the mouth through a pore.

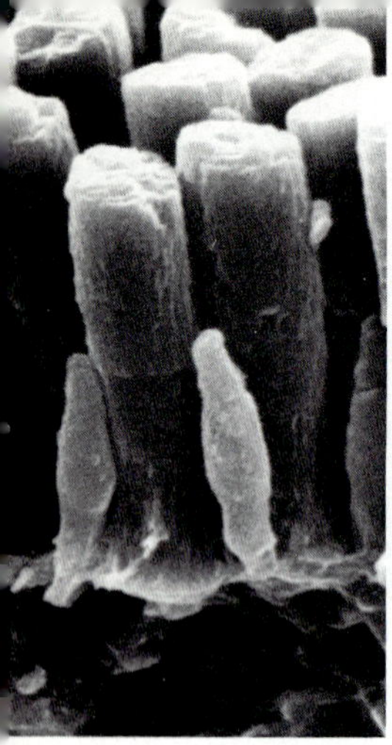

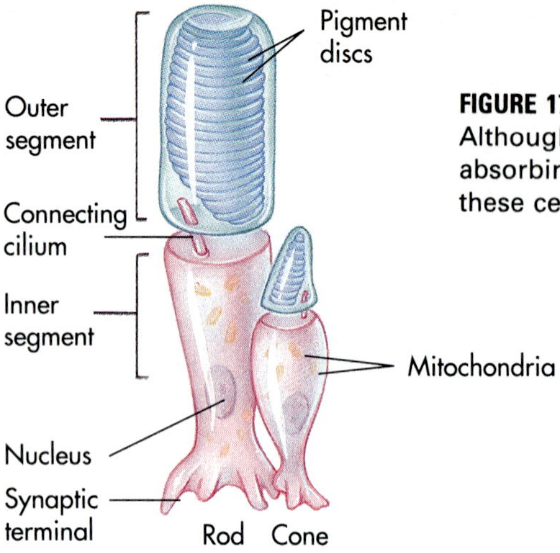

FIGURE 17-6 The structure of rods and cones.
Although not obvious from the electron micrograph, light-absorbing pigment is contained in the outer segments of these cells.

Sight

The eye is the only human sense organ that can detect wavelengths of electromagnetic energy known as *visible light.* (The concept of electromagnetic energy is described in Chapter 8.) The sensory receptors of the eye are called **rods** and **cones.** These receptors are located in the back of the eye and act somewhat like film in a camera. Light that falls on the eye is focused by a lens onto these receptors, just as the lens of a camera focuses light on film.

It is important to discuss how the eye's sensory receptors work before discussing the structure of the eye in greater detail. Just as in photosynthesis, a primary event is the absorption of a **photon** of light by a pigment. Photons, as you may recall from Chapter 8, are "units" or "packets" of electromagnetic energy. One complexity of electromagnetic energy is that it travels as waves but also behaves like individual units. In the eye, photons of light are absorbed by the pigment **rhodopsin.** This pigment is located in the tips of rod and cone cells that make up tissue at the back of the eye called the **retina** (Figure 17-6). Rhodopsin consists of **retinal,** a derivative of Vitamin A, coupled to a membrane protein called **opsin.** As each rhodopsin complex absorbs a photon of light, the molecular shapes of both retinal and opsin change. As a result, retinal separates from opsin. The "activated" opsin triggers a series of events that are responsible for *closing* sodium channels. As discussed in Chapter 15, stimuli usually lead to an *opening* of sodium channels, allowing sodium ions to rush into the cell, *depolarizing* the receptor membrane. Sodium channels close in rod and cone cells when they are stimulated. The result is **hyperpolarization**—the interior of the rod becomes even more negatively charged than before. Hyperpolarization of the cell in turn activates a mechanism that *lowers* the amount of neurotransmitter released by the cell. This decrease in neurotransmitter release lessens the frequency of firing in adjacent neurons—nerve signals that eventually reach the brain and are interpreted as patterns of light and dark.

Light receptors are rod and cone cells located within the retina at the back of the eye. The pigment within these cells absorbs photons of light, which causes a series of events leading to a hyperpolarization of the receptor cells and a subsequent firing of adjacent neurons.

Rod cells are able to detect low levels of light because even one photon can stimulate a rod cell. In addition, they detect only white light, so vision with rod cells alone is black, white, and shades of gray. Did you ever try to see color in extremely dim light? Your rods alone are at work, and they cannot "see" color. Color vision is achieved by your cone cells, which function in bright light.

There are three kinds of cone cells (although they all look the same under the electron microscope). Each type of cone cell possesses rhodopsin molecules that have slightly different shapes from one another because they have different types of opsin. These differences determine which wavelengths of light the pigment (and therefore the cone cell) will absorb. One type absorbs wavelengths of light in the 455 nanometer (blue-absorbing) range. Another type absorbs in the 530 nanometer (green-absorbing) range, and the third in the 625 nanometer (red-absorbing) range. The color you perceive depends on how strongly each group of cones is stimulated by a light source. The pigment in rod cells, on the other hand, absorbs in the 500 nanometer range. As shown in Figure 17-7, this range encompasses the three primary colors (blue, green, and red), which together make up white light.

Rod cells function in dim light and detect white light only, whereas cone cells function in bright light and detect color.

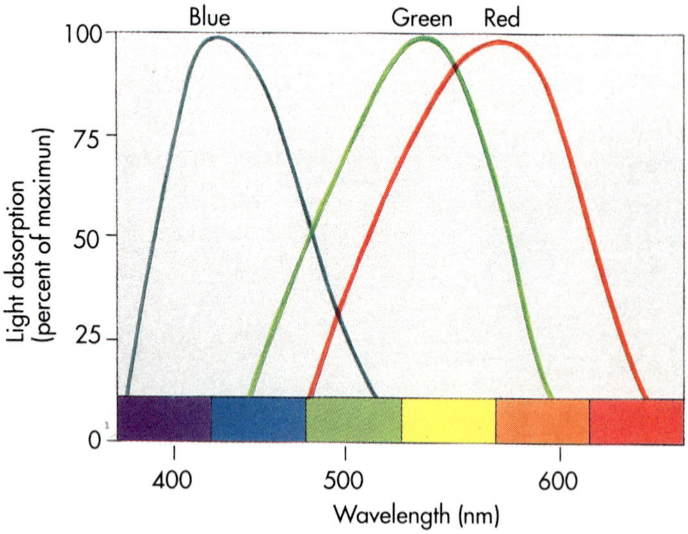

FIGURE 17-7 The absorption spectrum of human vision.
The wavelength of light absorbed by the visual pigment of cone cells depends on the opsin proteins to which the pigment is bound. There are three such proteins, producing cones that absorb at 455 nanometers (blue), 530 nanometers (green), and 625 nanometers (red).

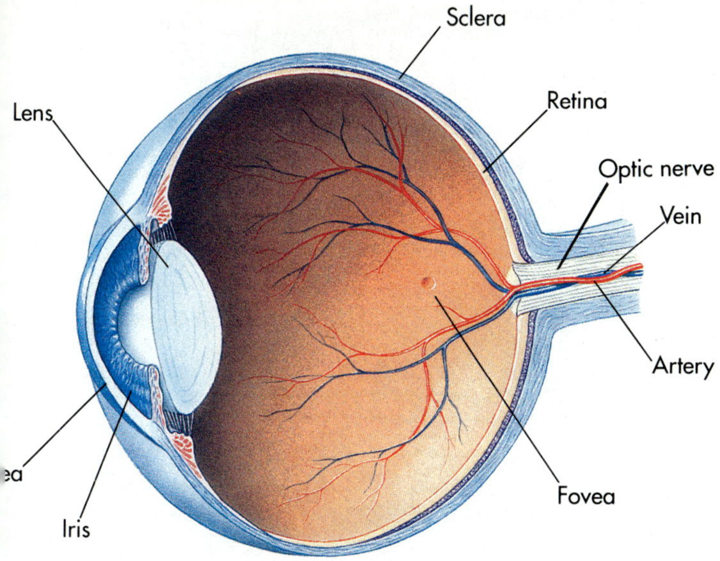

FIGURE 17-8 Structure of the human eye.
Light passes through the transparent cornea and is focused by the lens on the retina at a particular location called the *fovea.* The retina is rich in rods and cones.

The basic structure of the human eye is shown in Figure 17-8. Your eyes are each about 1 inch in diameter and are covered and protected by a tough outer layer of connective tissue called the **sclera.** The front of the eye is transparent, allowing light to enter the eye. This portion of the eye's outer layer is called the **cornea.** Because the cornea is rounded, it not only allows light to enter the eye but bends it as well. This bending, or refraction, of light occurs as the light waves slow down as they move from the air and pass through the tissue of the cornea. You can see this phenomenon if you put a straw in a glass of water. The parts of the straw both outside and inside the water do not appear connected. If the cornea were flat, the light waves or rays would exit the cornea traveling in a path parallel to their original path. However, the cornea is rounded; the light waves that enter above, below, or to the side of the center of the cornea are bent slightly toward one another. Therefore the cornea is the first structure of the eye that begins to focus incoming light onto a point at the rear of the eye.

A chamber behind the cornea is filled with a watery fluid called the **aqueous humor.** The word *humor* refers to "fluid within the body," and the word *aqueous* means "water." Fresh aqueous humor is continually produced as old fluid is drained into the bloodstream. This fluid nourishes both the cornea and the lens, since neither structure has a supply of blood vessels. In addition, the aqueous humor (along with fluid further back in the eye called the *vitreous* [Latin for "glass"] *humor*) creates a pressure within the eyeball that maintains the eyeball's shape and keeps the retina pressed properly against the back of the eyeball.

The **lens** lies just behind the aqueous humor and plays a major role in focusing the light that enters the eye onto the retina at the back of the eye. It looks much like a "lemon drop" candy or a somewhat flattened balloon. If you were to cut a lens in half, it would look similar to an onion, since it is made up of layer upon layer of protein fibers. The lens is encircled by ligaments that suspend it within the eye. These ligaments are attached to a tiny circular muscle called the **ciliary muscle** that by contracting or relaxing, slightly changes the shape of the lens. The greater the curve of the lens, the more sharply it bends light rays toward one another.

In persons with normal vision, this bending of the light results in its being focused on the retina. Some people, however, may have an elongated eyeball or a thickened lens. In either case, the light is focused on a spot in front of the retina. Such people are nearsighted. As Figure 17-9 shows, this condition can be corrected with a lens that spreads the light rays entering the eye so that they focus on the retina.

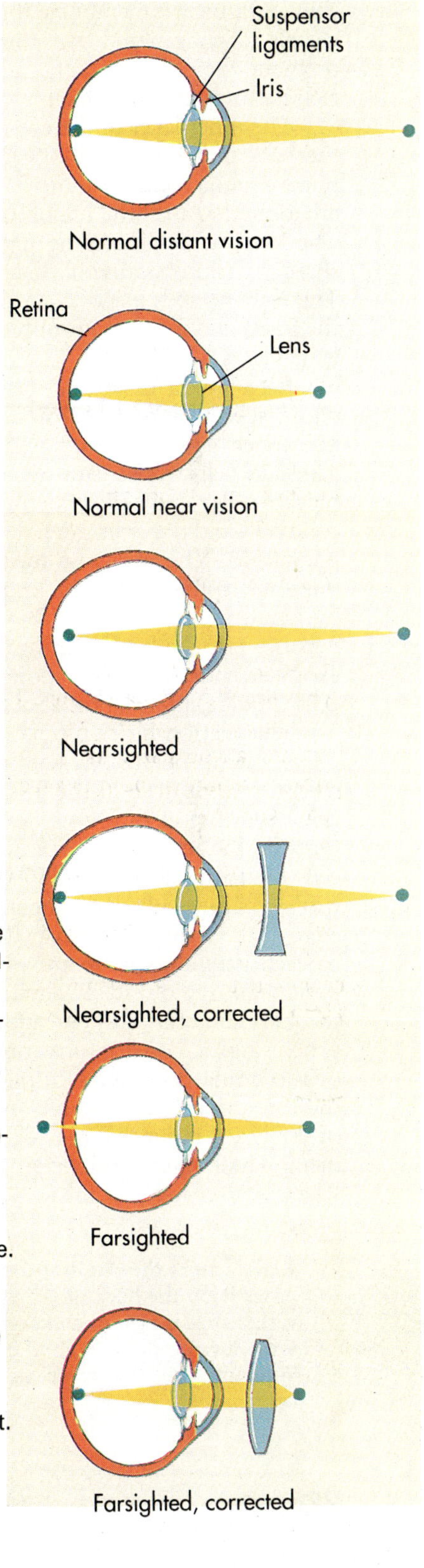

FIGURE 17-9 Focusing the human eye.
Contraction of ciliary muscles pulls on suspensory ligaments and changes the shape of the lens, which alters its point of focus forward or backward. In nearsighted people the ciliary muscles place the point of focus in front of the retina rather than on it. The problem can be corrected with glasses or contact lenses, which extend the focal point to where it should be. In farsighted people, the ciliary muscles make the opposite error, placing the point of focus behind the retina. Corrective lenses can shorten the focal point.

Some people are farsighted, having either a shortened eyeball or a thin lens. Light is focused on a spot behind the retina. This condition can be corrected with a lens that converges entering light rays so that they focus on the retina.

The amount of light entering the eye is controlled by a diaphragm called the **iris,** which lies between the cornea and the lens. The iris reduces the size of the transparent zone, or **pupil,** of the eye (what you see as a black "dot") through which the light passes. Together, the ciliary muscle and the iris make up two of the structures of the middle layer of the eye. The third structure is the **choroid.** As shown in Figure 17-8, the choroid extends toward the back of the eye as a thin, dark-brown membrane that lines the sclera. It contains vessels carrying blood that nourishes the retina and a dark pigment that absorbs light rays so that they will not be reflected within the eyeball.

The inner layer of the eye is the retina. The retina lines the back of the eye and contains the rods and cones. The human retina contains about 7 million cones, most of which are located at a central region of the retina called the **fovea.** The lens focuses incoming light on this spot, the area of sharpest vision because of this high concentration of cones. There are no rods in the fovea. Some are located just outside the fovea and increase in concentration as the distance from it increases.

Each foveal cone cell makes a one-to-one connection with a special kind of neuron called a *bipolar cell* (Figure 17-10). Each of the bipolar cells is connected in turn to a ganglion cell, whose axon is part of the **optic nerve.** The bipolar cells receive the hyperpolarization stimulus from the cone cells and transmit a depolarization stimulus to the ganglion cells. The axons of the ganglion cells transmit the impulses to the brain. The frequency of pulses transmitted by any one receptor provides information about light intensity. The pattern of firing among the different foveal axons provides a point-to-point image. The different cone cells provide information about the color of the image.

The relationship of receptors to bipolar cells to ganglion cells is one-to-one-to-one within the fovea. Put simply, each receptor (cone) cell in the fovea synapses with its own bipolar cell that in turn synapses with its own ganglion cell. Outside the fovea, however, the output of many receptor cells is channeled to one bipolar cell. Many bipolar cells, in turn, synapse with one ganglion cell. In fact, in the outer edge of the retina, more than 125 receptor cells feed stimuli to each ganglion cell in the optic nerve. In this outer, or peripheral, region, many additional neurons cross connect the ganglion cells with one another and carry out extensive processing of visual information. As a result, this portion of the retina does not transmit a point-to-point image as the fovea does but transmits instead a "processed" version of the visual input that may be interpreted simply as movement. Think of the periphery of your retina as a "detector," whereas your fovea is an "inspector."

Light enters the pupil of the eye and the lens focuses it on the back of the eye, on an area of the retina that is rich in cone cells. These cells, along with surrounding rod and cone cells, initiate nervous impulses that travel to the cerebrum via the optic nerve.

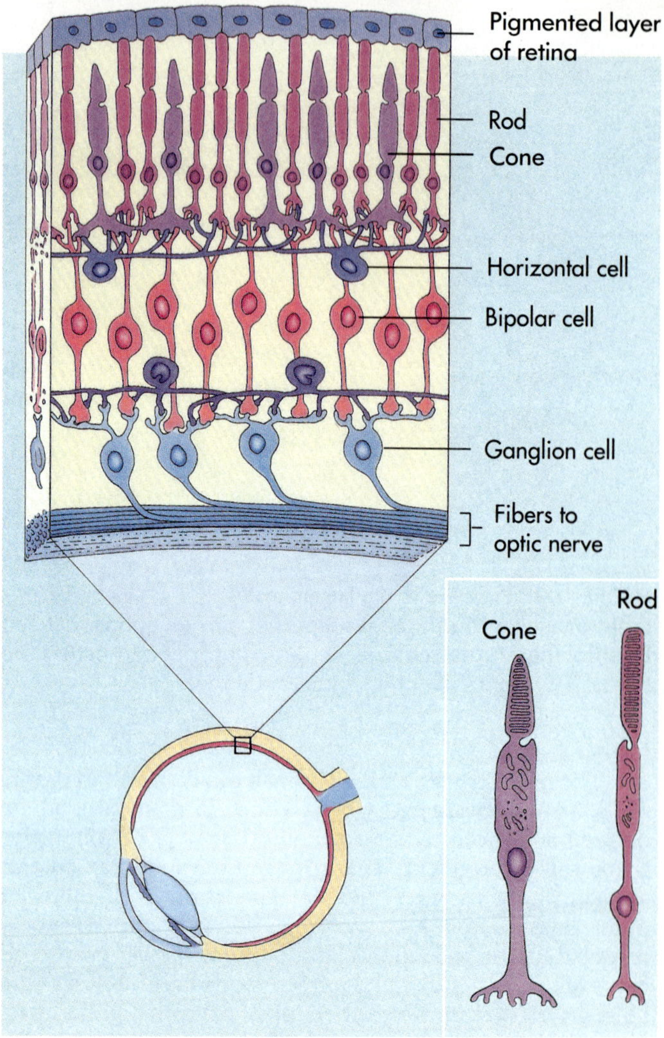

FIGURE 17-10 Structure of the retina.
Note that the rods and cones are at the rear of the retina, not the front. Bipolar cells receive the hyperpolarization stimulus from the cone cells and transmit a depolarization to the ganglion cells. The ganglion cells, whose axons are part of the optic nerve, transmit the impulse to the brain.

The positioning of your two eyes on each side of the head sends two slightly different sets of information to the brain about what you are seeing. Each set of information is slightly different because each eye views an object from a slightly different angle. This slight displacement of images gives **depth perception**—a three-dimensional quality—to your sight.

Hearing

Sound is a type of mechanical energy resulting from the vibration of an object. This vibration disturbs the air around the object, pushing the molecules in the air closer to one another, compressing them. These areas of compression are followed by areas of molecules that are not compressed. If you could visualize these disturbances, they would look much like the concentric circles of waves on the surface of water that occur when you throw a pebble into a still pond. Sound waves are these wave-like disturbances of molecules in the air (or in certain other substances such as water, glass, iron, or wood) caused by a vibrating object. Two parts of the ear—the outer and middle ear—work together to transmit sound waves to the inner ear, where sound stimuli are changed into nerve impulses.

The flaps of skin on the outside of the head that are called ears are only one part of the **outer ear.** Each flap, or **pinna,** funnels sound waves into an **auditory canal.** The auditory canal, which is about 1 inch long, leads directly to the eardrum. The eardrum, or **tympanic membrane,** is a thin piece of fibrous connective tissue that is stretched over the opening to the **middle ear.** On its external side the eardrum is covered with skin. On its internal side the eardrum is connected to one of the smallest bones in the body—the hammer, or **malleus.** The malleus is connected to another tiny bone called the *anvil,* or **incus.** The anvil, in turn, is connected to a third bone called the *stirrup,* or **stapes.** These three bones are the structures of the middle ear that "pick up" sound vibrations from the outer ear and transfer them to the inner ear.

Sound waves entering the outer ear beat against the tympanic membrane, causing it to vibrate like a drum. Vibrations of this membrane cause the malleus to move with a rocking motion, since it is attached to the internal surface of the tympanic membrane. This rocking is transferred, in turn, to the incus and stapes. These three bones are hinged to one another in a way that produces a lever system, a mechanism that causes them to act like an amplifier and increase the force of the vibrations.

Also located in the middle ear is an opening to the eustachian tube (a structure named after an Italian physician who lived in the 1500s). The **eustachian tube** connects the middle ear with the nasopharynx (the upper region of the throat). Its function is to equalize air pressure on both sides of the eardrum when the outside air pressure is not the same as the pressure in your middle ear. You have probably had the experience of your ears "popping" while you were driving up or down a mountain, taking off and landing in an airplane, or deep-sea diving; it is the result of the pressure equalization between these two sides of the eardrum.

The stirrup, the third in the series of middle ear bones, is attached to a membrane that separates the middle ear from the inner ear. This membrane covers the **oval window,** the entrance to the **inner ear,** the part of the ear in which hearing actually takes place. In addition, other parts of the inner ear detect motion and the effect of gravity on the body.

The oval window is the entrance to the fluid-filled **cochlea**—the part of the inner ear that contains the organ of hearing. To understand the cochlea's structure and function, imagine a tapering blind-ended tube (Figure 17-11) having two membrane-covered holes at its wider end. The "upper" hole is the oval window, and the "lower" hole is called the **round window.** Also imagine that another, smaller membranous tube runs down its center, serving as a partition between the oval and round windows and bisecting the large tube into upper and lower channels, or canals. The upper channel is called the **vestibular canal;** the lower channel is called the **tympanic canal.** The "inner tube" stops short of the blind, tapered end of the outer tube, however, so the two canals connect with one another there. Roll this tube up in your mind like a jellyroll, and you have the basic structure of the cochlea. In fact, the word *cochlea* comes from a Latin word meaning "snail" and describes the rolled-up shape of this structure extremely well.

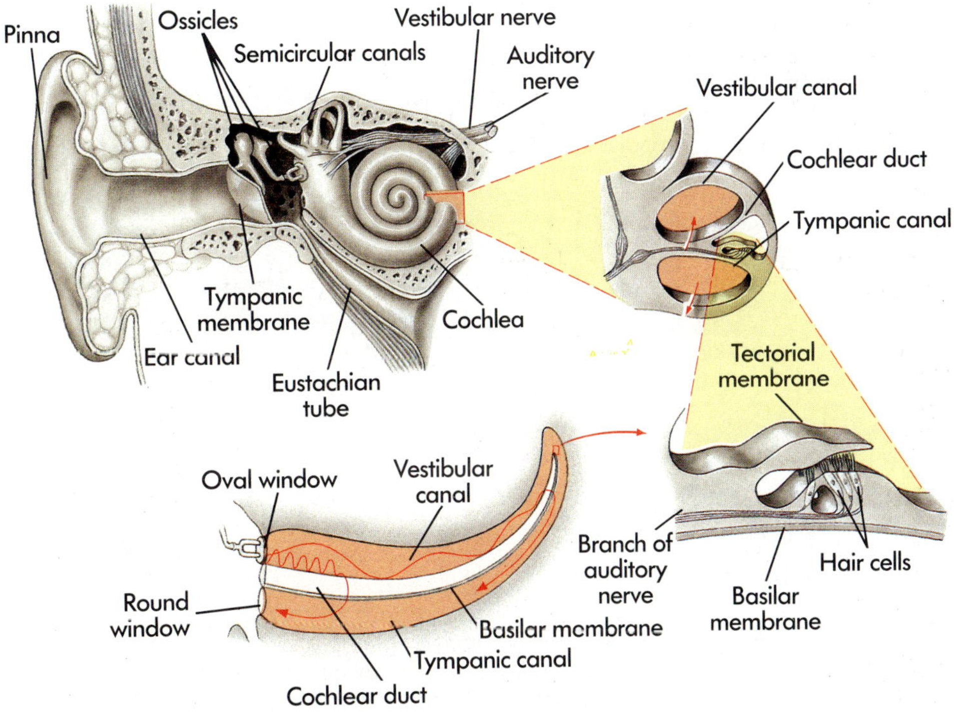

FIGURE 17-11 Structure of the human ear.
The human ear is composed of outer, middle, and inner sections. The outer ear extends from the pinna to the tympanic membrane. The middle ear contains the ear ossicles, or bones that transmit sound vibrations from the tympanic membrane to the cochlea. The cochlea contains the organ of hearing and makes up part of the inner ear.

The inner tube of the cochlea is called the **cochlear duct,** and it contains specialized cells that are the receptors of hearing. These cells are called *hair cells* and are embedded in the "floor" of the inner tube, the basilar membrane. The hairs that project from the hair cells stick up into the cochlear duct and are covered by a "roof" called the **tectorial membrane.** Together, the hair cells with supporting cells of the basilar membrane and the overhanging tectorial membrane are called the **organ of Corti**—the organ of hearing.

How does the organ of hearing detect sound? As the stapes rocks in the oval window, it sets the fluid within the vestibular canal in motion. As the oval window membrane rocks inward, it pushes on the fluid in the vestibular canal. This fluid moves forward in waves that pass into the tympanic canal. When the waves reach the membrane-covered round window, it bulges outward. Meanwhile, the stapes rocks backward in the oval window, setting up an opposite motion of the fluid. As the waves then flow "backward" through the tympanic canal on the way to the vestibular canal, they push on the basilar membrane. The basilar membrane bends, causing the hairs of the receptor cells to be pressed against the tectorial membrane. As the hairs bend, they develop generator potentials that fire adjacent neurons. These impulses then pass to the brain to be interpreted. The brain is able to interpret pitch (the "highness" or "lowness" of a sound) because sounds of different pitches produce sound waves of different frequencies (numbers of waves per second). These differences set up differing wave patterns within the fluid of the inner ear that cause specific regions of the basilar membrane to vibrate more intensely than others. In addition, louder sound waves cause greater vibrations of the basilar membrane than can be interpreted by the brain. Repeated exposure to extremely loud noises such as gunshots, jet engines, and loud music can damage the hairs of the receptor cells and cause partial but permanent hearing loss.

The ear has three parts: the outer, middle, and inner ear. The outer ear funnels sound waves in toward the eardrum, which changes these waves into mechanical energy. This energy is then transmitted to the bones of the middle ear, which increase its force. The inner ear, a complex of fluid-filled canals, contains the organ of hearing. The receptor cells of this organ change the mechanical "sound" energy into nerve impulses.

Balance

The inner ear has two other portions that are each structured as a fluid-filled "tube within a tube" like the cochlea, but their shapes are different from that of the cochlea. As shown in Figure 17-12, the cochlea makes up one side of the inner ear. The "bulge" in the midsection of the inner ear is called the **vestibule,** and it contains structures that sense whether you are upside down or right side up. In other words, it detects the effects of gravity on the body.

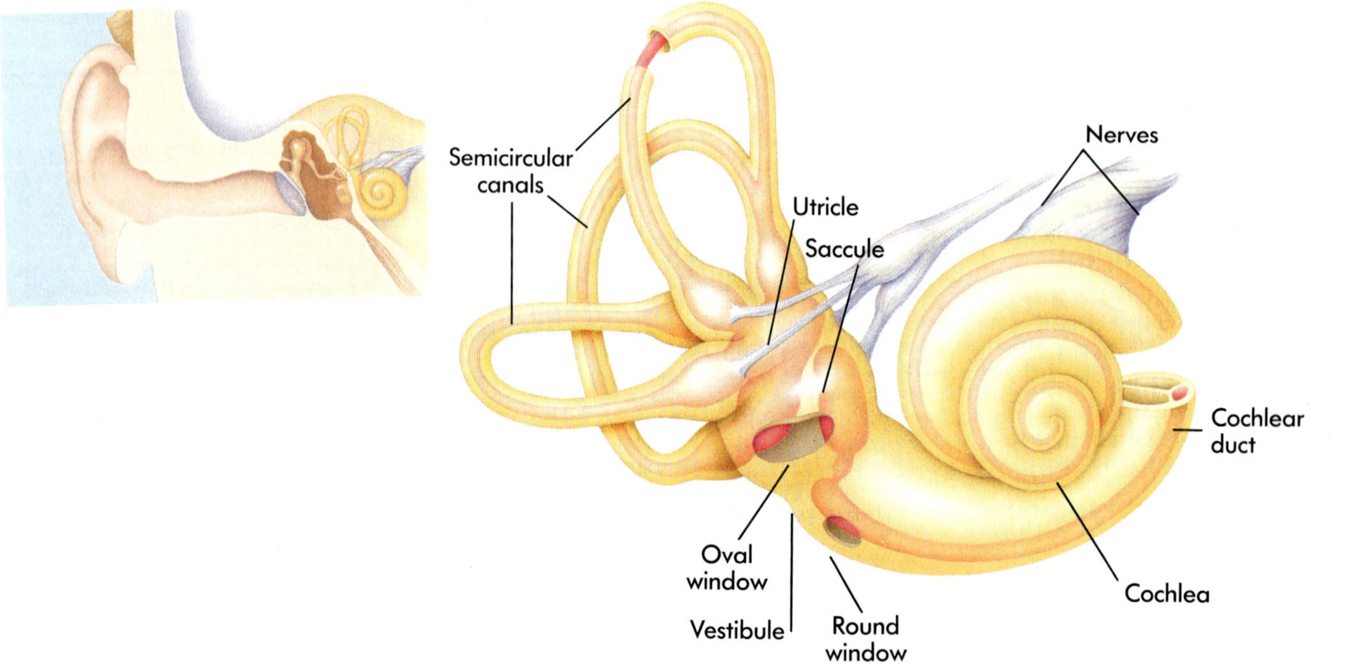

FIGURE 17-12 Structure of the inner ear.
The inner ear contains the organs of hearing and the organs that are responsible for maintaining the body's equilibrium.

Inside this bulge are two sacs called the **utricle** and the **saccule** (both words mean "a little bag or sac"). Within each of these sacs is a flat area composed of both ciliated and nonciliated cells. Spread over the surfaces of these cells is a layer of jelly-like material. Embedded within the jelly are small pebbles of calcium carbonate called **otoliths** (literally, "ear stones"). The ciliated cells are the receptor cells and have long cilia and thin cell extensions that stick up into the jelly like microscopic tufts of hair (Figure 17-13). When the head is moved, the otoliths slide within the jelly like a hockey puck slides on ice. As the otoliths move, they pull on the jelly, which pulls on the cilia, bending them. Any shift in the position of the otoliths results in different cilia being bent and cilia being bent to a specific degree and in a specific direction (Figure 17-14). These stimuli initiate generator potentials that are transmitted to neurons that make up a branch of cranial nerve VIII (see Chapter 16). The brain interprets these messages, resulting in your perception of "up" with respect to the pull of gravity.

Positioned above the saccule and utricle are three fluid-filled **semicircular canals** (see Figure 17-12). These "loops" are oriented at right angles to one another in three planes. At the base of each loop is a group of ciliated sensory cells that are connected to neurons. Lying above these cells is a mass of jelly-like material. When the head moves, the fluid within the semicircular canals moves and pushes the jelly (and the cilia) in a direction opposite to that of the motion. (You have experienced this phenomenon when you accelerate suddenly in your car and your head is thrown backward.) This movement initiates the depolarization of the cell membrane, which triggers a nerve impulse. Because the three canals are each oriented differently, movement in any plane is sensed by at least one of them. Complex movements are analyzed in the brain as it compares the sensory input from each canal.

The vestibule of the inner ear detects your position with respect to gravity, whereas the semicircular canals detect the direction of your movement.

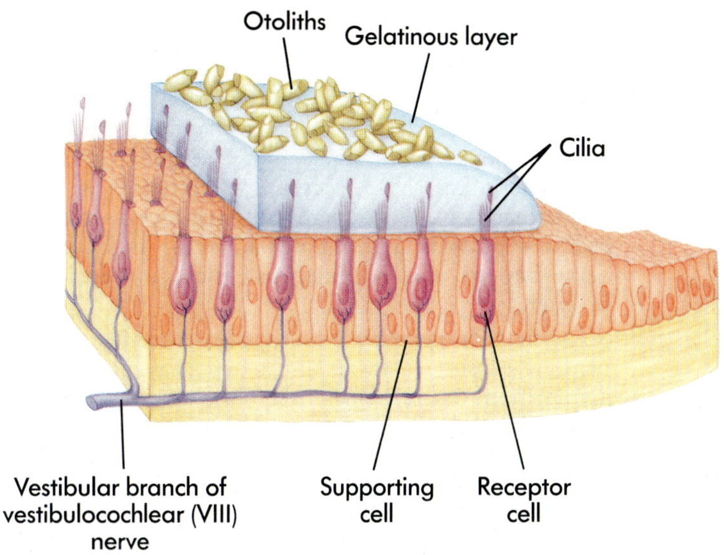

FIGURE 17-13 The structure of the interior of the utricle and saccule. Movement of the otoliths pulls on the cilia attached to the receptor cells, initiating a nerve impulse that is sent through a nerve to the brain. The utricle and saccule provide sensory information necessary to maintain your body's orientation relative to the ground and provide information regarding acceleration and deceleration.

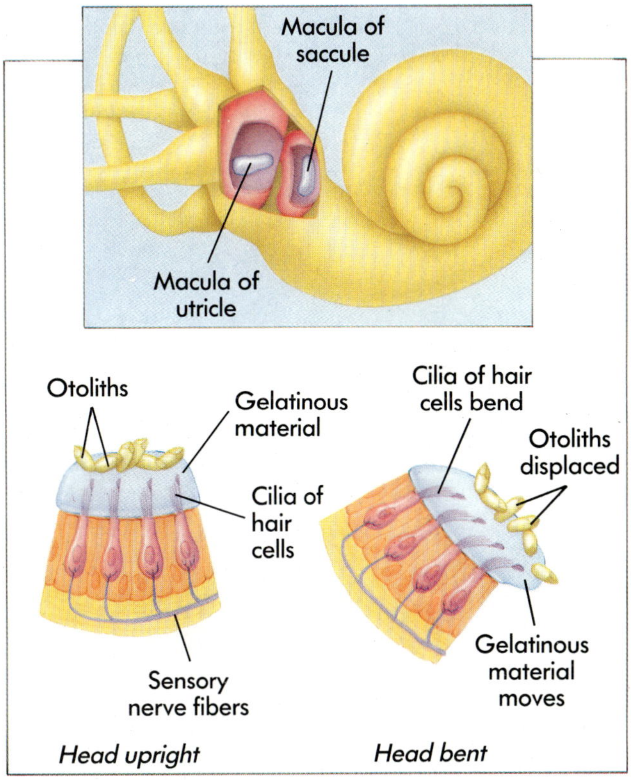

FIGURE 17-14 Balance is maintained by the semicircular canals. At the base of each semicircular canal is a group of ciliated cells connected to neurons. The cells are covered with a jelly-like material studded with otoliths. A sudden movement of the head causes the otoliths to be displaced, depolarizing the neurons and initiating a nerve impulse that is sent to the brain. Because each semicircular canal is oriented in a different plane in space, movement in any plane can be detected by at least one of the canals. The brain interprets complex movements by comparing information sent from each canal.

Summary

1. Sensory receptors change environmental stimuli into nerve impulses. These stimuli, or changes in the environment, must be of a sufficient magnitude to open ion channels within the membranes of receptor cells, depolarizing them. This initial depolarization is called a *generator potential* and initiates an action potential, or nerve impulse, that is carried to either the spinal cord or brain.
2. Some sensory receptors monitor body functions. Temperature-sensitive neurons in the hypothalamus, for example, monitor the temperature of the blood and trigger mechanisms to regulate this temperature. Other receptors embedded in the walls of certain arteries are sensitive to the carbon dioxide, oxygen, and pH levels of the blood and provide input to respiratory centers in the medulla and pons. Still others monitor the blood pressure and provide feedback to a cardiac center that regulates heartbeat.
3. Sensory receptors called *proprioceptors* gather information about the body's position in space. Proprioceptors buried deep within the muscles keep track of the degree to which they are contracted. Gravity is detected by structures in the inner ear called the *utricle* and *saccule*. Motion is sensed by the semicircular canals, also located in the inner ear. Together, these two structures help you maintain your balance and know which way is "up."
4. A variety of receptors located at or near the body surface senses changes in the external environment. Many simple receptors located in the skin detect "general" environmental stimuli such as touch, pressure, pain, and temperature. More specialized senses—smell, taste, sight, hearing, and balance—have more complex receptors located only at certain locations.
5. The olfactory receptors, which are located in the nasal epithelium just behind the bridge of the nose, detect various smells. Scientific evidence suggests that the olfactory receptors detect these different smells because of specific binding of airborne gases with receptor chemicals located within the cilia of the olfactory receptor cells.
6. Taste receptors, or taste buds, are located within the short projections on the tongue known as *papillae*. The receptor cells of the taste buds detect sweet, sour, salty, and bitter tastes in food as food molecules dissolve in the saliva and come into contact with cellular "hairs" that poke out of the taste pores.
7. The human eye detects wavelengths of electromagnetic energy known as *visible light*. Electromagnetic energy is generated by charged subatomic particles moving through space emitting waves of electrical and magnetic energy. The human eye contains receptor cells called *rods* and *cones* that contain pigments capable of absorbing "units" of electromagnetic energy of certain wavelengths. The absorption of light causes a change in the shape of the pigment, which in turn triggers a series of events resulting in a hyperpolarization of the receptor cell membrane. This hyperpolarization leads to an adjustment in the frequency of the firing of adjacent neurons—nerve signals that eventually reach the brain and are interpreted as patterns of light and dark.
8. Structurally, the human eye is similar to a camera. Light entering the eye is focused by a lens onto the receptors located at the back of the eye, which act like film in a camera.
9. Sound waves are disturbances of molecules in the air (and in certain other substances) that result from the vibration of an object. The ear collects sound waves and changes them into other forms of mechanical energy that stimulate the organ of hearing within the inner ear. The receptor cells of this organ propagate nerve impulses that are carried to the cerebral cortex.

KEY TERMS

aqueous humor 287
auditory canal 289
choroid 288
ciliary muscle 287
cochlea 289
cochlear duct 289
cones 286
cornea 287
depth perception 288
eustachian tube 289
fovea 288
hyperpolarization 286
incus 289
inner ear 289
iris 288
lens 287
malleus 289
middle ear 289
olfactory receptor 284
opsin 286
optic nerve 288
organ of Corti 290
otoliths 291
outer ear 289
oval window 289
pinna 289
proprioceptor 282
pupil 288
retina 286
retinal 286
rhodopsin 286
rods 286
round window 289
saccule 291
sclera 287
semicircular canals 291
stapes 289
stretch receptor 283
taste buds 285
tectorial membrane 290
tympanic canal 289
tympanic membrane 289
utricle 291
vestibular canal 289
vestibule 290

REVIEW QUESTIONS

1. What are sensory receptors? What three types of information do they give you?
2. Explain the events that occur at the cellular level that allow you to "sense" something.
3. What information is provided by receptors that monitor your body's internal environment? Why is this important?
4. What are proprioceptors? Explain their significance.
5. Distinguish between the general senses and the special senses. How do their receptors differ?
6. Explain how you can smell the difference between your favorite perfume and your least favorite food.
7. You may have noticed that certain smells are able to evoke specific memories and emotions—both good and bad. Explain why.
8. What are taste buds? Explain their role in allowing you to taste the flavors in your favorite ice cream. What other sense (besides taste) is involved?
9. What are rods and cones? Explain their role in your sense of vision.
10. How are you able to see different colors?
11. Explain the process that allows you to see objects.
12. Draw a diagram of the human eye. Label the sclera, cornea, aqueous humor, lens, iris, pupil, and fovea.
13. Mr. Garcia undergoes minor eye surgery and must wear a patch over his left eye for 3 days. During this time, he finds that he has trouble grasping things because he keeps underreaching and overreaching. Explain why.
14. Draw a picture of the human ear. Label the pinna, auditory canal, tympanic membrane, malleus, incus, stapes, oval and round windows, cochlea, cochlear duct, and the organ of Corti. Indicate the outer, middle, and inner ear.
15. You switch on the radio to your favorite program. Explain how your ears enable you to hear the sounds.
16. Even with your eyes closed, you can tell when your head tilts to one side. Explain how.

THOUGHT QUESTIONS

1. In *The Silence of the Lambs* and many other movies, you see people using "snooperscopes," devices that allow them to literally see in the dark. How might such a device work?
2. In a very interesting series of experiments, migrating song birds were placed in a cage lined with carbon paper similar to that used in credit card slips. The birds would peck at the wall of the cage, attempting to get out, but not at random—by far the greatest density of pecks is in the direction they were migrating when captured! Now when a large magnet is placed in the room, they peck in the direction of the magnet, even though they cannot see it. What sort of sensory system do you imagine the birds possess that lets them orient in this way?

FOR FURTHER READING

Miller, J. A. (1990). A matter of taste. *BioScience, 40*(2), pp. 78-82.
Innovative methods and materials have made possible new insights into the way your sense of taste works.

Finke, R. (1986, May). Mental imagery and the visual system. *Scientific American*, pp. 84-92.
The article explains the theory that the way in which visual information is perceived and carried to the brain has a strong influence on mental imagery.

Gibbons, B. (1986, September). The intimate sense of smell. *National Geographic*, pp. 324-361.
This detailed account of the human sense of smell is interesting and well illustrated.

18 Protection, support, and movement

Fashioning faces using computers is one type of medical miracle taking place at Johns Hopkins Children's Facial Rehabilitation Center in Baltimore, Maryland. Erin Williams, born with a genetic disorder called Treacher Collins syndrome, knows all about this medical miracle. She was born without a jaw, cheekbones, or fully formed ears. Although she still faced surgery at the time this photo was taken, Erin does not look like a child that had to struggle just to eat and breathe during her infancy.

OUTLINE

How skin, bones, and muscles work together

Skin

Bones

- The skeletal system
  - The axial skeleton
  - The appendicular skeleton
  - Joints

Muscles

- The muscular system
  - How skeletal muscles contract
  - How nerve impulses signal muscle fibers to contract

HIGHLIGHTS

Movement of the human body is possible because your tissues have a flexible outer covering and are supported by a scaffolding of bone that is powered by skeletal muscles.

The skin is a dynamic organ that flexes with body movement, protects the body from water loss and invasion by microbes, provides the body with a sensory surface, and helps control the body's internal temperature.

The rigid yet flexible bones of the body form an internal framework called the *skeletal system*, which supports the soft tissues of the body; provides places for the attachment of the "engines" of movement, the muscles; and acts as a lever system against which muscles can pull.

The skeletal muscles—the muscles of body movement—are made up of long, cylindrical cells packed with organized arrangements of thick and thin microfilaments capable of causing muscular contraction.

Doctors at the Children's Center performed "phantom surgery" to plan the reconstruction of Erin's face. Using three-dimensional computer images of her skull, doctors simulated Erin's surgery before they actually operated on her. Phantom surgery can be used to experiment with cutting and fitting bone pieces from other parts of a patient's body, or the computer can be used to design an artificial bone. This technology is called computer-aided design, or CAD for short. CAD has been used for years by engineers to design such products as cars and airplanes, but researchers are now applying this concept to medicine.

To help Erin, her skull was recreated as a computer graphic using a stack of computerized tomography (CT) scan images. CT scan images are "pictures" of a thin slice of the body. A CAD computer stacks these images to produce a three-dimensional graphic that can be rotated and viewed from any angle. Using this image, doctors designed prostheses—artificial bones—for Erin. The instructions to build these prostheses were put on tape and were then fed into a computer specialized in computer-aided manufacturing (CAM). CAM computers produce the mold necessary to make a prosthesis.

In Erin's case, one company used a CAM computer attached to a milling machine to make Erin a plastic jaw. During surgery, a real one was fashioned for Erin using the plastic model as a template. To do this, physicians used bits of Erin's ribs and pieces of a chemically treated South Pacific coral that mimics human bone. Another company used CAD/CAM to construct titanium cheekbones for her. This lightweight, durable, biocompatible metal will hold bone grafts that will be added to Erin's face as she grows. After her teenage years, Erin will no longer need reconstructive surgery. The growth and development of her face, fashioned with modern technology, will be complete.

How skin, bones, and muscles work together

Bones not only give shape to the body and protect its delicate inner structures but also help you move. Your skeletal muscles are attached to your bones, pulling on one while anchored to another. Rigid yet flexible, bones are able to bear a considerable amount of weight. If you have ever tried to lift a heavy object, you are familiar with this basic problem of movement: gravity exerts a force, or "pull," on objects. For any motion to take place, the force exerted on an object—including your body parts—must be greater than the opposing force of gravity. The manner in which muscles are attached to bones allows bones to be used as levers to increase the strength of a movement, similar to the way a claw hammer increases the force you can exert on a nail to pull it out of a piece of wood.

To supply the force needed for movement, your body uses the chemical energy of adenosine triphosphate, or ATP (see Chapter 6). By splitting an ATP molecule into adenosine diphosphate (ADP) and inorganic phosphate (P_i), energy is made available to do the work of movement. Your body uses this energy to move certain microfilaments within your muscle cells, resulting in the shortening of those cells. When a large group of muscle cells shortens all at once, the muscle cells exert a great deal of force. For a muscle to use this force to produce movement, it must direct its force against an object. In humans, muscles pull on bones, so, for example, muscle contraction results in the lifting of a leg or the bending of a finger. And although your movement is determined largely by your muscles and their attachments to bones, your skin plays an important role too. It is a soft and flexible covering, stretching to accommodate the myriad of movements of which your body is capable.

Movement results from the contraction of skeletal muscles anchored to bones. These muscles use bones like levers to direct force against an object. When a body part is pulled to a new position, the skin stretches to accommodate the change.

Skin

Your skin is far more than simply an elastic covering of epithelial cells encasing your body's muscles, blood, and bones. Instead, it is a dynamic organ that performs many functions:

1. Skin is a protective barrier. It keeps out microorganisms that would otherwise infect the body. Because skin is waterproof, it keeps the fluids of the body in and other fluids out. Skin cells also contain a pigment called *melanin*, which absorbs potentially damaging ultraviolet radiation from the sun.
2. Skin provides a sensory surface. Sensory nerve endings in skin act as your body's pressure gauge, telling you how gently to caress a loved one and how firmly to hold a pencil. Other sensors embedded in the skin detect pain, heat, and cold. Skin is the body's point of contact with the outside world (see Chapter 17).
3. Skin compensates for body movement. Skin stretches when you reach for something and contracts quickly when you stop reaching. It grows when you grow.
4. Skin helps control the body's internal temperature. When the temperature is cold, the blood vessels in the skin constrict, so less of the body's heat is lost to the surrounding air. When it is hot, these same vessels dilate, giving off heat. In addition, glands in the skin release sweat, which then absorbs body heat and evaporates, cooling the body surface.

Your skin is the largest organ of your body. In an adult human, 15% of the total body weight is skin. Much of the multifunctional role of skin reflects the fact that its tissues are made up of a myriad of specialized cells. In fact, one typical square centimeter of human skin contains 200 nerve endings, 10 hairs with accompanying microscopic muscles, 100 sweat glands, 15 oil glands, 3 blood vessels, 12 heat receptors, 2 cold receptors, and 25 pressure-sensing recep-

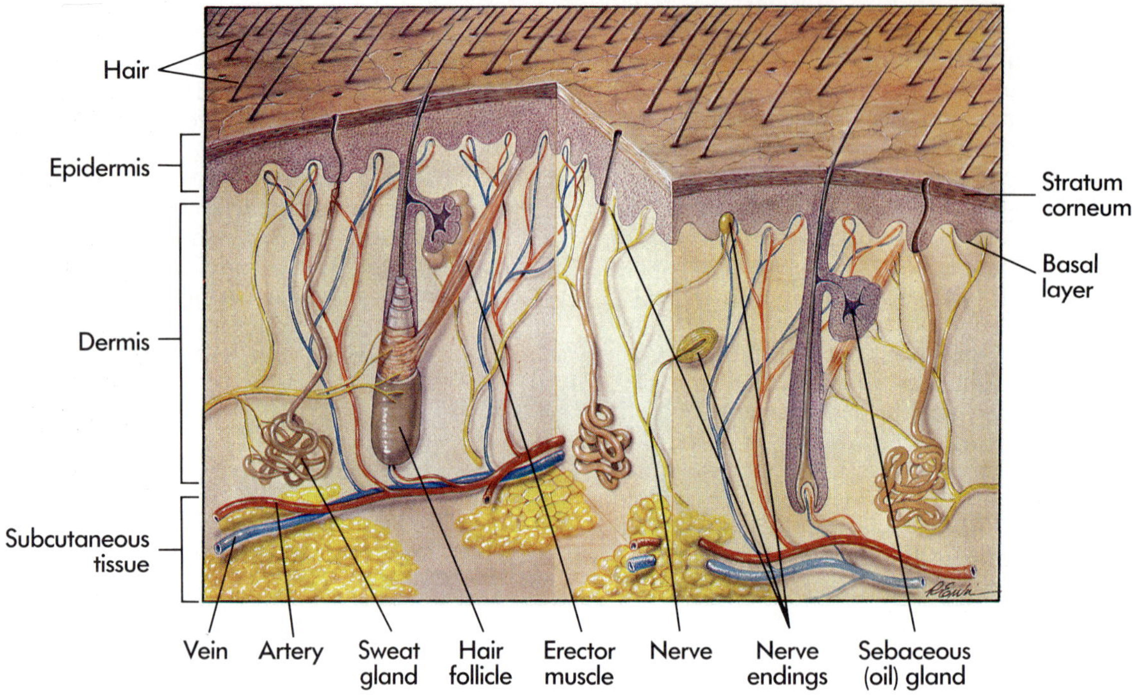

FIGURE 18-1 The structure of human skin.
Human skin is composed of three layers: the epidermis, the dermis, and the subcutaneous layer. Within these layers are specialized cells that perform many specialized functions.

tors (Figure 18-1). Together with the hair and nails, the skin is called the **integument** (meaning "outer covering") or the **integumentary system.**

> **The skin is the largest organ of the body and is made up of a myriad of specialized cells. It covers the surface of the body, protecting it while helping to control the body's internal temperature, provide a sensory surface, and compensate for body movement.**

Bones

Bone is a type of connective tissue (see Chapter 9) consisting of widely separated bone cells embedded in a matrix of collagen fibers and mineral salts. The bones' mineral salts are needle-shaped crystals of calcium phosphate and calcium carbonate. The collagen fibers (see Figure 9-6) are coated and surrounded by these mineral salts. Interestingly, these two components give bone a structure strikingly similar to that of fiberglass, producing bones that are rigid yet flexible. You can see this interplay of flexibility and hardness by soaking a chicken bone in vinegar overnight. The acetic acid of the vinegar dissolves the mineral salts in the bone. Without the hardness of the mineral salts, the collagen fibers leave the bone so flexible that you can tie it in a knot!

Living cells are as important a component of bone as the collagen fibers and mineral salts are. New bone is formed by cells called **osteoblasts.** These cells secrete the collagen fibers on which the body later deposits mineral salts. The osteoblasts lay down bone in thin, concentric layers called **lamellae** (Figure 18-2), like layers of insulation wrapped around an old pipe. You can think of the osteoblasts as being within the insulation and the open tube of the pipe as a narrow channel called the **haversian canal**. Haversian canals run parallel to the length of the bone and contain blood vessels and nerves. The blood vessels provide a lifeline to living bone-forming cells, whereas the nerves control the diameter of the blood vessels and thus the flow through them. Nutrients and oxygen diffuse from the bloodstream into thin cellular processes of the osteoblasts, and metabolic wastes diffuse out.

> **Bone is a type of connective tissue made up of widely separated living cells that secrete collagen fibers into their surrounding matrix. The body deposits mineral salts on these fibers. The salts give bone hardness, whereas the collagen fibers contribute flexibility. This bone tissue appears under the light microscope as concentric rings surrounding a central canal containing blood vessels and nerves.**

The bones of the human skeleton are composed of spongy bone and compact bone. Microscopically, **compact bone** has the concentric ring structure described previously. As shown in Figure 18-2, it runs the length of long bones, such as those in your arms and legs. However, if bones were

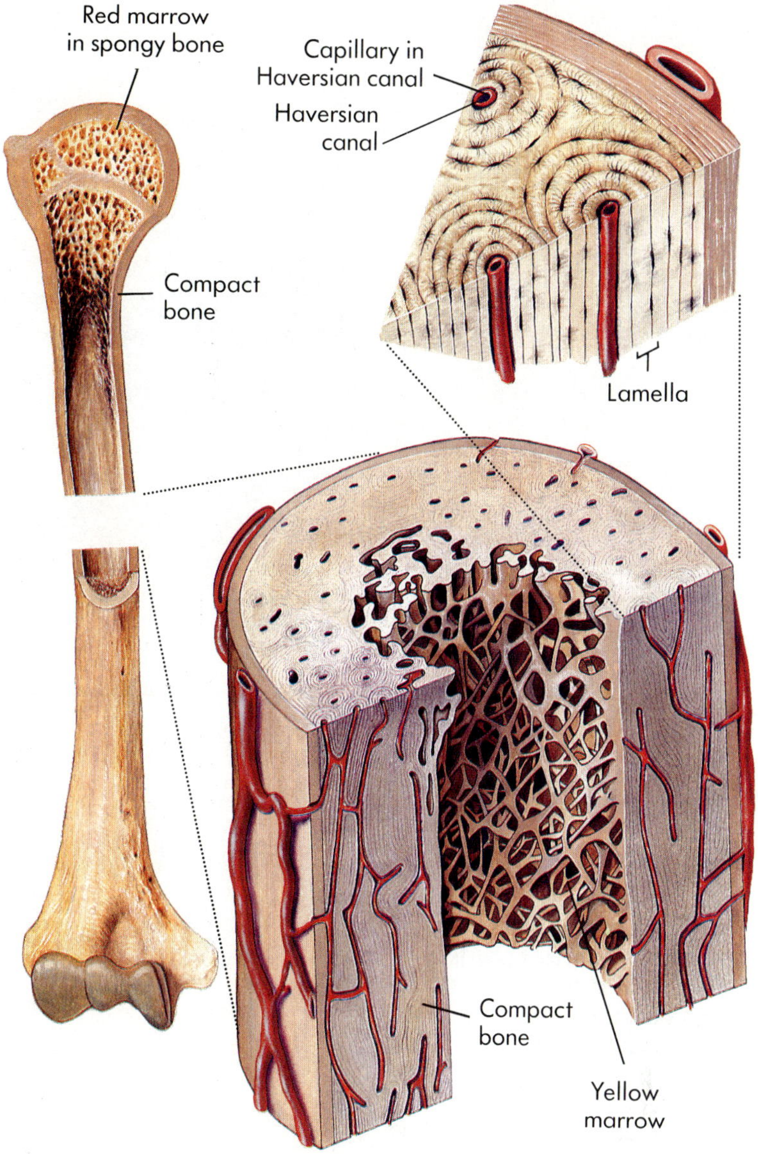

FIGURE 18-2 The organization of bone, shown at three levels of detail.
Some parts of bone are dense and compact, giving the bone strength. Other parts are spongy, with a more open lattice; it is here (within red bone marrow) that most red blood cells are formed. The microscopic structure of bone is shown in the upper right illustration.

completely made up of compact bone, your body would be very heavy, and your arms and legs would be nearly impossible to move! Instead, the central core of long bones is a hollow cylinder, and this cavity is filled with a soft, fatty connective tissue called **yellow marrow.**

Spongy bone makes up most of the ends of long bones and most of the bone tissue of short bones (like your wrist and ankle bones), flat bones (like your ribs), and irregularly shaped bones (like some of your facial bones). Spongy bone is an open latticework of thin plates, or bars, of bone. Microscopically, its structure does not show a regular concentric ring structure like compact bone; it has a somewhat more irregular organization. The spaces within its bony latticework are filled with **red bone marrow.** Here, most of the body's blood cells are formed. Surrounding the spongy bone tissue are layers of compact bone (see Figure 18-2). The compact bone gives bones the strength to withstand mechanical stress, and spongy bone provides some support and a storage place for the red bone marrow while helping to lighten bones.

The bones of the skeleton contain two kinds of tissue: spongy bone and compact bone. Compact bone runs the length of long bones and has no spaces within its structure visible to the naked eye. Spongy bone is found in the ends of long bones and within short, flat, and irregularly shaped bones. It looks like interweaving bars of bone with the intervening spaces filled with red bone marrow.

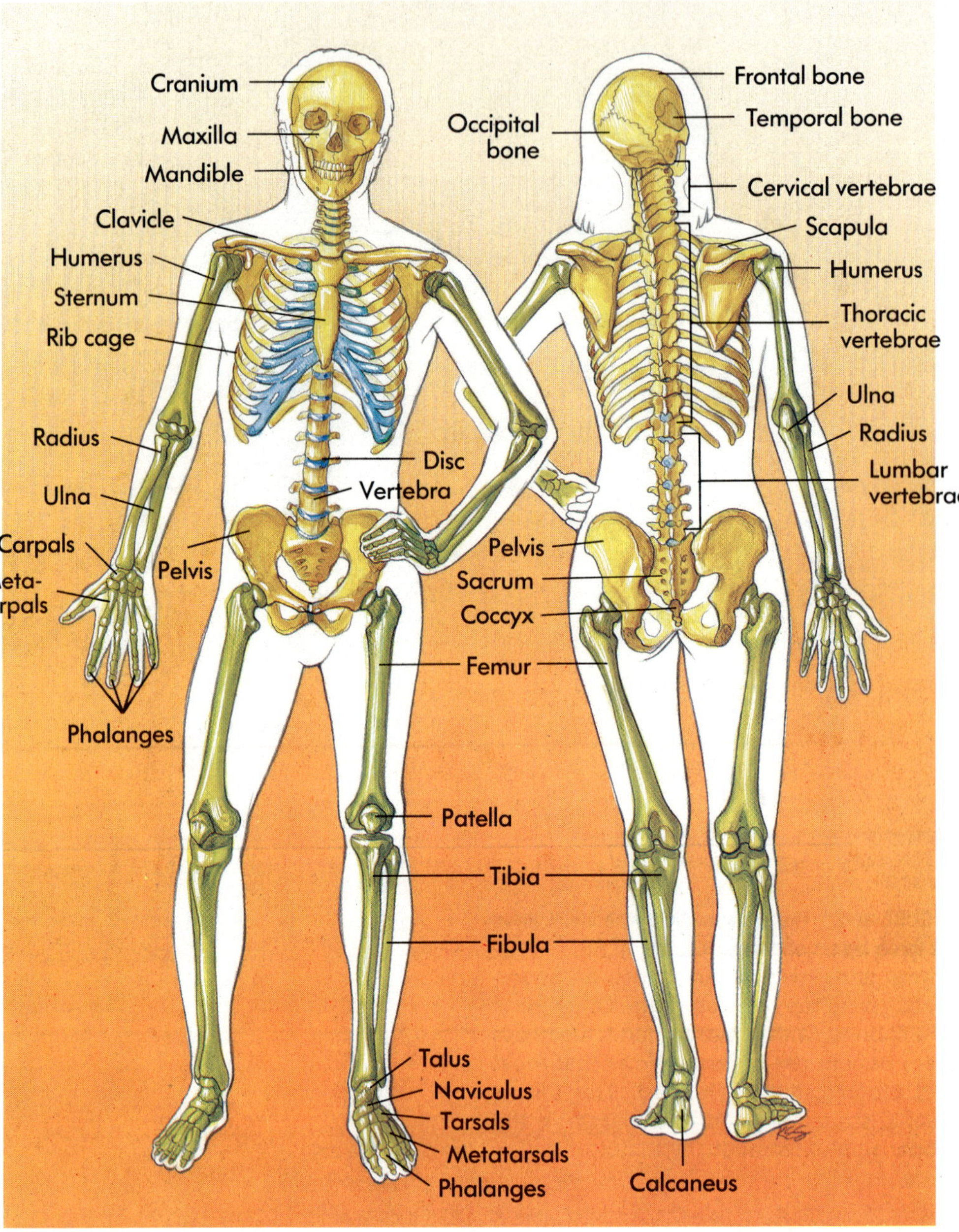

FIGURE 18-3 The human axial and appendicular skeletons. The axial skeleton is made up of the bones of the head, face, rib cage, and vertebral column. The appendicular skeleton includes the bones of the legs and feet, arms and hands, clavicles, scapula, and pelvis.

The skeletal system

The 206 bones of the body make up a working whole called the *skeleton,* or *skeletal system.* Together, the bones of the skeleton perform many important functions, some of which have already been mentioned. To summarize: (1) The skeletal system provides support for the body—otherwise you would be a shapeless blob of skin-covered tissues and organs. (2) It also provides for movement, with individual bones serving as points of attachment for the skeletal muscles and acting as levers against which muscles can pull. (3) Delicate internal structures are protected by the skeleton. The brain, for example, lies within a strong, bony casing called the *skull;* the rib cage surrounds and protects the heart and lungs. (4) The bones of the skeleton are a storehouse for minerals such as calcium and phosphorus. Amazingly, bone tissue is continually broken down and re-formed, the mineral salts being transported to other parts of the body on demand. In fact, scientists have estimated that your body completely "replaces" your skeleton over a period of 7 years! (5) Lastly, certain inner portions of bones contain red marrow and are your internal factory of blood cells. Without this factory you would die, since red blood cells have a short life span of approximately 120 days. Without these cells your body would have no efficient means to transport life-giving oxygen to your tissues. White blood cells are also produced in the red bone marrow. One of the main jobs of certain classes of these cells is to ingest bacteria and debris. White blood cells also form the cells of your immune system (see Chapter 13). It is essential to your survival that the red marrow

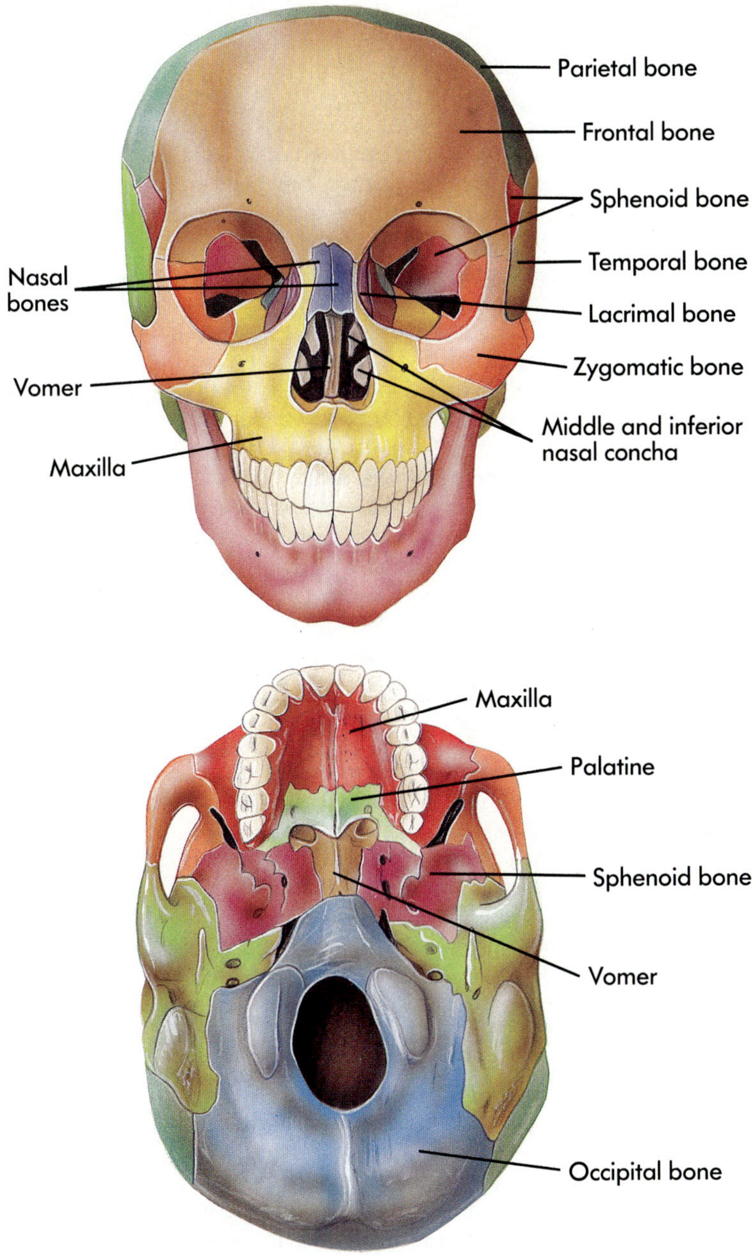

FIGURE 18-4 Frontal view of the human skull. The skull contains 8 cranial and 14 facial bones.

continue to manufacture these cells, which have varying life spans and help protect your body from infection and disease.

Scientists divide your internal scaffolding of bones into two parts: the axial skeleton and the appendicular skeleton (Figure 18-3). The **axial skeleton** is your axis, or central "column," of bones off which the appendages (arms and legs) of your **appendicular skeleton** hang. The appendicular skeleton also includes the bones that serve to attach the appendages to the axial skelton.

The axial skeleton

The 80 bones of the axial skeleton include the bones of the skull, vertebral column, and rib cage. The **skull** contains 8 cranial and 14 facial bones. Find the following skull bones in Figure 18-4: one frontal bone that forms the forehead, two parietal bones that form a large portion of the sides and top of the skull, two temporal bones that form the sides of the skull toward the back, and one occipital bone that forms the lower back portion of the skull. The irregularly shaped ethmoid and sphenoid bones are not easy to locate in a diagram because only a tiny portion of these bones lies on the surface of the skull. The ethmoid bone sits between the eyes; only a fraction of it can be seen forming a part of each orbit, or eyesocket. The rest of this bone forms part of the nasal cavity and the floor of the cranial cavity. The sphenoid bone makes up part of the back and sides of the eyesockets. Looking somewhat like wings, it extends from one side of the skull to the other, forming the middle portion of the base of the skull.

Some of the 14 facial bones are easy to see in Figure 18-4: 1 mandible, or lower jawbone; 2 maxillae, which unite to form the upper jawbone; 2 nasal bones, which form the bridge of the nose; and 2 zygomatic bones, or cheekbones. Look below the part of the ethmoid bone that forms part of the nasal septum to find the vomer. Along with the ethmoid and cartilage, the vomer helps divide the nose into right and left nostrils. The paired lacrimal bones are tiny, therefore difficult to see, and are located at the inner corners of the orbits near the nose. Each of these bones has a groove forming part of the tear ducts, canals that drain excess fluid bathing the eyes. The two L-shaped palatine bones are hidden from view because they form the back portion of the roof of your mouth and extend upward to form part of the floor and sides of the nasal cavity. The two inferior nasal conchae are "swirled" bones and project into the nasal cavity. These bones cause the air that enters the nose to be circulated and warmed before being breathed into the lungs.

The **vertebral column** is made up of 26 individual bones called *vertebrae* (Figure 18-5). Stacked one on top of the other, these bones act like a strong yet very flexible rod that supports the head. In addition, some of the vertebrae serve as points of attachment for the ribs. Each vertebra has a central hole, or foramen. Lined up, the vertebrae form a bony canal protecting the spinal cord, which runs down much of its length.

The 7 vertebrae closest to the head are the cervical (neck) vertebrae. Next are 12 thoracic (chest) vertebrae. Following these are 5 lumbar (lower back) vertebrae. The sacrum and the coccyx, or tailbone, are the last vertebrae in the column. Positioned between the vertebrae (except the sacrum and coccyx) are disks of fibrocartilage called **intervertebral disks.** These disks act as shock absorbers, provide the means of attachment between one vertebra and the next, and permit movement of the vertebral column.

Attached to the 12 thoracic vertebrae are 12 pairs of ribs. The ribs curve around to the front of the thoracic (chest) cavity, producing a bony cage that protects the heart and lungs. The upper 7 pairs of ribs directly connect to a flat bone that lies at the midline of the chest called the *breastbone,* or sternum. Each of these 14 ribs connects to the sternum by means of a strip of hyaline cartilage and are therefore called *true ribs.* The remaining five pairs of ribs, called *false ribs,* do not directly connect to the sternum. Instead the cartilages of ribs 8 through 10 attach to each other, and then to the cartilage of ribs 7. Ribs 11 and 12 do not attach to the sternum but hang free, supported by muscle tissue. These ribs are therefore called *floating ribs.*

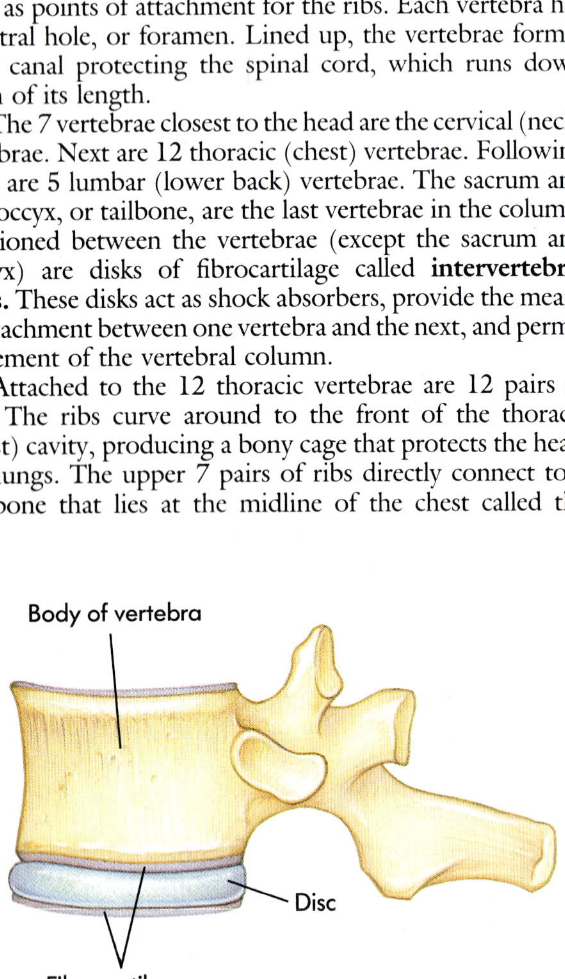

FIGURE 18-5 A vertebra with disk attached.
The vertebral column is composed of 26 vertebrae. Between each vertebra and the next is a disk composed of an outer layer of fibrocartilage and an inner, more elastic layer.

The remaining bone of the axial skeleton is the **hyoid bone,** a name that means "U shaped". This bone "floats" in the neck supported by ligaments. It is the only bone in the body that does not form a joint with other bones, and it serves as the point of attachment for the tongue. Since it is often broken when a person is strangled, the hyoid can provide important evidence in certain murder cases.

The appendicular skeleton

The appendicular skeleton is made up of the bones of the appendages (arms and legs), the **pectoral (shoulder) girdle,** and the **pelvic (hip) girdle.** The pectoral girdle is made up of two pairs of bones: the clavicles, or collarbones, and the scapulae, or shoulder blades. Find these bones in Figures 18-3 and 18-6, and then locate the edges of your own clavicles and scapulae. The bones of the pectoral girdle support and articulate with (form joints with) the arms. (Articulations, or joints, are discussed shortly.) Together, both arms, or upper extremities, contain 60 bones.

As shown in Figure 18-6, the bone of the upper arm is called the *humerus.* This bone articulates with both the clavicle and the scapula. The other end of this long bone articulates with the two bones of the forearm: the radius and the ulna. The wrist is made up of eight short bones called *carpals,* lined up in two rows of four. These bones are held together by ligaments and articulate with the bones of the hand, or metacarpals. The five metacarpals, in turn, articulate with the bones of the fingers, or phalanges (sing. phalanx). Each finger has three phalanges; the thumb has only two.

The pelvic girdle is made up of two bones called *coxal bones.* You know these bones as pelvic bones or hipbones. Find these bones in Figures 18-3 and 18-7. Male and female skeletons can be distinguished from one another by looking at their pelvic bones. Females have a wider pelvis with a large, oval opening in the center for giving birth. The male pelvic girdle, in contrast, is narrower and has a smaller, heart-shaped opening in the center.

The bones of the pelvis support and articulate with the legs. Together, both legs, or lower extremities, contain 60 bones as the arms do. As shown in Figure 18-7, the bone of the upper leg, or thigh, is called the *femur.* This bone has a rounded head, or "ball," that articulates with a depression, or "socket," in the hipbone. The femur, the longest and heaviest bone in the body, articulates at its lower end with the two bones of the lower leg: the tibia and the fibula. The tibia is the bone commonly referred to as the *shinbone* and can be felt at the front of the lower leg. If you have ever had shin splints, you are well aware of this bone. Shin splints are an inflammation of the outer covering of the tibia caused by repeated tugging by muscles and tendons. Vigorous walking, running, or other types of exercise can sometimes result in shin splints, which can be very painful.

The patella, or kneecap, is a small triangular bone that

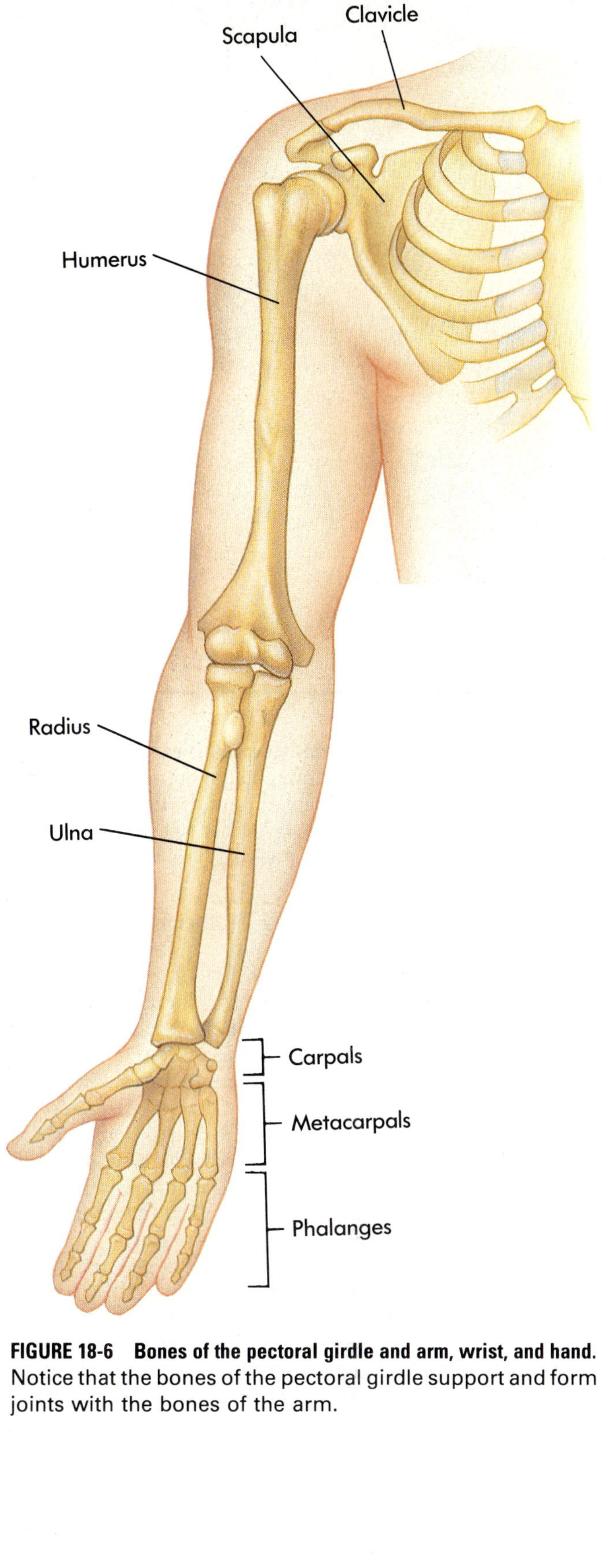

FIGURE 18-6 Bones of the pectoral girdle and arm, wrist, and hand. Notice that the bones of the pectoral girdle support and form joints with the bones of the arm.

FIGURE 18-7 Bones of the pelvic girdle and leg, ankle, and foot. The two pelvic bones (also called *coxal* or *hip bones*) articulate with the head of each femur, or thigh bone.

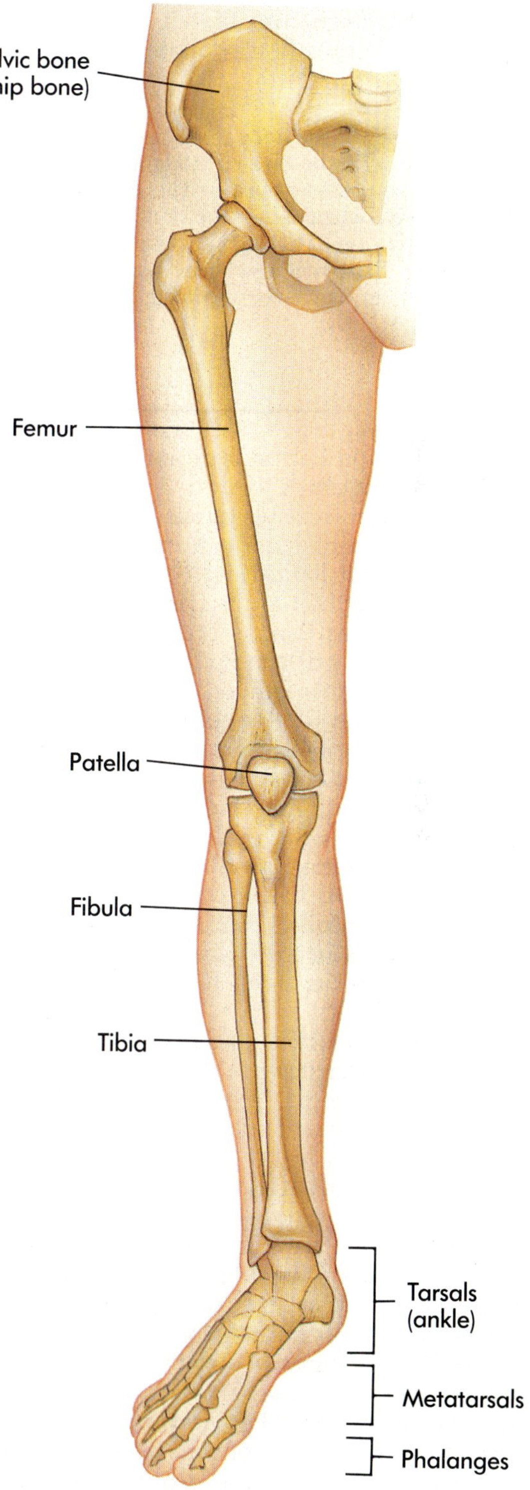

sits in front of the joint formed by the femur, tibia, and fibula. The patella is a sesamoid bone, one that is formed within tendons where pressure develops. Although everyone has two patellae, people develop a variable number of tiny sesamoid bones in their wrists (in addition to their eight carpals).

Each ankle is made up of seven short bones called *tarsals*. One of these bones forms the heel, and is the largest and strongest ankle bone. The tarsals are held together by ligaments and articulate with the bones of the foot, or metatarsals. The five metatarsals, in turn, articulate with the bones of the toes, or phalanges. Like the fingers, each toe has three phalanges; like the thumb, the "big toe" has only two.

The 206 bones of the body make up the skeletal system, which can be divided into the axial skeleton and the appendicular skeleton. The axial skeleton includes your central axis—the skull, vertebral column, and rib cage. The appendicular skeleton is made up of the appendages (arms and legs) and the bones that help attach the appendages to the axial skeleton.

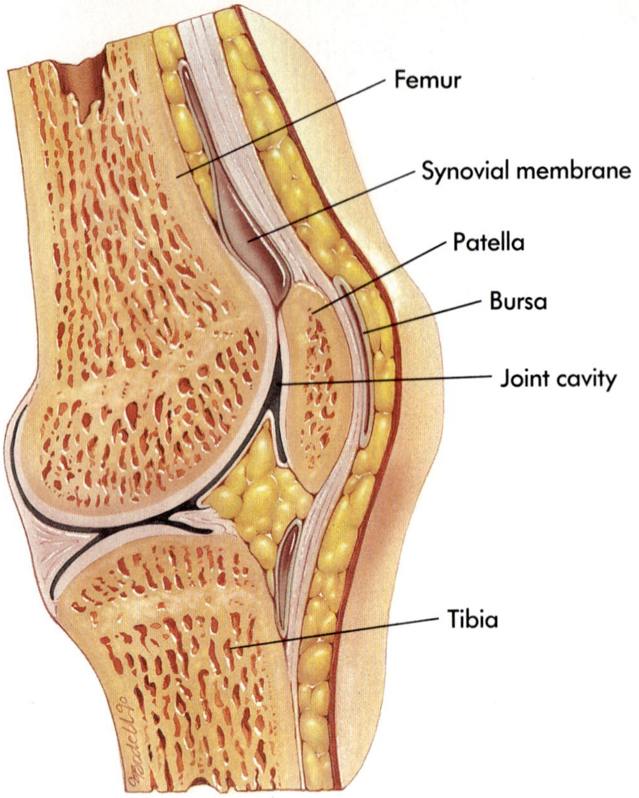

FIGURE 18-8 A synovial, or freely movable, joint. The articulation of the kneecap (patella) with the femur and tibia is an example of a synovial joint. In the knee joint a sac filled with synovial fluid (a bursa) lies on top of the kneecap, providing a cushion between the bone and skin.

Joints

If you have ever broken a leg or an arm and had a cast covering a joint, you learned firsthand just how important joints are to movement. A **joint,** or **articulation,** is a place where bones or bones and cartilage come together. All joints are not alike; some permit little or no movement (immovable joints); some permit limited movement (slightly movable joints); and some permit a considerable amount of movement (freely movable joints). The amount of movement afforded by a joint is a direct result of how tightly the bones are held together at that location.

The bones of the skull articulate with one another in a type of immovable joint called a **suture.** The edges of each skull bone are ragged but fit tightly with the ragged edges of adjoining bones—just like puzzle pieces fit together (see Figure 18-3). A layer of dense connective tissue helps keep the bones from separating (see Chapter 9). Another example of immovable joints is the articulation of your teeth with the mandible and maxillae. Here, peg-shaped roots fit into cone-shaped sockets in the jawbones. A ligament lies between each tooth and its socket, holding each tooth in place. **Ligaments** are bundles, or strips, of dense connective tissue that hold bones to bones. A third example of an immovable joint is the articulation between the first pair of ribs and the breastbone. Each of these two ribs is connected to the sternum by a strip of hyaline cartilage, connections that change to bone during adult life.

As you twist your forearm to the right and left or twist your lower leg in the same manner, imagine the movement taking place between the shafts of the radius and ulna or between the tibia and fibula at their lower ends. Dense connective tissue is present at these locations, holding these bones together while permitting some flexibility. These two articulations are examples of slightly movable joints. The articulation between one vertebra and the next is also an example of a slightly movable joint. Here, a broad, flat sheet of fibrocartilage covers the top and bottom of each intervertebral disk (see Figure 18-5), which is sandwiched between adjacent vertebrae creating a somewhat flexible connection.

In slightly movable and immovable joints, there is no space between the articulating bones. The bones forming these joints are held together by dense connective tissue or cartilage. In freely movable joints, there is a space between the articulating bones. Such joints are common in your body and are called **synovial joints.** Figure 18-8 diagrams the components of a synovial joint. Hyaline cartilage covers the ends of the articulating bones, which are separated by a fluid-filled space, or joint cavity. The entire joint is encapsulated in a double layer of connective tissue, with the joint cavity and its fluid lying in between these two layers. The outer layer of the capsule is made up of dense connective tissue that holds the bones of the joint together. Some of the connective tissue fibers are arranged in bundles forming strips of tissue, or ligaments. Although this outer capsule holds the bones of the joint in place, it permits a wide range of motion. The inner layer of the capsule is made up of loose connective tissue, including elastic fibers and adipose tissue, or fat. This inner layer of tissue secretes the fluid of the joint cavity, called *synovial fluid.* In fact, the synovial joint gets its name from this "egg white-like" fluid; the word *synovial* comes from a Greek word and a Latin word meaning "with" *(syn)* "an egg" *(ovum)*. Synovial fluid lubricates the joint, provides nourishment to the cartilage covering the bone, and contains white blood cells that battle infection. Table 18-1 summarizes the types of movements possible at synovial joints and gives other examples of joints.

TABLE 18-1

Locations and motions of various joints in the human body

TYPE OF JOINT	LOCATION	TYPE OF MOVEMENT
Hinge joint 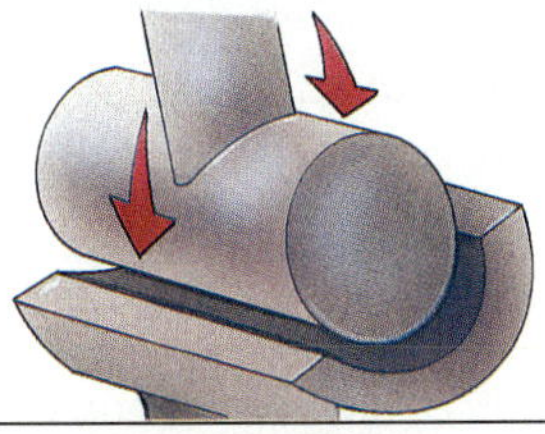	Elbow, knee 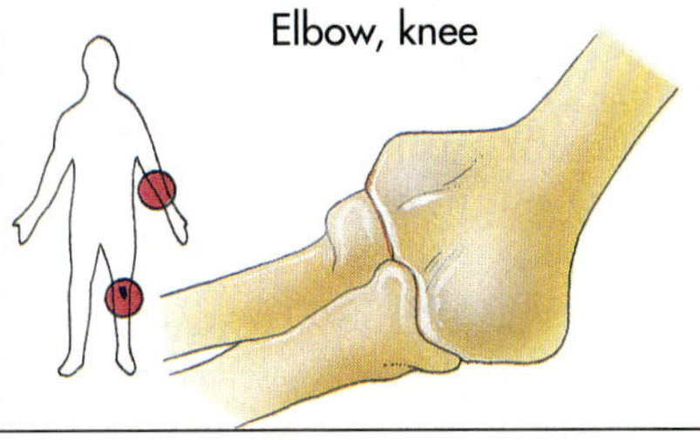	Hinge movement
Ball and socket joint 	Shoulder, hip 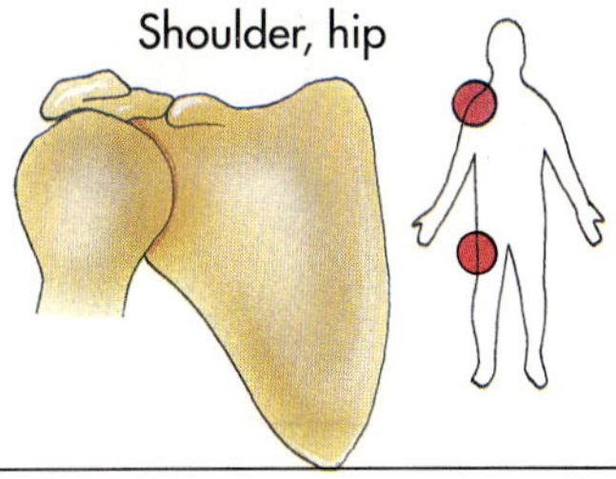	Wide range of movement in almost any direction
Pivot joint 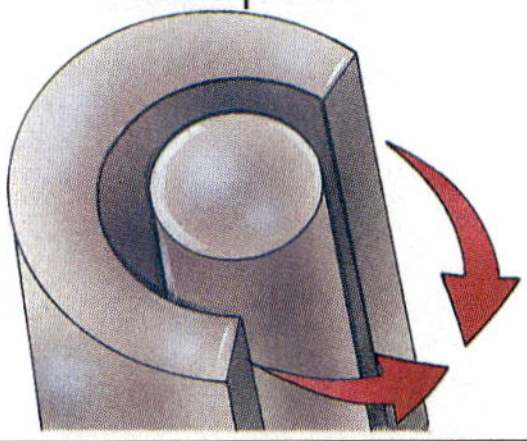	Joint between first two vertebrae 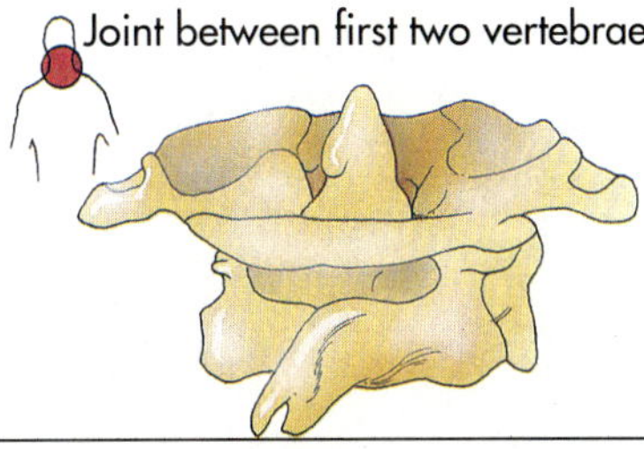	Rotational movement
Plane or gliding joint 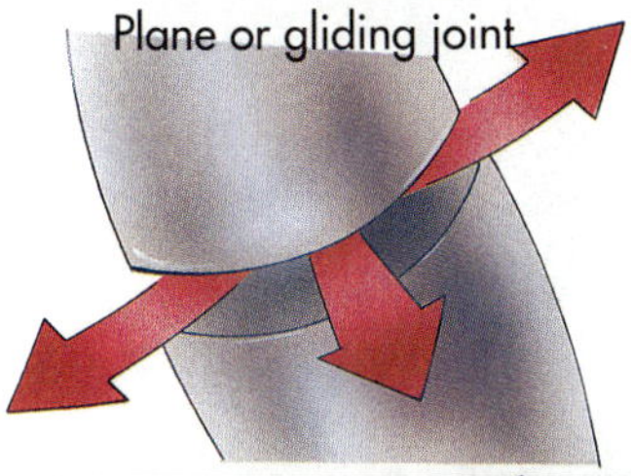	Processes (projections) between vertebrae 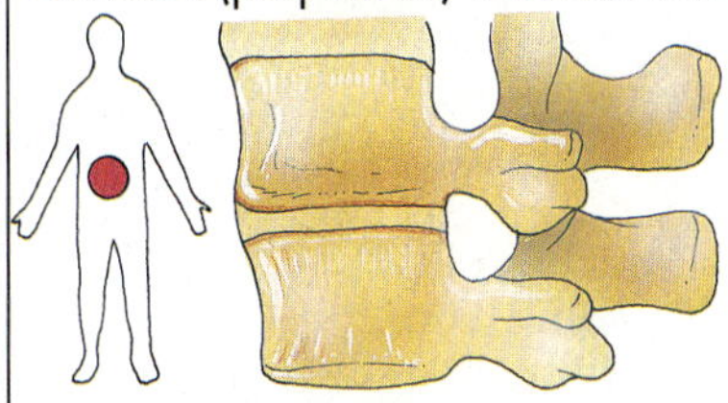	Sliding in many different directions
Saddle joint 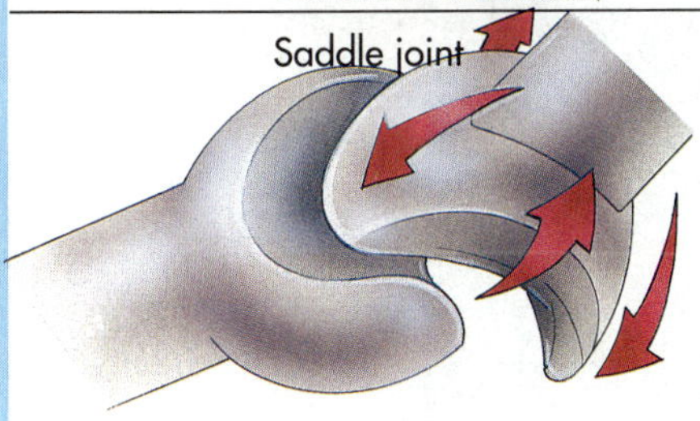	Thumb 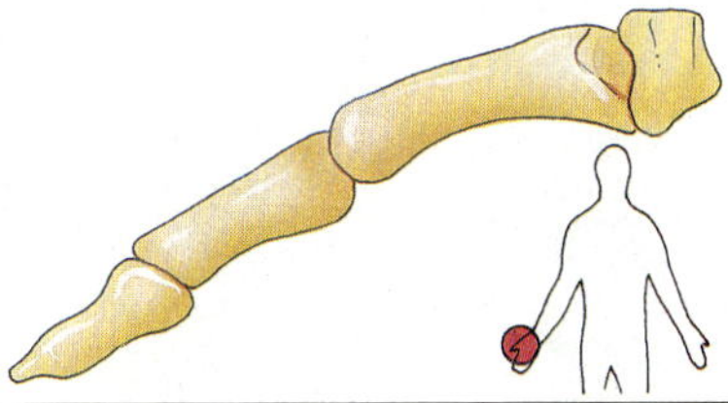	Movement in right angles
Ellipsoid joint 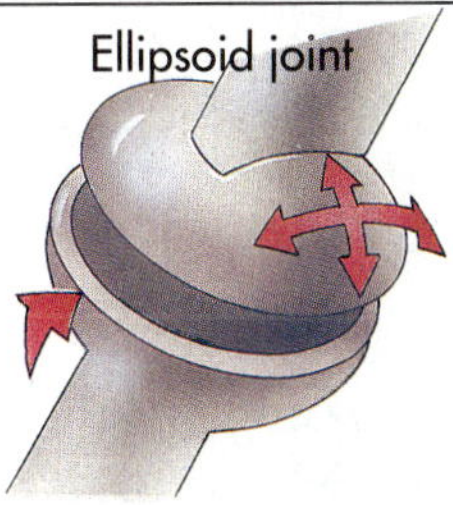	Joint between skull and first vertebra 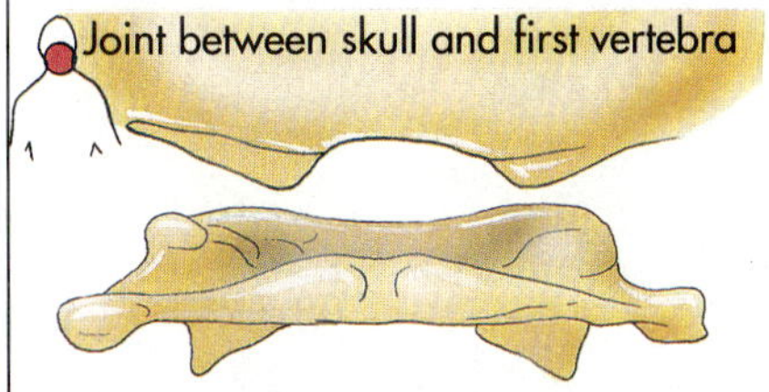	Nearly hinge movement; rotation restricted

Although synovial joints permit a wide range of movement, this movement is limited by several factors, including the tension, or "tightness," of the ligaments and muscles surrounding the joint and the structure of the articulating bones. You can see the various types of movement of synovial joints on your own body. For example, move your hand at the wrist or your foot at the ankle. During this movement the carpals or tarsals slide over one another to produce gliding movements. Other joints allow angular movements, which as the name suggests, increase or decrease the angle between two bones. When body-builders flex their biceps, for example, they bring their forearms closer to their upper arms, decreasing the angle between the radius/ulna and the humerus. The elbow joint permits this movement and is called a **hinge joint.** A hinge joint allows movement in one plane only, similar to how a door opens and closes by means of its hinges. Another type of movement is rotation, the movement of a bone around an axis. You see this type of movement when a pitcher "winds up" and throws a baseball or when you move your head from side to side. The type of joint the pitcher is using at the shoulder is called a **ball-and-socket** joint. In this joint the rounded "head" or "ball" of the humerus articulates with the concavity or "socket" formed by the ends of the clavicle and scapula. When you move your head from side to side, the type of joint you are using is a **pivot joint,** formed by the first two vertebrae, the atlas, and the axis. Your skull rests on the ring-like atlas (named after the Greek god who could lift the Earth) and pivots on a projection of the axis, rising from below.

A joint is a place where bones and cartilage come together. The amount of movement allowed between two bones depends on how tightly the bones are held together. Freely movable joints, those in which a space exists between articulating bones, are common in the body and include such joints as the elbow and knee.

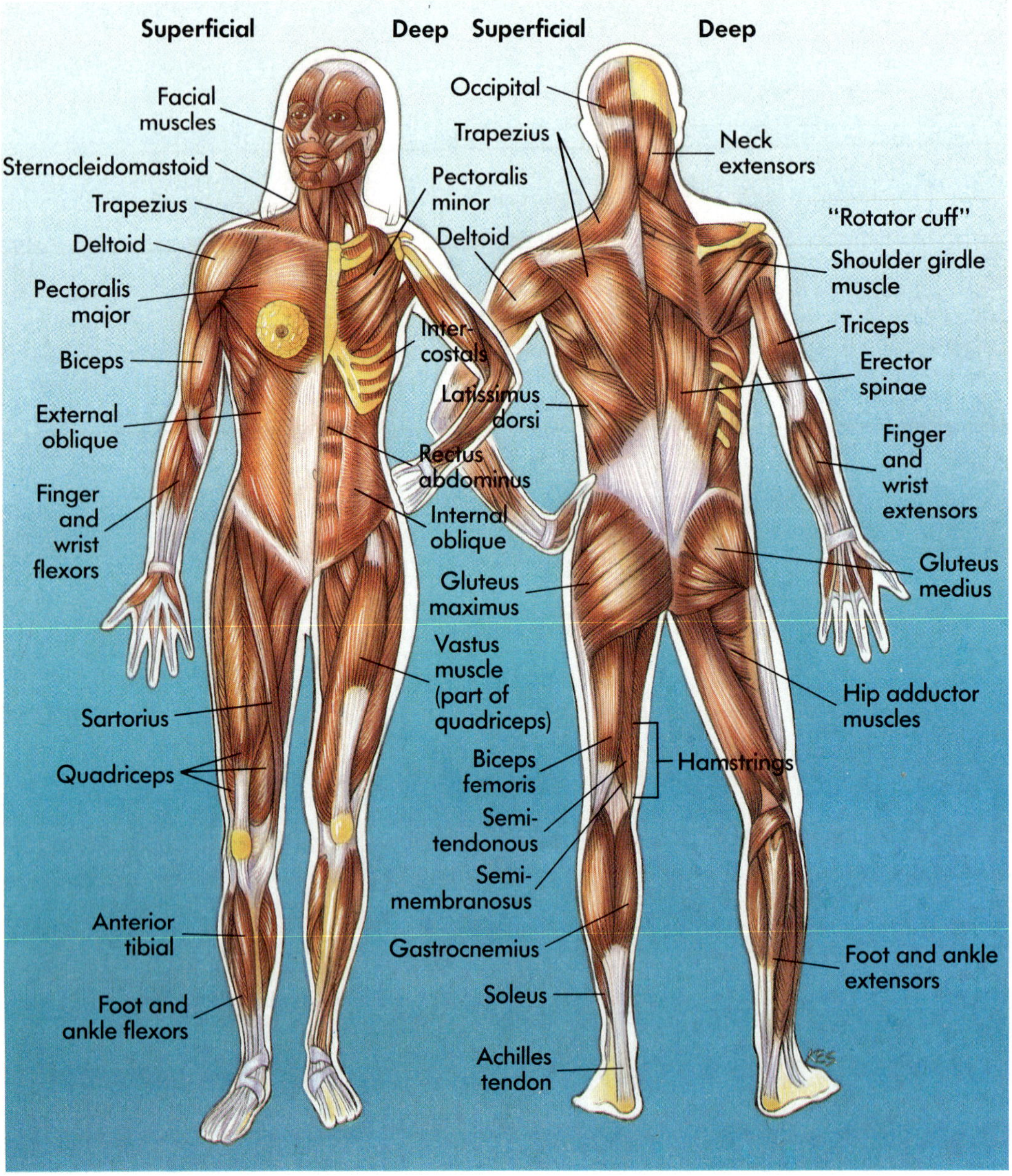

FIGURE 18-9 The human muscular system. Body movements are accomplished by the contraction of the skeletal muscles, which are attached to the bones.

Muscles

Bones and joints are of no use in movement unless muscles are attached to the bones. Muscles provide the power for movement and are made up of specialized cells packed with intracellular fibers capable of shortening. Your body has three different kinds of muscle tissue: smooth muscle, cardiac muscle, and skeletal muscle (see Table 9-4). Each type is found in certain locations within the body and performs specific functions. Smooth muscle tissue is made up of sheets of cells that are found in many organs, doing such jobs as mixing food in your stomach and narrowing the interior of your arteries to restrict blood flow. Cardiac muscle tissue makes up the heart—the pump of your circulatory system. Skeletal muscle tissue makes up the muscles that are attached to bones, allowing you to move your body; these muscles are the nearly 700 muscles of the muscular system.

The muscular system

The muscular system is shown in Figure 18-9. Its muscles are attached to bones by means of cords of dense connective tissue called **tendons.** One end of a skeletal muscle is attached to a stationary bone; this attachment is called the **origin.** The other end is attached to a bone that will move; this attachment is called the **insertion.** The origin serves to anchor the muscle as it pulls at the insertion. As an example, Figure 18-10 shows the biceps and triceps muscles and their origins and insertions. The biceps brachii is located on the front of the upper arm. In fact, the word *brachii* means "arm." This muscle has two upper ends, or heads (hence, *biceps,* meaning "double headed"). These two heads have origins on the edges of the scapula near the front side of the humerus. Their tendons pass from the scapula over the upper end of the humerus and then blend into the fattened midsection, or belly, of the biceps. The insertion of the biceps is by means of flattened tendons into the radius. When the biceps shortens, the insertion is brought closer to the origin of the muscle, an action that pulls the forearm to the upper arm. To lower the forearm, the triceps brachii goes into action. Skeletal muscles oppose each other in this way; such opposing muscle pairs are called **antagonists.** In this case the action of the triceps opposes the action of the biceps. The triceps sits on the back of the arm with one of its three heads originating on edges of the scapula near the back side of the humerus. Another head originates along the upper half of the back side of the humerus and the third along the lower half of the back side of the humerus. The triceps inserts on the ulna near the elbow. When the triceps shortens, its insertion is brought closer to its origin and the forearm is brought downward. The distance between the origin and insertion is measured as the length of the muscle and tendons, not as a straight line between the two points.

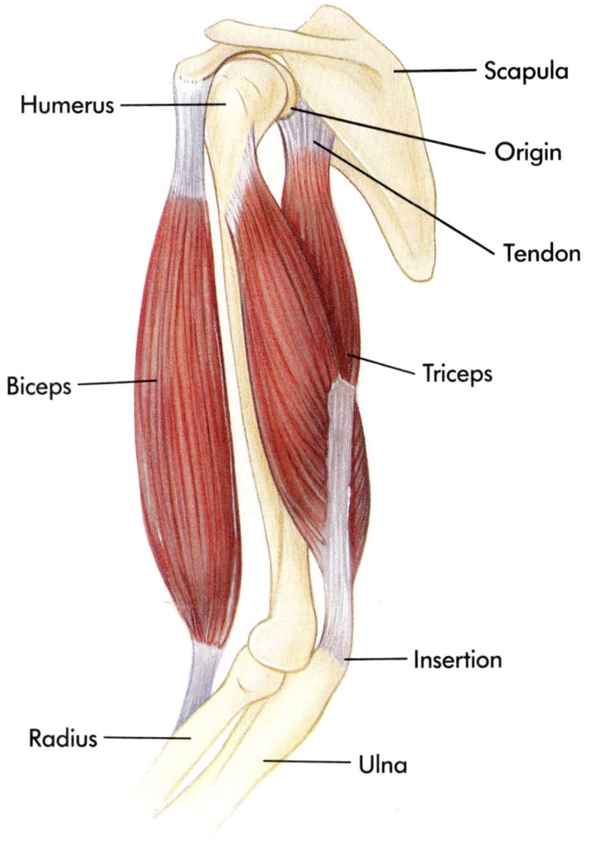

FIGURE 18-10 The attachments of skeletal muscles to bones. The bicep is attached to the stationary scapula at the origin and to the movable radius at the insertion with straps of tissue called *tendons.* The origin anchors the muscle as it pulls at the insertion.

The nearly 700 muscles of the skeletal system are attached to bones by means of connective tissue called *tendons.* One end of a muscle is attached to a stationary bone (the origin) and the other to a movable bone (the insertion). The origin anchors the bone as it pulls at the insertion.

How skeletal muscles contract

Skeletal muscles are also called **striated muscles** because their cells appear marked with striations, or lines, when viewed under the light microscope (Figure 18-12, *A*). These striations are due to an orderly arrangement of microfilaments within skeletal muscle cells. Special groupings of these microfilaments are the "contractile units" of muscle cells.

Skeletal muscle cells are extremely long cells formed by the end-to-end fusion of shorter cells during embryonic development. These long muscle cells are called **muscle fibers.** Each muscle fiber contains all the nuclei of the fused cells pushed out to the periphery of the cytoplasm (Figure 18-11, *D*). To see the relationship between muscle fibers and muscles, study Figure 18-11, noting that a muscle is made up of bundles of muscle fibers, bound together with connective tissue, and nourished by blood vessels. As you study the diagram, you will notice that the names of certain parts of muscle cells have the prefix *sarco,* which comes from a Greek word meaning "flesh." Table 18-2 lists these differences in terminology.

Each muscle fiber is packed with **myofibrils,** which are cylindrical, organized arrangements of special thick and thin microfilaments capable of shortening the muscle fiber (see Chapter 3). In muscle cells, microfilaments are called **myo-**

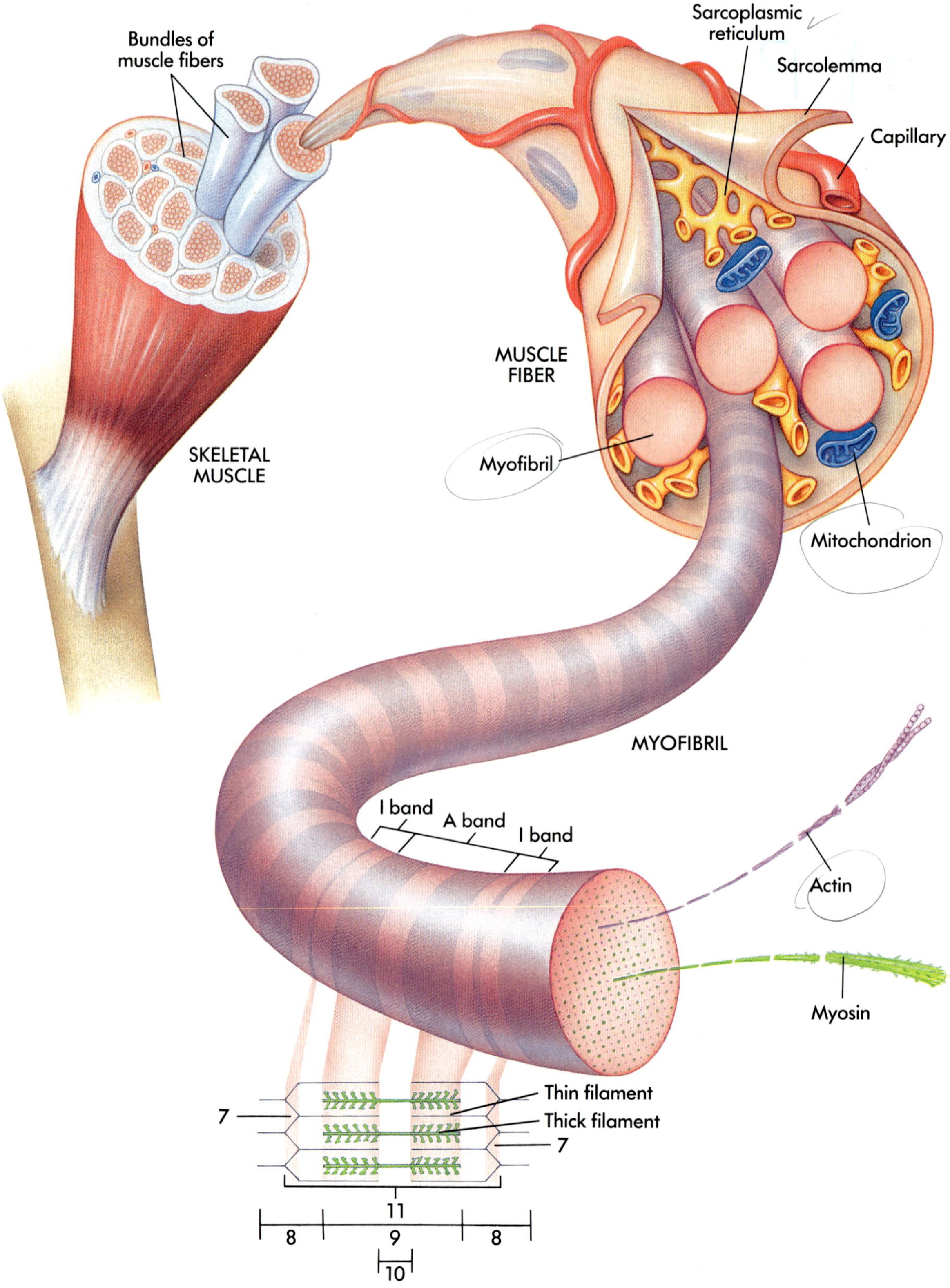

FIGURE 18-11 Structure of a skeletal muscle.
This diagram shows progressively magnified views of a skeletal muscle.

TABLE 18-2

Muscle cell terminology

CELL COMPONENT	MUSCLE CELL COMPONENT
Cell membrane (or plasmalemma)	Sarcolemma
Cytoplasm	Sarcoplasm
Endoplasmic reticulum	Sarcoplasmic reticulum

filaments (*myo* means "muscle"). The so-called thin myofilaments are made up of the protein **actin.** The thick myofilaments are made up of a much larger protein, **myosin.** Figure 18-11, *E* shows the arrangement of actin and myosin filaments in a myofibril. All the myofilaments within a muscle fiber are "lined up" in such a way that the cells appear to have interior bands of light and dark lines, a feature readily observable in Figure 18-11, *D*.

Figure 18-12 is a close-up view of the banding pattern in a muscle fiber as shown in an electron micrograph. This pattern is repeated throughout each muscle fiber. Below the electron micrograph in Figure 18-12, the diagram shows how the actin and myosin filaments are arranged, forming this banding pattern. The thin actin filaments are attached to plates of protein that appear as dark lines called *Z lines*. The actin filaments extend from the Z lines (plates) equally in two directions, perpendicular to the plates. In a resting muscle the actin filaments that extend from two sequential Z lines are not long enough to reach each other. Instead, they are joined to one another by interdigitating myosin filaments, similar to the way your fingers interlock when you fold your hands. This arrangement of protein plates and myofilaments produces the banding patterns shown in Figure 18-12: The I band contains the Z line and the thin actin filaments. The A band contains a portion of the thin filaments and the thick myosin filaments. The H zone appears as a light zone running down the center of the A band; it contains only thick filaments held in place by a series of fine threads referred to as the *M line*. The part of the myofibrils lying between adjacent Z lines is called the **sarcomere,** the contractile unit of muscles.

Skeletal muscle cells are long, multinucleated cells called *muscle fibers*. A muscle is made up of bundles of muscle fibers. Each muscle fiber is packed with organized arrangements of microfilaments that are capable of contracting.

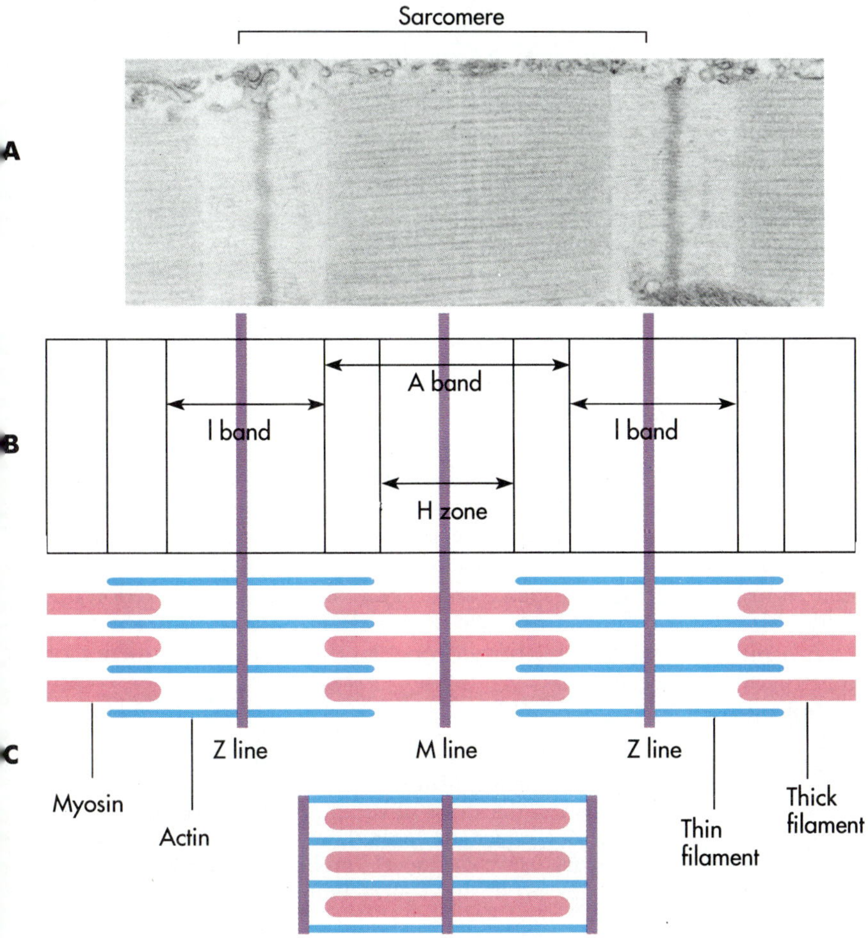

FIGURE 18-12 Structure of a sarcomere.
A The micrograph shows the banding patterns of the thick and thin filaments in a sarcomere, the contractile unit of muscles.
B Actin filaments (thin filaments) are attached to the Z line. Myosin filaments (thick filaments) interdigitate with actin filaments. The different zones of the muscle are also illustrated in this diagram; the I band contains thin filaments and the Z line, the A band contains some thin filaments and the thick filaments, and the H zone contains only thick filaments that have been anchored in place by the M line.
C In a contracting sarcomere, the thick and thin filaments move toward the center. As this movement occurs, the I band disappears.

BIOLOGY IN FOCUS

Can special shoes make you run faster?

In the 1963 Hollywood comedy *Son of Flubber,* an absent-minded professor invented a magical bouncy substance that, when inserted into the boots of the hapless football team players, gave them a lift and helped them to win their big game. What was fantasy then is the talk of runners today.

Shoes are the most important tool of any runner (Figure 18-A). They must both cushion against the impact of the foot with the road and prevent exaggerated sideways movements of the foot that can lead to injury. In modern shoes the foot is protected from unwanted sideways movements by a strong flexible plastic cup around the heel that holds it firmly in place. To cushion a shoe, manufacturers commonly use a midsole material made of springy ethyl vinyl acetate (EVA) encapsulated in various ways within polyurethane.

In the last 2 years, shoe manufacturers have added a new twist. Noting experiments carried out at Harvard University showing that runners ran 3% faster on tracks made slightly elastic, some shoe manufacturers reasoned that the same theory applying to the track should also apply to the shoes running on the track. Manufacturers therefore designed a more elastic shoe that used an "energy-return" system.

FIGURE 18-A Can the type of shoes you wear make you run faster? Shoe manufacturers say yes, but some researchers disagree.

In a typical "energy-return" shoe, the cushion is provided by hollow tubes of flexible thermoplastic that quickly spring back into their original shape—the idea is that the springback of the tubes imparts energy, aiding lift-off of the runner's foot. Touted by their manufacturers as a major advance, these shoes have rapidly become the rage of the jogging set.

Do they work? Skeptics argue that a functional energy-return system would have to capture the energy from heel strike beneath the rear foot and somehow transfer it to the forefoot to return any of the energy of running to the athlete, not simply send it back up the rear foot as energy-return shoes do. The heel recoils too early in the step for such energy return to be of any use to a runner. A high return of energy from the heel simply keeps the vibrations following impact going longer.

Perhaps more to the point, many energy-return shoes are less flexible and cushioned than their less technologically advanced counterparts, resulting in a more jarring run, just as a stiff suspension in a car produces a jarring ride. Peter Cavanagh of Penn State University summed up the problem succinctly. "A steel ball on a concrete surface would give an excellent score for energy return but no one would suggest that a shoe made from these materials would be acceptable." Anyway, for most runners the goal is usually to expend energy—to maximize energy expenditure, not minimize it.

Molecularly, each actin filament (that is, thin filament) consists of two strings of proteins wrapped around one another, like two strands of loosely wound pearls. The result is a long, thin, helical filament. Myosin has an unusual shape: one end of the molecule is a coil of two chains that forms a very long rod, whereas the other end consists of a double-headed globular region. In electron micrographs a myosin molecule looks like a two-headed snake. The contraction of myofilaments occurs when the heads of the myosin filaments change shape. This causes the myosin filaments to slide past the actin filaments, their globular heads "walking" step by step along the actin (Figure 18-13).

How does myofilament sliding lead to muscle contraction? Because actin myofilaments are anchored to the Z lines, the Z lines are pulled closer to one another when actin slides along the myosin, contracting the sarcomere. All the sarcomeres of a myofibril contract simultaneously, shortening the myofibril. In addition, all the myofibrils of a muscle fiber usually contract at the same time. However, all the muscle fibers within a muscle *do not* contract simultaneously. The forcefulness, speed, and degree of a muscle contraction depend on the number of muscle fibers that contract, their position in relation to one another, and the frequency of nerve stimuli.

Muscle cells contract as a result of the "sliding" of actin and myosin filaments past one another. Changes in the shape of the ends of the myosin molecules, which are located between adjacent actin filaments, cause the myosin molecule to move along the actin, producing contraction of the myofilament.

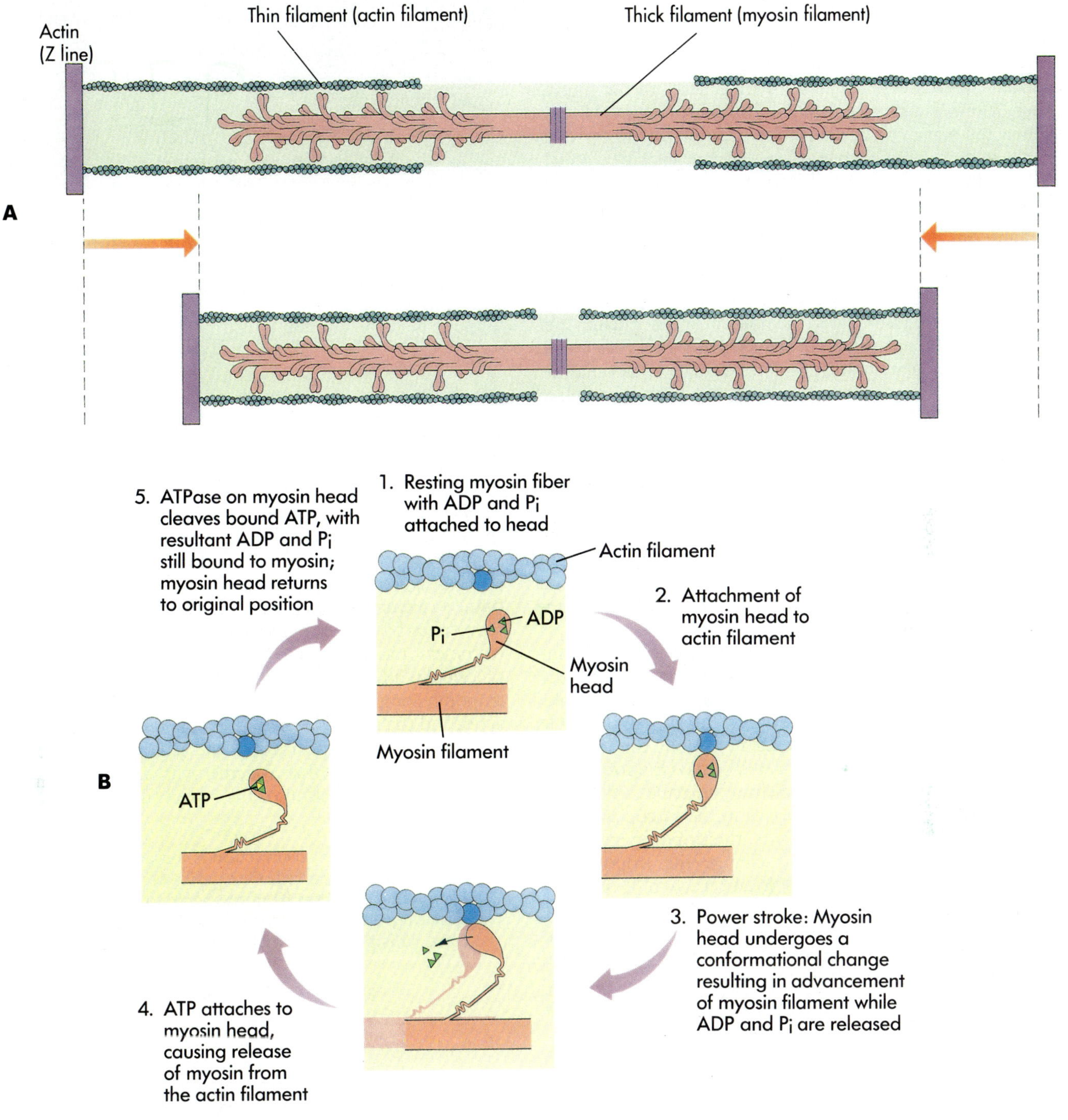

FIGURE 18-13 How the myofilaments move during muscle contraction.
A The sliding of myosin filaments past the actin filaments pulls the actin filaments toward the center.
B A microscopic view of how myosin slides past actin. Myosin moves along the actin filament by first binding to it and then moving forward as the result of a change in the shape of the myosin head.

How nerve impulses signal muscle fibers to contract

The contraction of skeletal muscles is initiated by a nerve impulse that causes a series of events ultimately resulting in the interaction of actin and myosin. The nerve impulse arrives as a wave of depolarization along the nerve fiber (see Chapter 15). The end of the nerve fiber, the **motor endplate,** is microscopically close to the surface of the muscle fiber, forming a **neuromuscular junction** (Figure 18-14). When the wave of depolarization reaches the motor endplate, it causes the release of the neurotransmitter acetylcholine from the nerve cell into the junction. The acetylcholine passes across to the muscle fiber membrane and opens the ion channels of that membrane, depolarizing it. Infoldings of the membrane of the muscle fiber called **transverse (T) tubules** carry this action potential deep into the muscle fiber, reaching each myofibril. In turn, the depolarization of the muscle fiber membrane opens calcium ion channels in the membrane of the **sarcoplasmic reticulum,** which is a tubular, branching latticework of endoplasmic reticulum that wraps around each myofibril like a sleeve (see Figure 18-11). In resting muscle, calcium ions are actively pumped through these channels, concentrating calcium ions within the spaces or sacs of the sarcoplasmic reticulum and thus polarizing the sarcoplasmic reticulum membrane with respect to calcium ions. When the ion channels of the sarcoplasmic reticulum membrane open, calcium ions stored within the sarcoplasmic reticulum move across its membrane and into the sarcoplasm (cytoplasm) (Figure 18-15). The presence of calcium ions triggers the series of chemical reactions of contraction.

In resting muscle, myosin filaments are not free to interact with actin because the myosin binding sites on the actin (called **cross bridge binding sites**) are not available. These binding sites are blocked by a thread-like molecule called **tropomyosin** (see Figure 18-15). Along this tropomyosin thread lie other globular proteins called **troponin.** The troponin molecules bind the calcium molecules when they are released from their storage sacs in the sarcoplasmic reticulum. When this binding occurs, the troponin molecules change their shape. With this change the tropomyosin thread is repositioned to a new location where it does not interfere with myosin interaction. The globular heads of the myosin filament bind to the actin, forming cross bridges that exert tension on the actin threads, pulling them toward the center of the sarcomere. As in a chain reaction, the myosin then cleaves a molecule of ATP (see Chapter 6) and uses the energy that is released to break the cross bridge. Spontaneously, the now-freed head of the myosin forms a new cross bridge at another actin binding site, once again pulling the actin thread closer to the center of the sarcomere. This cycle of the formation, breaking, and re-formation of the cross bridges results in the speedy and simultaneous contraction of the sarcomeres within a muscle fiber. After the passage of the stimulating nerve impulse, the membrane's calcium channels close. The sarcoplasmic reticulum repolarizes by the active transport of calcium ions back out of the cytoplasm and into the reticular spaces. As the level of calcium in the cytoplasm decreases, the tropomyosin-troponin complexes again block the myosin binding sites on actin, and the muscle relaxes. Thus the release of calcium by the nerve fiber's stimulation of the sarcoplasmic reticulum releases the troponin that, together with an input of energy, results in the contraction of the myofibril.

FIGURE 18-14 A neuromuscular junction.
The thick central portion is the body of an axon, with small branches each terminating in a motor end plate. This special synapse between a motor nerve fiber and a muscle fiber is the neuromuscular junction.

In skeletal muscle, contraction is initiated by a nerve impulse. Acetylcholine passes across the neuromuscular junction from the nerve to the muscle, initiating the process that causes the muscle to contract. This process is driven by the energy from ATP.

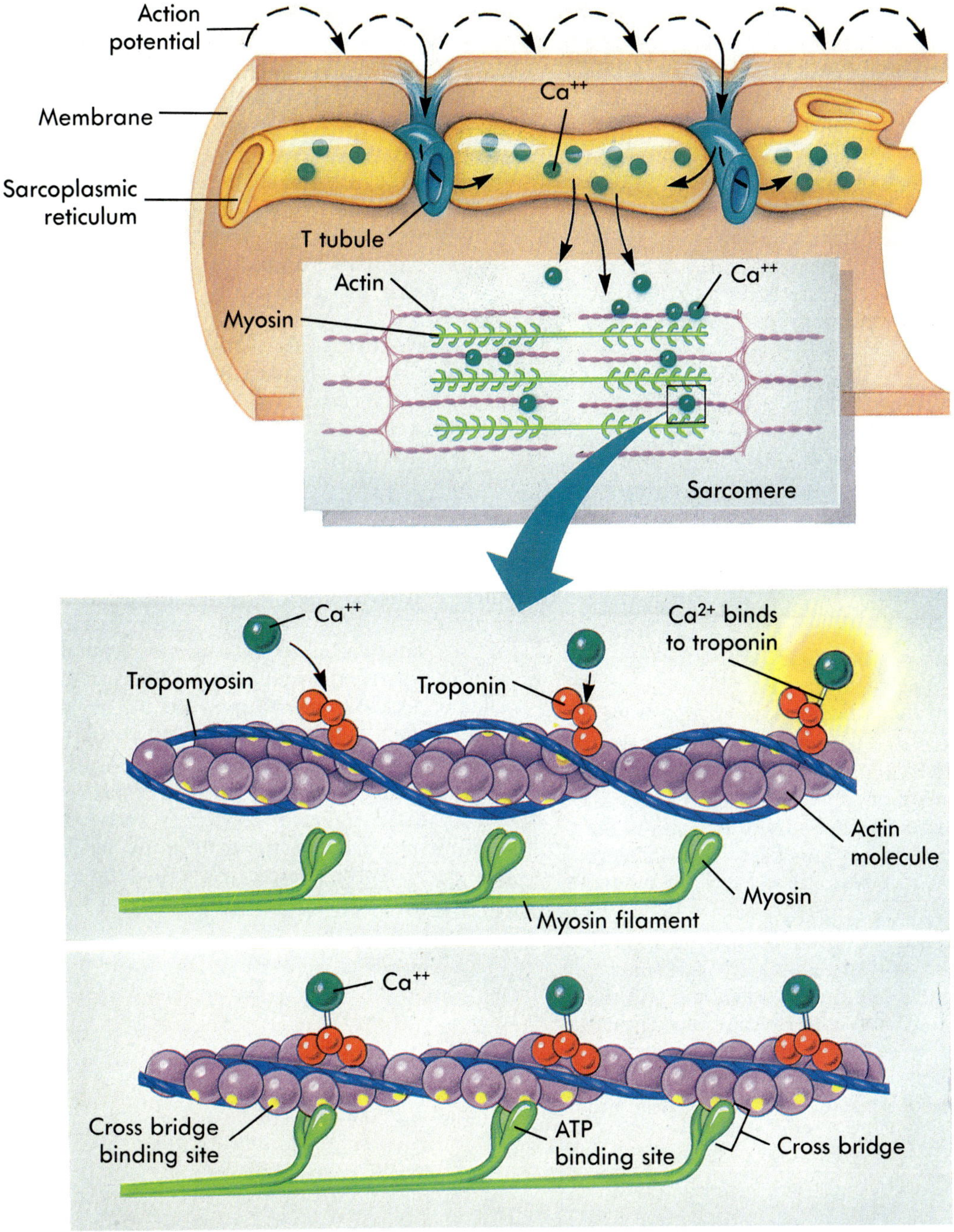

FIGURE 18-15 How nerve impulses signal muscle fibers to contract.
When a nerve impulse reaches a motor end plate, acetylcholine causes a depolarization in the muscle fiber membrane. The T tubules in the membrane transmit this action potential to each myofibril, causing calcium ions to be released from their storage sites in the sarcoplasmic reticulum. For muscle contraction to occur, myosin must be able to bind to the actin at specific sites called *cross bridge binding sites.* However, in resting muscle, these sites are obstructed by tropomyosin threads. When a muscle is signaled to contract, the calcium ions released from the sarcoplasmic reticulum bind to troponin, a molecule that studs the tropomyosin threads. This binding causes the tropomyosin to move out of the way so that myosin can bind to actin at the cross bridge binding sites.

Summary

1. The human body is able to move from place to place because it is supported by a rigid internal framework of bones powered by skeletal muscles. The skeletal muscles are attached to the bones and use them as levers to move the body itself or other objects. The skin provides a flexible covering for the body, accommodating each movement, no matter how large or small.
2. In addition to providing an elastic covering for the body, the skin serves as a protective barrier against water loss and invading microbes, provides a surface by which you sense your environment, and helps control the body's internal temperature.
3. Although bones may not look like living tissue, they contain living cells embedded in a matrix of collagen fibers and mineral salts. The collagen provides a flexibility to the bone, whereas the mineral salts provide hardness. Bone is nourished by blood that flows in blood vessels permeating its interior.
4. Not all bone is made up of "solid," or compact, bone. Compact type of bone is found along the length of long bones, surrounding a central core of fatty yellow marrow. A lighter, bony latticework called *spongy bone* is found within the ends of long bones and within short, flat, and irregularly shaped bones. Red bone marrow fills the spaces within the latticework of spongy bone.
5. Together, all the bones of the body make up the skeletal system. Along with providing support for the body and places for the attachment of skeletal muscles, the skeletal system protects delicate internal structures such as the heart, lungs, and brain; stores minerals such as calcium and phosphorus; and produces blood cells in its red marrow.
6. Scientists divide the skeletal system into two parts: the axial skeleton and the appendicular skeleton. The axial skeleton is your "central column" of bones and consists of the skull, vertebral column, and rib cage. The appendicular skeleton consists of the bones of the appendages (the arms and legs), and the bones off which the appendages hang (the collarbones, shoulder blades, and pelvic bones).
7. A joint, or articulation, is a place where bones or bones and cartilage come together. At places where bones are held tightly together, such as in the skull, little or no movement is possible at the joint. At places where bones are held together less tightly, such as between the two bones of the forearm, slight movement is possible at the joint. At places where a space exists between articulating bones, such as between your upper leg and your pelvic bone, somewhat free and varied movement is possible. Such joints are common in your body and are called *synovial joints*.
8. There are three kinds of muscle tissue: smooth muscle tissue, which is found in many body organs and contracts spontaneously; cardiac muscle tissue, which makes up the heart and may initiate contraction spontaneously; and skeletal muscle tissue, which is organized into trunks of long fibers and contracts only when stimulated by a nerve. Skeletal muscles are attached to bones and move the body. These are the muscles of the muscular system.
9. Skeletal muscles move bones by pulling on one while anchored to another. They are made up of bundles of extremely long muscle cells called *muscle fibers*. Each muscle fiber is packed with myofibrils, which are cylindrical, organized arrangements of special thick and thin filaments called *myofilaments,* capable of shortening the muscle fiber as they slide past one another.
10. The contraction of skeletal muscles is initiated by a nerve impulse, which causes a series of events that ultimately result in the interaction of actin and myosin. Acetylcholine passes across the neuromuscular junction from the nerve to the muscle, triggering the events leading to this interaction.

KEY TERMS

actin 307
antagonist 305
appendicular skeleton 299
axial skeleton 299
ball-and-socket joint 304
bone 296
compact bone 296
cross bridge binding sites 310
haversian canal 296
hinge joint 304
hyoid bone 300
insertion 305
integument 296
integumentary system 296
intervertebral disk 300
joint (articulation) 302
lamellae 296
ligaments 302
motor endplate 310
muscle fiber 305
myofibril 305
myofilament 307
myosin 307
neuromuscular junction 310
origin 305
osteoblasts 296
pectoral (shoulder) girdle 300
pelvic (hip) girdle 300
pivot joint 304
red bone marrow 297
sarcomere 307
sarcoplasmic reticulum 310
skull 299
spongy bone 297
striated muscle 305
suture 302
synovial joint 302
tendon 305
transverse (T) tubules 310
tropomyosin 310
troponin 310
vertebral column 300
yellow marrow 297

REVIEW QUESTIONS

1. You raise your hand to answer a question in class. Explain the roles played by your bones and skeletal muscles in this movement.
2. Fill in the blank: The largest organ in your body is your ________.
3. Explain the functions and importance of skin.
4. What is bone, and what are its functions? Explain its importance.
5. Distinguish between compact and spongy bone. What are the functions of each?
6. Distinguish between the axial and appendicular skeletons. What groups of bones are found in each?
7. Mrs. Gorman has injured her back; her doctor tells her she has a "slipped disk." What type of "disk" is involved, and what is its function?
8. Fill in the blanks: Your shoulder blades are part of your ________ ________; your hipbones make up your ________ ________. All of these bones are part of your ________ skeleton.
9. What is an articulation? Describe the three main categories of articulations, and give an example of each. On what factor are these categories based?
10. What do hinge, pivot, and ball-and-socket joints have in common? How are they different?
11. What is a muscle? List the three types of muscle tissue, and give an example of where each is found in the body.
12. Fill in the blanks: Skeletal muscles are attached to bones with cords of connective tissue called ________. At its ________ the muscle attaches to a stationary bone; at its ________ it attaches to a movable bone. The ________ anchors the muscle as it pulls at the ________.
13. Define actin, myosin, and myofilament. Explain their roles in the contraction of a muscle fiber.
14. Explain how a nerve impulse can lead to a muscle contraction.

THOUGHT QUESTIONS

1. When an adult begins to jog, doctors warn about the importance of building up slowly, starting with short distances and building up to longer ones only over periods of months. Individuals who ignore this warning often come down with painful stress fractures, tiny spiral cracks along the length of the leg bones. Why does building up slowly prevent such stress fractures?
2. For several hundred years people have been breeding race horses to run fast. However, when the track times of winning horses over the years are compared, there has been little if any improvement in winning times over the last century. Do you think this means there is a limit to how fast horses can run that has been reached?

FOR FURTHER READING

Carafoli, E., & Penniston, J. (1985, November). The calcium signal. *Scientific American,* pp. 70-78.
This article discusses how the release of calcium ions is the only known way in which the electricity of the nervous system is able to produce changes in the body. Nerves regulate all muscle contractions and hormone secretions by controlling the level of calcium ions.

Schmidt-Nielsen, K. (1987). *Animal physiology: adaptation and environment.* ed. 4, New York: Cambridge University Press.
Chapter 11 presents an outstanding treatment of muscles and bones from an evolutionary perspective.

19 Hormones

HIGHLIGHTS

▼

Hormones are chemical messengers produced by cells; they regulate other cells of the body.

▼

Hormones produced by glands and transported to cells via the bloodstream are known as *endocrine hormones,* whereas hormones that are secreted into the tissue fluid to affect nearby cells are known as *local hormones.*

▼

Hormones affect cells in one of four ways: by regulating their secretions and excretions, by helping them to respond to changes in the environment, by controlling activities related to reproductive processes, and by influencing their proper growth and development.

▼

The production of hormones is controlled by a feedback mechanism that acts much like a thermostat in a house.

OUTLINE

Endocrine glands and their hormones
- The pituitary gland
  - Controlling the pituitary: The hypothalamus
  - The anterior pituitary
  - The posterior pituitary
- The thyroid gland
- The parathyroid glands
- The adrenal glands
  - The adrenal cortex
  - The adrenal medulla
- The pancreas
- The pineal gland
- The thymus gland
- The ovaries and testes

Nonendocrine hormones

Shooting up in the locker room with anabolic steroids has caused the downfall of many winners. These controversial drugs are really synthetic hormones, chemicals that affect the activity of specific organs or tissues. The various hormones of the body all affect their "target" tissues in unique ways. Anabolic steroids affect the body in ways similar to the male sex hormone testosterone and stimulate the buildup of muscles. But along with building a championship body, anabolic steroids strikingly change the body's metabolism.

Female athletes on steroids experience side effects such as shrinking breasts, a deepening voice, and an increase in body hair. Male athletes find that their testicles shrink. Some users also experience life-threatening kidney and liver damage. Youngsters who take these drugs risk stunting their growth, since anabolic steroids cause bones to stop growing prematurely. And a great deal of controversy still surrounds claims that anabolic steroids can cause psychological effects such as "steroid rage," a state of mind in which users attack people and things around them.

Today, scientists still lack solid scientific data regarding all aspects and consequences of anabolic steroid use. However, it is clear that their use is risky at the least—and may put users in the cemetery rather than in the winner's circle.

Endocrine glands and their hormones

A **hormone** is a "chemical messenger" sent by a gland to other cells of the body. Traditionally, animal hormones have been described by scientists as the chemical products of glands that travel within the bloodstream to all parts of the body, causing an effect on specific cells, or target organs, far removed from that gland. Glands are individual cells or groups of cells that secrete substances. Their secretory portions are made up of specialized epithelial cells (see Chapter 9). The glands that secrete hormones spill these chemicals directly into the bloodstream and are called **endocrine,** or "ductless," **glands.** Glands that secrete other substances such as digestive enzymes or sweat route their secretions to specific destinations by means of ducts. In this way, for example, the digestive enzyme pancreatic amylase flows directly from the pancreas to the small intestine and goes nowhere else. Glands having associated "ductwork" are called **exocrine glands.**

Today, most scientists have expanded their definition of hormones to include any chemical produced by one cell that causes an effect in another. Included in this description, then, are substances such as neurotransmitters—chemicals produced by the axon end of a nerve cell that travel to and bind with the dendrite end of an adjacent nerve cell, contributing to the propagation of the nerve impulse along that neuron (see Chapter 15). Such chemicals are often called **local hormones** because they affect neighboring target cells. The human body produces many local hormones; they are described later in this chapter.

> **Hormones are "chemical messengers" secreted by cells that affect other cells. Hormones that travel within the bloodstream and affect cells in another part of the body are called *endocrine hormones*. Hormones that do not travel within the bloodstream but only affect cells lying near the secretory cells are called *local hormones*.**

The 10 different endocrine glands of the human body make over 30 different hormones. Together, these glands are called the **endocrine system** (Figure 19-1). The endocrine system works with the nervous system to integrate the functioning of the various tissues, organs, and organ systems of the body. The nervous system sends messages to muscles and glands, regulating muscular contraction and glandular secretion. The hormones of the endocrine system, on the other hand, carry messages to virtually any type of cell in the body. The messages of the endocrine hormones are varied but can be grouped into four categories:

1. **Regulation:** Hormones control the internal environment of the body by regulating the secretion and excretion of various chemicals in the blood, such as salts and acids.
2. **Response:** Hormones help the body respond to changes in the environment and cope with physical and psychological stress.
3. **Reproduction:** Hormones control the female reproductive cycle and other reproductive processes essential to conception and birth and control the development of sex cells, the reproductive organs, and secondary sexual characteristics (those that make men and women different) in both sexes.
4. **Growth and development:** Hormones are essential to the proper growth and development of the body from conception to adulthood.

Once molecules of a hormone are released into the bloodstream, they travel throughout the body. Although hormone molecules may pass billions of cells, specific hormones only affect specific cells called **target cells.** Hormones recognize target cells because they bind to receptor molecules embedded within the cell membrane or located within the cytoplasm of the cell. The binding of a hormone molecule to a receptor molecule activates a chain of events in the target cell that results in the effect of the hormone being expressed.

Two major classes of endocrine hormones work within the human body: **peptide hormones** and **steroid hormones.** Peptide hormones are made of amino acids, but the amino acid chain length varies greatly from hormone to hormone. The smallest are actually modifications of the single amino acid tyrosine. Somewhat larger are short peptide hormones that are several amino acids in length. Polypeptide hormones have chain lengths of several dozen or more amino acids, such as the hormone insulin. Even larger are protein hormones that may have over 200 amino acids with carbohydrates attached at several positions.

Unable to pass through the cell membrane, peptide hormones bind to receptor molecules embedded in the cell membrane of target cells. The binding of hormone to a receptor triggers an increase in that cell's production of a compound referred to as a **second messenger.** A second messenger triggers enzymes that cause the cell to alter its

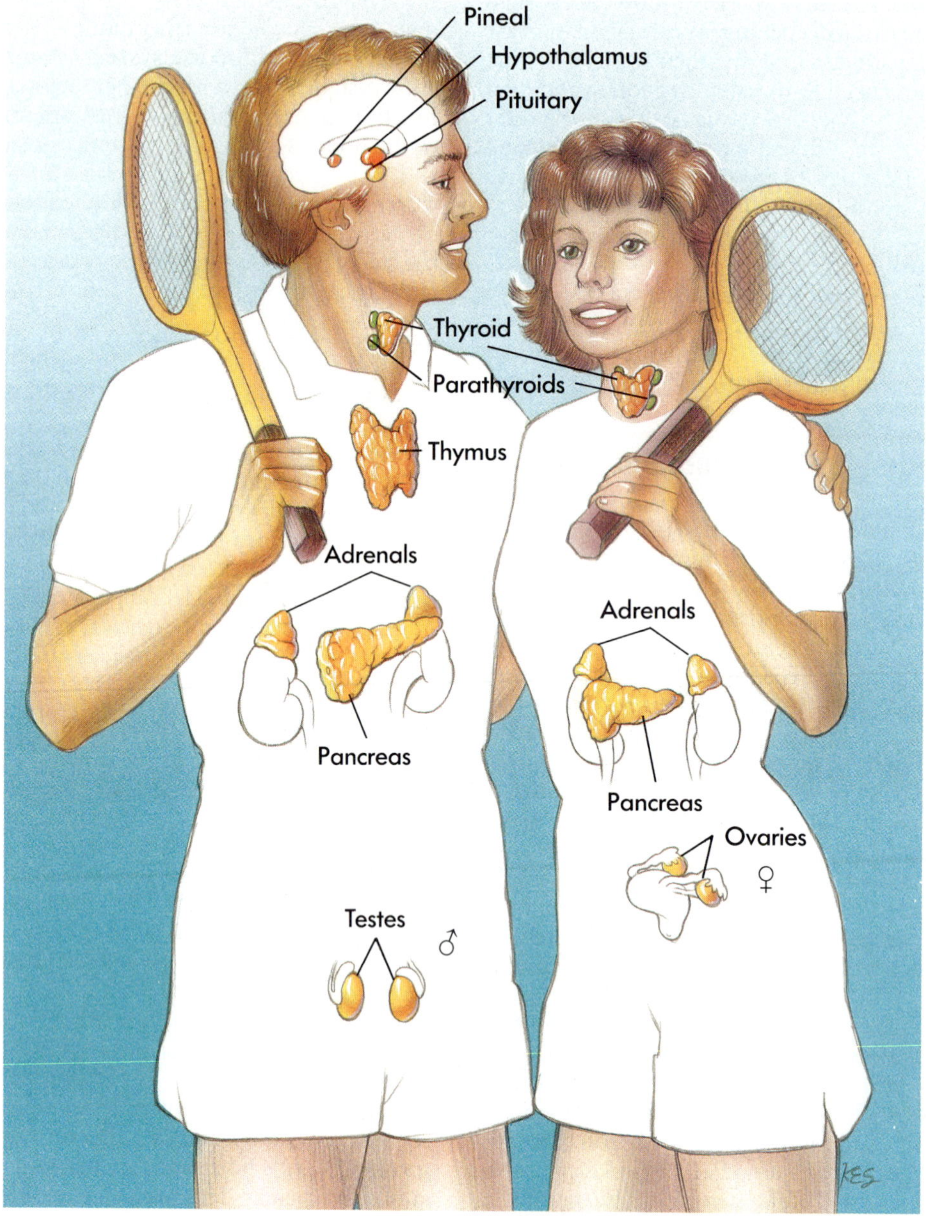

FIGURE 19-1 The human endocrine system.
The endocrine glands pictured in this diagram secrete chemical "messengers" that travel through the bloodstream to affect other cells in the body.

functioning in response to the hormone (Figure 19-2). For example, prolactin stimulates cells of the mammary glands to produce milk. Target cells respond as enzymes "go into action" catalyzing reactions that produce the components of mother's milk. Other types of hormone responses include the secretion of substances from target cells and the closing or opening of certain "protein doors" within target cell membranes. Cyclic adenosine monophosphate (cyclic AMP for short), a "cousin" of ATP (see Chapter 6), acts as a second messenger to many cells. Besides cyclic AMP, other second messenger molecules have been discovered.

Steroid hormones are all made from cholesterol, a lipid synthesized by the liver. You know cholesterol as that "dietary devil" present in certain foods such as eggs, dairy products, and beef. A characteristic of steroid hormones is their set of carbon rings. Steroid hormones, being lipid soluble, pass freely through the lipid bilayer of the cell membrane. Once inside a cell, these hormones bind to receptor molecules located within the cytoplasm of target cells. Together, the hormone-receptor complex moves into the nucleus of the cell, causing the cell's hereditary material, or DNA, to trigger the production of certain proteins (Figure 19-3). In

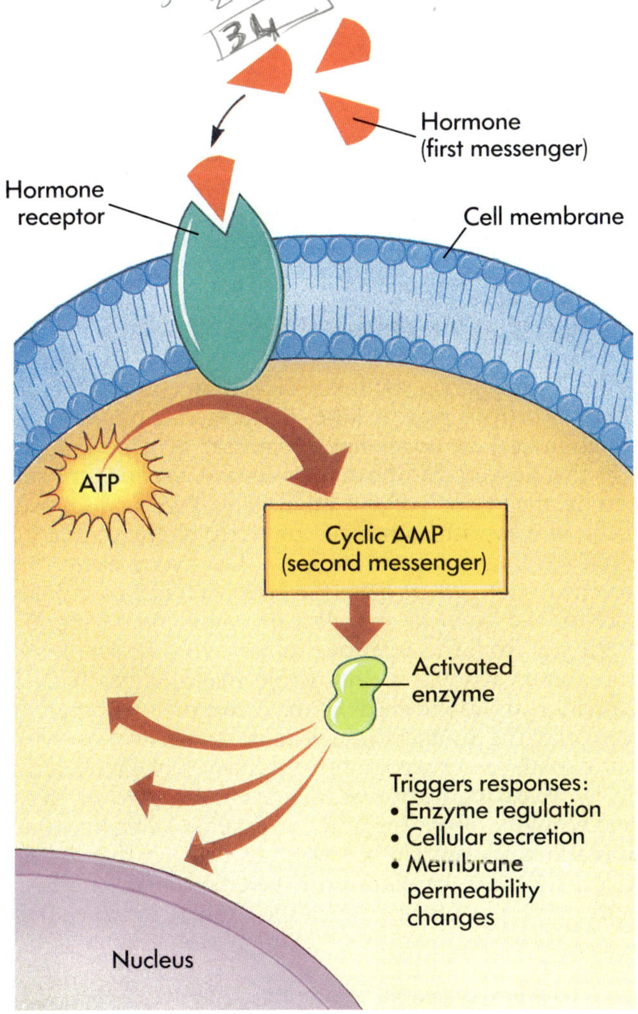

FIGURE 19-2 How peptide hormones work.
Once inside the cell, peptide hormones bind to the cell membrane and trigger an increase of second messenger compounds within the cell, such as cyclic AMP. The second messenger in turn activates enzymes that alter the cell's function in response to the hormonal message.

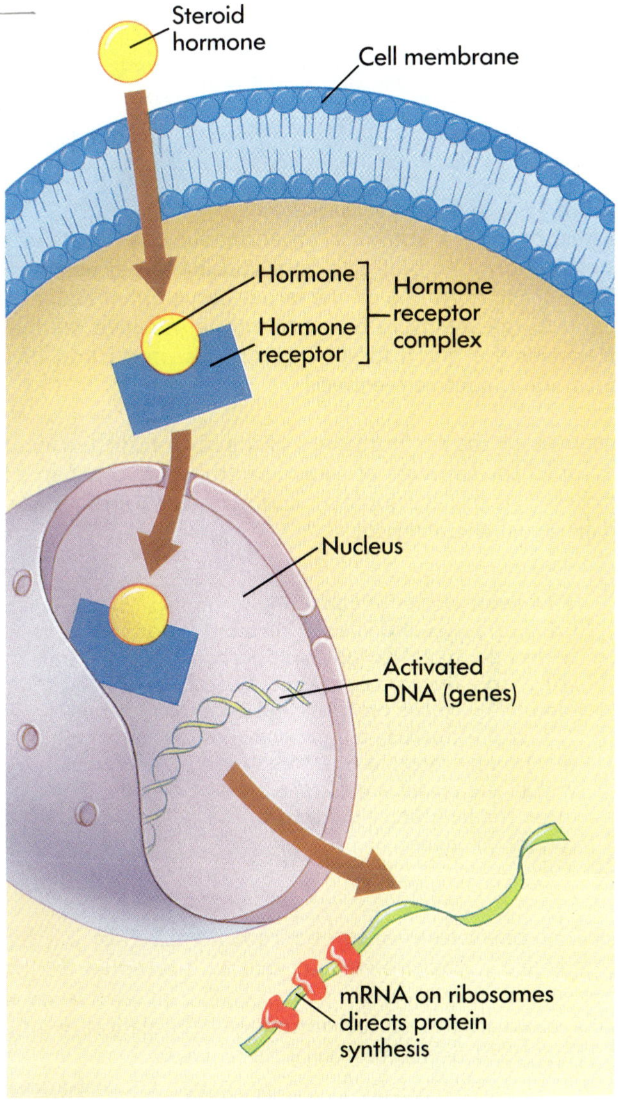

FIGURE 19-3 How steroid hormones work.
Steroid hormones are able to pass through the cell membrane without the aid of a receptor molecule. Inside the cell, they bind with receptor molecules. The hormone-receptor complex then enters the nucleus of the cell, where it acts on DNA to produce proteins. These proteins control physiological processes such as growth and development.

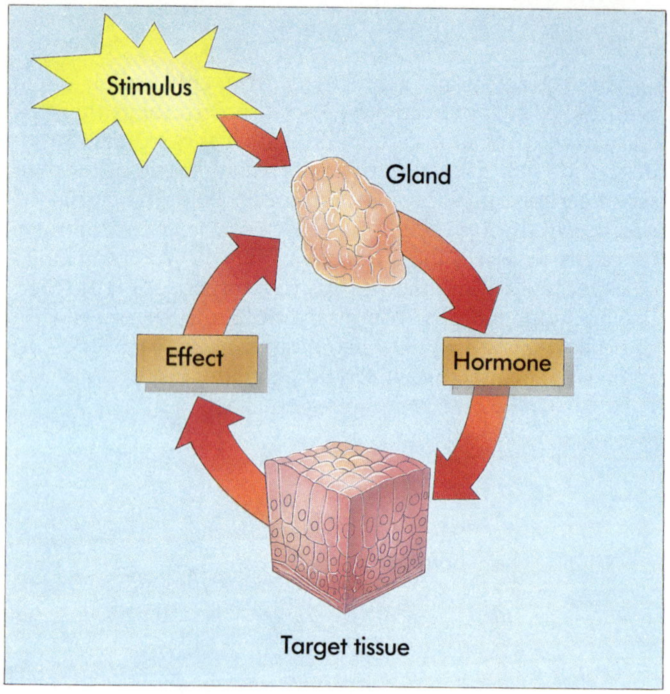

FIGURE 19-4 A simple feedback loop.
In response to a stimulus, an endocrine gland releases a specific hormone that acts on a specific target tissue. The effect of the hormone on the target tissue either causes the gland to release more of the hormone (positive feedback) or causes the gland to slow or stop its production of the hormone (negative feedback).

response to the sex hormones estrogen or testosterone, for example, the proteins produced are those involved in such processes as the development and maintenance of female or male sexual characteristics.

Two main classes of endocrine hormones are peptide hormones and steroid hormones. Both travel within the bloodstream to all parts of the body but affect only certain target cells. Peptide hormones bind to receptors on the cell membrane of target cells and ultimately trigger enzymes that alter cell functioning. Steroid hormones bind to receptors within the cytoplasm of target cells and ultimately cause the hereditary material of the cell to produce specific proteins.

The production of hormones is regulated by a mechanism called a *feedback loop.* In general, hormonal feedback loops work in the following way: endocrine glands are initially stimulated to release hormones. Stimulation of an endocrine gland occurs in one of three ways:

1. *Direct stimulation by the nervous system:* The sensation of fear, for example, can cause the autonomic nervous system to trigger the release of the hormone adrenaline from the adrenal medulla.
2. *Indirect stimulation by the nervous system by means of releasing hormones:* The hypothalamus is a specialized portion of the brain that produces and secretes releasing hormones. Some releasing hormones stimulate the release of other hormones; some prevent the release.
3. *The concentration of specific substances in the bloodstream:* The blood level of a substance such as glucose or calcium, for example, may signal an endocrine gland to "turn on" or "turn off."

After an endocrine gland secretes its hormone into the bloodstream, the hormone travels throughout the body via the circulatory system and interacts with target tissues. The target tissues cause the desired effect to be produced. This "effect" acts as a new stimulus to the endocrine gland (Figure 19-4). Put simply, the body "feeds back" information to each endocrine gland after it releases hormone. In a positive feedback loop, the information that is fed back causes the gland to produce more of its hormone. In a **negative feedback loop,** the feedback causes the gland to slow down or to stop the production of its hormone. Most hormones work by means of negative feedback loops. (Specific examples of feedback mechanisms and interactions are discussed throughout this chapter.)

The pituitary gland

The **pituitary** is a powerful gland that secretes nine different hormones. Although it secretes so many hormones, it is amazingly tiny—about the size of a marble. The pituitary "marble" hangs from the underside of the brain, supported and cradled within a bony depression of the sphenoid bone.

Controlling the pituitary: The hypothalamus

The pituitary secretes seven major hormones from its larger front portion, or lobe, the anterior pituitary. It secretes two from its rear lobe, the posterior pituitary. The secretion of these hormones is regulated by a mass of nerve cells that lies directly above the pituitary, making up a small part of the "floor" of the brain. This regulatory nervous tissue, the **hypothalamus,** is connected to the pituitary by a stalk of tissue (Figure 19-5, *A*). The hypothalamus uses information it gathers from the peripheral nerves and other parts of the brain to stimulate or inhibit the secretion of hormones from the anterior pituitary. In this way, the hypothalamus acts like a production manager, receiving information about the needs of the company's customers and regulating the production of products to satisfy those needs. The hypothalamus accomplishes its management job by producing **releasing hormones** that affect the secretion of specific hormones from the anterior pituitary. The hypothalamus also produces two hormones that do not regulate hormonal release in the pituitary. When they are needed by the body, the hypothalamus signals the pituitary to release them.

The pituitary is a tiny gland that hangs from the underside of the brain. The secretion of its many diverse hormones is controlled by a mass of nerve cells lying directly above it called the *hypothalamus.* The hypothalamus stimulates or inhibits the secretion of hormones from the pituitary by means of releasing hormones. In addition, the hypothalamus produces two hormones that it stores in the pituitary.

A

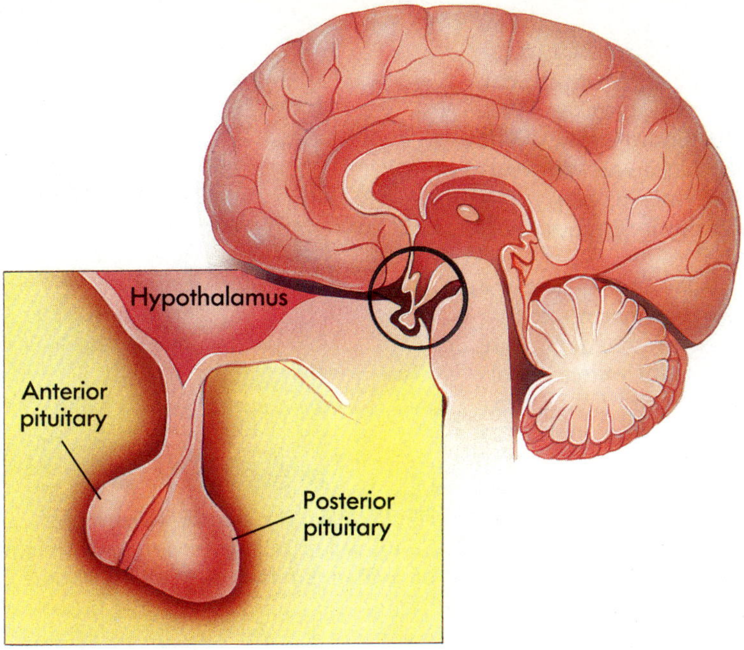

B

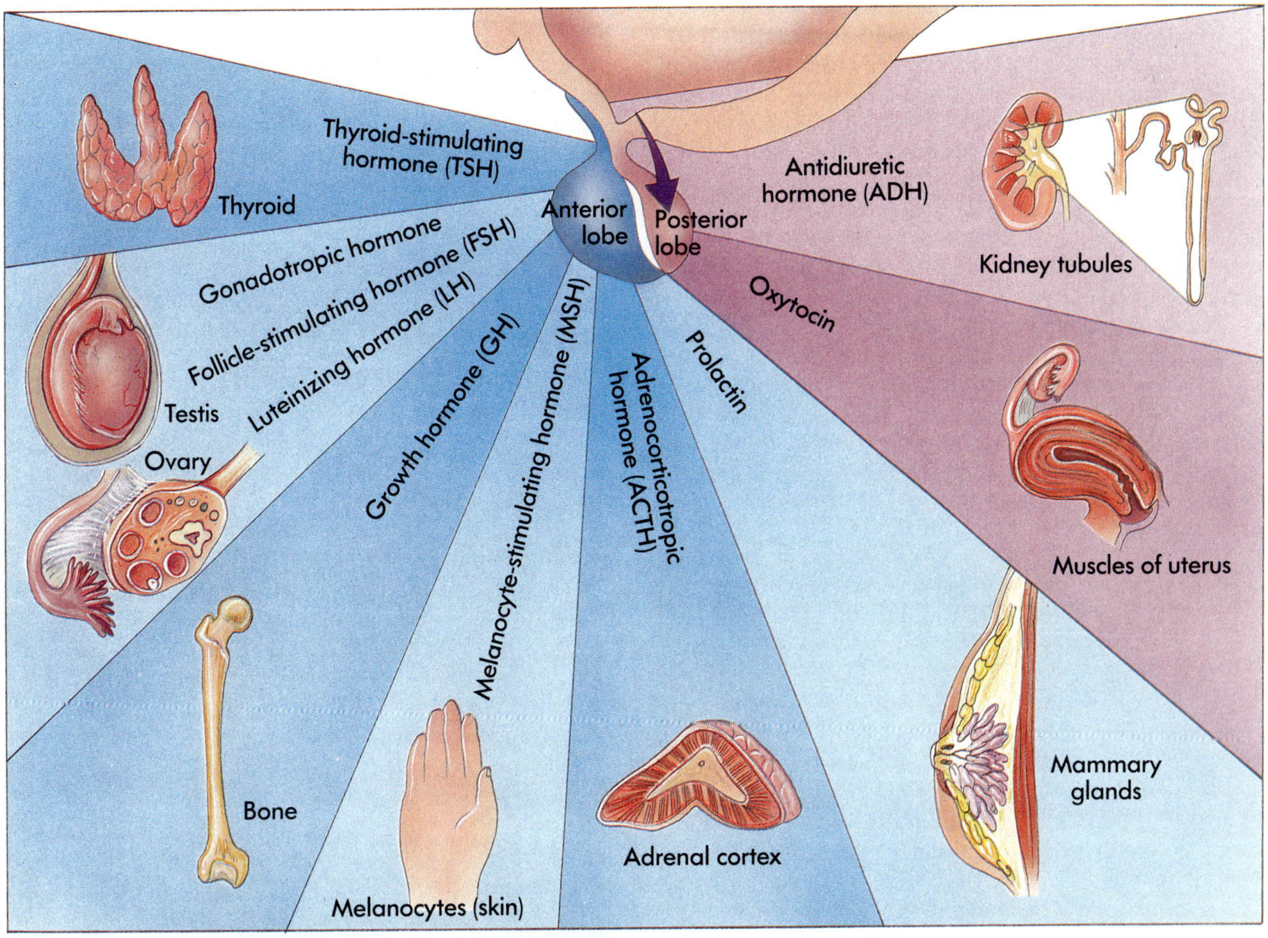

FIGURE 19-5 The role of the pituitary.
A In this diagram, the stalk of connecting tissue between the hypothalamus and the pituitary can be clearly seen.
B The hormones secreted by the pituitary can be divided into anterior pituitary hormones and posterior pituitary hormones.

The anterior pituitary

The seven hormones produced by the anterior pituitary regulate a wide range of bodily functions (Figure 19-5, *B*). Four of these hormones are called **tropic hormones.** The word *tropic* comes from a Greek word meaning "turning" and refers to the ability of tropic hormones to turn on or stimulate other endocrine glands. Of the four tropic hormones, two are **gonadotropins.**

The gonads are the male and female sex organs, the testes and the ovaries. The gonadotropins are hormones that affect these sex organs (considered endocrine glands because they secrete sex hormones). The two gonadotropins are **follicle-stimulating hormone (FSH)** and **luteinizing hormone (LH).** In females, FSH targets the ovaries and triggers the maturation of one egg each month. In addition, it stimulates cells in the ovaries to secrete female sex hormones called *estrogens*. In men, FSH targets the testes and triggers the production of sperm. LH stimulates cells in the testes to produce the male sex hormone testosterone. In females, a surge of LH near the middle of the menstrual cycle stimulates the release of an egg. In addition, LH triggers the development of cells within the ovaries that produce another female sex hormone—progesterone. (See Chapter 20 for the organs and processes of the reproductive system.)

The two other tropic hormones are **adrenocorticotropic hormone (ACTH)** and **thyroid-stimulating hormone (TSH).** ACTH triggers the adrenal cortex to produce certain steroid hormones. The adrenal glands are located on top of the kidneys (see Figure 19-1). Each of these two glands has two distinct parts: an outer cortex and an inner medulla. ACTH stimulates the adrenal cortex to produce hormones that regulate the production of glucose from "noncarbohydrates" such as fats and proteins. Others regulate the balance of sodium and potassium ions in the blood. Still others contribute to the development of the male secondary sexual characteristics. TSH triggers the thyroid gland to produce the three thyroid hormones. This endocrine gland is located on the front of the neck, just below the voice box (see Figure 19-1). Its hormones control normal growth and development and are essential to proper metabolism. (Further discussion of ACTH is on p. 326, and further discussion of TSH is on p. 322.)

The front portion of the pituitary, the anterior pituitary, secretes seven hormones. Of these seven, four stimulate other endocrine glands and are called *tropic hormones*.

Growth hormone (GH) is produced by the anterior pituitary and works with the thyroid hormones to control normal growth. GH increases the rate of growth of the skeleton by causing cartilage cells and bone cells to reproduce and lay down their intercellular matrix. In addition, GH stimulates the deposition of minerals within this matrix. GH also stimulates the skeletal muscles to grow in both size and number. In the past, children who did not produce enough GH did not grow to an average height; this condition is called *hypopituitary dwarfism*. However, in the past decade, scientists have been able to use the techniques of genetic engineering (see Chapter 28) to insert the human GH gene into bacteria to produce human GH. Currently, this laboratory-made hormone is being used successfully for treating growth disorders caused by hyposecretion (underproduction) of GH in children. The opposite problem may also occur: during the growth years, some children produce too much GH. This hypersecretion (overproduction) can cause the long bones to grow unusually long (Figure 19-6) and result in a condition known as *giantism*. In adults, hypersecretion of GH causes the bones of the hands and face to thicken, resulting in a condition known as *acromegaly* (Figure 19-7).

FIGURE 19-6 The world's tallest woman.
Sandy Allen is shown here with her mother, brother, and dog. Giantism is caused by the oversecretion of growth hormone.

FIGURE 19-7 Acromegaly.
This ancient carving of the Egyptian pharoah Akhenaton, who ruled from 1379-1362 BC, exhibits many characteristics of acromegaly.

BIOLOGY & you

Living with diabetes

When you think of dread diseases, the ones that come to mind are AIDS, cancer, Alzheimer's, multiple sclerosis, muscular dystrophy, and the like. Few would think to add diabetes to this list. Diabetes is a disorder in which the pancreas does not produce adequate amounts of the hormone insulin or the body becomes insensitive to it. As a result, muscles, fat, and liver cells are prevented from absorbing sugar from the blood; the sugar is excreted in urine, while the undernourished cells literally starve. Diabetes is treated by adding insulin directly to the bloodstream by injection (a protein, it would be destroyed in the stomach if administered as a pill), usually twice a day (Figure 19-A).

Over 100 million people worldwide have diabetes. Because initial symptoms are usually mild (fatigue and frequent urination), half the 12 million Americans with diabetes do not realize they have it. Untreated, the disease eventually leads to serious complications, doubling the risk of heart attack and stroke. It is the leading cause of blindness in adults and accounts for one third of all cases of kidney failure. Because of these many ill effects, diabetes is now the seventh leading cause of death in the United States.

There are two kinds of diabetes. In type I diabetes, which chiefly strikes in childhood, the body loses its ability to produce insulin. Affecting about 10% of diabetics, type I diabetes is a hereditary autoimmune disease in which the body's immune system attacks the pancreas, destroying its ability to produce insulin. Glucose builds up in the bloodstream to a three times normal level, while the cells starve. The high levels of blood glucose cause the thickening of capillary and artery walls, constricting blood flow and damaging critical organs. Affected individuals must take two insulin shots a day all their lives to avoid disaster. Researchers are working hard to uncover a way to prevent the immune system from attacking the pancreas.

FIGURE 19-A The father of this boy with diabetes is testing his son's blood-sugar level with a pocket-size machine.

In type II diabetes, which primarily affects adults over the age of 40, the body becomes resistant to the effects of insulin, requiring the pancreas to produce ever-larger amounts until it is unable to produce enough insulin to meet its needs and the same symptoms that ravage type I patients set in. In the United States, 90% of those who develop type II diabetes are obese. Cells, sated with food, adjust their appetite for glucose downward by reducing their sensitivity to insulin. To compensate, the pancreas pumps out ever more insulin. In most obese people, the pancreas is able to keep up with the work load, but in individuals carrying particular recessive genes for type II diabetes, obesity overloads the system and insulin-producing cells begin to stop functioning. These cells also make a protein called *amylin,* and some researchers believe that amylin deposits in the pancreas lead to type II diabetes, much as amyloid protein deposits in the brain lead to Alzheimer's disease. Fortunately, type II diabetes can usually be treated by diet and exercise, and most affected individuals do not need daily injections of insulin.

Prolactin is another hormone secreted by the anterior pituitary. Prolactin works with estrogen, progesterone, and other hormones to stimulate the mammary glands in the breasts to secrete milk after a woman has given birth to a child. During the menstrual cycle, milk is not produced and secreted because prolactin levels in the bloodstream are very low. Late in the menstrual cycle, however, as the levels of progesterone and estrogen fall, the pituitary is stimulated by the hypothalamus to secrete some prolactin. This rise in prolactin, although not sufficient to cause milk production, does cause the breasts of some women to feel sore before menstruation. After menstruation, estrogen levels begin to rise, and prolactin secretion is once again inhibited.

Melanocyte-stimulating hormone (MSH) acts on cells in the skin called *melanocytes,* which synthesize a pigment called *melanin.* This pigment is taken up by epidermal cells in the skin, producing skin colorations from pale yellow (in combination with another pigment called *carotene*) to black. Variations are caused by the amount of pigment the melanocytes produce; this variation is genetically determined and is an inherited characteristic.

The posterior pituitary

The posterior lobe of the pituitary stores and releases two hormones that are produced by the hypothalamus: **antidiuretic hormone (ADH)** and **oxytocin.** ADH helps control the volume of the blood by regulating the amount of water reabsorbed by the kidneys. For example, receptors in the hypothalamus can detect a low blood volume by detecting when the solute concentration of the blood is high. When the hypothalamus detects such a situation, it triggers its specialized neurosecretory cells to make ADH. This hormone is transported within axons to the posterior pituitary, which releases the hormone into the bloodstream. ADH binds to target cells in the collecting ducts of the nephrons of the kidneys, increasing their permeability. More water then moves out of these ducts and back into the blood, resulting in a more concentrated urine. ADH also acts on the smooth muscle surrounding arterioles. As these muscles tighten, they constrict the arterioles, an action that helps raise the blood pressure. Alcohol supresses ADH release, which is why excessive drinking leads to the production of excessive quantities of urine and eventually to dehydration.

Oxytocin is another hormone of the posterior pituitary: it is produced in the hypothalamus and transported within axons to the posterior pituitary for secretion. In women, oxytocin is secreted during the birth process, triggered by a stretching of the cervix of the uterus at the beginning of the birth process. Oxytocin binds to target cells of the uterus, enhancing the contractions already taking place. The mechanism of oxytocin secretion is an example of a **positive feedback loop** in which the effect produced by the hormone enhances the secretion of the hormone. For this reason, oxytocin is used by physicians to induce uterine contractions when labor must be brought on by external means. Oxytocin also targets muscle cells around the ducts of the mammary glands, allowing a new mother to nurse her child. The suckling of the infant triggers the production of more oxytocin, which aids in the nursing process and helps contract the uterus to its normal size.

The rear lobe of the pituitary, or posterior pituitary, stores and releases two hormones, ADH and oxytocin, which are produced by the hypothalamus.

The thyroid gland

Sitting like a large butterfly just below the level of the voice box, the thyroid gland can be thought of as your "metabolic switch." This gland secretes hormones that determine the rate of the chemical reactions of your body's cells. Put simply, thyroid hormones determine how fast bodily processes take place.

The thyroid hormones are **thyroxine (T_4)** and **triiodothyronine (T_3).** These hormones are called *amines:* single, modified amino acids. They are not considered to be "true" peptide hormones, however, because they act on the DNA of target cells as steroid hormones do. They are also unique because an inorganic ion—iodine—is part of their structures.

Your body uses iodine in the food you eat to help make the thyroid hormones; the *3* or *4* in each hormone name refers to the number of atoms of iodine in each hormone. Foods such as seafood and iodized salt are good sources of dietary iodine. If the diet contains an insufficient amount of iodine, the thyroid gland enlarges. This condition is called a **hypothyroid goiter** (Figure 19-8).

The hypothalamus and the thyroid gland work together to keep the proper level of thyroid hormone circulating in the bloodstream. This level is detected by the hypothalamus. A low level of thyroid hormones stimulates the hypothalamus to secrete a releasing factor—a chemical message—to the anterior pituitary. This message tells the pituitary to release more TSH. The thyroid responds, thereby raising the blood level of T_3 and T_4 back to normal (Figure 19-9). This mechanism of action is an example of a negative feedback loop in which the effect produced by stimulation of a gland "shuts down" the stimulus. Shutdown occurs when a sufficient effect has been produced, similar to the mechanism of a thermostat. In your home, your furnace is triggered to go on when the temperature goes below the thermostat setting. The furnace stays on until the house heats up to the desired level. The thermostat then signals the furnace to turn off.

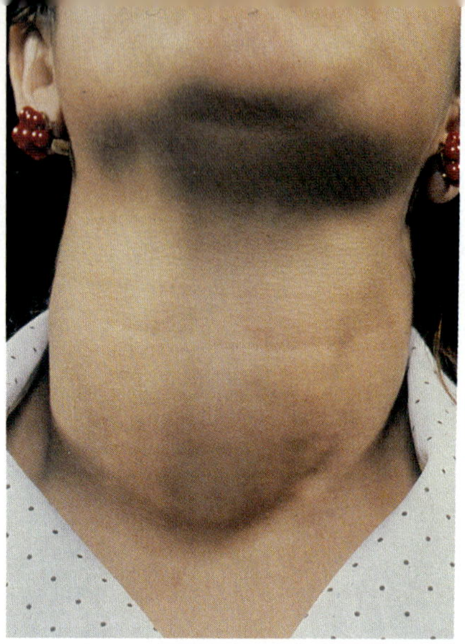

FIGURE 19-8 A goiter.
The thyroid gland becomes enlarged when the diet lacks iodine. Iodine is used in the production of thyroid hormones; when not enough iodine is available, the thyroid cannot produce thyroid hormones, and the thyroid swells from overstimulation by the anterior pituitary.

In certain disease conditions the amount of thyroid hormones in the bloodstream cannot be regulated properly. If the thyroid produces too much of the thyroid hormones, a person may feel as though the "engine is racing," with such symptoms as a rapid heartbeat, nervousness, weight loss, and protrusion of the eyes (Figure 19-10). This condition is called **hyperthyroidism.** On the other hand, if the thyroid produces too little of the thyroid hormones, a person may feel "run down," with such symptoms as weight gain and slow growth of the hair and fingernails. This condition is called **hypothyroidism.** Various factors can be the underlying cause of such problems; often medication or surgery can correct the situation.

In addition to secreting the thyroid hormones, the thyroid gland secretes a hormone called **calcitonin,** or **CT.** This hormone works to balance the effect of another hormone called **parathyroid hormone,** or **PTH.** PTH regulates the concentration of calcium in the bloodstream. Calcium is an important structural component in bones and teeth and aids in the proper functioning of nerves and muscles.

The thyroid gland, located in the neck near the voice box, produces hormones that regulate that body's metabolism.

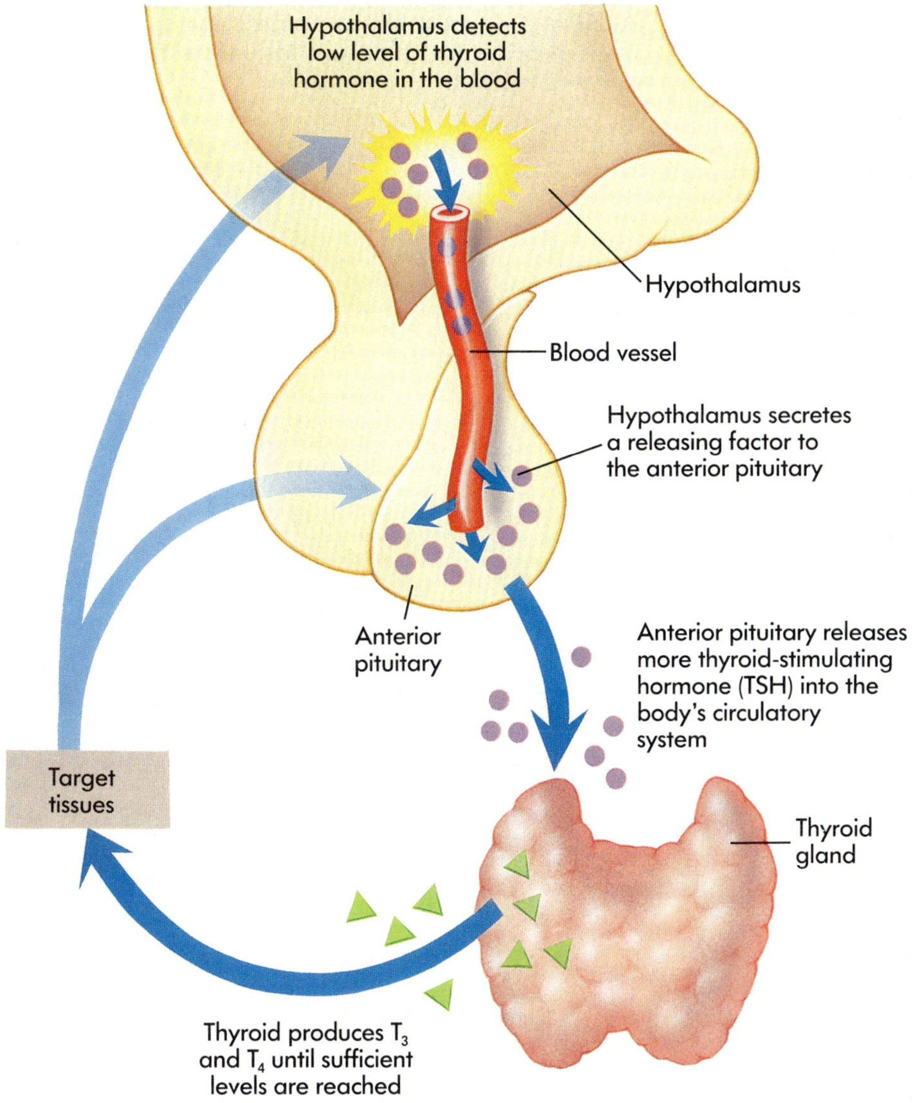

FIGURE 19-9 A negative feedback loop is used in the release of thyroid hormone. When enough thyroid hormone has been produced, the hypothalamus stops producing thyroid-releasing hormone.

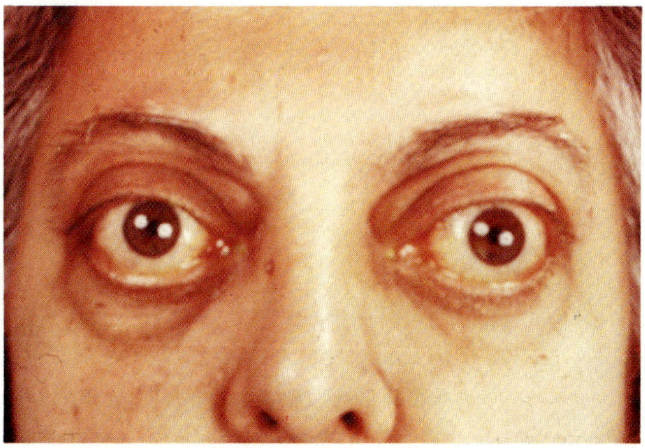

FIGURE 19-10 Hyperthyroidism.
One of the symptoms of an overactive thyroid is protrusion of the eyes.

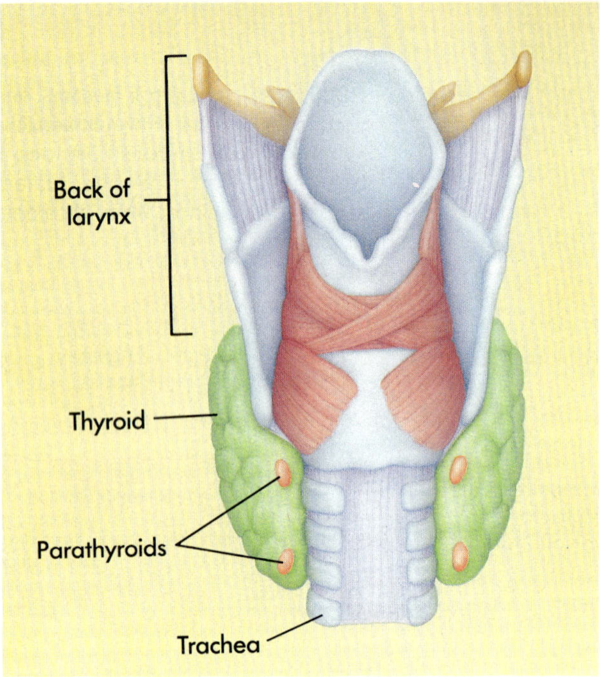

FIGURE 19-11 Location of the parathyroid glands.
These glands are embedded in the underside of the thyroid and secrete parathyroid hormone.

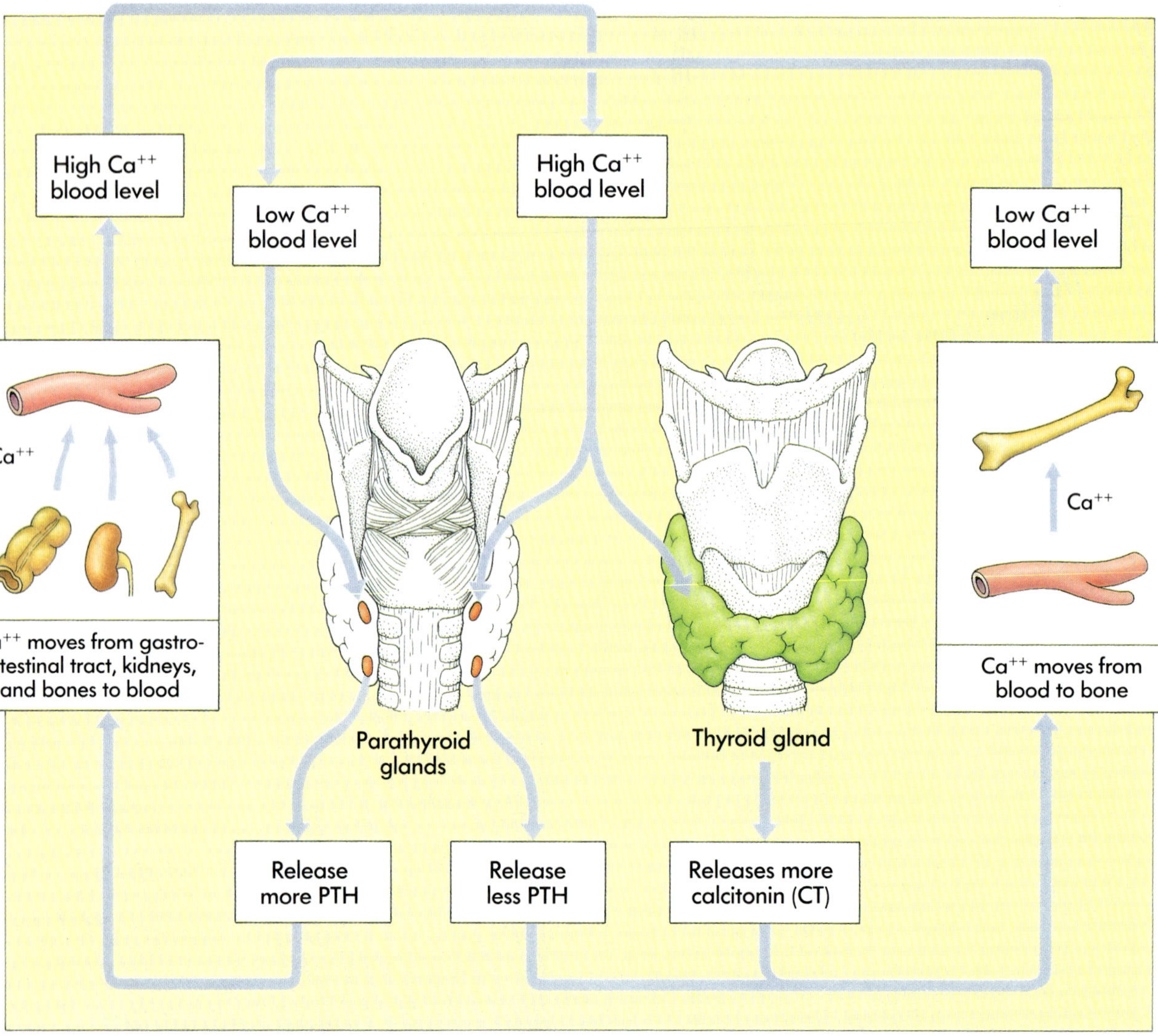

FIGURE 19-12 How parathyroid hormone and calcitonin work to maintain proper calcium levels in the blood.
Parathyroid hormone stimulates the removal of calcium from the bones, the reabsorption of calcium from the urine, and the absorption of calcium from digested food when calcium levels are low. When calcium levels in the blood are high, calcitonin works to inhibit calcium release from bone and speed up its absorption.

The parathyroid glands

Embedded in the posterior side of the thyroid are the **parathyroid glands.** Most people have two parathyroids on each of the two lobes of the thyroid, as shown in Figure 19-11. These are the glands that secrete PTH, which works antagonistically to CT to help maintain the proper blood levels of various ions, primarily calcium. Two of the many problems related to abnormal calcium levels in the blood are kidney stones and osteoporosis. If calcium levels in the blood remain high, tiny masses of calcium may develop in the kidneys. These masses, called *kidney stones,* can partially block the flow of the urine from a kidney. If calcium levels in the blood remain low, calcium may be removed from the bones, a disorder known as *osteoporosis*. Osteoporosis is most common in middle-aged and elderly women, who have stopped secreting estrogen at menopause (see Chapter 20). Estrogen stimulates bone cells to take calcium from the blood to build bone tissue.

PTH and CT work in the following way to keep calcium at an optimum level in the blood: If the calcium level is too low, PTH stimulates the activity of osteoclasts, or bone-destroying cells. These cells liberate calcium from the bones and put it into the bloodstream. PTH also stimulates the kidneys to reabsorb calcium from urine that is being formed and stimulates cells in the intestines to absorb an increased amount of calcium from digested food. CT acts antagonistically to PTH. When the level of calcium in the blood is too high, less PTH is secreted by the parathyroids and more CT is secreted by the thyroid. The CT inhibits the release of calcium from bone and speeds up its absorption, decreasing the levels of calcium in the blood. These interactions of PTH and CT are an example of a negative feedback loop that does not involve the hypothalamus or pituitary gland. The level of calcium in the blood directly stimulates the thyroid and parathyroid glands (Figure 19-12).

The parathyroid glands, embedded in the posterior side of the thyroid, secrete a hormone that helps maintain the proper levels of various ions such as calcium in the bloodstream.

The adrenal glands

The two **adrenal glands** are named for their position in the body: above (*ad* meaning "near") the kidneys (*renal* meaning "kidney"). Each of these triangular glands has two parts with two different functions (Figure 19-13). The **adrenal cortex** is the outer, yellowish portion of each adrenal gland. The word *cortex* comes from a Latin word meaning "bark" and is often used to refer to the outer covering of a tissue, organ, or gland. The **adrenal medulla** is the inner, reddish portion of the gland and is surrounded by the cortex. Not surprisingly, the word *medulla* comes from a Latin word meaning "marrow" or "middle."

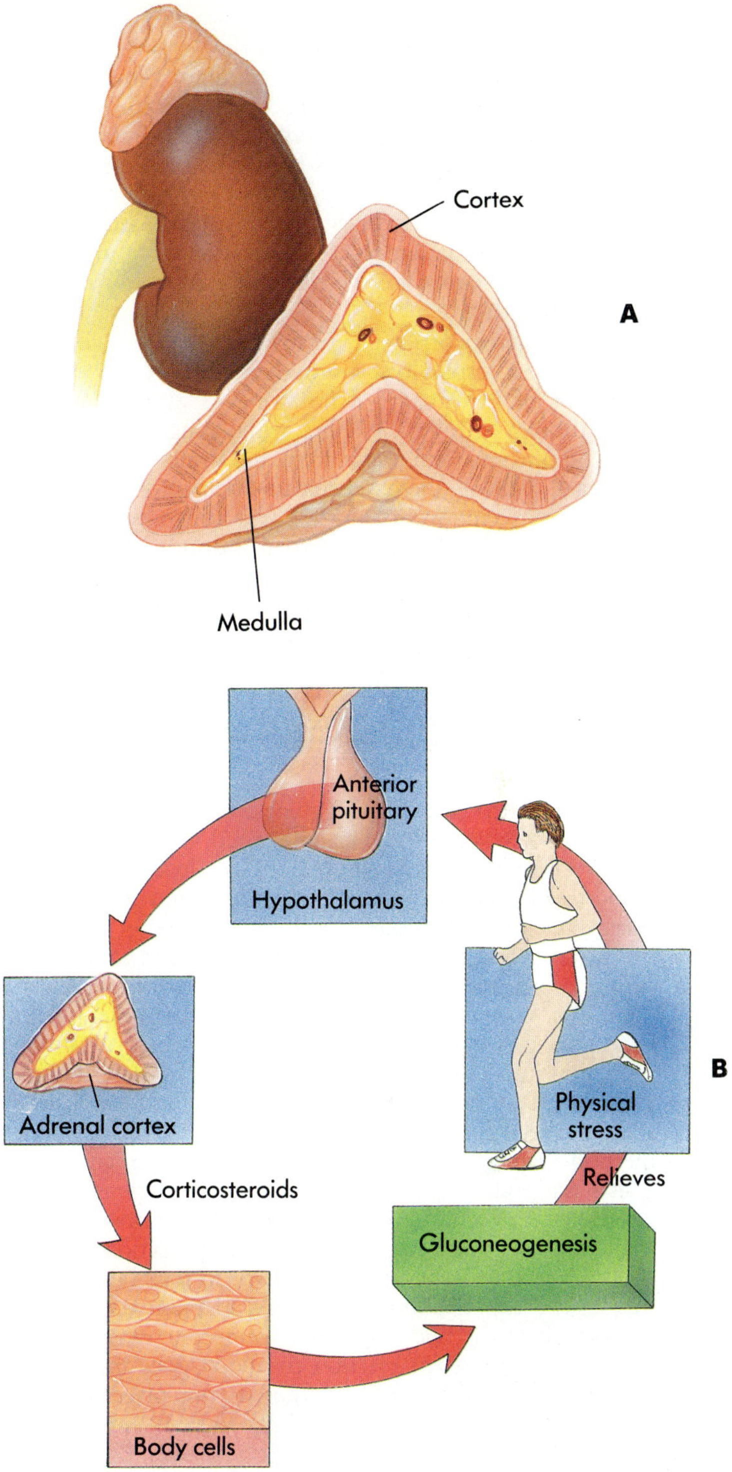

FIGURE 19-13 Structure and function of the adrenal glands.
A The adrenal glands are located on top of each kidney. Each gland has a cortex (the outer portion) and a medulla (the inner portion).
B The adrenal cortex plays a role in stress reduction. When the body is under stress, the hypothalamus, through the action of releasing hormone, induces the anterior pituitary to secrete ACTH. ACTH stimulates the adrenal cortex to produce corticosteroids. These corticosteroids cause the body to make glucose from noncarbohydrates such as protein, providing extra energy and thus reducing stress.

The adrenal cortex

As you may recall, the anterior pituitary gland secretes the hormone ACTH, adrenocorticotropic hormone. This hormone, as its name implies, stimulates the adrenal cortex to secrete a group of hormones known as **corticosteroids.** These steroid hormones act on the nucleus of target cells, triggering the cell's hereditary material to produce certain proteins. The two main types of corticosteroids produced by the adrenal cortex are the **mineralocorticoids** and the **glucocorticoids.**

The mineralocorticoids are involved in the regulation of the levels of certain ions within the body fluids. The most important of this group of hormones is **aldosterone.** It affects tubules within the kidneys, stimulating them to reabsorb sodium ions and water from the urine that is being produced, putting these substances back into the bloodstream. The secretion of aldosterone is triggered when the volume of the blood is too low, such as during dehydration or blood loss. Special cells in the kidneys "monitor" the blood pressure. When the blood pressure drops, these cells secrete an enzyme that begins a chain of reactions ending with the secretion of aldosterone. Conversely, when the blood pressure is within a normal range, the cell "detectors" in the kidneys are not stimulated, the release of aldosterone is not triggered, and the kidney tubules are not stimulated to conserve sodium and water.

The glucocorticoids affect glucose metabolism, causing molecules of glucose to be manufactured in the body from noncarbohydrates such as proteins. This glucose enters the bloodstream, is transported to the cells, and is used for energy as part of the body's reaction to **stress.**

Almost everyone is familiar with the term *stress*. And almost everyone can give examples of stressful situations: their boss "chewing them out" in front of co-workers, their kids fighting constantly with one another, or their sustaining a physical injury. The stress reaction was first described in 1936 by Hans Selye, a researcher who has since become the acknowledged authority on stress. Dr. Selye explained how the body typically reacted to stress—any disturbance that affects the body—and called this reaction the **general adaptation syndrome.** Over a prolonged period of stress, the body reacts in three stages: (1) the alarm reaction, (2) resistance, and (3) exhaustion. Contrary to maintaining homeostasis within the body, the general adaptation syndrome works to help the body "gear up" to meet an emergency.

During the alarm reaction the body goes into quick action. Imagine that you just entered your place of work and your boss confronted you, accusing you—in front of the office staff—of making a costly mistake. Your body reacts with a quickening pulse, increased blood flow, and an increased rate of chemical reactions within your body. Why does your body react in this way? Although the adrenal cortex is involved in the stress reaction, the beginning of the story lies in an understanding of the middle section of the adrenal glands, the adrenal medulla.

The adrenal medulla

The adrenal medulla is different from most other endocrine tissue in that its cells are derived from cells of the peripheral nervous system and are specialized to secrete hormones. These cells are triggered by the autonomic nervous system, which controls involuntary "automatic" responses. The other major nervous tissues with direct endocrine function are the secretory portion of the hypothalamus in the brain and the posterior pituitary just under the hypothalamus.

The two principal hormones made by the adrenal medulla are **adrenaline** and **noradrenaline** (also called *epinephrine* and *norepinephrine*). These two hormones are primarily responsible for the alarm reaction. The hypothalamus is responsible for sending the "alarm signal" (via the autonomic nervous system) to the adrenal medulla. The hypothalamus picks up the alarm signal as it monitors changes in the emotions and carries it as nerve signals on tracts of neurons that connect the hypothalamus with the emotional centers in the cerebral cortex. It can therefore sense when the body perceives an emotional stress. It can also sense physical stress, such as cold, bleeding, and poisons in the body. The hypothalamus reacts to stress by readying the body for "fight or flight," first triggering the adrenal medulla to dump adrenaline and noradrenaline into the bloodstream. These hormones cause the heart rate and breathing to quicken, the rate of chemical reactions to increase, and glucose (stored in the liver) to be dumped into the bloodstream. In general, the actions of adrenaline and noradrenaline increase the amounts of glucose and oxygen available to the organs and tissues most used for defense: the brain, heart, and skeletal muscles. (A summary of the stress reaction is shown in Figure 19-14.)

During the second stage of stress, the resistance stage, the hypothalamus triggers continuing responses by releasing regulating factors. These factors stimulate the pituitary to release ACTH, GH, and TSH. TSH stimulates the thyroid to secrete T_4 which stimulates the liver to break down stored carbohydrates (glycogen) to glucose. GH also stimulates the liver to produce glucose from glycogen, providing the body with an abundant energy source. And ACTH stimulates the adrenal cortex to secrete both mineralocorticoids and glucocorticoids. The mineralocorticoids cause the body to retain sodium and water, raising the blood pressure and providing more blood volume in the case of blood loss. The glucocorticoids also promote the production of glucose.

The adrenal glands, located on top of each kidney, are divided into two secretory portions: an outer cortex and an inner medulla. The hormones of the adrenal cortex, the corticosteroids, primarily regulate the level of sodium and consequently water in the bloodstream and stimulate the liver to produce glucose from stored carbohydrates. The hormones of the adrenal medulla, adrenaline and noradrenaline, ready the body to react to stress.

If a person continues to be highly stressed over a long time, the body may "lose the fight" and enter the third stage of the stress reaction: exhaustion. This stage is serious and can cause death. One cause of exhaustion is the loss of potassium ions, which are excreted when sodium ions are retained. This loss severely affects the cells' ability to function properly. Another cause is depletion of the glucocorticoids, resulting in a sharp drop in the blood glucose level. The organs also become weak and may cease to function. To

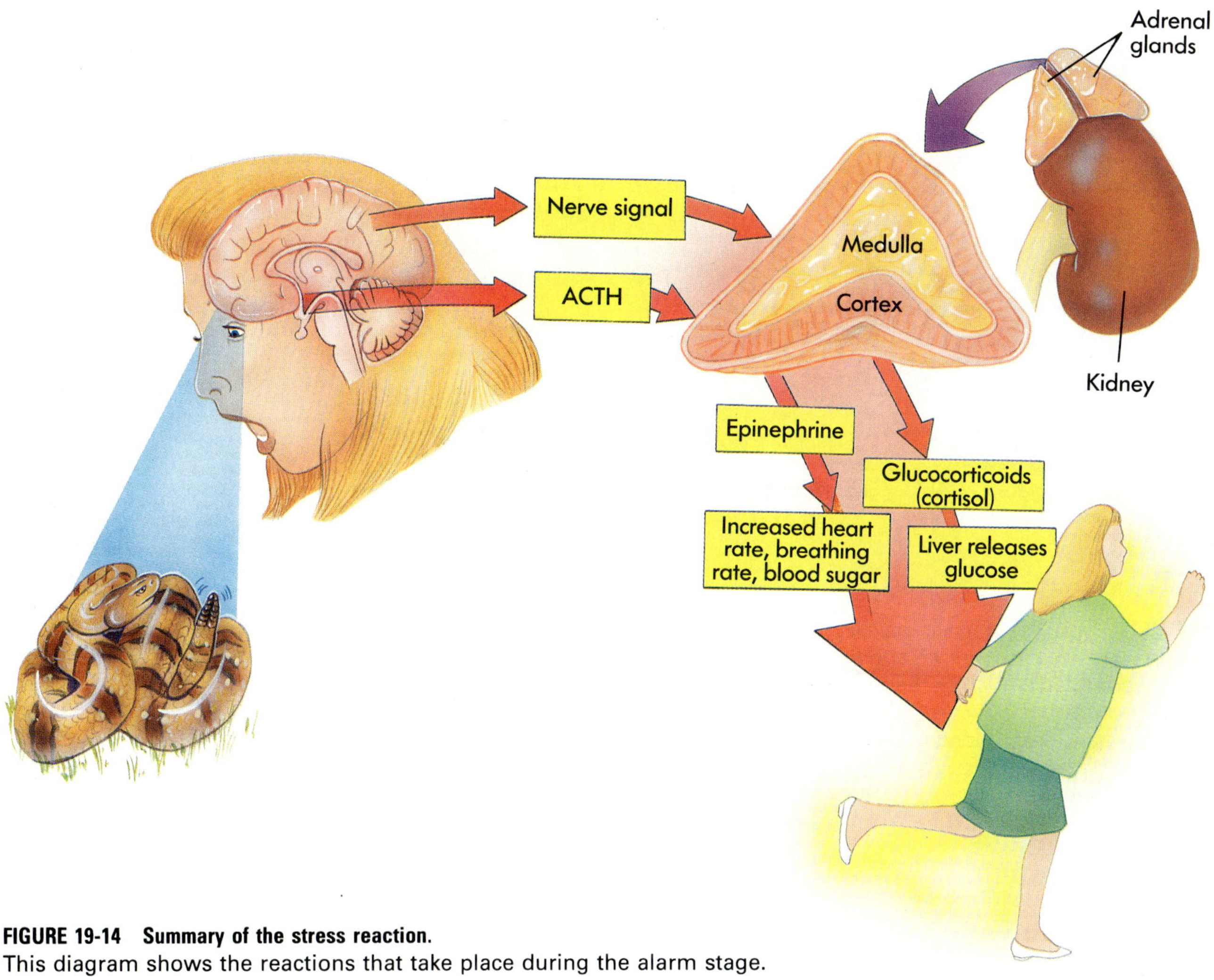

FIGURE 19-14 Summary of the stress reaction.
This diagram shows the reactions that take place during the alarm stage.

combat chronic stress, people can learn ways in which to psychologically "handle" their stress and can work toward a level of health and fitness that will help their bodies cope with the physical effects of stress.

The pancreas

The pancreas, located alongside the stomach (see Figure 19-1), is two glands in one: an exocrine gland and an endocrine gland. As an exocrine gland, it secretes the digestive enzymes discussed in Chapter 10. As an endocrine gland, it secretes the hormones **insulin** and **glucagon.**

The endocrine portion of the pancreas consists of separated clusters of cells that lie among the exocrine cells. For this reason, these cells are called *islets*—the **islets of Langerhans** (Figure 19-15). Separate types of cells within the islets produce insulin and glucagon. These hormones act antagonistically to one another to regulate the level of glucose in the bloodstream. Glucagon increases the blood glu-

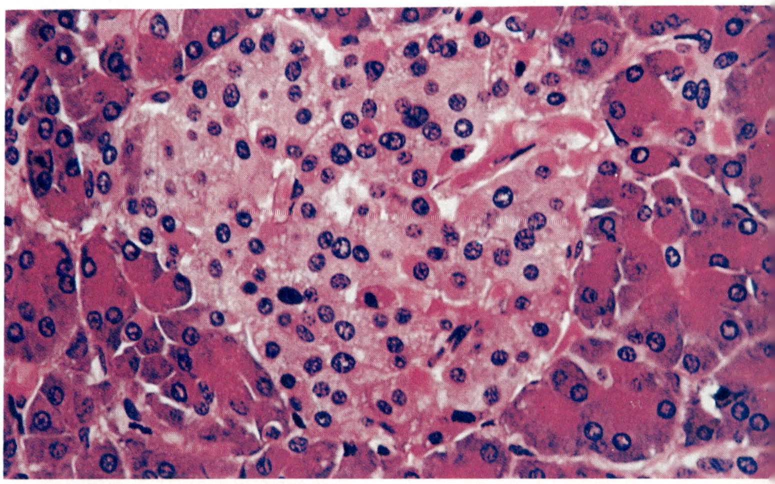

FIGURE 19-15 Islets of Langerhans.
Glucagon and insulin are produced by the islets of Langerhans, which are more lightly stained in this preparation.

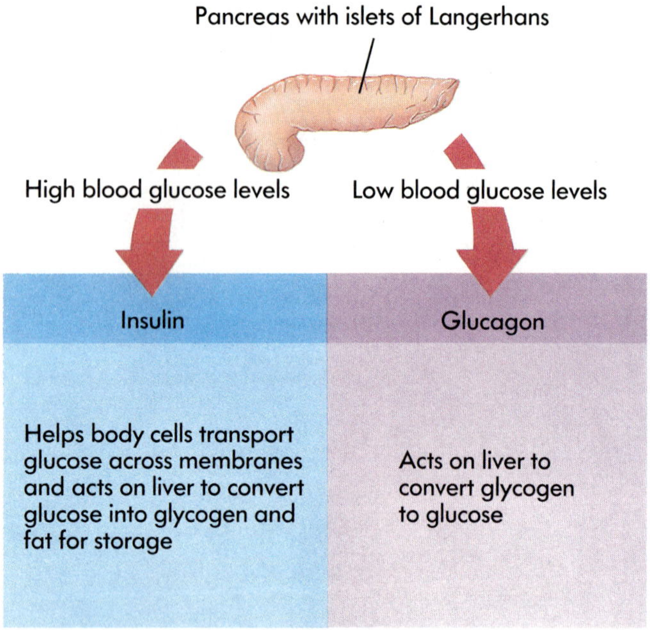

FIGURE 19-16 Functions of insulin and glucagon. Glucagon and insulin work antagonistically. Insulin helps cells transport glucose across their membranes and acts on the liver to convert the excess glucose to glycogen and fat. These activities lower blood glucose levels. When blood glucose levels are low, glucagon triggers the liver to convert glycogen to glucose.

cose level by triggering the liver to convert stored carbohydrates (glycogen) into glucose and to convert other nutrients such as amino acids into glucose. Insulin decreases the blood glucose level by helping body cells transport glucose across their membranes. In addition, insulin acts on the liver to convert glucose into glycogen and fat for storage (Figure 19-16).

Diabetes mellitus is a set of disorders in which a person tends to have a high level of glucose in the blood. There are many variations of the disorder and many causes. The underlying cause is the lack or partial lack of insulin, or the inability of tissues to respond to insulin, which leads to increased levels of glucose in the blood. The high levels of glucose in the blood result in water moving out of the body's tissues by osmosis. The kidneys remove this excess water through increased urine production, which causes excessive thirst and dehydration.

The primary types of diabetes mellitus are termed *type I* and *type II diabetes*. Persons with type I diabetes, or juvenile onset diabetes, have no effective insulin of their own. This type of diabetes usually develops in people younger than age 20. These persons must take injections of insulin as well as maintain an appropriate diet. Their deficiency of insulin causes the body to act as though it does not have enough glucose to fill its energy needs. Therefore it breaks down its fat reserves, resulting in the production of acids called *ketones*. These acids lower the pH of the blood (make it more acidic), a situation that can result in death.

Persons with type II diabetes, or maturity onset diabetes, have some effective insulin but not enough to meet their needs. Others have sufficient insulin, but the cells in many parts of the body lack enough insulin receptors. These persons can often treat their disorder with diet alone. Active research on the possibility of transplanting islets of Langerhans for persons who do not produce enough insulin holds much promise as a lasting treatment for this disease.

The pancreas secretes two hormones, insulin and glucagon, that act antagonistically to one another, regulating the level of glucose in the bloodstream.

The pineal gland

The **pineal gland** gets its name because it looks like a tiny pine cone embedded deep within the brain (see Figure 19-1). Although a great deal of research has been conducted regarding the workings of this gland, it still remains somewhat of a mystery. Interestingly, the pineal is the possible site of your "biological clock," or controlling center that regulates your daily rhythms, such as sleeping and waking. It may also stimulate such activity as the onset of puberty.

The thymus gland

The thymus gland is a small gland located in the neck a few inches below the thyroid (see Figure 19-1). In the thymus, certain immune system cells called *T lymphocytes* develop the ability to identify invading bacteria and viruses (see Chapter 13). The thymus produces a variety of hormones to promote the maturation of these cells. This gland is quite active during childhood but is replaced by fat and connective tissue by the time a person reaches adulthood.

The ovaries and testes

The ovaries produce female sex cells, or eggs, and the testes produce male sex cells, or sperm. In addition, the ovaries produce the hormones estrogen and progesterone, and the testes produce testosterone. The detailed structure of these endocrine glands and the specific roles their hormones play are discussed in Chapter 20.

Nonendocrine hormones

The principal endocrine glands and their hormones are listed in Table 19-1. However, certain cells in the body produce chemical messengers that regulate nearby cells without traveling in the bloodstream and are therefore not considered endocrine hormones. These intercellular chemical messengers are often called *local hormones* and include a wide variety of substances. Certain cells in the walls of the stomach and small intestine, for example, secrete hormones that regulate the release of digestive juices by various cells. The ends of axons secrete a variety of transmitter substances that stim-

TABLE 19-1

Endocrine glands and their hormones

ENDOCRINE GLAND AND HORMONE	TARGET TISSUE	PRINCIPAL ACTIONS
HYPOTHALAMUS		
Releasing hormones	Other endocrine glands	Stimulate the release of hormones by other endocrine glands
POSTERIOR PITUITARY		
Oxytocin	Uterus Mammary glands	Stimulates contraction of uterus and milk production
Antidiuretic hormone (ADH)	Kidneys	Stimulates reabsorption of water by the kidneys
ANTERIOR LOBE OF PITUITARY		
Follicle-stimulating hormone (FSH)	Sex organs	Stimulates ovarian follicle, spermatogenesis
Luteinizing hormone (LH)	Sex organs	Stimulates ovulation and corpus luteum formation in females
Adrenocorticotropic hormone (ACTH)	Adrenal cortex	Stimulates secretion of adrenal cortical hormones
Thyroid-stimulating hormone (TSH)	Thyroid	Stimulates secretion of T_3 and T_4
Growth hormone (GH)	Cartilage and bone cells, skeletal muscle cells	Stimulates division of cartilage and bone cells, growth of muscle cells, and deposition of minerals
Prolactin	Mammary glands	Stimulates milk production
Melanocyte-stimulating hormone (MSH)	Melanocytes	Stimulates production of melanin
THYROID GLAND		
Thyroxine (T_4) and triiodothyrone (T_3)	General	Regulates metabolism
Calcitonin	Bone	Regulates calcium levels in the blood
PARATHYROID GLAND		
Parathyroid hormone (PTH)	Bone, kidney, small intestine	Regulates calcium levels in the blood
ADRENAL CORTEX		
Aldosterone	Kidney	Increases sodium and water reabsorption and potassium excretion
Glucocorticoids	General	Stimulate manufacture of glucose
ADRENAL MEDULLA		
Adrenaline and noradrenaline	Heart, blood vessels, liver, fat cells	Regulate "fight or flight" response: increase cardiac output, blood flow to muscles and heart, conversion of glycogen to glucose
PANCREAS (ISLETS OF LANGERHANS)		
Insulin	Liver, skeletal muscle, fat	Decreases blood glucose levels by stimulating movement of glucose into cells
Glucagon	Liver	Increases blood glucose levels by converting glycogen into glucose
OVARY		
Estrogens	General, female reproductive organs	Stimulate development of secondary sex characteristics in females; control monthly preparation of uterus for pregnancy
Progesterone	Uterus	Completes preparation of uterus for pregnancy
	Breasts	Stimulates development
TESTIS		
Testosterone	General	Stimulates development of secondary sex characteristics in males and growth spurt at puberty
	Male reproductive structures	Stimulates development of sex organs, spermatogenesis

ulate the dendrites or cell body of a neighboring neuron, allowing a nerve impulse to be conducted from neuron to neuron. Cells throughout the body secrete substances called **prostaglandins** that are derived from the cell membranes. These local hormones accumulate in regions of tissue injury or disturbance. They stimulate smooth muscle contraction and the dilation and constriction of blood vessels. Overproduction of prostaglandins swells cerebral blood vessels, so their walls press against nerve tracts in the brain, causing pain. Aspirin relieves this pain, commonly known as a headache, because it inhibits prostaglandin production.

In recent years, a group of local hormones have been identified within the brain. The most commonly known local brain hormones are the **enkephalins** and **endorphins.** Both have pain-killing properties similar to the drug morphine. It has been suggested that these chemical messengers are the body's natural painkillers, and they have been linked to a feeling of well-being referred to as *runner's high* experienced by some joggers.

Certain cells in the body produce hormones such as prostaglandins, enkephalins, and endorphins that regulate nearby cells. Because these chemicals do not travel in the bloodstream, they are not considered endocrine hormones.

Summary

1. Hormones are the chemical products of cells, used as messengers to affect other cells within the body. Hormones that are secreted into the bloodstream are called *endocrine hormones*. The 10 different endocrine glands make over 30 different hormones. Together, the glands that secrete the endocrine hormones are known as the *endocrine system*.
2. Hormones that are not secreted into the bloodstream but move within the tissue fluid to affect cells near their sources are called *local hormones*. These intercellular chemical messengers include a wide variety of substances.
3. Hormones affect cells in one of four ways: by regulating their secretions and excretions, by helping them to respond to changes in the environment, by controlling activities related to reproductive processes, and by influencing their proper growth and development.
4. The two primary types of endocrine hormones are the peptide hormones and the steroid hormones. Peptide hormones work by binding to receptor molecules embedded in the cell membranes of target cells and ultimately triggering enzymes that cause cells to alter their functioning. Steroid hormones work by binding to receptor molecules within the cytoplasm of cells and ultimately causing the hereditary material to produce certain proteins.
5. The production of hormones is regulated by feedback mechanisms in the following way: an endocrine gland is initially stimulated directly or indirectly by the nervous system or by the concentration of various substances in the bloodstream. The gland secretes a hormone that interacts with target cells, and the effect produced by the target cells acts as a new stimulus to increase or decrease the amount of hormone produced.
6. The pituitary gland, which hangs from the underside of the brain, secretes nine different hormones: seven from its anterior lobe and two produced by the hypothalamus that are stored in its posterior lobe. The hypothalamus is a small mass of brain tissue lying above the pituitary. The hypothalamus controls the pituitary by means of releasing hormones.
7. Four of the hormones produced by the anterior pituitary control other endocrine glands. In addition, the anterior pituitary secretes growth hormone, which works with the thyroid hormones to control normal growth; prolactin, which works with other female sex hormones to stimulate the mammary glands to secrete milk after childbirth; and melanocyte-stimulating hormone, which affects certain pigment-producing cells of the body.
8. The posterior pituitary secretes two hormones produced by the hypothalamus. Antidiuretic hormone helps control the volume of the blood by regulating the amount of water reabsorbed by the kidneys. Oxytocin is secreted during the birth process and enhances uterine contractions.
9. The parathyroids, located on the underside of the thyroid gland in the neck, secrete parathyroid hormone. This hormone works with one of the thyroid hormones, calcitonin, to help maintain the proper blood levels of various ions, primarily calcium.
10. The adrenal glands, located on top of the kidneys, have two different secretory portions: the outer cortex and the inner medulla. The cortex secretes a group of hormones known as *corticosteroids* that regulate the levels of certain mineral ions, water, and glucose in the bloodstream. The adrenal medulla secretes the hormones adrenaline and noradrenaline, which ready the body for action during times of stress.

KEY TERMS

adrenal cortex 325
adrenal glands 325
adrenal medulla 325
adrenaline 326
adrenocorticotropic hormone (ACTH) 320
aldosterone 326
antidiuretic hormone (ADH) 321
calcitonin (CT) 322
corticosteroids 326
diabetes mellitus 328
endocrine gland 315
endorphins 330
enkephalins 330
exocrine gland 315
follicle-stimulating hormone (FSH) 320
general adaptation syndrome 326
glucagon 327
glucocorticoids 326
gonadotropins 320
growth hormone (GH) 320
hormone 315
hyperthyroidism 322
hypothalamus 318
hypothyroid goiter 322
hypothyroidism 322
insulin 327
islets of Langerhans 327
local hormone 315
luteinizing hormone (LH) 320
melanocyte-stimulating hormone (MSH) 321
mineralocorticoids 326
negative feedback loop 318
noradrenaline 326
oxytocin 321
parathyroid gland 325
parathyroid hormone (PTH) 322
peptide hormone 315
pineal gland 328
pituitary gland 318
positive feedback loop 322
prolactin 321
prostaglandins 330
releasing hormone 318
second messenger 315
steroid hormone 315
target cell 315
thyroid-stimulating hormone (TSH) 320
thyroxine (T_4) 322
triiodothyronine (T_3) 322
tropic hormone 320

REVIEW QUESTIONS

1. What are hormones, and why are they important?
2. Define *endocrine system* and summarize the functions of endocrine hormones.
3. How do the "messages" sent by the endocrine system differ from those carried by the nervous system?
4. Identify the two main classes of endocrine hormones. Describe their respective functions and effects.
5. Describe how a feedback loop regulates the production of a hormone. What is the difference between a negative feedback loop and a positive feedback loop.
6. Describe the relationship between the hypothalamus and the pituitary gland.
7. Explain the term *tropic hormone*. Identify the four tropic hormones produced by the anterior pituitary and summarize the function(s) of each.
8. Give a function for each of the following. Where are they produced?
   a. Growth hormone
   b. Oxytocin
   c. Aldosterone
   d. Prostaglandins
9. After playing softball on a hot sunny day, you are sweating hard and feeling very thirsty. A friend suggests getting a beer. Is this a good idea? Explain.
10. Which hormones are produced by the thyroid gland? Summarize their functions.
11. Describe two hormones that together regulate calcium levels in your blood and discuss how they do it.
12. Distinguish between the mineralocorticoids and glucocorticoids. What are their respective functions, and where are they produced?
13. Describe the three stages of the general adaptation syndrome. What hormones are involved?
14. Chronic emotional stress has been linked to a greater than normal risk of various diseases. Why do you think this would happen?
15. Which hormones regulate the level of glucose in your blood? Describe how they do this. Where are they produced?
16. Identify and give a function for each of the following:
    a. Pineal gland
    b. Thymus gland
    c. Ovaries
    d. Testes

THOUGHT QUESTIONS

1. There is no question that athletes using steroids gain a competitive advantage from the extra muscle the steroids induce. Why do you think the use of steroids is universally banned?
2. People who cannot manufacture enough insulin are called diabetics and are treated by the administration of insulin. Diabetics are common—diabetes is becoming a leading cause of death in America. Why are there not similarly large numbers of people who cannot manufacture glucagon? What do you imagine such a person would be like?
3. It has often been speculated that the pineal gland is involved with the body's rhythms. How would you test this hypothesis?

FOR FURTHER READING

Berridge, M. (1985, October). The molecular basis of communication within the cell. *Scientific American,* pp. 142-152.
This is an account of what is known about second messengers in the cell.

Cantin, M., & Genest, J. (1986, February). The heart as an endocrine gland. *Scientific American,* pp. 76-81.
This is a good example of how your body self-regulates its activities; in addition to pumping blood, the heart secretes a hormone that fine tunes the control of blood pressure.

Davis, J. (1984). *Endorphins: New waves in brain chemistry,* New York: Doubleday & Co., Inc.
This article provides a popular account of research on brain hormones.

PART FIVE

HOW HUMANS REPRODUCE AND PASS ON BIOLOGICAL INFORMATION

20 Sex and reproduction

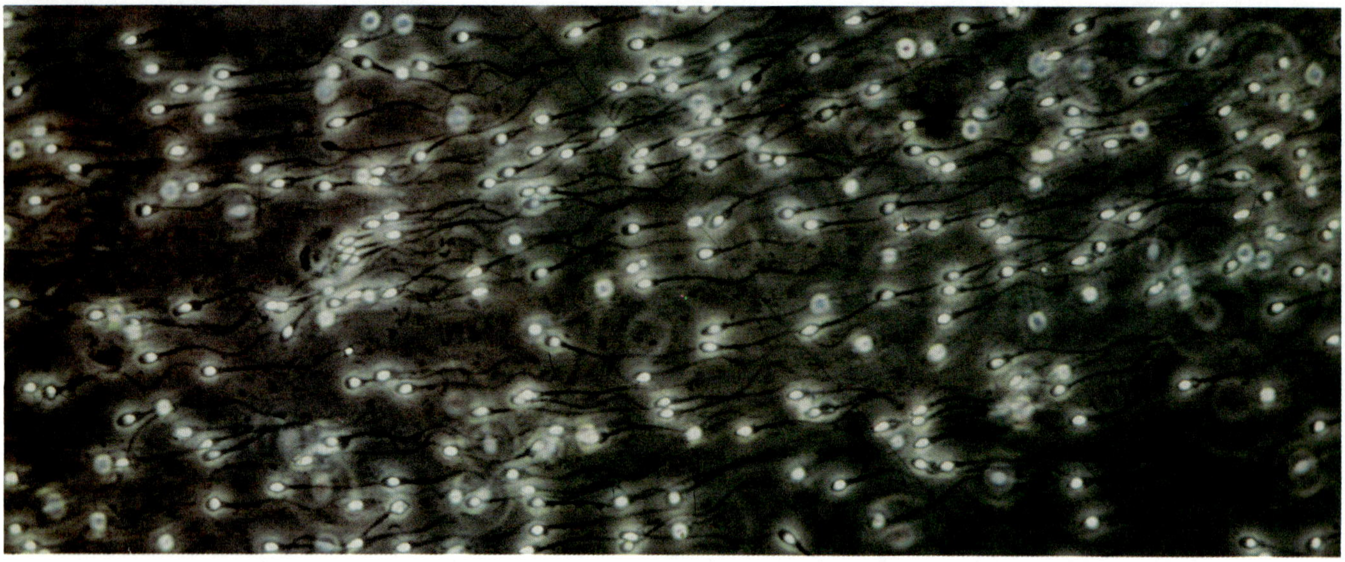

OUTLINE

Reproduction
The male reproductive system
- The production of sperm: The testes
- Maturation and storage: The epididymis and vas deferens
- Nourishment of the sperm: The accessory glands
- The penis

The female reproductive system
- The production of eggs: The ovaries
- Passage to the uterus: The uterine tubes
- The site of prenatal development: The uterus
- The reproductive cycle
- The vagina
- The external genitals
- The mammary glands

The sexual response
Contraception and birth control
- Abstinence
- Sperm blockage
- Sperm destruction
- Prevention of egg maturation
- Surgical intervention

Their nuclei glowing with a special fluorescent dye, these sperm are all moving in the same direction, propelled by their whip-like flagella. Only one of the hundreds of millions introduced into the vagina will be the first to encounter the female gamete—a developing egg. Enzymes in the sperm head will dissolve the jelly-like covering of the egg and pierce its membrane. Once the union of egg and sperm takes place, changes on the surface of the egg block the entry of additional sperm, ensuring that the fertilized egg will have only one complete set of hereditary material. Having more than one complete set could be fatal to a new embryo's existence. But with a set of genes from each of its parents and with conditions right for survival, a single human embryo has the potential to develop into the complex network of 100 trillion cells that make up the human body.

HIGHLIGHTS

▼

Sexual reproduction in humans, as well as in most sexually reproducing organisms, is a process in which a male and female sex cell unite to form the first cell of a new individual.

▼

The male reproductive system produces male sex cells called *sperm* and deposits them in the female reproductive tract during sexual intercourse.

▼

The female reproductive system receives the male sex cells, produces female sex cells called *eggs,* and provides an environment for a fertilized egg to grow and develop into a new individual.

▼

The human sexual response promotes sexual reproduction; techniques to avoid conception during sexual activity vary in their methods and success rates.

Reproduction

Each male sex cell, or **spermatozoon** (sperm), is a highly motile cell that can penetrate the membrane of a secondary oocyte (potential egg). After penetration, a sperm joins its half of the hereditary message—the "blueprints" for the development of a new individual—with the hereditary material of the oocyte. This process, whereby a male and a female sex cell combine to form the first cell of a new individual, is called *sexual reproduction*. The organs of the human male and female reproductive systems are the organs of sexual reproduction. Both the male and female reproductive systems have organs called **gonads** that produce the sex cells, or **gametes** (eggs and sperm). Other organs and ducts store, transport, and receive the sex cells. Accessory glands help nourish the sex cells.

The male reproductive system

The function of the male reproductive system is to produce sperm cells and transport them to the female reproductive tract. The production of sperm takes place in the male gonads, which are called the **testes.**

The production of sperm: The testes

The testes, or testicles (Figure 20-1), are located outside of the lower pelvic area of the male, housed within a sac of skin called the **scrotum** (Figure 20-2). The placement of

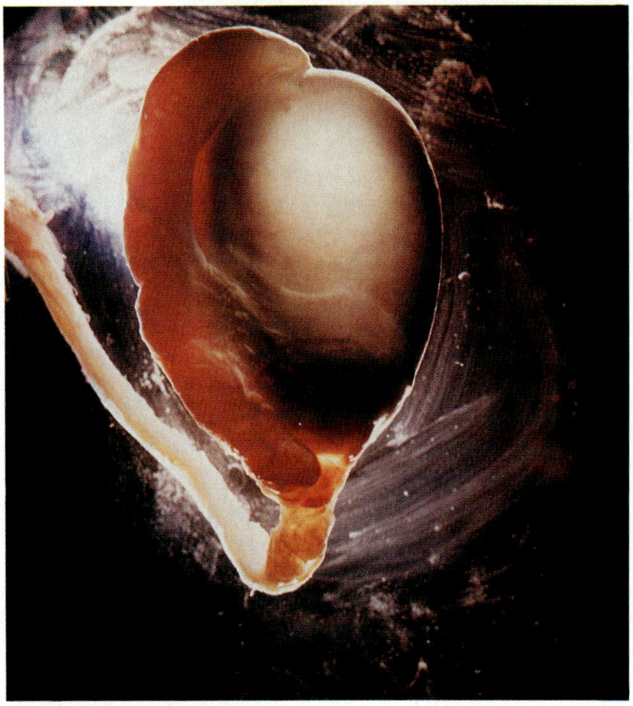

FIGURE 20-1 Human testis.
The testis is the ovoid structure in the center of the photograph; sperm are formed within it. Cupped along side of the testis is the epididymis, a highly coiled passageway within which sperm complete their maturation. Extending away from the epididymis is a long tube, the vas deferens.

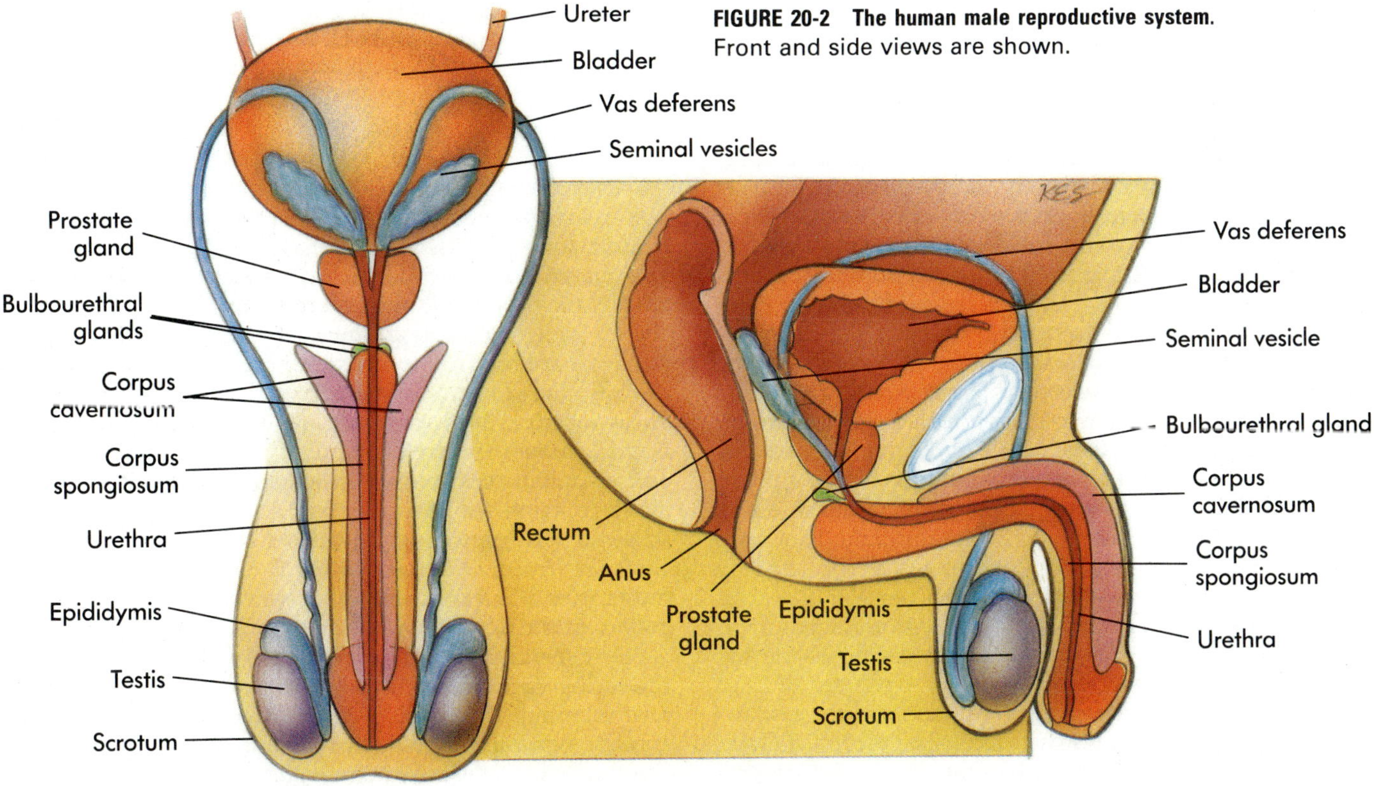

FIGURE 20-2 The human male reproductive system.
Front and side views are shown.

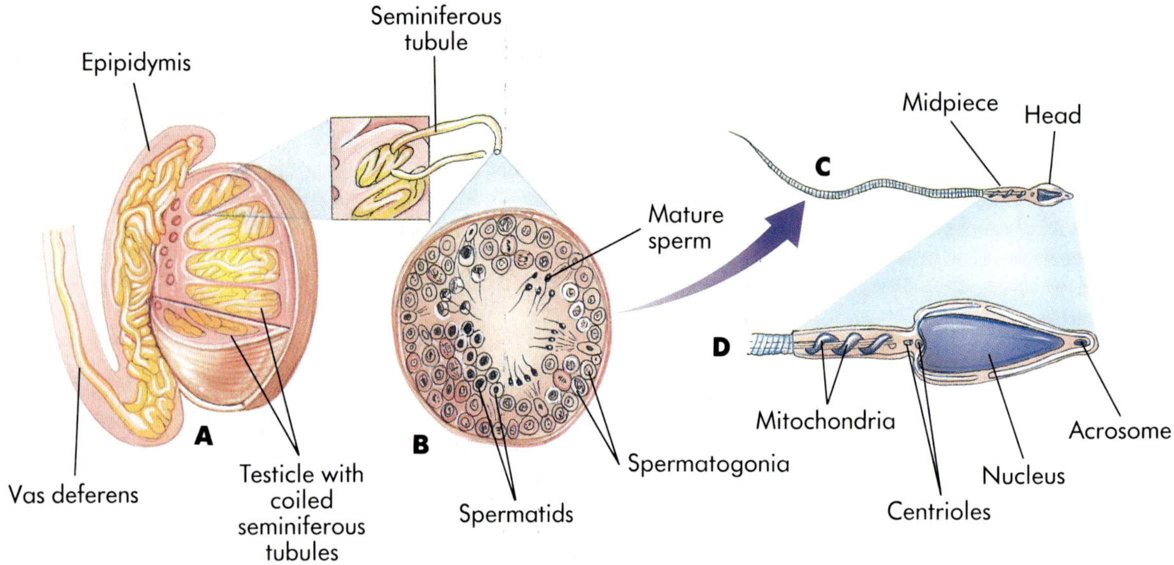

FIGURE 20-3 The interior of the testis, site of spermatogenesis.
Within the seminiferous tubules of the testis **(A)**, cells called *spermatogonia* **(B)** develop into sperm, passing through the spermatocyte and spermatid stages. Each sperm **(C)** possesses a long tail attached to a head **(D)**, which contains a haploid nucleus.

this organ allows the sperm to successfully complete their development at a temperature slightly lower than the 37° C (98.6° F) internal temperature of the human body. Tight jeans or other clothing that press the testes close to the body can have a detrimental effect on sperm development, lowering the number of viable sperm that a man produces.

Each testis is packed with approximately 750 feet of tightly coiled tubes called **seminiferous tubules** (Figure 20-3,*A*). Male gametes, or sperm cells, develop within the tissue lining these tubules during a process of meiosis (see Chapter 5) and development called **spermatogenesis** (literally, "the making of sperm"). Hundreds of millions of sperm are produced by this process each day! Spermatogenesis usually begins during the teenage years when a boy reaches sexual maturity, or **puberty,** and continues throughout his life. Sperm production is triggered by follicle-stimulating hormone, or FSH.

Luteinizing hormone, or LH, regulates the testes' secretion of testosterone, a sex hormone responsible for the development and maintenance of male secondary sexual characteristics. These characteristics, such as a deepening of the voice and the growth of facial hair, begin to develop at puberty and are signs that sexual maturation is taking place.

Both FSH and LH are secreted by the pituitary, a tiny gland that hangs from the underside of the brain (see Chapter 19). The secretion of these hormones is regulated by means of a negative feedback loop involving the hypothalamus, a tiny portion of the brain that controls the pituitary. A high level of testosterone in the blood inhibits the hypothalamus from triggering the pituitary to release FSH and LH. Without stimulation by these hormones, the testes produce less sperm and less testosterone. When the testosterone level drops too low, the hypothalamus once again triggers the release of FSH and LH from the pituitary, causing the testes to "step up" their production of sperm and testosterone.

Sperm production, or spermatogenesis, takes place within the coiled tubules of the testes and is triggered by the hormone FSH. The development of functionally mature sperm also requires LH and testosterone.

Packed within the walls of the seminiferous tubules are cells called **spermatogonia,** the cells that give rise to sperm (Figure 20-3,*B*). These cells have a full complement (diploid amount) of hereditary material and constantly produce new cells by the process of mitosis (see Chapter 5). Some of the "new" (daughter) cells produced by mitotic cell division move inward toward the interior of the tubule and undergo meiosis. The reduction-division of meiotic cell division produces four cells from each original cell, all having *half* the normal amount (or haploid amount) of hereditary material. These haploid cells are now called **spermatids.** Each spermatid contains one member from each pair of the 23 chromosome pairs present in the cells of the human body. Included in these 23 chromosomes is either an X sex chromosome or a Y sex chromosome, which will determine the gender of the new individual (Figure 20-4).

The spermatids then undergo a process of development, producing sperm cells, or spermatozoa. Spermatozoa are relatively simple cells (Figure 20-3, *C* and *D*). Each has an anterior portion, or head, that consists primarily of a cell membrane encasing hereditary material. A vesicle called an

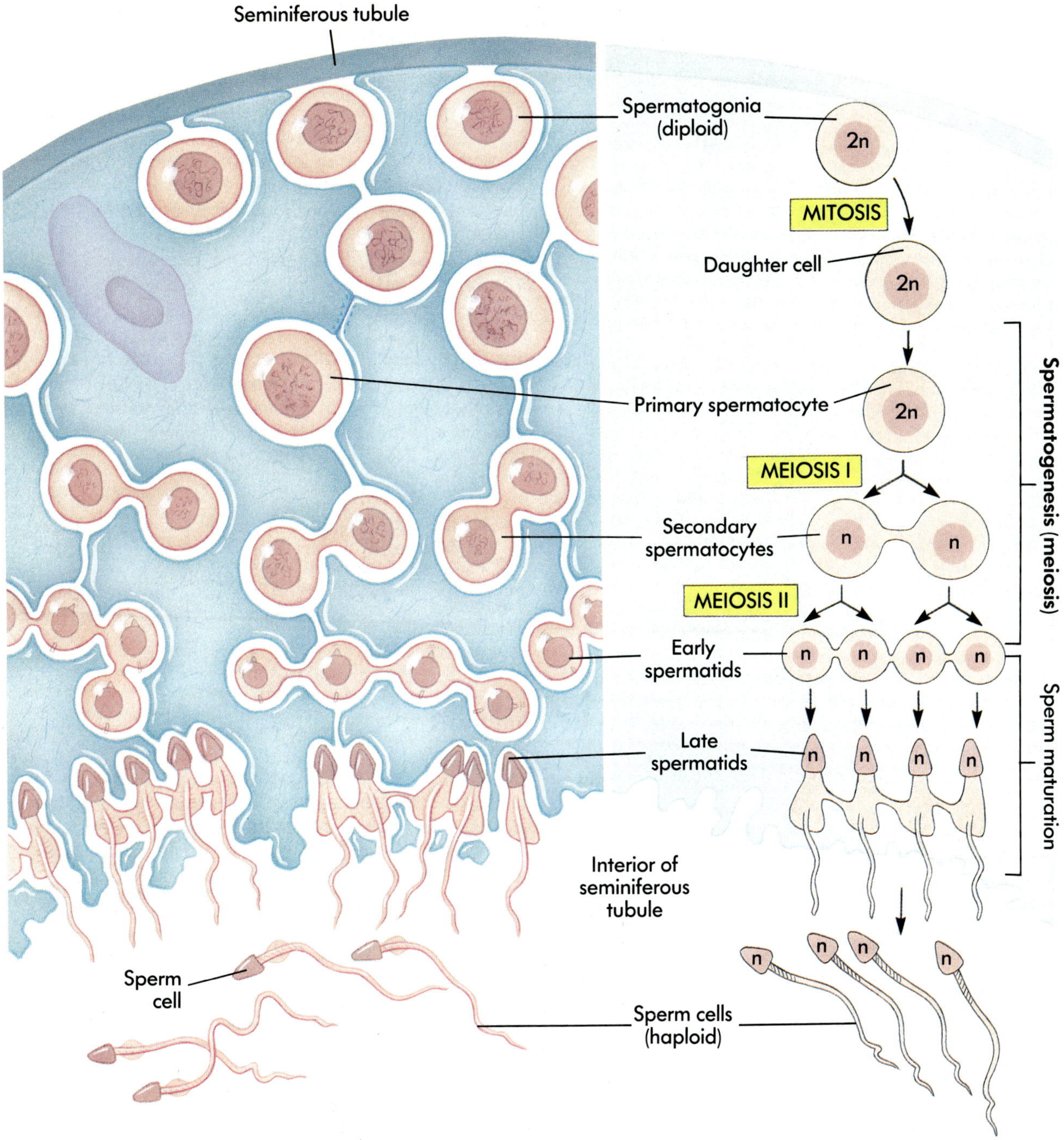

FIGURE 20-4 Spermatogenesis.
Spermatogonia undergo mitosis to produce more spermatogonia. Some of the spermatogonia, however move to the interior of the seminiferous tubule to undergo meiosis. Meiotic cell division produces four haploid spermatids that will eventually develop into spermatozoa.

acrosome, derived from the Golgi body, is located at the leading tip of the sperm cell. It contains enzymes that aid in the penetration of the protective layers surrounding the egg. In addition, each spermatozoon has a whip-like tail, or flagellum, that propels the cell and mitochondria, which produce ATP from which a sperm derives energy that powers its flagellum. Sperm development takes about 2 months from spermatogonia to mature spermatozoa. On average, 300 million mature sperm are produced per day.

In the testes, diploid cells called *spermatogonia* give rise to haploid cells called *spermatids.* These haploid cells contain half the normal amount of hereditary material of body cells. The spermatids then develop into sperm cells, or spermatozoa, which have a head containing the hereditary material, a vesicle containing enzymes to penetrate an egg, and a whip-like tail.

Maturation and storage: The epididymis and vas deferens

After its formation within the testes, the sperm move within its tubules to a nearby long, coiled tube called the **epididymis** (see Figure 20-3, *A*) that sits on the back side of the testes. Here the sperm undergo further development and may be stored for about 18 hours to 10 days, becoming capable of propelling themselves and penetrating an oocyte. The sperm then move toward a connecting tube, the **vas deferens,** a long tube that ascends from the epididymis into the pelvic cavity, looping over the side of the urinary bladder (see Figure 20-2). Stored in the epididymis and the very beginning of the vas deferens, sperm retain their ability to fertilize an oocyte for a short while. If they are not ejaculated, "old" sperm are reabsorbed by the body to make room for "new" sperm.

Spermatozoa mature in the epididymis, a long, coiled tube that sits on the back side of the testes and leads to the long ascending vas deferens.

Nourishment of the sperm: The accessory glands

When a male ejaculates during sexual intercourse (see p. 339), the sperm are propelled through the vas deferens to the **urethra,** where the reproductive and urinary tracts join. On the way, accessory glands add fluid to the sperm, producing an ejaculate called **semen.** The semen leaves the body through the urethral opening at the tip of the penis.

The first accessory glands to add fluid to the traveling sperm are the **seminal vesicles.** This pair of glands has short ducts that empty a thick, clear fluid into each vas deferens just before they join and flow into the urethra. The fluid secreted by the seminal vesicles is primarily composed of the sugar fructose, which serves as a source of energy for the sperm.

Near its beginning, the urethra is surrounded by a gland called the **prostate.** About the size of a chestnut, the prostate adds a milky alkaline fluid to the semen. This fluid neutralizes the acidity of any traces of urine left in the urethra and also neutralizes the acidity of the female vagina. Sperm are unable to swim in an acid environment, so the secretion of the prostate ensures their motility.

Just beneath the prostate lies a set of tiny, round accessory glands called the **bulbourethral glands.** The purpose of these glands is somewhat of a mystery. Like the prostate, the bulbourethral glands secrete an alkaline fluid into the semen. However, they secrete only a drop or two, which precedes the ejaculate. It does carry sperm to the outside before ejaculation, making the withdrawal method of birth control highly unreliable (see p. 347). Scientists think that the purpose of this secretion may be to neutralize the acidity of the urethra just before ejaculation as a further protection for sperm.

During ejaculation, sperm are propelled through the vas deferens to the urethra and out the penis. Fluid is added to the sperm from various accessory glands along the way. This fluid contains substances that nourish the sperm and neutralize the acidity of the vagina.

The penis

The scrotum (which encloses the testes and epididymi) and the penis are the two male external sexual organs, or **external genitals.** The **penis** is a cylindrical organ that transfers sperm from the male reproductive tract to the female reproductive tract. As shown in the cross-sectional view of the penis in Figure 20-5, this organ is composed of three cylinders of spongy erectile tissue. Two veins run along the top surface of the penis. Beneath these veins, two of the cylinders of

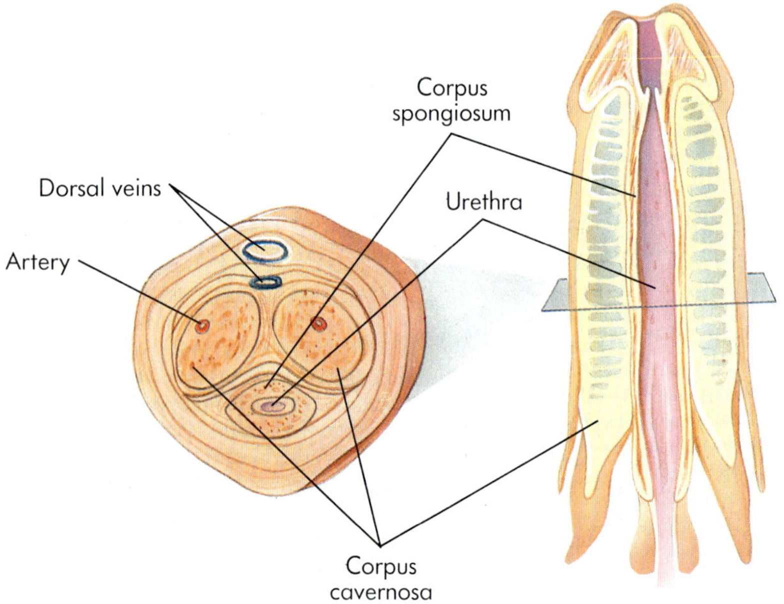

FIGURE 20-5 A penis in cross section.
The three cylinders of spongy tissue are the corpus spongiosum and the two sides of the corpus cavernosa.

erectile tissue sit side by side. And beneath them lies a third cylinder, surrounding the urethra.

The tissue that makes up the three erectile cylinders is riddled with spaces between its cells. These spaces normally contain a small amount of blood and are lined by a layer of flattened cells similar to the inner lining of blood vessels. During sexual stimulation, nerve impulses from the central nervous system cause the arterioles leading into this tissue to widen, causing additional blood to collect within the spaces. Pressure from the increased volume of blood within the spaces causes the erectile tissue to become distended and compresses the veins that normally drain blood from the penis, causing it to become erect and rigid. Continued stimulation by the central nervous system is required for this erection to be maintained.

Erection can be achieved without any physical stimulation of the penis. However, physical stimulation usually is required for the ejaculation of semen, which corresponds to the culmination of sexual excitement, or orgasm. Prolonged stimulation of the penis, as by repeated thrusts into the female's vagina, leads first to the mobilization of the sperm. In this process, muscles encircling the vas deferens contract, moving the sperm into the urethra. Then ejaculation takes place: muscles at the base of the penis and within the walls of the urethra contract repeatedly, ejecting approximately a teaspoon of semen containing about 200 million sperm out of the penis.

The female reproductive system

The part of the female reproductive tract that receives the sperm is called the **vagina,** a muscular tube 9 to 10 centimeters (3½ inches) long (Figure 20-6). But the female reproductive tract does more than receive sperm. It usually produces one secondary oocyte (potential egg) each month and transports it to where it can be fertilized and the resultant zygote nourished as it develops into an embryo and then a fetus (see Chapter 21).

The production of eggs: The ovaries

Secondary oocytes are produced in the female gonads, or **ovaries** (Figure 20-6). These two almond-shaped organs are about 3 centimeters long, slightly larger than the distance from the last knuckle on your thumb to the thumb's tip. They are located in the lower portion of the abdominal cavity in an area called the *pelvic cavity*. The ovaries contain **primary oocytes,** cells that have the potential to develop into eggs, surrounded by supporting cells called **follicular cells.** Together, the follicular cells and the oocyte are called a **follicle,** a word derived from a Latin word meaning "bag" or "sac."

The process of meiosis and development that produces mature female sex cells, or eggs, is called **oogenesis** (literally,

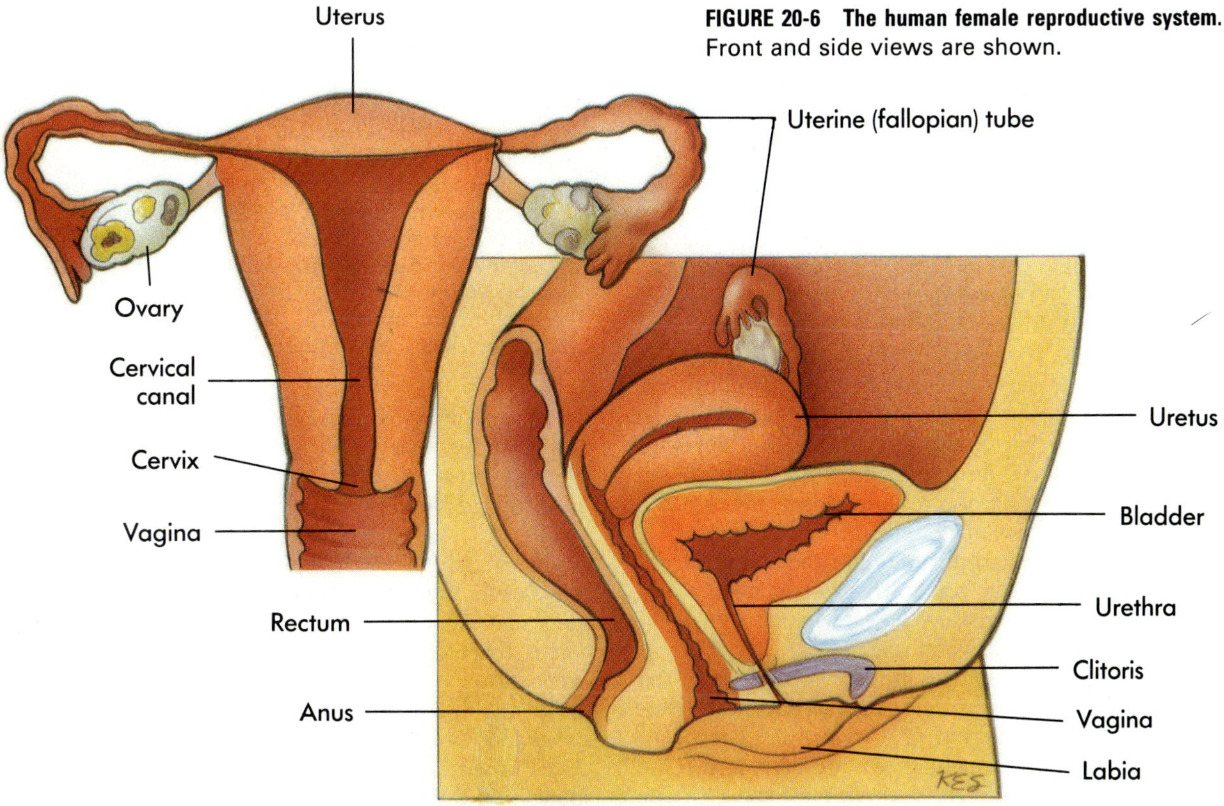

FIGURE 20-6 The human female reproductive system. Front and side views are shown.

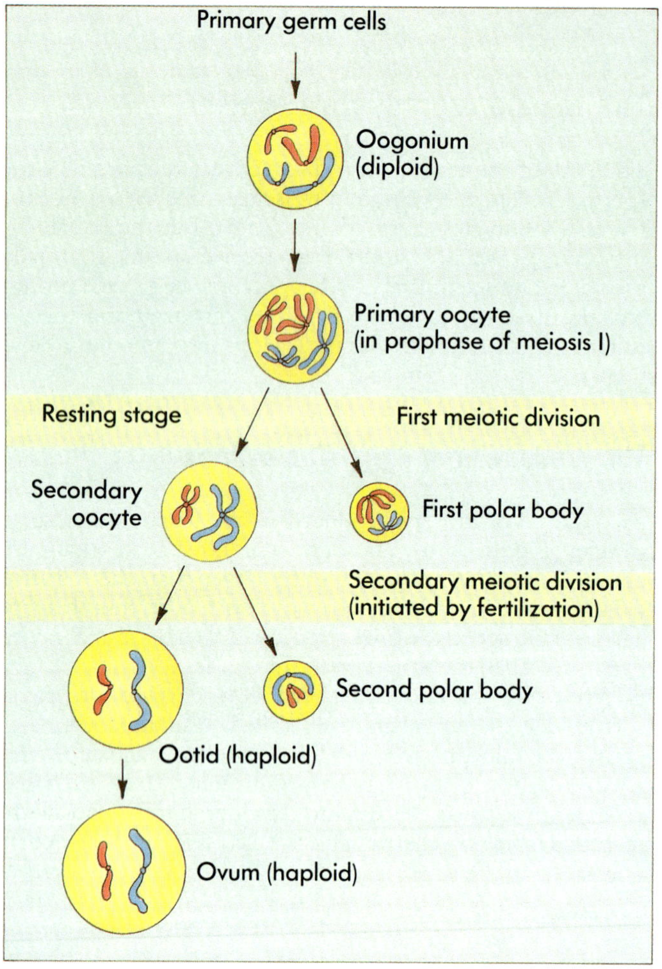

FIGURE 20-7 Oogenesis.
Notice that the oocytes have already completed part of the first meiotic division and are in a holding pattern until puberty, when one oocyte completes the first meiotic division approximately every 28 days.

"the making of eggs"). Unlike males, whose spermatogonia are constantly dividing to produce new spermatocytes, females have all of the primary oocytes at birth that they will ever produce—about 1 to 2 million of them! (By the age of puberty, more than half these oocytes have degenerated, leaving about 400,000 viable oocytes.) These primary oocytes were produced during fetal development from diploid cells called **oogonia.** The oocytes in the ovaries of a female began the process of meiosis before birth, but the process stopped during the first meiotic division (Figure 20-7). In other words, a female's primary oocytes are in a developmental "holding pattern" during her lifetime. As a woman ages, her oocytes age. The oocytes are continually exposed to chromosomal mutation, or change, throughout life. After 35 years, the odds of a harmful mutation having occurred are appreciable. This is one reason that developmental abnormalities occur with increasing frequency in pregnancies of women who are over 35 years old.

With the onset of puberty, one oocyte completes the first meiotic division approximately every 28 days. (The ovaries alternate in the production of oocytes). As a result of this division, two cells of unequal size are produced. One is called the **secondary oocyte** and contains most of the cytoplasm of the primary oocyte. The other cell, called a **polar body,** contains little cytoplasm and soon dies. The secondary oocyte breaks away from the ovary during **ovulation** and continues with the second meiotic division only if it is fertilized by a sperm. After the completion of the second meiotic division, the second polar body dies, and the egg, or **ovum,** is mature. The haploid nuclei of the sperm and the egg fuse, producing a **zygote,** the first cell of a new individual.

The ovaries are made up of primary oocytes—cells that have the potential to develop into eggs—surrounded by supporting cells called *follicular cells.* At birth, females have all of the primary oocytes that they will ever produce. With the onset of puberty, one primary oocyte matures into a secondary oocyte capable of breaking away from the ovary each month during a process called *ovulation.*

Passage to the uterus: The uterine tubes

At ovulation the secondary oocyte is released from the ovary. Lying close to each ovary is the entrance to a tube-like passageway that will take the egg to the **uterus,** the organ in which a fertilized egg can develop. Each passageway from an ovary to the uterus is called an *oviduct,* or **uterine tube.** In humans, the uterine tubes are commonly called the *fallopian tubes.*

Each uterine tube has an expanded "fringed" end near each ovary. The tube and its fringed ends are lined with ciliated cells that beat in a synchronized manner, creating a current in the surrounding fluid that sweeps the egg into the tube. Once in the uterine tube, the egg is swept along by the beating of the ciliated cells lining the tube and by rhythmic peristaltic contractions of the muscles within its walls.

he secondary oocyte journeys to the uterus, or womb, by means of a ciliated muscular tube called a *uterine tube.* One uterine tube extends from near each ovary to the uterus. If fertilized while in the tube, an oocyte will complete its maturation, join its hereditary material with that of a sperm, and begin its development as an embryo. A few days after reaching the uterus, the embryo will implant in the uterine lining to continue development.

The journey from the ovary to the uterus is a slow one, taking about 3 days to complete. Because unfertilized oocytes can live for only 24 hours, a secondary oocyte must be fertilized while in the uterine tube, or it is no longer viable (able to be fertilized). In addition, sperm live for approximately 48 to 72 hours; they cannot simply wait in the uterine tubes for an egg to come along. Therefore fertilization can only take place if sexual intercourse occurs within three days before to one day after ovulation. (However, knowing exactly when ovulation will occur is usually a difficult task unless specific testing procedures are used; see p. 347.) If fertilization occurs, the egg completes its development, its hereditary material joins with that of the sperm's, and development of the zygote begins as it moves along the uterine tube. By the time the new embryo reaches the uterus, it is a tiny ball of cells. After a few days, it implants in the lining of the uterus to receive nourishment. (Further development of the embryo is discussed in the next chapter.)

The site of prenatal development: The uterus

As shown in Figure 20-6, the uterine tubes lead directly to the uterus. The uterus (from a Latin word meaning "womb") sits above the urinary bladder and in front of the rectum and directly connects with the vagina. Its walls are made up of thick layers of muscle that contract during the birth process and during menstruation (see p. 342). The interior of the uterus is hollow and provides a cavity for the development of a fertilized egg. Amazingly, the uterus of a woman who has never been pregnant is only 7.5 centimeters (3 inches) long and 5 centimeters (2 inches) wide, yet it can stretch to accommodate the size of a newborn infant!

The inner lining of the uterus is called the **endometrium.** This lining has two layers: one functional, transient layer that is in contact with the uterine cavity and an underlying permanent layer. The functional layer grows and thickens each month, readying itself for the implantation of an embryo. If an embryo is not present, the functional layer is shed. The monthly development and shedding of the functional layer of the endometrium (called the *menstrual cycle*) and the monthly maturation of an egg and its release (called the *ovarian cycle*) are both governed by levels of various female sex hormones. Together, these two cycles are commonly referred to as the *reproductive cycle.*

The reproductive cycle

Each month, a small number of primary oocytes (20 to 25) begin to develop within an ovary. This initial development is triggered by follicle-stimulating hormone (FSH) released by the pituitary gland in the brain (see Chapter 19). After they are stimulated, these follicles start producing very low levels of a group of hormones called **estrogens.** Estrogens develop and maintain the female reproductive structures such as the ovarian follicles as well as the lining of the uterus and the breasts. Therefore as the follicles develop in the ovaries and secrete estrogens, the lining of the uterus begins to thicken, preparing for the implantation of a fertilized egg. As the developing follicles mature, a membrane forms around each potential egg, fluid builds up within each follicle (Figure 20-8), and each follicle's cells produce increasing amounts of estrogens. During the second week of the reproductive cycle, most of the developing follicles die; usually only one follicle continues to mature. The hypothalamus senses the rise in blood level of estrogens and triggers the pituitary to secrete a surge of luteinizing hormone, or LH. A smaller surge of FSH occurs also. The surge in LH triggers ovulation. During ovulation the follicle breaks open and releases its immature egg into the abdominal cavity. This rise in LH can be detected by home test kits, informing a woman one day before ovulation occurs. A couple can then use this information to help them conceive a child.

Ovulation usually (but not always) occurs at the midpoint of the reproductive (menstrual) cycle—around day 14 in a 28-day cycle (Figure 20-9). After ovulation the ciliated cells at the fringes of the nearby uterine tube "coax" the egg into the tube. It then begins its 3- to 4-day journey to the uterus. The ruptured follicle collapses, and then under the continued influence of LH, it begins to enlarge, forming a yellowish structure called the **corpus luteum,** or "yellow body." The corpus luteum plays an important role in the preparation of the endometrium for the implantation of the fertilized egg by secreting increasing quantities of estrogens and progesterone. The progesterone "steps up" the thickening of the lining of the uterus and dramatically increases its blood supply.

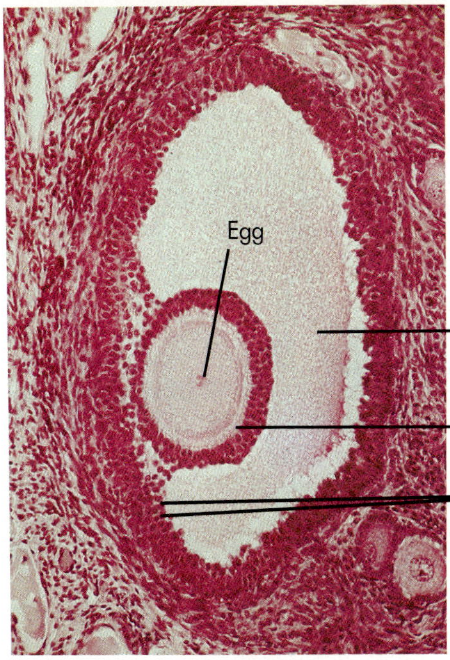

FIGURE 20-8 A mature egg within an ovarian follicle of a cat. The follicle produces estrogens, which develop and maintain other follicles within the ovary.

The rise in progesterone and estrogens also acts as negative feedback to the hypothalamus, ultimately inhibiting the secretion of FSH and LH from the pituitary. If the immature egg is fertilized while jouneying down the uterine tube, it completes a second meiotic division, joins its hereditary material with that of the sperm, and begins dividing by mitosis (see Chapters 5 and 21). A few days after reaching the uterus, the developing embryo nestles into the thickened uterine lining to continue its development. The placenta, a flat disk of tissue that provides an exchange of nutrients and wastes between the embryo and the mother, begins to form. Its embryonic tissues secrete a hormone called **human chorionic gonadotropin,** or **HCG.** This hormone maintains the corpus luteum so that it will continue to secrete progesterone and estrogen. Progesterone is needed for the maintenance of the uterine lining and the continued attachment of the fetus to this lining. Secretion of HCG occurs about 8 days after fertilization and is excreted in the urine. The detection of this hormone in the urine is the basis for both laboratory home pregnancy tests.

If the egg is not fertilized while in the uterine tube, it dies and cannot implant in the uterus and develop. Therefore no placenta develops and no HCG is produced. The corpus luteum begins degenerating on approximately day 24 of the

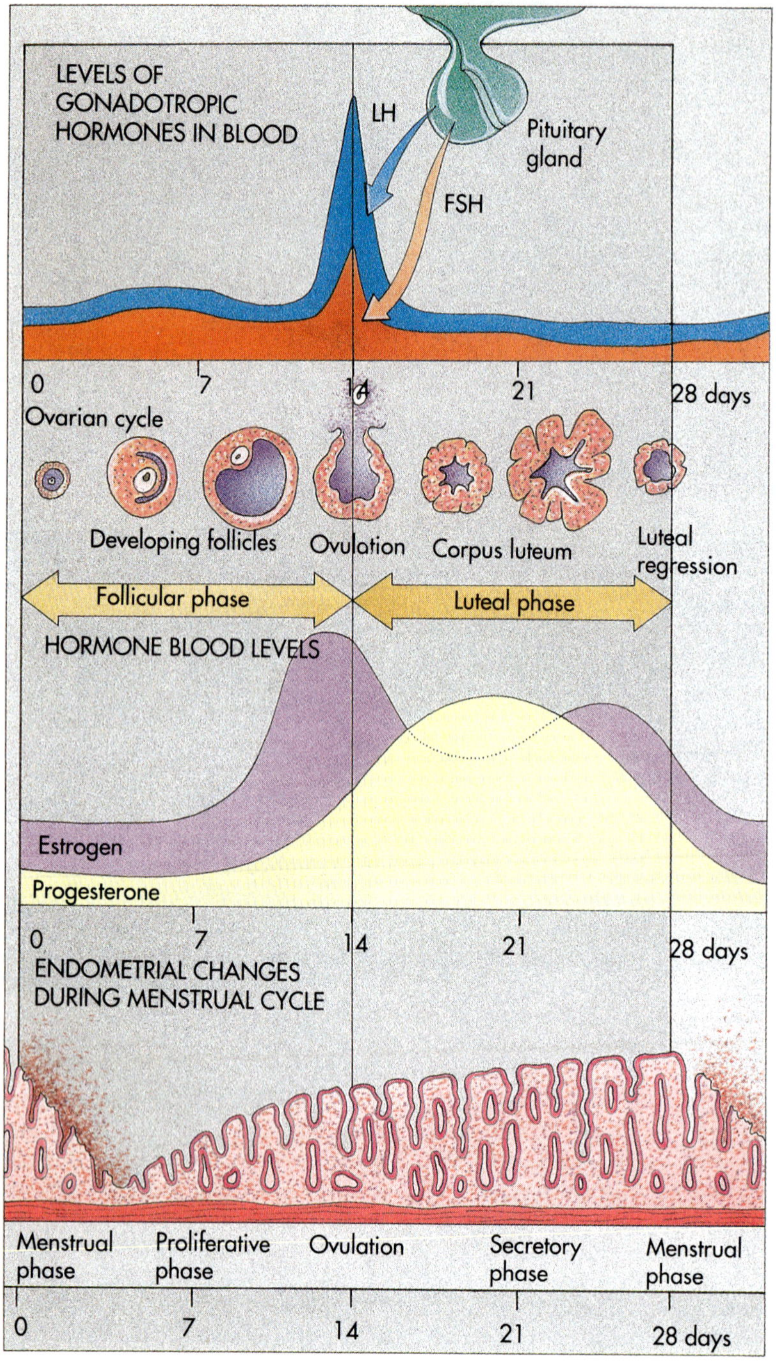

FIGURE 20-9 The human menstrual cycle. The growth and thickening of the uterine lining is governed by increasing levels of estrogens and progesterone; menstruation, the sloughing off of the blood-rich tissue, is initiated by decreasing levels of these hormones.

cycle. As it degenerates, it stops producing estrogens and progesterone, hormones necessary to maintain the buildup of the lining of the uterus. As the blood level of these hormones falls, the endometrial lining sloughs off in a process known as **menstruation,** or the menses. The lining degenerates over a period of approximately 5 days, causing a somewhat steady flow of blood, tissue, and mucus from the uterus and out through the vagina, usually heaviest during the first three days of the menses. The process of menstruation is traditionally considered to mark the *beginning* of the reproductive cycle.

The reproductive cycle of females occurs approximately every 28 days. This cycle includes the maturation of the primary oocyte, its release from the ovary, its journey to the uterus, and the development of the former follicular cells into the corpus luteum. In addition, it includes the thickening of the endometrial lining of the uterus in preparation for the implantation of a new embryo. The hormones FSH, LH, estrogen, and progesterone orchestrate these events.

Before the onset of a menstrual period, many women experience symptoms that are collectively referred to as *premenstrual tension syndrome,* or *PMS*. Premenstrual symptoms are thought to be related to the cutoff of progesterone and the retention of water. Symptoms vary from woman to woman and include breast swelling and tenderness, bloating, weight gain, headaches, fatigue, and backache. Physicians suggest that a diet high in protein and low in salt, alcohol, and carbohydrates can reduce water retention during this time and lessen the symptoms of PMS.

The reproductive cycle of women generally begins during adolescence and ends between the ages of 50 and 55 years. This cessation of the menses is termed **menopause.** As a woman ages, changes take place in the reproductive cycle that lead to menopause. These changes usually begin between the ages of 40 and 50 years. The menstrual cycles may become irregular, varying in length. Generally they become less frequent. These changes result from the failure of the ovaries to respond to stimulation by FSH and LH secreted by the pituitary. As a result of these changes, follicles do not develop; the ovaries therefore produce less estrogens. The drop in estrogen causes the two major symptoms associated with menopause: flushes, the reddening of the face and a feeling of warmth, and flashes, short periods of intense flushing. Although some people think that depression often accompanies menopause, this link does not exist. Recent psychiatric studies show that depression occurs no more often during this time of a woman's life than at any other.

The vagina

The uterus is shaped somewhat like a pear: wider at its top and narrower at its bottom. The narrower portion, called the **cervix** (or neck), opens into the vagina. The vagina, a muscular tube-like passageway to the exterior of the body, is an organ with three functions: it accepts the penis during sexual intercourse and is the place where sperm are deposited, it is the lower portion of the birth canal during childbirth, and it provides a passageway for the exit of the menstrual flow. Although only about 9 centimeters (3½ inches) in length, the vagina can stretch considerably during intercourse and childbirth because its walls are composed of elastic-like smooth muscle tissue. The mucous membrane that lines the inside of the vagina produces a viscous fluid that keeps the vagina moist and provides lubrication during intercourse.

Bacteria that normally reside in the vagina use the large amounts of glycogen in the vaginal secretion as a food source. As a by-product of the digestion of this glycogen, these "normal" vaginal bacteria produce acid, which accounts for the acidity of the vagina so harmful to sperm. However, the ejaculate contains substances that neutralize this acidity (see p. 338). The presence of these bacteria have a positive effect in that they discourage the growth of unwanted bacteria and yeasts. Often, when a woman takes antibiotics to kill bacteria infecting another part of her body, the antibiotic also kills the bacteria residing in the vagina. If these bacteria die, yeast cells or other bacteria can begin to grow. A vaginal infection can be a troublesome side effect of antibiotic therapy in women, since it can produce itching, discomfort, and discharges from the vagina.

The external genitals

The external genitals of a female are collectively called the **vulva.** Interestingly, parts of the vulva are homologous to the external genitals of the male, which means that the organs have a similar structure and anatomical position and originated from the same tissues during embryonic development. The most anterior structure is the **mons pubis** (Figure 20-10). The word *mons* comes from a Latin word meaning "prominence" and refers to the mound of fatty

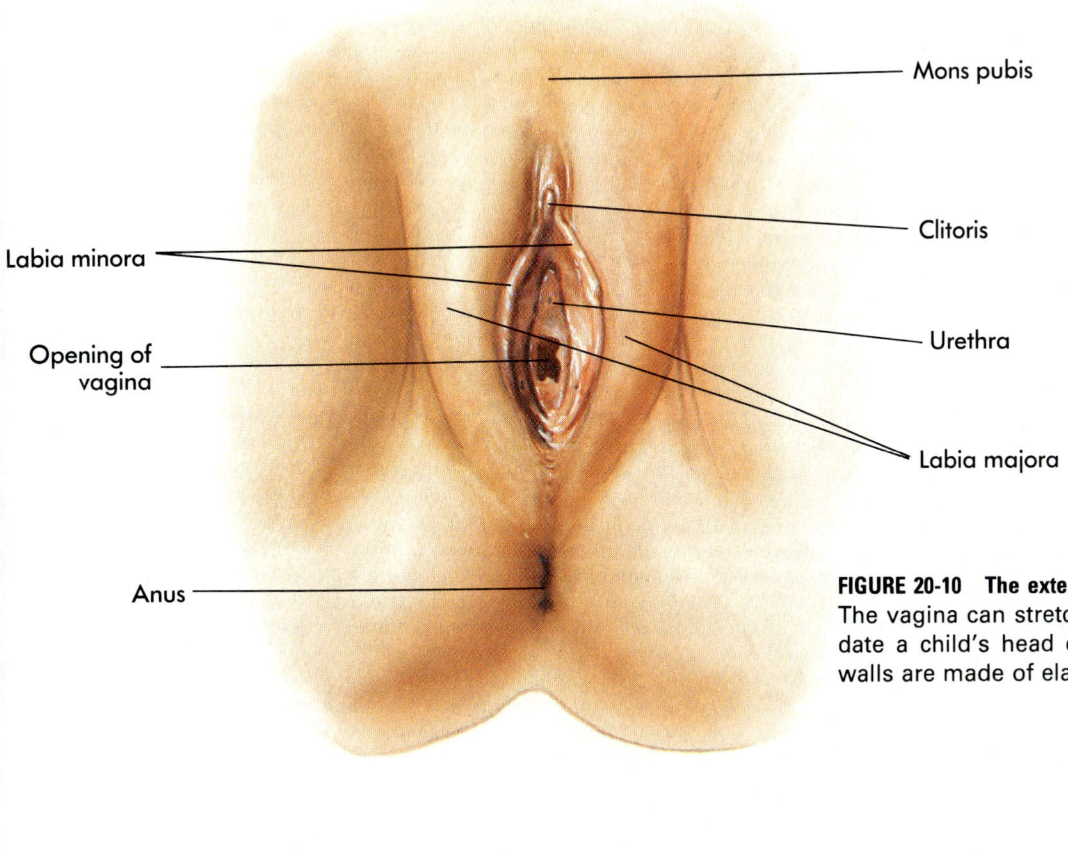

FIGURE 20-10 The external genitals of the female. The vagina can stretch considerably to accommodate a child's head during childbirth because its walls are made of elastic muscle tissue.

BIOLOGY & *you*

AIDS on the college campus

As most college students now know, AIDS is a serious disease that is rapidly becoming common in the United States and around the world. The disease, first reported in 1981, is transmitted by a virus and is always fatal. Over 200,000 individuals in the United States had contracted AIDS by 1991, and nearly 50 Americans will die each day of AIDS this year. The World Health Organization estimates that worldwide, over 5000 people become infected with the virus that causes AIDS *each day*. Few, if any, of these people will escape the disease.

The name *AIDS* is short for acquired (transmitted from another infected individual) immunodeficiency (a breakdown of the body's ability to defend itself against disease) syndrome (a spectrum of symptoms). The disease is caused by HIV, or human immunodeficiency virus, which attacks and destroys T cells, a key component in the body's immune system (See chapter 13). Without T cells, the immune system will not work. AIDS is fatal because no one can survive for long without an immune system to defend against viral and bacterial infections and to ward off cancer.

AIDS is not the only fatal disease to threaten humans, and it is not the most contagious. What makes AIDS an unusually serious threat is that the virus causing the disease does not show its effect immediately on infection. Recently infected people usually show no symptoms of the disease at all. Only much later, typically 2 to 9 years, does the virus begin to multiply and attack the immune system. During these years, however, the infected person is an unknowing carrier, able to transmit the virus to others. It is the large reservoir of undiagnosed, infected individuals that casts such a shadow over the future. In November of 1991, Magic Johnson, the Los Angeles Laker's star basketball player, revealed to the public that he had contracted the AIDS virus (Figure 20-A, *1*). He learned of his infection through a standard test administered for insurance purposes. Magic has stated that it frightens him to think what would have happened if he hadn't taken the physical required for his insurance and had been an unsuspecting carrier of HIV.

Current estimates of the number of individuals in their twenties and thirties who are infected with the virus in the United States vary, but most estimates are about 1 million. This statistic suggests both a staggering load of future suffering that will be difficult to avoid and a great danger that the infection will spread further. It is folly to assume that college campuses will escape infection—every student should face that fact squarely. A major step toward arming yourself against AIDS is through

1 **2**

AIDS cases by exposure category (percentages of total AIDS cases)

	Homosexual/ bisexual male	IV drug user	IV drug user/ homosexual male	Hemophilia	Heterosexual contact	Transfusion, blood components	Undetermined
1981	76.2	—	—	—	—	—	—
1982	74.6	—	—	—	—	—	—
1983	71.4	17.2	—	0.6	—	—	6.3
1984	73.0	17.0	—	1.0	1.0	1.0	4.0
1985	66.6	17.3	7.1	0.9	3.3	1.9	2.9
1986	65.7	16.4	7.3	0.9	3.8	2.2	3.6
1987	65.0	17	7.0	1.0	4.0	3.0	3.0
1988	57.0	24	7.0	1.0	5.0	3.0	3.0
1989	58.0	23	6.0	1.0	6.0	2.0	4.0
1990	56.0	24	5.0	1.0	6.0	2.0	6.0

FIGURE 20-A
1 Magic Johnson, former star player for the Los Angeles Lakers, learned in 1991 that he had been infected with HIV.
2 This table displays the percentages of total AIDS cases by exposure category for the years 1981 through 1990. Much of the information is incomplete for the years 1981 through 1984, which was before the exposure categories were made consistent by the Centers for Disease Control and methods for collecting data were formalized. Despite the gaps, it is easy to see that the percentages of AIDS cases among IV drug users and heterosexual persons have climbed steadily since 1984.

education about the disease and how it is transmitted. In the following paragraphs, some of the most common questions about AIDS will be answered, and important information about AIDS testing will be given.

How is AIDS transmitted? HIV is a fragile virus that does not survive outside body cells. It is present in the body fluids of infected individuals, notably in blood, semen, and vaginal fluid. HIV is transmitted from one individual to another when body fluid is transferred from an infected individual. You cannot catch AIDS from a toilet seat, from a hot tub shared with an AIDS victim, from kissing an AIDS carrier, or by being bitten by a mosquito that bit an AIDS victim. In studies of 619 households of AIDS victims, not one family member contracted the AIDS virus.

The *only* way you can become infected with HIV is to come into contact with the body fluids of an infected person. Among college students, there are two important routes of transmission:

1. Sexual intercourse: Both semen and vaginal fluid of infected individuals have high levels of HIV. This means that vaginal, anal, and oral sex with an infected individual can all transmit the virus successfully—and to either sex. Absorption across the vaginal wall or into the penis offers a ready means for the virus to enter the body; the tiny tears produced during anal sex facilitate entry even more. The microscopic abrasions in the mouth that everyone has from eating and chewing are a third easy means of entry, making oral sex also dangerous.

   A safer way to have sex with an infected individual is to use a condom and use it correctly. When used correctly, condoms offer the best available protection from infection. It is important, however, that they be used properly. It is not wise to mix alcohol or drugs with sexual encounters; they may cloud your judgment and lead you to do things you wouldn't do with a clearer head, such as forgetting to use a condom or using it carelessly. It is a mistake that could cost you your life.
2. Drug use: A needle used more than once by an infected individual typically harbors large quantities of HIV, both in the fluid that remains behind in the needle and in the body of the hypodermic syringe. Anyone else who reuses the needle may become infected. The use of intravenous drugs is itself dangerous—both illegal and life threatening—but if you engage in such dangerous practice, do not compound the damage by sharing needles or syringes.

Who is at risk? You are. AIDS is commonly perceived as a disease of homosexual men because the disease first appeared in the United States among the gay community. Because homosexuals tend to confine their sexual interactions to one another, the disease initially spread among homosexuals without entering the larger heterosexual community. That initial segregation appears to be ending: although only 5% of the AIDS cases diagnosed in 1990 were heterosexual nondrug users, the incidence of HIV among heterosexuals is now increasing (Figure 20-A, *2*). Estimates vary widely. the US Public Health Service estimates that between 1.5 million and 2 million people in the United States now harbor the virus. In Africa, where the virus first infected humans several years before it spread to this country, the epidemic has proceeded further than in the United States, even though homosexuality is rare there. In Africa, sexual transmission of AIDS is almost exclusively heterosexual. In some central African countries, as many as 10% of adults are thought to carry HIV.

Can people find out if they have been infected with HIV? Yes, easily. A simple test identifies infected individuals by detecting antibodies in their blood directed at HIV. Such antibodies are detectable only in the bodies of infected individuals. They are the remnants of the body's attempt to ward off the HIV infection.

How does an AIDS test work? The standard test for detecting the presence of antibody directed against HIV is called the *ELISA antibody test.* In the test, about 5 milliliters (roughly the amount of blood that would fit in a drinking straw) is drawn and checked to see whether any antibodies are present that will interact with bits of HIV attached to the surface of a plastic dish. If such antibodies are present, the test is said to be positive.

Who offers AIDS testing? The Red Cross offers an inexpensive walk-in AIDS antibody test at a variety of locations in most metropolitan areas. The Centers for Disease Control also maintains a national AIDS hotline that you can contact for information and advice by calling 1-800-342-AIDS.

Are the test results confidential? At the Red Cross the testing is completely confidential. The test is coded with a number that you select—the test is not associated with your name—and you telephone a few days later for the result of the test identified with that number. But maintaining the confidentiality of a positive result is very difficult, even though the test itself is completely confidential. When an individual does have a positive AIDS test, it is necessary to see a doctor frequently to monitor possibilities of disease progression. All this has to be documented in the physician's office, along with billing information. Because these files may be accessed when you sign a medical release form or pay bills with insurance, it is impossible to maintain confidentiality over a long period.

Can a person with AIDS survive? The only way to survive AIDS is not to contract it. The only way (short of contaminated needles) that any student will contract AIDS is by having sex with someone who has the virus. (Protected sex makes contact *safer* but not totally safe.) But the 2- to 9-year lag prevents anyone from knowing who is infected and who is not, and the disease continues to spread. Although widespread antibody testing on college campuses to identify infected individuals and so dampen the spread of the virus while it is still confined to relatively few individuals has been proposed by some, the proposal is controversial, and the associated dangers of invasion of privacy are a real concern. In the absence of such wide-scale on-campus testing, any student you have sex with might be a carrier and not know it. To avoid becoming part of the epidemic, you have to accept the responsibility of protecting your health.

tissue that lies over the place of attachment of the two pubic bones. Two longitudinal folds of skin called the **labia majora** (*labia* means "lips") run posteriorly from the mons. These folds are homologous to the scrotum in the male. Covered by the labia majora are additional folds of skin called the **labia minora.** Both sets of skin folds protect the vaginal and urethral openings beneath. Figure 20-10 shows that the urethra opens to the outside slightly in front of the vagina and is not connected to the reproductive system as in males. In addition to the vaginal and urethral openings, the labia minora also covers the openings of several tiny ducts that produce a mucus-like secretion during sexual stimulation. This secretion helps lubricate the vagina and penis during sexual intercourse. Slightly anterior to the convergence of the labia minora is the **clitoris,** a small mass of erectile and nervous tissue that responds to sexual stimulation. The clitoris is homologous to the tip of the penis in males.

The mammary glands

The **mammary glands,** or milk-producing glands, lie over the chest muscles. Each mammary gland is made up of 15 to 20 lobes that are separated by fat. The amount of fat determines the size of the breast and is not an indicator of a woman's ability to produce milk. Ducts extend from the glands to the nipple of the breast.

The mammary glands are secondary sexual characteristics that begin to develop in adolescent girls as the hormonal changes of puberty take place. These changes are initiated as the ovary begins to produce estrogens and accelerates as the menstrual cycle begins and the ovaries produce even higher blood levels of estrogens.

The function of the mammary glands is lactation, the production and secretion of milk. Prolactin, a hormone secreted by the anterior pituitary during pregnancy, is the hormone primarily responsible for stimulating the production of milk. As an infant nurses, its sucking sends nervous impulses to the hypothalamus. These "messages" stimulate the continued production of prolactin and therefore of milk. In addition, the nervous impulses trigger the hypothalamus to produce oxytocin, a hormone that is released by the posterior pituitary and that causes milk to be ejected into the ducts of the mammary glands. This process is called *milk letdown.*

The sexual response

The sexual act is referred to by a variety of names, including *intercourse, copulation,* and *coitus,* as well as a host of more informal ones. The physiological events associated with sexual intercourse can be described as having four phases, although the division of the sexual response into phases is somewhat arbitrary. These phases are excitement, plateau, orgasm, and resolution (Figure 20-11).

The *excitement phase* refers to the sexual activity that precedes intercourse. This phase therefore varies in length depending on the needs and desires of the sexual partners. During excitement the heartbeat, blood pressure, and rate of breathing increase. In addition, the blood vessels of the face, breasts, and genitals widen, producing a reddening of the skin in these areas called the *sex flush.* The nipples commonly harden and become more sensitive. In the genital area, increased circulation leads to an erection in the male and a swelling and parting of the labia in females. The vaginal walls become moister and the muscles of the vagina relax.

The act of sexual intercourse signals the beginning of the *plateau phase,* a period of intensifying physiological changes initiated in the excitement phase. During sexual intercourse the penetration of the vagina by the thrusting penis results in the repeated stimulation of nerve endings both in the tip of the penis and in the clitoris. The clitoris, which is now swollen, becomes very sensitive and withdraws up into a sheath, or "hood." Once it has withdrawn, the stimulation of the clitoris is indirect, with the thrusting movements of the penis rubbing the clitoral hood against the clitoris. This type of stimulation of the clitoris is insufficient for many women, however, who require a more direct stimulation of this organ.

The climax of intercourse (*orgasm*) is reached when the stimulation is sufficient to initiate a series of reflexive muscular contractions. The nerve signals producing these contractions are associated with other nervous activity within the central nervous system, activity that you experience as sexual pleasure. In females the contractions are initiated by impulses in the hypothalamus that cause the pituitary to release large amounts of the hormone oxytocin. This hormone, in turn, causes the muscles in the uterus and around the vaginal opening to contract and the cervix to be pulled upward. There may be one or several intense peaks of con-

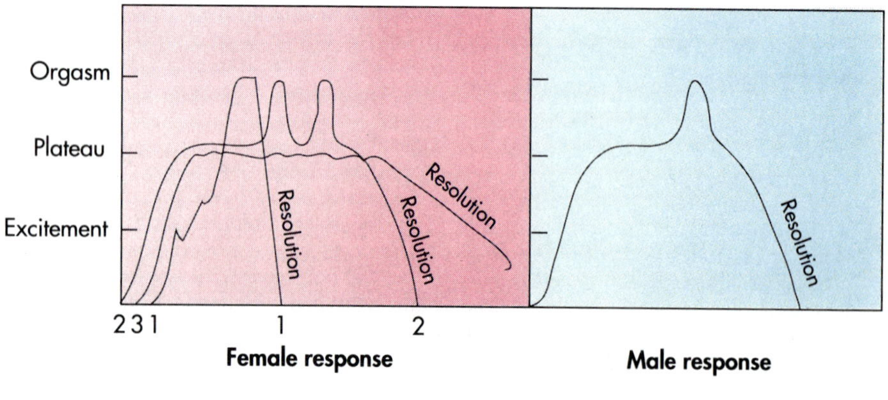

FIGURE 20-11 The human orgasmic response.
Among females, the response is highly variable. It may be typified by one of the three patterns illustrated here. Among males, the response is not as variable.

tractions (orgasm) or the peaks may be more numerous but less intense (see Figure 20-11).

Analogous contractions occur in the male, initiated by nerve signals from the brain. These signals first cause emission, in which rhythmic peristaltic contractions of the vas deferens and the prostate gland cause the sperm and seminal fluid to move to the base of the penis. Shortly after, nerve signals from the brain induce contractions of the muscles at the base of the penis, resulting in the ejaculation of the semen from the penis.

The *resolution phase* begins after orgasm. The bodies of both men and women return slowly over a period of several minutes to their normal physiological states. After ejaculation, males rapidly lose their erection and enter a period lasting 20 minutes or longer in which sexual arousal is difficult to achieve and ejaculation is almost impossible. By contrast, women can be aroused again almost immediately.

Contraception and birth control

In most vertebrates, sexual intercourse is associated solely with reproduction. In humans, however, sexual behavior serves a second important function—the reinforcement of pair bonding, the emotional relationship between two individuals. Couples that wish to limit the pregnancies resulting from sexual intercourse may use one of many methods of family planning, or birth control (Table 20-1). Several different approaches are commonly taken to achieve birth control. These methods differ from one another in their effectiveness and in their acceptability to different couples. Some of them are shown in Figure 20-12. In addition, some methods of birth control also protect against sexually transmitted diseases.

Abstinence

The most reliable way to avoid pregnancy and sexually transmitted diseases is not to have sex at all. A variant of this approach (for birth control purposes only) is to avoid sexual relations on the three days preceding and the day after ovulation, since this is the only time during which fertilization can occur. This approach, called the **rhythm method,** is satisfactory in principle but difficult in application, since ovulation is not easy to predict and may occur unexpectedly. The effectiveness of the rhythm method is low. However, a woman can use techniques that involve taking her basal body temperature or analyzing her cervical mucus to help determine when ovulation takes place. Home test kits are also available that detect the time of ovulation. (These methods can be used to help a couple conceive a child as well as to avoid conception.) As an added birth control measure, some couples using the rhythm method of birth control conservatively avoid sexual intercourse many days before and a few days after ovulation. The failure rate of the rhythm method is estimated to be 13% to 21% (13 to 21 pregnancies per 100 women practicing the rhythm method, per year).

Another variant of this approach is to withdraw the penis before ejaculation. This method is as unreliable as the

A
B
C
D
E

FIGURE 20-12 Five common methods of birth control.
A Condom.
B Foams.
C Diaphragm and spermicide.
D Oral contraceptives.
E Contraceptive sponge.

TABLE 20-1

Methods of birth control*

DEVICE	ACTION	FAILURE RATES (%)	ADVANTAGES	DISADVANTAGES
Rhythm method	Avoidance of sexual intercourse during woman's fertile period	13-21	Contraceptive devices unnecessary, agreeable to some religious groups	High failure rate due to difficulty of predicting ovulation, no protection against sexually transmitted diseases
Withdrawal	Removal of penis from vagina before ejaculation	9-25	Contraceptive devices unnecessary	High failure rate attributed to the secretion of sperm before ejaculation, self-control necessary
Condom	Thin rubber sheath for penis that collects semen	3-15	Easy to use, effective, and inexpensive; the most reliable protection against some sexually transmitted diseases with the exception of abstinence	Possible decrease in spontaneity, possible deterioration on the shelf
Diaphragm	Soft rubber cup that covers entrance to uterus, prevents sperm from reaching egg, and holds spermicide	4-25	No dangerous side effects, reliable if used properly; some protection against cervical cancer	Careful fitting required, some inconvenience associated with insertion and removal, possible dislodgment during sex
Cervical cap	Miniature diaphragm that covers cervix closely, prevents sperm from reaching egg, and holds spermicide	Probably comparable to diaphragm	No dangerous side effects, fairly effective, able to remain in place longer than diaphragm	Problems with fitting and insertion, limited number of sizes, not available in the United States
Sponge	Device acting as sperm barrier and releasing spermicide	15-30	Safe, easy to insert, possible to leave in place for 24 hours, no fitting needed	Relatively unreliable, only one size available, some sensitivity and removal problems, no use during menstruation

*Approximate effectiveness of these reversible methods of birth control is measured in pregnancies per 100 actual users per year.
†Due to recent lawsuits, no longer widely used in the United States.

DEVICE	ACTION	FAILURE RATES (%)	ADVANTAGES	DISADVANTAGES
Foams, creams, jellies, vaginal suppositories	Chemical spermicides inserted in vagina before intercourse that also prevent sperm from entering uterus	10-25	Possible use by anyone who is not allergic, no known side effects	Relatively unreliable, sometimes messy; necessary to use 5 to 10 minutes before each act of intercourse
Oral contraceptives (birth control pills)	Hormones, either in combination or progestin only, that primarily prevent release of egg	1-5, depending on type	Convenient and highly effective; significant noncontraceptive health benefits, such as protection against ovarian and endometrial cancers	Necessary to take regularly; possible minor side effects, which new formulations have reduced; not for women with cardiovascular risks, mostly those over 35 who smoke
Implant (Norplant)	Capsules surgically implanted under skin that slowly release a hormone that blocks release of eggs	0.3	Very safe, convenient and effective; very long lasting (5 years); possible nonreproductive health benefits like those of oral contraceptives	Irregular or absent periods, necessity of minor surgical procedure to insert and remove
Intrauterine devices	Small plastic or metal devices placed in the uterus that somehow prevent fertilization or implantation; some contain copper, others release hormones	1-5	Convenient, highly effective, infrequent replacement	Possible excess menstrual bleeding and pain; danger of perforation, infection, and expulsion; not recommended for those who are childless or not monogamous, risk of pelvic inflammatory disease or infertility; dangerous in pregnancy†

rhythm method, with a failure rate estimated between 9% and 25%. The reason for this high failure rate is that the penis can secrete prematurely released sperm within its lubricating fluid and a second sexual act may transfer sperm ejaculated earlier.

Sperm blockage

If sperm do not reach the uterine tubes, fertilization cannot take place. One way to prevent sperm from reaching the uterine tubes is for the male partner to use a **condom.** A condom is a sheath for the penis, constructed of thin rubber or other natural or synthetic materials. In principle, this method is easy to use and should be highly successful. However, in practice the failure fate of condoms is from 3% to 15%, primarily due to improper use or failure to use a condom consistently. Nevertheless, it is the most commonly used form of birth control in the United States. The use of latex condoms is also an excellent way to reduce the risk of contracting sexually transmitted diseases, including AIDS.

A second way to prevent the entry of sperm into the uterine tubes is to place a cover over the cervix. The cover may be a relatively tight-fitting **cervical cap** (not available in the United States at this time), which is worn for days at a time, or a rubber dome called a **diaphragm,** which is inserted immediately before intercourse. In addition, the vaginal side of a diaphragm is covered with a spermicide before insertion and is then kept in place for a minimum of six hours after intercourse. Because the dimensions of individual cervices vary, a cervical cap or diaphragm must be fitted by a physician. Failure rates average from 4% to 25% for diaphragms and are somewhat lower for cervical caps. These high failure rates are due, in part, to incorrect use of these birth control devices. Neither form of birth control is effective against sexually transmitted diseases.

Sperm destruction

The spermicidal (sperm-killing) sponge is inserted and used much like a diaphragm, but it is composed of a sponge-like material impregnated with a spermicide. The sponge provides somewhat of a barrier to the uterus but also absorbs the sperm into the sponge to be killed there. Sperm can also be destroyed in the vagina by the use of spermicidal jellies, creams, or foams applied immediately before intercourse. The failure rate of these methods varies widely from 10% to 25%, and they are not useful in the prevention of sexually transmitted diseases.

FIGURE 20-13 The contraceptive implant: a new method of birth control for women.
The tubes shown here are implanted in a woman's upper arm in a simple surgical procedure. Once implanted, the tubes release a constant, low-level flow of progestin that prevents ovulation. The implant is effective for about 5 years.

Prevention of egg maturation

Since about 1960, a widespread form of birth control has been the use of **birth control pills** by women. These pills contain estrogen and progesterone, either taken together in the same pill or in separate pills taken sequentially. In the normal reproductive cycle of a female, these hormones act to shut down the production of the pituitary hormones FSH and LH. The artificial maintenance of high levels of estrogen and progesterone in a woman's bloodstream fools the body into acting as if ovulation had already occurred, when in fact it has not. The ovarian follicles do not mature in the absence of FSH, and ovulation does not occur in the absence of LH. For these reasons, birth control pills provide a very effective means of birth control, with a failure rate of 0% to 10%. They play no role in the prevention of sexually transmitted diseases, but research shows that the synthetic progesterone (progestin) in the pill helps ward off cancer of the uterus and ovaries. Scientists are concerned, however, that the estrogen in the pill may increase the risk of breast cancer. Medical researchers are now experimenting with a pill that includes LH-releasing hormone. This pill is currently undergoing clinical trials. Another, more highly experimental pill, substitutes melatonin for the estrogen but is not yet approved for testing.

In 1990, a product that works in a manner similar to the pill but that prevents pregnancy for up to 5 years was introduced in the United States. This product is called Norplant (Figure 20-13). In a 15-minute operation in the doc-

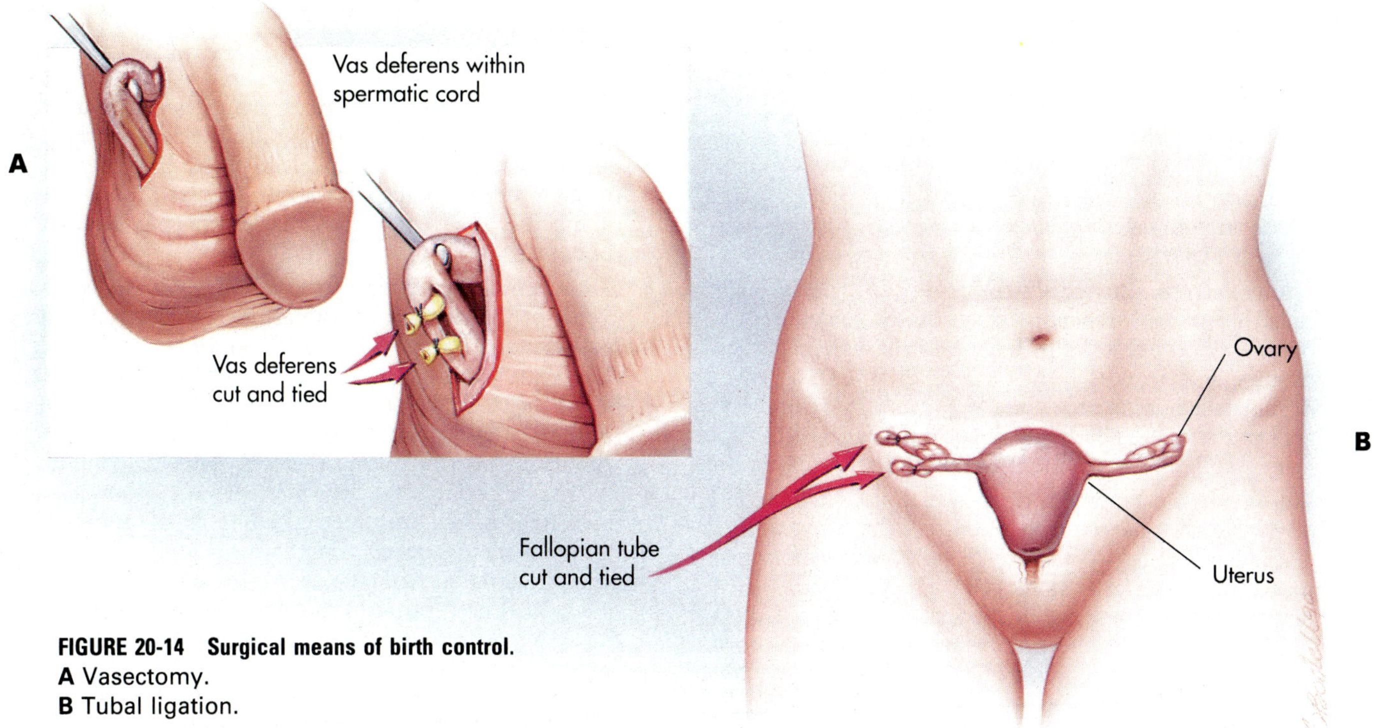

FIGURE 20-14 Surgical means of birth control.
A Vasectomy.
B Tubal ligation.

tor's office, the six matchstick-sized tubes are surgically implanted under the skin of the upper arm. These tubes contain the synthetic hormone progestin, the active ingredient in most birth control pills. Norplant prevents pregnancy by inhibiting ovulation as the pill does, but it also thickens the cervical mucus so that sperm cannot move through the female reproductive tract. Although 99.8% reliable, Norplant causes irregular menstrual bleeding in 75% of its users.

Surgical intervention

A completely effective means of birth control is the surgical removal of a portion of the tube through which gametes move from one reproductive structure to another (Figure 20-14). In males, such an operation involves the removal of a portion of the vas deferens, the tube through which sperm travel to the penis. This procedure, called a **vasectomy,** can be carried out in a physician's office. Reversing a vasectomy is difficult but can sometimes be done. In females the comparable operation involves the removal of a section of each of the two uterine tubes through which the oocyte travels to the uterus. Since these tubes are located within the abdomen, the operation, called a **tubal ligation,** is more difficult to perform than a vasectomy and is even more difficult to reverse.

Among methods of birth control, the rhythm method and withdrawal are not highly reliable. Condoms and diaphragms are effective when used correctly, but mistakes are common. Birth control pills and implanted hormonal tubes are very effective. Vasectomies and tubal ligations are completely effective, although usually permanent.

Summary

1. Sexual reproduction is the process whereby a male and a female sex cell combine to form the first cell of a new individual. The male and female reproductive systems are made up of organs and ducts that produce, store, transport, and receive the sex cells. Accessory glands help nourish the sex cells.
2. The male reproductive system produces male sex cells called *sperm* and transports them to the female reproductive tract. Sperm cells develop in the tightly packed tubules of the testes, which are covered with a pouch of skin called the *scrotum* and hang from the lower pelvic area of the male. Sperm production is triggered by follicle-stimulating hormone (FSH).
3. During the development of sperm, diploid cells called *spermatogonia* give rise to haploid cells called *spermatids*. These haploid cells contain half the normal amount of hereditary material of body cells. The spermatids then develop into sperm cells, or spermatozoa, which have a head containing the hereditary material, a vesicle containing enzymes to penetrate the oocyte membrane, and a whip-like tail. They mature in the epididymis, a long, coiled tube that sits on the back side of the testes.
4. The accessory glands of the male reproductive system secrete a fluid that combines with the sperm during ejaculation. This alkaline fluid nourishes the sperm and neutralizes the acidity of the female vagina. During ejaculation, sperm are propelled through the vas deferens to the urethra and out the penis.
5. The female reproductive tract produces one potential egg (secondary oocyte) each month. Each potential egg develops in the ovaries, travels to the uterus via one of the two uterine tubes, and implants in the inner lining of the uterus if fertilized while journeying down the tube.
6. The reproductive cycle of the female encompasses the events of the maturation of each potential egg, or primary oocyte, its release from the ovary as a secondary oocyte, its journey to the uterus, and the "healing" of its follicle. Certain primary oocytes in the ovaries continue the development they began before birth when stimulated by the hormone FSH. Their surrounding cells start producing estrogens, which promote the thickening of the endometrial lining of the uterus. Of the developing oocytes, one matures into a secondary oocyte that bursts from the ovary as a surge of LH occurs midcycle. The ruptured follicular cells that surrounded the oocyte secrete estrogens and progesterone. If fertilization of the oocyte does not take place, the progesterone and estrogens maintain the endometrial lining until diminished levels of these hormones cause it to degenerate. A sloughing off of the endometrial lining follows, which is referred to as *menstruation*.
7. The human sexual response has four physiological periods: excitement, plateau, orgasm, and resolution. Orgasm in women is variable and may be prolonged. Orgasm in men is more uniform and abrupt; it coincides with the ejaculation of sperm.
8. Humans practice a variety of birth control procedures; men using condoms and women using birth control pills are the most common.

KEY TERMS

acrosome 336
bulbourethral glands 338
cervical cap 350
cervix 343
clitoris 346
condom 350
corpus luteum 341
diaphragm 350
endometrium 341
epididymis 338
estrogens 341
follicle 339
follicular cell 339
gametes 335
gonads 335
human chorionic gondaotropin (HCG) 341
labia majora 346
labia minora 346
mammary glands 346
menopause 343
menstruation 342
mons pubis 343
oogenesis 339
oogonia 340
ovary 339
ovulation 340
ovum 340
penis 338
polar body 340
primary oocyte 339
prostate 338
puberty 336
rhythm method 347
scrotum 335
secondary oocyte 340
semen 338
seminal vesicles 338
seminiferous tubules 336
spermatids 336
spermatogenesis 336
spermatogonia 336
spermatozoon (spermatozoa) 335
testes 335
tubal ligation 351
urethra 338
uterine tube 340
uterus 340
vagina 339
vas deferens 338
vasectomy 351
vulva 343
zygote 340

REVIEW QUESTIONS

1. What is sexual reproduction?
2. Fill in the blanks: In humans, the male sex cells (__________) are produced in the gonads (__________), which are housed in a sac called the ______. Diploid cells called __________ produce daughter cells, which in turn produce haploid cells called __________ that develop into the sex cells.
3. Discuss the roles of the hormones involved in the development of male gametes and secondary sexual characteristics.
4. Relate the structure of a spermatozoon to its function.
5. What are the seminal vesicles, prostate gland, and bulbourethral glands? Describe the contribution made by each during ejaculation.
6. Explain why women over 35 years of age are more likely to bear children with developmental abnormalities.
7. Fill in the blanks: During ovulation the secondary __________ breaks away from the ovary and is carried in the __________ __________ to the __________. If fertilization occurs, the nuclei of the secondary oocyte and the __________ fuse to produce a __________, the first cell of a new individual.
8. Summarize the reproductive cycle of a human female when fertilization does not occur. What hormones are involved?
9. Summarize what happens during the reproductive cycle of the human female if fertilization occurs. What hormone is secreted?
10. Summarize the four phases of the sexual response.
11. Which methods of birth control can help prevent the spread of sexually transmitted diseases?
12. Explain how birth control pills prevent conception. What hormones are involved?
13. Compare the effectiveness of the various birth control methods discussed in the chapter. Which are most effective? Least effective?

THOUGHT QUESTIONS

1. Among all vertebrates, only primates are sexually receptive throughout the sexual cycle. Why do you think this is so?
2. Female gametes begin their development prenatally, so at age 20, a woman's gametes are all 20 years old. A male's gametes develop all during his life, so at age 20 a man's gametes are only a few days old. What are some implications inherent in this difference?

FOR FURTHER READING

Aitken, R. J., & Paterson, M. (1988, October 6). New horizons in contraception. *Nature,* pp. 492-493.
Research into contraceptive vaccines has yielded successful preliminary results.

Findlay, S. (1990, December 24). Birth control. *US News & World Report,* pp. 58-64.
This comprehensive article details the latest birth control options for Americans, with special attention paid to the new implant.

Leishman, K. (1987, February). Heterosexuals and AIDS. *The Atlantic,* pp. 39-58.
This is a chilling account of the difficulty of modifying sexual behavior, despite the knowledge of the dangers associated with AIDS.

Ulmann, A., Tuetsch, G., & Philibert, D. (1990, June). RU486. *Scientific American,* pp. 42-48.
This article covers the history of this controversial drug, which in conjunction with prostaglandins, can induce early abortion. The article also covers other uses of RU486, including the treatment of certain cancers.

21 Human development before birth

The image of a mother nursing, rocking, and cuddling her baby has been portrayed in art for hundreds of years. But scientists have only recently validated that these activities are not only emotionally good for moms and babies, they help babies grow properly. Cuddling and caressing an infant stimulates respiration, blood flow, and growth rate. The baby is healthier and cries less. For these reasons, volunteers at Duke University Medical Center cuddle hospital-bound premature infants whose parents must work and are unable to be with them continually. Interestingly, gentle rocking of an infant stimulates the baby's cerebellum—the part of the brain that controls coordination. Rocking hastens the maturation of this control center. And while the baby is being cuddled and rocked, the breast milk is providing nourishment and antibodies that will help stave off disease until the baby's own defenses are produced. In fact, this image of peaceful nurturing is providing the baby with medical benefits that even the most high-tech pediatric unit would find hard to match.

HIGHLIGHTS

Life before birth, or prenatal development, lasts approximately 8½ months in humans and consists of the growth and changes that take place from conception until birth.

Conception takes place in the uterine tube when a sperm penetrates an ovum and the nuclear material of both fuse to form the first cell of a new individual.

Most of the body systems of the human develop during the first 9 weeks of pregnancy; the ensuing 29 weeks are primarily a time of growth and refinement of these organ systems and of the body form of the developing individual.

Birth takes place 38 weeks from conception. Hormonal changes in the fetus and the mother trigger waves of uterine contractions that push the fetus through the birth canal.

OUTLINE

Development

Fertilization

The first and second weeks of development: The pre-embryo

Early development of the extraembryonic membranes

Development from the third to eighth week: The embryo

- The third week
- The fourth week
- The fifth through eighth weeks

Development from the ninth to thirty-eighth week: The fetus

- The third month
- The fourth through sixth months
- The seventh through ninth months

Birth

Physiological adjustments of the newborn

Development

What happens after the birth of an infant is crucial to its survival. However, what happens before birth may be even more crucial. In fact, if an infant does not develop properly before birth, normal growth after birth becomes difficult and sometimes impossible. This chapter focuses on life before birth, or **prenatal development,** the gradual growth and progressive changes in a developing human from conception until the time the fetus leaves the mother's womb. This time of development is also called **gestation** and lasts approximately 8½ months. (The 9-month pregnancy calculation is computed from the first day of a woman's last menstrual period, but conception usually takes place in the middle of her cycle.) The gestation period is commonly referred to as *pregnancy*.

Fertilization

The union of a male gamete (sperm) and a female gamete (egg) is called **fertilization,** or conception. In the human, fertilization usually takes place in the uterine tube of the female (see Chapter 20 for a discussion of gamete formation and the events leading up to fertilization). Therefore development begins in the uterine tube.

Figure 21-1 shows that a human egg, or **ovum,** is a relatively large cell. In fact, ova are among the largest cells within an animal's body. Making up most of the substance of an ovum is the **yolk,** the nutrient material that the developing individual lives on until nutrients can be derived from the mother. In humans, the yolk granules provide nutrition for a week or so. In the case of birds, however, the young develop totally within the egg. Therefore the yolk of these eggs must be substantial enough to maintain the animal throughout its entire course of development. The yolk of an ostrich egg, for example, is so large that it is approximately the size of a baseball!

As mentioned in Chapter 20, the secondary oocyte breaks away from the ovary during ovulation in placental mammals (which includes people) and is swept into the uterine tube. This cell is not yet a mature ovum, or egg. Only after penetration of the oocyte by a sperm will the second meiotic division be completed, yielding the mature ovum and a second polar body. And, although fertilization may sound like an easy job for sperm, only 50 to 100 sperm make it to the egg from an ejaculate that contains approximately 200 to 300 million!

Many things can go wrong from the time the sperm are manufactured in the testes, stored in the epididymis, and travel out of the man's body and into the vagina of a woman. For example, up to 20% of sperm are deformed in some way. A common abnormality is having two flagella instead of one. Sperm with this abnormality do not have normal motility and cannot make the lengthy journey from the vagina, through the constricted cervix, up through the uterus, and along the uterine tube.

Although the acid environment of the female vagina is treacherous territory for sperm and although they constantly swim "upstream" against the downward currents of female secretions, they are also aided by the female body. During the time of ovulation, a female produces strands of a special protein called **mucin.** This substance becomes a part of the cervical mucus her body produces and provides thread-like highways along which the sperm can travel. The sperm make their way up these strands of mucin to the uterus. As they move along, enzyme inhibitors located at the tips of their heads are slowly worn away, gradually uncovering enzymes that are capable of penetrating the egg's outer protective layers and its membrane. Figure 21-2, *A* pictures a sperm that has penetrated an egg's jelly-like covering, the **zona pellucida.**

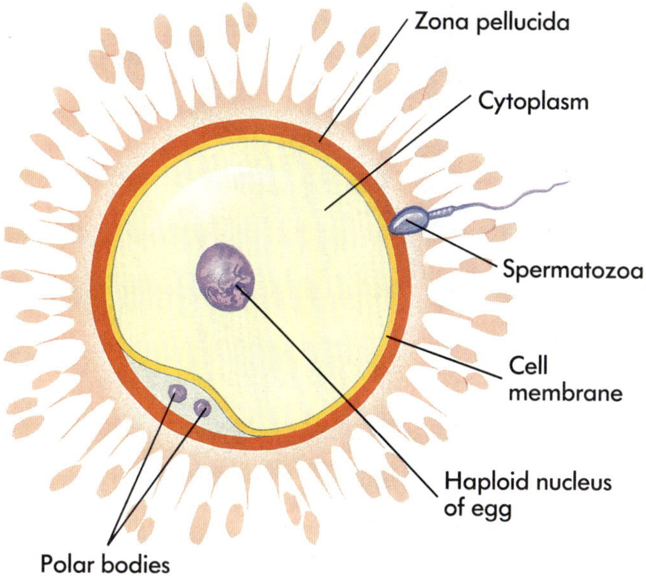

A

B

FIGURE 21-1 The egg.
A A human egg is surrounded by a membrane, is nourished by yolk, and contains a haploid nucleus.
B The membrane of this sea urchin egg has been penetrated by a sperm cell. Once the sperm cell enters the egg and reaches the nucleus, the fused cells begin to divide.

A

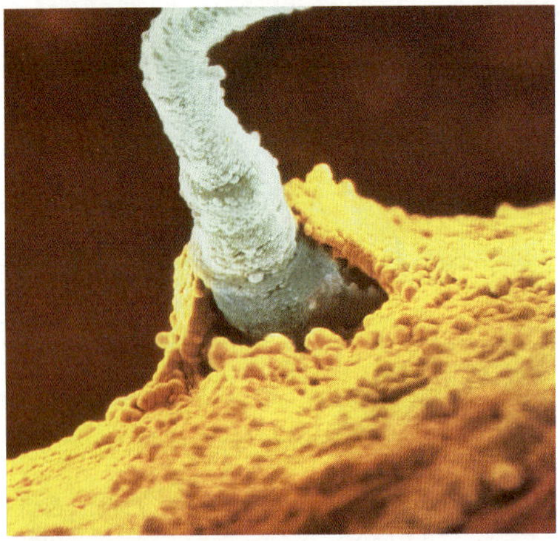

B

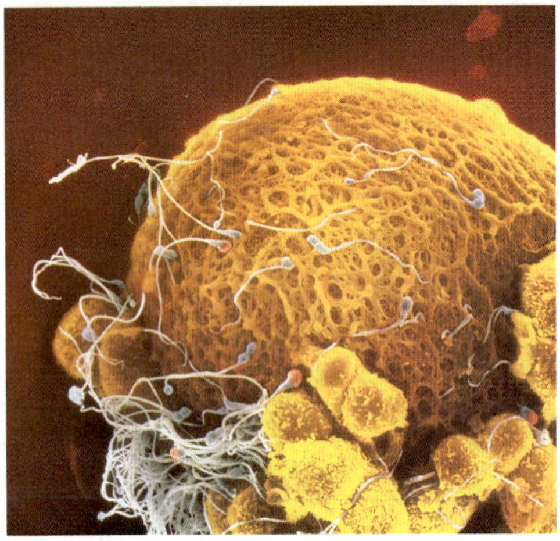

FIGURE 21-2 Fertilization.
The moment of penetration of a human egg by a sperm cell **(A)** initiates changes in the surface of the egg that prevent other sperm from entering **(B)**.

The moment of penetration of a human oocyte by a sperm is an event only recently captured on film. On entry of the cytoplasm of the oocyte, the sperm sheds its tail, and its head swells. Immediate changes in the surface of the egg allow no other sperm to penetrate, and oocyte meiosis is completed. Lastly, the sperm and ovum nuclei, clearly seen as separate from one another in Figure 21-3, *A,* fuse to form the nucleus of the first cell of a new individual (Figure 21-3, *B*). This new cell, which contains intermingling genetic material from both the mother and the father, is called a **zygote.**

The penetration of the secondary oocyte by the sperm is called *fertilization.* This penetration causes changes in the surface of the egg that allow no other sperm to enter and triggers the completion of meiosis. After the sperm sheds its tail, the nuclei of the egg and sperm fuse to form a zygote, the first cell of a new individual.

With the union of the chromosomal material of both gametes, the zygote begins to divide by mitotic cell division. As division proceeds during the next 2 weeks, the developing cell mass is referred to as a **pre-embryo.**

The first and second weeks of development: The pre-embryo

Within 30 hours, the one-celled zygote begins to divide rapidly: one cell into two, two into four, four into eight, and so forth, producing a cluster of cells. This process of cell division occurs without cell growth and is called **cleavage.** Occasionally, a single, fertilized ovum splits into two

A

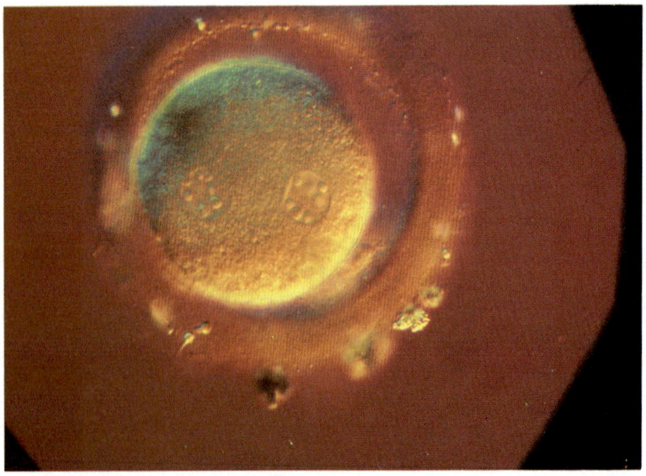

B

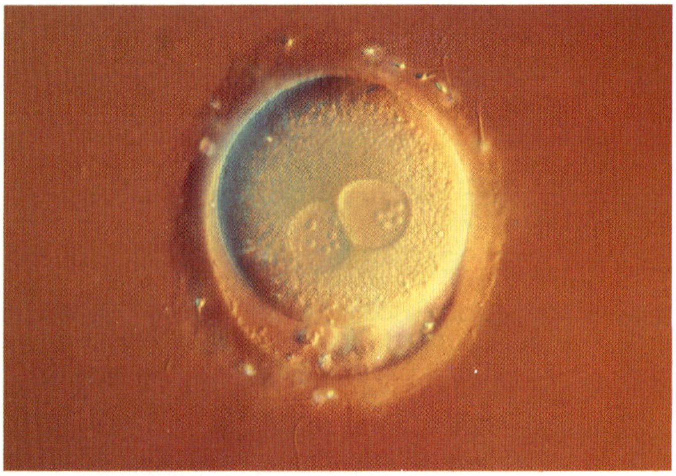

FIGURE 21-3 The fusion of nuclei.
A The nuclei of the sperm and ovum are clearly separate.
B The nuclei have merged to form the nucleus of the first cell of a new individual.

cell clusters during cleavage, and each cluster continues developing on its own. Since both cell clusters arose from the same cell, they have identical genetic information and will result in the development of identical twins. Identical twins are always the same sex. Fraternal twins arise from the release of two secondary oocytes and the fertilization of both. Fraternal twins are genetically different from one another because they are two separate oocytes fertilized by two different sperm cells. Therefore they may be different sexes.

As the cell cluster (pre-embryo) divides, it journeys along the remaining two thirds of the uterine tube. This trip takes approximately 3 days. Occasionally, a pre-embryo gets caught in the folded inner lining of the uterine tube and implants there, creating an **ectopic pregnancy.** The possibility of this situation occurring is increased if the tubes are scarred from previous infections. Acute pelvic pain usually signals this problem, which requires immediate medical attention. Normally, however, the pre-embryo reaches the uterus, or womb. Still the same size as a newly fertilized ovum, the pre-embryo now consists of about 16 densely clustered cells—a ball of cells called a **morula,** from a Latin word meaning "mulberry." This stage of development is common to all vertebrates and even invertebrates. In fact, the processes of development in all vertebrates are very similar to the course of human development before birth.

Cleavage, or cell division of the zygote without cell growth, is the first stage in the development of humans as well as other multicellular animals. It results in the formation of a mass of cells known as a *morula* approximately 3 days after fertilization.

The pre-embryo floats free in the uterus as its cells continue to divide. After 2 days in the uterus, the morula has developed into the **blastocyst,** a stage of development in which the pre-embryo is a hollow ball of cells. The center is filled with fluid from the uterine cavity. Like the term *morula,* the term *blastocyst* is descriptive, derived from two Greek words meaning "germ (germinal) sac."

One portion of the blastocyst contains a concentrated mass of cells destined to differentiate into the various body tissues of the new individual. It is referred to as the *inner cell mass.* Interestingly, each cell in this mass has the ability to develop into a complete individual. In fact, scientists have been able to produce test-tube mice using transplanted nuclei from inner cell mass cells. The outer ring of cells, called the **trophoblast** from a Greek word meaning "nutrition," will give rise to most of the extraembryonic membranes including much of the placenta, an organ that helps maintain the developing embryo (Figure 21-4).

The general term **blastula** is used to describe the saclike blastocyst of mammals and, in other animals, the stage that develops a similar fluid-filled cavity. This stage of development is significant for all vertebrates; it is the first time that cells begin to move, or migrate, to shape the new individual in a process called **morphogenesis.** However, the major morphogenetic events occur during the third to eighth weeks.

Approximately 1 week after fertilization, the blastocyst secretes enzymes that digest a microscopic portion of the uterus. It then nestles into this site, nourished by the digested uterine cells, in a process called **implantation** (Figure 21-5). Barely visible to the naked eye, the blastocyst most often attaches to the posterior wall of the uterus (Figure

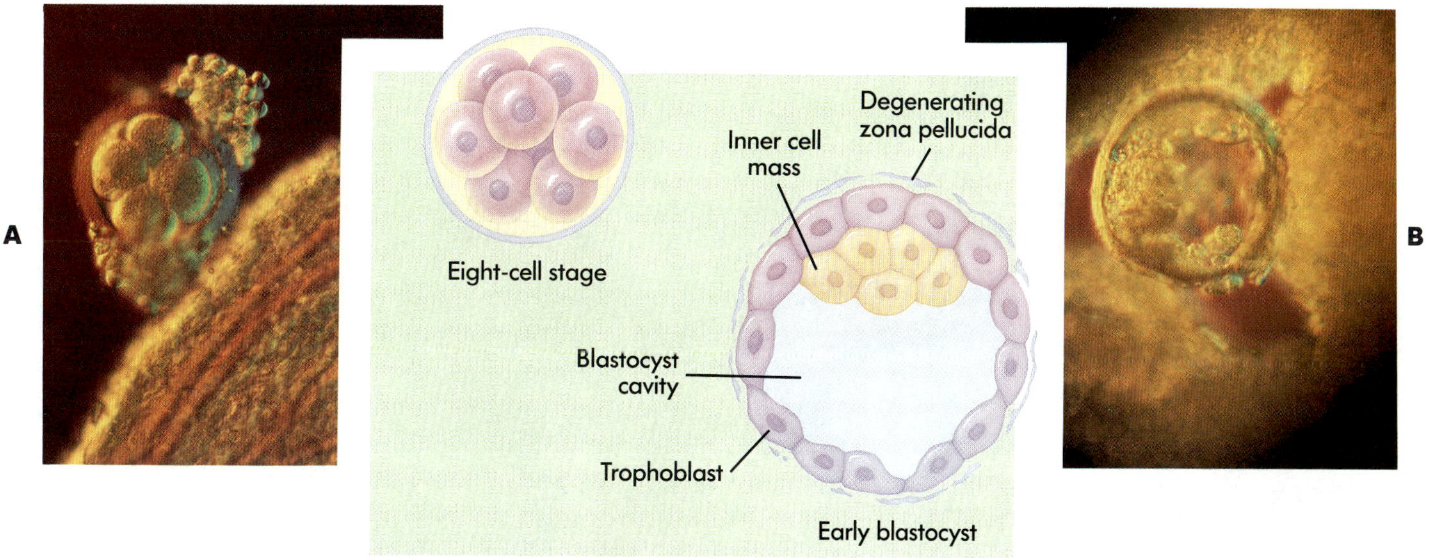

FIGURE 21-4 Development of the pre-embryo.
A View of the pre-embryo at the eight-cell stage. At this stage, the pre-embryo is floating free in the uterus as its cells continue to divide.
B After 2 days in the uterus, the ball of cells has developed into a blastocyst. The inner cell mass contains cells that will develop into the different body tissues of the new individual.

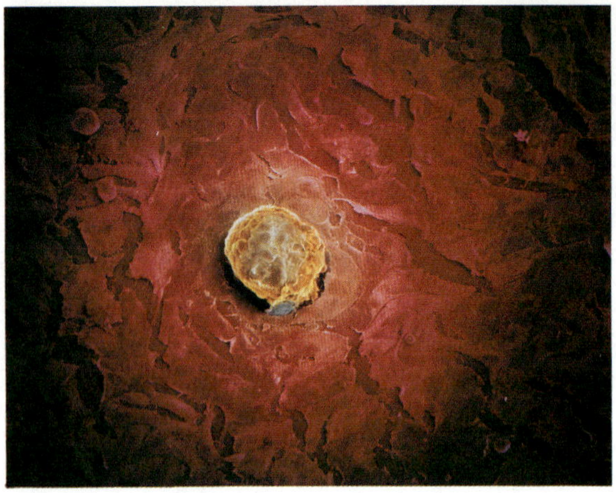

FIGURE 21-5 Implantation.
About 7 to 8 days after fertilization, the blastocyst imbeds itself into the wall of the uterus.

21-6). Figure 21-7 summarizes the events that take place from fertilization to implantation.

Cell migration, which begins to shape the pre-embryo, is the process that begins approximately 4 days after fertilization. Cell migration helps shape the developing individual in a process called *morphogenesis.* At this early stage, cell movement results in the formation of a hollow ball of cells known as a *blastocyst* in humans and in other mammals a stage more generally called the *blastula.*

By this time, a woman is nearing the end of her menstrual cycle (see Chapter 20), and the blastocyst is in danger of being swept away during menstruation. However, the blastocyst secretes a hormone called **human chorionic gonadotropin,** or **HCG,** which maintains the corpus luteum.

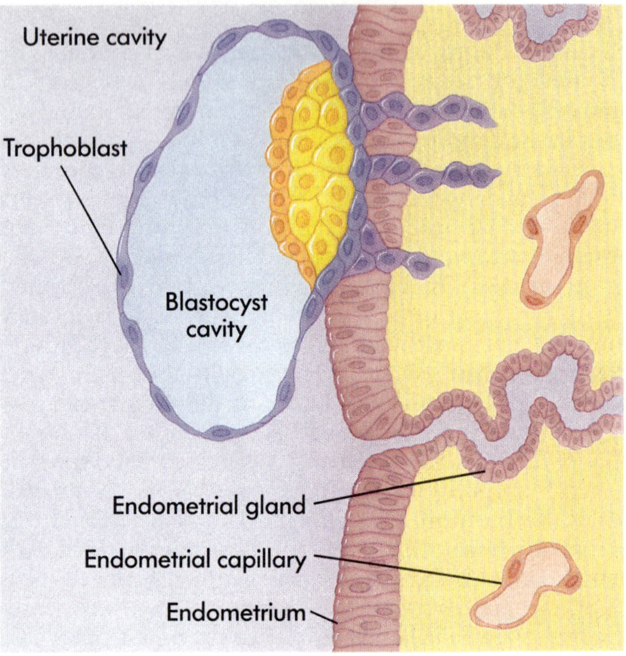

FIGURE 21-6 A detailed view of implantation.
Special finger-like projections produced by the trophoblast anchor the blastocyst to the wall of the uterus, or endometrium.

The corpus luteum is a structure derived from the ruptured follicle from which the egg was cast out of the ovary. Sustained by HCG, the corpus luteum continues to secrete progesterone and estrogen, hormones that maintain the uterine lining and allow the development of the pre-embryo to proceed. After the first 3 months of development, the placenta (see p. 360) begins to secrete the estrogens and progesterone that maintain the pregnancy.

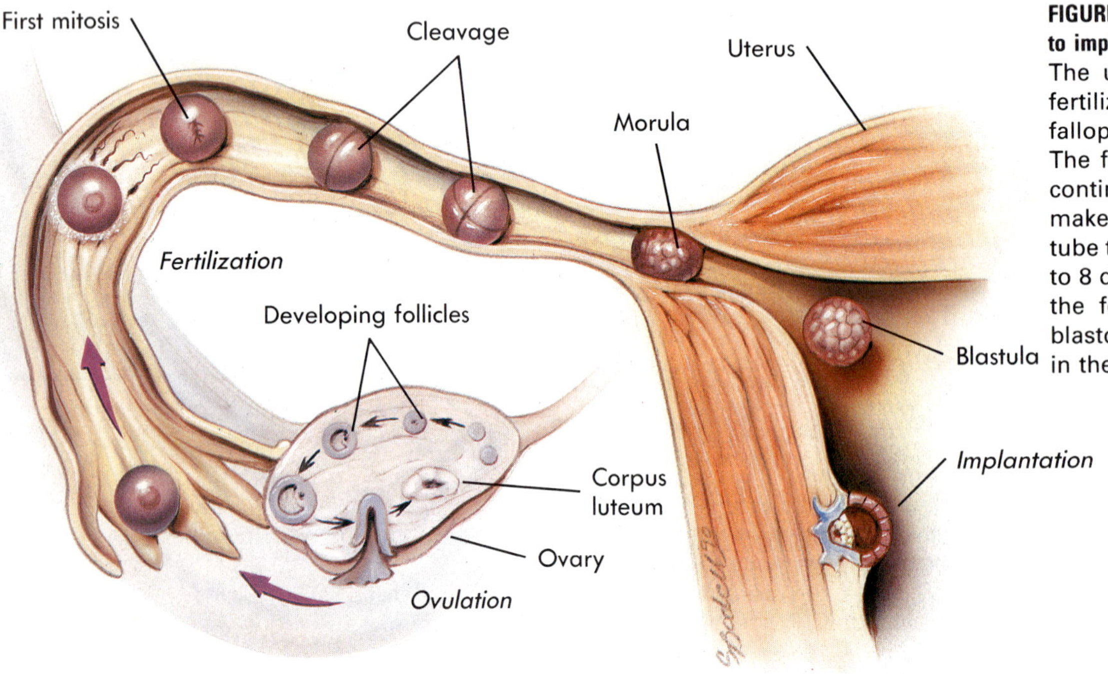

FIGURE 21-7 From fertilization to implantation.
The unfertilized oocyte is fertilized by a sperm in the fallopian (uterine) tube. The fertilized egg (zygote) continues to divide as it makes its way through the tube to the uterus. About 7 to 8 days after fertilization, the fertilized egg, now a blastocyst, implants itself in the wall of the uterus.

During its second week of development, the pre-embryo completes its implantation within the uterine wall, two of its three **primary germ layers** develop, and the extraembryonic membranes begin to form. The primary germ layers are three layers of cells that develop from the inner cell mass of the blastocyst and from which all the organs and tissues of the body develop. These three layers are called the *ectoderm* ("outside skin"), *endoderm* ("inside skin"), and the *mesoderm* ("middle skin"). The **ectoderm** forms the outer layer of skin, the nervous system, and portions of the sense organs. The **endoderm** gives rise to the lining of the digestive tract, the digestive organs, the respiratory tract, and the lungs; the urinary bladder; and the urethra. The **mesoderm** differentiates into the skeleton, muscles, blood, reproductive organs, connective tissue, and the innermost layer of the skin (Figure 21-8). Figure 21-9, *A* shows the completely implanted pre-embryo buried within the uterine lining. Figure 21-9, *B* diagrams this pre-embryo, visualizing two of the germ layers and the beginnings of the third. At the beginning of the second week only the endoderm and the ectoderm have formed.

Approximately 1 week after fertilization, the blastocyst implants in the lining of the uterus and secretes human chorionic gonadotropin (HCG). HCG acts on the corpus luteum in the ovary. The corpus luteum responds by continuing to produce estrogens and progesterone, hormones that sustain the implanted blastocyst as the placenta develops.

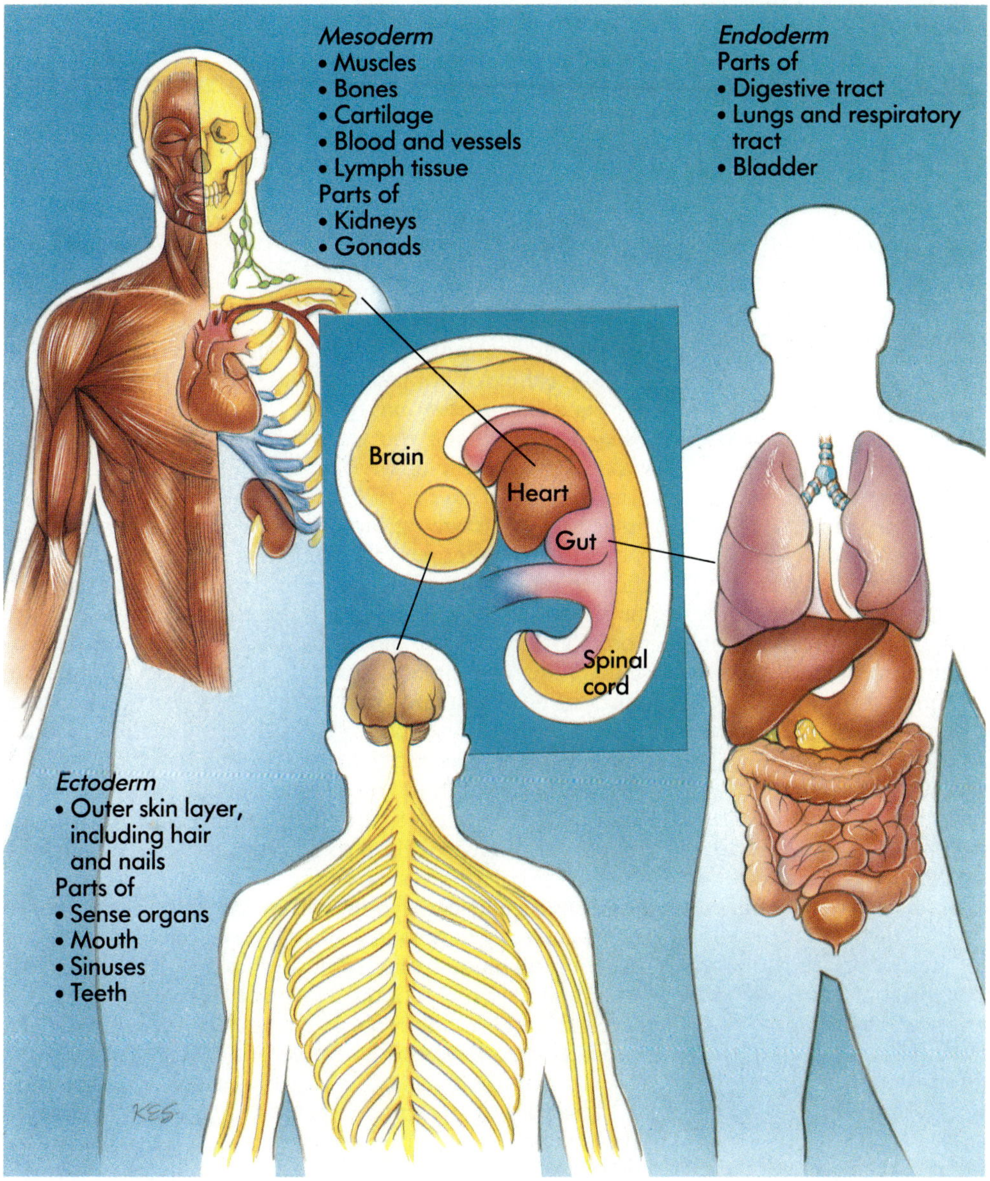

FIGURE 21-8 Structures produced by the three primary germ layers.
The diagram shows the body systems and tissues into which the germ layers develop.

FIGURE 21-9 **The implanted pre-embryo showing the three primary germ layers.**
A An implanted pre-embryo in the uterine wall.
B The endoderm and ectoderm have already formed at this stage, with the mesoderm just beginning its development.

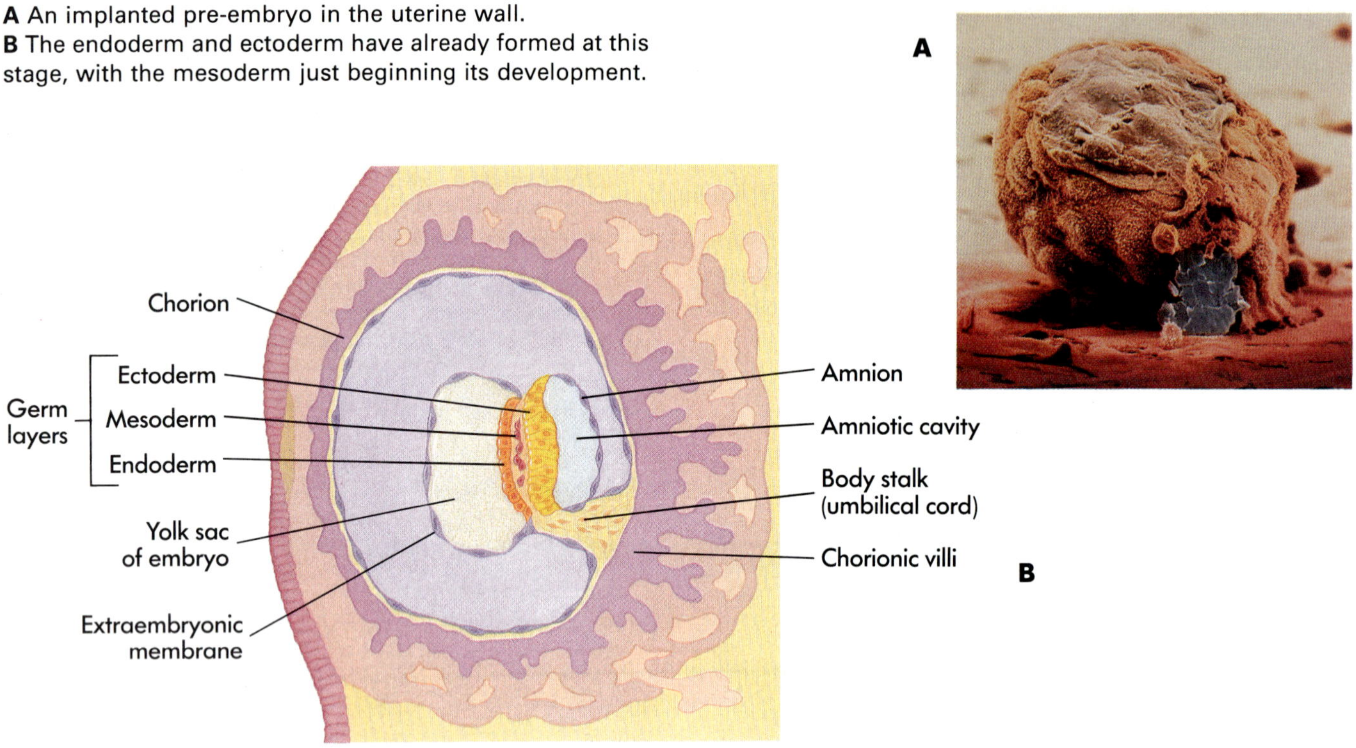

Early development of the extraembryonic membranes

In Figure 21-9, *B* the extraembryonic membranes have begun to form even at this early stage of development. The **extraembryonic membranes** all play some role in the life support of the pre-embryo, the embryo, and then the fetus. (The term **fetus** is used to describe the stage of development after 8 weeks.) Called *extraembryonic membranes* because they are not a part of the body of the embryo, these structures provide nourishment and protection. They form from the trophoblast, the ring of cells surrounding the inner cell mass. At the same time the pre-embryo is forming from the inner cell mass cells. The further development of these membranes continues into the fetal period.

The **amnion** is a thin, protective membrane that grows down around the embryo during the third and fourth weeks, fully enclosing the embryo in a membranous sac. The amniotic sac can be thought of as a shock absorber for the embryo. This thin membrane encloses a cavity that is filled with a fluid—**amniotic fluid**—in which the fetus floats and moves. This fluid also helps keep the temperature of the embryonic and fetal environment constant. The amniotic cavity is first seen at about 8 days as a slit-like space that appears between the inner cell mass and the underlying trophoblast cells that are invading the uterine wall. It can be seen as a better-developed structure in Figure 21-9, *B* at the end of 2 weeks of development.

Notice also in Figure 21-9, *B* that a structure called the **chorion** is beginning to develop as tissue outside of and including the trophoblast cells that ring the developing embryo and extraembryonic tissue. The chorion is highly specialized to facilitate the transfer of nutrients, gases, and wastes between the embryo and the mother's body. It is a primary part of an organ called the **placenta,** a flat disc of tissue about the size (at birth) of a large, thick pancake that grows into the uterine wall. The placenta is made up of both chorionic and maternal tissues. Digging deeply into the uterine lining, finger-like extensions of the chorion come into close contact with the blood-filled uterine tissues at the placenta. When the embryo reaches a stage of development in which its heart begins to beat, oxygen-poor blood filled with wastes is sent from the embryo's body through the **umbilical arteries** to the placenta.

The two umbilical arteries and a single vein are embedded in the connective tissue of the **umbilical cord,** the developing embryo's lifeline to the mother. This "highway" joins the circulatory system of the embryo with the placenta. At the placenta, embryonic wastes are exchanged for nutrients and oxygen through a thin layer of cells that separates the embryo's blood from the mother's blood. The embryonic blood and maternal blood do not mix. The fetal blood then travels through the umbilical vein back to the embryo. The umbilical cord develops from the body stalk, the yolk sac, and the allantois during the fourth week of development.

The **allantois** (from a Greek word meaning "sausage") gives rise to the umbilical arteries and vein as the umbilical cord develops. Although this extraembryonic membrane is

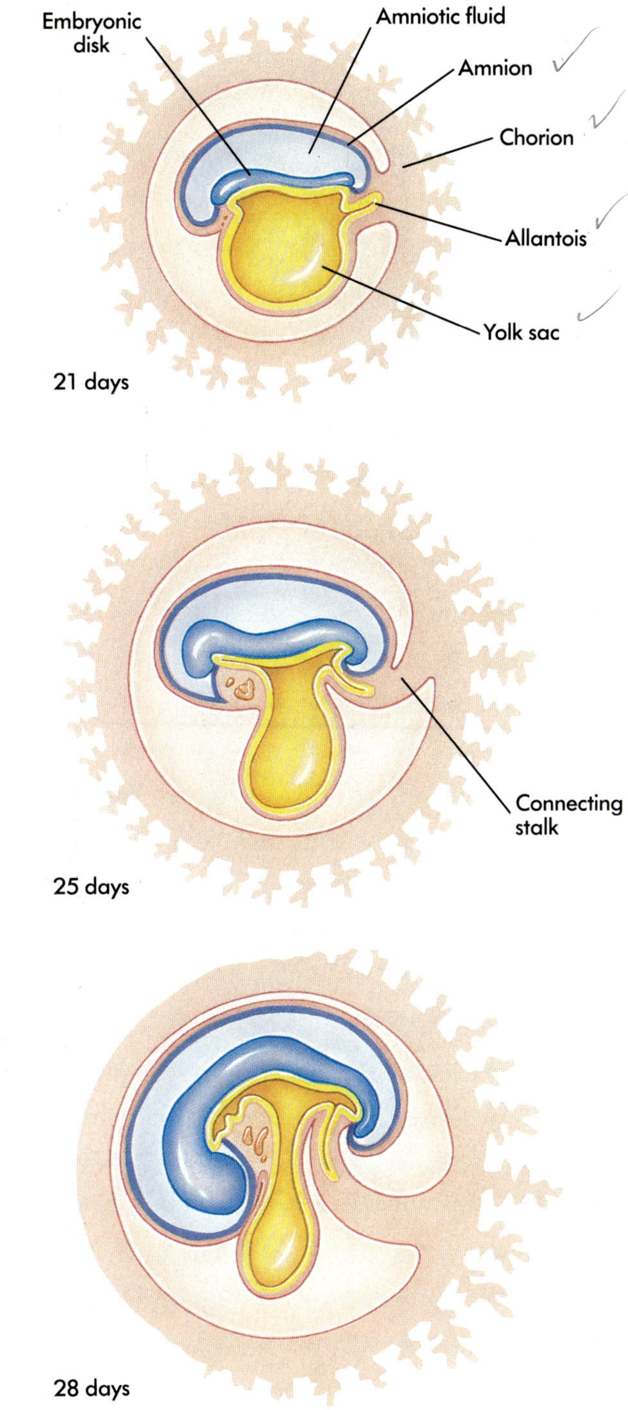

FIGURE 21-10 Extraembryonic membranes.
The embryo rapidly develops extraembryonic membranes during a single week. This diagram depicts the embryo and the development of these membranes at **(A)** 3 weeks, **(B)** 3½ weeks, and **(C)** 4 weeks.

small and cannot be seen in Figure 21-9, *B,* it is shown in Figure 21-10. Appearing during the third week of development as a tiny, sausage-shaped pouching on the yolk sac, the allantois is initially responsible for the formation of the embryo's blood cells, and later develops into the umbilical blood vessels. Notice that the **yolk sac,** a structure established during the end of the second week, also becomes a part of the umbilical cord. Before becoming a nonfunctional part of the cord, the yolk sac produces blood for the embryo until its liver becomes functional during the sixth week of development. In addition, part of the yolk sac becomes the lining of the developing digestive tract.

During the second week of development, the pre-embryo completes its implantation within the uterine wall, two of the three layers of cells develop from which its organs and tissues will arise, and the extraembryonic membranes—the amnion, chorion, yolk sac, and allantois—begin to form from the trophoblast.

Development from the third to eighth week: The embryo

During this crucial time period, the developing individual is termed an **embryo.** Throughout the next six weeks, the embryo takes on a human shape by means of the ongoing process called *morphogenesis* mentioned earlier. During morphogenesis, cells move to shape the developing embryo. In addition, the organs are established but need further development during the third through eighth weeks in a process called **organogenesis.** Because of the incredible array of developmental processes proceeding simultaneously during this period, the embryo is particularly sensitive to **teratogens,** certain agents such as alcohol that can induce malformations in the rapidly developing tissues and organs (see the boxed essay).

The third week

At the end of the second week and continuing into the third week, various cell groups of the inner cell mass begin to divide, move, and differentiate, changing the two layered pre-embryo into a three-layered embryo. This process is called **gastrulation.** The word *gastrulation* is derived from a Greek word *gastros* meaning "belly." In fact, the prefix *gastr-* is found in many words denoting parts of the human body, such as gastric, referring to the stomach. This term is descriptive of the fact that at this stage in many animals, a primitive gut is formed by the invagination (infolding) of the blastula, but the intestines develop differently in humans and many animals.

At the beginning of the third week, the embryo is an elongated mass of cells barely one tenth of an inch long. A streak (called the *primitive streak*) runs down the midline of what will be the "backside" of the embryo. Originally a mere spot of cells that lengthened over the period of a few days, the streak, along with the trophoblast, produces the mesoderm that develops during the gastrulation process. Cells at the "head" end of the streak grow forward to form the beginnings of the **notochord.** The notochord is a structure that forms the midline axis along which the vertebral column (backbone) develops. An embryonic notochord forms in all vertebrate animals. However, in humans and in most other vertebrates it degenerates and disappears long before birth.

BIOLOGY & *you*

Fetal alcohol syndrome

According to the Public Health Service, as many as 86% of women drink at least once during pregnancy, and between 20% and 35% of pregnant women drink regularly; that is, drinking is not confined to special occasions but is a daily or weekly event. These statistics are alarming because new studies now show that even moderate alcohol consumption during pregnancy can harm the unborn child.

When a pregnant woman drinks, the alcohol she consumes crosses the placenta and intoxicates the fetus. A recent study has demonstrated that women have less of a stomach enzyme to neutralize alcohol than men do, so women become intoxicated faster due to higher levels of alcohol in their bloodstreams. These high blood alcohol levels also mean that more alcohol is passed along to the fetus. To cause damage, a pregnant woman must drink above a certain threshold level of danger. But experts are not sure what the alcohol levels are that will cause damage from one day of the pregnancy to the next. Having two drinks on day 39 may not harm the fetus. But having two drinks on day 40 may cause fetal brain damage.

The damage that is inflicted on the fetus can be severe, or it can be more subtle. There seems to be a rough correlation between the amount of alcohol consumed and the severity of the birth defects that result. Full-blown **fetal alcohol syndrome** results in a child who has a number of physical and mental handicaps. These children require special education and sometimes physical therapy to help them contend with their disabilities. More subtle is **fetal alcohol effect** in which a child has learning difficulties, shows poor judgment, and is impulsive, unable to learn from mistakes, and undisciplined. These children are often placed in "regular" classrooms that are unsuited to their special needs. Often, these are the students that are doomed to fail in school and who may drop out of school.

Experts agree that the surest way to protect the unborn child from alcohol-related effects is not to drink at all during pregnancy. Some researchers believe that a woman even contemplating pregnancy should not drink. This position makes sense, since many women do not know they are pregnant until weeks after conception, and unwitting alcohol consumption may have already wreaked its damage on the fetus. Given the uncontrollable factors influencing the birth of a healthy baby (such as heredity), it makes sense to control those factors within our reach. Not drinking is one of the factors pregnant women can control that will contribute to a healthy baby.

FIGURE 21-A Drinking during pregnancy has been shown to cause harm to the unborn child.

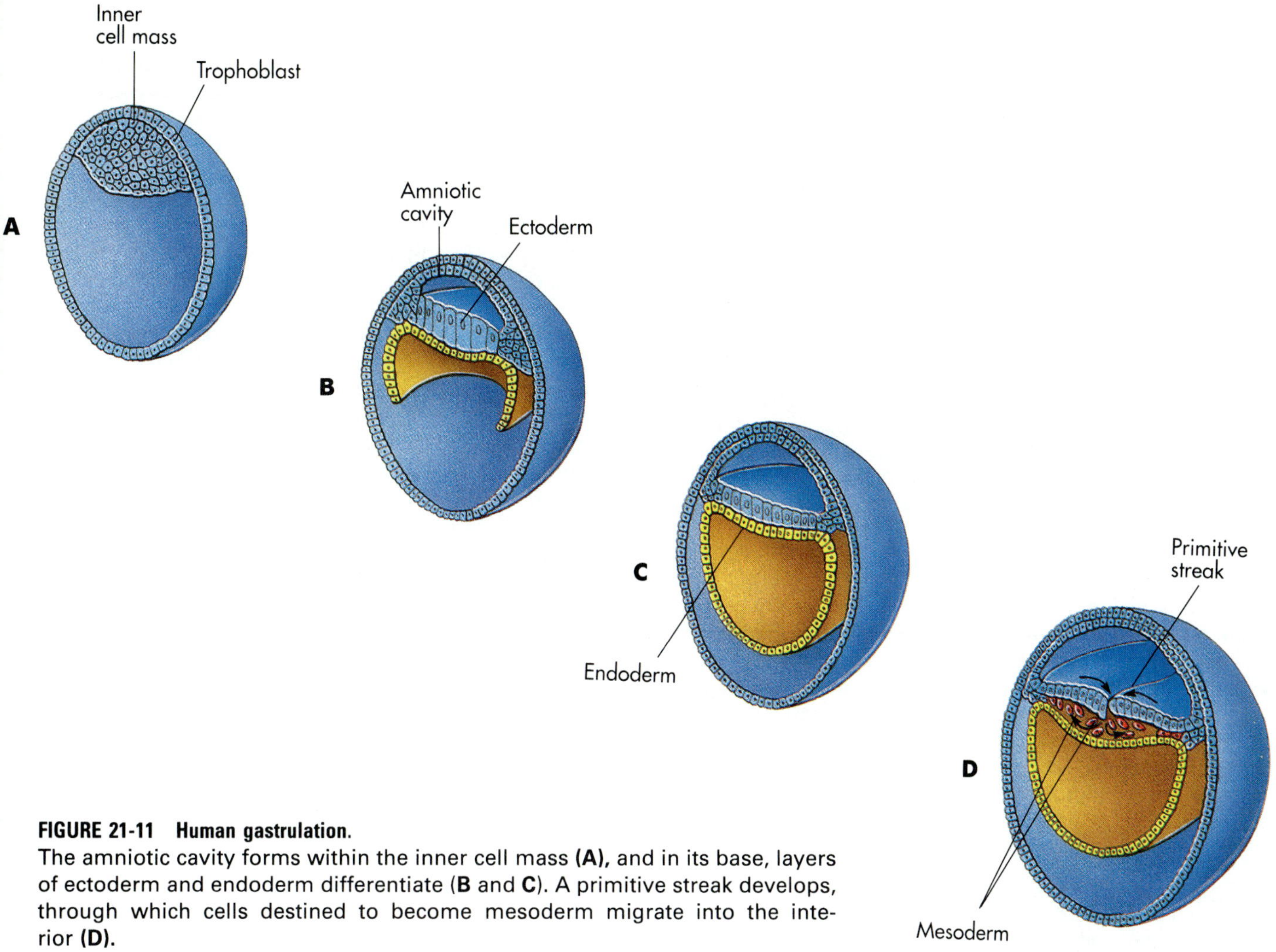

FIGURE 21-11 Human gastrulation.
The amniotic cavity forms within the inner cell mass **(A)**, and in its base, layers of ectoderm and endoderm differentiate (**B** and **C**). A primitive streak develops, through which cells destined to become mesoderm migrate into the interior **(D)**.

Gastrulation ends with the completion of the notochord, midway through the third week. By the end of gastrulation, a layer of ectoderm covers over the notochord tissue. Figure 21-11 diagrams the events of gastrulation.

> **Gastrulation is a process by which groups of inner cell mass cells migrate, divide, and differentiate into three primary germ layers from which all the organs and tissues of the body will develop. This process, in humans as well as all vertebrates, results in a developmental stage called the *gastrula*.**

In the photo of the embryo at 4½ weeks (Figure 21-12), its head can be seen to have the beginnings of eyes, a sign that **neurulation** has taken place. Neurulation is the development of a hollow nerve cord, which later develops into the brain, spinal cord, and related structures such as the eyes.

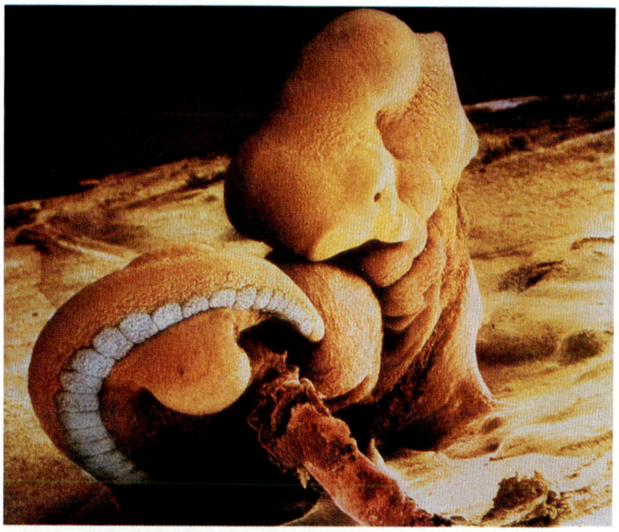

FIGURE 21-12 A human embryo at 4½ weeks.
Note that the eyes have begun to develop.

Neurulation begins in the third week with the folding of the ectoderm lying above the notochord, forming an indentation along the back of the embryo. This indentation is called the **neural groove.** On either side of the groove are areas of tissue called *neural folds*. In Figure 21-13, the neural folds of the 3-week embryo have come together at one spot and have fused. This spot is destined to be the neck region of the developing individual. The tissue above the fused region, looking somewhat like a pair of lips, will develop into the brain. The less broad area of tissue below the fused region will develop into the spinal cord. Eventually the edges of the groove will fuse along its length, forming a neural tube, the precursor to these structures.

During neurulation, the mass of ectodermal cells lying over the notochord curls up, forming a groove. The edges of the groove move together, eventually forming a tube that will develop into the brain and spinal cord—the central nervous system.

Neurulation results in an embryo called a *neurula* and signals the developmental process of **tissue differentiation.** During tissue differentiation, groups of cells become distinguished from other groups of cells by the jobs they will perform in the body. This differentiation appears to be accomplished by the interaction of cytoplasmic stimulators and inhibitors of certain cells with various genes of other cells. Put simply, some cells act like "on" and "off" switches for the genes of neighboring cells. This process is called **induction.**

Important processes of pre-embryonic and embryonic development are cleavage, cell migration, morphogenesis, and tissue differentiation. The four stages of development of the pre-embryo and the embryo are the morula, the blastula, the gastrula, and the neurula.

The embryo's heart also begins its development during the third week as a pair of microscopic tubes. In fact, the cardiovascular system is the very first system to become functional in the embryo. At the end of the third week of development, the heart tubes have fused and have linked with blood vessels in the embryo, body stalk, chorion, and yolk sac to provide a primitive circulation of blood (Figure 21-14), and the heart actually begins to beat! At the same time, finger-like structures called **villi** begin to protrude from the chorion. These blood-filled projections dig into the uterine tissues of the mother, increasing the surface area over which gases, nutrients, and wastes can be exchanged. At 8 to 10 weeks of development, doctors can take some of this tissue to detect whether genetic abnormalities exist in the embryo. Termed *chorionic villus sampling,* or *CVS,* this procedure is performed by inserting a suction tube through the vagina, into the uterus, and to the chorionic villi. Because chorion cells and fetal cells contain identical genetic information, doctors can use these cells for genetic studies.

Colorized with a blue-gray tint, paired segments of tissue are also prominent in Figure 21-12. These chunks of mesoderm are called **somites** (from a Greek word meaning "a body"). They will give rise to most of the axial skeleton (see Chapter 18) with its associated skeletal muscles and most of the dermis of the body—tissues that underlie the epidermis of the skin. The first somites appear during the third week of development and are added as the embryo grows. Eventually, 42 to 44 pairs develop by the middle of the fifth week.

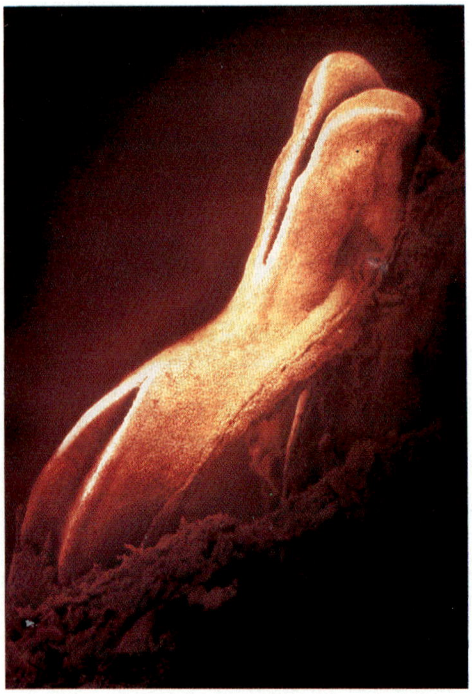

FIGURE 21-13 Neurulation.
The ectoderm folds inward to form the neural groove and, at 3 weeks gestation, fuses. The spot at which the fusion occurs will become the neck, and the tissue above this spot will become the brain. The area below the fused region will become the spinal cord.

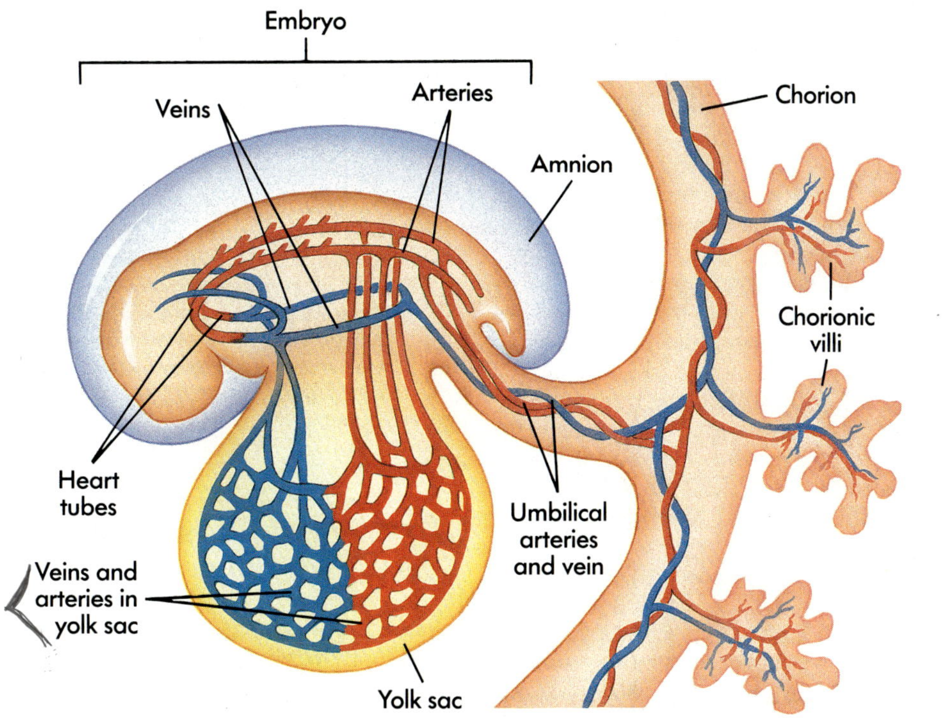

FIGURE 21-14 Cardiovascular system of a 3-week-old embryo. At the end of the third week, the heart begins to beat. Villi extending out from the chorion dig into the uterine wall and provide a surface for the exchange of gases, nutrients, and wastes.

The fourth week

During the fourth week of development, the embryo begins to curl (see Figure 21-12). The curling occurs as folds are formed at the head and tail end of the embryo. As the embryo curls, the yolk sac becomes squeezed into a narrow stem that fuses with the body stalk connecting the embryo to the placenta (see Figure 21-10). As this fused body stalk lengthens and as its surface is tightly covered by the growing amnion membrane, it is properly called the umbilical cord. Blood cells continue to be produced by the yolk sac until the liver completely takes over this job in the sixth week. The umbilical arteries and vein, already functional, arose from the pouch of the allantois in the body stalk and have been transporting blood between the embryo and the placenta since the third week.

Four visible sets of swellings called the **branchial arches** develop on either side of the head end of the embryo during the fourth week. The word *branchial* is derived from a Greek word meaning "gill" and better describes similar structures that develop into gill supports in fish. In fact, the head end of the human embryo at this stage somewhat resembles a fish embryo.

The presence of gill arches and a tail (see Figure 21-12) in the human embryo indicates a relationship between humans and the lower vertebrates (fish and amphibians). In the nineteenth century, the German scientist Ernst Haeckel described this link as "ontogeny recapitulates phylogeny." What Haeckel meant was that the embryological development of an individual organism (ontogeny) repeats (recapitulates) the evolutionary history of its ancestors (phylogeny). This statement, however, is untrue. Embryonic stages of particular vertebrates are not a "replay" of the succession of its adult ancestors. Rather, the embryonic stages of an individual organism often reflect embryonic stages of its ancestors. For example, gill arches form during the embryological development of humans. These arches develop into structures such as the middle ear, eustachian tube, tonsils, thymus, and parathyroids.

The embryo's heart begins development during the third week. Somites—chunks of mesoderm—also become prominent at this time. They will give rise to most of the axial skeleton. During the fourth week, the gill arches appear, which will later develop into such structures as the middle ear and the tonsils.

The fifth through eighth week

During the fifth week, the embryo doubles in length from 4 millimeters (3/16 inch) to about 8 millimeters (3/8 inch). A nose begins to take shape as tiny pits. Although **limb buds** were first visible during the fourth week, first the arms and then the legs, at this stage of development the limbs look like microscopic flippers. The brain grows rapidly this week, resulting in an embryo with a large head in proportion to the rest of the developing body.

Seen floating within the fluid-filled amniotic sac in Figure 21-15, the 6-week-old embryo now has arm buds with distinct wrists. Fingers are just beginning to form, as are the ears. The retina of the eyes is now darkly pigmented. Although the trunk of the body straightens out somewhat as the liver and digestive system grow, the neck area remains bent, forcing the head to rest on the chest above the red protruding heart, which now beats 150 times per minute. At the end of 6 weeks, the embryo is little more than half an inch long.

During the seventh week of development, eyelids begin to partially cover the eyes, and the ears develop more fully. The face begins to take on a human appearance. Each arm develops an elbow. Individual fingers can be distinguished, but they are connected with webs of skin that disappear during the eighth week. The legs, slower in their development than the arms, develop ankles and the suggestion of toes. During the eighth week—the last week as an embryo—the developing individual grows to 30 millimeters (slightly over an inch) in length.

From the fourth to eighth week of development, dramatic changes take place in the embryo. The embryo grows from a length of 2 millimeters to approximately 30 millimeters (slightly longer than an inch). A primitive circulation is established within the embryo, and a maternal-fetal exchange of nutrients, gases, and wastes begins. The central nervous system and its associated structures begin to develop. The body form is established, including the appendages.

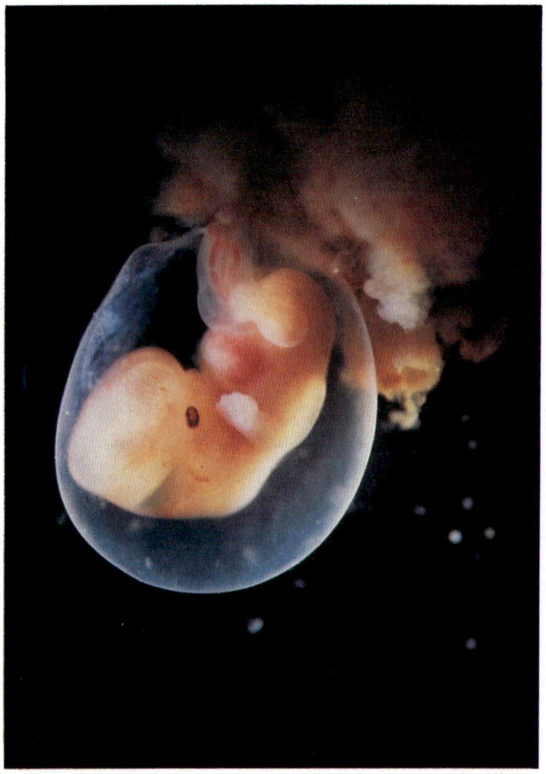

FIGURE 21-15 Embryo at 6 weeks of age.
At this stage, the arms and legs are beginning to form. No bones have formed in the skull, so the brain can be clearly seen above the eye.

Development from the ninth to thirty-eighth week: The fetus

The embryonic period is primarily one of development. By the ninth week, most of the body systems are functional. The primary job of the fetal period is the refinement, maturation, and growth of these organ systems and of the body form of the developing individual.

The third month

The **first trimester** ends with the completion of the third month of pregnancy. During the third month, the fetus grows rapidly, tripling its length to approximately 85 millimeters (3⅓ inches). Fine hair called **lanugo,** meaning "down," appears over its body, but this downy coat is lost before birth. Changes in the shape of the face cause the eyes to face forward and appear closer together, although the forehead remains prominent (Figure 21-16). The eyes also become well developed during this month. The eyelids have continued their development and now fuse shut. They will not open until the fetus is 7 months old. A suggestion of an ear can be seen on the side of the head in line with the jaw bone in the 12-week-old fetus in Figure 21-17.

During the embryonic period, male babies cannot be distinguished from female babies in their outward appearance. During early fetal development, outward differences between the sexes begin to take shape. The differentiation of male or female sex organs depends, of course, on the genetic makeup of the fetus but also on a complex interplay between hormones and tissues in the developing fetus. This interplay of factors results in the development of strikingly different reproductive systems by the end of the third month, with the penis in the male and the clitoris in the female arising from the same embryonic tissues. Likewise, the scrotum in the male is homologous to the labia majora in the female.

During the third month the fetus grows to 3⅓ inches in length. Eyes become well-developed, and ears begin to form.

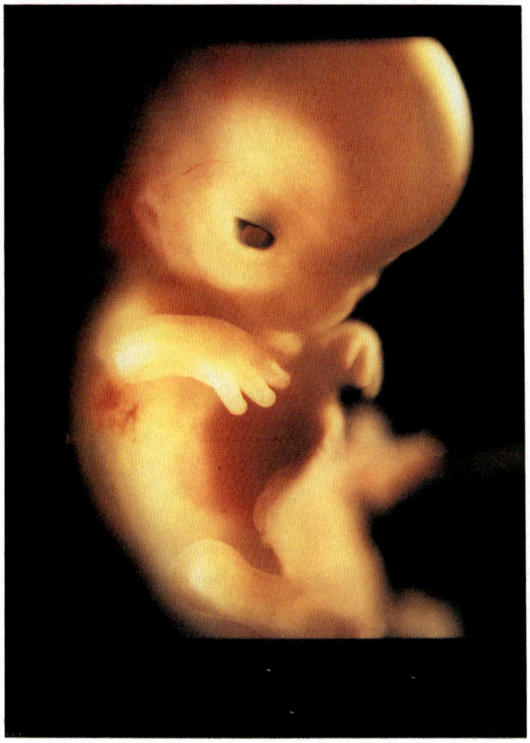

FIGURE 21-16 Fetus at 11 weeks of age.
Fingers have begun to develop at this stage. The eyes are well formed and are covered with eyelids.

The fourth through sixth month

During the **second trimester,** the fetus grows to about $\frac{3}{5}$ kilogram (about $1\frac{1}{2}$ pounds) and $\frac{3}{10}$ meter (1 foot) long. The 4-month-old fetus in Figure 21-18 looks quite human. Halfway through the fourth month, it can bring its hands together and suck its thumb. By 15 weeks, the sensory organs are almost completely developed, and by 16 weeks the fetus is actively turning inside the mother. Many of the bones are forming, replacing areas of cartilage. The process of bone formation, termed **ossification,** began at approximately 8 weeks of development and will continue beyond birth to the age of 18 or 19 years. By the end of the fifth month, the heartbeat of the fetus can be heard through a stethoscope. Although the body systems have been rapidly continuing their development, the fetus is still unable to survive outside of the mother's womb. At the end of the sixth month, the head is no longer quite as large compared to the rest of the body, the eyelids have separated, and the eyelashes have formed. The fetus is capable of independent survival at the sixth month but only with special medical intervention.

During the fourth through sixth months, the fetus grows to 30 centimeters (12 inches) in length and begins to look human. The sensory organs become well developed, and the bones begin to form, replacing cartilage.

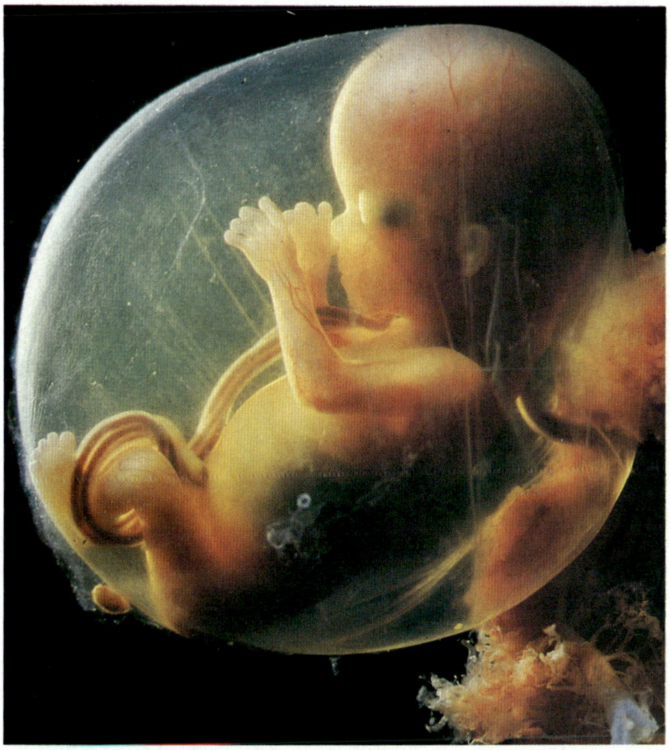

FIGURE 21-17 Fetus at 12 weeks of age.
Ears can be seen, as well as the fingers and toes.

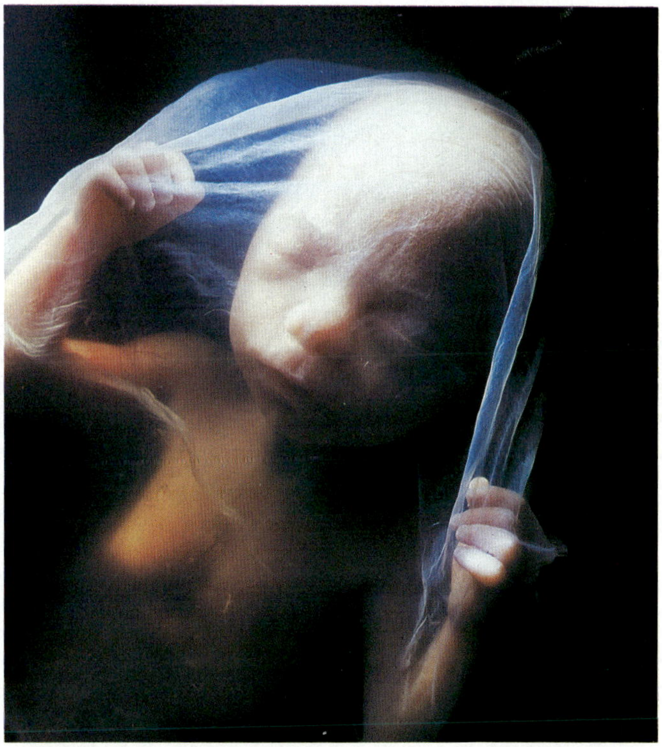

FIGURE 21-18 Fetus at 4 months of age.
The fetus at this point has a human appearance and can even bring its hands together and suck its thumb.

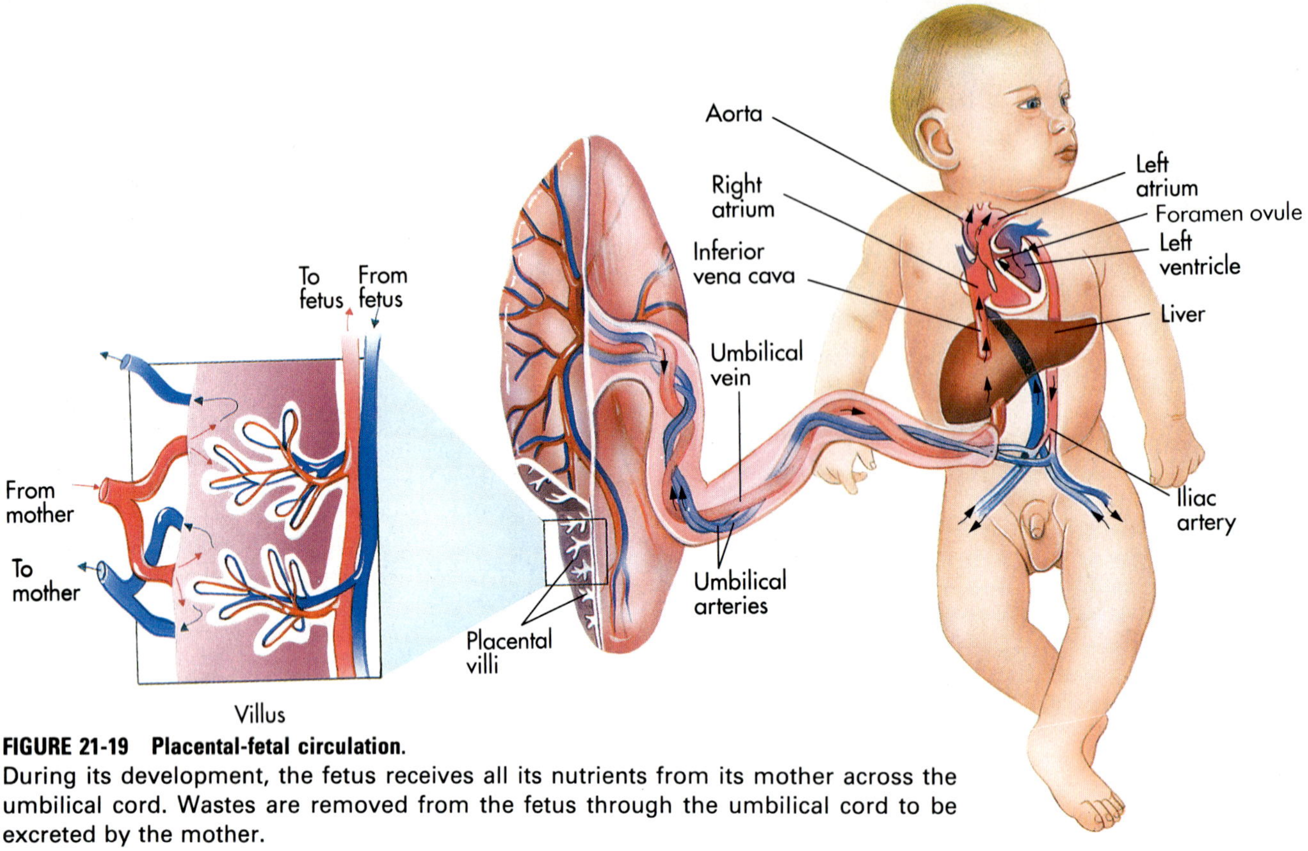

FIGURE 21-19 Placental-fetal circulation.
During its development, the fetus receives all its nutrients from its mother across the umbilical cord. Wastes are removed from the fetus through the umbilical cord to be excreted by the mother.

The seventh through ninth month

The **third trimester** is predominantly a period of growth rather than one of development. In the seventh, eighth, and ninth months of pregnancy, the weight of the fetus doubles several times because of growth but also because of fat that is laid down under its skin. This increase in bulk is not the only kind of growth that occurs. Most of the major nerve tracts in the brain, as well as many new brain cells, are formed during this period. The mother's bloodstream fuels all of this growth by the nutrients it provides. Within the placenta these nutrients pass into the fetal blood supply (Figure 21-19). The undernourishment of the fetus by a malnourished mother can adversely affect this growth and result in severe retardation of the infant. Retardation resulting from fetal malnourishment is a severe problem in many underdeveloped countries where poverty is common.

By the end of the third trimester, the neurological growth of the fetus is far from complete and, in fact, continues long after birth. By this time, however, the fetus is able to exist on its own.

The critical stages of human development take place quite early. All the major organs of the body have been established by the end of the third month. The following 6 months are essentially a period of growth.

Birth

Birth takes place at the end of the third trimester, 38 weeks from conception (40 weeks from the start of the last menstrual period). Although the exact mechanism of the onset of **labor,** the sequence of events that leads to birth, is not well understood, scientists know that changing hormone levels in the developing fetus initiates this process. These hormones induce placental cells of the mother to manufacture **prostaglandins** (see p. 330), hormone-like substances that cause the smooth muscle of the uterine wall to contract. In addition, the pressure of the baby's head against the cervix sends nerve impulses to the mother's brain that trigger the hypothalamus to release the hormone oxytocin from her pituitary. Working together, oxytocin and prostaglandins stimulate waves of contractions in the walls of the uterus, forcing the fetus downward (Figure 21-20). Initially, only a few contractions occur each hour, later increasing to one every 2 to 3 minutes. Eventually, strong contractions, aided by the mother's pushing, expel the fetus through the vagina, or **birth canal.** The fetus is now a newborn baby.

After birth, uterine contractions continue and expel the placenta and associated membranes, called the *afterbirth*. The umbilical cord is still attached to the baby, and to free the newborn, a doctor or midwife ties and cuts it; blood clots in the cord and contraction of its muscles prevent excessive bleeding.

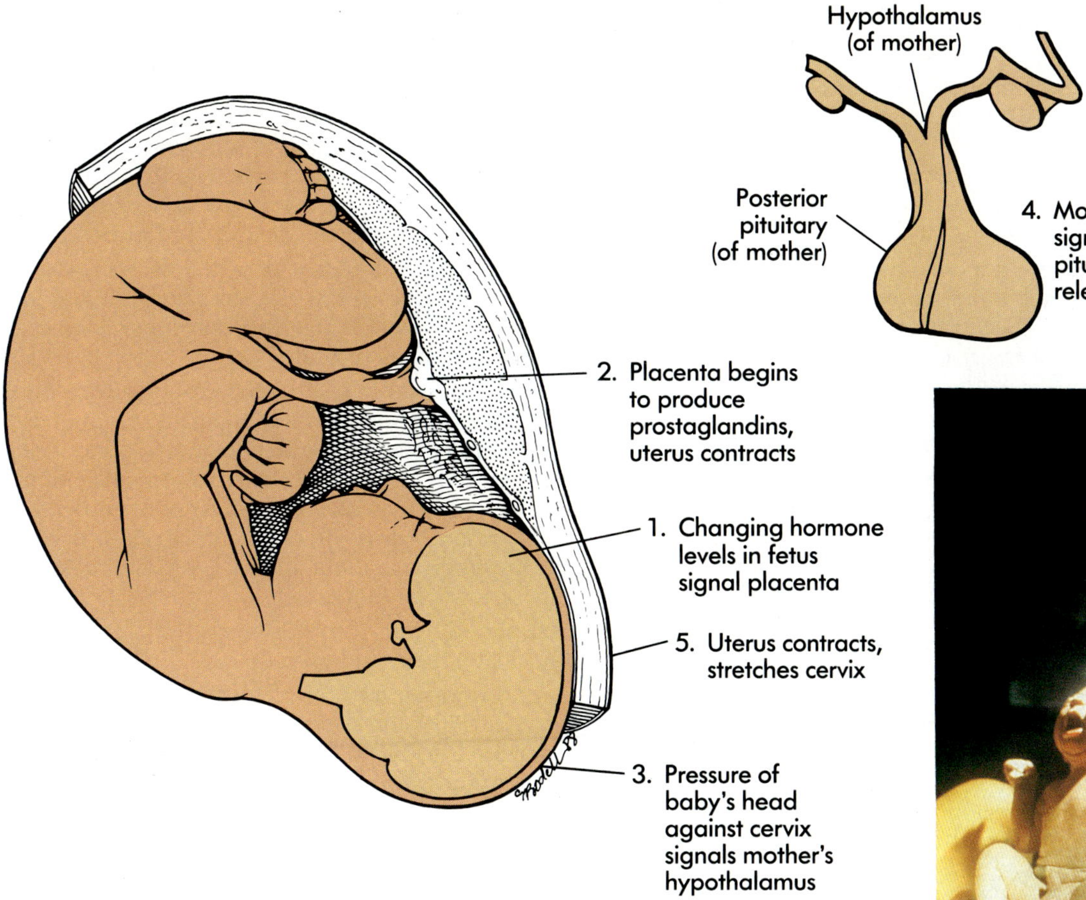

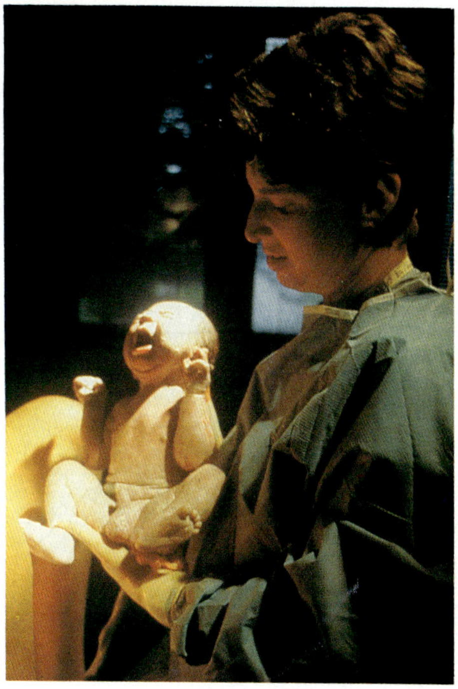

FIGURE 21-20 Birth is initiated and sustained by hormones in a positive feedback loop. Although the exact mechanism is not known, it is thought that birth is triggered by the following mechanism: changing hormone levels in the fetus *(1)* signal the placenta to produce prostaglandins, causing the uterus to contract *(2).* In addition, the fetus' head pushing against the cervix *(3)* signals the hypothalamus to release oxytocin from the pituitary *(4),* which causes the uterus to contract even harder *(5).* This feedback loop, in which the pressure of the fetus' head causes the release of oxytocin, continues until the baby is born.

Physiological adjustments of the newborn

At birth, the baby's lungs are not filled with air. Its first breath is therefore unusually deep. For the first time, its lungs are inflated; the baby cries. Because the baby is now obtaining its oxygen from its lungs rather than from the placenta, several major changes must take place in the circulation of the blood.

Until birth, the placenta was the source of nutrients and oxygen for the fetus; in addition, it was the site for the removal of waste products from the fetal circulation. The lungs were not functional as organs of gas exchange. The fetal body had two major adaptations to limit the flow of blood to the lungs. First, a hole between the two atria called the *foramen ovale* shunted most right atrial blood directly into the left atrium, thus avoiding the right ventrical and the pulmonary circulation (see Figure 21-19). In addition, the right ventricular blood that is pumped into the pulmonary artery was mostly shunted into the aorta rather than through the lungs by the *ductus arteriosus,* a direct connection between the pulmonary artery and the aorta. At birth, the foramen ovale is closed by two flaps of heart tissue that fold together and fuse. The ductus arteriosus is shut off by contractions of muscles in its walls. Complete closure may take several months. The umbilical arteries and vein also must close off.

At birth, the circulation of the newborn changes as the lungs rather than the placenta become the organ of gas exchange.

Summary

1. Human development before birth, or gestation, lasts aproximately 8½ months. Conception takes place when a sperm, or male sex cell, penetrates a secondary oocyte, or female sex cell. This penetration, or fertilization, usually takes place in the uterine tube.
2. The fertilized egg, or zygote, begins to divide after the genetic material of the egg and sperm unite. This process of cleavage is the first stage in the development of humans as well as other multicellular animals. Cleavage results in a ball of cells, or morula.
3. After 2 days in the uterus, the morula has developed into the blastocyst, a stage of development in which the embryo is a hollow ball of cells. This stage of development is significant; it is the first time that cells begin to move, or migrate, to shape the new individual. This process of cell movement to shape the developing individual is morphogenesis.
4. At approximately 1 week after fertilization, the blastocyst nestles into the lining of the uterine wall, which is the process of implantation. It also secretes a hormone that acts on the corpus luteum in the ovary, stimulating the body to produce estrogens and progesterone to maintain the uterine lining rather than having it slough off during a menstrual period.
5. During the first 2 weeks of development, the developing mass of cells is generally referred to as a pre-embryo; from 3 to 8 weeks, an embryo; and from 9 to 38 weeks, a fetus. Structures also develop that are not a part of the growing individual but that help sustain the pre-embryo, embryo, and fetus through development. These extraembryonic membranes are the amnion, chorion, yolk sac, and allantois.
6. At the end of the second week and continuing into the third week, various cell groups of the blastula begin to divide, move, and differentiate in a process called *gastrulation*. During this time a three-layered embryo is formed. Each of these three layers will give rise to specific tissues and organs of the developing individual.
7. Neurulation, the development of a hollow nerve cord that develops into the central nervous system, begins in the third week. However, the development of the nervous system continues even after birth.
8. Dramatic changes take place in the embryo from the fourth to the eighth weeks of development. Limb buds become visible in the fourth week and are quite well developed by the seventh week. A primitive circulation is established, and a maternal-fetal exchange of nutrients, gases, and wastes begins. By the eighth week of development, the embryo is about an inch long, and most of the body systems are functional.
9. The fetal period, from 9 to 38 weeks, is a time of refinement, maturation, and growth.
10. Changing hormone levels in the fetus at about the thirty-eighth week trigger the onset of labor, the sequence of events that leads to birth. Waves of contractions in the walls of the uterus force the fetus downward and out the birth canal.
11. At birth, the circulation of the newborn changes as the lungs rather than the placenta become the organ of gas exchange.

KEY TERMS

allantois 360
amnion 360
amniotic fluid 360
birth canal 368
blastocyst 357
blastula 357
branchial arches 365
chorion 360
cleavage 356
ectoderm 359
ectopic pregnancy 357
embryo 361
endoderm 359
extraembryonic membranes 360
fertilization 355
fetus 360
first trimester 366
gastrulation 361
gestation 355
human chorionic gonadotropin (HCG) 358
implantation 357
induction 364
labor 368
lanugo 366
limb buds 366
mesoderm 359
morphogenesis 357
morula 357
mucin 355
neural groove 364
neurulation 363
notochord 361
organogenesis 361
ossification 367
ovum 355
placenta 360
pre-embryo 356
prenatal development 355
primary germ layers 359
second trimester 367
somite 364
teratogen 361
third trimester 368
trophoblast 357
umbilical arteries 360
umbilical cord 360
villi 364
yolk 355
yolk sac 361
zona pellucida 355
zygote 356

REVIEW QUESTIONS

1. What is prenatal development, and how long does it last?
2. What happens to a sperm and a secondary oocyte immediately after fertilization occurs?
3. Fill in the blanks: After fertilization, the one-celled ______ begins to divide rapidly in a process called ______. Normally it travels through the rest of the ______ ______ to reach the ______. At this point it is a ball of cells called the ______.
4. What is implantation? Explain its significance.
5. Identify the three primary germ layers, and describe the structures into which they will develop.
6. What are the extraembryonic membranes? Explain their significance.
7. Summarize the events that occur during gastrulation.
8. Define neurulation, and explain its significance.
9. What is a notochord, and what role does it play in humans?
10. Put the following into the sequence they occur during development, and briefly describe each one and when it occurs.
    a. Gastrula
    b. Morula
    c. Neurula
    d. Blastula
11. Summarize the changes that occur in the embryo between the fourth and eighth weeks of development.
12. Summarize the changes that occur in the embryo/fetus during the first, second, and third trimesters. When do most body systems become functional?
13. Explain why good nutrition is so important during pregnancy.
14. Describe the physiological adjustments that a newborn baby must undergo to survive.

THOUGHT QUESTIONS

1. Can you suggest a mechanism that would enable an oocyte to stop the second meiotic division from completing after ovulation, and how do you think a sperm's penetration of the oocyte then triggers completion of the second meiotic division? What sort of machinery do you think this triggering requires?
2. What sort of physical changes might an egg use to prevent the penetration of a second sperm? What might be the consequences of a second sperm gaining entry?

FOR FURTHER READING

Blodel, C. P., et al. (1992, March 19). A potential fusion peptide and an integrin ligand domain in a protein-active sperm-egg fusion. *Nature.*
Scientists at the University of California—San Francisco are studying a protein that helps fuse egg and sperm during fertilization. This protein resembles some fusion proteins that help viruses fuse to cells.

Wassarman, P. M. (1988, December). Fertilization in mammals. *Scientific American,* p. 78.
This article describes the cellular mechanisms of sperm penetration and fertilization.

22 Patterns of inheritance

OUTLINE

Inheritance and variation within species
- Historical views of inheritance
- The birth of the study of inheritance

Gregor Mendel's experiments to determine inheritance patterns
- Conclusions Mendel drew from his experiments
- Analyzing Mendel's experiments
- How Mendel tested his conclusions
- Further questions Mendel asked
- Analyzing the results of Mendel's dihybrid crosses
  - Analysis of a dihybrid cross using probability theory
  - Analysis of a dihybrid cross using a Punnett square

The connection between Mendel's factors and chromosomes

Sex linkage

Solving the mysteries of Inheritance

The same yet different—How often have you heard that phrase? And what does it really mean? Scientists would answer the last question with a single word: variation.

Variation can be seen in the group of flamingos in the photograph. Although the flamingos are all the same—with long legs and beaks and pink feathers covering their bodies—they are all different. Look closely at the same characteristic in each. One has longer legs than the others. Another has a slightly different shade of pink. What other differences do you see?

HIGHLIGHTS

Over the centuries, the nature of heredity was misunderstood by scientists and philosophers until the work of Mendel, Sutton, Morgan, and others in the late 1800s began to uncover its mechanisms.

Working with pea plants that exhibited alternative forms of various characteristics (traits), Mendel established that traits are inherited as discrete packets of information, which scientists today call *genes.*

Through his experiments with pea plants, Mendel also discovered that alternative forms of genes (alleles) separate from one another during the formation of gametes and are distributed independently of other traits.

Mendel and other early geneticists took the first real steps toward solving the puzzles of inheritance and laid the foundation for one of the great revolutions in thinking of the twentieth century: an understanding of the nature of genetic material and how it is transmitted from generation to generation.

Clearly, each species of living things exhibits variation. Humans, for example, all have characteristics you recognize as human, but each person (except for identical twins) looks different from all others. What is the source of this variation? And how are these differences distributed among populations of living things that are all the same?

Inheritance and variation within species

Today, it is common knowledge that organisms inherit characteristics from their parents. During sexual reproduction, parents pass on traits to their offspring by means of genetic material within the eggs of the mother and sperm of the father. The intermingling of parental genes that takes place on fertilization, the union of the egg and sperm, is the material of variation. Organisms produced by asexual reproduction do not exhibit variation for just this reason. Their genetic makeup is derived from one parent only, so offspring are genetically identical to that parent. For example, a plant produced asexually by rooting a cutting will have the same genes as the "mother" plant, but a plant produced by cross-pollination—a type of sexual reproduction—will have certain characteristics of both its parents.

Sexual reproduction introduces variation within a species because offspring inherit characteristics from both their parents. Asexual reproduction does not introduce variation because the offspring is an exact duplicate of its parent.

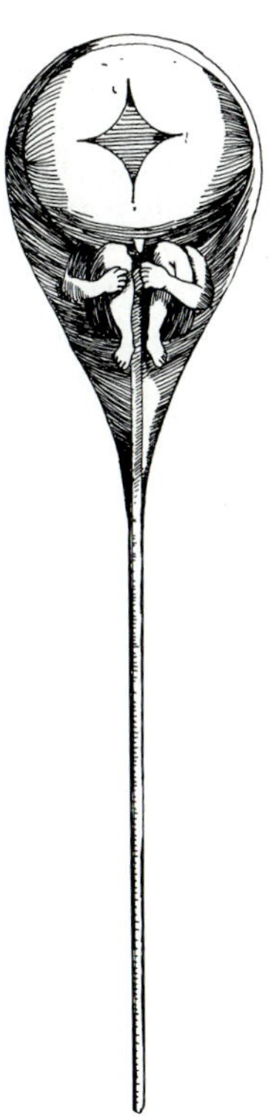

FIGURE 22-1 Drawing of a homunculus.
A widely held belief before the nineteenth century was that sperm contained a fully formed, miniature human being called a *homunculus.* It was supposedly implanted in the uterus during fertilization, where it grew into maturity.

Historical views of inheritance

Although genetic inheritance seems obvious today, this fact was not obvious to scientists and philosophers hundreds of years ago. Hippocrates (460-377 BC) believed that a child inherited traits from "particles" given off by all parts of the bodies of the father and mother. These particles, he suggested, travel to the sex organs. During intercourse, the father's particles merge with the mother's particles to form the child. This idea of inheritance was held by many until the mid-1800s. Most theories of the direct transmission of hereditary material assumed that the male and female traits blended in the offspring. Thus a parent with red hair and a parent with brown hair would be expected to produce children with reddish-brown hair, and a tall parent and a short parent would produce children of intermediate height. However, taken to its logical conclusion, this theory suggests that all individuals within a species would eventually look like one another as their traits continually blended together.

Other ideas regarding inheritance were formulated with the invention of a simple, hand-held microscope. Anton van Leeuwenhoek (1632-1723) observed sperm for the first time with the microscope, drew pictures of them, and developed hypotheses regarding inheritance based on his observations. A widely held notion at the time was that each sperm contained a tiny but whole human being (Figure 22-1). *Preformationists* (as this group was called) were separated into two camps. The spermists thought that sperm encased microscopic humans. Because the human was fully formed, the body contained sperm that each encased another preformed individual, and so forth . . . ad infinitum! Ovists, on the other hand, held that it was the eggs of the mother that contained minute humans.

Not until the mid-1800s with the work of Schleiden, Schwann, and Virchow (see Chapter 3) did scientists realize that new life arose from old life in the form of new cells arising from old cells. By means of the growth and division of these cells, new organisms developed from individual cells of parent organisms. Scientists of the late 1800s studied the nuclei of cells to uncover the mysteries of cell growth and division. They observed the complex process of mitosis (see Chapter 5) and wondered why cells went through this intricate process. Wouldn't it be more efficient for a cell to simply pinch in two along its middle? By 1883, scientists knew that the complex process of cell division ensured an equal distribution of the nuclear material to two daughter cells. And not only did each daughter cell receive the same amount of nuclear material, they each received a complete amount of the nuclear material.

At this same time, scientists also observed that an even more complex series of nuclear events (now known as *meiosis* [see Chapter 5]) preceded the formation of eggs and sperm. By 1885, several scientists independently concluded that this nuclear material was the physical bond that linked generations of organisms. But how nuclear material regulated the development of fertilized eggs and how it was related to heredity and variation was still a mystery. About 1900, scientists began to answer this question by piecing together research of the day with research that had long been ignored: the work of Gregor Mendel. Mendel was an Austrian monk trained in botany and mathematics at the University of Vienna.

The birth of the study of inheritance

Approximately 25 years before scientists had discovered a link between heredity and the complex processes of mitosis and meiosis, Mendel began his work with the garden pea. Mendel chose the pea plant because it was an annual plant that was small and easy to grow and had a short generation time. Therefore he could conduct experiments involving numerous plants and obtain results relatively quickly.

Pea plants are well suited to studies of inheritance because each pea flower contains both female parts (stigma, style, and ovary) and male parts (filaments that support anthers). Both are enclosed and protected by the petals (Figure 22-2). The gametes produced within each flower—pollen grains within the anthers and eggs within the ovary—are able to fuse and develop into seeds of new plants (see Chapter 31). Fertilization of this sort, called **self-fertilization,** takes place naturally within individual pea flowers if they are not disturbed. As a result, the offspring of self-fertilizing garden peas are derived from one pea plant, not two. After generations of self-fertilization, some plants produce offspring consistently identical to the parent with respect to certain defined characteristics; these plants are said to be **true-breeding.**

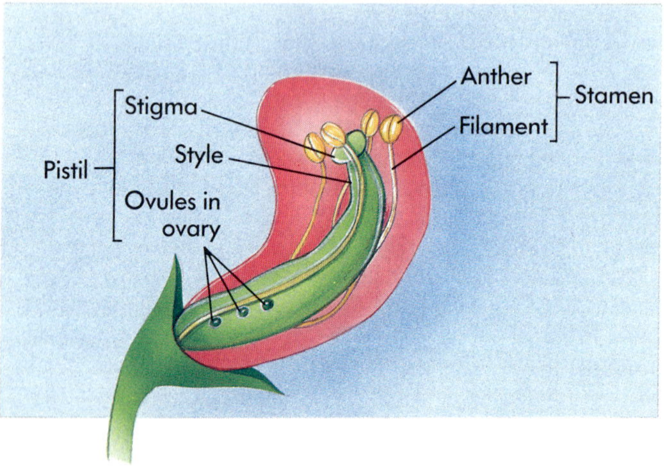

FIGURE 22-2 Anatomy of a pea plant.
In the flower of a pea plant the petals enclose the male (anther) and female (stigma) parts, ensuring that self-fertilization will take place.

Mendel selected a number of different true-breeding **varieties** of pea plants with which to work. Differing varieties, or strains, of an organism each belong to the same species, but each has one or more distinctive characteristics that are passed from parent to offspring. To study the inheritance patterns of these characteristics, Mendel took true-breeding plants and artificially cross-fertilized them. **Cross-fertilization** occurs when the pollen of one plant fertilizes the egg cells of another plant. To do this, Mendel removed the anthers of a flower before they shed pollen and then dusted this flower with pollen from another plant. In this way, Mendel was able to perform experimental crosses between two different true-breeding varieties of pea plants that exhibited differences regarding particular traits. The offspring, or progeny, of the cross between two different varieties of plants of the same species are called **hybrids.** (In other contexts, the word *hybrid* may also refer to the cross between two different species of organisms. A mule, for example, is the hybrid offspring of a horse and a donkey.)

Gregor Mendel's experiments to determine inheritance patterns

Mendel first designed a set of experiments that involved crossing varieties of pea plants that differed from one another in a single characteristic. Mendel chose seven different characteristics, or **traits,** to study (Figure 22-3). Although various other varieties of pea plants exist, Mendel chose only those varieties that differed from one another clearly and distinctly.

The purpose of Mendel's experiments was to observe the offspring from the crossing of each pair of plants and to look for patterns in the transmission of single traits. Although he was not the first to perform such experiments, he was the first to count and classify the peas that resulted from his crosses and compare the proportions with mathematical models. Mendel planned to use these data to try to deduce the laws by which these traits are passed on from generation to generation. First, Mendel crossed plants having contrasting forms of the single traits listed in Figure 22-3 by artificially fertilizing one with the other. Mendel called these plants the **parental (P) generation** and called their hybrid offspring the **first filial (F_1) generation.** (The word *filial* is from Latin words meaning "son" and "daughter.") These progeny are called **monohybrids** because they are the product of two plants that differ from one another in a single trait.

When Mendel crossed two contrasting varieties, such as purple-flowered plants with white-flowered plants, the hybrid offspring that he obtained were not intermediate in flower color, as the theory of blending inheritance predicted. Instead, all hybrid offspring in each case resembled *only one* of their parents. Thus in a cross of white-flowered plants with purple-flowered plants, *all* the F_1 offspring had purple flowers. In a cross of tall plants and short plants, *all* the F_1 offspring were tall plants. Mendel referred to the form of a trait that was expressed in the F_1 plants as *dominating,* or

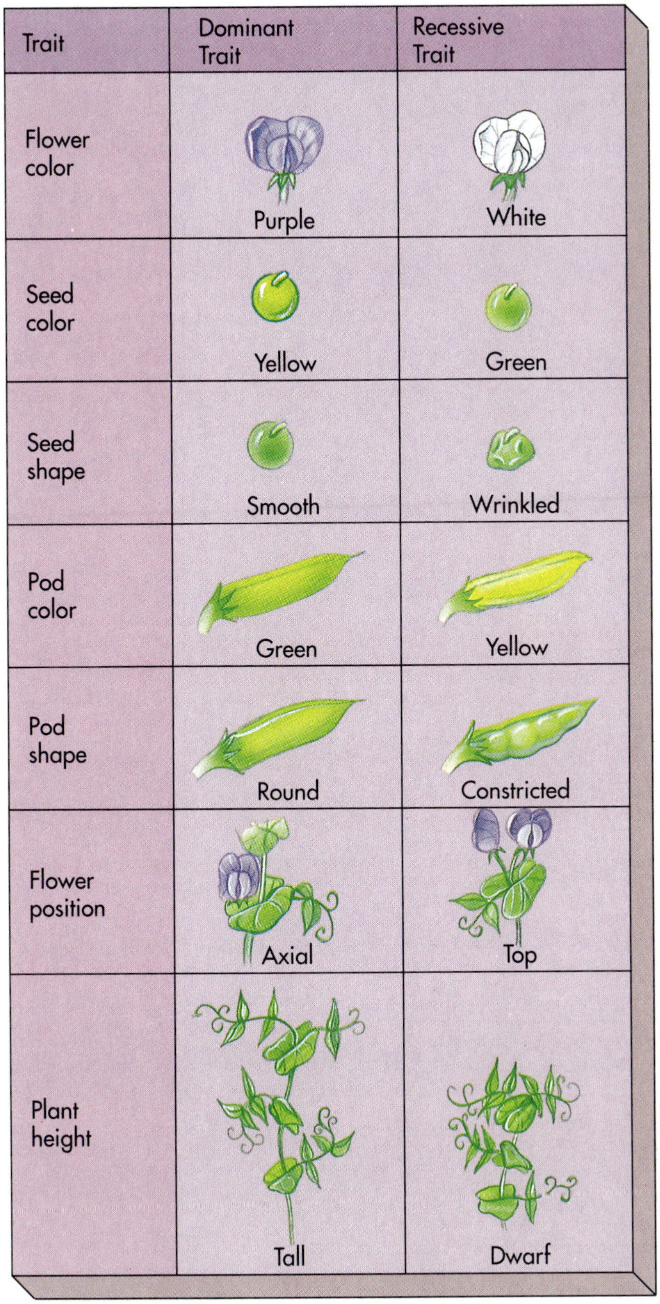

FIGURE 22-3 The seven pairs of contrasting traits in the garden pea studied by Mendel.

To determine inheritance patterns, Mendel crossed plants with contrasting forms of each of seven traits. The parent plants were the P (parental) generation, and their offspring were the F_1 (first filial) generation. Among the progeny of these crossing experiments, Mendel found that some traits dominated over other traits.

dominant, and to the alternative form, which was not expressed in the F_1 plants, as **recessive.** He chose the word *recessive* because this form of the trait receded, or disappeared entirely, in the hybrids. For each of the seven contrasting pairs of traits that Mendel examined, one member of each pair was dominant; the other was recessive. Figure 22-3 shows which traits Mendel found to be dominant and which traits he found to be recessive.

Then Mendel went a step further. He allowed each F_1 plant to mature and self-pollinate. He collected and planted seeds from each plant, which produced the **second filial (F_2) generation.** He found that most of the F_2 plants exhibited the dominating form of the trait and looked like the F_1 plants, but some exhibited the recessive form. None of the plants had blended characteristics. Therefore Mendel was able to count the numbers of each of the two contrasting varieties of F_2 progeny and compare these results.

Earlier in history, scientists had carried out hybridization experiments. However, the plants these scientists chose often produced hybrids that differed in appearance from both their parents because they exhibited blended traits. Scientists had also not quantified the results of their experiments. Mendel's change in experimental design—counting the progeny—was a key component of his ability to unravel the mystery of certain inheritance patterns.

When Mendel quantified the traits he observed in the F_2 generation, he discovered that for every three plants exhibiting the dominant form of the trait, one exhibited the recessive trait. Put another way, both contrasting forms of the parental characteristics reappeared in the F_2 generation in an approximate 3:1 ratio: three fourths of the plants exhibited the dominant form (determined in the F_1 generation), and one fourth exhibited the recessive form. Figure 22-4, *A* illustrates Mendel's experiments using the trait of flower color; Figure 22-4, *B* lists the results Mendel obtained in the F_2 generation for each of the contrasting forms of the seven traits. Notice that these numbers reflect a 3:1 ratio of dominant to recessive forms for each trait. Notice also that in his experiments Mendel used hundreds and sometimes thousands of plants. Why do you think that such a procedure is part of a good experimental design?

Mendel went on to examine what happened when each F_2 plant was allowed to self-pollinate. He found that the recessive one fourth were always true-breeding. For example, self-fertilizing white-flowered F_2 plants reliably produced only white-flowered offspring. By contrast, only one third of the dominant purple-flowered F_2 individuals (one fourth of the total offspring) proved true-breeding, whereas two thirds were not true-breeding. This last class of plants produced dominant and recessive F_3 individuals in a ratio of 3:1. The ratio of individuals in the entire F_2 population was the following (Figure 22-5):

1 true-breeding dominant:
2 not true-breeding dominant:
1 true-breeding recessive

A

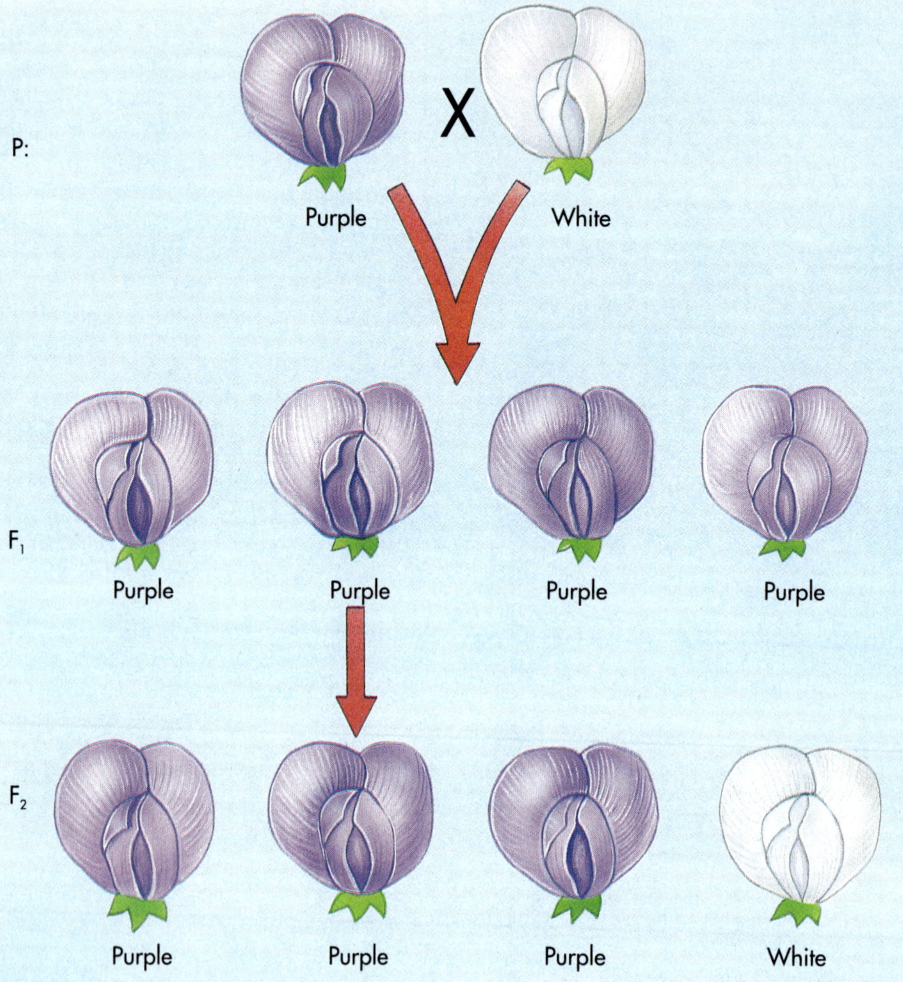

F_2	Stem length	Flower color	Seed color	Seed shape	Flower color	Pod shape	Flower position
Dominant allele	787	705	6022	5474	428	882	651
Recessive allele	277	224	2001	1850	152	299	207

B

FIGURE 22-4 Mendel's experiments refute the theory of blending inheritance.
A When crossing a purple flower with a white flower, Mendel observed that all the F_1 generation flowers were purple. When the F_1 plants were allowed to self-pollinate, three quarters of the F_2 generation exhibited the dominant trait (purple) and one quarter the recessive trait (white). There were no intermediate or blended flowers.
B Mendel found the approximate 3 : 1 ratio of dominant to recessive forms to be exhibited in each of the seven traits he studied.

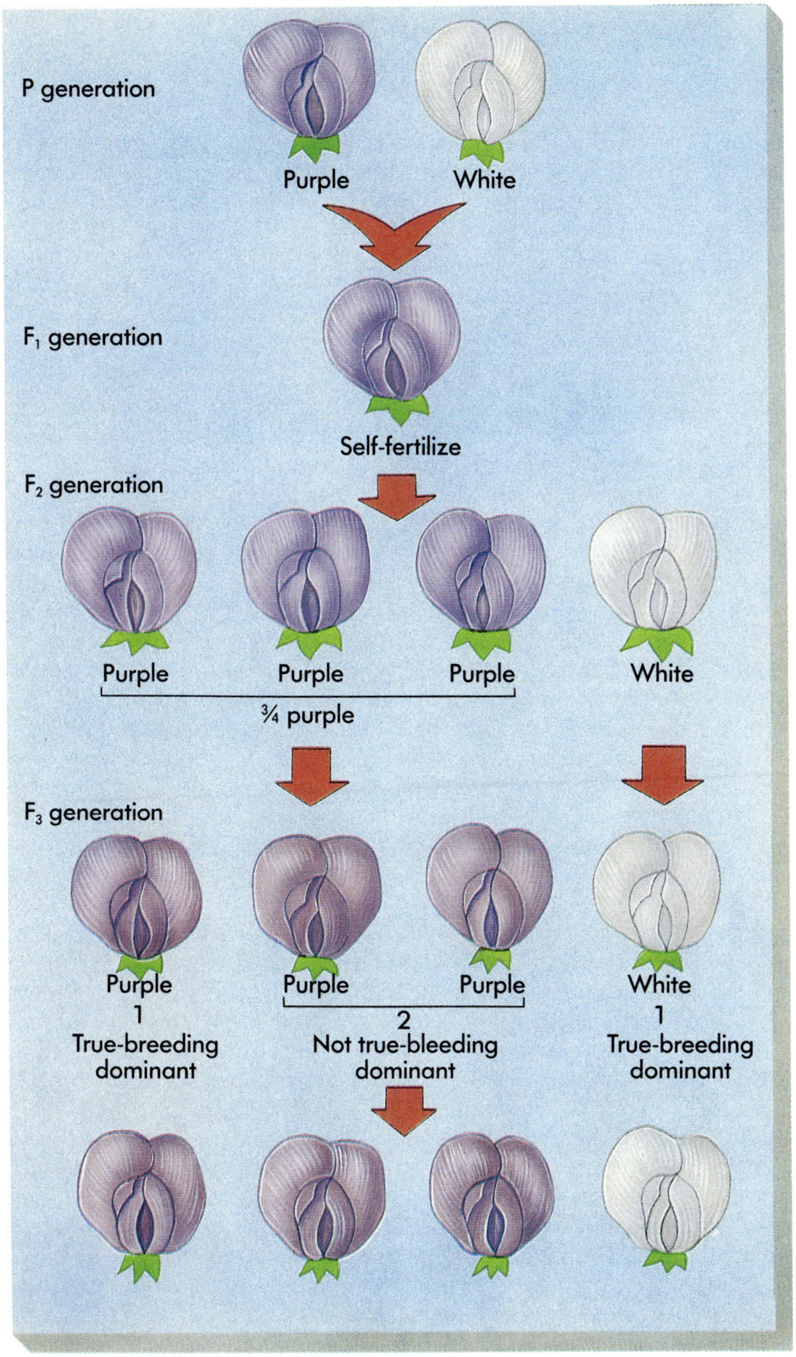

FIGURE 22-5 The F_3 generation allowed Mendel to determine which of the F_2 plants were true-breeding. By allowing the F_2 generation to self-pollinate, Mendel reasoned from the F_3 offspring that the F_2 generation exhibited the ratio of one true-breeding dominant, two not-true-breeding dominant, and one true-breeding recessive.

Conclusions Mendel drew from his experiments

From his experimental data, Mendel drew conclusions regarding the nature of heredity—conclusions that have withstood tests of time and further experimentation. In fact, Mendel's work is historically looked upon as the birth of **genetics,** the branch of biology dealing with the principles of heredity and variation in organisms. The statements that follow regarding Mendel's conclusions are considered to be the first established principles of genetics.

First, as mentioned previously, Mendel observed that the plants exhibiting the traits he studied did not produce progeny of intermediate appearance when crossed. These observations did not uphold the theory of blending inheritance but suggested instead that traits are inherited as discrete "packets" of information that are either present or absent in a particular generation. Mendel called these discrete bits of information *factors*. These factors, hypothesized Mendel, act later in the offspring to produce the trait. Today, scientists call these factors **genes,** the units of transmission of hereditary characteristics in an organism. In addition, scientists know that a gene is a segment, or piece, of DNA occupying a particular place on a particular chromosome. (These concepts regarding the molecular nature of genes and chromosomes are explained more fully in Chapters 23 and 24.)

Second, for each pair of traits that Mendel examined, one alternative form was not expressed in the F_1 hybrids, although it reappeared in some F_2 individuals. As mentioned earlier, Mendel referred to the form of the trait that was expressed in the F_1 plants as dominanting and referred to the alternative form, which was not expressed in the F_1 plants, as recessive. He inferred from these observations that each individual, with respect to each trait, contains two factors. Each pair of factors may contain information for (be a code for) the same form of a trait, or each member of the pair may code for an alternative form of a trait. Today, scientists call each member of a factor pair an **allele.** An allele is a particular form of a gene. Each human, for example, receives one allele for each gene from the mother's egg and one allele from the father's sperm. The 46 chromosomes of each human cell are actually 23 paired chromosomes, one member of each pair from the sperm and one from the egg.

Third, Mendel hypothesized that (1) because the two alleles that coded for a trait remained "uncontaminated," not blending with one another as the theory of blending inheritance predicted, and (2) since the results obtained from experiments on various traits showed similar results, then pea hybrids must form egg cells and pollen cells (gametes) in which the alleles for each trait separate from one another "in equal shares" during gamete formation. This concept is referred to as **Mendel's Law of Segregation.** Put in today's terms with today's understandings: Each gamete receives only one of an organism's pair of alleles. Chance determines which member of a pair of alleles becomes included in a gamete. This random segregating process takes place during the process of meiosis, or reduction-division of the nuclei of cells destined to be sex cells (see Chapter 5).

Fourth, Mendel realized that plants exhibiting the dominant trait in his monohybrid crosses of pea plants, when self-fertilized, would breed true or would produce plants exhibiting either the dominant or the recessive form of the characteristic in a 3:1 ratio, respectively. He observed that plants exhibiting the recessive trait, when self-fertilized, would always breed true. These data suggested to Mendel that true-breeding plants receive *only* the dominant factors or the recessive factors from each of their parents and that non-true-breeding plants were hybrids, which received the dominant and the recessive factors in equal shares. Today, scientists call an individual having two identical alleles for a trait **homozygous** for that trait. The prefix *homo* means "the same"; the suffix *zygous* refers to the zygote, or fertilized egg. An individual having two different alleles for a trait is said to be **heterozygous** (*hetero* means "different") for that trait.

Mendel's first law of inheritance, the Law of Segregation, states that each gamete receives only one allele of each pair of alleles in an organism's genetic makeup.

Scientists understand that Mendel's results were well-defined because he was studying traits that exhibited complete dominance. This is not always the case. Incomplete dominance is a situation in which neither member of a pair of alleles exhibits dominance over the other. In fact, the "blending" of genes observed by some early investigators may have been visible expressions of incomplete dominance. Since Mendel's time, many examples of incomplete dominance have been found for various traits in both plants and animals (see Chapter 23).

Analyzing Mendel's experiments

Looking back to Mendel's experiments, you can use the information from the conclusions he drew to further analyze and understand his experiments and data. Mendel used letters to represent alleles, with the dominant allele commonly denoted by an uppercase letter, and the recessive allele by the lowercase of the same letter. In the parental generation the cross of true-breeding (therefore homozygous) pea plants with purple flowers (the dominant trait) and true-breeding pea plants with white flowers (the recessive trait) can be represented as PP × pp (Figure 22-6). The *x* denotes a cross between two plants. Because the purple-flowered parent can produce only P gametes and the white-flowered parent can produce only p gametes, the union of an egg and a pollen grain from these parents can produce only heterozygous Pp offspring in the F_1 generation. But because the P allele is dominant, all of the F_1 individuals have purple flowers. The p allele, although present, is not visibly expressed.

To distinguish between the presence of an allele and its expression, scientists use the term **genotype** to refer to an organism's allelic makeup and the term **phenotype** to refer to the expression of those genes. The phenotype—the organism's outward appearance—is the end result of the functioning of the enzymes and the proteins coded by an organism's genotype.

The genotype of the F_1 generation of pea plants from the cross of true-breeding plants with purple flowers (PP) and true-breeding plants with white flowers (pp) is Pp, but the phenotype of the flowers of the hybrids is purple because P is the dominant allele. When these F_1 plants are allowed to self-fertilize, the P and p alleles segregate randomly during gamete formation. Their subsequent union at fertilization to form F_2 individuals is also random. Figure 22-6 shows the possible combinations of the gametes formed by the F_1 plants. Their random combination produces plants in an approximate ratio of 1 PP:2 Pp:1 pp. From these genotypes, can you determine the phenotypes of these plants?

Using a simple diagram called a **Punnett square** is another way to visualize the possible combinations of genes in a cross: Named after its originator, the English geneticist Sir Reginald Punnett, the Punnett square is used to align possible female gametes with possible male gametes in an orderly way. The male gametes are shown along one side of the square; the female gametes are shown along the other. As in Figure 22-7, the square is divided into smaller squares as columns are drawn vertically and rows are drawn horizontally, providing a "cell" for each possible combination of gene pair. Whether using a Punnett square or visualizing the gametes and their recombinations as in Figure 22-6, you can see the expected ratios of the three kinds of F_2 plants:

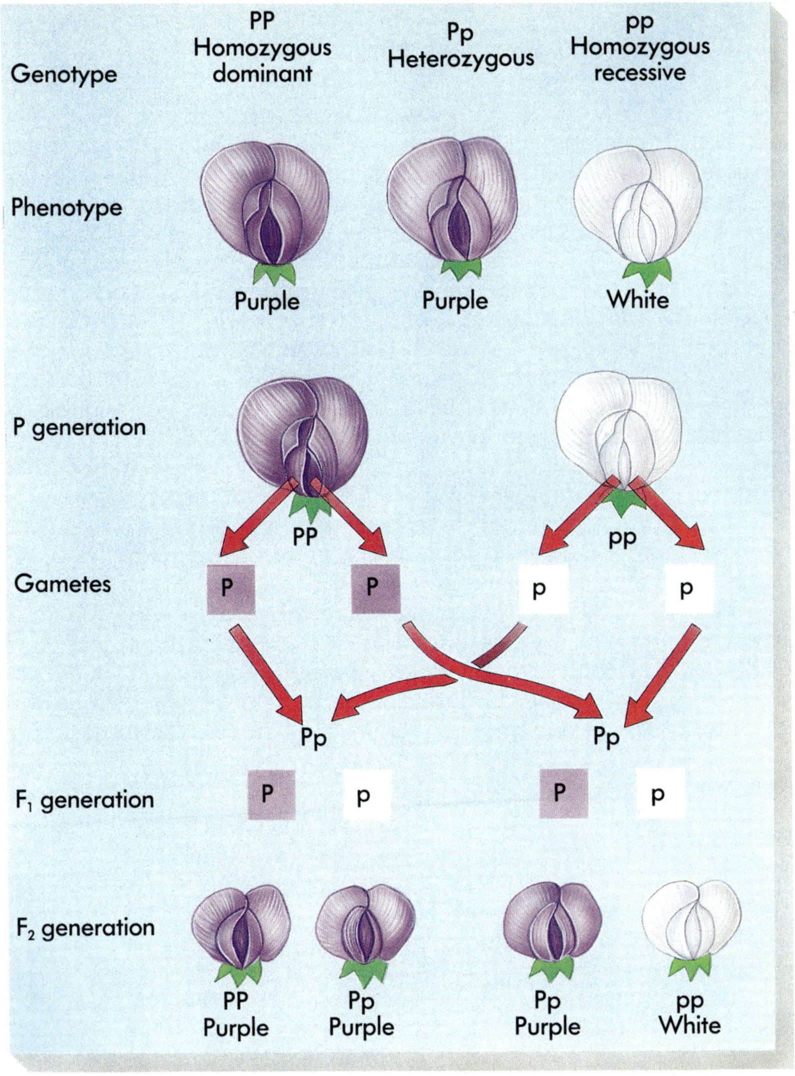

FIGURE 22-6 Analysis of Mendel's experiments. Mendel determined that the physical characteristics of his plants were a reflection of alleles, which he represented with letters. In this diagram, *P* represents purple flower color, and *p* represents white flower color. Because true-breeding purple flowers can produce only P gametes and white flowers only p gametes, a cross between these two types of flowers (the P generation) can yield only purple flowers (Pp) in the F_1 generation. The p allele is recessive and thus is not expressed in the F_1 generation. The P and p alleles segregate randomly during gamete formation and randomly recombine in the F_2 generation to produce plants in an approximate ratio of 1 PP : 2 Pp : 1 pp.

one fourth are true-breeding pp white-flowered plants, two fourths are heterozygous Pp purple-flowered plants, and one fourth are pure-breeding PP purple-flowered individuals. The 3 : 1 phenotypic ratio is really a 1 : 2 : 1 genotypic ratio.

> **The outward appearance of an individual (the expression of the genes) is referred to as its *phenotype.* The genetic makeup of an individual is referred to as its *genotype.* The genotype represents the alleles for given genes that are present. Using a Punnett square is one way to visualize the genotypes of progeny in simple Mendelian crosses and to illustrate their expected ratios.**

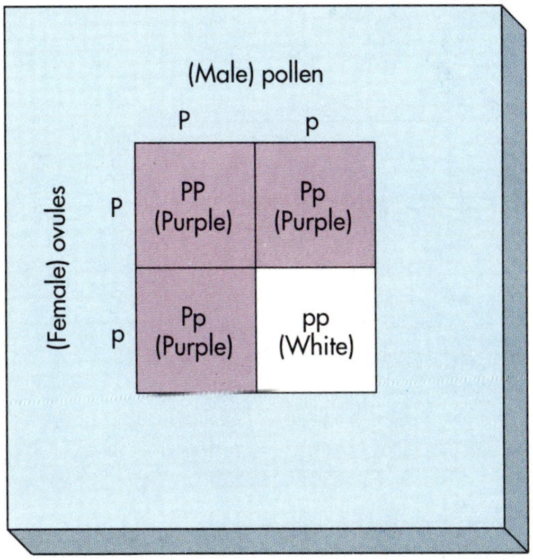

FIGURE 22-7 Punnett square analysis of Mendel's F_1 generation. Each smaller cell within the square contains the possible F_2 phenotypic combinations from this cross. The phenotypic ratio is 3 : 1, but the genotypic ratio is 1 : 2 : 1. Look back at Figure 22-5 and see that the genotypic ratio matches the expected F_2 ratio of true-breeding dominant to non-true-breeding dominant to true-breeding recessive.

How Mendel tested his conclusions

To test his conclusions further and to distinguish between homozygous dominant and heterozygous phenotypes, Mendel devised two procedures. The first test, already mentioned, was to self-fertilize the plant having a dominant phenotype. If the plant is homozygous dominant, it breeds true: purple-flowered plants produce progeny with purple flowers, tall plants produce tall plants, and so forth. If the plant is heterozygous, on the other hand, it produces dominant and recessive offspring in a 3 : 1 ratio when self-fertilized.

The other procedure Mendel used is called a **testcross.** In this procedure, Mendel crossed the phenotypically dominant test plant with a known homozygous recessive plant. He predicted that (1) if the test plant is homozygous for the dominant trait, the progeny will all be hybrids, Pp for example, and will therefore look like the test plant and that (2) if the test plant is heterozygous, then *half* the progeny will be heterozygous and look like the test plant, but half the progeny will be homozygous recessive and therefore will exhibit the recessive characteristic. Figure 22-8 shows the two possible testcross scenarios, using the characteristics that Mendel actually used in his first testcross experiments.

What did Mendel's data reveal? When he performed the testcross using hybrid F_1 plants having smooth seeds (Ss) and crossed them with a variety of plant having wrinkled seeds (ss), he obtained 208 plants: 106 with smooth seeds, and 102 with wrinkled seeds—a 1 : 1 ratio just as he had predicted. These data confirmed the primary conclusion Mendel drew from his earlier work: alternative alleles segregate from one another in the formation of gametes, coming together in the progeny in a random manner.

Further questions Mendel asked

Expanding on his Law of Segregation, Mendel asked a new question: Do the pairs of factors (alleles) that determine particular traits segregate independently of factor pairs that determine other traits? In other words, does the segregation of one factor pair influence the segregation of another?

To answer his question, Mendel first developed a series of true-breeding lines of peas that differed from one another with respect to two of the seven pairs of characteristics with which he had worked in his monohybrid studies. He then crossed pairs of plants that exhibited contrasting forms of the two characteristics and that bred true. For example, he crossed plants having round, yellow seeds with plants having wrinkled, green seeds. From his monohybrid studies, Mendel knew that the traits "round seeds" and "yellow seeds" are dominant to "wrinkled seeds" and "green seeds." Therefore the genotypes of the true-breeding parental (P) plants were RRYY and rryy.

Mendel's F_1 progeny are **dihybrids**—the product of two plants that differ from one another in two traits. As in his monohybrid crosses, all the F_1 progeny had round, yellow seeds—the dominant phenotype. Mendel then allowed the F_1 dihybrids to self-fertilize. The seeds from these self-

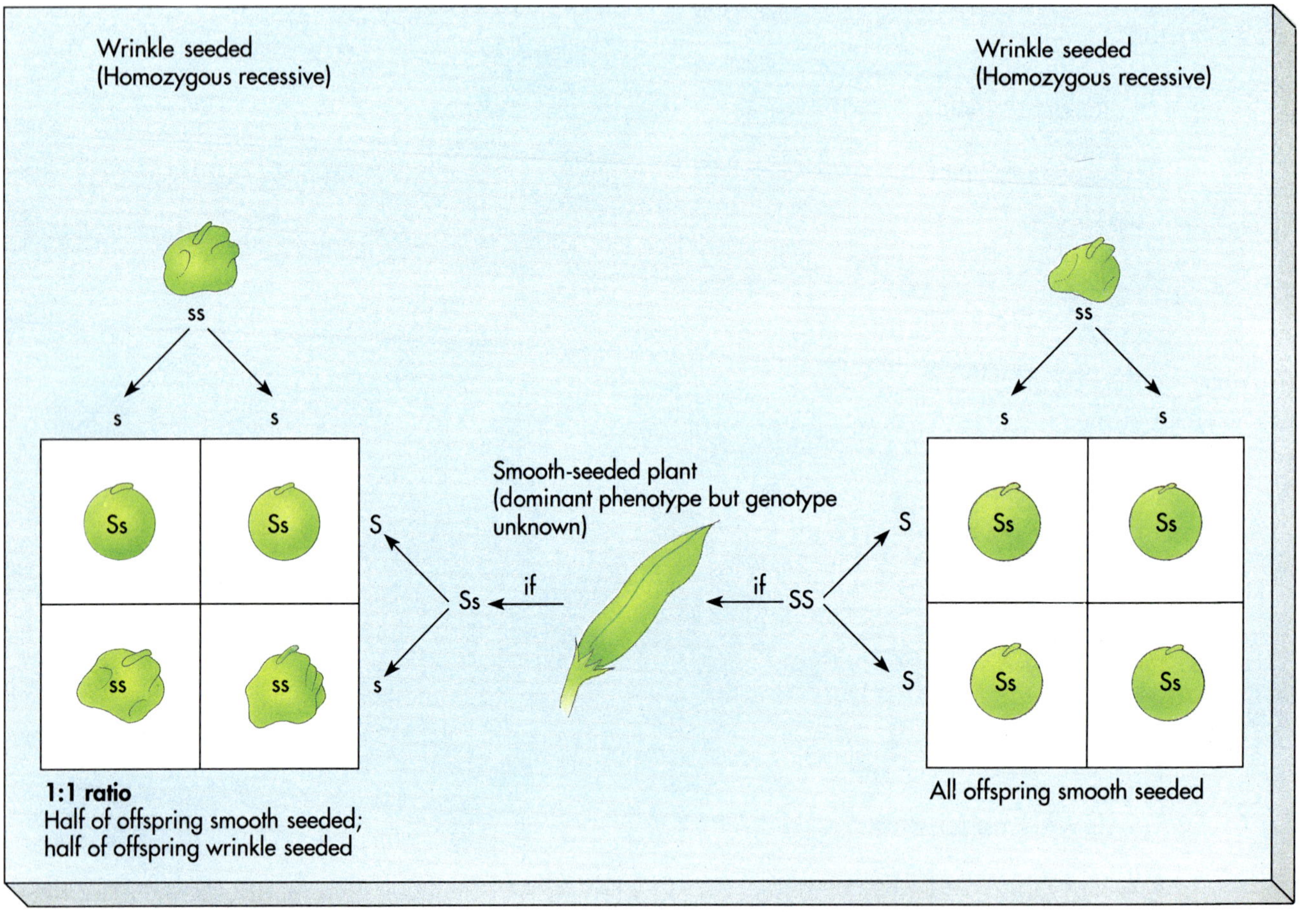

FIGURE 22-8 A testcross.
The phenotypically dominant test plant, which has smooth seeds but an unknown genotype, is in the center, crossed with a known homozygous recessive plant having wrinkled seeds. The Punnett square *(left)* shows the expected genotypes and phenotypes if the test plant is heterozygous dominant. The Punnett square on the right shows the expected genotypes and phenotypes if the test plant is homozygous dominant.

crosses grew into 315 plants having smooth, yellow seeds; 101 plants having wrinkled, yellow seeds; 108 plants having smooth, green seeds; and 32 plants having wrinkled, green seeds—an approximate ratio of 9:3:3:1. Other dihybrid crosses also produced offspring having the same approximate ratio.

Analyzing the results of Mendel's dihybrid crosses

Mendel reasoned that if the alleles for seed color and seed shape segregated into gametes independently of one another and were therefore inherited independently of one another, then the outcome for each trait would exhibit the 3:1 ratio of a monohybrid cross. Looking at Mendel's results, 315 + 108 (423) plants had smooth seeds and 101 + 32 (133) plants had wrinkled seeds. Put simply, *three times* as many plants had smooth seeds as had wrinkled seeds—a 3:1 ratio. Study these figures for the trait "seed color." How many plants produced yellow seeds? How many produced green seeds? What is the approximate ratio of plants having yellow seeds to plants having green seeds?

The contrasting alleles for the genes of seed shape and seed color assort independently from one another during gamete formation. This concept is referred to as **Mendel's Law of Independent Assortment.** Put in today's terms with today's understandings: The distribution of alleles for one trait into the gametes does not affect the distribution of alleles for other traits.

Even though you may understand the concept of the independent assortment of alleles, you may not understand how two 3:1 ratios combine to produce a 9:3:3:1 ratio. These ratios and the relationship between these ratios are governed by the laws of **probability,** or chance. Each ratio is not simply a statement of the comparative numbers of plants Mendel found but is also a predictive statement of events that could occur in the future under the same conditions.

Mendel's second law of inheritance, the Law of Independent Assortment, states that the distribution of alleles for one trait into the gametes does not affect the distribution of alleles for other traits.

Analysis of a dihybrid cross using probability theory

Figure 22-9 shows the self-cross of hybrid F_1 plants having smooth seeds. The genotype of these plants is Ss. The male gametes S and s are shown along one side of the Punnett square, and the female gametes S and s are shown along the other. These four gametes can combine in four ways. What is the probability that this cross will produce a

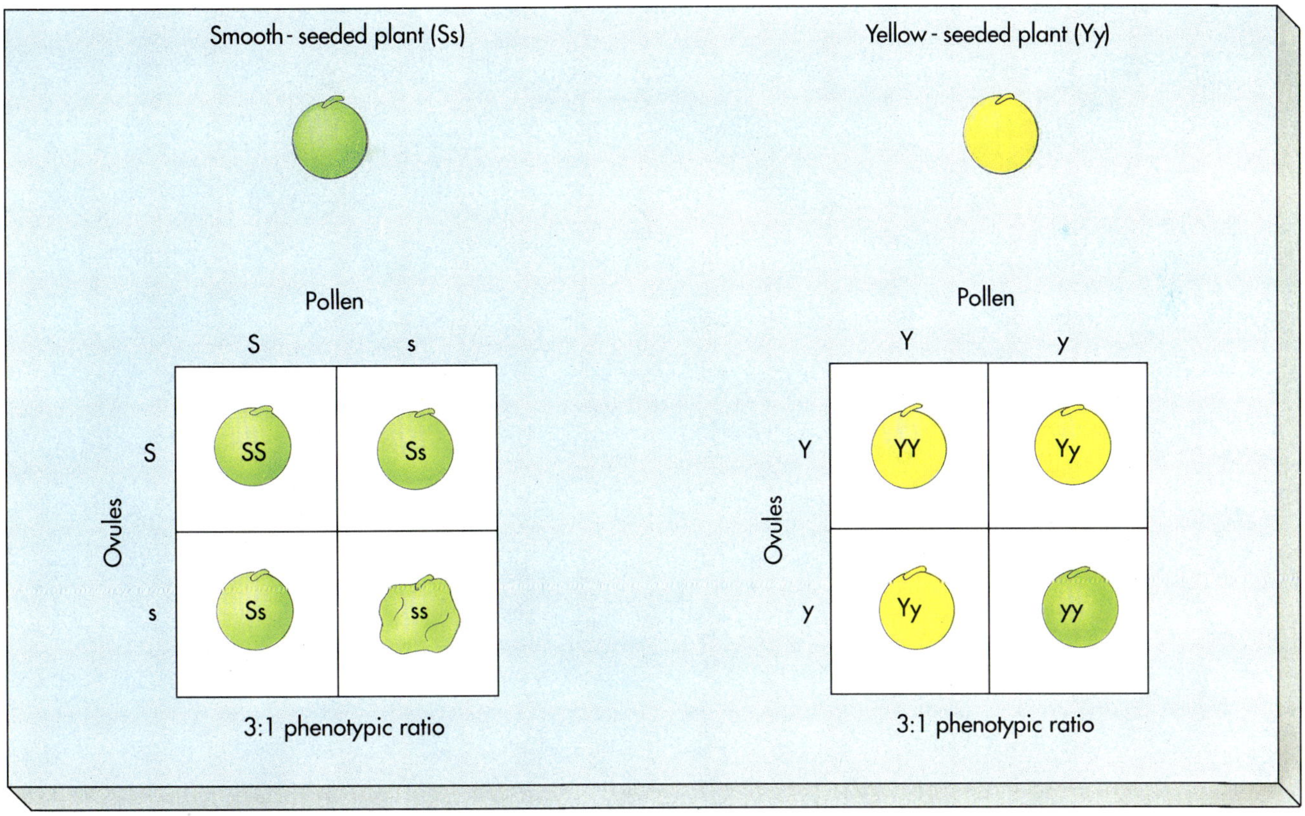

FIGURE 22-9 Mendel's law of independent assortment.
As this diagram demonstrates, the contrasting alleles for seed shape (S and s) and color (Y and y) assort independently from one another during gamete formation.

plant having wrinkled seeds? Because there are only four possible combinations of gametes and the alleles segregate randomly as the gametes are formed, there is only a one in four possibility that the ss gametes will combine. Likewise, there is a three out of four probability that gametes can combine in such a way that they will produce plants having smooth seeds: SS, Ss, and ss. Although Ss and sS are the same genotypes, the reversed position of the alleles represents the dominant and recessive alleles as contributed by each of two parent plants. Hence the 3:1 phenotypic ratio. The alleles for the trait seed color, when looked at independently of seed shape, segregate in the same manner to produce plants having yellow seeds and plants having green seeds in a 3:1 ratio, respectively (see Figure 22-9).

To illustrate further, you can simulate the combination of the gametes S and s of two parents by performing a simple activity. Take two pennies and tape the letter "S" on one side of each and "s" on the other. Flip the two pennies 100 times and record your results. The probability of flipping an S on one of the pennies is one out of two, or 1/2. The chance of flipping an S on the other penny is also 1/2. The probability that an event will occur at the same time as another independent event is simply the product of their individual probabilities. Therefore the probability of flipping SS with the two pennies is $1/2 \times 1/2$, or 1/4. (Did you flip SS approximately 25 times out of 100?) Likewise, the probability of flipping ss is 1/4, flipping Ss is 1/4, and flipping sS is 1/4. (Do your data match?) If the pennies were in fact gametes, 3/4 would produce plants having smooth seeds and 1/4 would produce plants having wrinkled seeds—a 3:1 ratio.

Now look at these traits as they occur together in a dihybrid cross. The probability that a plant with wrinkled, green seeds will appear in the F_2 generation is equal to the probability of observing a plant with wrinkled seeds (1/4) times the probability of observing a plant with green seeds (1/4), or 1/16. The probability that a plant with smooth, green seeds will appear in the F_2 is equal to the probability that the F_1 parents will produce a plant with smooth seeds (3/4) times the probability that they will produce a plant with green seeds (1/4), or 3/16. Can you figure out the probability that a plant with wrinkled, yellow seeds will appear in the F_2 or that a plant with smooth, yellow seeds will appear?

Analysis of a dihybrid cross using a Punnett square

Figure 22-10 shows the self-cross of hybrid F_1 plants having smooth seeds and yellow seeds, showing how the gametes segregate to produce the F_2 plants. The eight gametes of the two parents can combine in 16 different ways. What is the probability that this cross will produce a plant

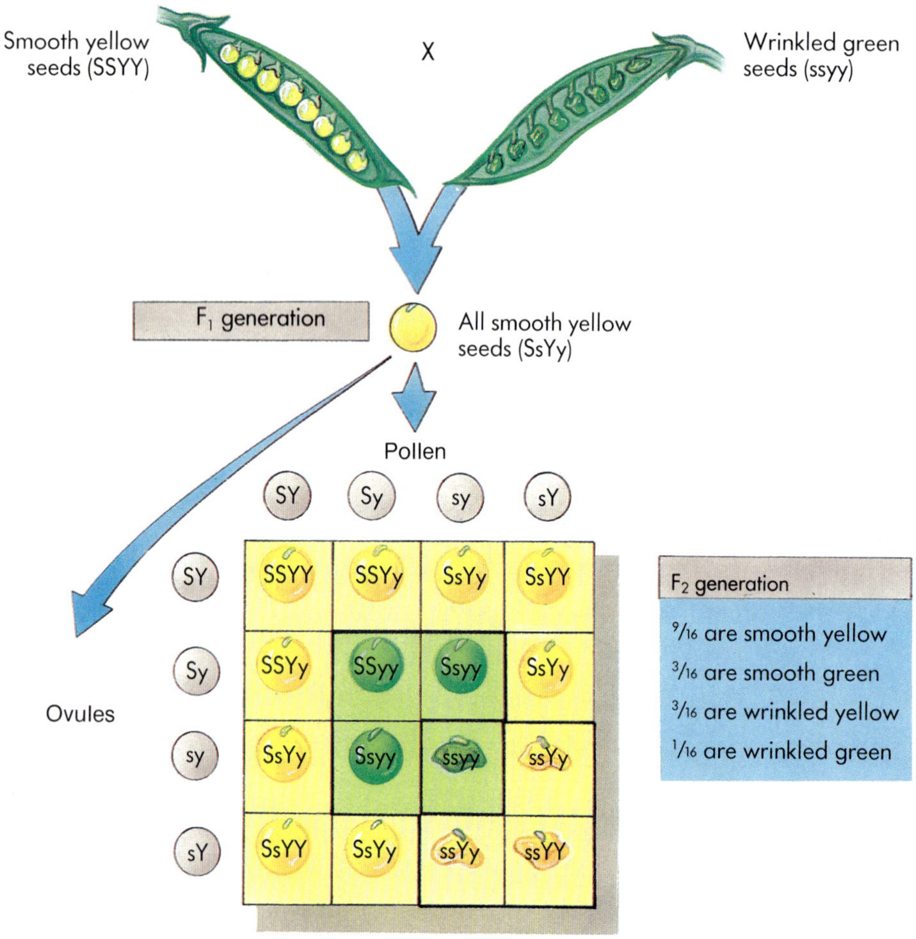

FIGURE 22-10 A Punnett square showing the results of Mendel's self-cross of dihybrid of smooth yellow-seeded plants. The approximate ratio of the four possible combinations of phenotypes is predicted to be 9:3:3:1, the ratio that Mendel found.

having wrinkled, green seeds? Because there are only 16 possible combinations of gametes and if you assume that the alleles segregate randomly and assort independently as the gametes are formed, there is only a 1 in 16 possibility that an sy egg will combine with an sy sperm (producing ssyy offspring). Likewise, there are only 3 combinations of gametes out of 16 possible combinations that will produce plants having wrinkled, yellow seeds: ssYY, ssYy, and ssyY. Following the same reasoning and using the Punnett square, notice that there are 3 possible combinations of gametes that will form plants producing smooth, green seeds and 9 possible combinations of gametes that will form plants producing smooth, yellow seeds. So whether you calculate the expected offspring in a dihybrid cross by using a Punnett square or whether you do this analysis by using the probability theory, the ratio of the F_2 progeny is predicted to be 9:3:3:1.

The connection between Mendel's factors and chromosomes

During the years Mendel was experimenting with pea plants to determine the principles of heredity, other scientists were studying the structure of cells. Although Mendel published his theories of inheritance in 1866, few scientists read his work or understood its significance. In the late 1870s, scientists had described the process of nuclear division now known as mitosis and the process of nuclear reduction-division that occurs as gametes are formed now known as meiosis (see Chapter 5). In fact, in 1888, scientists gave the name *chromosomes* to the discrete, threadlike bodies that form as the nuclear material condenses during these processes of cell division. But they still had no idea of the link between Mendel's *factors* of inheritance and the newly named chromosomes—both important keys to unlocking the mysteries of heredity.

In 1900, three biologists, Carl Correns, Hugo de Vries, and Eric von Tschermak, independently worked out Mendel's principles of heredity. However, they knew nothing of Mendel's work until they searched the literature before publishing their results. The rediscovery of Mendel's work—his hypotheses supported by the independent work of others—helped scientists begin to make connections between Mendel's ideas and chromosomes.

By the late 1800s, scientists had observed the process of fertilization and knew that sexual reproduction required the union of an egg and a sperm. However, they did not know how each contributed to the development of a new individual. Many scientists hypothesized that the sperm merely stimulated the egg to develop. An American graduate student, Walter Sutton, suggested that if Mendel's hypotheses were correct, then each gamete must make equal hereditary contributions. Sutton also suggested that because sperm contain little cytoplasm, the hereditary material must reside within the nuclei of the gametes. He noted that chromosomes are in pairs and segregate during meiosis, as did Mendel's factors. In fact, the behavior of chromosomes during the meiotic process paralleled the behavior of the hereditary factors. Using this line of reasoning, Sutton suggested that Mendel's factors were located on the chromosomes. However, Sutton had no experimental evidence to support his hypothesis. Years later, various scientists worked on this problem, but the most conclusive evidence to uphold Sutton's chromosomal theory of inheritance was provided by a single, small fly.

Sex linkage

In 1910, Thomas Hunt Morgan, studying the fruit fly *Drosophila melanogaster,* detected a male fly that differed strikingly from normal flies of the same species. This fly had white eyes (Figure 22-11) instead of the normal red eyes. Morgan quickly designed experiments to determine if this new trait was inherited in a Mendelian fashion.

Morgan first crossed the white-eyed male fly to a normal female to see if red or white eyes were dominant. All F_1 progeny had red eyes, and Morgan therefore concluded that

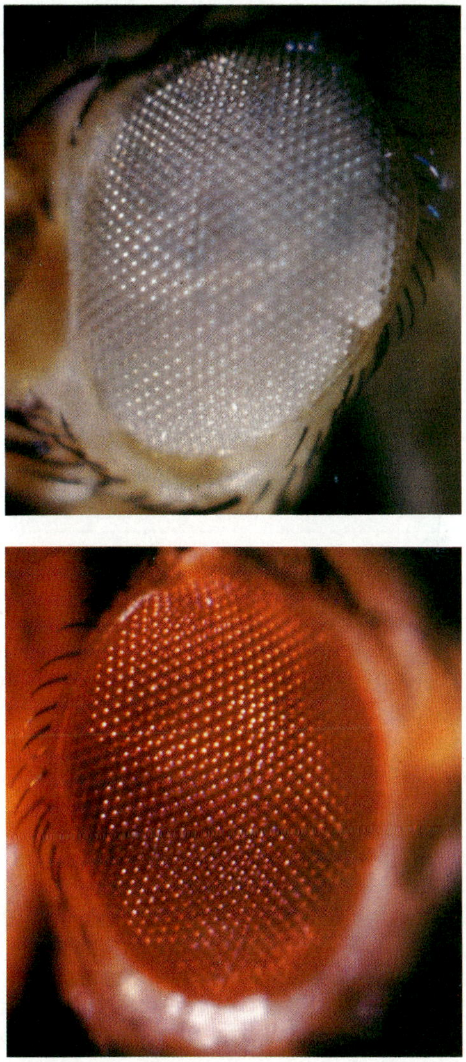

FIGURE 22-11 Red-eyed and white-eyed *Drosophila melanogaster.* The white-eyed defect in eye color is hereditary, the result of a mutation in a gene located on the sex-determining X chromosome. It was by studying this mutation that Morgan first demonstrated that genes are on chromosomes.

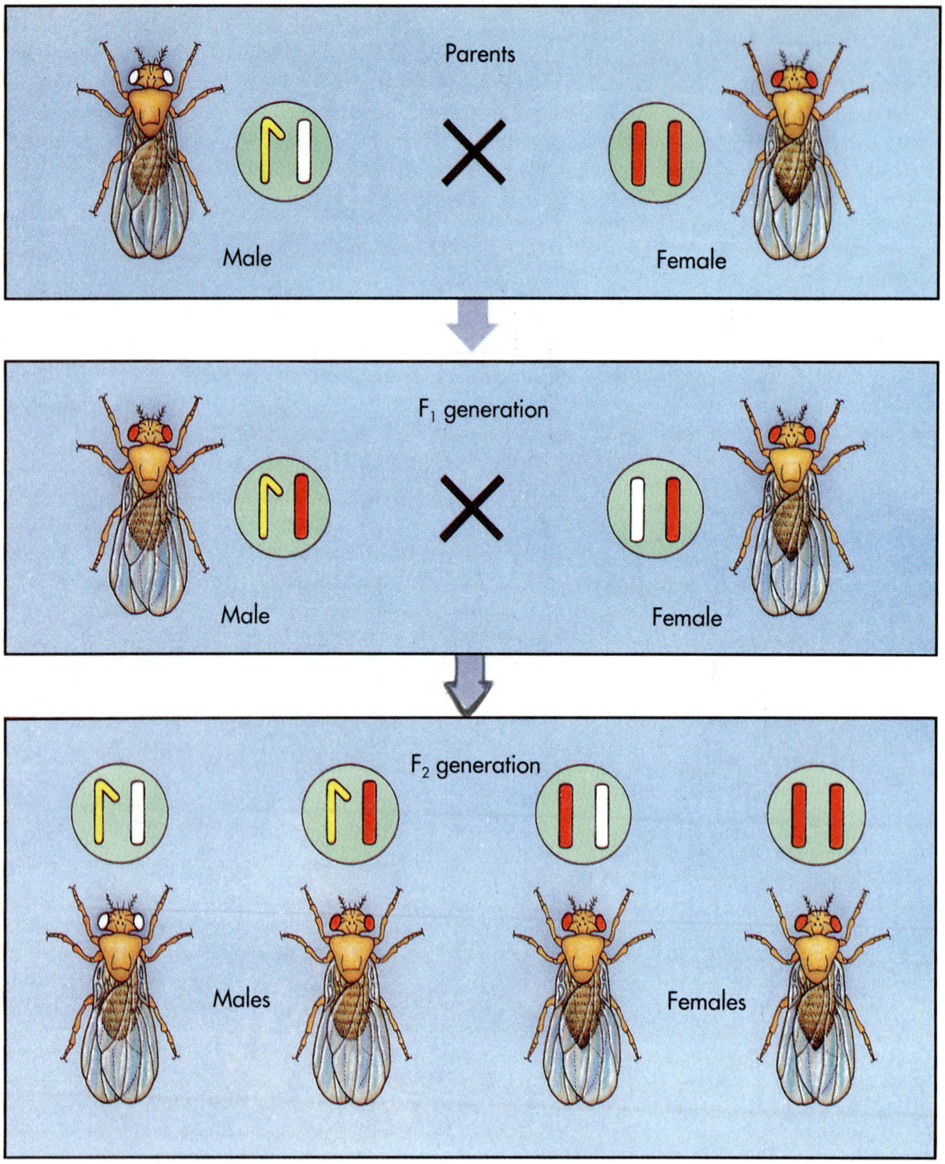

FIGURE 22-12 Morgan's experiment demonstrating the chromosomal basis of sex-linkage in *Drosophila melanogaster.*
White-eyed mutant male flies were crossed with normal red-eyed females. The F_1 generation flies all exhibited red eyes. The male flies carried the normal red-eyed allele along with a Y chromosome that contains no allele for eye color, whereas the female flies carried one red-eyed allele and one white-eyed allele. Therefore in the F_2 generation, all female progeny were red eyed, being either homozygous or heterozygous for this dominant trait. The F_2 male flies, however, exhibited the trait inherited on their single X chromosome. Because half of them bear the white allele (the F_1 mother is heterozygous), half the male progeny will exhibit white eyes. This is exactly the result that Morgan observed: All of the white-eyed generation flies were male.

red eye color was dominant over white. Following the experimental procedure that Mendel had established long ago, Morgan then crossed flies from the F_1 generation with each other. Eye color did indeed segregate among the F_2 progeny as predicted by Mendel's theory. Of 4252 F_2 progeny that Morgan examined, 782 had white eyes—an imperfect 3:1 ratio but one that nevertheless provided clear evidence of segregation. Something was strange about Morgan's result, however—something totally unpredicted by Mendel's theory: all of the white-eyed F_2 flies were males!

How could this strange result be explained? The solution to this puzzle involves gender. In drosophila (as in most animals), the gender of the fly is determined by specific chromosomes called *X* and *Y* chromosomes. Female flies have four pairs of chromosomes, with one of those pairs being two X chromosomes. Male flies, on the other hand, have three pairs plus an X and a Y chromosome. To explain his results, Morgan deduced that the white-eyed trait is located on the X chromosome but is absent from the Y chromosome. (Scientists now know that the Y chromosome carries relatively few functional genes.) Because the white-eye trait is recessive to the red-eye trait, Morgan's result was a natural consequence of the Mendelian segregation of alleles (Figure 22-12). Morgan's experiment is one of the most important in the history of genetics because it presented the first clear evidence upholding Sutton's theory that the factors determining Mendelian traits are located on the chromosomes. When Mendel observed the segregation of alternative traits in pea plants, he was observing the outward reflection of the meiotic segregation of homologous chromosomes.

From his observation of chromosomes during meiosis and his knowledge of Mendel's work, Walter Sutton hypothesized that the hereditary factors were on the chromosomes. Thomas Hunt Morgan provided experimental evidence to uphold this hypothesis, also revealing that certain traits may be located on the chromosomes that determine the gender of an organism. Such traits are said to be *sex-linked.*

Solving the mysteries of inheritance

Throughout human history, an understanding of the nature of heredity was highly speculative until the work of Mendel, Sutton, and Morgan. These early geneticists took the first real steps toward solving the puzzles of inheritance and laid the foundation for one of the great scientific advancements of the twentieth century: an understanding of the nature of genetic material and how it is transmitted from generation to generation. This basic outline of heredity led to a long chain of questions and was to be modified and refined as scientists provided additional experimental data and evidence. One line of questioning regards inheritance in humans—the topic of Chapter 23. And in Chapter 24, another path leads to genetics at the molecular level—the structure, natural functioning, and human manipulation of genes.

Summary

1. Scientists and philosophers hypothesized about the nature of inheritance over many centuries. Some suggested that parents gave off particles from their bodies that traveled to their sex organs during reproduction. Many assumed that the male and female contribution blended in the offspring. Others held that either sperm or eggs carried tiny preformed beings within them. Not until the mid-1800s did scientists realize that new life arose from old life in the form of new cells arising from old cells.
2. In the mid-1800s, Gregor Mendel, an Austrian monk trained in botany and mathematics, studied patterns of inheritance using garden pea plants. Using true-breeding varieties of pea plants that exhibited alternative forms of seven different traits, Mendel artificially cross-fertilized them. The purpose of his experiments was to observe the offspring from the crossing of each pair of plants and to look for patterns in the transmission of traits.
3. Mendel found that alternative traits segregate in crosses and may mask each other's presence. From counting progeny types, Mendel learned that the alternatives that were masked in hybrids appeared only 25% of the time when the hybrids were self-crossed. This finding, which led directly to Mendel's model of heredity, is usually referred to as the Mendelian ratio of 3:1 dominant to recessive traits.
4. Mendel deduced from the 3:1 ratio that traits are specified by discrete factors that do not blend in the offspring. Today, scientists refer to Mendel's factors as genes and to alternative forms of his factors as alleles.
5. Because Mendel observed that traits did not blend in the offspring and because the results he obtained from experiments on various traits showed similar results, Mendel hypothesized that alleles separate from one another during gamete formation. This concept is called *Mendel's Law of Segregation.*
6. Working with pea plants that differed from one another in two characteristics, Mendel discovered that alleles of different genes assort independently during gamete formation. This concept is referred to as *Mendel's Law of Independent Assortment.*
7. The first clear evidence that genes reside on chromosomes was provided by Thomas Hunt Morgan. Morgan demonstrated that the segregation of the white-eye trait in *Drosophila melanogaster* was associated with the segregation of the X chromosome, the one responsible for sex determination.

GENETICS PROBLEMS

1. The annual plant *Haplopappus gracilis* has two pairs of chromosomes 1 and 2. In this species, the probability that two traits *a* and *b* selected at random will be on the same chromosome of *H. gracilis* is the probability that they will both be on chromosome 1 (½), multiplied by the probability that they will both be on chormosome 2 (also ½): ½ × ½ = ¼, or 25%. This is often symbolized

$$1/2^{(\#\ \text{of pairs of chromosomes})}$$

Human beings have 23 pairs of chromosomes. What is the probability that any two human traits selected at random will be on the same chromosome?

2. Among Hereford cattle there is a dominant allele called ***polled;*** the individuals that have this allele lack horns. After college, you become a cattle baron and stock your spread entirely with polled cattle. You have many cows and few bulls. You personally make sure that each cow has no horns. Among the calves that year, however, some grow horns. Angrily you dispose of them and make certain that no horned adult has gotten into your pasture. The next year, however, more horned calves are born. What is the source of your problem? What should you do to rectify it?
3. Many animals and plants bear recessive alleles for albinism, a condition in which homozygous individuals completely lack any pigments. An albino plant lacks chlorophyll and is white. An albino person lacks any melanin pigment. If two normally pigmented persons heterozygous for the same albinism allele have children, what proportion of their children would be expected to be albino?
4. Your uncle dies and leaves you his race horse, Dingleberry. To obtain some money from your inheritance, you decide to put the horse out to stud. In looking over the stud book, however, you discover that Dingleberry's grandfather exhibited a rare clinical disorder that leads to brittle bones. The disorder is hereditary and results from homozygosity for a recessive allele. If Dingleberry is heterozygous for the allele, it will not be possible to use him for stud because the genetic defect may be passed on. How would you go about determining whether Dingleberry carries this allele?
5. In Drosophila, the allele for dumpy wings (d) is recessive to the normal long-wing allele (D). The allele for white eye (w) is recessive to the normal red-eye allele (W). In a cross of DDWw × Ddww, what proportion of the offspring are expected to be "normal" (long wing, red eye)? What proportion "dumpy, white"?
6. Your instructor presents you with a drosophila named Oscar. Oscar has red eyes, the same color that normal flies possess. Your add Oscar to your fly collection, which also contains Heidi and Siegfried, flies with white eyes, and Dominque and Ronald, which are from a long line of red-eyed flies. Your previous work has shown that the white-eyed trait exhibited by Heidi and Siegfried is caused by their being homozygous for a recessive allele. How would you determine whether Oscar was heterozygous for this allele?
7. In some families, children are born who exhibit recessive traits (and therefore must be homozygous for the recessive allele specifying the trait), even though one or both of the parents do not exhibit the trait. What can account for this occurrence?
8. You collect two drosophila in your backyard, one a young male and the other a young, unmated female. Both are normal in appearance, with the typical vivid red eyes of drosophila. You keep the two flies in the same vial, where they mate. Two weeks later, hundreds of little offspring are flying around in the vial. They all have normal red eyes. From among them, you select 100 individuals, some male and some female. You cross each individually to a fly you know to be homozygous for the recessive allele sepia, which leads to black eyes when homozygous (these flies thus have the genotype *se*/*se*). Examining the results of your 100 crosses, you observe that in about half of them only normal red-eyed progeny flies are produced. In the other half, however, the progeny are about 50% red eyed and 50% black eyed. What must have been the genotypes of your original backyard flies?

REVIEW QUESTIONS

1. How do offspring of sexual reproduction differ from those produced by asexual reproduction? Compare the amount of variation introduced within a species that reproduces sexually with a species that reproduces asexually.
2. Distinguish between self-fertilization and cross-fertilization.
3. Distinguish between a gene and an allele.
4. What is Mendel's Law of Segregation? When does this segregation occur?
5. Fill in the blanks: An individual possessing two identical alleles for a trait is ______ for that trait; an individual with two different alleles for a trait is ______.
6. Distinguish between genotype and phenotype.
7. Assume that "L" represents the dominant trait of having long leaves, and "l" represents the recessive short-leaved trait in a plant. In the parental generation, you cross a homozygous long-leaved plant with a homozygous short-leaved plant. Draw a Punnett square illustrating this cross, and give the genotypes and phenotypes of the F_1 generation.
8. Using a Punnett square, show the genotypes and phenotypes of the F_2 generation if the F_1 plants in question 7 are self-fertilized.
9. What was Mendel testing when he used a testcross? What procedure did he use, and what was the outcome?
10. Fill in the blanks: If two parent plants differ from each other in a single trait, their progeny are ______. If the parents differ from each other in two traits, their offspring are ______.
11. What is Mendel's Law of Independent Assortment?
12. In question 8, what is the probability that the F_2 plants will have short leaves?
13. In a dihybrid cross between organisms that are heterozygous for both traits, what is the probability that their offspring will exhibit the phenotype for (a) both recessive traits? (b) both dominant traits?
14. Fill in the blank: Traits that are located on the chromosomes that determine the gender of an organism are ______.

KEY TERMS

allele 378
cross-fertilization 374
dihybrid 380
dominant 375
first filial (F_1) generation 374
gene 377
genetics 377
genotype 378
heterozygous 378
homozygous 378
hybrid 374
Mendel's Law of Independent Assortment 381
Mendel's Law of Segregation 378
monohybrid 374
parental (P) generation 374
phenotype 378
probability 381
Punnett square 378
recessive 375
second filial (F_2) generation 375
self-fertilization 374
testcross 380
traits 374
true-breeding 374
varieties 374

THOUGHT QUESTIONS

1. Can you think of a distinctive human trait that is *not* inherited? How would you test this hypothesis?
2. Later work has shown that two of the seven traits Mendel used in his pea plant crosses (pod shape and plant height) are located near each other at the tip of chromosome 4. Is this consistent with Mendel's Law of Independent Assortment in light of Sutton's chromosomal theory of inheritance? Discuss.
3. What is wrong with explaining Morgan's failure to observe white-eyed female flies among the F_2 progeny of a cross between a white-eyed male and a red-eyed female reflecting the fact that homozygous recessive (ww) individuals do not survive (that is, that w is a recessive lethal, killing any homozygous embryos)? How would you test this hypothesis?

FOR FURTHER READING

Blixt, S. (1975). Why didn't Gregor Mendel find linkage? *Nature, 256,* 206.
This is an interesting examination of the pea strains first studied by Mendel.

Morgan, T. H. (1910). Sex-limited inheritance in *Drosophila. Science, 32,* 120-122.
This article describes Morgan's original account of his famous analysis of the inheritance of the white eye trait.

Sutton, W. S. (1903). The chromosomes of heredity. *Biological Bulletin, 4,* 213-251.
This is the original statement of the chromosomal theory of heredity.

23 Human genetics

OUTLINE

- **Studying inheritance patterns in humans using karyotypes**
  - The inheritance of abnormal numbers of autosomes
  - The inheritance of abnormal numbers of sex chromosomes
    - Triple X females
    - Klinefelter syndrome
    - Turner syndrome
    - XYY males
  - Changes in chromosome structure
  - Gene mutations
- **Studying inheritance patterns in humans using pedigrees**
- **Dominant and recessive genetic disorders**
- **Incomplete dominance and codominance**
- **Multiple alleles**
- **Genetic counseling**

About 100 years ago, this is what family photos looked like. But this family is not just any family—it is a royal family. Queen Victoria of England, matriarch of this family, is sitting in the center and is wearing a crown. At the time this photo was taken, Queen Victoria did not know that she would have an enormous impact on the lives of some of the family members surrounding her and on the lives of future members of the royal families of both Russia and Spain.

HIGHLIGHTS

▼

Genetic disorders can occur when a person inherits a fewer or greater number of chromosomes than is normal, chromosomes with pieces that have been added or removed, or chromosomes having changes in the molecular structure of their DNA.

▼

Geneticists use pictures of chromosomes arranged in pairs according to size, and pedigrees, diagrams of genetic relationships among family members over several generations, to study the inheritance of genetic disorders.

▼

Inherited traits do not always exhibit clear-cut dominance and recessiveness; certain traits exhibit incomplete dominance or codominance or may be coded by multiple alleles.

▼

Although most genetic disorders cannot yet be cured, research scientists are making continual progress in developing gene therapies that may, in the future, have a drastic impact on a physician's ability to treat genetic disorders before birth.

Queen Victoria affected many of her descendants because of a single allele—a mutant, or changed gene, which coded for a disorder known as *hemophilia.* Hemophilia is a hereditary condition in which the blood is slow to clot or does not clot at all. It is a recessive disorder, expressed only when an individual does not have a normal blood-clotting allele that masks the mutant's appearance. In addition to being recessive, the allele for this type of hemophilia is sex-linked: it is located on the X chromosome, one of the chromosomes that determines gender. This mutant gene probably arose in one of the sex cells from which Queen Victoria developed, since the disease was not manifested in earlier generations.

Genetically normal females have two X chromosomes; normal males have one X and one Y chromosome. (The genes that determine male characteristics are located on the Y chromosome.) Although the X and Y chromosomes act as pairing mates during meiosis (see Chapter 5), only a portion of these two chromosomes have "matching" genes. The Y chromosome lacks many (and in some organisms all) of the alleles that correspond to alleles on the X chromosome. A male who inherits an X chromosome with a mutant clotting allele will develop hemophilia because his Y chromosome does not have a corresponding allele to mask the allele on the X chromosome. A woman who inherits a mutant allele on one X chromosome and a normal allele on the other will not have the disorder but can pass the mutant allele along to her offspring. Such women are called *carriers.*

Of Queen Victoria's four daughters who lived to bear children, two daughters—Alice and Beatrice—were carriers of hemophilia. Two of Alice's daughters, Princess Irene of Prussia and Czarina Alexandra of Russia, were also carriers of hemophilia.

Studying inheritance patterns in humans using karyotypes

In addition to a pair of **sex chromosomes,** each human body cell also contains 22 pairs of **autosomes.** Unlike sex chromosomes, autosomes (or "body" chromosomes) are the same in both sexes. The 23 pairs of human chromosomes are shown in Figure 23-1. Arranged in this manner according to size, shape, and other characteristics, the chromosome pairs make up a **karyotype.**

Looking at a karyotype can often help researchers see genetic disorders if they are caused by the loss of all or part of a chromosome or by the addition of extra chromosomes or chromosome fragments. Changes in single genes *cannot* be seen. The differences between alleles that code for alternate forms of a trait lie in the chemical structure of the DNA, so they are invisible in a karyotype. Permanent changes in the genetic material, whether they affect single genes, pieces of chromosomes, whole chromosomes, or entire sets of chromosomes, are called **mutations.**

Each human cell contains 23 pairs of chromosomes. A karyotype is a picture of these chromosome pairs and shows whether a person has inherited a fewer or greater number of chromosomes than is normal or has lost or gained pieces of chromosomes.

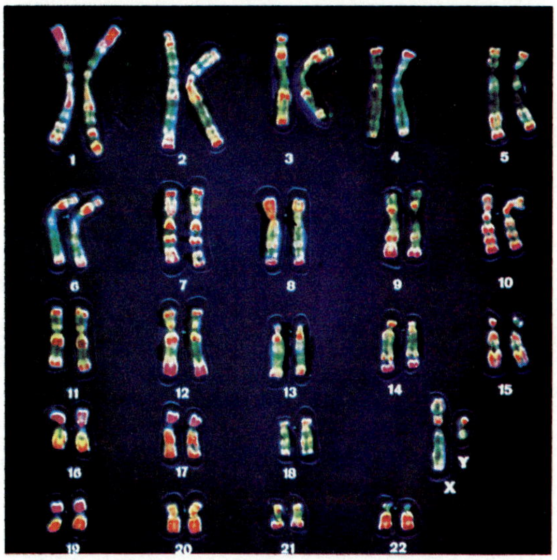

FIGURE 23-1 A normal karyotype of a human male. Humans have 23 pairs of chromosomes, including the two sex chromosomes. Notice the small Y chromosome in pair 23 that is characteristic of males.

The inheritance of abnormal numbers of autosomes

Look at the karyotype in Figure 23-2, *A* and compare it with that in Figure 23-1. What differences do you see? Carefully examine chromosome 21. One karyotype contains an extra copy of this chromosome, a situation called *trisomy 21.* The developmental defect produced by trisomy 21 was first described in 1866 by J. Langdon Down. For this reason, it is called **Down syndrome.** In these individuals the maturation of the skeletal system is delayed, so persons having Down syndrome are generally short and have poor muscle tone. In addition, they are mentally retarded.

Down syndrome is only one genetic disorder caused by the inheritance of an abnormal number of autosomal or sex chromosomes. How does such a situation arise? In humans, it comes about almost exclusively as a result of errors during the process of meiosis. Meiosis is the process of nuclear division in which the number of chromosomes in cells is halved during gamete formation (see Chapter 5).

Early in meiosis, pairs of chromosomes called *homologues* (the pairs of chromosomes shown in the karyotype) line up side-by-side in a process called *synapsis.* Because, at this time, each chromosome is made up of two sister chromatids, the paired homologous chromosomes together have four chromatids. Gemetes can gain or lose chromosomes at this point in the meiotic process if two homologous chromosomes fail to separate, or disjoin. If **nondisjunction** occurs, two of the resulting gametes carry a double of the chromosome; the other two gametes lack the chromosome entirely. Nondisjunction can occur later in the meiotic process as well, when two sister chromatids fail to separate (Figure 23-3).

The cause of nondisjunction is not known. However, the occurrence of nondisjunction of chromosome 21, which results in Down syndrome, increases as a woman's age in-

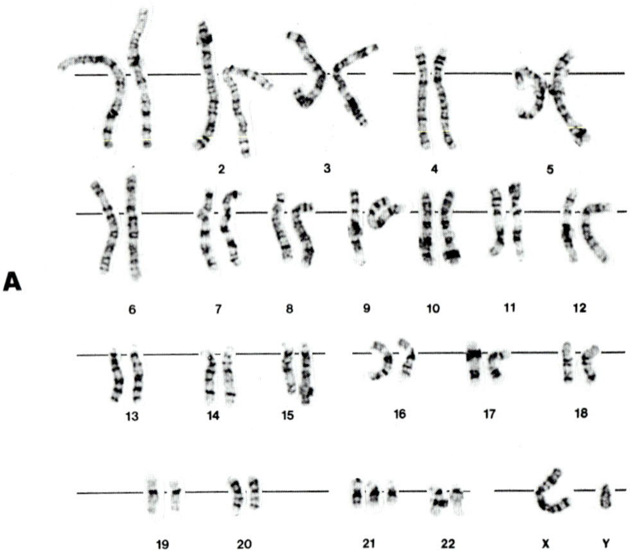

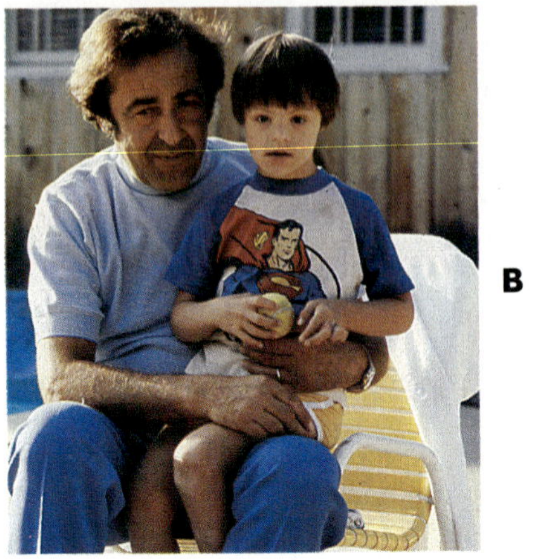

FIGURE 23-2 Down syndrome.
A As shown in this karyotype of a male, Down syndrome is usually associated with trisomy of chromosome 21.
B Down syndrome child sitting on his father's knee.

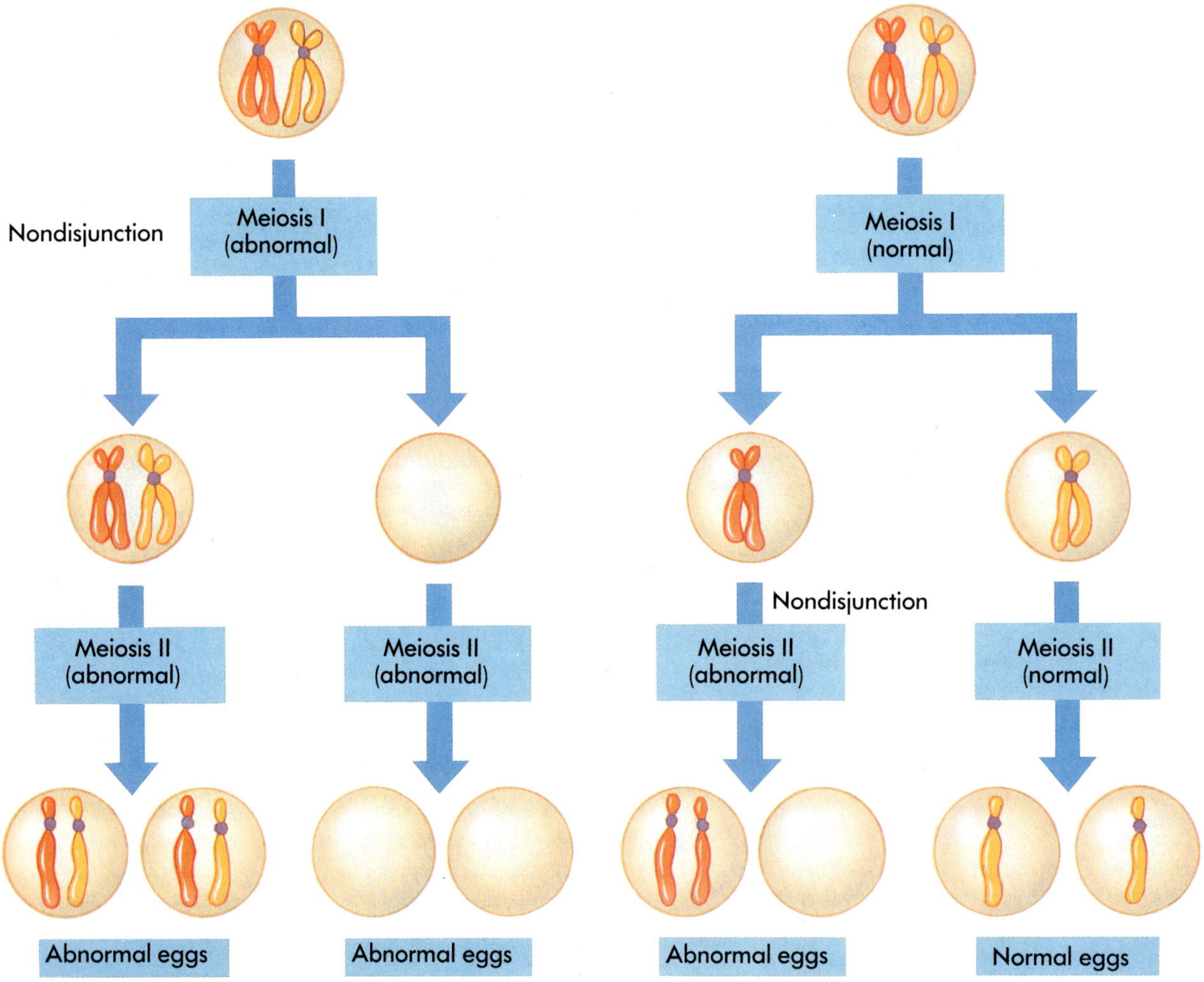

FIGURE 23-3 Nondisjunction.
Nondisjunction results from an error in meiosis in which homologous chromosomes fail to separate during meiosis I. Nondisjunction can also occur during meiosis II when two sister chromatids fail to separate.

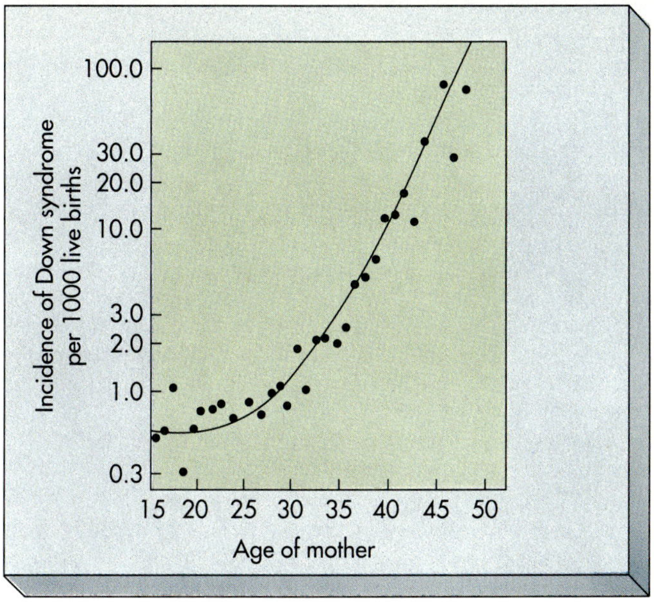

FIGURE 23-4 Incidence of Down syndrome vs. maternal age. The occurrence of this nondisjunction increases sharply as a woman grows older, thus the Y axis uses an exponential scale. The 10-fold increase from 1 to 10 is graphed using the same scale as the 10-fold increase from 10 to 100.

creases (Figure 23-4). In mothers less than 20 years of age, the occurrence of Down syndrome children is only about 1 per 1700 births. In mothers 20 to 30 years old, the incidence is only slightly greater—about 1 per 1400. However, in mothers 30 to 35 years old, the incidence almost doubles to 1 per 750. And in mothers over 45, the incidence of Down syndrome babies is as high as 1 in 16 births.

During gamete formation, homologous chromosomes occasionally fail to separate after synapsis. This occurrence, called *nondisjunction,* results in gametes with abnormal numbers of chromosomes. Down syndrome is a genetic disorder produced when an individual receives three (instead of two) 21 chromosomes.

The incidence of nondisjunction of other chromosomes also rises as a woman's age rises. Babies with other serious autosomal chromosome abnormalities are rare. Fertilized eggs with the improper number of chromosomes are almost always inviable; that is, they are unable to survive. These eggs do not begin normal development and implantation but are cast out of the body with the menstrual flow—a process called *spontaneous abortion*. This increase in the incidence of nondisjunction and its negative impact on the viability of zygotes is one reason older women often have a harder time conceiving than younger women.

The inheritance of abnormal numbers of sex chromosomes

Nondisjunction can also occur with the sex chromosomes. However, persons inheriting an extra X chromosome or inheriting one X too few do not have severe developmental abnormalities as persons with too many or too few autosomes do. However, persons who inherit abnormal numbers of sex chromosomes often have abnormal physical features.

Triple X females

When X chromosomes fail to separate in meiosis, some gametes are produced that possess both of the X chromosomes; the other gametes have no sex chromosome and are designated *O*. If an XX gamete joins an X gamete during fertilization, the result is an XXX (triple X) zygote. Triple X females usually have underdeveloped breasts and genital organs, but they can often bear children. In addition, a small number of XXX people have a lower-than-average intelligence. Although rare, a few individuals have been discovered to have tetra X (XXXX) and penta X (XXXXX) genotypes. Individuals having these genotypes are similar phenotypically to triple X individuals but are usually mentally retarded.

Klinefelter syndrome

If the XX gamete joins a Y gamete, the result is quite serious. The XXY zygote develops into a sterile male who has, in addition to male genitalia and characteristics, some female characteristics, such as breasts (Figure 23-5) and a high-pitched voice. In some cases, XXY individuals have a lower than average intelligence. This condition, called **Klinefelter syndrome,** occurs in about 1 out of every 600 male births.

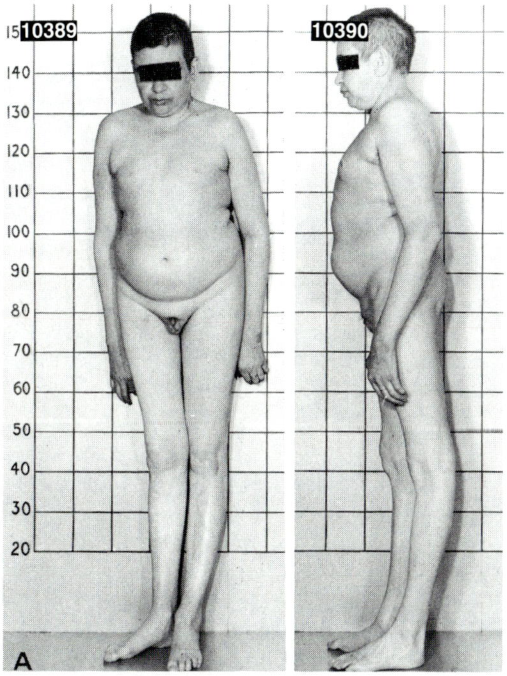

FIGURE 23-5 Klinefelter syndrome. A male with Klinefelter syndrome (XXY) exhibits some female characteristics, such as enlarged breasts and a high-pitched voice.

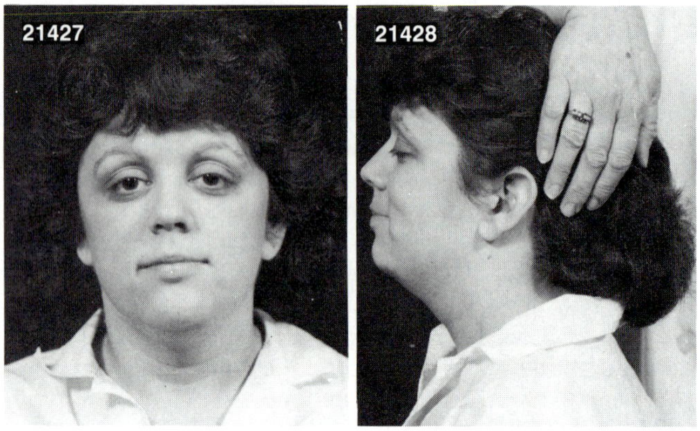

FIGURE 23-6 Turner syndrome.
A female with Turner syndrome (XO) exhibits short stature, webbed neck, low-set ears, broad chest, and immature sex organs.

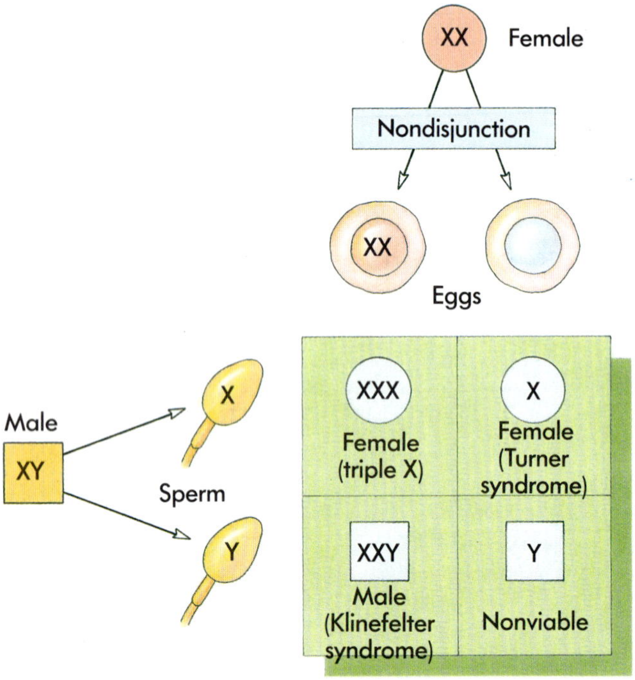

FIGURE 23-7 How nondisjunction can result in abnormalities in the number of sex chromosomes.
Both Klinefelter syndrome and Turner syndrome result from nondisjunction of either the male or female gamete. This diagram shows how these genetic disorders occur when nondisjunction takes place in female gametes. It also shows how nondisjunction in female gametes can result in a triple X female.

Turner syndrome

If an O gamete (no X) from the mother fuses with a Y gamete, the resulting OY zygote is inviable and fails to develop further. If, on the other hand, an O gamete from either the mother or the father fuses with an X gamete to form an XO zygote, the result is a sterile female of short stature, a "webbed" neck, low-set ears, a broad chest, and immature sex organs that do not undergo puberty changes (Figure 23-6). The mental abilities of an XO individual are slightly below normal. This condition, called **Turner syndrome,** occurs roughly once in every 3000 female births. The ways in which nondisjunction can result in abnormal numbers of sex chromosomes are shown in Figure 23-7.

XYY males

The Y chromosome occasionally fails to separate from its sister chromatid in meiosis II (see Chapter 5). Failure of the Y chromosome to separate leads to the formation of YY gametes and viable XYY zygotes that develop into males who are unusually tall but have normal fertility. The frequency of XYY among newborn males is about 1 per 1000. The XYY syndrome has some interesting history associated with it. In the 1960s, the frequency of XYY males in penal and mental institutions was reported to be approximately 2% (that is, 20 per 1000)—20 times higher than in the general population. This observation led to the suggestion that XYY males are inherently antisocial and violent. Further studies revealed that XYY males have a higher probability of coming into conflict with the law than XY (normal) males, but their crimes are usually nonviolent. Most XYY males, however, lead normal lives and cannot be distinguished from other males.

Persons inheriting abnormal numbers of sex chromosomes often have abnormal features and may be mentally retarded. Zygotes without an X chromosome are not viable.

Changes in chromosome structure

Although a person may have inherited the proper number of chromosomes, a chromosome that is structurally defective may occur if a chromosome breaks and the cell repairs the break incorrectly or does not repair the break at all. These broken and misrepaired chromosomes can then be passed on in the gametes of the parents to produce disorders in the offspring.

Chromosomes may break naturally or may be caused by outside agents, such as ionizing radiation and chemicals. **Ionizing radiation,** such as x-rays and nuclear radiation, is a form of energy known as *electromagnetic energy.* Sunlight is also a type of electromagnetic energy (see Chapter 8). Ionizing radiation, however, has a higher level of energy than ordinary light does. When ionizing radiation reaches a cell, it transfers energy to electrons in the outer shells of the atoms it encounters, raising them to a still higher energy level and ejecting them from their orbitals. The result is the breaking of covalent bonds and the formation of charged

Curing genetic disorders by gene transfer

BIOLOGY TECHNOLOGY & society

Genetic disorders create tragic situations for many families. Cystic fibrosis, multiple sclerosis, muscular dystrophy and many other inherited conditions strike thousands of families every year. Until recently, there was little one could do about such genetic diseases, but gene technology is rapidly providing hope. To understand how, it is necessary to appreciate that many genetic disorders arise because of a disabling mutation in a single gene. Cystic fibrosis, for example, is caused by a harmful mutation in a gene that codes a protein needed to transport chloride ions in and out of cells. In principle, all you would need to do to cure such gene disorders is add a working copy of the damaged gene to the body's cells. In practice, this idealized therapy was considered fantasy. On September 14, 1990, it was done successfully on a 4-year-old girl.

The girl suffered from a rare genetic disorder called *severe combined immune deficiency*. Certain critical white blood cells of her immune system called *T cells* were not working properly, leaving her body with no defense against infection. The symptoms are the same as AIDS, but this condition is not caused by a virus; it is inherited. The problem is in a gene coding the enzyme adenoside deaminase (ADA). T cells need this enzyme to function, and the girl's ADA genes were defective (that is, she had two inactive variants). Without the active ADA enzyme, her immune system was helpless to stave off infection, and she had little chance of living much longer.

Doctors removed some of her blood and from it cultured T cells. They then added to the cell culture viruses that possessed a good working copy of the human ADA gene the girl lacked. These viruses carried the ADA gene with them as they infected the T cells and spliced it along with their own genes into T-cell chromosomes. Doctors grew these altered T cells in the laboratory until they numbered into the billions and then injected a billion into a vein in the child's hand (Figure 23-A).

Over the next year the treatment was repeated several times as the genetically repaired cells multiplied in her body, populating the girl's immune system with gene-corrected T cells that produce the necessary enzyme. At this time, the girl is healthy and happy, her immune system fully restored. In time, it is hoped that the bone marrow cells from which her T cells are derived will be replaced by gene-corrected stem cells. Then she will be the first human cured of a genetic disorder. Medicine is alive with the hope that gene transfer therapy will provide a similar cure for many other genetic disorders.

FIGURE 23-A Dr. W. French Anderson, the scientist who performed the first successful gene therapy treatment on a human being.

fragments of molecules with unpaired electrons. The charged molecular fragments, or **free radicals,** are highly reactive. They may interact with DNA, producing chromosomal breaks or changes in the nucleotide structure.

When chromosomes break and are improperly repaired by the cell, some chromosomal information may be added, lost, or moved from one location on the chromosome to another. Chromosomal rearrangement changes the way that the genetic message is organized, interpreted, and expressed. Newly added chromosomal information can be caused by a **duplication** of a section of a chromosome. Seen in a karyotype, a chromosome having a duplication appears longer than its homologue. A **translocation** could also produce an abnormally long chromosome. In this situation, a section of a chromosome breaks off and then reattaches to another chromosome. The chromosome losing a section would then appear shorter due to this **deletion** of information. As gametes form, the progeny could receive one or both of these defective chromosomes. If a chromosomal **inversion** occurs, the broken piece of chromosome reattaches to the same chromosome but in a reversed direction (Figure 23-8).

A well-known disorder associated with a chromosomal deletion in humans is the **cri du chat syndrome.** Described in 1963 by a French geneticist, cri du chat means "cat cry" and describes the cat-like cry made by cri du chat babies. Other symptoms of this disorder include severe mental retardation and a round "moon face." (Table 23-1 summarizes the chromosomal aberrations discussed.)

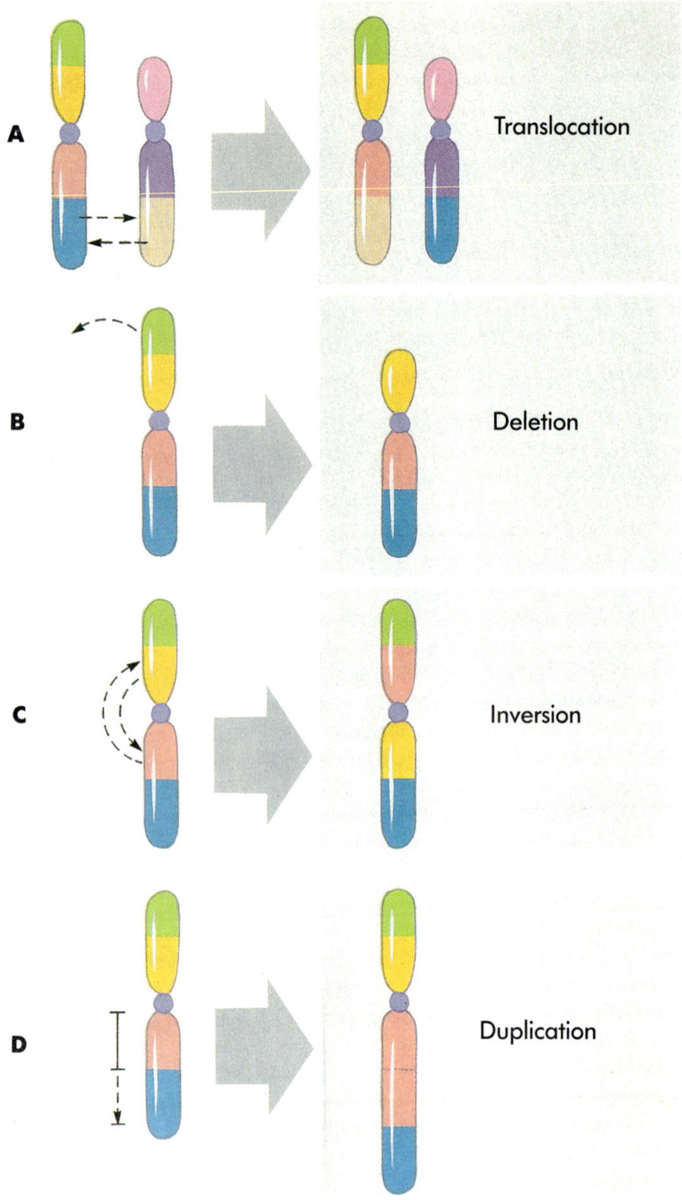

FIGURE 23-8 Types of chromosomal rearrangement.
A Translocation occurs when a section of a chromosome breaks off and attaches to another chromosome.
B A deletion occurs when a chromosome loses a section.
C Inversion occurs when a broken piece of a chromosome reattaches to the same chromosome but in a reversed direction.
D A duplication occurs when chromosomal information is duplicated within the chromosome.

Gene mutations

Sometimes changes take place in a single allele rather than entire sections of chromosomes. A change in the genetic message of a chromosome due to alterations of molecules within the structure of the chromosomal DNA is called a **point mutation** or **gene mutation.** Through mutation, a new allele of a gene is produced. Mutations may occur spontaneously (although their occurrence is rare) or may be caused by ionizing radiation, UV radiation, or chemicals.

TABLE 23-1 Summary of chromosomal aberrations

CONDITION	CHARACTERISTICS	ABERRATION
Trisomy 13	Multiple defects, including severe mental retardation and deafness Death before 6 months of age for 90% of those who survive birth	3 13 autosomes
Trisomy 18	Facial deformities Heart defects Death before 1 year of age for 90% of those who survive birth	3 18 autosomes
Down syndrome (trisomy 21)	Developmental delay of skeletal system Mental retardation	3 21 autosomes
Trisomy 22	Similar features to Down syndrome More severe skeletal deformities	3 22 autosomes
Cri du chat	Moon face Severe mental retardation	Deletion on 5
Triple X female	Underdeveloped female characteristics	XXX XXXX XXXXX
Turner syndrome	Sterile female Webbed neck Broad chest	XO
Klinefelter syndrome	Sterile male Male and female characteristics	XXY
XYY male	Fertile male Usually quite tall	XYY

Ultraviolet (UV) radiation, the component of sunlight that leads to suntan (and sunburn), is much lower in energy than x-rays but still higher in energy than ordinary light (see Figure 8-3). Certain molecules within the structure of chromosomes absorb UV radiation, developing chemical bonds among them that are not normally present. These unusual bonds produce a "kink" in the molecular structure of the DNA of chromosomes. Normally, a chromosome is able to repair itself by removing the affected molecules and synthesizing new, undamaged molecules. Sometimes, however, mistakes can take place in some part of the repair process and a change may occur in a gene.

Sometimes chemicals damage DNA directly. LSD (the hallucinogenic compound lysergide), marijuana (leaves of the *Cannabis sativa* plant), cyclamates (compounds widely used as nonnutritive sugar substitutes), and certain pesticides (compounds used to kill insects on crops and other plants) are a few examples of chemicals known to damage DNA. In general, chemicals alter the molecular structure of nucleotides resulting in a mispairing of bases. As with UV radiation, mistakes can take place as the cell works to repair the DNA and changes may then occur in a gene.

Three major sources of environmental damage to chromosomes are (1) high-energy radiation, such as x-rays; (2) low-energy radiation, such as UV light; and (3) chemicals, such as LSD, marijuana, cyclamates, and certain pesticides.

Studying inheritance patterns in humans using pedigrees

Karyotypes are useful in studying diseases or disorders caused by an abnormal number of chromosomes or the addition or deletion of chromosome pieces. However, most inherited disorders, such as the type of hemophilia discussed in the chapter opener, are caused by point mutations, which cannot be seen in a karyotype.

To study how this and other human traits are inherited, scientists study family histories—the records of relevant genetic features—to draw pedigrees. **Pedigrees** are diagrams of genetic relationships among family members over several generations. By studying which relatives exhibit a trait, it is often possible to say if the gene producing the trait is sex-linked (occurs on one of the sex-determining chromosomes) or autosomal (occurs on a non-sex-determining chromosome). Pedigree analysis also helps a geneticist determine whether the trait is a dominant or a recessive characteristic. In many cases, it is also possible to infer which individuals are homozygous or heterozygous for the allele carrying the trait.

Figure 23-9 is a pedigree that shows the history of red-green color blindness in a family. The circles in the diagram represent females, and the squares represent males. Solid-colored symbols stand for individuals that exhibit the trait being studied. In this case, those individuals would be color blind (Figure 23-10). Parents are represented by horizontal lines connecting a circle and a square. Vertical lines coming from two parents indicate their children, arranged along a horizontal line in order of their birth.

First, to determine whether a trait is sex-linked or autosomal, notice whether the trait is expressed more fre-

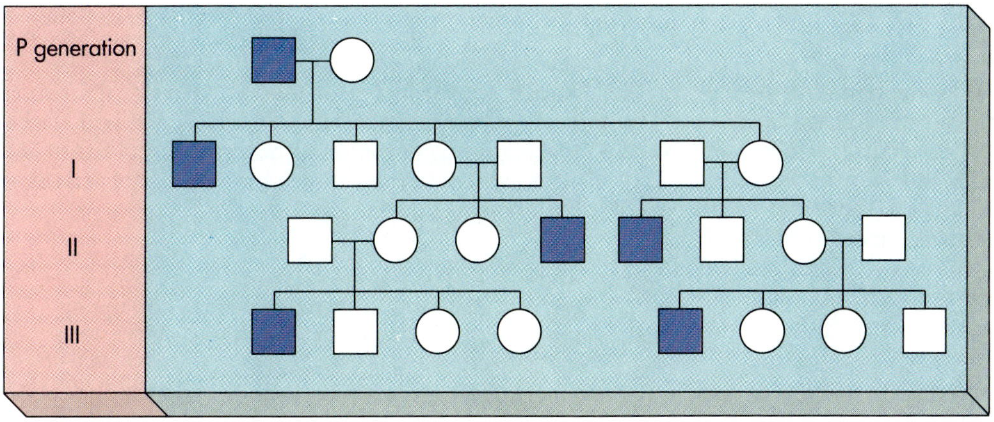

FIGURE 23-9 A pedigree showing red-green color blindness. *Squares* indicate males, *circles* indicate females. From this pedigree, it can be determined that the carriers of the color blindness gene are the female in the P generation, the three females in generation I, and the first and third females in generation II. The status of the other females is unknown.

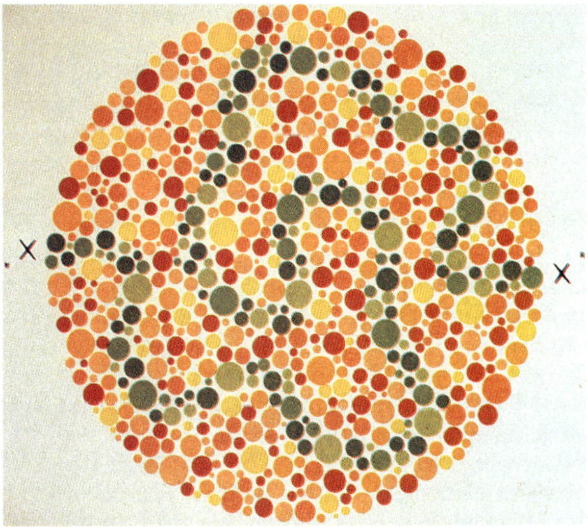

FIGURE 23-10 A test for red-green color blindness. About 8% of Caucasian males and 0.5% of Caucasian females have some loss of color vision, which is almost always caused by an X-linked recessive allele. This recessive allele causes a defect in one of three groups of color-sensitive cells in the eye. The most common form of color blindness is dichromatic vision, in which affected individuals cannot distinguish red from green or yellow from blue. The pattern tests for an inability to distinguish red from green. Those with normal vision will see a path of green dots between the points marked *X*. Those with red-green color blindness will not be able to distinguish the green dots from red and so will not see the green path.

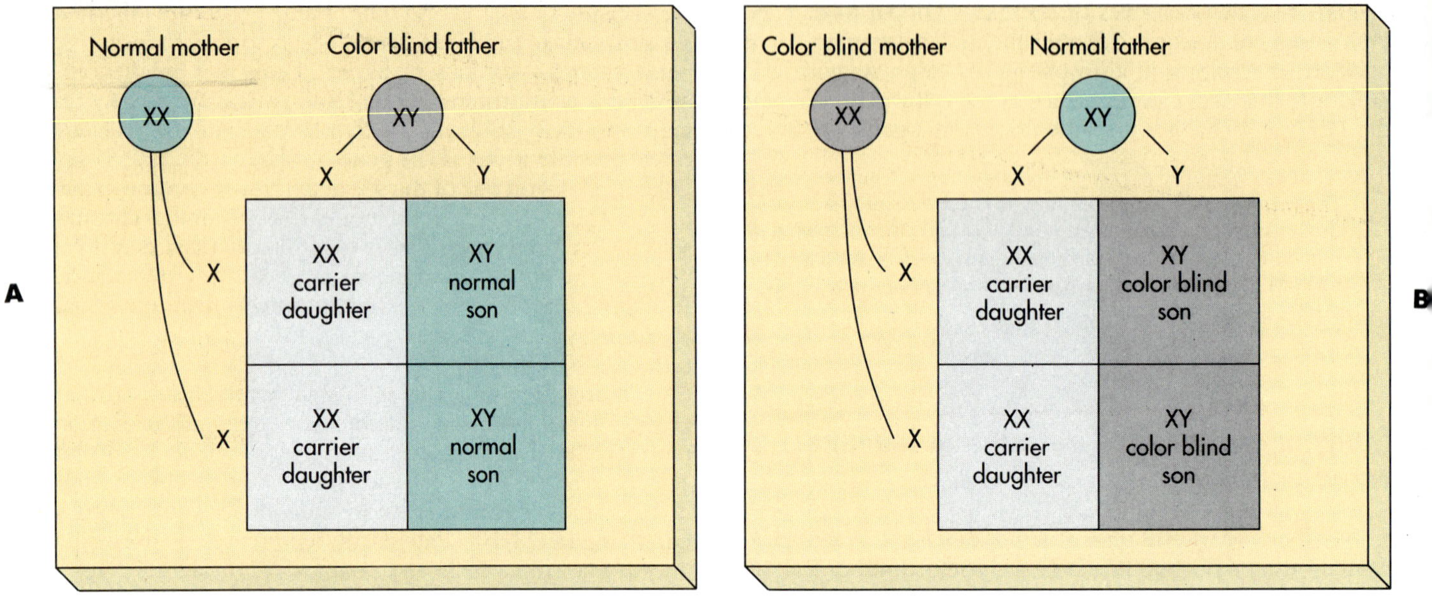

FIGURE 23-11 Punnett squares depicting the outcomes for the sex-linked disorder color blindness. **A** shows a Punnett square for a normal mother and color blind father. **B** shows a Punnett square for a color blind mother and normal father.

quently in males than in females. If so, the trait is most likely a sex-linked trait. In addition, the trait would then likely be recessive because few dominant sex-linked traits are known. An autosomal trait (whether dominant or recessive) is expressed equally in both males and females. A sex-linked trait, however, is always expressed in a male, since the relatively inert Y chromosome lacks alleles that correspond to alleles on the X chromosome; one deleterious gene on a male's X chromosome results in its phenotypic expression. Is red-green color blindness a sex-linked trait?

Second, to determine whether a trait is dominant or recessive, notice whether each person expressing the trait has a parent who expressed the trait. In this case, if color blindness is dominant to normal color vision, then each color blind person will have a color blind parent. However, if the trait is recessive, a person expressing the trait can have parents who do not express the trait. Both parents may be heterozygous (may carry the trait). If each parent passes on the recessive gene to a child, the child is then homozygous recessive and the trait will be expressed. Is color blindness a dominant or a recessive characteristic?

Next, determine which individuals are carriers of the color blindness gene. By now you probably realize that color blindness is a sex-linked recessive trait. A color blind male always contributes an X chromosome with the defective allele to his daughters; it is the only X he can contribute. His sons, however, get a Y chromosome from him and an unaffected X chromosome from their (normal) mother. A color blind female can only contribute a chromosome with the defective allele to her children because both her X chromosomes are affected. Women who are carriers of the trait, however, may contribute either a normal or a defective X. The Punnett square in Figure 23-11 illustrates the contributions of genes by a normal mother and a color blind father (23-11, *A*) and a color blind mother and normal father (23-11, *B*). Can you determine which individuals are carriers of color blindness in the pedigree?

Pedigrees, diagrams of genetic relationships among families over several generations, are useful in determining whether traits are sex-linked or autosomal and dominant or recessive. Carriers of recessive genes can also be identified.

Queen Victoria's family pedigree is shown in Figure 23-12. Persons carrying a mutant allele but not having the disorder are shown by partially colored symbols. Using the strategy outlined above, analyze the pedigree to determine whether hemophilia is sex-linked or autosomal and whether it is a dominant or recessive trait. Look back to the chapter opener to confirm your results.

As shown in the royal pedigree, in the six generations since Queen Victoria, 10 of her male descendants have had hemophilia. The British royal family escaped the disorder because Queen Victoria's son King Edward VII did not inherit the defective allele. Three of Victoria's nine children did receive the defective allele, however, and carried it by marriage into many of the royal families of Europe. It is still being transmitted to future generations among these family lines, except in Russia, where the five children of Alexandra, Victoria's granddaughter, were killed soon after the Russian revolution (Figure 23-13).

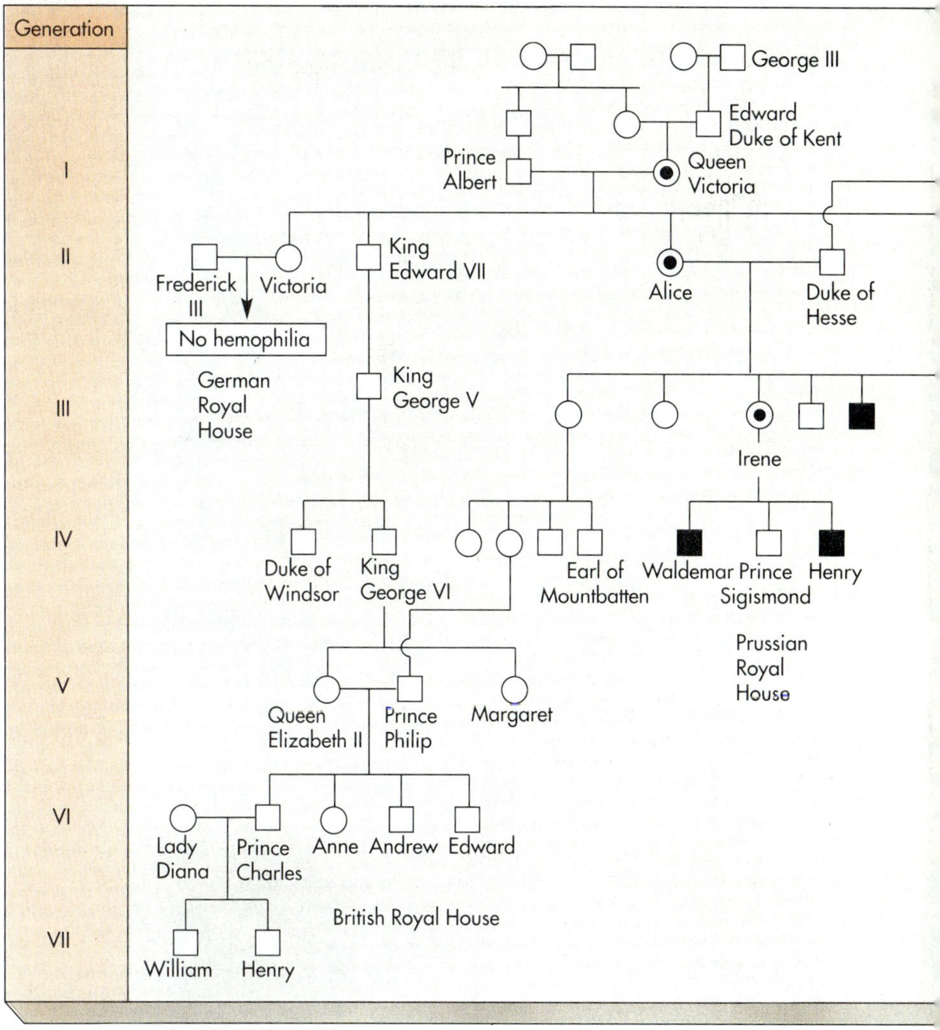

FIGURE 23-12 The royal hemophilia pedigree.
From Queen Victoria's daughter, Alice, the disorder was introduced into the Prussian and Russian royal houses, and from her daughter Beatrice, it was introduced into the Spanish royal house. Victoria's son Leopold, himself a victim, also transmitted the disorder in a third line of descent.

FIGURE 23-13 The last Russian royal family.
Shown here are Czar Nicholas II of Russia, the last Russian czar, and his wife Alexandra, with their five children: Olga, Tatiana, Maria, Anastasia, and Alexis. Alexandra, Queen Victoria's granddaughter, was a carrier of hemophilia and passed it on to her son Alexis. Because the kind of hemophilia they had is caused by a recessive mutant allele located on the X chromosome, it is not expressed in the heterozygous condition in women because women have two X chromosomes. It is not known which of Alexandra's four daughters might have received the hemophilia allele from her; none of the four daughters lived to bear children.

Dominant and recessive genetic disorders

Most genetic disorders are recessive; these mutant genes are able to persist in the population among carriers. Unfortunately, persons receiving a mutant allele from each of their parents are affected with the disease and, in the case of some diseases, may die before reaching adulthood. Likewise, persons having certain dominant disorders may die before reaching reproductive age, so lethal dominant disorders are less likely to persist in the population.

Huntington's disease, the disorder that killed folksinger and songwriter Woody Guthrie, is caused by a mutant dominant allele that does not show up until individuals are more than 30 years old—after they may have had children. This disorder causes progressive deterioration of brain cells. Other dominant genetic disorders include the following:

- **Marfan's syndrome,** which results in skeletal, eye, and cardiovascular defects
- **Polydactyly,** which results in extra fingers or toes
- **Achondroplasia,** which results in a form of dwarfism
- **Hypercholesterolemia,** which results in high blood cholesterol levels and a higher likelihood of developing coronary artery disease

Recessive genetic disorders are often seen primarily within specific populations or races of people unless many members of the population marry and have children outside of their population or race. Among the Caucasian population, for example, the most common fatal genetic disorder is **cystic fibrosis.** Affected individuals secrete a thick mucus that clogs the airways of their lungs and the passages of their pancreas and liver. Among Caucasians, about 1 in 20 individuals has a copy of the defective gene but shows no symptoms. Approximately 1 in 1800 has two copies of the gene and therefore has the disease. These individuals inevitably die from complications that result from their disease. It was learned only recently that the cause of cystic fibrosis is a defect in the way cells regulate the transport of chloride ions across their membranes.

Sickle-cell anemia is a recessive disorder most common among African blacks and their descendants. In the United States, for example, about 9% of African Americans are heterozygous for this allele; about 0.2% are homozygous and therefore have sickle-cell anemia. In some groups of people in Africa, up to 45% of the individuals are heterozygous for this allele. Heterozygous carriers of the sickle-cell gene do not have the disease but can pass the gene along to their children.

Individuals afflicted with sickle-cell anemia are unable to transport oxygen to their tissues properly because the molecules within red blood cells that carry oxygen—molecules of the protein hemoglobin—are defective. Red blood cells that contain large proportions of such defective molecules become sickle shaped and stiff; normal red blood cells are disk shaped and much more flexible (Figure 23-14). As a result of their stiffness and irregular shape, the sickle-shaped red blood cells are unable to move easily through capillaries. Therefore they tend to accumulate in blood vessels, reducing the blood supply to the organs they serve, causing pain, tissue destruction, and an early death.

Interestingly, the gene for sickle-cell anemia is most prevalent in the regions of Africa where malaria is prevalent. Malaria is a disease caused by microorganisms that live in a person's red blood cells. These microbes are injected into a person's bloodstream by the bite of a female *Anopheles* mosquito. A long-lasting disease, malaria affects the physical and mental development of its victims, causing damage to many body organs. Scientists have discovered that the defective hemoglobin molecules of the person with sickle-cell anemia produce conditions that are unfavorable to the growth of the malaria organism. However, these persons eventually die from their anemia. But persons heterozygous for the sickle-cell gene, although not afflicted with the disease, are more resistant to malaria. Heterozygotes, then, have a survival advantage with respect to malaria and sickle-cell anemia (although their offspring may be afflicted with sickle-cell anemia if they reproduce with persons also carrying the defective gene).

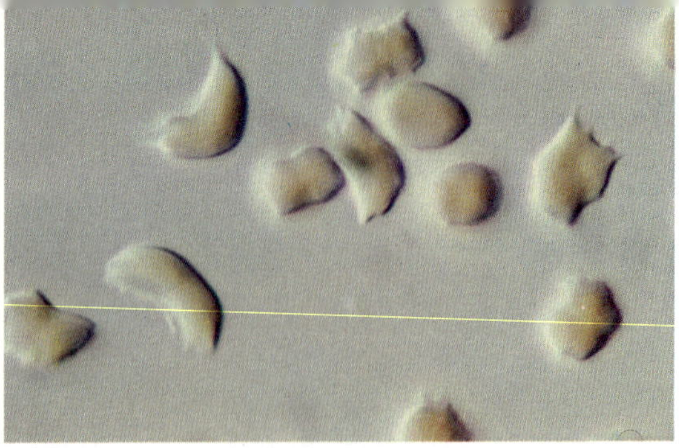

FIGURE 23-14 Sickle-shaped red blood cells.
In individuals who are homozygous for the recessive sickle cell trait, many of their red blood cells are sickle shaped and stiff. These irregularities are caused by defective hemoglobin molecules within the red blood cells. Because of their irregular shape, sickle-shaped red blood cells are unable to move easily through capillaries and become lodged in these blood vessels.

Most genetic disorders are recessive because mutant genes are able to persist in the population among carriers. Some more common recessive genetic disorders among humans are color blindness, hemophilia, cystic fibrosis, and sickle-cell anemia.

Tay-Sachs disease is an incurable, fatal recessive hereditary disorder. Although rare in most human populations, Tay-Sachs has a high incidence among Jews of Eastern and Central Europe and among American Jews (90% of whom are descendants of Eastern and Central European ancestors). In fact, geneticists estimate that 1 in 28 individuals in these Jewish populations carries the allele for this disease, and that approximately 1 in 3600 infants within this population is born with this genetic disorder. Affected children appear normal at birth but begin to show signs of mental deterioration at about 8 months of age. As the brain begins to deteriorate, affected children become blind; they usually die by the age of 5 years.

Table 23-2 summarizes the dominant and recessive disorders discussed. Many common human traits are also coded by either dominant or recessive genes. Table 23-3 lists some of these characteristics.

TABLE 23-2 Dominant and recessive disorders

NAME	CHARACTERISTICS
DOMINANT DISORDERS	
Huntington's disease	Progressive deterioration of brain cells
Marfan's syndrome	Skeletal, eye, and cardiovascular defects
Polydactyly	Extra fingers or toes
Achondroplasia	Dwarfism
Hypercholesterolemia	High blood cholesterol levels
RECESSIVE DISORDERS	
Cystic fibrosis	Production of thick mucus that clogs airways
Sickle-cell anemia	Defective hemoglobin in red blood cells, effect on oxygen transport
	Misshapen red blood cells, tendency to form clots in vessels
Tay-Sachs disease	Brain deterioration beginning at 8 months of age
SEX-LINKED RECESSIVE DISORDERS	
Color blindness	Inability to discern certain colors
Hemophilia	Improper clotting of blood
Albinism	Lack of pigment in skin, hair, and eyes
Duchenne muscular dystrophy	Degeneration of muscle
Fragile X syndrome	Easy breakage of tip of X chromosome
	Long face, squared forehead, large ears
	Mental retardation

TABLE 23-3 Common dominant and recessive human traits

DOMINANT TRAITS
- Widow's peak
- Hair on middle segment of fingers
- Short fingers
- Inability to straighten little finger
- Brown eyes
- A or B blood factor

RECESSIVE TRAITS
- Common baldness
- Blue or gray eyes
- O blood factor
- Attached ear lobes

Incomplete dominance and codominance

Alleles do not always exhibit clear-cut dominance or recessiveness. The pea plants that Gregor Mendel worked with exhibited complete dominance, and for this reason his work disputed the concept of blended inheritance (see Chapter 22). However, since Mendel's time, researchers have discovered many cases of **incomplete dominance** in which alternative alleles are *not* dominant over or recessive to other alleles governing a particular trait. Instead, heterozygotes are phenotypic "intermediates."

Incomplete dominance is found in various plant and animal traits. For example, a cross between snapdragons having red flowers and those having white flowers yields plants having pink flowers. Crossing black Andalusian chickens with white Andalusian chickens yields an intermediate hybrid that is slate blue. In humans, a curly haired Caucasian and a straight haired Caucasian will have children with wavy hair. And persons with wavy hair will have children that are either straight haired, wavy haired, or curly haired. Figure 23-15 shows a cross between two wavy haired people. Notice that an upper case H denotes the "hair" allele. No lower case letters are used because neither allele (curly or straight) is recessive. Instead, one allele is designated as H and the other as H'.

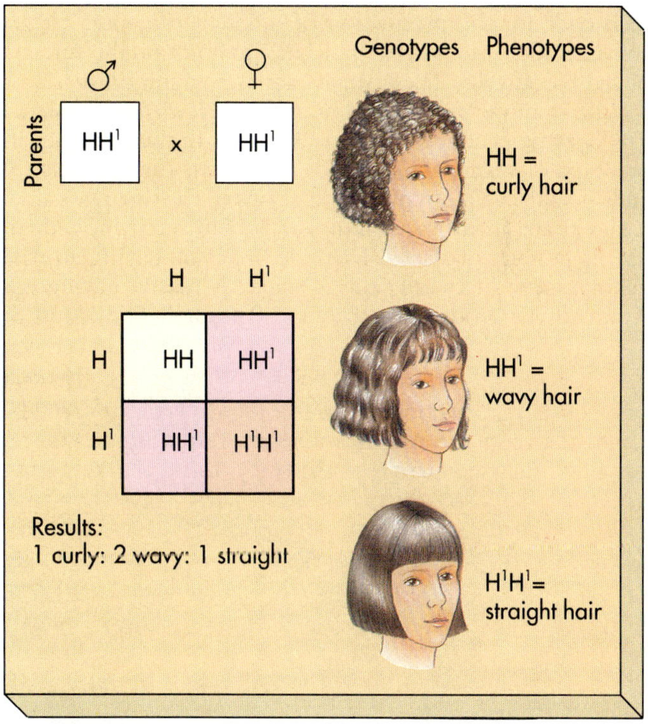

FIGURE 23-15 Incomplete dominance.
Sometimes, alternative alleles are not dominant or recessive. For example, neither straight nor curly hair is dominant in Caucasians. The predicted ratio of the offspring of two wavy haired people will be one curly haired individual, one straight haired individual, and two wavy haired individuals. Wavy hair is the phenotypic intermediate.

A slightly different situation occurs with alleles that are codominant. **Codominant** alleles are *both* dominant; both characteristics are exhibited in the phenotype. In humans, the alleles that code the A, B, or AB blood types are a good example of codominance. For example, a person having allele A and allele B has blood type AB. In addition to being coded by codominant alleles, blood types are coded by more than two alleles. So far, genes that consist of only two—a pair of—alleles have been discussed. But a gene may be represented by more than two alleles within the population. Some genes, such as the ABO blood type genes, consist of a system of alleles, or **multiple alleles.** In this system of multiple alleles, the alleles A and B exhibit codominance, and the O allele is recessive.

In traits exhibiting incomplete dominance, alternative forms of an allele are neither dominant nor recessive; heterozygotes are phenotypic intermediates. In traits exhibiting codominance, alternative forms of an allele are both dominant; heterozygotes exhibit both phenotypes.

Multiple alleles

As mentioned, human ABO blood types are coded by multiple alleles. In fact, four phenotypes (A, B, AB, and O blood types) are determined by the presence of two of three possible alleles (A, B, or O) in an individual. The A and B alleles each code for the production of different enzymes. These enzymes add certain sugar molecules to lipids on the surface of red blood cells. These sugars act as recognition markers for the immune system and are called *cell surface antigens.* The enzyme produced by allele A results in the addition of one type of sugar; allele B adds a different sugar. Allele O adds no sugar. Different combinations of the three possible alleles occur in different individuals, with each individual having *one pair* of alleles. Therefore a person having an AA genotype produces only the A sugar. A person having an AB genotype produces both A and B sugars. (The A and B alleles are codominant and both are expressed.) A person having an AO phenotype will produce the A sugar (so is said to have type A blood). The A allele is expressed because it is dominant to the O allele.

Some genes consist of a system of alleles, or multiple alleles, which are usually codominant to one another. The ABO system of human blood types is an example of codominant multiple alleles.

Genetic counseling

Although most genetic disorders cannot yet be cured, research scientists and physicians are continually learning a great deal about them. Exciting methods of genetic therapy (see Chapter 24) are being studied. Geneticists can identify couples at risk of having children with genetic defects and can help them have healthy children. This process is called **genetic counseling.**

By analyzing family pedigrees and applying the laws of statistical probability, genetic counselors can often determine the probability that a person is a carrier of a recessive disorder. When a couple is expecting a child and both parents have a significant probability of being carriers of a serious recessive genetic disorder, the pregnancy is said to be a high-risk pregnancy. In such a pregnancy, there is a significant probability that the child will exhibit the clinical disorder. Another class of high-risk pregnancies occurs with mothers who are more than 35 years old. Remember, the frequency of birth of infants with Down syndrome increases dramatically among the pregnancies of older women (see Figure 23-4).

When a pregnancy is diagnosed as being high risk, many women elect to undergo **amniocentesis,** a procedure that permits the prenatal diagnosis of many genetic disorders (Figure 23-16). In the fourth month of pregnancy, a small sample of amniotic fluid is taken from the mother using a sterile hypodermic needle. The amniotic fluid, which bathes the fetus, contains free-floating cells derived from the fetus.

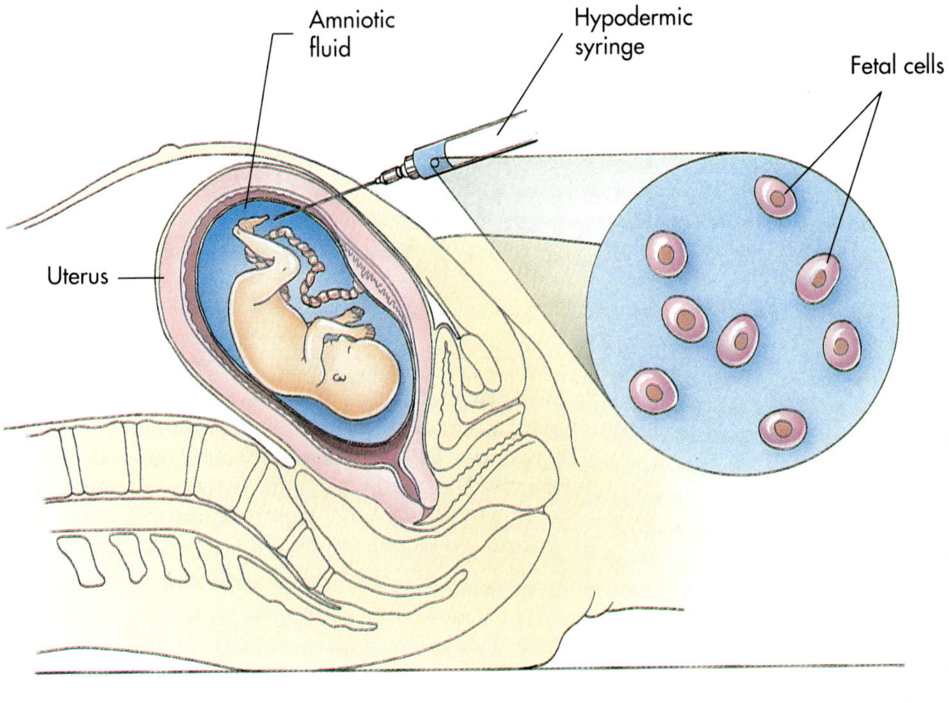

FIGURE 23-16 Amniocentesis.
A needle is inserted into the amniotic cavity, and a sample of amniotic fluid, containing some free cells derived from the fetus, is drawn into a syringe. The fetal cells are then grown in tissue culture so that their karyotype and many of their metabolic functions can be examined.

BIOLOGY & you

Using blood typing to test for paternity

Paternity testing is not just the stuff of soap operas. Disputes involving adoption, divorce, and legal heirs can require a determination of a person's true father. Although no test can definitely state that a suspect is the true father of a child, the paternity tests being performed today can yield a probability of paternity in excess of 99%.

Paternity testing involves the analysis of genetic information inherited by a child from his or her mother and true father. Certain antigens (see Chapter 13) are attached to the surface of each person's red blood cells. The type of antigens present on your red blood cells is inherited from your parents. Testing a child's mother and suspected father for these antigens and then using the principles of genetics to determine whether the child has received its complement of antigens from the hypothetical parents is the basis of paternity testing.

There are several types of antigens that are analyzed in paternity testing. The most familiar are the antigens associated with the ABO system. Each person is a particular blood type: A, B, AB, or O. Type A persons make antibodies against the B antigen. If an A person needed a blood transfusion and was given B blood, the B antibodies in the recipient's blood would attack the B antigen in the transfused blood. The blood cells would clump together, and death could occur. Similarly, B individuals have A antibodies, AB individuals have no antibodies (and both A and B antigens), and O individuals have both antibodies (and neither antigen). The presence or absence of these antigens is governed by alleles inherited from one's parents. The alleles for each blood type are shown in Figure 23-B, *1*.

How can you rule out a suspected father using ABO blood groups? Say a child's mother is type O, and the suspected father is AB. The child is type AB. Is the suspect the true father? Examine the Punnett square in Figure 23-B, *2*. The only blood types possible from O and AB parents are types A and B. An AB child could not be born to such parents, and thus the suspect is not the true father.

Paternity testing, however, is based on more than ABO blood groups. DNA analysis is now used in conjunction with antigen analysis to test paternity. It is this combination of tests that contributes to the high accuracy of the results.

1

Type	Possible alleles
A	I^AI^A, I^Ai
B	I^BI^B, I^Bi
AB	I^AI^B
O	ii

2

O mother \ AB father	I^A	I^B
i	I^Ai	I^Bi
i	I^Ai	I^Bi

FIGURE 23-B 1 The alleles for the ABO blood group system.
2 A Punnett square showing the possible offspring from a type O mother and a type AB father.

Once removed, these cells can be grown as tissue cultures in the laboratory. Studying these tissue cultures, genetic counselors can test for many of the most common genetic disorders.

During amniocentesis, the position of the needle in relationship to the fetus is observed by means of a technique called **ultrasound.** The term *ultrasound* refers to high-frequency sound waves. Pulses of these waves are sent into the body and are reflected back in various patterns depending on the tissues or fluids the waves hit. These patterns of sound-wave reflections are then mapped to produce a picture of inner tissues as in Figure 23-17. Because ultrasound allows the position of the fetus to be determined, the person performing the amniocentesis can avoid damaging the fetus. Ultrasound also allows the fetus to be examined for the presence of major abnormalities.

At this time, most of the problems that can be diagnosed by pedigree analysis, amniocentesis, chorionic villus sampling (see Chapter 22), and ultrasound techniques cannot be treated. Sometimes the only options available to a couple are to continue the pregnancy and deal with the problems after birth or to have a therapeutic abortion. However, continual progress in the developing gene therapies discussed in Chapter 24 may, in the future, have a drastic impact on a physician's ability to treat genetic disorders before birth.

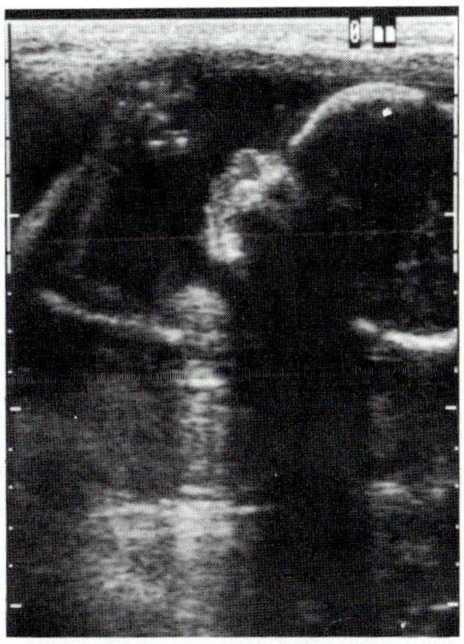

FIGURE 23-17 Ultrasound.
Before amniocentesis, the fetus is observed with ultrasound to determine its position. Imaging the fetus using ultrasound before amniocentesis is important, since the amniocentesis needle must be inserted precisely without damaging the fetus.

Summary

1. Human cells contain 46 chromosomes: 44 autosomes and 2 sex chromosomes. The autosomes form 22 pairs of homologous (matched) chromosomes. In females, the two sex chromosomes are similar in size and appearance and are designated XX. In males, the two sex chromosomes are not similar in size and appearance. The smaller of the two is designated Y and the larger X. The genes that determine male characteristics are located on the Y chromosome.
2. Researchers often use karyotypes to study the inheritance of genetic disorders caused by the loss of all or part of a chromosome or by the addition of extra chromosomes or chromosome fragments. A karyotype is a picture of chromosomes paired with their homologues and arranged according to size.
3. In humans, the inheritance of one autosome too few usually results in an inviable zygote. The inheritance of one autosome too many, or a trisomy, also results in an inviable zygote with the exception of chromosome 21. Individuals with an extra copy of this chromosome are retarded and have Down syndrome. Down syndrome is much more frequent among pregnant mothers over 35 years of age, and it occurs when two sister chromatids fail to separate, or disjoin, during the meiotic process.
4. Nondisjunction can occur with the sex chromosomes. Persons who inherit abnormal numbers of sex chromosomes often have abnormal features and may be mentally retarded. Examples of genetic disorders caused by abnormal numbers of sex chromosomes are poly-X females, XXY males (Klinefelter syndrome), XO females (Turner syndrome), and XYY males.
5. Persons may inherit chromosomes that are structurally defective due to chromosomal breaks and misrepairs or changes that may take place in the molecular structure of the chromosomal DNA. The three major sources of environmental damage to chromosomes are (1) high-energy radiation, such as x-rays; (2) low-energy radiation, such as UV light; and (3) chemicals, such as LSD, marijuana, cyclamates, and certain pesticides.
6. Patterns of inheritance observed in family histories, or pedigrees, can be used to determine the mode of inheritance of a particular trait. By such analysis, it can often be determined whether a trait is associated with a dominant or a recessive allele and whether the gene determining the trait is sex linked.
7. Some genetic disorders are relatively common in human populations, whereas others are rare. Many of the most common genetic disorders are associated with recessive alleles, the functioning of which may lead to the production of defective versions of enzymes that normally perform critical functions. Because such traits are determined by recessive alleles and therefore expressed only in homozygotes, the alleles are not eliminated from the human population. Dominant alleles that lead to severe genetic disorders are less common; in some of the more frequent ones, the expression of the alleles does not occur until after the individuals have reached their reproductive years.
8. Alleles do not always exhibit clear-cut dominance or recessiveness but may instead exhibit incomplete dominance or codominance. In traits exhibiting incomplete dominance, alternative forms of a trait are neither dominant nor recessive; heterozygotes are phenotypic intermediates. In traits exhibiting codominance, alternative forms of an allele are both dominant; heterozygotes exhibit both phenotypes. Some genes consist of a system of alleles, or multiple alleles, which are usually codominant to one another. The ABO system of human blood types is an example of multiple alleles with two of the alleles (A and B) codominant to one another and one allele (O) recessive to the other two.
9. Although most genetic disorders cannot yet be cured, research scientists are making progress in developing gene therapies that may, in the future, have an important impact on a physician's ability to treat genetic disorders before birth.

REVIEW QUESTIONS

1. Fill in the blanks: Each human cell contains one pair of ______ ______ and ______ pairs of ______.
2. What is nondisjunction? Explain what causes it and what the results can be.
3. What is Down syndrome and what causes it? Give another term for this condition, and summarize its symptoms.
4. What happens when human offspring inherit abnormal numbers of sex chromosomes? Summarize the four examples discussed in the chapter and give the genotype of each.
5. Fill in the blank: A human must have at least one ______ chromosome to survive.
6. Distinguish among a chromosomal duplication, translocation, and inversion.
7. What are the three major sources of damage to chromosomes?
8. Describe how heavy use of drugs could affect a woman's ova or a man's sperm.
9. What is a point mutation? What results from it?
10. What would you conclude about a specific human trait in each of the following situations?:
    a. The trait is expressed more frequently in males than in females.
    b. Offspring who exhibit this trait have at least one parent who exhibits the same trait.
    c. Offspring can exhibit this trait even though the parents do not.
    d. The trait is expressed equally in both males and females.
11. Why do most genetic disorders in humans result from recessive genes? Name several examples.
12. Distinguish between incomplete dominance and codominance. Describe the phenotype of a heterozygote in each case.
13. A woman whose blood type is AB marries a man with the same blood type. Draw a Punnett square to illustrate the possible genotypes of their children. What blood type will each genotype have?
14. If a couple wants to have a child but suspects that they may be at risk for a genetic disorder, what can they do? If a pregnancy turns out to be high risk, what options are available?

HUMAN GENETICS PROBLEMS

1. George has Royal hemophilia and marries his mother's sister's daughter Patricia. His maternal grandfather also had hemophilia. George and Patricia have five children: two daughters are normal, and two sons and one daughter develop hemophilia. Draw the pedigree.
2. A couple with a newborn baby are troubled that the child does not appear to resemble either of them. Suspecting that a mix-up occurred at the hospital, they check the blood type of the infant. It is type O. Because the father is type A and the mother is type B, they conclude that a mistake must have been made. Are they correct?
3. How many chromosomes would you expect to find in the karyotype of a person with Turner syndrome?
4. A woman is married for the second time. Her first husband was ABO blood type A, and her child by that marriage was type O. Her new husband is type B, and their child is type AB. What is the woman's ABO genotype and blood type?
5. Total color blindness is a rare hereditary disorder among humans in which no color is seen, only shades of gray. It occurs in individuals homozygous for a recessive allele and is not sex linked. A man whose father is totally color blind intends to marry a woman whose mother was totally color blind. What are the chances that they will produce offspring who are totally color blind?
6. This pedigree is of a rare trait in which children have extra fingers and toes. Which if any of the following patterns of inheritance is consistent with this pedigree?
   a. Autosomal recessive
   b. Autosomal dominant
   c. Sex-linked recessive
   d. Sex-linked dominant
   e. Y-linkage

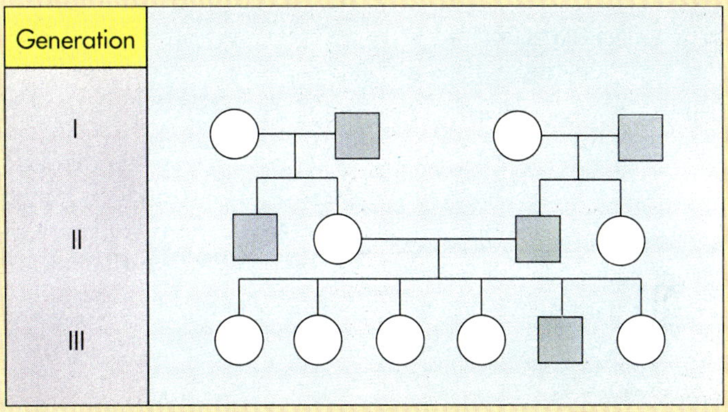

KEY TERMS

achondroplasia 398
amniocentesis 400
autosome 389
codominant 400
cri du chat syndrome 393
cystic fibrosis 398
deletion 393
Down syndrome 389
duplication 393
genetic counseling 400
Huntington's disease 398
hypercholesterolemia 398
incomplete dominance 399
inversion 393
ionizing radiation 392
karyotype 389
Klinefelter syndrome 391
Marfan's syndrome 398
multiple alleles 400
mutation 389
nondisjunction 389
pedigree 395
point mutation (gene mutation) 394
polydactyly 398
sex chromosome 389
sickle-cell anemia 398
Tay-Sachs disease 398
translocation 393
Turner syndrome 392
ultrasound 401
ultraviolet (UV) radiation 394

THOUGHT QUESTIONS

1. Women are advised to have babies before they are 35 years old if possible, but no such advice is usually offered to men. Why the difference?
2. Some people smoke all their lives and never develop lung cancer. Do you think the propensity to develop lung cancer on exposure to the mutagens in cigarette smoke is hereditary? Discuss.

FOR FURTHER READING

Anderson, W. French. (1992, May). Human gene therapy. *Science.*
The author of this article supervised the first use of gene therapy to treat human patients. Here he summarizes 11 new uses for gene therapy currently underway. The author also discusses the safety, social, and ethical concerns that gene therapy poses.

Collins, F. S. (1992, May). Cystic fibrosis: molecular biology and therapeutic implications. *Science.*
This challenging article presents the latest information revealed since the identification of the defective gene in patients with cystic fibrosis. Prospects for gene therapy are also discussed.

Diamond, J. (1989, February). Blood, genes, and malaria. *Natural History.*
This is a lucid account of the evolutionary history of sickle-cell anemia.

24 The molecular basis of inheritance

HIGHLIGHTS

DNA is located in the nucleus of eukaryotic cells and in the nuclear region of prokaryotic cells.

RNA plays key roles in the translation of the hereditary message of DNA.

Most traits reflect the action of enzymes; the hereditary message is a molecular code in DNA that results in the production of these enzymes.

Today, scientists are trying to cure certain disorders by inserting normal genes—units of DNA—into the cells of persons having damaged genes.

OUTLINE

The location of the hereditary material

The chemical nature of the hereditary material: Nucleic acids

The nucleic acids: DNA and RNA

A major scientific breakthrough: The structure of DNA

How DNA replicates

Genes: The units of hereditary information

Gene expression: How DNA directs the synthesis of polypeptides

An overview

Transcribing the DNA message to RNA

Translating the transcribed DNA message into a polypeptide

Differences in prokaryotic and eukaryotic genes

Differences in prokaryotic and eukaryotic gene expression

Prokaryotic gene expression

Eukaryotic gene expression

Gene therapy: Changing genes to cure disease

The final piece to the puzzle is almost in place. Scientists around the world have been working on the human genome project, pooling their data to create a complete molecular picture of the substance of human heredity: deoxyribonucleic acid, or DNA. It is the DNA within each cell of your body that carries information regarding your eye color, gender, and height. It is this incredible molecule that holds the key to life.

FIGURE 24-1 The marine green alga *Acetabularia* has been the subject of many experiments in biology.
Although *Acetabularia* is a large organism with clearly differentiated parts, individuals are actually single cells.

Amazingly simple in design, the "message" of DNA is constructed of only four molecular building blocks that together act like a code. The sequence of molecular units is actually a cellular message that is ultimately interpreted into human traits. But although the message of heredity uses a simple coding system, the code is lengthy. In fact, approximately *five million* molecular "code letters" are located on a single human chromosome!

Deciphering the code of all 46 human chromosomes is the task of those involved in the Human Genome Project. The term **genome** refers to the complete set of genes in an individual organism, and these scientific pioneers are planning to decipher the code hidden within all 46. Their reward will be tremendous—the ability to detect flaws in the code of a person having a genetic disorder and the potential ability to fix those flaws with the groundbreaking techniques of gene therapy.

The location of the hereditary material

Today, scientists understand (in considerable detail) the way that information in the DNA of a developing organism is translated into its eyes, arms, and brain. However, this understanding is the result of over a half century of work by a succession of scientific investigators who asked and answered questions and whose answers helped formulate new questions. Mendel's work suggested that traits are inherited as discrete packets of information (see Chapter 22). At the turn of the century, Walter Sutton suggested that these packets of information were on the chromosomes, but he had no direct evidence to support this hypothesis. In 1910, however, Thomas Hunt Morgan's experiments provided the first clear evidence upholding Sutton's theory. Beginning in the 1930s, experiments by biologists like Joachim Hammerling further probed the question of where the hereditary information is stored within the cell.

Hammerling chose a single-celled organism to focus his experiments on—the long, slender alga *Acetabularia* (Figure 24-1). Although unicellular, *Acetabularia* is visible to the naked eye—about half as long as your little finger. This huge cell has three distinct parts: a foot, a stalk, and a cap.

Hammerling conducted a preliminary experiment to gather data he would use to develop a hypothesis regarding the role of the nucleus. He amputated the caps from some *Acetabularia* cells and the feet from others. He found that when the cap was amputated, a new cap regenerated from the remaining portions (foot and stalk) of the cell. When the foot was amputated, however, a new foot was *not* regenerated from the cap or the stalk. After analyzing these data, Hammerling hypothesized that the hereditary information resides in the nucleus within the foot of *Acetabularia* and is necessary for the regeneration of amputated parts of the cell.

To test this hypothesis, Hammerling used two different species of *Acetabularia: A. mediterranea,* which has a disk-shaped cap, and *A. crenulata,* which has a branched, flower-like cap. He cut the stalk and cap away from a disk-capped cell, and grafted a stalk from a flower-capped cell to its remaining foot (Figure 24-2). Interestingly, the new cap that formed had characteristics of both types of caps. However, when Hammerling cut off this regenerated cap, a disk-shaped one—exactly like that of *A. mediterranea*—grew in its place and in every regeneration thereafter.

From this experiment and other experiments producing similar results, Hammerling concluded that the instructions specifying the type of cap are stored in the foot of the cell—probably in the nucleus. The data suggested that these instructions must pass from the foot through the stalk to the cap. In his regeneration experiments, the initial "blended" cap was formed as a result of flower-shape instructions remaining in the transplanted stalk and disk-shape instructions transported from the nucleus. All the subsequent caps used "new" information from the nucleus.

Gregor Mendel's work in the late 1800s suggested that traits are inherited as discrete packets of information. Walter Sutton's and Thomas Hunt Morgan's work suggested that these packets might be located on the chromosomes. The experiments of Joachim Hammerling beginning in the 1930s provided firm evidence that the hereditary information is located within the nucleus of the cell.

Identifying the nucleus as the most likely source of hereditary information again focused attention on the chromosomes, which are located within the nucleus. Scientists had been studying the chromosomes since the late 1800s and already suspected them to be the vehicles of inheritance. In 1869—only 4 years after the publication of Gregor Mendel's work—a German chemist named Friedrich Miescher isolated the DNA contained within the chromosomes of various types of cells. He did not call it DNA (this designation would not come until the 1920s). Miescher called this chromosomal material **nucleic acid** because it seemed to be specifically associated with the cell nucleus and was slightly acidic. In addition, it contained a proportion of nitrogen and phosphorus that was very different from any other molecules isolated from cells: Miescher suspected that he had discovered a new biological substance.

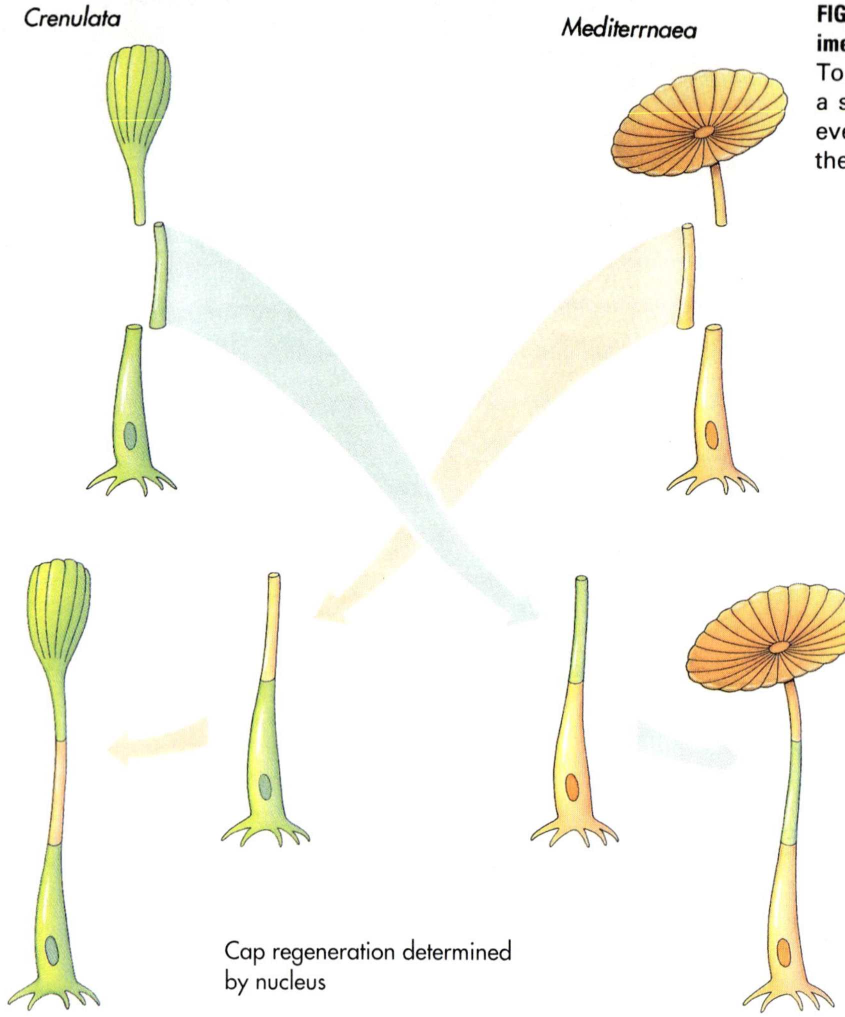

FIGURE 24-2 Hammerling's *Acetabularia* graft experiment.
To the foot of each species, Hammerling grafted a stalk of the other. In each case, the cap that eventually developed was dictated by the foot, the location of the nucleus, and not the stalk.

The chemical nature of the hereditary material: Nucleic acids

In the 1920s, the American biochemist P.A. Levene discovered that two types of nucleic acid were located within cells. Today, scientists understand that one type of nucleic acid contains the hereditary message, whereas the other type works to express that message. Levene found that both types were nearly identical in their structure and contained three molecular parts (Figure 24-3): (1) phosphate (PO_4) groups, (2) 5-carbon sugars, and (3) four types of nitrogen-containing bases. Although nucleic acids as a whole are acidic and tend to form hydrogen ions when in solution, the nitrogen-containing bases tend to accept hydrogen ions. For this reason, these portions of nucleic acids are called bases.

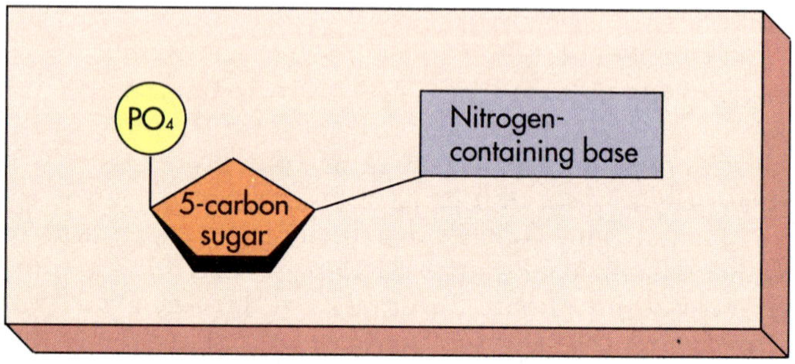

FIGURE 24-3 Structure of a nucleotide.
A nucleotide contains three different molecular components: a phosphate group (PO_4), a five-carbon sugar, and a nitrogen-containing base.

The nucleic acids: DNA and RNA

Levene found that nucleic acids are composed of roughly equal portions of these three molecular parts. He concluded (correctly) from this information that these three molecular parts, bonded together, form the unit of structure of nucleic acids. Each unit—a five-carbon sugar bonded to a phosphate group and a nitrogen-containing base—is called a **nucleotide** (see Figure 24-3). One of the two types of nucleic acids found in cells is composed of units containing the 5-carbon sugar, ribose. It is therefore called **ribonucleic acid,** or **RNA** for short. The other type of nucleic acid is composed of units containing a 5-carbon sugar similar to ribose but having one less oxygen atom. This sugar is called

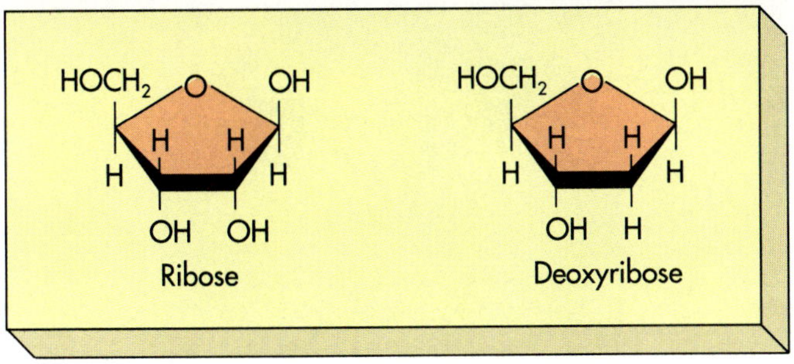

FIGURE 24-4 Comparison of ribose and deoxyribose.
Both sugars have five carbons, but deoxyribose has one less oxygen atom. Ribose is found in RNA and deoxyribose is found in DNA.

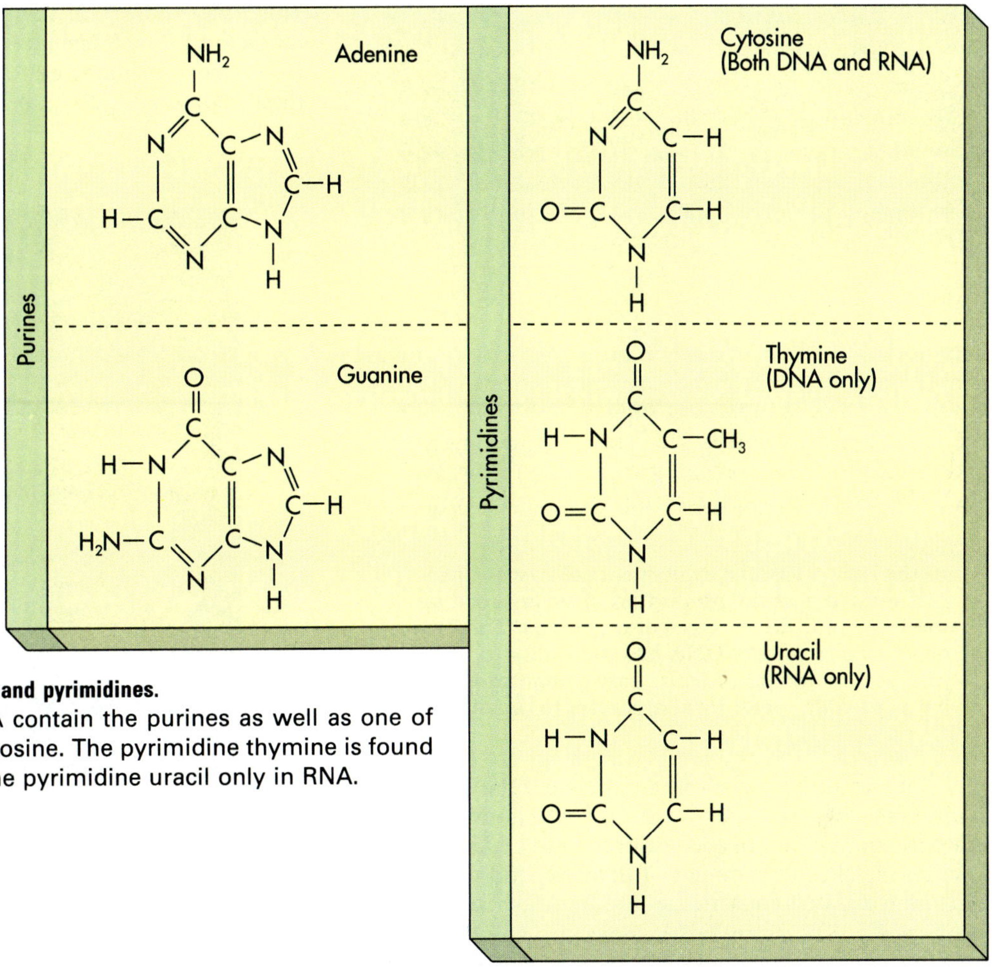

FIGURE 24-5 Purines and pyrimidines.
Both DNA and RNA contain the purines as well as one of the pyrimidines, cytosine. The pyrimidine thymine is found only in DNA, and the pyrimidine uracil only in RNA.

deoxyribose, and the nucleic acid is called **deoxyribonucleic acid,** or **DNA.** Figure 24-4 shows the structure of both ribose and deoxyribose sugars. Can you find the difference between them? How do their names reflect this difference?

There are two types of nucleic acids located within cells: deoxyribonucleic acid (DNA) and ribonucleic acid (RNA). Both types of nucleic acid are composed of units called *nucleotides,* which are made up of three molecular parts: a sugar, a phosphate group, and a base.

Each nucleotide making up the structure of DNA and RNA contains one of four different bases. DNA and RNA both contain the bases adenine and guanine, which are double-ring compounds called **purines** (Figure 24-5). DNA also contains the bases thymine and cytosine, which are single-ring compounds called **pyrimidines.** RNA contains the pyrimidine cytosine as well but contains the single-ring base uracil instead of thymine.

A key advance to understanding how these parts are put together to form DNA and RNA molecules came after World War II. A study carried out by Erwin Chargaff showed that the nucleotide composition of DNA molecules varied in complex ways depending on the source of the DNA. As shown in Table 24-1, the amount of adenine in the DNA of yeast cells exists in a much higher proportion than it does in the DNA of bacteria. The proportion of thymine in human DNA is greater than the proportion of

TABLE 24-1 Chargaff's analysis of DNA nucleotide base compositions

ORGANISM	BASE COMPOSITION (MOLE PERCENT) A	T	G	C
Escherichia coli	26.0	23.9	24.9	25.2
Streptococcus pneumoniae	29.8	31.6	20.5	18.0
Mycobacterium tuberculosis	15.1	14.6	34.9	35.4
Yeast	31.3	32.9	18.7	17.1
Sea urchin	32.8	32.1	17.7	18.4
Herring	27.8	27.5	22.2	22.6
Rat	28.6	28.4	21.4	21.5
Human	30.9	29.4	19.9	19.8

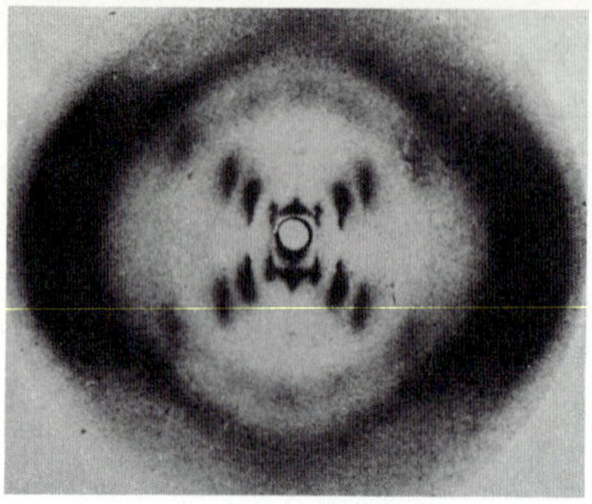

FIGURE 24-6 Evidence for the helical structure of DNA. This x-ray diffraction photograph of crystals of DNA was made in 1953 by Rosalind Franklin in the laboratory of Maurice Wilkins. It suggested to Watson and Crick that the DNA molecule was a helix, like a winding staircase.

thymine in a rat but is lower than its proportion in a sea urchin. These variations in the base composition of the DNA of different types of organisms suggest that DNA might be able to code for (scientists say "encode") the information used by cells for development, growth, and repair. Put simply, DNA had the properties required of hereditary material.

In the late 1940s, the experiments of Erwin Chargaff showed that the proportion of bases varied in the DNA of different types of organisms. This evidence suggested that DNA had the ability to be used as a molecular code. Its base composition varied as its code varied from organism to organism.

Along with the variations among the DNA molecules of different organisms, Chargaff also noted an important similarity: the amount of adenine present in DNA molecules is always equal to the amount of thymine, and the amount of guanine is always equal to the amount of cytosine (A = T and G = C). Therefore DNA molecules always have an equal proportion of purines (A and G) and pyrimidines (C and T).

FIGURE 24-7 Watson and Crick and the double helix. Watson and Crick examine their first model of the DNA double helix.

A major scientific breakthrough: The structure of DNA

In 1952, Alfred Hershey and Martha Chase performed now-famous experiments showing that DNA, not protein, directs virally infected cells to produce more virus particles. As the boxed essay describes, the experiments of Hershey and Chase showed clearly that DNA *is* the hereditary material. The next puzzle to solve was the structure of DNA. How were its building blocks—its nucleotides—put together? The significance of the regularities in the proportion of pyrimidines and the proportion of purines in DNA pointed out by Chargaff became clear through the work of two later experiments.

Two British biophysicists, Rosalind Franklin and Maurice Wilkins, carried out x-ray diffraction analysis of fibers of DNA. In this process, the DNA molecule is bombarded with an x-ray beam. When individual rays encounter atoms, each ray's path is bent or diffracted; the pattern created by all of these diffractions can be captured on a piece of photographic film. The pattern resembles the ripples created by tossing a rock into a smooth lake. With careful analysis of the diffraction pattern, it is possible to develop a three-dimensional image of the molecule causing the diffractions of the x-rays. The diffraction patterns that Franklin and Wilkins obtained (Figure 24-6) suggested that the DNA molecule was a helical coil. Put simply, this complex molecule was probably shaped like a spring.

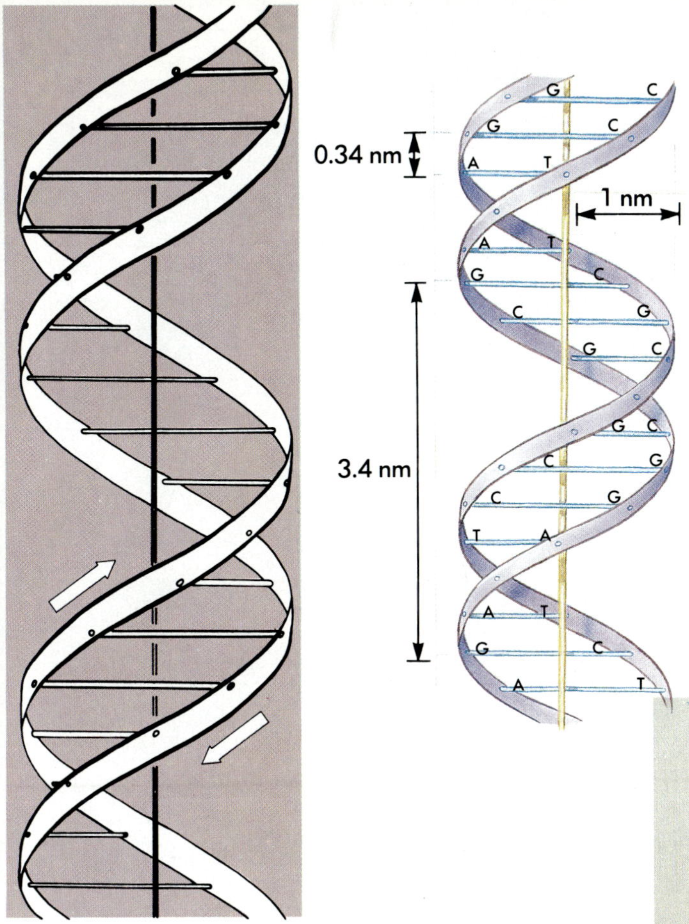

FIGURE 24-8 Diagram of DNA.
On the left is a reproduction of the original diagram of the double-helical structure of DNA presented in the 1953 paper by Watson and Crick. The diagram on the right shows the location of the nucleotide bases. This diagram also points out that the helix has a radius of 1 nanometer (nm) and completes a 360-degree turn every 3.4 nm. The base pairs are 0.34 nm apart. The arrows indicate the direction of replication of the DNA strands, which run in opposite directions. The vertical line shown in both diagrams is not part of the molecule—it is the imaginary axis.

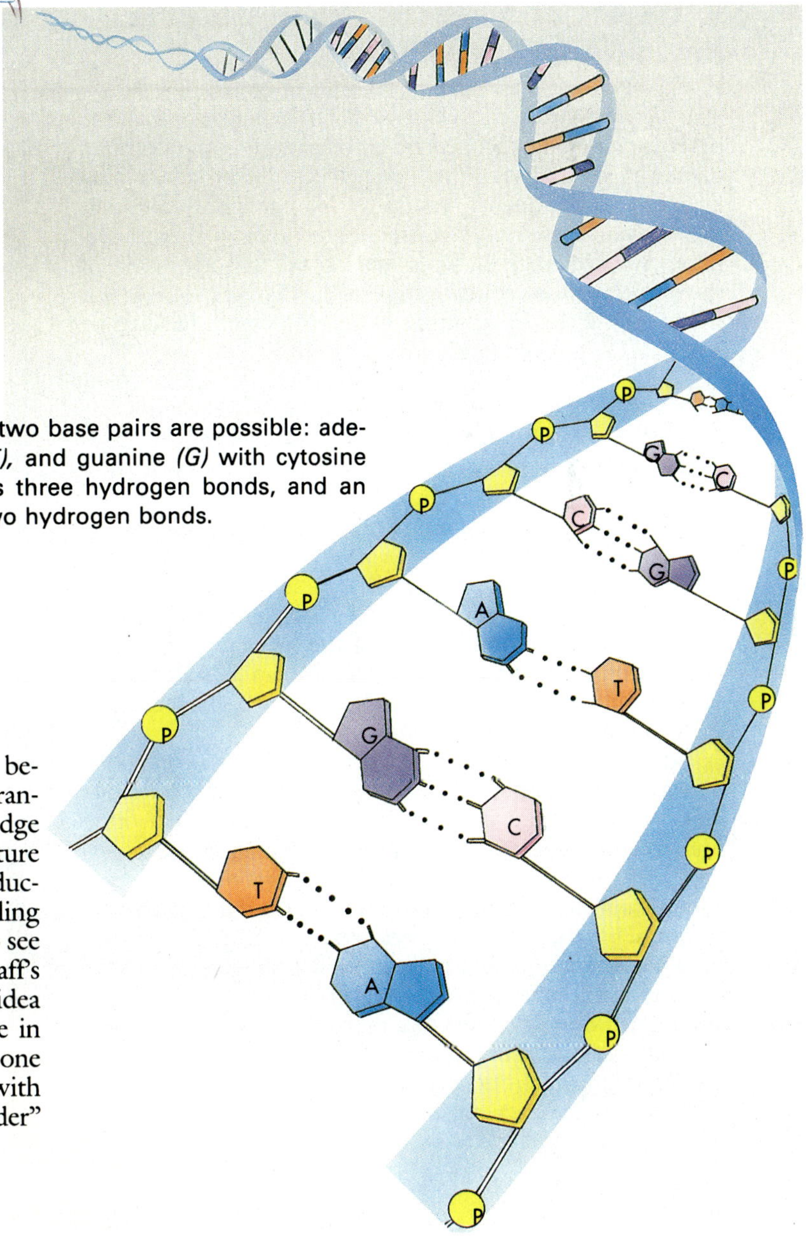

FIGURE 24-9 Base pairing.
In a DNA molecule, only two base pairs are possible: adenine *(A)* with thymine *(T)*, and guanine *(G)* with cytosine *(C)*. A G-C base pair has three hydrogen bonds, and an A-T base pair has only two hydrogen bonds.

Learning informally of Franklin and Wilkins' results before they were published in 1953, James Watson and Francis Crick (Figure 24-7), two young scientists at Cambridge University in England, quickly worked out a likely structure of the DNA molecule. They analyzed the problem deductively by building models of the nucleotides, assembling them into molecular structures, and then testing each to see whether its structure fit with what they knew from Chargaff's and Franklin and Wilkins' work. They finally hit on the idea that the molecule might be a simple double helix, one in which the bases of two strands pointed inward toward one another (Figure 24-8). Pairing a purine (which is large) with a pyrimidine (which is small) resulted in a helical "ladder" with "rungs" of uniform length (Figure 24-9).

In fact, always pairing adenine with thymine, and cytosine with guanine yielded a molecule in which A = T and C = G—a molecule consistent with Chargaff's observations. In their model, the sugar and phosphate units linked in an alternating fashion to form the sides of this twisted ladder. For their groundbreaking work, Watson and Crick shared the Nobel Prize in 1962.

Chargaff determined that although the proportion of the bases in DNA varied among organisms, the amount of adenine always equalled the amount of thymine and guanine always equalled cytosine. This information, coupled with the work of Rosalind Franklin, Maurice Wilkins, James Watson, and Francis Crick, led to the determination that the DNA molecule is a double-stranded helical "ladder." The bases of each strand of DNA form the rungs of uniform length, and alternating sugar-phosphate units form the uprights.

How DNA replicates

The Watson-Crick model immediately suggested that the basis for copying the genetic information is the complementarity of its bases. One side of the DNA ladder may have any base sequence, resulting in a particular code that the body understands. This sequence of bases then determines the sequence of bases making up the other side of the ladder. If the sequence on one side is ATTGCAT, for example, the sequence of the other side would have to be TAACGTA—its complementary image.

Scientists have also learned how DNA replicates, or makes more DNA. Before prophase of cell division (see Chapter 5), the bonds between the complementary bases break in short sections of the double-stranded DNA molecules, and the complementary strands separate from one another. This process is commonly referred to as *unwinding* or *unzipping*. Each split in the molecule is called a **replication fork.** At these forks, each separated strand serves as a template for the synthesis of a new complementary strand. Figure 24-10 shows a diagram of the replication fork and an electron micrograph of the DNA from a bacterial cell during the replication process. Notice the Y-shaped replication point.

At the replication fork, in a process carefully controlled by a complex of enzymes, each of the separated single strands is now free to bond with new nucleotides. Free nucleotide units that accumulate in the nucleus are linked to complementary bases on each of the old DNA strands. The sugars and phosphates of the new nucleotides bond together to form the backbones of both new strands. Thus the process of DNA replication begins with one double-stranded DNA molecule and ends with two double-stranded DNA molecules. Each double strand contains one strand from the old "parent" molecule plus a new complementary strand assembled from free nucleotides. Each of the new double-stranded molecules is identical to the other and is also identical to the original parent molecule.

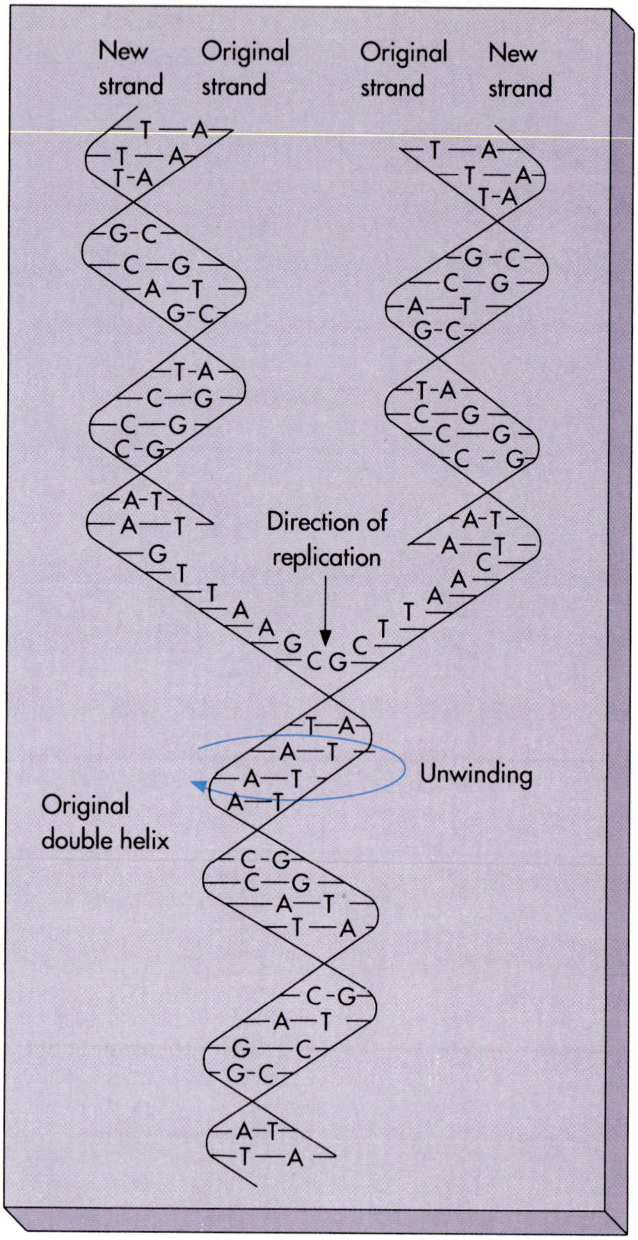

FIGURE 24-10 A replication fork.
The double-stranded DNA molecule separates at its bases, forming a split, or fork. At this fork, each separated strand acts as a template for the synthesis of a new complementary strand.

DNA begins replication by unwinding at intervals along its helix before cell division. In a process initiated and coordinated by enzymes, free nucleotides bond to the exposed bases, producing two new DNA strands, each identical to one another and to the original double strand from which they were replicated.

BIOLOGY TECHNOLOGY & society

The human genome project

More than 4000 diseases are known to result from failure or abnormality of a gene, ranging from cystic fibrosis to epilepsy. However, only 164 of these disease-related defects have been pinpointed to specific locations on the human chromosomes. In fact, only 1700 genes of any sort have been located precisely on the human chromosomes, out of an estimated total of 100,000 genes. This lack of a precise map of genes is a matter of some importance now that gene transfer therapy offers hope of curing genetic defects. The problem is that doctors cannot transfer a healthy copy of a gene into an individual lacking it unless researchers have first isolated that gene from a normal individual and located the disease-related gene's position on the chromosomes. If gene transfer therapy is going to be applied to a wide range of genetic defects, a far more detailed map of the human chromosomes is needed.

The preparation of this map is now well underway. The initial stage involved identifying the positions of a variety of short nucleotide sequences. Used as recognition sites by DNA-cutting enzymes, these sequences are scattered around the chromosomes (much as the sequence of letters "en" is scattered about at various locations on this page). Each kind of enzyme has a different recognition site, called a *restriction site*. By looking at many different enzymes, researchers can identify restriction sites all over the chromosome. By looking to see how the sites are inherited in families with hereditary disorders, it becomes possible to quickly associate a genetic defect with a particular restriction site and thus to locate the gene's position. Scientists hope to have such "signposts" positioned every million nucleotides along the chromosomes by 1995 (Figure 24-A).

FIGURE 24-A Maynard Olson (left) and Eric Green review DNA sequences. Olson was instrumental in developing key technologies that have made the Human Genome Project possible.

At the same time, efforts are underway to determine the full nucleotide sequence of all the chromosomes—the Human Genome Project. This is a mammoth task because there are some 3×10^9 nucleotide base pairs in the human chromosomes—a complete map, printed in the same size type as this text, would stretch from St. Louis to Anchorage! Two recent key developments have made this gargantuan project possible. The first is the polymerase chain reaction (PCR), which makes it possible for a scientist to obtain millions of copies of any small segment of DNA. Researchers can use PCR to test for the presence of a particular gene in large numbers of chromosome fragments. The second key development is a method of combining human DNA with pieces of yeast chromosomes. The yeast artificial chromosomes, or YAKs, enable scientists to isolate and clone far larger pieces of DNA than had been possible before. This is greatly speeding the sequencing of the human genome because it eliminates the need to sort through tiny bits of DNA that often contain only parts of a gene. Although scientists have a long way to go in meeting their goal of sequencing the entire human genome, no one believes it an impossible task.

Genes: The units of hereditary information

Just before the scientific discovery of Watson and Crick, scientists were also working to determine how DNA directs the growth and development of an organism. They asked the question, "What is the hereditary message that DNA carries?" This question was answered by means of experiments conducted in the 1940s and early 1950s by two geneticists—George Beadle and Edward Tatum.

In the early 1900s, before Beadle and Tatum's work, the British physician Archibald Garrod showed some insight into these questions. He suggested that inherited diseases reflect the inability of a person to manufacture a necessary enzyme. He therefore called inherited diseases *inborn errors in metabolism*. Garrod was ahead of his time; his contemporaries looked at physical characteristics as the phenotypic expression of genes. However, a few decades after Garrod's work, Beadle and Tatum supported his theory and officially marked the beginning of molecular genetics as they studied biochemical characteristics as the phenotypic expression of genes. They worked with the common red bread mold *Neurospora*.

First, Beadle and Tatum exposed cultures of the mold to x-rays, hoping to induce changes, or mutations, in its DNA (see Chapter 23). This technique produced mutants that each needed a specific amino acid supplied in their medium, or food. Before treatment with x-rays, the normal *Neurospora* were able to make the amino acids they needed from other substances such as glucose and a nitrogen source, so they could grow on media lacking amino acids. The mutant cultures could not grow on such media.

At the same time Beadle and Tatum were doing this work, various biochemists were also studying the manufacture and breakdown of organic molecules within cells. They determined that cells synthesize and degrade molecules by means of biochemical pathways. These pathways are sequences of steps in the process of synthesis or breakdown; each step is catalyzed by a specific enzyme.

As Beadle and Tatum studied different amino acid mutants and the known biochemical pathways of *Neurospora,* they hypothesized that the synthesis of any particular organic molecule was determined by the action of a specific enzyme. This enzyme directed the manufacture of a specific compound. Beadle and Tatum hypothesized that because their mutants had lost the ability to manufacture certain amino acids, specific enzymes must not be doing their jobs because of mutational changes.

To test their hypothesis, Beadle and Tatum looked for the presence of specific enzymes in their mutants that directed the synthesis of particular amino acids. They concentrated on one group of mutants that required the amino acid arginine for growth. First, they found that the mutants were actually different from one another. One group of arginine-requiring mutants could grow on a medium containing the organic compounds ornithine and citrulline. Another mutant group could not grow on that medium but would only grow if given arginine. A third group was able to grow on a medium containing ornithine alone. By supplying the mutants with these various compounds intermediate in the pathway to synthesize arginine and observing whether or not the mutant was able to synthesize arginine, Beadle and Tatum were able to infer the presence of defective enzymes (Figure 24-11).

From these studies, Beadle and Tatum proposed the **one gene–one enzyme theory,** which states that the production of a given enzyme is under the control of a specific gene. If the gene mutates, the enzyme will not be synthesized properly or will not be made at all. Therefore the reaction it catalyzes will not take place, and the product of the reaction will not be produced. Today, scientists have refined this hypothesis to say that the production of a given polypeptide (a portion of an enzyme or other protein) is under the control of a single gene. In addition, their conclusions have been upheld by other biochemical evidence. For their groundbreaking work, Beadle and Tatum were awarded the Nobel Prize in Physiology and Medicine in 1958.

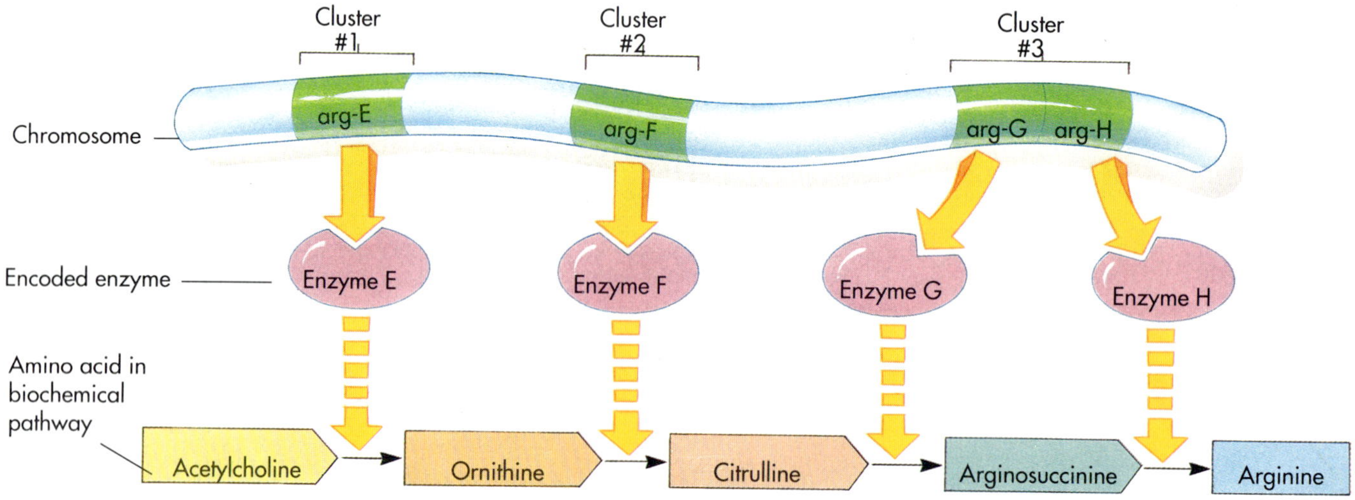

FIGURE 24-11 The one gene-one enzyme theory.
This theory states that the production of a given enzyme is under the control of a specific gene. This theory was postulated by Beadle and Tatum after their work with *Neurospora.*

These experiments and other related ones have finally brought a clear understanding of what the unit of heredity is. A unit of heredity—a **gene**—is a sequence of nucleotides that codes for the amino acid sequence of an enzyme or other protein. Although most genes code for a string of amino acids (polypeptides), there are also genes devoted to the production of special forms of RNA, which play important roles in protein synthesis.

DNA carries a code that directs the synthesis of polypeptides—pieces of enzymes and other proteins that orchestrate and regulate the growth, development, and daily functioning of cells. Each hereditary unit, or gene, codes for one polypeptide.

Gene expression: How DNA directs the synthesis of polypeptides

An overview

Interestingly, polypeptides are not made in the nucleus where DNA is located. Instead, they are made in the cytoplasm at the **ribosomes.** These complex polypeptide-making factories contain over 50 different proteins in their structure. Along with proteins, ribosomes are also made up of RNA—a hint that RNA molecules play an important role in polypeptide synthesis. In fact, cells contain three types of RNA, and each plays a special role in the manufacture of polypeptides.

The type of RNA found in ribosomes is called **ribosomal RNA,** or **rRNA.** A second type of RNA, called **transfer RNA,** or **tRNA,** is found in the cytoplasm. During polypeptide synthesis, tRNA molecules transport amino acids (used to build the polypeptide) to the ribosomes. In addition, tRNA molecules position each amino acid at the correct place on the elongating polypeptide chain. A third type of RNA, called **messenger RNA,** or **mRNA,** brings information from the DNA (within the nucleus) to the ribosomes (in the cytoplasm) to direct which polypeptide is assembled.

Transcribing the DNA message to RNA

The first step in the process of polypeptide synthesis and gene expression is the copying of the gene into a strand of messenger RNA. Scientists call this copying process **transcription.** Transcription begins when a special enzyme, called an **RNA polymerase,** binds to a particular sequence of nucleotides on a single DNA strand. This sequence of nucleotides is located at the beginning of the gene that is being expressed. You can think of such a sequence of nucleotides as an enzyme code that says, "start here." Scientists call this place on a DNA strand a **promoter site.**

First, the DNA base-pair bonds break, no longer hold-

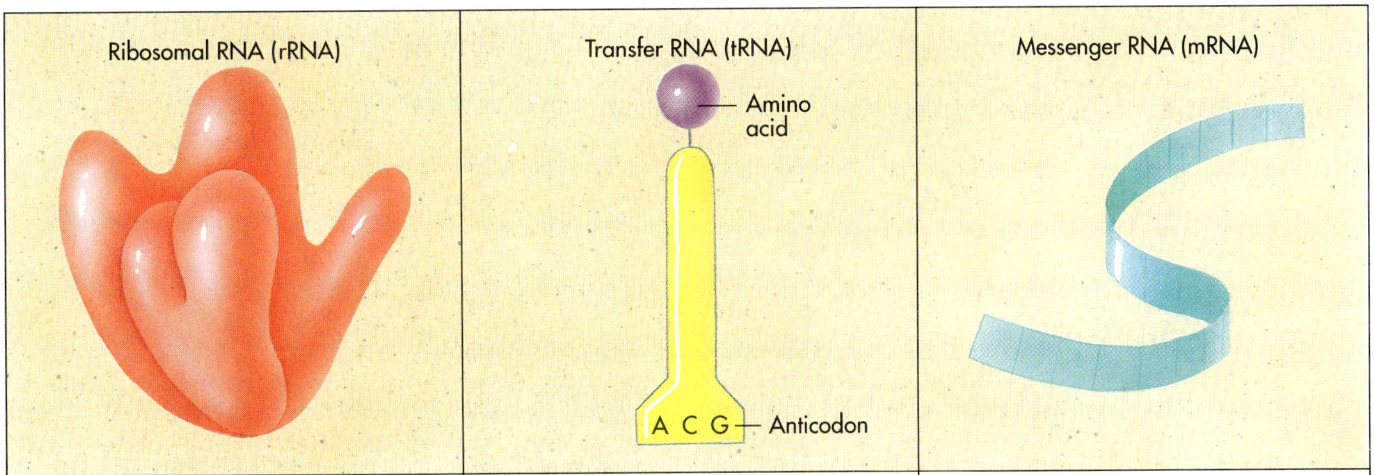

Location: Ribosomes	**Location:** Cytoplasm	**Location:** Made in nucleus; transported through nuclear pores to cytoplasm
Function: not involved in coding; directs protein synthesis	**Function:** During polypetide synthesis, tRNA molecules transport amino acids to the cytoplasm; also positions each amino acid at the correct place on the elongating polypeptide chain.	**Function:** Brings information from the DNA (within the nucleus) to the ribosomes (in the cytoplasm) to direct which polypeptide is assembled

BIOLOGY IN FOCUS

The Hershey-Chase experiment

Students studying biology today sometimes take the information presented in textbooks for granted. For instance, it seems a *given* that all the genetic information about you—your appearance, how your cells function, your sex—is contained in your DNA. But as little as 40 years ago, many scientists thought that the genetic information was contained in and passed along to offspring by proteins located in chromosomes. A series of important experiments that helped prove DNA and not protein was the genetic material was performed in 1952 by Alfred Hershey and Martha Chase. Building on an increasing body of experimental evidence that pointed to DNA, Hershey and Chase used viruses to carry out a clear either/or test and proved that an infection by a virus involved an injection of the virus DNA, not protein, into a cell.

Hershey and Chase reasoned that if DNA was indeed the genetic carrier, then bacteria infected with a virus would contain DNA from the virus inside their cells. They reached this conclusion because they knew that when viruses infect cells, they somehow direct the cell to manufacture more virus particles. If, on the other hand, protein is the material of which genes are made, then virus-infected bacteria would contain virus proteins within their cells.

The challenge for Hershey and Chase was to find a way to look inside an infected cell and see what parts of the infecting virus could be found. Hershey and Chase chose to work with a bacteriophage, a special kind of virus that infects bacteria. The bacteriophage has a nucleic acid core surrounded by a protein coat. "Legs" are attached to the protein coat that allow the virus to dock with a cell.

The key to the success of Hershey and Chase's experiments was the way they told DNA and protein apart—they used radioactive "name tags." Proteins contain sulfur, but DNA does not, so radioactive sulfur served as a label for protein. DNA contains phosphorus, but protein does not, so radioactive phosphorus served as a label for DNA.

Hershey and Chase performed two experiments (Figure 24-B). In experiment 1, they labeled the protein coat of the bacteriophage with 35S, an isotope of sulfur (see the box in Chapter 2). Then they mixed the labeled bacteriophage with bacteria (a kind called *E. coli*, which are found in abundance in your digestive tract) and allowed the viruses to infect the bacteria. To examine the bacteria after infection, Hershey and Chase had to separate the part of the bacteriophages that did not enter the bacteria from the infected bacteria. They did this by pouring their mixture into a blender and blending it very fast. The mixture was then centrifuged (spun rapidly), which pulled the heavier bacteria to the bottom of a test tube, leaving the lighter virus particles in the solution on top, called the *supernatant*. Hershey and Chase found all the 35S-labeled protein in the supernatant, not in the bacteria. Thus the bacteriophage protein did not enter the bacteria during infection.

In experiment 2, Hershey and Chase labeled the bacteriophage DNA with an isotope of phosphorus, 32P, and performed the same mixing, blending, and centrifuge procedure as they did in experiment 1. The 32P was found in the bottom of the test tube, with the bacteria. Clearly, 32P-labeled DNA had entered the bacteria.

These experiments went a long way toward convincing the scientific community that DNA was the genetic material—the logic was straightforward and the result was unambiguous. Within the year, Watson and Crick's discovery of DNA's structure hammered the final nail into the coffin of the protein hypothesis, showing how easily DNA could fulfill its suggested genetic role.

ing the double-stranded helix together. As the DNA strands separate, one strand begins to function as a template as RNA polymerase binds to a promoter site on the DNA molecule. RNA nucleotides bind with the now-exposed DNA bases in a sequence complementary to that of the DNA. For example, RNA nucleotides having the base adenine pair with DNA nucleotides having the base thymine. Likewise, RNA nucleotides having the base uracil pair with DNA nucleotides having the base adenine. Figure 24-12 shows a representation of the process of RNA transcription. Which DNA nucleotides bond with RNA nucleotides having the base guanine?

As the RNA polymerase moves along the DNA strand encountering each DNA nucleotide in turn, it adds the corresponding complementary RNA nucleotide to the growing single strand of mRNA. When the enzyme arrives at a special "stop" sequence (frequently a loop in the DNA) located at the end of the gene, it disengages from the DNA and releases the newly assembled mRNA strand. A processing step removes segments of RNA that are not used in polypeptide synthesis from the original transcript. The remaining pieces are connected to produce each messenger RNA strand. (This process of "cutting" and "splicing" the mRNA of eukaryotes is discussed more fully later in this chapter.) The mRNA strands leave the nucleus through the nuclear pores and travel to the ribosomes in the cytoplasm of the cell.

In the first step of polypeptide synthesis, the DNA message is copied onto a strand of messenger RNA. In a process similar to DNA replication, mRNA is constructed from free RNA nucleotides by using a single strand of DNA as a template. Messenger RNA leaves the nucleus and travels to the ribosomes, where polypeptide synthesis takes place.

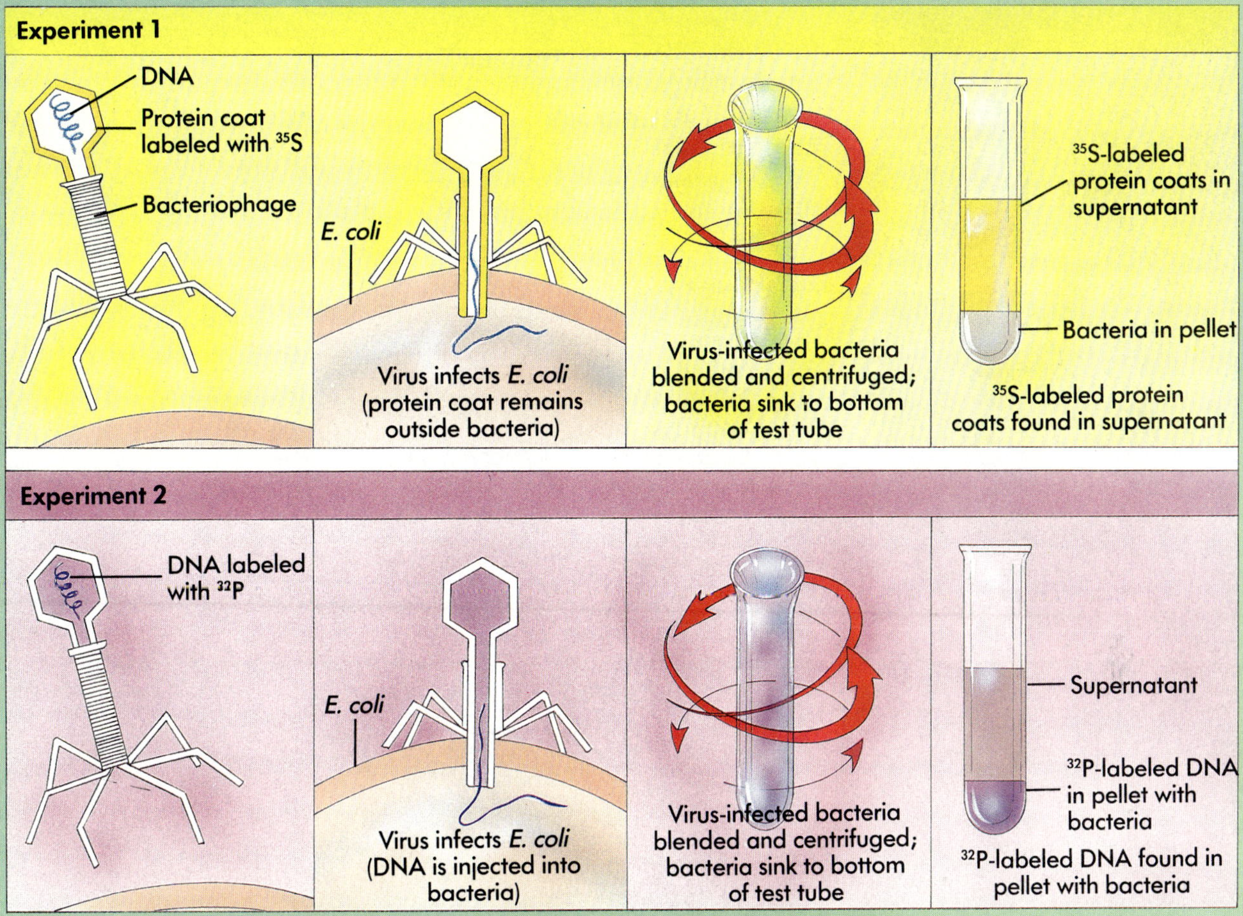

FIGURE 24-B How the Hershey-Chase experiments were performed.

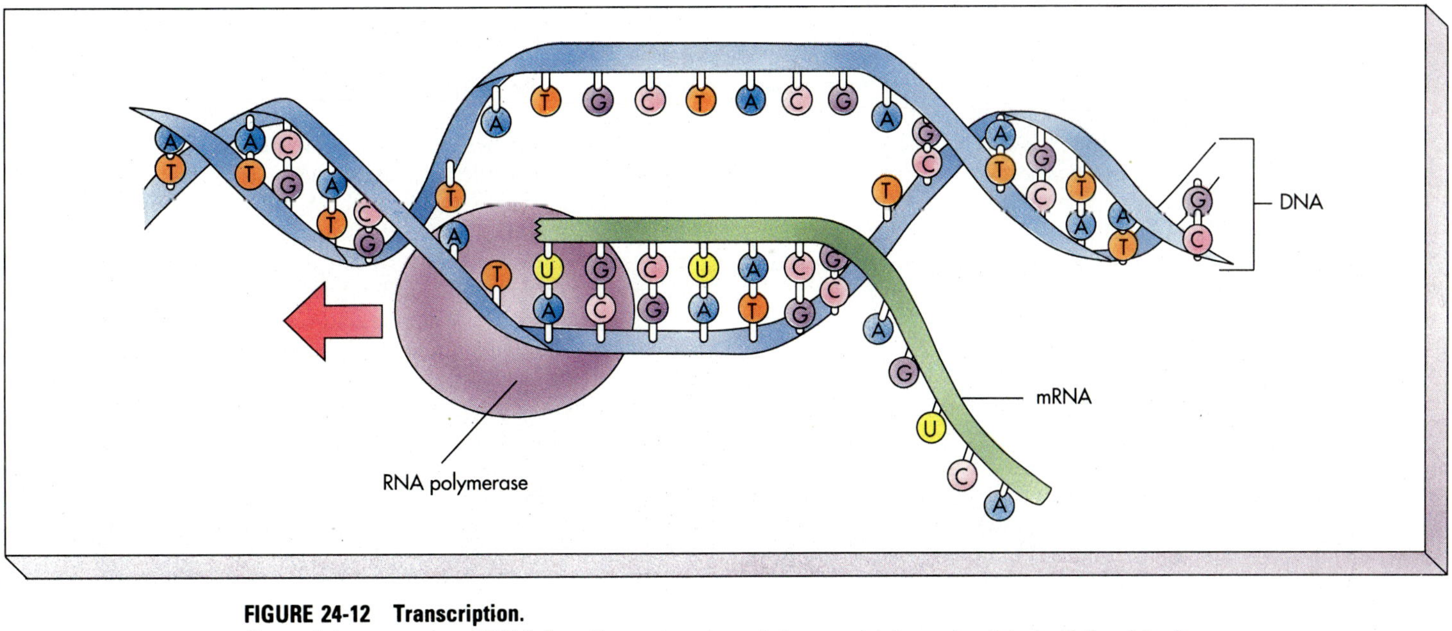

FIGURE 24-12 Transcription.
One of the strands of DNA functions as a template on which nucleotide building blocks are assembled by RNA polymerase into RNA.

TABLE 24-2

The genetic code

FIRST LETTER	SECOND LETTER U	C	A	G	THIRD LETTER
U	Phenylalanine	Serine	Tyrosine	Cysteine	U
	Phenylalanine	Serine	Tyrosine	Cysteine	C
	Leucine	Serine	Stop	Stop	A
	Leucine	Serine	Stop	Tryptophan	G
C	Leucine	Proline	Histidine	Arginine	U
	Leucine	Proline	Histidine	Arginine	C
	Leucine	Proline	Glutamine	Arginine	A
	Leucine	Proline	Glutamine	Arginine	G
A	Isoleucine	Threonine	Asparagine	Serine	U
	Isoleucine	Threonine	Asparagine	Serine	C
	Isoleucine	Threonine	Lysine	Arginine	A
	(Start); Methionine	Threonine	Lysine	Arginine	G
G	Valine	Alanine	Aspartate	Glycine	U
	Valine	Alanine	Aspartate	Glycine	C
	Valine	Alanine	Glutamate	Glycine	A
	Valine	Alanine	Glutamate	Glycine	G

A codon consists of three nucleotides on mRNA read in the sequence shown above. For example ACU codes threonine. The first letter, A, is read in the first column; the second letter, C, from the second letter column; and the third letter, U, from the third letter column. Each of the mRNA codons is recognized by a corresponding anticodon sequence on a tRNA molecule. Most amino acids are encoded by more than one codon. For example, threonine is encoded by four codons (ACU, ACC, ACA, and ACG), which differ from one another only in the third position.

Translating the transcribed DNA message into a polypeptide

In the second step of gene expression, the mRNA—using its *copied* DNA code—directs the synthesis of a polypeptide. Scientists call this "decoding" process **translation.** Understanding the code of DNA is the first step to understanding the process of translation.

In 1961, as a result of experiments led by Francis Crick (one of the researchers who worked out the structure of DNA), scientists learned that the DNA code is made up of sequences of three nucleotide bases. These triplet-nucleotide "code words"—as they appear on the transcribed mRNA—are called **codons.** Researchers soon "broke" the code and learned which codons stand for which amino acids, the "building blocks" of polypeptides.

Table 24-2 is a list of the mRNA codons and the amino acids for which they stand. The table is structured so that every triplet-base sequence of the four mRNA bases is listed. As you can see from the table, using a three-base sequence of four different bases produces 64 possible codon combinations ($4 \times 4 \times 4$ or 4^3). Approximately 20 amino acids are used in polypeptide production, so "extra" codons exist, providing alternate code words for many of the amino acids. Also notice that certain codons do not code for amino acids but act as stop signals for the process of translation.

During the process of translation the genetic code—the sequence of codons—is deciphered. First, the initial portion of the mRNA transcribed from DNA in the nucleus binds to an rRNA molecule interwoven in the ribosome (Figure 24-13). The mRNA lies on the ribosome in such a way that only the three-nucleotide portion of the mRNA molecule—the codon—is exposed at the polypeptide-making site. As each bit of the mRNA message is exposed in turn, a molecule of transfer RNA, or tRNA, binds to the mRNA.

tRNA molecules each have an **anticodon** loop—a portion of the molecule with a sequence of three base pairs complementary to a specific mRNA codon (Figure 24-14). On the opposite end of the anticodon loop, each transfer RNA molecule carries an amino acid specific to its anticodon sequence. For example, a tRNA molecule having the anticodon AGA carries the amino acid serine. A special "family" of enzymes links the amino acids to the tRNA molecules. Using the information in Table 24-2 and your understanding of the relationship between codons and anticodons, can you determine which amino acid a tRNA molecule with the anticodon GGG carries?

One by one, as tRNA molecules bind to codons at the mRNA, amino acids are "lined up" in an order determined by the sequence of codons. As each amino acid is added to the chain, a bond is formed. When an amino acid bonds to

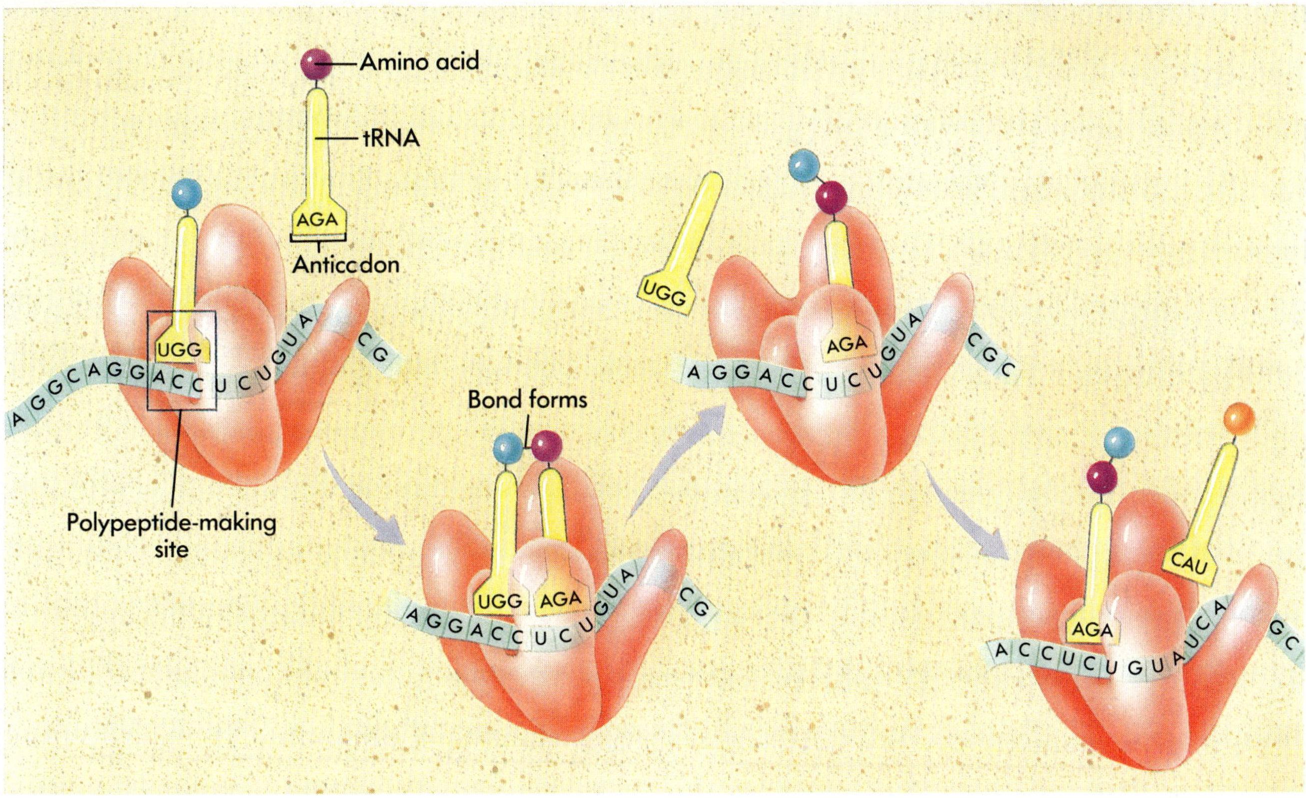

FIGURE 24-13 Translation.

The steps of translation are as follows:

1. mRNA binds to rRNA.
2. tRNA with an attached amino acid approaches the mRNA-rRNA complex.
3. The tRNA attaches to the mRNA at a special binding site. The amino acid at the end opposite to the anticodon loop forms a bond with the amino acid attached to the adjacent tRNA molecule.
4. The adjacent tRNA molecule breaks its bond with the mRNA and falls from its site.
5. The rRNA moves along the mRNA, exposing the next codon of mRNA.

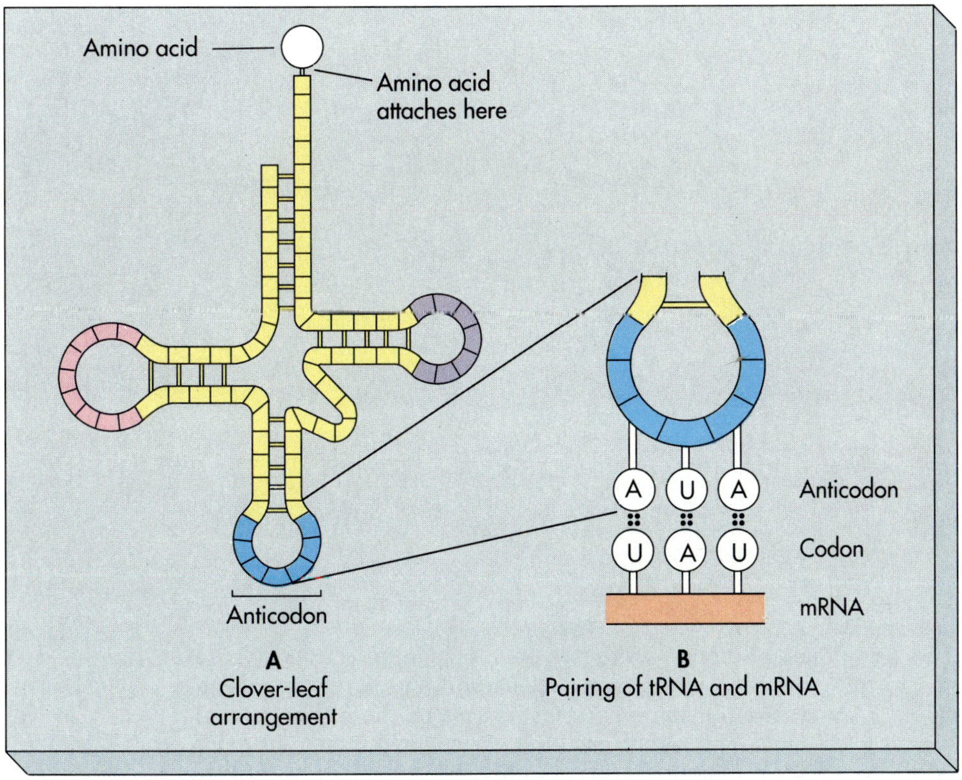

FIGURE 24-14 Structure of a tRNA molecule.

A A tRNA molecule is shaped like a cloverleaf. The anticodon is at the bottom of the cloverleaf, and the amino acid specific to the anticodon is at the top.

B The anticodon pairs with a complementary codon on the mRNA molecule.

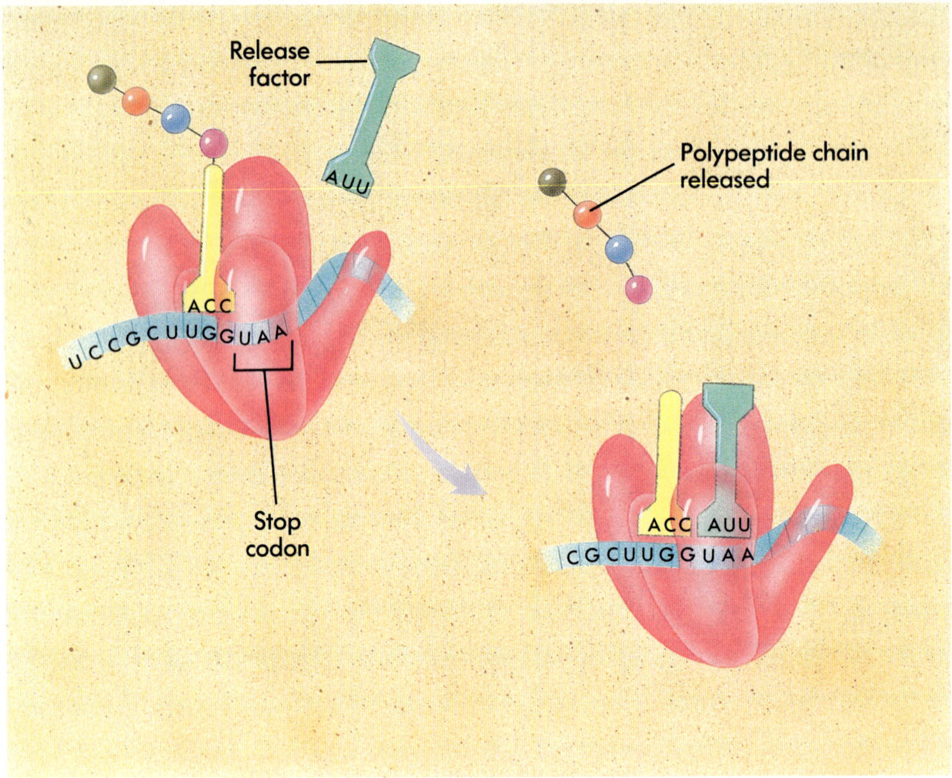

FIGURE 24-15 Termination of protein synthesis.
There is no tRNA with an anticodon complementary to any of the three stop codons, such as UAA. When a ribosome encounters a stop codon, it stops moving along the mRNA. The anticodon, however, *is* recognized by a special release factor that brings about the release of the newly made polypeptide from the ribosome.

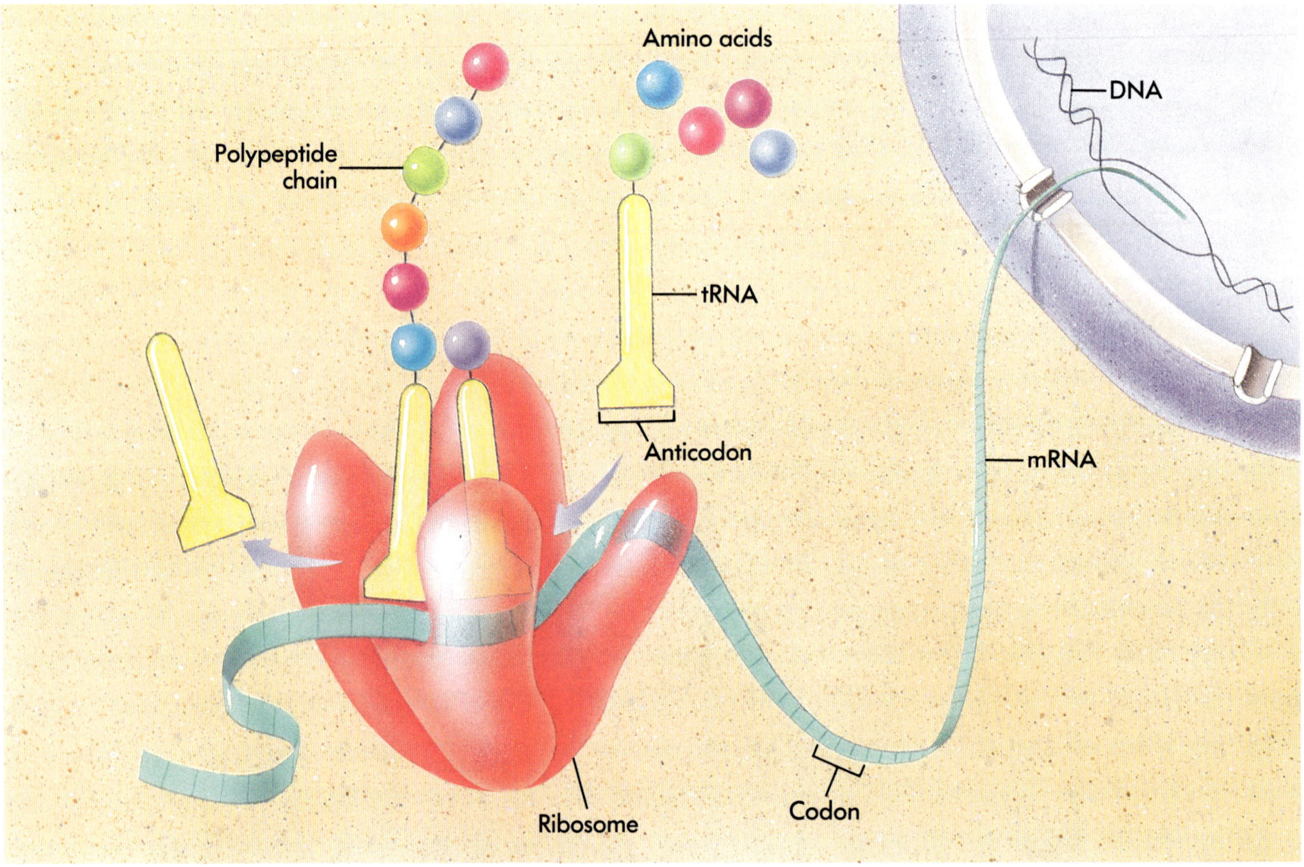

FIGURE 24-16 Overview of protein synthesis.
This diagram presents a summary of protein synthesis, beginning in the nucleus where DNA is transcribed into mRNA. mRNA leaves the nucleus and enters the cytoplasm where it binds with a ribosome. The ribosome exposes codons on the mRNA so that tRNA molecules with complementary anticodons can bind. These tRNA molecules carry amino acids. In this manner, a chain of amino acids—a polypeptide—is formed.

the forming polypeptide, its bond to its "carrier" tRNA molecule breaks. The tRNA falls from its site on the ribosome, leaving that site vacant. The ribosome then moves along the mRNA strand, exposing the next codon of the mRNA. When a tRNA molecule that "recognizes" this next codon appears, its anticodon bonds to the codon, adding a new amino acid to the growing chain. When a stop codon is encountered (Figure 24-15), no tRNA exists to bind to it. Instead, it is recognized by special release factors, proteins that bring about the release of the newly made polypeptide from the ribosome. Figure 24-16 shows an overview of protein synthesis.

Polypeptide synthesis takes place at the ribosomes. Ribosomes bind to sites at one end of a mRNA strand and then move down the strand, exposing the codons one-by-one. At each step of a ribosome's progress, it exposes a codon to binding by a tRNA molecule having a three-base sequence complementary to the exposed mRNA codon. The amino acid carried by that particular tRNA molecule is then added to the end of the elongating polypeptide chain.

Figure 24-17 shows that more than one ribosome at a time "reads" and "translates" the mRNA message to synthesize a polypeptide. Here, a group of many ribosomes (called a *polyribosome*), are reading along an mRNA molecule taken from a cell of a fly. The polypeptides being made can be seen dangling behind each ribosome.

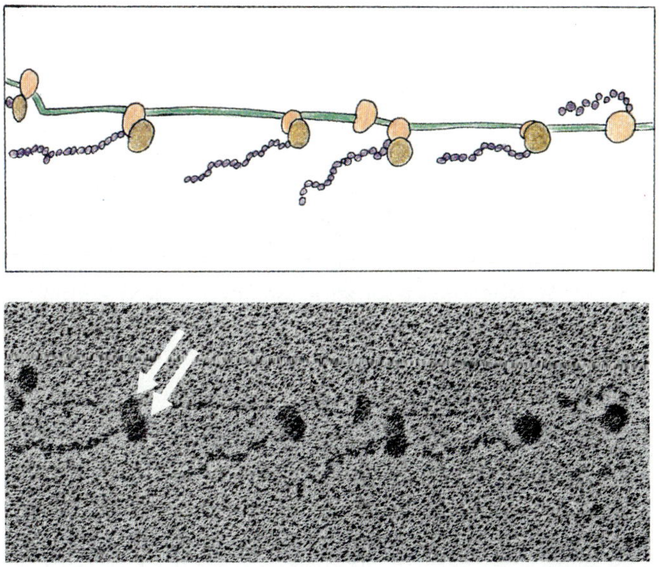

Differences in prokaryotic and eukaryotic genes

Polypeptide synthesis occurs in a similar way in prokaryotes, such as bacteria, and in eukaryotes, such as amebas, worms, plants, and humans. However, a few differences exist. One relates to cell structure. Prokaryotic cells have no membrane-bound nucleus as eukaryotic cells do. Their single, circular strand of DNA lies within the cytoplasm of the cell close to the ribosomes. Therefore bacteria often begin the translation of mRNA molecules before their transcription is complete—they can transcribe and translate simultaneously.

Although other small differences exist, one difference is of particular importance. Unlike bacterial genes, most eukaryotic genes are much larger than they need to be. In fact, eukaryotic genes contain long stretches of nucleotides that are not used in polypeptide synthesis. These "extra" sequences of nucleotides are actually cut out of the mRNA before it leaves the nucleus. These sequences that intervene between the polypeptide-specifying portions of the gene are called **introns** (they *intrude* into the gene but are not expressed) (Figure 24-18). The remaining segments of the gene—the nucleotide sequences that encode the amino-acid sequence of the polypeptide—are called **exons** (expressed portions of a gene). Exons are typically much shorter than introns and are scattered among the larger noncoding sequences. In a typical human gene the nontranslated intron portion of a gene can be 10 to 30 times larger than the coding exon portion! For example, even though only 432 nucleotides are required to encode the 144 amino acids of one of four polypeptide chains of hemoglobin, there are actually 1356 nucleotides in this hemoglobin gene. Scientists theorize that introns have evolutionary significance. Such intervening sequences within a gene may allow easier rearrangement of its exons than if they were located one next to the other.

FIGURE 24-17 Protein synthesis in a fly.
These ribosomes are reading along an mRNA molecule of the fly *Chironomus tentans* from left to right, assembling polypeptides that dangle behind them like a tail. Clearly visible are the two subunits *(arrows)* of each ribosome translating the mRNA.

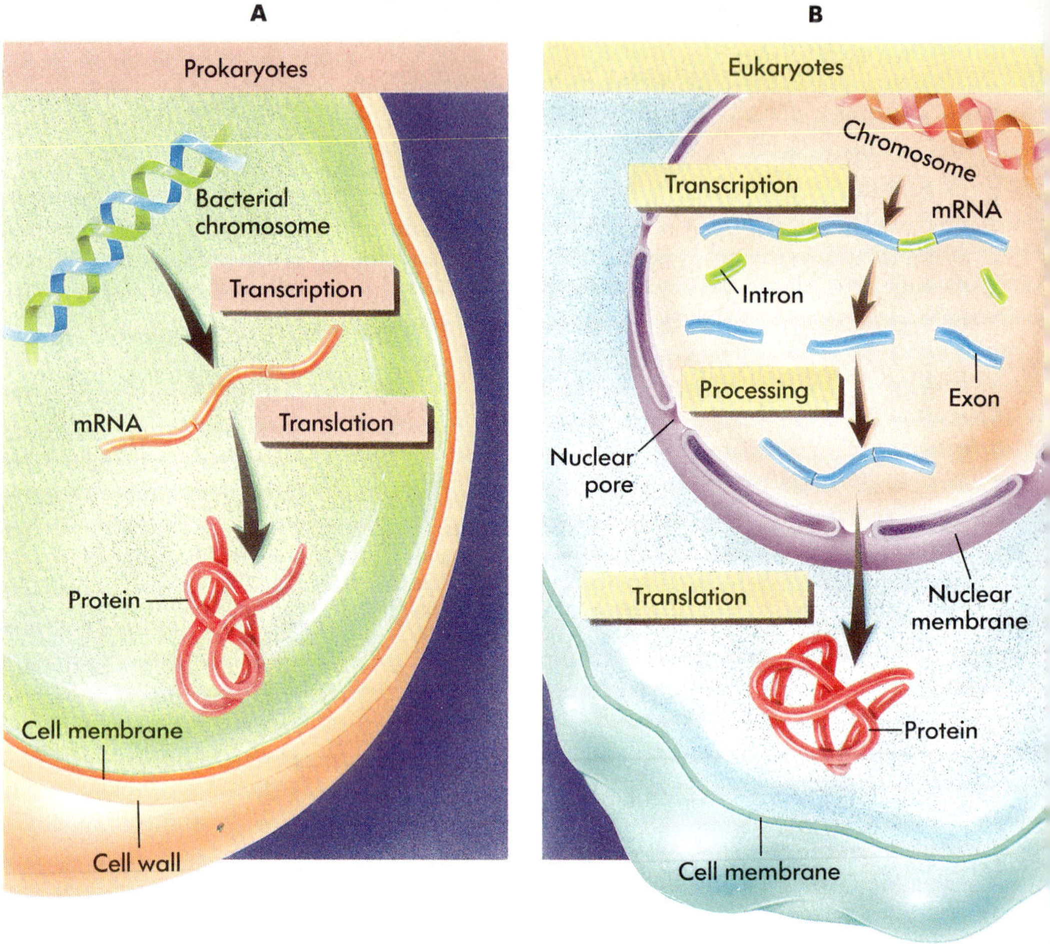

FIGURE 24-18 Differences in polypeptide synthesis in prokaryotes and eukaryotes.
A In prokaryotes the DNA that makes up the chromosomes lies within the cytoplasm of the cell. Bacterial genes are transcribed into mRNA, which is translated directly. Therefore the sequence of DNA nucleotides in bacteria corresponds to the sequence of amino acids in the encoded polypeptide.
B Genes in eukaryotes are long and contain extra sequences of nucleotides called *introns,* which are not used in protein synthesis. These introns must be cut out of the mRNA transcript of the gene, leaving exons, before protein synthesis can proceed.

Differences in prokaryotic and eukaryotic gene expression

Cells not only know how to make particular polypeptides but also when to make them. Interestingly, cells have "gene switches." These switches are really proteins that interact with specific nucleotide sequences called **regulatory sites.** Regulatory sites (also called *regulatory genes*) control the transcription of structural genes—those genes that code for the production of polypeptides. Control can be positive and result in turning a gene on, or it can be negative and result in turning a gene off. Eukaryotic regulatory genes act primarily as "on" switches because eukaryotic genes are usually inactive. Prokaryotic cells, such as bacteria, turn their genes on and off (regulate their genes) according to environmental conditions.

Prokaryotic gene expression

The environments that bacteria live in vary tremendously. Although you cannot see the bacteria, look around where you are right now and you will see some of the many places they live. Bacteria are on your desk top, in the air, and on the floor. They are on your hands, on your clothes, and in your body. The conditions under which bacteria live change often. As a bacterium's environment changes, many of its genes may be turned on or off in response to varying conditions.

Geneticists and molecular biologists have learned a great deal about gene expression by studying bacterial genes. In bacteria and viruses (see Chapter 28), a cluster of genes having related functions that are regulated as a unit is called an **operon.** These genes are: (1) **structural genes,** which direct the synthesis of enzymes having related functions; (2) a **promoter,** which marks the starting point for transcription of the genes; and (3) an **operator,** which acts as an on or off switch. The operator region is either located within the promoter or overlaps it and is the region that serves as a binding site for RNA polymerase (Figure 24-19).

As mentioned previously, bacteria live almost everywhere—including within your body. One common inhabitant of your intestines is a species of bacterium called *Escherichia coli.* This bacterium is one of the species scientists study to learn about gene expression in prokaryotes. Interestingly, scientists discovered that if the sugar lactose is not present in the environment of a population of *E. coli,* the bacteria have very few enzymes present within their cells that can take lactose into the cell and break it down. When *E. coli* is "fed" lactose, such as when you drink a glass of milk, the bacteria quickly manufacture the enzymes that do these jobs. Scientists discovered that the metabolism of lactose in *E. coli* is programmed by a cluster of genes they named the ***lac* operon.** The *lac* operon has three genes that

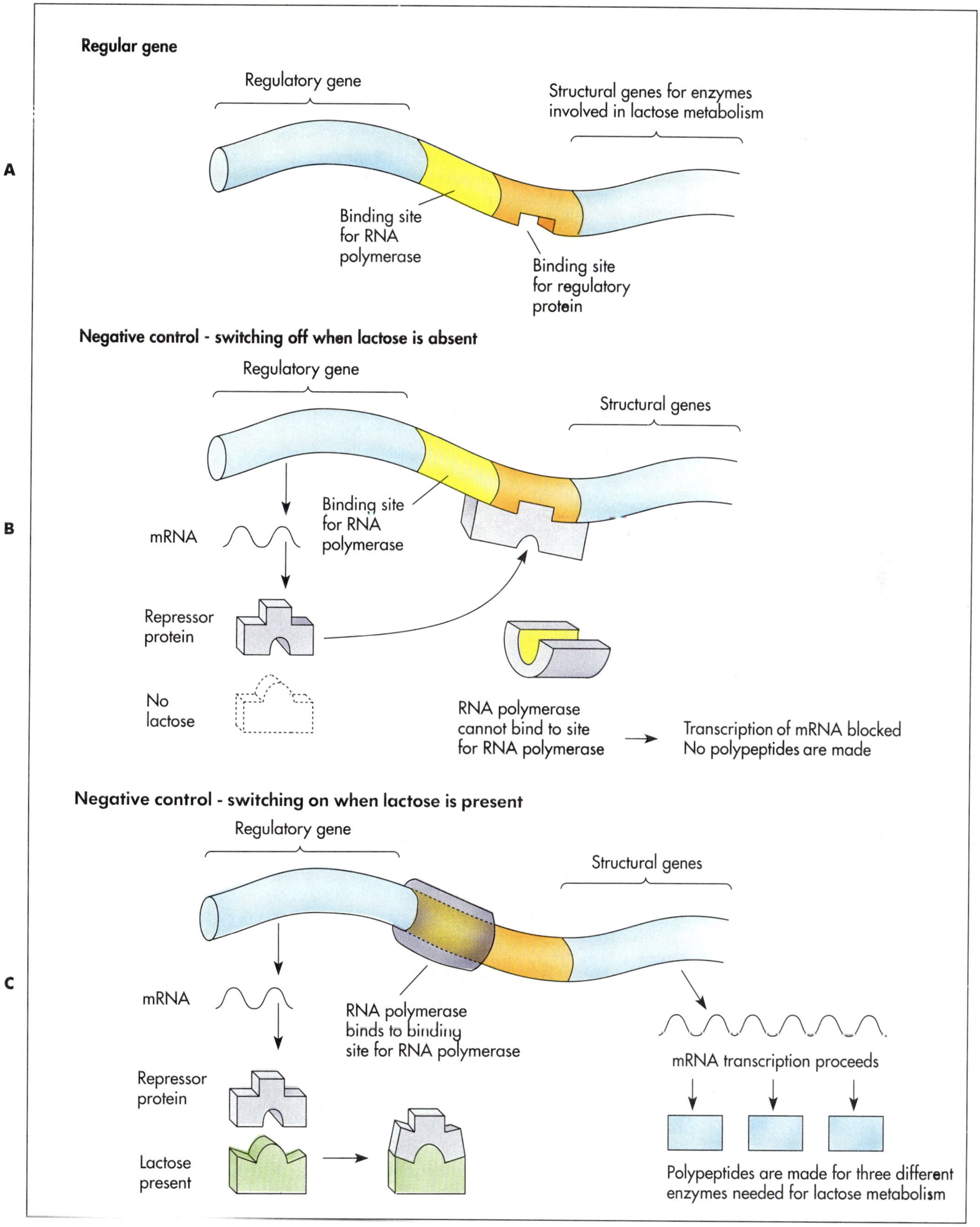

FIGURE 24-19 The *lac* operon.
A Structure of the *lac* operon.
B How the operon is switched off.
C How the operon is switched on.

direct the synthesis of enzymes that metabolize milk sugar, or lactose. It has become one of the most widely studied operons in bacteria and is often used as an example of gene expression in prokaryotes.

The *lac* operon is on (meaning it is transcribed) until transcription is switched off, or blocked. This type of system operates, then, by **negative control.** The off switch is a protein called the ***lac* repressor,** which is produced by a regulatory site. As shown in Figure 24-19, the *lac* repressor fits into the operator region of the *lac* operon like a key fits into a lock, similar to the binding of enzymes and substrates (see Chapter 6). The binding of the repressor blocks the attachment of RNA polymerase to the promoter. Therefore RNA polymerase cannot transcribe the structural genes. When lactose is present in the cell's environment, however, it binds to the repressor, changing its shape. The repressor is no longer able to bind to the operator, and the cell is able to produce the enzymes that direct the metabolism of lactose.

Although the *lac* operon is on unless it is switched off by a repressor, some operons must be switched on by proteins called **activators.** This type of control is called **positive control.** Activators bind to a specific nucleotide sequence at or within the promoter and stimulate transcription. Operons regulated by both activators and repressors have both positive and negative control. Regulatory sites code for activator proteins as well as for repressor proteins.

Eukaryotic gene expression

Gene expression in multicellular, eukaryotic organisms is less understood than in prokaryotic organisms because eukaryotes have much more genetic material in each cell and in their chromosome structures. However, the amount of genetic material in a species does not increase as the complexity of the organism increases as you might expect. For example, amphibians have more DNA in their cells than humans do. This fact suggests that eukaryotic cells contain extra DNA. Research suggests that as much as 90% percent of the DNA in certain eukaryotes does not code for proteins.

What is this extra DNA? One kind of untranslated DNA is intervening sequences, or introns, mentioned earlier. Introns interrupt the coding sequences of genes. After transcription but before translation, eukaryotic mRNA is "processed," removing introns in a type of splicing operation, much like a film is cut and spliced together to make the final version of a movie (see Figure 24-18). Scientists now know that this process allows eukaryotes to use the same genes to code for different proteins by employing changes in the cutting and splicing procedure.

Another source of extra DNA in eukaryotes is multiple copies of genes. Sometimes these gene copies are identical and allow a cell to produce more of a particular protein product in a certain length of time. In other gene copies, the nucleotide sequences are related but not identical. They produce related but differing products. Some copies are nonfunctional. Such multiple copies of genes in eukaryotes, called **gene families,** are derived from a common ancestral gene and are a reflection of the evolutionary process.

Another difference between prokaryotic and eukaryotic gene expression regards the switching on and off of genes in the various cell types of the eukaryote. Each cell in a multicellular eukaryote does not produce the same enzymes at the same time as in prokaryotes. For example, during your fetal development, specific enzymes played crucial roles at certain times, directing the series of biochemical reactions that resulted in your growth. As this growth and development took place, your genes were transcribed in a specific order, each gene for a specified time. Likewise, the cells in your body that produce digestive enzymes know when to manufacture these enzymes. Your red blood cells—not other types of cells—synthesize hemoglobin to carry oxygen to all parts of your body. How are these specific genes activated? Molecular geneticists are only now beginning to understand these mechanisms. One interesting answer appears to be that steroid hormones (see Chapter 19), along with protein receptor molecules from the cytoplasm, activate transcription by binding to promoter sites on the DNA. Scientists are still probing the details of this regulatory process.

Organisms control the expression of their genes by selectively inhibiting the transcription of some genes and facilitating the transcription of others. Regulatory proteins control the transcription process.

Gene therapy: Changing genes to cure disease

There are more than 4000 known inherited genetic disorders. Persons with genetic disorders are sick simply because one or more of their genes has a mistake in its code—its sequence of nucleotides is faulty. Today, scientists are trying to cure these disorders by inserting genes with the proper genetic message into patients having defective genes. This technique is called **gene therapy.**

The first federally approved clinical trial of gene therapy in the United States took place at the National Institutes of Health (NIH) in the fall of 1990. The disease targeted for treatment was one form of severe combined immunodeficiency, or SCID. This disorder attracted national attention in the 1970s when a film was made telling the story of a boy with SCID who lived his life in a sterile "bubble" (Figure 24-20). Like all SCID patients, his immune system was impaired because of his disorder, so he was unable to combat even the most common microbe.

Scientists have not yet been able to cure SCID, and the experimental gene therapy being tried at NIH will require repeated treatments through the patients' lives (see the boxed essay in Chapter 23). Ideally, scientists are hoping to provide real cures to genetic disorders by actually replacing damaged copies of genes with undamaged copies in a single treatment. This process involves removing cells from a patient, introducing copies of undamaged or healthy genes, and returning the cells to the patient.

In late 1990, scientists carried out gene therapy on a patient with a fatal type of skin cancer called melanoma.

FIGURE 24-20 Severe combined immunodeficiency syndrome (SCID).
David, shown here in his plastic "bubble," was born with a genetic disorder that leaves the body defenseless against infection. David was confined to a sterile "bubble" his entire life. In 1983, David received a transfusion of his sister's bone marrow in an attempt to stimulate the production of immune cells. Tragically, his sister's bone marrow harbored the Epstein-Barr virus, which infected the few B cells that David had (see Chapter 13). The infected B cells grew into tumors, and David died of cancer on February 22, 1984. He was 12 years old. In the fall of 1990, gene therapy—a new technique—offered hope to another sufferer of SCID.

Researchers removed a portion of a tumor and extracted certain white blood cells from it called tumor-infiltrating lymphocytes, or TILs. After inserting a gene that produces a tumor-killing protein into the patient's TILs, the cells were then injected back into the patient's body. The hope is that the cells will multiply and produce the tumor-killing protein, and then the melanoma will be cured.

Both these cases of gene therapy are types of **somatic cell gene therapy** in which body (somatic) cells are taken, treated, and then reintroduced to the patient. The gene therapy affects that patient alone. Because the treated cells are body cells, not sex cells, the new genes are not passed on to the recipient's offspring.

Another type of gene therapy being discussed is **germ line gene therapy.** In this process, scientists would extract eggs or sperm from an individual carrying genes for an inherited disorder. Healthy genes would be inserted into the germ cells (sex cells), and they would be placed back into the person's body. In this way, the genetically engineered traits would be passed on to a person's children. This type of gene therapy also raises serious bioethical concerns.

The story of gene therapy is just beginning to unfold. Maybe, in the future, gene therapy will be used not only to repair genetic defects but also to enhance a person's ability to combat disease. Today, only one thing is certain: genetic therapy holds not only amazing promise but also the potential for tremendous controversy as scientists attempt to change the human genome.

Summary

1. Eukaryotic cells store hereditary information within their nuclei in nucleic acids. There are two types of nucleic acids in cells: deoxyribonucleic acid, or DNA, and ribonucleic acid, or RNA. DNA carries the hereditary message.
2. The nucleic acids of all organisms are made up of nucleotides. Each nucleotide contains three molecular parts: one phosphate group, one 5-carbon sugar, and one nitrogen-containing base. The relative proportion of the bases found in DNA varies from species to species.
3. DNA has the structure of a double helix, two molecular "chains" held to each other by hydrogen bonds between its nucleotides. Its four bases are adenine (A), thymine (T), cytosine (C), and guanine (G). The nucleotide A always bonds with T and G with C. These bases are complementary to one another.
4. Before cell division, DNA replication is orchestrated by a battery of enzymes that "unzip" the DNA, cause it to "unwind," and then use each of the single strands as a template to assemble a complementary new strand.
5. The hereditary message of DNA is actually a code hidden within its sequence of bases. A gene, or unit of heredity, is a sequence of bases that codes for a specific polypeptide, or portion of an enzyme. Most hereditary traits reflect the actions of enzymes.
6. The expression of hereditary information in all organisms takes place in two stages. First, in the process of transcription, a portion of the DNA message is copied onto a single strand of mRNA by using the DNA as a template. Second, in the process of translation, an amino acid chain is assembled by a ribosome and tRNA, using the mRNA base sequence to direct the sequence of amino acids.
7. Most eukaryotic genes contain additional base sequences embedded within their coding sequences. These sequences, or introns, are removed from the mRNA before it is translated.
8. The control of gene expression is exercised largely by regulating transcription. Cells have "on" and "off" gene transcription "switches." Control is positive, in which cellular signals are required before transcription can start, and negative, in which other cellular signals prevent transcription except under appropriate circumstances.
9. Today, scientists are trying to cure certain disorders by inserting genes with the proper genetic message into patients having defective genes. This technique is gene therapy.

REVIEW QUESTIONS

1. Summarize the contributions of Mendel, Sutton, Morgan, and Hammerling to the present-day understanding of genetic inheritance.
2. What are the two types of nucleic acids found in your cells? Describe the structures of each.
3. What characteristics of DNA did Chargaff's experiments reveal? Why was this significant?
4. Explain the term *double helix,* and describe its structure.
5. Summarize the DNA replication process.
6. Fill in the blanks: A unit of heredity (a ______) is a sequence of ______ that codes for the ______ ______ sequence of a polypeptide.
7. Distinguish among rRNA, tRNA, and mRNA. What does each abbreviation stand for, and what are their respective functions?
8. Summarize the process of transcription.
9. Place the following steps in the correct sequence. What process is being described?:
   a. Ribosomes "move along" an mRNA strand and expose codons.
   b. Amino acids bind to the new polypeptide.
   c. Ribosomes bind to sites at one end of an mRNA strand.
   d. A tRNA molecule binds in turn to each exposed codon.
10. Distinguish among introns, exons, and codons. What is the significance of each?
11. What are regulatory genes? Explain why they are important.
12. Fill in the blanks: The process that blocks transcription is called ______, and the protein that is responsible is called the ______ ______. "Turning on" the transcription of specific genes is ______, and the protein that is responsible is called the ______ ______.
13. What is gene therapy?
14. The chapter notes that gene therapy is controversial. Some people feel it could be greatly abused; others feel the potential benefits outweigh the risks. What is your opinion? Explain why you feel this way.

KEY TERMS

activator 422
anticodon 416
codon 416
deoxyribonucleic acid (DNA) 407
exon 419
gene 413
gene family 422
gene therapy 422
genome 405
germ line gene therapy 423
intron 419
lac operon 420
messenger RNA (mRNA) 413
negative control 422
nucleic acid 405
nucleotide 406
one gene–one enzyme theory 412
operator 420
operon 420
positive control 422
promoter 420
promoter site 413
purine 407
pyrimidine 407
regulatory site 420
replication fork 410
ribonucleic acid (RNA) 406
ribosomal RNA (rRNA) 413
ribosome 413
RNA polymerase 413
somatic cell gene therapy 423
structural gene 420
transcription 413
transfer RNA (tRNA) 413
translation 416

THOUGHT QUESTIONS

1. There is much excitement about the possiblity of curing cystic fibrosis with somatic cell gene therapy by using a virus as the gene carrier. Doctors will attempt a cure by placing a "healthy" copy of the cf gene into the DNA of a common cold virus (adenovirus) and having the patient inhale the engineered virus. Infecting the cells of the patient's lungs, the adenovirus will introduce the working copy of the cf gene into these cells, curing the disorder. What do you think the limitations of such an approach might be? What other human genetic disorders might be candidates for similar treatment?
2. The bulk of a typical eukaryotic gene is untranslated introns; the sequences encoding the protein are scattered within this noncoding background like aces shuffled within a deck of playing cards. Why do you think evolution has favored this design over the straightforward design found in most bacteria?

FOR FURTHER READING

Franklin-Barbajosa, C. (1992, May). DNA profiling: the new science of identity. *National Geographic,* pp. 112-124.
Since the discovery of DNA's role in transferring hereditary information in 1944, the practical uses of DNA in science have grown steadily. Scientists use DNA to investigate crimes, match organ donors, study fossils, and even reunite families. This article, written for a nonscientific audience, describes the array of practical DNA applications.

Ptashne, M. (1989, January). How gene activators work. *Scientific American,* pp. 41-47.
This is a lucid overview of how genes are turned on and off.

Ross, J. (1989, April). The turnover of messenger RNA. *Scientific American,* pp. 48-55.
The level of many proteins in the body is determined by how fast the messenger RNA encoding them is broken down rather than by how speedily new transcripts are churned out.

PART SIX

EVOLUTION

HOW LIVING THINGS CHANGE OVER TIME

25 The scientific evidence for evolution

Brought back to life with computer chips and fleshlike polymers, the eighteenth century English naturalist Charles Darwin speaks to visitors from his library at the St. Louis Zoo in St. Louis, Missouri. Darwin's library is located in The Hall of Animals, one of the indoor exhibit areas at The Living World, the Zoo's educational complex. Darwin explains:

> A finch, not very different from this one, led my mind to an idea of which you may have heard—that the great diversity of creatures on Earth, the wealth of form and color, is the result of evolution by natural selection.

Darwin then continues to explain the experiences that affected his thinking and led to the development of his theory:

> As a young man, I sailed around the world on Her Majesty's ship *Beagle,* and on the little Galapagos Islands—a thousand miles at sea—I saw finches. Some had big bills with which they cracked seeds like sparrows. Others darted around catching insects as a warbler might. There was even one that used a twig to dig for grubs, trying to be a woodpecker. And I noticed a peculiar thing . . . the birds that I saw on the Galapagos were not at all like our English Finches. Instead, they looked like finches I had seen before, earlier in my voyage in South America. This was the clue. An ancestor of these Galapagos finches must have come from South America to the islands, and these finches are its descendants. But how could Galapagos finches be so different from one another if they all have the same ancestor? Somehow, they must have changed. And that is the heart of it . . . descent with modification . . . evolution.

HIGHLIGHTS

▼ One of the central theories of biology is the theory of evolution, which states that living things change over time.

▼ Charles Darwin, a nineteenth century English naturalist, developed this theory based on observations of plants and animals of oceanic islands, geological evidence that the Earth has changed over time, the results of artificial breeding, and Malthus' ideas on population growth and decline.

▼ Darwin proposed that evolution occurs as a result of natural selection.

▼ Although proposed over 100 years ago, the theory of evolution is still upheld and is further supported by the fossil record, the molecular record, progressive changes in homologous structures, and the presence of vestigial structures.

OUTLINE

The development of Darwin's theory of evolution
- Darwin's observations
- Factors that influenced Darwin's thinking
  - The influence of geology
  - The results of artificial breeding
  - The study of populations
- Natural selection: A mechanism of evolution
- An example of natural selection at work: Darwin's finches

The publication of Darwin's theory

Testing the theory
- The fossil record
- The age of the Earth: Rock and fossil dating
- Comparative anatomy
- Comparative embryology
- Molecular biology

Visitors at the St. Louis Zoo have a chance to better understand Darwin's statements. Likewise, this chapter describes the history of evolutionary thought and explains its scientific evidence. Then you, too, can better understand Darwin's statements and your own evolutionary past.

The development of Darwin's theory of evolution

The story of Darwin and his theory begins in 1831, when he was 22 years old. On the recommendation of one of his professors at Cambridge University, he was selected to serve as naturalist for a 5-year voyage (from 1831 to 1836) around the coasts of South America on the H.M.S. *Beagle* (Figure 25-1). Darwin had the chance to study plants and animals on continents, islands, and seas distant from his native England. He was able to experience firsthand the remarkable diversity of living things on the Galapagos Islands off the west coast of South America. Such an opportunity clearly played an important role in the development of his thought about the nature of life on Earth.

When the *Beagle* set sail, Darwin was fully convinced that species were unchanging. Indeed, he wrote that it was not until 2 or 3 years after his return that he began to consider seriously the possibility that species could change. And it was not until years after that time that Darwin began to formulate a theory integrating his observations of the trip and his understanding of geology, population biology, and the fossil record. Beginning in 1842, Darwin began to write his explanation of the diversity of life on earth and the ways in which living things are related to one another.

Darwin's observations

During his 5 years on the ship, Darwin observed many phenomena that were of central importance to the development of his theory of evolution. While in southern South America, for example, Darwin observed fossils (the preserved remains) of extinct armadillos that were similar to armadillos still living in that area (Figure 25-2). He found it interesting that such similar yet distinct living and fossil organisms were found in this same small geographical area. This observation suggested to Darwin that the fossilized armadillos were related to the present-day armadillos; that they were "distant" relatives.

Another observation made by Darwin was that geographical areas having similar climates, such as Australia, South Africa, California, and Chile, are each populated by different species of plants and animals. A **species** is a group of related organisms that share common characteristics and are able to interbreed and produce viable offspring. Organ-

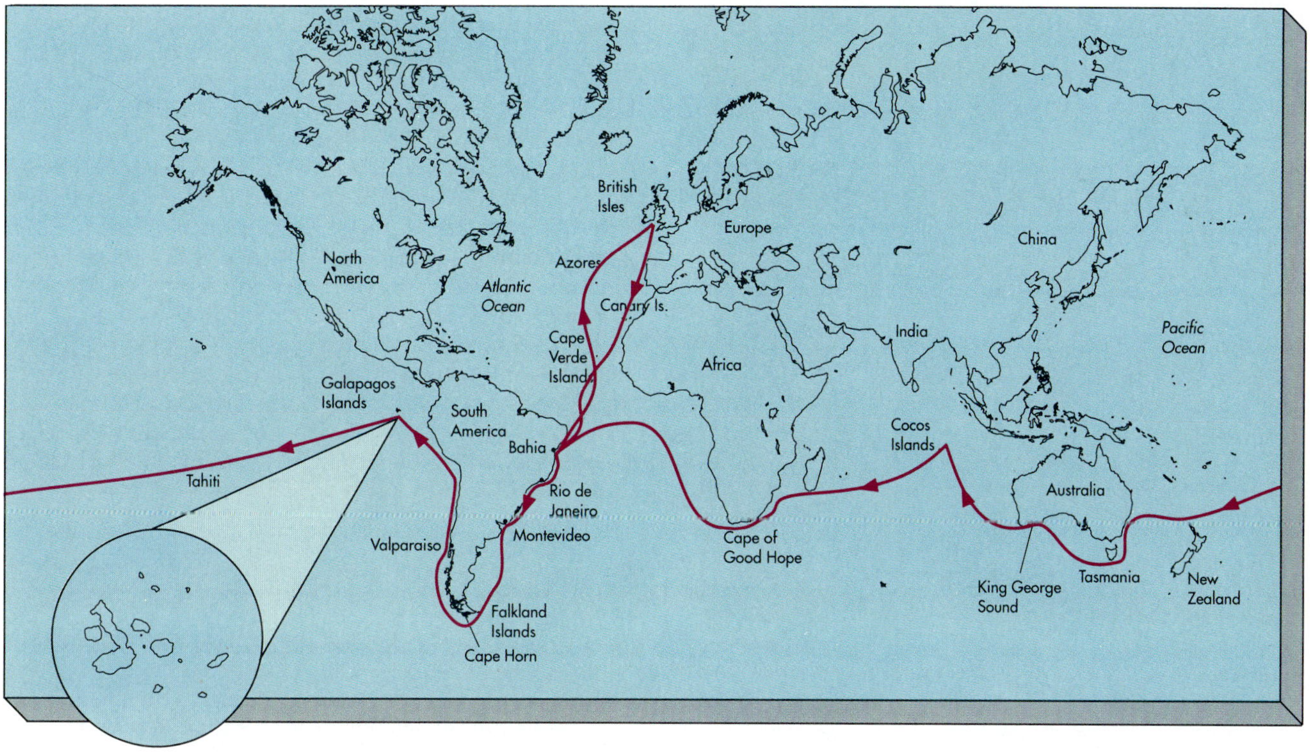

FIGURE 25-1 Voyage of the H.M.S. *Beagle*.
Most of Darwin's time was spent exploring the coasts and coastal islands of South America, such as the Galapagos Islands. Darwin's studies of the animals of the Galapagos Islands played a key role in his development of the theory of evolution by natural selection.

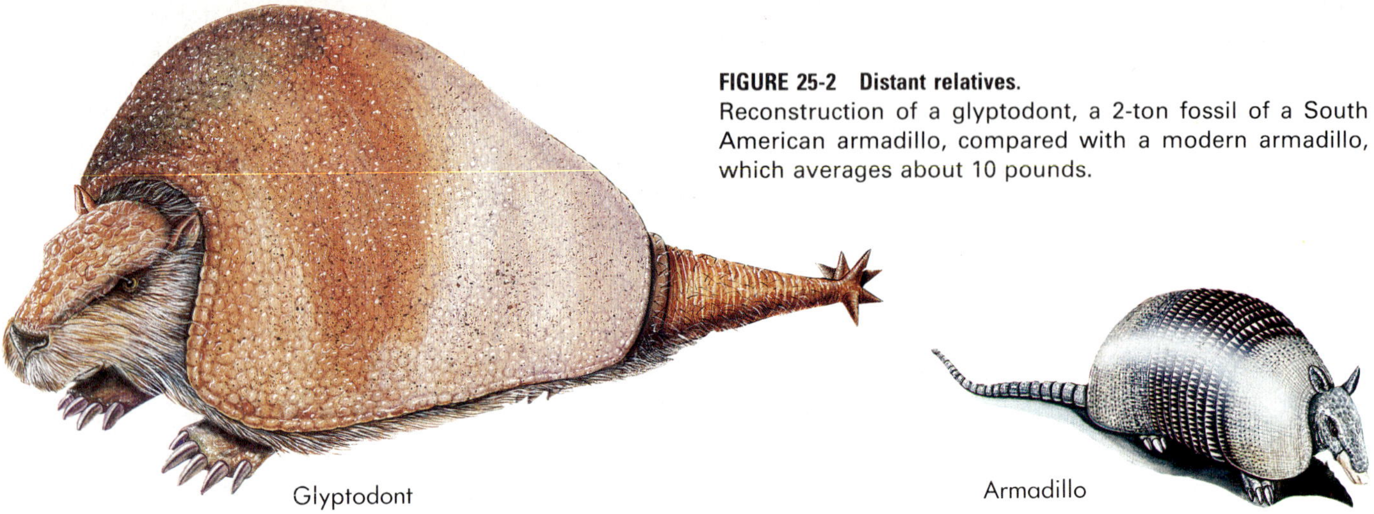

FIGURE 25-2 Distant relatives.
Reconstruction of a glyptodont, a 2-ton fossil of a South American armadillo, compared with a modern armadillo, which averages about 10 pounds.

isms found in different geographical areas having similar climates are often similar to one another, "shaped" by environmental similarities, but are, in fact, different and often unrelated species. These differences suggested to Darwin that factors other than or in addition to climate must play a role in plant and animal diversity. Otherwise, all lands having the same climate would have the same species of animals and plants.

On the Galapagos Islands, Darwin encountered giant land tortoises. The tortoises on the various islands were similar yet different from one another. In fact, local residents and sailors who captured the tortoises for food could tell which island a particular animal had come from just by looking at its shell (Figure 25-3). This pattern of physical variation suggested that all of the tortoises were related, but that they had changed slightly in appearance after becoming isolated on the different islands.

Darwin was also struck by the fact that the relatively young Galapagos Islands (formed by undersea volcanoes) were home to a profusion of living organisms resembling plants and animals that lived on the nearby coast of South America. Notice, for example, the similarity of the two birds in Figure 25-4. The bird on the left is a medium ground finch and is found on the Galapagos. The blue-black grass-quit, shown on the right, is found in grasslands along the Pacific Coast from Mexico to Chile. These observations suggested to Darwin that the Galapagos organisms were related to ancestors who, long ago, flew, swam, or "hitchhiked" (were transported by other organisms) to the islands from the mainland.

Darwin made observations during his 5-year voyage that suggested to him:

1. **Organisms of the past and present are related to one another.**
2. **Factors other than or in addition to climate play a role in the development of plant and animal diversity.**
3. **Members of the same species often change slightly in appearance after becoming geographically isolated from one another.**
4. **Organisms living on oceanic islands often resembled organisms found living on a close mainland.**

A

B

FIGURE 25-3 Galapagos tortoises.
The tortoises of the Galapagos Islands exhibit variation. Tortoises with large, domed shells **(A)**, are found in relatively moist habitats, whereas those with lower, saddleback-type shells with the front bent up **(B)** are found in drier habitats. This pattern of physical variation suggested to Darwin that the island tortoises were related but had changed slightly after being isolated in different habitats.

A B

FIGURE 25-4 Evidence from the finches.
A One of Darwin's finches, the medium ground finch.
B The blue-black grassquit, which is found in grasslands along the Pacific coast from Mexico to Chile. This bird may have a common ancestor with Darwin's finches.

A B C

FIGURE 25-5 Artificial selection: another clue in the natural selection puzzle.
The differences that have been obtained by artificial selection of the wild European rock pigeon **(A)** and domestic races such as the red fantail **(B)** and the fairy swallow **(C)** are so great that birds probably would, if wild, be classified in entirely different major groups. In a way similar to that in which these races were derived, widely different species have originated in nature by means of natural selection.

Factors that influenced Darwin's thinking

As Darwin studied the data he collected during his voyage, he reflected on its significance in the context of what was known about geology, the breeding of domesticated animals, and population biology.

The influence of geology

In the late eighteenth and early nineteenth centuries, scientists studying the rock layers of the Earth noticed two things. First, scientists saw evidence that the Earth had changed over time, acted on by natural forces such as the winds, rain, heat, cold, and volcanic eruptions. Geologists began to hypothesize that the Earth was much older than the 6000 years originally suggested by a literal translation of the Bible. Second, they noticed that the fossils found within the Earth's rock layers were similar to but different in many ways from living organisms—an observation Darwin himself had made on his voyage. Not only had the Earth changed, thought scientists, but evidence existed that the organisms living on its surface had changed also.

Geological evidence suggests that the Earth is much older than the 6000 years originally thought and that the Earth and organisms living on it have changed over time.

The results of artificial breeding

As he pondered these ideas that the earth and its organisms may have changed over time, Darwin reflected on the results of a process called **artificial selection.** In artificial selection a breeder selects for desired characteristics, such as those of the pigeons shown in Figure 25-5. At one time these pigeons came from the same stock, but through artificial breeding over successive generations, their offspring have changed dramatically.

Artificial selection is based on the *natural variation* all organisms exhibit. For example, looking back at p. 372, you can see that although individuals within this population of flamingos are similar, they possess characteristics that vary from individual to individual. Farmers and animal breeders, both today and in Darwin's time, take advantage of the natural variation within a population to select for characteristics they find valuable or useful. By choosing organisms

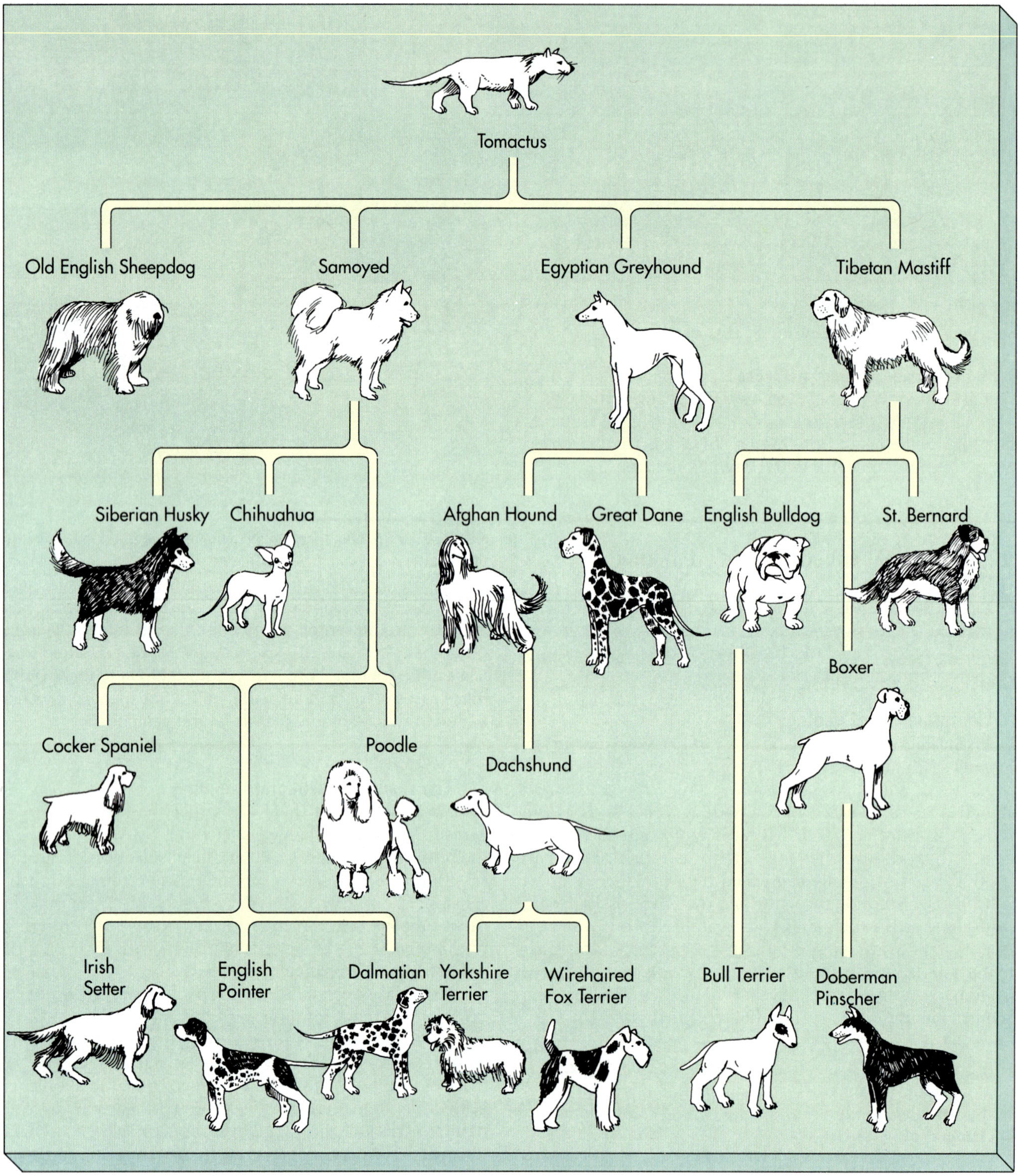

FIGURE 25-6 Artificial selection leads to different breeds of the same species. All dogs belong to the species *Canis familiaris,* but through artificial selection, breeders have been able to choose which traits each breed should retain. The result is a myriad of different breeds within one species.

that naturally exhibit a particular trait and then breeding that organism with another exhibiting the same trait, breeders are able (over successive breedings) to produce animals or plants having a desired, inherited trait. This trait will breed true in successive generations when these organisms are bred with one another. For example, dogs have been artificially bred for centuries. Although your collie may look much different from your neighbor's terrier, both animals belong to the same species but have been artificially bred to retain traits that are characteristic of their breeds (Figure 25-6). Even the turkey you eat on Thanksgiving has been artificially bred for large cavities for stuffing.

Breeders of plants and animals are able to alter the characteristics of organisms by selecting those with desired, inheritable traits and breeding them. After successive breedings, these inherited traits will consistently appear in offspring. Darwin hypothesized that a similar type of selection might take place in nature and result in changes within populations of organisms over time.

The study of populations

Pondering his observations, Darwin began to study Thomas Malthus' *Essay on the Principles of Population*. Malthus, an economist who lived from 1766 to 1834, pointed out that populations of plants and animals (including human beings) tend to increase geometrically. In a geometric progression, a population (for example) increases as its number is multiplied by a constant factor. In the geometric progression 2, 6, 18, 54, and so forth, each number is three times the preceding one. Figure 25-7 shows how the numbers in a geometric progression increase quickly! Malthus suggested that although populations grow geometrically, food supplies increase only arithmetically. An arithmetic progression, in contrast, is one in which the elements increase by a constant difference, as the progression 2, 6, 10, 14, and so forth. In this progression, each number is 4 greater than the preceding one. Figure 25-7 shows graphically how each type of progression increases.

Although Malthus suggested that populations grow at a geometric rate, he realized that factors existed to limit this astounding growth. If populations grew unchecked, organisms would cover the entire surface of the Earth within a surprisingly short time. But the world is not covered in ants, spiders, or poison ivy. Instead, populations of organisms vary in number within a certain limited range. Space and food are limiting factors of population growth; death limits infinite population growth. Malthus noted that in human populations, death was caused by famine, disease, and war. Sparked by Malthus' ideas, Darwin saw that in nature, although every organism has the potential to produce many offspring thereby contributing to a geometric growth rate of its population, only a limited number of organisms actually survive to reproductive age. Darwin realized that factors similar to those limiting human populations must also act to limit plant and animal populations in nature.

A key contribution to Darwin's thinking was Malthus' concept of geometric population growth. Real populations do not expand at this rate, and this implies that nature acts to limit population numbers.

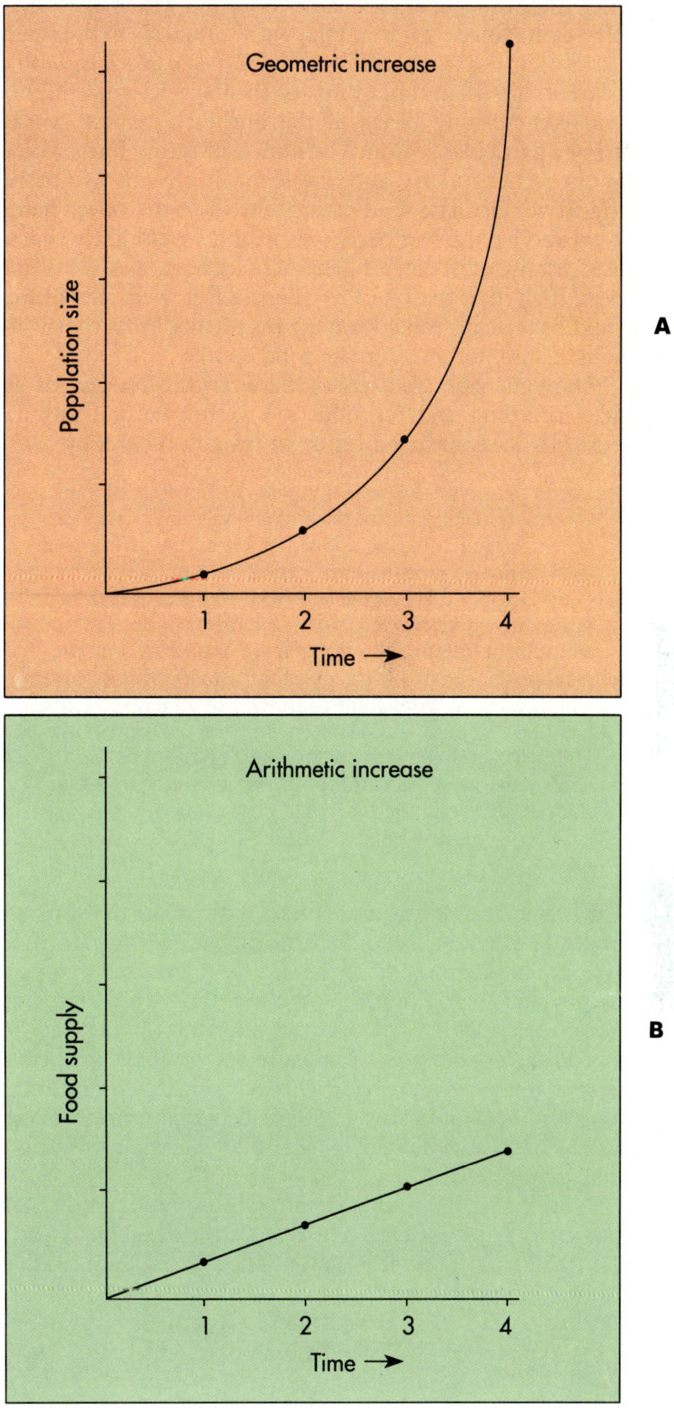

FIGURE 25-7 Types of mathematical progressions.
A In a geometric progression, a population increases as its number is multiplied by a constant factor. The numbers in a geometric progression increase rapidly.
B In an arithmetic progression, the numbers increase by a constant difference.

Natural selection: A mechanism of evolution

Darwin realized that environmental factors could influence which organisms in a population lived and which ones died. Berry-eating birds, for example, with variations in the structure of their beaks that allowed them to crush seeds, would survive longer than other birds of their species who did not have seed-crushing beaks, if the bushes bore few berries during a particular season. The non-seed-eating birds would die out rather quickly and would probably not live to reproductive age. The seed-eating birds, on the other hand, would survive the bad berry season and would likely reproduce. Many of their progeny would have seed-crushing beaks. If another season of few berries followed, these birds would have a survival advantage over other birds that were unable to live on an alternate food source.

Darwin made associations between the process of artificial breeding and reproduction within natural populations. His ideas were expressed in his autobiography:

> I soon perceived that selection was the keystone of man's success in making useful races of animals and plants. But how selection could be applied to organisms living in a state of nature remained for some time a mystery to me.
>
> In October 1838, that is, fifteen months after I had begun my systematic enquiry, I happened to read for amusement 'Malthus on Population' and being well prepared to appreciate the struggle for existence which everywhere goes on from long-continued observation of the habits of animals and plants, it at once struck me that under these circumstances favourable variations would tend to be preserved, and unfavourable ones to be destroyed. The result of this would be the formation of new species. Here then I had at last got a theory by which to work . . . .

Darwin was saying that those individuals that possess physical, behavioral, or other attributes well-suited to their environment are more likely to survive than those that possess physical, behavioral, or other attributes less suited to their environment. The survivors have the opportunity to pass on their favorable characteristics to their offspring. These characteristics are naturally occurring inheritable traits found within populations and are called **adaptations.** (Populations are individuals of a particular species inhabiting a locale or region.)

Notice that the term *adaptation* is used differently than in its everyday sense. Here, it refers to naturally occurring inheritable traits present in a population of organisms rather than noninheritable traits in individuals. Adaptive traits are inherited characteristics that confer a reproductive advantage to the portion of the population possessing them. In its everyday sense, an *adaptation* refers to something a single individual does to change how it responds to the environment. For example, you may adapt to getting up early for an 8:00 class. But this is not an inherited trait in the entire population that confers a reproductive advantage!

As adaptive, or reproductively advantageous, traits are passed on from surviving individuals to their offspring, the individuals carrying these traits will increase in numbers within the population, and *the nature of the population as a whole will gradually change.* Darwin called this process, in which organisms having adaptive traits survive in greater numbers than those without such traits, **natural selection.** Change in populations of organisms therefore occurs over time because of natural selection: the environment imposes conditions that determine the results of the selection and thus the direction of change. The driving force of change—natural selection—is often referred to as *survival of the fittest.* Again, the term *fittest* does not have the everyday meaning of the healthiest, strongest, or most intelligent. You may be fit if you work out at the local health club regularly. But fitness in the context of natural selection refers to reproductive fitness—the ability of an organism to survive to reproductive age in a particular environment and produce viable offspring.

Natural selection provides a simple and direct explanation of biological diversity—why animals are different in different places. Environments differ, so organisms are "favored" by natural selection differently in different places. The nature of a population gradually changes as more individuals are born that possess the "selected" traits. **Evolution** by means of natural selection is this process of change over time by which existing populations of organisms develop from ancestral forms through modification of their characteristics.

An example of natural selection at work: Darwin's finches

Interestingly, the results of evolution by natural selection can actually be seen if the process takes place relatively quickly (over a period of years to several thousand years), resulting in the existence of groups of closely related species from an original ancestral species. The results that can be seen are clusters of these closely related species found living nearby one another. Such clusters of species are often found on a group of islands, in a series of lakes, or in other environments that are close to but separated from one another. Organisms living in such sharply discontinuous habitats are said to be **geographically isolated** from one another.

The Galapagos Islands are a particularly striking example of sharply discontinuous habitats, providing a "natural laboratory" to view the results of natural selection. The islands are all relatively young in geological terms (several million years) and have never been connected with the adjacent mainland of South America or with any other area. Made up of 13 major islands (and some very tiny islands), the Galapagos are separated from one another by distances of up to 100 miles and are 600 miles from the South American mainland (see Figure 25-1). As a group, they exhibit diverse habitats. For example, the lowlands of the Galapagos are covered with thorn scrub. At higher elevations, attained only on the larger islands, there are moist, dense forests.

Formed by undersea volcanoes, the Galapagos Islands were uninhabited when they appeared above the surface of the water. The ancestors of all the organisms found on the Galapagos today reached these islands by crossing the sea by water or wind or on the bodies of other organisms. Only eight species of land birds reached the islands. One of these species was the finch, which fascinated Darwin.

Presumably, the ancestor of Darwin's finches reached these islands earlier than any of the other birds. If so, all the types of habitats where birds occur on the mainland were unoccupied on the Galapagos—and the ancestral finches were able to take advantage of them all! As the finches moved into these vacant habitats, the ones best suited to each particular habitat were selected for by nature. In other words, those birds possessing naturally occurring variations in their characteristics that were beneficial to survival lived to reproduce. Their offspring also possessed these inheritable traits. Over time, the population of finches occupying each habitat changed, and the ancestral finches split into a series of diverse populations. This phenomenon, by which a population of a species changes as it is dispersed within a series of different habitats within a region, is referred to as **adaptive radiation.** Some of these populations became so changed from the others that interbreeding was no longer possible: new species of finches were formed. This process, by which new species are formed during the process of evolution, is termed **speciation.**

The evolution of Darwin's finches on the Galapagos Islands provides one of the classic examples of speciation. The descendants of the original finches that reached the Galapagos Islands now occupy many different kinds of habitats on the islands (Figure 25-8) and are found nowhere else in the world. Among the 13 species of Darwin's finches that inhabit the Galapagos, there are three main groups: ground finches, tree finches, and warbler finches. The ground finches feed on seeds of different sizes. The size of their bills is related to the size of the seeds on which the birds feed. The tree finches, as their name suggests, eat insects, buds, or fruit found in the trees. Again, the size and shape of their bills is related to their food. The most unusual member of this group is the woodpecker finch. This bird carries around a twig or a cactus spine, which it uses to probe for insects in deep crevices. It is an extraordinary example of a bird that uses a tool. And lastly, the warbler finches, named for their beautiful singing, search continually with their slender beaks over leaves and branches for insects.

The evolution of Darwin's finches illustrates the same kinds of processes by which species are originating continuously in all groups of organisms. Isolated populations subjected to unique combinations of selective pressures (the conditions imposed by nature) diverge from one another and may ultimately become so different that they are distinct species.

The publication of Darwin's theory

Darwin drafted the overall argument for evolution by natural selection in 1842 and continued to refine it for many years. The stimulus that finally brought it into print was an essay that he received in 1858. A young English naturalist named Alfred Russel Wallace (1823-1913) sent the essay to Darwin from Malaysia; it concisely set forth the theory of evolution by means of natural selection (Figure 25-9). Like Darwin, Wallace had been influenced greatly in his development of this theory by reading Malthus' 1798 essay. After

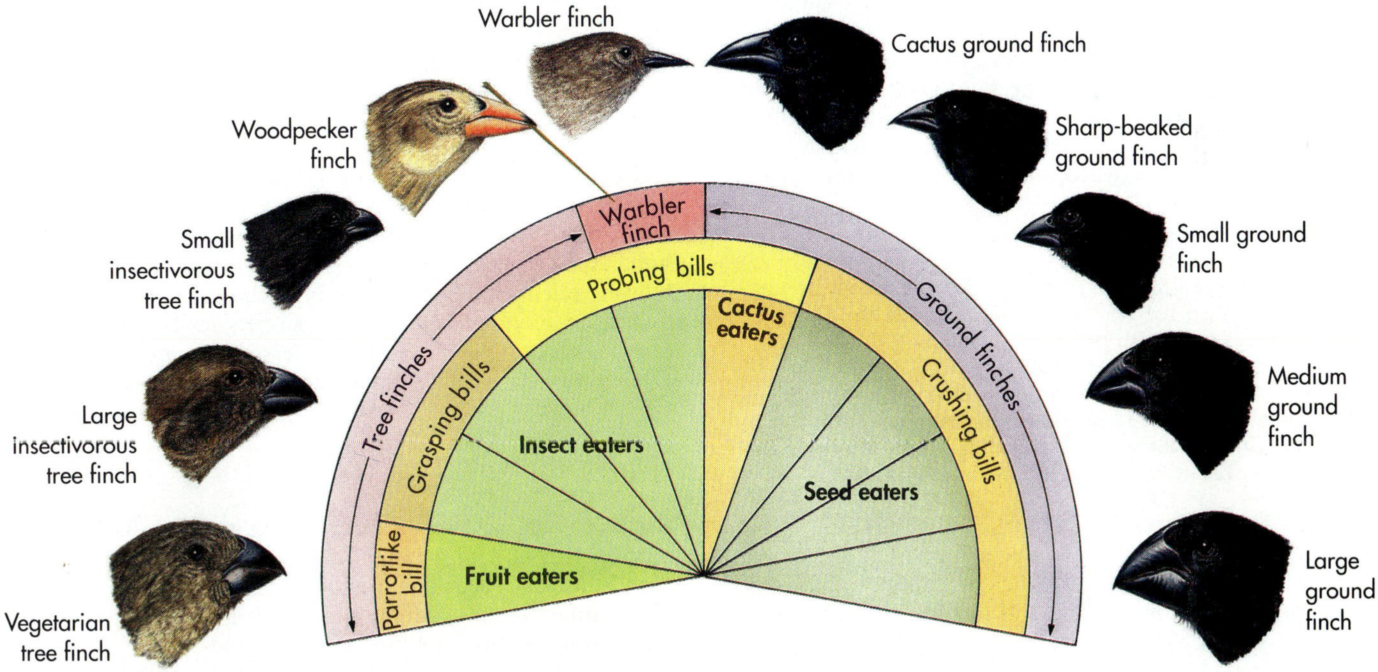

FIGURE 25-8 Darwin's finches.
Ten species of Darwin's finches from Indefatigable Island, one of the Galapagos Islands, showing differences in bills and feeding habits. The bills of several of these species resemble those of different, distinct families of birds on the mainland. All of these birds are thought to have been derived from a single common ancestor.

FIGURE 25-9 Alfred Russel Wallace. Wallace arrived at the theory of natural selection independently of Darwin, but the ideas of the two were presented together at a seminar in 1859.

receiving Wallace's essay, Darwin arranged for a joint presentation of their ideas at a seminar in London. Darwin then proceeded to complete his own book, which he had been working on for some time, and submitted it for publication in what he considered an abbreviated version.

Darwin's book, *On the Origin of Species,* appeared in November 1859 and caused an immediate sensation. Some called the book "glorious" and were in complete agreement with Darwin's theories. Others criticized the book, admiring some parts and asserting that other parts were "totally false." Still others attacked Darwin on religious grounds because Darwin's ideas were different from the exact word of the Bible. Many clergy, however, openly agreed with Darwin, saying that the theory of evolution did not deny the existence of God.

Darwin published his argument for the theory of evolution in 1859, in a book entitled *On the Origin of Species.* He also presented his ideas at a scholarly meeting with another scientist, Alfred Russel Wallace, who had independently developed a theory of evolution. These ideas were hotly debated at that time: the mechanism of evolution—natural selection—was not well accepted nor understood.

At the end of June 1860, a debate was held at a meeting of the British Association for the Advancement of Science. The debate lasted many days and attracted huge crowds of people. The debate was heated, but the outcome pleased Darwin: people read his book and gave his arguments serious consideration. Within 20 years or so after the publication of *On the Origin of Species,* the concept that species have changed over time was well accepted. The mechanism of this change—natural selection—was not. At that time, no one had any concept of genes or of how heredity works, and so it was impossible for Darwin to explain completely how evolution occurs. Gregor Mendel (see Chapter 22) had not yet begun his ground-breaking work in the study of inheritance. In fact, the science of genetics was not established until the beginning of the twentieth century, 40 years after the publication of Darwin's book. An understanding of the laws of inheritance and the mechanism by which inheritable traits are passed on from one generation to the next helped scientists understand the process of natural selection.

Testing the theory

More than a century has elapsed since Charles Darwin's death in 1882. During this period the evidence supporting his theory has grown progressively stronger. In fact, evolution is no longer considered a theory but is accepted as a **scientific law,** a theory that has been upheld countless times as it is tested and retested. By convention, the phrase *the theory of evolution* is still used, but its use does not suggest that evolution by means of natural selection is a highly tentative concept. Evolution is, rather, a scientific observation that is as well accepted in the scientific community as is the theory that the Earth revolves around the sun. Scientists are continuously learning more, however, about the intricacies of the mechanism of natural selection and the history of the evolution of life on Earth.

The concept of evolution by means of natural selection is established as valid within the scientific community today and scientists are continually developing their understanding of natural selection.

What is the scientific evidence that upholds the theory of evolution? Scientists find evidence in the fossil record, using widely accepted techniques to assess the age of the rocks in which fossils are often found, while gathering a picture of the history of the earth and its organisms. In addition, the tools of comparative anatomy help researchers understand relationships among organisms alive today. Significant scientific advances, such as those in genetics and molecular biology, have given scientists tools Darwin did not have, so today scientists understand more fully than Darwin ever could how organisms change over time.

The fossil record

A **fossil** is any record of a dead organism. Fossils may be nearly complete impressions of organisms or merely burrows, tracks, molecules, or other traces of their existence. Unfortunately, only a minute fraction of the organisms living at any one time are preserved as fossils.

Most fossils are preserved in **sedimentary** rocks. Sedimentary rocks are made up of particles of other rocks, cast off as they weather and disintegrate. Running water, such as a river or stream, picks up these pieces of rock and carries them to lakes or oceans where they are deposited as **sediment,** better known as mud, sand, or gravel. Over time, the

FIGURE 25-10 The Grand Canyon.
In this photo, the layers of sedimentary rock can be clearly seen.

sediment hardens into rock. But while some sediment is hardening, other sediment is still being deposited, creating layers of rock formed one on top of the other. Therefore most sedimentary rock has a stratified appearance, like that seen in the Grand Canyon (Figure 25-10).

During the formation of sedimentary rock, dead organisms are sometimes washed along with the mud or sand and eventually reach the bottom of a pond or lake. Dead marine organisms fall to the bottom of the ocean. As the sediments harden into rock, they harden around the bodies of these dead organisms (Figure 25-11). The hard parts of these organisms, such as their skeletons, may become preserved or may be broken down and replaced with other minerals. Fossils of organisms having hard parts are the type most often found (Figure 25-12) rather than fossils formed from soft body parts, which usually decay quickly and leave no

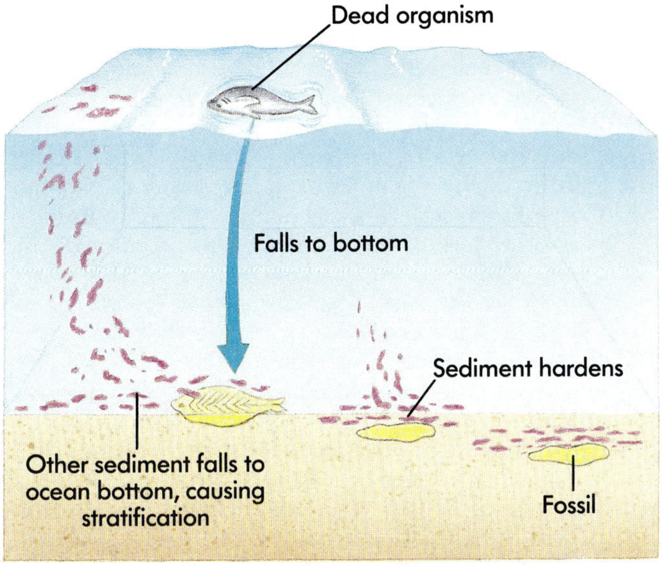

FIGURE 25-11 The process of fossil formation.
Fossils are formed when dead organisms are trapped in hardening sediment.

FIGURE 25-12 *Archaeopteryx*, a prehistoric ancestor of modern birds.
A well-preserved fossil of this bird, about 150 million years old, was discovered within 2 years of the publication of *On the Origin of Species.* This particular specimen is an excellent example of a fossil formed from the hard parts of the organism. The skeleton can be clearly seen.

FIGURE 25-13 A fossil "mold."
Shown here is *Mawsonites spriggi,* a fossil jellyfish from South Australia. This fossil illustrates the remarkable preservation of soft-bodied fossils in fine-grained sedimentary rocks.

trace of their existence. Sometimes, however, soft-bodied animals are preserved in exceptionally fine-grained muds, in conditions where the supply of oxygen was poor while the muds were being deposited, thus slowing the decomposition of the organism. Eventually, the soft parts of an organism decay completely, leaving behind a **mold,** or impression, of its body. Molds may become filled with minerals, such as lime or silica found in underground water, forming **casts,** which resemble the original organism or body part (Figure 25-13). In general, however, fossils of soft-bodied organisms, such as worms, are rare. Even though soft-bodied animals undoubtedly evolved before their hard-bodied counterparts, there is comparatively little evidence of their history in the fossil record.

Fossils provide an actual record of organisms that once lived, an accurate understanding of where and when they lived, and some appreciation of the environment in which they lived. Limestone that contains corals, for example, would have been deposited when the location was an ocean. Oak leaves found in sandstone suggest that a location was once a continent. In this way, fossils and the rock in which they are embedded provide information about the history of an area, which gives scientists clues to the location of the continents, ponds, lakes, and oceans and how their positions have changed over time.

Scientists can determine the age of fossils and use this information to establish the broad patterns of the progression of life on Earth (Figure 25-14). One of the ways of determining the age of particular fossils is to compare the sequences in which they appear in different layers, or strata, of sedimentary rock. Sedimentation deposits new layers mostly on older ones, so the fossils in upper layers mostly represent younger species than the fossils in lower layers. Fossils found in the same strata are assumed to be of the same age. Such correlations, in fact, were known at the time of Darwin and were used by him to formulate the theory of evolution.

Fossils, impressions of organisms that once lived, provide a record of the past.

The age of the Earth: Rock and fossil dating

Direct methods of dating rocks and fossils first became available in the late 1940s. This process depends on naturally occurring isotopes of certain elements that are found in rock. Isotopes are atoms of an element that have the same number of protons but different numbers of neutrons in their nuclei (see Chapter 2). They therefore differ from one another in their atomic numbers. **Radioactive isotopes** are unstable; their nuclei decay, or break apart, at a steady rate, producing other isotopes and emitting energy. After decay some radioactive isotopes may give rise to elements. Many different isotopes are used in radioactive dating. Some methods give scientists information about the age of rocks; others measure the length of time since the death of an organism.

One of the most widely used methods of dating, the carbon-14 (^{14}C) method, estimates the relative amount of the different isotopes of carbon present in a fossil (or other organic material). Most carbon atoms have an atomic weight of 12 (6 protons and 6 neutrons); the symbol of this particular isotope of carbon is ^{12}C. A fixed proportion of the atoms in a given sample of carbon, however, consists of carbon with an atomic weight of 14 (^{14}C), an isotope that has two more neutrons than ^{12}C. Interestingly, ^{14}C is produced from ^{14}N (nitrogen-14) as these atoms are bombarded by cosmic rays—high-energy particles from space. The cosmic rays (usually protons) bump a proton from the nucleus of a nitrogen atom, leaving it with 6 protons (the atomic number of carbon) and 7 neutrons. As a result of the collision, the atom also captures a neutron and becomes an atom of ^{14}C. This newly created ^{14}C reacts with oxygen, becoming carbon dioxide, a gas commonly found in the air. Plants use this carbon during photosynthesis, incorporating it into the sugars and starches they make. By eating plants, animals incorporate ^{14}C into their bodies as well.

The carbon that is incorporated into the bodies of living organisms consists of the same fixed proportion of ^{14}C and ^{12}C that occurs in the atmosphere. After an organism dies, however, and is no longer incorporating carbon, the ^{14}C in it gradually decays back to nitrogen by emitting a beta particle. (A beta particle is an electron discharged from the nucleus when a neutron splits into a proton and an electron.) It takes 5730 years for half of the ^{14}C present in a sample to be converted by this process; this length of time is called the **half-life** of the ^{14}C isotope. By measuring the amount of ^{14}C in a fossil, scientists can estimate the proportion of ^{14}C to all other carbon that is still present and compare that to the ratio of these isotopes as they occur in the atmosphere. In this way, scientists can then estimate the length of time over which the ^{14}C has been decaying, which is the same as the length of time since the organism died.

For fossils older than 50,000 years, the amount of ^{14}C remaining is so small that it is not possible to measure it precisely enough to provide accurate estimates of age. These fossils may be dated using the isotope thorium-230, which has a half-life of 75,000 years, and decays from uranium-238. These techniques have been most useful to scientists who study deep sea sediments too old to be dated with ^{14}C.

With the use of radioactive dating methods, knowledge of the ages of various rocks has become more precise. The oldest rocks on Earth that have been dated include rocks

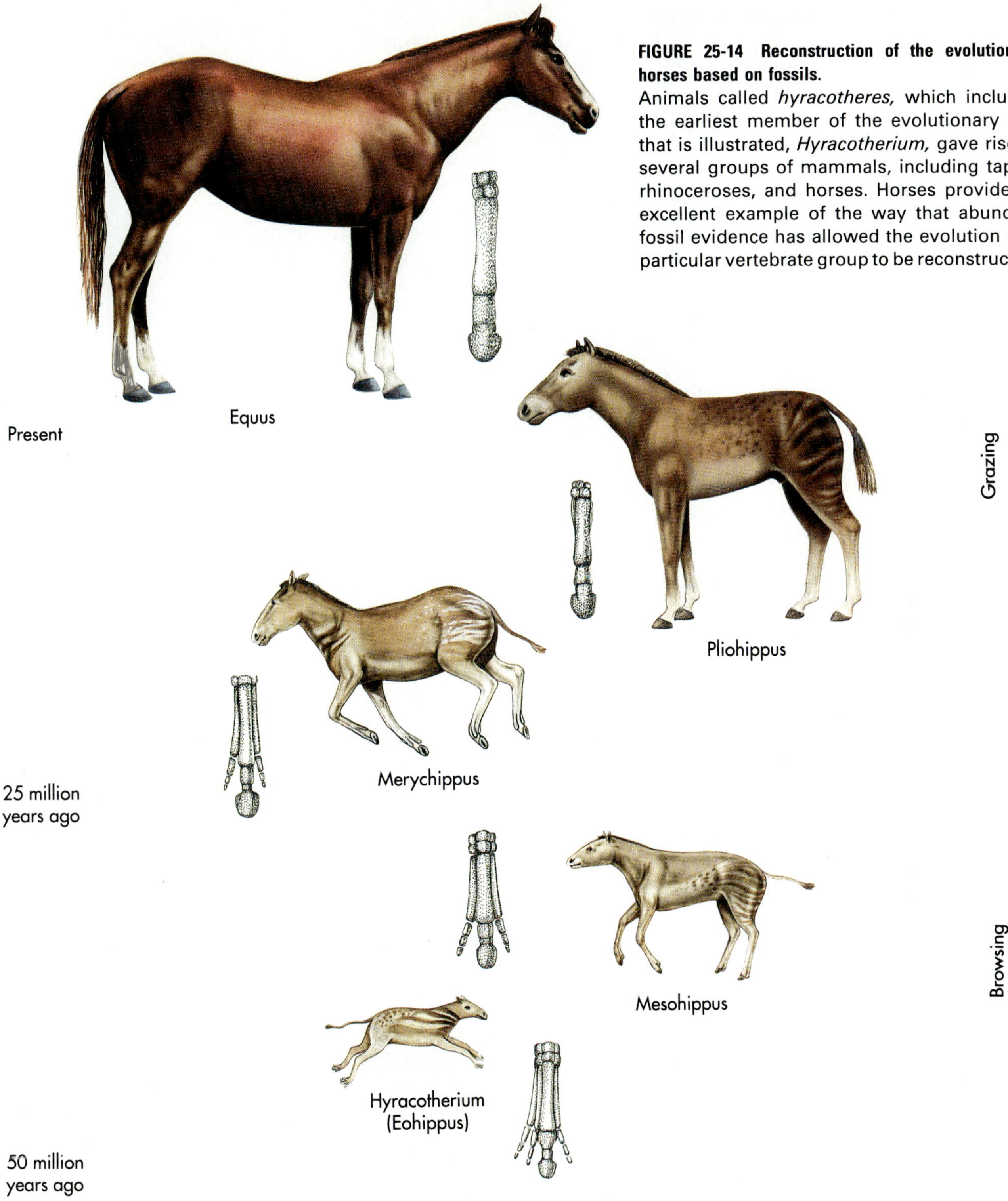

FIGURE 25-14 Reconstruction of the evolution of horses based on fossils.
Animals called *hyracotheres,* which included the earliest member of the evolutionary line that is illustrated, *Hyracotherium,* gave rise to several groups of mammals, including tapirs, rhinoceroses, and horses. Horses provide an excellent example of the way that abundant fossil evidence has allowed the evolution of a particular vertebrate group to be reconstructed.

from South Africa, southwestern Greenland, and Minnesota that are approximately 3.9 billion years old. Meteorites have been dated at about 4.6 billion years. Recently, rocks brought back to Earth from the moon have been dated from 3.3 to 4.6 billion years old. These pieces of evidence suggest that the Earth and the moon, most likely formed from the same processes at the same time, are about 4.6 billion years old.

What significance does the age of the Earth hold for the theory of evolution? The "accumulation" of adaptations and the development of new species usually takes thousands and probably millions of years. Until the time of Darwin, most held the belief that the Earth was approximately 6000 years old as described in the Judeo-Christian Bible. This time frame would not allow enough time for the process of evolution to take place. In fact, Sir Isaac Newton (1642-1727), an English physicist, calculated that it would take 50,000 years just for the Earth, after its formation (see Chapter 26), to cool to a temperature that would sustain life. Although scientists such as William Thomson, Lord Kelvin

FIGURE 25-15 Homology among vertebrate limbs.
Homologies among the forelimbs of four mammals and a frog show the ways that the proportions of the bones have changed in relation to the particular way of life of the organism and that the forelimb of each animal has the same basic bone structure.

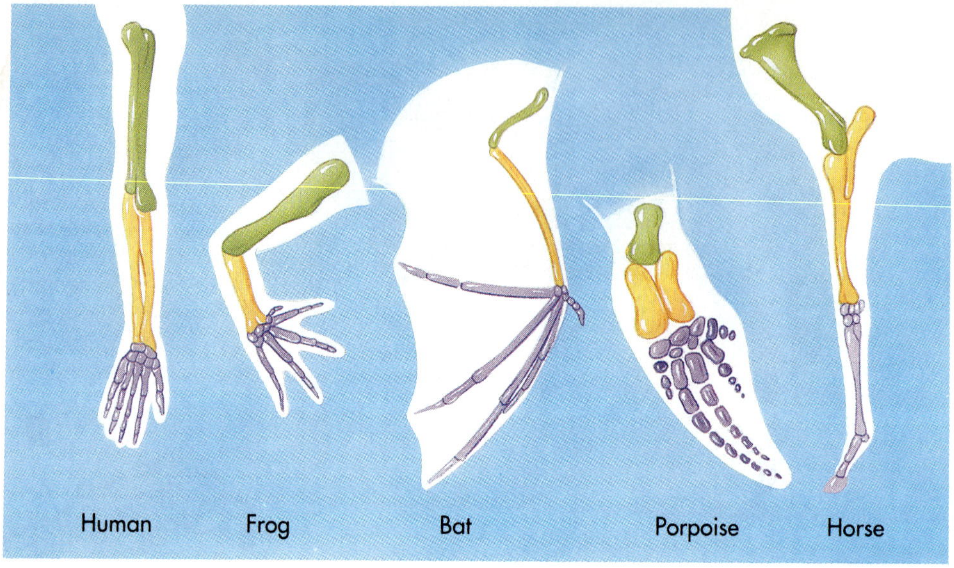

(1824-1907) used the techniques available to him to date the Earth at about 100 million years, this time span still seemed too short to allow for evolution. Not until the present techniques of radioactive dating were developed could scientists begin to solve the "time problem" of evolution and accurately measure the geological age of the Earth, its rocks, and its fossils.

By studying the comparative amounts of certain radioactive isotopes found within rocks and fossils, scientists can determine the age of fossils and use this information to establish patterns of life's progression. Studies using radioactive dating methods provide evidence that the Earth is 4.6 billion years old. This time is sufficient for organisms to have developed and evolved from "original" forms.

Comparative anatomy

Comparative studies of animal anatomy provide strong evidence for evolution. After Darwin proposed his theory, scientists began looking for evolutionary relationships in the anatomical structures of organisms. If derived from the same ancestor, organisms should possess similar structures with modifications reflecting adaptations to their environments. Such relationships have been shown most clearly in vertebrate animals.

Within the subphylum Vertebrata, the classes of organisms, such as birds, mammals, and amphibians, have the same basic anatomical plan of groups of bones (as well as nerves, muscles, and other organs and systems), but these bones are put to different uses among the classes. For example, the forelimbs seen in Figure 25-15 are all constructed from the same basic array of bones, modified in one way in the wing of a bat, in another way in the fin of a porpoise, and in yet another way in the leg of a horse. The bones are said to be **homologous** in the different vertebrates—that is, of the same evolutionary origin, now differing in structure and function. Although these vertebrates deviated from one another in their evolution, they all use the same bones in the same relative positions to do different jobs.

In some cases, homologous structures exist among related organisms but are no longer useful. These structures have diminished in size over time. Figure 25-16 shows tiny leg bones of the python that no longer serve a purpose. Such organs or structures that are present in an organism in a diminished size but are no longer useful are called **vestigial organs.** Humans have a tiny pouch called the appendix that hangs like a little worm at the junction of the small and large intestines. It serves no useful purpose in humans, but helps in digestion of plant material in organisms that are evolutionarily related to humans. And if you can wiggle your

FIGURE 25-16 A vestigial organ in a python.
Pythons possess tiny leg bones that serve no purpose in locomotion. Humans also possess a vestigial organ—the appendix—that is not needed for digestion or any other purpose.

FIGURE 25-17 Examples of convergent evolution.
The North American wolf, a placental animal **(A)** and the Tasmanian wolf, a marsupial animal **(B)** exhibit convergent evolution. The Tasmanian wolf has been extinct for approximately 110 years. Euphorbiaceae **(C)** found in Africa and the North American cactus **(D)** bear a striking resemblance to each other but belong to quite different families.

ears, you are using external ear muscles, which are also no longer useful except to amuse your friends, inherited from a distant ancestor.

> **Comparative studies of animal anatomy show that many organisms have groups of bones, nerves, muscles, and organs with the same anatomical plan but with different functions. These homologous structures provide evidence of evolutionary relatedness.**

In contrast to homologous structures, similar structures often evolve within organisms that have developed from different ancestors. Such body parts or organs, called **analogous** structures, have a similar form and function but have different evolutionary origins. The eyes of vertebrates and octopuses, which evolved independently but are similar in design, are analogous structures, as well as the wings of birds and insects.

Plants also show analogous structures. For example, three different families of flowering plants—the cacti, euphorbs, and milkweeds—have all developed thick, barrel-like fleshy stems that store water as an adaptation to a desert environment (Figure 25-17). In fact, these plants look so much like one another the casual observer might think they were all cacti. However, they evolved independently of one another in different parts of the world (southwest North America, Africa, and the Mediterranean, respectively).

To show that structures are indeed analogous, comparative anatomists do detailed dissections and study the embryological development of organisms. Analogous structures arise developmentally from different tissues. The presence of analogous structures shows how, in similar habitats, natural selection can lead to similar but not identical anatomical structures. Changes over time among different species of organisms having different ancestors that result in similar structures and adaptations is called **convergent evolution** (Figure 25-17).

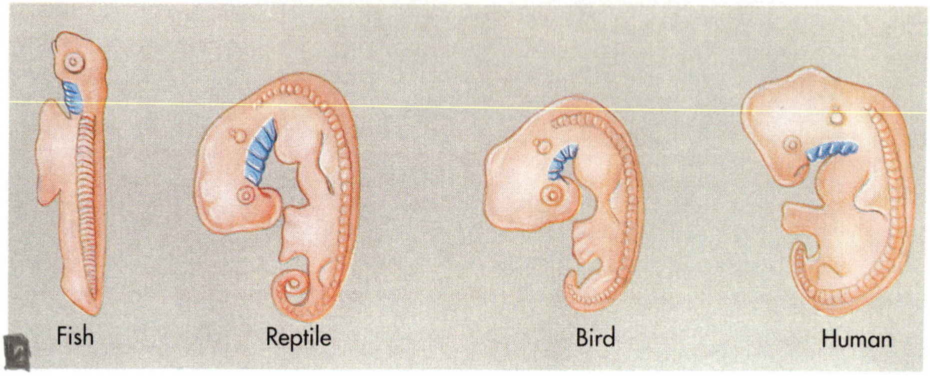

FIGURE 25-18 Embryos show our evolutionary history.
The embryos of various groups of vertebrate animals show the primitive features that all vertebrate animals share early in development, such as gill arches and a tail.

Comparative embryology

Embryologists, scientists who study the development of organisms from conception to birth, noticed as early as the nineteeth century (around the time of Darwin) that various groups of organisms, although different as adults, possessed early developmental stages that were quite similar. For example, the embryological development of vertebrate animals is similar in that all vertebrate embryos have a similar number of gill arches, seen as pouches below the head. Only fish, however, actually develop gills. Likewise, the embryological development of the backbone of vertebrates is similar, but some organisms develop a tail, and others such as humans do not (Figure 25-18). "Tailbones"—the fused coccyx bones at the end of the spine—are actually vestigial structures.

These similar developmental forms tell scientists that similar genes are at work during the early developmental stages of related organisms. The genes active during development have been passed on to distantly related organisms from a common ancestor. Over time, new instructions are added to the old, but both are expressed at different times resulting in similar embryos that develop into organisms quite different from one another.

Comparative embryological studies show that many organisms have early developmental stages that are quite similar. These similar developmental forms provide evidence of evolutionary relatedness.

Molecular biology

Today, biochemical tools provide additional evidence for evolution and give scientists new insights into the evolutionary relationships among organisms. Molecular biologists study the progressive evolution of organisms by looking at their hereditary material, DNA. According to evolutionary theory, every evolutionary change involves the formation of new alleles from the old by mutation; favorable new alleles persist because of natural selection. In this way, a series of evolutionary changes in a species involves a progressive accumulation of genetic change in its DNA. Organisms that are more distantly related will therefore have accumulated a greater number of changes in their DNA than organisms that more recently evolved from a common ancestor.

Within the last decade, molecular biologists have learned to study DNA by "reading" genes, much as you read this page. They have learned to recognize the order of the nitrogenous bases—the "letters"—of the long DNA molecules. By comparing the sequences of bases in the DNA of different groups of animals or plants, scientists can show the degree of relatedness among groups of organisms and develop detailed "family trees," called **phylogenetic trees** (Figure 25-19).

Interestingly, phylogenetic trees constructed from analyses of molecular differences are the same as those built using anatomical studies. For example, whales, dolphins, and porpoises cluster together using both approaches, as do the primates and the hoofed animals. By studying and interpreting the evidence from the fossil record, comparative anatomy, and genetic studies, it is often possible for scientists to estimate the rates at which evolution is occurring in different groups of organisms.

The pattern of progressive change seen in the molecular record supports other scientific evidence for evolution and provides strong, direct evidence for change over time.

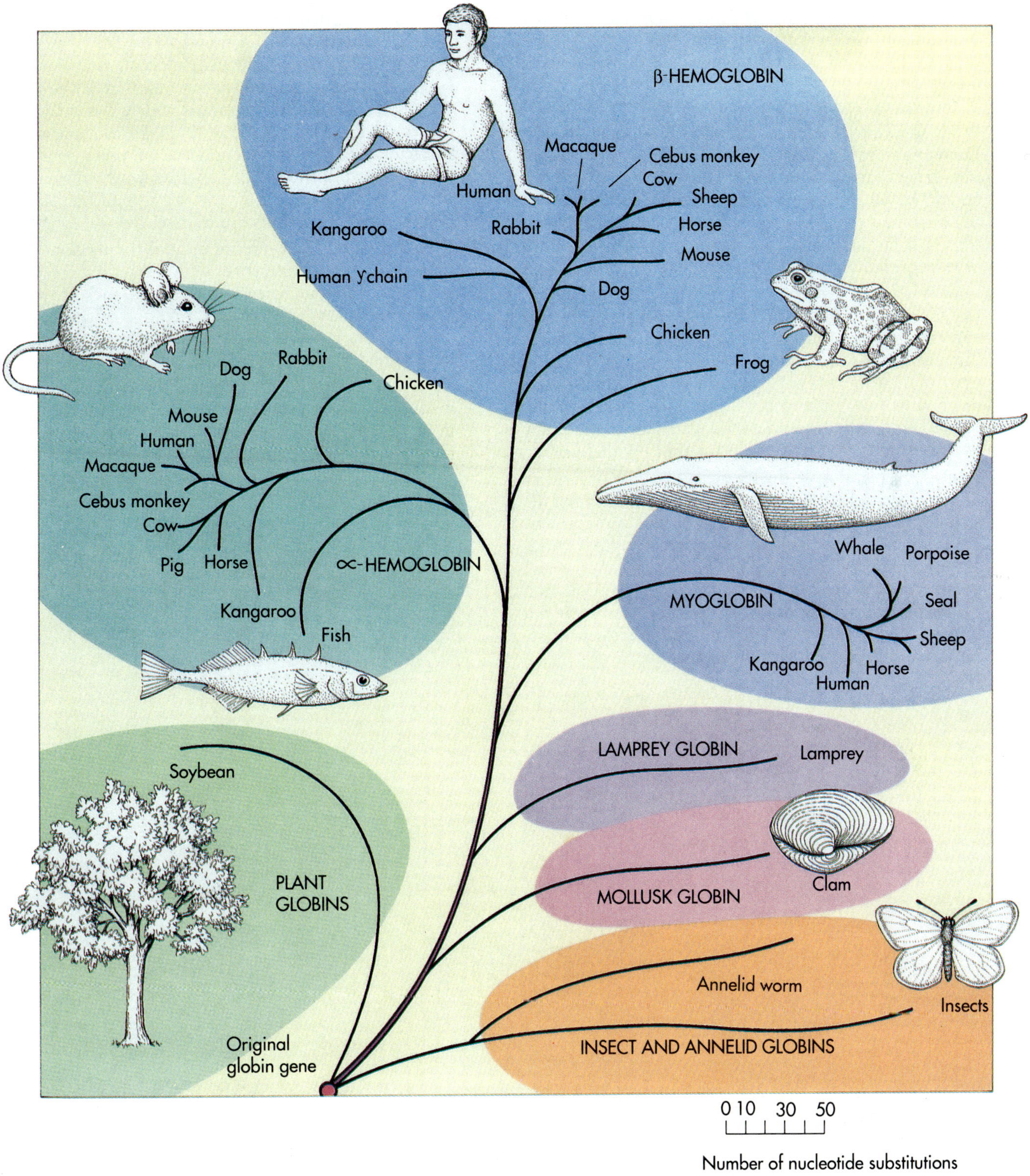

FIGURE 25-19 Evolution of the globin gene.
The length of the various lines is proportional to the number of nucleotide substitutions in the gene and measures the relatedness of the species.

Summary

1. One of the central theories of biology is Darwin's theory of evolution, which states that living things change over time by means of natural selection. Proposed over a hundred years ago, this theory has stood up to a century of testing and questioning.
2. While studying the animals and plants of oceanic islands, Darwin accumulated a wealth of evidence that organisms have changed over time. Other factors that influenced Darwin's thinking were geological evidence that the earth had changed over time, the changes in organisms that farmers were able to attain by using artificial breeding methods, and Malthus' ideas on population dynamics.
3. Darwin proposed that evolution occurs as a result of natural selection: some individuals have traits that make them better-suited to a particular environment, allowing more of them to survive to reproductive age and produce more offspring than other individuals lacking these traits. These characteristics, or adaptations, are naturally occurring inherited traits found within populations.
4. Fitness is a measure of the tendency of some organisms to leave more offspring than competing members of the same population. Therefore genetic traits possessed by the fitter individuals will appear in greater proportions among members of succeeding generations. As a result of this process, the traits allowing greater reproduction will increase in frequency over time. The environment imposes conditions that determine the direction of selection and thus the direction of change.
5. Darwin published his theory in 1859 in a book titled *On the Origin of Species*. A wealth of evidence since Darwin's time has supported his proposals that evolution occurs and that its mechanism is natural selection. By the 1860s, natural selection was widely accepted as the correct explanation for the process of evolution, but the mechanism of natural selection was not understood. The field of evolution did not progress much further until the 1920s because of the lack of a suitable explanation of how hereditary traits are transmitted.
6. Two direct lines of evidence uphold the theory of evolution: (1) the fossil record, which exhibits a record of progressive change correlated with age, and (2) the molecular record, which exhibits a record of accumulated changes, the amount of change correlated with age as determined in the fossil record.
7. Several indirect lines of evidence uphold the theory of evolution, including progressive changes in homologous structures, the existence of vestigial structures, and changes in DNA sequences.

REVIEW QUESTIONS

1. Summarize four of Darwin's conclusions that were inspired by his observations during his voyage.
2. What three factors, in addition to his voyage on the H.M.S. *Beagle*, influenced Darwin's thinking on evolution?
3. A dachshund and a Siberian Husky are both dogs, but they look very different from each other. By what process did the two breeds come to look so different? What observable genetic principle is this process based on?
4. Distinguish between a geometric progression and an arithmetic progression. Which increases more quickly? Relate these concepts to the development of Darwin's theory of evolution.
5. What are adaptations? Explain their significance.
6. Explain the phrase *survival of the fittest*. What does "fit" mean in this context?
7. Most species of bears are black, brown, or gray. Why are polar bears white?
8. Distinguish between adaptive radiation and speciation.
9. Fill in the blanks: A ______ is a record of a dead organism, often preserved in ______ rocks. It can be dated using naturally occurring ______ ______ of certain elements found in rocks.
10. Explain how scientists can use the ^{14}C method to date a fossil.
11. How old is the Earth according to radioactive dating methods? Why is this significant?
12. What are vestigial organs? Give an example from the human body.
13. Distinguish between homologous and analogous structures.
14. Explain how studies of comparative anatomy and comparative embryology support the theory of evolution.
15. What is a phylogenetic tree? How does it support the concept of evolution?

KEY TERMS

adaptation 432
adaptive radiation 433
analogous 439
artificial selection 429
cast 436
convergent evolution 439
fossil 434
geographical isolation 432
half-life 436
homologous 438
mold 436
natural selection 432
phylogenetic tree 440
radioactive isotopes 436
scientific law 434
sediment 434
sedimentary 434
speciation 433
species 427
vestigial organs 438

THOUGHT QUESTIONS

1. Imagine you were attempting to date an old skull by the ^{14}C method. Assume the atmospheric ratio of ^{14}C to ^{12}C is one molecule of ^{14}C to every 999 molecules of ^{12}C (that is, the proportion of ^{14}C is 1 in a 1000, or 0.001), and the half-life of ^{14}C is 6000 years. How old is the skull if its proportion of ^{14}C is 0.00025?

FOR FURTHER READING

Carson, H. L. (1987). The process whereby species originate. *BioScience, 37,* 715-720.
One of the modern masters of evolution discusses a fascinating aspect of the origin of species.

Gould S. J. (1987, January). Darwinism defined: The difference between fact and theory. *Discover,* pp. 64-70.
This is a clear account of what biologists do and do not mean when they refer to the theory of evolution by natural selection.

Sibley, C., & Alquist, J. (1986, February). Reconstructing bird phylogenies by comparing DNAs. *Scientific American,* pp. 82-92.
This is a good introduction to the way in which biologists are beginning to use the tools of molecular biology to answer questions about evolution.

26 The evolution of the five kingdoms of life

HIGHLIGHTS

▼ Approximately 15 billion years ago, the universe condensed and then exploded in the "big bang." Earth and the other planets of the solar system were formed approximately 4.6 billion years ago.

▼ The long-held "primordial soup" theory states that life arose in the sea as elements and simple compounds reacted to form organic molecules about 3.5 billion years ago.

▼ The evolution of oxygen-producing bacteria was a key event, making possible an "explosion of life" during the Cambrian Period.

▼ The evolution of the five kingdoms of life has been influenced by the physical conditions of the Earth that have changed over time.

OUTLINE

Theories about the origin of organic molecules

Theories about the origin of cells

The history of life

The Archean Era: Oxygen-producing cells appear

The Proterozoic Era: The first eukaryotes appear

The Paleozoic Era: The occupation of the land

Mass extinctions

The Mesozoic Era: The age of reptiles

Changes in the Earth: The impact on evolution

The Cenozoic Era: The age of mammals

Vast areas of gas and dust particles swirl in the Orion nebula M42. Orion is a specific constellation, or group of stars, that can be seen in the night sky of the northern hemisphere. Surrounding some of its stars are concentrations of gases and "stardust" called *nebulae,* a name derived from a Latin word meaning "cloud." In fact, nebulae are found in many sections of the Milky Way galaxy and are part of a cosmic life cycle—the birth, life, and death of stars.

Nebulae are "born" from the dust and gases hurled from unstable stars as they explode. Over time, this material condenses, giving birth to second- and later-generation stars. Then, in turn, these stars become unstable and explode, spewing dust and gases into space. And so the nebula life cycle continues.

The American astronomer, Edwin P. Hubble (1889-1953) observed that the galaxies in the universe are moving away from one another. Put simply, the universe is expanding. This movement suggests that at one time all the matter of the universe was in one spot, an assumption that leads scientists to hypothesize that the Earth (and the solar system) originated in the same way as nebulae do. Approximately 15 billion years ago, scientists theorize, *all* the material of the universe was condensed into a single small space. It then exploded in an event called the big bang. The universe—as it is known today—did not exist before that time. The gases and dust from the big bang produced an early generation of stars. Then, over billions of years, these stars exploded and their "space debris" formed other stars and planets. The solar system formed in this way some 4.6 billion years ago. For almost a billion years or so the molten Earth cooled, eventually forming a hardened, outer crust. Approximately 3.8 billion years ago, the oldest rocks on Earth formed. After that time, some 3.5 billion years ago, life began.

Theories about the origin of organic molecules

Primitive Earth was an incubator for life, but scientists know very little about what that incubator was like. Various hypotheses have been proposed that describe what the atmospheric and environmental conditions of "incubator Earth" might have been. But because scientists have no record of these conditions, their hypotheses are speculative and therefore highly controversial.

In trying to understand what the Earth was like 3.8 billion years ago, scientists often use information from events that are observable today. They use these pieces of information as starting points to build plausible hypotheses. For example, scientists have determined that the dust and gases ejected from unstable stars contain mostly hydrogen, as well as helium and varying amounts of other elements such as nitrogen, sodium, sulfur, and carbon. Some of these elements combine to form compounds such as hydrogen sulfide, methane, water, and ammonia. As the Earth coalesced, or came together into one mass, it may have contained some of these gases and vapors produced by the planet as it cooled. As the water vapor in this mix condensed, it could have caused millions of years of torrential rains extensive enough to form oceans. Other scientists suggest that water and gas deep within the earth were vented to the surface by volcanoes.

Scientists agree that the environment of primitive Earth was harsh and violent, bathed in ultraviolet radiation from the sun and subjected to violent electrical storms and constant volcanic eruptions. Under these conditions, many scientists hypothesize that the elements and simple compounds of the primitive atmosphere reacted with one another. New and more complex molecules were formed, capturing the surrounding energy within their bonds. In 1953, Stanley L. Miller and Harold C. Urey, working at the University of Chicago, designed an experiment that modeled an environment that included methane (CH_4), ammonia (NH_3), water vapor (H_2O), and hydrogen gas (H_2). They wondered if organic molecules—the molecules that make up the structure of all living things—could have spontaneously formed under such conditions.

Using an apparatus similar to that in Figure 26-1, Miller and Urey filled a glass chamber with the four gases. Electrodes within the chamber shot sparks of electricity through the mixture, while condensers cooled it. Miller and Urey wondered if complex molecules formed in their "atmosphere," dissolved in the water vapor, and fell into their "ocean." Within a week, they had their answer. A total of 15% of the carbon that was originally present as methane gas had been converted into other more complex compounds of carbon!

Miller and Urey, joined later by many other scientists in laboratories across the country, performed experiment after experiment using various combinations of simple compounds that might have made up the primitive atmosphere. In addition, scientists experimented with a variety of energy sources, such as ultraviolet light, heat, and radioactivity. The outcome was always the same: more complex organic molecules formed from simpler ones. These newly formed molecules included amino acids—the building blocks of protein. Sugars, including glucose, ribose, and deoxyribose, were also shown to form under the conditions of the primitive atmosphere. As scientists experimented further, they discovered that many of the molecules important to life—and important to the structure and function of cells—could be synthesized under certain primordial conditions. Not only were ribose and deoxyribose, components of nucleic acids, synthesized in their methane environment but so were purine and pyrimidine bases. Fatty acids, used in membranes and storage tissues of living organisms, were also made. From this array of experiments, scientists developed the primordial soup theory, which states that life arose in the primitive seas or in smaller lakes as complex organic molecules formed from the simple molecules in the ancient atmosphere.

During the past decade, some scientists have begun to doubt the primordial soup theory as described by the experiments of Miller, Urey, and others. Using computerized reconstructions of the atmosphere, James C. G. Walker of the University of Michigan at Ann Arbor suggests that the major components of the primitive atmosphere were carbon dioxide and nitrogen gases, spewed forth from volcanoes. But these gases could not have formed complex organic molecules—the precursors to life—in the seas or tidal pools of the Earth's surface.

Recently, some scientists have shifted from adherence to the primordial soup theory to thinking that life might

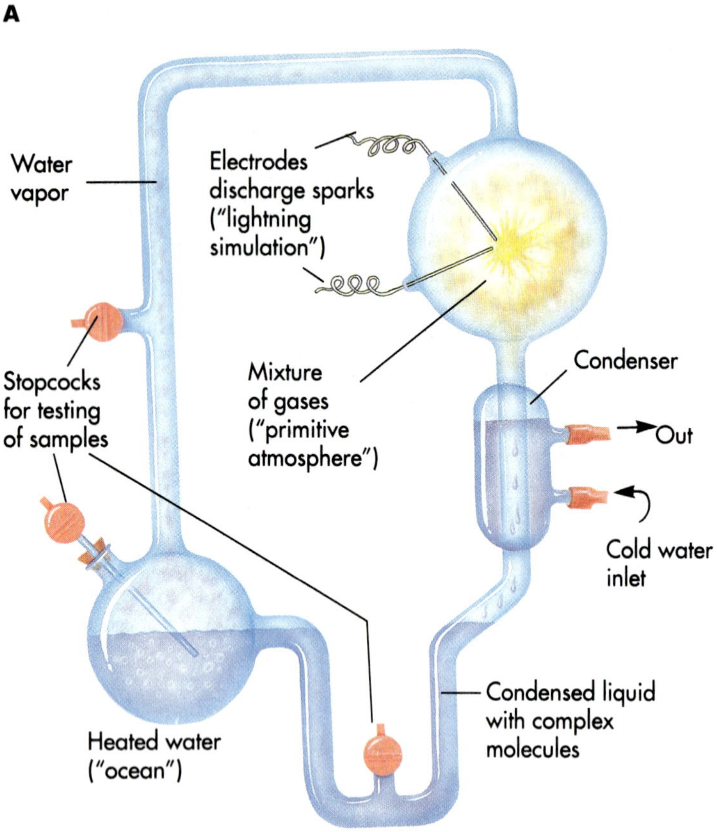

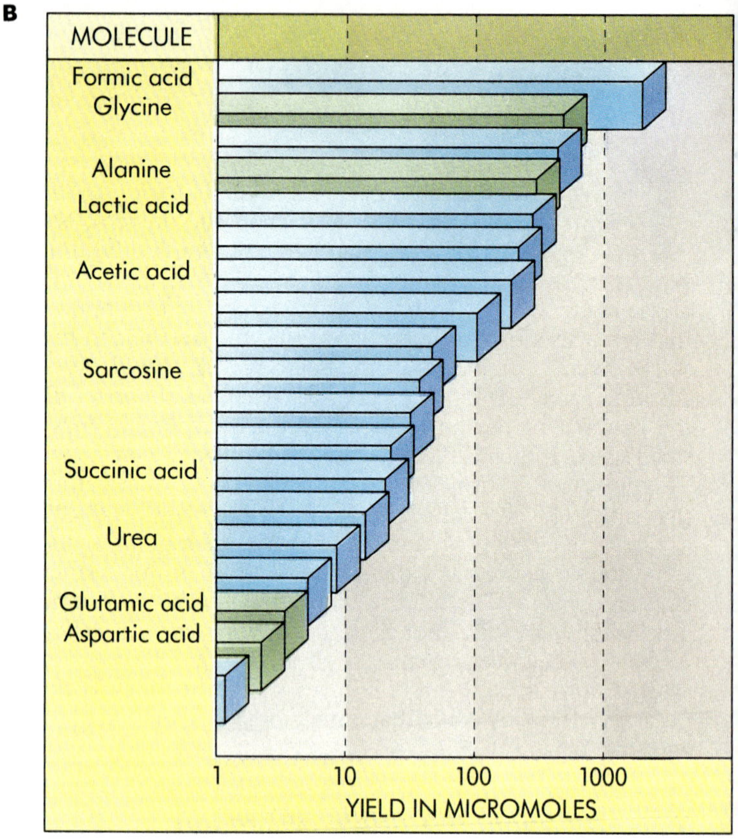

FIGURE 26-1 The Miller-Urey experiment.
A Miller and Urey's apparatus consisted of a closed tube connecting two chambers. The upper chamber contained a mixture of gases thought to resemble the Earth's atmosphere. Any complex molecules formed in the atmosphere chamber would be carried and dissolved in these droplets to the lower "ocean" chamber, from which samples were withdrawn for analysis.
B Some of the 20 most common complex molecules detected in the original Miller-Urey experiments are indicated. Among these 20 molecules are four amino acids (shown by green bars).
C Dr. Stanley Miller standing in front of the apparatus in 1991.

have arisen in hydrothermal vents, spots deep in the oceans where hot gases and sulfur compounds shoot from cracks in the Earth's crust (Figure 26-2). Some scientists hypothesize that the vents could have supplied the energy and nutrients needed for life to arise and be sustained. In fact, the hydrothermal vents in existence today support extensive communities of living organisms, such as tube worms, clams, and bacteria.

The long-held primordial soup theory states that life arose in the sea as elements and simple compounds of the primitive atmosphere reacted with one another to form simple organic molecules such as amino acids and sugars. Recently, scientists have begun to reexamine the premises of this theory and have proposed alternate hypotheses.

FIGURE 26-2 Hydrothermal vents.
Located under deep oceans, hydrothermal vents are cracks in the Earth's crust that release hot gases and sulfur compounds. Some researchers think that these vents could have provided enough energy to start and sustain life.

Theories about the origin of cells

Although scientists can replicate various plausible conditions of the primitive Earth, they cannot duplicate the millions of years that passed. Although scientists can observe the "birth" of simple organic molecules, they cannot generate life from nonlife. But at some point in the history of the Earth, organisms appeared. Scientists are still hard at work trying to unravel this mystery of life's beginnings. The work of many scientists over many years has led to various lines of reasoning regarding the evolution of the precursors to life, and life itself. However, most of these hypotheses have been based on the existence of a primordial soup.

The first step toward the evolution of life must have been the synthesis of even more complex organic molecules than the ones previously described. However, scientists know that complex molecules such as **polymers** (for example, polysaccharides and proteins) do not spontaneously develop from a mixture of their simpler building blocks, or **monomers** (in this case, sugars and amino acids). Most synthesis reactions are dehydrations and depend on the removal of water and the input of energy to chemically link one molecule with another. This process, called **dehydration synthesis** (see Chapter 2), could have taken place if condensing agents were present with the monomers in a primordial soup. Condensing agents are molecules that combine with water and release energy. If condensing agents were not present, heat and evaporation could also promote dehydration synthesis and the synthesis of polymers.

In the 1950s, the American scientist Sidney Fox and his co-workers developed a technique in which they used heat to produce polymers, which Fox called **proteinoids,** from dry mixtures of amino acids. Perhaps such complex molecules formed over 3.5 billion years ago as pools and puddles of a primordial soup evaporated under the heat of the sun or were heated near areas of volcanic activity. Interestingly, the proteinoid molecules synthesized by Fox act like enzymes because they are able to increase the rate of various organic reactions. Metal ions and clays (composed chiefly of minerals) could also have served as "early enzymes." When combined into sequences, enzymatic reactions can be thought of as the beginning of metabolic systems. To be considered "life" as scientists define it, metabolic systems must be organized within a cellular structure, be able to "carry" information about themselves, and be able to pass this information on by the process of replication.

The first step toward the evolution of life is the synthesis of highly complex organic molecules. Such complex molecules might have formed as pools of a primordial soup evaporated under the heat of the sun, promoting the linking of small molecules. Another theory suggests that metal ions and clays might have served as early enzymes promoting such reactions.

Recently, scientists have begun to suspect that self-replicating systems of RNA molecules may have started the process of evolution. RNA molecules have the ability, as do all polynucleotides, of acting as a **template,** or guide, for the synthesis of a second polynucleotide based on the complementarity of its bases. DNA and RNA synthesis occurs in this fashion—DNA serves as a template for its own synthesis and for the synthesis of RNA (see Chapter 24). As mentioned previously, the building blocks of nucleotides were probably present in a primordial soup: deoxyribose and ribose sugars, purine and pyrimidine bases, and molecules bearing phosphate groups (PO_4^{2-}). These compounds could have joined to form nucleotides, which then could have linked to form polynucleotides. These polynucleotides could have served as templates for the synthesis of others—a process that many scientists think played a crucial role in the origin of life. In fact, theoretical work by Eigen and his co-workers suggests that short strands of self-replicating RNA eventually "worked" together to assemble enzymes from the amino acids present in the primordial soup.

For replicating RNA to be considered life, however, it must have become enclosed by some sort of a membrane forming the first cells. Chemical reactions could then take place in a closed environment—not the open environment of a primordial soup. Within the boundaries of a membrane, enzymes could be organized to carry on life functions. Chemicals could be selectively allowed into or kept out of the cell's interior. Wastes could be eliminated from the cell, and hereditary material could be passed on from cell to cell.

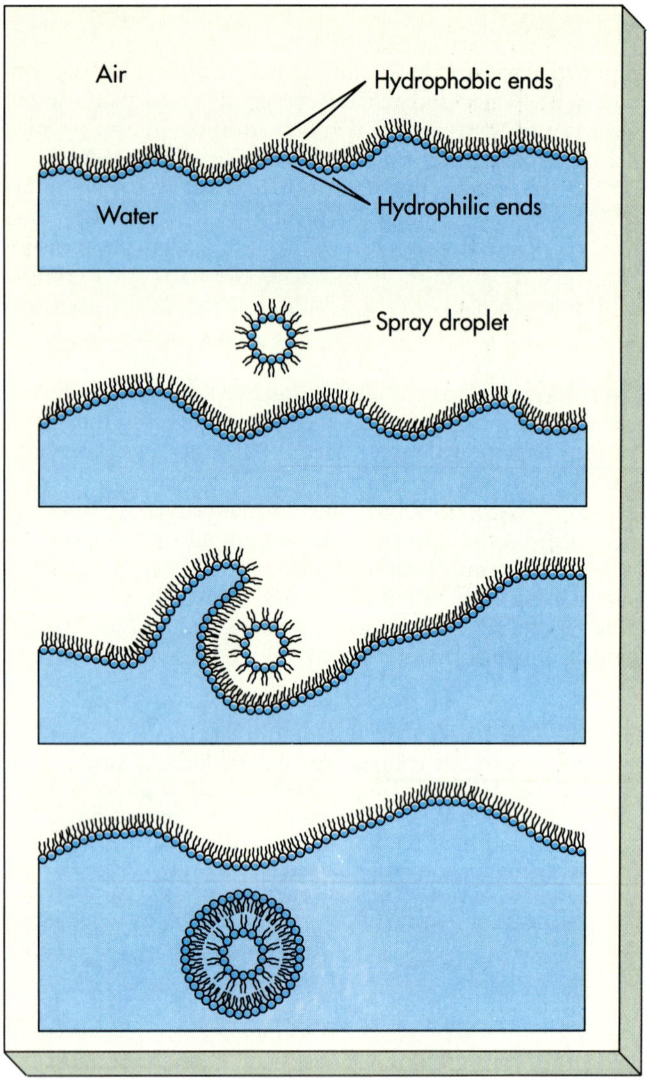

FIGURE 26-3 How membranous droplets can form.
The top diagram shows a single layer of lipid molecules floating on water. The sequence shows how wave action could cause these molecules, which have hydrophilic (water-loving) and hydrophobic (water-hating) ends, to form bilayered vesicles.

No one knows how much time passed before biochemical evolution became biological evolution. And no one knows how this evolution took place. Scientists do know, however, that single-celled life existed in the shallow seas of the primitive earth. How did these cells form? In the recent past, scientists thought that early precursors to cells—called **protocells**—were membranous droplets that had formed spontaneously. Scientists know (and can demonstrate) that many different kinds of molecules are attracted to one another when they are in solution because of forces such as electrical charge and surface tension. Sometimes these aggregates form microscopic, membranous droplets. Figure 26-3 shows how membranous droplets can form from molecules that have hydrophilic (water-loving) and hydrophobic (water-hating) portions, much like droplets of oil form in water when the two are mixed together. This droplet level of organization was thought to have been an important step in the origin of life. But although scientists have been able to produce two different types of spontaneously forming droplets in the laboratory, they recognize today that such droplets do not carry any genetic information and lack the internal organization of cells. Because of these facts, most scientists have abandoned the hypothesis that these droplets are the precursors to life.

After the synthesis of complex organic molecules had taken place, self-replicating systems of RNA molecules may have begun the process of evolution. To be considered life, however, this "early" genetic material must have been organized within a cellular structure. And although scientists have shown that membrane-bounded droplets can form spontaneously in solution, they recognize that such droplets are devoid of cellular components and therefore doubt their role as the "first cells."

The history of life

When discussing the history of life on Earth, scientists divide the time from the formation of the Earth until the present day into five major time periods or **eras:**

- The **Archean Era,** which extends from the formation of the Earth 4.6 billion to 2.5 billion (2500 million) years ago
- The **Proterozoic Era,** which extends from 2500 to 590 million years ago
- The **Paleozoic Era,** which extends from 590 to 250 million years ago
- The **Mesozoic Era,** which extends from 250 to 65 million years ago
- The **Cenozoic Era,** which extends from 65 million years ago to the present

The eras are subdivided into shorter time units called **periods,** as shown in the table of geologic time (Table 26-1). Additionally, the periods of the Cenozoic era are further subdivided into **epochs.** However, each time unit (such as an era or a period) does not stand for a consistent length of time because early geologists defined and named these time units as they discovered them, and because the distinctive events that mark the beginning and end of a time unit occurred over varying amounts of time. A 1-year geological "calendar" is shown on the left of Table 26-1 on p. 450, which relates the geological time scale to a single year, making it easier to understand how long each geological time unit is in relation to the others.

The Archean Era: Oxygen-producing cells appear

Before the Archean Era (meaning "ancient life"), the Earth was coalescing from an original cloud of dust formed as the Earth was "born" 4.6 billion years ago. Approximately 800 million (0.8 billion) years passed before the first rocks were formed. Approximately 3.8 billion years ago, chemical evolution began, resulting, somehow, in the formation of the first cells 3.5 billion years ago. (Interestingly, some scientists have recently suggested that this 300 million year time span is *too short* a time for cells to have evolved from the chemicals of the primordial Earth!)

TABLE 26-1

Table of geological time

Era	Period	Epoch	(MYA)	Major biological and geological events
Cenozoic	Quaternary	Pleistocene	1-2	
	Tertiary	Pliocene	7	First humans.
		Miocene	26	Origin of first human-like forms.
		Oligocene	38	Monkey-like primates appear.
		Eocene	54	Origin of *Eohippus*.
		Paleocene	65	Small mammals undergo adaptive radiation.
Mesozoic	Cretaceous		130	Major extinction of the dinosaurs and many marine organisms. Flowering plants appear, insects become more diverse.
	Jurassic		210	Large dinosaurs dominate Earth. First birds.
	Triassic		250	Small dinosaurs appear. First mammals.
Paleozoic	Permian		285	Major extinction occurs. Most species disapppear. Conifers appear.
	Carboniferous		370	First reptiles and arthropods, coal deposits formed, horsetails, ferns, and seed-bearing plants abundant. Fungi appears.
	Devonian		410	"Age of the fishes." Fishes with bones and jaws appear. Amphibians also appear. Major extinction of marine invertebrates and fishes.
	Silurian		430	Notochord becomes flexible as single rod is replaced with separate pieces, as seen in the Ostracoderms (armored fish without bones, jaws, or teeth). Plants invade the land.
	Ordovician		500	First vertebrates appear. Major extinction of marine species.
	Cambrian		590	Major extinction of the trilobites. Origin of the main invertebrate phyla.
Proterozoic			2500	Multicellular eukeryotic animals appear. First eukaryotic cells appear. Oxygen-producing bacteria present; atmosphere and oceans oxygenated.
Archean				Earth is born. Chemical evolution resulting in formation of first cells. Stromatolites formed. First rock formed.

TABLE 26-1

Table of geological time—cont'd

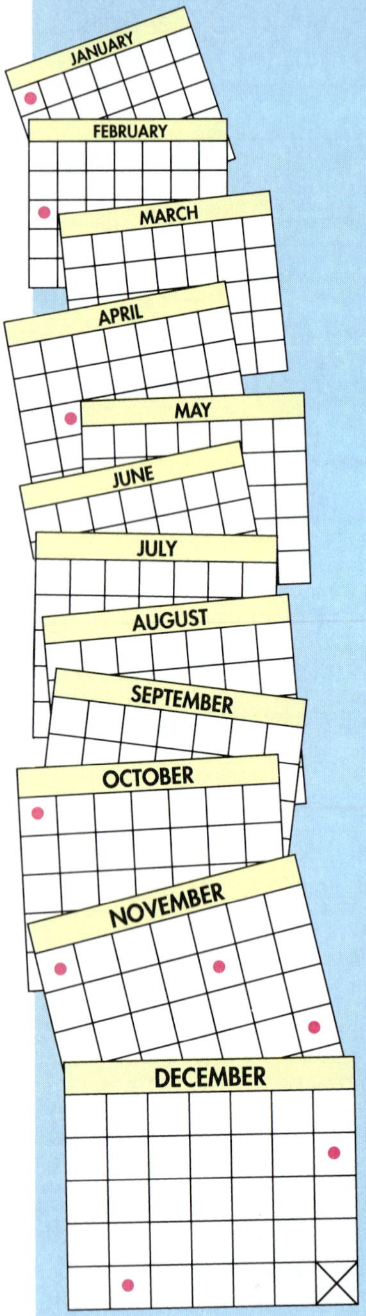

January 1 (4600 mya)	Origin of Earth
February 10 (3800 mya)	Oldest dated rocks
April 16 (3500 mya)	First known life
October 1 (1000 mya)	Normal oceans and atmospheres
November 1 (800 mya)	Primitive higher animals
November 12 (600 mya)	Beginnings of well-known geology
November 28 (410 mya)	First land plants
December 10 (200 mya)	Supercontinent
December 26 (65 mya)	End of dinosaurs
December 31, 6 PM (2 mya)	Humans appear
December 31, 80 sec to midnight (15000 ya)	End of last Glacial Age
December 31, 14 sec to midnight (2000 ya)	Birth of Christ

Scientists speculate that the first cells to evolve fed on organic materials in the environment. Because there would have been a very limited amount of suitable organic food available, a type of nutrition must have soon evolved in which cells were able to capture energy from inorganic chemicals. The evolution of a pigment system that could capture the energy from sunlight and store it in chemical bonds most likely evolved after that. Then, oxygen-producing bacteria evolved. Probably similar to present-day **cyanobacteria** (formerly called blue-green algae), these single-celled organisms became key figures in the evolution of life as it is known today. Their photosynthesis gradually oxygenated the atmosphere and the oceans around 2 billion years ago as evidenced by Archean and Proterozoic sediments. Scientists have also discovered fossils of the cyanobacteria.

Although fossils of single-celled organisms are usually difficult to find, the cyanobacteria sometimes grew in "piles," creating fossilized columns of organisms and the sediments that collected around them. These ancient columns of cyanobacteria are called **stromatolites** and date back 3.5 billion years. In Figures 26-4, *A* and *C*, the tops of stromatolites are visible above the surface of the water in which they form. Like icebergs, much of the stromatolite is hidden under water, as shown in Figure 26-4, *B*. Actively developing stromatolite formations can be observed even today in the warm, shallow waters of places such as the Gulf of California, western Australia, and San Salvador, Bahamas. Interestingly, the shapes and sizes of the bacteria within present-day stromatolites look very much like the bacteria found within fossil stromatolites (Figure 26-5). In the absence of competing organisms, cyanobacteria produced stromatolites abundantly in all fresh water and marine communities until about 1.6 billion years ago.

Scientists speculate that the first cells to evolve fed on organic materials in the environment. Because of limited organic material, the evolution of a pigment system that could capture the energy from sunlight and store it in chemical bonds most likely evolved relatively quickly. Then, oxygen-producing bacteria evolved, gradually oxygenating the atmosphere.

Until recently, all prokaryotes were considered to be genetically and evolutionarily similar. However, a small group of bacteria have been found to have biochemical differences from all other bacteria—differences that are as significant as the differences between prokaryotes and eukaryotes. This group of bacteria are fundamentally different from most bacteria in their ribosomal structure, RNA polymerase structure, cell membrane and cell wall composition, and other characteristics. Some of the "biochemically different" bacteria can multiply at higher temperatures than any other forms of life and require sulfur-containing compounds to survive. These unusual metabolic requirements remind many biologists of some of the habitats on Earth over 3 billion years ago. Such differences suggest an ancient evolutionary separation of these two "types" of bacteria. Speculation about the possible ancient origins of these cells led to naming them the **archaebacteria,** or "ancient bacteria." The larger group of more common bacteria are called

FIGURE 26-4 Stromatolites.

A These stromatolites are located in Shark Bay, Western Australia. The largest structures are about 1.5 meters (approximately 4½ feet) across. These stromatolites formed about 4000 years ago.

B This diagram shows that much of the stromatolite formation is underwater.

C Fossilized stromatolites (about 1.9 billion years old) in Great Slave Lake, Canada, showing some of the internal lamination typical of stromatolites.

D The internal, laminated structure of a fossilized stromatolite in Great Slave Lake, Canada.

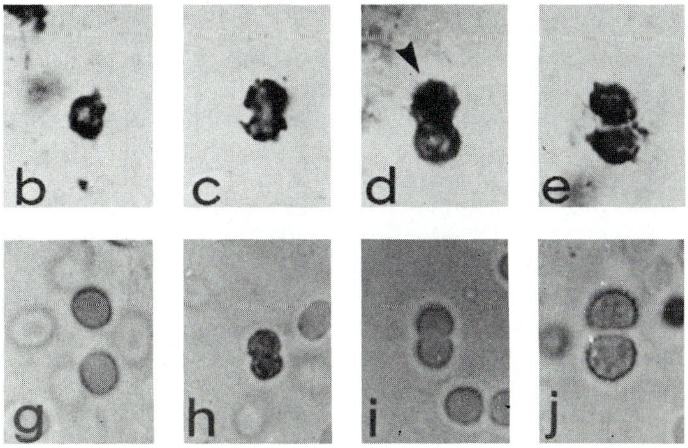

FIGURE 26-5 Fossilized bacteria.
The fossilized bacteria in **B** through **E** were found in stromatolites located in South Africa. These stromatolites are about 3.4 billion years old. The bacteria in **G** through **J** show living, similar bacteria. The fossilized bacteria bear a striking resemblance to the living bacteria.

eubacteria, or "true bacteria." Little is known about the evolutionary history of the archaebacteria, but scientists think that archaebacteria and eubacteria most likely had a common ancestor.

Most scientists still classify the archaebacteria and the eubacteria in a single kingdom, usually called *Monera*. However, others place the archaebacteria in a kingdom separate from the eubacteria, creating two prokaryotic kingdoms, or six kingdoms in all. Still other scientists propose a three kingdom system of classification: archaebacteria, **eubacteria,** and eukaryotes. They then further subdivide the kingdom of eukaryotic organisms into familiar subkingdoms: the plants, animals, protists, and fungi.

The Proterozoic Era: The first eukaryotes appear

The beginning of Proterozoic time, which extends from 2.5 billion to 590 million years ago, is marked by the formation of a stable oxygen-containing atmosphere. The term *Proterozoic* is derived from two Greek words meaning "prior" (proteros) "life" (zoe) and is descriptive of the fact that, until a few decades ago, scientists did not find evidence of life during this time (or during Archean time) because they were not looking for microfossils.

During the Proterozoic Era, the first eukaryotic organisms appeared on Earth. A fossil stromatolite found in California dating back 1.4 to 1.2 billion years ago contains larger and more complex microfossils than the fossil cyanobacteria previously found. Scientists think that these microfossils may have been eukaryotes, organisms separate from the cyanobacteria that may have lived in harmony with stromatolites or have fossilized with them. Other fossils in rocks of the same time add additional evidence to the theory that eukaryotes appeared around 1.4 billion years ago.

How did the eukaryotes arise? The most widely accepted hypothesis at this time is called the **endosymbiotic theory.** The term *symbiosis* refers to a relationship between two different organisms in which both organisms benefit. According to this theory, bacteria became attached to or were engulfed by "pre-eukaryotic" (host prokaryotic) cells. The bacteria carried out chemical reactions necessary for living in an atmosphere that was increasing in its amounts of oxygen, which benefited their host cells. The bacteria, or endosymbionts, benefited by living within the protective environments of their hosts. The cells that engulfed the bacteria had a reproductive advantage over those cells not associated with bacteria and, therefore, flourished.

The cell membranes of the pre-eukaryotes are thought to have pouched inward during these symbiotic relationships, loosely surrounding the bacteria. These membranous invaginations are thought to have been precursors to nuclear envelopes and endoplasmic reticula. The bacteria are thought to be "early" mitochondria. Symbiotic events similar to those postulated for the origin of mitochondria also seem to have been involved in the origin of chloroplasts, which are thought to be derived from symbiotic photosynthetic bacteria (Figure 26-6).

Multicellular eukaryotic animals date back 630 million years. Scientists have found fossils of a diverse array of unusual multicellular eukaryotes dating from that time. They regard these soft-bodied organisms as ancestors of the jellyfish, hydra, soft coral, and worms. However, other scientists suggest that they are not related to any currently existing species.

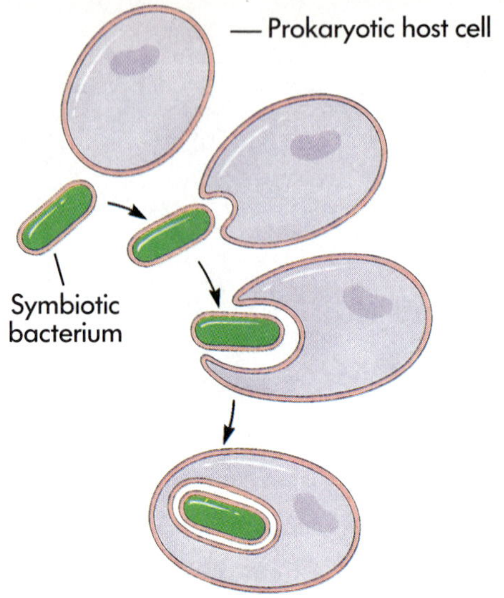

FIGURE 26-6 Endosymbiosis.
How pre-eukaryotes are thought to have engulfed bacteria, which may have become mitochondria (heterotrophic bacteria) and chloroplasts (photosynthetic bacteria) as eukaryotic cells evolved.

Fossil evidence suggests that the first eukaryotes, larger and more complex than the cyanobacteria, appeared around 1.4 billion years ago. The chloroplasts and mitochondria of eukaryotic cells are thought to be derived from bacteria (referred to as *endosymbionts*) that came to live within pre-eukaryotic cells. Multicellular eukaryotes appeared approximately 630 million years ago.

The Paleozoic Era: The occupation of the land

The Paleozoic Era (590 to 250 million years ago) is marked by an abundance of easily visible, multicellular fossils. In fact, the earliest such fossils are found in rocks of southern Australia and are approximately 630 million years old. For many years such fossils were the oldest known, which led scientists to name the era in which these fossils became plentiful as the *Paleozoic,* meaning "old" (*paleos*) "life" (*zoos*). However, of the roughly 250,000 different kinds of fossils that have been identified, described, and named, only a few dozen are more than 630 million years old.

The Paleozoic Era is divided into six shorter time spans called *periods*. (These periods are shown in Table 26-1.) The **Cambrian Period,** which ended roughly 500 million years ago, is the oldest period within the Paleozoic Era. It represents an important point in the evolution of life; all of the main phyla and divisions of organisms that exist today (except for the chordates and land plants) evolved by the end of the Cambrian Period. Because so many new kinds of

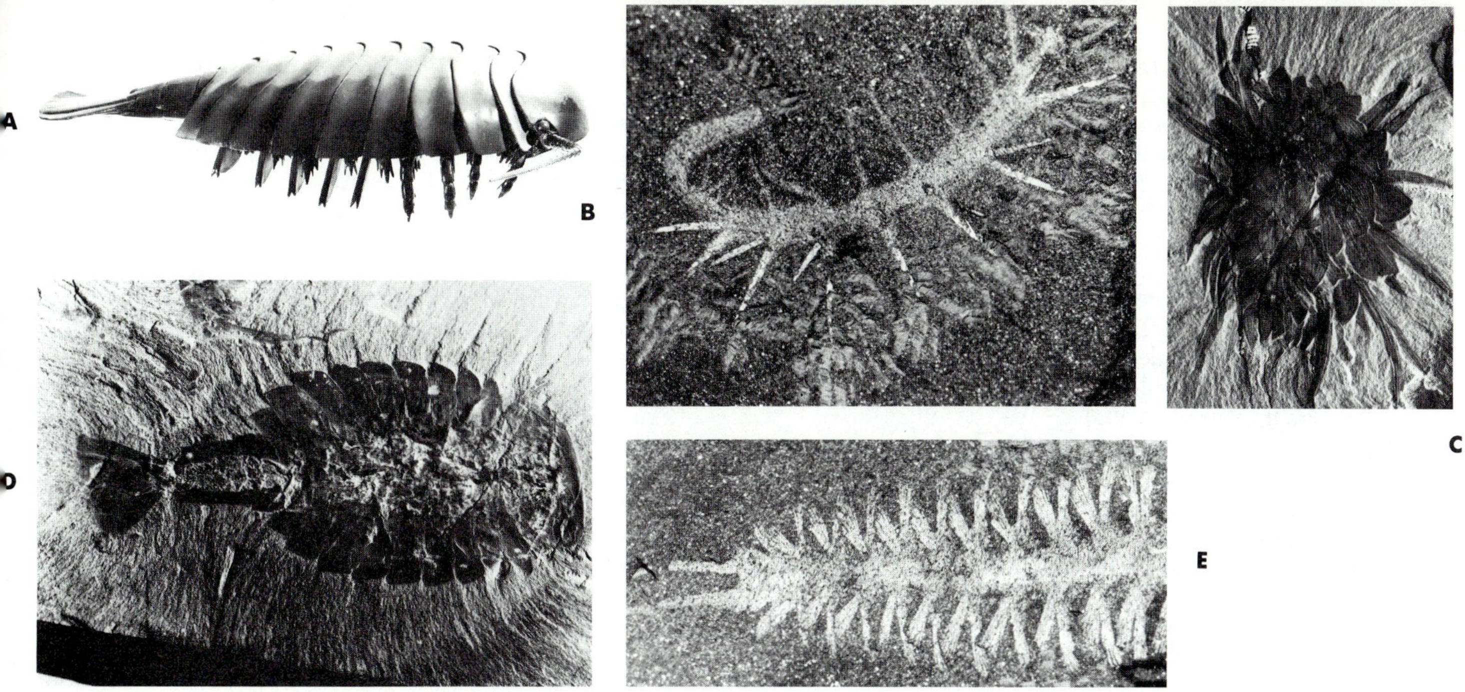

FIGURE 26-7 Cambrian Period fossils from the Burgess Shale, British Columbia, Canada, about 530 million years old.
A and **B** *Sidneyia inexpectans* and a model of the organism. This creature was an arthropod.
C *Hallucigenia sparsa.* This specimen is 12.5 millimeters (approximately $\frac{1}{2}$ inch) long. This animal seems to have been supported on seven pairs of spines; its trunk bore seven long tentacles and an additional group of tentacles near the rear end. *H. sparsa* may have been a scavenger on the bottom of the sea.
D *Wiwaxia corrugata.* The body of this animal was covered with scales and also bore spines. This specimen is 30.5 millimeters (approximately $1\frac{1}{4}$ inches) across, excluding the spines. It may have been a distant relative of the mollusks.
E *Burgessochaeta setigera,* a segmented worm. The photograph shows the front end of the worm, which had a pair of tentacles. Foot-like structures are attached to the worm in pairs along the sides of the body. The specimen is 16.5 millimeters (approximately $\frac{5}{8}$ inch) long.

organisms appeared in such a relatively short time span, paleontologists (scientists who study fossil life) speak of a Cambrian "explosion" of living forms and often refer to geological strata older than Cambrian time as **Precambrian.**

The evolution of Cambrian organisms took place in the sea. Figure 26-7 shows fossils of the organisms that lived on the seafloor during the Cambrian Period—quite unusual by today's standards! These fossils, together with well over 100 other species, have been found in the Burgess Shale, a geological strata formed from a fine-grained mud (Figure 26-8). During past movements of the Earth's crust, the Burgess Shale was uplifted to high elevations in the Rocky Mountains of British Columbia, Canada.

FIGURE 26-8 The Burgess Shale, located in the Rocky Mountains of British Columbia.
The stratification of this formation can be clearly seen.

Many kinds of multicellular organisms that thrived in the early Paleozoic Era have no relatives living today. As multicellular organisms arose that were capable of moving from place to place, the forces of natural selection resulted in an array of organisms capable of filling previously unoccupied ecological niches, or roles, within their communities (see Chapter 37). These possibilities led to diversification along new evolutionary pathways; some resulted in "dead ends," and others ultimately led to the contemporary phyla of organisms. The trilobites, for example (Figure 26-9, *A*), appear to be derived from the same evolutionary line that gave rise to one living group of arthropods, the horseshoe crabs (Figure 26-9, *B*).

During the Cambrian Period of the early Paleozoic Era, an array of species evolved in the sea. Some of these species still exist today.

As the Paleozoic Era continued into the **Ordovician Period** (500 to 430 million years ago), worm-like aquatic animals similar to lancelets and lampreys began to evolve. Characteristic of these organisms was the stiff, internal rod that ran down the back, parallel to the central nerve. This stiffening rod is called a **notochord** (meaning "back cord"); these organisms are therefore called **chordates.** As the **Silurian Period** began 430 million years ago, this rod was replaced by a sequence of pieces that provided more flexibility for the animal and protected the nerve cord. Eventually these organisms developed a bony, protective armor and cartilaginous internal skeletons. They are considered the first vertebrates—the **ostracoderms.** During this time, some groups of fishes developed jaws, bony internal skeletons, and scales, much like those of the fishes of today. During the **Devonian Period** (410 to 370 million years ago), these fishes became abundant and diverse, giving the Devonian the nickname "the age of fishes."

At the interface of the Silurian and Devonian Periods, the first terrestrial plants evolved. The earliest known fossil plants are about 410 million years old. The ancestors to these plants are believed to be the green algae, since their chloroplasts are biochemically similar to those of the plants. The green algae are an extremely varied group of more than 7000 protist species, found growing in the water as well as on tree trunks or in the soil. Many of these algae are microscopic and unicellular. However, the plants that appeared during the Silurian were multicellular and had developed mechanisms for transporting water within their bodies as well as for conserving it. Within the next 100 million years, plants became abundant and diverse on the land and eventually formed extensive forests.

Amphibians, the descendants of fishes, appear in the fossil record by the end of this period and subsequently became abundant in the great swamps that formed at that time. Leading "double lives," they spent part of their time on land and part of their time in the water. Then, about 300 million years ago in the late **Carboniferous Period,** the amphibians gave rise to the first reptiles. The reptiles, being better suited for a terrestrial existence than their am-

FIGURE 26-9 Trilobites and horseshoe crabs.
The trilobites flourished in the seas of the Cambrian Period, 500 to 550 million years ago. The example in **A** was found in the Burgess Shale. Trilobites appear to be directly related to modern-day horseshoe crabs **(B)**.

A

B

phibian ancestors, became the dominant organisms over the next 50 million years.

The Carboniferous Period (370 to 285 million years ago) is named for the great coal deposits formed during this time. Much of the land was low and swampy due to a worldwide moist, warm climate—conditions that contributed to the fossil preservation of the still-prevalent forests. These coal deposits provide a relatively complete record of the horsetail plants, ferns, and primitive seed-bearing plants of the Carboniferous Period.

The abundant plant life of that time provided the opportunity for invasion of the land by insects, a type of arthropod. Arthropods are invertebrate animals having segmented bodies, jointed appendages, and a tough, outer "skin" called a *cuticle*. Arthropods evolved as a phylum during the Cambrian from segmented "worm-like" animals. During the subsequent history of this group, only a few kinds of insects have moved back into fresh water habitats, and a few have returned to marine habitats, living on the surface of the ocean.

The fungi, a kingdom of organisms that live on dead and decaying matter, also invaded the land in the late Carboniferous Period. One component of the success of the fungi on land has been related to the structure of their chitin-rich cell walls. Chitin, a stiff, hard substance that also characterizes the outer skeletons of the arthropods, helped both insects and fungi to remain "drought resistant." In addition, fungi are successful land organisms because of their roles (along with the bacteria) as decomposers.

As the Paleozoic Era progressed, chordate and then vertebrate animals evolved in the sea. The first vertebrates appeared approximately 430 million years ago. Plants and insects colonized the land about 410 million years ago; amphibians did so about 50 million years later. Fungi may also have colonized the land at about the same time as plants.

The **Permian Period** (285 to 250 million years ago) was a period of drought and extensive glaciation, unlike the swamps of the preceding Carboniferous Period. The conifers—a group of seed-bearing plants that is represented today by pines, spruces, firs, and similar trees and shrubs—originated then. A major extinction event occurred at the end of the Permian, wiping out many of the species living at that time.

Mass extinctions

One of the most prominent features of the history of life on earth has been periodic **mass extinctions,** the global death of whole groups of organisms. There have been five such events. Four of these events occurred during the Paleozoic Era, the first occurring near the end of the Cambrian Period (about 500 million years ago). At that time, most of the existing families of trilobites became extinct. Additional mass extinctions occurred about 440 and about 360 million years ago. The fourth and most drastic mass extinction happened at the close of the Permian Period. It is estimated that approximately 96% of all species of marine animals living at that time died out! All of the trilobites, and many other groups of organisms as well, disappeared forever. The fifth major extinction event occurred at the close of the Mesozoic Era, 65 million years ago. This was the famous event when dinosaurs became extinct.

Major episodes of extinction produce conditions that can lead to the rapid evolution of those relatively few plants, animals, and microorganisms that survive. Habitats, or places to live, formerly occupied suddenly open up, providing the "raw material" for **adaptive radiation,** a phenomenon by which populations of a species change as they disperse within a series of different habitats within a region (see Chapter 25). In fact, many paleontologists suggest that extinction episodes have been key events in the "pattern" of evolution.

The Mesozoic Era: The age of reptiles

The Mesozoic Era, or "middle life" (250 to 65 million years ago), was a time of adaptive radiation of the terrestrial plants and animals that had been established during the mid-Paleozoic Era but that had not been "wiped out" by the mass extinction of the Permian Period. The radiation of these organisms led to the establishment of the major groups of organisms living today (Figure 26-10). This era was dom-

FIGURE 26-10 Basic body plans were established long ago. The fossil dragonfly, which is about 170 million years old, closely resembles its modern counterpart. It dramatically illustrates the establishment of modern groups of organisms in the Mesozoic Era; many of these have persisted to the present day.

inated by the adaptive radiation of the reptiles. During the course of the Mesozoic, one group of reptiles evolved into mammal-like reptiles, precursors to mammals; another group led to the evolution of the birds; and yet another developed into the dinosaurs, or "terrible lizards." Reptiles also invaded the sea and the air, but these kinds of reptiles eventually died out.

The Mesozoic is divided into three periods: the **Triassic,** the **Jurassic,** and the **Cretaceous.** These names refer to geological events or formations of the eras and do not chronicle the evolution of life during this time. During the Triassic Period, the earliest period of the Mesozoic, small dinosaurs and primitive mammals appeared. Using fossil evidence regarding changes in the jaws and teeth of these organisms, some scientists think that mammals evolved from reptiles of the Permian Period (Paleozoic Era) that were distinguished by "sails" on their backs. Scientists speculate that their upright fans of tissue collected the sun's heat when it was cold and radiated excess body heat when it was hot. The small dinosaurs of the Triassic are pictured to the far right of these organisms.

During the Jurassic Period, large dinosaurs dominated the Earth, and birds first appeared. The earliest known bird, *Protoavis* ("first bird") (Figure 26-11), was reported in 1986 by Sankar Chaterjee of Texas Tech University. The fossil bones of this organism were found in rocks dated to be 225 million years old, very early in the Mesozoic Era. Until this find, the oldest known fossil bird was *Archaeopteryx,* estimated to be about 150 million years old (Figure 26-12).

FIGURE 26-11 Protoavis, a 225 million-year-old fossil bird, from west Texas.
This remarkable find, which, if it proves to be a true ancestor of birds, will extend the history of birds back 75 million years, was announced by Sarkar Chaterjee of Texas Technological University in 1986.

FIGURE 26-12 Archaeopteryx.
About the size of the crow, *Archaeopteryx* lived in the forests of central Europe 150 million years ago. The teeth and long, jointed tail are features not found in any modern birds. Discovered in 1862, *Archaeopteryx* was cited by Darwin in later editions of *On the Origin of Species* in support of his theory of evolution.

BIOLOGY & EVOLUTION

Was *Protoavis* the first bird?

Those who study the history of life on Earth by examining fossils are paleontologists, and in June 1991 a paleontologist named Sankar Chaterjee published a detailed description of a fossil that has created a furor among other paleontologists. The fossil he described was 225 million years old, from the late Triassic Period, found by Chaterjee and colleagues 5 years earlier as a jumble of bones in an ancient west Texas mudbed (Figure 26-A). After 5 years of careful study, Chaterjee had assembled the bones into a nearly complete skeleton—and he claimed it was the skeleton of a bird! He gave it the name *Protoavis* ("first bird").

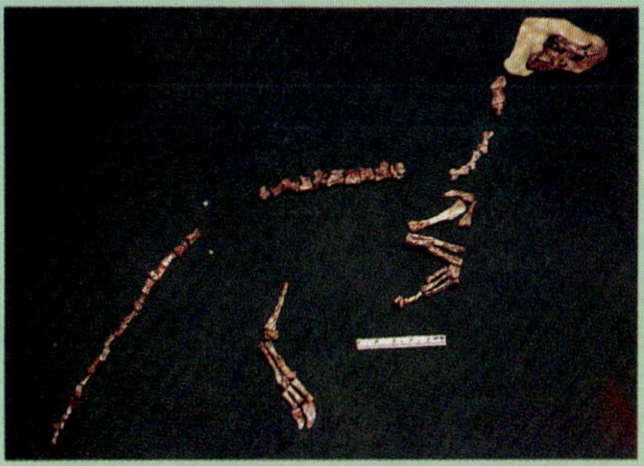

FIGURE 26-A The bones of *Protoavis.* Was *Protoavis* the first bird?

Why all the fuss? Chaterjee's claim is in effect a frontal assault on one of the most widely known animals in all paleontology—the fossil bird *Archaeopteryx* ("ancient wing"). *Archaeopteryx* dates back to the late Jurassic Period, about 150 million years ago, during the middle of the dinosaurs' reign on earth, whereas Chaterjee's fossil is as old as the oldest dinosaur, from the time when they first appeared on Earth. Most paleontologists believe that birds evolved from dinosaurs in the mid-Jurassic Period (in effect, that birds are dinosaurs with a feather coat) and that *Archaeopteryx* represents a transitional form, having the feathers and breastbone of a bird and the teeth and tail of a dinosaur.

Chaterjee bases his claim on the fact that the skeleton shows bird-like elements, such as a wishbone, a shoulder modified for flying, and a keeled sternum (which serves as an attachment point for flight muscles in modern birds). He claims the skull has no fewer then 23 features that are fundamentally birdlike. In particular, the creature's skull appears to lack holes that are present on the skulls of dinosaurs of the period; in *Protoavis,* these holes have merged with the eye sockets, making it similar to modern birds. The jaws of *Protoavis* are able to slide forward, and the upper jaw can elevate. Modern birds possess both these features, but ancient reptiles had neither.

The mud in which *Protoavis* fossilized was incapable of preserving any feathers, however, and indeed most of the *Protoavis* bones are poorly preserved. Other paleontologists are not convinced by Chaterjee's reconstruction, complaining that the poor state of the fossils makes it extremely difficult to identify many of the features important to Chaterjee's argument. Also, because the fossil was not articulated (that is, its bones were not attached to one another but were found lying in a jumble), they worry that Chaterjee's skeleton may in fact contain the bones of several different kinds of animals. If Chaterjee is right, birds are not the descendants of dinosaurs but their cousins. If a vote were taken today among paleontologists, his claim would probably lose. The world of paleontology awaits a better fossil.

Whether *Archaeopteryx* arose from the same evolutionary line as *Protoavis* or independently from another dinosaur is not known. Some scientists question the validity of Chaterjee's "find" (see the boxed essay).

Archaeopteryx had feathers but apparently not enough of them to fly very effectively. Feathers, thought to have evolved from reptilian scales, probably helped "early" birds glide from tree to tree and provided them with insulation. Ultimately, feathers made possible the subsequent evolution of birds that could fly well. This ability allowed birds to inhabit unoccupied habitats and resulted in the evolution of a large and diverse class of organisms. Today, there are about 9000 different species of birds.

During the Mesozoic Era, the reptiles, which had evolved earlier from amphibians, became dominant and in turn gave rise to mammals (about 200 million years ago) and birds (at least 150 million years ago).

During the Cretaceous Period, about 127 million years ago, flowering plants (angiosperms) began to appear, becoming the dominant form of plant life about 100 million years ago. Seed-bearing plants with fern-like leaves, similar to the living cycads, were abundant at that time. As flowering plants became more diverse, so did insects, which had feeding habits that were closely linked with the characteristics of flowering plants. Insects and flowering plants have **coevolved** with one another. Indeed, all groups of terrestrial organisms, including mammals, birds, and fungi, have evolved characteristics largely in relation to those of the flowering plants. Today, with about 240,000 species of flowering plants, the angiosperms still dominate the plant kingdom, greatly outnumbering all other kinds of plants.

Around 65 million years ago, as the Cretaceous Period (and the Mesozoic Era) ended, sudden shifts occurred in the kinds of marine organisms that existed. Some became extinct and others began to flourish. Scientists can infer these changes in the populations of marine organisms living at that time by studying marine fossils exposed in certain European geological strata. For example, many of the larger plankton (free-drifting microscopic protists) disappeared about 65 million years ago, and a fewer number of smaller ones took their place. The same rapid changes occurred in at least some nonplanktonic marine animal groups, such as the bivalve mollusks (clams and their relatives). The ammonites, a large and diverse group related to octopuses but with shells, abruptly disappeared. And, of course, a major extinction occurred on land: the dinosaurs disappeared, although at a pace much slower than that of the plankton.

Scientists have long discussed hypotheses regarding the extinction event of the late Cretaceous. Some propose that the climate cooled. Others have suggested that the sea level changed or that the vegetation cover of the earth changed. These reasons, however, would not account for the change in the plankton community. In 1980 a group of scientists headed by Luis W. Alvarez of the University of California, Berkeley, presented a dramatic hypothesis about the reasons for these changes. Alvarez and his associates observed that the usually rare element iridium was abundant in a thin layer in the geological strata that marked the end of the Cretaceous Period. Alvarez and his colleagues proposed that if a large, iridium-rich meteorite or asteroid had struck the surface of the earth, a dense particulate cloud would have been thrown up. This cloud would have darkened the earth for a time, greatly slowing or temporarily halting photosynthesis and driving many kinds of organisms (particularly photosynthetic plankton) to extinction. By disrupting and killing plant life, other organisms dependent on the plants would die out as well. Then, as its particles settled, the iridium in the cloud would have been incorporated in the layers of sedimentary rock that were being deposited at that time. Although the "meterorite hypothesis" is the most plausible one put forth to date, scientists are not certain that the impact happened, nor that it caused any of the extinctions of that time.

Changes in the Earth: The impact on evolution

During the Mesozoic Era, the continents did not exist as they do today. Instead, they were all joined in one giant continent scientists call **Pangea,** meaning "all Earth." Pangea began to break up into smaller pieces during the Mesozoic Era, with this movement continuing during the following era, the Cenozoic. These changes greatly affected evolution.

After the Earth coalesced approximately 4.5 billion years ago, its surface was hot and violent for about 600 to 700 million years. At first, these conditions were too inhospitable for living things or even for biochemical evolution to take place. Scientists hypothesize that geological activity eventually resulted in the formation of fragments of continents. Some geologists think that these land masses were formed as melted rock material rose to the surface of the Earth, pushed upward by other heavier, sinking, melted metals such as iron and nickel. Others hold that the highly active volcanoes prevalent at that time spewed so much lava into the seas that it eventually accumulated, forming land masses.

Approximately 200 million years ago, small land masses united to form the single, large "supercontinent" Pangea. Its northern half was called **Laurasia** and consisted of the present-day North America, Europe, and Asia. Its southern half was called **Gondwana** and was a combination of present-day South America, Africa, India, Antarctica and Australia. Pangea remained as a supercontinent for approximately 100 million years. However, forces were at work beneath it that eventually divided it—once again—into smaller land masses.

Scientists now have some insight into the movement of land masses over time, such as those that resulted in the formation of Pangea and its subsequent division. Today, the outer shell of the Earth is made up of six large "plates" and several smaller ones (Figure 26-13). According to the **plate tectonics theory,** these plates are rigid pieces of the Earth's crust, which are from 75 to 150 kilometers (approximately 50 to 100 miles) thick. The rocks beneath these plates are less rigid than the plates they support. The surface plates move, or drift, over this underlying semiliquid rock. Scientists think that this "liquid" rock flows slowly, rising as it heats, and sinking as it cools (Figure 26-14). At the mid-oceanic ridges—long, narrow mountain ranges under the sea—the plates move away from one another in a process called *seafloor spreading.* Molten rock wells up from deep within the Earth, pushing the plates apart and filling in the space between. At other places the plates move toward each other. As they meet, one plate sinks below the other, pushing up mountain ranges in the process. The Himalayas, for example, have been thrust up to the highest elevations on Earth as a result of the grinding, prolonged collision of the Indian subcontinent with Asia. Other types of violent geological activity are also associated with the movements of the crustal plates. The earthquakes that destroyed San Francisco in 1906 and the one that occurred in the same region in 1989 resulted from two plates sliding past each other.

Figure 26-15 summarizes the movements of the continents as they gradually moved apart from their positions as parts of Pangea 200 million years ago. The South Atlantic Ocean opened about 125 to 130 million years ago as Africa and South America moved apart from one another. South America then moved slowly toward North America and, approximately 3.1 to 3.6 million years ago, became connected to it by the Isthmus of Panama. Changes occurred in the positions of all the continents, which played a major role in the distribution of organisms that are seen today.

As land masses move away from one another, populations inhabiting those lands become separated from one another. Geographically isolated, populations of the same species often undergo speciation as differing pressures of natural selection act on them. Similarly, as land masses collide, species are thrust together and must compete with one another to survive. Shorelines disappear and with them go a variety of habitats and species. And as continents move, ocean currents change, causing climatic changes and yet new selection pressures.

The Cenozoic Era: The age of mammals

The mass extinctions of the Cretaceous marked the end of the Mesozoic Era and heralded in the Cenozoic Era about 65 million years ago. Extending to the present, the Cenozoic Era has two periods—the **Tertiary** and the **Quaternary**—

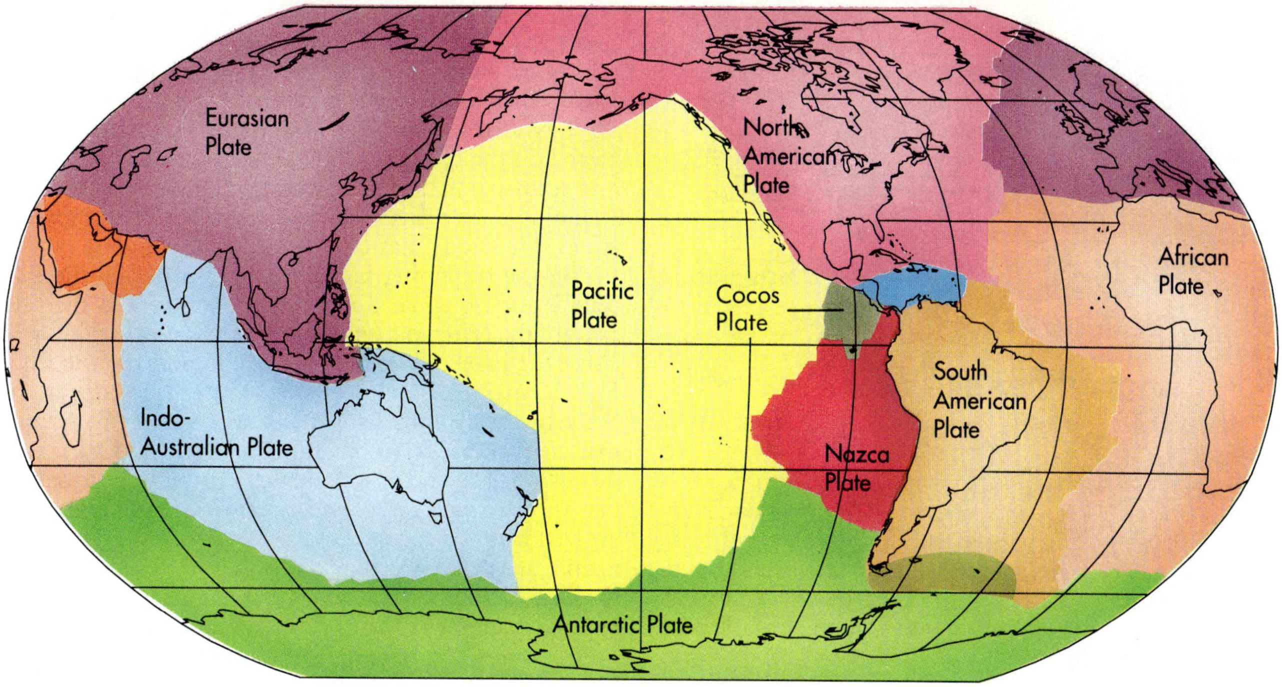

FIGURE 26-13 Plate tectonics theory.
Plate tectonics describes and explains the movement of the continents. According to this theory, Earth is made up of six large plates and several smaller plates. The rocks beneath these plates are semiliquid. The plates drift over this semiliquid rock, occasionally colliding. When this occurs, one plate is usually pushed below the other, a process called *subduction.* When plates move apart, the semiliquid rock wells up in the space created, a process called *seafloor spreading.*

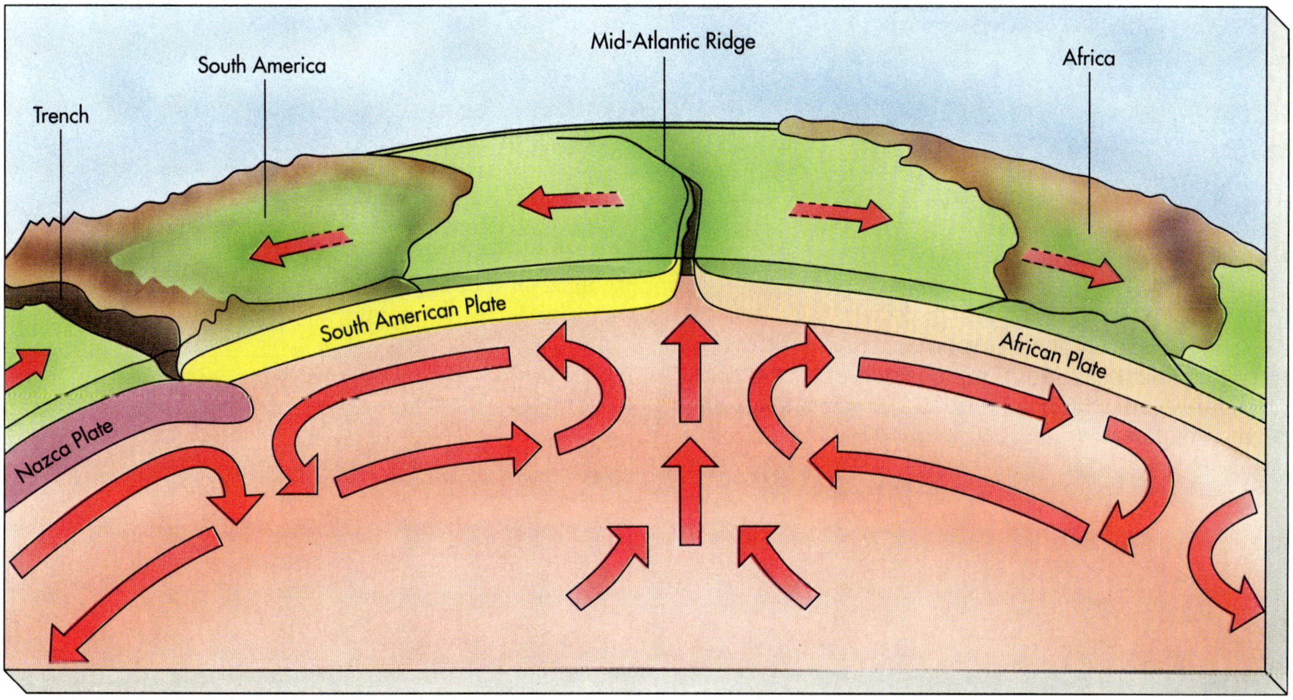

FIGURE 26-14 Why do the plates move?
The semiliquid rock on which the plates rest rises as it heats and sinks as it cools, causing movement.

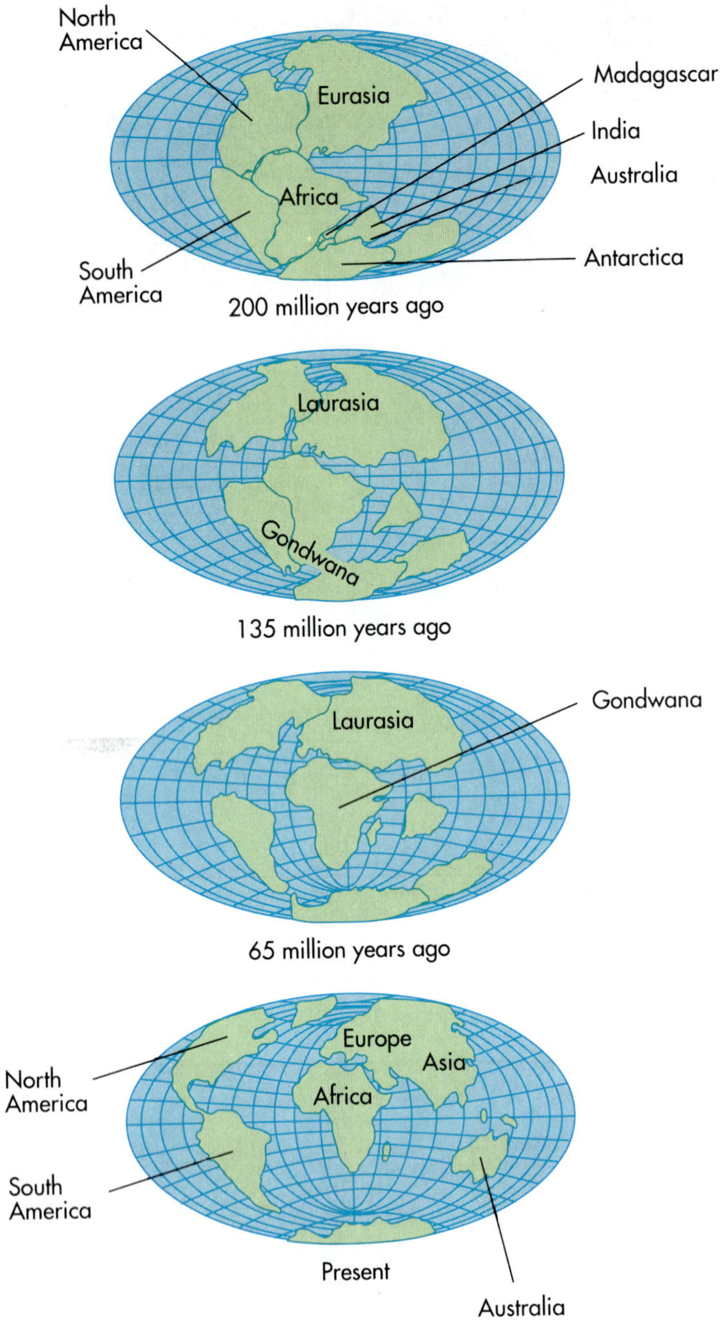

FIGURE 26-15 Movement of the continents.
The changes that occurred in the position of all the continents as a result of shifting plates have played a major role in the distribution of organisms seen today.

which are subdivided into many smaller time units called *epochs*. The epochs of the Tertiary Period are listed in Table 26-1. Notice also in Table 26-1 that most of the "year" has passed on the geologic calendar. Although 65 million years may seem like an incredibly long time, in terms of the history of life, it is incredibly short. During this very short time, life as it is known has evolved.

At the beginning of the Cretaceous during the **Paleocene Epoch,** the small mammals that survived the extinctions of the Mesozoic underwent adaptive radiation, quickly filling the habitats vacated by the dinosaurs. By the **Eocene Epoch,** many species of mammals were no longer small. Some of the mammals appearing at this time are organisms that were ancestors to present-day whales and bats. Eohippus, a small mammal (about the size of a dog) that gave rise to the horse, also appeared at that time. Eohippus evolved from an ancestor that gave rise to several groups of mammals, including tapirs and rhinoceroses, in addition to the horses. Monkey-like primates, the highest order of mammals, first appear as fossils about 36 million years ago in the early **Oligocene Epoch.** These **anthropoid primates**—monkeys, apes, and human beings and their direct ancestors—are the focus of Chapter 27, a continuation of the story of the evolution of life on Earth.

Summary

1. Approximately 15 billion years ago, scientists theorize, all the material of the universe condensed into a single small space and then exploded in an event called the "big bang," forming an early generation of stars. Then, approximately 4.6 billion years ago, these stars exploded, forming the Earth and the other planets of the solar system.
2. When the Earth cooled, a hardened outer crust formed. This time is marked by the formation of the rocks, the oldest of which have been dated at about 3.8 billion years old. Fossil bacteria about 3.5 billion years old are the oldest direct evidence of life on Earth. These fossil bacteria, or cyanobacteria, are found in the massive limestone deposits of stromatolites.
3. Somehow, in the 300 million years between the the time when the Earth cooled and the first bacteria appeared, biochemical evolution became biological evolution. The long-held "primordial soup" theory states that life arose in the sea as elements and simple compounds of the primitive atmosphere reacted with one another to form simple organic molecules such as amino acids and sugars. Recently, scientists have begun to reexamine the premises of this theory and have proposed alternate hypotheses.
4. After the synthesis of complex organic molecules had taken place, self-replicating systems of RNA molecules may have begun the process of biological evolution. To be considered life, however, this "early" genetic material must have been organized within a cellular structure. Although scientists have shown that membrane-bounded droplets can form spontaneously in solutions, they recognize that such droplets are devoid of cellular components and doubt their role as the "first cells."
5. Scientists speculate that the first cells to evolve fed on organic materials in the environment. Because of limited organic material, the evolution of a pigment system that could capture the energy from sunlight and store it in chemical bonds most likely evolved relatively quickly. Then, oxygen-producing bacteria evolved, gradually oxygenating the atmosphere.
6. The first unicellular eukaryotes appeared about 1.4 billion years ago during the Proterozoic Era. Eukaryotic cells likely arose as "pre-eukaryotic" cells became hosts to endosymbiotic prokaryotes. Multicellular organisms first appeared about 630 million years ago; the earliest, soft-bodied forms are poorly represented in the fossil record.
7. Many of the phyla of organisms in existence today, except the chordates and the land plants, appear to have evolved during the Cambrian Period (590 to 500 million years ago) of the Paleozoic Era. These organisms evolved in the sea. Chordates and then vertebrates evolved later in the Paleozoic Era, with the first vertebrates appearing about 430 million years ago.
8. Terrestrial plants appeared about 410 million years ago and flourished during the Carboniferous Period (370 to 285 million years ago). Fishes became abundant and diverse during the Devonian Period (410 to 370 million years ago). The amphibians, organisms that spend part of their time on land and part in the water, evolved from the fishes, appearing about 360 million years ago. Then, about 300 million years ago, the amphibians gave rise to the first reptiles.
9. Insects evolved on land during the late Carboniferous Period. The abundant plant life of that time provided them with food and shelter. The fungi, a kingdom of organisms that live on dead and decaying matter, also evolved on land during the late Carboniferous Period.
10. A major extinction event occurred during the Permian Period, wiping out many of the species living at that time. The Mesozoic Era (250 to 65 million years ago) was a time of the adaptive radiation of the organisms surviving the Permian extinction, dominated by the evolution of the reptiles. The reptiles gave rise to the mammals (about 200 million years ago) and the birds (at least 150 million years ago).
11. The evolution of the five kingdoms of life has been influenced by movement of the land masses that has occurred over time.

REVIEW QUESTIONS

1. What is the "primordial soup" theory, and what does it attempt to explain? Describe another theory that explains the same event.
2. What is dehydration synthesis? What role might it have played in the evolution of life?
3. Explain the importance of a cell membrane to the evolution of early cells.
4. Create a table that lists the major eras and periods of geological time.
5. What are cyanobacteria, and when did they first appear? Explain their evolutionary significance.
6. Is the term *Proterozoic* accurate? Explain your answer.
7. When was the Cambrian Period, and why was it important?
8. Which period has been called *the age of fishes,* and why? What other evolutinary event occurred during this period?
9. Describe the major events of the Carboniferous Period. What were the living conditions on Earth during this time?
10. Summarize the major extinction events that have occurred on Earth. When did each happen?
11. Fill in the blanks: During the Mesozoic Era, one group of ______ evolved into precursors of ______, another led to the evolution of ______, and a third group developed into the large ______.
12. What might have caused the extinction event of the late Cretaceous Period? What organism(s) became extinct at that time?
13. What is the plate tectonics theory? Relate it to the evolution of living organisms.
14. Summarize the events of the Cenozoic Era. When did it begin?
15. Fill in the blanks: Monkeys, apes, and human beings are all examples of ______ ______, a particular order of ______.
16. Many movies have shown early human beings battling ferocious dinosaurs. Is this accurate? Explain.

KEY TERMS

adaptive radiation 455
anthropoid primate 460
archaebacteria 450
Archean Era 448
Cambrian Period 452
Carboniferous Period 454
Cenozoic Era 448
chordate 454
coevolution 457
Cretaceous Period 456
cyanobacteria 450
Devonian Period 454
endosymbiotic theory 452
Eocene Epoch 460
epoch 448
era 448
eubacteria 452
Gondwana 458
Jurassic Period 456
Laurasia 458
mass extinction 455
Mesozoic Era 448
notocord 454
Oligocene Epoch 460
Ordovician Period 454
ostracoderm 454
Paleocene Epoch 460
Paleozoic Era 448
Pangea 458
period 452
Permian Period 455
plate tectonics theory 458
Precambrian 453
proteinoids 447
Proterozoic Era 448
protocell 448
Quaternary Period 458
Silurian Period 454
stromatolite 450
template 447
Tertiary Period 458
Triassic Period 456

THOUGHT QUESTIONS

1. The Miller-Urey experiments suggest that the key building blocks of biological macromolecules could have been present in a "primordial soup," but the experiments do not suggest a way that the building blocks could have assembled themselves. The problem is that RNA monomers join in a dehydration synthesis to form RNA chains (that is, water is a *product* of the reversible reaction). So how could the first RNA molecules have formed in a watery primordial soup?
2. The most diverse time in the history of life was the pre-Cambrian Era, when creatures with many different body plans lived. Only a few of these basic body plans survived into the post-Cambrian Era, the other body plans were weeded out by evolution. Stephen Gould has speculated that evolution does not foster ever-greater complexity, more complex forms evolving from simpler primitive ones, but rather acts in the opposite fashion, tending to eliminate more and more options. Discuss his suggestion in light of what you have learned about evolution.
3. Archaebacteria have introns and in many other ways seem to be the direct ancestors of eukaryotes. Eubacteria, on the other hand, seem the direct ancestors of mitochondria. Which transition must have happened first?

FOR FURTHER READING

Alvarez, W., & Asaro, F. (1990, October). What caused the mass extinction?—an extra-terrestrial impact. *Scientific American,* pp. 78-84.
This article features the debate of the cause of the dinosaurs' extinction. The discoverers of iridium deposited at the time argue that a giant meteorite collided with the Earth.

Cortillot, V. (1990, October). What caused the mass extinction?—a volcanic eruption. *Scientific American,* pp. 85-92.
In a companion article, geologists argue that the iridium came not from a meteorite but from within the Earth, blown into the sky by enormous volcanoes produced when plumes in the Earth's mantle reached the surface 65 million years ago.

Silk, J. (1992, April). Cosmology back to the beginning. *Nature,* pp. 741-742.
Using data drom NASA's COBE satellite, scientists have detected temperature ripples in the cosmological microwave background, which they believe are remnants of the big bang that spawned the universe. This short article describes the findings in a simple, straightforward way.

Zimmer, C. (1992, May). Ruffled feathers. *Discover,* pp. 44-54.
Sakar Chaterjee, the Texas Tech University professor who claims he has discovered the fossil of the first bird, has been hit with charges from other scientists that his work is misleading and his conduct unprofessional. This article delineates the Protoavis *controversy and provides a look into the sometimes contentious process of scientific discovery.*

Human evolution

27

Footprints in the sand of a parent and child walking hand in hand on the wet edge of a beach might leave a permanent imprint of their presence. But the story behind these footprints is more complex than it might appear. Not surprisingly, these footprints were made near a lake. However, they are not embedded in sand but in volcanic ash that has preserved them for over 3½ million years. Although human in appearance, these footprints are not human. They were made by non-human ancestors, *Australopithecus afarensis*, and give scientists important clues to our heritage. One clue, for example, is the size of the footprints. Some scientists suspect that they were made by a male and a female rather than by a parent and child and therefore may reflect our ancestors' sexual dimorphism, or size difference with gender. Another clue to our heritage is that these organisms are walking erect on two legs instead of four, a characteristic called *bipedalism*. Bipedalism is only one of the many evolutionary changes that led to the appearance of modern humans.

HIGHLIGHTS

▼ The story of human evolution began over 500 million years ago with the appearance of the first chordates, organisms with a stiff internal rod that runs down the back beneath a central nerve.

▼ Vertebrates evolved from the chordates and include a class of organisms called *mammals*, animals that are warm blooded and have hair and whose females secrete milk from mammary glands to feed their young.

▼ The primates, the order of mammals to which humans belong, reflect an arboreal heritage having excellent depth perception and flexible, grasping hands; the earliest primates arose about 75 million years ago and lived on the ground before moving into the trees.

▼ The first human fossils date about 2 million years ago and are of the extinct species *Homo habilis*, a species that is considered human because they exhibited an intelligence far greater than their ancestors by making tools and clothing.

OUTLINE

Evolution of the vertebrates
- Evolution of the fishes
- Evolution of the amphibians, reptiles, and birds
- Evolution of the mammals

Evolution of the primates
- The prosimians
- The anthropoids

Evolution of the anthropoids

Evolution of the hominids
- The first hominids: *Australopithecus*
- The first humans: *Homo habilis*
- Human evolution continues: *Homo erectus*
- Modern humans: *Homo sapiens*

Evolution of the vertebrates

The beginning of the human evolution story can be traced all the way back to Cambrian times (see Chapter 26) and the appearance of the first chordates. This phylum of animals is named for the presence of a notochord: a stiff, internal rod that runs down the back beneath the central nerve cord. One of the earliest known chordate fossils is shown in Figure 27-1, *A*. This photo is of *Pikaia*, a small fishlike organism that resembles today's lancelets, shown in Figure 27-1, *B*.

Fossil evidence suggests that ancient flatworms are the ancestors to all the more complex animals, including the chordates (and *Pikaia*). As the flatworms exploited the variety of habitats in the sea, the processes of adaptive radiation resulted in the evolution of various phyla. Remarkably well-preserved fossils exist from this period of 100 million years ago, known as the Precambrian and early Cambrian Periods. The large numbers suggest an "explosion" of animal diversity during which time all the major phyla of animals originated with the exception of chordates. Included in this diversity are the roundworms, the arthropods, the mollusks, and the echinoderms (Figure 27-2). Chordate fossils became abundant in the Ordovician Period.

The approximately 42,500 species of chordates include fishes, amphibians, reptiles, birds, and mammals (and therefore humans). Chordates are distinguished by three principal features, two of which have already been mentioned: (1) a single hollow **nerve cord** located along the back; (2) a rod-shaped **notochord,** which forms between the nerve cord and the developing gut (stomach and intestines); and (3) **pharyngeal (gill) arches and slits,** which are located at the

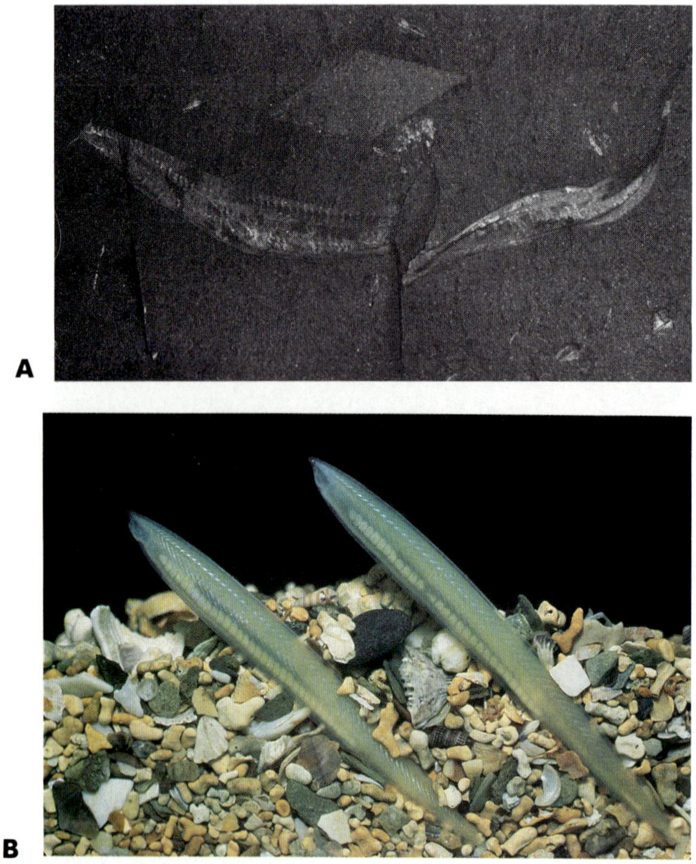

FIGURE 27-1 The earliest known chordate.
A *Pikaia gracilens* is a small fish-like animal, and it is one of the first organisms with a notochord. This particular fossil was found in the Burgess Shale.
B A lancelet. Note the resemblance to *Pikaia.*

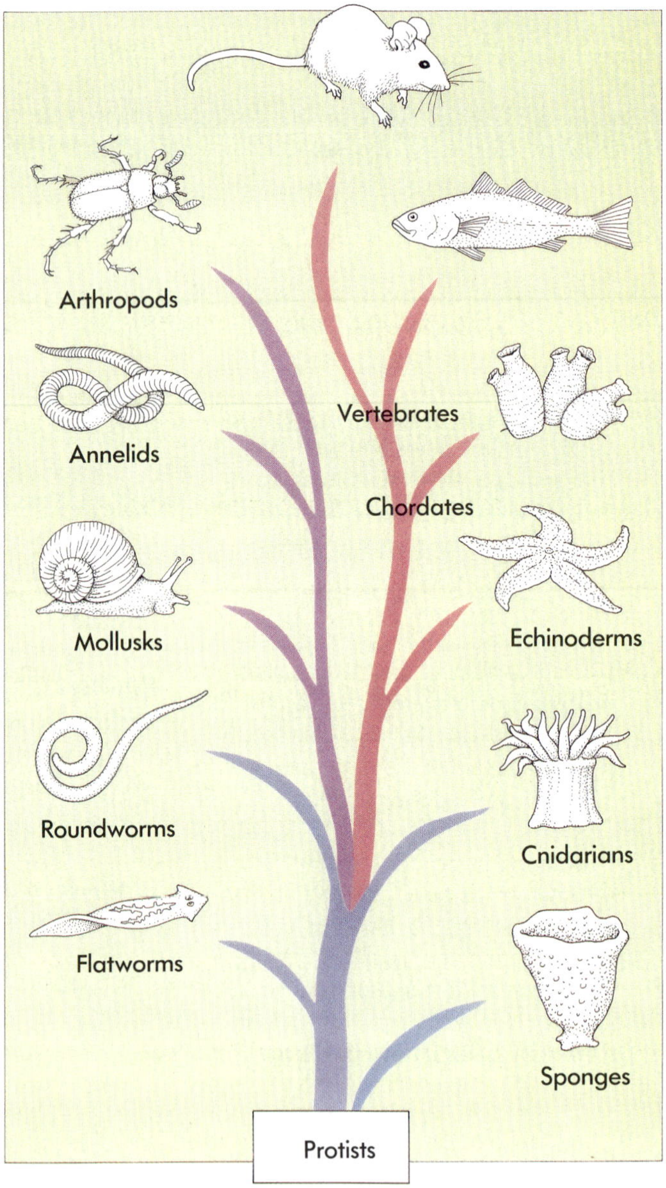

FIGURE 27-2 Evolution of the major phyla.
During the Precambrian and Cambrian Periods, an explosion of animal diversity took place, resulting in the origination of all the major phyla with the exception of the chordates.

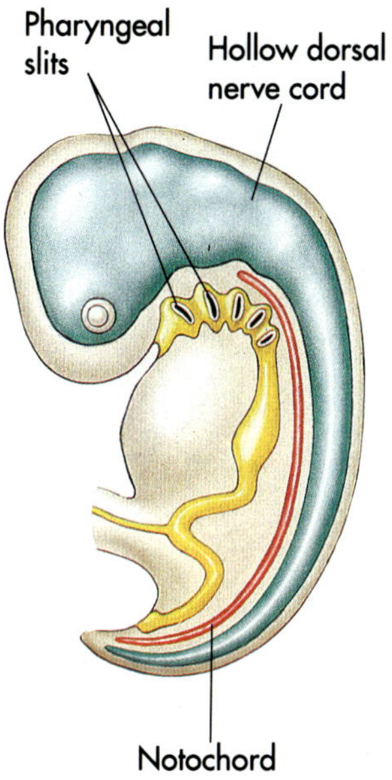

FIGURE 27-3 Chordate features. Embryos reveal the three principal features of the chordates: a nerve cord, a notochord, and pharyngeal (gill) arches and slits.

FIGURE 27-5 A lamprey, member of class Agnatha. Lampreys attach themselves to fish by means of their sucker-like mouths, rasp a hole in the body cavity, and suck out blood and other fluids from within. Lampreys do not have a distinct jaw.

throat (pharynx). (These three chordate features are shown in Figure 27-3 as they appear in the embryo.) All the features of chordates are evident in their embryos, even if they are not present in the adult form of the organism.

Human evolution can be traced back to the appearance of the first chordates, which were small, fish-like organisms. Today, chordates include fishes, amphibians, reptiles, birds, and mammals.

Each of the three chordate characteristics played an important role in the evolution of the chordates and the vertebrates. The dorsal nerve cord increased the animals' responsiveness to the environment. In the more advanced vertebrates, it became differentiated into the brain and spinal cord. The notochord was the starting point for the development of an internal stabilizing framework, the backbone and skeleton, which provided support for locomotion. The bony structures that support the pharyngeal arches evolved into jaws with teeth, allowing these organisms to feed differently than their ancestors (Figure 27-4).

With the exception of two groups of relatively small marine animals, the tunicates and lancelets (see Figure 27-1, *B*), all chordates are vertebrates. Vertebrates (a subphylum of the chordates) differ from these two chordate groups in that the adult organisms have a **vertebral column,** or backbone, that develops around and replaces the embryological notochord. In addition, most vertebrates have a distinct head and a bony skeleton, although the living members of two classes of fishes, Agnatha (lampreys and hagfishes) (Figure 27-5) and Chondrichthyes (sharks and rays), have a cartilaginous skeleton.

Vertebrates evolved from chordates. Vertebrates are characterized by a vertebral column surrounding a nerve cord located along the back. In addition, most vertebrates have a distinct head and a bony skeleton.

Evolution of the fishes

The first vertebrates to evolve were jawless fishes, members of the class Agnatha, about 500 million years ago (see Figures 26-4 and 27-5). Although traces of agnathan fossils are found in Cambrian strata, most date back to the Ordovician and Silurian Periods, about 400 to 500 million years ago.

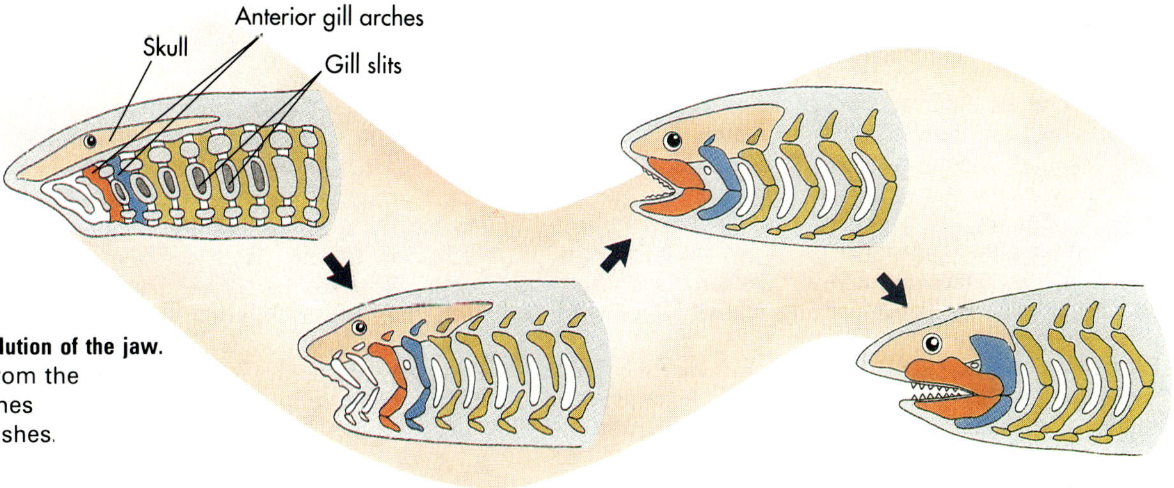

FIGURE 27-4 Evolution of the jaw. Jaws evolved from the anterior gill arches of the jawless fishes.

A

B

C

FIGURE 27-6 Sharks and rays.
A Blue shark.
B Manta ray.
C Diamond sting ray.

Jaws first developed among vertebrates that lived about 410 million years ago, toward the end of the Silurian and beginning of the Devonian Periods. The first jawed fishes that evolved were the placoderms, ancestors to today's bony fishes and cartilaginous fishes—the sharks, skates, and rays (Figure 27-6). They were also one of the first groups of fishes to have paired fins. During the Devonian Period, fishes having a bony skeleton appeared: the "ray-finned" fishes and the "fleshy-finned" fishes.

The ray-finned fishes are the ancestors of the variety of bony fishes that exists today, such as the salmon, trout, cod, and tuna. (Others in this group are shown in Figure 27-7.) These fishes are the most successful class of vertebrates, accounting for about half of all living vertebrate species. One of the reasons for the success of the bony fishes is the development of the swim bladder, a gas-filled sac that allows these fishes to regulate their buoyant density; for this reason they can remain suspended at any depth in the water.

In the late Devonian Period, the "lungfishes" and the "lobe-finned" fishes appear in the fossil record. Fleshy-finned fishes are ancestors to both. Only a few species of lungfishes exist today (Figure 27-8). The lungs of these fishes are supplementary to their gills but allow them to breathe air when they are buried in the mud of dried-up lakes and streams. Most of the Devonian Period lobe-finned fishes also had primitive lungs similar to the lungfishes. Interestingly, these fishes were thought to be extinct until the discovery of a living member of this group in 1938. Since then, other specimens of this same species have been found.

The first vertebrates to evolve were the jawless fishes, about 500 million years ago. Jawed fishes evolved toward the end of the Silurian Period, about 410 million years ago.

Evolution of the amphibians, reptiles, and birds

One group of the lobe-finned fishes is believed to be the ancestor of the land-living **tetrapods,** or four-limbed vertebrates. The first land vertebrates were **amphibians,** animals able to live both on land and in the water. Amphibians are a class of vertebrates that lay their eggs in water or moist places and live in water during their early stages of development. Although a mature amphibian spends time on land, it must be able to return to the water frequently to keep its thin skin moist. The early amphibians had fish-like bodies,

A

B

C

D

FIGURE 27-7 The bony fishes.
The bony fishes (class Osteichthyes) are extremely diverse.
A Koran angelfish. *Pomacanthus semicircularis,* in Fiji.
B Puffer. Puffers avoid being eaten by inflating themselves when attacked, thus projecting their spines outward.
C A sea horse. Sea horses move slowly and are difficult to see among marine algae and sea grasses.
D Moray eel, in Rangiroa, French Polynesia.

FIGURE 27-8 A lungfish.
Lungfishes have supplemental lungs that allow them to breathe air when they are buried in mud.

FIGURE 27-9 Reconstruction of *Ichthyostega*, one of the early amphibians.
Ichthyostega had efficient limbs for crawling on land, an improved olfactory sense associated with a lengthened snout, and a relatively advanced ear structure for picking up airborne sounds. Despite these features, *Ichthyostega,* which lived about 350 million years ago, was still quite fish-like in overall appearance.

short stubby legs, and lungs (Figure 27-9). Abundant about 300 million years ago during the Carboniferous Period, these amphibians probably gave rise to the other tetrapod groups.

The moist climate of the Carboniferous Period fostered the dominance of the amphibians during this time. However, following the Carboniferous Period, the climate changed dramatically as the Permian Period began. The amphibians, the dominant land vertebrates for about 100 million years, were gradually replaced by the reptiles as the dominant organisms. In contrast to amphibians, reptiles have water-resistant skins and more efficient lungs. These adaptations to dry climates gave reptiles a survival advantage over the amphibians. In addition, reptiles developed a reproductive advantage—the amniotic egg. An **amniotic egg** has a thick shell that encloses the developing embryo and a nutrient source within a watery sac (see Chapter 33). This type of egg protects the embryo from drying out, nourishes it, and enables it to develop away from water. Birds, too, have amniotic eggs.

One group of jawed, bony fishes, the lobe-finned fishes, are thought to be the ancestors of the amphibians. Amphibians, animals that live on land but return to water to breed, arose over 300 million years ago and became dominant during the warm, moist climate of the Carboniferous Period. As the climate changed during the Permian Period, however, reptiles took over the land.

Birds (class Aves) evolved from reptiles during the Mesozoic Era, although the entire story of the evolution of birds is somewhat of a mystery due to gaps in the fossil record. Scientists have fossils of *Protoavis* and *Archaeopteryx,* two of the earliest known birds (see Chapter 26). No fossils have been found of birds that lived after *Archaeopteryx* until about 10 million years later in the Cretaceous Period. However, scientists think that all modern birds evolved from animals like *Archaeopteryx.*

Evolution of the mammals

The fossil record clearly shows that **mammals,** too, evolved from reptiles. Mammals are warm-blooded vertebrates that have hair and whose females secrete milk from mammary glands to feed their young. Organisms having *both* mammalian and reptilian characteristics, called **transitional forms,** first appear in the fossil record approximately 245 million years ago. Then, about 200 million years ago in the early Mesozoic Era, the first known mammals appear. These early mammals resemble today's shrews (Figure 27-10). They were small, fed on insects, and were probably **nocturnal,** or active at night. About 65 million years ago, following the extinction of the dinosaurs (see Chapter 26), mammals became abundant.

Various natural selection pressures resulted in the emergence of a number of significant anatomical and physiological changes in the mammals as they evolved from their reptilian ancestors. These changes resulted in a class of organisms that not only survived but flourished in a wide variety of habitats. Changes occurred in the reptilian arrangement of limbs as the mammals evolved, raising them high off the ground and allowing them to walk quickly and to run. A hinge developed between the lower jaw and the

FIGURE 27-10 A shrew.
Shrews are small nocturnal animals that feed on insects. They resemble the first known mammals that appeared about 200 million years ago.

BIOLOGY & EVOLUTION

All about Eve, or how scientists are attempting to construct the human family tree

Have you ever wanted to trace your "roots" and discover who your long-lost relatives were? Scientists, too, have this same curiosity but are much more ambitious. Ever since Darwin published his great works *On the Origin of Species* in 1860 and *The Descent of Man* in 1871, scientists have been searching for the "missing links" that link modern humans to their ancestors.

This search for missing links officially began when Dutch anatomist Eugene Dubois travelled to Java, a small island situated off the coast of Indonesia, and unearthed the first evidence of a hominid species called *Homo erectus* in 1877. Other fossil finds followed, and paleontologists, such as Richard and Mary Leakey, their son Louis Leakey, and Donald Johansen, built their careers out of the elusive hunt for hominid remains. Based on this fossil evidence, most scientists agree that *H. sapiens* evolved from *H. erectus* about 500,000 years ago. Fossil remains of *H. erectus* have been found throughout Europe, Asia, and Africa, leading many scientists to conclude that the races of *H. sapiens* evolved independently from separate isolated groups of *H. erectus* living in these different locations.

Recently, however, new molecular data has challenged this view, using DNA within mitochondria of living humans rather than fossils. Mitochondria are the cellular organelles that provide energy for the cell. Like tiny bacteria, mitochondria also contain their own DNA. Because mitochondrial DNA is passed from the mother only, looking at mitochondrial DNA made the genetic sleuthing much easier.

Scientists examined samples of blood from 189 people of different races and origins, including the !Kung people of Botswana, the Eastern pygmies of Zaire, and the Yorubans of Nigeria, as well as Europeans, Asians, and African-Americans. Feeding this information on DNA into a computer, they then constructed a family tree and reported that there was a single point of origin—in Africa (Figure 27-A)! The oldest fragment of mitochondrial DNA present in all humans today was first carried by an early human female in Africa, perhaps as recently as 200,000 years ago. This report argues strongly that our species has a single origin in Africa and that this African group of humans emigrated out of Africa and completely replaced the groups of *H. erectus* living in Europe, Asia, and other locations.

However, the methodology used to generate this now-famous family tree has been questioned, and it appears that many other family trees are as likely—the data do not allow a clear conclusion about human origins. Because African mitochondrial DNA is far more diverse, scientists are still betting on an African origin for *H. sapiens*, but the matter is far from proven.

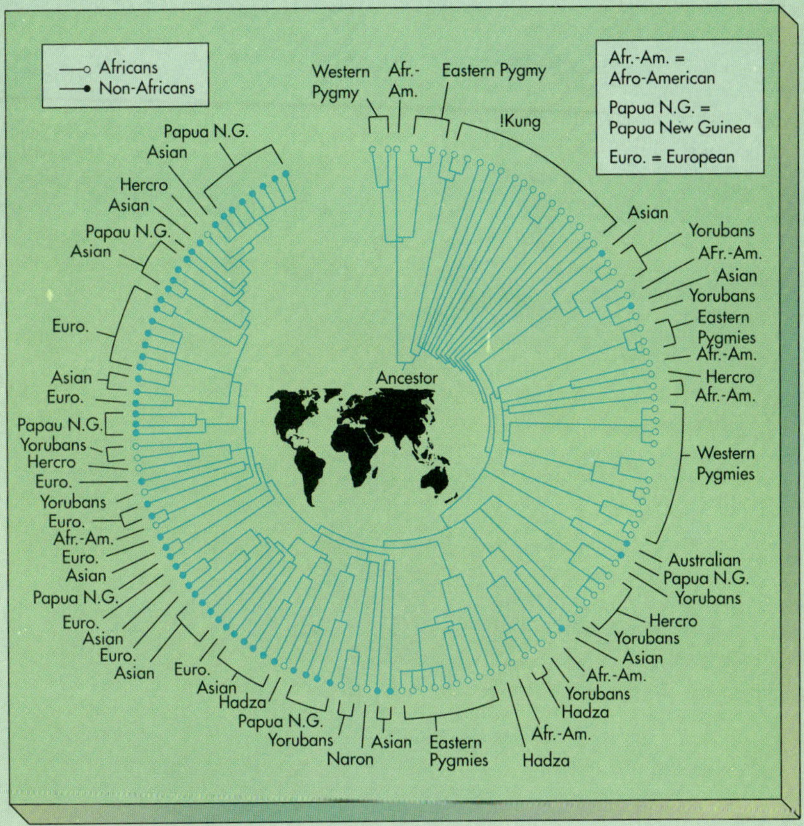

FIGURE 27-A This genetic tree is based on analysis of genetic sequences in 189 people of diverse races and shows that a common female ancestor existed in Africa.

FIGURE 27-11 Convergent evolution.
This illustration demonstrates the convergent evolution of marsupials in Australia and placental mammals in the rest of the world.

skull, and the teeth became differentiated, allowing mammals to eat a wide variety of food. The reptilian heart, having two ventricles with an incomplete separation, developed a complete wall in the mammals, preventing the mixing of oxygenated and deoxygenated blood. This change was significant in that mammals had also developed **warm bloodedness:** a constant internal body temperature. Warm-blooded animals are also called **endotherms,** meaning "within" (*endo-*) "temperature" (*-therm*). Their ancestors were **cold blooded,** having internal body temperatures that followed the temperature of their environments. Cold-blooded animals are also called **ectotherms,** meaning "outside" (*ecto-*) "temperature" (*-therm*). Endotherms have a higher rate of metabolism than do ectotherms and need more oxygen for the increase in the rate of cellular respiration.

Important changes also occurred in reproduction as the mammals evolved from the reptiles. The most "reptilian" type of reproduction is seen in the **monotremes,** one of the three subclasses of mammals living today. Monotremes lay eggs with leathery shells and incubate these eggs in a nest. The underdeveloped young that hatch from these eggs feed on their mother's milk until they mature. The ability of a mother to feed her young with milk she produces in mammary glands is another characteristic that has been advantageous to the adaptive radiation of the mammals. Present-day examples of monotremes are platypuses and spiny anteaters of Australia. These organisms are thought to have arisen on Gondwana, which is today's South America, Africa, and Australia (see Chapter 26), and were then isolated as the continents drifted apart.

The **marsupials,** a second subclass of mammals, do not lay eggs but give birth to immature young. These blind, embryonic-looking creatures crawl to the mother's pouch and nurse until they are mature enough to venture out on their own. Like the monotremes, marsupials were present in Gondwana; today most marsupials, such as kangaroos, wombats, and koalas, are found in Australia. Interestingly, the marsupials of Australia resemble the placental mammals that are present on the other continents. Figure 27-11 compares individual members of these two sets of mammals. The members of each pair have similar habitats and find their food in similar ways. These characteristics, along with their strikingly similar anatomy, suggest that the evolution of these two subclasses of mammals is a product of **convergent evolution:** the development of similar structures having similar functions in different species as the result of the same kinds of selection pressures.

Mammals evolved from the reptiles approximately 200 million years ago. Changes in the structure and placement of their limbs, the structure of the heart, and reproductive strategies resulted in their ability to survive in a variety of climates and to eventually become abundant.

In **placental mammals,** the young develop to maturity within the mother. They are named for an organ formed during the course of their embryonic development, the placenta. The placenta is located within the walls of the uterus, or womb. Composed of both maternal and fetal tissues, the placenta is connected to the fetus by the umbilical cord. At the placenta, fetal wastes pass into the bloodstream of the mother, and oxygen and nutrients in the mother's bloodstream pass into the bloodstream of the fetus, a mechanism that allows the fetus to develop within the mother until it reaches a certain stage of maturity.

Evolution of the primates

There are 14 orders of placental mammals, which include many animals familiar to you, such as dogs, cats, horses, whales, squirrels, rabbits, bats, and a variety of others. The order of placental mammals that humans belong to is the **primates.** Primates are mammals that have characteristics reflecting an arboreal, or tree-dwelling, lifestyle. Among these characteristics are hands and feet that are able to grasp objects (such as tree branches), flexible limbs, and a flexible spine. Table 27-1 lists primate characteristics. (Figure 27-12 diagrammatically represents the taxonomic relationships of organisms living today.)

The earliest fossils that resemble any living primates belong to a group of organisms that look very much like today's insect-eating shrews (Figure 27-13). (Once classified as primates, today's tree shrews are now considered a separate order of placental mammals.) These primate ancestors are thought to have evolved in the late Mesozoic Era, about 75 million years ago, filling new habitats created by the appearance of the flowering plants, shrubs, trees, and the insects that fed on these plants. These early primates lived in the underbrush, hiding as they competed with the dinosaurs and other reptiles that were dominant at that time. As thousands of years passed, humans' shrew-like ancestors probably moved into the trees, eating insects that fed on the fruits and flowers growing on tree branches.

TABLE 27–1 Primate characteristics

- Ability to spread toes and fingers apart
- Opposable thumb (thumb can touch the tip of each finger)
- Nails instead of claws
- Omnivorous diet (teeth and digestive tract adapted to eating both plant and animal food)
- A semi-erect to an erect posture
- Binocular vision (overlap of visual fields of both eyes) resulting in depth perception
- Well-developed eye-hand coordination
- Bony sockets protecting eyes
- Flattened face without a snout
- A complex brain that is large in relation to body size

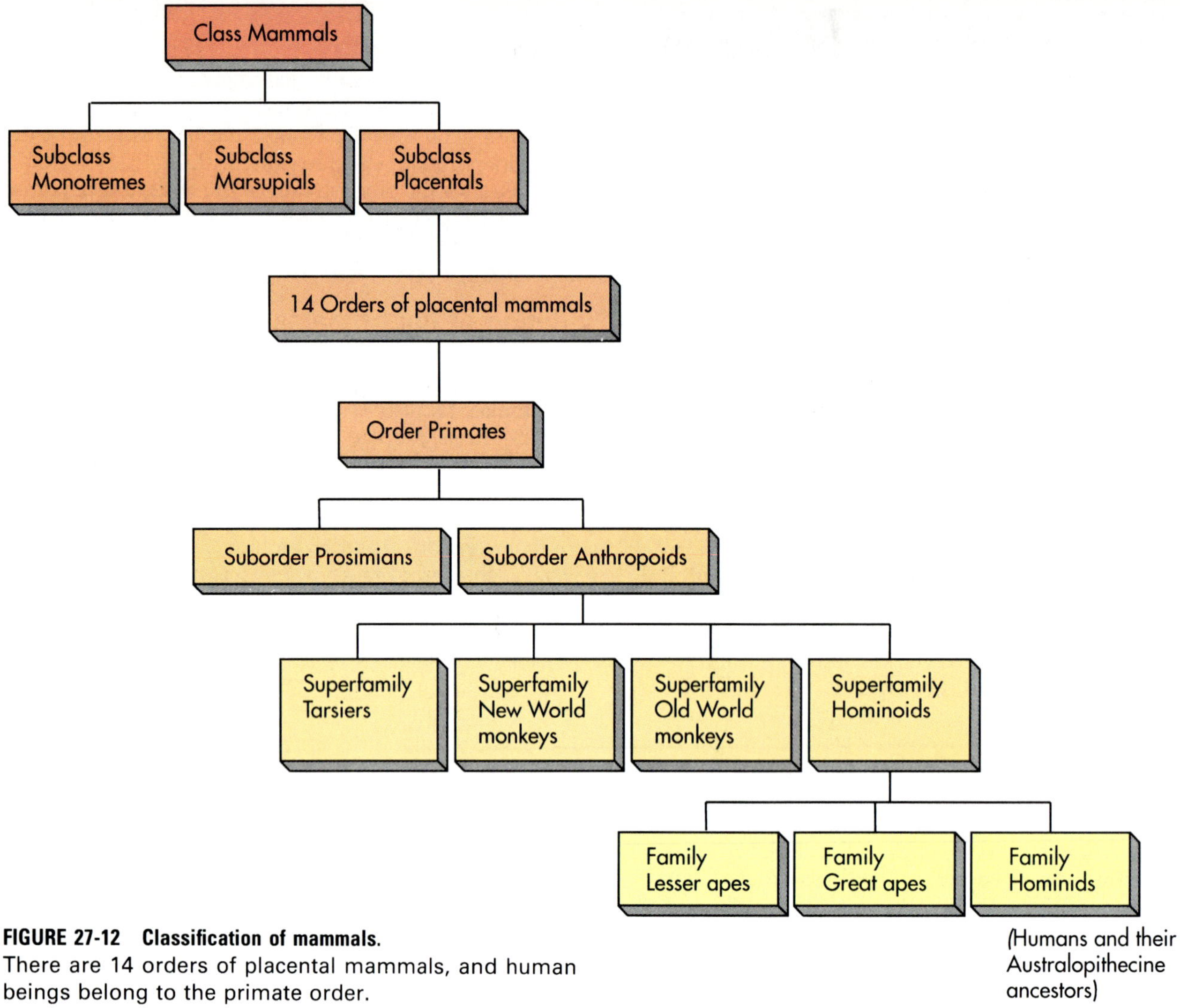

FIGURE 27-12 Classification of mammals.
There are 14 orders of placental mammals, and human beings belong to the primate order.

FIGURE 27-13 Where primates came from.
Reconstruction of an insect-eating mammal from the late Mesozoic Era, a mouse-sized animal that probably resembled the common ancestor of the primates and insectivores, two contemporary orders of mammals.

Selection pressures must have been great for these arboreal creatures. Those that survived in this habitat developed excellent **depth perception,** the ability to see in more than one plane as they moved from tree to tree. Depth perception is a function of **stereoscopic vision,** vision created by two eyes focusing on the same object. As primates evolved from their shrew-like ancestors, the eyes moved closer together from their placement on either side of the head (Figure 27-14). With this new placement, each eye can focus on the same object but from a slightly different angle. The brain puts both sets of messages together, resulting in the perception of three-dimensional form and shape. For example, eye placement that results in stereoscopic vision is apparent in predatory birds, such as owls, but is absent in nonpredatory birds, such as robins.

The primates also developed long limbs with flexible hands and feet adapted to grasping and swinging from branch to branch. Primates have two bones in the lower part of a limb that enable the wrists and ankles to rotate. In addition, the hands and feet of primates have digits that can be spread apart from one another, helping primates balance themselves when walking or running or helping

FIGURE 27-14 Evolution of depth perception.
As primates evolved, their eyes moved closer together from their placement on either side of the head. Because the eyes are closer together, each eye focuses on the same object but with a slightly different angle, resulting in greater depth perception.

them grasp objects. An **opposable thumb,** one that can touch the tip of each finger, helps in grasping. Most primates also developed flattened nails at the end of the digits, replacing the claws of their mammalian relatives.

> **The primates, an order of mammals that humans belong to, reflect an arboreal heritage. The earliest primates arose about 75 million years ago and lived on the ground, feeding on plants and insects. Over time, the primates moved into the trees, developing excellent depth perception and flexible, grasping hands.**

The prosimians

The primates are divided into two suborders: the **prosimians** and the **anthropoids** (the suborder that includes humans). The prosimians (meaning "before ape") are small animals, such as lemurs, indris, aye-ayes, and lorises; they range in size from less than a pound to approximately 14 pounds—about the size of a cat or a small dog. Most are nocturnal and have large ears and eyes to help them see and hear in the night. Prosimians have elongated snouts, reflecting their highly developed sense of smell. These characteristics are clearly seen in the two prosimians pictured in Figure 27-15, in addition to their elongated rear limbs, which help them leap from tree to tree in their tropical rain forest habitat. By the end of the Eocene Epoch (38 million years ago), prosimians were abundant in North America and Eurasia and were probably present in Africa also. However, their descendants now live only in the tropics of Asia, in tropical Africa, and on the island of Madagascar.

A

B

FIGURE 27-15 Prosimians.
A Ringtail lemur, *Lemur catta.* All living lemurs are restricted to the island of Madagascar, and all are in danger of extinction as the rain forests are destroyed.
B Tarsier, *Tarsius syrichta,* tropical Asia. Note the large eyes of the tarsier, an adaptation to nocturnal living.

A

B

FIGURE 27-16 Facial differences between an Old World monkey and a New World monkey.

A This macaque has a nose in which the nostrils are next to each other and point downward. Such noses are typical of Old World monkeys.

B This marmoset has a nose in which the nostrils face outward. This trait is characteristic of New World monkeys.

The anthropoids

The *anthropoids* (meaning "human-like") include the monkeys, apes, gorillas, chimpanzees and humans. These primates differ from the prosimians in the structure of the teeth, brain, skull, and limbs. The prosimians have pointed molars and horizontal lower front teeth. These horizontal teeth are used to comb the coat or to get at food. The anthropoids have more rounded molars and no horizontal lower front teeth. The brain of the prosimian is much smaller in relation to body size than the brain of an anthropoid. The face of an anthropoid is somewhat flat. In addition, the eyes of an anthropoid are closer together than those of a prosimian. Lastly, a prosimian's front limbs are short in relation to its long hind limbs, whereas both the front and hind limbs of an anthropoid are long. Compare these features in the prosimians (see Figure 27-15) and the anthropoids (see Figures 27-17 and 27-18).

In addition to these structural differences, prosimians and anthropoids also exhibit behavioral differences. Anthropoids are **diurnal,** that is, active during the day, whereas the prosimians are nocturnal. The anthropoids have evolved color vision, probably in relationship to their diurnal existence. Also, the anthropoids live in groups in which complex social interaction occurs. In addition, they tend to care for their young for prolonged periods.

The **hominoids** (also meaning "human-like") are one of four superfamilies of the anthropoid suborder. This superfamily includes apes and humans. The other three anthropoid superfamilies include (1) tarsiers (formerly classified with the prosimians, see Figure 27-14), (2) New World monkeys, and (3) Old World monkeys. In general, New World monkeys are found in the New World—South and Central America and they have flat noses whose nostrils face outward (Figure 27-16, *B*). Old World monkeys have noses similar to humans in which the nostrils are next to each other and point downward (Figure 27-16, *A*); these monkeys are found in the Old World—Africa, southern Asia, Japan, and Indonesia.

The hominoid superfamily is divided into three families: the lesser apes, the great apes, and the hominids, or humans. A variety of characteristics put the apes in a different superfamily from the monkeys: most apes are bigger than monkeys, have larger brains than monkeys, and lack tails.

The **lesser apes** include the gibbons, which are the smallest hominoids and closest in size to monkeys. They are about the size of a small dog, weighing 4 to 8 kilograms (9 to 18 pounds). As all monkeys and apes do, they live in tropical rain forests. Here, they leap and swing from tree to tree with their long arms.

The **great apes** include the orangutans, gorillas, and chimpanzees. These apes are much larger than the gibbons (Figure 27-17). The orangutans are about the size of humans, weighing between 50 and 100 kilograms (110 to 220 pounds). These apes exhibit sexual dimorphism; the females weigh about half that of the males. Like the gibbons, the orangutans have long arms, but they walk—they do not swing—from branch to branch of the rain forest trees. The

FIGURE 27-17 The apes.
A Gorilla, *Gorilla gorilla.*
B Mueller gibbon, *Hylobates muelleri.*
C Chimpanzee, *Pan troglodytes.*
D Orangutan, *Pongo pygmaeus.*

gorillas are the largest of the apes; they weigh about 160 kilograms, or 350 pounds. The females are smaller, ranging in weight from 165 to 240 pounds. The gorillas spend most of their time on the ground. When alarmed, male gorillas beat on their chests in a behavioral display. However, they are not the fierce animals humans think they are. In fact, gorillas are quite peaceable. The smallest of the great apes are the chimpanzees—humans' closest relatives. These animals weigh between 40 and 50 kilograms, or 90 to 110 pounds; females are slightly lighter. Like the other apes (except the gorillas), they spend their time in the trees.

The **hominids,** the family that includes humans of today, are the most intelligent of the hominoids. They are distinguished from the other families of hominoids in that they walk upright on two legs; they are said to be **bipedal.** In addition, hominids communicate by language and exhibit culture—a way of life that is passed on from one generation to another. The only living hominids are humans—***Homo sapiens.***

Evolution of the anthropoids

Scientists have found jaw fragments that suggest that the ancestors of monkeys, apes, and humans began their evolution approximately 50 million years ago. The tarsioids may have begun their evolution even earlier—about 54 million years ago. Scientists do not know what adaptive pressures resulted in the appearance of these early anthropoid primates, nor do they know which prosimian was their ancestor.

Although the only fossil evidence of the emergence of the anthropoids consists of pieces of jaw, biochemical studies complement the knowledge gained from the fossil record. Together, these techniques tell scientists a great deal about the evolution of the anthropoids and the relationships among humans, apes, and monkeys. The basis for biochemical studies lies in the concept that as organisms change over time, their genetic material, or DNA, also changes. Each evolutionary change involves the formation of "new" genes from "old" by mutation; favorable new genes endure because of natural selection. Thus evolutionary change involves a progressive accumulation of genetic changes between organisms and their ancestors. Using biochemical techniques, scientists can determine the relatedness of species that have evolved from a common ancestor by comparing their DNA molecules. Scientists compare DNA indirectly by determining the similarity between comparable protein molecules or directly by measuring the degree of "mismatch" between two comparable stretches of DNA. Organisms that are more distantly related will have accumulated a greater number of genetic differences than organisms that more recently evolved from a common ancestor. In this way, scientists can estimate the time since two species diverged from a single ancestor, thereby determining their evolutionary relatedness. Relatedness can then be depicted in a **phylogenetic tree,** or family tree.

The phylogenetic tree of the anthropoids is shown in Figure 27-18. Fossils and biochemical studies show that the New World monkeys branched from the line leading to the Old World monkeys and the hominoids about 45 million years ago, in the mid-Eocene Epoch. In addition, scientists have discovered many fossils in North Africa that date from 25 to 30 million years ago, in the Oligocene Epoch. One of the earliest of these fossils, which scientists have named *Parapithecus,* probably led to the line of Old World monkeys,

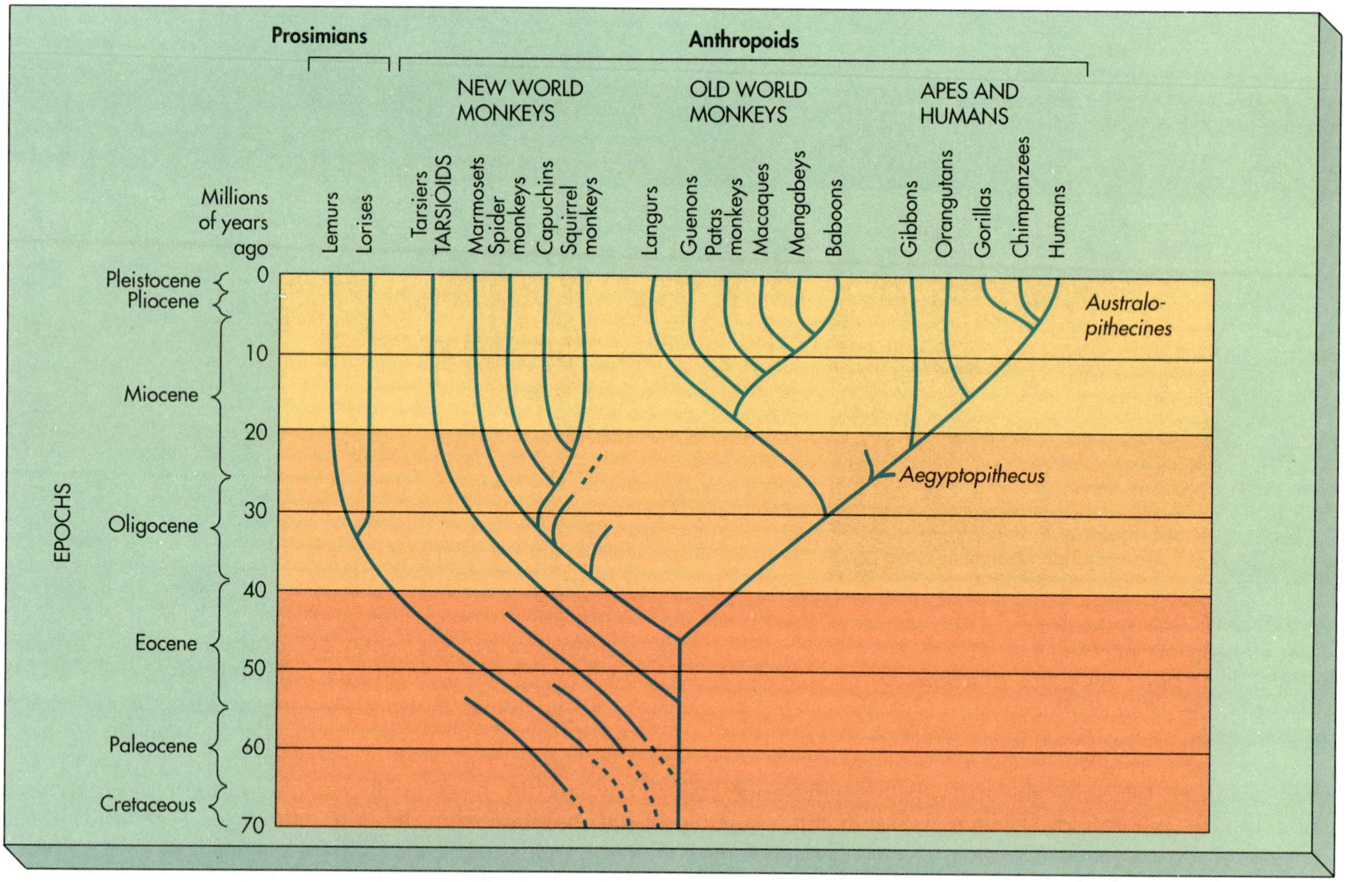

FIGURE 27-18 The primate family tree.
This tree is based on DNA comparisons between living species. The dating of the branches is based on the ages of fossils.

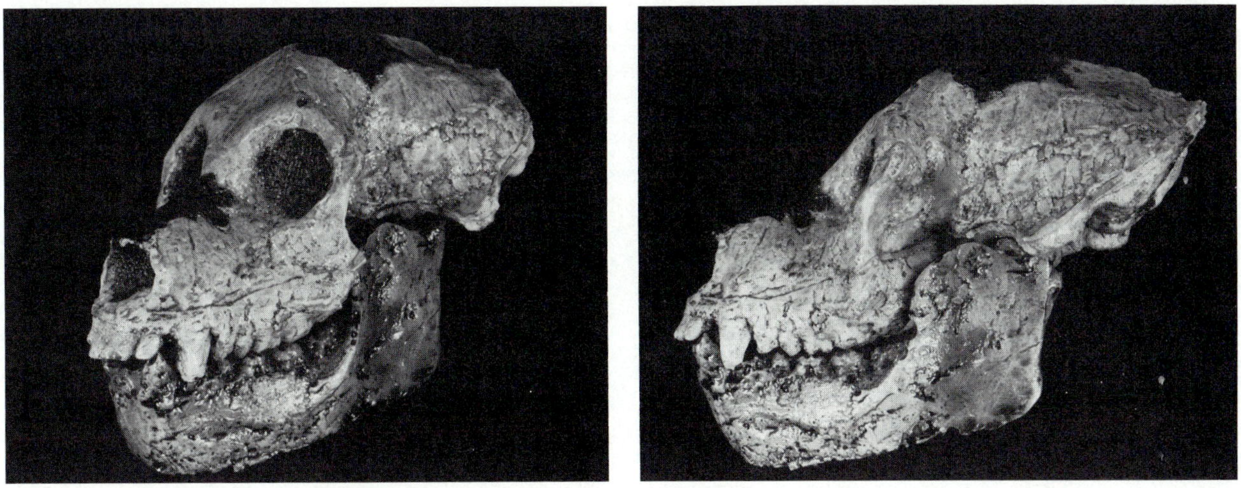

FIGURE 27-19 ***Aegyptopithecus zeuxis.***
This primate fossil dates to the late Oligocene Era and is thought to be the ancestor of the hominoids that lived in Africa during the early Miocene Era.

splitting from the hominoids approximately 30 million years ago. The others, scientists think, are probably members of the evolutionary line that leads to the hominoids.

The most well-known of the Oligocene fossils found in North Africa is called *Aegyptopithecus.* This fossil is from the late Oligocene Epoch and is thought to be the ancestor to the early Miocene hominoids of Africa. Looking at the partially restored skull in Figure 27-19, you can see that *Aegyptopithecus* had some prosimian characteristics. It had a pronounced snout, leading scientists to believe that its sense of smell was still highly specialized, much like that of the prosimians. In addition, like the prosimians, *Aegyptopithecus* lived singly rather than in social groups.

Fossil evidence and biochemical studies show that the anthropoids, a suborder of primates that includes humans, monkeys, and apes, first appeared about 50 million years ago. The New World monkeys branched from the evolutionary line leading to the Old World monkeys and the hominoids about 45 million years ago. The Old World monkeys split from the hominoids approximately 30 million years ago.

Fossil evidence is still being accumulated that will help tell the story of hominoid evolution. From mid-Miocene times on, the hominoid fossil record is quite extensive but consists primarily of skull and teeth fragments. Investigators have calculated that the evolutionary line leading to gibbons diverged from the line leading to the other apes about 18 to 22 million years ago, the line leading to orangutans split off roughly 13 to 16 million years ago, the line leading to gorillas diverged 8 to 10 million years ago, and the split between hominids and chimps occurred approximately 5 to 8 million years ago. This last statement suggests that chimpanzees and gorillas are humans' closest relatives, with a common relative alive 5 to 8 million years ago.

Evolution of the hominids

The first hominids: *Australopithecus*

The two critical steps in the evolution of humans were the evolution of bipedalism (walking on two feet) and the enlargement of the brain. For many years, scientists have hypothesized how and when bipedalism arose. The current theory is that bipedalism arose as a *preadaptation* because of the way humans' arboreal, ape-like ancestors moved through the trees. In other words, the skeletons and muscles of human ancestors were structured in a way that these hominoids were able to walk bipedally even though they lived in the trees. Interestingly, these structural adaptations developed as a part of their arboreal life and the types of locomotion they exhibited in the trees. Figure 27-20 shows the movements of gibbons and other related brachiators—those hominoids that move through trees by hanging from the branches with their arms, "walking" themselves along. As the weather in Africa cooled somewhat and became seasonal, patches of savannah grasslands began to invade former areas of tropical forest. The hominoids best able to walk efficiently on the ground survived, and bipedalism evolved.

Although hominids may have first appeared as long ago as 5 million years, the oldest undisputed evidence of the hominids is 3.6 to 3.8 million years old. This find was made relatively recently, in 1976, by Mary D. Leakey and an international team of scientists. The team discovered the fossil footprints shown on p. 463. Anthropologists think that the individuals who made the footprints are or are very closely related to human ancestors. Fossil bones discovered a few years earlier, at a site in Ethiopia about 2000 kilometers (1250 miles) north of the footprints, support this hypothesis. At 3.5 million years old, these hominid bones are somewhat younger than the footprints but are the oldest hominid bones ever found. These fossils were named ***Australopithe-***

FIGURE 27-20 Brachiation and bipedalism.
Brachiators, such as gibbons and siamangs, locomote by hanging from branches with their arms and reaching from hold to hold. Brachiation has preadapted these animals to bipedal walking, which they do on broad branches (as the siamang in the illustration is doing) or on the ground.

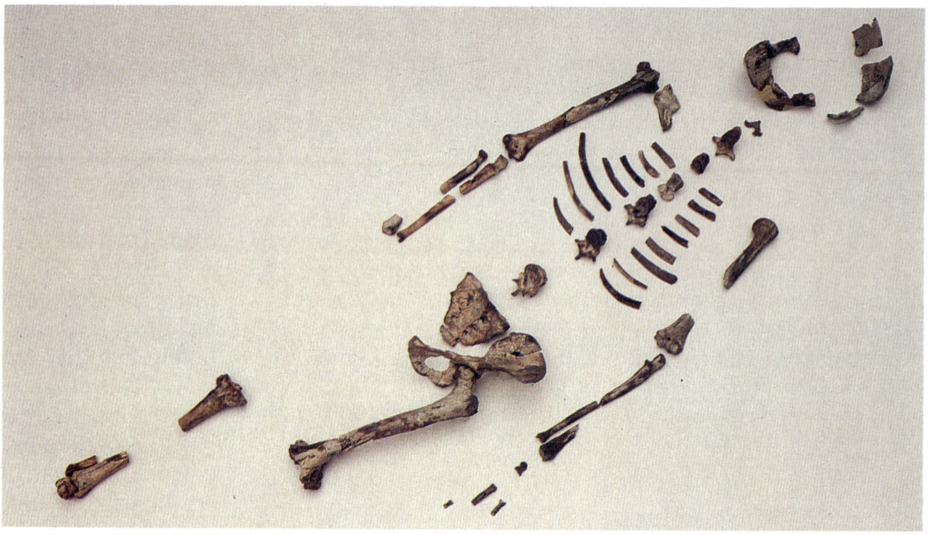

FIGURE 27-21 Lucy, from Ethiopia, is the most complete skeleton of *Australopithecus* discovered so far.
The reconstruction was made by a careful study of muscle attachments to the skull.

cus afarensis, meaning "southern ape of Afar," by Donald C. Johanson and an international team of scientists who found them in the Afar region of Ethiopia. In 1974, this team also found and pieced together one of the most complete fossil skeletons of *A. afarensis,* which has since become famous (Figure 27-21). This "first" hominid was named Lucy after the Beatles song "Lucy in the Sky with Diamonds," which was popular at that time.

Although the australopithecines are considered to be the first hominids, they are not humans (members of the genus *Homo*). Their brains were still small in comparison to present-day human brains, and they had long, monkey-like arms. In addition, their faces were ape-like as shown in the photo of a reconstruction of Lucy. Lucy probably weighed about 25 kilograms (55 to 60 pounds) and was about 1 meter tall (slightly over 3 feet). Males were larger than females, however; they stood up to 1.2 meters tall (4 feet) and weighed as much as 45 kilograms (100 pounds).

A. afarensis evolved into two (and possibly more) lineages, including the species *A. africanus, A. robustus,* and *A. boisei*. These australopithecines lived on the ground in the open savannah of eastern and southern Africa. Their diets consisted primarily of plants. At night they probably slept in the few trees that existed in these grasslands, much like the savannah baboons do today to protect themselves from predators.

The oldest evidence of the first appearance of the hominids is 3.6 to 3.8 million year old footprints. The oldest hominid fossil skeleton is 3.5 million years old and is not classified as human. This hominid is of the species *Australopithecus afarensis* and is nicknamed Lucy.

The phylogenetic relationships among these australopithecines is not clear; a family tree that is widely accepted at this time is shown in Figure 27-22. Notice from the diagram that no australopithecines are alive today; the last ones disappeared about 1 million years ago.

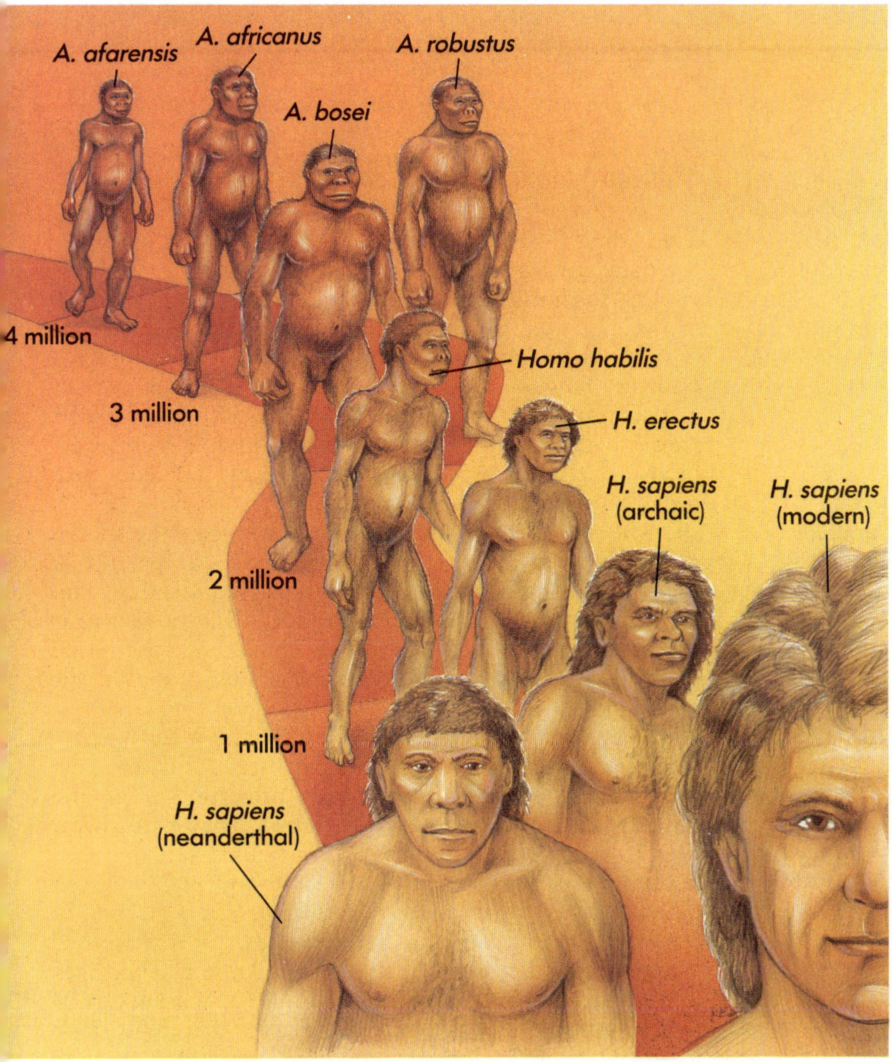

FIGURE 27-22 The path of human evolution. *Australopithecus robustus* and *A. boisei* seem to represent evolutionary dead ends, with no living descendants. Whether *A. africanus* was ancestral to *Homo habilis* or to *A. robustus* is currently in dispute, as is whether Neanderthals are the ancestors of modern humans or a parallel line that did not survive.

The first humans: *Homo habilis*

Climatic changes during the Pleistocene Epoch (2 million to 10,000 years ago) may have contributed to the disappearance of the australopithecines and the evolution of a new, more intelligent genus of hominids: the human (genus *Homo*). During the Pleistocene Epoch, the earth cooled, repeatedly undergoing **ice ages** during which vast regions of the earth (from the poles to a latitude of about 40 degrees) were covered by massive glaciers. For example, ice covered the land from the north pole southward over Canada and parts of the northern United States. Organisms such as the australopithecines had a low level of intelligence and no innate cold weather survival behaviors, and they died. Apparently, however, a species of hominid more intelligent than the australopithecines, was evolving from (most likely) *A. afarensis*. These hominids protected themselves from the cold by building shelters and wearing clothes. They are thought to have been the first humans and were given a name to emphasize their intelligence—***Homo habilis,*** or "skillful human." Scientists have fossil bones and tools of *H. habilis* dating back about 2 million years. The first *H. habilis* fossils were discovered in 1964 by Louis Leakey, Philip Tobias, and John Napier in the Olduvai Gorge of eastern Africa.

Judging from the structure of the hands, *H. habilis* regularly climbed trees as did their australopithecine ancestors, although the *H. habilis* people spent much of their time on the ground and walked erect on two legs. Skeletons found in 1987 indicate that the *H. habilis* people were small in stature like the australopithecines, but fossil skulls and teeth reveal that the diet of *H. habilis* was more diverse than that of the australopithecines, including meat as well as plants. The tools found with *H. habilis* were made from stones fashioned into implements for chopping, cutting, and pounding food. The use of stone tools by these early humans marks the beginning of the **Stone Age,** a time which spans approximately 2 million to 35,000 years ago.

The first human fossils date back about 2 million years and are of the extinct species *H. habilis,* meaning "skillful human." This species is considered human because they exhibited an intelligence far greater than their ancestors by making tools and clothing.

Human evolution continues: *Homo erectus*

All of the early evolution of the genus *Homo* seems to have taken place in Africa. There, fossils belonging to the second, also extinct species of *Homo*—***Homo erectus***—are widespread and abundant from 1.6 million to about 300,000 years ago. By 1 million years ago, however, *H. erectus* had migrated into Asia and Europe.

The *H. erectus* people were about the size of modern-day humans, were fully adapted to upright walking, and had brains that were roughly twice as large as those of their ancestors. However, they still retained prominent brow ridges, rounded jaws, and large teeth. The tools of *H. erectus* were much more sophisticated than those of *H. habilis* and were used for hunting, skinning, and butchering animals. These people were hunter gatherers, which means that they collected plants, small animals, and insects for food, while occasionally hunting large mammals. Researchers have found the first evidence of the use of fire by humans at 1.4 million-year-old campsites of *H. erectus* in the Rift Valley of Kenya, Africa. Fire is characteristically associated with populations of this species from that time onward. All of these activities: tool making, hunting, using fire, and building shelters are signs of **culture,** or a way of life that depends on intelligence and the ability to communicate knowledge of the culture to succeeding generations. Inherent in the concept of culture, then, is the concept of language. The ability to communicate enhances the ability of a species to survive, especially in harsh conditions, such as an ice age, by sharing survival tactics and warning each other of danger. Anthropologists think that the development of language was probably one of the most important factors in the appearance of *Homo sapiens.*

Fossils of a second extinct species of human, *H. erectus,* date back to 1.6 million years ago. These humans made sophisticated tools, built shelters, used fire, and probably communicated with language.

Modern humans: *Homo sapiens*

The earliest fossils of *Homo sapiens* (meaning "wise humans") are dated to be about 200,000 years old and most likely evolved from the *H. erectus* species in Africa. The oldest *H. sapiens,* however, are not considered to be anatomically modern, that is, to have the same anatomical features of today's humans. This species is therefore referred to as an early or archaic form (see Figure 27-22). In general, these early *H. sapiens* had larger brains, flatter heads, more sloping foreheads, and more protruding brow ridges and faces than today's humans.

The fossil record shows gradual change of the species *H. sapiens,* with the early form evolving over a 75,000 year span to a subspecies of *H. sapiens* called **Neanderthal.** This subspecies was named after the Neander Valley in Germany where their fossils were first found. The Neanderthals lived from about 125,000 to 35,000 years ago in Europe and the Middle East.

Compared with modern humans, the Neanderthal people were powerfully built, short, and stocky. Their skulls were massive, with protruding faces, projecting noses, and rather heavy bony ridges over the brows. Their brains were even larger than those of modern humans, a fact that may have been related to their heavy, large bodies. The Neanderthals made diverse tools, including scrapers, borers, spearheads, and hand axes. Some of these tools were used for scraping hides, which they used for clothing. They lived in hutlike structures or in caves. Neanderthals took care of their injured and sick and commonly buried their dead, often placing food and weapons and perhaps even flowers with the bodies. Such attention to the dead strongly suggests that they believed in a life after death. For the first time, the

FIGURE 27-23 Cave painting.
Cave paintings, almost always showing animals and sometimes hunters, were made by Cro-Magnon people, our immediate ancestors. These paintings are found primarily in Europe and were made for about 20,000 years, until 8000 to 10,000 years ago.

kinds of thought processes characteristic of modern *H. sapiens,* including symbolic thought, are evident in these acts.

Approximately 10,000 years before the Neanderthal subspecies died out, modern *H. sapiens* made their appearance. This "modern" subspecies (our subspecies) is called ***Homo sapiens sapiens.*** The early members of this subspecies are the Cro-Magnons and are named after a cave in southwestern France where scientists found some of their fossils.

The **Cro-Magnons** had a stocky build, much like the Neanderthals, but their heads, brow ridges, teeth, jaws, and faces were much smaller than the Neanderthals and were more similar to today's humans. However, just as modern humans show variation among races, so, too, did the Cro-Magnons. The Cro-Magnons used sophisticated tools that were made not only from stone but from bone, ivory, and antler—materials that were not used by earlier peoples. Hunting was an important activity for the Cro-Magnons, evidenced by the abundance of animal bones found with human bones and elaborate cave paintings of animals and hunt scenes (Figure 27-23). The paintings appear to have been part of a ritual to ensure the success of the hunt.

The subspecies of *H. sapiens* that preceded us showed a gradual development of culture and society, which was the foundation for the development of "modern" culture and society. About 10,000 years ago, the last ice age came to a close, the global climate began to warm, and various groups of *H. sapiens sapiens* began to cultivate crops and breed animals for food. Archeologists have uncovered the remains of small, ancient cities, such as those of Jericho shown in Figure 27-24, which give evidence that by 9000 years ago humans had developed complex social structures. By 5000 years ago the first large cities and "great civilizations" appeared, such as those in Egypt (3100-1090 BC) and Mesopotamia (3100-1200 BC), and the final break occurred with the hunter-gatherer way of life.

The rest of the story of human development is one of history and is best left to the historians, anthropologists, and sociologists to tell. From a biologist's perspective, however, the spread of modern humans throughout the world has a dark side. As people inhabited new lands and increased their numbers, they destroyed many populations of animals and plants. Today, the growing human population is hastening global pollution, destruction, and devastation. Although humans have made incredible progress in the ability to survive, they still need to make progress in living in harmony with the rest of the living world and its resources.

FIGURE 27-24 The beginnings of civilization.
This photo shows the remains of a house in Jericho, which dates to about 7000 BC. The ruins of Jericho also contain the remnants of city walls and towers, demonstrating that by about 9000 years ago, many of our ancestors had moved away from the hunting-gathering life-style into an agricultural life-style.

Summary

1. The story of human evolution can be traced back to the appearance of the first chordates about 500 million years ago. Chordates are characterized by a single, dorsal, hollow nerve cord; a flexible rod, the notochord, which forms between the nerve cord and the developing gut in the early embryo; and pharyngeal (gill) arches located at the throat.
2. The vertebrates evolved from the chordates. Today, most chordates are vertebrates. Vertebrates differ from other chordates in that they usually possess a vertebral column, a distinct and well-differentiated head, and a bony skeleton.
3. The first vertebrates to evolve were the jawless fishes. Once abundant and diverse, they are represented among the living vertebrates only by the lampreys and the hagfishes. Jaws first developed among vertebrates that lived about 410 million years ago. Jawed fishes make up about half of the species of vertebrates and are dominant in fresh waters and saltwaters everywhere. The two classes of jawed fishes are the cartilaginous fishes, such as the sharks, rays, and skates, and the bony fishes, such as salmon, cod and tuna.
4. One ancient group of bony fishes, the lobe-finned fishes, is believed to be the ancestors of the first land vertebrates, the amphibians. Amphibians depend on water and lay their eggs in moist places. The amphibians arose over 300 million years ago. They gave rise to the reptiles.
5. The reptiles were the first vertebrates that were fully adapted to life on land. Amniotic eggs, which evolved in this group but are also characteristic of the birds and the few egg-laying mammals, represent a significant adaptation to the dry conditions that are widespread on land. The birds and mammals evolved from the reptiles.
6. Mammals, vertebrates that have hair and whose females secrete milk from mammary glands to feed their young, evolved from the reptiles approximately 200 million years ago. Mammals are warm blooded, or able to maintain a constant internal body temperature. Other than a few organisms such as the birds, all other living animals are cold blooded; their body temperatures vary with the temperature of the environment. Changes in the structure and placement of the mammalian limbs, the structure of the heart, and reproductive strategies resulted in mammals' ability to survive in a variety of climates and to eventually become abundant.
7. Primates, one of the 14 orders of mammals, first appeared 75 million years ago. Primates have large brains in proportion to their bodies, binocular vision, and five digits, including an opposable thumb; they exhibit complex social interactions.
8. The primates are divided into two suborders: the prosimians (small animals such as lemurs, indris, and aye-ayes) and the anthropoids (a group that includes monkeys, apes, and humans).
9. The hominoids (meaning "human-like") are one of four superfamilies of the anthropoid suborder. This superfamily includes apes and humans. The other three anthropoid superfamilies include tarsiers, New World monkeys, and Old World monkeys. The New World monkeys branched from the evolutionary line leading to the Old World monkeys and the hominoids about 45 million years ago. The Old World monkeys split from the hominoids approximately 30 million years ago. Ancestors to the apes gave rise to the gibbons, orangutans, chimpanzees, gorillas, and hominids.
10. The two critical steps in the evolution of humans were the evolution of bipedalism (walking on two feet) and the enlargement of the brain. The earliest hominids and the direct ancestors of humans belong to the genus *Australopithecus.* They appeared in Africa about 5 million years ago. They were small hominids, standing 3 to 4 feet tall and weighing from 50 to 100 pounds.
11. The genus *Australopithecus* gave rise to humans belonging to the genus *Homo.* The first species of this genus, *H. habilis,* appeared in Africa about 2 million years ago. Now extinct, the people of this species are considered human because they exhibited an intelligence far greater than their ancestors by making tools and clothing.
12. The second species of *Homo, H. erectus,* appeared in Africa approximately 1.6 million years ago. These people used fire, built shelters, fashioned sophisticated tools, and exhibited culture. *H. sapiens* probably evolved from *H. erectus* about 200,000 years ago.

REVIEW QUESTIONS

1. Describe the three features that distinguish chordates from other organisms.
2. Distinguish between chordates and vertebrates.
3. Fill in the blanks: The first land vertebrates were ______, organisms, which can live both on land and in water. They were gradually replaced as dominant organisms by the ______, animals with more efficient lungs and water-resistant skins.
4. Choose the correct answer: Birds evolved from: a) winged insects; b) flying mammals; c) reptiles; d) amphibians with webbed feet.
5. Fill in the blanks: Mammals are ______ -blooded vertebrates with body ______. Female mammals secrete ______ from ______ ______ to feed their young.
6. What is warm bloodedness, and why is it important?
7. What do monotremes, placental mammals, and marsupials have in common? How do they differ?
8. What are primates? What two characteristics have helped them to be successful?
9. Distinguish between the prosimians and the anthropoids. Give an example of each.
10. Match each family with the appropriate animal(s):

1. Hominids	a. Gibbon
2. Hominoids	b. Chimpanzee
3. Lesser apes	c. You, the chimp, and the gibbon
4. Greater apes	d. You

11. Which statement is true? (Or are they both false)? Explain your answer:
    a. All hominids are hominoids.
    b. All hominoids are hominids.
12. Explain the significance of each of the following: *Australopithecus afarensis*, *Homo habilis*, and *Homo erectus*. Which is most like you?
13. "Neanderthal" is sometimes used as a derogatory term, meaning someone who is brutish and unsophisticated. Is this accurate? Explain your answer.
14. How did *Homo sapiens sapiens* differ from earlier *Homo* species?

KEY TERMS

amniotic egg 468
amphibian 466
anthropoid 473
Australopithecus afarensis 477
bipedal 475
convergent evolution 471
Cro-Magnon 481
culture 480
depth perception 472
diurnal 474
ectotherm 471
endotherm 471
great ape 474
hominid 475
hominoid 474
Homo erectus 480
Homo habilis 480
Homo sapiens 475
Homo sapiens sapiens 481
ice age 480
lesser ape 474
mammal 468
marsupial 471
monotreme 471
Neanderthal 480
nerve cord 464
nocturnal 468
notochord 464
opposable thumb 473
pharyngeal (gill) arches 464
phylogenetic tree 476
placental mammal 471
primate 471
prosimian 473
stereoscopic vision 472
Stone Age 480
tetrapod 466
vertebral column 465

THOUGHT QUESTIONS

1. If the oldest fossils of *Homo sapiens* are from Africa, then *Homo* must have migrated out of Africa twice, first as *Homo erectus* and later as *H. sapiens*. *H. erectus* survived in Africa until 500,000 years ago, in Europe until 400,000 years ago, and in Asia until about 250,000 years ago. Do you think *H. erectus* was driven to extinction by *H. sapiens*? Discuss.

FOR FURTHER READING

Thorne, A. G., & Wolpoff, M. (1992, April). The multiregional evolution of humans. *Scientific American*, pp. 66-83.

Wilson, A. C., & Cann, R. (1992, April). The recent African genesis of humans. *Scientific American*, pp. 66-83.

These two articles outline polar positions on the origins of humans: are humans descended from a single African ancestor, as postulated by researchers investigating DNA, or are modern humans related to a number of interconnected lineages scattered throughout the world, as suggested by fossil remains?

Walker, A., & Teaford, M. (1989, January). The hunt for proconsul. *Scientific American*, pp. 76-82.

This is an exciting account of the last common ancestor of great apes and human beings.

PART SEVEN

THE DIVERSITY AND UNITY OF LIVING THINGS

28 Viruses, bacteria, and genetic engineering

HIGHLIGHTS

Viruses are noncellular infectious agents capable of causing certain diseases and cancers.

Bacteria have a prokaryotic cell structure, occur in many habitats, play key ecological roles, are important in producing food, and can cause disease.

Some bacteria can transfer genetic material, naturally recombining genes.

Gene transfer is used to improve the characteristics of plants and animals and to produce medically important products. Scientists are beginning to treat genetic disorders in humans.

OUTLINE

Viruses
- The discovery of viruses
- The structure of viruses
- Viral replication
  - The lytic cycle
  - The lysogenic cycle
- Viruses and cancer
- The classification of viruses

Bacteria
- Bacterial reproduction
- Bacterial diversity and classification
  - Archaebacteria
  - Eubacteria
- Bacteria as disease producers

Genetic engineering
- Natural gene transfer among bacteria
- Human-engineered gene transfer among bacteria

Applications of recombinant DNA technology
- Medically important products
- Gene therapy
- Agriculture

The statistics are astounding! A recent study revealed that nearly one out of four college women tested in Seattle, Washington are harboring a sexually transmitted virus that is strongly associated with cancer of the cervix. Computer color-enhanced, these deadly virus particles (formally known as *human papillomavirus 16*) are shown in the photo, as the green spheres speckled with red. The human cell they are destroying is yellow.

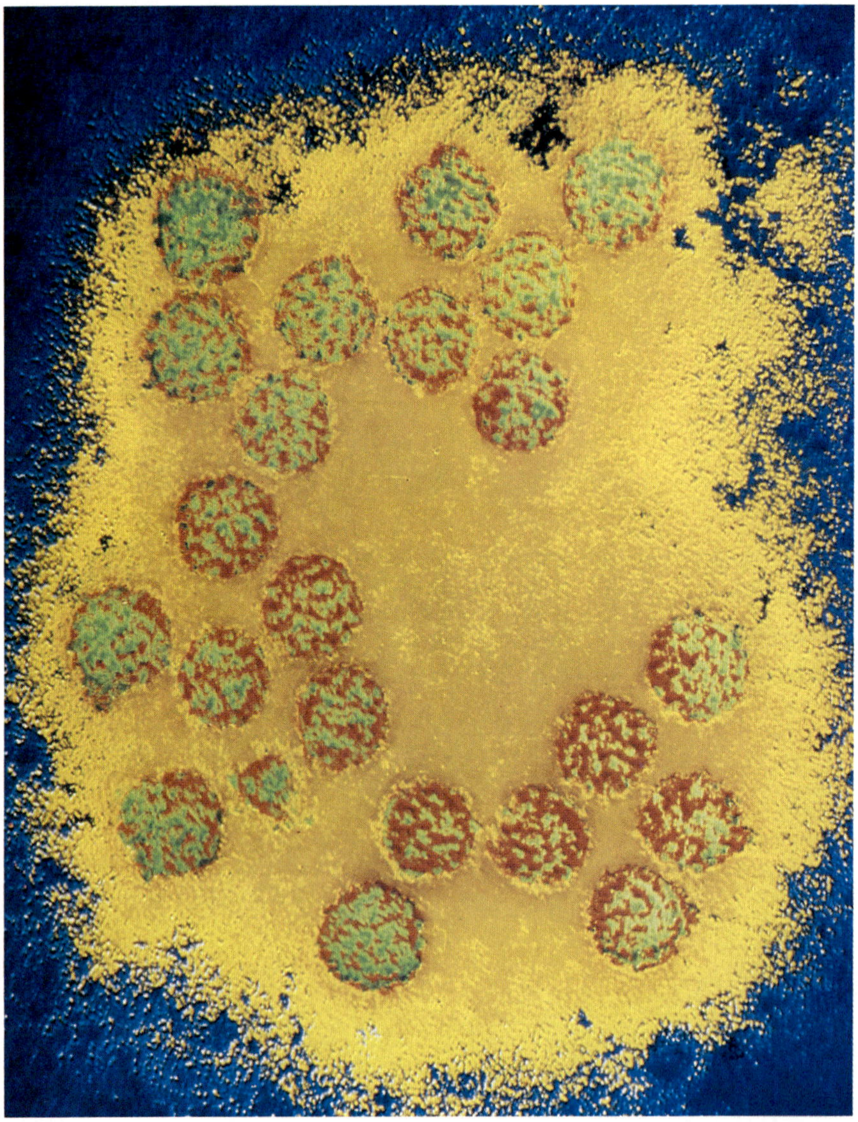

The human papillomavirus (HPV) causes a sexually transmitted disease commonly known as *genital warts.* HPV comes in 60 different varieties. Infection with any one of the 60 types can lead to an outbreak of warts, although some are more likely to cause warts than others. Ironically, the types that most often result in warts are the least dangerous. HPV types such as HPV 16 rarely cause warts on infection but are those most likely to be associated with cancer and the least likely to be detected.

Two other viral diseases transmitted by sexual contact are genital herpes and AIDS. Unlike sexually transmitted diseases (STDs) spread by bacteria, viral STDs cannot be cured. The drugs available to treat these diseases only help ease the symptoms but cannot destroy the viruses. Why are these diseases different in that regard? What makes a virus so hard to kill?

Viruses

Viruses are **infectious agents.** They enter living organisms and cause disease. But although viruses invade living things and cause cells to make more viruses, the viruses themselves are not living! They do not have a cellular structure, which is the basis of all life. They are nonliving **obligate parasites,** which means that viruses cannot reproduce outside of a living system. They must exist in association with and at the expense of other organisms. Unfortunately, that "other organism" may be you!

The discovery of viruses

At the end of the nineteenth century, several groups of European scientists working independently first realized that viruses existed. As they filtered fluids derived from plants with tobacco mosaic disease and cattle with hoof-and-mouth disease, the scientists discovered that the infectious agents passed right through the fine-pored filters they used, which were designed to hold back bacteria. They concluded that the infectious agents associated with these diseases were *not* bacteria—they were too small. As they studied the filtrate containing these mysterious agents, the scientists also discovered that the disease-causing agents could multiply only within living cells. These infection agents, they hypothesized, must lack some of the critical "machinery" cells use to reproduce.

For many years after their discovery, viruses were regarded as very primitive forms of life, perhaps the ancestors of bacteria. Today, scientists know that this view is incorrect—viruses are not living organisms. The true nature of viruses became evident in the 1930s after the ground-breaking work of an American scientist, Wendell Stanley. Stanley prepared an extract of tobacco mosaic virus (TMV), purified it, and studied its chemical composition. His conclusion: TMV was a protein—and he was partially right. Scientists later discovered that TMV also contains ribonucleic acid, or RNA. In the late 1930s, with the development of the electron microscope, scientists were able to see the virus that Stanley purified (Figure 28-1, *A*).

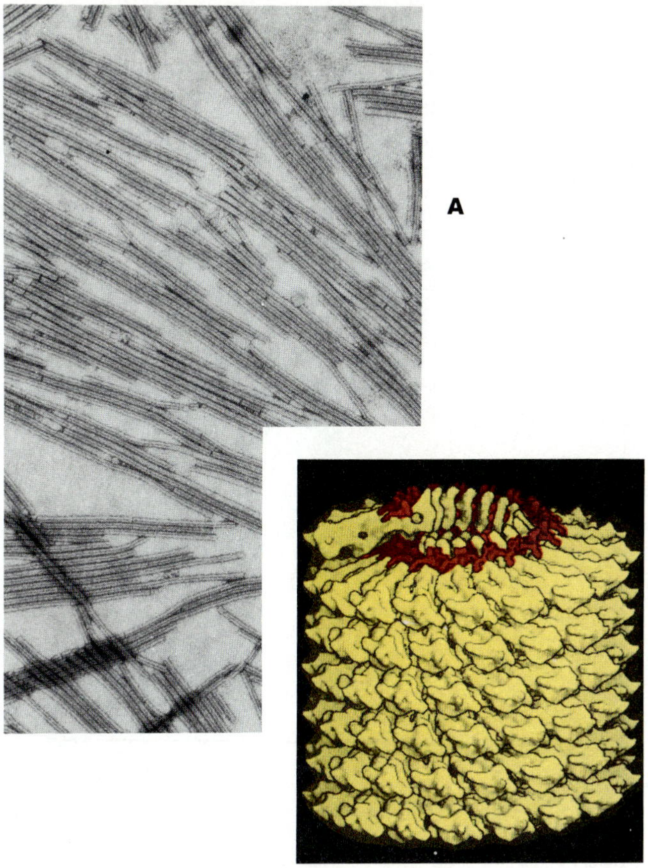

FIGURE 28-1 Tobacco mosaic virus.
A An electron micrograph of purified tobacco mosaic virus.
B Computer-generated model of a portion of tobacco mosaic virus. An entire virus consists of 2130 identical protein molecules—the yellow knobs—which form a cylindrical coat around a single strand of RNA (colored *red*).

The structure of viruses

Viruses primarily infect plants, animals, and bacteria. A specific virus can only infect a certain species of organisms. So you cannot be infected by a bacterial virus, nor can your dog catch your cold. Some viruses, however, can infect more than a single species of organism; for example, the human immunodeficiency virus (HIV virus that causes acquired immunodeficiency syndrome [AIDS]) is thought to have been introduced to humans from African monkeys.

Each virus has its own unique shape (Figure 28-2), but all contain the same basic parts: a nucleic acid **core** (either DNA or RNA) and a protein "overcoat" called a **capsid.** The structure of the TMV is shown in Figure 28-1, *B* and illustrates one way that a virus is put together. This virus is helical, with its single strand of RNA coiled like a spring, surrounded by a spiraling capsid of protein molecules. Many viruses have another chemical layer over the capsid called the **envelope,** which is rich in proteins, lipids, and carbohydrate molecules. Figure 28-3, *A* is an electron micrograph of a typical enveloped virus, the causative agent of herpes. Its structure is illustrated in Figure 28-3, *B*.

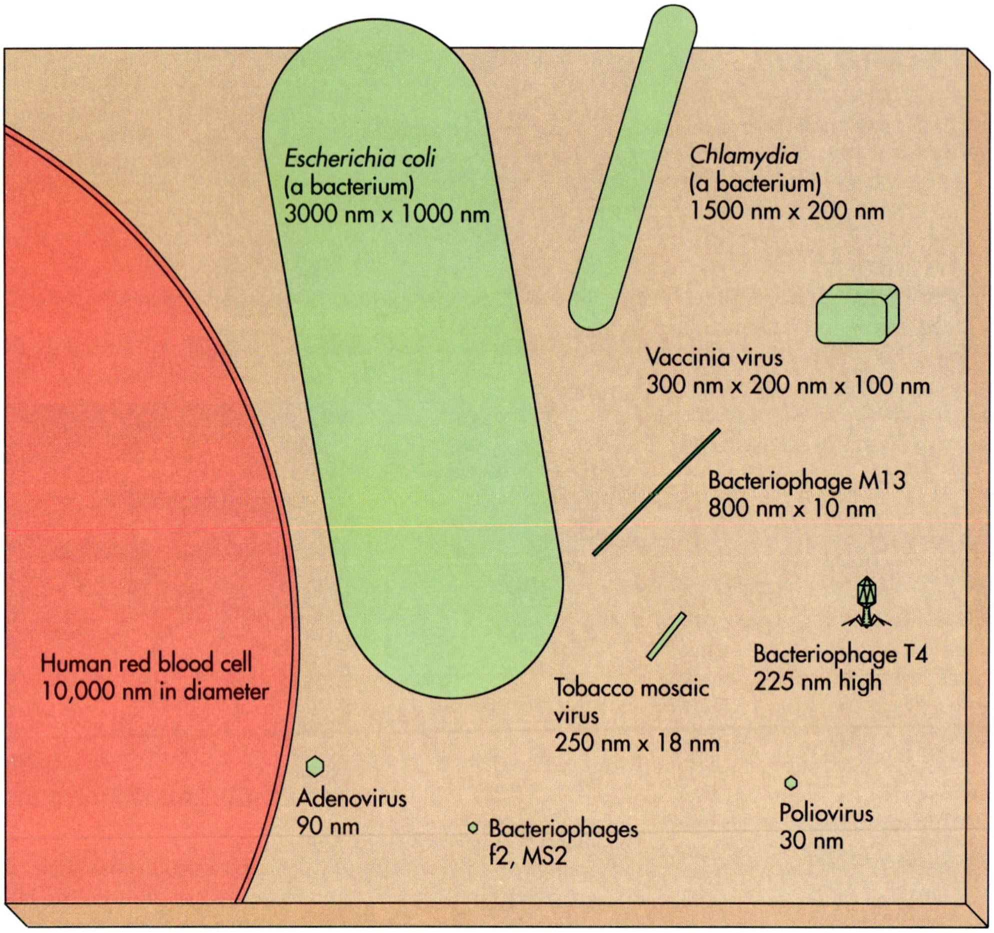

FIGURE 28-2 The shapes and sizes of viruses.
The size of various viruses are shown here in relation to a bacterium and a human red blood cell. Dimensions are given in nanometers.

A

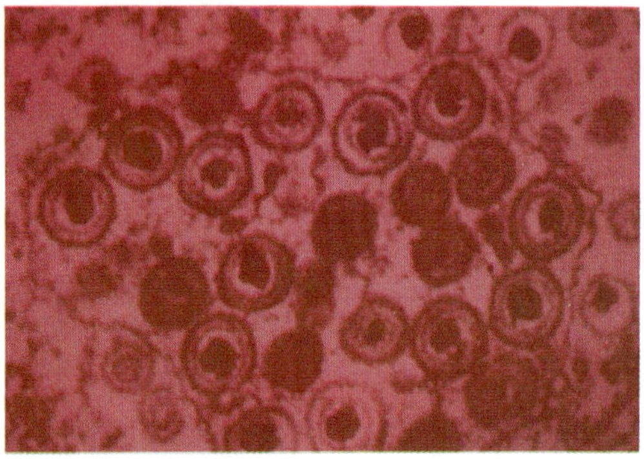

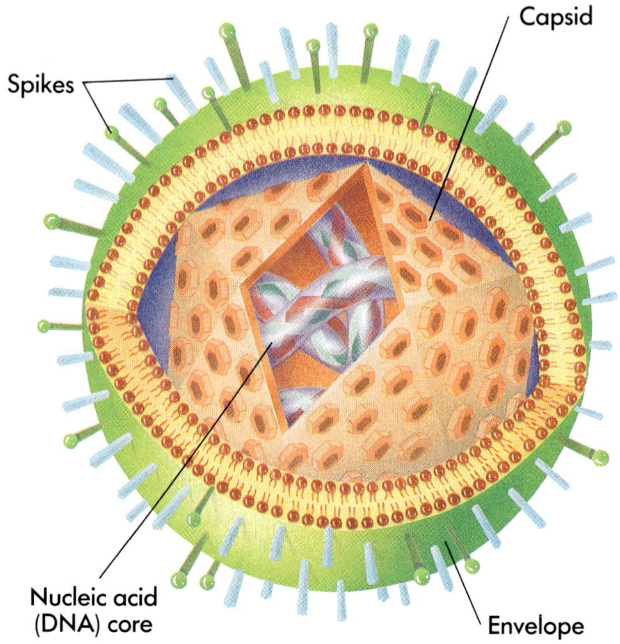

B

FIGURE 28-3 Structure of a typical virus.
A Electron micrograph of the herpesvirus.
B Structure of the herpesvirus.

It is hard to conceptualize how small viruses are, but Figure 28-2 helps by showing the size of a few viruses relative to the size of a bacterium and a human red blood cell. Some viruses, such as the poliovirus, are as small as the width of the plasma membrane on a human cell. The largest viruses are barely visible with a light microscope. Most viruses can be seen only by using an electron microscope.

Viruses are made up of a nucleic acid core surrounded by a protein covering called a *capsid.* Some viruses also have an additional covering, or envelope.

Viral replication

Viruses cannot multiply on their own. They must enter a cell and use the cell's enzymes and ribosomes to make more viruses. This process of viral multiplication within cells is called **replication.** Various patterns of viral replication exist. Some viruses enter a cell, replicate, and then cause the cell to burst, releasing new viruses. This pattern of viral replication is called the **lytic cycle.** Other types of viruses enter into a long-term relationship with the cells they infect, their nucleic acid replicating as the cells multiply. This pattern of viral replication is called the **lysogenic cycle.**

The lytic cycle

The process of viral replication has been studied most extensively in bacteria because bacteria are easier to grow in the laboratory and infect with viruses than plant or animal cells. Many bacterial viruses (usually called **bacteriophages** or simply **phages**) follow a lytic cycle pattern of replication. As shown in Figure 28-4, a bacteriophage first attaches to a receptor site on a bacterium. The virus then injects its nucleic acid into the host cell, while its protein capsid is left outside the cell. Next, the viral genes take over cellular processes and direct the bacterium to produce viral "parts" that will be used to assemble whole viruses. After their manufacture, these strands of nucleic acid and proteins are assembled into mature viruses that **lyse,** or break open, the host cell. Each bacterium releases many virus particles. Each "new" virus is capable of infecting another bacterial cell.

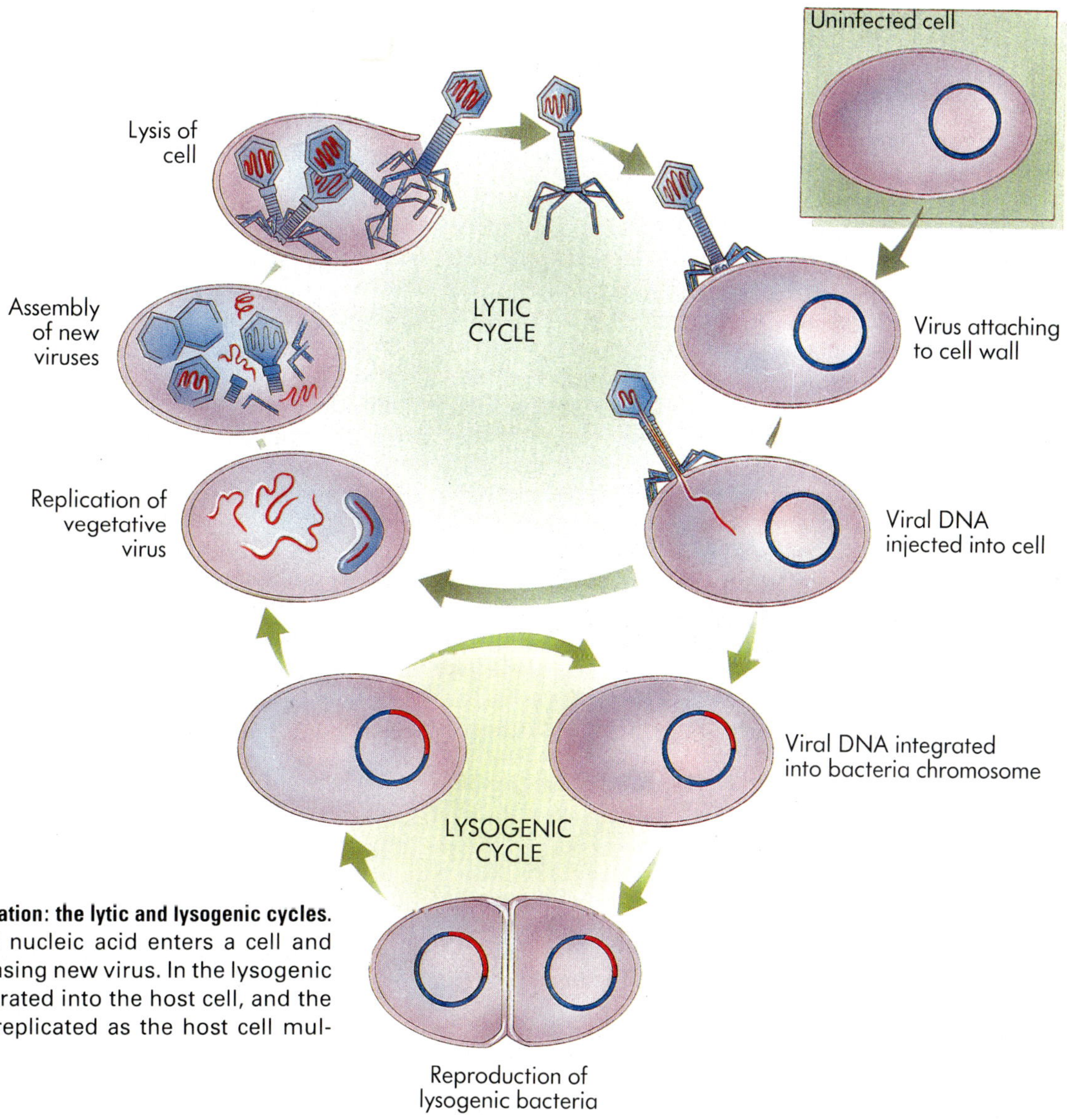

FIGURE 28-4 Viral replication: the lytic and lysogenic cycles. In the lytic cycle viral nucleic acid enters a cell and causes it to burst, releasing new virus. In the lysogenic cycle the virus is integrated into the host cell, and the virus' nucleic acid is replicated as the host cell multiplies.

Some animal viruses infect cells in a manner similar to bacterial virus infection, but they enter animal cells by endocytosis and must be uncoated before they can cause the cell to manufacture viruses. These cells may die as new virus particles are released in a lytic infection, or they may survive if virus particles are slowly budded from the cell by a process similar to exocytosis. This "slow budding" causes a persistent infection.

During a lytic pattern of viral infection, a bacterial virus injects its nucleic acid into a host cell, whereas an animal virus enters by endocytosis. The viral nucleic acid directs the cell to produce "new" viral nucleic acid and protein coats. After assembly of these parts, the bacterial virus particles cause the cell to burst open, releasing them. Animal viruses often leave the cell by slow budding.

Plant cells are somewhat protected from viral infection by their rigid cell walls and protective outer, waxy cuticles. Viruses can only enter plants if they are damaged or if other organisms, such as sucking insects or fungi, assist them.

The lysogenic cycle

Some viruses, instead of killing host cells, integrate their genetic material with that of the host. Then, each time the host cell reproduces, the viral nucleic acid is replicated as if it were a part of the cell's genetic makeup (see Figure 28-4). In this way, the virus is passed on from cell to cell. Infection of this sort is called a **latent infection.** These integrated latent genes may not cause any change in the host for a long time. Then, triggered by an appropriate stimulus, the virus may enter a lytic cycle and produce symptoms.

The herpes simplex virus causes latent infections of the skin. Herpes nucleic acid remains in nerve tissue (sensory ganglia) without damaging the host until a cold, a fever, or other factor such as ultraviolet radiation from the sun acts as a trigger, and the cycle of cell damage begins. This "damage" is manifested as cold sores or fever blisters. The herpes zoster virus (the chickenpox virus) can also act in the same way. This virus may remain latent in the nerve tissue of a person having had chickenpox, only to be triggered at a later time to cause the painful nerve disorder shingles.

During a lysogenic pattern of viral infection, a virus integrates its genetic material with that of a host and is replicated each time the host cell reproduces.

Viruses and cancer

Depending on which genes a virus carries, a virus can often seriously disrupt the normal functioning of the cells it infects. For thousands of years, diseases caused by viruses have been known and feared. Among the diseases caused by viruses are smallpox, chickenpox, measles, German measles (rubella), mumps, influenza, colds, infectious hepatitis, yellow fever, polio, rabies, and AIDS. Additionally, certain animal viruses can disrupt the normal functioning of cells so much that they change normal cells into cancer cells.

TABLE 28-1 Human cancers that may be caused by viruses

CANCER	VIRUS	FAMILY
Adult leukemia (cancer of the white blood cells)	Human T cell leukemia virus I and II	Retrovirus
Burkitt's syndrome (cancer of the white blood cells, often accompanied by a large facial tumor)	Epstein-Barr virus	Herpes
Cervical cancer	Herpes simplex II virus	Herpes
Liver cancer	Hepatitis B virus	Not classified
Nose and throat cancer	Epstein-Barr virus	Herpes
Skin and cervical cancer	Papillomavirus	Papova

Infection with certain viruses can lessen the growth requirements of the invaded cells, resulting in their growing faster than other cells—often uncontrollably. Such uncontrolled growth leads to the formation of large masses called **tumors.** If the body successfully walls off the tumor so that it is unable to spread, it is called a **benign** tumor and its cells are not considered cancerous. The human papillomavirus, for example, causes the growth of small benign tumors commonly known as warts.

Certain viruses can also transform normal cells into rapidly-growing but highly invasive cells. Along with viruses, other stimuli, such as chemicals called **carcinogens** and certain types of radiation such as ultraviolet rays or x-rays, can transform cells in this way. These highly invasive cells are cancer cells. Cancer cells grow out of control, forming tumors that invade and destroy body tissues. Such tumors are called **malignant** (cancerous) tumors. Human cancers that may be induced by viruses are listed in Table 28-1.

Infection with certain viruses can transform normal cells into cancer cells.

The classification of viruses

Because viruses are not living things, they are not included in the five kingdoms of life. Scientists have, however, devised a classification scheme for viruses based on the host they infect. Viruses are first grouped according to whether they infect plants, animals, or bacteria. Further classification usually focuses on differences in morphology (shape and structure), type of nucleic acid, and manner of replication.

Bacteria

Bacteria and viruses both have "reputations" as being agents of disease. But the role of bacteria in the world of living things is much broader than that of a pathogen. Bacteria make life on Earth possible because they perform integral functions as decomposers of organic material and are natural recyclers of nitrogen and other inorganic compounds in ecosystems (see Chapter 38). They are also used to produce certain foods, such as yogurt, sauerkraut, dill pickles, and olives.

Bacteria have a cell structure different from other organisms: the cytoplasm contains no internal compartments or organelles, the hereditary material is not enclosed by a membrane to form a nucleus, and the cell is bounded by a membrane encased within a cell wall. For these reasons, bacteria are classified as a separate kingdom of organisms, Monera.

Bacteria are the oldest, most abundant, and simplest organisms. Bacteria were abundant for well over 2 billion years before eukaryotes appeared in the world. They were largely responsible for creating the properties of the atmosphere and the soil during the long ages in which they were the only form of life on Earth (see Chapter 26). Bacteria are present on and in virtually everything you eat and you touch. Bacteria are also the only organisms with a prokaryotic cellular organization. (This type of cellular organization is described in Chapter 3.) It differs from eukaryotic cellular organization in primarily two ways: (1) the prokaryotic cell has no membrane-bounded nucleus, and (2) it contains no membrane-bounded organelles that compartmentalize the cell. This structural uniqueness places bacteria in a kingdom all their own: **Monera,** meaning "alone."

Bacterial reproduction

Reproduction among bacteria is asexual; one cell divides into two with no exchange of genetic material among cells. This process is called **binary fission.** Before fission, or division of the cell, the genetic material replicates and divides.

Bacterial DNA exists as a single, circular molecule that is attached at one point to the interior surface of the cell membrane. As eukaryotic cells do, bacteria make a copy of their genetic material before cell division. The bacterium also grows in size and manufactures sufficient ribosomes, membranes, and macromolecules for two cells before dividing. When the cell reaches an appropriate size and the synthesis of cellular components is complete, binary fission begins.

The first step of binary fission is the formation of a new cell membrane and cell wall between the attachment sites of the two DNA molecules (Figure 28-5). As the new membrane and wall are added, the cell is progressively constricted in two. Eventually the invaginating membrane and wall reaches all the way into the cell center, forming two cells from one.

Most bacteria reproduce every 1 to 3 hours. Some bacteria take a great deal longer. However, bacteria having conditions favoring their growth could produce a population of billions in little more than a day! This type of growth

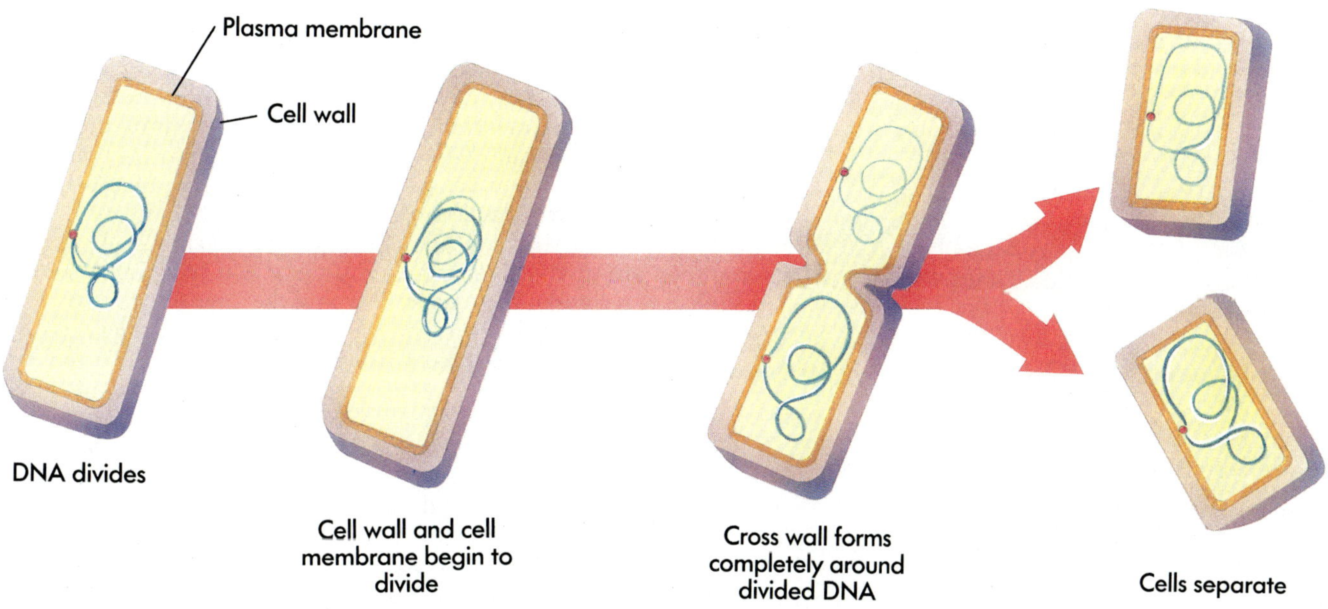

FIGURE 28-5 Binary fission in bacteria.
During binary fission, one bacterium divides into two bacteria.

is referred to as **exponential growth,** a period of rapid doubling of cells. However, the population cannot grow unchecked indefinitely. Many cells compete for food and other growth factors they need to survive. With resources limited, the entire population cannot be maintained, and cells begin to die. Wastes also accumulate, poisoning some of the cells. The growth of the population "levels out." Eventually, if the growth requirements of the bacterial population are no longer met, the population may begin to die as rapidly as it once grew.

Bacteria reproduce asexually by binary fission. First they produce sufficient cell parts for two cells and replicate their single, circular molecule of DNA. Then the cell wall and membrane grow inward, between the attachment sites of the two DNA molecules, and literally split the cell in two.

Bacterial diversity and classification

Bacteria occur in a wide range of habitats and play key ecological roles in each of them. Some thrive in hot springs where the water temperature can be as high as 78° C (172° F). Others live more than a quarter of a mile beneath the surface of the ice in Antarctica. Still other bacteria, capable of dividing only under high pressures, exist around deep-sea vents formed by undersea volcanoes. Bacteria live practically everywhere, even in ground water where they were once thought to be absent. They are able to play many ecological roles because they make up a kingdom of organisms that is extremely diverse in its physiology.

This diverse kingdom of organisms is classified according to criteria listed in Table 28-2. They are not grouped into phyla but into four divisions that reflect differences in the chemistry of their cell walls. Each division is further split into sections rather than classes of organisms. Classification from this point on is similar to the classification of the eukaryotes in that further subdivisions use the titles of order, family, and genus. Within this classification scheme, most of the bacteria are considered **eubacteria,** or "true" bacteria.

TABLE 28-2

Criteria for classifying bacteria

- Shape of individual cells
- Arrangements of groups of cells
- Presence of flagella
- Staining characteristics
- Nutritional characteristics
- Temperature, pH, and oxygen requirements
- Biochemical nature of cellular components, such as RNA and ribosomes
- Genetic characteristics, such as percentages of DNA bases

But one section of organisms is different from all the rest. These are the **archaebacteria,** or "ancient bacteria." Many taxonomists consider archaebacteria to be a separate, sixth kingdom (see Chapter 26).

The kingdom of bacteria, Monera, is divided into four divisions, which are further subdivided into sections, orders, families, genera, and species.

Archaebacteria

The cell walls and the cell membranes of the archaebacteria are chemically different from those of the eubacteria, as are certain of their physiological processes. These bacteria have such unusual chemical processes that they live in quite unusual places!

One group of archaebacteria produces methane (also known as marsh gas) from carbon dioxide and hydrogen. These bacteria can be found in places such as the gut of cattle and the depths of landfills. The methane they produce unfortunately adds to the blanket of greenhouse gases surrounding the Earth and therefore to the problem of global warming (see Chapter 40). Another group of archaebacteria live only in areas having high concentrations of salt, such as in salt marshes in the intertidal zone, where fresh water meets seawater in stagnant, concentrated salt pools. A third group of archaebacteria were already mentioned: those that live in the incredible heat and pressure of the deep-sea vents.

The archaebacteria are a taxonomic section of bacteria that are chemically different in certain structures and metabolic processes from all other bacteria.

Eubacteria

The eubacteria make up all the rest of the bacteria—quite a diverse collection. However, these thousands of species can be placed into one of three groups according to their mode of nutrition.

One of these groups is the photoautotrophic (photosynthetic) bacteria. Like plants, photosynthetic bacteria use energy from the sun to produce "food" in the form of carbohydrates. They are therefore called **photoautotrophs,** which means *light* (photo) *self* (auto) *feeders* (trophs). Photosynthesis takes place in green bacteria, purple bacteria, and cyanobacteria (formerly known as blue-green algae). These bacteria contain chlorophyll (chemically different from the chlorophyll in plants), which is found within a system of membranes that ring the interior periphery of the cell. In the green and purple bacteria, photosynthesis does not take place exactly as it does in plants, and oxygen is not a by-product. In fact, oxygen tends to break down their chlorophyll, so these species usually live in polluted water that contains little oxygen. The cyanobacteria, however, release oxygen during photosynthesis as plants do and can be found living near the surface of lakes and ponds (Figure 28-6). The cyanobacteria were among the first cells to evolve and probably oxygenated Earth's primitive environment.

A second group of eubacteria are the chemoautotrophic bacteria. "Chemical self-feeders" or **chemoautotrophs,** derive the energy they need from inorganic molecules such as ammonia (NH_3), methane (CH_4), and hydrogen sulfide

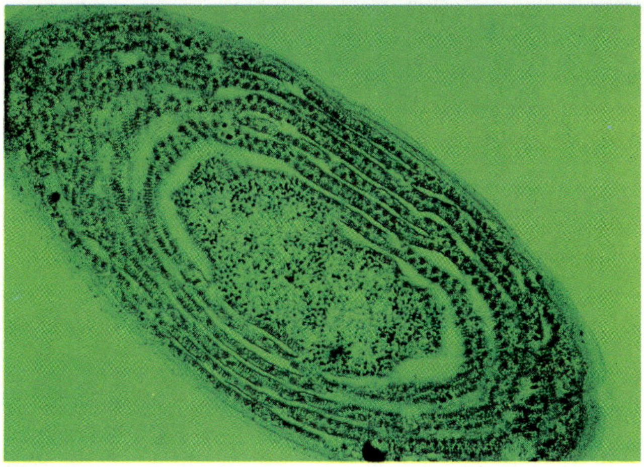

FIGURE 28-6 A cyanobacterium.
The outer regions of the cell are filled with photosynthetic membranes. The dark spots between the membranes are storage areas for the carbohydrates produced by photosynthesis.

(H_2S) gases. With this energy and carbon dioxide (CO_2) as a carbon source, they can manufacture all their own carbohydrates, fats, proteins, nucleic acids, and other growth factors. The archaebacteria that live near deep-sea vents are chemoautotrophs, as are eubacteria called **nitrifying bacteria.** Nitrifying bacteria play an important role in the cycling of nitrogen between organisms and the environment. These bacteria live in nodules (spherical swellings) in the roots of legumes, such as beans, peas, and clover (Figure 28-7). They convert ammonia to nitrates, a form of nitrogen used by plants. Other species of chemoautotrophs play a key role in the sulfur cycle, using H_2S or elemental sulfur for energy and converting it in the process to sulfates, other plant nutrients.

The third group of eubacteria are heterotrophic bacteria. Most bacteria are "other feeders" or **heterotrophs,** obtaining their energy from organic material that enters these cells by diffusion and active transport. Humans are heterotrophs, too, eating plants and animals—organic material that once lived. Heterotrophic bacteria are considered **decomposers** and play a key role in the carbon cycle. They break down large organic compounds, such as proteins and carbohydrates, into small compounds, such as CO_2, which is released into the atmosphere to be recycled as it is "fixed" by plants during the process of photosynthesis.

FIGURE 28-7 Soybean root nodules.
The nodules growing on the roots of this soybean plant contain nitrogen-fixing bacteria. These bacteria convert atmospheric nitrogen (N_2) to ammonia (NH_3), which can then be used as a nitrogen source by the plants.

Bacteria as disease producers

Most plant diseases are not caused by bacteria but by fungi (see Chapter 29). However, a few genera of bacteria do infect plants, primarily causing types of plant rot and wilt. Wilt occurs when bacteria block water from moving up the xylem vessels in the plant. In addition, some genera of bacteria cause tumorlike growths called *galls* in plants. Figure 28-8 is a photograph of a plant with crown gall disease caused by bacteria of the genus *Agrobacterium*. Interestingly, scientists have learned ways to use these disease-causing bacteria in productive ways in the genetic engineering of plants.

FIGURE 28-8 Crown gall disease on a tobacco plant.
This disease is caused by bacteria of the genus *Agrobacterium*.

In contrast to plant diseases, many human diseases are caused by bacteria, including the diseases listed in Table 28-3 with which you may be familiar. The bacteria that cause disease are all heterotrophic. Most disease-producing bacteria use their hosts for food, but some poison their hosts. To cause disease, bacteria or the poisons they produce must first get into the body. Usually this happens if you eat contaminated food or drink contaminated water. You may inhale bacteria present in the air after an infected person coughs or sneezes, touch a contaminated object, or have sexual intercourse with an infected partner. Sometimes bacteria enter the body through broken skin as the result of an injury or injection with a contaminated needle.

After entering the body, bacteria then attach to body cells and cause various types of tissue damage. The chemicals bacteria produce digest the tissues so that the breakdown products can be taken into the bacteria and metabolized. In addition, certain bacteria, such as the bacteria that cause staphylococcal food poisoning, gas gangrene, tetanus, or cholera, produce **toxins,** or poisons. These toxins can cause powerful effects, such as high fevers, violent muscle spasms, vomiting, diarrhea, heart damage, and respiratory failure. All the while, your body combats these cells using its nonspecific and specific defense systems (see Chapter 13). The ability of your body to wage this war affects the degree and length of the illness and its eventual outcome.

TABLE 28-3 Some diseases caused by bacteria

RESPIRATORY DISEASES	Bacterial pneumonia Bacterial meningitis Strep throat Tuberculosis Legionnaire's disease
GASTROINTESTINAL DISEASES	Cholera Dysentery Typhoid fever
GASTROINTESTINAL POISONING	Staph food poisoning Botulism
SEXUALLY TRANSMITTED DISEASES	Syphilis Gonorrhea Chancroid
SKIN DISEASES	Staph infections Leprosy Yaws

Genetic engineering

Producing disease is only one of the varied roles that bacteria play in the web-like interactions of living things. In recent decades, scientists have added a new role—bacteria are now being used as microscopic factories, manufacturing hormones and a variety of medically useful products and helping to produce crop plants and feed animals with desired characteristics. These new techniques of molecular biology that involve the manipulation of genes are called **genetic engineering** or **recombinant DNA technology.**

Natural gene transfer among bacteria

Scientists have long known that certain bacteria can transfer genetic material from one cell to another. Genes move from one prokaryote to another by three methods: **transformation, transduction,** and **conjugation** (Figure 28-9).

In transformation, free pieces of DNA move from a donor cell to a recipient cell. The pieces of DNA come from bacteria that have lysed and released their DNA. Certain cells, called **competent cells,** are able to "take up" this DNA and incorporate pieces into their genomes, thereby becoming transformed, or changed. Not all bacteria are able to be transformed; the ability to take up DNA fragments is an inherited characteristic.

During transduction, DNA from a donor cell is transferred to a recipient cell by a virus. For this transfer to occur, the virus must first merge its genetic material with that of a bacterium it has infected. Only certain lysogenic viruses or damaged viruses are capable of incorporating bacterial genes with their genome. Also, not all bacteria are capable of being infected by a virus carrying bacterial genes. But this process does occur in certain viral-bacterial interactions and results in genes from one bacterium being transferred to another bacterium.

During conjugation, a donor and a recipient cell make contact, and the DNA from the donor is injected into the recipient cell. This transfer takes place in bacteria having extrachromosomal pieces of DNA called **plasmids** that have genes for chromosome transfer. Plasmids are fragments of DNA separate from the main circular molecule that occurs in bacteria. They replicate independently of the main chromosome and make up about 5% of the DNA of many bacteria. (Figure 28-10 shows these small circles of DNA.) Certain plasmids are called **F plasmids** because they have a fertility factor—several special genes that promote the transfer of the plasmid to other cells. These genes code for proteins that form a tube called a **pilus** on the surface of the cell. When the pilus of one cell makes contact with the surface of another cell that lacks pili (and therefore does not contain the F plasmid), the replication and then transfer of the plasmid occurs (Figure 28-11). If the F plasmid has been incorporated into the larger bacterial chromosome, as sometimes happens, the cell will copy and transfer the entire bacterial chromosome to the recipient cell.

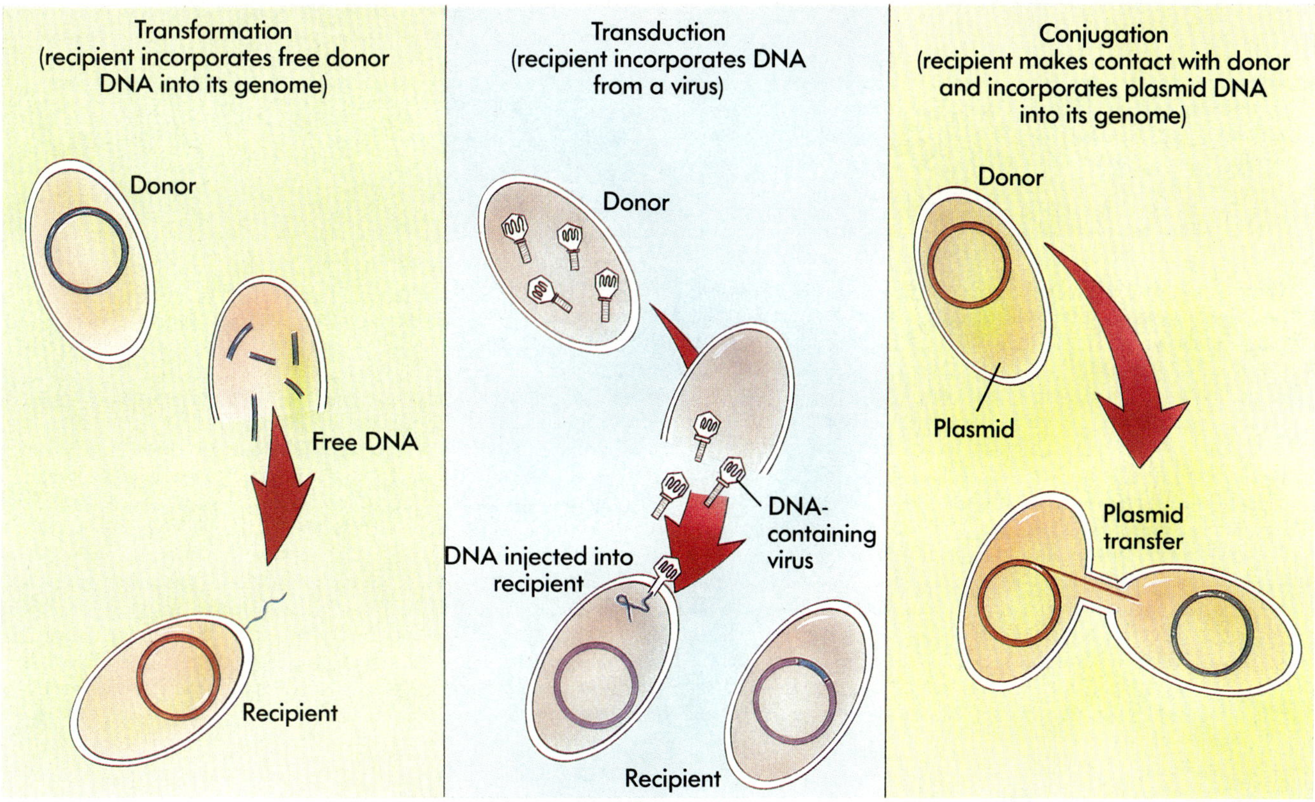

FIGURE 28-9 Natural gene transfer among bacteria.
Transformation, transduction, and conjugation.

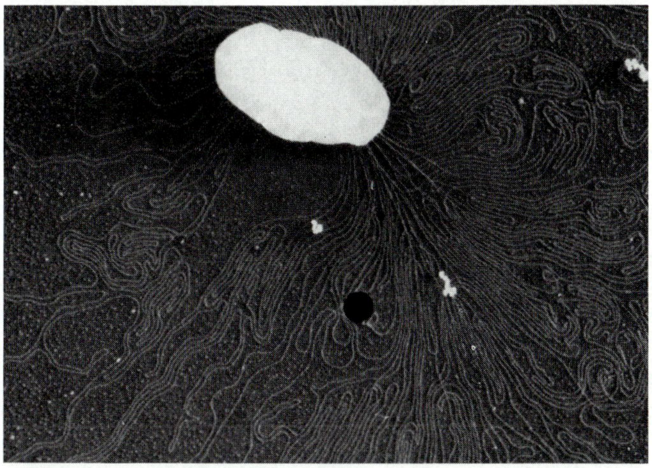

FIGURE 28-10 A ruptured bacterial cell *(Escherichia coli)* has released its chromosomes and plasmids.
Plasmids are the small circles of "extra" DNA that replicate independently of the main chromosome.

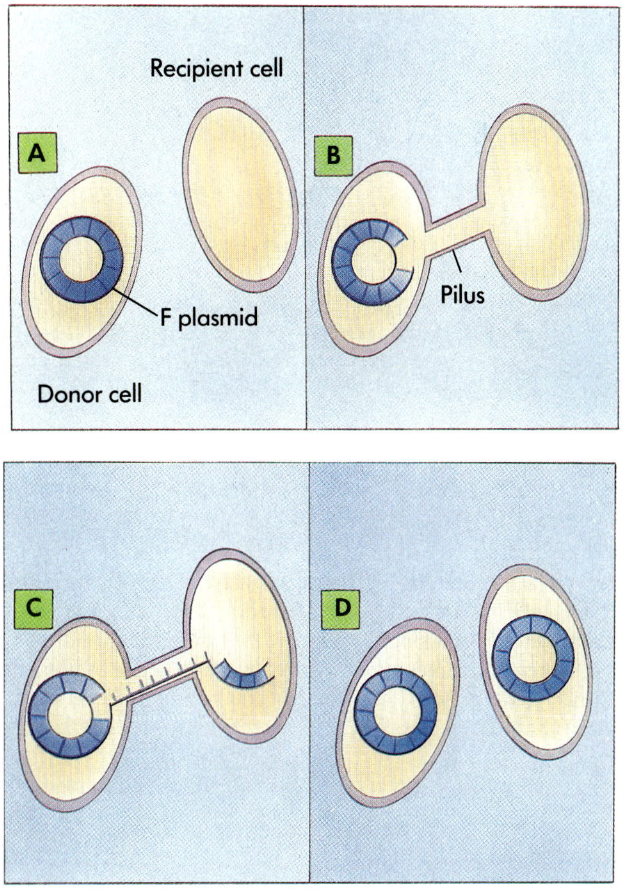

FIGURE 28-11 Conjugation.
A-C During conjugation, a pilus is formed between two bacterial cells across which the F plasmid is transferred. **D** If the F plasmid becomes incorporated into the larger bacterial chromosome, the cell will copy and transfer the entire bacterial chromosome to the recipient cell.

BIOLOGY TECHNOLOGY & society

The new agriculture

Genetic engineering is revolutionizing how farmers grow their crops. In the past, much of a farmer's efforts were devoted to cultivating soil to eliminate weeds and spraying crops with pesticides to rid them of crop-destroying insects. Over 40% of the chemical insecticides used in the world today, for example, are sprayed on cotton plants to kill plant-eating insects.

Now weeds can be eliminated from fields of many commercially important broadleaf plants in a new way. These plants have been genetically engineered to be resistant to the herbicide glyphosate, a powerful biodegradable chemical that kills most growing plants. The herbicide blocks an enzyme that plants need to make aromatic amino acids. (Humans do not make them but get them from their diet, so the herbicide is harmless to humans.) Genetic engineers have added a gene encoding a glyphosate-resistant enzyme to the crops. Now when fields are treated with glyphosate, all the plants in the field die—except the genetically engineered crop. Instead of cultivating and applying many different kinds of herbicides (most herbicides kill only a few kinds of weeds), only a single herbicide is required, one that readily breaks down in the environment.

Genetic engineering is also making many crops immune to attack by insects. In one approach, tomatoes have been made resistant to the caterpillar-like tomato hornworm. Scientists inserted a gene isolated from soil bacteria (a kind called *Bacillus thuringiensis*) into the tomato's chromosomes. This gene encodes a protein that converts enzymes in the caterpillar's stomach into a toxin causing paralysis and death in insects. Because the necessary enzyme is not found in mammals, the protein is harmless to humans. The genetically engineered tomato plants are resistant to the hornworm, without the application of any pesticide (Figure 28-A).

Many plant pests attack roots, and to counter this threat biologists have introduced the gene encoding the *B. thuringiensis* protein into root-colonizing bacteria. Insects that eat the roots convert the protein into toxin and die.

A major effort is underway to transfer into crop plants the genes that enable soybeans and other legume plants to "fix" atmospheric nitrogen, which all other plants must obtain from the soil. Worldwide, farmers applied over 60 million tons of nitrogen fertilizer in 1990, an expensive undertaking. The genes needed to obtain the necessary nitrogen directly from the air are present only in certain symbiotic bacteria living in the legume plant's roots. These genes have been isolated and introduced into crop plants, but they do not seem to function properly in their new eukaryotic environment. Experiments are being pursued actively because perfecting the ability to grow crops without nitrogen fertilizers would greatly expand the world's ability to feed a hungry population.

FIGURE 28-A A successful experiment in plant resistance. The two tomato plants were exposed to destructive caterpillars under laboratory conditions. The nonengineered plant on the left has been completely eaten, but the engineered plant shows no signs of damage.

Human-engineered gene transfer among bacteria

Within the past couple of decades, scientists have learned how to manipulate these natural mechanisms of gene transfer to produce bacteria with desired genomes. One use of this technology has been to insert *human genes* into bacteria that instruct the bacteria to produce human proteins. For example, the genes that code for the production of human insulin have been inserted into bacteria, causing the bacteria to produce this hormone. Bacteria grown in large numbers, such as shown in Figure 28-12, produce large quantities of this hormone, which can then be purified and used in the treatment of diabetes.

How do scientists put genes into bacteria? Continuing with the example of the production of human insulin, scientists first cut the pieces of DNA they need—in this case the gene that codes for insulin—from human DNA. To do this job, researchers extract human DNA from cells and treat it with "chemical scissors" called **restriction enzymes.** Restriction enzymes work by "recognizing" certain base sequences in DNA molecules and breaking the bonds between the bases at that point. Such a sequence is called a **recog-**

FIGURE 28-12 A pharmaceutical engineer checks on the purification process for Humulin, a human insulin created from recombinant DNA technology.
Humulin is used to manage diabetes in humans.

nition sequence. In nature, restriction enzymes are produced by bacteria to protect themselves from bacteriophage invasion. They protect bacteria by "cutting" viral DNA into pieces before it can take over the cell. And now these naturally occurring bacterial enzymes have become one of the many tools of the genetic engineer.

A myriad of restriction enzymes exist. Different restriction enzymes recognize different recognition sequences, so scientists can pick the enzyme that will cut DNA in a spot they choose. The DNA sequences that restriction enzymes recognize are typically four to six nucleotides long. Because the nucleotides on double-stranded DNA are complementary to one another, the restriction enzyme recognizes and cuts both strands. However, the bases of the recognition sequence, being complementary to one another, run in opposite directions on the two strands. Therefore the restriction enzyme cuts both strands but cuts one at each end of the recognition segment as shown in Figure 28-13. The result is that the cut piece of DNA has ends in which one strand is longer than the other. These ends are "sticky ends" because they can bond with another piece of DNA cut with the same restriction enzyme. Therefore fragments of human DNA and bacterial DNA cleaved by the same restriction enzyme have the same complementary nucleotide sequences at their ends and can be joined to one another. A "sealing" enzyme called a **ligase** helps re-form the bonds. In this example, insulin genes and bacterial plasmids that have been cut with the same restriction enzyme can be joined.

The next step is to insert "hybrid" plasmids into bacteria. If these plasmids are mixed under the proper conditions with competent bacteria, the bacteria will take up the plasmids from the mixture. These bacteria now have the ability to produce human insulin. (This procedure of genetically engineering bacteria is summarized in Figure 28-14.)

Inserting fragments of foreign DNA into bacterial cells using plasmids is a technique that scientists use frequently. Scientists also use certain viruses to insert DNA fragments into bacteria. First, they insert the DNA into the genome of a virus instead of into a plasmid. The virus then carries the DNA fragment into a cell it is capable of infecting. Such viruses are called **vectors.** Animal viruses can also be used as vectors. In fact, scientists have used animal viruses to carry bacterial genes into monkey cells and animal genes into plant cells.

There has been considerable discussion about the potential danger of inadvertently creating an undesirable life form in the course of a recombinant DNA experiment. What if someone fragmented the DNA of a cancer cell and then incorporated the fragments at random into viruses that are propagated within bacterial cells? Might there not be a danger that one of the resulting bacteria could be capable of infecting humans and causing an infective form of cancer?

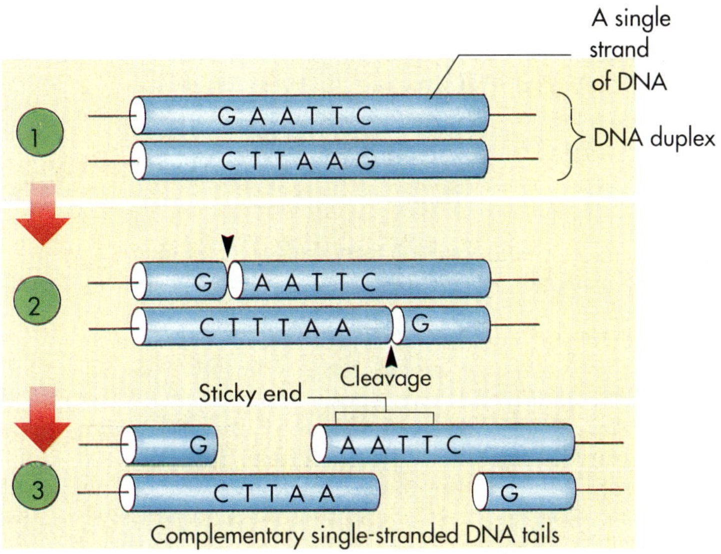

FIGURE 28-13 The method in which restriction enzymes produce DNA fragments with "sticky ends."
The restriction enzyme EcoRI always cleaves sequence GAATTC at the same spot, after the first G. Because the same sequence occurs on both strands, both are cut. But the position of the G is not the same on the two strands; the sequence runs the opposite way on the other strand. As a result, single-stranded tails are produced. The single-stranded tails are complementary to each other, or sticky.

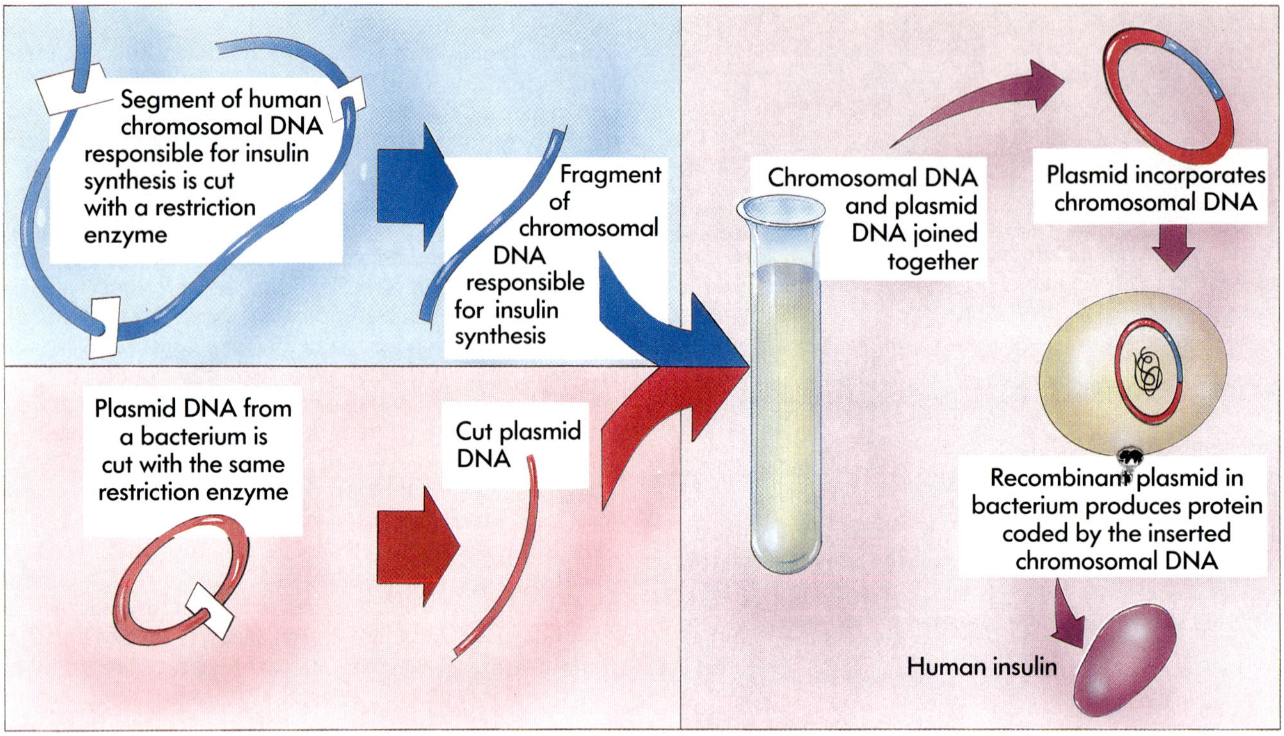

FIGURE 28-14 How insulin is made through genetic engineering.
DNA is inserted into plasmids, and the plasmids are mixed with bacteria. The bacteria will take up the plasmids through transformation and will produce the proteins directed by the inserted DNA.

Even though most recombinant DNA experiments are not dangerous, such concerns should be taken seriously. Both scientists and governments monitor experiments in genetic engineering to detect and forestall any such hazard. In addition, researchers have established appropriate experimental safeguards. For example, the bacteria used in many recombinant DNA experiments are unable to live outside of laboratory conditions; many of them can live only in an atmosphere that is free of oxygen and therefore cannot survive in the Earth's atmosphere. Experiments that are clearly dangerous are prohibited.

Applications of recombinant DNA technology

The newly found ability to isolate individual genes and transfer them from one kind of organism to another has revolutionized scientist's ability to improve the characteristics of useful plants and animals and to avoid disease. These techniques offer enormous potential for the agriculture and medicine of the future and have already produced important applications. Genetic engineering is also a useful tool that helps scientists learn about gene structure, function, and regulation. So these powerful techniques are used not only to produce new products but also to generate new knowledge.

Medically important products

Using gene technology to produce proteins useful in treating human disorders or diseases involves the process described earlier: introducing human genes that code for clinically important proteins into bacteria. Along with producing human insulin, this technique has been used to produce more than 150 medically important products. Some of them are human growth hormone, used to treat growth disorders in children; interferon, used to fight cancer, viral diseases, and rheumatoid arthritis; and interleukin-2, used to treat certain deficiencies of the human immune system. Previously, these medically important products were unavailable or were obtained from animals at great expense. Humans often had allergic reactions to proteins obtained from other animals, but the genetically engineered products do not produce side effects because they have the human protein structure.

Genetic engineering techniques have also been used to produce vaccines, such as hepatitis B vaccine that helps protect against the leading cause of liver cancer. In one approach, scientists use gene-splicing techniques to produce a noninfective virus whose protein coat will stimulate the body's immune system to produce antibodies against the "real" virus. Scientists use the DNA from a harmless virus and add to it the genes that code for the protein coat of the infective virus. When injected into humans, these viruses do not cause disease but stimulate the immune system to produce antibodies specific for the infective form of the virus.

Gene therapy

Recombinant DNA technology entered a new phase of application on September 14, 1990, when a 4-year-old girl became the first person to undergo **gene therapy.** Gene therapy is the treatment of a genetic disorder by the insertion of "normal" genes into the cells of a patient. The young girl that was treated suffered from the rare genetic disorder severe combined immune deficiency. Key immune system cells called *T cells* (see Chapter 13) were not working because they lacked the enzyme ADA (adenosine deaminase). Without her T cells, this girl had no defense against infection.

To treat her, doctors removed some of her blood and cultured her defective T cells. They then added viruses that contained working copies of the ADA gene to the cell culture. When the viruses infected the T cells, they inserted these ADA genes into the cells. Doctors grew these altered T cells in the laboratory until they numbered in the billions and then injected them into the child's blood. The genetically repaired cells multiplied, populating the girl's immune system with gene-corrected T cells that produce the necessary enzyme.

Agriculture

Another important application of genetic engineering is the manipulation of the genes of crop plants. In broadleaf plants, such as tomatoes, tobacco, and soybeans, a plasmid of the bacterium *Agrobacterium tumefaciens,* which causes crown gall in plants, is the main vehicle that has been used to introduce foreign genes. A part of this plasmid integrates into the plant DNA, carrying its spliced genes. (Figure 28-15 diagrams the steps of this process.) Through the use of this technique, scientists are working to improve crops and forest trees by making them more resistant to disease, frost, and herbicides (chemicals that kill weeds).

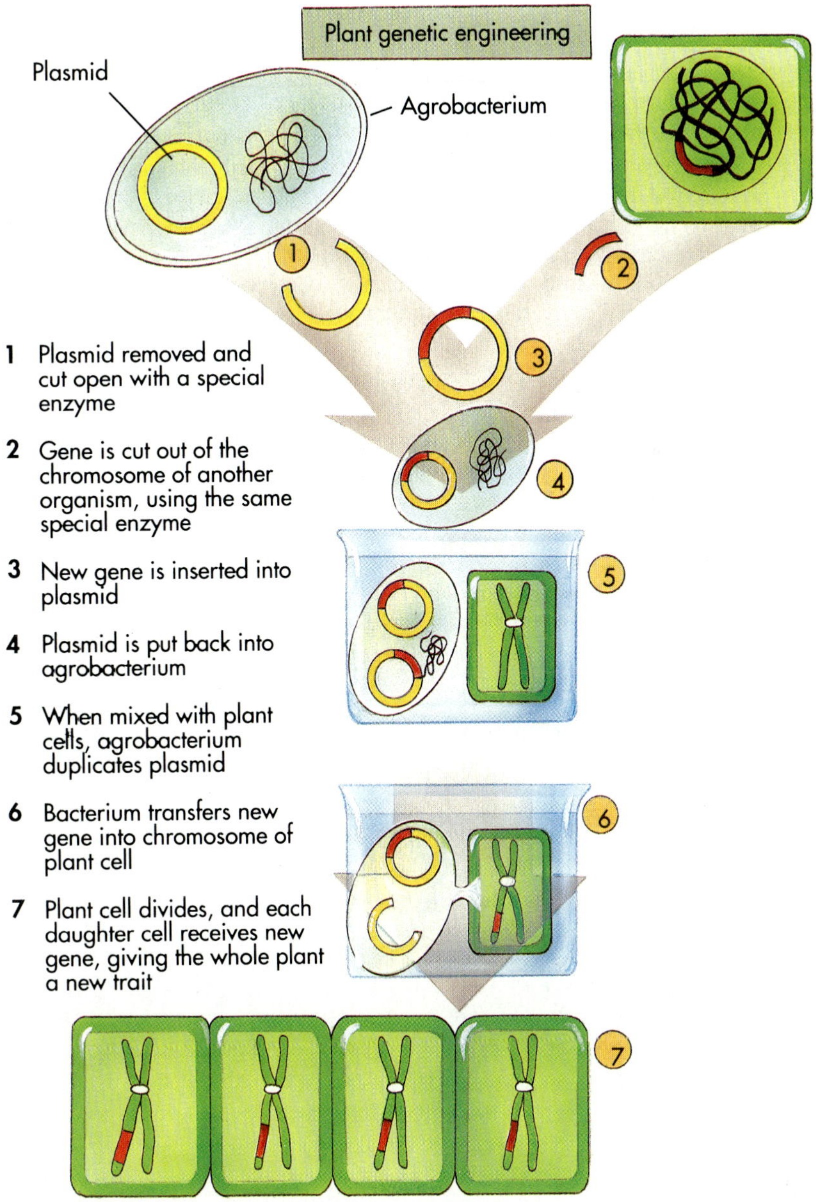

FIGURE 28-15 Use of a plasmid of *Agrobacterium tumefaciens* in genetic engineering.
These genetic engineering procedures induce resistance to viruses, herbicides, and frost in broadleaf plants.

Summary

1. Viruses were discovered at the end of the nineteenth century. For many years after their discovery, viruses were regarded as primitive forms of life. Today, scientists know that viruses are not living organisms because they are not cells. Viruses are protein-coated nucleic acids that replicate (multiply) within living cells.
2. Viruses primarily infect plants, animals, and bacteria, replicating within their cells. Various patterns of viral replication exist. In the lytic cycle a virus enters a cell and causes it to produce viral nucleic acid and protein coats. After these viral parts are assembled, the new virus particles may burst from the host cell or may leave the host cell by budding. In the lysogenic cycle, viruses enter into a long-term relationship with the cells they infect, their nucleic acid replicating as the cells multiply.
3. Some viruses can seriously disrupt the normal functioning of the cells they infect, transforming them into rapidly growing, invasive cells. These cells grow out of control, forming cancerous tumors that destroy body tissues.
4. Bacteria (kingdom Monera) have a prokaryotic cell structure. They differ from eukaryotic cells in many ways but primarily in that they have no membrane-bounded nucleus or membrane-bounded cellular organelles.
5. Bacteria reproduce asexually by binary fission, a splitting in two, after replication of the gentic material takes place. The numbers of bacteria within a population increase rapidly when growth conditions are favorable. These growth conditions vary among bacteria, since they occur in a wide range of habitats.
6. Most bacteria are considered eubacteria, meaning "true bacteria." These thousands of genera of bacteria can be placed into one of three groups according to their mode of nutrition: photoautotrophs, which make their own food by photosynthesis using the energy of the sun; chemoautotrophs, which make their own food by deriving energy from inorganic molecules; and heterotrophs, which obtain energy from organic material.
7. The archaebacteria differ from the eubacteria in their structural and physiological chemistry. Archaebacteria inhabit harsh environments, such as land fills, salt marshes, and deep-sea vents.
8. Bacteria play diverse ecological roles and are extremely important as decomposers. Bacteria are also important in the manufacture of certain foods. A few genera of bacteria cause diseases in plants, and many genera cause diseases in animals, including humans.
9. Scientists have learned to use the techniques of natural gene transfer in bacteria to direct gene transfer in ways that are useful to humans. Bacteria are now being used to manufacture hormones and a variety of medically useful products and to help produce crop plants and animals with desired characteristics. Scientists are also beginning to learn to use these techniques to treat genetic disorders.

REVIEW QUESTIONS

1. What are viruses? What do scientists mean when they say that viruses are not alive?
2. Diagram the structure of a generalized virus. Label the core, capsid, and envelope.
3. Distinguish between the lytic cycle and the lysogenic cycle.
4. One of your friends usually develops a cold sore on his mouth when he gets a cold. Why? What pattern of viral replication is involved?
5. What is the difference between a benign tumor and a malignant tumor? Which is more dangerous?
6. Why are bacteria classified into a kingdom that is unique from all other organisms?
7. Although many bacteria can cause dangerous diseases, in general, bacteria make life on earth possible. Why?
8. Fill in the blanks: Bacteria reproduce by a process called ______ ______. This can lead to a period of rapid doubling of cells called ______ ______.
9. What group(s) of bacteria would you expect to find living in each of the following locations: a) in nodules in the roots of bean plants; b) on a fallen log on the forest floor; c) in a shallow freshwater pond; d) in a highly concentrated salt pool?
10. What is meant by the terms *genetic engineering* and *recombinant DNA technology*?
11. What do transformation, transduction, and conjugation have in common? How are they different?
12. Summarize how researchers insert human genes into bacteria. Why is this technique useful?
13. What is gene therapy?
14. Summarize some of the potential risks and benefits of genetic engineering experiments.

KEY TERMS

archaebacteria 492
bacteriophage 489
benign 490
binary fission 491
capsid 487
carcinogen 490
chemoautotroph 492
competent cells 494
conjugation 494
core 487
decomposer 493
envelope 487
eubacteria 492
exponential growth 492
F plasmid 494
gene therapy 499
genetic engineering 494
heterotroph 493
infectious agent 487
latent infection 490
ligase 497
lyse 489
lysogenic cycle 489
lytic cycle 489
malignant 490
Monera 491
nitrifying bacteria 493
obligate parasite 487
photoautotroph 492
pilus 494
plasmids 494
recognition sequence 496
recombinant DNA technology 494
replication 489
restriction enzyme 496
toxin 494
transduction 494
transformation 494
tumor 490
vector 497
virus 487

THOUGHT QUESTIONS

1. Most plant diseases are caused by fungi and few by bacteria; most human diseases are caused by bacteria and few by fungi. Why the difference?
2. Why would you use genetic engineering to make a crop resistant to a herbicide?

FOR FURTHER READING

Mee, C.L., Jr. (1990, February). How a mysterious disease laid low Europe's masses. *Smithsonian*, pp 66-79.
This is the story of the bubonic plague, the Black Death, which changed the character of Europe in the fourteenth century.

Pennisi, E. (1992, February 22). Making sense of the disorder inside viruses. *Science News*, 141(8):116-117.
Scientists peer into the interior of tumor viruses using a new, unusual technique.

Radetsky, P. (1989, April). Taming the wily rhinovirus. *Discover*, pp. 38-43.
This article examines contemporary approaches to preventing colds, which affect about 1 in 20 people at any given time.

Resenberg, Z., & Fauci, A. (1990, February 10). Inside the AIDS virus. *New Scientist*, pp. 51-54.
Understanding the mechanisms by which the AIDS virus destroys the immune system is of central importance in developing a cure.

29 Protists and fungi

OUTLINE

Protists
- Animal-like protists: Protozoans
  - Amebas
  - Flagellates
  - Ciliates
  - Sporozoans
- Plant-like protists: Algae
  - Dinoflagellates
  - Golden algae
  - Brown algae
  - Green algae
  - Red algae
- Fungus-like protists
  - Cellular slime molds
  - Plasmodial (acellular) slime molds
  - Water molds

Fungi
- Zygote-forming fungi
- Sac fungi
- Club fungi
- Imperfect fungi
- Lichens

"Please don't drink the water!" may seem like an unnecessary warning if you were camping along this pristine-looking mountain stream. But after an invigorating hiking trip, you could return home with nausea, cramps, bloating, and diarrhea—all symptoms of giardiasis, or "hiker's diarrhea." The culprit of this uncomfortable ailment is a single-celled organism no larger than a red blood cell: *Giardia lamblia*.

HIGHLIGHTS

▼

Protists make up an extremely diverse kingdom of eukaryotic organisms that are primarily single-celled but that include certain phyla containing multicellular forms.

▼

Protists can be grouped according to their mode of nutrition into those that are animal-like heterotrophs, plant-like photosynthetic autotrophs, and fungus-like saprophytes, although taxonomists differ in the classification of many forms.

▼

Fungi make up a kingdom of multicellular eukaryotic organisms that are primarily saprophytic; that is, they feed on dead or decaying organic material.

▼

Lichens are symbiotic associations between fungi and algae that allow both to survive in extremely harsh environments.

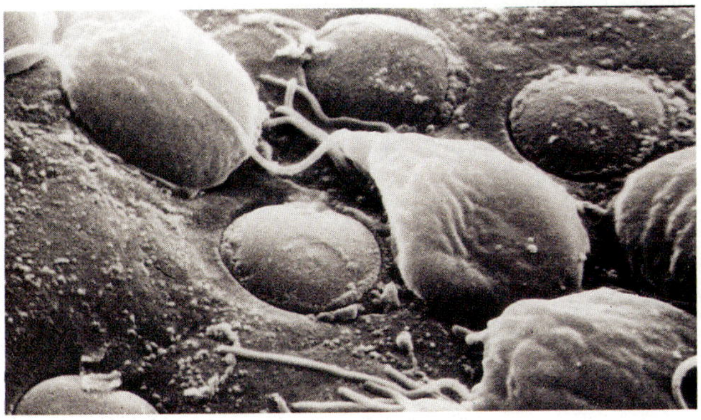

FIGURE 29-1 ***Giardia lamblia.***
These single-celled organisms live in water. When humans drink water infected with *Giardia* organisms, they can experience nausea, cramps, bloating, vomiting, and diarrhea. It is important not to drink *any* untreated water, even though it may look safe.

The organism *G. lamblia* is found throughout the world, including all parts of the United States and Canada. It occurs in water, including the clear water of mountain streams and the water supplies of some cities. In addition to humans, it infects at least 40 species of wild and domesticated animals. These animals can transmit *G. lamblia* to humans by contaminating water with their feces.

Flagella protude from one end of *G. lamblia,* allowing it to move along the intestinal wall of its host. When feeding, it attaches to the wall, sucking blood for nutrition and leaving the suction-cup–shaped marks shown in Figure 29-1. This motile form of the protist exists only while inside the body of its victim. Dormant, football-shaped cysts are expelled in the feces, which can survive for long periods of time outside of their hosts—especially in the cool water of mountain streams. When ingested by other hosts, the cysts develop into their motile, feeding form.

What should you do to prevent infection by *G. lamblia* when hiking or camping? First, *never* drink untreated water, no matter how clean it looks. Because *G. lamblia* is resistant to usual water treatment agents, such as chlorine and iodine, you should boil the water you drink for at least a minute. Better still, *do not* drink the water—bring your own bottled water to be safe.

Protists

The **protists** (kingdom Protista) are a varied group of eukaryotic organisms. Many are single celled, although some phyla of protists include multicellular or colonial forms (single cells that live together as a unit). Within this kingdom are animal-like, plant-like, and fungus-like organisms (Figure 29-2)—quite a diverse array! The animal-like protists are called **protozoans** and are considered animal-like because they are heterotrophs: they take in and use organic matter for energy. *G. lamblia,* for example, is a protozoan. The plant-like protists are the **algae** (including **diatoms**).

A

B

C

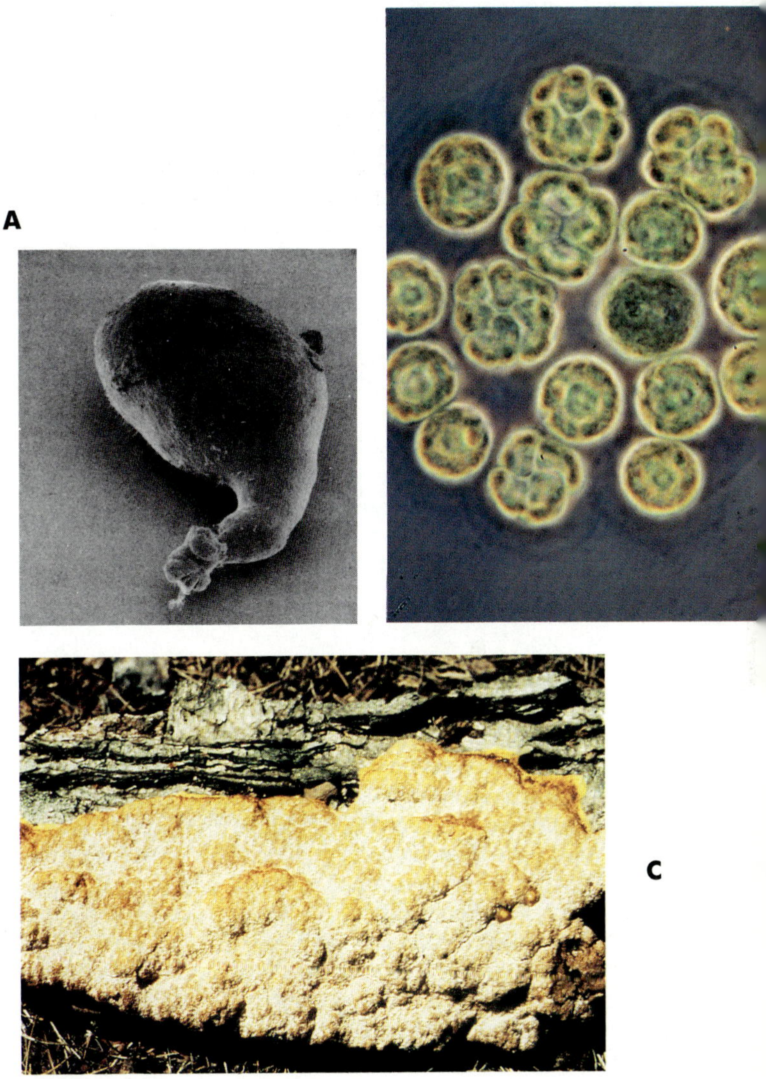

FIGURE 29-2 Single-celled protists.
A Animal-like protist *Pelomyxa palustris,* a unique, single-celled organism that lacks mitochondria and does not divide mitotically.
B Plant-like protist *Eudorina elegans,* a type of green alga.
C Fungus-like protist, the plasmodial slime mold. The individual cells cannot be distinguished in such a plasmodium.

These organisms are plant-like because they are eukaryotic photosynthetic autotrophs: they manufacture their own food using energy from sunlight. The fungus-like protists consist of two phyla of **slime molds** and one of **water molds.** They secrete enzymes onto food sources to predigest the food before they absorb it as the fungi do.

Although many protists are single celled, these organisms are incredibly different from prokaryotic single-celled organisms, the bacteria. The single-celled protists are much larger than bacteria, having approximately 1000 times the volume, and they contain typical eukaryotic cellular organelles. This compartmentalization of eukaryotic cells by membrane-bounded organelles increases their organization to a level much more complex than that of bacteria and allows single-celled eukaryotic organisms to carry out cell functions that support their larger cell volume.

Most of the protist phyla are unicellular eukaryotes, although some phyla contain multicellular forms. Protists can be grouped as animal-like (protozoans), plant-like (algae), or fungus-like according to their mode of nutrition.

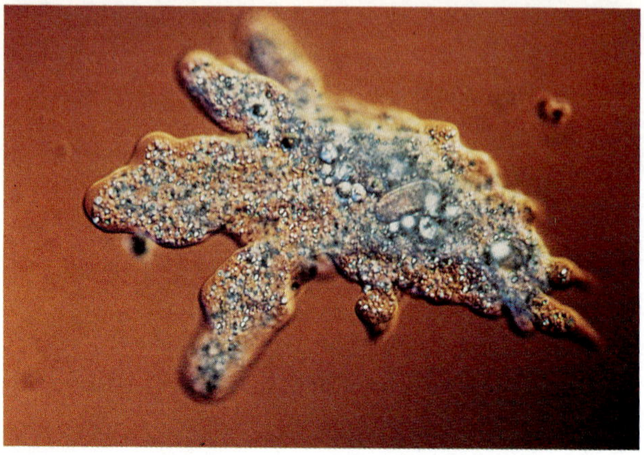

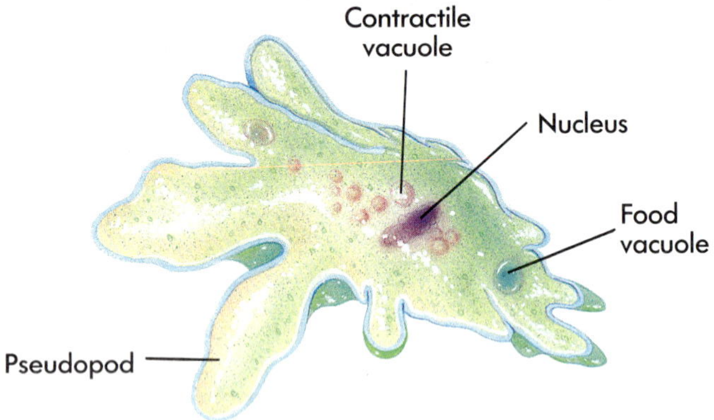

FIGURE 29-3 Structure of an ameba.
The pseudopods, literally "false feet," allow the ameba to move and also trap food particles.

Animal-like protists: Protozoans

The animal-like protists, or protozoans, obtain their food in diverse ways. These heterotrophic characteristics are linked to the ways that these organisms move and provide a means by which they can be grouped. There are four protozoan phyla: the amebas, flagellates, ciliates, and sporozoans.

Amebas

Amebas appear as soft, shapeless masses of cytoplasm. Within the cytoplasm lies a nucleus and other typical eukaryotic organelles. The cytoplasm streams, pushing out certain parts of the cell while retracting others. These cytoplasmic extensions are called **pseudopods** (from the Greek meaning "false feet") and are a means of both locomotion and food procurement. In fact, these cell extensions give the amebas their phylum name, *Rhizopoda,* which means "rootlike feet." Shown in Figure 29-3, the pseudopods simply stream around the ameba's prey, engulfing it within a vacuole. Enzymes digest the contents of the vacuole, which are then absorbed into the cytoplasm to be further broken down for energy. Amebas reproduce by binary fission: the nucleus reproduces by mitosis and then the cell splits in two.

Amebas are abundant throughout the world in fresh water and saltwater, as well as in the soil. Many species are parasites of animals, including humans, and can cause diseases such as amebic dysentery, an infection of the digestive system that produces a diarrhea containing blood and mucus. Although amebic dysentery is a disease associated with poor sanitation and is found primarily in the tropics, medical researchers estimate that about 2 million Americans are infected with the causative agent, *Entamoeba histolytica.*

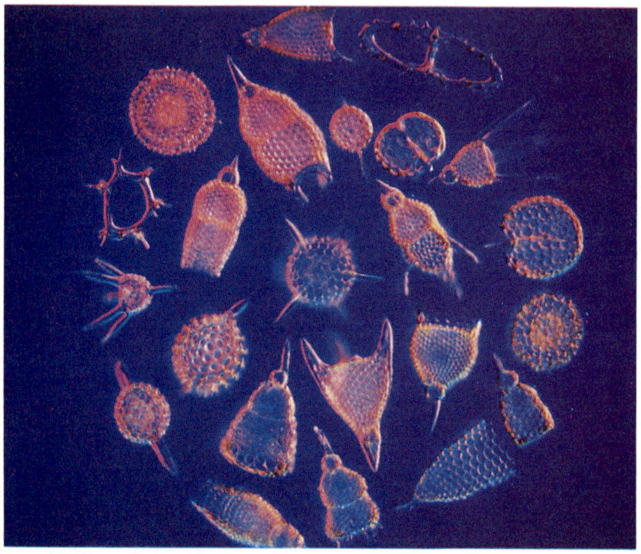

FIGURE 29-4 Radiolarian shells.
The shells of radiolarians resemble delicate glass sculptures.

A

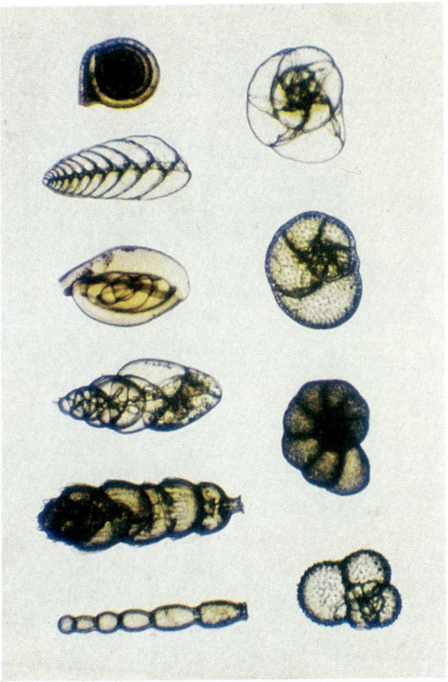

B

FIGURE 29-5 Foraminiferan shells.
A These beautiful foraminiferan shells are made out of limestone.
B The shells of foraminifera are punctured with tiny holes through which the foraminiferan pseudopods poke.

Certain groups of ameba secrete shells that cover and protect their cells. The **radiolarians** secrete shells made of silica that are glass-like and delicate (Figure 29-4). The **foraminifera** (or forams) secrete beautifully sculpted shells made out of calcium carbonate ($CaCO_3$), or limestone (Figure 29-5, *A*). The name *foraminifera* means "hole bearers" and refers to the microscopic holes in their shells through which their pseudopods poke (Figure 29-5, *B*). Food particles stick to these cellular extensions and are then absorbed into the cell.

Forams are abundant in the sea—so abundant, in fact, that their shells litter the sea floor. When studying geological strata, scientists often use the forams as indicators of geological age by noting the types of forams present in ancient rock. Interestingly, the white cliffs of Dover (Figure 29-6) are actually masses of foram shells, uplifted millions of years ago with the sea floor in an ancient geological event.

Amebas are protozoans that have changing shapes brought about by cytoplasmic streaming, which forms cell extensions called *pseudopodia*.

FIGURE 29-6 The white cliffs of Dover.
The picturesque white cliffs are actually composed of masses of foram shells, which were uplifted from the sea floor millions of years ago.

FIGURE 29-7 The flagellates.
A *Codosiga,* a colonial flagellate that remains attached to its substrate.
B *Trichonympha,* one of the flagellates that inhabit the gut of termites and wood-feeding cockroaches. *Trichonympha* ingests wood cellulose, which it finds in abundance in the digestive tracts of its hosts.
C *Trypanosoma* causes sleeping sickness in humans. It has a single flagellum attached to its head.

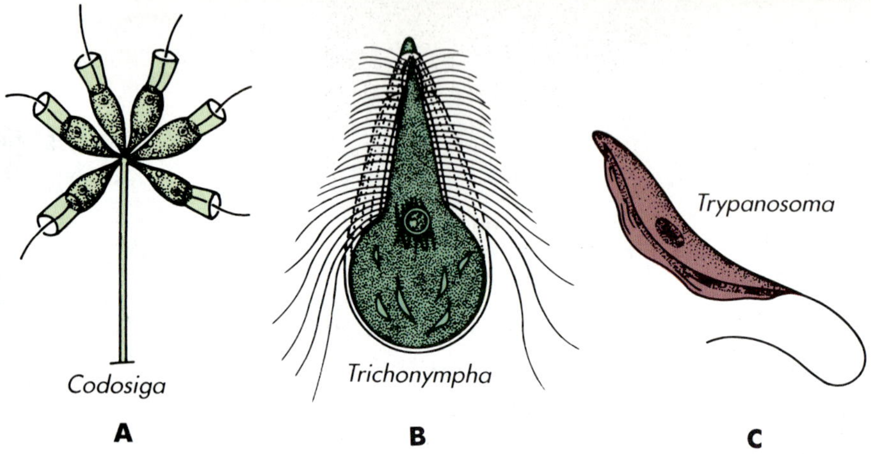

Flagellates

Flagellates are an interesting group because they are so diverse; a few representative genera of flagellates are shown in Figure 29-7. Although they all have at least one **flagellum** (a long, whip-like organelle of motility), some members of this phylum have many flagella, and some have thousands! Their phylum name is *Zoomastigina,* which means "animal whip."

All flagellates have a relatively simple cell structure. They do not have cell walls or protective outer shells as some of the amebas or ciliates do. They also have no complex internal digestive system of organelles as the ciliates do. A flagellate simply absorbs food through its cell membrane, sometimes using its flagella to ensnare food particles.

The flagellates are generally found in lakes, ponds, or moist soil where they can absorb nutrients from their surroundings. The colonial flagellate *Codosiga* shown in Figure 29-7, *A,* for example, consists of groups of cells that are often found anchored to the bottom of a lake or pond by a cellular stalk. The flagella create currents in the water that draw food toward the cells. The flagellate *Trichonympha,* shown in Figure 29-7, *B,* lives a protected life in the gut of termites, digesting wood particles the termite eats. Many flagellates are found living within other organisms; some of these relationships are not harmful to the hosts but other relationships are. Figure 29-7, *C* shows the flagellate *Trypanosoma* that lives in a parasitic relationship with certain mammals, including humans, causing the disease sleeping sickness.

Flagellates are protozoans characterized by fine, long hair-like cellular extensions called *flagella.*

One class of flagellates, the **euglenoids,** generally have chloroplasts and make their own food by photosynthesis. Each euglenoid has two flagella, a short one and a long one that are located on the anterior end of the cell. They move by whipping the long flagellum (Figure 29-8). The euglenoids reproduce asexually by transverse fission, a process in which the parent cell divides across its short axis (see Figure 29-11, *A*). No sexual reproduction is known among this group, which is named after its most well-known member: *Euglena.*

Euglena lives in ponds and lakes and can withstand stagnant water. It has a hard yet flexible covering beneath its plasma membrane called a *pellicle,* with ridges spiraling around its body. These ridges can be seen clearly in Figure 29-8. Two organelles, an **eyespot** and a **photoreceptor,** help *Euglena* stay near the light. (The name *Euglena,* in fact, means "true eye.") The photoreceptor, located near the base of its longer flagellum, is shaded by the nearby eye spot. As light filters through the pigment of the eyespot, the receptor senses the direction and intensity of the light source. Information from the receptor assists the movement of *Euglena* toward the light, a behavior known as **positive phototaxis.**

Interestingly, *Euglena* can survive without light—the chloroplasts simply become small and nonfunctional, and the organism begins to absorb food like a heterotroph. *Euglena* is a good example of an organism that is difficult to classify because of its ability to change its mode of nutrition. In fact, some euglenoids never have chloroplasts and live a totally heterotrophic existence.

Euglenoids are flagellates, most of which have chloroplasts.

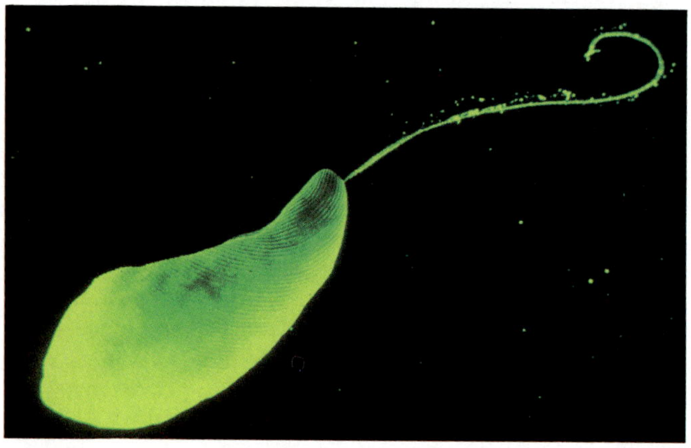

FIGURE 29-8 *Euglena,* a euglenoid.
Euglenoids make their own food by photosynthesis and swim by means of flagella.

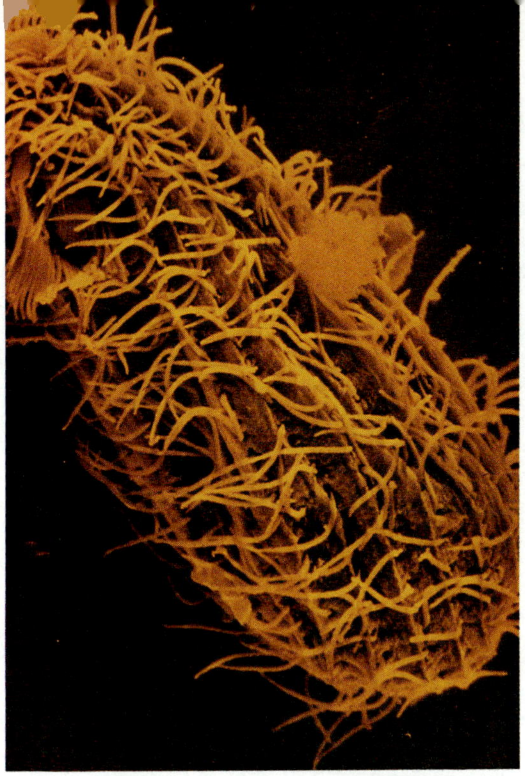

FIGURE 29-9 The ciliates.
Stentor, a funnel-shaped ciliate, showing spirally arranged cilia.

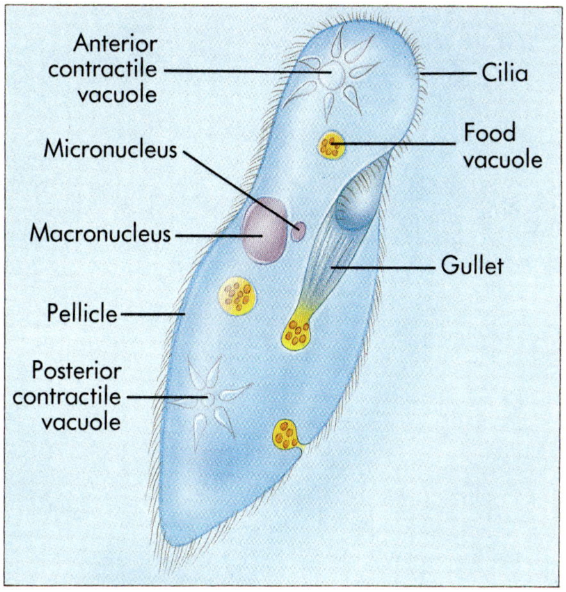

FIGURE 29-10 ***Paramecium.***
This diagram shows the location of the contractile vacuoles, the pellicle, micronucleus and macronucleus, cilia, and food vacuole.

Ciliates

Ciliates (phylum Ciliophora) get their name from a Latin word meaning "eyelash"—a name that is descriptive of the fact that all or parts of these cells are covered with hair-like extensions called *cilia* (Figure 29-9). These cilia beat in unison, moving the cell about or creating currents that move food particles toward the gullet of the cell.

Ciliates possess a wide array of cellular organelles that perform functions similar to the organs of multicellular organisms. An example of this interesting cellular organization is shown in the diagram of the paramecium in Figure 29-10. The paramecium is a ciliate that is classically used as one example of this group. The beating cilia of the paramecium sweep food into its gullet. From the gullet, food passes into a vacuole where enzymes and hydrochloric acid aid in digestion. After absorption of the digested material is complete, the vacuole empties its waste contents into the **anal pore,** located in a special region of the pellicle. The waste then leaves the cell by a process similar to exocytosis.

Paramecia reproduce asexually by **transverse fission** (Figure 29-11, *A*.) In addition, paramecia undergo a type of sexual reproduction shown in Figure 29-11, *B* called **conjugation.** However, conjugation is not really a reproductive process. Instead, two cells exchange genetic material. So although conjugation does not produce offspring cells, it does promote genetic variability among cells that normally produce clones of identical cells when they reproduce. Genetic variability enhances the ability of the population to survive. Some algae, fungi, and bacteria also exchange genetic material in similar processes, which are also termed conjugation.

Most ciliates live in fresh water or salt water and do not infect other organisms. However, one species called *Balantidium coli* inhabits the intestinal tracts of pigs and rats.

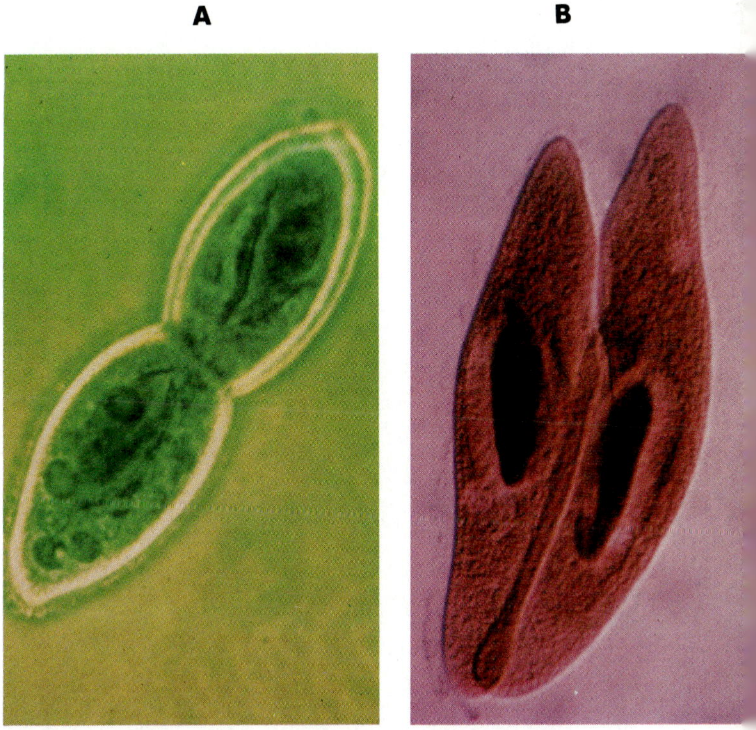

FIGURE 29-11 Reproduction in ***Paramecium.***
A Transverse fission. When a mature *Paramecium* divides, two complete individuals result.
B Conjugation. In this process, two *Paramecia* exchange genetic material to promote genetic variability.

BIOLOGY & *you*

Malaria today

Malaria, caused by the sporozoan *Plasmodium,* is one of the most serious diseases in the world. As many as 250 million people are infected with it at any one time, and approximately 1 million of them, mostly children, die of malaria each year.

Malaria occurs primarily in the tropics, where the female *Anopheles* mosquito, the carrier of the disease, is prevalent. When an infected mosquito bites a person, *Plasmodium* sporozoans are injected into the bloodstream. As Figure 29-A shows, these sporozoans are carried to the liver where they undergo several cycles of asexual reproduction. In a week or two, these cells emerge from the liver, invade red blood cells, and undergo additional stages of asexual division and development. After these stages of development, the sporozoans are called *merozoites*. These cycles of development cause cycles of symptoms. The host often feels well during development of the parasite within the blood cells. Then the infected red blood cells burst, releasing the merozoites, which enter other red blood cells. This event produces chills and fever, accompanied by nausea, vomiting, and severe headache. These symptoms cycle every few days as cells are infected and then burst. During their development with the blood cells, some of the merozoites develop into gametes (sex cells). If the host is again bitten by an *Anopheles* mosquito, these gametes are taken up by the mosquito, and the sexual stages of reproduction take place.

Scientists are currently focusing on developing a vaccine against malaria and are starting to produce promising results. Each of the three different stages of the life cycle of *Plasmodium* sporozoans (the asexual stages in the liver, the stages in the red blood cells, and the sexual stages) produces different antigens and is sensitive to different antibodies. Because the numbers of sporozoans injected by the *Anopheles* mosquito are so high (about a thousand are injected!) and because they can travel to the liver within a few minutes and increase eightfold within 24 hours, a vaccine against these sporozoans is not likely to be effective. Scientists are therefore searching for a compound vaccine that will be effective against the sporozoans, merozoites, and gametes. Human trials are underway and promise to eventually control the disease, despite the great complexity of the *Plasmodium* sporozoan life cycle.

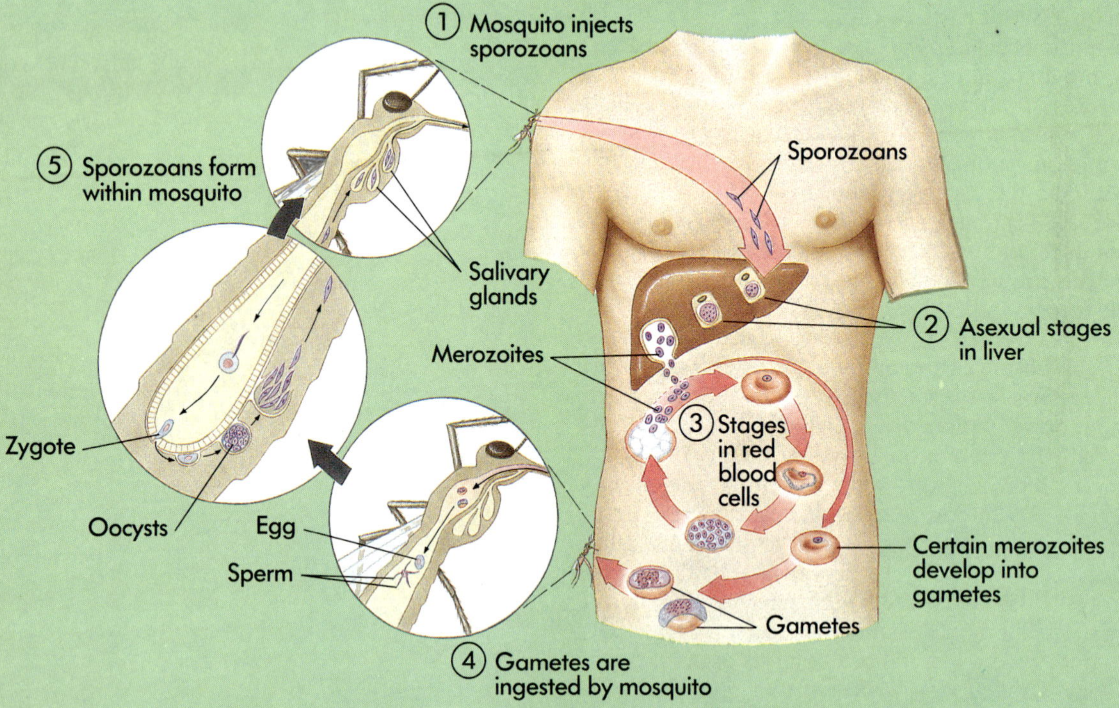

FIGURE 29-A The life cycle of *Plasmodium,* a parasite that causes malaria.

People who come in contact with this protozoan on farms or in slaughterhouses can become infected. The ciliates embed themselves into the lining of the intestines, producing sores and causing dysentery, similar to amebic dysentery. Occasionally, epidemics of balantidiasis occur in areas having poor sanitation.

Ciliates are protozoans characterized by fine, short, hair-like cellular extensions called *cilia.*

Sporozoans

All **sporozoans** (phylum Sporozoa) are nonmotile, spore-forming parasites of vertebrates, including humans. Sporozoans have complex life cycles that involve both asexual and sexual phases. Their spores are small, infective bodies that are transmitted from host to host by various species of insects.

The sporozoan used classically to represent this phylum of protists is *Plasmodium,* the causative agent of malaria. Approximately 1 million people die from this disease each year, so it is considered one of the most serious diseases in the world. The boxed essay describes this disease and the life cycle of *Plasmodium* in more detail.

Sporozoans are nonmotile protozoans that are parasites of vertebrates, including humans. They undergo complex life cycles in which they are passed from host to host by various species of insects.

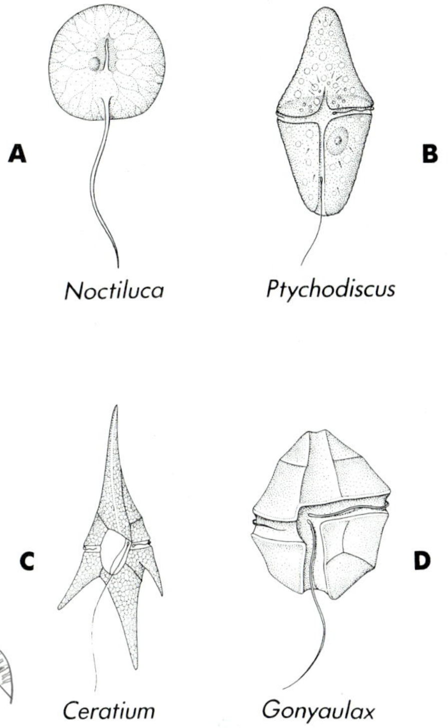

FIGURE 29-12 The dinoflagellates.
A *Noctiluca* lacks the heavy cellulose armor characteristic of most dinoflagellates.
B *Ptychodiscus.*
C *Ceratium.*
D *Gonyaulax.*

Plant-like protists: Algae

Algae are eukaryotic organisms that contain chlorophyll and carry out photosynthesis, so they can be thought of as plant like. However, algae lack true roots, stems, leaves, and vascular tissue (an internal water-carrying system). Algae are widely distributed in the oceans and lakes of the world, floating on or near the surface of the water no lower than the sun's rays can reach.

Some systems of classification place the unicellular algae in the protist kingdom and the multicellular algae in the plant kingdom. However, current thinking favors placing all the algae with the protists, broadening the scope of this kingdom. Formerly, the protist kingdom included only the protozoans and the unicellular algae. It now includes multicellular algae and the fungus-like protists, each of which lacks some of the characteristics of either plants or fungi. These changes reflect the ongoing debate among scientists regarding the evolutionary relationships among organisms and how organisms should be classified to reflect these relationships. Classifying all algae as protists suggests closer evolutionary relationships among the various phyla of algae and other protists than between the algae and the plants. The green algae, thought to be the ancestors to the land plants, are sometimes still placed in the plant kingdom.

There are three phyla of multicellular algae: the **brown algae,** the **green algae** (which includes many unicellular forms), and the **red algae.** These are the phyla that were formerly classified with the plants. There are two phyla of unicellular algae: the dinoflagellates and the **golden algae.**

Dinoflagellates

Although the **dinoflagellates** (phylum Dinoflagellata) have flagella, they look nothing like the flagellates or euglenoids. Many dinoflagellates have outer coverings of stiff cellulose plates, which give them very unusual appearances (Figure 29-12). Their flagella beat in two grooves, one encircling the cell like a belt and the other perpendicular to it. As they beat, the encircling flagellum causes the dinoflagellate to spin like a top; the perpendicular flagellum causes movement in a particular direction. This spinning is a characteristic for which this phylum was named. The word *dinoflagellate* comes from Latin words meaning "whirling swimmer."

Most dinoflagellates live in the sea and carry on photosynthesis. Their photosynthetic pigments are usually golden brown, but some are green, blue, or red. The red dinoflagellates are also called *fire algae.* In coastal areas, these organisms often experience population explosions or "blooms," causing the water to take on a reddish hue referred to as a *red tide.* Red tides destroy other living things because many species of dinoflagellates produce powerful toxins.

FIGURE 29-13 Red tide.
Red tides are caused by a population explosion of certain dinoflagellates. The release of toxins from these dinoflagellates can poison marine life.

These poisons kill fishes, birds, and marine mammals (Figure 29-13). In addition, shellfish strain these dinoflagellates from the water and store them in their bodies. Although the shellfish are not harmed, they are poisonous to humans and other animals that eat them.

Dinoflagellates reproduce primarily by longitudinal cell division, but sexual reproduction has also been shown in more than 10 genera of dinoflagellates.

Dinoflagellates, a type of flagellated unicellular red algae, are characterized by stiff outer coverings. Their flagella beat in two grooves, one encircling the cell like a belt and the other perpendicular to it.

Golden algae

The three types of organisms found in phylum Chrysophyta are the **yellow-green algae,** the **golden-brown algae,** and the diatoms. These organisms are named after the gold-green color of the photosynthetic pigments in their chloroplasts. (The prefix *chryso-* means "color of gold.") An unusual characteristic of these organisms is that they store food as oil. Some forms smell "fishy," giving unpleasant odors to the fresh water lakes they inhabit.

The most well-known members of this group, abundant in both the ocean and in fresh water, are the diatoms—microscopic aquatic pillboxes. Diatoms are made up of top and bottom shells that fit together snugly (Figure 29-14, *A*). These organisms reproduce asexually by separating their top from their bottom, each half then regenerating another top or bottom shell within itself. The shells are composed of silica, so like the radiolarians, the diatoms look somewhat glass like. However, the shells of the diatoms are so characteristically striking and intricate (Figure 29-14, *B*) that it would be hard to confuse them with any other group of protists. Interestingly, the shells of fossil diatoms often form very thick deposits on the sea floor, which are sometimes mined commercially. The resulting "diatomaceous earth" is used in water filters, as an abrasive, and to add the sparkling quality to products, such as the paint used on roads and frosted fingernail polish.

The golden algae have gold-green photosynthetic pigments and store food as oil. They are often represented by diatoms, organisms that look like microscopic pillboxes.

A

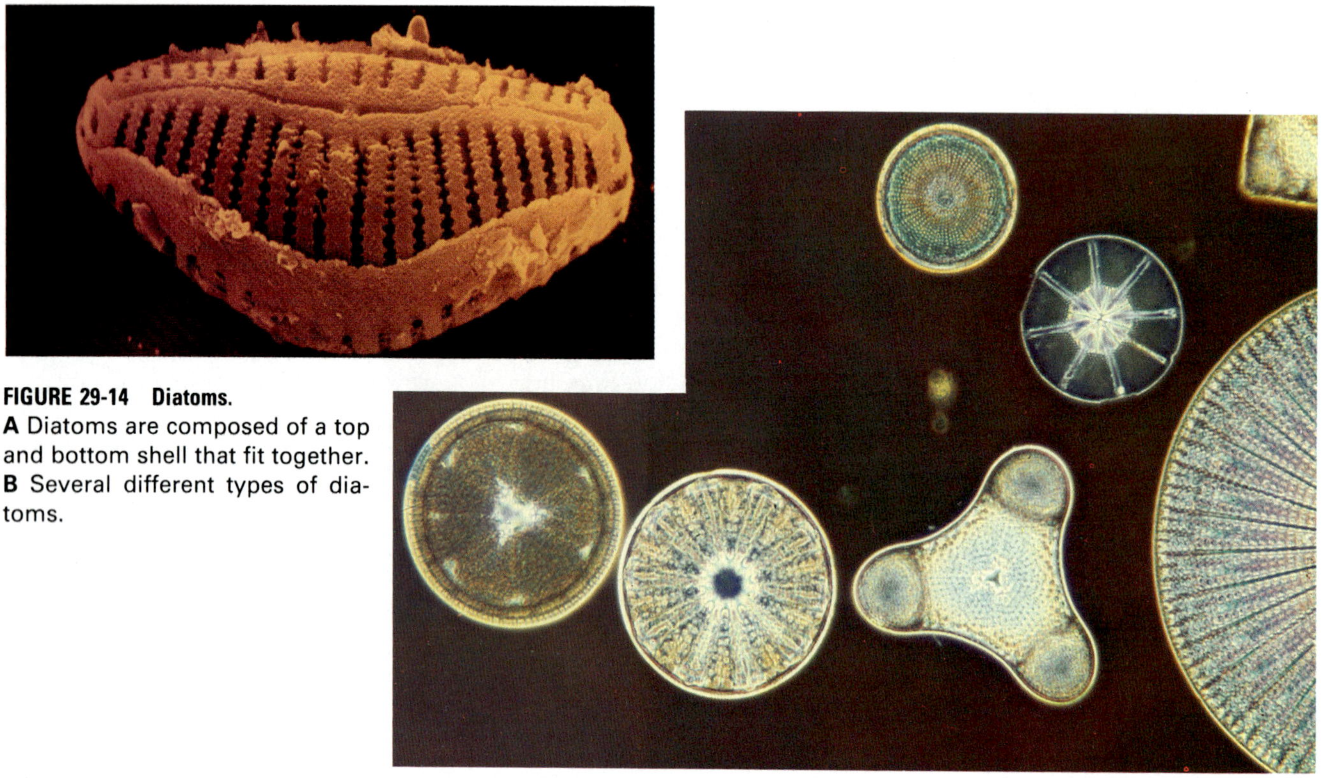

B

FIGURE 29-14 Diatoms.
A Diatoms are composed of a top and bottom shell that fit together. **B** Several different types of diatoms.

Brown algae

The brown algae (phylum Phaeophyta) are the dominant algae of the rocky, northern shores of the world. The types of brown algae that grow attached to rocks at the shoreline are known as *rockweed* (Figure 29-15, *A*). Their puffy air bladders keep the plant afloat during high tide. One type of rockweed is also called *Sargasso weed* and gave the Sargasso Sea its name. The Sargasso Sea is an area of ocean in the mid-Atlantic, east of Bermuda, with unusual water and current patterns that cause it to be quite calm. This calmness allows floating species of the Sargasso weed to proliferate and dominate the area.

The large brown alga with enormous leaf-like structures is kelp (Figure 29-15, *B*). These algae are an important source of food for fish and invertebrates, as well as some marine mammals and birds that live among these seaweeds. Some genera of the kelps are among the longest organisms in the world (rivaling the height of the giant sequoia trees), reaching lengths of up to 100 meters (328 feet)! These algae usually have a structure in which a root-like portion, descriptively termed a **holdfast,** anchors the seaweed to the ocean floor or to rocks. A stem-like **stipe** carries the leaf-like **blades** of the seaweed, which float on or in the water, capturing the sun's rays (Figure 29-15, *C*).

Most brown algae have life cycles that parallel the generalized life cycle of plants (see Chapter 30 and Figure 30-1). The other multicellular algae (brown algae and red algae) also have plant-like life cycles. However, the life cycles of the red algae are quite complex.

Brown algae are large, multicellular algae found predominantly on northern, rocky shores. Some grow to enormous sizes.

A

B

C

FIGURE 29-15 The brown algae.

A Rockweed. The air bladders attached to the algae help keep the plant afloat.

B Kelp. Kelp have large blades (leaf-like structures) and are an important food source for marine fishes, birds, mammals, and invertebrates.

C A kelp forest. The blades of kelp capture sunlight, which they use for photosynthesis. Kelp forests can be quite large.

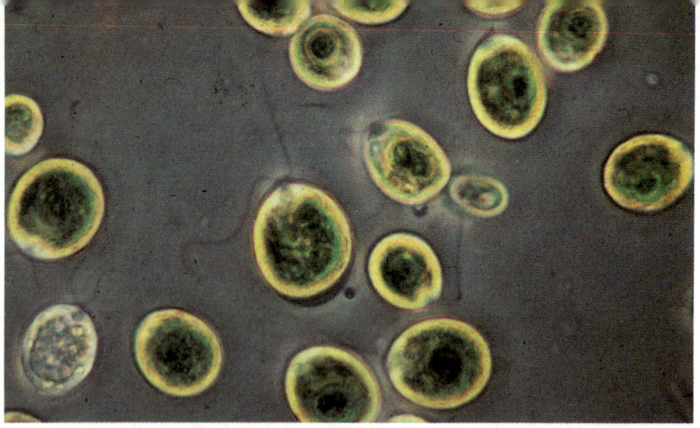

A

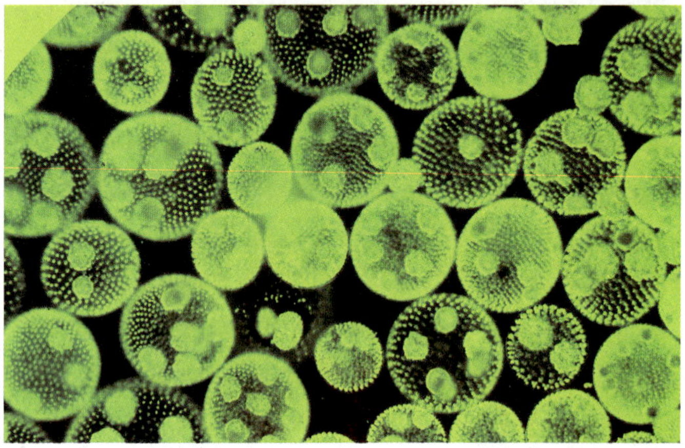

B

C

D

Green algae

The green algae, or Chlorophyta (literally, "green plants") are an extremely varied phylum of protists. In fact, more than 7000 species exist! Of these species, most are aquatic (as are other algae), but some are **semiterrestrial.** Semiterrestrial algae live in moist places on land, such as on tree trunks, on snow, or in the soil. These algae are primarily unicellular microscopic forms, but some are multicellular. Green algae show many similarities to land plants: they store food as starch, they have cell walls composed of cellulose, they have a similar chloroplast structure, and their chloroplasts contain chlorophylls *a* and *b*. For these reasons, scientists think the green algae were ancestors to the plant kingdom.

Well known among the unicellular green algae is the genus *Chlamydomonas* (Figure 29-16, *A*). Individuals are microscopic, green, and rounded and have two flagella at their anterior ends. They are aquatic and move rapidly in the water as a result of the beating of their flagella in opposite directions. They have eyespots that help direct the algae to the sunlight. The life cycle of *Chlamydomonas* is very simple. This haploid organism reproduces asexually by cell division and sexually by the functioning of some of its cells as gametes. After gametes fuse, the diploid zygote divides by meiosis, restoring the haploid state.

Some genera of green algae live together in groups. *Volvox* is one of the most familiar of these colonial green algae. Each colony is a hollow sphere made up of a single layer of individual, biflagellated cells (Figure 29-16, *B*). The flagella of all of the cells beat in such a way as to rotate the colony in a clockwise direction as it moves through the water. In some species of *Volvox* there is a division of labor among the different types of cells, making them truly multicellular organisms.

The multicellular forms of green algae grow in either fresh water or saltwater, such as the sea lettuce *Ulva* shown in Figure 29-16, *C*. Sea lettuce is extremely plentiful in the ocean and is often found clinging to rocks or pilings. Interestingly, this alga consists of sheets of tissue only two cells thick. Another familiar multicellular green alga is the filamentous *Spirogyra*. This alga is interesting because it has spiral chloroplasts (Figure 29-16, *D*).

Green algae include both unicellular and multicellular forms. Most forms are aquatic, as are other algae, but some species live in moist places on land.

FIGURE 29-16 The green algae.

A *Chlamydomonas* is a unicellular green alga that has two flagella and an eyespot sensitive to sunlight.

B *Volvox* is a colonial green alga. Each colony is made up of individual cells.

C *Ulva,* or sea lettuce.

D *Spirogyra,* a filamentous green alga.

Red algae

Almost all red algae, or Rhodophyta ("red plants"), are multicellular, and the majority of their species are marine. Their color comes from the types and amount of photosynthetic pigments present in their chloroplasts. Many species have a predominance of red pigments in addition to chlorophyll and so are red (as their name suggests). Some red algae have a predominance of other pigments so they look green, purple, or greenish black (Figure 29-17). The red pigment, however, is especially efficient in absorbing the green, violet, and blue light that penetrates into the deepest water. (It *reflects* red light.) This enables some of the red algae to grow at greater depths than other algae and inhabit areas in which most algae cannot exist.

The red algae produce substances that make them interesting both ecologically and economically. The coralline algae, for example, deposit calcium carbonate (limestone) in their cell walls. Along with coral animals, these red algae play a major role in the formation of coral reefs. Also, all red algae have glue-like substances in their cell walls: agar and carrageenan. Agar is used to make gelatin capsules, the material for making dental impressions and a base for cosmetics. It is also the main ingredient in the laboratory media on which bacteria, fungi, and other organisms are often grown. Carrageenan is used mainly as a stabilizer and thickener in dairy products, such as creamed soups, ice cream, puddings, and whipped cream, and as a stabilizer in paints and cosmetics. Some of the red algae are used as food in certain parts of the world, such as in Japan.

The red algae play a major role in the formation of coral reefs and produce glue-like substances that make them commercially useful.

A

B

C

FIGURE 29-17 The red algae.

Red algae do not always appear red. Some algae show different colors depending on their daily exposure to light.

A *Ahnfeltia plicata* growing on rocks. This type of alga is edible and is considered a delicacy in many Asian cultures.

B *Bossiella* is a coralline red alga, which means that it secretes the hard substance calcium carbonate.

C *Ahnfeltia concinna,* a common red alga.

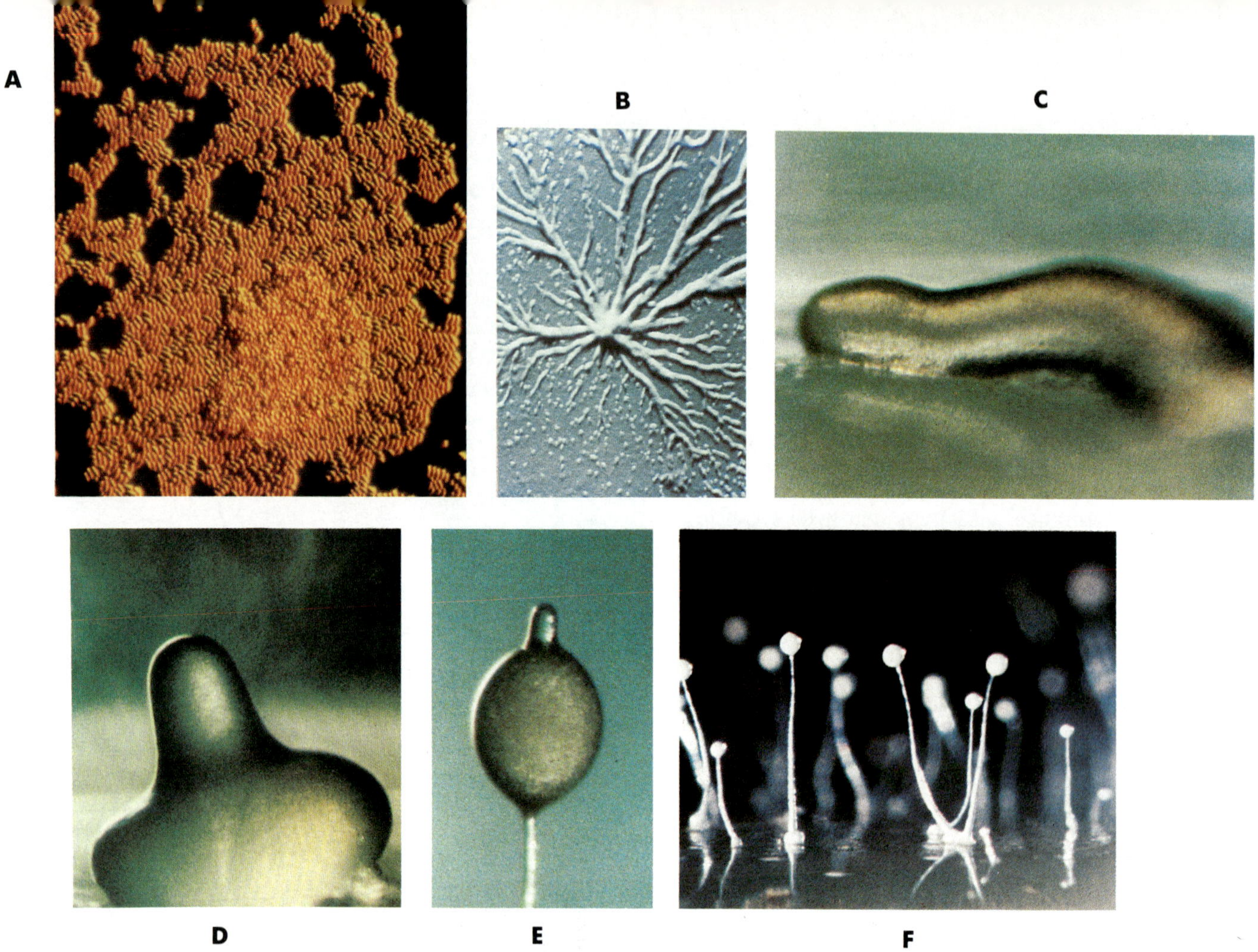

FIGURE 29-18 Development in a cellular slime mold.
A The ameba stage, in which the mold looks and behaves like an ameba.
B The amebas aggregate and move toward a fixed center.
C They form a slug that migrates towards light.
D The slug stops moving and begins to form into a fruiting body.
E The head of a fruiting body.
F Many fruiting bodies together.

Fungus-like protists

The slime molds make up two unique and interesting phyla of protists. They are called molds because they give rise to mold-like (fungus-like) spore-bearing stalks during one stage of their life (see p. 518). However, slime molds look very much like amebas at other times in their life cycles, forming visible "slimy" masses. In fact, scientists now think slime molds may be most closely related to the amebas.

Slime molds are heterotrophs and feed on bacteria, which they find in damp places rich in nutrients, such as rotting vegetation (especially rotting logs), damp soil, moist animal feces, and water. The slime molds are subdivided into two classes: the **cellular slime molds** and the **plasmodial (acellular) slime molds.**

> **Slime molds are organisms that are fungus-like in one phase of their life cycles and ameba-like in another phase of their life cycles.**

Cellular slime molds

Most of their lives, the cellular slime molds look and behave like amebas, moving along and capturing bacteria by means of pseudopods (Figure 29-18, *A*). At a certain phase of their life cycle, which is often triggered by a lack of food, the individual organisms move toward one another (Figure 29-18, *B*) and form a large, moving mass called a *slug* (Figure 29-18, *C*). (Their phylum name, Acrasiomycota, comes from the name of the chemical attractant, acrasin, the cellular form produces, which causes the cells to aggregate.) The slug eventually transforms into a fruiting body (Figure 29-18, *D, E, F*). The tips of the fruiting bodies contain dormant, cyst-like forms of the ameba-like cells and are called *spores.* Some of these spores fuse and undergo a type of sexual reproduction before being released; others do not. Each spore becomes a new "ameba" if it falls onto a suitably moist habitat. The amebas begin to feed and continue the life cycle.

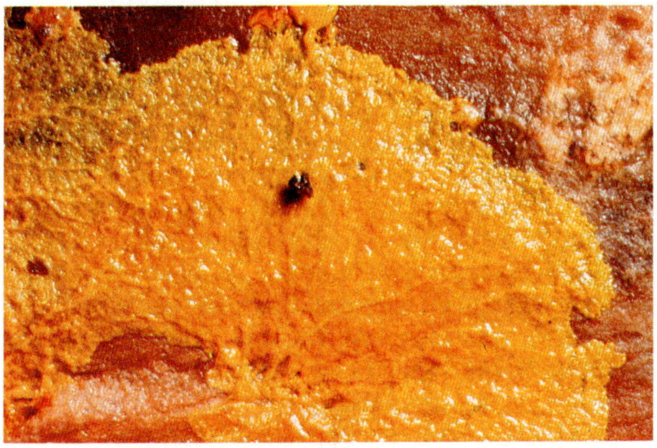

FIGURE 29-19 The plasmodial slime mold.
Plasmodial slime molds move about as a plasmodium, a multinucleated mass of cytoplasm.

Plasmodial (acellular) slime molds

The plasmodial slime molds also have an ameboid stage to their life cycle, but these "amebas" are quite unusual. These bizarre organisms stream along as a **plasmodium**—a nonwalled, multinucleate mass of cytoplasm—which resembles a moving mass of slime (Figure 29-19). Their phylum name, Myxomycota, literally means "mucous mold." The plasmodia engulf and digest bacteria, yeasts, and other small particles of organic matter as they move along. At this stage of its life cycle, a plasmodium may reproduce asexually; the nuclei undergo mitosis simultaneously, and the entire mass grows larger.

When food or moisture is in short supply, the plasmodium moves to a new area and forms spores. These spores are held in spore cases, which have a characteristic look for each genera of acellular slime mold. Figure 29-20 shows three different types of spore cases. The spores are resistant to unfavorable environmental influences and may last for years if dry. Meiosis occurs in the spores. When conditions are favorable, the spore cases open and release spores that germinate into flagellated, haploid cells called *swarm cells.* The swarm cells can divide, producing more swarm cells, or can act as gametes. Gametes can fuse and form a new plasmodium by repeated mitotic divisions.

Water molds

Taxonomists disagree as to whether the water molds should be considered protists or fungi. A primary issue is that water molds have flagellated spores, which are not characteristic of fungi. In this textbook, these organisms are classified with the fungus-like protists.

If you have an aquarium and have seen white fuzz on any of your fish, you have been introduced to the water molds. They live not only in fresh water and saltwater and on aquatic animals but also in moist soil and on plants. Some of the plant diseases caused by this group are late blight and downy mildew. Although this group gets the name *water molds* because many species thrive in moisture, they are sometimes called *egg fungi* because of the large egg cells present during their sexual reproduction.

Water molds thrive in moist places and aquatic environments, parasitizing plants and animals. During sexual reproduction, they produce large egg cells.

A

B

C

FIGURE 29-20 Spore cases of three types of plasmodial slime molds.
A *Arcyria.*
B *Fuligo.*
C *Tubifera.*

Fungi

Fungi are a separate kingdom of mostly multicellular eukaryotic organisms that are **saprophytic;** that is, they feed on dead or decaying organic material. To do this, fungi secrete enzymes onto a food source to break it down and then absorb the breakdown products. Some fungi are **parasites** and feed off living organisms in the same way (as happens in ringworm and athlete's foot). Most fungi are multicellular. Unlike plants, fungi have no chloroplasts and do not produce their own food by photosynthesis. They are composed of slender filaments that may form cottony masses or that may be packed together to form complex structures, such as mushrooms.

The slender filaments of fungi are barely visible to the naked eye. Termed **hyphae** (singular, hypha), they may be divided into cells by cross walls called **septa** (sing., septum) or may have no septa at all. The septa rarely form a complete barrier, however, so cytoplasm streams freely throughout either type of hyphae. Because of this streaming, proteins made throughout the hyphae may be carried to their actively growing tips. As a result, the growth of fungal hyphae may be very rapid when food and water are abundant and the temperature is optimum.

A mass of hyphae is called a **mycelium** (plural, mycelia). Part of the mycelial mass grows above the food source (substrate) and bears reproductive structures. The rest grows into the substrate (Figure 29-21). The part of the mycelium embedded in the food source secretes enzymes that digest the food. For this reason, many fungi are harmful because their mycelia grow into and decay, rot, and spoil foods (and sometimes other organic substances, such as leathers). In addition, some fungi cause serious diseases of plants and animals, including humans.

In their roles as decomposers, fungi may seem troublesome to humans, but they are essential to the cycling of materials in ecosystems (see Chapter 38). In addition, many antibiotics (drugs that act against bacteria) are produced by fungi. Yeasts, which are single-celled fungi, are also useful in the production of foods such as bread, beer, wine, cheese, soy sauce, and yogurt. These foods all depend on the biochemical activities of yeasts, in which they produce certain acids, alcohols, and gases as by-products of their metabolism of sugar. Yeasts can also cause infections in humans, such as vaginal and oral yeast infections.

A

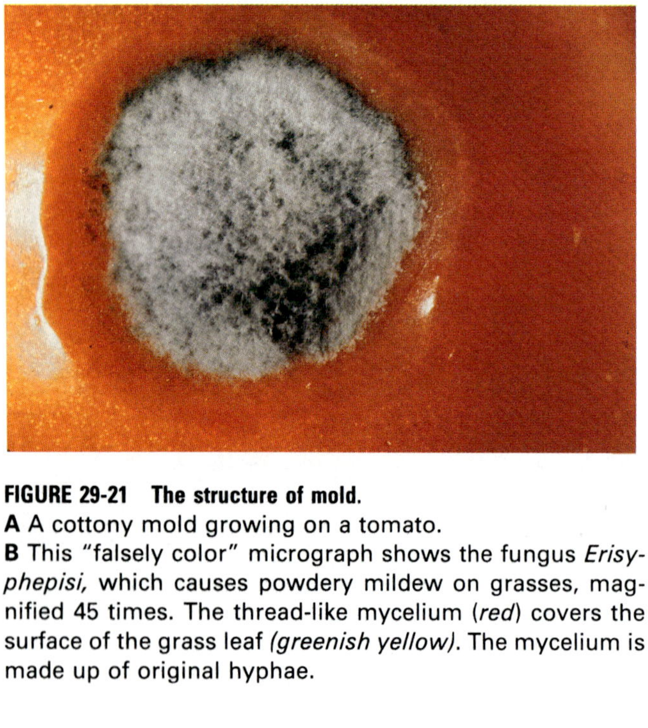

B

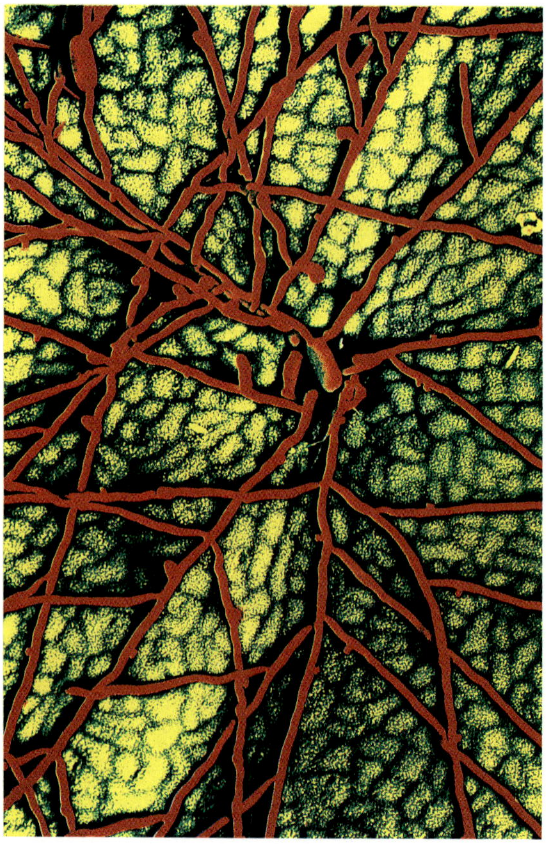

FIGURE 29-21 The structure of mold.
A A cottony mold growing on a tomato.
B This "falsely color" micrograph shows the fungus *Erisyphepisi,* which causes powdery mildew on grasses, magnified 45 times. The thread-like mycelium (*red*) covers the surface of the grass leaf *(greenish yellow)*. The mycelium is made up of original hyphae.

Most fungi reproduce both asexually and sexually. (Figure 29-22 shows a generalized life cycle for many fungi.) During sexual reproduction, hyphae of two genetically different mating types (called + and - strains) come together, and their haploid nuclei fuse and produce a diploid zygote. This zygote germinates into diploid hyphae, which bear **spores.** Spores are reproductive bodies formed by cell division (mitosis or meiosis) in the parent organism. These spores are formed by meiosis in a diploid parent and by mitosis in a haploid parent; the spores are always haploid. They germinate into haploid hyphae after being released and finding appropriate growth conditions. The cycle comes full circle when their hyphae fuse with those of a genetically different mating type, forming zygotes. Some fungi may reproduce asexually by budding, by growing new hyphae from fragments of parent hyphae, or by producing spores by mitosis.

Fungi are classified into divisions rather than into phyla as plants are. Three divisions of fungi include those organisms that have a distinct sexual phase of reproduction: the **zygote-forming fungi,** the **sac fungi,** and the **club fungi.** A fourth group is a "catch-all" division in which organisms are placed because they appear to have no sexual stage of reproduction. This division is termed the **imperfect fungi.**

Fungi are eukaryotic organisms that feed on dead or decaying organic material or parasitize living organisms. Although some fungi cause disease in plants and animals, some are important in the production of food, and most are ecologically important decomposers.

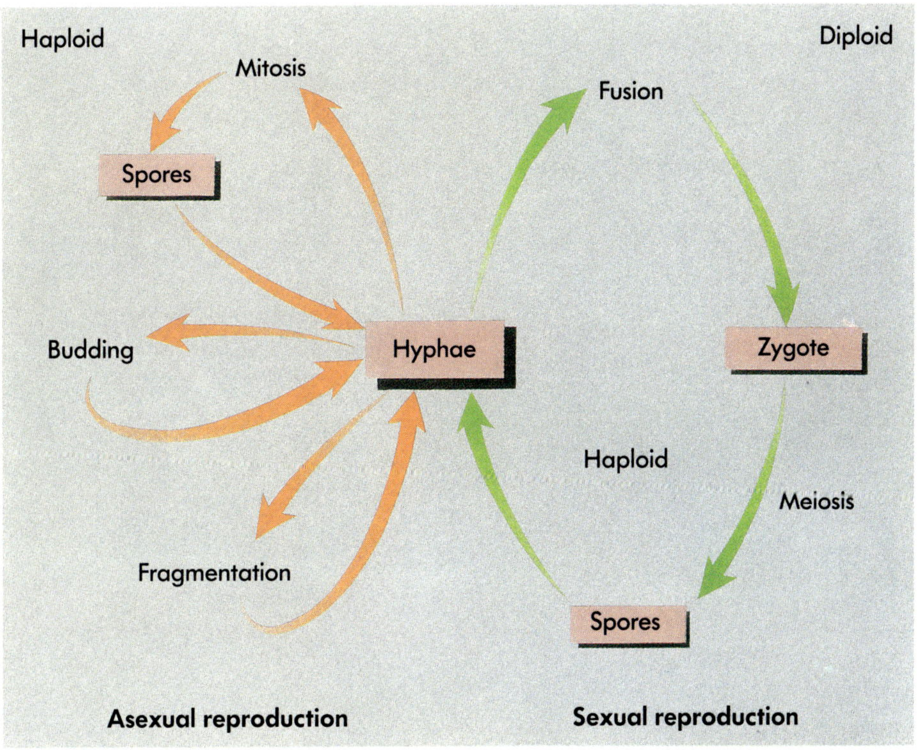

FIGURE 29-22 A generalized life cycle for fungi.
Fungi alternate between sexual and asexual reproductive stages.

Zygote-forming fungi

You may have seen black mold growing on bread or other food (Figure 29-23, *A*), but you may not have seen this mold under a microscope (Figure 29-23, *B*). Members of this group are found on decaying food and other organic material and are characterized by their formation of sexual spores called **zygospores.** For this reason, this group of fungi is called the *zygote-forming fungi* (division Zygomycota). The life cycle of these fungi (Figure 29-23, *C*) parallels the generalized life cycle shown in Figure 29-22 quite closely.

> **Zygote-forming fungi, such as black bread mold, live on decaying organic material and are characterized by their formation of sexual spores called *zygospores.***

A

B

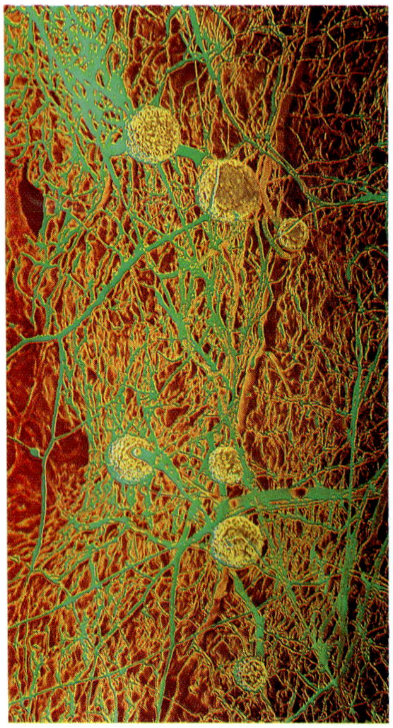

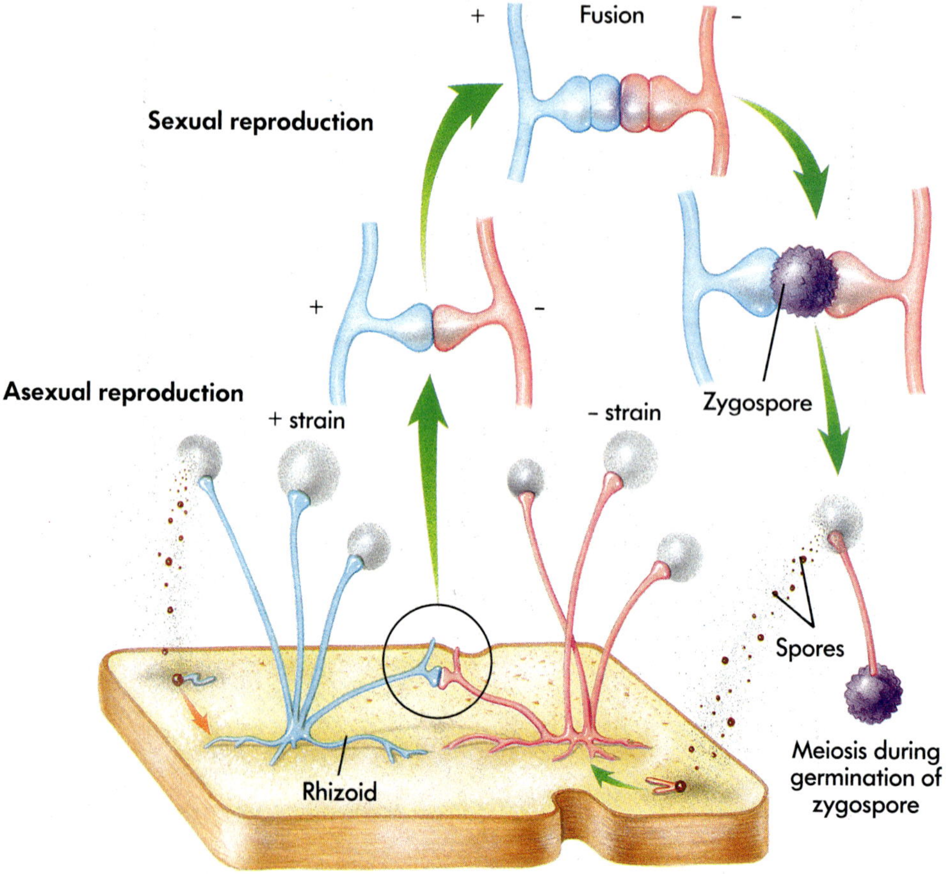

C

FIGURE 29-23 Black bread mold.
A Black bread mold growing on food.
B A microscopic view of black bread mold. This scanning electron micrograph has been color-enhanced to bring out contrasts. The hyphae (the *green strands*) and the sporangia *(yellow balls)* can be clearly seen.
C The life cycle of black bread mold.

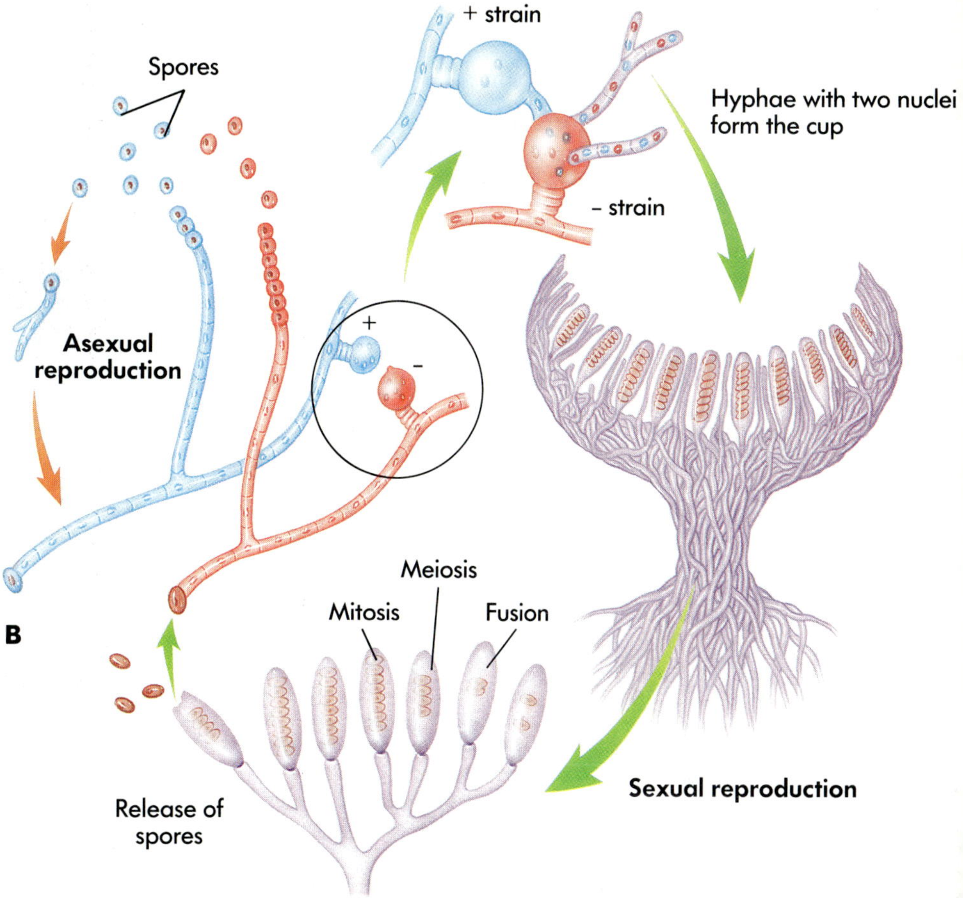

FIGURE 29-24 Sac fungi (division Ascomycota).
A A cup fungus, *Cookeina tricholoma,* from the rain forest of Costa Rica.
B The life cycle of a cup fungus.

Sac fungi

The sac fungi are the largest division of fungi, having at least 30,000 named species. They live in a wide variety of places, such as in the soil, in saltwater and fresh water, on dead plants and animals, and on animal feces. They are called sac fungi because their sexual spores, which occur in groups of eight, are enclosed in sac-like structures called **asci** (sing., ascus). Their life cycles are similar to the club fungi.

Among the sac fungi (division Ascomycota) are such familiar and economically important fungi as yeasts, cup fungi (Figure 29-24), and truffles. Unfortunately, this class of fungi also includes many of the most serious plant pathogens, including the causative agent of Dutch elm disease, which has killed millions of elms in North America and Europe (Figure 29-25). The spores of the mold causing this disease are carried from tree to tree by a bark beetle.

FIGURE 29-25 Dutch elm disease.
Millions of elms in North America and Europe have been killed by Dutch elm disease, which is caused by an ascomycete, *Ceratocystis ulmi.* The spread of Dutch elm disease can be controlled by a combination of sanitation (removing dead and dying trees promptly), killing beetles in traps baited with chemicals, and treating the trees in advance with fungicides and bacteria that inhibit the growth of the fungus.

Although single cells, the yeasts are also sac fungi and are one of the most economically important of the class because of their use in the production of various foods. Most of the reproduction of the yeasts is asexual and takes place by binary fission or by budding (the formation of a smaller cell from a larger one). Sometimes, however, whole yeast cells may fuse, forming sacs with zygotes. These cells divide by meiosis then mitosis, forming eight spores within each sac. When the spores are released, each functions as a new cell.

The sac fungi live in both aquatic and terrestrial environments and are characterized by sexual spores borne in sac-like structures. Familiar examples of this group are cup fungi and yeasts.

Club fungi

This division of fungi includes the mushrooms, toadstools, puffballs, jelly fungi, and shelf fungi (Figure 29-26). Some species of club fungi are commonly cultivated; for example, the button mushroom—commonly served in salad bars—is grown in more than 70 countries, producing a crop with a value of over 15 billion dollars. Other kinds of club fungi are represented by the rusts and smuts, which are devastating plant pathogens. Wheat rust, for example, causes enormous economic losses to wheat wherever it is grown.

B

C

FIGURE 29-26 Club fungi (division Basidiomycota).
A An earth star, *Geastrum saccatus.* Earth stars are a kind of puffball. Basidia form within puffballs, which may have spores. They release hundreds of thousands—or even millions—of basidiospores when mature.
B A shelf fungus, *Trametes versicolor,* growing on a tree trunk.
C A jelly fungus growing in the Amazonian rain forest of Brazil. Although many familiar basidiomycetes form their basidia within mushroom-like basiocarps, many do not.

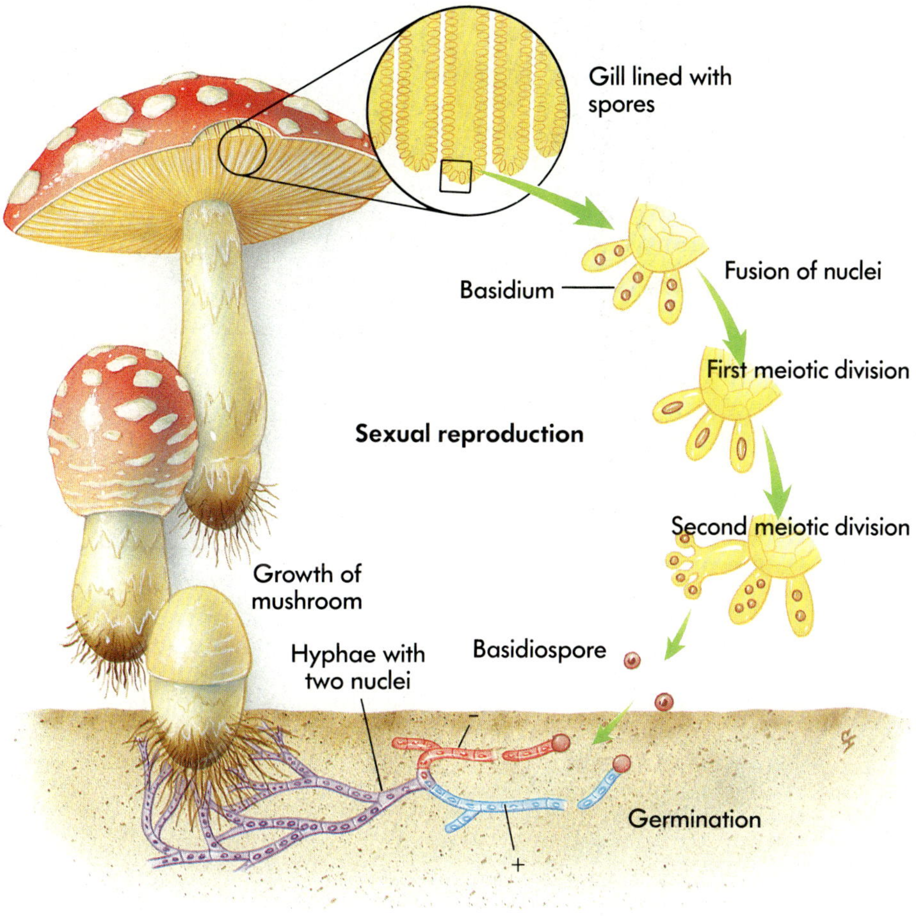

FIGURE 29-27 Life cycle of a common club fungus—a mushroom.
This class of fungi rarely undergoes an asexual stage of reproduction.

The club fungi are so-named because they have club-shaped structures from which unenclosed spores are produced. These structures are called **basidia** (sing., basidium) and give the division name of Basidiomycota. In mushrooms, basidia line the gills found under the cap (Figure 29-27). In rusts and smuts, basidia arise from hyphae at the surface of the plant.

As shown in Figure 29-27, a mushroom is formed when pairs of hyphae (often of two different mating strains) fuse and intermingle their cytoplasm and haploid nuclei. The fused hyphae develop into the mushroom, including the gills on the underside of its cap. The basidia develop as single cells on the free edges of the gills. Within the cells that develop into basidia, the nuclei present from the fusion of hyphae now fuse to form zygotes. Each zygote undergoes meiosis, producing four haploid spores in each basidia. The zygotes within some basidia undergo mitosis, producing basidia having eight haploid spores. Turgor pressure (water build-up within the basidia) bursts the basidia and hurls the spores from mushroom—at an average rate of 40 million per hour! These spores germinate to produce hyphae, which fuse to begin a sexual cycle of reproduction once again. Asexual reproduction is rare in this class of fungi.

The club fungi, named because they have club-shaped structures from which unenclosed spores are produced, include the mushrooms, puffballs, and shelf fungi.

A

B

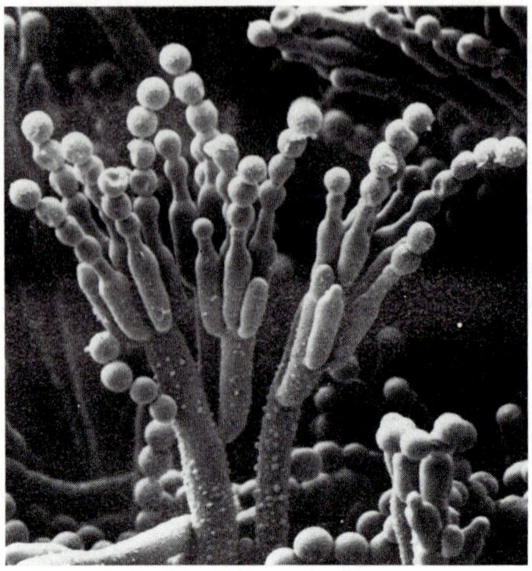

C

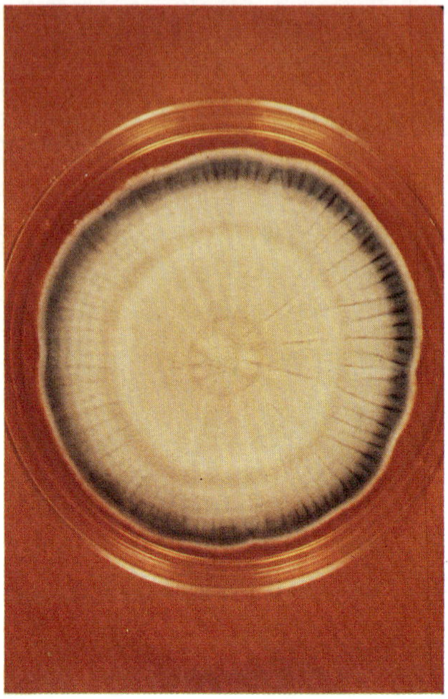

FIGURE 29-28 Imperfect fungi.
A *Aspergillis.*
B *Penicillium* growing on an orange.
C *Penicillium* growing in a petri dish.

Imperfect fungi

This division of fungi includes members whose hyphal structure and asexual reproduction suggest that they are zygote-forming fungi, sac fungi, or club fungi. However, the fungi in this group have lost the ability to reproduce sexually, or their sexual stages of reproduction have not been observed. Therefore these genera cannot be placed in their appropriate division because placement is based on features related to sexual reproduction. They are therefore referred to as *imperfect*—the Fungi Imperfecti. Although these fungi do not reproduce sexually, hyphae of different mating types often fuse, providing some genetic recombination. This characteristic is important to the survival of each species, especially in the plant pathogens because mutation and recombination produce new strains that can still infect plants bred for resistance to these fungal diseases.

The imperfect fungi reproduce asexually by spores. An enormous range of diversity occurs in the structure of the spore cases found at the tips of their hyphae. These variations represent adaptations of the fungi to dispersal of their spores. For example, spores distributed by insects usually have sticky, slimy spore cases with odors attractive to insects. Those dispersed by the wind have dry spore cases.

Among the economically important genera of imperfect fungi are *Penicillium* and *Aspergillus* (Figure 29-28). Some species of *Penicillium* are sources of the well-known antibiotic penicillin, and other species of the genus give the characteristic flavors and aromas to cheeses such as Roquefort and Camembert. Species of *Aspergillus* are used for fermenting soy sauce and soy paste, processes in which certain bacteria and yeasts also play important roles.

Most of the fungi that cause diseases in humans are members of the imperfect fungi. Some of these fungi cause infections of the skin, such as athlete's foot (Figure 29-29) and other forms of ringworm. The term *ringworm* refers to any fungal infection of the skin, including the scalp and nails. There is also a small group of fungi that cause diseases of various organs, including the brain. Some of these diseases are spread by birds, bats, or contaminated soil. Such diseases are usually contracted by breathing air heavily contaminated with spores of the causative agent. Other fungal diseases occur only in persons weakened by other diseases, making them more suseptible to infection by pathogenic fungi.

The imperfect fungi have no sexual stage of reproduction and reproduce asexually by spores. Some members of this division are sources of antibiotics or are important in food production, but many cause disease in humans.

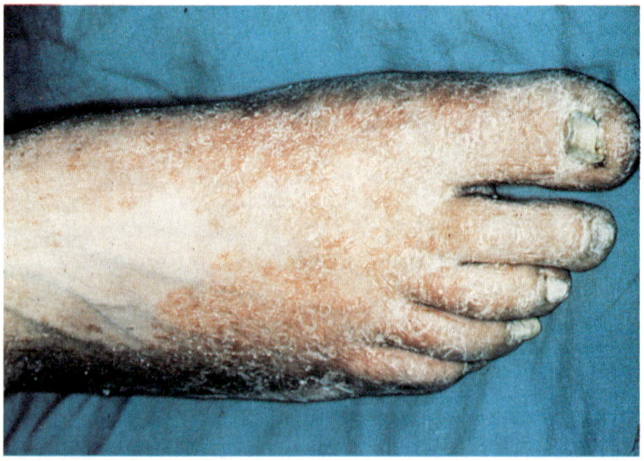

FIGURE 29-29 Athlete's foot.
This itchy, painful condition is caused by a member of the imperfect fungi.

A

B

C

Lichens

Lichens (Figure 29-30) are associations between fungi and photosynthetic partners and are classified with the fungi. They provide an example of mutualism, a living arrangement in which both partners benefit. Most of the visible body of a lichen consists of fungus, but cyanobacteria or green algae (or sometimes both) live within the fungal tissues (Figure 29-31). The photosynthetic organism provides nutrients for itself and the fungus. Specialized fungal hyphae penetrate or envelop the photosynthetic cells and transfer water and minerals to them, although this part of the mutualistic relationship has been questioned recently. Researchers now suspect that the fungus acts more like a parasite than a mutualistic partner.

Lichens are able to invade the harshest of habitats at the tops of mountains, in the farthest northern and southern latitudes, and on dry, bare rock faces in the desert. In such harsh, exposed areas, lichens are often the first colonists, breaking down the rocks and setting the stage for the growth of other plants. These amazing algae-fungus partnerships are able to dry or freeze and then recover quickly and resume their normal metabolic activities. The growth of lichens may be extremely slow in harsh environments—so slow, in fact, that many small lichens appear to be thousands of years old and are among the oldest living things on Earth.

FIGURE 29-30 Three types of lichens.
A Crustose (encrusting) lichens growing on a rock in California.
B A fruticose (shrubby) lichen, *Cladina evansti,* growing on the ground in Florida. Fruticose lichens predominate in deserts because they are more efficient in capturing water from moist air than either of the other two types.
C A foliose (leafy) lichen, *Parmotrema gardneri,* growing on the bark of a tree in the mountain forest in Panama.

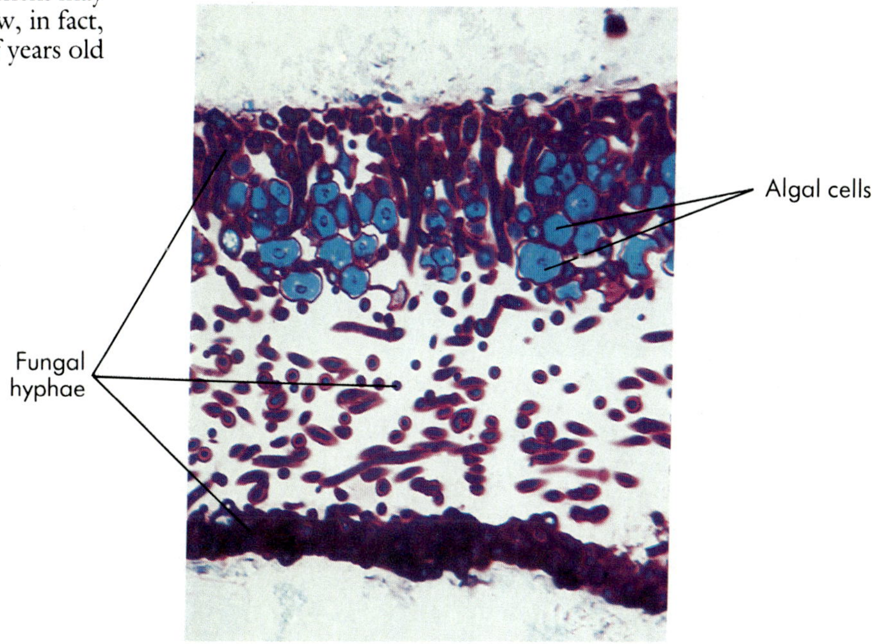

FIGURE 29-31 Section of a lichen.
The fungal hyphae are more densely packed into a protective layer on the top and bottom layers of the lichen. The cells (shown as blue) near the surface of the lichen are a green alga. This alga supplies carbohydrates to the fungus.

Summary

1. The protists (kingdom Protista) are a varied group of eukaryotic organisms. Many are single celled, although some phyla of protists include multicellular or colonial forms. Within this kingdom of organisms are animal-like cells, which take in organic matter for nutrition; plant-like cells, which manufacture their own food; and fungus-like cells, which live on dead or decaying organic matter.
2. Animal-like protists, the heterotrophs, are usually referred to as protozoans. The protozoans are classified according to the means by which they move and feed. The amebas move and eat by means of cell extensions, or pseudopods. The flagellates move by means of long whip-like cellular extensions, their flagella. The ciliates move and sweep food toward themselves by means of short hair-like cellular processes, their cilia. And the sporozoans are nonmotile parasites of vertebrate animals, carried from one host to another by insects.
3. The algae contain chlorophyll, carry out photosynthesis, but lack true roots, stems, leaves, and vascular tissue. There are five phyla of plant-like protists: the dinoflagellates, the golden algae, the brown algae, the green algae, and the red algae. The dinoflagellates and the golden algae are unicellular organisms, whereas brown algae, green algae, and red algae are multicellular. However, the green algae has certain unicellular species.
4. The slime molds have unique life cycles that have fungus-like spore-bearing stages and "slimy" ameba-like stages. The water molds grow in fresh water and saltwater, on aquatic animals, in moist soil, and on plants. These protists produce large egg cells during sexual reproduction.
5. Fungi are multicellular, eukaryotic organisms, most of which feed on dead or decaying organic material. They are important decomposers that are essential to the cycling of materials in ecosystems, are important in the production of certain foods, and also cause certain diseases. Most fungi are composed of slender microscopic filaments, or hyphae, that form cottony masses or compact plant-like structures.
6. Three divisions of fungi have sexual stages of reproduction. These are the zygote-forming fungi (such as black bread mold), the sac fungi (such as cup fungi and yeasts), and the club fungi (such as mushrooms, puffballs, and shelf fungi).
7. The imperfect fungi are a fourth division of fungi. Members of this division have no known sexual stages of reproduction. The imperfect fungi include organisms that produce antibiotics and are used in the production of certain foods but that also cause certain diseases in humans.
8. Lichens are associations of fungi with green algae and/or cyanobacteria. The photosynthetic organism provides nutrients for itself and the fungus. Researchers now question whether the fungus helps the photosynthetic organism; it has been thought that fungal hyphae penetrated or enveloped the photosynthetic cells and transferred water and minerals to them. Lichens are able to exist in harsh environments and live for long periods of time. Some lichens may be among the oldest organisms on Earth.

KEY TERMS

algae 503
ameba 504
anal pore 507
asci 519
basidium (pl., basidia) 521
blade 511
brown algae 509
cellular slime mold 514
ciliate 507
club fungi 517
conjugation 507
diatom 503
dinoflagellate 509
euglenoid 506
eyespot 506
flagellum 506
foraminiferan (pl., foraminifera) 505
golden algae 509
green algae 509
holdfast 511
hyphae 516
imperfect fungi 517
mycelium (pl., mycelia) 516
plasmodial (acellular) slime mold 514
plasmodium 515
positive phototaxis 506
protist 503
protozoan 503
pseudopod 504
radiolarian 505
red algae 509
sac fungi 517
saprophytic 516
semiterrestrial 512
septum (pl., septa) 516
slime mold 504
spore 517
sporozoan 509
stipe 511
swarm cell 525
transverse fission 507
water mold 504
zygospore 518
zygote-forming fungi 517

REVIEW QUESTIONS

1. Explain how the following terms are related: protists, protozoa, algae, diatoms, slime molds.
2. Describe an ameba. How does it move and take in food?
3. What do flagellates and ciliates have in common? How do they differ?
4. What are sporozoans? Briefly summarize a sporozoan's life cycle.
5. What do euglenoids, dinoflagellates, and golden algae have in common? Describe each one.
6. Identify the phylum or phyla to which the following belong: yellow-green algae, diatoms, golden-brown algae.
7. Fill in the blanks: The ________ ________ are protists that are ________ -like in one phase of their life cycles, and ________ -like in another phase of their life cycles.
8. Fill in the blanks: A fungus that feeds on a fallen tree in a forest is ________. A fungus that invades the skin between your toes, creating the uncomfortable condition known as athlete's foot, is a(n) ________.
9. Fungi are both helpful and harmful to human beings. Give some examples of each.
10. Match each term with the most appropriate statement:
    ________ Club fungi
    ________ Zygote-forming fungi
    ________ Sac fungi
    ________ Water molds
    ________ Imperfect fungi
    a. Sexual spores borne in sac-like structures
    b. Have no sexual stage of reproduction
    c. Sexual spores are produced by basidia
    d. Sexual spores are called zygospores
    e. During sexual reproduction, they produce large egg cells
11. What are lichens, and where are they found?
12. Fill in the blanks: Amebas that secrete delicate shells made of silica are called ________; amebas that secrete shells made of calcium carbonate or limestone are called ________.

THOUGHT QUESTIONS

1. Some single-celled protists like *Volvox* form spherical colonies of hundreds of cells. Within the colony, some of the cells become specialized for sexual reproduction. Why is such a colony not considered to be a multicellular organism?
2. If the protist *Euglema* can act "plant-like" by photosynthesizing in the presence of light, and "animal-like" by injesting its food in the absence of light, in which group does it belong? What does this say about the usefulness of this approach to classifying protists?
3. Many fresh water lakes, when polluted by high-phosphate detergents, become overgrown with mats of photosynthetic green algae. If the pollution continues, all other life in the lake soon dies. Why? Why can you not avoid this lethal result by simply poisoning the algae?

FOR FURTHER READING

Barron, G. (1992, March). Jekyll-Hyde mushrooms, *Natural History*, pp. 47-52.
Many fungi belie their benign appearance and actively trap and consume a variety of organisms.

Cochran, M.F. (1990, February). Back from the brink: Chestnuts. *National Geographic*, pp. 128-140.
Nearly destroyed by the chestnut blight, the American chestnut is being rescued by dedicated scientists and volunteers.

McKnight, K.H., & McKnight, V. (1987). *A field guide to mushrooms of North America.* Peterson Field Guide Series. New York: Houghton Mifflin.
This is an excellent guide that includes most of the common and edible species of North America.

Saffo, M.B. (1987). New light on seaweeds. *BioScience, 37*, 654-664.
Seaweeds occur at different depths in the ocean, their photosynthetic pigments being efficient in harvesting energy for the particular light waves that reach that depth.

30 Plants

Reproductive patterns and diversity

Although this does not look like a family photo of two generations of individuals—it is! The leaf-like greenery are plants of one generation and the stalks are of another generation. But interestingly, both are the same plant—the hairy-cap moss *Polytrichum*.

The green plants are the gametophyte generation. Gametophytes do as their name suggests: they produce gametes, or sex cells. These male and female gametes fuse during fertilization to form zygotes.

HIGHLIGHTS

▼

Plants are multicellular, eukaryotic organisms that typically live on land and make their own food by the process of photosynthesis.

▼

Plants, unlike animals, have two different multicellular phases that occur alternately within their sexual life cycles.

▼

The differences between the life cycles of the lower, nonvascular plants and the higher, vascular plants reflects an adaptation of the vascular plants for life on land.

▼

Plants are also able to reproduce asexually by vegetative propagation, a process in which a new plant develops from a portion of a parent plant.

OUTLINE

Characteristics of plants

The general pattern of reproduction in plants

Patterns of reproduction in nonvascular plants

- The bryophytes
  - Mosses
  - Liverworts
  - Hornworts

Patterns of reproduction in vascular plants

- Seedless vascular plants
- Vascular plants with naked seeds
- Vascular plants with protected seeds
- Mechanisms of pollination
- Fruits and their significance in sexual reproduction
- Seed formation and germination
- Types of vegetative propagation in plants

Regulating plant growth: Plant hormones

The generation of plants that develops from the zygotes, the stalk-like plants called *sporophytes*, looks very different from its gametophyte parents. The sporophyte generation is named because it produces spores. And the gametophyte generation is in turn produced from these spores, resulting in a cycling of two very different generations of the same plant.

Characteristics of plants

Because plants have the ability to reproduce sexually, a characteristic they have in common with most living things, other characteristics differentiate them from other organisms and place them in their own kingdom. Plants are multicellular, eukaryotic, photosynthetic autotrophs. That is, they produce their own food (which most plants store as starch) by using energy from the sun and carbon dioxide from the atmosphere (a process described in Chapter 8). However, all organisms fitting this description are *not* plants. The multicellular algae, at one time classified as plants and now classified by most systematists (taxonomists) as protists, also fit this description. Today, plants are often defined as multicellular, photosynthetic eukaryotes that *live on land*, although some botanists disagree with this definition. The few species of plants that live in the water have returned to the water after evolving on land.

TABLE 30-1 The ten divisions of plants

GROUP	DIVISION	EXAMPLES
Nonvascular plants	Bryophyta	"True" mosses, liverworts, hornworts
Seedless vascular plants	Psilophyta	Whisk ferns
	Lycophyta	Club mosses
	Sphenophyta	Horsetails
	Pterophyta	Ferns
Vascular plants with naked seeds	Coniferophyta	Conifers
	Cycadophyta	Cycads
	Ginkgophyta	Ginkgos
	Gnetophyta	Gnetae
Vascular plants with protected seeds	Anthophyta	Flowering plants
	Class Monocotyledons	Grasses, irises
	Class Dicotyledons	Flowering trees, shrubs, roses

Another characteristic that differentiates the multicellular algae from the plants is their type of chlorophyll, the pigment that absorbs energy from the sun during the process of photosynthesis. Plants have the pigments chlorophyll *a* and chlorophyll *b*. The red and brown algae do not have chlorophyll *b* and so are thought not to have evolved from the same ancestors as the plants. The green algae do have chlorophyll *a* and *b* but, because they have many single-celled forms, are also classified as protists. The green algae are thought by most scientists to be the evolutionary predecessors of the land plants.

All plants can be placed into one of two groups: the nonvascular plants and the vascular plants. Most plants (over 80% of all living plant species) are **vascular plants,** those having specialized tissues to transport fluids. There are three major groups of vascular plants: seedless vascular plants (such as ferns), vascular plants with naked seeds (such as pine trees), and vascular plants with protected seeds (such as flowering plants). Plants lacking these specialized transport tissues are called **nonvascular plants.** There is one major group of nonvascular plants: the bryophytes (such as "true" mosses).

Traditionally, plants are taxonomically separated into **divisions** rather than into phyla, but these categories are basically equivalent to one another. In this classification scheme, the plant kingdom has ten divisions. Table 30-1 lists these divisions, including examples. It shows how the divisions fit into the groupings of plants previously described.

The general pattern of reproduction in plants

It is necessary to take a broader look at the patterns of reproduction of all living things to investigate patterns of reproduction in specific divisions of plants. The series of events that take place from one stage during the life span of an organism, through a reproductive phase, and until a stage similar to the original is reached in the next generation is called a **life cycle.** For example, a bacterium's life cycle can be described as beginning when it (along with another daughter cell) splits from a mother cell during binary fission. This life cycle is complete when the bacterium itself splits into two new cells. This is an example of an asexual life cycle.

Animals, plants, some protists, and some fungi have sexual life cycles, which are characterized by the alternation of meiosis and fertilization. For example, your life cycle could be described as all of the events of growth and development that began at fertilization and continued after your birth through the time of your sexual maturation. This is the time you began producing sex cells by the process of meiosis. The cycle would begin again, or have come "full circle," when you have a child—the next generation.

Figure 30-1, *A* diagrams this type of sexual life cycle, which is common to most animals. Notice in the diagram that during the animal life cycle, the multicellular organism has a double set of hereditary material; it is diploid. (One set was contributed by a father and one set was contributed by a mother.) The gametes, however, are haploid single cells.

Animal life cycles are characterized by a multicellular diploid phase and a unicellular haploid phase. The multicellular diploid organisms produce haploid gametes by meiosis. These gametes fuse during fertilization, forming the first cell of a new diploid multicellular organism.

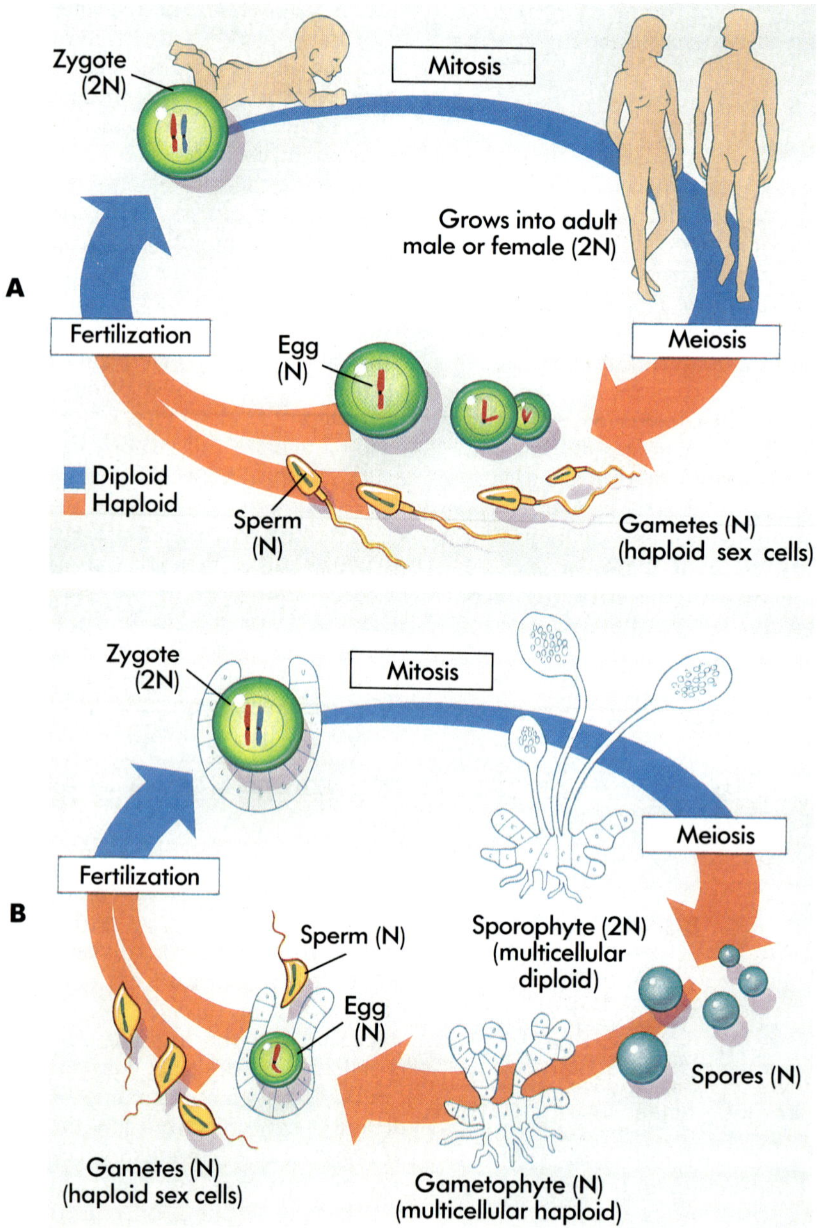

FIGURE 30-1 Sexual life cycles.
A The animal sexual life cycle. The organism that results from the union of haploid gametes is diploid.
B A generalized plant life cycle showing the alternation of generations. The sporophyte generation (diploid) alternates with the gametophyte generation (haploid). Haploid spores are produced by the sporophyte generation by meiosis. When dispersed, these haploid spores grow into gametophytes.

In some protists and many fungi, the haploid phase of the life cycle is multicellular. Gametes are produced from these organisms by mitosis rather than by meiosis. Only the zygotes are diploid (see Chapter 29).

Plants and some species of algae have life cycles that differ from the two sexual life cycles already described. They have both a multicellular haploid phase and a multicellular diploid phase. Because both phases of the life cycle are multicellular, this type of life cycle is called **alternation of generations** (Figure 30-1, *B*). The alternating generations of plants are called the **sporophyte** (spore-plant) generation and the **gametophyte** (gamete-plant) generation. As previously mentioned, gametophytes form gametes, or sex cells. Because they are **haploid** individuals, having *half* the usual number of chromosomes for that species of organism, they form gametes (which are haploid) by the process of mitotic cell division. These gametes fuse during fertilization, a process that depends on the presence of water in which sperm swim to the egg. The plants produced from fertilized eggs are **diploid** plants, having a full complement of genetic material. These diploid plants are the sporophytes. Because sporophytes are diploid, they use the process of meiosis to produce haploid spores. When dispersed, spores grow into gametophyte plants.

Plant life cycles are marked by an alternation of generations of diploid sporophytes with haploid gametophytes. As a result of meiosis, sporophytes produce spores, which grow into gametophytes. Gametophytes produce gametes as a result of mitosis. These gametes fuse during fertilization and grow into sporophytes.

Looking at general characteristics of the bryophytes and the vascular plants and their patterns of reproduction, you will notice that each group has its own variation of a life cycle. However, the gametophyte (haploid) generation often dominates the life cycles of the nonvascular plants, whereas the sporophyte (diploid) generation dominates the life cycles of the vascular plants. This difference reflects an adaptation of the vascular plants for life on land as they evolved from earlier, nonvascular forms. These adaptations included the development of spores with protective walls able to tolerate dry conditions, efficient water and food-conducting systems, and gametophyte generations that became protected by and nutritionally dependent on the sporophyte generation.

Plants produce gametes and spores in specialized structures. In the bryophytes and several divisions of vascular plants, eggs are formed in structures called **archegonia** (singular, **archegonium**). Sperm are produced in structures called **antheridia** (sing., **antheridium**). In the more specialized vascular plants, except in a few species of seed-bearing plants, these specialized gamete-producing structures have been reduced in size during the course of evolution as the sporophyte generation began to dominate the life cycle. Eggs and sperm differentiate from a small number of haploid cells within the sporophyte. These haploid cells are actually the gametophyte generation.

A

B

FIGURE 30-2 Bryophytes.
A A liverwort, *Marchantia.* The sporophytes are carried within the tissues of the umbrella-shaped structures that arise from the surface of the flat, green, creeping gametophyte.
B *Antheros,* a hornwort. This photo shows the long, slender sporophyte with the gametophyte below.

Patterns of reproduction in nonvascular plants

The bryophytes

Bryophytes are a single division of small, low-growing plants that are commonly found in moist places. The three classes of bryophytes are mosses (such as the hairy-cap moss shown on p. 526), liverworts (Figure 30-2, *A*), and hornworts (Figure 30-2, *B*).

A

B

FIGURE 30-3 Peat mosses, *Sphagnum*.
A The glistening black, round objects are the spore cases, which contain spores. The spore cases of *Sphagnum* have a lid that blows off explosively, releasing the spores. **B** A peat bog being drained by the construction of a ditch. In many parts of the world, accumulations of peat, which may be many meters thick, are used as fuel.

Mosses

The largest class of bryophytes and probably the one most familiar to you is the **mosses.** The gametophytes of most mosses have small, simple leaf-like structures often arranged in a spiral around stem-like structures. These bryophyte structures are different from the stems and leaves of vascular plants; see Chapter 31. Many small, carpet-like plants are mistakenly called mosses. For example, *Spanish moss* is actually a flowering plant, a relative of the pineapple. Even eliminating these impostors, however, there are still some 10,000 species of true mosses, and they are found almost everywhere on Earth.

One kind of moss that is most important economically is *Sphagnum.* (Figure 30-3). This moss grows in boggy places (low-lying, wet, spongy ground), forming dense and deep masses that are often dried and sold as peat moss. Peat moss is used in gardening as a mulch, layered around trees or plants to protect the roots from temperature fluctuations, retain moisture, control weeds, and enrich the soil. It functions well as a mulch and a soil additive because its tissues have special water storage cells. These cells allow the peat to absorb and retain up to 90% of its dry weight in water, whether or not the moss is alive.

Figure 30-4 shows the life cycle of the hairy-cap moss. Flask-shaped archegonia are found among the top leaves of the female gametophytes. Each archegonium produces one egg. Antheridia are found in a similar place in the male gametophytes. Each antheridium produces many sperm. The flagellated sperm swim through drops of water from rain, dew, or other sources into the neck of the archegonium and then to the egg. After fertilization takes place, the zygote develops into a young sporophyte within the archegonium. It then grows out of the archegonium and differentiates into a slender stalk. This stalk is initially green, but its chlorophyll disintegrates as it matures, leaving the stalk yellow or brown. At this stage the sporophyte stalk derives its nourishment from the gametophyte. Each sporophyte stalk bears a spore capsule near its tip. Haploid spores are produced by meiosis within this capsule. When the top of the spore capsule pops off, the spores are freed. Under the proper conditions, these spores germinate into thread-like filaments; the characteristic leafy gametophytes arise from buds that form on these filaments.

Liverworts

Liverworts were given their name in medieval times when people believed that plants resembling particular body parts were good for treating disease of those organs. Some liverworts are shaped like a liver and were thought to be useful in treating liver ailments. The ending *-wort* simply means "herb."

A well-known example of a liverwort is *Marchantia*, shown in Figure 30-2, *A*. It has green, leafy gametophytes that grow close to the ground. Antheridia and archegonia develop within the umbrella-like portion of the stalks that grow up from the leaf-like gametophyte. The sporophytes develop encased within these tissues; spores are freed from these structures.

Hornworts

The **hornwort** is named because it has elongated sporangia that protrude like horns from the surface of the creeping gametophytes (Figure 30-2, *B*). Its life cycle parallels quite closely those of the mosses and liverworts.

The life cycles of all three classes of bryophytes—the mosses, liverworts, and hornworts—are somewhat uniform, all having two distinct phases. The sporophyte generation grows out of or is embedded in the tissues of the gametophyte generation and depends on the gametophyte generation for nutrition. Although the gametophyte plant is the most familiar, neither the sporophyte nor gametophyte generation tends to dominate the life cycle.

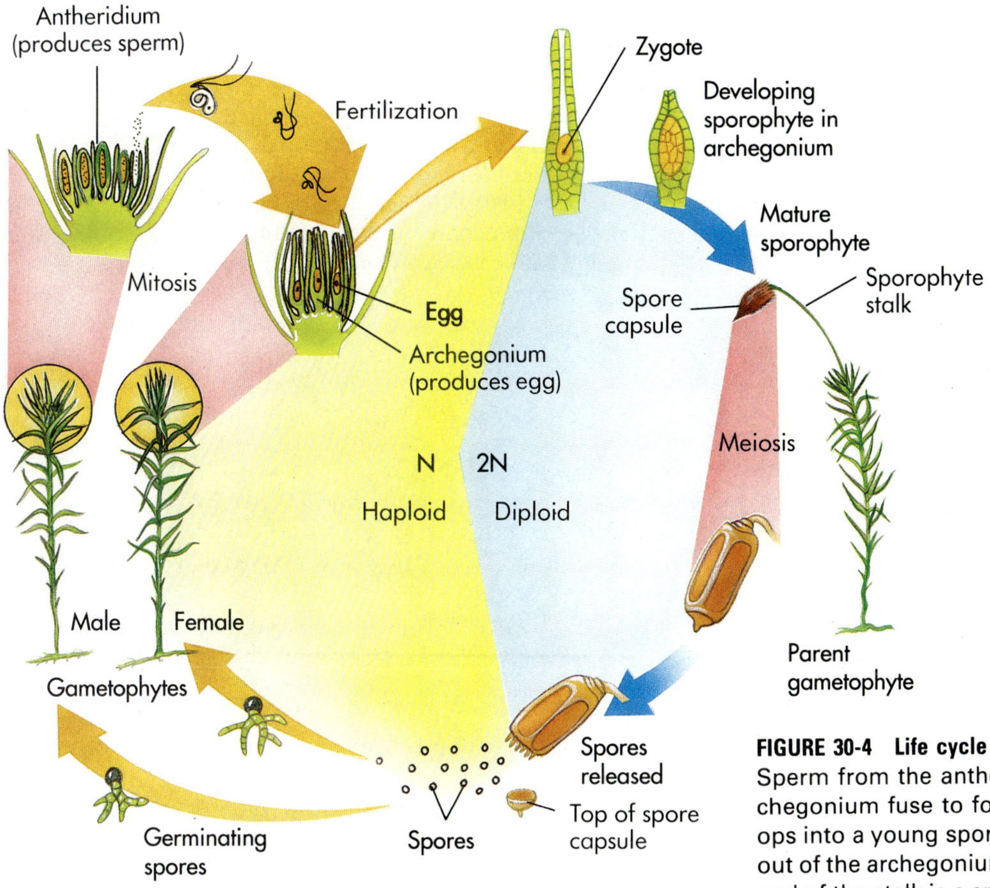

FIGURE 30-4 Life cycle of a moss.
Sperm from the antheridium and an egg from an archegonium fuse to form a zygote. This zygote develops into a young sporophyte, which eventually grows out of the archegonium and into a slender stalk. At the end of the stalk is a spore case. The spore case breaks open, releasing spores, which germinate and produce gametophytes.

Patterns of reproduction in vascular plants

Seedless vascular plants

As listed in Table 30-1, the members of four divisions of vascular plants do not form seeds. **Seeds** are structures from which new sporophyte plants grow; they protect the embryonic plant from drying out or being eaten when it is at its most vulnerable stage. Seeds also contain stored food for the new plant. The seedless plants overcome these problems in interesting ways. Figure 30-5 diagrams the life cycle of a fern—a familiar member of the **seedless vascular plants.** Its life cycle is representative of this group.

When a fern plant is mature, it produces spores by meiosis. The spore cases of ferns look like dots on the underside of the fern leaf, or **frond.** Because it produces spores, the fern is the sporophyte generation—the dominating form in the life cycle of the seedless vascular plants. After its spores are dispersed, those that settle in a moist environment will germinate into haploid plants that look very *unlike* ferns. These plants are small, ground-hugging, heart-shaped gametophytes. Their antheridia and archegonia are protected somewhat by being located on the underside of the plant. Sperm, when released, swim through moisture collected on the underside of the leaf to the archegonia. Each fertilized egg (zygote) starts to grow within the protection of the archegonium. After this initial protected phase of growth, the fern sporophyte is able to grow on its own and becomes much larger than the gametophyte.

The life cycles of seedless vascular plants are similar to one another. The sporophyte (diploid) generation is dominant and lives separately from the gametophyte. The gametophytes produce motile sperm that need water to swim to the eggs. The fertilized eggs produce young sporophytes that grow protected within gametophyte tissues but eventually become free living.

Many of the seedless vascular plants that lived about 300 million years ago were converted long ago to a fuel used today—coal. Club mosses that grew on trees, horsetails, ferns, and tree ferns made up great swamp forests during the Carboniferous Period (see Chapter 26). Areas of New York State, Pennsylvania, and West Virginia, for example, were lying near the equator at that time (see Chapter 26). Dead plants did not completely decay in the stagnant, swampy waters, and they accumulated. These swamps were later covered by ocean waters. Marine sediments piled on top of the plant remains. Pressure and heat acted on the layers of dead plant material beneath the ocean floor and converted the remains to coal. When you burn fossil fuels such as coal, you are burning a resource that was formed under special conditions that have not been repeated in the last 300 million years. Coal is therefore called a *nonrenewable resource*. Burning coal adds carbon dioxide to the atmosphere that had been locked in this fossil fuel for millions of years. This gas adds to the amount of carbon dioxide in the atmosphere and contributes to the problems of global warming and acid rain (see Chapter 40).

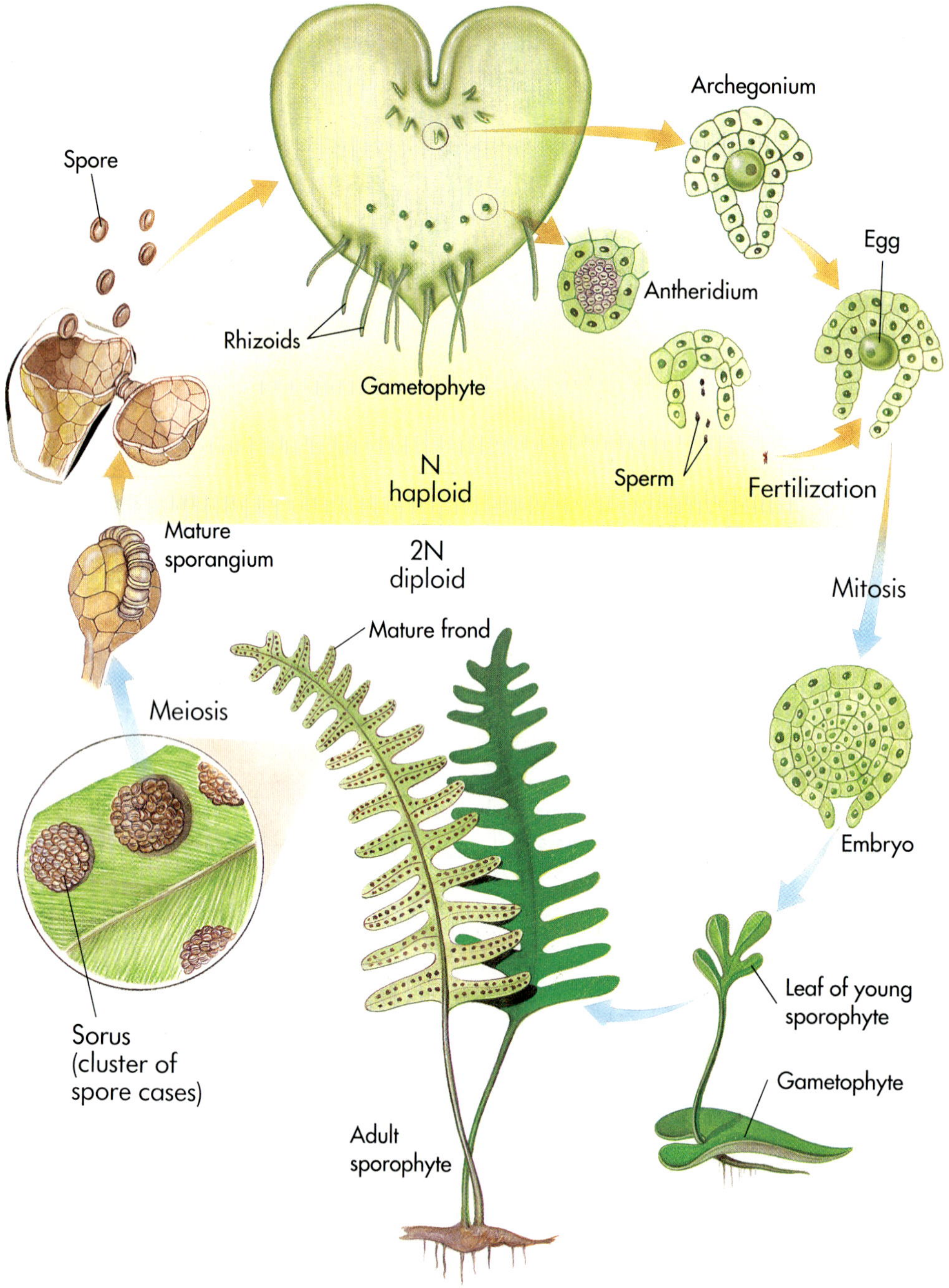

FIGURE 30-5 Life cycle of a fern.
The dark dots that you have seen on the underside of a fern frond are actually clusters of spore cases. The spore case breaks open, releasing spores, which develop into a heart-shaped gametophyte. The archegonia and antheridia produce eggs and sperm, which fuse to form a zygote. The zygote develops into a sporophyte, which becomes much larger than the gametophyte.

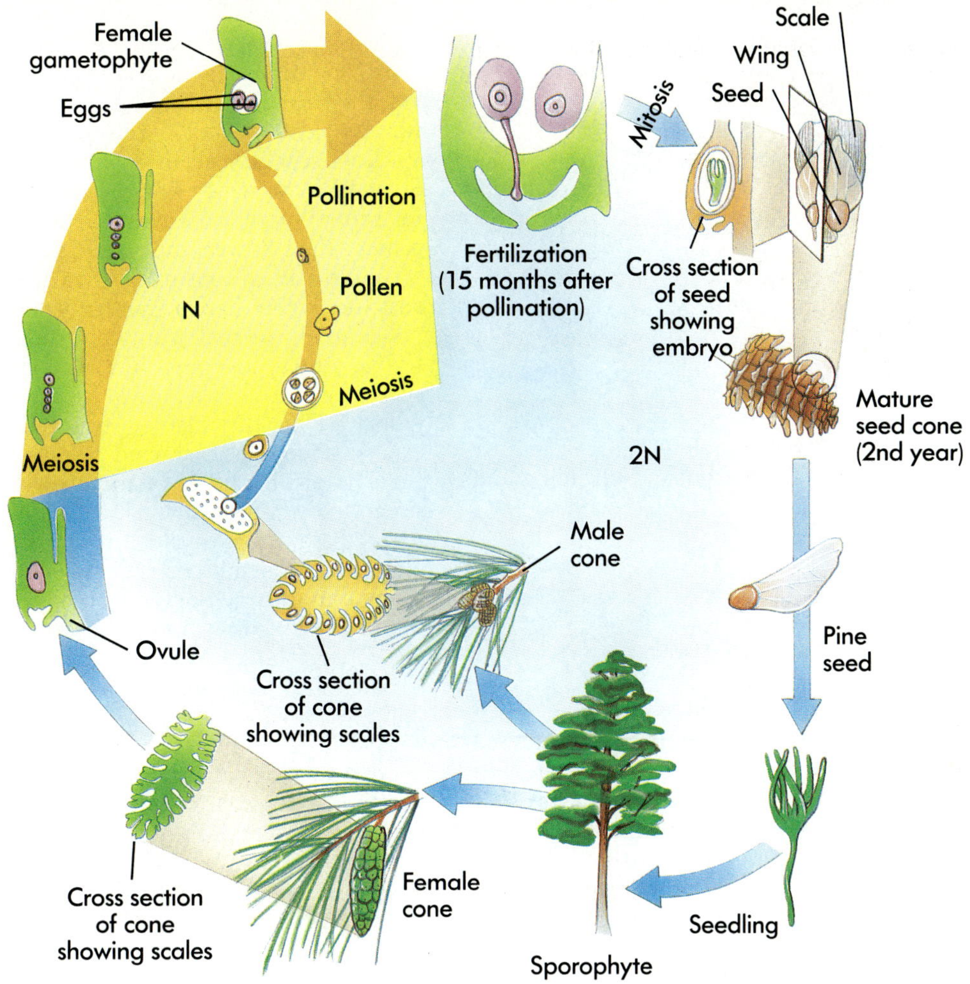

FIGURE 30-6 Life cycle of a pine.
Start at the sporophyte, which is a full-grown pine tree. The gametophyte generation of a pine is so small that it is not visible to the naked eye! The male gametophyte is the pollen. The female gametophyte (archegonium) is multicellular and contains two eggs, of which only one will develop into an embryo. The female gametophyte is depicted as the white area around the eggs and is multicellular tissue.

Vascular plants with naked seeds

Four divisions of vascular plants fall in the category of vascular plants with naked seeds: the conifers, cycads, ginkgo, and gnetophytes. This group is called the **gymnosperms,** a name derived from Greek words meaning "naked seed." The term *naked* refers to the seeds of gymnosperms, which are not completely enclosed by the tissues of the parent at the time it is pollinated.

The most familiar division of gymnosperm are the conifers, or cone-bearing trees. The conifers include pines, spruces, firs, redwoods, and cedars. Figure 30-6 diagrams the life cycle of a pine. The pine tree is the sporophyte (diploid) generation. Interestingly, the gametophyte generation is so reduced that it is not visible to the naked eye and is comparatively short lived.

Pines and most other conifers bear both male cones and female cones on the same tree. You can tell them apart because the female cones are the larger of the two. These cones produce spores that undergo meiosis, producing the male and female gametophytes. Male gametophytes are pollen grains, each consisting of four cells. The male gametophytes produce sperm, which remain located in the pollen grains. The multicellular female gametophytes each produce two to three eggs. The eggs develop within protective structures called **ovules** within the scales of the female pine cones. In the spring, the male cones release their pollen, which is blown about by the wind. (Those allergic to pine pollen know this fact quite well!) As some of this pollen passes by female cones, it gets trapped there by a sticky fluid produced by the now open female cones. As this fluid evaporates, the pollen is drawn further into the cone. When the pollen comes into contact with the outer portion of the ovule, it germinates and forms a pollen tube that slowly makes its way into the ovule—to the egg. After 15 months the tube reaches its destination and discharges its sperm. Fertilization takes place, producing a zygote. The development of the zygote into an embryo takes place within the ovule, which matures into a seed. Eventually the seed falls from the cone and germinates, and the embryo resumes growing and becomes a new pine tree.

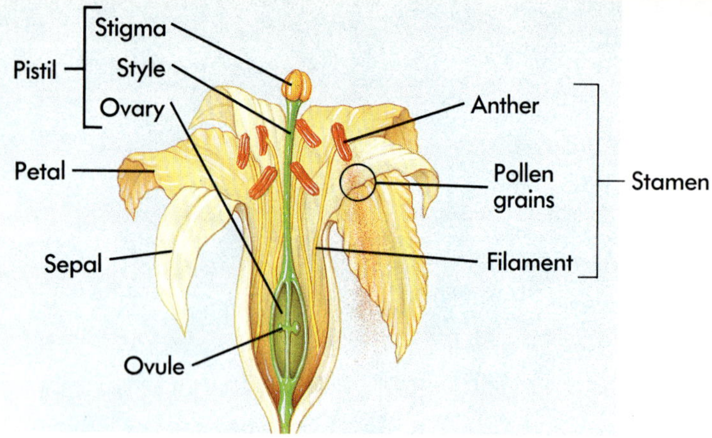

FIGURE 30-7 An angiosperm flower.
The main parts of a flower, which contain both male and female structures.

Vascular plants with protected seeds

Flowers are the organs of sexual reproduction in the vascular plants with protected seeds—the **angiosperms.** These plants bear seeds in a **fruit.** The flower is structured to promote sexual reproduction—the union of gametes that ultimately develops into an embryo within the seed.

Developmentally, flower parts are actually modified leaves. The outermost whorl, or ring, of modified leaves are the **sepals** (Figure 30-7). Sepals enclose and protect the growing flower bud. In some flowers such as roses, the sepals remain small, green, and somewhat leaflike. In other flowers, such as tulips, the sepals become colored and look like the **petals,** the next whorl of flower parts.

Flower petals are frequently prominent and colorful and attract pollinating animals, especially insects. Although the sepals and petals are the dominant outward features of flowers, they are not the organs of sexual reproduction. Many flowers either do not have sepals or petals or have inconspicuous sepals or petals—particularly flowers that are pollinated by the wind. The sex organs of the flower are the innermost modified leaves, the male **stamens** and, at the center of the flower, the female **pistil.**

Each stamen consists of an **anther,** a compartmentalized structure where haploid **pollen grains** are produced. Each pollen grain is a male gametophyte, enclosed within a protective outer covering. This gametophyte will produce sperm by mitosis, which will be enclosed within the pollen grain. A long, thin filament bears and supports the anther, exposing the pollen to wind or pollinating animals.

The pistil consists of three parts. At its tip is a sticky surface called the **stigma** to which pollen grains can adhere. At the base of the pistil is the **ovary,** a chamber that completely encloses and protects the ovules. Within each ovule, a single mother cell develops and then divides meiotically, producing four cells. One of these cells develops into a female gametophyte. When mature, the female gametophyte is called an *embryo sac.* Within this sac are typically eight cells, one of which is the egg. The ovary will become a seed when its eggs are fertilized by the male gamete. The **style** is a narrow stalk arising from the top of the ovary that bears the stigma. The style may be either long or short to facilitate the best exposure of the stigma to the method of pollination that characterizes the plant (Figure 30-8). Pollination in the flowering plants refers to the transfer of pollen from the anther to the stigma.

Figure 30-9 diagrams the life cycle of an angiosperm. After a pollen grain lands on the stigma, it produces a long tube that grows from the pollen grain down the **style** and penetrates the ovary, entering an ovule. One of the two haploid nuclei (sperm) in the pollen tube fertilizes the haploid egg nucleus in the ovule, producing a diploid zygote.

A

B

FIGURE 30-8 Different modes of pollination.
A Pollination by a bumblebee. As this bumblebee, *Bombus,* collects nectar from the flame azalea, the stigma contacts its back and picks up any pollen that the bee might have acquired there during a visit to a previous flower.
B Pollination by a butterfly. This copper butterfly, *Lycaena gorgon,* is probing a flower with its proboscis, a coiled, tongue-like organ, to extract its nectar.
C Pollination by a hummingbird. Long-tailed hermit hummingbird extracting nectar from the flowers of *Heliconia imbricata* in the forests of Costa Rica. Notice the pollen grains on the bird's beak.

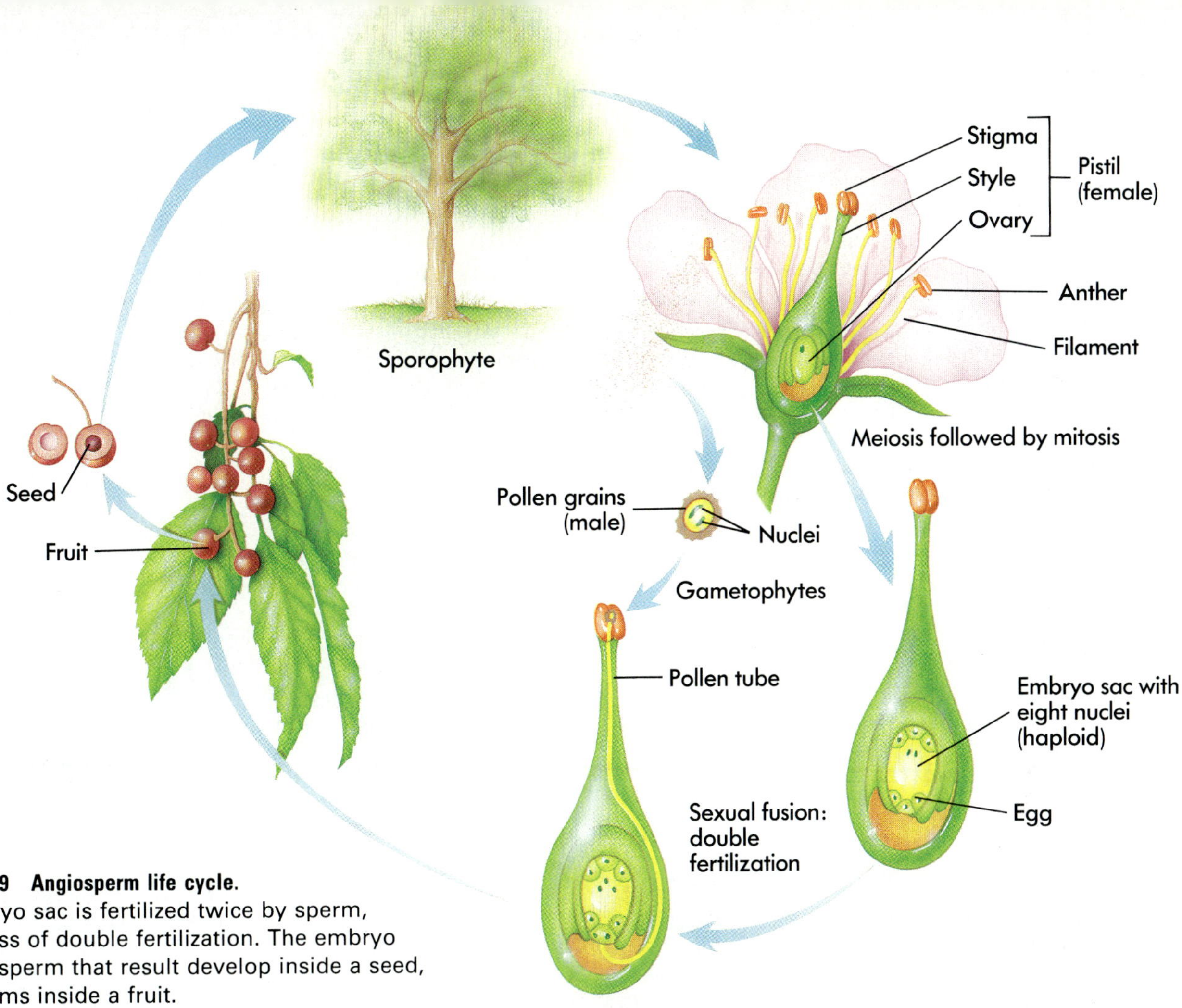

FIGURE 30-9 Angiosperm life cycle.
The embryo sac is fertilized twice by sperm, the process of double fertilization. The embryo and endosperm that result develop inside a seed, which forms inside a fruit.

The zygote will become a new plant embryo. The other sperm nucleus fuses with two other nuclei of the embryo sac, producing a *triploid* endosperm nucleus. The endosperm nucleus develops into tissue that will feed the embryo as it grows into a plant. The ovule becomes the seed within which the embryo develops, and the ovary ripens into a fruit. The fruit is a food that attracts animals, which play a role in seed dispersal.

Sexual reproduction in gymnosperms and angiosperms produces seeds by the union of male gametes released from pollen grains and female gametes within ovules. Germinating pollen tubes convey sperm to eggs without a watery environment. The seeds, formed within the ovary after the union of sperm and eggs, protect and nourish the young sporophytes as they begin to develop into new plants.

Mechanisms of pollination

Pollination takes place in many ways, such as by insects, animals, and the wind. Insects and other animals often visit the flowers of angiosperms for a liquid called **nectar.** Nectar is a food rich in sugars, amino acids, and other substances.

In certain angiosperms and in most gymnosperms, pollen is blown about by the wind and reaches the stigmas passively. However, wind does not carry pollen far or precisely compared with insects or other animals. Therefore plants pollinated by the wind usually grow close together. These plants typically produce large quantities of pollen.

In some angiosperms, the pollen does not reach other individuals at all. Instead, it is shed directly onto the stigma of the same flower, sometimes in bud. This process is termed **self-pollination.**

The relationship between animals known as **pollinators** and flowering plants has been one of the most important features in the evolution of both groups. For plants to be effectively pollinated by animals, a particular insect or other animal must visit many plants of the same species. Flowers have evolved various colors and forms that attract certain pollinators, thereby promoting effective pollination. Yellow flowers, for example, are particularly attractive to bees, whereas red flowers attract birds but not insects. Insects in turn have evolved a number of special traits that enable them to obtain food efficiently from the flowers of the plants they visit.

Plants can be pollinated by animals, insects, or the wind. The evolution of both flowering plants and their animal pollinators is closely linked.

Fruits and their significance in sexual reproduction

Parallel to the evolution of the angiosperms' flowers and nearly as spectacular has been the evolution of their fruits. Fruits have evolved a diverse array of shapes, textures, and tastes and exhibit many differing modes of dispersal.

Fruits that have fleshy coverings—often black, bright blue, or red—are normally dispersed by birds and other vertebrates. Just as red flowers attract birds, the red fruits signal an abundant food supply. By feeding on these fruits, birds and other animals carry seeds from place to place and thus transfer the plants from one suitable habitat to another (Figure 30-10, *B*). Other fruits, such as snakeroot, beggar ticks, and burdock (Figure 30-10, *A*), have evolved hooked spines and are often spread from place to place because they stick to the fur of mammals or the clothes of humans. Others, like the coconut and those that occur on or near beaches, are regularly spread by water (Figure 30-10, *C*). Some fruits have wings and are blown about by the wind. The dandelion is a familiar example (Figure 30-10, *D*).

Seeds are dispersed by sticking to or being eaten by animals, by being blown by the wind, or by floating to new environments across bodies of water.

A

B

C

D

FIGURE 30-10 How seeds are dispersed.
A Animal dispersing seeds. This English setter is covered with hound's tongue weeds seeds that have stuck to his coat.
B Bird dispersing seeds. This cedar waxwing feeds berries to the waiting young in the nest. Birds that eat seeds digest them rapidly so that much of the seed is left intact. What is excreted from the bird can grow into a mature plant.
C The seeds of a coconut, *Cocos nucifera,* sprouting on a sandy beach. One of the most useful plants for humans in the tropics, coconuts have become established even on the most distant islands by drifting in the waves.
D The seeds of a dandelion, *Pyropappus caroliniana,* are dispersed by the wind. The "parachutes" disperse the fruits of dandelions widely in the wind, much to the gardener's despair.

Finding new food plants

Just three species of plants—rice, wheat, and corn—supply half of all human energy requirements, and only about 150 kinds of plants are used extensively. Among the 250,000 known species of plants, there may be tens of thousands of additional kinds of plants that could be used for human consumption if their properties were known and they were cultivated. Finding new food plants in the tropics is especially crucial, since the population in these areas is expected to climb steadily. Making these tropical countries less dependent on imported food is a goal that researchers are striving to fulfill. Unfortunately, efforts to find new food plants in the tropics are hampered by deforestation. Cutting tropical rain forests to make way for more traditional food crops will greatly reduce the opportunity to uncover new food plants that may reside in the rain forest.

One of the most fertile areas for biological exploration is the study of plant use by people who live in direct contact with natural plant communities. These people know a great deal about the plants with which they come into contact, and the screening that they and their ancestors have performed over many generations can, if the results are understood and cataloged, help scientists recognize new plants. For example, the herbal curer in the jungles of southern Surinam has a profound knowledge of uses of plants that grow in his region, but this knowledge is being lost as civilization and modern medicine move into the area.

Some recent triumphs in the search for new, useful plants are the following:

Jojoba *(Simmondsia chinensis)* (Figure 30-A, *1*) is native to the deserts of northwestern Mexico and the southwestern United States. It has been an important crop in several parts of the world since the late 1970s because the liquid waxes in its seeds have characteristics similar to sperm whale oil, which is important for certain kinds of fine lubrication. Jojoba has the added advantage of being easy to grow: it can be cultivated in lands too dry for most other crops to grow. Industries once dependent on sperm whale oil for lubrication have found the jojoba oils cheaper to use. This new availability of jojoba oils has made the hunting of sperm whales uneconomical, helping to save them from extinction.

Grain amaranths (*Amaranthus* species) were important grain crops of the Latin American highlands in the days of the Incas and Aztecs, but are little-used now (Figure 30-A, *2*). Their use was suppressed because they played a role in pagan ceremonies of which the Spanish conquerors disapproved. The grain amaranths, fast-growing plants that produce abundant grain, are rich in lysine, an amino acid rare in most plant proteins but essential in animal nutrition. These ancient crop plants are currently being investigated for widespread development.

Winged bean *(Psophocarpus tetragonolobus)* is a tropical vine that produces highly nutritious seeds, pods, and leaves. Its tubers are eaten like potatoes, and its seeds produce large quantities of an edible oil much like safflower oil. First cultivated locally in New Guinea and Southeast Asia, the winged bean has spread since the 1970s throughout the tropics, where it holds great promise as a source of food. (Figure 30-A, *3*).

Perhaps the strangest of all the new food crops is not a plant at all, but a bacterium. *Spirulina* is a photosynthesizing bacterium that is used as a traditional food source in Africa, Mexico, and other regions. It is currently being studied as a possible food source for other countries. *Spirulina* has a higher protein content than soybeans and is easy to grow—the bacterium thrives in very alkaline water and is cultivated easily in large ponds. However, people must overcome certain psychological barriers to eating this kind of food if the efforts to transplant *Spirulina* out of its traditional cultures are to be successful.

FIGURE 30-A
1 Jojoba.
2 Grain amaranth.
3 Winged bean.

1

2

3

Seed formation and germination

The long chain of events between fertilization and maturity is **development**. During development, cells become progressively more specialized, or differentiated. Development in seed plants results first in the production of an embryo, which remains dormant within a seed until the seed germinates, or sprouts.

Seed **germination** depends on a variety of environmental factors, the most important of which is water. However, the availability of oxygen (for aerobic respiration in the germinating seed), a suitable temperature, and sometimes the presence of light are also necessary. The first step in germination of a seed occurs when it imbibes, or takes up, water. Once this has taken place, metabolism within the embryo resumes.

Germination and early seedling growth require the mobilization of food storage reserves within the seed. A major portion of almost every seed consists of food reserves. Angiosperms fall into two groups regarding the placement of stored food in their seeds: the **monocots** and the **dicots.** In dicots, most of the stored food is in the **cotyledons**, or seed leaves. In monocots, most of the food is stored in extraembryonic tissue called **endosperm** (Figure 30-11). (Although dicots store some food in endosperm, the amount of endosperm varies among species of plants.) Its single cotyledon absorbs food from the endosperm and shuttles it to the embryo. Monocots and dicots also differ from one another in a number of features. Monocots usually have parallel veins (fluid-carrying tissues) in their leaves, and their flower parts are often in threes. Among the monocots are the lilies, grasses, cattails, orchids, and irises (Figure 30-12, *A* to *C*). Dicots usually have netlike veins in their leaves and their flower parts are in fours or fives. The dicots include the great majority of familiar plants: almost all kinds of flowering trees and shrubs and most garden plants, such as snapdragons, chrysanthemums, roses, and sunflowers (Figure 30-12, *D* to *F*).

Usually the first portion of the embryo to emerge from the germinating seed is the young root, or **radicle,** which anchors the seed and absorbs water and minerals from the soil. Then, the shoot of the young seedling elongates and emerges from the ground.

As the shoot emerges from the soil, the first true leaves are protected by a straight sheath or a curved stem. Multiplication of the cells in the tips of the stem and roots, along with their elongation and differentiation, initiates and continues the growth of the young seedling.

The embryo of a plant remains dormant within the seed until the seed germinates. Germination begins when the seed takes up water and begins to sprout.

FIGURE 30-11 Monocots and dicots.
A In monocots, most of the food for the embryo is stored in endosperm. Other characteristics of monocots are the parallel veins in their leaves and the occurrence of their flower parts in threes.
B Dicots store their food in cotyledons, or seed leaves. They have net-like veins in their leaves, and their flower parts occur in fours or fives.

A

B

D

E

F

FIGURE 30-12 Examples of monocots and dicots.
Monocots include the cattail **(A)**, the crested dwarf iris **(B)**, and the pink lady's slipper **(C)**. All of these plants show the parallel leaf veination typical of monocots. Dicots include the rose **(D)**, the sea grape **(E)**, and the bloodroot **(F)**. These examples show the net-like veination typical of dicots.

FIGURE 30-13 Clones of aspen.
The contrast between the golden quaking aspens *(Populus tremuloides)* and the dark green Engelmann spruces *(Picea engelmannii)* evident in this autumn scene near Durango, Colorado, makes it possible to see their clones, large colonies produced by single individuals spreading underground and sending up new shoots periodically.

Types of vegetative propagation in plants

Vegetative propagation is an asexual reproductive process in which a new plant develops from a portion of a parent plant. Some plants, such as irises and grasses, produce new plants along underground stems called **rhizomes.** Other plants, such as strawberries, have horizontal stems that grow above the ground called **runners** or **stolons.** These plants produce new roots and shoots at nodes (places where one or more leaves are attached) along these stems. New plants can also arise vegetatively from specialized underground storage stems called **tubers.** A white potato, for example, is a tuber and can grow a new plant from each of its eyes.

In some plants, new shoots can arise from roots and grow up through the surface of the soil. For example, a group of aspen trees often consists of a single individual that has given rise to a colony of genetically identical trees by producing new shoots from its horizontal roots (Figure 30-13).

In a few species of plants, even the leaves are reproductive. The plant in Figure 30-14 is commonly called the *maternity plant* because small plants arise in the notches along the margins of the leaf. When mature, they drop to the soil and take root. Gardeners commonly propagate African violets from leaf cuttings and many other plants from stem cuttings.

A major breakthrough in the asexual propagation of plants has been the development of **cell culture techniques** (Figure 30-15). Using these techniques, scientists are able to remove individual cells from a parent plant and grow these cells into new individuals. In this way, botanists are able to produce virtually unlimited numbers of genetically identical offspring. Cell culture techniques have been particularly useful in propagating plants that are slow to multiply on their own, such as coconut palms and redwoods, and in cultivating varieties of individual plants with special characteristics, such as large flowers. Award-winning varieties of orchids, for example, are often produced in this way. Cell culture is also used for the commercial production of tremendous numbers of plants in a short period of time—such as the chrysanthemums you may buy at the grocery store. The timber industry uses cell culture techniques in developing rapidly growing conifers such as the Douglas fir.

The use of cell culture techniques in agriculture will revolutionize the way in which farmers produce some crops. In the future, farmers will be able to grow crop plants that have been mass-produced by cell culture and are genetically superior in some way, such as disease-resistance, high yield per plant, or desirable taste. In addition, the techniques of genetic engineering allow scientists to insert genes that express favorable characteristics into the chromosomes of culture cells—another potential revolution in worldwide agriculture.

Many plants can reproduce vegetatively by growing new plants from their roots, stem, or leaves. Scientists, using cell culture techniques, can also propagate many plants from a single cell of the parent. Use of these techniques along with the genetic engineering of plants will revolutionize the way that farmers produce some crops.

FIGURE 30-14 ***Kalanchoe daigremontiana.***
The small plants growing in the notches along the leaf margins will drop to the soil and take root. For this reason, *Kalanchoe daigremontiana* is called the "maternity plant."

FIGURE 30-15 A plant growing in a cell culture.
The small shoots growing in the petri dish are genetic clones of carrots. Researchers use clones and tissue cultures to try to find hardier and more insect-resistant varieties of plants.

Regulating plant growth: Plant hormones

As a plant grows, it is influenced by environmental factors such as the amount of water and light it receives. However, a plant's growth, differentiation, maturation, flowering, and many other activities are also regulated by chemicals called **hormones.** Hormones are chemical substances produced in small, often minute, quantities in one part of an organism and then transported to another part of the organism where they bring about physiological responses.

There are at least five major kinds of hormones in plants. **Auxins, gibberellins,** and **cytokinins** promote and regulate growth. Cytokinins stimulate cell division, and auxins and gibberellins promote growth through cell elongation (Figure 30-16). The differences in concentration of auxin from one side of the stem or root to another, for example, controls the bending of plants toward or away from light (phototropism) (Figure 30-17) and toward or away from gravity (gravitropism). In contrast, **abscisic acid** is a growth inhibitor that induces and maintains dormancy or otherwise opposes the three growth-promoting hormones. **Ethylene** is released by plants as a gas and effects the ripening of fruit and leaf drop in nearby plants.

Other chemicals, such as plant pigments called **phytochromes,** also affect a plant's response to its environment. Phytochromes change form in response to day length and thereby stimulate or inhibit flowering. The study of plant hormones and other regulatory chemicals, especially how they produce their effects, is an active and important field of research today.

FIGURE 30-16 The effect of the plant hormone gibberellin.
This cabbage plant has been treated with gibberellin, which promotes growth through cell elongation.

FIGURE 30-17 Phototropism.
Plants grow toward the light because auxin, a plant hormone that controls plant growth, is produced in higher levels in darkness. Thus the side of the plant away from the light will grow faster, causing the plant to bend toward the light. The stems of this blood sorrel are oriented toward the window, which lets in plenty of sunlight.

Summary

1. Plants are multicellular, eukaryotic, photosynthetic autotrophs that live on land: the bryophytes (nonvascular plants) and the vascular plants. Current classification schemes usually classify the multicellular algae, once classified as plants, with the protists.
2. Animals, plants, some protists, and some fungi have sexual life cycles that are characterized by the alternation of the processes of meiosis and fertilization. Plants and some species of algae have life cycles that have both a multicellular haploid phase and a multicellular diploid phase. Because both phases of the life cycle are multicellular, this type of life cycle is called *alternation of generations*.
3. In plant life cycles, multicellular haploid plants produce gametes by mitosis and are gametophytes. These gametes fuse during fertilization and grow into the multicellular, diploid spore-producing plants, or sporophytes. These individuals form spores by meiosis, which grow into gametophytes.
4. In general, the gametophyte generation often dominates the life cycles of the nonvascular plants, whereas the sporophyte generation dominates the life cycles of vascular plants. This difference reflects an adaptation of the vascular plants for life on land.
5. The bryophytes include three divisions of nonvascular plants: the mosses, liverworts, and hornworts. The life cycles of most have two distinct phases, with neither the sporophyte nor the gametophyte dominating the life cycle, although the gametophyte plants are the ones usually more familiar. The sporophyte plants live on and derive nutrients from the gametophyte plants.
6. The sporophyte phase of the life cycle is the dominant phase of the life cycle of vascular plants. The seedless vascular plants produce motile sperm that swim through moisture on the gametophyte to fertilize the egg. The sporophytes that develop are protected initially by the gametophyte tissues. The gametophytes in seed-bearing plants consist of only a few cells. These cells produce eggs and sperm. After fertilization the embryonic sporophyte is protected and nourished within a seed.
7. The organs of sexual reproduction in the flowering plants, or angiosperms, are in the flower. Pollen grains are male gamete–producing cells. The female gamete–producing cells are in the ovules in the ovary.
8. The flowers of angiosperms make possible the precise transfer of pollen from the anther to the stigma, an event that precedes fertilization and formation of a seed. Pollen is transferred by insects, particularly bees; birds and other animals; and the wind.
9. The seeds of angiosperms remain within the ovary, which develops into a fruit. Many fruits are fleshy and often sweet; animals that consume fruit may carry the seeds for long distances before excreting them as waste. The seeds can then germinate, or sprout, in the new location.
10. Seeds germinate only when they receive water and appropriate environmental cues. In the germination of seeds, mobilization of the food reserves stored in the cotyledons and in the endosperm is critical.
11. The change of a zygote into a mature individual, initiated immediately after fertilization, is development. Development is a process of progressive specialization that results in differentiation, the production of highly individual tissues and structures.
12. Plants can also reproduce asexually by vegetative propagation; a new plant grows from a portion of another plant.

KEY TERMS

abscisic acid 541
alternation of generations 529
angiosperm 534
anther 534
antheridium (pl., antheridia) 529
archegonium (pl., archegonia) 529
auxin 541
cell culture techniques 540
cotyledon 538
cytokinin 541
dicot 538
divisions 527
endosperm 538
ethylene 541
frond 531
fruit 534
gametophyte 529
germination 538
gibberellin 541
gymnosperm 533
hornwort 530
liverwort 530
monocot 538
moss 530
nectar 535
nonvascular plant 527
ovary 534
ovule 533
petal 534
phytochrome 541
pistil 534
pollen grain 534
pollinator 535
radicle 538
rhizome 540
runner/stolon 540
seed 531
seedless vascular plant 531
self-pollination 535
sepal 534
sporophyte 529
stamen 534
stigma 534
style 534
tuber 540
vascular plant 527
vegetative propagation 540

REVIEW QUESTIONS

1. Fill in the blanks: __________ plants contain specialized tissues within them to transport fluids. __________ plants lack these specialized tissues.
2. Fill in the blanks: __________ produce spores, which grow into __________. These __________ produce gametes, which fuse during fertilization and grow into __________.
3. Name the three classes of bryophytes and summarize their life cycles.
4. What are seeds? Explain their function(s).
5. Summarize the life cycle of seedless vascular plants. Give an example of this type of plant.
6. Match each of the following with the most appropriate term:
   Gymnosperms
   Angiosperms
   Bryophytes
   Ferns
   a. Seedless vascular plants
   b. Nonvascular plants
   c. Vascular plants with naked seeds
   d. Vascular plants with protected seeds
7. Draw a generalized diagram of a flower. Label the following: sepals, petals, stamens, pistil, anther, pollen grains, stigma, ovary, ovules, and style.
8. Summarize the sexual reproduction of angiosperms and gymnosperms.
9. The evolution of flowering plants and their animal pollinators is closely linked. Explain why.
10. Summarize the significance of fruits in sexual reproduction.
11. Place the following events in the correct sequence:
    a. The radicle emerges from the seed.
    b. Fertilization occurs.
    c. The shoot emerges from the ground.
    d. An embryo develops.
    e. The seed germinates.
12. What do monocots and dicots have in common? How do they differ?
13. What is vegetative propagation? Give an example.
14. What are hormones? Summarize the function(s) of the five major kinds of plant hormones mentioned.

THOUGHT QUESTIONS

1. The incredible success of flowering plants argues that increasing the efficiency of sexual reproduction has been of great evolutionary advantage to the angiosperms. Why then are so many angiosperms self-pollinators?
2. You probably know what a dandelion looks like, a delicate white sphere at the tip of a slender stalk—blowing on dandelion flowers is something almost all children do. Dandelions are obligatorily asexual, which is to say they never practice sexual reproduction. So why do they have flowers?
3. Why do you think angiosperm seed tissue (endosperm) has evolved to be triploid when there is no triploid tissue in the seeds of gymnosperms?

FOR FURTHER READING

Batra, S.W.T. (1984, February). Solitary bees. *Scientific American*, pp. 120-127.
This is an excellent article on these diverse and fascinating pollinators, some of which are of great commercial importance.

Cook, R.E. (1983). Clonal plant populations. *American Scientist, 71*, 244-253.
This is an excellent discussion on the role of asexual reproduction among plants in natural populations.

Schaller, G. B., Jinchu, H., Wenshi, P., & Jing, Z. (1985). *The giant pandas of Wolong*. Chicago: University of Chicago Press.
A fascinating account of the mutual adaptations of pandas and bamboo, this book offers the first glimpse of the life of pandas in their remote mountain home in western China.

31

Plants

Patterns of structure and function

HIGHLIGHTS

Vascular plants are organized along a vertical axis with underground roots and aboveground shoots made up of stems and leaves.

Vascular plants have specialized tissues that make up the structure of roots, stems, and leaves, enabling them to absorb nutrients, transport fluids, store food, carry on photosynthesis, and produce new cells for plant growth.

Water and dissolved minerals move upward from root to stem and leaves by transpirational pull; water and dissolved sugar are transported throughout the plant by mass flow.

The nonvascular plants have various levels of organization within their groups; all have a much simpler organization than the vascular plants.

OUTLINE

The organization of vascular plants

Tissues of vascular plants

- Fluid movement: Vascular tissue
- Food storage: Ground tissue
- Protection: Dermal tissue
- Growth: Meristematic tissue

Organs of vascular plants

- Roots
- Shoots
  - Stems
  - Leaves

Movement of water and dissolved substances in vascular plants

The organization of nonvascular plants

Although the photo looks like the discarded shell of a huge turtle, it is actually a plant. In fact, one nickname for this plant is the "turtleback." The turtleback's "shell" is really a modified stem that stores water for the plant. This helps the turtleback survive the dry season in its tropical environment. During the rainy season, a system of green branches grows out its top, manufacturing food and bearing leaves, flowers, and fruits. Through the years these branches die, covering the turtleback with an intertwining mass.

Turtlebacks commonly grow in Africa. The native people, however, call this plant *Hottentot's bread* because its fleshy storage organ can be eaten and has a texture similar to a giant turnip. This organ commonly grows a meter (3 feet) in diameter. Some grow as tall as 2⅓ meters (approximately 7 feet) and weigh up to 318 kilograms (700 pounds)—rather large for the produce section of the grocery store!

The organization of vascular plants

For any plant to grow as tall as the turtleback, it must have a means to transport substances to and from its various parts. Most plants (over 80% of all living species) have a system of specialized tissues that performs this function. Plants with these tissues are called **vascular plants** (see Chapter 30 for discussion of their reproductive cycles). Despite their reproductive diversity, however, vascular plants all have the same basic architecture.

A vascular plant is organized along a vertical axis (Figure 31-1). The part below the ground is called the **root.** The part above the ground is called the **shoot.** The root penetrates the soil and absorbs water and various ions crucial for plant nutrition. It also anchors the plant. The shoot consists of a **stem** and **leaves.** The stem serves as a framework for the positioning of the leaves, where most photosynthesis takes place. The arrangement, size, and other characteristics of the leaves are critically important in the plant's production of food. Flowers and ultimately fruits and seeds are also formed on the shoot.

Vascular plants are those having a system of tissues that transports water and nutrients. Vascular plants have the same basic architecture: they are made up of underground roots and aboveground shoots. The shoot consists of a stem and leaves.

Tissues of vascular plants

The organs of a vascular plant—the leaves, roots, and stem—are made up of different mixtures of tissues, just as your legs are composed of different tissues such as bone and muscle. A tissue is a group of cells that works together to carry out a specialized function. Vascular plants have three

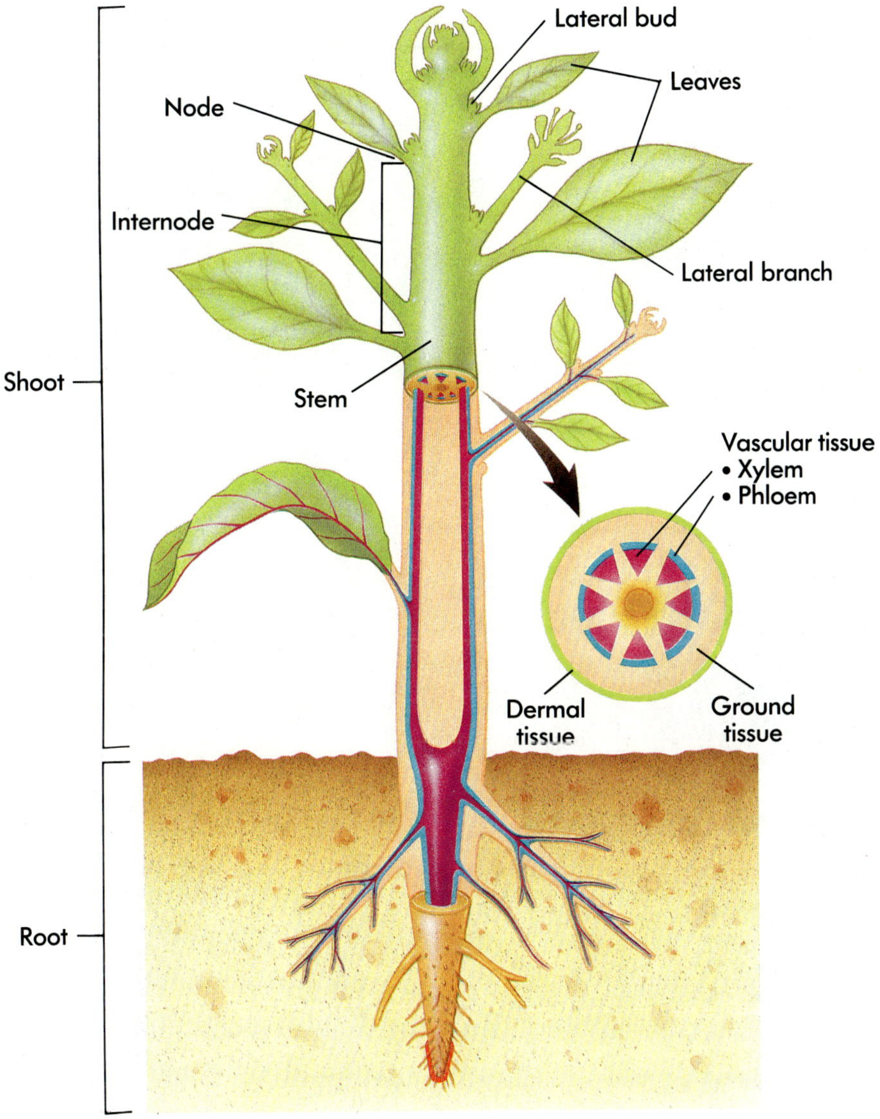

FIGURE 31-1 Structure of a typical plant. The root anchors the plant while the shoot supports the leaves, structures in which photosynthesis takes place. The vascular tissue is the circulatory system of plants and transports water and nutrients.

types of differentiated tissues, groups of cells having specific structures to perform specific functions: **vascular tissue, ground tissue,** and **dermal tissue** (see Figure 31-1).

The word *vascular* comes from a Latin word meaning "vessel." Thus vascular tissue forms the "circulatory system" of a plant, conducting water and dissolved minerals up the plant and the products of photosynthesis throughout the plant. Ground tissue stores the carbohydrates the plant produces. It is the tissue in which the vascular tissue is embedded and forms the substance of the plant. Dermal tissue covers the plant, protecting it—like skin—from water loss and injury to its internal structures. Plants also contain **meristematic tissue,** or growth tissue, an undifferentiated tissue in which cell division occurs. This tissue, often just called *meristem,* is considered undifferentiated because the cells it produces will eventually become one of the other three cell types.

Fluid movement: Vascular tissue

Vascular tissue is of two principal types: **xylem,** which conducts water and dissolved minerals, and **phloem,** which conducts carbohydrates the plant uses as food along with other needed substances. Each type of vascular tissue is made up of different, specialized conducting cells.

Figure 31-2, *A* and 31-2, *B* show two different types of xylem cells. Both types of cells conduct water and dissolved minerals only after they die and lose their cytoplasm, becoming hollow and thick walled. Stacked end to end, they form pipelines that extend throughout the plant. The cells pictured in Figure 31-2, *A,* called *tracheids,* show the pores these two cell types have along their length, like soaking hoses used in a garden. These pores allow some water to move laterally to cells surrounding the xylem pipelines. Figure 31-2, *C* shows a sieve tube member and companion cell.

The three types of differentiated tissues in vascular plants are (1) dermal tissue, which protects the plant; (2) vascular tissue, which conducts water, minerals, carbohydrates, and other substances throughout the plant; and (3) ground tissue, which stores food the plant manufactures. In addition, an undifferentiated type of tissue called *meristem* produces new plant cells during growth.

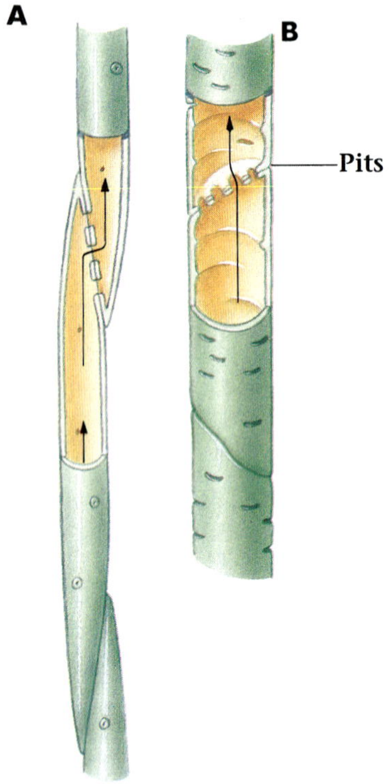

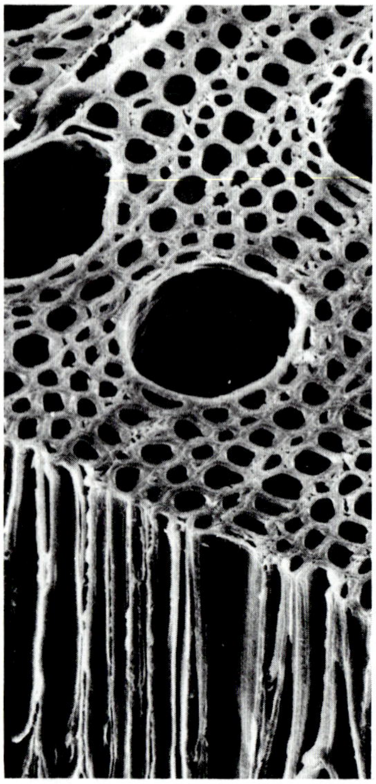

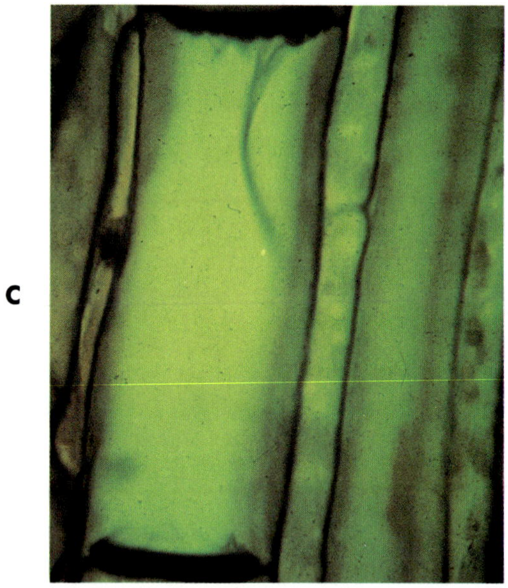

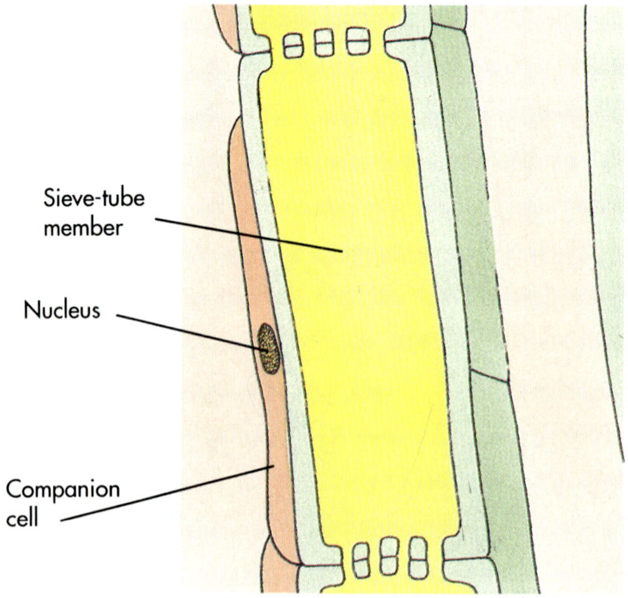

FIGURE 31-2 Vascular tissues of a plant.
A Tracheid. In tracheids, the water passes from cell to cell by means of pits.
B Vessel elements. In vessel elements, water passes from cell to cell by means of pits. Both tracheids and vessel elements make up xylem.
C Seive tube. Individual cells, or seive tube members, are stacked end to end to form a tube. Seive tube members are a part of the phloem and transport nutrients.

The conducting cells (sieve tube members) are alive and contain cytoplasm but are not typical cells. They lack nuclei, for example. The ends of these elongated cells have pits that allow the easy passage of sugar-filled water from where it is produced—in the leaves, for example—to where it is used rapidly—at the reproductive structures, for example. Companion cells, which contain all of the organelles commonly found in plant cells (including nuclei), secrete substances into and remove substances from the sieve tube members. These substances include sugars produced during photosynthesis.

Food storage: Ground tissue

Ground tissue contains **parenchyma cells,** which function in photosynthesis and storage. These thin-walled cells contain large vacuoles and may also be packed with chloroplasts. They are the most common of all plant cell types and form masses in leaves, stems, and roots. Fleshy storage roots, such as carrots or sweet potatoes, contain a predominance of storage parenchyma. The flesh of most fruits is also made up of these cells. **Sclerenchyma cells,** also found in the ground tissue, are hollow cells with strong walls. These cells help support and strengthen the ground tissue. The gritty texture of a pear is due to schlerenchyma cells that are dispersed among the softer parenchyma cells. Both types are shown in Figure 31-3.

A

B

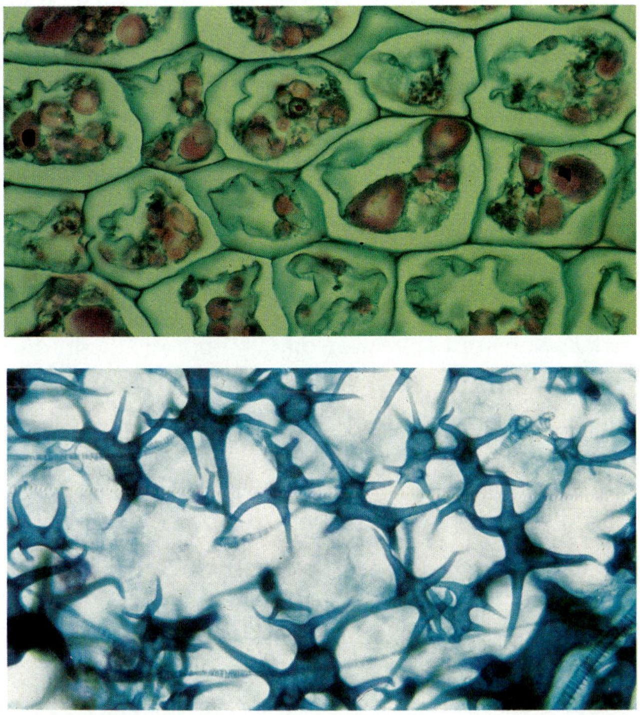

FIGURE 31-3 Ground tissue.
A Parenchyma cells. Parenchyma cells store starch and contain chloroplasts.
B Sclerenchyma cells. These cells are hollow, with strong walls. Sclerenchyma cells help support the plant.

Protection: Dermal tissue

Dermal tissue covers the outside of a plant with the exception of woody shrubs and trees, which have protective bark in its place. Bark is made up of other tissue types (see p. 552). **Epidermal cells** are the most abundant type of cell found in the dermal tissue. These cells are often covered with a thick, waxy layer called a **cuticle** that protects the plant and provides an effective barrier against water loss (Figure 31-4, *A*). Other types of cells found in the dermis

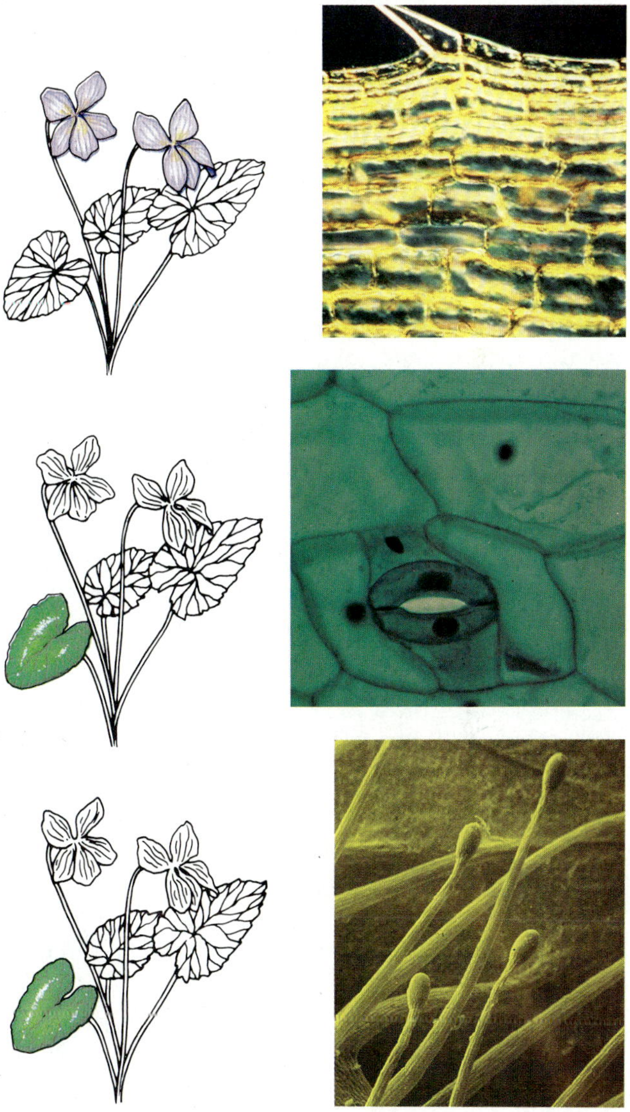

A

B

C

FIGURE 31-4 Dermal tissue.
A Epidermal cells. These cells are covered with a waxy substance and thus protect the plant from water loss.
B Guard cells. These cells regulate the movement of water vapor and gases into and out of the plant.
C Trichomes. Trichomes perform various functions: they increase surface area, defend the plants against insects and other predators, and reflect the sunlight away from the plant, reducing water loss.

are guard cells, which surround openings in the leaves through which gases and water vapor enter and leave (Figure 31-4, *B*), and trichomes (Figure 31-4, *C*), which are outgrowths of the epidermis (much like the hairs on your body) that have various functions. For example, on "air plants," trichomes help provide a large surface area for the absorption of water and minerals. On leaves or fruits, fuzz-like trichomes reflect sunlight, which helps control water loss. In some desert plants, white multicellular trichomes reflect enough sunlight to reduce temperatures in internal tissues. Some trichomes defend a plant against insects or larger animals with their sharpness or by secreting chemicals.

Growth: Meristematic tissue

Plants contain meristems, areas of undifferentiated cells that are centers of plant growth. Every time one of these cells divides, one of the two resulting cells remains in the meristem. In this way, meristem cells remain "forever young," capable of repeated cell division. The other cell goes on to differentiate into one of the three kinds of plant tissue, ultimately becoming part of the plant body.

Plants can grow only in relationship to where their meristematic tissue is located. Located at the tips of the roots and the tips of the shoots, this tissue is called **apical meristem** (*apex* meaning "tip"). Apical meristem allows plants to grow taller and their roots to grow deeper into the ground. This type of plant growth, which occurs mainly at the tips of the roots and shoots, is called **primary growth.**

Some plants not only grow taller but grow thicker as well. This type of growth is called **secondary growth** and occurs in all woody trees and shrubs such as pines, oaks, and rhododendrons. These plants have a cylinder of meristematic tissue along the length of their stems and branches, which is called **lateral meristem.** Herbaceous (nonwoody) plants, such as tulips, have only primary growth. However, their stems do grow somewhat thicker as they develop. One reason for this growth is that plant development involves both cell division *and* cell growth; as cells enlarge, plant parts enlarge. Additionally, some cell division occurs in the cells differentiating from meristematic tissue.

Organs of vascular plants

Roots, stems, and leaves are the organs of vascular plants. Together, the stems and leaves make up the shoot. The roots and shoots of all vascular plants share the same basic architecture, but there are differences. This chapter discusses the structure of the plants that dominate the plant world today: the flowering plants, or **angiosperms.** The principal differences in the structures of the roots, stems, and leaves in the seedless vascular plants and those with naked seeds lie primarily in the relative distribution of the vascular and ground tissue systems.

Roots

The function of a root system is to anchor a plant in the ground and to absorb water and minerals from the soil.

TABLE 31-1 Inorganic nutrients important to plants

NUTRIENT	RELATIVE ABUNDANCE IN PLANT TISSUE (ppm)
MACRONUTRIENTS (nutrients required in greater concentration)	
Hydrogen	60,000,000
Carbon	35,000,000
Oxygen	30,000,000
Nitrogen	1,000,000
Potassium	250,000
Calcium	125,000
Magnesium	80,000
Phosphorus	60,000
Sulfur	30,000
MICRONUTRIENTS (nutrients required in minute quantities)	
Chlorine	3000
Iron	2000
Boron	2000
Manganese	1000
Zinc	300
Copper	100
Molybdenum	1

ppm, Parts per million. Parts per million equals units of an element by weight per million units of oven-dried plant material.

(The minerals [essential inorganic nutrients] that plants use are listed in Table 31-1.) During the process of photosynthesis, plants use the water they absorb along with carbon dioxide they capture from the air to produce carbohydrates (see Chapter 8). But primarily, water absorbed at the roots replaces the water released by the plant into the air in a process called *transpiration* (see p. 555). Leaves are the principal organs of transpiration. But why do plants need to absorb minerals? The answer is simple. Although plants manufacture carbohydrates during photosynthesis, these sugars are not the only substances that plants need to live. Plants need proteins, fats, and vitamins. These substances are formed from the carbohydrates plants manufacture and from the minerals that plants take in from the soil and concentrate. Many plants therefore are an important source of minerals in the human diet. Broccoli and cabbage, for example, are excellent sources of calcium. Bananas provide you with potassium.

Roots, the part of a plant usually found below ground, absorb water and minerals as well as anchor the plant.

Structurally, the roots of dicots (plants having netlike veins in their leaves, see Chapter 30) have a central column of xylem with radiating arms. Between these arms are strands of phloem (Figure 31-5). Ringing this column of vascular

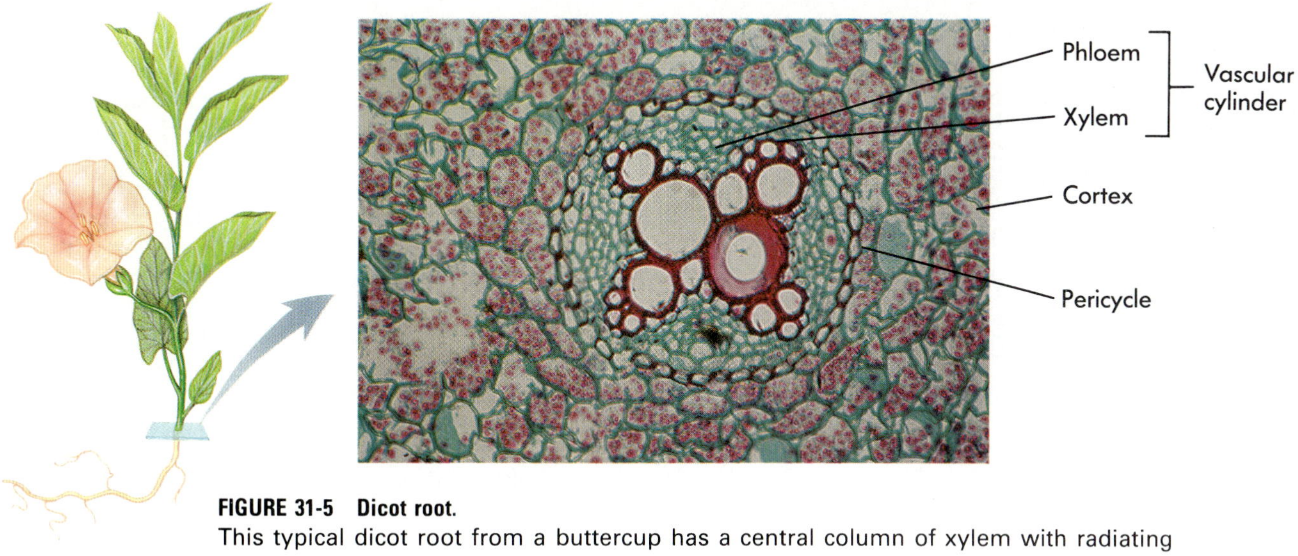

FIGURE 31-5 Dicot root.
This typical dicot root from a buttercup has a central column of xylem with radiating arms.

tissue (often called the *vascular cylinder*) and forming its outer boundary is another cylinder of cells called the **pericycle.** This tissue is made up of parenchyma cells able to undergo cell division to produce branch roots (roots that arise from other older roots). Surrounding the pericycle is a mass of parenchyma called the **cortex.** These cells store food for the growth and metabolism of the root cells. The innermost layer of the cortex is called the **endodermis,** which consists of specialized cells that regulate the flow of water between the vascular tissues and the outer portion of the root. The outer layer of the root is the **epidermis,** which absorbs water and minerals. These cells have extensions called root hairs that provide the epidermal cells with a larger surface area over which absorption can take place—like the function of microvilli in the intestinal walls of animals.

Monocot roots (plants having parallel veins in their leaves, see Chapter 30) are similar to dicot roots with one important exception: monocot roots often have centrally located parenchyma (storage) tissue called **pith.** The xylem and phloem are arranged in rings around the pith (Figure 31-6).

Most dicot roots have a central column of xylem with radiating arms and strands of phloem between these arms. Surrounding this vascular tissue is a layer of cells called the *pericycle* that is capable of cell division. Parenchyma (storage) cells of the cortex surround the pericycle. The entire root is covered with a protective epidermis. Monocot roots are structured similarly but often have an additional storage tissue called *pith,* which is located in the center of the root.

FIGURE 31-6 Monocot root.
The important difference between a dicot and monocot root is that the monocot root has a centrally located storage tissue, or pith. The xylem and phloem are arranged in rings around the pith.

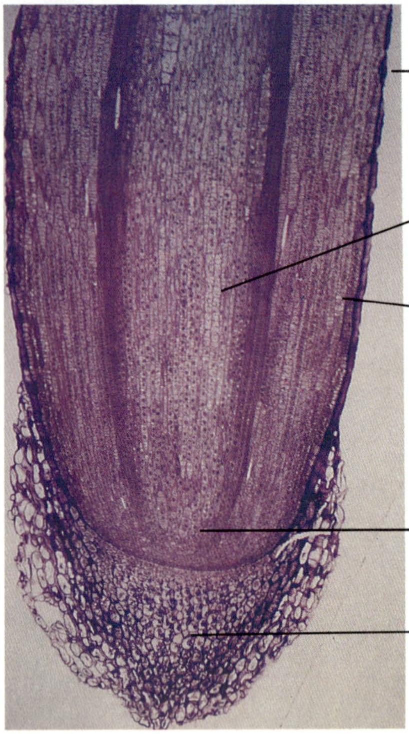

FIGURE 31-7 A root tip.
Section of a root tip in corn, *Zea mays,* showing the differentiation of epidermis, cortex, and column of vascular tissues.

The end of a root is tipped with apical meristem. These growth cells divide and produce cells inwardly, back toward the body of the plant, and outwardly. Outward cell division results in the formation of a thimble-like mass of relatively unorganized cells, the **root cap,** which covers and protects the root's apical meristem as it grows through the soil. Just behind its tip, the root cells elongate. Velvety root hairs, the tiny projections from the epidermis, form above this area of elongation (Figure 31-7). Here, too, the cells differentiate to produce specialized cell types.

If you have ever pulled a plant up by its roots, you will have noticed that the roots of one type of plant may look different from those of another type of plant. These differences in appearance are linked to differences in function. Many dicots, like dandelions, for example, have a single root called a **taproot,** which is sometimes the only major underground structure (Figure 31-8, *A*). Taproots grow deep into the soil, firmly anchoring the plant. Some taproots, like carrots and radishes, are fleshy because they are modified for food storage. The plant draws on these food reserves when it flowers or produces fruit, thus taproot crops are harvested before that time.

A B C

FIGURE 31-8 Types of roots.
A Taproots in a dandelion, *Taraxacum officinale.* Even a small section of these taproots can regenerate a new plant, which is one reason why dandelions are so difficult to eliminate from lawns and gardens.
B Prop roots in corn, *Zea mays,* are adventitious—they arise from stem tissue and take over the function of the main root.
C Fibrous roots in grass. These roots have no central, predominant root. Fibrous roots are especially efficient in achoring the plant and absorbing nutrients.

The taproot that develops in monocots, on the other hand, often dies during the early growth of the plant and new roots develop from the lower part of the stem. These roots are called **adventitious roots,** roots that develop from an aboveground structure. Often, adventitious roots help anchor a plant, such as "prop" roots in corn (Figure 31-8, *B*). Certain dicots, such as ivy plants, also develop adventitious roots. The adventitious roots of ivy plants help them cling to walls.

Have you ever pulled up a clump of grass and looked at its roots? Grass has **fibrous roots,** a type of root system that has no predominant root (Figure 31-8, *C*). Most monocots have fibrous roots. They work well to anchor the plant in the ground and absorb nutrients efficiently because of their large surface area. They also help prevent soil erosion by holding soil particles together.

Shoots

Plant shoots grow aboveground and are made up of stems and leaves. As with roots, the growing end of a shoot is tipped with apical meristematic tissue. Young leaves cluster around the apical meristem, unfolding and growing as the stem itself elongates (see Figure 31-1).

Leaves form on the stem at locations called **nodes.** The portions of the stem between the nodes are called the **internodes**. As the leaves grow, tiny undeveloped side shoots called **lateral buds** develop at the angles between the leaves and the stem. These buds, which contain their own embryonic leaves, may elongate and form lateral branches given the proper environmental conditions. However, most of these branches remain small and dormant.

Stems

One purpose of stems is to support the parts of plants that carry out photosynthesis. This process takes place primarily in the leaves, which are arranged on the stem so that light will fall on them. In addition, stems conduct water and minerals from the roots to all plant parts and bring the products of photosynthesis to where they are needed or stored (Figure 31-9).

The stem "transportation system" is made up of strands of xylem and phloem tissue that are positioned next to each other, forming cylinders of tissue called **vascular bundles** (often referred to as the *veins* of leaves). The xylem tissue

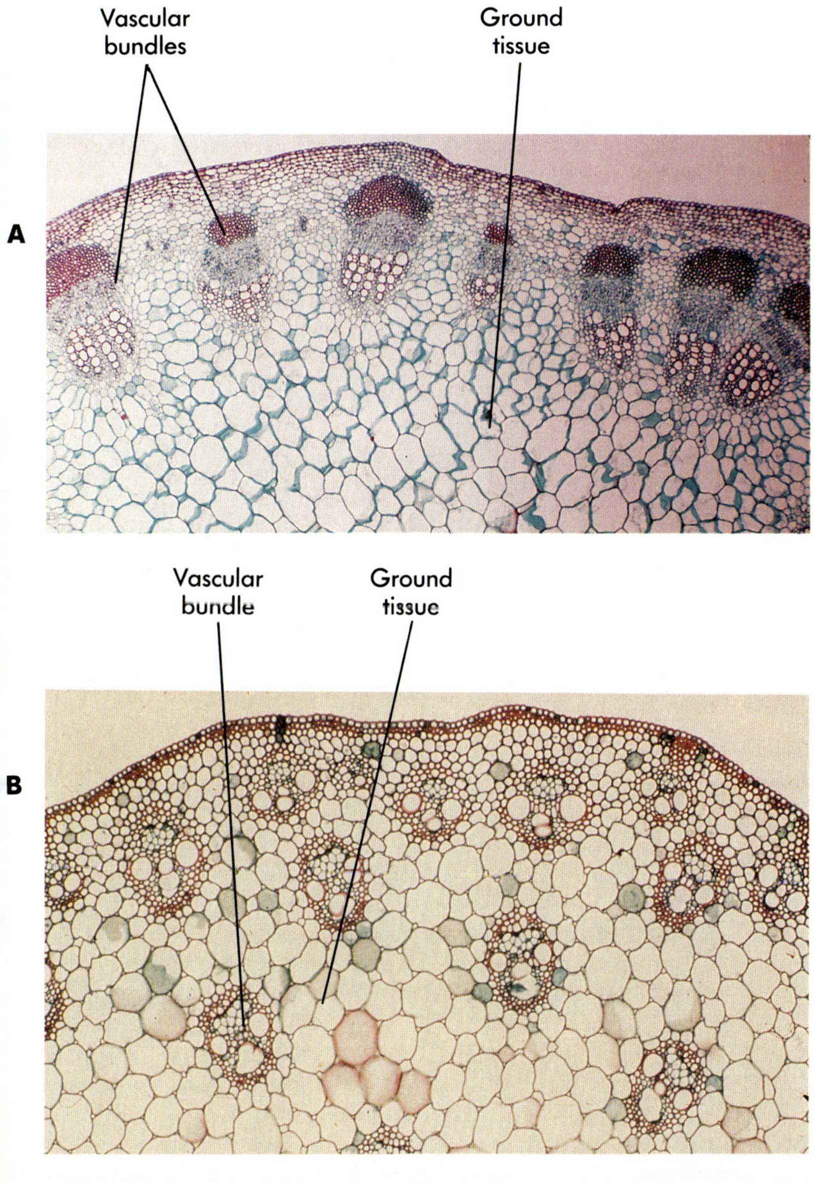

FIGURE 31-9 Stems.
A A dicot stem from the common sunflower, in which the vascular bundles are arranged around the outside of the stem. **B** A monocot stem from corn, showing the scattered vascular bundles characteristic of monocots.

characteristically forms the part of each bundle closer to the interior of the stem. The phloem lies closer to the epidermis. In herbaceous dicots (those with soft stems rather than woody, tree-like stems), the vascular bundles are arranged in a ring near the periphery of the stem (Figure 31-9, *A*). In monocots, the vascular bundles are scattered throughout the stem (Figure 31-9, *B*).

Because of the arrangement of their vascular bundles, dicot stems have a mass of ground tissue in the center of the stem called the pith and a ring of ground tissue between the epidermis and the vascular bundles called cortex. Monocot stems also have ground tissue, but because it surrounds scattered vascular bundles, ground tissue does not form areas of pith or cortex. The epidermis of both monocot and herbaceous dicot stems is covered with a protective, waxy coating called the *cuticle*.

Stems support the photosynthetic structures—the leaves—and transport water and dissolved substances throughout plants. Herbaceous dicot stems are characterized by an inner cylinder of ground tissue called *pith* surrounded by a ring of vascular bundles. Encircling the ring of vascular bundles is additional ground tissue called *cortex*. Monocot stems are characterized by scattered vascular bundles embedded in ground tissue.

Dicots and gymnosperms (such as pine trees) with woody stems (or trunks), such as flowering trees, have lateral meristems called **cambia.** One type of cambium in woody stems is called **vascular cambium.** As the name suggests, this growth tissue lies between the vascular tissue—the xylem and phloem—connecting the bundles to form a ring (Figure 31-10). As the cambial cells divide during secondary growth, one of the resulting daughter cells remains as a cambial cell, and the other differentiates into either a xylem or a phloem cell. This new xylem and phloem is called *secondary xylem* and *secondary phloem* (31-11).

The wood of trees is actually accumulated secondary xylem. The wood of dicot trees (such as cherry, hickory, oak, and walnut) is commonly referred to as *hardwood,* whereas the wood of conifers (such as fir, cedar, pine, and spruce) is called *softwood.* However, these names are not accurate descriptions of each group. Each contains trees having woods of varying hardness. The hardness of wood relates to its density, which depends on its proportion of wall substance to the space bounded by the cell wall. The denser a wood, the more wall substance it has in relation to the space it bounds and the stronger it is. When used for building, denser woods are harder to nail and machine, but they generally shrink and swell less than less dense, softer woods. Denser woods are also better fuel woods.

When growth conditions are favorable, as in the spring and early summer in most temperate regions, the cambium divides most actively, producing large, relatively thin-walled cells. During the rest of the year, the cambium divides more slowly, producing small, thick-walled cells. This pattern of growth results in the formation of rings in the wood. These rings are called **annual rings** and can be used to calculate the age of a tree (Figure 31-12).

Other lateral meristem tissue called **cork cambium** lies just under the epidermis. The outermost cells of this growth tissue produce densely packed cork cells. Cork cells have thick cell walls that contain fatty substances, making these cells waterproof and resistant to decay. When mature, the cork cells lose their cytoplasm and become hardened in a manner similar to the epidermal cells on your skin. The innermost cells of the cork cambium produce a dense layer of parenchyma cells. The cork, the cork cambium, and the parenchyma cells make up the outer protective covering of the plant called the **bark.**

Woody stems have lateral meristem tissue that provides for secondary growth, which increases the diameter of the stem. Other lateral meristem tissue produces a dense layer of cells called *bark* that protects the plant.

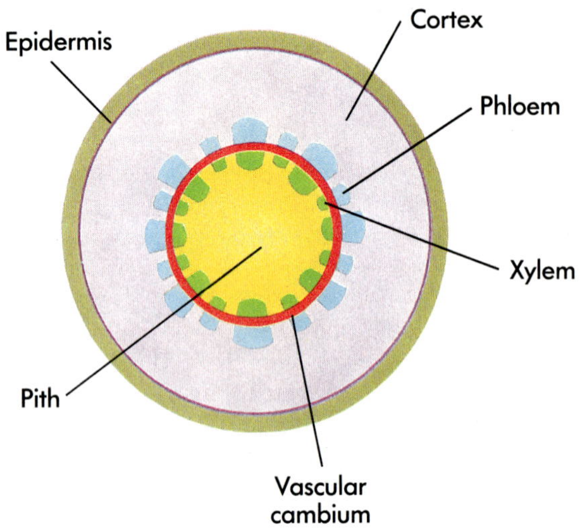

FIGURE 31-10 The vascular cambium in a stem.
The vascular cambium is the growth tissue in woody stems and lies between the xylem *(green)* and the phloem *(blue).*

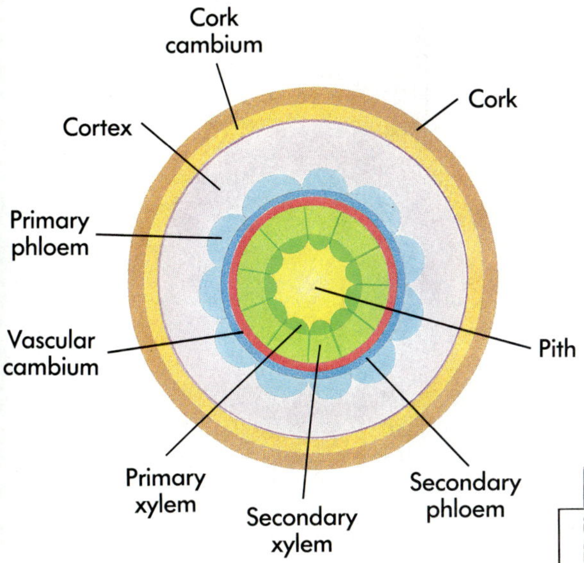

FIGURE 31-11 Secondary growth.
During secondary growth, cambial cells divide. One of the resulting daughter cells remains a cambial cell, and the other daughter cell becomes either a new xylem or phloem cell. These new cells are secondary xylem and secondary phloem.

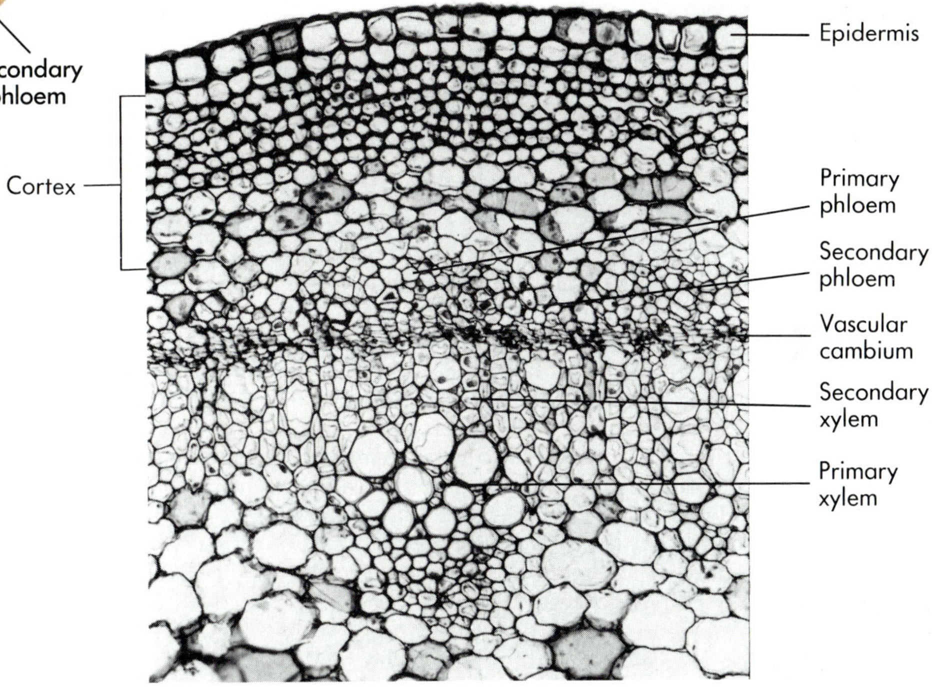

FIGURE 31-12 The structure of a tree trunk.
The yearly growth of the vascular cambium produces growth rings.

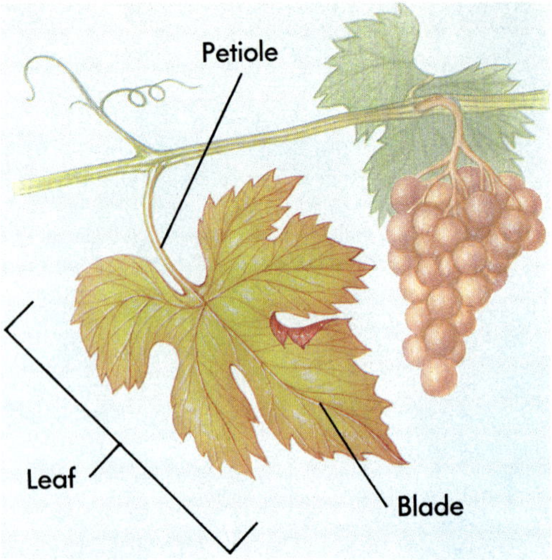

FIGURE 31-13 Structure of a leaf.
A leaf consists of a petiole and blade.

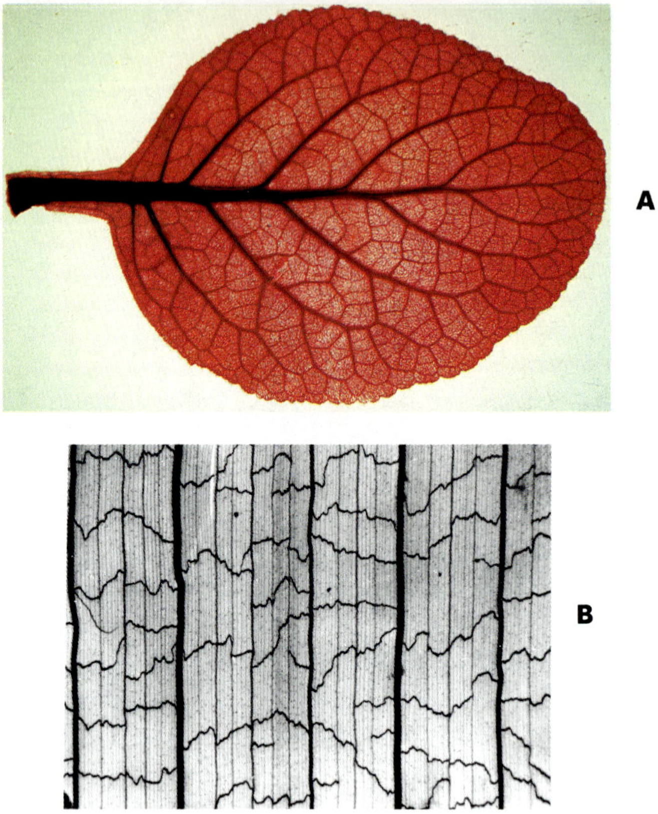

FIGURE 31-14 Dicot versus monocot leaves.
The leaves of dicots, such as this African violet relative **(A)**, have net veination; those of monocots, like this palm **(B)**, have parallel veination.

Leaves

Leaves, outgrowths of the shoot apex, are the light-capturing photosynthetic organs of most plants. Some exceptions exist; a major exception is found in most cacti, in which the stems are green and have largely taken over the function of photosynthesis for the plants.

Most leaves have a flattened portion, the **blade,** and a slender stalk, the **petiole** (Figure 31-13). Veins, consisting of both xylem and phloem, run through the leaves. In monocots, veins are usually parallel, and in dicots, they are usually net-like (Figure 31-14). Many conifers (vascular plants with naked seeds) have needle-like leaves suited for growth under dry and cold conditions. These modified leaves have thick, waxy coverings beneath, which are compactly arranged, thick-walled cells.

Microscopically, a cross section of a typical leaf looks somewhat like a sandwich: parenchyma cells in the middle, bounded by epidermis. The vascular bundles, or veins, run through the parenchyma. The leaf parenchyma is appropriately called the **mesophyll,** or "middle-leaf" (Figure 31-15).

The mesophyll of most dicot leaves is divided into two layers: the **palisade layer** and the **spongy layer.** The palisade layer lies beneath the upper epidermis of the leaf and consists of one or more layers of closely packed, column-like cells. These cells and the cells of the spongy layer contain chloroplasts, the organelles in which photosynthesis takes place. The spongy layer lies beneath the palisade layer and is made up of irregularly shaped cells. These cells are not closely packed like those in the palisade layer but instead have many air spaces between them. These spaces are connected, directly or indirectly, with openings to the outside called **stomata** (singular, **stoma**). Each stoma is bracketed by two **guard cells** that regulate its opening and closing (Figure 31-16).

The stomata open and close because of changes in the water pressure of their guard cells. As Figure 31-16 shows, guard cells are thicker on the side next to the stomatal opening and thinner on their other sides and ends. When the guard cells are plump and swollen with water, the cells bow. This change in the shape of the guard cells opens the stomata. When photosynthesis is taking place, water enters the guard cells, opening the stomata because the guard cells actively transport potassium ions to their interior. The water follows the potassium ions by osmosis because of the osmotic gradient the ions create. The oxygen produced by photosynthesis diffuses into the atmosphere through the stomatal openings, while the carbon dioxide needed for pho-

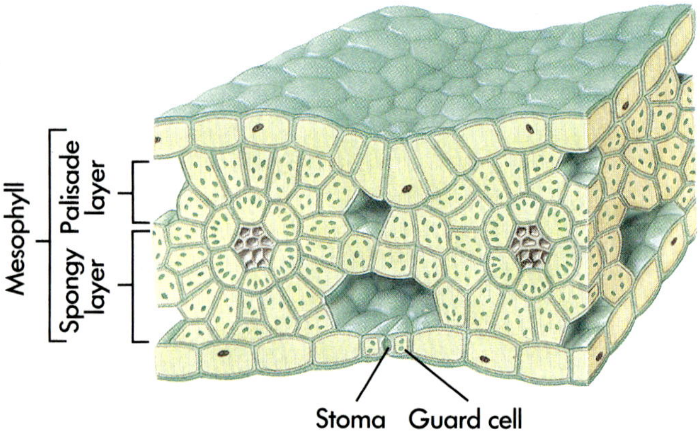

FIGURE 31-15 Internal structure of a leaf.
This diagram shows the internal structure of a typical leaf.

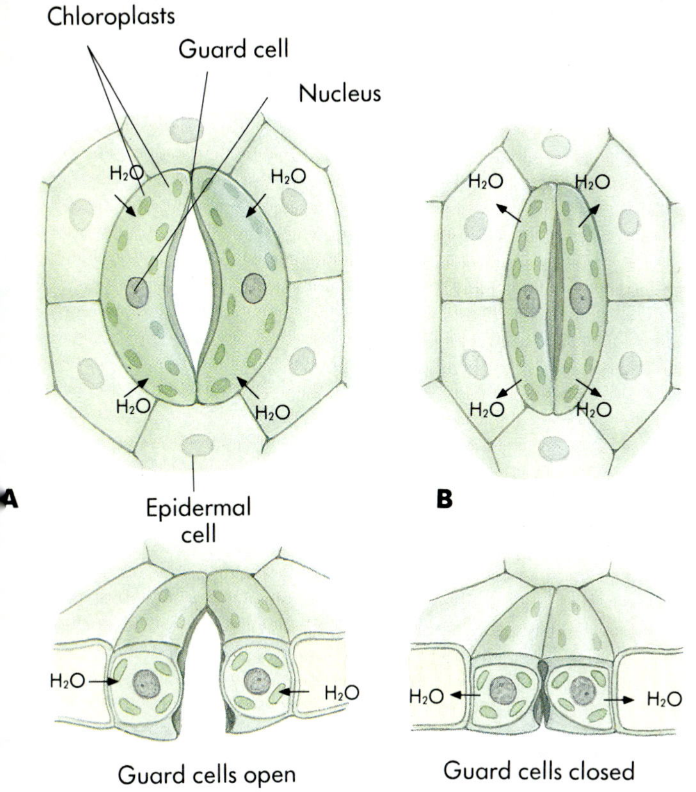

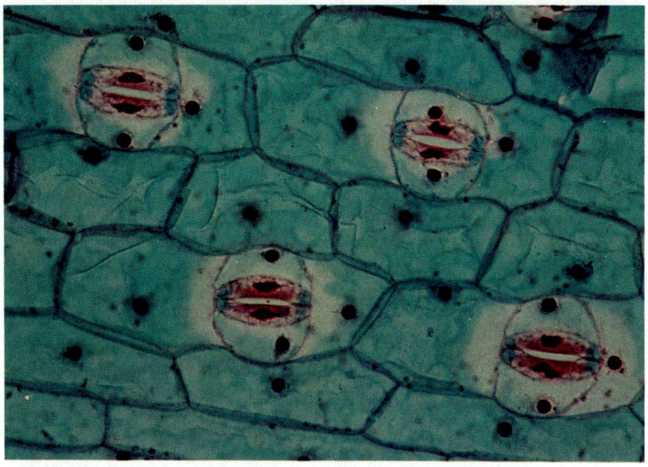

FIGURE 31-16 Guard cells.
A When solute pressure is high within the guard cells, the two guard cells bow outward, opening the stomata. **B** When solute pressure is low, the guard cells become limp and close the stomata.

tosynthesis to take place diffuses in. In addition, water leaves the leaf through these openings in the form of water vapor, a process called **transpiration.**

> **Leaves, the light-capturing organs of most vascular plants, are made up of parenchyma cells bounded by epidermis. The parenchyma cells contain chloroplasts, the organelles of photosynthesis.**

Movement of water and dissolved substances in vascular plants

Did you ever wonder how trees manage to move water to their uppermost leaves? Osmosis alone cannot account for this amazing feat. In fact, many forces are at work. To understand these forces and their interactions, try a little experiment. Fill a glass jar with water—a jar that apple or cranberry juice comes in, for example. Now fill a dishpan about three-fourths full with water. Put a piece of cardboard over the mouth of the jar, quickly turn it upside down as you place the mouth of the jar under water, and then remove the cardboard. What happens? Unless some water escaped from the jar before the mouth was submerged, the water stayed in the jar well above the level of the water in the dishpan. But why? The answer to this question is one of the reasons plants can move water up their vascular system from the roots to the leaves. Interestingly, the answer is air pressure.

The force of gravity pulls on the air that is over the water in the dishpan creating a force called *air pressure*. The air pushing down on the water in the dishpan causes water to be pushed up into the jar and helps it stay there. At the same time, however, gravity pulls down the water within the jar. The interaction of these two forces (the push up and the pull down) are both dependent on gravity and determine the level of water in the jar. At sea level, air pressure overcomes the pull of gravity on a column of water in a microscopically thin tube (such as xylem in plants), pushing it to a height of about 10.4 meters (about 31 feet or 3½ stories)! Two other forces, adhesion and cohesion, help prevent the collapse of this column. The term **adhesion** means that water molecules adhere, or stick to, the walls of the very narrow xylem of plants. **Cohesion** means that water molecules tend to stick to one another. (See Chapter 2 for a more complete discussion of the properties of water molecules.) These two forces link water molecules together and to the sides of the xylem with weak chemical bonds called *hydrogen bonds*.

Where does the water come from that is in the xylem of plants? Most of the water absorbed by a plant comes in through its root hairs. Minerals and other nutrients also pass into the cells of the root hairs by means of special cellular "pumps." In this way, root cells maintain a higher concentration of dissolved minerals than the concentration of minerals in the water of the soil. Therefore water tends to steadily move into the root hair cells from an area of higher concentration (in the soil) to an area of lower concentration (in the root hairs) by osmosis,

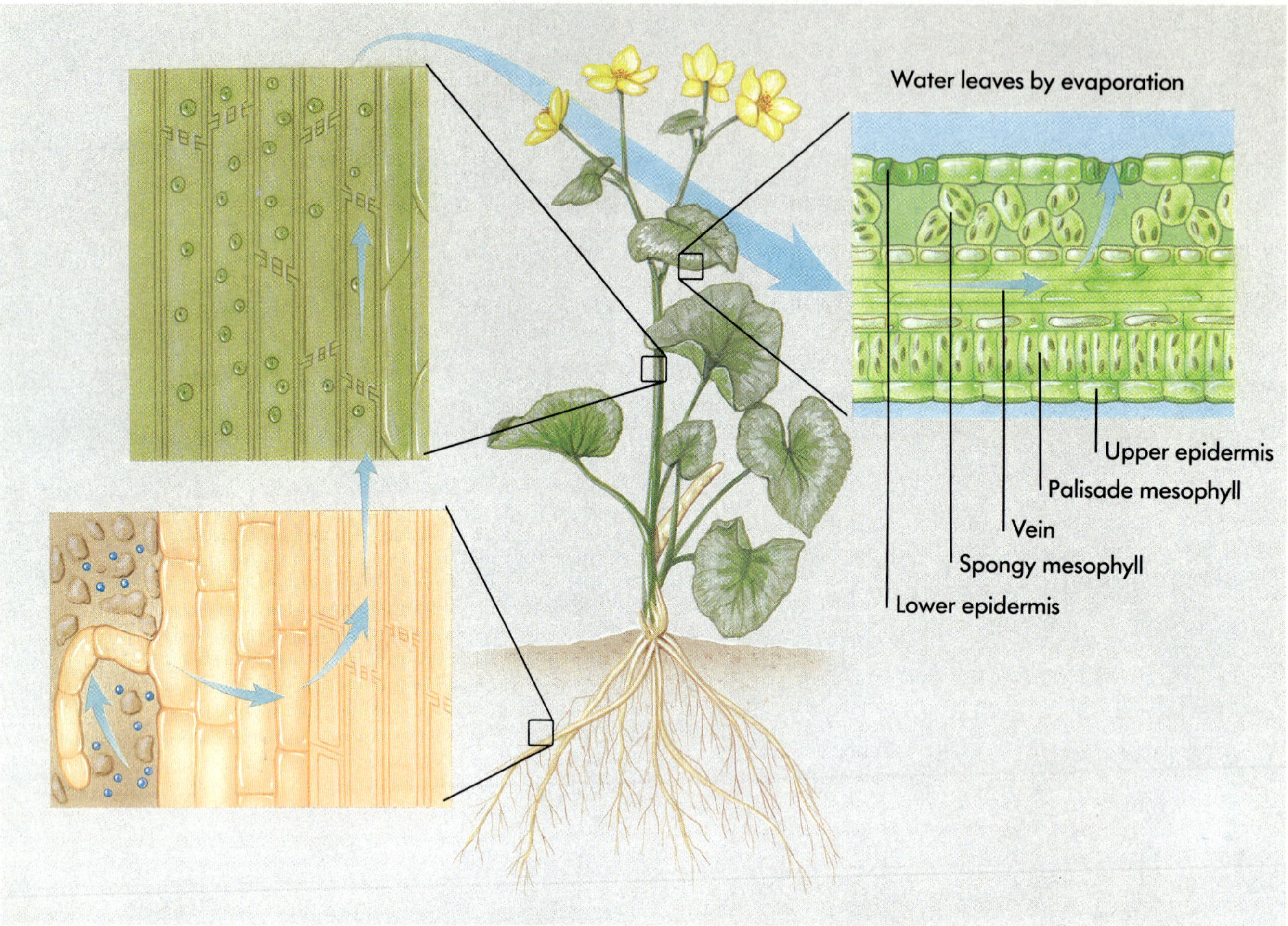

FIGURE 31-17 Water movement through a plant (transpiration). Water is pulled up through the plant by a combination of forces.

developing a force called **root pressure.** Once inside the roots, the water and dissolved minerals pass inward to the conducting elements of the xylem.

In addition to air pressure, adhesion, cohesion, and root pressure, the water in the xylem is pulled upward by forces associated with transpiration. Transpiration is a process by which water vapor passes out of a plant through the stomatal openings in its leaves. The water moves into the leaves after travelling up the plant through the xylem. When it reaches the end of the xylem, it diffuses into the photosynthetic mesophyll cells. Many air spaces surround the mesophyll cells, and water molecules easily pass through the membranes of the cells and into the spaces. Notice the close placement of the xylem, the mesophyll cells, and the air spaces in the cross-sectional view of the leaf in Figure 31-15. From the air spaces, the water evaporates through open stomata. The evaporation of water from the surfaces of the leaves creates a concentration gradient of water molecules, which causes the water molecules in areas of higher concentration (the xylem and mesophyll cells) to move to areas of lower concentration (the air spaces), creating a tension, or pull, on the column of water. Figure 31-17 shows how the forces of transpiration, adhesion, cohesion, and root pressure work together to move water and minerals from the roots to the leaves of a plant.

Water rises in a vascular plant beyond the point at which it would be supported by air pressure because evaporation from its leaves produces a force that pulls up on the entire water column all the way down to the roots. The forces of cohesion and adhesion work to maintain an unbroken column of water.

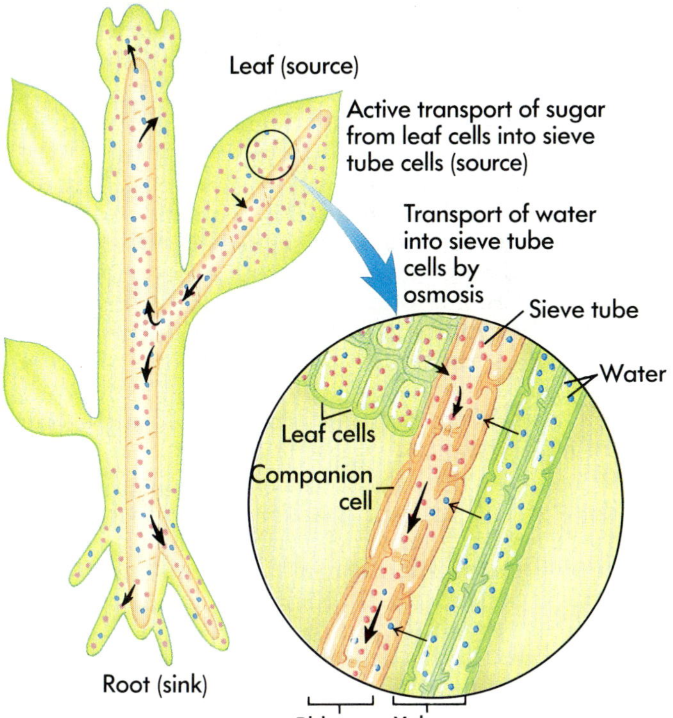

FIGURE 31-18 Mass flow.
Sucrose from the source is transported into sieve tube members by companion cells, and water follows by osmosis. At a sink—a place where sucrose is needed—sucrose is actively transported from the phloem, and water again follows by osmosis. The high water pressure in the phloem near the source and the low pressure near the sink cause this flow of sucrose, which can be very rapid.

Fluid is also transported in plants by the phloem. As mentioned previously, these tissues transport the products of photosynthesis—sugars—dissolved in water. This sugary solution is commonly called *sap*. Have you ever seen the collection of sap from maple trees? If so, you realize that plants can rapidly transport large volumes of fluid very quickly. Phloem sap contains 10% to 25% sucrose in addition to minerals, amino acids, and plant hormones and may travel as fast as one meter per hour!

The forces of diffusion and osmosis alone cannot account for this rapid movement of phloem sap. Instead, a pressure-flow or **mass flow** system performs this function. The mass flow system is shown in Figure 31-18 and works in the following way: Sucrose is produced at a **source,** such as a photosynthesizing leaf, and is actively transported into sieve tube members by companion cells. As the concentration of sucrose increases in the phloem, water follows by osmosis. In the roots below or at some other **sink** where sucrose is needed, companion cells actively transport sucrose out of the phloem. Water again follows by osmosis. The high hydrostatic (water) pressure in the phloem near the source and low pressure near the sink causes the rapid flow of the sap. After the sap reaches its destination in the plant, the water can be recycled by moving back to the source through the xylem.

The organization of nonvascular plants

The nonvascular plants, the bryophytes, are made up of three groups of plants: the liverworts, the hornworts, and the mosses (see Chapter 30). These plants are not organized the same as the vascular plants. Only certain genera have specialized vascular-like tissue. None have true roots, stems, or leaves. Bryophytes range from those forms that look like filamentous algae to those that look somewhat like certain vascular plants. But as a consequence of having less sophisticated transport systems than the vascular plants do, the bryophytes do not grow very tall. Some look as though they are creeping over their substrate, or food source.

Some bryophytes have distinct stems and leaf-like structures, such as most mosses (see Chapter 30), whereas other bryophytes do not. Bryophyte stems and leaf-like structures look outwardly different from the stems and leaves of vascular plants and are anatomically similar but simpler. These organs have an outer layer of epidermis made up of protective cells and growth cells. The cortex is made of parenchyma cells, much like those found in vascular plants. In addition, these stems and leaf-like structures have a central area of water-conducting tissue.

Bryophytes have no roots, but some have slender, usually colorless projections called **rhizoids** that anchor these simple plants to their substrate. Unlike roots, however, rhizoids consist of only a few cells and do not play a major role in the absorption of water or minerals. These substances often enter a bryophyte directly through its stems or leaves.

In general, the bryophytes have no specialized vascular tissues. The sporophytes of many moss species and the gametophytes of some moss species do have a central strand of somewhat specialized water-conducting tissue in their stems, and food-conducting tissue has been identified in a few genera. Even when such tissues are present, however, their structures are much less complex than those found in the vascular plants.

The bryophytes—liverworts, hornworts, and mosses—are primarily low-growing plants. Some have stems and leaf-like structures, but these structures are anatomically much simpler than those of vascular plants. Although most bryophytes have no specialized vascular tissues, a few species have somewhat specialized water-conducting tissues. Food-conducting tissues are rare.

Summary

1. The body of a vascular plant has two parts, a root and a shoot. These parts are made up of three principal tissues: vascular tissue, ground tissue, and dermal tissue. Vascular tissue conducts water and dissolved minerals up the plant, ground tissue stores the carbohydrates the plant produces and forms the substance of the plant, and dermal tissue covers the plant, protecting it. In addition, plants contain meristematic tissue, an undifferentiated tissue in which cell division occurs.
2. Vascular tissue is of two types: xylem and phloem. The xylem conducts water and dissolved minerals from the roots through the stem and to the leaves. Phloem conducts water and dissolved sugars (sap).
3. Plants grow in length by cell elongation and apical meristems, zones of active cell division at the ends of the roots and the shoots. This type of growth is primary growth. Secondary growth in both stems and roots takes place in woody trees and shrubs by means of lateral meristems along the length of their stems and branches.
4. Roots, stems, and leaves are the organs of vascular plants. The root system anchors a plant in the ground and absorbs water and minerals from the soil. Stems support the leaves—the primary organs of photosynthesis. In addition, stems conduct water and minerals from the roots to all plant parts and bring the products of photosynthesis to where they are needed or stored.
5. Together, the xylem and phloem are called *vascular bundles*. They form different, distinct organizational patterns in the roots, stems, and leaves of monocots (plants with parallel vascular bundles, or veins, in their leaves) and the dicots (plants with a web-like pattern of veins in their leaves).
6. Dicots with woody stems have lateral meristem tissue, or cambium. The vascular cambium lies between the xylem and phloem; its dividing cells form xylem toward the interior (secondary xylem) and phloem (secondary phloem) toward the exterior. As a result, the diameter of a plant increases.
7. Wood is accumulated secondary xylem; it often displays rings because it exhibits different rates of growth during different seasons.
8. Leaves, the photosynthetic organs of most vascular plants, are made up of specialized ground tissue cells, or parenchyma, bounded by epidermis. Vascular bundles run through the parenchyma, and its cells contain chloroplasts, the organelles in which photosynthesis takes place. Openings in the epidermis—stomata—allow the carbon dioxide needed for photosynthesis to enter and allow the oxygen produced by photosynthesis to escape. Water vapor also evaporates from the plant through the stomata.
9. Water flows through plants in a continuous column, driven mainly by the evaporation of water vapor from the stomata and osmotic pressure at the roots. The force of air pressue helps maintain the height of the column of water in the xylem. In addition, the cohesion of water molecules and their adhesion to the walls of the narrow xylem through which they pass are important factors in maintaining the flow of water to the tops of plants.
10. Sucrose moves from where it is produced in a plant to where it is used by the process of mass flow. First, it is actively transported into phloem cells where it is produced and is actively transported out of the phloem cells where it is used. These active transport processes produce a sugar gradient and a water pressure gradient, which cause their movement.
11. The bryophytes are not organized the same as the vascular plants. Only large, multicellular forms of these plants have tissue differentiation and specialized transport cells. However, the organization and transport systems of these genera are much less complex than that of the vascular plants.

KEY TERMS

adhesion 555
adventitious root 551
annual ring 552
apical meristem 548
bark 552
blade 554
cambium 552
cohesion 555
cork cambium 552
cortex 549
cuticle 547
dermal tissue 546
endodermis 549
epidermal cell 547
epidermis 549
fibrous root 551
ground tissue 546
guard cell 554
internode 551
lateral bud 551
lateral meristem 548
leaf 545
mass flow 557
meristematic tissue 546
mesophyll 554
node 551
palisade layer 554
parenchyma cell 547
pericycle 549
petiole 554
phloem 546
pith 549
primary growth 548
rhizoid 557
root 545
root cap 550
root pressure 556
sclerenchyma cell 547
secondary growth 548
shoot 545
sink 557
source 557
spongy layer 554
stem 545
stomata (sing., stoma) 554
taproot 550
transpiration 555
vascular bundle 551
vascular cambium 552
vascular plant 545
vascular tissue 546
xylem 546

REVIEW QUESTIONS

1. What are vascular plants?
2. Fill in the blanks: In a vascular plant, the part below ground is called the __________. The part aboveground, called the __________, consists of __________(structures in which most photosynthesis takes place) and a(n) __________(which serves as a framework).
3. Match each type of tissue to its function:
   1. Meristem
   2. Dermal tissue
   3. Ground tissue
   4. Vascular tissue

   a. Stores food manufactured by the plant
   b. Protects the plant
   c. Produces new plant cells during growth
   d. Conducts water, minerals, carbohydrates, and other substances throughout the plant
4. What do xylem and phloem have in common? How do they differ?
5. Identify and give a function for each of the following: a) parenchyma cells; b) epidermal cells; c) stomata; d) root cap.
6. Distinguish between primary and secondary growth.
7. Parents are always telling their children to eat lots of vegetables. Using your knowledge of plant physiology, give one benefit of eating vegetables.
8. Draw two diagrams, one showing the root of a "typical" monocot, the other the root of a "typical" dicot. Label the pericycle, xylem, cortex, endodermis, epidermis, and pith.
9. What type of root system would you expect to find in a dandelion, an ivy plant, and a clump of grass? What are the advantages of each type of root?
10. How can annual rings help you estimate the age of a tree? What type of tissue is involved? In which type of growth does this result?
11. Describe the process by which water rises in a vascular plant. What forces are involved?
12. Diagram the movement of phloem sap in a mass flow system. Label the source and sink, and show the direction of flow.
13. In general, how does the appearance of nonvascular plants differ from that of vascular plants? Why?

THOUGHT QUESTIONS

1. In tropical climates, many tall plants are CAM plants—they shut their stomates during the hot days and open them at night (see the boxed essay in Chapter 8). If their stomates are closed during the day so that transpiration cannot suck water up the xylem, why doesn't the water within the plant fall down the stem?
2. The roots of many plants have permanent symbiotic associations with the fungi mycorrhizae. Among them are the most ancient fossil plants known. What might be the advantage of this association to the plant?

FOR FURTHER READING

Sandved, K. B., & Prance, G. T. (1984). *Leaves.* New York: Crown Publishers, Inc.
This book provides a beautiful introduction to the diversity of leaves.

Swain, R. B. (1988, November). Notes from the radical underground. *Discover*, pp. 16-18.
Cooperation and competition between tree roots—this article offers a new view of the relationships between plants.

Woodward, L. (1989, February 18). Plants, water and climate. Part I. *New Scientist*, (supplement), pp. 1-4.
This is an excellent account of the way water moves through plants.

32 Invertebrate animals

Patterns of structure, function, and reproduction

HIGHLIGHTS

Animals are a diverse group of multicellular, heterotrophic, eukaryotic organisms that most likely evolved from protist ancestors.

Other than the sponges, all animals exhibit either radial symmetry or bilateral symmetry.

All bilaterally symmetrical organisms (except roundworms and flatworms) have coeloms, which are lined, fluid-filled enclosures that contain body organs suspended by thin sheets of connective tissue.

The echinoderms have embryological developmental patterns suggesting that they are more closely related to the vertebrates than to the other invertebrates.

OUTLINE

Characteristics of animals
- Patterns in symmetry
- Patterns in body cavity structure
- Patterns in embryological development

Invertebrates: Diversity in symmetry and coelom
- Asymmetry: Sponges
- Radial symmetry: Hydra and jellyfish
- Bilateral symmetry: Variation on a theme in many phyla
  - Flatworms
  - Roundworms
  - Clams, snails, and octopuses
  - Earthworms and leeches
  - Lobsters, insects, and spiders
  - Sea urchins and starfish

Coral reefs are one of the most beautiful, exotic, and colorful living landscapes. They are only partially made up of corals, which are animals that grow on rocks in shallow, tropical waters (from 30 degrees north to 30 degrees south of the equator). In fact, the organisms that make up a coral reef are quite diverse. Many organisms such as red algae help build and strengthen coral reefs. Even the corals themselves show incredible diversity.

The term *coral* is a general term that refers to a variety of invertebrate animals (those without a backbone) of the phylum Cnidaria. Some corals are hard corals—those that have a hard, outer shell. Others are soft corals—those that lay down fragments of shell-like material within their tissues. The tissues of corals contain pigments that produce a wide range of hues, as do the microscopic algae that live within the coral animals. Corals therefore exist in many colors: brown, blue, green, yellow, pink, purple, red, and even black. Later in this chapter you will learn about the patterns of structure and function that corals and other invertebrate animals have in common despite their incredible diversity. But first you will look at the "bigger picture" of the animal kingdom to discover its unifying themes.

Characteristics of animals

How are animals different from the other kingdoms of living things? Like the plants, fungi, and protists, animals are eukaryotic organisms, having a cellular structure different from the prokaryotic structure of bacteria (see Chapter 28). Also like the plants, most fungi, and some protists, animals are multicellular. No single-celled animals exist. Only their gametes are single-celled, but these cells are not independently living organisms. As soon as fertilization takes place, the development of a new multicellular individual begins. (See Figure 30-1, which diagrams the sexual life cycle of animals.)

Animals are heterotrophs, unable to make their own food. Therefore animals must eat plants, other organisms, or organic matter for food. Some simple animals, such as the sponges, take organic matter directly into their cells. Most animals digest food within a body cavity. The resulting molecules are then taken into the body cells to be broken down further by the chemical reactions of cellular respiration (see Chapter 7). The end product of cellular respiration is energy, which is used to drive the activities of life, including growth, maintenance, reproduction, and response to the external environment. As part of this response, most animals are capable of movement to capture food or to protect themselves from injury.

The cells of all animals (except the sponges) are organized into tissues, which are groups of cells combined into structural and functional units (see Chapter 9). In most animals, the tissues are organized into organs, complex structures made up of two or more kinds of tissues. Organs that work together to perform a function are organ systems.

Animals are extraordinarily diverse in their forms and how they function. This diverse group is often divided into two subgroups: the invertebrates and the vertebrates. The invertebrates are animals without a backbone, a series of bones that surrounds and protects a dorsal (back) nerve cord. In some invertebrates, the dorsal nerve, as well as the backbone, does not exist. Examples of invertebrates are spiders, sponges, jellyfish, snails, and worms. The vertebrates (a group that includes you) have both a backbone and a dorsal nerve cord. Examples of vertebrates are bears, fish, dogs, cats, birds, and frogs. Interestingly, although the vertebrates are usually larger and more commonly known, the invertebrates make up over 95% of all animal species.

Most zoologists think that the animals arose from protist ancestors (see Chapters 26 and 27). (The evolutionary relationships among the animals and their ancestors are shown in Figure 32-1.) Each animal on the lower branches of the tree represents a present-day phylum in the animal kingdom. The animals on the upper, right branches—the mammals, birds, reptiles, amphibians, and fishes—are all classes of animals in the phylum Chordata, subphylum Vertebrata. The chordate phylum includes three subphyla: the lancelets, the tunicates, and the vertebrates. Lancelets and tunicates (sea squirts) are both groups of marine animals that have a cartilaginous notochord instead of a bony vertebral column. In addition, the lancelets and tunicates lack a brain, an organ common to all vertebrates. The animals on the upper, left branches—the crustaceans, insects (with the centipedes and millipedes), and spiders—are the three subphyla of the phylum Arthropoda. Arthropods are animals that have a hard outer shell, jointed legs, and a segmented body.

Animals are a diverse group of eukaryotic, multicellular, heterotrophic organisms. This diverse group is often divided into two subgroups: the invertebrates and the vertebrates. The invertebrates are animals without backbones; the vertebrates are animals with a backbone that surrounds and protects a dorsal nerve cord. Most zoologists think that the animals evolved from protist ancestors.

Figure 32-1 shows that the sponges, cnidarians (jellyfish, corals), and flatworms are the present-day phyla of animals most closely related to the evolutionary ancestors of the animals. Likewise, animals higher on the tree represent present-day phyla, subphyla, or classes of organisms that are less closely related to the ancestors of the animals. Each branch of this phylogenetic tree represents an evolutionary pathway that diverged from an ancestral pathway.

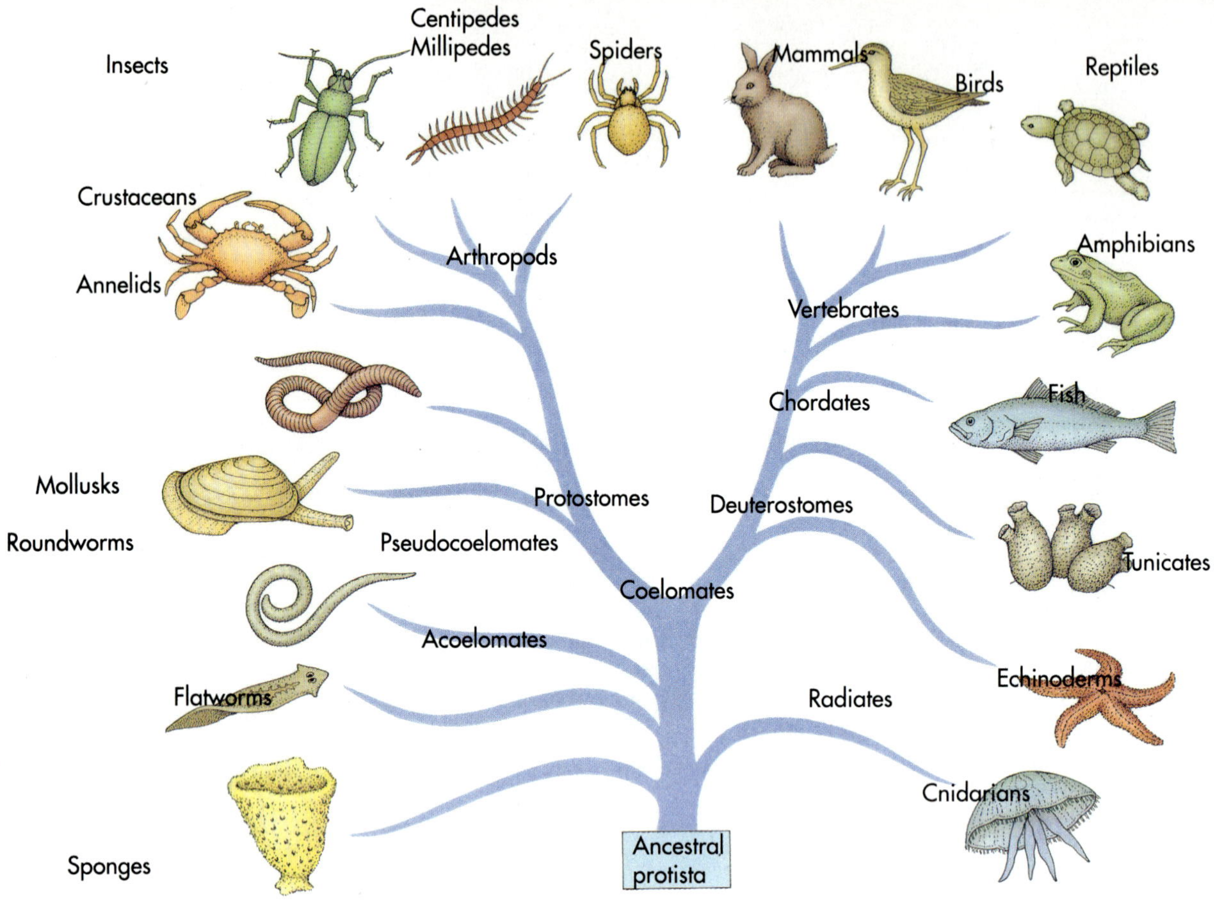

FIGURE 32-1 The animal ancestral tree.
All animals on the lower branches of the tree are phyla. The animals on the upper right branches are classes of phylum Chordata, subphylum Vertebrata. The animals on the upper left branches of the tree are the subphyla of phylum Arthropoda.

Patterns in symmetry

The word *symmetry* refers to the distribution of the parts of an object or living thing. All animals (except the sponges) exhibit either **radial symmetry** or **bilateral symmetry.** Sponges are asymmetrical, or without symmetry.

Radial symmetry means that body parts emerge, or radiate, from a central point, much like spokes on a wheel (Figure 32-2). As noted in Figure 32-1, the cnidarians and their relatives are radially symmetrical. These organisms have a top and a bottom, but they have neither a dorsal (back) side nor a ventral (belly) side (see Figure 32-6).

Animals exhibiting radial symmetry are aquatic organisms. Many are sessile; that is, they are anchored in one place. Their radially symmetrical body allows them to interact with the watery environment in all directions. Other organisms such as starfish (phylum Echinodermata) are also radially symmetrical but are not grouped with the cnidarians because their embryonic development and internal anatomy suggest that they are related to organisms having bilateral symmetry. In addition, their larvae are bilaterally symmetrical.

Bilateral symmetry means that the right side of an object or an organism is a mirror image of the left side. Animals with bilateral symmetry have a dorsal and a ventral side and a cephalic (head) end and a caudal (tail) end. All phyla of animals other than the sponges, jellyfish, and echinoderms exhibit bilateral symmetry.

Patterns in body cavity structure

A **coelom,** or body cavity, is a fluid-filled enclosure lined with connective tissue within most bilaterally symmetrical organisms and the echinoderms. Organs are located within this body cavity. For example, you have an abdominal cavity lined with a thin, nearly transparent sheet of connective tissue (peritoneum). Suspended by thin sheets of connective tissue arising from the peritoneum, the stomach and intestines hang in this coelom. All bilaterally symmetrical organisms have coeloms, except roundworms and flatworms.

Organisms having a coelom are called **coelomates.** Most of the organisms in Figure 32-1 are coelomates. Notice that the roundworms are labeled **pseudocoelomates.** These organisms (along with tiny aquatic organisms called *rotifers*) have a false body cavity. Although a fluid-filled cavity houses their organs, it is not lined nor are the organs suspended by thin sheets of connective tissue. This difference arises during the embryological development of the pseudocoelomates.

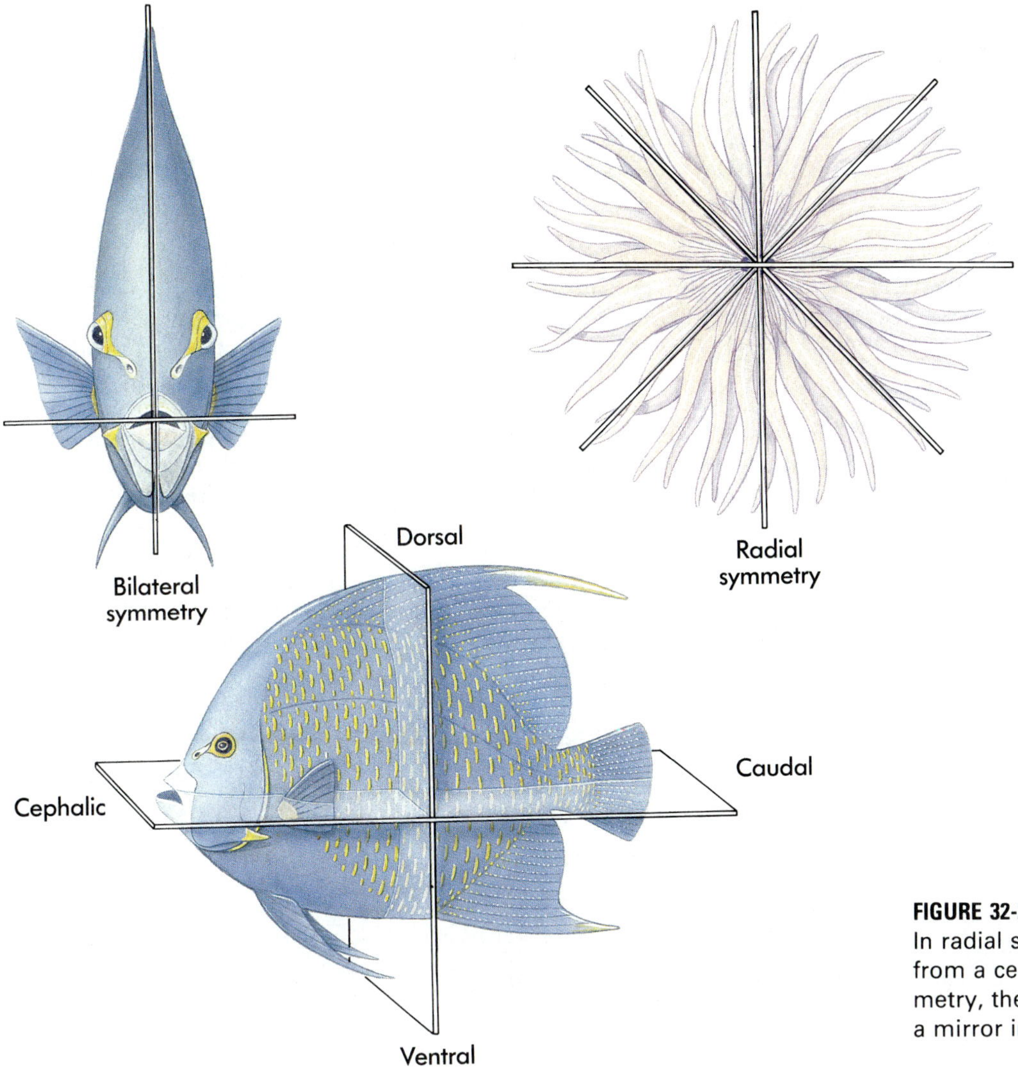

FIGURE 32-2 Patterns in symmetry. In radial symmetry, body parts emerge from a central point. In bilateral symmetry, the right side of an organism is a mirror image of the left side.

The **acoelomates,** the flatworms, have no body cavity. Embedded within the other tissues of the body, their organs are not protected from the movements and compressions of other body tissues.

The three patterns of body cavity structure are shown in Figure 32-3.

Coelomates have lined, fluid-filled body cavities in which organs are suspended by thin sheets of connective tissue. Pseudocoelomates have body cavities but they are not lined and do not have organs suspended in them. Acoelomates have no body cavities.

FIGURE 32-3 Three plans for construction of animal bodies. Animals can be coelomate (having a coelom), pseudocoelomate (having a false coelom), and acoelomate (having no coelom).

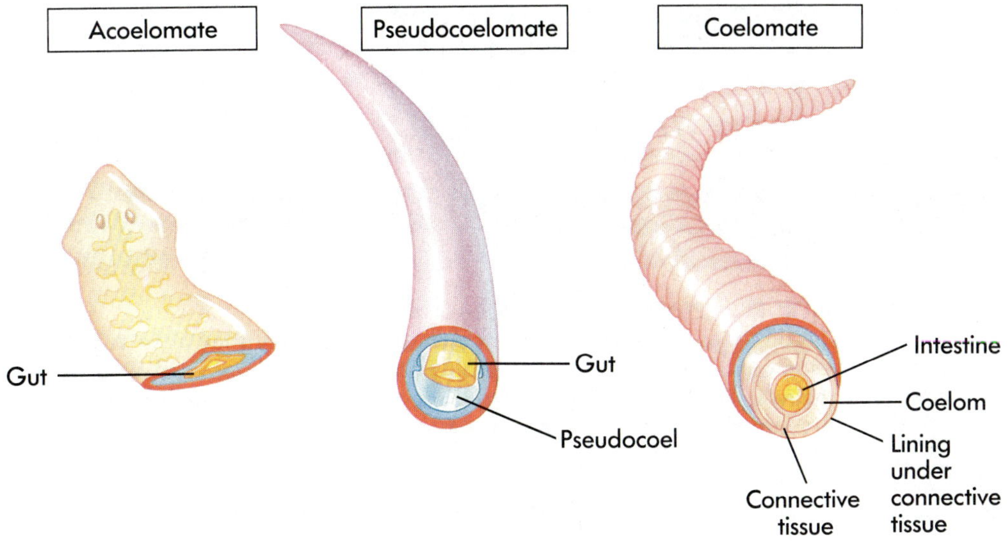

Patterns in embryological development

Two main branches arise from the trunk of the animal family tree in Figure 32-1. These two branches of coelomate animals represent two distinct evolutionary lines. One includes the mollusks (clams), annelids (segmented worms), and arthropods (lobsters, insects, and spiders). These organisms are called *protostomes,* meaning "first" (*proto*) "mouth" (*stome*). This strange name refers to events in the embryological development of these organisms.

During early embryological development (see Chapter 21), all organisms consist of a solid ball of cells usually called a *morula.* As development proceeds, the cells secrete fluid that fills the interior of the ball and pushes the cells to the edges of the sphere. This stage is called the *blastula.* This fluid-filled ball of cells then forms an indentation, assuming the shape of a blown-up balloon with a fist pushing in one side. This stage of development, the *gastrula,* gives rise to a two-layered embryo and begins the formation of the gut. In the **protostomes,** this first indentation becomes the mouth of the organism. In another group of organisms called **deuterostomes,** which includes the echinoderms (starfish) and chordates, the first indentation becomes the anus and a second one becomes the mouth. The term *deuterostome* means "second mouth." Figure 32-4 shows these differences.

The protostomes and the deuterostomes also show different cleavage patterns during embryological development. During the morula stage of development, the cells of the protostomes divide in a way that forms a spiral pattern. The cells of the deuterostomes cleave in a radial pattern. These patterns of embryological development (along with other developmental similarities) suggest close evolutionary relationships among the organisms within each group. In addition, because the protostome developmental pattern occurs in all acoelomates, scientists think that it was the pattern of development in the ancestors to the animals.

The protostomes (mollusks, annelids, and arthropods) differ from the deuterostomes (echinoderms and chordates) with regard to certain embryological events. These differences suggest close evolutionary relationships among the organisms within each group.

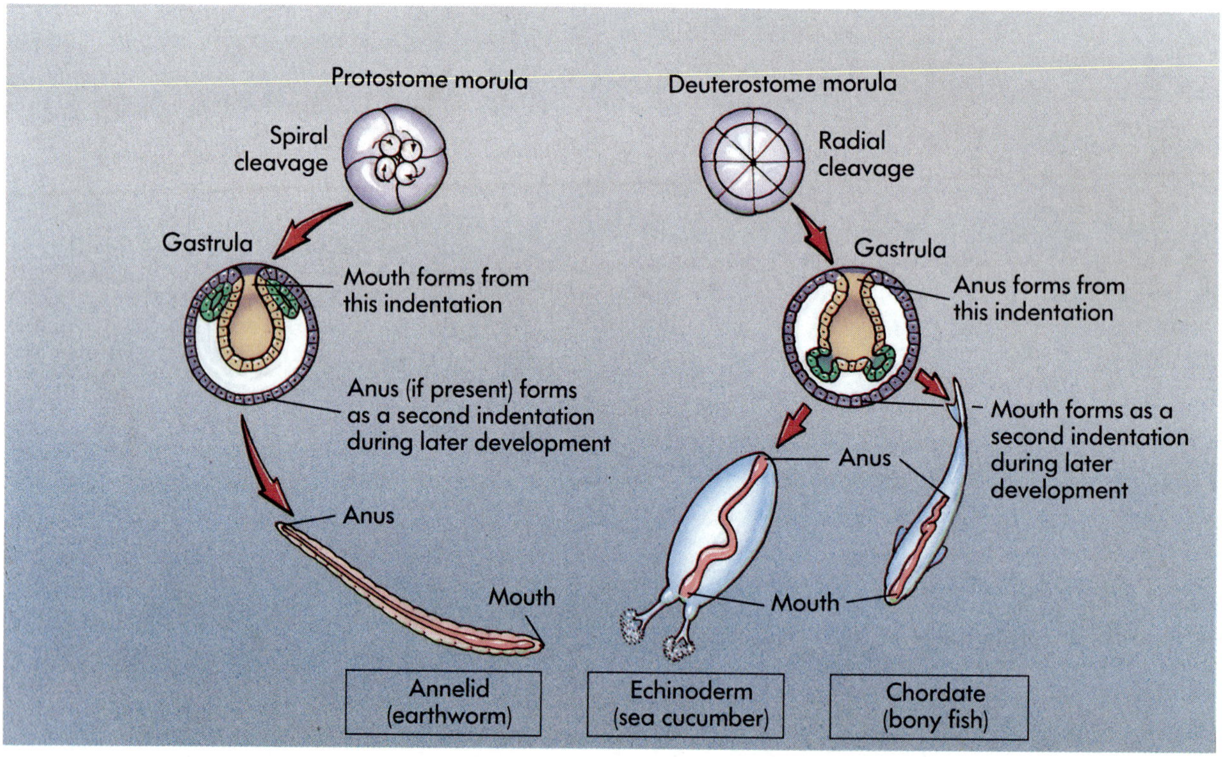

FIGURE 32-4 Protostomes and deuterostomes.
The differences between protostomes and deuterostomes arise during embryonic development. The mouth of the protostome develops from the first indentation of the gastrula. The mouth of a deuterostome develops from a second indentation. The deuterostome anus develops from the first indentation.

Invertebrates: Diversity in symmetry and coelom

Asymmetry: Sponges

Sponges (phylum Porifera) are aquatic organisms; most species live in the ocean rather than in fresh water. These sessile creatures are considered simpler than other animals in their organization because they have no tissues, no organs, and no coelom. The asymmetrical bodies of sponges consist of little more than masses of cells embedded in a gelatinous material, or matrix (Figure 32-5, *A*).

The body of a sponge is shaped like a sac or vase (Figure 32-5, *B, C*). The body wall is covered on the outside by a layer of flattened cells called the **epithelial wall.** Lining the inside cavity of the sponge are specialized, flagellated cells called **collar cells.** The matrix makes up the substance of the sponge, sandwiched between the outer, epithelial layer and the inner layer of collar cells. Within the matrix are ameboid-type cells, needle-like crystals of calcium carbonate or silica, and tough protein fibers. Pores, channel-like openings that span the matrix, are dispersed throughout the sponge. These pores are integral to the movement of water, dissolved substances, and particulate matter to the interior of the sponge. They give the sponges their phylum name *Porifera,* which means "pore bearers."

As they beat, the flagella of the collar cells create a current of water that flows from the outside of the sponge, through pores in the matrix, to the internal cavity of the

A

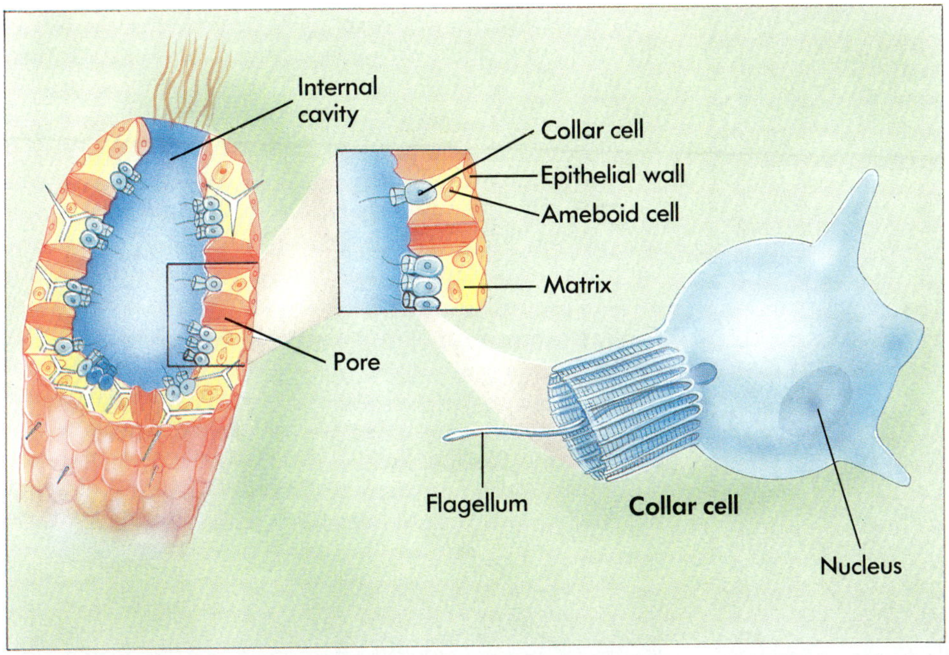

FIGURE 32-5 The sponge (phylum Porifera).
A Diagram of a sponge, with detail of a collar cell.
B A barrel sponge.
C Red boring sponge.

B

C

sponge, and then out again through the large opening at the top of the sponge. The circulation of water in this way brings the nutrients in the water to the collar cells.

Sponges frequently reproduce asexually by fragmentation; groups of cells become separated from the body of the sponge and develop into new individuals. In addition, sponges may develop branches that grow over the rocks on the sea floor, much like a plant develops underground runners. Colonies of sponges grow along these branches.

Sponges reproduce sexually and asexually. Most species of sponges produce both female sex cells (eggs) and male sex cells (sperm), which arise from cells in the matrix. Both types of sex cells are produced within the same organism. Such individuals are called **hermaphrodites,** after the Greek male god Hermes and the female goddess Aphrodite. The sperm are released into the cavity of the sponge and are carried out of the sponge with water currents and into neighboring sponges through their pores. Fertilization occurs in the gelatinous matrix where the eggs are held. There, the fertilized eggs develop into flagellated, free-swimming larvae that are released into the sponge's cavity. After the larvae leave the interior of the sponge, they settle on rocks and develop into adults.

Sponges (phylum Porifera) are aquatic, asymmetrical, acoelomate organisms shaped somewhat like a vase. They are considered the simplest of animals because they have no tissues or organs.

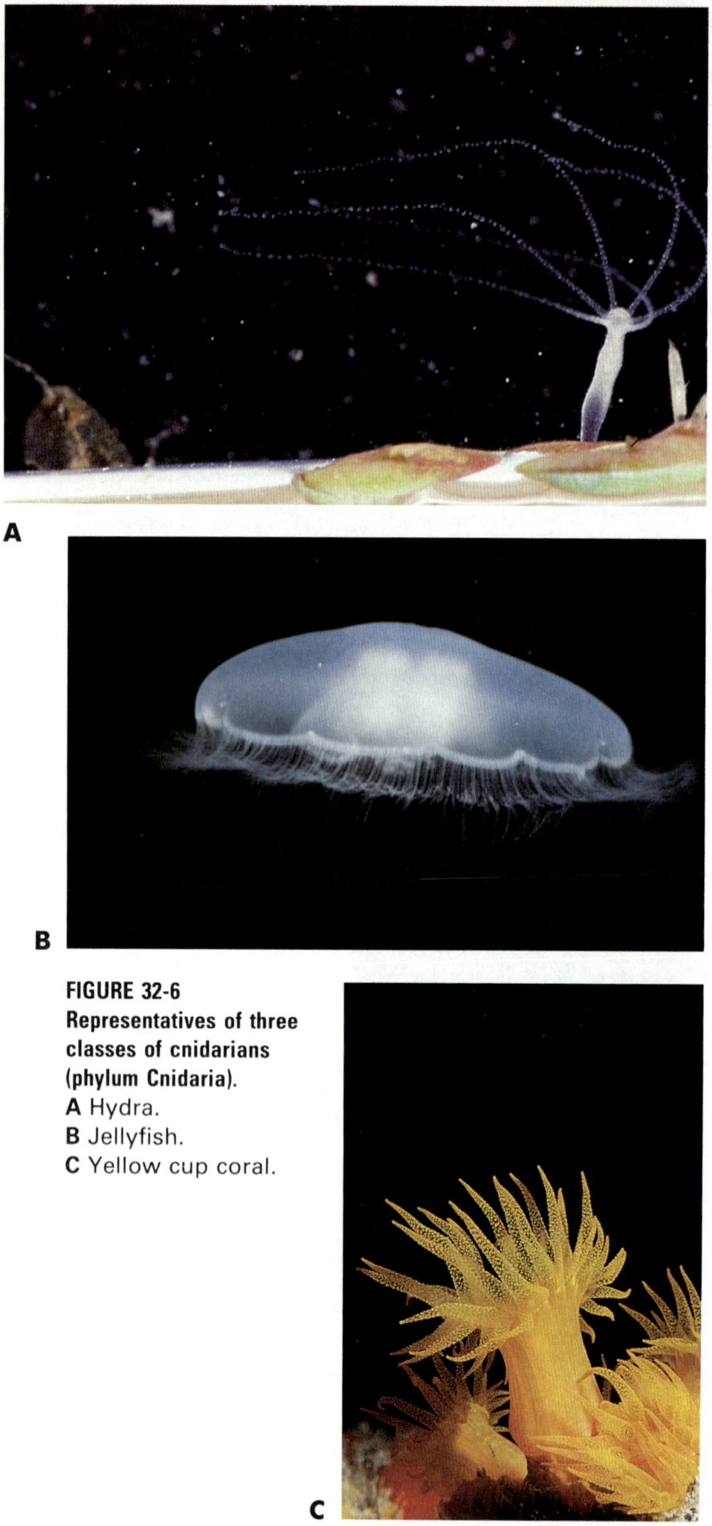

FIGURE 32-6
Representatives of three classes of cnidarians (phylum Cnidaria).
A Hydra.
B Jellyfish.
C Yellow cup coral.

Radial symmetry: Hydra and jellyfish

Animals other than the sponges have a definite shape and symmetry. Only two phyla are classified as radially symmetrical: the Cnidaria, which includes jellyfish, hydra, sea anemones, and corals (Figure 32-6), and Ctenophora, a minor phylum that includes the comb jellies (Figure 32-7).

FIGURE 32-7 A comb jelly.
Note the comb-like plates and two tentacles.

There are two basic body plans exhibited by the cnidarians: **polyps** and **medusae** (Figures 32-6 and 32-8). Polyps are aquatic, cylindrical animals with a mouth at one end that is ringed with tentacles. (The name *polyp* actually means "many feet.") Polyps such as the sea anemones and corals live attached to rocks. Like the corals, many polyps build up a hard outer shell, an internal skeleton, or both. Some polyps are free floating, such as the fresh water *Hydra.* In contrast, most medusae are free floating and are often umbrella shaped. Commonly known as jellyfish, medusae have a thick, gelatinous interior. The mouth of a medusa is usually located on the underside of its umbrella shape, with its tentacles hanging down around the umbrella's edge.

Structurally, epithelial tissue covers the outside of cnidarians, and an inner tissue layer, the gastrodermis, lines the gut cavity (see Figure 32-8). The mesoglea (literally, "middle glue") lies between. This layer is quite thick within medusae and gives them their jelly-like appearance. A network of nerve cells extends through cnidarians, but they have no brain-like controlling center.

Some cnidarians such as the hydra, sea anemones, and corals occur only as polyps. Simple polyps such as hydra usually reproduce asexually by budding (see Figure 5-6). However, they do have some of their tissue organized into primitive ovaries that produce eggs and testes producing sperm, as do the more complex polyps such as sea anemones and corals. (Some species of these organisms *never* reproduce by budding.) Some hydra, like sponges, are hermaphrodites. Others exist as males and females. Eggs remain attached to the hydra but exposed to the water. Sperm are discharged from the testes and swim to the egg. After fertilization, developing hydras grow while attached to the parent.

Some cnidarians exist only as medusae. Medusae reproduce sexually. Ovaries hang from the underside of female medusae, and testes hang from the males. Eggs and sperm are shed into the water, where fertilization takes place. The fertilized egg develops into a larva that never settles down to become a sessile polyp but develops directly into a medusa. However, most species have medusae that have a life cycle in which the larvae develop into polyps. These polyps develop into medusae. This type of alternating life cycle is shown in Figure 32-9.

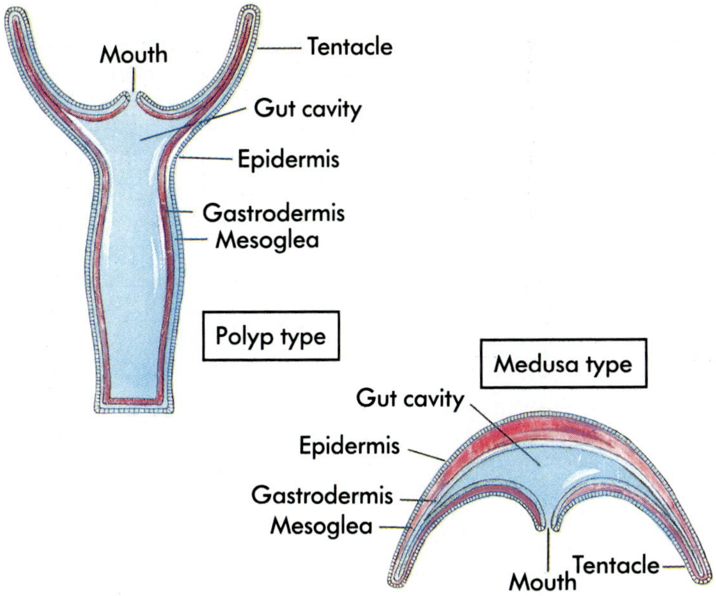

FIGURE 32-8 Body forms of cnidarians: the medusa and the polyp.
These two phases alternate in the life cycles of many cnidarians, but a number—including the corals and sea anemones—exist only as polyps.

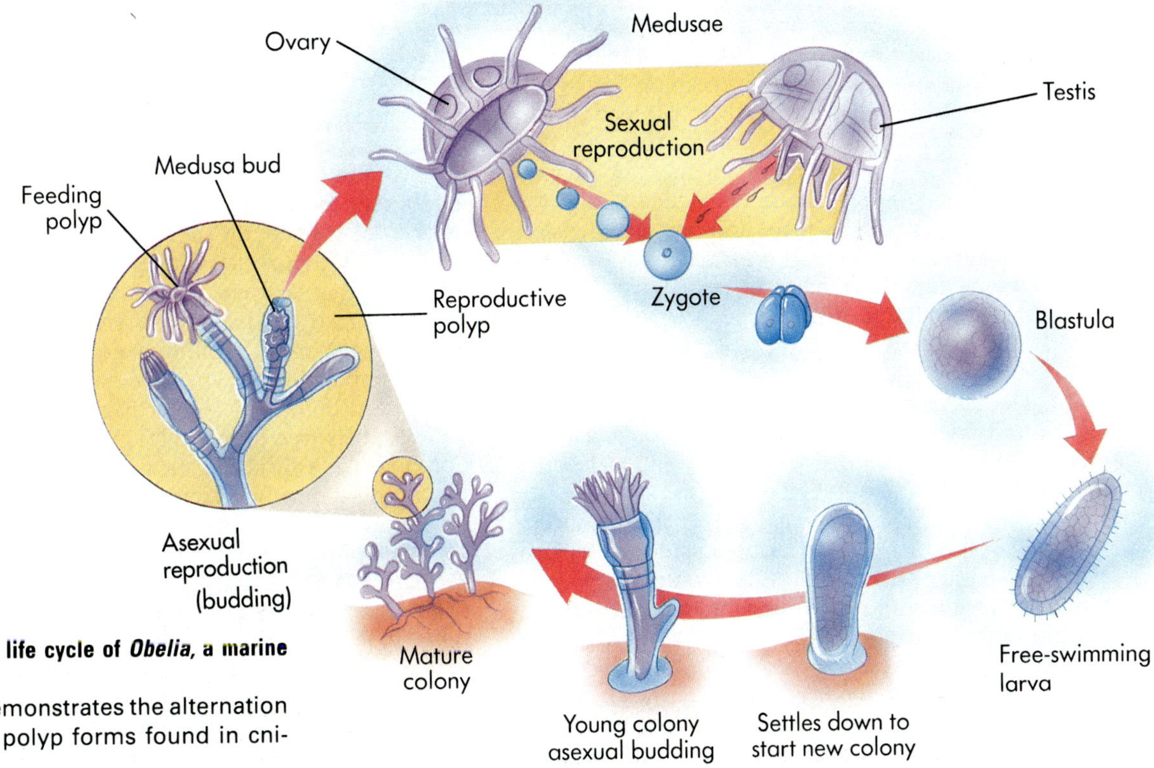

FIGURE 32-9 The life cycle of *Obelia*, a marine cnidarian.
This life cycle demonstrates the alternation of medusa and polyp forms found in cnidarians.

The tentacles of a cnidarian help it capture prey—other animals such as small fishes, shrimp, and aquatic worms. The tentacles bear stinging cells called *cnidocytes,* which give the phylum its name. You can see these cells as tiny dots in the tentacles of the yellow cup coral shown in Figure 32-6, *B*. If you have ever been stung by a jellyfish, you know how powerful its sting can be. These stinging cells work much like harpoons. Powered by water pressure, the stinging cells are jettisoned out of the animal, spearing and immobilizing the prey. The tentacles then draw the prey back to the mouth.

Cnidarians have two layers of tissues and a nerve net that coordinates cell activities. They exist either as polyps (corals and sea anemones), which are cylindrically shaped animals that anchor to rocks, or as medusae (jellyfish), which are free-floating, umbrella-shaped animals. In some cnidarians, these two forms alternate during the life cycle of the organism.

Bilateral symmetry: Variation on a theme in many phyla

Flatworms

The remaining phyla of animals are all bilaterally symmetrical or (in the case of the echinoderms) are thought to have evolved from bilaterally symmetrical forms. In addition, they all develop embryologically from three layers of tissue: an inner layer or **endoderm,** an outer layer or **ectoderm,** and a middle layer or **mesoderm.** Although cnidarians have a middle layer between their two tissue layers, it consists of a jelly-like material with only widely dispersed cells.

The bilaterally symmetrical animals with the simplest body plan are the flatworms (phylum Platyhelminthes). This phylum name comes from Greek words meaning "flat" (*platys*) "worm" (*helminthos*) and describes their flattened ribbon or leaf-like shapes. Although simple in structure, the flatworms have organs and some organ systems, but because they have no coelom, the organs are embedded within the body tissues (see Figure 32-3). Flatworms are also the simplest animals to have a distinct head, a characteristic common to many of the bilaterally symmetrical animals.

There are three classes of flatworms: the turbellarians, the flukes, and the tapeworms (Figure 32-10). The turbellarians are free living and found in fresh or salt water or in damp soil. The flukes and tapeworms are parasites and live on or in other animals, deriving nutrition from their hosts.

Free-living flatworms move from place to place, feeding on a variety of small animals and bits of organic debris. They move by means of ciliated epithelial cells that are concentrated on their ventral surfaces. In fact, the name *turbellarian* comes from a Latin word meaning "to bustle or stir" and refers to the water turbulence created by their movement. Sensory pits or tentacles along the sides of their heads detect food, chemicals, and movements of the fluid in which they are moving. They also have eyespots on their heads, which contain light-sensitive cells that enable the worms to distinguish light from dark. These organs are part of the flatworm nervous system and connect to a ladder-like paired nerve cord than extends down the length of the animal. Tiny swellings at the cephalic, or head end, of the organism are considered a primitive brain.

FIGURE 32-10 Flatworms (phylum Platyhelminthes).
A A marine, free-living turbellarian.
B The human liver fluke, *Clonorchis sinensis.*

The flatworm has a digestive system consisting of a digestive sac, or gut, open only at one end (Figure 32-11). Muscular contractions in the upper end of the gut of flatworms cause a strong sucking force by which the flatworms ingest their food and tear it into small bits. The cells making up the gut wall engulf these particles; most digestion takes place within these cells. Wastes from within the cells diffuse into the digestive tract and are expelled through the mouth. In addition, the flatworm excretes excess water and some wastes by means of a primitive excretory system: a network of fine tubules that runs along the length of the worm. Specialized bulb-like cells, or flame cells, are located along these tubules (see Figure 32-11). As cilia within them beat (looking like a flickering flame), they move the water and wastes into the tubules and out excretory pores.

Reproduction in flatworms is much more complicated than in sponges or cnidarians. Although most flatworms are hermaphroditic, a characteristic exhibited by the sponges and cnidarians, the organs of reproduction are better developed. When flatworms mate, each partner deposits sperm in a copulatory sac of the other. The sperm travel along special tubes to reach the egg. In free-living flatworms, the fertilized eggs are laid in cocoons and hatch into miniature adults. In some parasitic flatworms, there is a complex

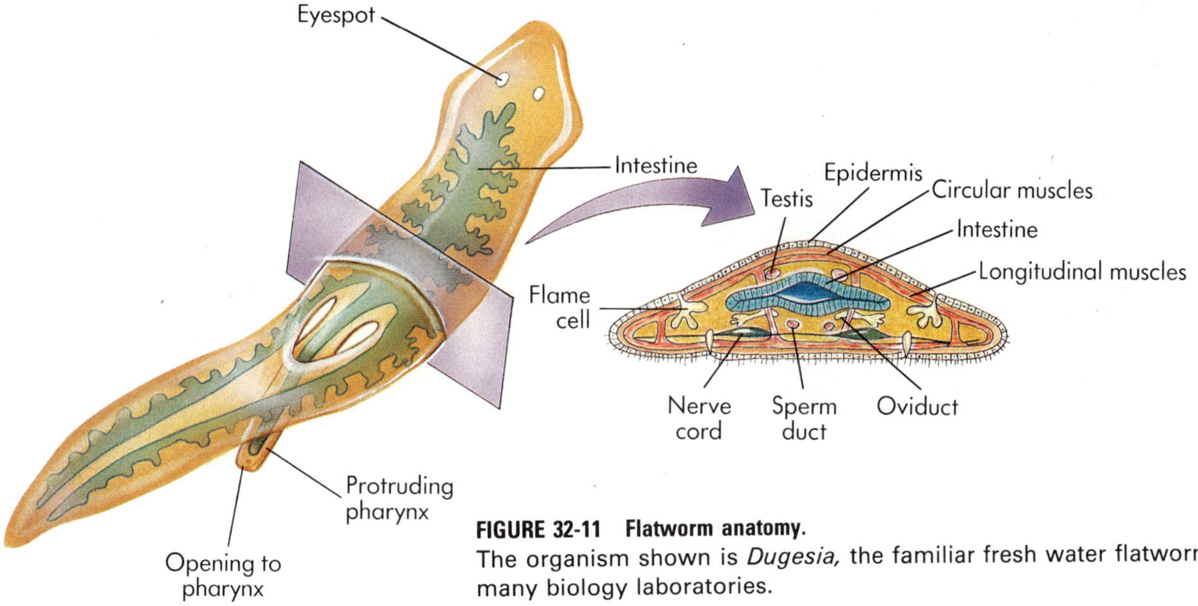

FIGURE 32-11 Flatworm anatomy.
The organism shown is *Dugesia,* the familiar fresh water flatworm of many biology laboratories.

succession of distinct larval forms. Flatworms are also capable of asexual reproduction. In some genera, when a single individual is divided into two or more parts, each part can regenerate an entire new flatworm.

The flukes and the tapeworms are two classes of flatworms that live within the bodies of other animals. The adult form of both classes parasitize humans. The life cycles of these organisms are discussed in the boxed essay.

The acoelomates, typified by the flatworms, are the most primitive bilaterally symmetrical animals. Although simple in structure, the flatworms have organs and some organ systems, including a primitive brain. Because flatworms do not have a coelom, their organs are embedded within the body tissues.

Roundworms

Seven phyla have a pseudocoelomate body plan (see Figure 32-3)—a body cavity with no lining. Only one of the seven, the roundworms, includes a large number of species.

Roundworms are classified in the phylum Nematoda, a word that comes from a Greek word meaning "thread." These cylindrical worms are diverse in size, but some of them are so small and slender that they look like fine threads. In fact, they can be so microscopically small that a spadeful of fertile soil may contain millions of these worms (Figure 32-12). Some species are abundant in fresh or salt water.

Many members of this phylum are parasites of vertebrates. About 50 species of roundworms parasitize human beings, causing problems such as blockage of the lymphatic vessels (Figure 32-13) or intestines and infections of the

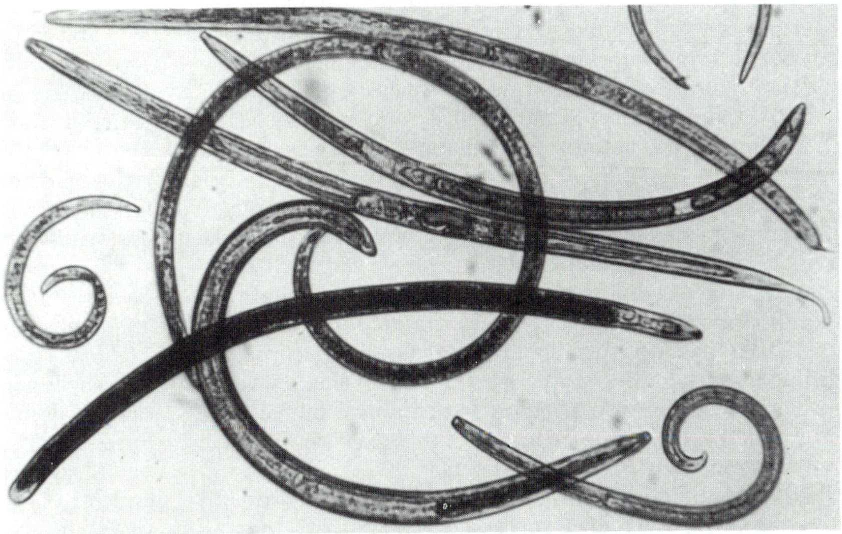

FIGURE 32-12 Living dirt.
One square meter of ordinary lawn soil teems with 2 to 4 million roundworms. Although most are similar in form, they range in length from about 0.2 millimeters to about 6 millimeters.

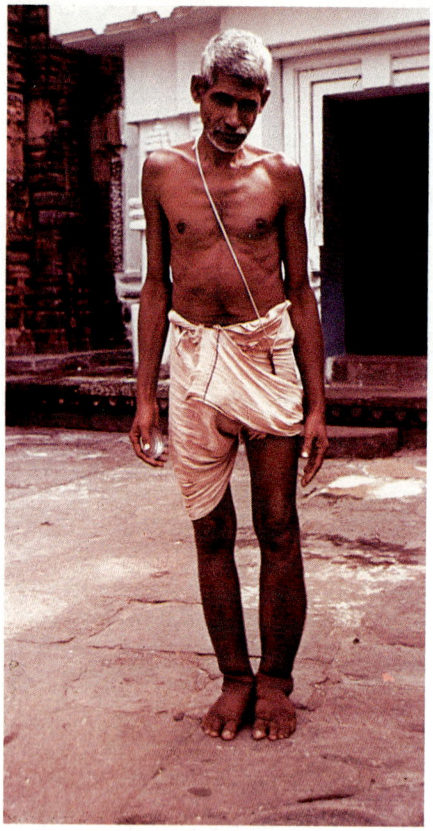

FIGURE 32-13 Elephantiasis.
This condition is caused by roundworms that live in the lymphatic passages and block the flow of lymph. As a result, fluids cannot drain, and swelling occurs.

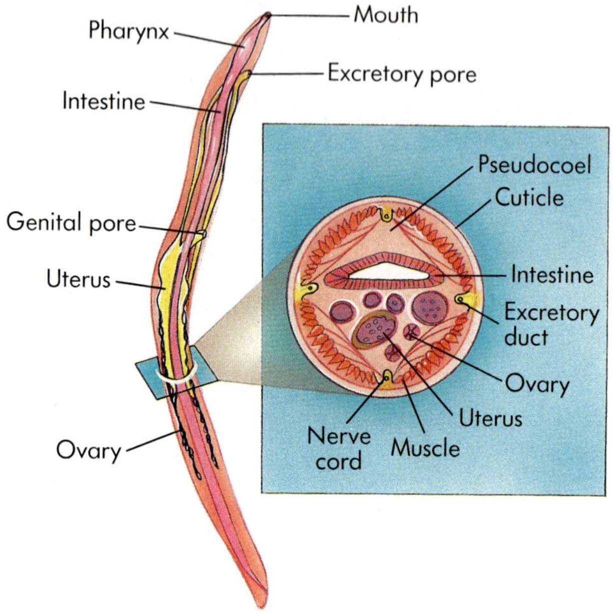

FIGURE 32-14 Anatomy of *Ascaris,* a parasitic roundworm of humans.
Note the pseudocoel.

muscles or lungs. Roundworms also parasitize invertebrates and plants. For this reason, some nematodes are being investigated as agents of biological control of insects and other agricultural pests.

Roundworms are covered by a flexible, tough, transparent multilayered tissue (cuticle), which is shed as they grow. A layer of muscle lies beneath this epidermal layer and extends lengthwise (Figure 32-14). These longitudinal muscles pull against both the cuticle and the firm, fluid-filled pseudocoelom, similar to how your muscles pull against your bones. All this effort gets them nowhere in clear water, but they can move in muddy water or soil, which provides surfaces against which the worm's body pushes. The movements do, however, push on the fluid-filled pseudocoelom and aid in the distribution of food and oxygen throughout the worm.

The roundworm digestive system has two openings—a mouth and anus. Most roundworms have raised, hair-like sensory organs near their mouths. The mouth itself often has piercing organs, or stylets. Food passes through the mouth as a result of the sucking action of a muscular pharynx. After passing through these organs, food continues through the digestive tract where it is broken down and then absorbed. The roundworms that parasitize animals take in digested food of the host; the cells lining the digestive system simply absorb these nutrients.

The roundworms also contain primitive excretory and nervous systems. The nervous system consists of a ring of tissue surrounding the pharynx and a solid dorsal and a ventral nerve cord (unlike the hollow nerve cord of a chordate [see p. 581]). The excretory system consists of two lateral canals that unite near the anterior end to form a single tube ending in an excretory pore.

The pseudocoelomates are typified by the roundworms. Muscles attached to their thick, outer cuticle push against the fluid-filled pseudocoelom, resulting in a whip-like movement. Many of the roundworms are parasites of invertebrates, vertebrates (including humans), and plants.

Clams, snails, and octopuses

The rest of the invertebrate animals have a "true" coelom. The next three phyla of coelomates: the mollusks (clams, snails, and octopuses), the annelids (earthworms and leeches), and the arthropods (lobsters, insects, and spiders) are all protostomes. The acoelomate animals already discussed are not protostomes but exhibit similar developmental patterns to the protostomes. These similarities suggest an evolutionary closeness among these groups.

The mollusks are a large phylum of invertebrate animals having a muscular foot and a soft body covered by a mantle and are usually covered with a hard shell. The phylum name *Mollusca* comes from a Latin word meaning "soft bodied." The shelled mollusks include the snails, clams, scallops, and oysters. Unshelled mollusks are represented by the octopuses, squids, and slugs.

Mollusks are widespread and often abundant in marine and fresh water environments, and some, such as certain snails and the slugs, live on land. They range in size from

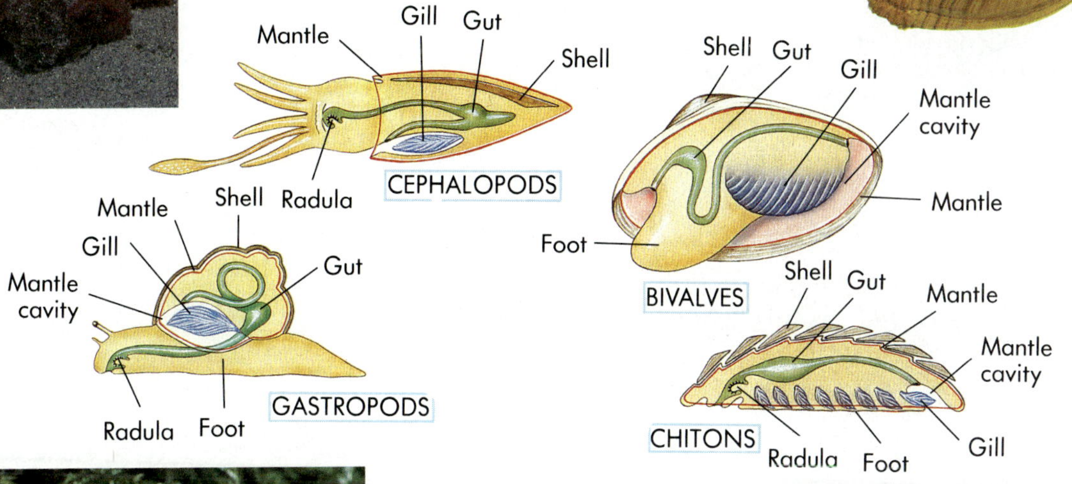

FIGURE 32-15 Body plans among the mollusks.
The name of each group describes its prominent features. The photos of each group are: slug, *Arion ater* (gastropod); a mussel of a nearly extinct species from the Wabash and upper Maumee river in Indiana (bivalve); octopus (cephalopod); and a chiton, *Tonicella lineata.*

being near microscopic to having the huge proportions of giant squid, which may be as long as 21 meters (almost 70 feet) and weigh up to 250 kilograms (550 pounds).

Mollusks exhibit four body plans: gastropod, cephalopod, bivalve, and chiton (Figure 32-15). The name of each group describes its prominent features. Although each group (with the exception of the bivalves) has a head end, the cephalopods (literally, "head-foot") have the most well-differentiated head and the most well-developed nervous system. The bivalves ("two-shelled") gastropods have the least well-developed nervous sytem of these groups. All four groups have a **visceral mass,** or group of organs, consisting of the digestive, excretory, and reproductive organs. The visceral mass is covered with a soft epithelium called the **mantle,** which arises from the dorsal body wall and encloses a cavity between itself and the visceral mass. This cavity is *not* the coelom; the coelom surrounds the heart only. The mollusk's gills, the organs of respiration, lie within the mantle cavity. **Gills** are a system of filamentous projections of the mantle tissue that are rich in blood vessels. These projections greatly increase the surface area available for gas exchange. In land-dwelling mollusks, a network of blood vessels within the mantle cavity serves as a primitive lung.

Mollusks exhibit both open circulatory systems (in clams, for example) and closed circulatory systems (in squid, for example). Both types of circulatory system have a heart to pump blood. In a **closed circulatory system,** blood is enclosed within vessels as it travels throughout the body of the organism. In an **open circulatory system,** blood flows in vessels leading to and from the heart but through irregular channels called **blood sinuses** in many parts of the body.

In most coelomate animals (most of which have a closed circulatory system), the blood vessels are intimately associ-

BIOLOGY & you

Parasitic flatworms

Parasitic flatworms are an unusual and frightening group. Although tiny, their effects on humans and animals on which humans depend can be devastating.

Most parasitic flatworms have two hosts. Eggs or larvae grow and develop within an animal called the *secondary host*. Once the appropriate stage of development is reached (different for each species of flatworm), the parasites then migrate into a *primary host,* where they complete their maturation into adults and begin to reproduce. Humans most often serve as primary hosts to parasitic flatworms, and animals such as snails, fish, cattle, and pigs serve as secondary hosts. It is easy to see how humans can become infected with parasitic worms—the secondary hosts are sources of food.

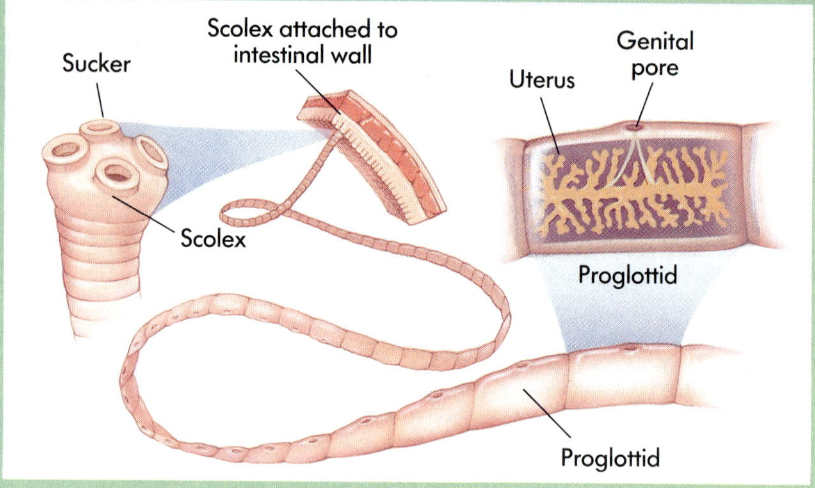

FIGURE 32-A Anatomy of the beef tapeworm.

Parasitic flatworms are of two types: the flukes and the tapeworms. Several species of the parasitic fluke *Schistosoma* are found in Africa, South America, the West Indies, the Middle East, and the Far East. Infection with *Schistosoma* organism ranks as one of the major infectious diseases of the world. The life cycle of *Schistosoma* flatworms begins when the eggs are shed into the water in the feces or urine of an infected individual. The eggs hatch in the water. The larvae must come into contact with a specific genus of snail that serves as their secondary host within a few hours, or they die. Each species of *Schistosoma* parasitizes a specific genus of snail. Once within the snail, the *Schistosoma* larvae develop into a form that is equipped to swim. They then migrate out of the snail and into the water. When they come into contact with humans, they penetrate the skin and burrow until they find a blood vessel. The young schistosomes migrate first to the liver, where they complete their development into adults, before migrating to various other parts of the body. The females produce prodigious amounts of eggs, and these eggs are shed in the feces and urine of the infected individual, thus completing the life cycle.

Other parasitic flukes include the liver fluke (*Clonorchis*), which uses the snail *Parafossarulus* and fishes of the family Cyprinidae as secondary hosts. Humans become infected when they eat raw or undercooked fish. The lung fluke uses fresh water crabs as secondary hosts and migrates to the lungs of its primary host, the human, after being ingested. Infection with the lung fluke causes severe respiratory problems.

Tapeworms are a second type of parasitic flatworm. The body of a tapeworm is long and flat (Figure 32-A). The head region is equipped with a scolex, an organ that contains suckers or hooks with which the tapeworm attaches itself to the intestinal wall of its primary host. The rest of the body is made up of repeating sections called *proglottids*. Each proglottid contains a full set of reproductive organs. Each proglottid, then, is capable of producing hundreds of offspring. No wonder tapeworms have been referred to as "reproduction factories!"

The most common tapeworm that affects humans is the beef tapeworm (*Taeniarhynchus saginatus*). The life cycle of the beef tapeworm begins when a few proglottids are broken off in the feces of an infected animal. The proglottids produce eggs as they lie in the vegetation or soil. The eggs are extremely hardy—they have been known to live up to 5 months as they lie on grass. When the eggs are ingested by grazing cattle, they hatch within the intestine, and the larvae burrow through the intestinal wall until they reach muscle. Here, the larvae encyst—they become encapsulated and quiescent. When a human eats raw or undercooked beef, the cyst dissolves, and the tapeworm attaches to the intestine with its scolex. New proglottids develop, and the tapeworm matures in 2 to 3 weeks. After maturation, the tapeworm produces numerous offspring. Proglottids are broken off in the feces of the affected individual, and the life cycle begins again.

Pork tapeworm (*Taenia solium*) has a life cycle similar to that of the beef tapeworm, except that humans can also serve as the secondary host in which the tapeworm encysts. Common sites for encystment are the brain and the eye, resulting in blindness or severe neurological problems.

While the thought of sharing your body with such creatures is quite unpleasant, the use of efficient sanitation procedures has greatly reduced the rate of parasitic worm infection in the United States. A further precaution against these parasites is to avoid eating raw or undercooked meat or fish—remember that cattle, fish, and other animals serve as these parasites' secondary hosts, and high cooking temperatures will kill the parasites.

ated with the excretory organs, making the direct exchange of materials between these two systems possible. Mollusks were one of the earliest evolutionary lines to develop an efficient excretory system. Wastes are removed from the mollusk by tubular structures called **nephridia.** In mollusks with open circulatory systems, wastes move from the coelom into the nephridia and are discharged into the mantle cavity. From there they are expelled by the continuous pumping of the gills. In animals with closed circulatory systems, such as annelids, some mollusks, and the vertebrates, the coiled tubule of a nephridium is surrounded by a network of capillaries. Wastes move from the circulatory system to the nephridium for removal from the body. All coelomates (except for arthropods and chordates) have basically similar excretory systems.

Mollusks, widespread in marine and fresh water environments, exhibit four body plans as represented by snails, clams, squid, and chitons. These animals use gills for respiration in water; terrestrial species have adaptations for breathing on land. They exhibit both open and closed circulatory systems and well-developed excretory systems.

Earthworms and leeches

The other two major phyla of protostomes, the annelids and arthropods, have segmented bodies, whereas the mollusks do not. Segmentation underlies the organization of all of the more complex phyla of animals: the annelids (earthworms and leeches), arthropods (lobsters, insects, and spiders), echinoderms (sea urchins and starfish), and chordates (tunicates, fishes, amphibians, reptiles, birds, and mammals). In some adult arthropods the segments are fused; their segmentation is only apparent in their embryological development. So, too, embryological development reveals segmentation in the chordates. Segmentation in this phylum is exhibited only by the vertebrates in the repeating units of their backbones. Because segmentation in animals is different among phyla, scientists think it arose independently in more than one line of evolution.

The annelids (phylum Annelida) are worms characterized by a soft, elongated body composed of a series of ringlike segments. In fact, the word *annelid* means "tiny rings." Annelids are abundant in the soil and in both marine and fresh water environments throughout the world. Internally, their segments are divided from one another by partitions called **septa.** Digestive, excretory, and neural structures are repeated in each segment.

There are three classes of annelids: marine worms, fresh water and terrestrial worms, and leeches (Figure 32-16). Exhibiting unusual forms and sometimes iridescent colors, the marine worms live in burrows, under rocks, inside shells, and in tubes of hardened mucus they manufacture. Leeches occur mostly in fresh water, although a few are marine and some tropical leeches occur in terrestrial habitats. Most leeches are predators or scavengers, and some suck blood from mammals, including humans. The best-known leech is the medicinal leech, which was used for centuries to remove

A

B

C

FIGURE 32-16 Representative annelids.
A Shiny bristleworm.
B Earthworms mating.
C A fresh water leech, with young leeches visible in the brood pouch.

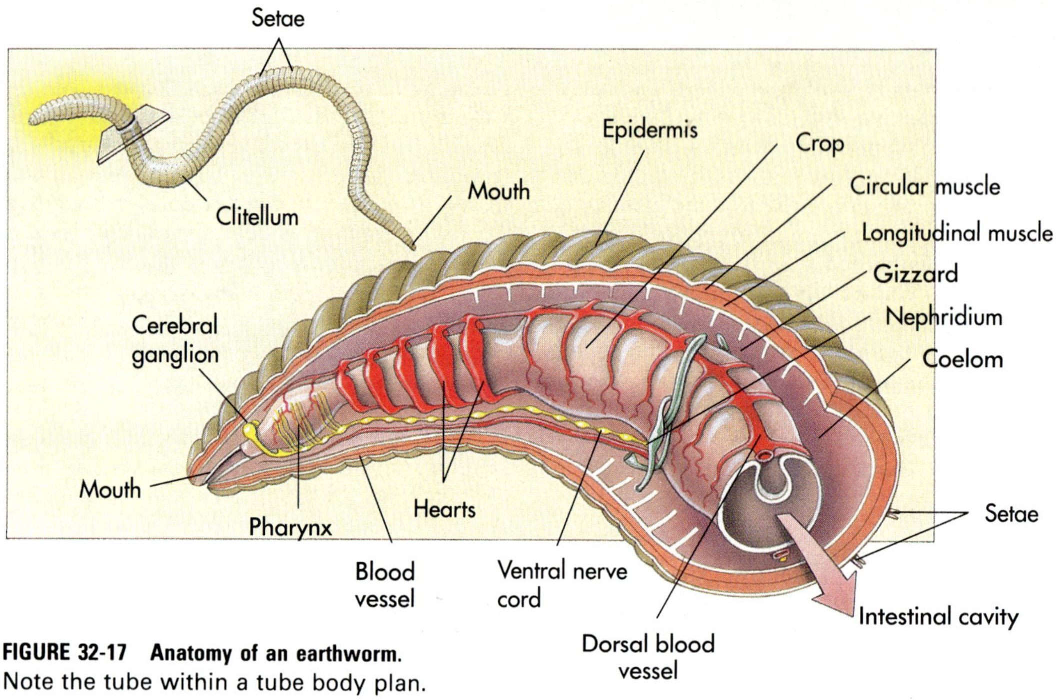

FIGURE 32-17 Anatomy of an earthworm. Note the tube within a tube body plan.

what was thought to be excess blood responsible for certain illnesses. The medicinal leech is now used as a source of anticoagulant, which is used in research focusing on blood clotting. This animal is also used by some physicians to remove excess blood after surgery and to help restore circulation to severed body parts (such as fingers) after reattachment.

The earthworm exhibits the generalized body plan of this phylum: a tube within a tube (Figure 32-17). The digestive tract, a straight tube running from mouth to anus, is suspended within the coelom. An earthworm sucks in organic material by contracting its strong pharynx. It grinds this material in its muscular gizzard, aided by the presence of soil particles it takes in with its food.

The anterior segments of an earthworm contain a well-developed **cerebral ganglion,** or brain, and a few muscular blood vessels that act like hearts, pumping the blood through the closed circulatory system. Sensory organs are also concentrated near the anterior end of the worm. Some of these organs are sensitive to light, and elaborate eyes with lenses and retinas have evolved in certain members of the phylum. Separate nerve centers, or ganglia, are located in each segment and are connected by nerve cords. Each segment also contains both circular and longitudinal muscles, which annelids use to crawl, burrow, and swim. **Setae,** or bristles, help anchor the worms during locomotion or when they are in their burrows.

The annelids are characterized by serial segmentation and a tube within a tube body plan. The body is composed of numerous similar segments, each with its own circulatory, excretory, muscular, and neural structures.

Reproduction differs among the annelid classes. In the marine worms, the sexes are usually separate, and fertilization is often external, occurring in the water and away from both parents. The earthworms and leeches, on the other hand, are hermaphroditic. When they mate, their anterior ends point in opposite directions, and their ventral surfaces touch. The **clitellum,** a thickened band on an earthworm's body, secretes a mucus that holds the worms together as they exchange sperm. Ultimately, the worms release the fertilized eggs into cocoons also formed by mucous secretions of the clitellum.

Lobsters, insects, and spiders

Lobsters (crustaceans), insects, and spiders (arachnids) are representatives of the diverse phylum Arthropoda (Figure 32-18). The name *arthropod* comes from the two Greek words "arthros" (jointed) and "podes" (feet) and describes the characteristic jointed appendages of all arthropods. The nature of the appendages differs greatly in different subgroups; appendages may take the form of antennae, mouthparts of various kinds, or legs.

The arthropods have a rigid external skeleton, or **exoskeleton,** which varies greatly in toughness and thickness among arthropods (Figure 32-19). The exoskeleton provides places for muscle attachment, protects the animal from predators and injury, and most importantly, protects arthropods from water loss. As an individual outgrows its exoskeleton, that exoskeleton splits open and is shed. A new soft exoskeleton lies underneath, which subsequently hardens. The animal then grows into its new "shell."

All arthropods can be placed into one of two groups: those with jaws and those without jaws. The crustaceans, insects, centipedes, millipedes, and a few other small groups of arthropods have jaws, or **mandibles.** These jaws are formed by the modification of one of the pairs of anterior appendages (but *not* the first pair). The appendages nearest the anterior end are sensory antennae (Figure 32-20, *A*). The remaining arthropods, which include the spiders, mites, scorpions, and a few other groups, lack mandibles. Their mouthparts usually take the form of pincers or fangs, which evolved from the appendages nearest the anterior end of the animal (Figure 32-20, *B*).

A

B

FIGURE 32-18 **Representatives of phylum Arthropoda.**
A A freshwater crayfish, *Procambarus.*
B Copulating grasshoppers.
C The arrowhead spider.

C

A

B

FIGURE 32-19 Exoskeletons.
Some arthropods have a tough exoskeleton, like this South American scarab beetle **(A)**; others have a fragile exoskeleton, like the green darner dragonfly **(B)**.

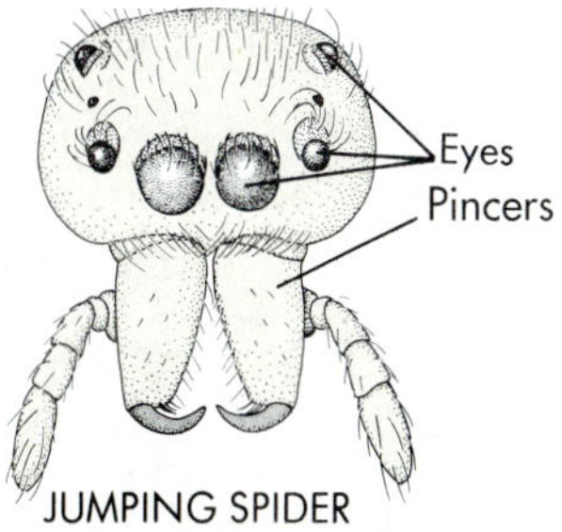

A

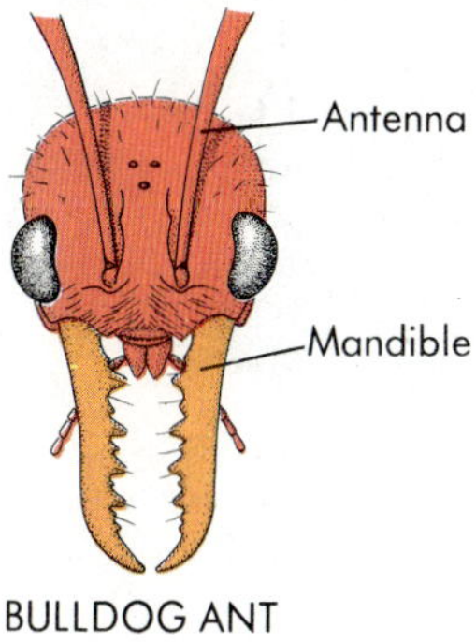

B

FIGURE 32-20 Arthropod mouthparts.
A The bulldog ant does have a jaw.
B The jumping spider lacks a jaw but does have pincers with which it catches prey.

FIGURE 32-21 The compound eye. This compound eye is found on a robberfly.

An important structure of many arthropods such as bees, flies, moths, and grasshoppers is the **compound eye** (Figure 32-21). Compound eyes are composed of many independent visual units, each containing a lens. **Simple eyes,** composed of a single visual unit having one lens, are found in many arthropods with compound eyes and function in distinguishing light and darkness. In some flying insects, such as locusts and dragonflies, simple eyes function as horizon detectors and help stabilize the insects during flight.

In the course of arthropod evolution, the coelom has become greatly reduced, consisting only of the cavities that house the reproductive organs and some glands. Figure 32-22 illustrates the major structural features of a grasshopper as a representative of the arthropods. Like the annelids, the arthropods have a tubular gut that extends from the mouth to the anus. The circulatory system of arthropods is open; their blood flows through cavities between the organs. One longitudinal dorsal vessel functions as a heart, helping move the blood along.

Most aquatic arthropods breathe by means of gills. Their feathery-looking structure provides a large surface area over which gas exchange takes place between the surrounding water and the animal's blood. The respiratory systems of terrestrial arthropods generally have internal surfaces over which gas exchange takes place. The respiratory systems of insects, for example, consist of small, branched air ducts called **tracheae.** These tracheae, which ultimately branch into very small **tracheoles,** are a series of tubes that transmits oxygen throughout the body. The tracheoles are in direct contact with the individual cells, and oxygen diffuses from them to other cells directly across the cell membranes. Air passes into the trachea by way of specialized openings called **spiracles,** which can be closed and opened by valves in most insects.

Although there are various kinds of excretory systems in different groups of arthropods, a unique excretory system evolved in terrestrial arthropods in relation to their open circulatory system. The principal structural element is the **malpighian tubules,** which are slender projections from the digestive tract. Fluid is passed through the walls of the malpighian tubules to and from the blood in which the tubules are bathed. The nitrogenous wastes in it are precipitated and then emptied into the hindgut (the posterior part of the digestive tract) and eliminated. Most of the water and salts in the fluid is reabsorbed by the hindgut, thus conserving water.

Arthropods are a diverse phylum of organisms having jointed appendages and rigid exoskeletons.

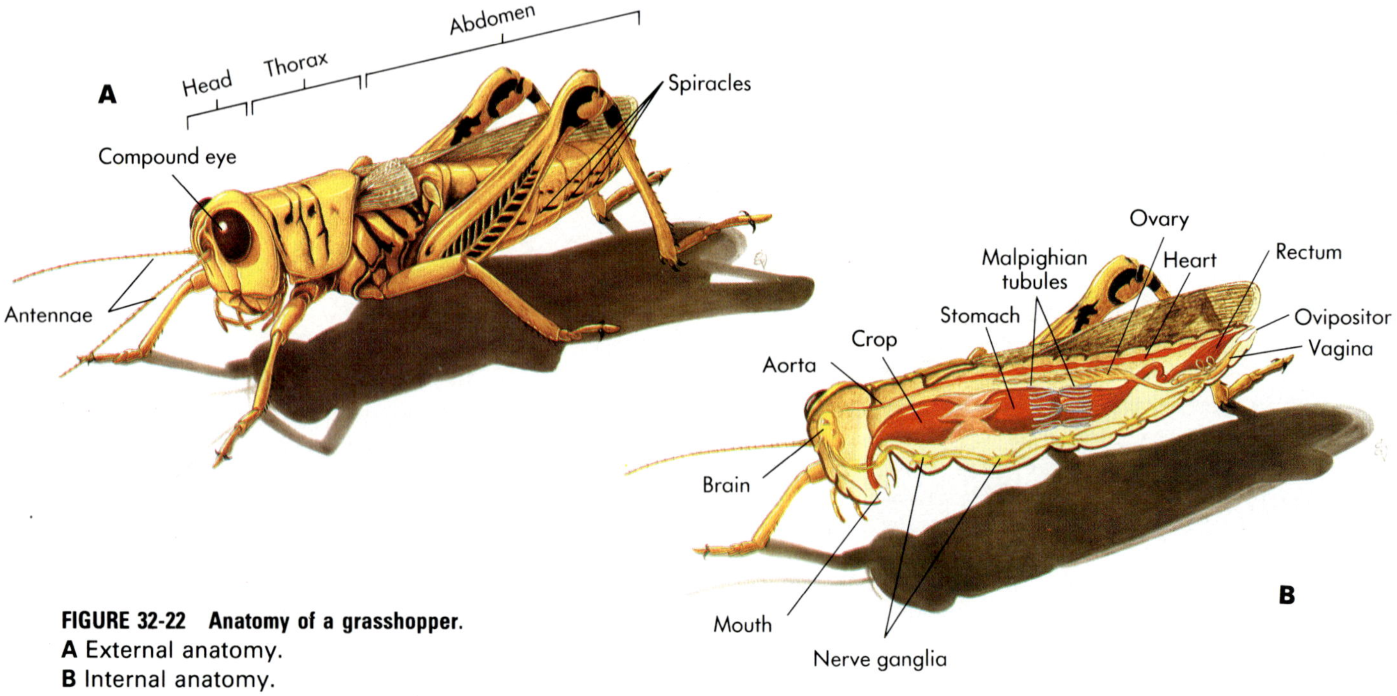

FIGURE 32-22 Anatomy of a grasshopper.
A External anatomy.
B Internal anatomy.

FIGURE 32-23
Representatives of the echinoderms.
A Sea star.
B Sea cucumber.
C Sea lilies.
D Brittle star.
E Sand dollar.

Sea urchins and starfish

Sea urchins and starfish are representatives of the phylum Echinodermata. The term *echinoderm* means "spine skin," an appropriate name for many members of this phylum (Figure 32-23). The five living classes of echinoderms are shown: the sea lilies, sea stars (starfish), brittle stars, sea urchins and sand dollars, and sea cucumbers. These marine animals live on the sea floor, with the exception of a few swimming sea cucumbers.

The echinoderms are different from the other invertebrates in that they are deuterostomes. The other invertebrates are protostomes. The embryological differences that distance them evolutionarily from the other invertebrates connect them with the chordates. It is thought that the echinoderms and the chordates, along with two smaller phyla not mentioned here, evolved from a common ancestor.

Echinoderms are bilaterally symmetrical as larvae but radially symmetrical as adults. They are evolutionary related to and grouped with the bilaterally symmetrical animals, however. Adult echinoderms have a five-part body plan corresponding to the arms of a sea star or the design on the "shell" of a sand dollar. Five **radial canals,** the positions of which are determined early in the development of the embryo, extend into each of the five parts of the body (Figure 32-24). This water vascular system is used for locomotion and is unique to echinoderms.

FIGURE 32-24 Anatomy of an echinoderm. Notice the water vascular system of a sea star *(blue).*

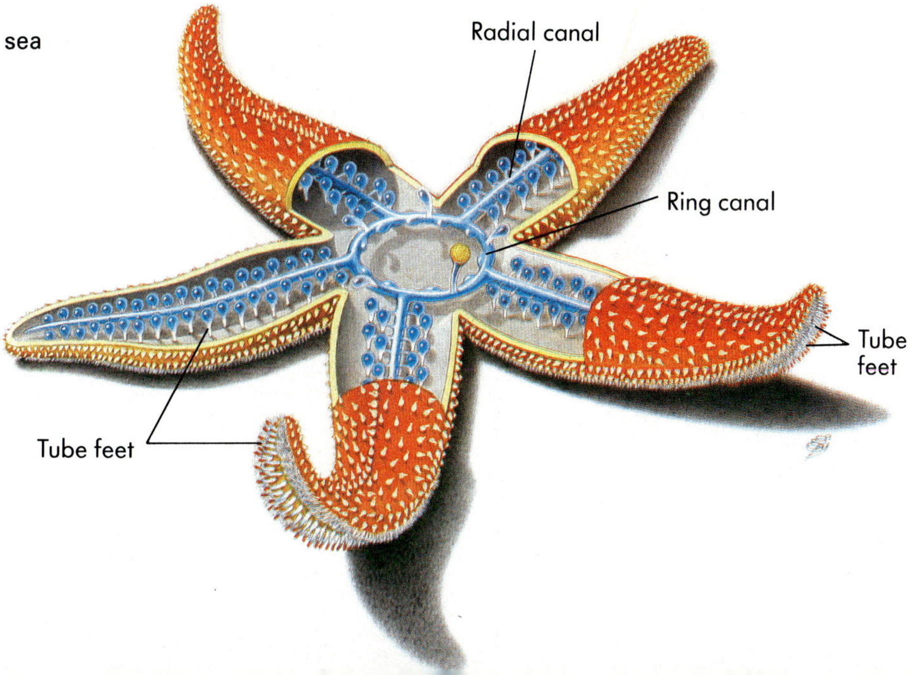

As adults, these animals have no head or brain. Their nervous systems consist of central **nerve rings** from which branches arise. The animals are capable of complex response patterns, but there is no centralization of function.

There is no well-organized circulatory system in echinoderms. Food from the digestive tract is distributed to all the cells of the body in the fluid that lies within the coelom. In many echinoderms, respiration takes place by means of skin gills, which are small, finger-like projections that occur near the spines. Waste removal also takes place through these skin gills. The digestive system is simple but usually complete, consisting of a mouth, gut, and anus.

Although the echinoderms are well-known for regenerating lost parts, most reproduction is sexual and external. The sexes in most echinoderms are separate, although usually little external difference exists between them. The fertilized eggs of echinoderms usually develop into free-swimming bilaterally symmetrical larvae that look quite different from the adults. They eventually change through a series of stages, or metamorphose, to adult forms.

The echinoderms, or spiny-skinned animals, are marine organisms represented by sea urchins and starfish. They are closely related to the chordates.

Summary

1. Animals are a diverse group of multicellular, eukaryotic, heterotrophic organisms. The cells of all animals (with the exception of sponges) are organized into tissues, and in most animals the tissues are organized into organs. Most zoologists think that the animals arose from protist ancestors.
2. Other than the sponges, all animals exhibit either radial symmetry (the jellyfish and starfish) or bilateral symmetry (all other animals). Radial symmetry means that body parts emerge from a central point. Bilateral symmetry means that the right side of an organism is a mirror image of the left side.
3. A body cavity, or coelom, is a lined, fluid-filled enclosure, which contains body organs that hang suspended by thin sheets of connective tissue. All bilaterally symmetrical organisms and the echinoderms (starfish) have coeloms except roundworms and flatworms. Roundworms have pseudocoeloms, or false body cavities, which are not lined; roundworms' organs are not suspended by thin sheets of connective tissue. Flatworms have no coeloms.
4. Two main branches of coelomate animals, representing two distinct evolutionary lines, are the protostomes and the deuterostomes. The protostomes (mollusks, annelids, and arthropods) differ with regard to certain embryological events from the deuterostomes (echinoderms and chordates).
5. Sponges (phylum Porifera) are acoelomate, asymmetrical, aquatic animals. They are considered simple animals because they have no tissues or organs.
6. The hydra and jellyfish (phylum Cnidaria) are acoelomate, radially symmetrical, aquatic animals. They have two layers of tissues and exist either as cylindrically shaped polyps (corals and sea anemones) or medusae (jellyfish). In some cnidarians, these two forms alternate during the life cycle of the organism.
7. All animals other than the sponges and the cnidarians are bilaterally symmetrical or, in the case of the echinoderms, evolved from bilaterally symmetrical ancestors. In addition, they all develop embryologically from three tissue layers: endoderm, ecotoderm, and mesoderm.
8. The flatworms (phylum Platyhelminthes) are ribbon-like worms that live either in the soil or water or within other organisms. Having no coelom, they are considered to have the simplest body plan of the bilaterally symmetrical animals. However, the flatworms have organs and some organ systems, including a primitive brain.
9. The roundworms (phylum Nematoda) are pseudocoelomate, cylindrical worms that live in the soil, in water, or within other organisms. They have simple digestive, excretory, and nervous systems.
10. The mollusks (phylum Mollusca) are a large phylum of coelomates that exhibit four body plans. All four groups have a visceral mass, or group of organs, consisting of the digestive, excretory, and reproductive organs. They exhibit both closed and open circulatory systems.
11. The annelids (phylum Annelida) are worms characterized by a soft, elongated body composed of a series of ring-like segments, each with its own circulatory, excretory, muscular, and neural structures. They have a tube within a tube body plan.
12. The arthropods (phylum Arthropoda) are an extremely diverse phylum that includes organisms such as lobsters, insects, and spiders. Arthropods are characterized by a rigid external skeleton and jointed appendages.
13. The phylum Echinodermata, represented by sea urchins and starfish, are spiny-skinned marine animals that live on the sea floor. They are the only deuterostome invertebrates, a characteristic that shows relatedness with the chordates. Although adult echinoderms are radially symmetrical, their larvae are bilaterally symmetrical.

REVIEW QUESTIONS

1. What do you and jellyfish have in common? What is an important taxonomic difference between the two of you?
2. "Animals are a diverse group of eukaryotic, multicellular, heterotrophic organisms." Explain this statement.
3. Fill in the blanks: If an organism's body parts all emerge from a central point, it exhibits __________ __________. If an organism has two sides that are a mirror image of one another, it shows ______ ______.
4. Distinguish among coelomates, pseudocoelomates, and acoelomates. Give an example of each.
5. Which organisms are considered the "simplest animals"? Why?
6. What two basic body plans are shown by cnidarians? Summarize the differences between these two plans.
7. Which organisms are the most primitive bilaterally symmetrical animals? Briefly describe these animals' structure.
8. People can become very ill if infested by roundworms. Describe these organisms and why they can be a health risk to humans.
9. Distinguish between open and closed circulatory systems. Which do you have?
10. Summarize the characteristics of mollusks.
11. If you've ever tried to swat a fly, you know that flies are quick to respond to even the slightest movements. What feature of a fly's body allows it to perceive motion so rapidly?
12. Briefly summarize the characteristics of arthropods.
13. In casual conversation, you refer to a spider on the wall as an insect. A friend who's studied biology informs you that spiders are not insects. Explain what she means.
14. How do echinoderms differ from all other invertebrates? What is the significance of this fact?

KEY TERMS

acoelomate 563
bilateral symmetry 562
blood sinus 571
cerebral ganglion 574
clitellum 574
closed circulatory system 571
coelom 562
coelomate 562
collar cell 565
compound eye 576
deuterostome 564
ectoderm 568
endoderm 568
epithelial wall 565
exoskeleton 574
gills 571
hermaphrodite 566
malpighian tubule 576
mandible 574
mantle 571
medusa 567
mesoderm 568
nephridium (pl., nephridia) 573
nerve ring 578
open circulatory system 571
polyp 567
protostome 564
pseudocoelomate 562
radial canal 577
radial symmetry 562
septa 573
setae 574
simple eye 576
spiracle 576
tracheae 576
tracheole 576
visceral mass 571

THOUGHT QUESTIONS

1. Coral reefs are among the most diverse communities of organisms on Earth. Why do you think this is so? Why aren't free-living marine communities as diverse?
2. Pseudocoelomates evolved before coelomates. Why do you think the coelomates have been so much more successful?

FOR FURTHER READING

Bavendam, F. (1989, August). Even for ethereal phantasms, it's a dog-eat-dog world. *Smithsonian*, pp. 94-101.
Aspects of the lives of some of the incredibly beautiful sea slugs, or nudibranchs, are explained.

Wilson, E.O. (1990, March). Empire of the ants. *Discover*, pp. 45-50.
This is an outstanding essay on the fascinating ways of ants.

33 Vertebrate animals

Patterns of structure, function, and reproduction

Although bats often conjure up images of blood-sucking vampires, these bats would be totally disinterested in your neck. Nicknamed "little red flying-foxes," they find the eucalyptus blossoms of their Australian homeland a delicacy. When this favorite food is in short supply, they often raid orchards. Although these bats can become pests, you can leave the vampire-fighting garlic and pointed stakes at home if you venture on a trip "Down Under." Little red flying foxes only bother the fruit-growers.

HIGHLIGHTS

- Chordates are organisms that, at some time during their development, have a single, hollow nerve cord along the back, a rod-shaped notochord between the nerve cord and the gut, and gill arches located at the throat.
- There are three subphyla of chordates: the tunicates, sac-like sessile marine organisms; the lancelets, tiny scaleless fish-like organisms, and the vertebrates.
- A vertebrate has a vertebral column surrounding the dorsal nerve cord, a distinct head with a skull that encases the brain, a closed circulatory system with a heart to pump the blood, and many other complex organs and organ systems.
- There are seven classes of living vertebrates: three classes are fishes and four classes are land-dwelling, four-footed animals.

OUTLINE

Chordates and vertebrates: Unity in symmetry and coelom
- Tunicates
- Lancelets
- Vertebrates
  - Jawless fishes
  - Cartilaginous fishes
  - Bony fishes
  - Amphibians
  - Reptiles
  - Birds
  - Mammals

Interestingly, bats are mammals—the same class of animals to which humans belong. Bats are more like you than they are like birds! As this chapter progresses, you will come to understand those similarities as you explore the patterns of structure, function, and reproduction of the vertebrates.

Chordates and vertebrates: Unity in symmetry and coelom

The chordates and vertebrates are characterized by three principal features (see Chapter 27): (1) a single, hollow **nerve cord** located along the back; (2) a rod-shaped **notochord,** which forms between the nerve cord and the gut (stomach and intestines) during development; and (3) **pharyngeal (gill) arches,** which are located at the throat (pharynx). (These three chordate features are shown in Figure 27-3 as they appear in the embryo because they are present in the embryos of all chordates.) In addition, lancelets exhibit these characteristics as adults, and tunicates exhibit them as larvae. In vertebrates, the nerve cord differentiates into a brain and spinal cord. The notochord develops into a vertebral column that encases the nerve cord, protecting it. The pharyngeal arches develop into the gill structures of the fishes and into ear, jaw, and throat structures of the terrestrial vertebrates. The presence of the gill arches in all vertebrate embryos provides a clue to the aquatic ancestry of the subphylum Vertebrata.

Along with having these three traits, chordates have many other characteristics in common. All have a true coelom and bilateral symmetry. The embryos of chordates exhibit segmentation. Figure 33-1 shows segments of tissue called *somites* in the human embryo, which develop into the skeletal muscles. Most chordates have an internal skeleton to which their muscles are attached and work against, providing movement. (Larval tunicates and adult lancelets do not have internal skeletons; their muscles are attached to their notochords.) Finally, chordates have tails that extend beyond the anus, at least during embryonic development. Nearly all other animals have a terminal anus.

Chordates are characterized by a single, hollow nerve cord located along the back, a rod-shaped notochord, and gill arches located at the throat. The vertebrates are an important subphylum of the chordates.

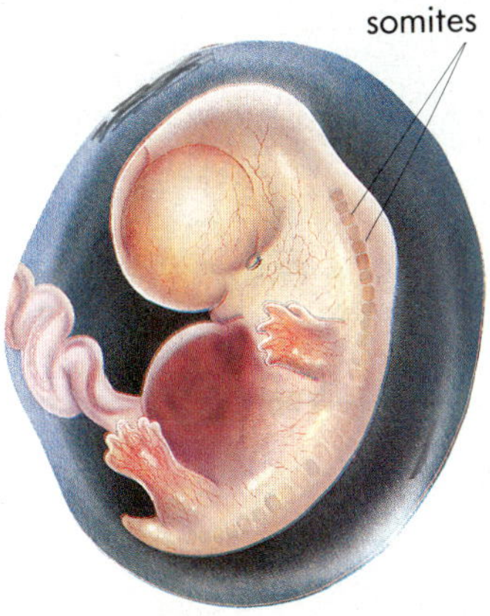

FIGURE 33-1 A chordate embryo.
All chordate embryos have segments of tissue called *somites.* The somites eventually develop into the skeletal muscles.

Tunicates

The **tunicates** are a group of about 2500 species of marine animals, most of which look like living sacs attached to the floor of the ocean (Figure 33-2, *A, B*). As shown in Figure 33-2, *C,* the tunicates are not much more than a large pharynx covered with a protective **tunic.** The tunic is a tough outer "skin" composed mainly of cellulose, a substance found in the cell walls of plants and algae but rarely found in animals. Some colonial tunicates live in masses on the ocean floor and have a common sac and a common opening to the outside (Figure 33-3). Colonial tunicates reproduce asexually by budding. Individual tunicates are hermaphodites, with each organism having both male and female sex organs.

The pharynx of a tunicate is lined with numerous cilia. As these cilia beat, they draw a stream of water through the incurrent siphon into the pharynx, which is lined with a sticky mucus. Food particles are trapped within the pharynx, and the filtered water flows out of the animal through the excurrent siphon. Because 90% of all species of tunicate have this structure and forcefully squirt water out their excurrent siphons when disturbed, they are also called *sea squirts.*

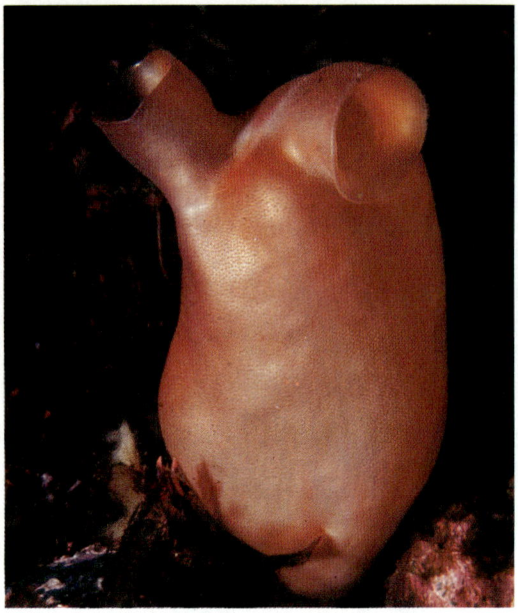

A

B

FIGURE 33-2 Tunicates (subphylum Urochordata).
A The sea perch *Halocynthia auranthium.*
B A beautiful blue and gold tunicate.
C The structure of a tunicate.

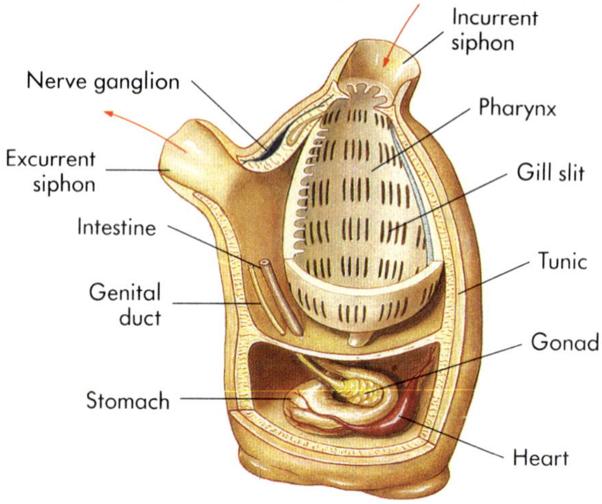

C

As adults, tunicates lack a notochord and a nerve cord. The gill slits (which develop from the gill arches) are the only clue that adult tunicates are chordates. Only the larvae, which look like tadpoles (Figure 33-4), have notochords and nerve cords. The notochords are in their tails, and their nerve cords run dorsal to their notochords almost the entire length of their bodies. The subphylum name *Urochordate* comes from the placement of the larval notochord and literally means "tail chordate." The larvae remain free swimming for no more than a few days. Then they settle to the bottom and become sessile, attaching themselves to a suitable substrate by means of a sucker. As they mature, they adjust to a filter-feeding existence. Some tunicates that live in the world's warmer regions retain the ability to swim and never develop into a sessile form.

FIGURE 33-3 Colonial tunicates.
This colony was found at a depth of about 2 meters. The colony on the right is in the process of dividing.

Adult tunicates are sac-like, sessile, marine filter-feeders. Their only chordate characteristic is their gill slits. Their larvae, however, have notochords and nerve cords in the tails of their tadpole-like bodies.

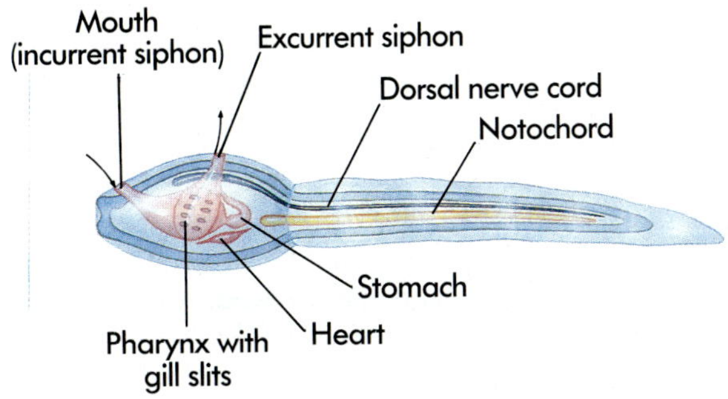

FIGURE 33-4 The structure of a larval tunicate.
The larvae of tunicates resemble tadpoles and contain notochords and nerve cords.

Lancelets

The **lancelets** are tiny, scaleless, fish-like marine chordates that are just a few centimeters long and pointed at both ends (Figure 33-5). They look very much like tiny surgical blades called *lancets,* from which they get their name. You may have had a few drops of blood taken in the doctor's office from a "fingerstick" done with a lancet.

Lancelets have a segmented appearance because of blocks of muscle tissue that are easily seen through their thin, unpigmented skin. Although they have pigmented light receptors, lancelets have no real head, eyes, nose, or ears. Unlike the the tunicates, the lancelet's notochord runs the entire length of its dorsal nerve cord. For this reason, lancelets are called *cephalochodates,* or *head chordates.* The lancelet retains its notochord throughout its lifespan.

The 23 species of lancelets live in the shallow waters of oceans all over the world. They spend most of their time partly buried in the sandy or muddy bottom with only their anterior ends protruding, feeding on plankton, floating microscopic plants and animals. In a manner similar to the tunicates, lancelets filter plankton from the water. Cilia line the anterior ends of their alimentary canals, and these beating cilia create an incoming current of water. The filtered water exits at an excurrent siphon. An oral hood projects beyond the mouth, or incurrent siphon, and bears sensory tentacles. The males and females are separate, but no obvious external differences exist between them.

The lancelets are scaleless, fish-like marine chordates. The adult forms exhibit chordate characteristics.

Vertebrates

The **vertebrates** differ from the other chordates in that most have vertebral columns in place of notochords. A vertebral column, or backbone, is a stack of bones, each with a hole in its center, that forms a cylinder surrounding and protecting the dorsal nerve cord. (Each bone in the column is a vertebra.) One class of vertebrates that does not have a vertebral column is the jawless fishes. Present-day jawless

FIGURE 33-5 Lancelets (subphylum Cephalochordata).
A Two lancelets partly buried in shell gravel, with their heads protruding. The muscle segments are clearly visible.
B The structure of a lancelet, showing the path of water flow through the animal.

A

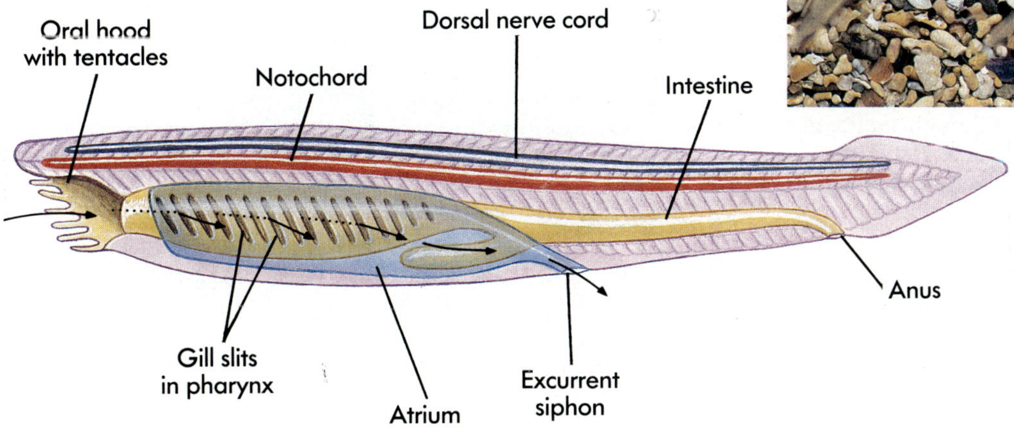

B

fishes have notochords, but their ancestors had bony skeletons and vertebral columns. Another class, the cartilaginous fishes, have skeletons and vertebral columns composed of cartilage (a tough yet elastic type of connective tissue) rather than bone.

In addition to having vertebral columns, vertebrates have distinct heads with skulls that encase their brains. They have closed circulatory systems (their blood flows within vessels) and a heart to pump the blood. Most vertebrates also have livers, kidneys, and endocrine glands. (Endocrine glands are ductless glands that secrete hormones, which play a critical role in controlling the functions of the vertebrate body.)

The human body plan is representative of the vertebrate body plan. Even though vertebrates have many similarities, they are a diverse group, consisting of animals adapted to life in the sea, on land, and in the air. There are seven classes of living vertebrates: three classes are fishes, and four classes are land-dwelling **tetrapods,** or four-footed animals.

Vertebrates are a subphylum of chordates characterized by a vertebral column surrounding a dorsal nerve cord.

Jawless fishes

Other than the lamprey eels and hagfishes, the major groups of the **jawless fishes,** or agnatha (*a* meaning "not," *gnatha* meaning "jaws"), have been extinct for hundreds of millions of years. Only about 20 to 30 species of each of these two groups are alive today. Both groups are long, tube-like aquatic animals that usually live in the sea or in brackish (somewhat salty) water where the fresh water of a river meets the ocean. In addition to their lack of jaws, they have no paired fins to help them swim, and they have no scales. They do have notochords, however, and portions of cartilaginous skeletons that are remnants from their extinct ancestors.

Lampreys parasitize other fish. In fact, they are the only parasitic vertebrates. They have round mouths that function like suction cups (see Figure 27-5), attaching them to their prey—the bony fishes. When a lamprey attaches to a fish, it uses its spine-covered tongue like a grater, rasping a hole through the skin of the fish and then sucking out its body fluids. Sometimes lampreys can be so abundant that they are a serious threat to commercial fisheries, preying on salmon, trout, and other commercially valuable fishes. Entering the Great Lakes from the sea, they have become important pests there; millions of dollars are spent annually on their control. The hagfishes, although similar in size and shape to lampreys, are not parasitic. They are scavengers, often feeding on the insides of dead or dying fishes or large invertebrates.

Lampreys and hagfish are tubular, scaleless, finless, jawless organisms that live in the sea or in brackish water. Lampreys parasitize bony fish and pose a serious threat to some commercial fisheries. Hagfish are scavengers.

To reproduce, jawless fishes **spawn** as do most fishes, amphibians, and shellfish. During spawning, the males and females deposit eggs and sperm directly into the water. Fertilization that takes place outside the body of the female is called **external fertilization.** Lampreys swim upstream to spawn as salmon do and create nests in which they deposit eggs and sperm. The fertilized eggs develop into larvae that feed on plankton. Over a period of years, the larvae mature and metamorphose (change) into parasitic adults. In contrast, the eggs of hagfishes do not develop into larvae. Completely formed hagfish hatch directly from fertilized eggs.

Most fishes, amphibians, and shellfish have an external type of fertilization. Eggs are fertilized by sperm outside the body of the organism.

Cartilaginous fishes

The chondrichthyes (*chondri* meaning "cartilage," *ichthyes* meaning "fishes") include the sharks, skates, and rays (Figure 33-6). Hundreds of extinct species of **cartilaginous fishes** are known from the fossil record, but less than 800 species exist today.

FIGURE 33-6 Cartilaginous fishes (class Chondrichthyes). Members of this class spend most of their time in graceful motion.
A Blue shark.
B Diamond sting ray.
C Manta ray.

Sharks, skates, and rays are all marine fishes. The skin of sharks (as well as the skates and rays) is covered with small, pointed, tooth-like scales called **denticles,** which give the skin a sandpaper texture. They have streamlined bodies and two pairs of fins: pectoral fins just behind the gills and pelvic fins just in front of the anal region. The dorsal (back) fins provide stability while motions of the other fins (including the asymmetrical tail fin) and sinuous motions of the whole body give the shark lift and propel it through the water.

Many sharks are predators and eat large fishes and marine mammals. Their sharp, triangular teeth saw and rip off pieces of flesh as the shark thrashes its head from side to side. Some sharks feed on plankton rather than prey on other animals. These sharks swim with their mouths open, and the plankton is strained from the water by specialized denticles on the inner surfaces of the gill arches.

Skates and rays are generally smaller than sharks and have flattened bodies with enlarged pectoral fins (see Figure 33-7, *B, C*) that undulate when these fishes move. Their tails, which are not the principal means of locomotion that they are in skarks, are thin and whip-like. They are sometimes armed with poisonous spines that are used as defense mechanisms rather than for predation. These animals have a mouth on their underside and feed mainly on invertebrates on or near the ocean floor.

The cartilaginous fishes are all marine fishes and include the sharks, skates, and rays. The skin is covered with small, pointed, tooth-like scales that result in a sandpaper texture. Many sharks are predators, but skates and rays feed on invertebrates on the ocean floor.

The sensory systems of sharks, skates, and rays are quite sophisticated and diverse. Sharks have a lateral line system as all fishes and amphibians do. A **lateral line system** is a complex system of *mechanoreceptors* that lies in a single row along the sides of the body and in patterns on the head. These receptors can detect mechanical stimuli such as sound, pressure, and movement. Sharks often detect prey via their lateral line systems, but they can also detect electrical fields emitted by other fishes using their electroreceptors. Sharks have these receptors on their heads, and rays have them on their pectoral fins. Scientists think these fishes may use their electroreceptors for navigation as well. Chemoreception is another important sense in the chondricthyes. In fact, sharks have been described as "swimming noses" because of their acute sense of smell. Vision is also important to the feeding behavior of sharks; they have well-developed mechanisms for vision at low-light intensities.

All fishes and amphibians have a complex system of mechanoreceptors called a *lateral line system.* These receptors, which lie in a row on the sides of the body and on the head, detect mechanical stimuli such as sound, pressure, and movement.

Another important characteristic of fishes is their ability to regulate their buoyancy, or ability to float at various depths in the water. Bony fishes have gas-filled *swim bladders* that regulate their buoyancy. Cartilaginous fishes do not have swim bladders but can adjust the size and oil content of their livers. Because oil is less dense than water and the liver has a high oil content, adjusting the oil content and size of the liver can regulate buoyancy to a certain degree. Most species of sharks, however, swim continually to keep from sinking and to keep water flowing over their gills.

Aquatic organisms show various adaptations regarding **osmoregulation,** the control of water movement into and out of their bodies. Marine organisms tend to lose water to their surroundings because their body fluids usually have a solute concentration (concentration of dissolved substances) lower than that of seawater. The cartilaginous fishes maintain solute concentrations close to that of seawater, however, because they change potentially toxic nitrogen-containing wastes (such as ammonia) into a less toxic compound called *urea* and retain it in their bodies rather than excreting it (Figure 33-7). (In fact, shark meat must be soaked in fresh water before it is eaten to remove most of the urea.) With similar solute concentrations both inside and outside the fish, water movement remains relatively equal in both directions.

Although most aquatic animals reproduce using external fertilization, the cartilaginous fishes have developed a method of internal fertilization. These fishes have *pelvic claspers,* which are rod-like projections between the pelvic fins of

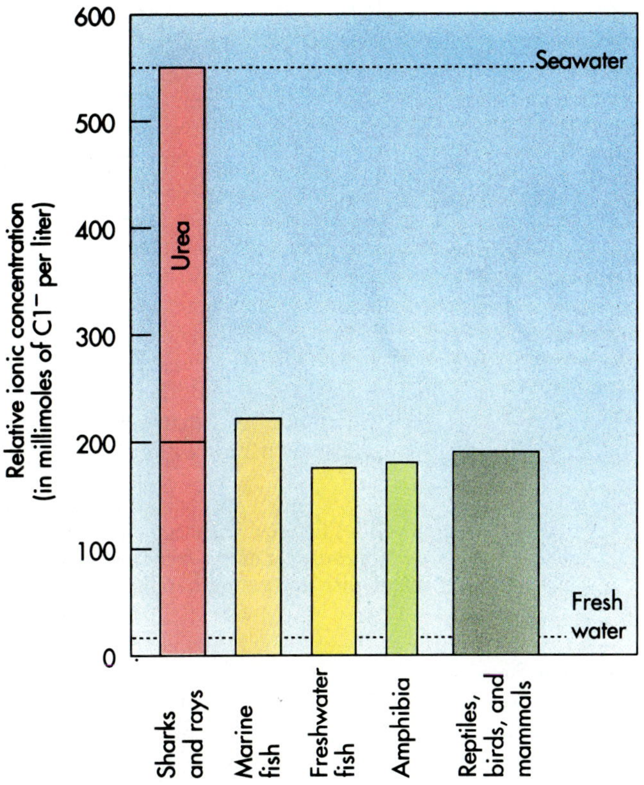

FIGURE 33-7 Ion concentrations for different classes of vertebrates. Ion concentrations in the bodies of different classes of vertebrates are roughly similar, with the exceptions of sharks and rays. Sharks and rays keep their ion concentrations close to that of seawater by adding urea to the bloodstream.

FIGURE 33-8 A viviparous fish giving birth.
Viviparous fishes such as this lemon shark carry live young within their bodies. The young complete their development inside their mother's body and are then released as small but competent adults.

the male fish. During *copulation,* or coupling of the male and female animals, the male inserts his clasper into the female's **cloaca.** The cloaca is the terminal part of the gut into which ducts from the kidney and reproductive systems open. The fish either swim side-by-side, or the male wraps around the female or holds on to her with his jaws. He then ejaculates into a groove in the clasper, which directs the sperm into the female's cloaca. The sperm then swim up the female's reproductive tract.

In addition to having a mechanism of internal fertilization, the cartilaginous fishes are ovoviviparous. **Ovoviviparous** organisms retain fertilized eggs within their oviducts until the young hatch (*ovum* meaning "egg," *vivus* meaning "alive," *pario* meaning "to bring forth"). However, although the young are born alive, they do not receive nutrition from the mother but from the yolk of the egg. The mother acts like an internal "nest" for the eggs until the young hatch. Certain fishes, some reptiles, and many insects are ovoviviparous. In contrast, **oviparous** organisms lay their eggs, and the young hatch from the eggs outside of the mother. Many skates and rays, for example, release eggs within protective egg cases. Birds and most reptiles, along with some of the cartilaginous fishes, are oviparous. **Viviparous** organisms such as mammals bear young alive as do ovoviviparous organisms, but the developing embryos primarily derive nourishment from the mother and not the egg. A few sharks, such as hammerhead sharks, blue sharks, and lemon sharks, are viviparous fishes (Figure 33-8).

Most cartilaginous fishes fertilize their eggs internally and are ovoviviparous, retaining fertilized eggs within their oviducts until the young hatch.

Bony fishes

The vast majority of known species of fishes belong to the class Osteichthyes (*oste* meaning "bone"). **Bony fishes** live in both salt and fresh water, with many species spending a portion of their lives in fresh water and another portion in the sea.

The osteichthyes get their name from their bony internal skeletons. However, the skin of osteichthyes is also covered with thin, overlapping bony scales. Sometimes these scales have spiny edges (see Figure 27-7, *A*, *B*), which provide some protection for the animal. Bony fishes also have a protective flap that extends posteriorly from the head and protects the gills. This flap is called the **operculum.** Along with protecting the gills, the movement of the operculum enhances the flow of water over the gills, bringing more oxygen in contact with the gas-exchanging surfaces and allowing a fish to breathe while stationary. Unlike most of the cartilaginous fishes, bony fishes can remain motionless at various depths because of their ability to more finely regulate their buoyancy with their swim bladders.

The swim bladders of most bony fishes, the *ray-finned fishes* (the largest subclass of bony fishes), evolved from lung-like sacs of their ancestors. These sacs aided in respiration. In the ray-finned fishes, gas exchange between the swim bladder and the blood regulates its density and allows it to remain suspended in water without sinking to the bottom. The ray-finned fishes are the most familiar fish and include perch, cod, trout, tuna, herring, and salmon. Another subclass of bony fishes continues to use lungs to aid the gills in breathing: the *lobe-finned fishes.* Only four genera of lobe-finned fishes exist today; all are found in the southern hemisphere. Ancestors of this subclass of fishes gave rise to land-dwelling tetrapods.

Bony fishes live in both salt and fresh water and get their name from their bony internal skeletons. By means of gas exchange between their swim bladders and their blood, bony fishes can regulate their buoyancy better than the cartilaginous fishes do. Cartilaginous fishes regulate buoyancy by controlling the amount of oil in their livers.

All chordates have closed circulatory systems with a system of blood vessels and a pump to push the blood through these vessels. The cartilaginous fishes and the bony fishes have tube-like hearts with four chambers, one right after the other (see Figure 33-12, *A*), but these four chambers really function as a two-chambered heart. The first two chambers (**sinus venosus** and **atrium**) collect blood from the organs. The second two (**ventricle** and **conus arteriosus**) pump blood to the gills. From the gills the blood moves to the rest of the body. Because of the great resistance in the narrow passageways of the capillaries at the gills, the movement of blood to the rest of the body is sluggish.

Although the cartilaginous fishes maintain a solute concentration of their body fluids near that of seawater, bony fishes do not. Only some bony fishes live in the sea; many groups live in fresh water. Marine and fresh water bony fishes have opposite situations regarding osmoregulation. The body fluids of marine fishes are *hypoosmotic* with respect to seawater. That is, their body fluids contain a lower concentration of dissolved substances than seawater does (see Figure 33-7). Consequently, water tends to leave these fishes (at the gill epithelium) by osmosis. To regulate water balance, marine fishes drink seawater and then excrete the salt by means of active transport at the gill epithelium. The fish kidney is not able to get rid of excess salt in the urine because it has no loop of Henle (see Chapter 14). This part of the kidney nephron works to concentrate urine and enable an organism to produce urine with a high solute concentration. Only birds and mammals have loops of Henle.

In contrast to marine fishes, the body fluids of fresh water fishes are *hyperosmotic* with respect to their fresh water environment. That is, their body fluids contain a higher concentration of dissolved substances than fresh water does. Consequently, water tends to enter fresh water fishes by osmosis. To regulate water balance, fresh water fishes do not drink water and excrete large amounts of very dilute urine. They reclaim some of the ions they lose in their urine by the uptake of sodium and chlorine ions by the gills and in the food they eat.

Because they tend to lose water by osmosis, marine fishes drink seawater and excrete salt at the gill epithelium. Because they tend to take on water, fresh water fishes do not drink water and excrete large amounts of very dilute urine.

Most osteichtheyes are oviparous, fertilizing their eggs externally. These eggs are often food for other marine organisms, however, and must survive other risks such as drying out. Most fishes (as well as other organisms that externally fertilize their eggs) lay large numbers of eggs, with some surviving these dangers. Many species of fishes build nests for their eggs and watch over them, which also enhances the chances of survival.

Amphibians

The word *amphibian* means "two lives" (*amphi* meaning "both," *bios* meaning "life") and refers to both the aquatic and terrestrial existence of this class of animals (Figure 33-

A B C D E

FIGURE 33-9 Amphibians (class Amphibia).
Examples of amphibians that have tails are the spotted salamander **(A)** and the Tennessee cave salamander **(B)**. Examples of amphibians without tails, the frogs and toads, are the pickeral frog, shown here in tadpole form **(C)**; the giant toad **(D)**; and the red-eyed tree frog **(E)**.

9). **Amphibians** (unlike reptiles, birds, and mammals) depend on water during their early stages of development. Many amphibians live in moist places like swamps and in tropical areas even when they are mature, which lessens the constant loss of water through their thin skins. The two most familiar orders of amphibians are those that have tails—the salamanders, mudpuppies, and newts—and those that do not have tails—the frogs and toads.

Most frogs and toads fertilize their eggs externally. The male grasps the female and sheds sperm over the eggs as they are expelled from the female. Most salamanders use internal fertilization but are still oviparous (lay their eggs). Because amphibian eggs have no shells or membranes to keep them from drying out, amphibians lay their eggs directly in water or in moist places. Figure 33-10, *A* shows the "nests" of foam that tropical tree frogs create to incubate their eggs. When the tadpoles develop, they drop from the tree branches into the water (Figure 33-10, *B*). Some amphibians protect their eggs by incubating them in their mouths, on their backs, or even in their stomachs! A few amphibian species are ovoviviparous and incubate the eggs within their reproductive organs until they hatch, and a few are viviparous.

The young of frogs and toads undergo *metamorphosis,* or change, during development from a larval to an adult form. The larvae are immature forms that do not look like the adult. The larvae of frogs and toads are tadpoles, which usually live in the water and have internal gills and a lateral line system like that of fishes. They feed on minute algae. These fish-like forms develop into carnivorous adults having legs and lungs; their gills and lateral line system disappear. The lungs of the adults are inefficient, however. Much of the gas exchange takes place across the skin and on the surfaces of the mouth. The skin of amphibians must therefore remain damp to allow gases to diffuse in and out.

The adults of certain salamanders (such as mudpuppies) live permanently in the water and retain gills and other larval features as adults. Other salamanders are terrestrial but return to water to breed. They, like frogs and toads, usually live in moist places such as under stones or logs or among the leaves of certain tropical plants.

Amphibians live both aquatic and terrestrial existences. Because they fertilize their eggs externally and their eggs have no shells or membranes, amphibians lay their eggs in water or in moist places. The young of frogs and toads change from a larval form to an adult form during their development.

Along with having lungs rather than gills (in most cases), amphibians have a pattern of blood circulation different from that of the fishes. After the blood is pumped through the fine network of capillaries in the amphibian lungs, it does not flow directly to the body as it does in fishes. Instead, it returns to the heart. It is then pumped out to the body at a much higher pressure than if it were not returned to the heart. However, the oxygenated blood that returns to the heart from the lungs mixes with the deoxygenated blood returning to the heart from the rest of the body. Consequently, the heart pumps out a mixture of oxygenated and deoxygenated blood rather than fully oxygenated blood. Figure 33-12, *B* shows this pathway of blood flow. The blood from the lungs and the blood from the body enter the right and left atria of the heart, respectively. The blood in both these chambers flows into the single ventricle of the three-chambered amphibian heart and is pumped through two large vessels to both the lungs and the body.

Amphibians have a three-chambered heart rather than the two-chambered heart of fishes. However, oxygenated and deoxygenated blood mix in the heart.

A

B

FIGURE 33-10 An amphibian egg-protection strategy. To keep their eggs from drying out, tropical tree frogs lay their eggs in nests of foam **(A)**. When the tadpoles hatch, they drop from their foam nest into a pond of water below **(B)**.

Reptiles

The three major orders of **reptiles** are the crocodiles and alligators, the turtles and tortoises, and the lizards and snakes (Figure 33-11). Reptiles have dry skins covered with scales that help retard water loss. As a result, reptiles can live in a wider variety of environments on land than amphibians can, but the crocodiles, alligators, and turtles are aquatic organisms.

The hearts of reptiles differ from amphibian hearts in that a partition called a *septum* subdivides the ventricle, the pumping chamber of the heart. The septum reduces the mixing of oxygenated and deoxygenated blood in the heart.

FIGURE 33-11 Reptiles (class Reptilia).
A The river crocodile. Crocodiles, like birds, are related to the dinosaurs.
B Red-bellied turtles. This attractive turtle is found frequently in the northeastern United States.
C An Australian skink. Although this species of burrowing lizard has legs, some species lack legs. Snakes evolved from one line of legless lizard.
D Smooth green snake.

In most crocodiles, the separation is complete. Figure 33-12, *C* shows that the reptilian heart closely resembles the four-chambered heart of birds and mammals shown in Figure 33-12, *D*. It is still considered to be a three-chambered heart, however.

One of the most critical adaptations of reptiles to life on land is the evolution of the shelled **amniotic egg.** Amniotic eggs are also characteristic of birds and egg-laying mammals (monotremes). The amniotic egg protects the embryo from drying out, nourishes it, and enables it to develop outside of water (Figure 33-13). This type of egg contains a yolk and albumin (egg white). The yolk is the primary food supply for the embryo, and the albumin provides additional nutrients and water. The embryo's nitrogenous wastes are excreted into the allantois, a sac that grows out of the embryonic gut. Blood vessels grow out of the embryo through the sac surrounding the yolk and through the allantois to the egg's surface, where gas exchange takes place. The amnion surrounds the developing embryo, enclosing a liquid-filled space within which the embryo develops and is protected. Lying just within the shell, a membrane called the *chorion* surrounds the embryo, amnion, yolk sac, and allantois and, along with the shell, controls the movement of gases into and out of the egg. In most reptiles, the eggshell is leathery, unlike the hard shell of bird eggs. Because of this difference, reptile eggs are somewhat permeable to water, whereas bird eggs are not.

Reptiles, like amphibians and fishes (but unlike birds and mammals), are **ectothermic** (*ectos* meaning "outside," *thermos* meaning "heat"). Ectothermic animals regulate body temperature by taking in heat from the environment. Even though ecotothermic animals are often called "cold blooded" (a misleading term), they often maintain body temperatures much warmer than their surroundings. Did you ever wonder why fish do not freeze in water near the freezing temperature? Interestingly, fishes (as well as certain lizards, invertebrates, and plants) produce their own internal antifreeze—chemical compounds that lower the freezing temperature of the body fluids of an organism. Along with such physiological adaptations, ectothermic animals protect themselves against the cold in behavioral ways. For example, frogs help protect themselves against freezing by spending the winter buried in the soil or in the mud at the bottom of ponds. Ectothermic animals also protect themselves from high heat by burrowing under rocks or remaining in shady, somewhat cooler areas. Desert tortoises, for example, construct shallow burrows to stay in during the summer and deeper burrows for hibernation in winter. Reptiles often

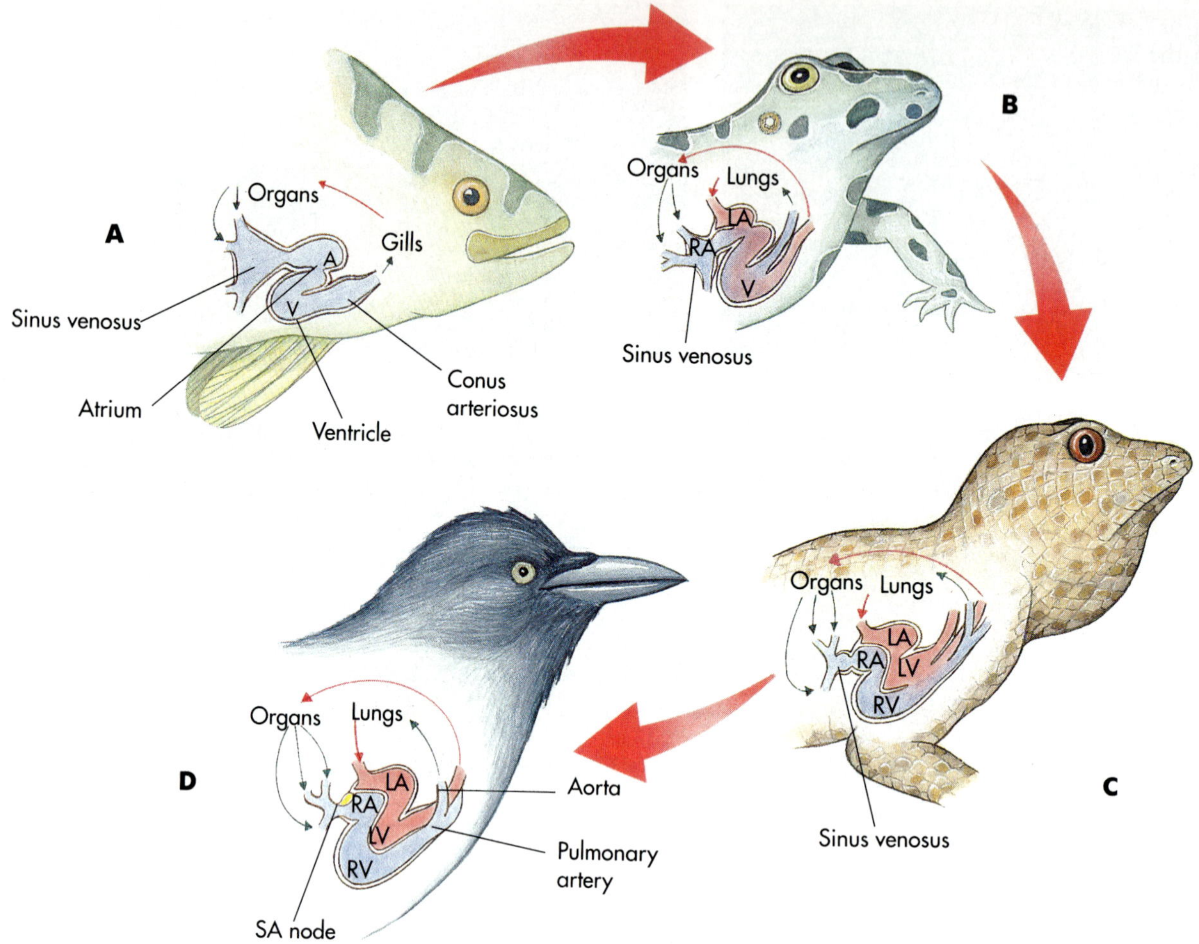

FIGURE 33-12 Evolution of the vertebrate heart.
A Fishes (two chambers).
B Amphibians (two chambers).
C Reptiles (three chambers).
D Birds and mammals (four chambers).

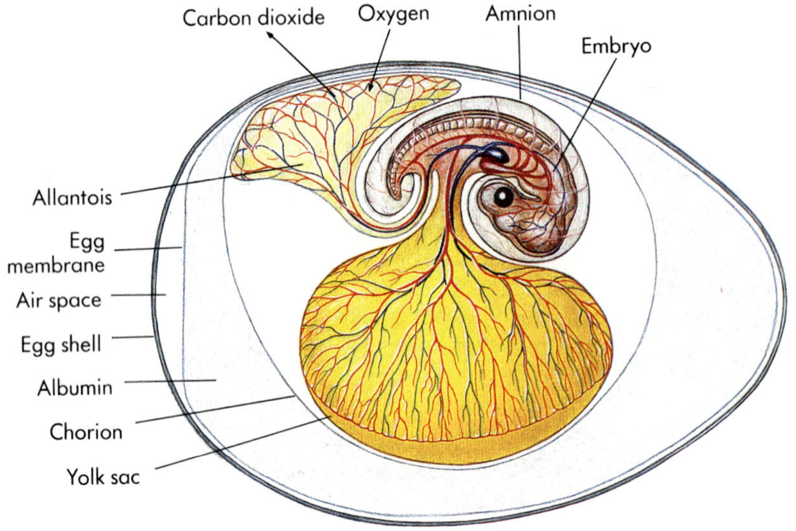

FIGURE 33-13 The amniotic egg.
The amniotic egg is an important adaptation that allows reptiles to live in a wide variety of terrestrial habitats.

bask in the sun, which raises their body temperature and their metabolic rate. When cold-blooded animals are cold, the metabolic rate slows down and they are unable to hunt for food or move about very quickly.

Reptiles have dry skins covered with scales that help retard water loss. Another critical adaptation to their life on land (although some reptiles are aquatic) is the development of the amniotic egg, which protects the embryo from drying out, nourishes it, and enables it to develop outside of water. Amniotic eggs are also characteristic of birds and egg-laying mammals.

Birds

There are approximately 9000 species of birds living today (Figure 33-14). In birds, the wings are homologous to the forearms of other vertebrates. That is, they are derived from the same evolutionary origin but have been modified in the course of evolution. Birds have reptilian-like scales on their legs and lay amniotic eggs as reptiles do. They have hard, horny extensions of the mouth called **beaks** that tear, chisel, or crush their food. They also have digestive organs called **gizzards,** often filled with grit, that grind food. Beaks are not limited to birds. Many reptiles have beaks, including turtles, and so do some fishes, including the parrot fish, which uses its beak to rip away fragments of coral reef. Birds, however, are the only animals that have feathers.

A

B

C

D

E

FIGURE 33-14 Birds (class Aves).
The birds are a large and successful group of about 9000 species, more than any other class of vertebrates except the bony fishes.
A Ostrich. The group of flightless birds includes the ostrich (Africa), rhea (South America), emu (Australia), and kiwi (New Zealand). Flightless birds seem to represent an evolutionary line distinct from all other birds.
B A bee-eater in Tanzania. Bee-eaters nest communally, with many adult individuals participating in the care of the young.
C Northern saw-whet owl. Owls, like hawks, hunt other animals (prey) for food.
D A pair of wood ducks, showing the marked difference between males and females.
E Tufted puffin, one of many unique groups of sea birds.

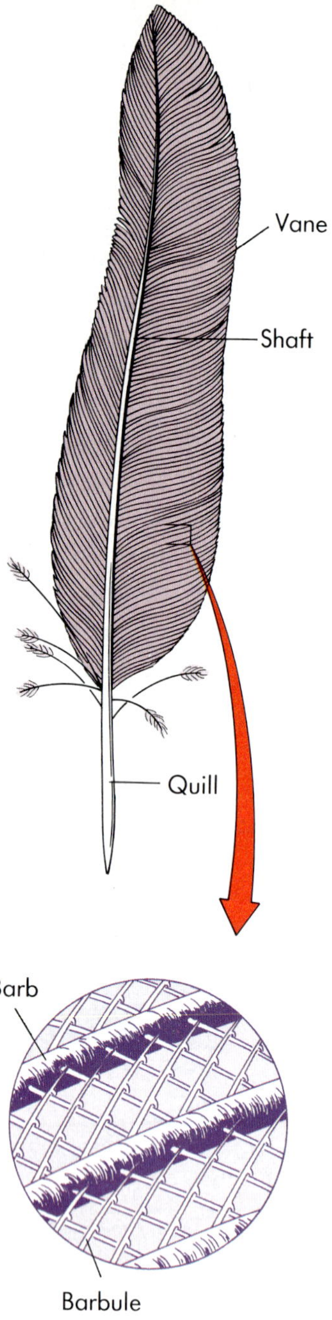

FIGURE 33-15 A feather.
The enlarged view shows the way in which the vanes are linked together by the microscopic barbules.

Feathers are flexible, light, waterproof epidermal structures. Several types of feathers form the body covering of birds, including contour feathers and down feathers. Contour feathers are flat (except for a fluffy, downy portion at the base) and are held together by tiny barbules (Figure 33-15). These feathers provide a streamlined surface for flight, and some are modified to reduce drag on the wings or act like individual propeller blades. In addition, birds can alter the area and shape of their wings by altering the positions of their feathers. Feathers also provide birds with waterproof coats and play an important role in insulating birds against temperature changes.

Along with feathers, birds have light, hollow bones adapted to flight. Birds also have highly efficient lungs that supply the large amounts of oxygen necessary to sustain muscle contraction during prolonged flight. Unlike fishes, amphibians, and reptiles (except for the crocodiles), birds and mammals have hearts that act as a double pump. The sides of the heart are completely separated with a septum (see Figure 33-12). The right side of the heart pumps blood to the lungs. The left side of the heart pumps blood to the body. Oxygenated blood and deoxygenated blood do not mix. As you compare the pathway of blood in the hearts of fishes, amphibians, reptiles, and birds in Figure 33-12, note that the four-chambered heart of birds and mammals evolved from only two of the chambers of the fish heart.

Birds and mammals are **endothermic** (*endo* meaning "within"); they regulate body temperature internally. The evolution of the four-chambered heart with separate pathways to the lungs and the body is thought to have been important in the evolution of endothermy in the birds and mammals. More efficient circulation is necessary to support the great increase in metabolic rate that is required to generate body heat internally. In addition, blood is the carrier of heat in the body, and an efficient circulatory system is required to distribute heat evenly throughout the body. Although endothermic animals are sometimes called "warm blooded," body temperature may often be cooler than that of the surroundings. Because endotherms maintain high internal body temperatures (37° C [98.6° F] in humans, for example), their internal temperatures are usually higher than that of the environment.

Endotherms maintain a constant high body temperature by adjusting heat production to equal heat loss from their bodies under various environmental conditions. The high metabolic rate of endotherms and the energy released during these chemical reactions produces much of this body heat. Increasing the action of skeletal muscles increases the metabolic rate and the amount of heat produced. Shivering in the cold, for example, is an action that produces body heat. But because endotherms usually live in environments cooler than body temperature, restricting heat loss is usually the concern. Animals that live in extremely cold temperatures, such as arctic birds and polar bears, are well-insulated with either feathers or hair that traps air. Raising or lowering the feathers or hair adjusts the insulating capacity. Getting "goosebumps" when you are cold is your body's reaction to raise your hairs and increase your insulation. Although humans no longer have substantial body hair, this mechanism is important in reducing heat loss in other endothermic vertebrates.

Birds are winged vertebrates that are covered with feathers and have scales on their legs. They have beaks, which are horny extensions of the mouth. Their light, hollow bones are an adaptation to flight. They have a four-chambered heart as mammals do. Unlike the reptiles, amphibians, and fishes, birds can regulate body temperature internally.

Mammals

There are about 4500 species of living **mammals,** including human beings. Mammals are endothermic vertebrates that have hair and whose females secrete milk from mammary glands to feed their young. Mammals, like birds and most crocodiles, have a four-chambered heart with circulation to the lungs and separate circulation to the body. The locomotion of mammals is advanced over that of the reptiles, which in turn is advanced over that of the amphibians; the legs of mammals are positioned much farther under the body than those of reptiles and are suspended from limb girdles, which permit greater leg mobility.

The evolution of specialized teeth in mammals represents a major evolutionary advance. In fishes, amphibians, and reptiles, all teeth are essentially the same size and shape; in mammals, evolutionary specialization has resulted in incisors, which are chisel-like teeth used for cutting; canines, which are used for gripping and tearing; and molars, which are used for crushing and breaking (see Figure 10-2, *A*).

There are three subclasses of mammals: monotremes, marsupials, and placental mammals. The only **monotremes** that exist are the duckbilled platypus and two genera of spiny anteaters (Figure 33-16). Monotremes lay eggs with leathery shells similar to those of reptiles. The platypus generally lays one egg and incubates it in a nest. The spiny anteater generally lays two eggs and incubates them in a pouch. When the young hatch, they feed on milk produced by specialized sweat glands of the mother. **Marsupials** are mammals in which the young are born early in their development and are retained in a pouch. After birth, the embryos crawl to the pouch and nurse there until they mature. The kangaroo and koala are familiar examples of marsupials (Figure 33-17). In **placental mammals,** the young develop to maturity within the mother. They are named for the first organ to form during the course of their embryonic development, the placenta. (The development of the human as an example of the development of a placental mammal is described in detail in Chapter 21.)

Placental mammals are extraordinarily diverse (Figure 33-18). There are 14 orders of mammals (see Figure 27-12), of which one is the primates, the order that includes monkeys, apes, and humans. The bats in the opening photograph (as well as all bats) belong to the order Chiroptera, a name that means the forelimbs are modified to form wings (*chiro* meaning "hand," *pteron* meaning "wing"). So although humans and bats are quite different, they have remarkable similarities in their circulatory, thermoregulatory, osmoregulatory, and reproductive patterns.

Mammals are endothermic vertebrates that have hair and whose females secrete milk from mammary glands to feed their young. There are three subclasses of mammals: monotremes, marsupials, and placentals.

A

B

FIGURE 33-16 Monotremes (class Mammalia).
A Duckbilled platypus at the edge of a stream in Australia.
B Echidna.

A

B

FIGURE 33-17 Marsupials (class Mammalia).
A Kangaroo with young in its pouch.
B Koala in a eucalyptus tree.

A

B

C

D

E

FIGURE 33-18 Placental mammals (class Mammalia).
A Snow leopard, a cat.
B Starnosed mole, a burrowing insectivore.
C White-tailed deer, abundant in eastern and central temperate North America.
D Greater horseshoe bat in flight.
E Orca, a carnivorous whale.

Summary

1. The chordates are characterized by three main features: (1) a single, hollow nerve cord located along the back; (2) a rod-shaped notochord, which forms between the nerve cord and the gut (stomach and intestines) during development; and (3) pharyngeal (gill) arches, which are located at the throat (pharynx).
2. There are three subphyla of chordates: the tunicates (subphylum Urochordata), the lancelets (subphylum Cephalochordata), and the vertebrates (subphylum Vertebrata). The tunicates are sessile, sac-like, marine organisms that filter food from the surrounding water. Although the adults have gill slits, only their larvae have notochords and nerve cords. The lancelets are tiny, scaleless, fish-like marine chordates that are just a few centimeters long.
3. Vertebrates are a subphylum of chordates characterized by a vertebral column surrounding a dorsal nerve cord. Vertebrates have distinct heads with skulls that encase their brains, closed circulatory systems, and a heart to pump the blood. Most vertebrates also have livers, kidneys, and endocrine glands. In spite of these similarities, the vertebrates are an extremely diverse subphylum of organisms. Three classes of vertebrates are fishes; four are tetrapods—animals with four limbs.
4. The jawless fishes—lampreys and hagfishes—are long, tube-like animals that live in the sea or brackish water. They lack fins and scales. Lampreys are parasites of bony fishes, and hagfishes are scavengers.
5. The cartilaginous fishes—sharks, skates, and rays—are all marine fishes and are covered with tooth-like scales. They have sophisticated and diverse sensory systems. They fertilize their eggs internally, and most are ovoviparous, rataining fertilized eggs within their oviducts until the young hatch.
6. The vast majority of fishes are the bony fishes. Along with having bony internal skeletons, these fishes have thin, bony, plate-like scales. Both the bony fishes and the cartilaginous fishes have two-chambered hearts that pump blood to the gills. From there, the blood moves sluggishly around the body.
7. The amphibians live both in water and on land. Because their eggs have no shells or membranes to keep them from drying out, amphibians lay their eggs directly in water or moist places. The young of frogs and toads undergo change from larval to adult forms during development. The adults of certain salamanders live permanently in the water and retain gills and other larval features as adults. Although amphibians have a three-chambered heart, oxygenated and deoxygenated blood mix in the heart.
8. Reptiles are better adapted to life on land than the amphibians due to their dry, scaly skin that retards water loss and their shelled (amniotic) egg. The amniotic egg retains a watery environment within the egg while protecting and nourishing the developing embryo.
9. Fishes, amphibians, and reptiles are ectothermic; that is, they regulate body temperature by taking in heat from the environment. Ectothermic animals protect themselves from the cold and high heat in behavioral ways.
10. Birds are winged vertebrates that are covered with feathers and adapted to flight. They lay amniotic eggs like the reptiles but have a four-chambered heart like the mammals. Birds, like mammals and unlike reptiles, amphibians, and fishes, are endothermic; that is, they regulate body temperature internally.
11. Mammals are endothermic vertebrates that have hair and whose females secrete milk from mammary glands to feed their young.

REVIEW QUESTIONS

1. Fill in the blanks: The three major features that are characteristic of all vertebrates are: a(n) ________ ________, a(n) ________ ________, and ________ (________) ________.
2. Into what structure(s) does each of the features in question 1 develop?
3. Match each term to the most appropriate description:
   Terms
   (1) Jawless fishes
   (2) Tunicates
   (3) Lancelets
   Descriptions
   a. Scaleless, fish-like marine chordates that retain a notochord throughout life
   b. Tubular, scaleless, finless organisms; most members of this class have been extinct for millions of years
   c. Sac-like, sessile, marine filter-feeders
4. The "Jaws" movies starred members of which two classes of vertebrates?
5. What is a lateral line system? Which animals have one, and why is it important?
6. What is osmoregulation, and what is its significance?
7. Distinguish among ovoviviparous, oviparous, and viviparous.
8. What does the term *amphibian* mean?
9. Distinguish between ectothermic and endothermic. Give an example of an ectotherm and an endotherm.
10. Describe two important adaptations of reptiles to life on land.
12. What are feathers, and what functions do they perform for birds?
13. What characteristics differentiate mammals from other vertebrates?

KEY TERMS

amniotic egg 589
amphibian 588
atrium 586
beak 591
bony fishes 586
cartilaginous fishes 584
cloaca 586
conus arteriosus 586
denticle 585
ectothermic 589
endothermic 592
external fertilization 584
gizzard 591
jawless fishes 584
lancelet 583
lateral line system 585
mammal 593
marsupial 593
monotreme 593
nerve cord 581
notochord 581
operculum 586
osmoregulation 585
oviparous 586
ovoviviparous 586
pharyngeal (gill) arches 581
placental mammal 593
reptile 588
sinus venosus 586
spawn 584
tetrapod 584
tunic 581
tunicate 581
ventricle 586
vertebrate 583
viviparous 586

THOUGHT QUESTIONS

1. How are the characteristics of the three subphyla of chordates related to the way of life of each group?
2. What limits the ability of amphibians to occupy the full range of terrestrial habitats and allows other terrestrial vertebrates to live in them successfully?
3. When you were an embryo, you had a notochord, just as a lancelet does. What do you think happened to it?

FOR FURTHER READING

Ehrlich, P. R., Dobkin, D. S., & Wheye, D. (1988). *The birder's handbook.* Simon and Schuster, Inc., New York.
A field guide to the natural history of North American birds, this book will greatly deepen your insight into the lives of these familiar animals.

Martin, J. (1990, June). The engaging habits of chameleons suggest mirth more than menace. *Smithsonian,* pp. 44-53.
The wonderful variety in a single group of tropical lizards is shown in this article.

Rismiller, P. O., & Seymour, R. S. (1991, February). The echidna. *Scientific American.*
This article outlines the natural history of this unusual and fascinating monotreme.

Waldvogel, J. A. (1990, July/August). The bird's eye view. *American Scientist.*
Scientists are trying to reconstruct the bird's excellent visual acuity in the laboratory.

PART EIGHT

HOW LIVING THINGS INTERACT WITH EACH OTHER AND WITH THEIR ENVIRONMENT

34 Innate behavior and learning in animals

HIGHLIGHTS

▼

Ethology, the study of animal behavior in the natural environment, examines the biological basis of the patterns of movement, sounds, and body positions of animals.

▼

All behaviors depend on nerve impulses, hormones, and other physiological mechanisms; behaviors are therefore directed in part by the genes that control the development of these systems and mechanisms.

▼

Certain animal behaviors are unchangeable and are performed correctly the first time they are attempted; these innate, or instinctive, behaviors are directed by nerve pathways developed before birth.

▼

Animals that can change their behaviors based on experience are capable of learning, a process that helps animals adapt to changes in the environment.

OUTLINE

Ethology: The biology of behavior

The link between genetics and behavior

Innate (instinctive) behaviors

- Kineses
- Taxes
- Reflexes
- Fixed action patterns

Learning

- Imprinting
- Habituation
- Trial-and-error learning (operant conditioning)
- Classical conditioning
- Insight

A case of mistaken identity? It would certainly seem to be the case. Scientists, however, would describe this "identity crisis" as a case of misfiring. Misfiring refers to a behavior that is normally appropriate being exhibited in an inappropriate circumstance.

Female birds like this cardinal are stimulated to feed their young when signaled by their tiny, gaping mouths. Such a signal is called a *releaser.* Releasers are structures or behavior patterns exhibited by one member of a species that trigger a response—a series of precise physical movements—in another member of that species. The behaviors triggered by releasers are also influenced by hormones or other internal stimuli. This cardinal recently lost her nest of young but is still hormonally "driven" to feed them. Therefore the tiny, wide-open mouth triggered a response in the cardinal—a response that will be quite a surprise for the fish!

Ethology: The biology of behavior

Animals (including humans) continually exhibit a range of behaviors. **Behaviors** include the patterns of movement, sounds (vocalizations), and body positions (postures) exhibited by an animal. In addition, behaviors include any type of change in an animal, such as a change in coloration or the releasing of a scent, that can trigger certain behaviors in another animal.

Some behaviors are simple, automatic responses to environmental stimuli. A bacterium "behaves" when it moves toward higher concentrations of sugar. You behave when you slam on the brakes to avoid a car accident. Other types of behaviors, such as the eating behavior of the sea otter, are quite complex (Figure 34-1). Complex behaviors are limited to multicellular animals with a neural network that can sense stimuli, process these stimuli in a central nervous system (brain), and send out appropriate motor impulses (Figure 34-2). You exhibit complex behavior when you walk and talk, drive a car, interact with your family, and do your job.

The study of the behavior of animals in their natural environments is called **ethology.** Ethologists observe and interpret the behavior of animals from a physiological rather than psychological perspective. They attempt to uncover the biological significance of animal behavior—the importance of a behavior for a particular species in its natural environment. Ethologists bring order to their observations, breaking down behavior patterns into recognizable units, naming these patterns, and categorizing them.

One of the most famous ethologists is the Austrian scientist Konrad Z. Lorenz (see Figure 34-8), referred to by some as the father of modern ethology. In 1973, Lorenz, Dutch ethologist Nikolaas Tinbergen, and Austrian zoologist Karl von Frisch won the Nobel Prize for their contributions to the study of animal behavior. Lorenz based his work on the premise that animal behaviors are evolutionary adaptations, as are physical adaptations. He referred to behaviors as being part of an animal's "equipment for survival." Today, this concept is a fundamental premise in the study of animal behavior.

FIGURE 34-2 Complex behaviors are characteristic of multicellular animals with a central nervous system and a neural network. These two penguins are exploring their interpersonal space. Most vertebrates exhibit complex behaviors.

FIGURE 34-1 Complex behavior in a sea otter. This sea otter is having dinner while swimming on its back. It is using the rock as a hard surface against which to hit the clam and break it open. Often a sea otter will keep a favorite rock for a long time, suggesting that it has a clear idea of what it is going to use the rock for. The sea otter may learn this pattern of eating behavior from others while young, but the capacity to use tools and consciously foresee their future use certainly depends on inherited abilities.

Scientists who study the behaviors of animals in their natural environments are ethologists. Behaviors include the patterns of movement, sounds, and body positions exhibited by animals.

The link between genetics and behavior

All behaviors depend on nerve impulses, hormones, and other physiological mechanisms such as sensory receptors. Therefore genes play a role in the development of behaviors because they direct the development of the nervous system. In addition, automatic responses depend on specific nerve pathways within the central nervous system of an organism. These pathways are neural programs and are genetically determined. Neural programs are part of the nervous system at the time of birth or develop at an appropriate point in maturation, resulting in behaviors that are called **instinctive, innate,** or **inborn.** Interestingly, innate behaviors are performed in a reasonably complete form the first time they are exhibited. A newborn baby, for example, will turn to suckle when touched on the cheek near the mouth. The baby performs this innate behavior without **learning,** or altering its behavior based on experience. Innate behaviors are important to the survival of an animal because they help an animal stay alive in certain situations and provide adaptive advantages that contribute to its fitness, or ability to achieve reproductive success.

The development of the nervous system and certain automatic nervous responses that are "preprogrammed" within the nervous system are directed by genes.

The neural programs of innate behavior patterns can be likened to the "hard wiring" of the circuitry within a calculator. Certain functions within a calculator are preprogrammed and unchangeable; the range of hard-wired functions depends on the calculator. Basic functions such as addition, subtraction, multiplication, and division are always hard wired, meaning that the program a calculator needs to follow is permanently set within its electronic circuitry. Likewise, some calculators are hard wired for other functions such as determining square roots. In a similar manner, innate behaviors are preprogrammed into the nerve circuitry of an animal. These "nerve programs" are passed from parent to offspring by means of genetic (hereditary) material. In general, such hard-wired behaviors cannot be changed, but scientists have observed that some can be gradually perfected during the course of an animal's life.

In animals, some nerve circuits are not completely set as they are in the neural programs that govern innate behaviors. These neural programs can accept the input of growth and experience, resulting in learned behavior. Learned behavior can become automatic; you learned as a small child, for example, how to drink from a cup. This behavior quickly becomes set in your neural circuitry resulting in behavior that does not require conscious thought. Learned behavior is the area in which the effect of environment can be most dramatically seen on the inherited characteristics of the neural circuitry.

Innate (instinctive) behaviors

Ethologists group innate behaviors into four categories: kineses, taxes, reflexes, and fixed action patterns.

Kineses

A **kinesis** is the change in the speed of the random movements of an animal with respect to changes in certain environmental stimuli. Put simply, movement slows down in an environment favorable to the animal's survival and speeds up in an unfavorable one. Have you ever picked up a rotting log or a clump of damp leaves in a wooded area? The pillbugs that you may have seen living under the leaves or the log (Figure 34-3) stay in this favorable environment because of their low levels of activity and then scurry when the log is rolled over.

Taxes

Kineses are *nondirected* types of movements. In other words, an animal is not attracted by a favorable environment; it tends to "blunder" there as it moves about quickly and then stays there as it moves about slowly. A **taxis,** however, is a *directed* movement toward or away from a stimulus, such as light, chemicals, or heat. Animals having preprogrammed taxes also have receptors that sense the particular stimuli to which the animal can orient. Female mosquitos and ticks, for example, have sensory receptors that detect warmth, moisture, and certain chemicals emitted by mammals. Sensing these stimuli helps the insects orient to their victims. In

FIGURE 34-3 Kinesis.
These pillbugs scurry when the log they are living under is moved. This change in the level of activity that occurs with a change in the environmental stimulus is called *kinesis.*

FIGURE 34-4 Phototaxis and gravitaxis in fish.
The normal fish in the left column orient to both light and gravity. The fish in the right column, which have had their gravity-detecting organ removed, orient to light only and have difficulty staying upright. The lower left diagram points out that orientation to gravity overrides orientation to light.

fact, some mosquito repellents work by "blocking" the insect's receptors so that it cannot sense and then locate its victim. The crowding of flying insects about outdoor lights is another familiar example of a taxis called *phototaxis*. Other insects, like the common cockroach, avoid light (are negatively phototactic). Fish swim upright by orienting to gravity and light. If the gravity-detecting organ in the inner ear is removed or if a light is shone into the sides or bottom of a fish tank, the fish become disoriented and do not swim upright (Figure 34-4). Certain species of fish such as trout and salmon also automatically orient against a current and therefore face and swim upstream (Figure 34-5). Even organisms without a nervous system such as protozoans and bacteria exhibit taxes. These organisms have cellular organelles or inclusions that act as receptors or that react to specific environmental stimuli.

FIGURE 34-5 Taxis in a salmon.
Salmon automatically orient themselves against a current and therefore always swim upstream.

FIGURE 34-6 Fixed action pattern in a graylag goose.
The egg retrieval movement in the graylag goose is an entirely programmed behavior, completely instinctive.

Reflexes

A **reflex** is an automatic response to nerve stimulation. The knee jerk is one of the simplest types of reflexes in the human body. If the tendon just below the kneecap is struck lightly, the lower leg automatically "kicks" or extends. This behavior occurs as the tendon is stretched and pulls on the muscles of the upper leg that are attached to it. Stretch receptors in these muscles immediately send an impulse along sensory nerve fibers to the spinal cord. At the spinal cord these fibers synapse directly with motor neurons that extend back to upper leg muscles. The upper leg muscles are stimulated to contract and jerk the leg upward while opposing muscles simultaneously relax (see Figure 17-14). These reflexes play an important role in maintaining posture. More complex reflexes involve a relay of information from a sensory neuron through one or more interneurons to a motor neuron. The outcomes of these reflexes are modulated by the interneurons. In complex organisms, reflexes play a role in survival, such as when you jerk your hand away from a hot stove before you consciously realize that your hand hurts. In animals with extremely simple nervous systems such as the cnidarians (hydra, jellyfish, sea anemones, and corals), most behaviors are the result of reflexes, although some simple learning can take place (see p. 567). In these animals a stimulus is detected by sensory neurons, and the impulse is passed on to other neurons in the animal's nerve net, eventually reaching the body muscles and causing them to contract. There is no associative activity in which other neurons can influence the outcome, no control of complex actions, and little or no coordination. The nerve net of the cnidarians possesses only the barest essentials of nervous reaction.

Fixed action patterns

More complex than other innate behaviors are **fixed action patterns.** Fixed action patterns are sequences of innate behaviors in which the actions follow an unchanging order of muscular movements, such as the mother cardinal popping an insect into the mouth of her young. This sequence is a recognizable "unit" of behavior. When a releaser triggers the behavior, the sequence of activity begins and is carried through to completion. Fixed action patterns are often seen in body maintenance behaviors (such as a cat washing its face), courtship behavior, nest building, and attainment of food.

The classic example used to illustrate a fixed action pattern behavior is the retrieval of an egg that rolls out of the nest by the graylag goose (Figure 34- 6). If a goose notices an egg has been knocked out of the nest, it will extend its neck toward the egg, reach up, and roll the egg back into the nest with its bill. Because this behavior seems so logical, it is tempting to believe that the goose saw the problem and figured out what to do. But, in fact, the entire behavior is totally instinctive. During experimentation, ethologists have discovered that any rounded object, regardless of size or color, acts as a releaser and triggers the response. Beer bottles, for example, are an effective releaser. And, as a fixed action pattern behavior, the goose completes the action even if the egg rolls away from its retrieving bill. Instead of stopping in the middle of the behavior, the goose will continually repeat the entire behavior until the egg is brought back to the nest.

Innate behaviors are those that are preprogrammed within the nervous system before birth and are demonstrated in a somewhat complete form the first time they are exhibited. Four categories of innate behaviors are kineses, taxes, reflexes, and fixed action patterns.

Learning

Innate behaviors are certainly important to the survival of an animal. An animal with protective coloration, for example, does not have time to learn to "freeze" when it detects a predator; its survival depends on its instinct to do so (Figure 34-7). Many social behaviors, such as certain mating or food-attainment behaviors, also depend on each individual's performance of instinctive behaviors. In addition, organisms with simple nervous systems, such as the cnidarians, have an extremely limited capacity for learning and must rely primarily on hard-wired innate behavior patterns for their survival.

Although important in many respects to the survival and fitness of animals, innate behaviors can become a liability if environmental conditions change, and an animal's behavior cannot change to adapt to new conditions. For this reason, behavior patterns that can change in response to experience have adaptive advantages over the firmly set programs of instinct. In most animals, only some behaviors are innate and immutable; many behaviors can be changed or modified by an individual's experiences during the process of learning. Learned behaviors can help an animal become better suited to a particular environment or set of conditions.

FIGURE 34-7 Protection against predation is instinctive. This tiny tropical frog has markings that resemble a bird dropping. The effectiveness of this protective coloration depends on the frog remaining completely still. This stillness is *not* learned but is an instinctive behavior on which the animal depends for survival.

Behaviors based on experience are learned behaviors. Learning helps an animal change its behavior to adapt to changes in environmental conditions.

Ethologists distinguish between types of learning, grouping them into five categories: imprinting, habituation, trial-and-error learning, classical conditioning, and insight.

Imprinting

Innate behavior and learning interact closely in a time-dependent form of learning called **imprinting.** Imprinting is a rapid and irreversible type of learning that takes place during an early developmental stage of some animals. Various types of imprinting exist. One type is object imprinting and has been observed in birds such as ducks, geese, and chickens. During a short time early in the bird's life (optimally 13 to 16 hours after hatching in mallard ducks, for example), the young animal forms a learned attachment to a moving object. Usually this object is its mother. However, animals can imprint on various objects regardless of size or color, such as balloons, clocks, and people.

Konrad Lorenz performed classic experiments regarding object imprinting during the 1930s. In one of his most famous experiments, Lorenz divided a clutch of graylag goose eggs in half. He left one half of the eggs with the mother goose and put the other half in an incubator. The half that hatched with their mother displayed normal behavior, following her as she moved. As adults, these geese also exhibited normal behavior and mated with other graylag geese. The other half of the clutch, however, hatched in the incubator and then spent time with Lorenz. These goslings did not behave in the same manner as their siblings; they followed Lorenz around as if he were their mother (Figure 34-8). After their initial time with him, Lorenz introduced the geese to their mother, but they still preferred to follow Lorenz. As adults, these geese tried to "court" adult humans!

Other types of imprinting also exist. Many species of birds hone their singing skills by motoric imprinting. A young North American white-crowned sparrow, for example, listens to its father sing long before it has matured and is ready to do the same. The father's song forms a "mental template" for the young bird, which it uses to compare its song later in life (Figure 34-9).

Many migrating birds and fishes learn to recognize their birthplace by locality imprinting. Pacific salmon, for example, are imprinted with the odor of the stream or lake in which they were born. Amazingly, 2 to 5 years later when they return from the sea to spawn, they are able to find their birthplace by its odor. Such long-range two-way movements are called **migrations.** The physiological and behavioral mechanisms that interact to help migrating animals navigate have interested and puzzled biologists for centuries.

FIGURE 34-8 Imprinting is a time-dependent form of learning. The eager goslings following Konrad Lorenz think he is their mother. He is the first animal they saw when they hatched, and they have become imprinted with his image. The goslings will always recognize Lorenz as their parent.

FIGURE 34-9 Song learning in birds. The singing of a sparrow, which sounds so spontaneous and free, is actually a carefully orchestrated performance that the immature bird learns from its father. The bird perfects its song by listening to others sing and by months of practice. A bird raised in isolation lacks singing partners and thus sings only a rough approximation of its song.

In many animals, migrations occur once a year and result from interactions among various environmental factors (such as day length) with animals' physiological and hormonal changes. Ducks and geese, for example, migrate down flyways from Canada across the United States each fall and return each spring. Caribou migrate across the top of North America each year (Figure 34-10). Monarch butterflies migrate from the eastern United States to Mexico and back, a journey of over 3000 kilometers, which takes from two to five generations of butterflies to complete. Perhaps the longest migration is that of the golden plover, a bird that flies from Arctic breeding grounds to wintering areas in southeastern South America, a distance of approximately 13,000 kilometers.

Migrating animals have the ability to orient themselves in relation to an environmental cue, such as the sun, the stars, or the Earth's magnetic field. In addition, they can navigate, or set a course and follow it. Day migrators such as some birds, ants, and bees use the sun's position to chart a course. Night migrators, a group that includes birds such as the indigo bunting, use the stars to chart a course. All these animals also have a biological clock, some sort of internal "timepiece" (of which scientists have limited understanding) that interacts with information from the environment to help migrating animals find their way. It helps animals compensate for the daily movements of the sun.

Imprinting is a rapid and irreversible type of learning that takes place during an early developmental stage of some animals. Imprinting helps animals recognize kin, hone singing skills, or recognize their "home" or birthplace. The last type, locality imprinting, plays an important role in the migrations of some animals.

FIGURE 34-10 Caribou migration. In caribou the seasonal migration across northern Canada may cover 1500 kilometers, although hunting and development are disrupting the long-used migration routes.

FIGURE 34-11 Habituation in gull chicks.
At first, a chick will crouch at any object that flies overhead. Gradually, the chick stops crouching at the familiar objects that fly overhead—the chick has gotten used to them. However, the chick will crouch when an unfamiliar flying object poses a potential threat.

Habituation

Habituation is the ability of animals to "get used to" certain types of stimuli. People get used to stimuli all the time. A single shotgun blast would startle most people, but this response would fade at the end of a day in a firing range. Not only would you get used to the noise level (and wear ear protection), but you would perceive the gunshots as nonthreatening. Likewise, you quickly get used to the feel of your clothing after dressing in the morning. Animals also stop responding to stimuli that they learn is neither harmful nor helpful. Learning to ignore unimportant stimuli is a critical ability in an animal confronting a barrage of stimuli in a complex environment and can help an animal conserve its energy. Hydra, for example, stop contracting if they are disturbed too often by water currents. Sea anemones stop withdrawing if they are touched repeatedly. Young black-headed gull chicks stop crouching as familiar-shaped birds fly overhead. However, they still crouch for the occasional unfamiliar shape that may pose a threat (Figure 34-11).

Trial-and-error learning (operant conditioning)

More complex than imprinting or habituation, **operant conditioning** is a form of learning in which an animal associates something that it does with a reward or punishment. (The word *operant* means "having the power to produce an effect" and is a form of the verb "to operate.") In operant conditioning an animal must make the proper association between its response (such as pressing a lever) and a reward (the appearance of a food pellet) before it receives this reinforcing stimulus. It may also learn to avoid a behavior when the stimulus is negative. Animal trainers use the techniques of operant conditioning.

The American psychologist B.F. Skinner studied such conditioning in rats by placing each in a specially designed box (today called a *Skinner box*) fitted with levers and other experimental devices (Figure 34-12). Once inside, the rat would explore the box feverishly, running this way and that. Occasionally, it would accidentally press a lever, and a pellet of food would appear. At first, a rat would ignore the lever and continue to move about, but soon it learned to press the lever to obtain food.

This sort of trial-and-error learning is of major importance to most vertebrates in nature. The toad in Figure 34-13 learned quickly, for example, not to eat a bumblebee.

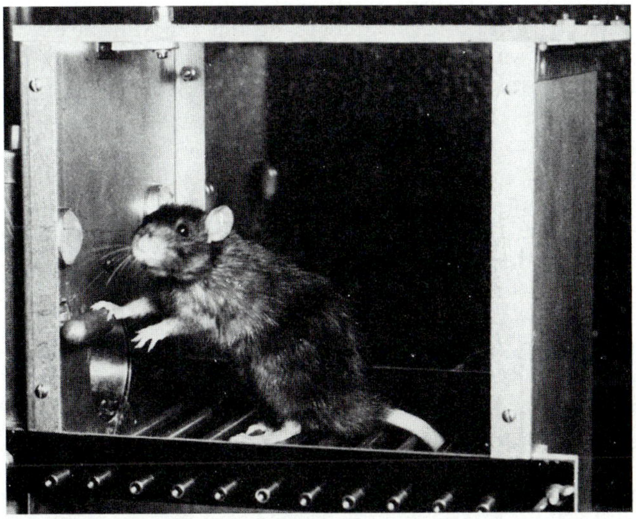

FIGURE 34-12 Operant conditioning.
This rat in a Skinner box rapidly learns that pressing the lever results in the appearance of a food pellet. This kind of learning is based on trial and error with a reward for success.

FIGURE 34-13 A toad is taught a lesson through trial and error.
A toad gobbles a bumblebee, but as it does, it gets a violent sting on its tongue and spits out the bee.

The frog did not use reasoning to determine that the bumblebee was not good to eat—it merely became conditioned by experience to avoid a response that caused it pain.

Behavioral psychologists used to think that animals could be conditioned in the laboratory to perform *any* learnable behavior in response to *any* stimulus by operant conditioning. However, through experimentation, researchers have discovered that animals only tend to learn in ways that are compatible with their neural programs. Put simply, operant conditioning only works for stimuli and responses that have meaning for animals in nature. Rats can be conditioned, for example, to press levers with their paws to obtain food because in nature they obtain food with their paws. They cannot, however, be conditioned to obtain food by jumping, an unnatural food-attaining behavior for a rat.

Some animals can learn by trial-and-error; by associating a specific behavior with a reward or punishment, they can learn to apply (or avoid) certain behaviors to affect outcomes in particular situations. Using the trial-and-error method, scientists can condition animals to perform specific behaviors in the laboratory. This type of learning is operant conditioning.

Classical conditioning

Classical conditioning is a form of learning in which an animal is taught to associate a new stimulus with a natural stimulus that normally evokes a response in the animal. Repeatedly presenting an animal with the new stimulus in association with the natural stimulus can cause the animal's brain to form an association between the two stimuli. Eventually, the animal will respond to the new stimulus alone; it will act as a substitute for the natural stimulus. The connection between the new stimulus and the natural stimulus is the result of a learning process but must be reinforced periodically with the presence of the natural stimulus or the animal will stop responding to the substitute.

In his famous study of classical conditioning, the Russian psychologist Ivan Pavlov worked with dogs to condition them to salivate in response to a stimulus normally unrelated to salivation. A natural stimulus to trigger the response of salivation in a dog is meat. In his experiments, Pavlov also presented the dog with a second, unrelated stimulus. He shone a light on the dog at the same time that meat powder was blown into its mouth. As expected, the dog salivated. After repeated trials, the dog eventually salivated in response to the light alone. The dog had learned to associate the unrelated light stimulus with the meat stimulus (Figure 34-14).

The natural stimulus-response connection in an animal is an inborn reflex, or an *unconditioned* response. Such innate responses are important to animals; many of them are protective, such as blinking, sneezing, vomiting, and coughing. Even the knee-jerk reflex mentioned earlier is part of the mechanism by which humans (and other vertebrates) maintain their posture. The work of researchers such as Pavlov contributed to the understanding that animal behavior depends on innate neural circuitry but that these neural programs can most often be modified and directed by the processes of learning.

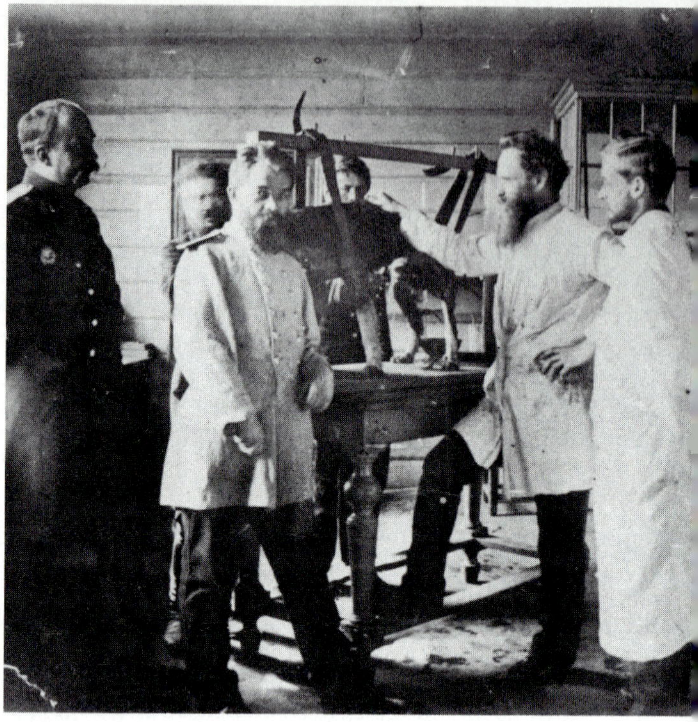

FIGURE 34-14 Classical conditioning.
The Russian psychologist Ivan Pavlov (second from right) in his laboratory. In the background a dog is suspended in a harness that Pavlov used to condition salivation in response to light.

Insight

Best developed in primates such as chimpanzees and humans, **insight,** or **reasoning,** is the most complex form of learning. An animal capable of insight can recognize a problem and solve it mentally before ever trying out a solution. Therefore the animal is able to perform a correct or appropriate behavior the first time it tries, without having been exposed to the specific situation.

German psychologist Wolfgang Kohler was the first to describe learning by insight, performing extensive experiments on chimpanzees in the 1920s. Kohler showed that animals must perceive relationships and manipulate concepts in its mind to solve a problem on the first try. In his classic experiments, he placed chimpanzees in a room with a few crates, poles that could be joined together, and a banana hung high above the grasp of the animals. The chimpanzees were able to stack the crates and join the poles appropriately to retrieve the banana (Figure 34-15, *A*). Unlike chimpanzees, most animals are unable to use insight to solve problems. Your dog or cat, for example, must use trial-and-error to "try out" solutions to problems. Your dog has no insight into the problem that having its leash wrapped around a tree is keeping it from reaching its food (Figure 34-15, *B*). The dog can merely continue walking in various directions and ultimately free itself by chance. This experience, however, may help the dog perform the appropriate behavior in the future, having learned by trial-and-error.

FIGURE 34-15 Insight is the most complex form of learning.

A A chimpanzee is able to see a problem and develop a solution, even before taking any action. To reach the bananas, the chimpanzee sees a solution in stacking the boxes one on top of the other.

B A dog, on the other hand, lacks insight and cannot develop a solution. A dog will eventually learn by trial and error to go around the post to reach the food.

B I O L O G Y & *you*

Yawn!

Everyone yawns. Most people yawn within 5 minutes of birth and keep doing it at odd moments every day as long as they live. Did you ever wonder why?

First, there is nothing particularly human about yawning. Most carnivores yawn. Few herbivores seem to yawn, although there are exceptions—rhinoceroses have enormous yawns. Frogs yawn. Even fish appear to yawn. Whatever is going on, it must be a basic behavior, like eating or sleeping, that is exhibited in many animals.

It used to be felt that a yawn was a silent scream for oxygen that usually occurs when people are tired or bored, a type of deep breath to increase oxygen in the blood or get rid of excess carbon dioxide. Not so. When a group of freshman psychology students at the University of Maryland inhaled air containing different mixtures of oxygen and carbon dioxide and counted their yawns, only breathing rates went up or down to compensate for changing levels of oxygen and carbon dioxide. Yawning rates did not change.

Some researchers believe that yawning is the body's way of promoting arousal in situations where you have to stay awake. Most humans yawn when stimulation is lacking. When a team of yawn counters observed people engaged in various activities, they found that people riding subway cars yawned far more often when the cars were empty. Students yawned a lot studying in the library. But the highest yawn rates they recorded were in a college calculus class, where, on the average, each student racked up an astonishing 24.6 yawns per hour!

FIGURE 34-A Yawn!

The arousal hypothesis helps explain why people driving late at night on the highway yawn a lot and why very few people yawn when they are actually in bed—they don't need to because it's okay to go to sleep. Whether this hypothesis is plausible or not, one thing is sure: yawning is highly contagious. Seeing another person yawn releases a powerful urge for you to yawn yourself. This trait seems to be restricted to humans; no other animal responds in this way.

Yawning is not a hotbed of research, but studies are ongoing in the laboratories of Robert Provine at the University of Maryland and Ronald Baenninger at Temple University, where the work described here was carried out. Provine points out the research may be more important than people realize. "I think a lot of times yawning is trivialized," he says, "because people think it's a little behavior in a compartment off to the side of all other human experiences and actions." To the contrary, yawning seems associated with many diseases in ways not yet understood—brain lesions and epilepsy often lead to excessive yawning, but schizophrenics yawn very little. There is clearly much to learn.

Summary

1. The study of animal behavior encompasses the patterns of movement, sounds, and body positions of animals. Ethologists study the behavior of animals in their natural environments, observing and interpreting their behavior in the context of physiology while attempting to uncover the biological significance of the behavior.
2. Some patterns of behavior are inborn, or innate, resulting from neural pathways developed before birth. Such instinctive behaviors are important to animals, providing them with fixed patterns of survival, such as those used to elude predators, to mate, and to care for young.
3. Ethologists group innate behaviors into four categories: kineses, taxes, reflexes, and fixed action patterns. Certain animals slow down in suitable environments and speed up in unsuitable ones by means of kineses, move toward or away from certain stimuli by means of taxes, automatically respond to nerve stimulation by means of reflexes, and perform sequences of movements by means of fixed action patterns. These behaviors can be modified only slightly or not at all.
4. Animals can learn, or change their behaviors based on experience. Behaviors are changed within certain limits by means of various types of learning: imprinting, habituation, trial-and-error, and conditioning.
5. Certain animals learn to recognize their kin or birthplace by means of imprinting, a rapid and irreversible type of learning that takes place during an early developmental stage. Habituation helps animals adapt to certain types of stimuli. Some animals learn to associate a behavior with a reward or punishment by means of trial-and-error learning. During classical conditioning, animals learn to associate a new stimulus with a natural stimulus that normally evokes a response in the animal.
6. Insight, or reasoning, is the most complex form of learning. An animal capable of insight can recognize a problem and solve it mentally by using experience, thereby performing an appropriate behavior the first time it tries. Insight is best developed in primates such as chimpanzees and humans.

REVIEW QUESTIONS

1. What are behaviors?
2. Explain this statement: Behaviors are part of an animal's equipment for survival.
3. Fill in the blanks: Behaviors that are performed in a fairly complete form the first time they are exhibited are __________ __________. Behaviors that are based on experience are __________ __________.
4. Fill in the blanks: Ethologists group innate behaviors into four categories: __________, __________, __________, and __________ __________ __________.
5. Define and give an example of each of the categories you listed for question 4.
6. Explain the importance of innate and learned behaviors. How do they complement each other?
7. Fill in the blanks: Ethologists distinguish between five types of learning: __________, __________, __________ __________, __________ __________, and __________.
8. Identify the type of behavior involved in each of the following situations:
   a. Fireflies are attracted to the flashing luminescence of other fireflies.
   b. While studying your biology text, you hear a faucet dripping. For a while it bothers you, but then you no longer notice it.
   c. You praise your puppy every time it sits down when you say "sit." Soon you've trained it to sit on command.
9. Distinguish between operant conditioning and classical conditioning.
10. What is insight? Explain its significance.
11. Identify the category of behavior involved in each of the following situations:
    a. Every year, purple martins fly to Brazil and return to the United States around April.
    b. When a bright light flashes near your eyes, you automatically blink.
    c. Your cat frequently grooms itself by licking its fur and rubbing its paws over its face.

KEY TERMS

behavior 599
classical conditioning 606
ethology 599
fixed action pattern 602
habituation 605
imprinting 603
insight (reasoning) 606
instinctive (innate, inborn) 600
kinesis 600
learning 600
migration 603
operant conditioning 605
reflex 602
taxis 600

THOUGHT QUESTIONS

1. Rats can be taught to press levers to get food but can never be taught to obtain food by jumping. Why not? Can you think of any behaviors dogs cannot be taught to do?
2. Monarch butterfly migrations take from two to five generations to complete, each way. How do you imagine a butterfly born midway along the journey knows which way to go?
3. Pacific salmon are born at the headwaters of rivers, then swim downstream hundreds of miles to the sea where they spend their adult lives. Years later, when it is time to spawn (that is, to lay and fertilize the eggs that will be the next generation), the adults swim up the same rivers to the precise location where they were born. Swimming up the river, how do they know which way to go when they come to a fork in the river?

FOR FURTHER READING

Dawkins, R. (1989). *The selfish gene.* ed. 2. New York: Oxford University Press.
This is an entertaining account of the sociobiologist's view of behavior.

Gould, J., & Marler, P. (1987, January). Learning by instinct. *Scientific American,* pp. 74-85.
This is a clear and interesting account of the relative roles of instinct and learning in behavior. The authors argue that learning is often limited or controlled by instinct.

Huber F., & Thorson, J. (1985, December). Cricket auditory communication. *Scientific American,* pp. 60-68.
This gives an unusually clear example of how nervous system activity underlies animal behavior.

35 Social behavior in animals

These kissing fish only appear to be kissing. In fact, their behavior is one of aggression, not affection. Engaged in a territorial war, each wants to win the rights to its own "space" within the water. Neither fish will die during this bloodless battle, but one will back down, allowing the other to lay claim to the win.

Their fight begins as they encounter one another at a territorial boundary. First, each tries to intimidate the other, puffing up its body with air to appear as big and threatening as possible. Then they swim side-by-side, pushing currents of water past one another with their fins. These currents provide clues to each contender regarding the size of the opponent. If neither retreats, the tug-of-war begins, mouths locked in battle. Soon, one fish emerges the victor. The loser signals submission, and the tournament is over.

OUTLINE

Types of social behavior
- Competitive behaviors
  - Threat displays
  - Submissive behavior
  - Territorial behavior
- Reproductive behaviors
- Parenting behaviors
- Group behaviors
  - Insect societies
  - Rank order in vertebrate groups

Human behavior

HIGHLIGHTS

Sociobiology, the biology of social behavior, applies the knowledge of evolution to the study of animal behavior.

Social behaviors, those that help members of the same species communicate and interact with one another, are valuable mechanisms that ultimately aid the reproductive fitness of a species.

Many species of animals live in social groups; some insects, in fact, live in societies having a division of labor.

Many researchers question the scientific validity of applying sociobiology to human behavior and propose that the application may lead to racism, sexism, or other types of group stereotyping.

Types of social behavior

The aggressive behaviors exhibited by the kissing gouramis are types of **social behaviors,** which help members of the same species communicate and interact with one another. The advantages of social behaviors are numerous, and these advantages differ from species to species. In general, however, all species that reproduce sexually exhibit interactive patterns of behavior for the purposes of reproduction, the care of offspring, and the defense of a territory. In addition, some animals use social behaviors to hunt for and share food and to warn and defend against predators.

The biology of social behavior is called **sociobiology.** This science applies the knowledge of evolutionary biology to social behavior. Its purpose is to develop general laws of the biology and evolution of social behavior.

Social behaviors are interactive patterns of animal behavior that help members of the same species communicate and interact with one another. The biology of social behavior is sociobiology.

Competitive behaviors

When two or more individuals strive to obtain the same needed resource, such as food, water, nesting sites, or mates, they are exhibiting **competitive behavior.** This type of behavior occurs when resources are scarce.

Threat displays

How do animals compete? Many first engage in **threat,** or **intimidation, displays,** a form of aggressive behavior. The purpose of these displays is to do as the name suggests: scare other animals away or cause them to "back down" before fighting takes place. Exact forms of this behavior vary widely among species but usually involve such behaviors as showing fangs or claws, making noises such as growls or roars, changing body color to one that is a releaser of aggression in an opponent, and making the body appear larger by standing upright, making the fur or hair stand on end, or inflating a throat sac (Figure 35-1). Threat displays are important social signals and communicate the intent to fight. Interestingly, some of the movements and body postures (body language) of threat displays that repel competitors also attract members of the opposite sex. Biologically, this makes sense because the competition may be over a mate.

Submissive behavior

Animals use other behaviors to avoid fighting. One type of behavior is called **submissive behavior** and is usually a behavior opposite to a threat display. The behavior might include making the body appear smaller, "putting away weapons" of fangs or claws, turning a vulnerable part of the body to an opponent, or removing body colors that are releasers of aggression. As Figure 35-2 shows, an animal may, in fact, possess an array of color "signals." If an animal is losing a fight, it might also display submissive behaviors to stop the fight. Contrary to popular belief, animals of the same species rarely fight to the death. One reason is that most animals do not have the means to do so—they do not have sufficiently dangerous fangs, claws, or horns, for example. Species with dangerous weapons usually have defenses against those weapons, too, such as a strong hide, long hair, or a thick layer of body fat. Scientists theorize that aggression within a species is meant to chase off rather than kill the rival. Such behaviors are adaptive for the species as a whole because they tend to disperse the population among the available resources rather than decrease the size of the population and conserve the energy of all the members of the species.

FIGURE 35-1 Defense by deception.
This common toad, confronted by a snake, inflates its body and sways from side to side in an attempt to make itself appear larger.

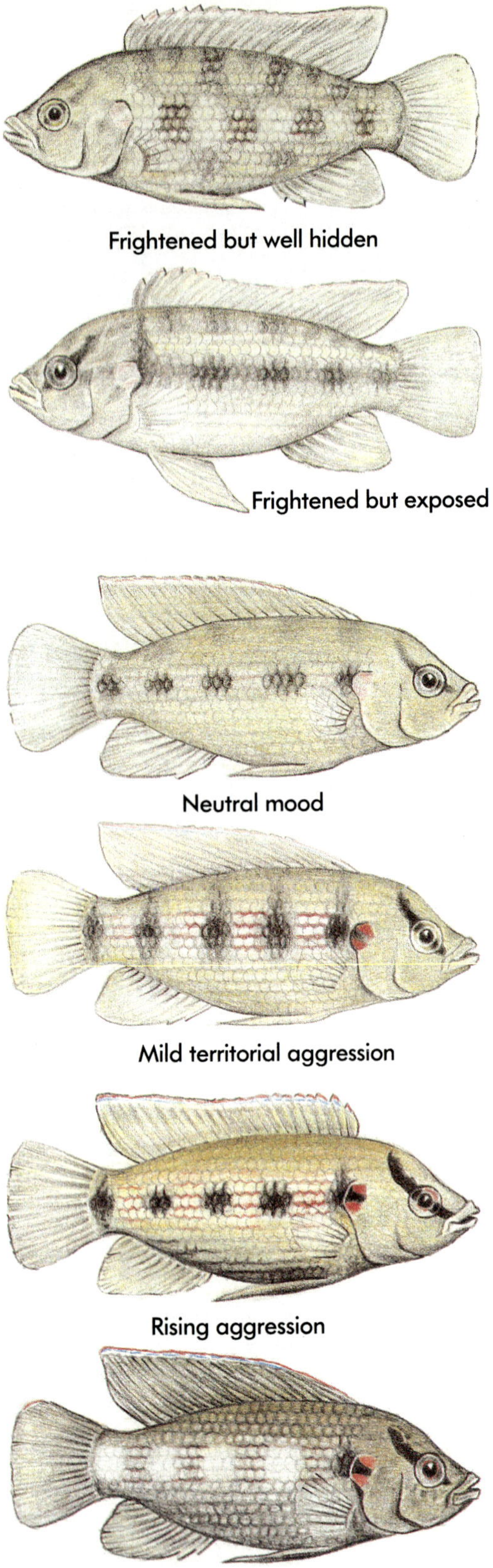

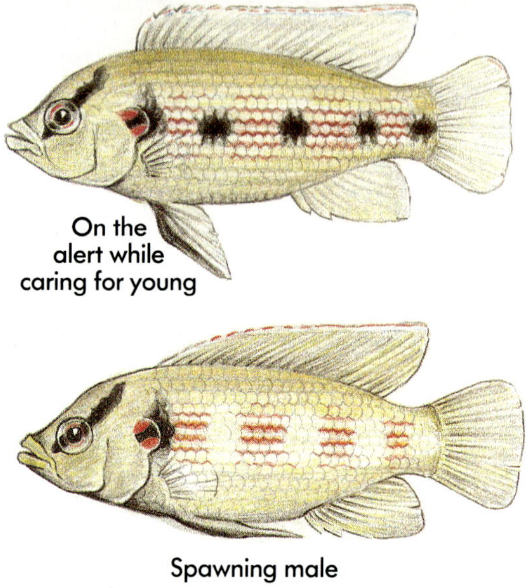

FIGURE 35-2 The technicolor fish.
This freshwater fish, a member of the cichlid family, changes its color patterns in different situations. It does this by expanding and contracting special red and black color cells in its skin.

Territorial behavior

Another means by which animals limit competition and aggression within their species is by the use of **territorial behavior.** A territory is an area that an animal marks off as its own, defending it against the same sex members of its species. Members of the opposite sex are always allowed into the territory; in fact, the territory is their meeting and mating place.

Territoriality is very common in all classes of vertebrates and is even found in some invertebrates such as crickets, wasps, and praying mantises. Some animals such as dragonflies mark their territories by conspicuously patrolling their borders, resting and then moving from prominent landmarks. As an animal demonstrates the borders of its territory (Figure 35-3), it may exhibit specific movements or body postures as signals. Some animals, such as song birds, monkeys, apes, frogs, and lizards, mark their territory by producing sounds that announce their ownership. Animals with a well-developed sense of smell, such as wolves, hippopotamuses, rhinoceroses, some rodents, and even domestic dogs, mark their territories with substances that have an odor. Many species, in fact, possess special scent glands that secrete substances just for this purpose. The odor marking helps keep out members of the same sex and helps owners of large territories orient themselves to its borders.

Territorial behavior has several adaptively important consequences. In many mammalian societies, territoriality often has an important influence on the establishment of new populations, with the dominant male defeating the lesser males and leaving them to found their own populations. Individuals surviving in the surrounding marginal areas repopulate any vacant territories. Territoriality is also a means to ration resources among members of a species.

FIGURE 35-3 **Territory of a male Demoiselle dragonfly.**
The male dragonfly sits on a perch above the egg deposit site. He periodically flies around his territory *(red arrows)* or circles within the territory.

When resources such as nesting sites and food are limited (Figure 35-4), each member of a species is in danger of the resources being spread too thin, so that no member gets an adequate amount of what it needs. In addition, breeding pairs of animals will not reproduce unless they have adequate resources; territorial behavior ensures that at least some members of the species will have their own space, food, and nesting sites so that they will survive and reproduce. Territorial behaviors also enhance reproductive capabilities by placing the peak time of competition and aggressive behavior at the time of the marking of the territory, *before* the time of reproduction and raising of young.

Individuals within the same species compete with each other for various resources. Competition among animals includes a repertoire of behaviors such as threat displays, submissive behavior, and territorial behavior. Such competitive behaviors disperse the population among the available resources and enhance the reproductive capabilities of the species.

FIGURE 35-4 **The emperor penguins in Antarctica nest in rookery sites.**
There are only a limited number of such sites available, which may reflect the limited amount of resources available to the breeding penguins.

BIOLOGY IN FOCUS

Elephant talk

How do male and female elephants find each other to reproduce? Consider the obstacles: females spend 2 years in gestation (the time from fertilization and implantation of the embryo until birth) and another 2 years nursing their young. During these periods, female elephants will not mate. The period of estrus, in which the female elephant is receptive to mating, occurs only once every 4 years—and lasts only 2 days! Male elephants have their work cut out for them. For half the year, male elephants are solitary creatures and travel long distances in a search for willing female partners. It seems a miracle that elephants reproduce at all.

But is it a miracle that elephants find each other? The answer is not at all. In 1989, researchers investigated the possibility that a female elephant in estrus sends out a signal that informs males for miles around of her condition. Using sophisticated microphones, the researchers recorded a sequence of intense, low-frequency calls emitted from an elephant in estrus. The estrus call is the same in all female elephants and is a song. The low frequency of the song places it in the infrasound frequency range—the same range that includes thunderstorms and earthquakes. Humans can barely detect infrasound—you "feel" more than hear the low, dull throbbing associated with these low-frequency sounds. The female elephant sings her song for only a half an hour, and within a day she is surrounded by male elephants, some of whom have heard the song from several miles away.

Infrasound is used by elephants for purposes other then mating as well. Warnings are issued by infrasound, and the whereabouts of groups of elephants are transmitted to other groups by infrasound. Observers have noted for years that entire groups of elephants will suddenly freeze, raising and spreading their ears. Researchers surmise that these elephants are listening to faint, distant calls and are remaining as silent as possible to do so. Moments later, the elephants will come out of their trance and rush off together to follow the signal.

The elephant's "talk" is a factor that enhances elephants' survival by communicating useful information and facilitating the reproductive process. Researchers are currently looking into the possibility that elephants use their infrasound calls to maintain their elaborate hierarchical society over distances of several miles.

FIGURE 35-A A bull elephant can locate a female elephant in estrus by listening for her infrasound mating call.

FIGURE 35-5 Asexual reproduction. Offspring of sea anemones can be produced by budding. A bud grows from a few cells of a single parent and, when developed, breaks off and becomes completely separate individuals. Budding results in genetically identical parents and offspring. Buds can be seen emerging at the sea anemone's midsection.

Reproductive behaviors

Sexual reproduction, the union of a male sex cell (sperm) and a female sex cell (egg) to form the first cell of a new individual (zygote), takes place in most animals. Even some animals that can reproduce asexually, such as sponges that can bud off bits of themselves that grow into new individuals, reproduce sexually at certain times. Sexual reproduction is biologically significant because it provides genetic variability among organisms of the same species. Put simply, the combining of genes from two parents produces offspring that are similar but different from their parents. In addition, because the genes of each parent become assorted independently during the meiotic process that produces sex cells (see Chapters 5 and 22), the offspring are different not only from their parents, but from one another. Differences among organisms within a species are the "raw materials" needed for adaptation to environmental change. The ability of some organisms to survive and reproduce in an environment in which others of their species cannot is, in turn, the basis of natural selection.

Genetic variability does not exist in organisms that reproduce asexually; these organisms are clones of one parent. That is, organisms produced by means of asexual reproduction develop from one or a few cells of a single parent (Figure 35-5) and are therefore genetically identical to that parent. Because organisms within a population of clones do not vary from one another, the entire population can be at risk if environmental conditions change.

Behaviors that promote successful sexual reproduction, then, are highly adaptive behaviors. Organisms that reproduce sexually must have patterns of male-female interactions that lead to fertilization—the penetration of an egg by a sperm. Fertilization can be preceded by copulation, the joining of male and female reproductive organs. However, some organisms such as many fish and frogs do not copulate; instead the male releases sperm over the eggs after they have been deposited by the female. The term **mating** refers to male-female behaviors that result in fertilization, regardless of whether copulation occurs. The term **courtship** refers to the behavior patterns that lead to mating.

FIGURE 35-6 A male peacock.
Males of certain species, such as the peacock, use fantastic displays of color or markings to attract females.

For courtship to take place, males and females must first find each other. In some species, males and females live together in pairs or groups, but many animal species do not. Most often the job of attracting a mate goes to the male of the species. Notable exceptions, however, are often found within insect populations.

Males often attract a mate by marking a territory and defending it against intrusion by other males. As mentioned previously, aggressive behaviors against other males are often the same behaviors that attract a female. In addition, males can attract females by displaying body colors or markings like the male peacock shown in Figure 35-6. Many birds sing songs to attract a mate, and certain frogs and many insects use other sounds or mating calls for this purpose. Many mammals and insects produce odors that are attractive to females. In some species, the males congregate and perform various dances or songs as a group to attract their mates (Figure 35-7).

FIGURE 35-7 Courtship dance in male prairie chickens. Male prairie chickens perform a courtship dance to attract females. They produce sounds as part of their behaviors, puffing up the orange air sacs on the sides of their necks.

After the formation of a male-female pair, a **courtship ritual,** unique to each species, leads to mating. Courtship behaviors usually consist of a series of fixed action patterns of movement (see Chapter 34), each triggered by some action of the partner that acts as a releaser and triggers in turn a preprogrammed movement pattern by the potential mate (Figure 35-8). For example, cranes dance around one another, peacocks strut with their tail feathers extended in an elaborate fan, and hummingbirds participate in a courtship flight (Figure 35-9). In addition, some species, such as various types of geese, exhibit postcopulatory behaviors. Both birds rise out of the water, wings extended, facing one another (Figure 35-10). Such behaviors appear to strengthen the pair bond, a behavior that is necessary when both parents care for the young.

Reproductive behaviors within a sexually reproducing species serve to bring males and females together in a manner that leads to the fertilization of eggs by sperm.

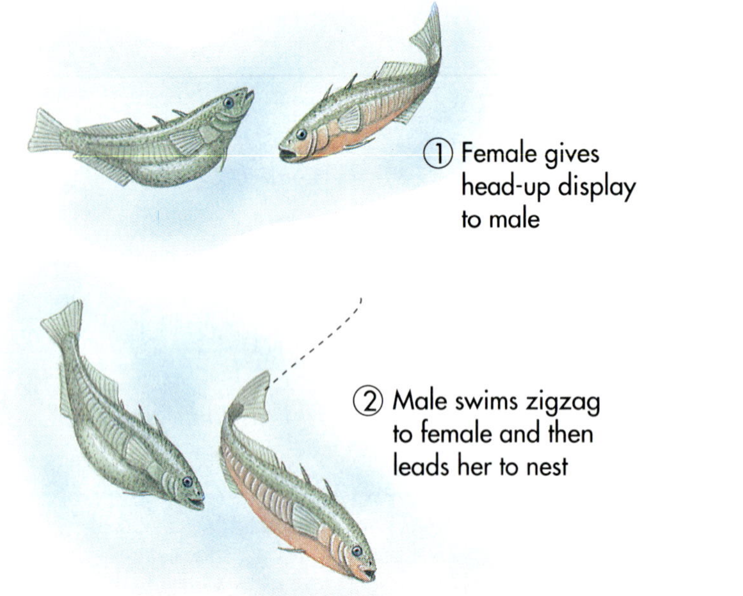

FIGURE 35-8 Stickleback courtship. Each set of movements in the courtship ritual is a fixed action pattern that triggers the next set of movements by the partner.

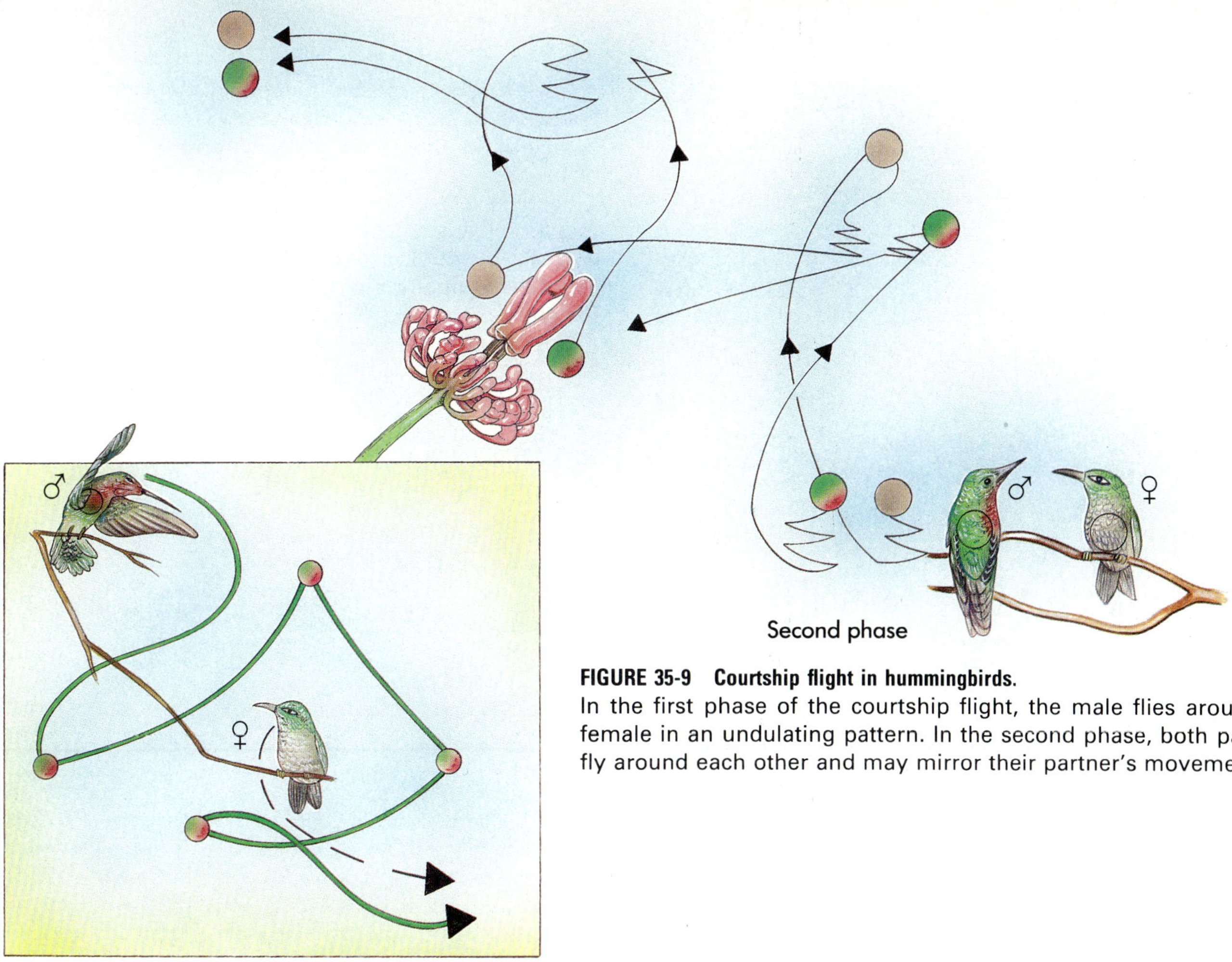

FIGURE 35-9 Courtship flight in hummingbirds.
In the first phase of the courtship flight, the male flies around the female in an undulating pattern. In the second phase, both partners fly around each other and may mirror their partner's movements.

FIGURE 35-10 Geese exhibiting postcopulatory behavior.
Here, a goose rises out of the water, wings extended. These postcopulatory behaviors seem to strengthen the pair bond between the mating geese.

FIGURE 35-11 A marsupial with young.
The kangaroo protects its young in its pouch. The young are nourished by mammary glands located in the pouch.

Parenting behaviors

For the offspring to survive, grow, and eventually reproduce, parents must either make preparations for the care of their young or care for them themselves. Parenting behaviors are found in almost all animal groups, but the participation of the parents varies. In many species of birds and fishes, both parents care for the young. In these species, a division of labor usually exists, with each parent carrying out a specific role. Because female mammals produce milk to nourish the young, most often the mother assumes the parenting role in this class of animals. In only a few species, such as certain fishes, sea horses, and birds, does the male alone take care of the offspring.

To prepare for young, animals build protective structures such as nests or cocoons. Usually a food supply is gathered, and the eggs are deposited in areas that are protected and that may also lie near a ready food source. Marsupials, mammals in which the young are born developmentally early, keep their young safe and nourish them as they complete their development by storing them in a pouch, conveniently situated at the mammary glands (Figure 35-11). Some animals simply carry their young to a safe place when they sense danger.

Some species exhibit quite interesting and unique parenting behaviors. Very early in his studies of behavior, Tinbergen noted that certain birds remove the shell fragments of broken eggs from their nests. Any shell fragment that Tinbergen added was removed quickly. The adaptive advantage of this parenting behavior, Tinbergen discovered, was that the odor from shell fragments attracted predators and their prompt removal helped protect the other eggs in the nest.

Parenting behaviors lead to the reproductive fitness of a species by helping offspring survive and grow to reproductive age.

Group behaviors

Individuals within many species of animals form pair bonds for the purposes of reproduction and parenting. In addition, the individuals of many species form temporary or permanent associations with many other members of their species, forming social groups. Animals within social groups work together for common purposes. Individuals within a group, for example, have more protection against predators than single individuals or pairs. They can warn each other of danger by using calls or chemical signals or cluster together when a predator appears, forcing the predator to either separate individuals from the group or fight the entire group. In addition, some animals such as musk-oxen and bison place their young within the center of the group (Figure 35-12), forming a circle around them much like the early American settlers did with their covered wagons.

FIGURE 35-12 Defensive formation of a herd of musk-oxen.
Like the pioneers who circled their wagons against danger, these musk-oxen form a defensive ring with horns pointing outward. The young musk-oxen are inside the ring.

FIGURE 35-13 Hunting behaviors of pelicans.
Eastern white pelicans hunt in groups **(A)** and then circle their prey **(B)**.

Many animals living within social groups search for food together and, scientists have observed, are more successful as group hunters than as single individuals. Food-gathering strategies become group strategies; often animals such as those that prey on fish will drive the prey to one location, encircle them, and feast. The eastern white pelicans shown in Figure 35-13 hunt for fish in this way.

In groups, animals can also ration the "work load" so that each animal does not have to perform an array of daily tasks but may focus on one task (Figure 35-14). Many species of insects, particularly bees and ants, are organized into social groups having a division of labor. These social groups are called **insect societies.**

FIGURE 35-14 Ant society.
This worker ant, a member of the genus *Polyergus,* is responsible for capturing slaves for the nest. The slave, a member of the genus *Formica,* is in its pupa stage. When it emerges as an adult, it will be a nonreproductive worker slave.

Insect societies

In insect societies, individuals are organized into highly integrated groups in which each member of the society performs one special task or a series of tasks that contributes to the survival of the group. The role an individual plays depends on its body structure which, in turn depends on its sex or age. In some animal societies the same animal performs different roles during the course of its life. In addition, a common feature among social insects is a queen that outlives other members of the society. The role of the queen is to produce offspring and to promote cooperation among members of the group.

A honeybee colony, for example, is made up of three different types of bees called **castes:** the queen, the workers, and the drones. The queen is the focus of the colony; she lays the eggs from which all the other bees develop (Figure 35-15). The drones are the male bees that fertilize the eggs of the queen. Interestingly, they develop from unfertilized eggs and so have no male parent themselves—and only one set of chromosomes! The workers are females that develop from fertilized eggs. Genetically, a queen and a worker are no different; any fertilized egg can develop into either a queen or a worker depending on how the egg is housed and fed.

In one hive, there may be up to 50,000 workers, 5000 drones, and of course 1 queen. The workers perform all the tasks of the hive except for mating and egg laying. In the winter months, the average length of the worker bee's life is 6 months. In the summer, workers live only 38 days! During the first 20 days of a worker's life, she performs various duties within the hive sequentially as she matures: feeding larvae, producing wax, building the honeycomb, passing out food, and guarding the hive. Although a worker may go on short play flights around the hive during this time, she does not venture very far. It is only after the first 20 days of her life that the worker takes long flights away from the hive to forage for food.

FIGURE 35-15 A bee colony.
The queen bee has a red spot on her back.

An interesting means of communication among the workers has evolved that enables them to communicate the location of flowers at which they can collect pollen and nectar. This communication is a form of body language called the **waggle dance.** The manner in which a worker shakes, or waggles, her abdomen communicates the distance to the flowers. The angle at which the worker dances indicates the direction her co-workers must fly (Figure 35-16).

The drones play almost no role in the hive other than to mate with the queen. They help keep the larvae warm by congregating near them in the morning while it is still cool. Drones also help distribute food. At the time of mating, many drones follow the queen out of the hive, following her pheromone trail, and mate in flight. After mating, the drones die. Drones that do not mate have no better fate, since the workers pinch them, sting them, and then throw them out of the hive as the winter approaches.

The queen bee controls the worker bees with a chemical she produces known as *queen substance.* This substance is a **pheromone,** a chemical produced by one individual that alters the physiology or behavior of other individuals of the same species. The bees ingest the queen substance by licking the queen and then pass the substance around from one to the other as they pass food. Queen substance renders the worker bees sterile and also inhibits the workers from making queen cells, compartments in which fertilized eggs can develop into queens. If the hive becomes too congested, however, the queen may begin producing less queen substance, resulting in the removal of the inhibition to produce queen cells. Workers make a half-dozen or more new queen cells in which replacement queens begin to develop. The old queen and a swarm of females and male drones leave to establish a new hive (Figure 35-17). The first new queen to emerge may kill the other candidate queens and assume rule or may create another swarm and leave to establish another hive.

The life-styles of social insects can be so unusual as to be bizarre, none more so than the little reddish ants—the leafcutters (Figure 35-18). Leafcutters live in the tropics, organized into colonies of up to several million individuals. These ants are farmers, growing crops of fungi beneath the ground. Their mound-like nests look like tiny underground cities covering more than a hundred square yards, with hundreds of entrances and chambers as deep as 16 feet underground. Long lines of leafcutters march daily from the mound to a tree or bush, hack its leaves into small pieces, and carry the pieces back to the mound. Each ant finds its way by following a trail of secretions left by those who came before it. Although these ants are nearly blind, they follow the scent by holding their antennae close to the ground. At the underground site, worker ants chew the leaf fragments into a mulch that they spread like a carpet in underground chambers. They wet the leaf mulch with their saliva and fertilize it with their feces. Soon a luxuriant lawn of fungi is growing, which serves as the sole food for all the ants,

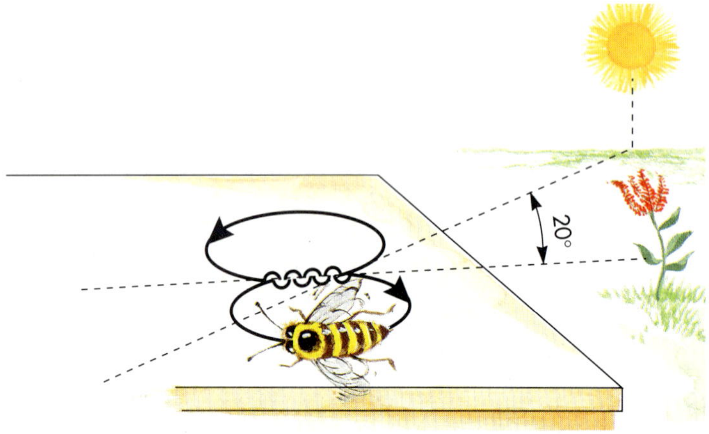

FIGURE 35-16 The waggle dance.
This dance is performed by a forager bee who, on spotting a food supply (a flower), goes back to the hive and communicates to the other bees through dance the exact location of the flower. The dance tells how far away the flower is and also the angle of the flower to the sun. The part of the dance that yields both kinds of information is the straight run, *(zigzag line between two circles).* The forager shakes its abdomen back and forth and buzzes with its wings while moving in the straight run. Then it makes a semicircle and returns to the starting point of the straight run. After completing this run, it circles in the opposite direction to land again at the beginning of the straight run and so on. How fast the bee performs the straight run determines how far away the flower is—a distance of 500 meters equals a straight run duration of about 1 second. The direction of the flower is indicated by the angle that the straight run deviates from the vertical. This angle equals the angle of the flower to the sun.

no matter what their age. Nurse ants even feed this fungus to the larvae, carrying them around to browse on choice spots.

The complex caste systems of bees and ants apparently evolved a long time ago. Ant fossils 80 million years old exhibit three castes (males, queens, and workers), indicating that their complex social system had already evolved.

Many animals not only interact with other members of their species but form social groups and aid one another in various ways. Many species of insects, particularly bees and ants, form highly organized social groups exhibiting a division of labor.

FIGURE 35-17 A swarm resting on a horse chestnut branch. These bees are in the process of establishing a new hive.

Rank order in vertebrate groups

A social hierarchy, or **rank order,** exists in groups of fishes, reptiles, birds, and mammals. In chickens, the rank order is often called a **peck order** because these animals establish the order by pecking at one another. In many species of animals, the rank order is linear, with the highest-ranking individual dominant over all the others and the lowest-ranking individual submissive to all others. Some rank orders show certain complexities within the ranking, such as a rank order in which animal "A," for example, may be dominant over animal "B" and "B" dominant over "C," but "C" may be dominant over "A." In some rankings, the sex of the animal may play no role. In other rankings, males and females are ranked separately, and in still other vertebrate groups, the female takes on the rank of the male. Such rankings help reduce aggression and fighting within social groups, focusing it on the time that the ranking is developed. This behavior is only one of a repertoire of behaviors, including territoriality and submissive behaviors, that help contribute to the stability of the social relationships among animals and within groups.

Human behavior

FIGURE 35-18 Leafcutter ants.
These leafcutter ants are carrying their day's harvest back to the nest. The ants do not live on the leaf material—they use it as a substrate on which to grow the fungi that they cultivate in underground farms.

Humans are unique animals. Because of this uniqueness, many scientists disagree with applying the principles of sociobiology to human behavior. Many scientists question the scientific validity of sociobiological research on human activity. Others question the social and political implications of human sociobiology, proposing that such endeavors might lead to yet another form of racism, sexism, or other types of group stereotyping. In addition, the study of human sociobiology has inherent limitations: scientists cannot experimentally manipulate the genes or vary the environments of humans for scientific studies. It is therefore extremely difficult to learn whether observed behaviors are a product of heredity or a person's environment. This "nature or nurture" dichotomy continually raises questions that scientists have a difficult time researching. Interesting observations have been made, however, in the study of identical twins raised by different families in different environments. From these studies and others, most biologists hold that much of the behavioral *capacity* of humans is genetically determined but that the neural circuits specified by genes can be shaped and molded—within certain limits—by learning.

Summary

1. Social behaviors help members of the same species communicate and interact with one another. Scientists who study the social behaviors of animals are sociobiologists. They seek to determine general laws of the biology and evolution of behavior.
2. Animals of the same species naturally compete for resources they need to survive. Within their repertoire of behaviors are aggressive behaviors that threaten and intimidate but do not kill rivals. Other behaviors also avoid fights to the death, such as submissive behaviors—signals of "backing down."
3. Animals also use territorial behavior to limit aggression within their species. Territoriality is very common in all classes of vertebrates and in some invertebrates. A territory is an area that an animal marks off as its own, defending it against same sex members of its species. This behavior helps disperse a population among limited resources and provides a breeding ground.
4. Reproductive behaviors include patterns of male-female interactions that lead to successful fertilization—the penetration of an egg by a sperm. These behaviors include methods by which mates attract each other and then interact with one another in types of courtships that lead to mating and fertilization.
5. Parenting behaviors include preparations for the care of young and their direct care to aid their survival and ultimate reproduction.
6. Individuals within species form temporary or permanent associations within social groups. Animals within social groups work together for a variety of common purposes. Some species of insects, particularly various species of bees and ants, are organized into social groups having a division of labor.
7. A social hierarchy, or rank order, exists in groups of fishes, reptiles, birds, and mammals. Behaviors relating to rank order help contribute to the stability of the social relationships among animals and within groups.
8. Some scientists question the scientific validity of applying sociobiological principles to human behavior.

REVIEW QUESTIONS

1. Distinguish between behaviors and social behaviors.
2. Explain the functions of social behaviors.
3. What is competitive behavior? Explain its significance.
4. Identify the type of behavior involved in the following situation:
   a. The hair on a dog's back stands on end as the dog growls at an approaching stranger.
   b. When you scold the dog for growling, it rolls over on its back and exposes its belly.
5. Define *territorial behavior*. Why is it important?
6. What is sexual reproduction? Explain the advantages that it can offer to a species.
7. Distinguish between mating and courtship.
8. Identify the type of behavior shown in each of the following situations:
   a. A robin builds its nest.
   b. A male peacock extends its tail feathers into a colorful fan.
   c. A cat chases another cat away from its food bowl.
9. Why do some animals form social groups?
10. Summarize the social structure of a honeybee colony.
11. Explain *rank order*. What is its significance, and in what animal(s) does it appear?
12. Do you feel the principles of sociobiology can be applied to human behavior? Explain your answer.

KEY TERMS

caste 619
competitive behavior 611
courtship 615
courtship ritual 616
insect societies 619
mating 615
peck order 621
pheromone 620
rank order 621
social behavior 611
sociobiology 611
submissive behavior 611
territorial behavior 612
threat (intimidation) displays 611
waggle dance 620

THOUGHT QUESTION

1. Very few mammals exhibit the complex societies seen among bees and ants—but a few do. Naked mole rats of the Middle East, for example, maintain large colonies with queens and special worker castes, organized remarkably like the colonies of bees. Why do you think such societies are so much rarer among mammals?

FOR FURTHER READING

Borgia, G. (1986, June). Sexual selection in bowers birds. *Scientific American*, pp. 92-100.
Behavior among birds is often bizarre, but none is more so than that of the Australasian bower birds, in which the females choose their mates depending on how well they adorn their bowers or ritualized nests.

Bull, J. J., et al. (1992, April 3). Selfish genes. *Science*, pp. 65-66.
An analysis of a new class of selfish genes—genes that are successful at propagating themselves while being detrimental to the organisms that carry them—is presented in this article. The authors offer a lucid explanation for the paradox of selfish genes, implicating the more controversial aspects of sociobiology.

Emlen, S. T., & Wrege, P. (1992, March 26). Parent-offspring conflict and recruitment of helpers among bee-eaters. *Nature*, pp. 331-332.
This short article describes the conflict between parent and offspring bee-eaters, in which fathers disrupt their sons' attempts to breed so that the sons can help raise their father's other offspring. Surprisingly, the sons do not resist their father's harassment.

Heinrich, B. (1989). *Ravens in winter: A zoological detective story*. New York: Summit Books.
What do ravens do in winter, and how are their social systems organized? Vermont zoologist Bernd Heinrich talks about his outstanding field investigations of these intelligent animals.

36 Population ecology

HIGHLIGHTS

Under ideal conditions, populations grow at an exponential rate, leveling off at the carrying capacity of the environment.

Very small populations are less able to survive than large populations; in addition, inbreeding in small populations leads to a loss of genetic diversity and increases the probability of the extinction of that species.

Large populations face threats to survival such as competition for resources such as food, light, and shelter and predation by other organisms.

In 1992, the global human population of more than 5.3 billion people was growing at approximately 1.8% per year; at that rate, the human population will reach well over 10 billion people by 2029.

OUTLINE

An introduction to ecology
Population growth
- Exponential growth
- Carrying capacity

Population size
Population density and dispersion
Regulation of population size
- Density-independent limiting factors
- Density-dependent limiting factors

Mortality and survivorship
Demography
The human population explosion

Looking like homes for oversized mud wasps, these cliff swallow nests hang precariously from the face of a rock outcropping. Any rough, vertical surface will serve as a nesting site for these birds. In fact, cliff swallows have recently discovered that the sides of bridges work well for this purpose . . . and come complete with protective overhangs! As a result, cliff swallows, once found mainly in the western part of the United States, can now be found inhabiting the prairie, nesting on the sides of bridges that span the major prairie rivers.

The cliff swallows that inhabit these nests are a population of organisms. A **population** consists of the individuals of a given species that occur together at one place and at one time. This flexible definition allows the use of this term in many contexts, such as the world's human population, the population of protozoans in the gut of an individual termite, or the population of blood-sucking swallow bugs living in the feathers of a cliff swallow.

An introduction to ecology

Population ecologists study how populations grow and interact. The science of ecology is much more broad, however, and includes the study of interactions between organisms and the environment. Scientists usually classify the study of ecological interactions into four levels: populations, communities, ecosystems, and the biosphere. This chapter discusses the first level of this hierarchy: populations.

Population growth

Most populations will grow rapidly if the ideal conditions for growth and reproduction of its individuals exist. But why, then, is the Earth not completely covered in bacteria, cockroaches, or houseflies? Why do some populations change from season to season or year to year?

To answer these questions, you need to understand how the size of a population is determined. The size of a population at any given time is the result of additions to the population from births and from **immigration,** the movement of organisms into a population, and deletions from the population from deaths and **emigration,** movement of organisms out of a population. Put simply, (births + immigrants) - (deaths + emigrants) = population change.

These statistics (births and deaths) are often expressed as a *rate:* numbers of individuals per thousand per year. For example, the population of the United States at the beginning of 1989 was approximately 250 million people. During 1989 there were the following:

- 3,975,000 live births: The birth rate was 3,975,000 per 250,000,000 people, or 15.9 births per 1000.
- 2,175,000 deaths: The death rate was 2,175,000 per 250,000,000 people, or 8.7 deaths per 1000.
- 570,009 immigrations: The immigration rate was 570,009 per 250,000,000 people, or 2.4 immigrants per 1000.
- 177,600 emigrations: The emigration rate was 177,600 per 250,000,000 people, or 0.7 emigrants per 1000.

The population change in the United States in 1989 can be calculated as follows: (15.9 births/1000/year + 2.4 immigrants/1000/year) - (8.7 deaths/1000/year - 0.7 emigrants/1000/year) = 8.9 people/1000/year. This figure can also be expressed as a population change of 0.89%—an increase of nearly 1%.

The size of a population is the result of additions to the base population due to births and immigrations, and deletions from the population due to deaths and emigrations.

In natural populations of plants and animals, immigration and emigration are often minimal. Therefore, a determination of the **growth rate** of a population does not include these two factors. Growth rate (r) is determined by subtracting the death rate (d) from the birth rate (b):

$$r = b - d$$

Using the figures from our previous example:

r = 15.9 births/1000/year - 8.7 deaths/1000/year
r = 7.2/1000 or 0.0072

To figure out the number of individuals added to a population of a specific size (N) in a given time *without* regard to immigration and emigration, r is multiplied by N:

$$\text{population growth} = rN$$

Therefore the population growth in the United States in 1989 solely from births and deaths was 0.0072 x 250,000,000 people = 1,800,000 people

The determination of the growth rate of a population does not include the factors of immigration and emigration. It is determined by subtracting the death rate from the birth rate.

Exponential growth

Although the *rate* of increase in population size *may stay the same,* the *actual increase* in the number of individuals *grows.* This sort of growth pattern is similar to the growth pattern of money in the bank as interest is earned and compounded. If you put $1000 in the bank at 8% per year, the first year you will earn $80. The second year you will earn 8% interest on $1080, or $86.40. Although your interest rate has stayed the same, the amount of money you earn grows as your money grows. Actually, in a bank your earnings would grow even more quickly because interest would be posted and compounded more often than once a year.

Time (Hours)	Number of bacteria
10	1,048,576
$9\frac{1}{2}$	524,288
9	262,144
$8\frac{1}{2}$	131,072
8	65,536
$7\frac{1}{2}$	32,768
7	16,384
$6\frac{1}{2}$	8,192
6	4,096
$5\frac{1}{2}$	2,048
5	1,024
$4\frac{1}{2}$	512
4	256
$3\frac{1}{2}$	128
3	64
$2\frac{1}{2}$	32
2	16
$1\frac{1}{2}$	8
1	4
$\frac{1}{2}$	2
0	1

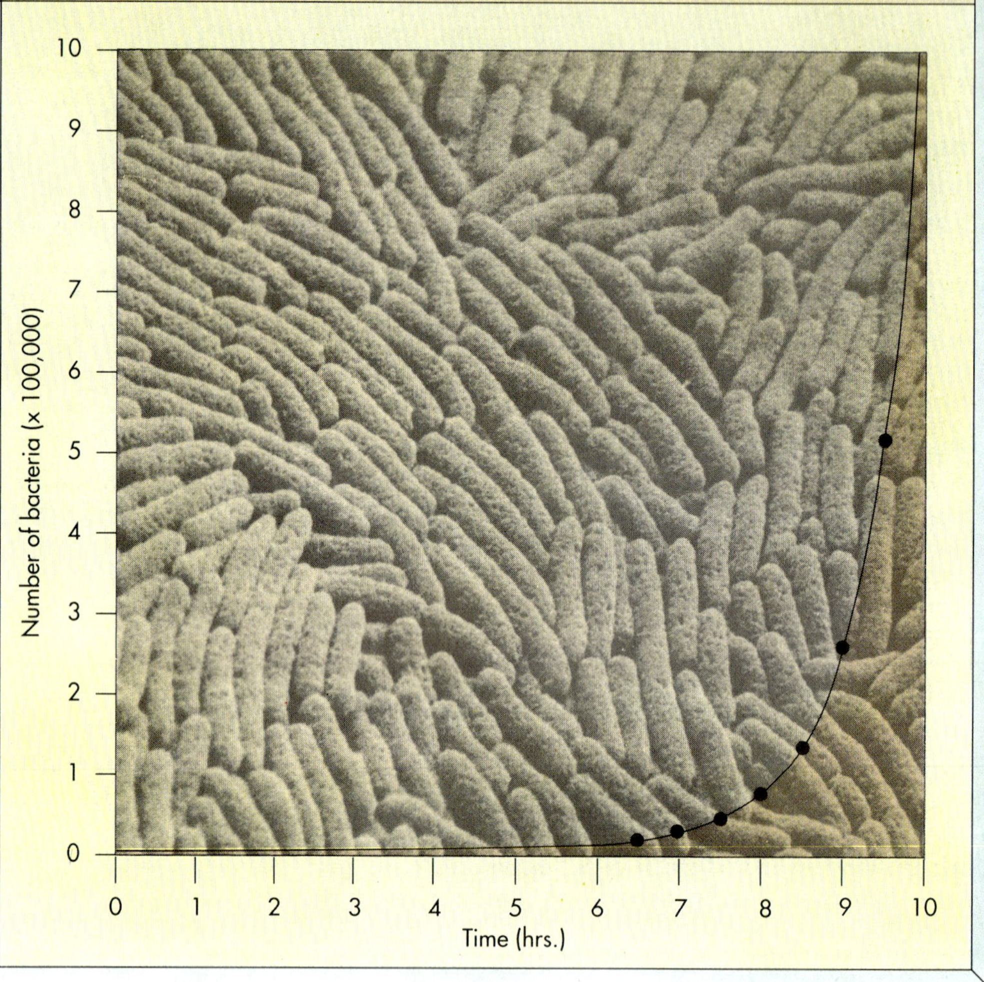

FIGURE 36-1 Exponential growth in a population of bacteria.
This period of rapid exponential growth can only be sustained as long as resources are abundant.

Figure 36-1 illustrates this principle with a population of bacteria in which each individual divides into two every half hour. The rate of increase remains constant, but the actual increase in the number of individuals accelerates rapidly as the size of the population grows. This type of mathematical progression found in the growth pattern of bacteria is termed **exponential growth.** (For example, 2 cells split to form 2^2 or 4, 4 become 2^3 or 8, and so on. The number, or power, to which 2 is raised is called an *exponent.*) Exponential growth refers to the rapid growth in numbers of a population of any species of organism, even though most organisms reproduce sexually and one organism does not split into two "new" organisms.

A period of exponential growth can occur only as long as the conditions for growth are ideal. In nature, exponential growth often takes place when an organism begins to grow in a new location having abundant resources. Such a situation occurred when the prickly pear cactus was introduced into Australia from Latin America. The species flourished, overrunning the ranges. In fact, the cactus became so abundant that cattle were unable to graze (Figure 36-2, *A*). Scientists regulated the population by introducing a cactus-eating moth to the area. The larvae of the moth fed on the pads of the cactus and rapidly destroyed the plants. Within relatively few years, the moth had reduced the population; the prickly pear cactus became rare in many regions where it was formerly abundant (Figure 36-2, *B*).

A B

FIGURE 36-2 A cactus takes over Australia. After an initial period in which prickly pear cacti, introduced from Latin America, choked many of the pastures of Australia with their rampant growth, they were controlled by the introduction of a cactus-feeding moth from the areas where the cacti were native. **A** An infestation of prickly pear cacti in scrub in Queensland, Australia, in October 1926. **B** The same view in October 1929, after the introduction of the cactus-feeding moth.

Carrying capacity

No matter how rapidly a population may grow under ideal conditions, however, it cannot grow at an exponential rate indefinitely. As a population grows, each individual takes up space, uses resources such as food and water, and produces wastes. Eventually, shortages of important growth factors will limit the size of the population. In some populations such as bacteria, a buildup of toxic wastes may also limit population growth. Ultimately, a population stabilizes at a certain size, called the **carrying capacity** of the particular place where it lives. The carrying capacity is the number of individuals within a population that can be supported within a particular environment for an indefinite period. A population actually rises and falls in numbers at the level of the carrying capacity but tends to be maintained at an average number of individuals (Figure 36-3). The exponential growth of a population and its subsequent stabilization at the level of the carrying capacity is represented by an S-shaped **sigmoid growth curve** (after the Greek letter *sigma*).

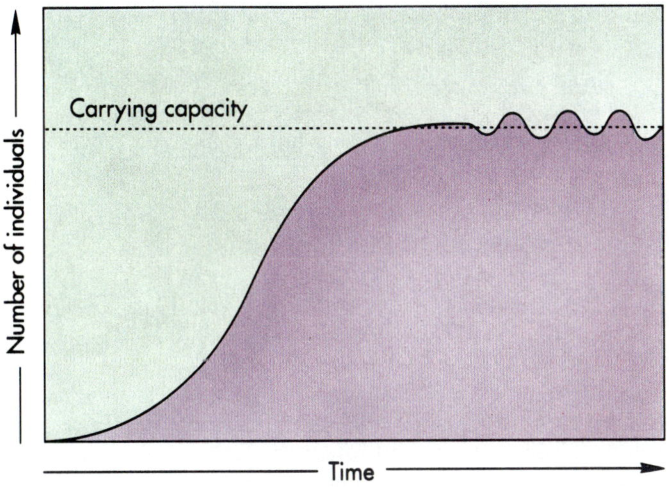

FIGURE 36-3 The sigmoid growth curve.
The sigmoid growth curve, characteristic of biological populations, begins with a slow period of growth quickly followed by a period of exponential growth. When the population approaches its environmental limits, the growth begins to slow down, and it finally stabilizes, fluctuating around the maximum number of individuals that the environment will hold. This maximum number is the carrying capacity.

Under ideal conditions, populations grow at an exponential rate and show some stability in size at the carrying capacity of that place for that species.

Population size

The size of a population has a direct bearing on its ability to survive. Very small populations are less able to survive than large populations and are more likely to become extinct. Random events or natural disturbances can wipe out a small population, whereas a large population—simply due to its larger numbers—is more likely to have survivors. Inbreeding, reproduction between closely related individuals, is also a negative factor in the survival of small populations. Inbreeding tends to produce many homozygous offspring (see Chapter 22), which results in the expression of many recessive deleterious traits that are usually masked by dominant genes. In addition, inbreeding reduces the level of variability in the gene pool (the genes of all breeding individuals) of the population, detracting from the population's ability to adjust to changing conditions. Loss of genetic diversity therefore increases the probability of extinction of that species.

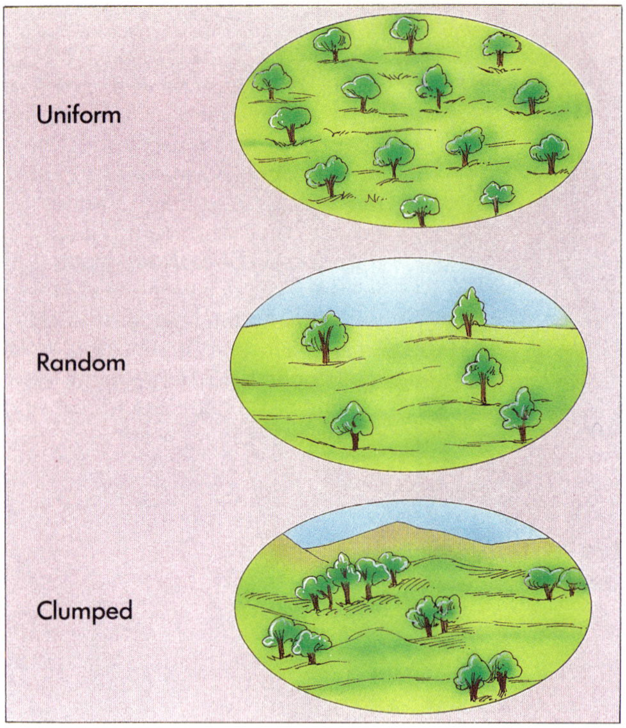

FIGURE 36-4 Distribution patterns in populations.
Uniform distribution patterns are rare in nature but are sometimes caused by allelopathy, in which plants secrete toxic chemicals that discourage the growth of other plants near them. Random patterns in plants are often caused by random seed dispersal, such as by wind. Clumped patterns, the most common type, result in an environment in which resources are unevenly distributed.

Population density and dispersion

In addition to a population's size, its **density**—the number of organisms per unit of area—influences its survival. For example, if the individuals of a population are spaced far from one another, they may rarely come into contact. Sexually reproducing animals cannot produce offspring if they do not mate. Therefore the future of such a population may be limited even if the absolute numbers of individuals over a wide area are relatively high.

A factor related to population density is **dispersion,** the way in which the individuals of a population are arranged. In nature, organisms within a population may be distributed in one of three different patterns: uniform, random, and clumped (Figure 36-4). Each of these patterns reflects the interactions between a given population and its environment, including the other species that are present.

Uniform, or evenly spaced, distributions are rare in nature. Populations of plants exhibiting allelopathy, the secretion of toxic chemicals that harm other plants, often show a uniform distribution. For example, the creosote bush, often the dominant vegetation covering wide areas of the deserts of Mexico and the southwestern United States, grows well spaced and evenly dispersed. This uniform pattern of distribution is probably due to chemicals secreted by the bush that retard the establishment of other individuals near established ones.

Random distributions occur if individuals within a population do not influence each other's growth and if environmental conditions are uniform—that is, if the resources necessary for growth are distributed equally throughout the area. Random distributions are often seen in plants as the result of certain types of seed dispersal, such as scattering by the wind.

Clumped distributions are by far the most frequent in nature. Organisms that show a clumped distribution are close to one another but far from others within the population. Clumping occurs as a result of the interactions among animals, plants, microorganisms, and unevenly distributed resources in an environment. Organisms are found grouped in areas of the environment that have the resources they need. Furthermore, animals often congregate for a variety of other reasons, such as for hunting, mating, or caring for their young; territorial behavior (see Chapter 35) is a means by which some animals group themselves for such purposes.

Regulation of population size

As a population grows and its density increases, competition among organisms for resources such as food, shelter, light, and mating sites increases and toxic waste products accumulate. Factors resulting from the growth of a population *regulate its subsequent growth,* and are "nature's way" of keeping the population size of every species in check. Such factors increase in effectiveness as population density increases and are appropriately termed **density-dependent limiting factors.** Other factors such as the weather, availability of soil nutrients, and physical disruptions of an area (such as volcanoes or earthquakes) can also limit the growth of a population. Because these factors operate regardless of the density of a population, they are called **density-independent limiting factors.**

Density-dependent limiting factors come into play when a population size increases in a given area; density-independent limiting factors operate regardless of population size.

Density-independent limiting factors

A variety of environmental conditions can limit populations. For example, freak snowstorms in the Rocky Mountains of Colorado in the summer can kill butterfly populations there. The size of insect populations that feed on pollen and flower tissues vary seasonally with the blooming of flowering plants. Humans, too, can affect the sizes of populations. Poachers have killed so many African elephants for their ivory, for example, that the species may become extinct.

Density-dependent limiting factors

Individuals within a species and individuals of differing species **compete** for the same limited resources. (Chapter 37 discusses interspecific competition in detail and points out that competition among different species of organisms is

The urban explosion

BIOLOGY TECHNOLOGY & society

Cities are a relatively new phenomenon. Villages and towns were first organized only about 5000 years ago, long after the development of agriculture. The first great cities of Mesopotamia date soon after that, but relative to a modern metropolis, the cities of ancient times were small. Only in the last thousand years have people lived in cities containing hundreds of thousands of people. Until the Industrial Revolution, only one person in five lived in a town of over 10,000 people. In the past hundred years, however, there has been a mass influx of people into cities; 14% of the world's population was urbanized in 1900. Starting in the industrial north, this mass migration has since spread to the less developed nations of the tropics. The cities of the Third World have swollen to hold one third of its total population. In 1920 the world's urban population was 360 million. By the year 2000, it will be near 3 billion—over 50% of the world's population will be urbanized. Two thirds of all people will be in cities of 100,000 or more, and over 20% in cities with more than a million people.

This massive movement of people is alarming because it is difficult to manage so many people crowded together. Mexico City, the largest city in the world today, is plagued by smog, traffic, waste disposal, and other problems, all worsened by the incredible congestion of over 20 million inhabitants (Figure 36-A). The prospects of supplying adequate food, water, and sanitation to these people, in a city whose population will increase to over 30 million by the end of the century, are almost unimaginable.

Nor is Mexico City's astonishing urban growth unique. Only 7 cities had a population larger than 5 million in 1950; by 2000 there will be 57 such megacities, 42 of them in the Third World. New York City has almost 16 million inhabitants, and Tokyo and São Paulo are bigger than New York! These four cities alone contain over 70 million inhabitants—half as many as the entire human population in the year 1 AD. In 1992 there were 16 cities with populations greater than 10 million. By 2000 there will be 25 such cities—and all but 3 of these will be in the northern hemisphere.

Most major Third World cities are really two cities: Shanty towns ring the outskirts of most megacities—in Calcutta they house 67% of the city's population. What draws the people to the city? Urban incomes in the northern hemisphere average three times rural incomes, and modern services such as doctors, teachers, sanitation, clean water, and electricity are at least within reach. Appalling though conditions are, they can be even worse in the countryside. As population numbers swell and resources become even more stretched, the difference between urban and rural becomes greater and the mass movement to the cities intensifies even more.

FIGURE 36-A Mexico City, the world's largest city.

Management of cities will be a priority for governments in the twenty-first century. Environmentally sound ways of disposing of waste, provision of livable housing, and the supply of safe food and water to the multitudes living in large urban areas are just some of the concerns that will be at the forefront of city management.

often greatest between those that obtain their food in similar ways.) In fact, if two species are competing with one another for the same limited resource in a specific location, the species able to use that resource most efficiently will eventually eliminate the other species in that location. This concept is called the principle of **competitive exclusion.** In fact, competition among organisms of the same and differing species was described by Charles Darwin as resulting in natural selection and survival of the fittest or most well-adapted organisms. Competition, therefore, not only limits the sizes of populations but is one of the driving forces of evolutionary change.

Predation is another factor that limits the size of populations and works most effectively as the density of a population increases. Predators are organisms of one species that kill and eat organisms of another—their prey. The intricate interactions between predators and prey are an essential factor in the maintenance of diverse species living in the same area. By controlling the levels of some species, the predators make the continued existence of other species in that same community possible. In other words, by keeping the numbers of individuals of some of the competing species low, the predators prevent or greatly reduce competitive exclusion. In fact, a given predator may very often feed on two or more different kinds of plants or animals, switching from one to the other as their relative abundance changes. Similarly, a given prey species may be a primary source of food for increasing species of predators as it becomes more abundant, a factor that will limit the size of its population automatically.

Parasitism also limits the size of populations by weakening or killing host organisms. Parasites live on or in larger species of organisms and derive nourishment from them. As a population increases in density, parasites such as bacteria, viruses, and a variety of invertebrates can more easily move from one organism to another, infecting an increasing proportion of a population. Once again, this limiting factor to population size acts in negative feedback fashion, becoming more effective as the density of the population increases.

Three density-dependent limiting factors are competition, predation, and parasitism.

Mortality and survivorship

A population's growth rate depends not only on the availability of needed resources and on the ability of its individuals to survive and compete effectively for those resources but also on the ages of the organisms in it. Interestingly, when a population lives in a constant environment for a few generations, its **age distribution**—the proportion of individuals in the different age categories—becomes stable. This distribution, however, differs greatly from species to species and even to some extent within a given species from place to place.

Scientists express the age distribution characteristics of a population by means of a survivorship curve. **Survivorship** refers to the proportion of an original population that survives to a certain age. The curve is developed by graphing the number of individuals within a population that survive through various stages of the life span. Samples of survivorship curves that represent certain types of populations are shown in Figure 36-5.

In the hydra, individuals are equally likely to die at any age, as indicated by the straight survivorship curve (type II). This type of survivorship curve is characteristic of organisms that reproduce asexually, such as hydra, bacteria, and asexually reproducing protists. Oysters, on the other hand, produce vast numbers of offspring, but few of these offspring live to reproduce. The rate of death, or **mortality,** of organisms that reach reproductive age is extremely low (type III survivorship curve). This type of survivorship curve is characteristic of organisms producing offspring that must survive on their own and therefore die in large numbers when young because of predation or their inability to acquire the resources they need. The survivorship curve for humans and other large vertebrates is much different from that for hydra and oysters. Humans, for example, produce few offspring but protect and nurture them; therefore most humans (except in areas of great poverty, hunger, and disease) survive past their reproductive years (type I survivorship curve).

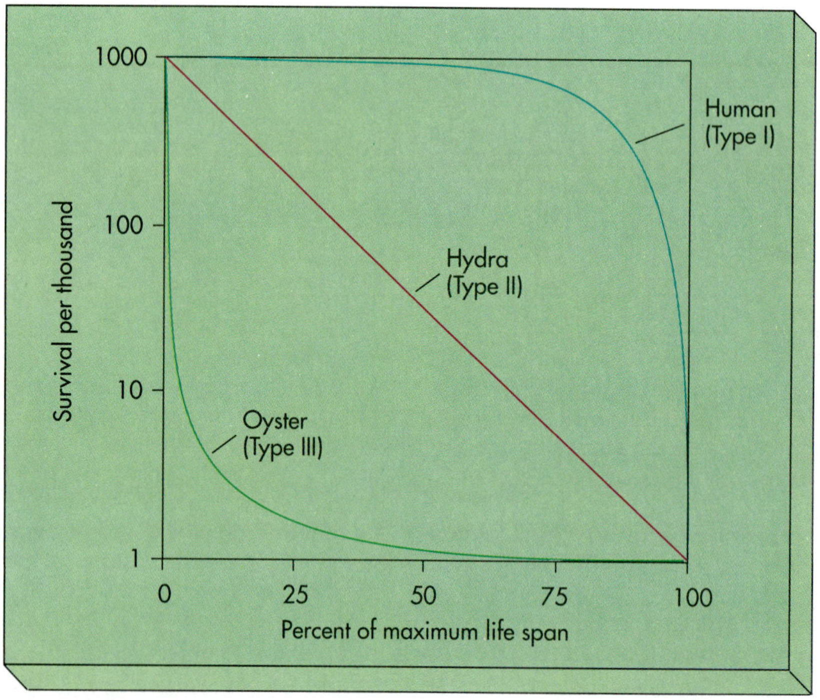

FIGURE 36-5 Survivorship curves.
The shapes of the respective curves are determined by the percentages of individuals in populations that die at different ages.

Many animal and protist populations in nature have survivorship curves that lie somewhere between those characteristic of type II and type III. Many plant populations, with high mortality at the seed and seedling stages, have survivorship curves close to type III. Humans have probably approached type I more and more closely through the years, with the life span being extended because of better health care and new medical technology.

When a population lives in a constant environment for a few generations, the proportion of individuals in various age categories becomes stable. Characteristic age distributions occur among species and are closely linked in animal populations to parental care for offspring.

Demography

Demography is the statistical study of populations. The term comes from two Greek words: *demos,* "the people" (the same root in the word *democracy*), and *graphos,* "to write." It therefore means the description of peoples and the characteristics of populations. Demographers predict the ways in which the sizes of populations will alter in the future, taking into account the age distribution of the population and its changing size through time.

A population whose size remains the same through time is called a **stable population.** In such a population, births plus immigration exactly balances deaths plus emigration. In addition, the number of females of each age group within the population is similar. If this were not the case, the population would not remain stable. For example, if there were many more females entering their reproductive years than older females leaving the population, the population would grow.

The age distribution of males and females in human populations of Kenya, the United States, and Austria is shown in Figure 36-6. A **population pyramid** is a bar graph that shows the composition of a population by age and sex. Males are conventionally enumerated to the left of the vertical age axis and females to the right. By using population pyramids, scientists can predict the future size of a population. First, the number of females in each age group is multiplied by the average number of female babies women in that age group bear. These numbers are added for each age group to see whether the new number will exceed, equal, or be less than the number of females in the population being studied. By such means the future growth trends of the human population as a whole and of individual countries and regions can be determined.

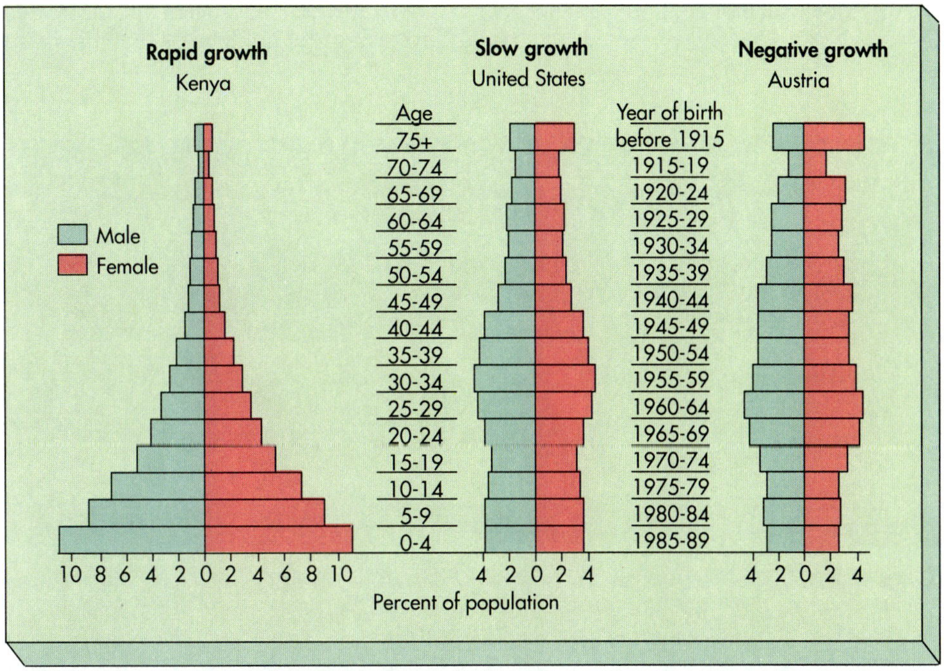

FIGURE 36-6 Three patterns of population change.
Kenya shows a rapid growth pattern—its pyramid has a broad base because of high fertility. The United States and Austria both show declining fertility and mortality, but the United States' "baby boom" has resulted in higher fertility rates than Austria, whose population is declining.

The population pyramids in Figure 36-6 show the differences in the pattern of a rapidly growing population (Kenya), a slowly growing population (the United States), and a country experiencing negative growth (Austria). The population pyramid of Kenya is characteristic of **developing countries,** those that have not yet become industrialized, such as countries in Africa, Asia, and Latin America. Each of these countries has a population pyramid with a broad base, reflecting the large numbers of individuals yet to enter their reproductive years. In Kenya, for example, 50% of the population is under the age of 15. These children will reach reproductive age in the near future, leading to a population explosion in Kenya over the next decade.

In the United States, birth rates are higher than death rates at present, producing a growth rate of approximately 0.7%. The high birth rate is not due to couples having large families but to the large size of the "baby boom" generation who are at the peak of their reproductive years. Individuals in this age group were born within the 20 years or so following World War II. The large number of women in this group causes the births to still outnumber the deaths. Austria and the United States are both experiencing a decline in fertility and mortality. However, Austria's population does not include as high a percentage of women in their childbearing years as the United States, so that deaths are outnumbering births.

Developing countries, those that have not yet become industrialized, have proportionately young populations and are experiencing rapid growth. Developed countries have populations with similar proportions of their populations in each age group, and are growing very slowly or not at all.

The human population explosion

Although some countries have populations that are no longer growing, such as Denmark, West Germany, Hungary, and Italy, and some countries such as Austria are declining in numbers, the population of the world as a whole is growing at the rate of 1.8% a year. This growth rate may sound low, but with the world population numbering over 5.3 billion, it adds over 95 *million* people to the population each year. Scientists estimate that the world population will double between 1990 and 2029!

How did the human population reach its present-day size? With the development of agriculture 11,000 years ago (Figure 36-7), human populations began to grow steadily (Figure 36-8). Villages and towns were first organized about 5000 years ago, and human effects on the environment began to intensify. In these centers of civilization, however, the specialization of professions such as metallurgy became possible; technology advanced. By 1650, the world population totaled approximately 500 million people. The Renaissance in Europe, with its renewed interest in science, ultimately led to the establishment of industry in the seventeenth century and to the Industrial Revolution of the late eighteenth and early nineteenth centuries. By the mid-nineteenth century, Louis Pasteur put forth the germ theory of disease, the understanding that microbes caused infection. With this understanding came new medical technology and discoveries. In the 1920s, Alexander Fleming accidentally discovered penicillin and opened the door to antibiotic therapy—medicine's "magic bullets" against bacterial infection. These medical advancements decreased the death rate by increasing the number of individuals surviving infection.

FIGURE 36-7 The development of agriculture was a key step in the growth of human populations. By producing abundant supplies of food, agriculture made the growth of cities and the future development of human culture possible. These workers are using a primitive method to thresh sorghum in southern India.

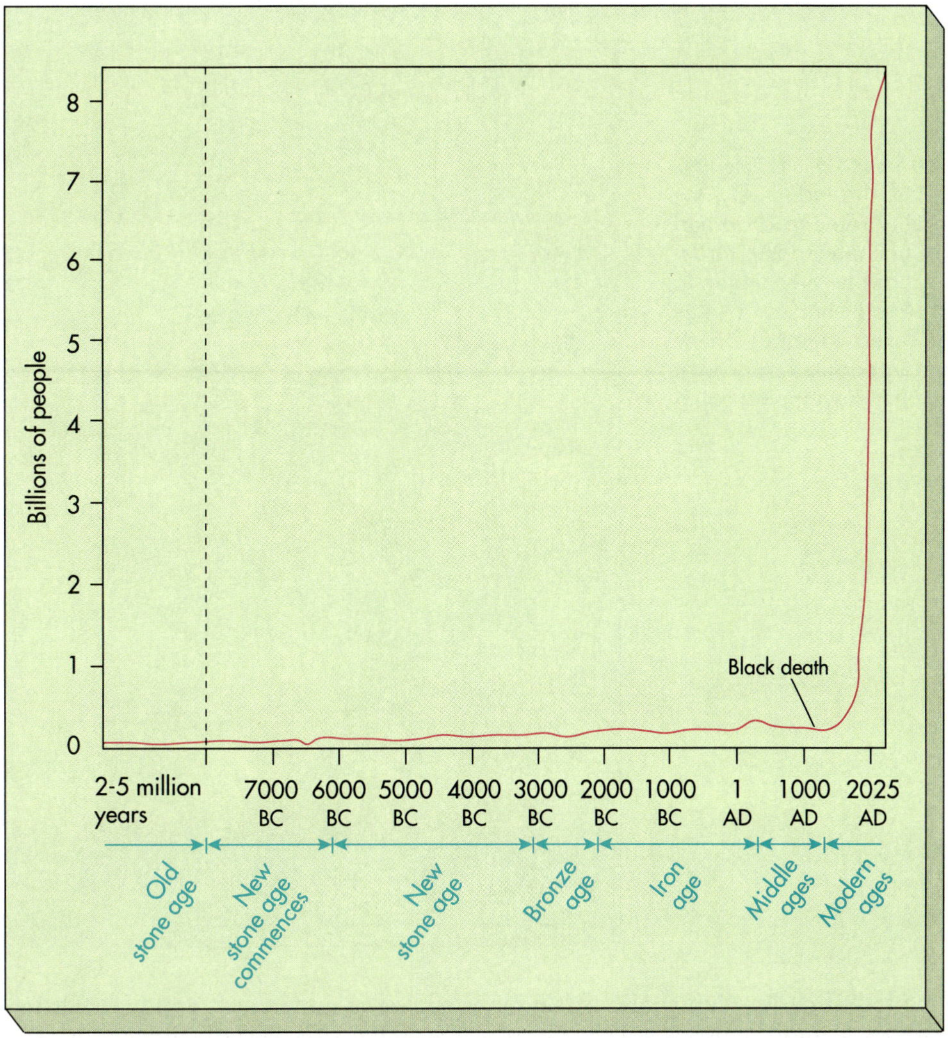

FIGURE 36-8 World population growth through history.
Except for the blip caused by the Black Death (25 to 35 million people died in Europe alone), the human population has grown steadily since the Stone Age.

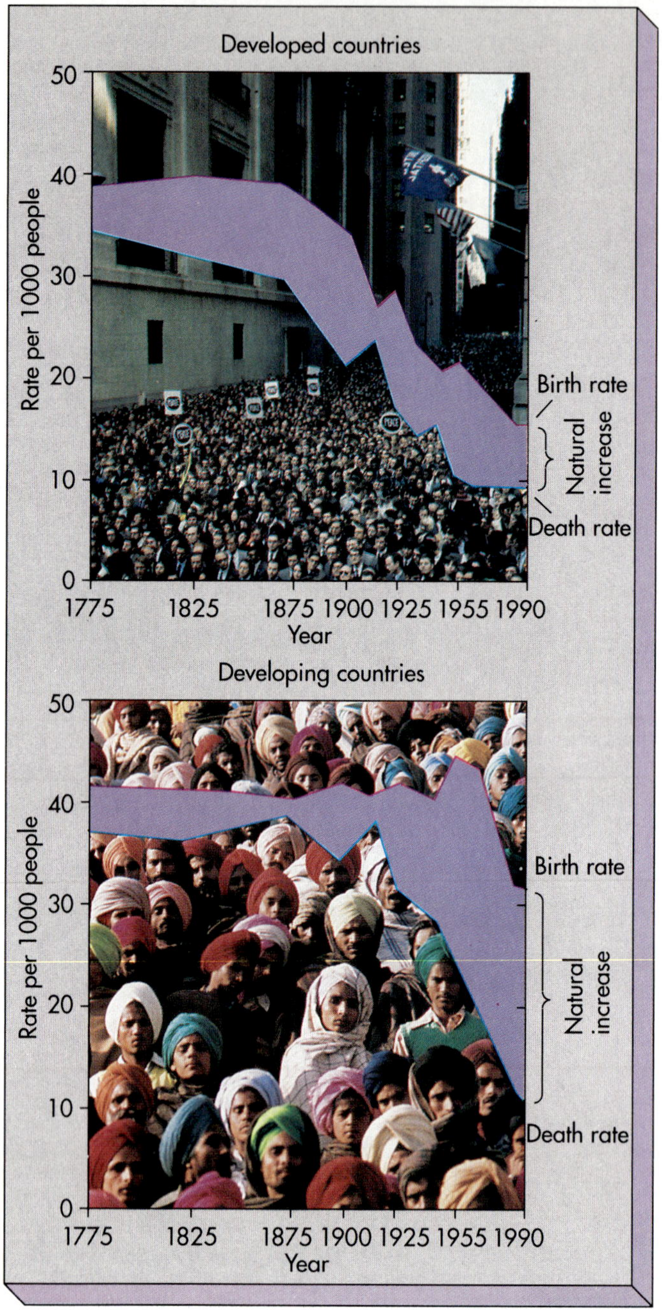

FIGURE 36-9 Population growth from 1775 to 1990. In developed countries, the Industrial Revolution caused mortality rates to drop and birth rates to remain unchanged **(A)**. In developing countries, mortality rates began to decline after World War II, but birth rates remained high **(B)**. Taken together, these changes caused human population growth rates to soar, especially in developing countries.

The advent of the Industrial Revolution also heralded new farming and transportation technology, which helped provide better nutrition for many people, especially those in industrialized countries. With better nutrition and increased medical understanding and technology, the death rate fell steadily and dramatically from the mid-nineteenth century on (Figure 36-9, *A*). In developing countries, international foreign aid imported this new technology along with food aid after World War II. The mortality rate plunged in a matter of years (Figure 36-9, *B*). However, birth rates remained largely unchanged, and as a result, the world's population growth rate soared. In fact, at their present rates of growth, the population of tropical South America will double in 31 years and that of tropical Africa will double in only 24 years. By the year 2000, about 60% of the people in the world will be living in countries that are at least partly tropical or subtropical, 20% will be living in China, and the remaining 20% will be in the developed countries of Europe, the Commonwealth of Independent States (formerly the Soviet Union), Japan, the United States, Canada, Australia, and New Zealand together (Figure 36-10). Putting it another way, for every person living in an industrialized country like the United States in 1950, there were two people living elsewhere; by 2020, just 70 years later, there will be five.

The world population rose sharply and dramatically after the Industrial Revolution because of new technology in agriculture, transportation, industry, and medicine.

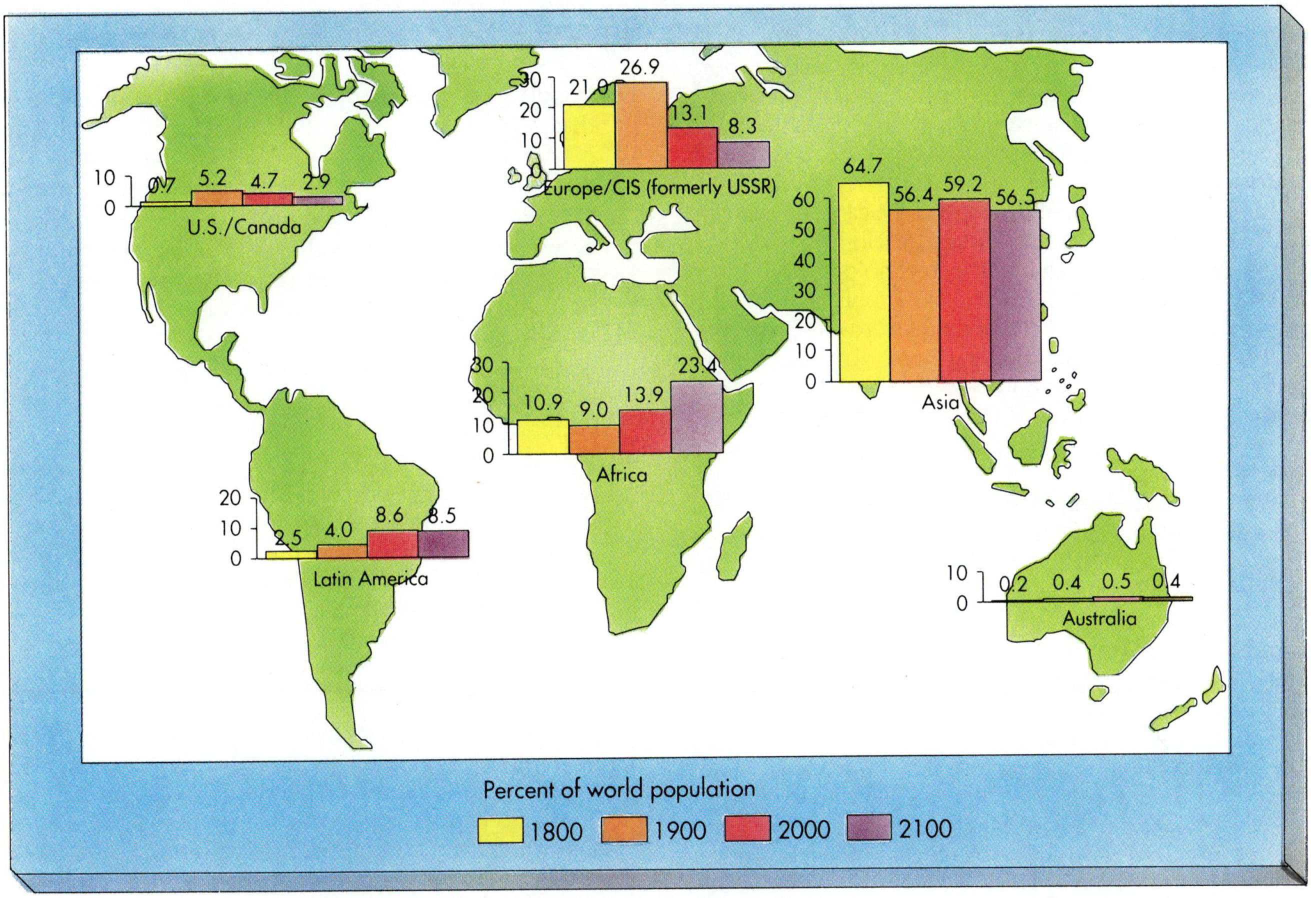

FIGURE 36-10 World population distribution by region, 1880 to 2100.
If current trends continue, Asia will have 57% of the total world population in 2100, Africa nearly a quarter, and Europe's share will drop to less than 10%.

Of the estimated 2.8 billion people living in the tropics in the late 1980s, the World Bank estimated that about 1.2 billion people were living in absolute poverty. (The World Bank is an international bank that provides loans and technical assistance for economic development projects in developing member countries.) These people cannot reasonably expect to be able to consistently provide adequate food for themselves and their children. Even though experts estimate that enough food is produced in the world to provide an adequate diet for everyone in it, the distribution is so unequal that large numbers of people live in hunger. The United Nations Children's Fund (UNICEF) estimates that in the developing world, about 14 million children under the age of 5—40,000 per day—starve to death each year, mainly of malnutrition and the complications associated with it (Figure 36-11).

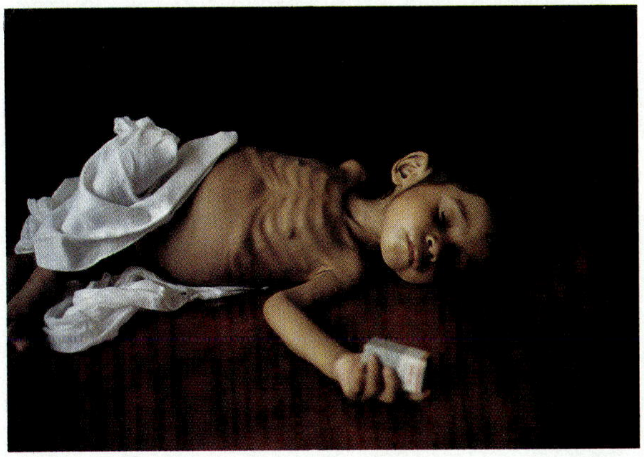

FIGURE 36-11 The face of starvation.
This Colombian child suffers from malnutrition. UNICEF estimates that 14 million children die each year as a consequence of malnutrition.

The size of human populations, like those of other organisms, is or will be controlled by the environment. Early in its history, human populations were regulated by both density-dependent and density-independent limiting factors, including food supply, disease, and predators; there was also ample room on Earth for migration to new areas to relieve overcrowding in specific regions. In the past century, however, humans have been able to expand the carrying capacity of the Earth because of their ability to develop technological innovations. Gradually, changes in technology have given humans more control over their food supply and enabled them to develop superior weapons to ward off predators, as well as the means to cure diseases. Improvements in transportation and housing have increased the efficiency of migration. At the same time, improvements in shelter and storage capabilities have made humans less vulnerable to climatic uncertainties.

As a result of the ability to manipulate these factors, the human population has been able to grow explosively to its present level of more than 5.3 billion people. Both the current human population level and the projected rate of growth have potential consequences for the future that are extremely grave. Pressures are placed on the land, water, forests, and other natural resources. Industrialization, although raising the standard of living by increasing the availability of goods and services, adds to air and water pollution. In the developed countries, nations of consumers have developed into "throwaway" societies, adding billions of tons of solid waste to landfills every year.

The most effective means of dealing with the population explosion has been the support of governments to encourage small families, the establishment of family planning clinics, improvement in education, and socioeconomic development. The developed countries of Western Europe and North America, Japan, and Australia have very low birth rates at this time. Mexico, Indonesia, Thailand, South Korea, Hong Kong, and Singapore have had considerable success with lowering their birth rates. Countries such as China and India have had some success in reducing their birth rates but are still striving toward this goal. The countries in sub-Sahara Africa have the highest birth rate of any countries in the world—some as high as 4.1%. Family planning programs are now being implemented across the continent.

Summary

1. Populations consist of the individuals of a given species that occur together at one place and at one time. They may be dispersed in an evenly spaced, clumped, or random manner. Clumped patterns are the most frequent.
2. The rate of growth of any population is the difference between the birth rate and the death rate per individual per unit of time. The actual rate is affected by emigration from the population and immigration into it.
3. Most populations exhibit a sigmoid growth curve, which implies a relatively slow start in growth, a rapid increase, and then a leveling off when the carrying capacity of the species' environment is reached.
4. Survivorship curves are used to describe the characteristics of growth in different kinds of populations. Type I populations are those in which a large proportion of the individuals approach their physiologically determined limits of age. Type II populations have a constant mortality throughout their lives. Type III populations have very high mortality in their early stages of growth, but an individual surviving beyond that point is likely to live a very long time.
5. Each population grows in size until it eventually reaches the limits of its environment to support it; resources are always limiting. Some of the limits to the growth of a population are related to the density of that population, but others are not. Competition within the populations and between populations of any two species limits their coexistence. Predation and other forms of interaction among populations also play an important role in limiting population size.
6. Agriculture was developed in several centers about 11,000 years ago when about 5 million people lived throughout the world. By 1990, the global population was more than 5.3 billion people and was growing at a rate of 1.8% per year, a rate that will double the population in approximately 39 years.
7. In the 1980s, the World Bank estimated that about 1.2 billion of the more than 2.75 billion people living in the tropics and subtropics existed in absolute poverty. Some 4 million children under 5 years of age were starving to death each year. With the populations of tropical countries growing from 1.8% to 4.1% per year, the task of feeding these people will be extraordinarily difficult.

REVIEW QUESTIONS

1. What is ecology?
2. Fill in the blanks, and explain what this equation means: __________ __________ = (births + immigrants) − (deaths + emigrants)
3. Fill in the blanks: Under ideal conditions, a population grows rapidly (__________ __________) until it stabilizes at a certain size called the __________ __________ of its particular environment.
4. Which is more likely to survive, a small population or a large one? Why?
5. Distinguish between density and dispersion. How does each affect a population's chances for survival?
6. Identify the three patterns of population dispersion found in nature. Into which pattern do human populations fall?
7. Distinguish between density-dependent and density-independent limiting factors. Give an example of each.
8. What do predation and parasitism have in common? How do they differ?
9. Draw type I, type II, and type III survivorship curves. Summarize the types of organisms that are characteristic of each curve, and give an example of each.
10. Compare the typical population pyramids of a developing country and industrialized country.
11. Fill in the blanks: The population of the world as a whole is growing at an annual rate of __________%, adding over __________ people to the population each year. Scientists estimate the world population will __________ between 1990 and 2029.
12. Both the current human population level and the projected rate of growth worldwide have potential consequences for the future that are extremely grave. Explain why.
13. Suppose that you were given the political power to deal with the world's population explosion. What steps would you take?

KEY TERMS

age distribution 630
carrying capacity 627
competitive exclusion 629
demography 631
density 628
density-dependent limiting factors 628
density-independent limiting factors 628
developing countries 632
dispersion 628
emigration 625
exponential growth 626
growth rate 625
immigration 625
mortality 630
parasitism 630
population ecologist 625
population pyramid 631
predation 629
sigmoid growth curve 627
stable population 631
survivorship 630

THOUGHT QUESTIONS

1. No other primate species has undergone a population explosion like humans. Why do you think humans alone have not controlled their population growth?
2. If female elephants have a single baby every 5 years and live on the average 60 years, how long would it take under ideal conditions for an elephant population to grow from 1000 individuals to 1 million individuals? Why does this not happen?
3. Much of the world's future seems to depend on bringing the world's human population growth rate under control. What do you think ought to be done?

FOR FURTHER READING

Ehrlich, P. R., & Roughgarden, J. (1987). *The science of ecology.* New York: Macmillan Publishing Company.
An excellent account of the entire field of ecology, this book is especially useful in defining the dynamics of populations.

Keyfitz, N. (1989, September). The growing human population. *Scientific American,* pp. 119-126.
This article discusses the general issues surrounding the burgeoning human population.

Population Reference Bureau. (Annual). *World population data sheet.* Washington, DC: Population Reference Bureau.
This annual data sheet provides the latest population statistics from around the world.

37 Interactions within communities of organisms

HIGHLIGHTS

A community is made up of populations of species that interact with one another.

Each organism plays a special role within a community; this role is its niche.

Members of the various species within a community interact by competing with one another for resources, killing one another for food, and living in close associations that both benefit and harm one another; these interactions affect the niche that a species can occupy in that community.

Communities, like species, change over time in the dynamic process of succession.

OUTLINE

Ecosystems
Communities
Types of interactions within communities
- Competition
  - The study of competition in the laboratory
  - The study of competition in nature
- Predation
  - Plant-herbivore coevolution
  - Protective coloration
- Symbiosis
  - Commensalism
  - Mutualism
  - Parasitism

Changes in communities over time: Succession

Although this photograph may look like an aerial view of a volcano-riddled landscape, these "volcanoes" will never compete with Krakatoa or Mount St. Helens. In fact, they are not volcanoes but tiny marine organisms living on a seashore rock. Each mound is an individual whose cone-like shell is toughened by lime and whose underside is firmly cemented to the rock. Together they form a population of acorn barnacles—organisms that are relatives of lobsters, crayfish, crabs, and shrimp. The acorn barnacles compete for food and space with a population of colonial invertebrates commonly called *moss animals*. These low-growing organisms look like dimpled areas near some of the barnacles. In addition, one member of a population of snails can be seen at the edge of the photograph. The snails prey on the barnacles. Together, these interacting populations form a community of organisms that transform a lifeless seashore rock into a dynamic ecosystem.

Ecosystems

Scientists have long known that the nonliving or **abiotic** factors within the environment—such as air, water, and even rocks—affect an organism's survival, as do the living or **biotic** factors—such as surrounding plants, animals, and microorganisms. All of the biotic and abiotic factors together within a certain area are called an **ecosystem.** Therefore the rock is as important to the makeup of the ecosystem as the barnacles.

Within an ecosystem, each living thing has a home, an actual area in which it resides. This space, including the factors within it, is an organism's **habitat.** Organisms not only reside in their habitats, they interact with the biotic and abiotic factors within it and use them to survive. Each organism also plays a special role within an ecosystem; this role is called a **niche.** A niche may be described in terms of space, food, temperature, appropriate conditions for mating, and requirements for moisture. A full portrait of an organism's niche also includes the organism's behavior and the ways in which this behavior changes at different seasons and at different times of the day. These concepts are important to the understanding of the concept of communities.

Communities

The interactions among organisms within ecosystems are varied. Individuals of the same species make up a **population** of organisms. (The individuals within a population interact with one another in ways described in Chapters 35 and 36.) Populations also interact with one another, forming **communities.** Thus, an ecosystem can be thought of as a community of organisms, along with the abiotic factors with which the community interacts.

The magnificent redwood forest that extends along the coast of central and northern California and into the southwestern corner of Oregon is an example of a community. Within it, the most obvious organisms are redwood trees (Figure 37-1). However, populations of other organisms, such as the sword fern and the ground beetle in Figure 37-2, make up this community too. The coexistence of these various populations is made possible in part because of the special conditions that are created by the redwoods: shade, water (dripping from the branches), and relatively cool temperatures. For this reason and because the redwoods visually dominate the area, this distinctive group of populations is the redwood community.

Individuals of a species compose a population; many populations make up a community. These populations of living things interact with one another in many complex ways.

FIGURE 37-1 The redwood community.
The redwood forests in coastal California and southwestern Oregon are dominated by the redwoods themselves.

FIGURE 37-2 Plants that occur regularly in the redwood community.
A The redwood sorrel.
B A ground beetle is feeding on a slug on a leaf of sword fern.

Types of interactions within communities

Interactions within communities can be grouped into the following three categories:

1. **Competition:** Organisms of different species that live near one another strive to obtain the same limited resources.
2. **Predation:** An organism of one species kills and eats an organism of another. (Animals that kill and eat members of their own species are cannibals.)
3. **Symbiosis:** Two or more organisms of different species live together in close association over a prolonged time (Figure 37-3).

Scientists study the interactions among organisms and between organisms and their environments in the laboratory and in nature. This specialized field of biology is called **ecology;** the scientists who work within this field are ecologists. Ecologists study the physical and biological variables governing the distribution and growth of living things. Ecologists also study the theoretical bases of these interactions, and some use computers to develop mathematical models of ecological systems. The knowledge gained by ecologists is essential to the basic understanding of the world and provides a foundation to find solutions to the many environmental problems created by humans.

Competition

When two kinds of organisms use the same limited resource for survival, they compete for that resource. Complex animals such as vertebrates compete by using innate behaviors such as threat displays and territorial behavior (see Chapter 35). Lower forms of animals and plants do not exhibit complex behaviors; they compete with one another simply by their adaptive fitness—by reproducing more and better-adapted offspring that crowd out opponents.

Competition among different species of organisms is often greatest between organisms that obtain their food in similar ways. Thus green plants compete mainly with other green plants, meat-eating animals with other meat-eating animals, and so forth. Competition within a genus or between individuals of the same species occurs as well.

> **Organisms of different species compete with one another when they need the same limited resource for survival. This competition is greatest among organisms that obtain their food in similar ways.**

Competition among organisms has been observed by scientists for a long time. Over 50 years ago, the Russian scientist G.F. Gause formulated the principle of **competitive**

FIGURE 37-3 Living—and dying—together.
Organisms that live within a community interact in a number of ways.
A Competition. The kudzu vine in the southern United States competes with other plants in the area for resources. However, the kudzu vine seems to be winning out in many locations.
B Predation. These lions have successfully ambushed their dinner.
C Symbiosis. This moray eel shelters a smaller fish in its mouth. The eel benefits from a thorough teeth cleaning, and the smaller fish receives an undisturbed meal.

exclusion, which was based on his experimental work. This principle states that if two species are competing with one another for the same limited resource in a specific location, the species able to use that resource most efficiently will eventually eliminate the other species in that location.

The study of competition in the laboratory

Scientists sometimes study competition between species in the laboratory so that they can control the environmental conditions. John Harper and his colleagues at the University College of North Wales, Australia, for example, performed competition experiments with two species of clover: white clover and strawberry clover. Each species was sown with the other at one of two densities: 36 or 64 plants per square foot. Various plots of the two species were planted, using all the possible combinations of the two densities of plants. One plot contained 36 white clover plants and 36 strawberry clover plants per square foot. Another contained 64 white and 36 strawberry per square foot, and so forth. The white clover initially formed a dense canopy of leaves in each experimental plot. However, the slower-growing strawberry clover, whose leaf stalks are taller, eventually produced leaves that grew above the white clover leaves. In competing more effectively for light, the strawberry clover overcame the white clover, causing it to die out. The outcome was the same, regardless of the initial densities at which the seeds of the plants were sown.

In a competitive situation and under controlled laboratory conditions, the species able to use a particular resource more efficiently will be the species to survive.

The study of competition in nature

As well as competing for sunlight, plants compete for soil nutrients. The roots of one species, for example, may outcompete another species by using up minerals in the soil essential to both species. In addition, one species may secrete poisonous substances that depress the growth of other species. Sage plants, for example, inhibit the establishment of other plant species nearby, producing bare zones around populations of these plants (Figure 37-4). These interactions are poorly understood, but experimental studies are beginning to resolve their complexity.

Another interesting view of competition in nature is demonstrated by the acorn barnacles. Highly adapted to their environment, acorn barnacles are typically found in the intertidal zone of rocky shores—the narrow strip of land exposed during low tide and covered during high tide. When submerged, an acorn barnacle feeds by extending appendages from the hole in its shell. Spread out, these appendages act like a net, sweeping the water and collecting food that it then brings into its shell and eats. When exposed to the air, an acorn barnacle pulls in its feeding appendages and shuts down, actually using much less oxygen than when underwater. Interestingly, barnacles of the genus *Balanus* have been kept out of the water as long as 6 weeks without detectable ill effects. However, a relative of *Balanus* organisms, barnacles of the genus *Chthamalus* (pronounced with the first two letters silent), have been kept out of water for 3 years, being submerged only 1 or 2 days a month—and they survived! Although both organisms have adaptations that make them well suited to the intertidal environment, their differences play an important role in determining where each genus lives. Of the two, *Chthamalus* barnacles live in shallower water, where they are often exposed to air as the tide rolls in and out. *Balanus* barnacles live deeper in the intertidal zone and are covered by water most of the time (Figure 37-5).

In studying these two genera of barnacles, J.H. Connell of the University of California, Santa Barbara, found that in this deeper zone, *Balanus* barnacles could always outcom-

FIGURE 37-4 Competition among plants.
The bare zones around these colonies of sage are plainly visible in this aerial photograph taken in the mountains above Santa Barbara, California. The plants secrete chemicals that suppress the growth of other plants nearby.

FIGURE 37-5 *Chthamalus* and *Balanus* barnacles growing together on a rock.
Competing species belonging to these two genera were studied by J.H. Connell along the coast of Scotland. *Balanus* are the larger organisms of the two genera.

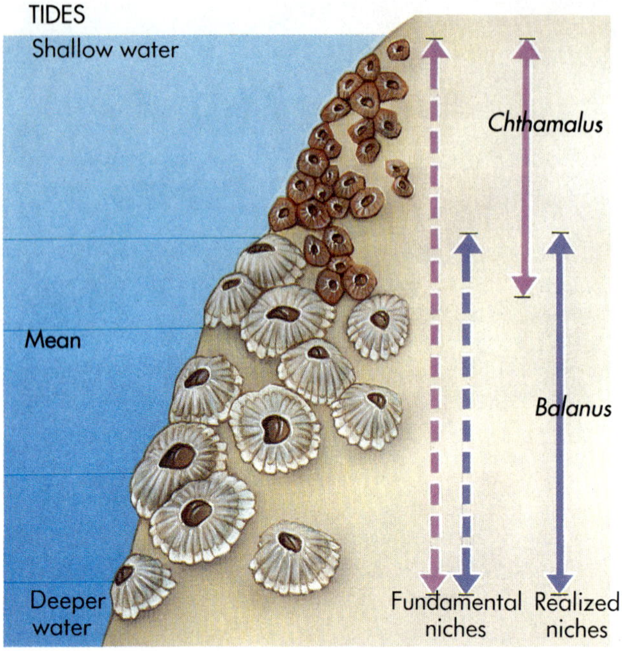

FIGURE 37-6 Competition can limit niche use.
The distribution of *Chthamalus* and *Balanus* barnacles with respect to different water levels is shown. In theory, *Chthamalus* barnacles can live in both deep and shallow zones, whereas *Balanus* organisms can live only in the deeper zone. These are the barnacles' fundamental niches. In reality, however, *Balanus* organisms outcompete *Chthamalus* barnacles in the deeper zone because *Balanus* barnacles use the resources of the deeper zone more efficiently. These are the barnacles' realized niches.

pete *Chthamalus* barnacles. *Balanus* organisms would crowd *Chthamalus* barnacles off the rocks, replacing it even where it had begun to grow. When Connell removed *Balanus* barnacles from the area, however, *Chthamalus* organisms were easily able to occupy the deeper zone, indicating that no physiological or other general obstacles prevented it from becoming established there. *Balanus* barnacles, however, must use the resources of the deeper zone more efficiently than *Chthamalus* organisms do, even though *Chthamalus* barnacles are able to survive there in the absence of its competitor. In contrast, *Balanus* barnacles cannot survive in the shallow-water where *Chthamalus* organisms normally occur. It evidently does not have the special physiological and morphological adaptations that allow *Chthamalus* barnacles to occupy this zone.

Along with illustrating the principle of competitive exclusion, these experiments with *Balanus* and *Chthamalus* barnacles illustrate that the role an organism plays in an ecosystem—its niche—can vary depending on the biotic and abiotic factors in the ecosystem. In this example, the niche occupied by *Chthamalus* barnacles is its **realized niche**—the role it *actually plays* in the ecosystem. It is distinguished from its **fundamental niche**—the niche that it *might* occupy if competitors were not present. Thus the fundamental niche of the barnacle *Chthamalus* in Connell's experiments included that of *Balanus* barnacles, but its realized niche was much narrower because *Chthamalus* organisms were outcompeted by *Balanus* organisms. However, the realized and fundamental niches of *Balanus* barnacles are the same (Figure 37-6).

Gause's principle of competitive exclusion can be restated in terms of niches as follows: no two species can occupy exactly the same niche indefinitely. Certainly, species can and do coexist while competing for the same resources. Nevertheless, Gause's theory predicts that when two species do coexist on a long-term basis, one or more features of their niches will always differ; otherwise the extinction of one species will inevitably result. The factors that are important in defining a niche are often difficult to determine, however, so Gause's theory can sometimes be difficult to apply or investigate.

The role an organism plays in an ecosystem can vary depending on the biotic and abiotic factors in the ecosystem.

Predation

A **predator** is an organism that kills and eats other organisms, or **prey.** Predation includes one kind of animal capturing and eating another, an animal feeding on plants, and even a plant, such as the Venus flytrap shown in Figure 37-7, capturing and eating insects! In a broad sense, parasitism is also considered a form of predation. Parasitism is a close association between two organisms in which the parasite (predator) is much smaller than the prey but feeds on the prey, harming it and benefiting the parasite.

How do predator populations affect prey populations? When experimental populations are set up under very simple conditions in the laboratory, the predator often exterminates its prey and then becomes extinct itself because it has nothing to eat. This fact was illustrated nicely in experiments performed by Gause. In his experiments, Gause used populations of the two protozoans shown in Figure 37-8: *Didinium,* and its prey *Paramecium.* As shown in Figure 37-9, *A,* when *Didinium* protozoans are introduced into a growing population of *Paramecium* protozoans, the population instantly begins to decline and quickly dies out. The *Didinium* population lives on for a short while, then dies out itself.

If refuges are provided for the prey, however, its population can be driven to low levels but can recover. In another of Gause's experiments, he provided sediment in the bottom of the test tubes in which he was growing *Didinium* and *Paramecium* protozoans. Interestingly, as *Didinium* began to prey on *Paramecium,* only those organisms in the clear fluid of the test tubes were killed. Those in the sediment were not eaten. Eventually, *Didinium* protozoans died from lack of food; meanwhile, the *Paramecium* prey multiplied (Figure 37-9, *B*) and overtook the culture!

In another series of experiments, Gause discovered that when he introduced new prey at successive intervals (Figure 37-9, *C*), the decline and rise in the numbers of the predator-prey populations followed a cyclical pattern. As the number of prey increased, the number of predators in-

A

B

C

FIGURE 37-7 Different forms of predation.
A A cheetah eats its kill, an impala, while vultures wait patiently in the background.
B A gazelle feeds on the grasses of the African plain.
C This predatory plant, the Venus flytrap, inhabits the low, boggy ground of North and South Carolina. The Venus flytrap traps insects within its snapping leaves.

A

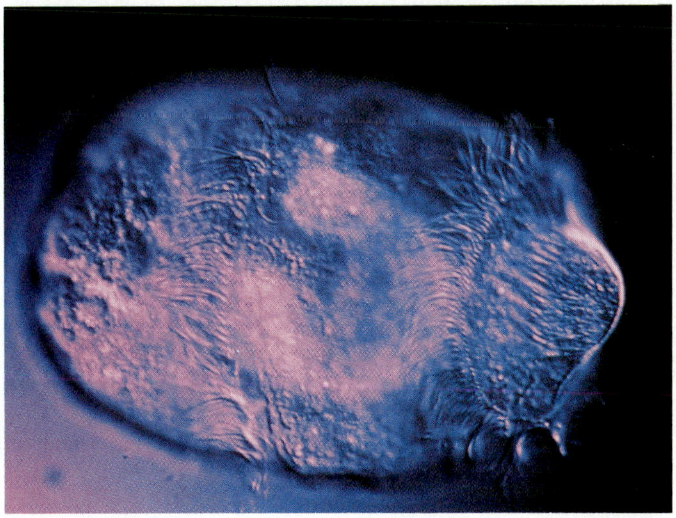

B

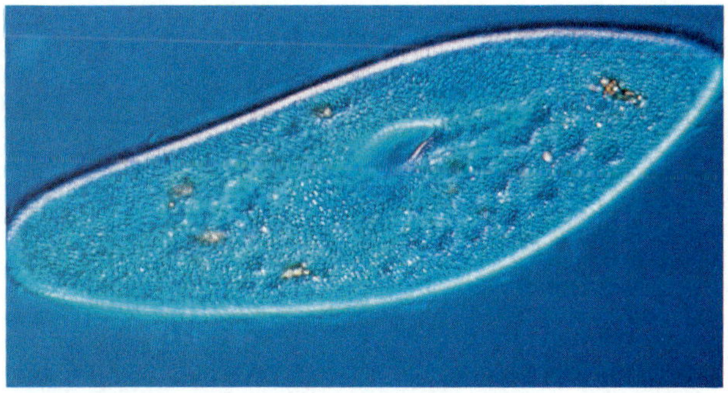

FIGURE 37-8 Gause's experimental subjects.
A *Didinium.*
B *Didinium's* prey, *Paramecium.*

A
B
C

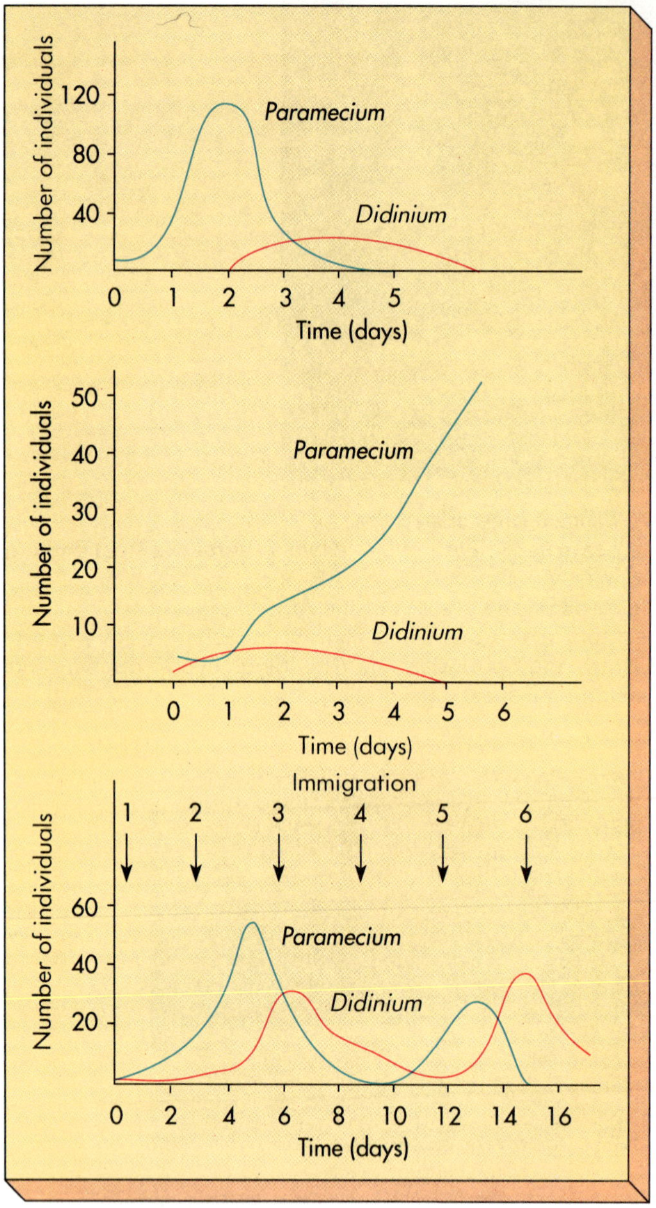

FIGURE 37-9 Outcome of Gause's experiments with *Paramecium* and *Didinium*.

A Results obtained when *Didinium* is introduced into a population of *Paramecium* protozoans.

B Results of the same experiment, only this time sediment was added to the test tubes, providing a place for *Paramecium* protozoans to hide.

C Results from third experiment in which new prey were introduced at intervals. As the number of prey increased, so did the number of predators.

creased. As the numbers of the prey were lowered by predation, the large predator population did not have enough food to eat; some died and the predator population declined. As this decline occurred, the prey recovered (aided by the addition of new organisms) and again became abundant, starting the cycle once again.

At one time, scientists thought that predator-prey populations always cycled in this manner. However, they have come to realize that in nature, conditions for survival are complex and do not always lead to such a cycling of populations. From experiments such as those described, scientists know that predators cannot survive when the prey population is low. Immigration of prey (movement of new prey into the community) may be necessary to sustain the predator population. From other experiments, scientists have come to learn that changes in predator-prey populations also depend on how the prey are dispersed in an area and the manner in which the predator searches for the prey. Factors other than the relationship between a single predator population and a single prey population also influence the survival and abundance of both predator and prey. For example, adverse weather conditions may result in the death of the predator and/or prey species, the predator may eat more than one type of prey and/or the prey may be eaten by more than one predator, and fluctuations may occur in the food source of the prey, limiting the survival of this population.

The effect of predator populations on prey populations is difficult to predict because their complex interactions depend on their interactions with other organisms in the community, movement of new organisms into and out of the community, and the abiotic factors that influence their survival.

The intricate interactions between predators and prey often affect the populations of other organisms in a community. By controlling the levels of some species, for example, predators help species survive that may compete with their prey. In other words, predators sometimes prevent or greatly reduce competitive exclusion by limiting the population of one of the competing species. Such interactions among organisms involving predator-prey relationships are key factors in determining the balance among populations of organisms in natural communities.

Plant-herbivore coevolution

Plants, animals, protists, fungi, and bacteria that live together in communities have changed and adjusted to one another continually over millions of years. Such interactions, which involve the long-term, mutual evolutionary adjustment of the characteristics of the members of biological communities in relation to one another, are examples of **coevolution.**

Plants and plant-eating predators called **herbivores** are a group of organisms that change and adjust to one another over time. Natural selection favors plants that have developed some means of protection against herbivores. In the dynamic equation of coevolution, however, natural selection also favors adaptations that enable animals to prey on plants in spite of their protective mechanisms. To avoid being eaten, for example, some plants have developed hard parts that are difficult to eat or unpalatable. In fact, certain grasses defend themselves by incorporating silica (a component of glass) in their structure (Figure 37-10). If enough silica is present, the plants may simply be too tough to eat. Some groups of herbivores, however, have developed strong, grinding teeth and powerful jaws. In addition, herbivores such as cattle have developed adaptations of their digestive systems. One such adaptation allows them to store the grass they have eaten in a digestive pouch called a *rumen*. Bacteria that live in the rumen attack the grass chemically, aiding in

FIGURE 37-10 Plants can defend themselves.
These zebras grazing on the East African savanna dislike eating grasses with leaves reinforced with silica. The more silica in the leaf cells, the less likely zebras and other grazing animals are to eat that particular kind of grass.

the digestive process. This stored food is then regurgitated and rechewed at a later time, providing a better breakdown of the cell walls within the grass.

Some plants have developed chemical defenses against herbivores. The best known plant groups with toxic effects are the poison ivy, poison oak, and poison sumac plants. All contain the contact poison urushiol. Castor bean seeds are also toxic to a wide variety of animals, producing a protein that attaches to ribosomes, blocking protein synthesis. Other plants produce toxins that inhibit the growth of bacteria, fungi, and roundworms. Still others produce chemicals having odors that act as a warning or as a repellent to a predator. Today, using the techniques of genetic engineering, scientists have been able to grow plants that chemically repel certain predators, thereby reducing the need for artificial pesticides (Figure 37-11).

However, associated with each family or other group of plants naturally protected by a particular kind of chemical compound are certain groups of herbivores that are adapted to feed on these plants, often as their exclusive food source. For example, the larvae of cabbage butterflies feed almost exclusively on plants of the mustard and caper families, which are characterized by the presence of protective chemicals—the mustard oils. Although these plants are protected against most potential herbivores, the cabbage butterfly caterpillars have developed the ability to break down the mustard oils, rendering them harmless. In a similar example of coevolution, the larvae of monarch butterflies are able to harmlessly feed on the toxic plants of the milkweed and dogbane families (Figure 37-12).

Over time, plants have developed various morphological and chemical adaptations that help protect them against plant eaters, or herbivores. In turn, however, herbivores have changed and adjusted to the plants and their adaptations. Such interactions, involving long-term mutual evolutionary adjustment, are coevolution.

FIGURE 37-11 Defense through genetic engineering.
Recently, scientists have developed plants that through genetic engineering, chemically repel predators. The tobacco plant on the *right* is a nonengineered plant and shows the effects of insect predation. The tobacco plant on the *left,* however, has been engineered to produce an insect toxin that deters insects and protects the plant from insects.

FIGURE 37-12 Monarch butterflies make themselves poisonous.
All stages of the life cycle of the monarch butterfly are protected from predators by the poisonous chemicals that occur in the milkweeds and dogbanes on which they feed as larvae. Both caterpillars and adult butterflies advertise their poisonous nature with warning coloration.

BIOLOGY IN FOCUS

Competition and killer bees

In the fall of 1956 a prominent Brazilian bee geneticist named Warwick Kerr traveled to Africa and brought back, at his government's request, a small number of queen bees. These bees were reported to be high honey producers, and the government was hoping to improve local bee varieties by breeding this desirable trait into them. On their return, the bees were placed in quarantine at a research station 200 miles inland from Kerr's university and crossed to local bees. Although most of the African queens were commercial bees obtained by Kerr in South Africa, one that was used much more than the others in crosses was a wild bee from Tanzania. Workers quickly confirmed that the African bees produced a great deal of honey and were very fast breeders. They also noticed that they were unusually sensitive to disturbance, becoming quite aggressive (Figure 37-A).

FIGURE 37-A A killer bee.

In the fall of 1957 a visting beekeeper, not knowing of the quarantine, removed the queen-excluder doors (holes too small for the more massive queen to pass through) from the hives, normal practice in a nonquarantine colony if a queen has started laying. That day 26 queens left their hives, each with her swarm of 1000 to 2000 bees.

The swarms traveled long distances and quickly supplanted the populations of wild bees in the surrounding countryside. In these new surroundings, they were freed from the competition and predation that had kept them in check in their African home and armed with evolutionary advantages: aggressiveness, fast reproduction, far traveling ability, and frequent founding of new hives.

Within months, reports began to come in of farm animals killed, followed by the first human fatalities. The African bees had evolved in the highly competitive environment of their homeland an effective defense against attack—at the slightest hint of provocation, they attacked first, in overwhelming numbers. A human not sensitive to bee stings can survive some 200 to 300 stings but not the number these attacks inflicted. A University of Miami graduate student killed in Costa Rica in 1986 by a swarm of African bees had 8000 stings.

By 1970 the African bees had overrun all of Brazil. Local geneticists suspected that they were breeding with local bees, but molecular studies by American geneticists suggest that they simply replaced them. By 1979 the bees had reached Central America. They first appeared in Mexico in 1986. The first African bee was detected in Texas in October of 1990. In 35 years the bees have conquered over 5 million square miles, killing hundreds of thousands of farm animals and over 1000 people. When they eventually reach their climatic limits in the next 20 years, where winters are too cold for them to survive, they will extend in an arc from mid-California to North Carolina.

FIGURE 37-13 Warning coloration.
The coloring of all these animals is meant to warn other animals to stay away.
A The skunk ejects a foul-smelling liquid when threatened.
B The poisonous gila monster is a member of the only genus of poisonous lizards in the world.
C The red-and-black African grasshopper feeds on highly poisonous *Euphorbia* plants.
D The tiny tropical tree frog is so poisonous that Indians in western Colombia use their venom to poison their blow darts.

Protective coloration

Some groups of animals that feed on toxic plants receive an extra benefit one of great ecological importance. When the caterpillars of monarch butterflies feed on plants of the milkweed family, for example, they do not break down the chemicals that protect these plants from most herbivores. Instead, they store them in fat within their bodies. As a result, the caterpillars and all developmental stages of the monarch butterfly are protected against predators by this "plant" poison. A bird that eats a monarch butterfly quickly regurgitates it. Although this is no help to the eaten insect, the bird will soon learn not to eat another butterfly with the bright orange and black pattern that characterizes the adult monarch. Such conspicuous coloration, which "advertises" an insect's toxicity, is called **warning coloration.** Warning coloration is characteristic of animals that have effective defense systems, such as poisons, stings, or bites. Other examples of animals that exhibit warning coloration are shown in Figure 37-13.

During the course of their evolution, many unprotected species have come to resemble distasteful ones that exhibit warning coloration. Provided that the unprotected animals are present in low numbers relative to the species they resemble, they too will be avoided by predators. If the unprotected animals are too frequent, of course, many of them will be eaten by predators that have not yet learned to avoid individuals with a particular set of characteristics. Such a pattern of resemblance is called **batesian mimicry,** after the British naturalist H.W. Bates, who first described this concept in the 1860s. Many of the best-known examples of batesian mimicry occur among butterflies and moths (Figure 37-14, *A* and *B*).

A

B

C

FIGURE 37-14 Mimicry.

A The viceroy butterfly is a müllerian mimic of the poisonous monarch. Larvae of the viceroy feed on willows or other nontoxic plants.

B The red-spotted purple is a batesian mimic of another poisonous butterfly, the pipevine swallowtail **(C)**, which obtains its toxic chemicals from plants of the pipevine family on which its larvae feed.

FIGURE 37-15 Insect herbivores are well suited to their hosts. The green caterpillars of the cabbage butterfly are camouflaged on the leaves of cabbage and other plants on which they feed. These caterpillars are able to break down the toxic mustard oils that prevent most insects from eating cabbage.

Another kind of mimicry, **müllerian mimicry,** was named for the German biologist Fritz Müller, a contemporary of Bates. Interestingly, in müllerian mimicry, the protective colorations of different animal species come to resemble one another as in batesian mimicry. However, unlike batesian mimicry, the organism and its mimic *do* possess similar defenses (Figure 37-14, *C*).

Some organisms are colored so as to blend in with their surroundings—a protective coloration called **camouflage.** Both cabbage caterpillars and cabbage butterflies have evolved a green coloration, allowing them to hide while feeding (Figure 37-15). Unlike monarchs, these insects do not store the toxic chemical in the plants they eat for use against predators. Instead, the cabbage butterflies break down the toxin. Similarly, insects that eat plants lacking specific chemical defenses are seldom brightly colored. Figure 37-16 shows examples of other animals who are camouflaged from their predators.

Some organisms within communities exhibit conspicuous coloration, or warning coloration, that advertises their ability to poison, sting, or bite. Organisms without specific defenses are often colored, or camouflaged, so as to blend in with their surroundings.

A

B

FIGURE 37-16 **A few striking examples of camouflage.**
A A young jackrabbit in the desert near Tucson displays both camouflage and stillness.
B Protective coloration in a flatfish on the sea floor.
C A tropical hawkmoth at rest, blending in perfectly with the moss-covered and lichen-covered tree bark.
D The green-brown coloring of the unicorn caterpillar gives it the appearance of a dried leaf edge.
E An inchworm caterpillar, which closely resembles a twig.

C

D

E

Symbiosis

Symbiotic relationships are those in which two different species of organisms live together in close association. All symbiotic relationships provide the potential for coevolution between the organisms involved, and in many instances the results of this coevolution are fascinating. The major kinds of symbiotic relationships include the following:

1. **Commensalism:** One species benefits and the other species neither benefits nor is harmed.
2. **Mutualism:** Both participating species benefit.
3. **Parasitism:** One species benefits but the other (the host) is harmed.

Parasitism is sometimes considered a form of predation. However, unlike a true predator, the successful parasite does not kill its host.

Commensalism

Many examples of the "one-sided" relationship of commensalism exist in nature. Often, the individuals deriving benefit are physically attached to the other species in the relationship. For example, plants called *epiphytes* grow on the branches of other plants. The epiphytes derive their nourishment from the air and the rain—not from the plants to which they attach for support (Figure 37-17). Similarly, various marine animals such as barnacles, grow on other,

FIGURE 37-17 **Commensalism: epiphytes and trees.**
The brownish growths attached to the tree limb are epiphytes. These plants derive their nourishment from the rain and air and use the tree limb for support only.

A

FIGURE 37-18 Commensalism: barnacles and whales.
A This breaching whale displays the barnacles growing on its skin.
B A close-up of a gray whale's skin reveals hitchhikers—lice and barnacles. The lice are actually parasites, whereas the barnacles cause no harm to the whale.

B

often actively moving, sea animals (Figure 37-18). These "hitchhikers" gain more protection from predation than if they were fixed in one place, and they continually reach new sources of food. They do not, however, harm the organisms to which they are attached.

Possibly one of the best-known examples of commensalism involves the relationship between certain small tropical fish—the clownfish—and sea anemones, marine animals that have stinging tentacles. The fish have developed an adaptation that allows them to live among the deadly tentacles of the anemones (Figure 37-19). These tentacles quickly paralyze other species of fish, protecting the clownfish against predators.

FIGURE 37-19 Commensalism: clownfish and sea anemones.
Two clownfish peer out from the tentacles of a large red sea anemone off the coast of Australia.

Mutualism

When both species benefit in a close relationship, it becomes one of mutualism. A particularly striking example of mutualism involves one genus of stinging ants and a Latin American plant of the genus *Acacia*. The modified leaves of acacia plants appear as paired, hollowed thorns. These thorns provide a home for the ants, protecting them and their larvae. In addition, the ants eat nectar the plants produce. In turn, the ants attack any herbivore that lands on the branches or leaves of an acacia and clear away vegetation that comes in contact with their host shrub, increasing the plant's ability to survive.

Many other interesting examples of mutualism exist in nature. Certain birds, for example, spend most of their time clinging to grazing animals (such as cattle), picking insects from their hides. In fact, the birds carry out their entire life cycles in close association with the cattle. The birds are provided with food and the cattle benefit by having their parasites removed. In another similar mutualistic relationship, ants use the tiny insects aphids, or greenflies, as a provider of food. The aphids suck fluids from the phloem of plants, extracting a certain amount of sucrose and other nutrients. However, many of these nutrients are not absorbed within the digestive tract of the aphid. A substantial portion runs out (somewhat altered) through the anus. The ants use this nutritional excrement as a food source and, in turn, actually carry the aphids to new plants so that they can continue eating!

Parasitism

Parasites include viruses, many bacteria, fungi, and an array of invertebrates. A different species of organism, larger than the parasite itself, is "home" for a parasite. During the intimate relationship between parasite and host, the parasite derives nourishment and the host is harmed.

Many instances of parasitism are well known. Intestinal hookworms, for example, are parasites (Figure 37-20). A person is infected when walking barefoot in soil containing hookworm larvae. These larvae are able to penetrate the skin, entering the bloodstream. The blood carries the larvae to the lungs. From there, they are able to migrate up the windpipe to the esophagus. The larvae are then swallowed and reach the intestines. After growing into adult worms, they attach to the inner lining of the intestines. They remain attached here, feeding on the blood of the host.

Some parasites do not live within an organism as hookworms do but attach to the outer surface of a plant or animal. The attachment may be fleeting, as with the bite of a mosquito, or may take place over a longer period, such as the burrowing of mites.

FIGURE 37-20 Parasitism: hookworms and humans.
These parasites of humans live in the intestine, feeding on the blood of the host.

FIGURE 37-21 Parasitism: dodder and plants.
Dodder has lost its chlorophyll, along with its leaves, in the course of evolution. Since it is unable to make its own food, it obtains food from host plants on which it grows.

Many fungi and some flowering plants are parasitic. The dodder plant, for example, has lost its chlorophyll and leaves in the course of its evolution and is unable to manufacture food. Instead, it obtains food from the host plants on which it grows (Figure 37-21).

The more closely the life of a parasite is linked with that of its host, the more its morphology and behavior are likely to have been modified during the course of its evolution. The human flea, for example, is flattened from side to side and slips easily through hair. The ancestors of this species of flea were brightly colored, large, winged insects. The structural and behavioral modifications of the human flea have come about in relation to a parasitic way of life.

Changes in communities over time: Succession

Communities, like species, change over time. Even when the climate of a given area remains stable year after year, the composition of the species making up a community, as well as the interactions within the community, shows a dynamic process of change known as **succession.** During the process of succession, a sequence of communities replaces one another in an orderly and predictable way. This process is familiar to anyone who has seen a vacant lot or cleared woods slowly become occupied with plants and animals or has seen a pond become filled with vegetation (Figure 37-22).

Primary succession takes place in areas not previously changed by organisms such as humans. Primary succession occurs in lakes left behind after the retreat of glaciers, for example, or on volcanic islands that may rise above the sea.

FIGURE 37-22 Succession of a pond.
This pond in central Maine is beginning to fill with aquatic vegetation **(A)**. As the process of succession proceeds, the pond is slowly filled **(B)**. Gradually, the area where the pond once existed will become indistinguishable from the surrounding vegetation **(C)**.

In the latter case, succession may begin as lichens take hold on bare rock. Lichens are made up of an alga and a fungus living together in a symbiotic relationship. As lichens grow, they produce acids that can break down rock, forming small pockets of soil. When enough soil has accumulated, mosses may begin to grow. This first community, a pioneer community, consists of plants that are able to grow under harsh conditions. The pioneer community paves the way for the growth and development of vegetation native to that climate. Over many thousands of years or even longer, the rocks may be completely broken down, and vegetation may cover a once-rocky area. As plants take hold, the area becomes able to sustain other forms of life. And as the plant community changes, so too do the other living things. Eventually, the mix of plants and animals becomes somewhat stable, forming what is termed a **climax community** (Figure 37-23). However, with an increasing realization that (1) climates may change, (2) the process of succession is often

FIGURE 37-23 Climax community.
A stage in succession toward a climax community on the west side of the San Francisco Peaks in northern Arizona. Here, Ponderosa pine *(Pinus ponderosa)* is replacing aspen *(Populus tremuloides)*.

very slow, and (3) the nature of a region's vegetation is determined to a great extent by human activities, ecologists do not consider the concept of a climax community as useful as they once did.

Succession occurs not only within terrestrial communities but in aquatic communities as well. A lake poor in nutrients, for example, may gradually become rich in nutrients as organic materials accumulate (see Figure 37-22). Plants growing along the edges of the lake, such as cattails and rushes, and those growing submerged, such as pondweeds, may contribute to the formation of a rich organic soil as they die and are decomposed by bacteria. As this process of soil formation continues, the pond may become filled in with terrestrial vegetation. Eventually, the area where the pond once stood may become an indistinguishable part of the surrounding vegetation.

Secondary succession occurs in areas that have been disturbed and that were originally occupied by living organisms. Humans are often responsible for initiating secondary succession throughout portions of the world that they inhabit. Abandoned farm fields, for example, undergo secondary succession as they revert to forest. Secondary succession may also take place after natural disasters, such as a forest fire or eruption of a volcano (Figure 37-24).

Communities change over time by means of the dynamic process of succession. During this process, a sequence of communities replaces one another in an orderly and predictable way.

A

B

C

FIGURE 37-24 Secondary succession.
Mount St. Helens in the state of Washington erupted violently on May 18, 1980. The lateral blast devastated more than 600 square kilometers of forest and recreation lands within 15 minutes. **A** shows an area near Lang Ridge 4 months after the blast; 4 years later **(B)**, succession was underway at the same spot, with shrubs, blueberries, and dogwoods following the first plants that became established immediately after the blast. In 1989, more plants have become established and stable communities have formed **(C)**.

Summary

1. All of the living (biotic) factors and nonliving (abiotic) factors together within a certain area are an ecosystem. Within an ecosystem, each living thing has a habitat, an area in which it lives. In addition, each organism plays a special role within an ecosystem, which is its niche.
2. Of the biotic factors within an ecosystem, groups of organisms of the same species are populations. Various populations of organisms make up a community. Thus an ecosystem can be thought of as a community of organisms, along with the abiotic factors with which the community interacts.
3. The interactions within communities fall into three categories: competition, predation, and symbiosis. Competition involves organisms of different species striving to obtain the same needed resource. During predation, one species (the prey) becomes a resource, being killed and eaten by another species (the predator). During a symbiotic relationship, two or more organisms of different species live together in close association over a prolonged time.
4. Symbiotic relationships also fall into three categories: commensalism, mutualism, and parasitism. Commensalism is a one-sided relationship: one species benefits whereas the other species neither benefits nor is harmed. In mutualism, both species benefit. In a parasitic relationship, one species (the parasite) benefits but the other (the host) is harmed.
5. Organisms living together in communities continually change and adjust to one another. Such interactions, which involve the long-term, mutual evolutionary adjustment of the characteristics of the members of biological communities in relation to one another, are forms of coevolution.
6. Communities, like species, change over time. This dynamic process of change, during which a sequence of communities replaces one another in an orderly and predictable way, is succession. Primary succession takes place in areas that are originally bare, like rocks or open water. Secondary succession takes place in areas where the communities of organisms that existed initially have been disturbed.

REVIEW QUESTIONS

1. Distinguish among population, community, and ecosystem.
2. Fill in the blanks: Within an ecosystem, each organism has a(n) __________, an area in which it resides. Each organism plays a role within the ecosystem, which is its __________.
3. Interactions within communities can be grouped into three categories. Name and describe each category.
4. True or false? (If it's false, change the statement so that it's correct): Competition among different species is often greatest between organisms that obtain their food in different ways.
5. State the principle of competitive exclusion in your own words.
6. Explain the terms *realized niche* and *fundamental niche*.
7. Fill in the blanks: A(n) ______ is an organism that kills and eats another organism called its ______. A(n) ______ also feeds on another organism but does not necessarily kill it.
8. Summarize the pattern of the predator-prey relationship as shown by Gause's experiments.
9. What other factors, in addition to those revealed in Gause's work, can affect the balance between predators and prey?
10. What is coevolution? Give an example of coevolution involving a herbivore.
11. What type of adaptation is shown in each of the following:
    a. The bright yellow and black stripes of a bee's body.
    b. The unobtrusive color of some lizards that make them difficult to see against surrounding rocks.
12. Distinguish among commensalism, mutualism, and parasitism. What do they have in common? Give an example of each.
13. What is primary succession? Summarize the process.
14. Fill in the blanks: Forests that have been cut down for logging can often no longer support the same animals and birds that lived in that ecosystem previously. Over time, the changes that take place in the communities of organisms in this logged area are examples of __________ __________.

KEY TERMS

abiotic 639
batesian mimicry 647
biotic 639
camouflage 648
climax community 652
coevolution 644
commensalism 649
community 639
competition 640
competitive exclusion 640
ecology 640
ecosystem 639
fundamental niche 642
habitat 639
herbivore 644
müllerian mimicry 648
mutualism 649
niche 639
parasitism 649
population 639
predation 640
predator 642
prey 642
primary succession 651
realized niche 642
secondary succession 653
succession 651
symbiosis 640
warning coloration 647

THOUGHT QUESTIONS

1. Can you think of a case of symbiosis that involves arthropods and humans? What type of symbiosis is involved? How did you make this determination?
2. How could you determine whether a resemblance between two butterfly species is mimicry or not?

FOR FURTHER READING

Breitwisch, R. (1992, March). Tickling for ticks. *Natural History*, pp. 57-63.
Red-billed oxpeckers spend most of their time preening mammals for parasites; the mammals seem to actively accommodate the birds' activities.

Creeping through the crinoids. (1992, May). *International Wildlife*, pp. 4-51.
This spectacular photo essay documents the symbiosis between the crinoids (relatives of seastars) and a variety of camouflaged animals that live among the crinoids' feathery arms.

Robinson, M. H. (1992, April). An ancient arms race shows no signs of letting up. *Smithsonian*, pp. 75-82.
In this engaging article, the author compares human military tactics to defensive tactics used by plants and animals.

38 Ecosystems

HIGHLIGHTS

Ecosystems are communities of organisms and the nonliving environment with which they interact.

Energy flows through ecosystems, initially captured from the sun by green plants and then passed from organism to organism as they feed on one another.

The elements essential to life, primarily hydrogen, carbon, oxygen, and nitrogen, are cycled from the atmosphere, through living things and back to the atmosphere once again.

Other elements such as phosphorus, potassium, calcium, and sodium are held in the soil in small amounts, incorporated in the tissues of living things, and cycled to the soil once again.

OUTLINE

Populations, communities, and ecosystems

The flow of energy through ecosystems

Food chains and webs

Food pyramids

The cycling of chemicals within ecosystems

The water cycle

The carbon cycle

The nitrogen cycle

The phosphorus cycle

A whole, raw mouse might not be your idea of a gourmet meal, but to this bullfrog it is a five-star dinner. Located in a rural village in South Africa, this "frog restaurant" is part of a food chain, a series of organisms that feed on one another. The bullfrog crushed the mouse with its powerful jaws. Its lower jaw has bony projections that help pull the prey into the frog's cavernous mouth and will soon help push it down into the stomach. Minutes before this picture was taken, it was the mouse that was feasting—on the seeds of nearby plants. And, unknown to the frog, a snake lies in wait behind the next branch, ready to add yet another link to this chain of gourmet relationships among the plants and animals of this community.

Populations, communities, and ecosystems

Individuals of a species, such as bullfrogs and mice, are each part of an individual population of organisms. Together, interacting populations are communities (see Chapters 36 and 37). The living organisms in a community interact not only with each other but with the nonliving substances in their environment, such as the soil, water, and air, to form an ecological system, or ecosystem.

An **ecosystem** is a community consisting of plants, animals, and microorganisms that interact with one another and with their environments and are interdependent on one another for survival. The living, or **biotic,** components of an ecosystem are made up of two types of organisms: those that can make their own food, or **producers,** and those that eat other organisms for food, or **consumers.** Many consumers kill and eat their food. A special group of consumers, called **decomposers,** obtains nourishment from dead matter such as fallen leaves or the bodies of dead animals. You know the decomposers as bacteria and fungi (although some species of bacteria can manufacture their own food and are therefore producers).

Figure 38-1 is a diagram of an ecosystem. Notice that the **abiotic,** or nonliving, components of the environment contribute substances needed for the ecosystem to function. In addition, notice that the exchanges of nutrients and other chemical substances among the organisms within the ecosystem form a cyclic pattern. Energy, however, does not cycle through the ecosystem but flows *through* the ecosystem, first captured from the sun by the producers, then used by herbivores that eat the producers, and ultimately used by all the consumers living in the ecosystem. Ecosystems are therefore systems in which there is a regulated transfer of energy and an orderly, controlled cycling of nutrients. The individual organisms and populations of organisms in an ecosystem act as parts of an integrated whole, adjust over time to their roles in the ecosystem, and relate to one another in complex ways that are only partly understood.

An ecosystem is made up of communities of organisms living within a defined area and the nonliving environmental factors with which they interact. The organisms of an ecosystem—the producers, consumers, and decomposers—each play a specific role within it, contributing to the flow of energy and the cycling of nutrients.

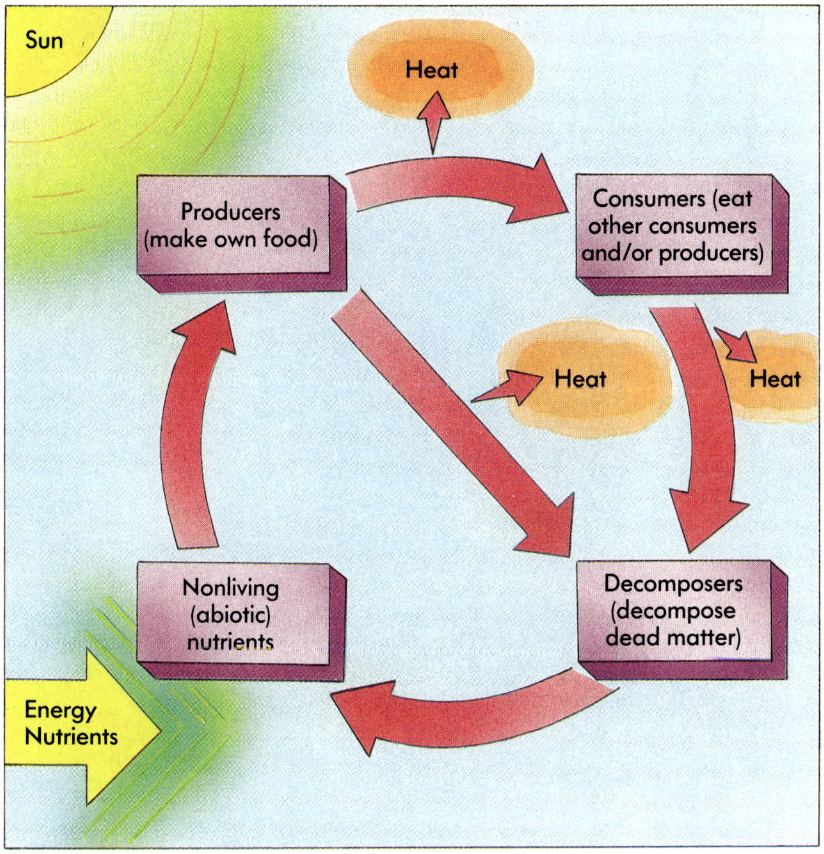

FIGURE 38-1 An ecosystem.
Nutrients flow through an ecosystem in a cyclic pattern, whereas energy flows through in one direction only.

FIGURE 38-2 Two distinct ecosystems.
In the coast ranges of California, the boundary between the evergreen shrub ecosystem known as chaparral and the grassland ecosystem is often sharp, as shown in this photograph taken along the western edge of the Santa Clara Valley near Morgan Hill. These two ecosystems each include a characteristic set of nonliving factors.

Where does one ecosystem begin and another end? Some ecosystems have clearly recognizable boundaries, such as that of a pond or a puddle or those found within the coastal ranges of California (Figure 38-2). Sometimes humans produce artifical ecosystems with human-made boundaries, such as the glass walls of an aquarium or the fencing surrounding a cultivated field. But the boundaries of many natural ecosystems blend with one another, sometimes almost imperceptibly. Ecosystems also change over time and with climate changes, slowly becoming modified into new ecosystems whose characteristics differ increasingly from those that preceded them.

The flow of energy through ecosystems

The energy that flows through an ecosystem comes from the sun. Green plants, the primary producers of terrestrial ecosystems, are able to capture some of the sun's radiant energy that falls on their leaves and convert it to chemical energy during the process of photosynthesis (see Chapter 8). Producers, then, are the key to life on Earth, since no other organisms can capture this energy for use in living systems. **Primary consumers,** or herbivores, feed directly on the green plants, incorporating some of this energy into molecules that make up their bodies and using the rest to perform the activities of life. **Secondary consumers** are meat-eaters, or carnivores, that feed in turn on the herbivores. And so the chain continues, with one living thing feeding on another, passing energy along that was once captured from the sun.

FIGURE 38-3 A decomposer doing its job.
Fungi are breaking down this hardwood stump, converting the organic materials contained within it into nutrients that can be reused by plants.

The refuse or waste material of an ecosystem is known as **detritus.** Organisms that are decomposers break down the organic materials of detritus into inorganic nutrients that can be reused by plants (Figure 38-3). Some of the energy still held in the tissues of once-living things is used by the decomposers, but they are the last link in this transfer of energy among organisms.

The producers of an ecosystem capture energy from the sun and convert it to chemical energy usable by themselves and consumers. Consumers feed on the producers and other consumers in the ecosystem, passing energy along that was once captured from the sun. Decomposers break down the organic molecules of dead organisms, serving as the last link in the flow of energy through an ecosystem and contributing to the recycling of nutrients to the environment.

Food chains and webs

All of the feeding levels previously described and additional levels such as tertiary consumers are represented in any fairly complicated ecosystem. These feeding levels are called **trophic levels,** from the Greek word *trophos,* which means "feeder." (In fact, this Greek word is the root for the words **heterotroph,** or "other feeder"—another word for consumer, and for **autotroph,** or "self-feeder"—another word for producer.) Organisms from each of these levels, feeding on one another, make up a series of organisms called a **food chain.** An example of a food chain can be seen in a pond ecosystem in which water fleas (primary consumers) feed on green algae (producers). Sunfish (secondary consumers) eat the water fleas but in turn are eaten by green heron (tertiary consumers) (Figure 38-4). The length and complexity of food chains vary greatly.

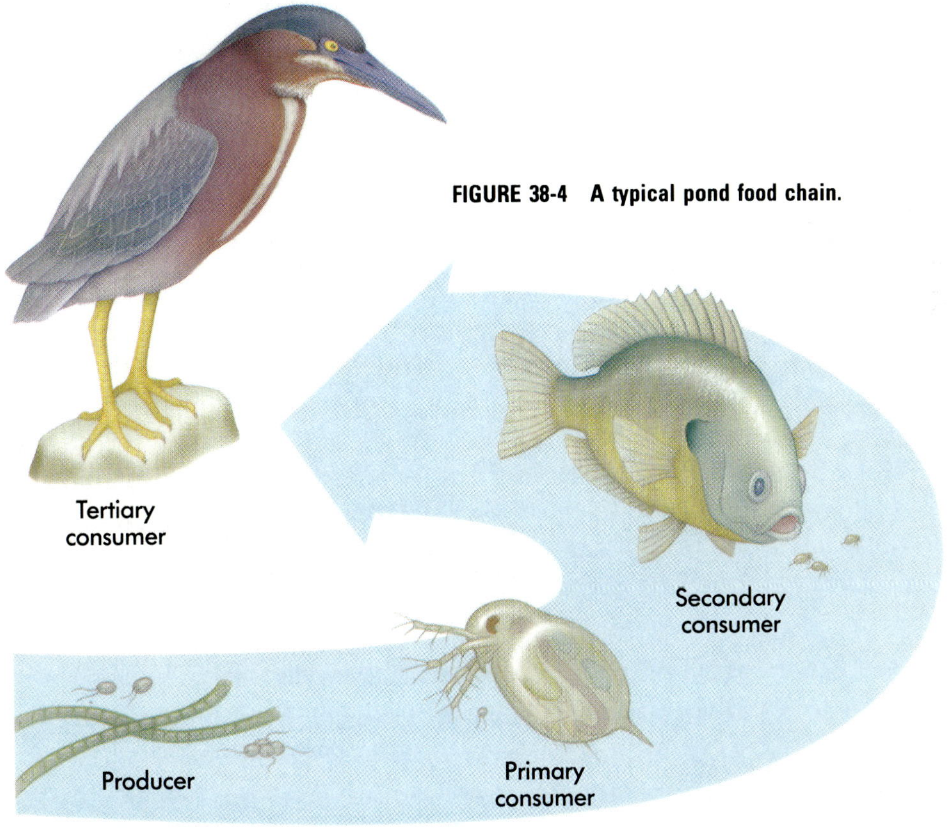

FIGURE 38-4 **A typical pond food chain.**

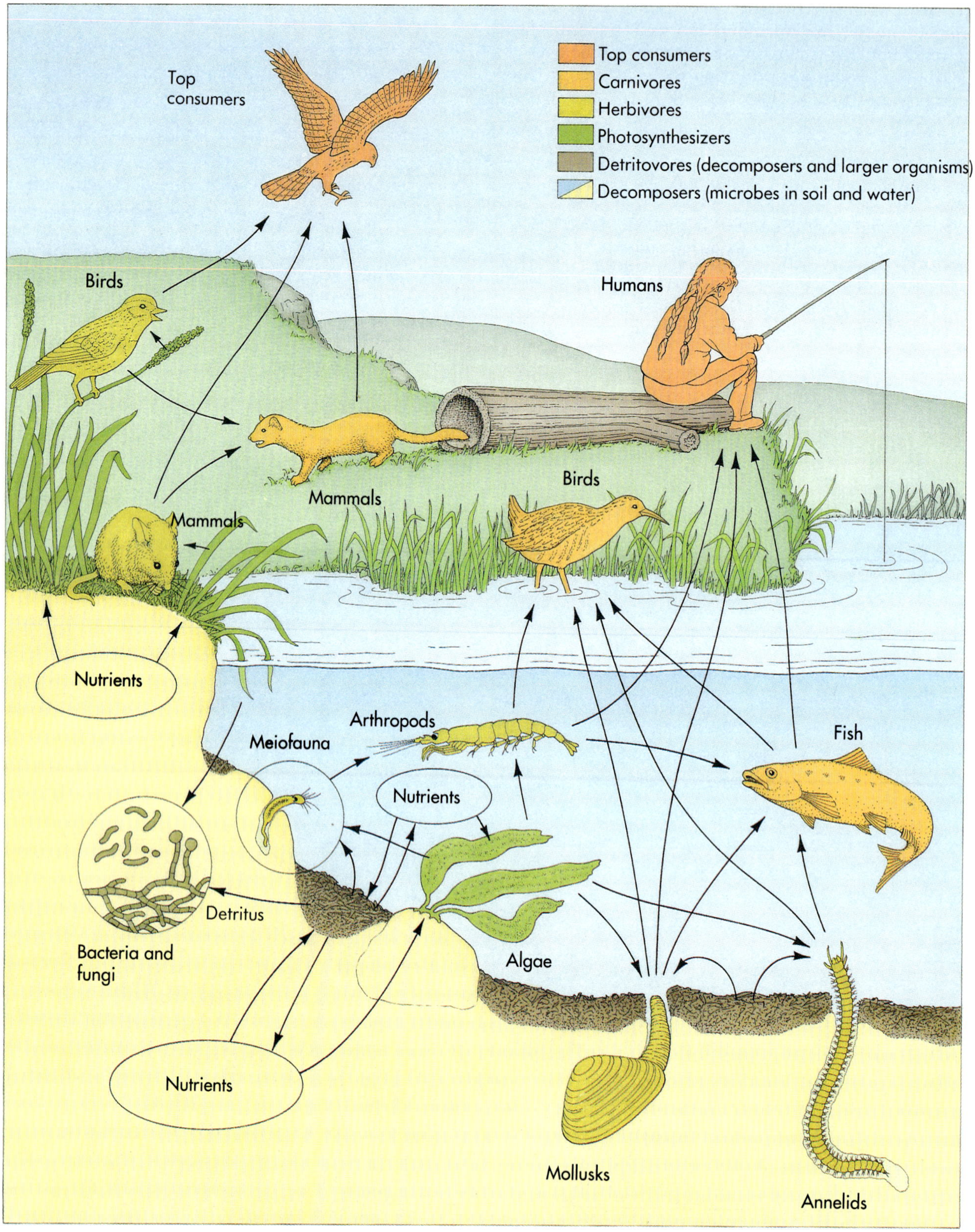

FIGURE 38-5 The food web in a salt marsh.
Each level of rectangles represents a trophic level. Each trophic level feeds on, or gains energy from, the layer below.

In reality, it is rare for any species of organism to feed on only one other species. Organisms feed on many different species and types of organisms and are, in turn, food for two or more other kinds. These relationships appear as a series of branching and overlapping lines rather than as one straight line. The organisms in an ecosystem that have such interconnected and interwoven feeding relationships make up a **food web.** Figure 38-5 shows a food web in a salt marsh ecosystem.

A linear relationship among organisms that feed on one another is a food chain. Food chains that interweave are food webs.

Food pyramids

In any ecosystem the number of organisms and the amount of energy making up each successive trophic level are less than the level that preceded. Lamont Cole of Cornell University illustrated this concept in his study of the flow of energy in a fresh water ecosystem in Cayuga Lake, New York. He calculated that approximately 150 calories of each 1000 calories of energy "fixed" by producers during photosynthesis was transferred into the bodies of small heterotrophs that feed on these plants and bacteria (Figure 38-6). Smelt, which are tiny fish, eat the heterotrophs; these secondary consumers obtain about 30 calories of each original 1000. If human beings eat the smelt, they gain about 6 calories from each 1000 calories that originally entered the system. If trout eat the smelt and humans eat the trout, humans gain only about 1.2 calories from each original 1000.

These types of calculations show that, on average, only 10% of plants' accumulated energy is actually converted into the bodies of the organisms that consume them. What happens to the rest? A certain amount of the energy that is ingested by organisms goes toward heat production. A great deal of energy is used for digestion and work, and usually 40% or less goes toward growth and reproduction. An invertebrate, for example, typically uses about a quarter of this 40% for growth. In other words, about 10% of the food that an invertebrate eats is turned into new body tissue. This figure varies from approximately 5% for carnivores to nearly

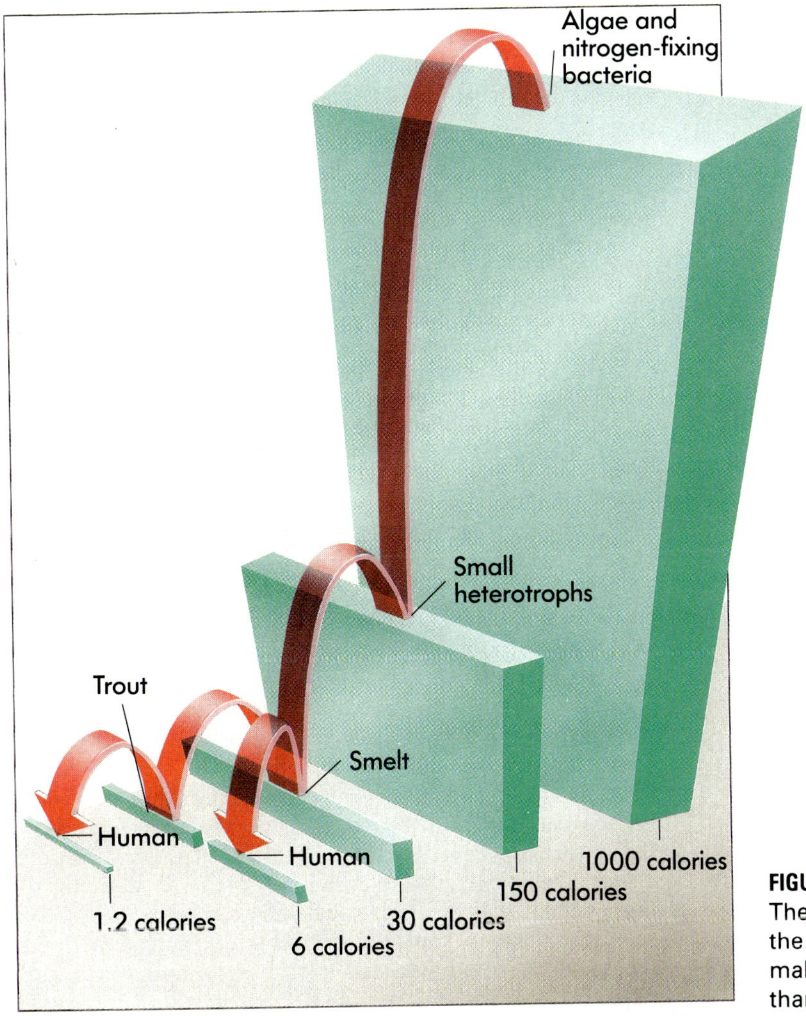

FIGURE 38-6 The flow of energy in Lake Cayuga. The experiments in Lake Cayuga demonstrated that the number of organisms and the amount of energy making up each successive trophic level is smaller than the preceding level.

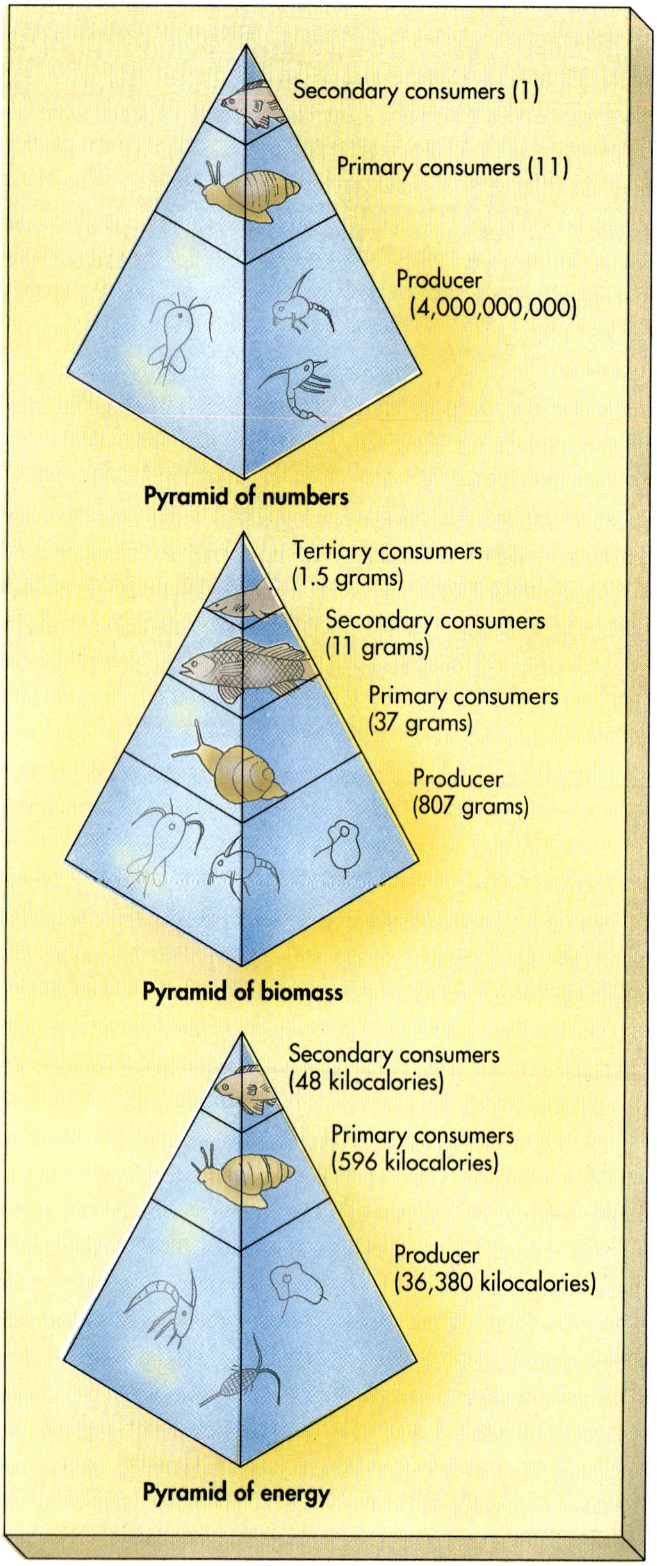

FIGURE 38-7 Pyramids of numbers, biomass, and energy for an aquatic ecosystem.
The pyramid of numbers shows the number of organisms at each trophic level in the ecosystem. The pyramid of biomass depicts the total weight of organisms supported at each level. The pyramid of energy depicts the amounts of energy available at each level.

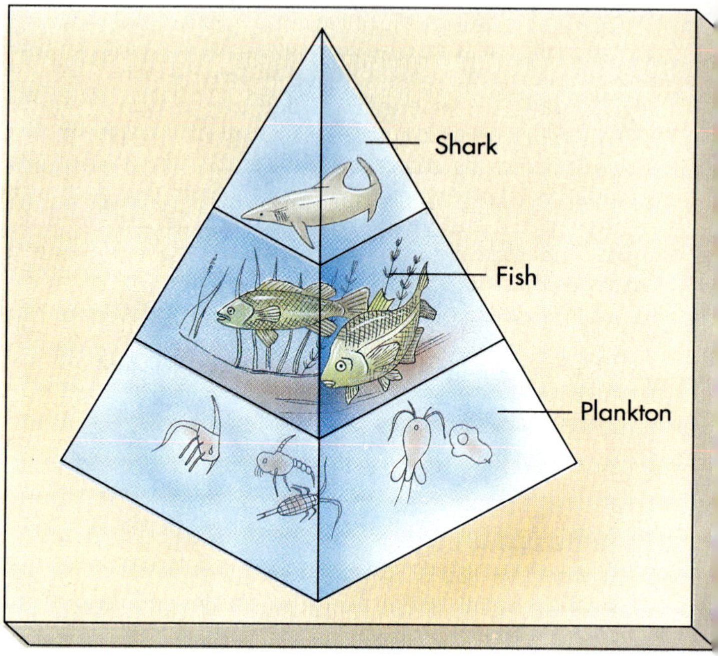

FIGURE 38-8 Food pyramid.
All food pyramids are basically the same—each successive trophic level is smaller than the preceding level (see Figure 38-7).

20% for herbivores, but 10% is an average value for the amount of energy (or organic matter) that organisms incorporate into their bodies from the energy available in the previous trophic level.

A plant captures some of the sun's energy that falls on its green parts. The successive members of a food chain process about 10% of the energy available in the organisms on which they feed into their own bodies. The rest is released as heat (and is lost from the food chain) or is used during various metabolic activities.

If shown diagrammatically, the relationships between trophic levels appear as pyramids. Diagrams that depict the energy flow through an ecosystem are called *pyramids of energy* (Figure 38-7). Those that depict the total weight of organisms supported at each trophic level in an ecosystem are referred to as *pyramids of biomass* (Figure 38-7). *Pyramids of number* depict, as the name suggests, the total number of organisms at each feeding level (Figure 38-7). Figure 38-8 shows a food pyramid that incorporates all three concepts.

The relationships between organisms at the various trophic levels of a food pyramid make it clear that herbivores have more food available to them than carnivores. In other words, the lower a population eats in the food chain, the higher the number of individuals that can be fed. Such considerations are increasingly important as humans work to maximize the food available for a hungry and increasingly overcrowded world.

BIOLOGY IN FOCUS

Hubbard Brook and the cycling of nutrients in ecosystems

In 1963, scientists at the Yale School of Forestry and Environmental Studies began an interesting experiment designed to show the effects of the cycling of nutrients within ecosystems. Their studies have yielded much of the information that is now known about nutrient cycles, and their ingenious experimental design has provided the basis for the development of the experimental methods used in the study of other ecosystems.

Hubbard Brook is the central stream located in a temperate deciduous forest in New Hampshire. For measurement of the flow of water and nutrients within the Hubbard Brook ecosystem, concrete weirs with V-shaped notches were built across six tributary streams that were selected for study. All of the water that flowed out of the six valleys (the areas surrounding the six streams) had to pass through the notch. The precipitation that fell in the six streams was measured, and the amounts of nutrients present in the water flowing in the six streams were also determined. By these methods the scientists demonstrated that the undisturbed forests in this area were very efficient in retaining nutrients. The small amounts of nutrients that fell from the atmosphere with the rain and snow were approximately equal to the amounts of nutrients that ran out of the six valleys. There was a net loss of calcium—about 0.3% of the total calcium in the system per year—and small net gains of nitrogen and potassium.

Then the researchers disturbed the ecosystem. In 1965, the investigators felled all of the trees and shrubs in one of the six valleys and then prevented their regrowth by spraying the area with herbicides. The effects of these activities were dramatic. The amount of water running out of the valley was increased by 40%, indicating that water normally would have been taken up by the trees and shrubs and then evaporated into the atmosphere from their leaves but was now running off. The amounts of nutrients running out of the system also increased. The loss of calcium was 10 times higher than it had been previously. The change in the status of nitrogen was particularly striking. The undisturbed ecosystem in this valley had been accumulating nitrogen at a rate of 2 kilograms per hectare per year, but the cut-down ecosystem lost it at a rate of about 120 kilograms per hectare per year! The nitrate level of the water rapidly increased to a level exceeding that judged safe for human consumption, and the stream that drained the area generated massive blooms of cyanobacteria and algae.

The Hubbard Brook experiment demonstrated that nutrient cycling depends on the vegetation present in the ecosystem among other things. The fertility of the deforested valley decreased rapidly, and at the same time the danger of flooding greatly increased. The Hubbard Brook experiment is particularly instructive in the 1990s, since large areas of tropical rain forest are being destroyed to make way for cropland.

FIGURE 38-A Experimental weir at Hubbard Brook. Water is forced over the concrete, and samples of it are representative of the flow from the valley where the stream is located.

The cycling of chemicals within ecosystems

Although energy flows through ecosystems and most is lost at each successive level in food pyramids, the matter making up the organisms at each level is not lost. All of these substances are recycled and are only used temporarily by living things. Hydrogen, carbon, nitrogen, and oxygen—the principal elements that make up all living things—are primarily held in the atmosphere in molecules of water, carbon dioxide, nitrogen gas, and oxygen gas. Other recycled substances necessary for life such as phosphorus, potassium, sulfur, magnesium, calcium, sodium, iron, and cobalt are held in rocks and, after weathering, enter the soil. The atmosphere and rocks are therefore referred to as the **reservoirs** of inorganic substances that cycle within ecosystems.

The cycling of materials in ecosystems is usually described as beginning at the reservoirs. Living things incorporate substances into their bodies from their reservoirs or from other living things, passing these materials along the food chain. Ultimately these substances, with the help of decomposers, move from the living world back to the nonliving world, becoming part of the soil or the atmosphere once again.

Within ecosystems, matter cycles from its reservoir in the environment, to the bodies of living organisms, and back to the environment.

The water cycle

Heated by the sun, water evaporates into the atmosphere from the surfaces of oceans, lakes, and streams (Figure 38-9). In terrestrial (land-based) ecosystems, as much as 90% of the water that reaches the atmosphere comes from plants as they release water vapor into the air during the process of transpiration (see Chapter 31). But because oceans cover three fourths of the Earth's surface, these bodies contribute most of the water to the atmosphere worldwide.

Atmospheric water condenses in clouds and eventually falls back to the Earth as precipitation. Most of it falls directly into the oceans. But some falls onto the land, flowing into surface bodies of water or trickling through layers of soil and rock to form subsurface bodies of fresh water called **ground water.** Plants take up water as it trickles through the soil, almost in a continuous stream. Crop plants, for example, use about 1000 kilograms of water just to produce 1 kilogram of biomass. Animals obtain water directly from surface water or from the plants or other animals they eat.

FIGURE 38-9 The water cycle.
In terrestrial ecosystems, 90% of the water that reaches the atmosphere comes from plant transpiration. In general, however, oceans contribute the most water to the Earth's atmosphere.

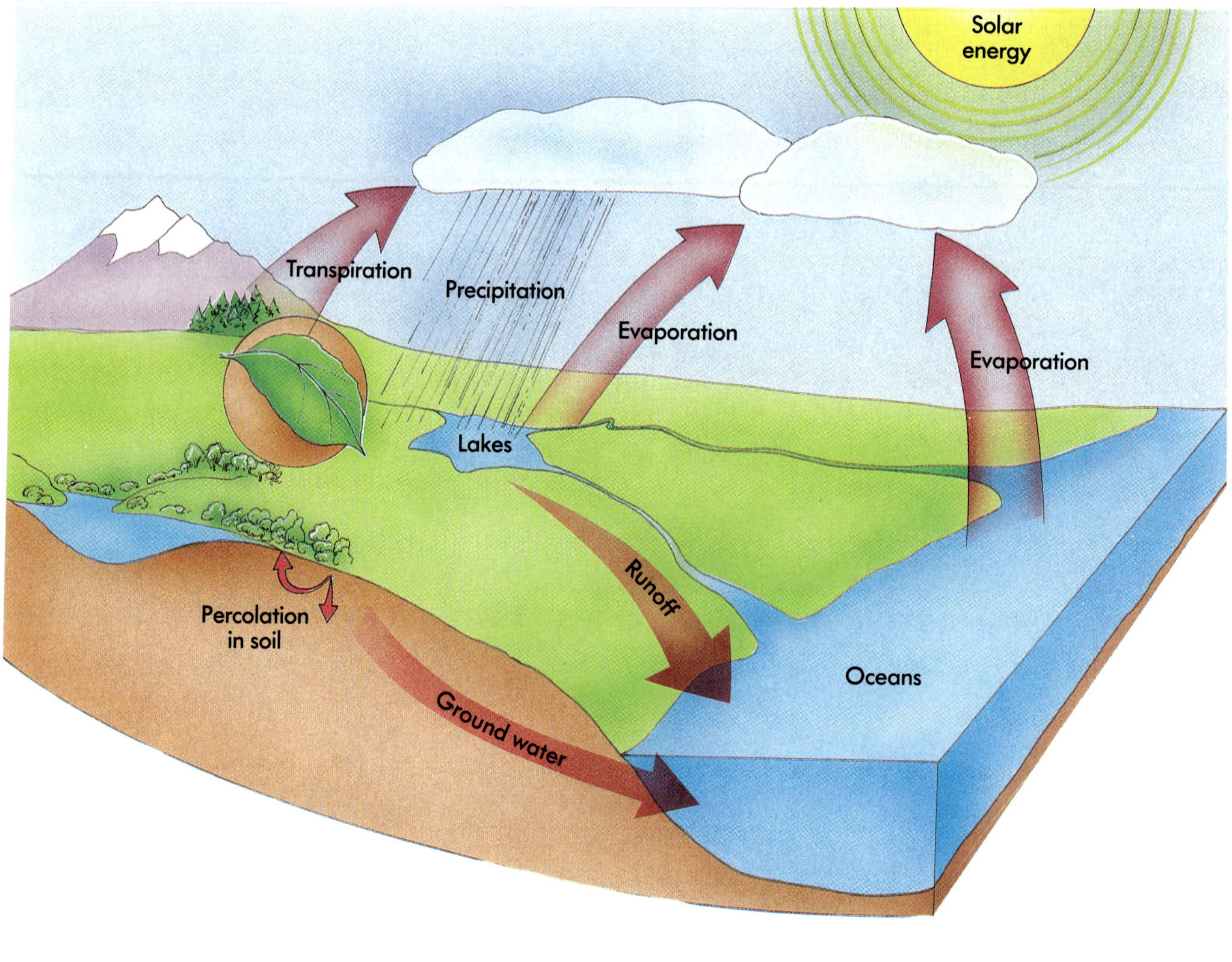

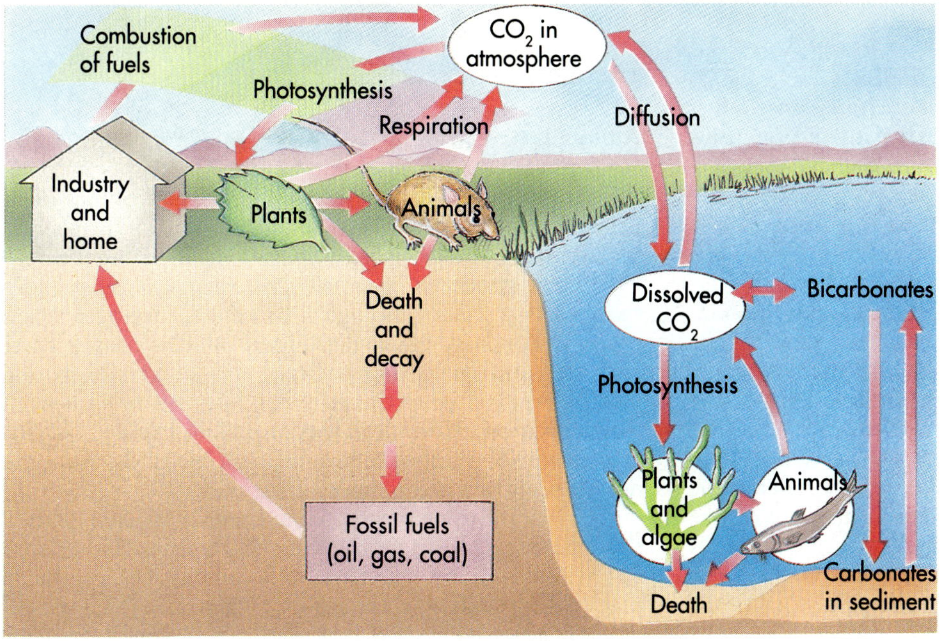

FIGURE 38-10 The carbon cycle. Carbon dioxide is found in the oceans and in the atmosphere. Producers and some consumers incorporate this carbon into substances necessary for life.

In the United States, ground water provides about a quarter of the water used by humans for all purposes and provides about half of the population with drinking water.

Water in the atmosphere condenses in clouds and falls back to the Earth as precipitation. Plants take up water from the soil, and animals obtain water from surface water or from the plants or other animals they eat. Water returns to the atmosphere through the evaporation of surface water and transpiration by plants.

Unfortunately, about 2% of the ground water in the United States is polluted, and the situation is worsening. Pesticides are one source, being carried to **aquifers,** underground reservoirs in which the ground water lies within porous rock as rain washes the chemicals from the surfaces of leaves and the topsoil. Chemical wastes, stored in surface pits, ponds, and lagoons, are another key source of groundwater pollution. Scientists have no technology that will remove pollutants from underground aquifers.

The carbon cycle

The carbon cycle is based on carbon dioxide (CO_2), which makes up about 0.03% of the atmosphere and is found dissolved in the oceans (Figure 38-10). Terrestrial as well as marine producers use CO_2—along with energy from the sun—to build carbon compounds such as glucose during the process of photosynthesis (see Chapter 8). The producers and the consumers that eat them break down these carbon compounds during cellular respiration and use the energy locked in their chemical bonds to carry on the metabolic processes of life. Consumers use some of the carbon atoms and compounds from the food they eat to produce needed substances. However, most of this carbon is waste and is released to the atmosphere (or to the oceans) as CO_2.

Intimately linked to the cycling of carbon is the cycling of oxygen (O_2). As plants use CO_2 during the process of photosynthesis, they produce O_2 as a waste product. This O_2 is released into the atmosphere and becomes available to organisms for the process of cellular respiration (Figure 38-11).

FIGURE 38-11 How carbon dioxide is linked to oxygen cycling. Plants use carbon dioxide in photosynthesis and give off oxygen. This oxygen is available to animals for the process of cellular respiration. Animals breathe out carbon dioxide; some carbon dioxide is liberated from the decomposition of dead organisms.

Some aquatic organisms such as mollusks use the CO_2 dissolved in water and combine it with calcium to form their calcium carbonate ($CaCO_3$) shells. When these organisms die, their shells collect on the sea floor. Years of exposure to water slowly dissolves the $CaCO_2$ releasing the CO_2 and making its carbon once again available to aquatic producers to use in the process of photosynthesis.

When organisms die, decomposers break down the carbon compounds making up their bodies. Certain carbon-containing compounds, such as the cellulose found in the cell walls of plants, are more resistant to breakdown than others, but certain bacteria, fungi, and protozoans are able to accomplish this feat. Some cellulose, however, accumulates as undecomposed organic matter. Over time and with heat and pressure, this undecomposed matter results in the formation of fossil fuels such as oil and coal. When these fuels are burned, as when wood is burned, the CO_2 is returned to the atmosphere. The release of this carbon as CO_2, a process that is proceeding rapidly as a result of human activities, may change global climates.

Carbon is held in the atmosphere in the form of carbon dioxide. This gas is taken in by photosynthetic organisms and is used to build the carbon compounds that plants manufacture during the process of photosynthesis. Both producers and consumers use these compounds as an energy source, metabolizing them during cellular respiration and releasing carbon dioxide to the atmosphere.

The nitrogen cycle

Although nitrogen gas (N_2) makes up 78% of the Earth's atmosphere, only a minute amount is incorporated into chemical compounds in the soil, oceans, and bodies of organisms. However, this N_2 is an essential part of the proteins within living things. Relatively few kinds of organisms—only a few genera of bacteria—can convert N_2 into a form that can be used for biological processes, playing a crucial role in the cycling of nitrogen (Figure 38-12). These bacteria are called nitrogen-fixing bacteria, and convert N_2 to ammonia (NH_3). Living things depend on this process of nitrogen fixation. Without it, they would ultimately be unable to continue to synthesize proteins, nucleic acids, and other necessary nitrogen-containing compounds.

Certain of the nitrogen-fixing bacteria are free living in the soil. Others form mutualistic relationships (see Chapter 37) with plants by living within swellings, or nodules, of plant roots. Some of these plants are legumes—plants such as soybeans, alfalfa, and clover. Plants having mutualistic associations with nitrogen-fixing bacteria can grow in soils having such low amounts of available nitrogen that they are unsuitable for most other plants. Growth of a leguminous crop can enrich the nitrate level of poor soil enough to benefit the next year's nonleguminous crop. This is the basis for crop rotation in which, for example, a field may be planted with soybeans (a legume) and corn (a nonlegume) in alternating years.

The roots of a few other kinds of plants form associations with nitrogen-fixing bacteria of the group actinomy-

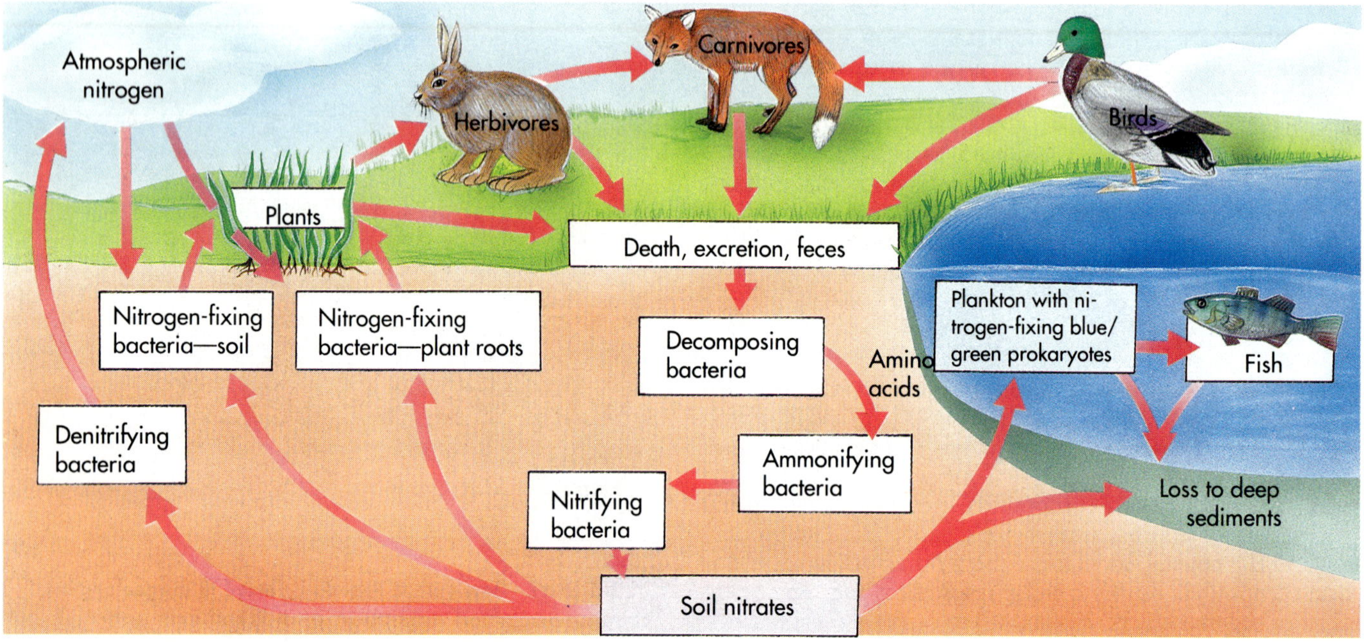

FIGURE 38-12 The nitrogen cycle.
Bacteria that can convert nitrogen into a usable form (nitrogen-fixing bacteria) are extremely important in the nitrogen cycle.

A

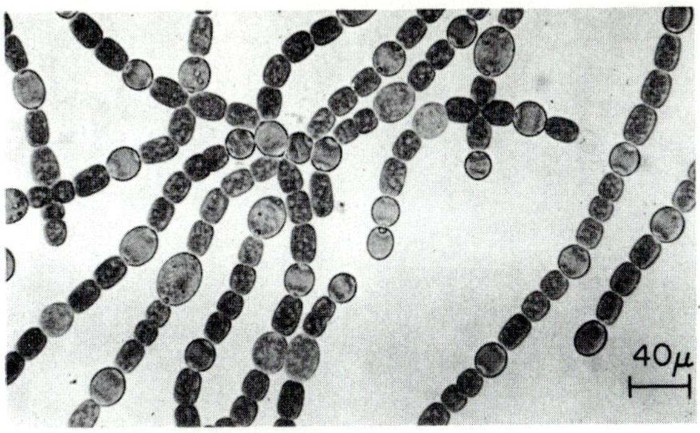

FIGURE 38-13 The use of nitrogen-fixing bacteria in agriculture.
The nitrogen-fixing bacteria *Anabaena azollae* **(A)** lives in the spaces between the leaves of the floating water fern *Azolla* **(B)**, which is deliberately introduced into the rice paddies of the warmer parts of Asia. Rice, here cultivated in Sri Lanka **(C)**, is the major food for well over one fourth of the human race.

B

C

cetes. Some of the plants involved are alders and mountain lilac. In addition, nitrogen-fixing bacteria of the genus *Anabaena* contribute large amounts of nitrogen to the rice paddies of China and Southeast Asia (Figure 38-13).

Other bacteria—certain decomposers—play another key role in the nitrogen cycle, producing NH_3 from the amino acids that make up the proteins and wastes of dead organisms. Still other bacteria convert NH_3 to nitrates (NO_3^-). NO_3^- is also produced by lightning, which causes N_2 to react with O_2 in the atmosphere. Humans add NO_3^- to the soil by spreading chemical fertilizers. Plants are able to use the nitrogen within molecules of NH_3 and NO_3^- to build their own proteins, nucleic acids, and vitamins. The nitrogen cycle comes full circle as nitrogen is continuously returned to the environment by bacteria that break down NO_3^- liberating N_2 to the atmosphere.

Although nitrogen gas constitutes about 78% of the Earth's atmosphere, it becomes available to organisms only through the metabolic activities of a few genera of bacteria, some of which are free living and others of which live symbiotically on the roots of legumes and some other plants. This bound nitrogen is released back to the atmosphere by other microorganisms capable of breaking down certain nitrogen compounds.

The phosphorus cycle

The reservoir of the nutrients phosphorus, potassium, sulfur, magnesium, calcium, sodium, iron, and cobalt is in rocks and minerals rather than in the atmosphere.

Phosphorus, more than any of the other required plant nutrients except nitrogen, is apt to be so scarce that it limits plant growth. In the soil, phosphorus is found as relatively insoluble phosphate (PO_4^{-3}) compounds, present only in certain kinds of rocks. For this reason, phosphates exist in the soil only in small amounts. Therefore humans add millions of tons of phosphates to agricultural lands every year. Plants take up the phosphates from the soil, and animals obtain the phosphorus they need by eating plants or other plant-eating animals. When these organisms die, decomposers release the phosphorus incorporated in their tissues, making it again available for plant use (Figure 38-14).

Phosphorus is held in the soil in the form of relatively insoluble phosphate compounds, found in most soils in small amounts. Therefore humans add phosphate compounds to the soil for plants to take up, and animals obtain the phosphates they need by eating plants or plant-eating animals. When organisms die, decomposers release phosphorus back to the soil.

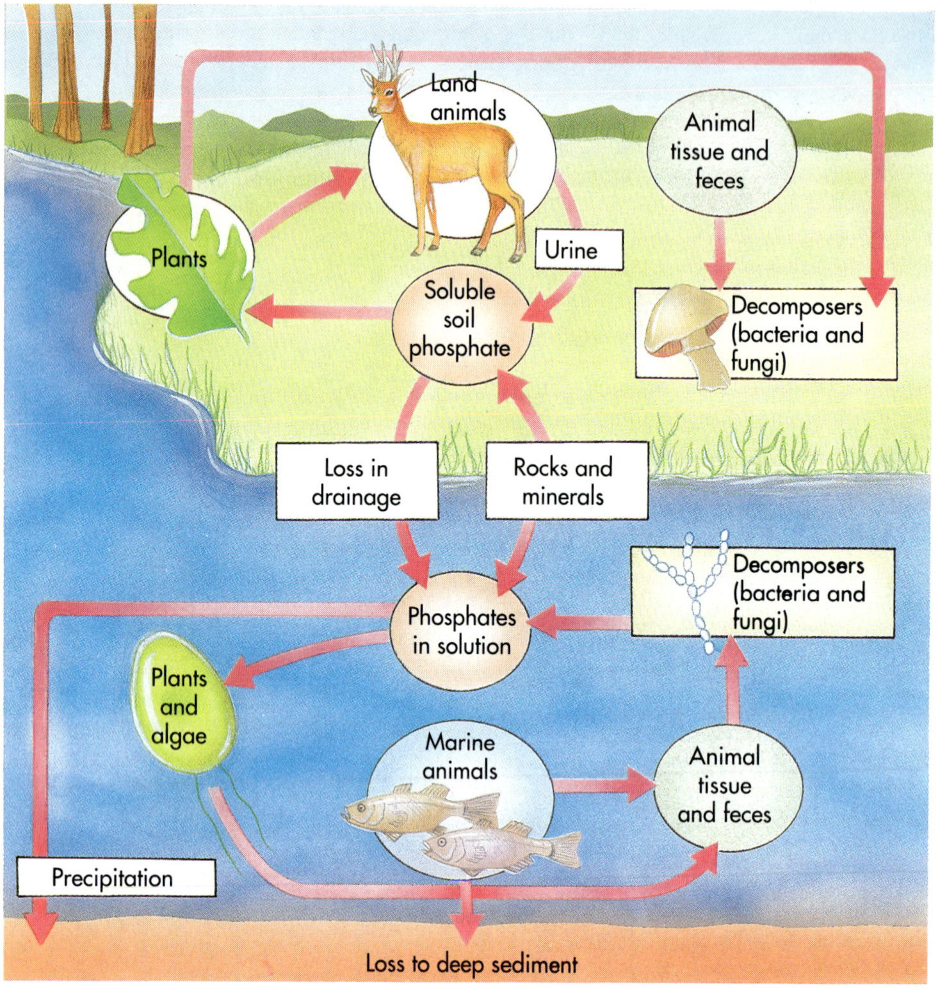

FIGURE 38-14 The phosphorus cycle. Phosphorus is found in only very small amounts in the soil; to make up for this lack, humans add phosphates to the soil. Plants take up phosphate from the soil, and animals obtain the necessary levels of phosphate by eating plants.

Humans unnecessarily alter the phosphorus cycle by adding up to four times as much phosphate as a crop requires each year. Much of this phosphate runs off the land or is eroded into rivers, streams, and the oceans. In addition, phosphates in sewage and waste water from homes and industry find their way into surface water. Rivers and streams become **eutrophic,** or "well fed," causing algae and other aquatic plants to overgrow. As they die, the decomposers feed on them, using up much of the oxygen in the water and choking out other forms of life.

Rivers and streams carry phosphates to the oceans. Some of these phosphates become incorporated into the bodies of fishes and other marine animals. The sea birds that eat these animals deposit enormous amounts of guano (feces) rich in phosphorus along certain coasts (these deposits have traditionally been used for fertilizer) (Figure 38-15). Phosphates not incorporated into the bodies of animals precipitate out of the water and become part of the bottom sediment. These phosphates become available again only if the sea floor rises up during eras of climatic change, such as along the Pacific coast of North and South America.

FIGURE 38-15 A guano coast.
Marine animals incorporate phosphates from runoff into their bodies. The seabirds that feast on these marine animals deposit feces rich in phosphorus on certain coasts.

Summary

1. Ecosystems are communities of organisms and the nonliving factors of their environments through which energy flows and nutrients cycle.
2. Through photosynthesis, plants growing under favorable circumstances capture and lock up some of the sun's energy that falls on their green parts. They may then be eaten by herbivores (primary consumers), which in turn may be eaten by secondary consumers (carnivores). The dead remains of all organisms are broken down by the decomposers. This sequence of organisms, one feeding on another, constitutes a food chain.
3. Each link in a food chain is a trophic, or feeding, level. Each trophic level includes organisms that can transfer about 10% of the energy that exists at each level to the next level. The rest is lost as heat or is used for various metabolic activities.
4. Carbon dioxide, nitrogen gas, oxygen gas, and water are the atmospheric reservoirs of the carbon, nitrogen, oxygen, and hydrogen used in biological processes. All of the other elements that organisms incorporate into their bodies come from the Earth's rocks.
5. Water in the atmosphere condenses in clouds and falls to the Earth as precipitation. Plants take up water from the soil, and animals obtain water from surface water or from the plants or other animals they eat. Water returns to the atmosphere through the evaporation of surface water and transpiration by plants.
6. Carbon is held in the atmosphere in the form of carbon dioxide. This gas is taken in by photosynthetic organisms and is used to build the carbon compounds that plants manufacture during the process of photosynthesis. Both producers and consumers use these compounds as an energy source, metabolizing them during cellular respiration and releasing carbon dioxide to the atmosphere.
7. Atmospheric nitrogen is converted to ammonia by several genera of symbiotic and free-living bacteria. The ammonia, in turn, is assimilated into amino groups in proteins of cells or is converted to nitrites and then to nitrates by other bacteria. Nitrates are incorporated into the bodies of plants and are converted back into ammonium ions, which are used in the manufacture of many kinds of molecules in the bodies of living organisms. The breakdown of these molecules either converts them to recyclable forms or results in the release of atmospheric nitrogen.
8. Phosphorus is a key component of many biological molecules; it weathers out of soils and is transported to the world's oceans where it tends to be lost. Phosphorus is relatively scarce in rocks; this scarcity often limits or excludes the growth of certain kinds of plants. Therefore humans add phosphate compounds to the soil for plants to take up, and animals obtain the phosphates they need by eating plants or plant-eating animals. When organisms die, decomposers release phosphorus back to the soil.

REVIEW QUESTIONS

1. What is an ecosystem? Briefly describe three of the living groups it contains with respect to mode of nutrition.
2. To which of the groups in question 1 do you belong?
3. Draw a diagram that illustrates the relationships between the biotic and abiotic components of a generalized ecosystem.
4. Fill in the blanks: __________ __________ feed directly on green plants. __________ __________ feed on herbivores. Other organisms known as __________ live on the refuse or waste material of an ecosystem, called __________.
5. Distinguish among trophic level, food chain, and food web.
6. Summarize what happens to the sun's energy as it travels through a food chain.
7. Explain how chemicals are cycled through an ecosytem.
8. Describe how water is cycled within ecosystems.
9. What is ground water, and how can it become polluted? Explain why this is a serious problem.
10. Summarize the carbon cycle.
11. Why can crop rotation make soil more fertile? To what chemical cycle does this relate?
12. Many farmers fertilize their crops heavily with phosphates, believing that this will improve the soil. Explain how this practice can affect an ecosystem.

KEY TERMS

abiotic 657
aquifer 665
autotroph 659
biotic 657
consumer 657
decomposer 657
detritus 659
ecosystem 657
eutrophic 668
food chain 659
food web 661
ground water 664
heterotroph 659
primary consumer 658
producer 657
reservoir 664
secondary consumer 658
trophic level 659

THOUGHT QUESTIONS

1. Imagine you had a hollow glass sphere the size of a basketball and you wanted to create within it a self-sustaining stable ecosystem that needed only sunlight and moderate temperature to persist indefinitely. What would you put in this ecosphere?
2. Most terrestrial food chains have only three or rarely four links. Why do you imagine this is so? Why not forest food chains with ten links?
3. Extensive cutting and burning of tropical rain forests often results in a drastic and permanent lowering of rainfall in the cleared area (an effect noted by Alexander von Humboldt over 100 years ago). Why?

FOR FURTHER READING

Colinvaux, P. A. (1989, May). The past and future Amazon. *Scientific American*, pp. 102-108.
This is a very interesting article about historical changes in the Amazon rain forest ecosystem, based on research.

Houghton, R. A., & Woodwell, G. M. (1989, April). Global climatic change. *Scientific American*, pp. 36-44.
This is a good summary of research on global warming.

Biomes and life zones of the world

39

Does your backyard look like the photo? Probably not, unless you happen to live near the equator. In this steamy climate, the vegetation is lush and the trees are tall, "topping out" at about 160 feet! Very little light penetrates this screen of foliage and finds its way to the forest floor beneath. Vines and other plants cling high up on the trunks of trees, competing for the light that filters through their leaves. Animal and plant species abound; so many species inhabit the tropical rain forest, in fact, that thousands remain unknown, unclassified, and unstudied.

Interestingly, not all regions of the Earth lying near the equator are tropical rain forests. Latitude is only one factor having an impact on the growth and distribution of living things. This chapter will describe other factors, explaining why organisms live where they do . . . and why your backyard is probably not a tropical rain forest.

HIGHLIGHTS

Biomes are distinctive, broad, terrestrial ecosystems of plants and animals that occur over wide areas within specific climatic regions.

Seven categories of biomes, listed by their distance from the equator, are tropical rain forests, savannas, deserts, temperate grasslands, temperate deciduous forests, taiga, and tundra.

The fresh water of ponds, lakes, streams, and rivers also provides various habitats for organisms to live; estuaries, places where fresh water meets salt water as rivers open into oceans, provide nutrient-rich environments intermediate between fresh and salt water.

Marine organisms occur most plentifully in the intertidal zone and the photic zone of the neritic and open sea; here, they have abundant light and oxygen.

OUTLINE

Biomes and climate
- The sun and its effects on climate
- Atmospheric circulation and its effects on climate

Life on land: The biomes of the world
- Tropical rain forests
- Savannas
- Deserts
- Temperate grasslands
- Temperate deciduous forests
- Taiga
- Tundra

Life in fresh water

Estuaries: Life between rivers and oceans

Life in the oceans
- The intertidal zone
- The neritic zone
- The open-sea zone

Biomes and climate

Biomes are ecosystems of plants and animals that occur over wide areas of land within specific climatic regions and are easily recognized by their overall appearance. Each biome is similar in its structure and appearance wherever it occurs on Earth and differs significantly from other biomes. Biomes are sometimes named by the climax vegetation (stable plant communities) of the region (see Chapter 37), such as the tropical rain forest.

The characteristics of biomes are a direct result of their temperature and rainfall patterns. These patterns result from the interaction of the features of the Earth itself (such as the presence of mountains and valleys) with two physical factors:

1. The amounts of *heat from the sun* that reach different parts of the Earth and the seasonal variations in that heat
2. *Global atmospheric circulation* and the resulting patterns of oceanic circulation

Together these factors determine the local climate, including the amounts and distribution of precipitation.

Biomes are broad, distinctive climatic regions containing specific types of plants and animals. Temperature and rainfall patterns are the primary determinants of the characteristics of these huge ecosystems.

The sun and its effects on climate

Because the Earth is a sphere, some parts receive more energy from the sun than others. The tropics are warmer than the temperate regions because the sun's rays arrive almost perpendicular to regions near the equator, whereas near the poles the rays' angle of incidence spreads them over a much greater area (Figure 39-1). Therefore the greater the latitude (distance from the equator), the colder the climate.

The northern and southern hemispheres also experience a change of seasons as well as the gradations in temperature that vary with latitude. Seasons occur because the Earth is tilted on its axis, so the northern and southern hemispheres receive unequal amounts of sunlight at various times during the Earth's year-long journey around the sun. One of the poles is closer to the sun than the other, except during the spring and autumn equinoxes (Figure 39-2).

Atmospheric circulation and its effects on climate

Near the equator, warm air rises and flows toward the poles. If the Earth were a stationary sphere, this air would reach the poles, cool, fall to the ground, and flow back toward the equator. The Earth is not stationary, however; as it spins, its rotary movement breaks the air into six "coils" of rising and falling air that surround the Earth (Figure 39-3, *A*). With increasing latitude, each air mass is cooler than the one before but warmer than the next. Therefore the air in each mass rises at the region of its lowest latitude, moves toward the poles, sinks to the ground at the region of its highest latitude, and flows back toward its lowest latitude. As it warms, it rises again, completing the cycle.

These moving masses of air that encircle the earth set up the prevailing wind patterns (Figure 39-3, *B*) and also have an effect on precipitation. The moisture-holding capacity of air increases when it is warmed and decreases when it is cooled. Therefore precipitation is relatively high near 60 degrees north and south latitude, where it is rising and being cooled. Conversely, precipitation is generally low near 30 degrees north and south latitude, where air is falling and being warmed (Figure 39-3, *A*).

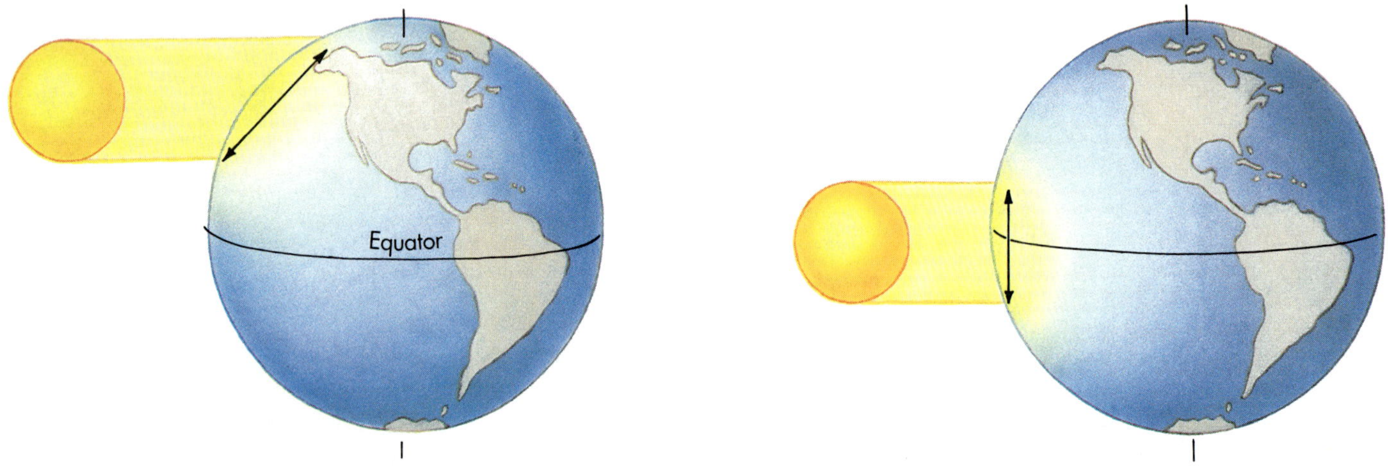

FIGURE 39-1 How the angle of solar energy striking the Earth affects climate.
The angle of a beam of solar energy striking the Earth at the equator is perpendicular to the surface of the Earth. The angle of this same beam striking an area at the middle latitude is larger, and the solar energy is more spread out over a larger area. In general, the greater the latitude, the colder the climate.

FIGURE 39-2 The rotation of the Earth around the sun has a profound effect on climate. In the northern and southern hemispheres, temperatures change in an annual cycle because the Earth is slightly tilted on its axis in relation to its pathway around the sun.

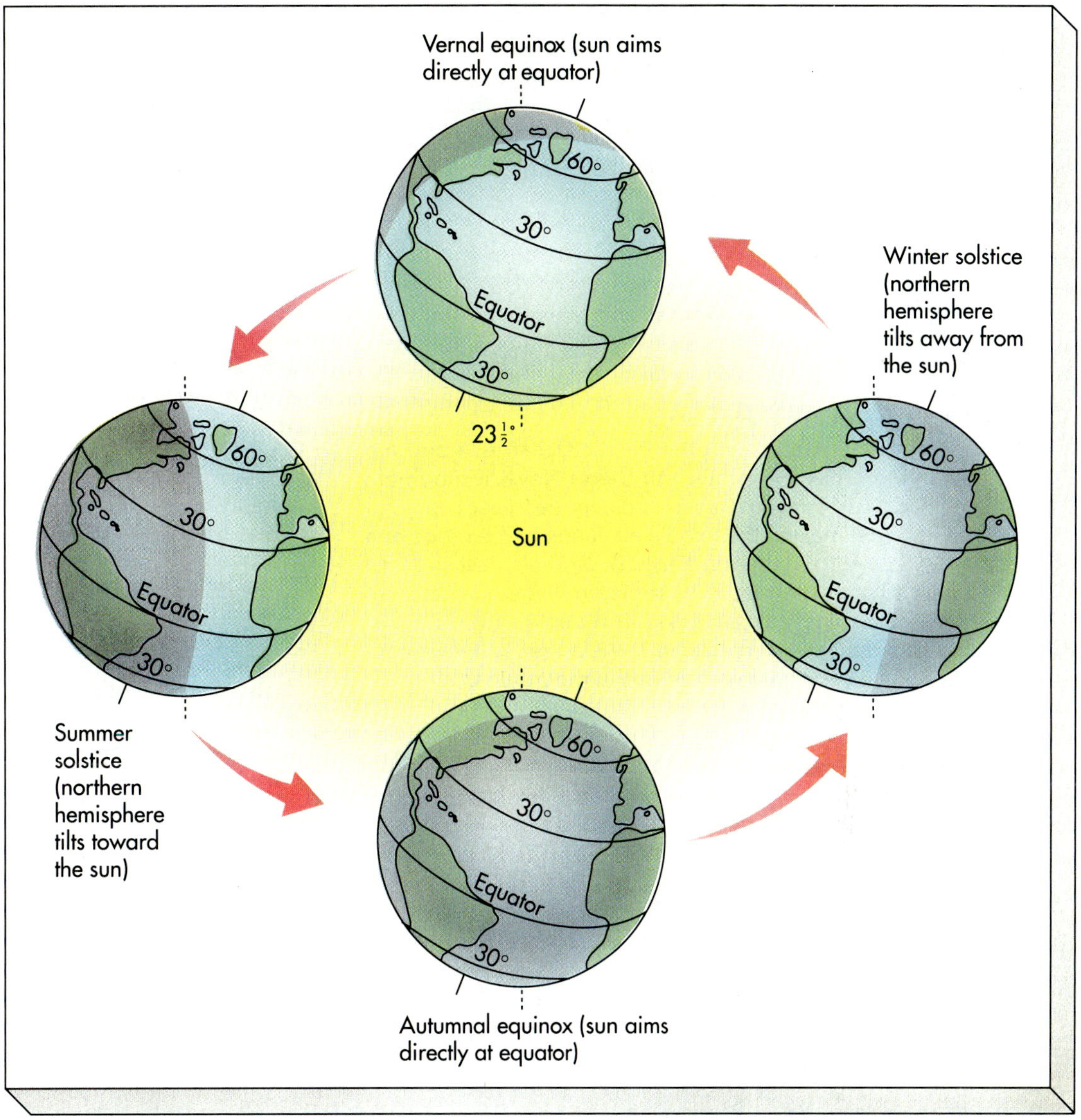

FIGURE 39-3 How the Earth's atmospheric circulation influences climate.
A Air surrounding the rotating Earth is broken into six coils of rising and falling air. These masses of air are associated with certain climates where moist air rises.
B The wind patterns around the Earth are set up by the six moving air masses.

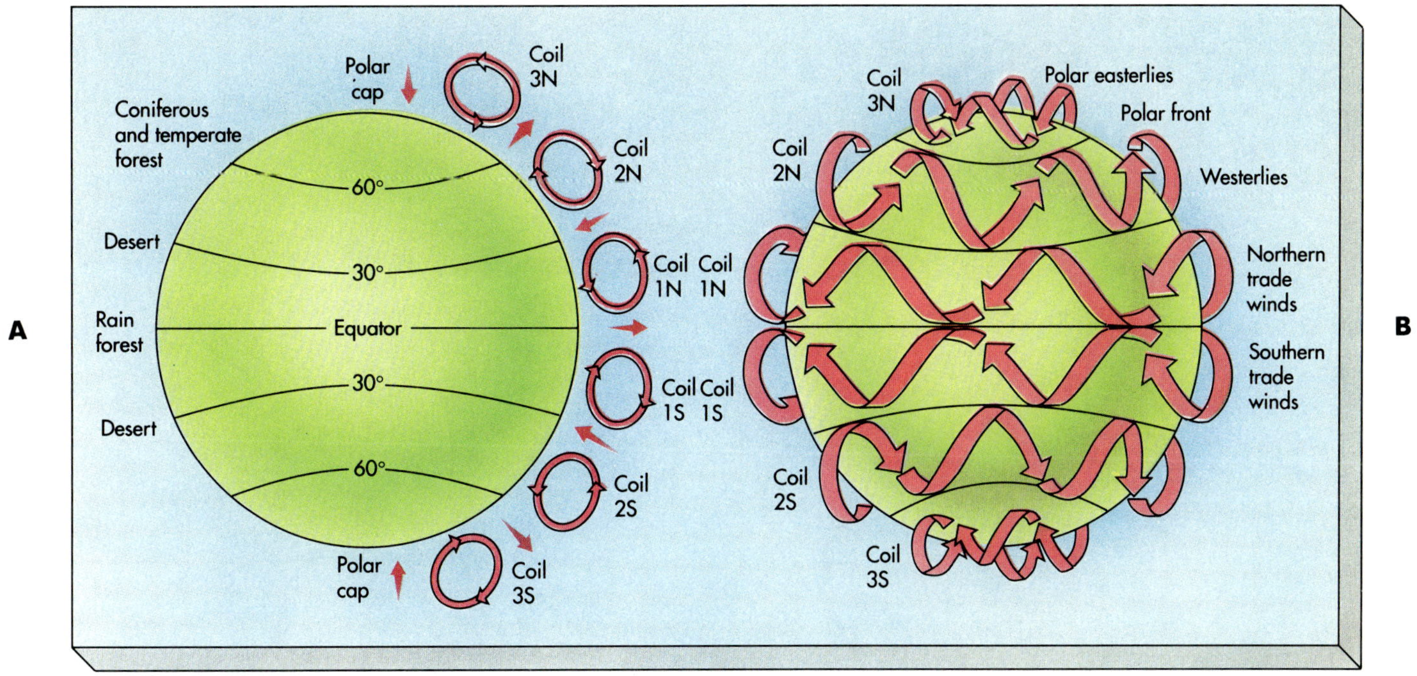

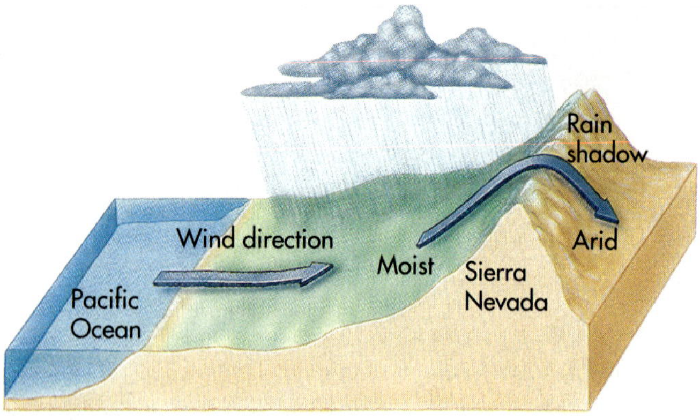

FIGURE 39-4 The rain shadow effect.
Moisture-laden winds from the Pacific Ocean rise and are cooled when they encounter the Sierra Nevada mountain range. As their moisture-laden capacity decreases, precipitation occurs, making the middle elevation of the range one of the snowiest regions on Earth. As the air descends on the east side of the range, its moisture-laden capacity increases again, and the air picks up moisture from its surroundings rather than releasing it. As a result, desert conditions prevail on the east side of the mountains.

Partly as a result of these factors, all the great deserts of the world lie near 30 degrees north or 30 degrees south latitude, and some of the great temperate forests are near 60 degrees north or south latitude. But other factors also come into play. Some major deserts are formed in the interiors of the large continents; these areas have limited precipitation because of their distance from the sea, the ultimate source of most precipitation. Other deserts sometimes occur because mountain ranges intercept the moisture-laden winds from the sea (Figure 39-4). As the air travels up a mountain (its windward side), it is cooled, and precipitation forms. As the air descends the other side of the mountain (its leeward side), it is warmed; its moisture-holding capacity increases. For these reasons, the windward sides of mountains are much wetter than the leeward sides, and the vegetation is often very different. This phenomenon is the rain shadow effect. Seattle, Washington, for example, lies on the windward side of the Cascade Mountain range in the northwestern United States. This city receives 99 centimeters (39 inches) of rainfall per year. Yakima, Washington, slightly south from Seattle on the leeward side of this mountain range, receives only 20 centimeters (8 inches) of rain per year.

The climate of a region is determined primarily by its latitude and wind patterns. These factors, interacting with surface features of the Earth such as mountains and distance from the ocean, result in the particular rainfall patterns. The temperature, rainfall, and altitude of an area, in turn, provide conditions that result in the growth of vegetation characteristic of that area.

The patterns of atmospheric circulation influence patterns of circulation in the ocean, modified by the location of the land masses around and against which the ocean currents must flow. Oceanic circulation is dominated by huge surface gyrals (Figure 39-5), which move around the subtropical zones of high pressure between approximately 30 degrees north and 30 degrees south latitude. These gyrals move clockwise in the northern hemisphere and counterclockwise in the southern hemisphere. They profoundly affect life not only in the oceans but also on coastal lands

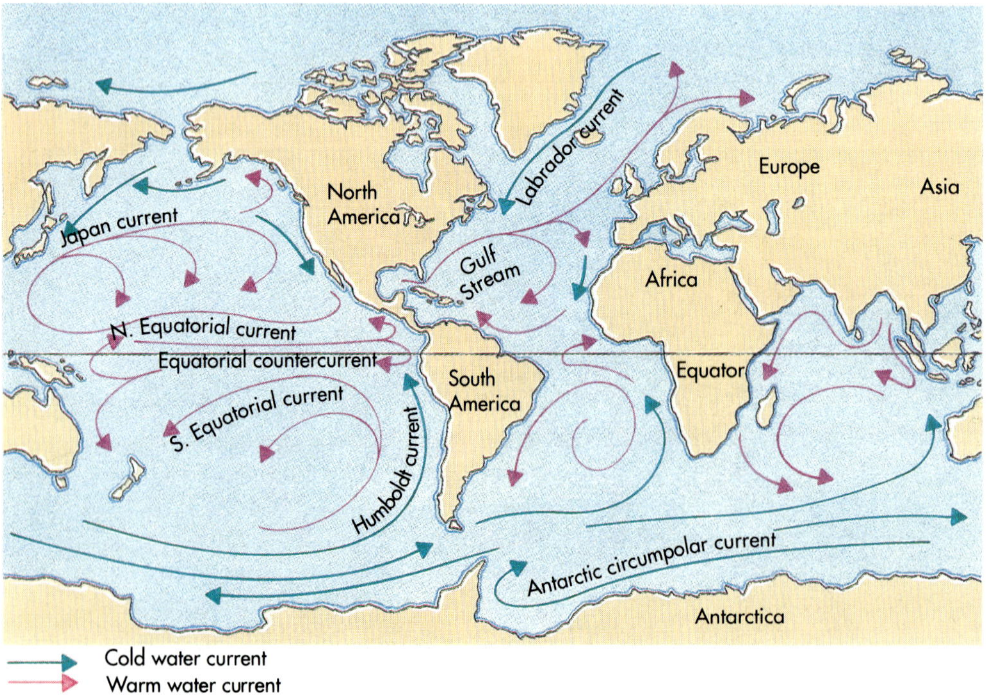

FIGURE 39-5 Ocean circulation. The circulation in the oceans moves in great surface spiral patterns, or gyrals, and affects the climate on adjacent lands.

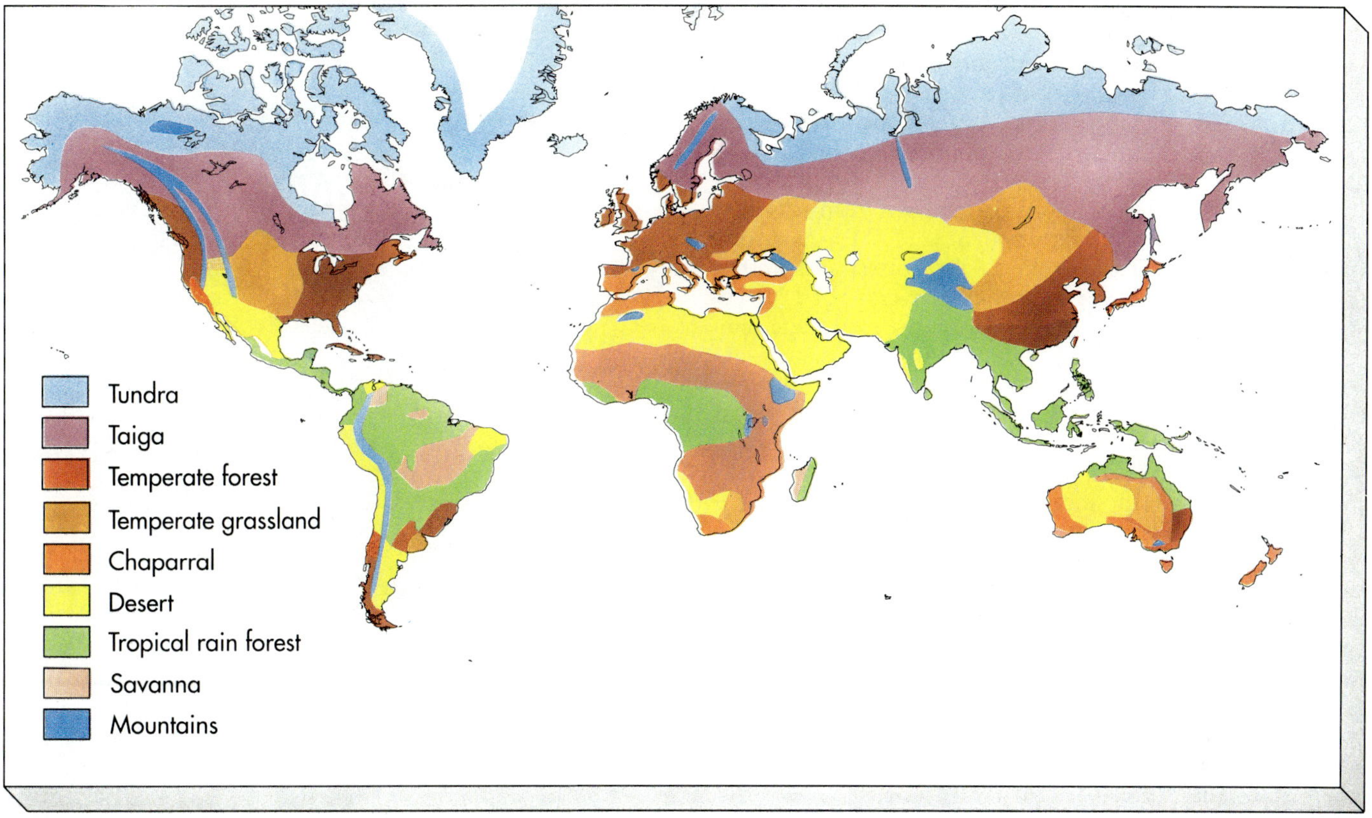

FIGURE 39-6 Distribution of biomes.
Mountain ranges and chaparrals are also shown on this map. Chaparrals are temperate scrublands that border grasslands and deserts in certain parts of the world.

because they redistribute heat. For example, the Gulf Stream in the North Atlantic swings away from North America near Cape Hatteras, North Carolina, and reaches Europe near the southern British Isles. Because of the Gulf Stream, western Europe is much warmer and thus more temperate than eastern North America at similar latitudes. As a general principle, the western sides of continents in the temperate zones of the northern hemisphere are warmer than their eastern sides; the opposite is true in the southern hemisphere.

Life on land: The biomes of the world

Biomes are often classified in seven categories: (1) tropical rain forests, (2) savannas, (3) deserts, (4) temperate grasslands, (5) temperate deciduous forests, (6) taiga, and (7) tundra (Figure 39-6). This list is arranged by distance from the equator, but the biomes do not encircle the Earth in neat bands. Their distribution is greatly affected by the climatic effects caused by the presence of mountains, the irregular outlines of the continents, and the temperature of the surrounding sea. In addition, the climate (and vegetation) changes with the elevation of land similar to changes with increasing latitude (Figure 39-7). Thus tundra-like vegetation occurs near the top of a tall mountain in the tropics, as well as near the North and South Pole.

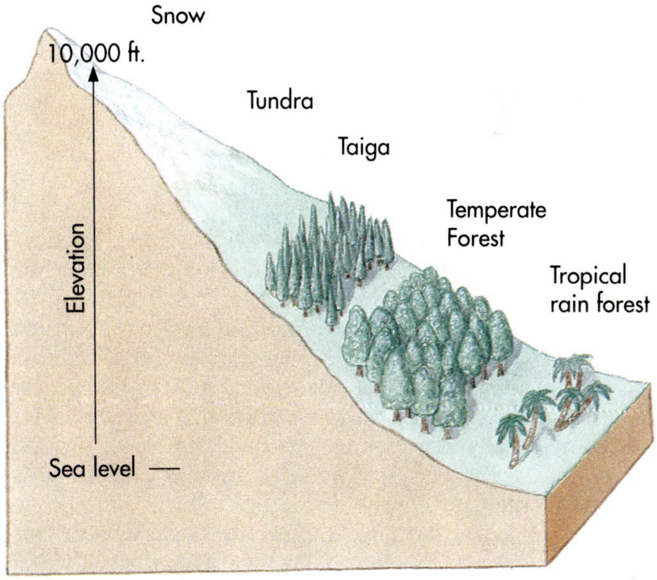

FIGURE 39-7 Elevation and biomes.
Biomes that normally occur far north and far south of the equator at sea level occur also in the tropics but at high mountain elevations. Thus on a tall mountain in southern Mexico or Guatemala, you might see a sequence of biomes like the one illustrated here.

Tropical rain forests

Tropical rain forests occur in Central America; in parts of South America, particularly in and around the Amazon Basin; in Africa, particularly in central and west Africa; and in southeast Asia. As the name suggests, the tropical rain forests occur in regions of high temperature and rainfall, generally 200 to 450 centimeters (80 to 175 inches) per year, with little difference in its distribution from season to season. The temperature averages 25° C (77° F). As a comparison, Houston, Texas—one of the hottest and wettest cities in the United States—receives an average of 121 centimeters (48 inches) of rainfall per year; the average temperature is 20.5° C (69° F). New Orleans, Louisiana—another hot, wet city—receives an average of 145 centimeters (57 inches) of rainfall per year; the average temperature is 20° C (68° F). These cities almost seem dry and cool in comparison to the tropical forest!

You may have the idea that the tropical rain forest is thick with lush vegetation and creeping vines, creating a network too dense to penetrate without a machete. These forests *are* thick and lush—but not at the forest floor. Little can grow on the ground far beneath the canopy of trees whose branches and leaves form an overlapping roof to the forest. In fact, only 2% of the light shining on the forest canopy reaches its floor! Plants that do grow there have large, dark green leaves adapted to conducting photosynthesis at low light levels. Other types of vegetation have interesting adaptations that enable them to compete successfully with the large trees for sunlight. Vines, for example, have their roots anchored in the soil but climb up the trees with their long stems. They reach the canopy where leaves grow to capture sunlight. Other interesting plants are the epiphytes, or "air plants." **Epiphytes** grow on the trees or other plants for support but draw their nourishment mostly from rain water. Some epiphytes catch moisture with modified leaves or flower parts; others have roots that hang free in the air and absorb water (with its dissolved minerals) from the rain (Figure 39-8).

The giant trees of the tropical forest support a rich and diverse community of animals on their branches. Figure 39-9 provides examples of just a few of these interesting organisms. The roots of the trees are interesting also, spreading out from thickened trunks into a thin layer of soil, often no more than a few centimeters deep. These roots transfer the nutrients from fallen leaves and other organic debris quickly and efficiently back to the trees after bacteria and fungi break it down. Very few nutrients remain in the soil. Therefore when humans cut down and then burn these trees to clear the land for agriculture, they are "burning away" the nutrients held in the trees as well as breaking down organic matter to carbon dioxide. The small amount of ash that is left provides few nutrients for the crops farmers try to grow. In 2 to 3 years of farming, these few remaining nutrients are depleted from the soil, and the land remains barren (Figure 39-10).

The tropical rain forest, a biome that occurs in regions of high temperature and rainfall, is characterized by tall trees that support a variety of plant and animal life on their branches.

A

B

FIGURE 39-8 Tropical rain forests: lush equatorial forests.
A Epiphytes in the Monteverde Cloud Forest Reserve in Costa Rica. **B** Blooming only for 24 hours, the flower of a passion vine is a specialist in rain forest competition. Sweet spots reward vigilant guard ants, and poison leaves deter unwanted intruders.

FIGURE 39-9 Ecological specializations in the tropical rain forest.
A Orange and red flower clusters of *Heliconia irrasa* stand out like beacons in the Costa Rican rain forest. They are visited and pollinated by the hummingbirds that they attract.
B Army ants live as huge, mobile, foraging communities in the tropical rain forest. Here a small column is transporting a wasp larva to be used as food.
C Some species of birds, such as the bicolored antbird, fly above the columns of army ants and feed on the insects that they flush from the foliage.
D A mother and baby three-toed sloth hang from a tree limb. Sloths reside in the canopy of the tropical rain forest, as shown here in Panama.
E Sloth moths carry out their entire life cycle in the fur of the three-toed sloth, where their larvae feed on the green algae that grow luxuriantly there.
F A dead leaf butterfly in the forests of Sumatra is superbly camouflaged.

FIGURE 39-10 The destruction of the rain forest.
To clear land for agriculture, farmers first cut down the trees and then burn the stumps in the technique of slash-and-burn agriculture. Here, a rain forest is being destroyed in the Amazon basin of Peru.

BIOLOGY IN FOCUS

The El Niño southern oscillation

The west coast of South America is normally a highly productive area, cooled by the Humboldt current, which sweeps up from the south and makes it possible for penguins (cold-blooded birds) to live on the Galapagos Islands near the equator. The cool Humboldt current is also responsible for bringing the cool, nutrient-rich waters up from the depths and replacing the depleted waters at the surface. Every year, usually in December or January, this cool current begins to warm as the trade winds slack off and warm water flows down the coast to southern Peru and northern Chile. Local fishermen are familiar with this event, since it signals the end of the fishing season, and have named it *El Niño* for "the Christ Child" because the warm current occurs around the Christmas season.

Every few years, however, the El Niño current is much warmer than usual. Accompanying the warmer waters, scientists have also noted changes in atmospheric pressure systems around the world. The term *southern oscillation* is used to describe these atmospheric results of a severe El Niño southern oscillation (ENSO) event such as the one that occurred in 1972 can be devastating on fisheries. Commercial fish stocks disappeared from the waters of Peru and northern Chile. The anchovy fishery was especially hard hit—the commercially valuable anchovy fisheries of Peru were destroyed.

Other populations of animals are affected as well by a severe ENSO. In the ENSO of 1982 to 1983, many birds starved to death on islands throughout the central Pacific, and a huge number of bird colonies simply gave up breeding. Christmas Island, located about 2000 kilometers (1243 miles) south of Honolulu, is an important breeding site for many species of birds. During the spring and summer of 1983, 95% of the 14 million birds that normally nest on the island, representing 18 species, simply left, abandoning their eggs; by June 1983, there were only 150,000 birds on the island.

The effects of an ENSO on the weather are more well known. The severe ENSO of 1986 to 1988 correlated with unusual drought in the states of Washington and Oregon in the summer of 1987. The ENSO event of 1991 to 1992 brought heavy flooding to southern California and parts of Texas.

Not all the effects of an ENSO are negative and catastrophic. The moist, rainy conditions that accompany an ENSO are favorable for crops on land. The Peruvian rice crop, for example, benefits from the ENSO rains. Land birds, such as Darwin's finch on the Galapagos Islands, breed abundantly during an ENSO, and their population sizes increased remarkably during the ENSO of 1982 to 1983.

Scientists still do not know what causes an ENSO event; they do not know, for instance, whether the warm current causes the southern oscillation, or vice versa. Because the cause is still unknown, an ENSO cannot be predicted. Nevertheless, scientists have used the knowledge they have gained to recognize ENSO events in their early stages. Further study of the ENSO may hold the key to understanding the complex interplay of atmospheric and water currents around the globe.

Savannas

Not all areas near the equator are wet; some areas experience prolonged dry seasons or have a lower annual rainfall than the tropical forests (generally, about 90 to 150 centimeters per year, or 35 to 60 inches). The heat, periodic dryness, and poor soils cannot support a forest but have led to the evolution of savannas: open grasslands with scattered shrubs and trees. These areas, situated between the tropical forests and deserts, cover much of central and southern Africa, western India, northern Australia, large areas of northern and east-central South America, and some of Malaysia.

The vegetation of the savanna supports large grazing herbivores such as buffalo and zebra (Figure 39-11); these animals, in turn, are food for carnivores such as lions. The savanna also supports a large number of plant-eating invertebrates, such as mites, grasshoppers, ants, beetles, and termites. The termites, in fact, are one of the most important soil organisms, breaking down dried twigs, leaves, and grass to usable nutrients. In addition, their huge, complex mounds (Figure 39-12) provide passageways for rain water to deeply penetrate the ground rather than just running off or evaporating from the surface.

Like tropical forests, savannas are found near the equator but in areas having less annual rainfall. This climate supports grasslands with only scattered trees and shrubs.

FIGURE 39-11 Savannas: dry, tropical grasslands.
These zebras are grazing on a savanna in Tanzania.

FIGURE 39-12 Termite mound in southern Africa.
Termite mounds allow rainwater to penetrate the ground, thwarting run-off and evaporation.

Deserts

Deserts are biomes that have 25 centimeters (10 inches) or less of precipitation annually. For this reason, the vegetation in deserts is characteristically sparse. The higher the annual rainfall a desert has, however, the greater the amount of vegetation it will be able to support. In fact, ecologists classify deserts based on their annual rainfall: semideserts receive about 25 centimeters (10 inches) per year (Phoenix, Arizona, and San Diego, California, for example), true deserts receive less than 12 centimeters (4.7 inches) per year (Las Vegas, Nevada), and extreme deserts average below 7 centimeters (2.8 inches) per year (Namib Desert in southwestern Africa). The photos of these three types of deserts shown in Figure 39-13 point out the differences in their patterns of vegetation.

A

B

C

FIGURE 39-13 Deserts: arid and hot.
The semidesert in **A** is in South Mountain Park in Phoenix, Arizona. The true desert in **B** is the Great Basin desert near Baker, Nevada. The extreme desert in **C** is the Namib desert in Namibia.

Major deserts occur around 20 to 30 degrees north and south latitude, where the warm air that rose from the equator falls. As previously mentioned, the air at the equator rises, cooling and releasing its moisture, which falls on the tropical forests. The dry air then falls over desert regions, resulting in little precipitation. Deserts also occur in the interiors of continents far from the moist sea air, especially in Africa (the Sahara Desert), Eurasia, and Australia. Some deserts, such as the Baja region of California, are near the ocean yet dry; the winds blow from the north, carrying little moisture because they are cool. High pressure areas off the West Coast of the United States also deflect storms moving down from the north. In addition, some deserts form on the leeward side of mountain ranges, such as in the Great Basin of Nevada and Utah in the United States.

Because desert vegetation is sparse and the skies are usually clear, deserts radiate heat rapidly at night. This situation results in substantial daily changes in temperature, sometimes more than 30 degrees between day and night. Although both hot deserts (the Sahara, for example) and cool deserts (the Great Basin of North America) exist, summer daytime temperatures in all deserts are extremely high, frequently exceeding 40° C (104° F). In fact, temperatures of 58° C (136.4° F) have been recorded both in Libya and in San Luis Potosi, Mexico—the highest that have been recorded on Earth.

Plants have developed a wide variety of adaptations to survive in this difficult environment. Annual plants are often abundant in deserts and simply bypass the unfavorable dry season in the form of seeds. After sufficient rainfall, many germinate and grow rapidly, sometimes forming spectacular natural displays. Characteristics of deserts, of course, are the many species of succulent plants, those with tissues adapted to store water, such as cacti. The trees and shrubs that live in deserts often have deep roots that reach sources of water far below the surface of the ground. The woody plants that grow in deserts may be either deciduous, losing their leaves during the hot, dry seasons of the year, or evergreen, with hard, reduced leaves. The creosote bush of the deserts of North and South America is an example of an evergreen desert shrub. Near the coasts in areas where there are cold waters offshore, deserts may be foggy, and the water that the plants obtain from the fog may allow them to grow quite luxuriantly.

A

B

FIGURE 39-14 How two desert animals conserve water.
A Spadefoot toads, which live in the deserts of North America, can burrow nearly a meter below the surface and remain there for as much as 9 months of each year. Under such circumstances, their metabolic rate is greatly reduced, and they depend largely on their fat reserves. When moist, cool conditions return to the desert, they emerge and breed rapidly. The young toads mature quickly and burrow back underground, using the horny projections on their feet that gave them their name.
B On the very dry sand dunes of the Namib desert in southwestern Africa, the beetle *Onymacris unguicularis* collects fog water by holding up its abdomen at the crest of a dune, thus gathering condensed water on its body.

Desert animals, too, have fascinating adaptations that enable them to cope with the limited water of the deserts (Figure 39-14). Many limit their activity to a relatively short period of the year when water is available or even plentiful; they resemble annual plants in this respect. Many desert vertebrates live in deep, cool, and sometimes moist burrows. Organisms that are active for much of the year emerge from their burrows only at night, when temperatures are relatively cool. Other organisms, such as camels, can drink large quantities of water when it is available and store it for use when water is unavailable. A few animals simply migrate to or through the desert and exploit food that may be abundant seasonally; when the food disappears, the animals move on to more favorable areas.

Deserts occur around 20 to 30 degrees north and south latitude and in other areas that have 25 centimeters (10 inches) or less of precipitation annually. Desert life is somewhat sparse but exhibits fascinating adaptations to life in a dry environment.

FIGURE 39-15 Temperate grasslands: seas of grass.
Among the other names for these grasslands are prairie, steppe, veld, and pampa.

Temperate grasslands

Temperate grasslands have various names in different parts of the world: the prairies of North America, steppes of Russia, pusztas of Hungary, veld of South Africa, and pampas of South America. All temperate grasslands have 25 to 75 centimeters (10 to 30 inches) of rainfall annually, much less than that of savannas but more than that of deserts. Temperate grasslands also occur at higher latitudes than savannas but are often found bordering deserts as savannas do.

Temperate grasslands are characterized by large quantities of perennial grasses; the rainfall is insufficient to support forests or shrublands. Grasslands are often populated by herds of grazing mammals, such as the North American bison (Figure 39-15). The soil in temperate grasslands is rich; in fact, much of the temperate grasslands are farmed for this reason. Grasslands are often highly productive when they are converted to agriculture, and many of the rich agricultural lands in the United States and southern Canada were originally occupied by prairies.

Temperate grasslands experience a greater amount of rainfall than deserts but a lesser amount than savannas. They occur at higher latitudes than savannas, but, like savannas, are characterized by perennial grasses and herds of grazing mammals.

FIGURE 39-16 Temperate deciduous forests: rich hardwood forests.
The leaves of the trees in temperate deciduous forests often change color in the autumn before they fall from the trees as winter approaches.

Temperate deciduous forests

The climate in areas of the northern hemisphere such as the eastern United States and Canada and an extensive region in Eurasia supports the growth of trees that lose their leaves during the winter. Such trees are called *deciduous* (from a Latin word meaning "to fall"), dropping their leaves and remaining dormant throughout the winter. These vast areas of trees are therefore temperate deciduous forests (Figure 39-16) and thrive in climates where summers are warm, winters are cold, and the precipitation is moderate, generally from 75 to 150 centimeters (30 to 60 inches) annually. Precipitation is well distributed throughout the year, but water is generally unavailable during the winter because it is frozen.

The temperate forest differs from the tropical forest in that more vegetation grows near the forest floor than in the tropical forest. Although the temperate forest does have an upper canopy of dominant trees such as beech, oak, birch, and maple, there is a lower tree canopy and a layer of shrubs beneath. On the ground, herbs, ferns, and mosses abound (Figure 39-17). In addition, animal life in the temperate forest is abundant on the ground as well as in the trees; in the tropical forest, animal life is primarily arboreal, with the exception of large mammals, a few bird species, and soil invertebrates.

In areas having less than 75 centimeters (30 inches) of precipitation annually, temperate deciduous forests are replaced by grassland, as in the prairies of North America and the steppes of Eurasia. Where conditions are more limiting—restricted, for example, by intense cold—these forests may be replaced by coniferous forests.

Temperate deciduous forests occur in areas having warm summers, cold winters, and moderate amounts of precipitation. The trees of this forest lose their leaves and remain dormant throughout the winter.

FIGURE 39-17 The floor of the temperate deciduous forest is rich in plant growth.
Plants such as herbs, ferns, and mosses inhabit the forest floor.

Taiga

The northern coniferous forest is called **taiga.** The cone-bearing trees of this forest are primarily spruce, hemlock, and fir and extend across vast areas of Eurasia and North America (Figure 39-18). The taiga is characterized by long, cold winters with little precipitation; most of the precipitation falls in the summers. Because of the latitude where taiga occurs, the days are short in winter (as little as 6 hours) and correspondingly long in summer. The light, warmth, and rainfall of the summer allows plants to grow rapidly, and crops often attain a large size in a surprisingly short time.

The trees of the taiga occur in dense stands of one or a few species of cone-bearing trees. Alders, a common species, harbor nitrogen-fixing bacteria in nodules on their roots; for this reason, they are able to colonize the infertile soils of the taiga. Marshes, lakes, and ponds also characterize

FIGURE 39-18 Taiga: great evergreen forests of the north.
The taiga located in Alaska is dominated by spruces and alders.

the taiga; they are often fringed by willows or birches. Many large mammals can also be found there, such as the elk (Figure 39-19). Other herbivores, including moose and deer, are stalked by carnivores such as wolves, bear, lynx, and wolverines.

To the south, taiga grades into temperate forests or grasslands, depending on the amount of precipitation. Coniferous forests also occur in the mountains to the south, but these are often richer and more diverse in species than those of the taiga. Northward, the taiga gives way to open tundra.

The taiga, or northern coniferous forest, consists of evergreen, cone-bearing trees. The climate of this biome is characterized by long, cold winters with little precipitation.

FIGURE 39-19 Bull elk crossing a mountain stream.
Elk, which are herbivores, are hunted by carnivores such as wolves, bears, lynx, and humans.

FIGURE 39-20 Tundra: cold boggy plains of the north. Mount McKinley National Park in Alaska.

Tundra

Farthest north in Eurasia, North America, and their associated islands, between the taiga and the permanent ice, is the open, often boggy community known as the **tundra** (Figure 39-20). Dotted with lakes and streams, this enormous biome encircles the top of the world, covering one fifth of the Earth's land surface. (A well-developed tundra does not occur in the Antarctic because there is no land at the right latitude.) The tundra is amazingly uniform in appearance, dominated by scattered patches of grasses and sedges (grass-like plants), heathers, and lichens. Some small trees do grow but are primarily confined to the margins of streams and lakes.

Annual precipitation in the tundra is very low, similar to desert-like precipitation of less than 25 centimeters (10 inches) annually. In addition, the precipitation that falls remains unavailable to plants for most of the year because it freezes. During the brief Arctic summers, some of the ice melts. The **permafrost,** or permanent ice, found about a meter down from the surface never melts, however, and is impenetrable to both water and roots. When the surface ice melts in the summer, it has nowhere to go and forms puddles on the land. In contrast, the alpine tundra found at high elevations in temperate or tropical regions does not have this layer of permafrost.

The tundra teems with life during its short summers. As in the taiga, perennial herbs grow rapidly then, along with various grasses and sedges. Large grazing mammals, including musk-oxen, caribou, and reindeer, migrate from the taiga. Many species of birds and waterfowl nest in the tundra in the summer and then return to warmer climates for the winter. Populations of lemmings, small rodents that breed throughout the year beneath the snow, rise rapidly and then crash on a long-term cycle, influencing the populations of the carnivores that prey on them, such as wolves, foxes, and lynx.

The tundra encircles the top of the world. This biome is characterized by desert-like levels of precipitation, extremely long and cold winters, and short, warmer summers.

Life in fresh water

Only 2% of the Earth is covered by fresh water, found standing in lakes and ponds or moving in rivers and streams. Fresh water ecosystems lie near and are intertwined with terrestrial ecosystems. For example, some organisms such as amphibians may move from one ecosystem to another. In addition, organic and inorganic material continuously enters bodies of fresh water from terrestrial communities. Often, the wet, spongy land of marshes and swamps provides habitats intermediate between the two.

Ponds and lakes have three life zones, or regions, in which organisms live: the shore zone, the open-water zone, and the deep zone. The **shore zone** is the shallow water near edges of a lake or pond in which plants with roots, such as cattails and water lilies, may grow. Consumers such as frogs, snails, dragonflies, and tiny shrimp-like organisms live among these producers. The **open-water zone** is the main body of water through which light penetrates. Microscopic floating algae, or phytoplankton, grow here. Microscopic floating animals (zooplankton) feed on the phytoplankton in this aquatic food chain. They, in turn, are eaten by small fish, which are eaten by larger fish. The **deep zone,** the water into which light does not penetrate, is devoid of producers. This dark zone is inhabited mainly by decomposers and other organisms such as clams that feed on the organic material that filters down to them (Figure 39-21). Ponds differ from lakes in that they are smaller and shallower. Therefore light usually reaches to the bottom of all levels of a pond; it has no deep zone.

Rivers and streams differ from ponds and lakes primarily in that their water flows rather than remains stationary. The nature of this ecosystem is therefore different from that of a pond or lake. One difference is that the level of dissolved oxygen is usually much higher in a river or stream than in a standing body of water because moving water mixes with

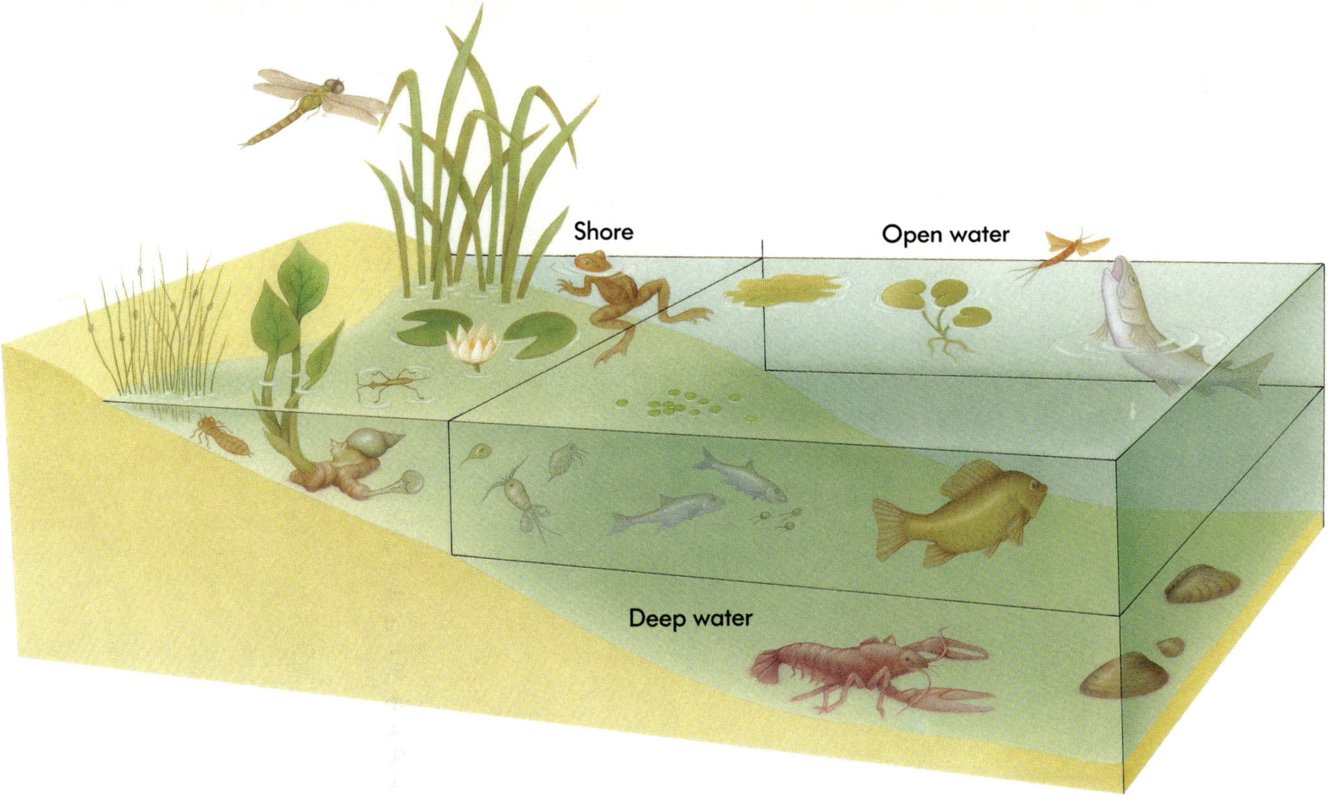

FIGURE 39-21 The three life zones of a lake.
Organisms live in the shore zone, the open-water zone, and the deep-water zone.

the air as it churns and bubbles along. A high level of dissolved oxygen allows an abundance of fish and invertebrates to survive. In addition, only a few types of producers inhabit rivers and streams: various species of algae that grow on rocks and a few types of rooted plants such as water moss.

A river or stream is characterized as an open ecosystem; that is, it derives most of its organic material from sources other than itself. Detritus (debris or decomposing material) flows from upstream or enters from the land. Leaves and woody material drop into the stream from vegetation bordering its banks. Rainwater washes organic material from overhanging leaves. In addition, water seeps into a river or stream from below the surface of adjoining land, carrying with it organic materials and, in some cases, fertilizers and other chemicals. These nutrients feed the producers and small consumers. As commonly occurs in pond and lake food chains, large fish feed on smaller fish that feed on tiny invertebrates. The river/stream ecosystem is largely heterotrophic and is strongly tied to terrestrial ecosystems that surround it.

Fresh water ecosystems lie near and are intertwined with terrestrial ecosystems. From them, organic and inorganic material continuously enters fresh water ecosystems.

As the water from a stream or river flows into a lake or the sea, the velocity of the water decreases. As the water slows, sediment carried along by faster water now sinks to the bottom. These deposits form fan-shaped areas called **deltas.** Accumulated sediment breaks the path of the water into many small channels that course through the delta. Occasionally, the path of the water is actually stopped as sediment accumulates. For this reason a delta created by a large river, such as the Mississippi River delta, consists of a great deal of swampy, marshy land.

Estuaries: Life between rivers and oceans

As rivers and streams flow into the sea, an environment called an **estuary** is created where fresh water joins salt water. In the shallow water of the delta areas, rooted grasses often grow. Other producers of the estuary are various types of algae and phytoplankton. Consumers are primarily mollusks, crustaceans, and fish. All organisms inhabiting estuaries, however, have adaptations that allow them to survive in an area of moving water and changing salinity. (*Salinity* refers to the concentration of dissolved salts in the water.) Therefore many estuarine organisms are bottom-dwellers and attach themselves to bottom material or burrow in the mud. These organisms are found living in regions of optimum salinity for their survival.

Oysters are one of the most important bottom-dwellers in estuaries, providing a habitat for many other species of organisms. These mollusks either bury themselves in the mud, forming oyster beds, or cement themselves in clusters to the partially buried shells of dead oysters, forming oyster reefs. Many other invertebrates, such as sponges and bar-

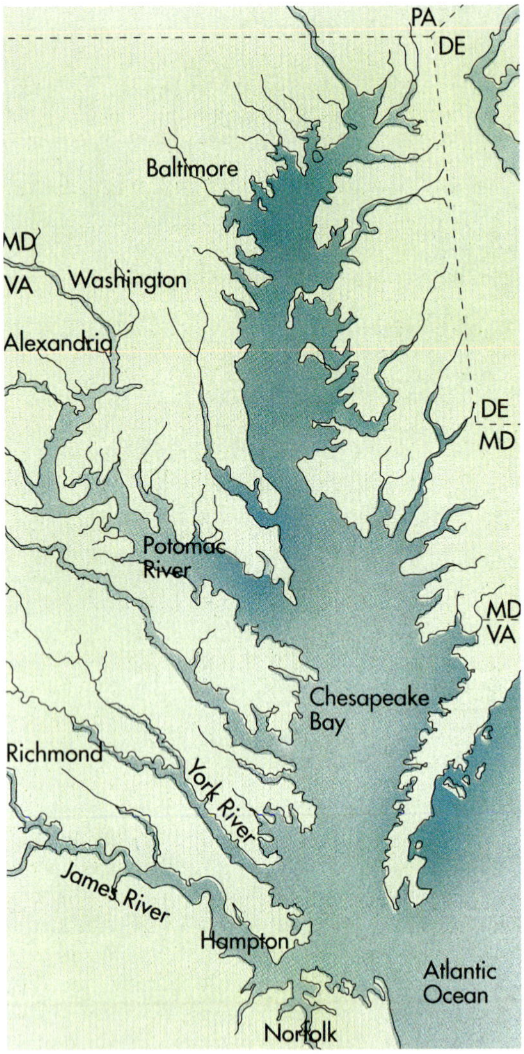
PA
DE
Baltimore
MD
VA
Washington
Alexandria
DE
MD
Potomac River
MD
VA
Chesapeake Bay
Richmond
York River
James River
Hampton
Atlantic Ocean
Norfolk

A

B

C

D

E

FIGURE 39-22 Chesapeake Bay.
Chesapeake Bay has more than 11,300 kilometers (7020 miles) of shoreline and drains more than 166,000 square kilometers (64,100 square miles) in one of the most densely populated and heavily industrialized areas in North America. The body of open water is about 320 kilometers (199 miles) long and, at some points, nearly 50 kilometers (31 miles) wide.
A Large metropolitan areas and shipping facilities make the bay one of the busiest natural harbors anywhere.
B Early settlers of the bay area tell tales of scooping fish out of the bay with their hands. Today, rockfish, once the backbone of bay fishery, are so scarce that a rockfish ban was imposed in 1989 in an effort to increase the population.
C One of the most biologically productive bodies of water in the world, the bay yielded an annual average of about 275,000 kilograms (600,000 pounds) of fish in the 1960s but only a tenth as much in the 1980s. The human population grew 50% during the same period.
D Oil transport and commercial shipping is expected to double by the year 2020. This grebe is coated from a recent oil spill off the mouth of the Potomac River.
E Uncontrolled erosion from certain agricultural practices, pesticides, and increases in nutrients block the light needed for photosynthesis and upset the delicate ecological balance on which the productivity of the bay depends. The states that border the bay are cooperating, with the assistance of the Environmental Protection Agency, to try to bring the bay back to its former productivity.

nacles, attach themselves to the oysters and feed on algae. Other species such as crabs, snails, and worms live on, beneath, and between the oysters, feeding on the oysters themselves or on detritus trapped in the oyster reef. One researcher, in fact, has documented over 300 species of organisms living in association with a single oyster reef!

The motile organisms of the estuary are primarily crustaceans such as crabs, lobsters, and shrimp and various species of fish. Fish exhibit interesting reproductive adaptations to the varying salinities of the estuary. Some species, such as the striped bass, spawn upstream from the estuary where the salinity of the water is low. The larvae and young fish move downstream through increasing concentrations of salt as they develop, moving into the ocean in adulthood. In a similar manner, shad spawn upstream in fresh water, and the young spend their first summer in the estuary before swimming to the sea.

> **Estuaries are places where the fresh water of rivers and streams meets the salt water of oceans. These ecosystems are made up of plentiful communities of organisms exhibiting behaviors and growth characteristics adapted to changing salinity.**

Nutrients are more abundant in estuaries than in the open ocean because they are close to terrestrial ecosystems and derive much of their nutrients from them as rivers and streams do. Unfortunately, estuaries are also easily polluted from these sources. In Chesapeake Bay (Figure 39-22), for example, complex systems of rivers enter the Atlantic Ocean, forming one of the most biologically productive bodies of water in the world. In the 1960s, the bay yielded an annual average of about 275,000 kilograms (600,000 pounds) of fish. As the human population in this area increased, however, along with oil transport and commercial shipping, pollution also increased. More than 290 oil spills were reported in the bay in 1983 alone. In addition, uncontrolled erosion of the land into the water and leaching of pesticides has caused a 90% decrease in the yield of fish. Maryland and Virginia, the states that border the bay, along with the Environmental Protection Agency, are working to bring the Chesapeake Bay estuary back to its former productivity.

Life in the oceans

Although only 2% of the Earth is covered by fresh water, nearly three quarters of the Earth's surface is covered by ocean. These seas have an average depth of 4 kilometers (approximately 2½ miles), and they are, for the most part, cold and dark. The concentration of oxygen, as well as the availability of light and food, is a factor that limits life in the ocean. Although cold water is able to "hold" more oxygen than warm water, the warmer sea water near the surface of the ocean mixes with the oxygen in the atmosphere. Therefore oxygen is present in its highest concentrations in the upper 200 meters (650 feet) or so of the sea. Light is most abundant in the top 100 meters (325 feet).

The marine environment provides a variety of habitats, but it can be divided into three major life zones:

1. The **intertidal zone,** the area between the high tide mark and the low tide mark
2. The **neritic zone,** the area of shallow waters along the coasts of the continents, which extends from the low tide mark to waters up to 200 meters deep
3. The **open-sea zone,** composing the remainder of the ocean

The intertidal zone

The wind-swept shoreline is a harsh place for organisms to live. As the tide rolls in and out, environmental conditions change from hour to hour: wet to dry, sun protected to sun parched, and wave battered to calm. Nevertheless, life abounds in the intertidal zone, exhibiting interesting adaptations and characteristics necessary for survival.

Figure 39-23 shows the rocky shore at low tide at the Pacific Grove Monterey Bay in California. The exposed rocks are teeming with life, but these organisms vary along a continuum from the driest areas (those least often covered with water) to the wettest areas (those most often covered with water). Highest up on the rocks, in areas that the high tide sometimes does not reach, grow certain lichens and algae. Somewhat lower on the rocks grow barnacles. Then oysters, blue mussels, and limpets (mollusks that have conical shells) take over, followed by brown algae and red algae. "Forests" of large brown algae, or kelp, take over in areas that are exposed for only short periods of time. All of these organisms have adaptations such as hard shells (Figure 39-24) or

FIGURE 39-23 A rocky shore in Pacific Grove in Monterey Bay, California.
The different areas on the rock all receive different amounts of moisture, providing a variety of habitats for a variety of organisms.

gelatinous coverings that keep them from drying out and are either anchored within the sand or stick to the surfaces of rocks so that they will not be washed away.

In contrast to the rocky shore, the sandy shore (Figure 39-25) looks as though no life is present. However, life abounds beneath the sand and mud. Because organisms have no large surfaces to which they can attach, they are adapted to burrowing under the sand during low tide. Copepods, tiny "micro" crustaceans, are predominant organisms. In addition, worms, crabs, and mollusks such as clams burrow to safety when the tides roll out. In areas near the low tide mark, sea anemones, sea urchins, and sea stars make their home.

FIGURE 39-24 Rocky shore organisms.
A rocky shore in Maine shows a few of the organisms that make this shore their home. The pink sea star clings to the rocks, as do barnacles, which use special filaments to anchor themselves.

FIGURE 39-25 A sandy shore.
In contrast to the rocky shore, there seems to be no life present on this sandy shore in Hawaii. However, microscopic animals make their homes in the spaces between grains of sand, and larger animals burrow beneath the sand during low tide.

The intertidal zone has plentiful light and is home to a variety of producers. Along with the algae, phytoplankton float in the water and are used for food by the zooplankton and many other consumers. In addition, the heterotrophs of the intertidal zone have the waves to thank for bringing fresh organic material to them and for washing away their wastes.

The neritic zone

Surrounding the continents of the world is a shelf of land that extends out from the intertidal zone, sloping to a depth of approximately 200 meters (approximately 650 feet) beneath the sea. This margin of land is called the **continental shelf.** The waters lying above it make up the neritic zone, which is derived from a Greek root referring to the sea. Because light reaches the waters of most of this zone, it supports an abundant array of plant and therefore animal life.

One outstanding community in the neritic zone is the **coral reef.** The term *reef* refers to a mass of rocks in the ocean lying at or near the surface of the water. Coral reefs are built by marine animals, or corals (phylum Cnidaria), that secrete carbonate, a hard, shell-like substance. With the help of algae that reside in their bodies, the coral build on already shallow portions of the continental shelf or on submerged volcanoes in the ocean. Complex and fascinating ecosystems, coral reefs provide habitats for a variety of invertebrates and fish (Figure 39-26). Along with the tropical rain forests, coral reefs are the most highly productive ecosystems in terms of biomass (see Chapter 38).

A

B

C

D

E

F

FIGURE 39-26 The coral reef: a delicate biome in danger. The living corals in coral reefs depend on constant conditions for survival. Increasingly, the conditions in which coral reefs live are being disrupted by pollution and human carelessness. Preventative measures and a sensitivity to the precarious balance necessary for the coral reefs' survival will be instrumental in saving the reefs from destruction.

A Coral reef lagoon, Fiji.

B Coral reef in Fiji, showing hard and soft corals.

C Coral reef in Borneo, showing the brilliant colors of the different coral formations.

D Green turtle hovering over a soft and hard coral-encrusted reef off Sipadan Island.

E Scallop attached to a limestone reef in the Indo-Pacific.

F Diver destroying a reef in Key Largo, Florida. When exploring a coral reef, divers must never touch or stand on the corals.

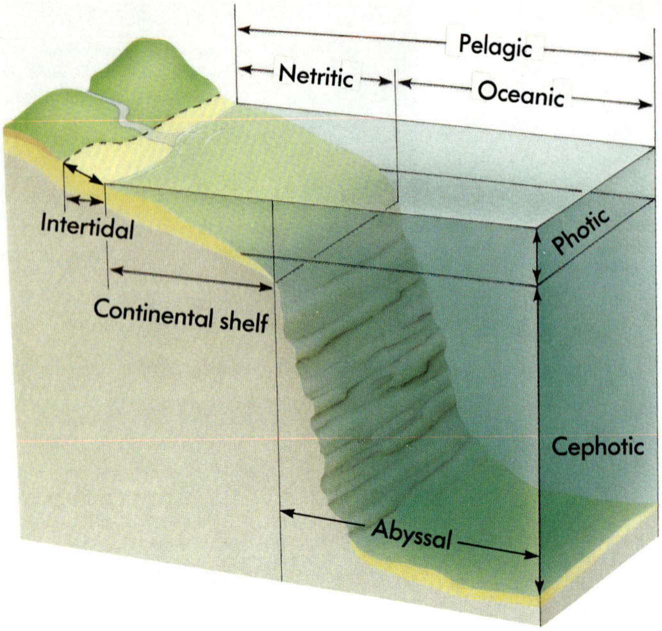

FIGURE 39-27 Zones of the ocean.
The pelagic zone encompasses the oceanic and neritic zones. Nearer to shore is the continental shelf and the intertidal zone. The photic zone and aphotic zones are measured by the amount of light that penetrates the water.

The open-sea zone

Beyond the continental shelf lies the great expanse of the open ocean. This open-sea zone is often referred to as the **pelagic zone,** a term derived from another Greek word meaning "ocean" (Figure 39-27). Within this huge ecosystem exists many diverse forms of life—some with which you are familiar, such as the floating plankton and various species of fishes. But other forms of life are unfamiliar—even bizarre—such as the anglerfish shown in Figure 39-28, *D*.

Organisms live in the vast expanse of the ocean based on the availability of light and food. Temperature, salinity, and water pressure also play roles in creating the various habitats of the ocean. Light is only available to organisms from the water's surface to an approximate depth of 200 meters (650 feet). In fact, this area of the open ocean is called the **photic zone** for this reason. Phytoplankton thrives in this well-lighted layer of the ocean (especially within the better-lighted upper 100 meters [325 feet]), drifting freely with the ocean currents and serving as the base of oceanic food webs. Zooplankton float with the phytoplankton and are first-order consumers in many photic food webs. Other typical heterotrophs of this zone are most air-breathing mammals such as whales, porpoises, dolphins, seals, and sea lions and fishes such as herring, tuna, and sharks. These fishes and mammals of the sea are called **nekton** and feed on the plankton and on one another. Together, the organisms that make up the plankton and the nekton provide all of the food for those that live below.

Little light penetrates the ocean from 200 to 1000 meters (650 to 3250 feet). Therefore no photosynthetic organisms live in this zone known as the **mesopelagic,** or middle ocean. At this depth, temperatures remain somewhat steady throughout the year but are cooler than the water above; water pressure increases steadily with depth. Under these conditions, many bizarre organisms have evolved, such as those that exhibit bioluminescence (see Figure 26-12), which they use to communicate with one another or attract prey. The organisms common to middle ocean life are fish with descriptive names such as swordfish, lanternfish, and hatchetfish; certain sharks and whales; and cephalopods such as octopi and squid.

Peculiar creatures also live in the ocean depths of 1000 or more meters—a mile or so (and more) beneath the surface. This region of the ocean is called the **abyssal zone,** meaning "bottomless." Although the ocean is not really bottomless, the water at this tremendous depth contains high concentrations of salt, is under immense pressure from the water above, and is very cold. The organisms living at this level cannot make trips into the photic zone to capture food as some mesopelagic organisms do, but must feed on material that settles from above.

Organisms collectively referred to as **benthos,** or bottom-dwellers, live on the ocean floor. Sea cucumbers and sea urchins crawl around eating detritus. Various species of clams and worms burrow in the mud, feeding on a similar array of decaying organic material. Bacteria are also rather common in the deeper layers of the sea, playing important roles as decomposers as they do on land and in fresh water habitats.

Of the three main life zones in the ocean, the intertidal zone, the neritic zone and the photic area of the open-sea zone support the most life because of the presence of light and the availability of oxygen. Organisms that live in waters beneath the photic zone must feed on organic material and detritus that falls from above and be adapted to increased water pressure and salinity.

A

B

C

D

E

FIGURE 39-28 The open ocean.
A What mysteries lie beneath the ocean's deepest waters?
B Flying fish in the eastern Pacific ocean.
C School of yellowfin tuna swimming around a coral reef.
D A deep-sea anglerfish showing the lure used by the fish to attract prey. The lure is actually part of the fish, but to prey, the lure looks like a tempting bit of food. When the prey comes close to investigate, the anglerfish quickly eats the prey.
E Hydrothermal vent community located in the deep waters off the Galapagos Islands.

Summary

1. Biomes are wide-ranging, distinctive ecosystems of plants and animals that occur together within certain climatic regions and have definite characteristics. These characteristics are a direct result of temperature and rainfall patterns.
2. The climate of a region is determined primarily by its latitude and wind patterns. These factors interact with the surface features of the Earth, resulting in particular rainfall patterns. The temperature, rainfall, and altitude of an area, in turn, provide conditions that result in the growth of vegetation characteristic of that area.
3. Seven categories of biomes, arranged by distance from the equator, are (1) tropical rain forests, (2) savannas, (3) deserts, (4) temperate grasslands, (5) temperate deciduous forests, (6) taiga, and (7) tundra.
4. The tropical rain forests and savannas lie nearest the equator. The tropical rain forest receives an enormous amount of rain year round, whereas the savanna experiences less rain and prolonged dry spells. These differences promote the growth of tall trees and other lush vegetation in the tropical rain forest and open grasslands with scattered trees and shrubs in the savanna.
5. Deserts are extremely dry biomes. Hot deserts are hot year round, whereas cool deserts are hot only in the summer. Deserts are of great biological interest due to the extreme behavioral, morphological, and physiological adaptations of the plants and animals that live there.
6. Temperate grasslands receive less rainfall than savannas but more than deserts. The soil in these grasslands is rich, so they are well suited to agriculture.
7. The temperate deciduous forests receive moderate precipitation that is well distributed throughout the year. The climate of these forests differs from tropical forests in that they receive less rainfall, are found at a higher and cooler latitude, and experience cold winters. The trees are therefore deciduous, losing their leaves and remaining dormant throughout the winter.
8. The taiga is the coniferous forest of the north. It consists primarily of cone-bearing evergreen trees, which are able to survive long, cold winters and low levels of precipitation. Even farther north is the tundra, which covers about 20% of the Earth's land surface and consists largely of open grassland, often boggy in summer, which lies over a layer of permafrost.
9. Fresh water ecosystems make up only about 2% of the Earth's surface; most of them are ponds and lakes. Ponds and lakes have a shore zone and an open-water zone. Lakes have a deep zone. Rivers and streams differ from these bodies of water because they contain moving water that mixes with the air to provide high levels of oxygen for its fish and invertebrate inhabitants.
10. Estuaries are places where fresh water meets salt water as rivers empty into the ocean. These ecosystems receive nutrients from the surrounding land and support a large number of living things.
11. The marine environment consists of three major life zones: the intertidal zone, between the high and low tide marks; the neritic zone, the water from the low tide mark that lies over the continental shelf, and the open-sea zone. Within these zones, the ocean supports the most life in areas that have light and sufficient quantities of dissolved oxygen.

REVIEW QUESTIONS

1. What is a biome?
2. January is a winter month in the United States but part of summer in Australia. Why?
3. Summarize how atmospheric circulation affects climates around the world.
4. What factors determine the climate of a region?
5. Identify the terrestrial biome associated with each of the following:
   a. Animals in these areas live mostly in trees.
   b. Dominant trees such as oak and beech allow enough light so that herbs, ferns, and mosses grow on the ground.
   c. Because of permafrost, when surface ice melts, the ground cannot absorb the water.
   d. Although located near the equator, these regions are mostly open grasslands.
   e. These coniferous forests receive most of their precipitation during the summer.
   f. Herds of North American bison once roamed these areas, grazing on large expanses of perennial grasses.
   g. These regions radiate heat rapidly at night, causing wide temperature differences between day and night.
6. About 100,000 square kilometers of tropical rain forest are being destroyed each year (more than 1.5 acres per second), often to make room for planting crops. Do you think these areas will make good farmland? Explain your answer.
7. What factors limit rainfall in a desert?
8. Distinguish between taiga and tundra.
9. Describe the three life zones found in fresh water. What types of organisms live in each?
10. What is an estuary? Summarize how its inhabitants are adapted to its conditions.
11. Why are estuaries and fresh water regions so vulnerable to human pollution?
12. Identify and briefly describe the three major life zones of the ocean.
13. The ocean is divided into three zones based on depth. Describe what types of organisms are found in each zone.

KEY TERMS

abyssal zone 690
benthos 690
biome 672
continental shelf 688
coral reef 688
deep zone 684
delta 685
epiphyte 676
estuary 685
intertidal zone 687
mesopelagic zone 690
nekton 690
neritic zone 687
open-sea zone 687
open-water zone 684
pelagic zone 690
permafrost 684
photic zone 690
shore zone 684
taiga 682
tundra 684

THOUGHT QUESTIONS

1. Why do you think there are so many more species in the tropical rain forests of Brazil than in the temperate forests of Michigan? How would you go about testing your idea?
2. Temperatures over the antarctic ice cap (South Pole) are markedly colder than over the arctic ice cap (North Pole). Why do you think this is so?
3. Much of the world's most productive agriculture is carried out on soil of temperate grasslands. Agriculture in the tropics is far less productive. Why do you think temperate grassland soil is so much richer than soil in a tropical rain forest?

FOR FURTHER READING

Attenborough, D. (1984). *The living planet: A portrait of the Earth.* London: William Collins, Sons & Co., Ltd., and British Broadcasting Corporation.
This is a beautifully written and illustrated account of the biomes.

Gore, R. (1990, Februrary). Between Monterey tides. *National Geographic,* pp. 2-43.
This is a beautifully illustrated account of life along the California coast.

Ropelewski, C. F. (1992, April 9). Predicting El Ninõ events. *Nature,* vol. 38.
Using statistical correlations of patterns of circulation in the upper atmoshpere, it may be possible to predict the onset of an El Ninõ event.

Swan, L. (1992, April). The aeolian biome. *BioScience,* 42(4):262-270.
The aeolian biome is a region around and beyond the limits of flowering plants, such as a high altitude spot on Mt. Everest. The animals in these biomes are nourished by airborne nutrients, and studies of aeolian zones can reveal much about how life began on Earth billions of years ago.

Whitehead, J.A. (1989, February). Giant ocean cataracts. *Scientific American,* pp. 50-57.
Enormous undersea cataracts play a crucial role in determining the chemistry and climate of the deep ocean.

40

The biosphere

Today and tomorrow

HIGHLIGHTS

The biosphere extends from the tops of the mountains to the depths of the seas; it is all the parts of the Earth where biological activity occurs.

Humans need to work toward a sustainable society that uses nonfinite and renewable sources of fuel, recycles resources to the fullest extent, and manages forests to meet global needs while not compromising the ability of future generations to survive.

Water is being polluted by multiple sources, causing physical or chemical changes that harm living and nonliving things; prevention of this contamination is essential to ensure a safe water supply.

A primary cause of air pollution is the combustion of fossil fuels; energy conservation measures and new energy technology are needed.

OUTLINE

The biosphere

The land

- Diminishing natural resources
  - Fossil fuels
  - Mineral resources
  - Deforestation
- Species extinction
- Solid waste

The water

- Surface water pollution
- Ground water pollution
- Acid rain

The atmosphere

- Air pollution
- Ozone depletion

Overpopulation and environmental problems

Just imagine that you could hold the world in your hands and make a difference in its future. Well, imagining isn't even necessary. All people do hold the world in their hands in a figurative way. Everyone's actions have a direct impact on the Earth and on the quality of life that you and others will experience for generations to come.

However, you must ask yourself some crucial questions. How *are* you affecting the Earth? What environmental problems do people face today? And what can you do to deal with those problems to be sure that the quality of life on Earth will be enhanced for yourself and your children? The answers to these questions are the focus of this chapter and will help you understand more about how you shape this fragile planet on which you live.

The biosphere

Life on Earth is confined to a region called the **biosphere,** the global ecosystem in which all others exist. The biosphere extends from approximately 9000 meters (30,000 feet) above sea level to about 11,000 meters (36,000 feet) below sea level. You can think of it as extending from the tops of the highest mountains (such as Mount Everest or some of the Himalayas) to the depths of the deepest oceans (such as the Mariana Trench of the Pacific Ocean)—the part of our Earth in which the land, air and water come together to help sustain life (Figure 40-1).

The biosphere is often spoken of as the **environment.** This general term refers to everything around you—not only the land, air, and water but other living things as well. You can speak, for example, of the environment of an ant, a water lily, or all the peoples of the world. Your particular environment can change during the day from a home environment, to a classroom environment, and then to an office environment. The environment can include a great deal—or very little—of the total biosphere and its living things.

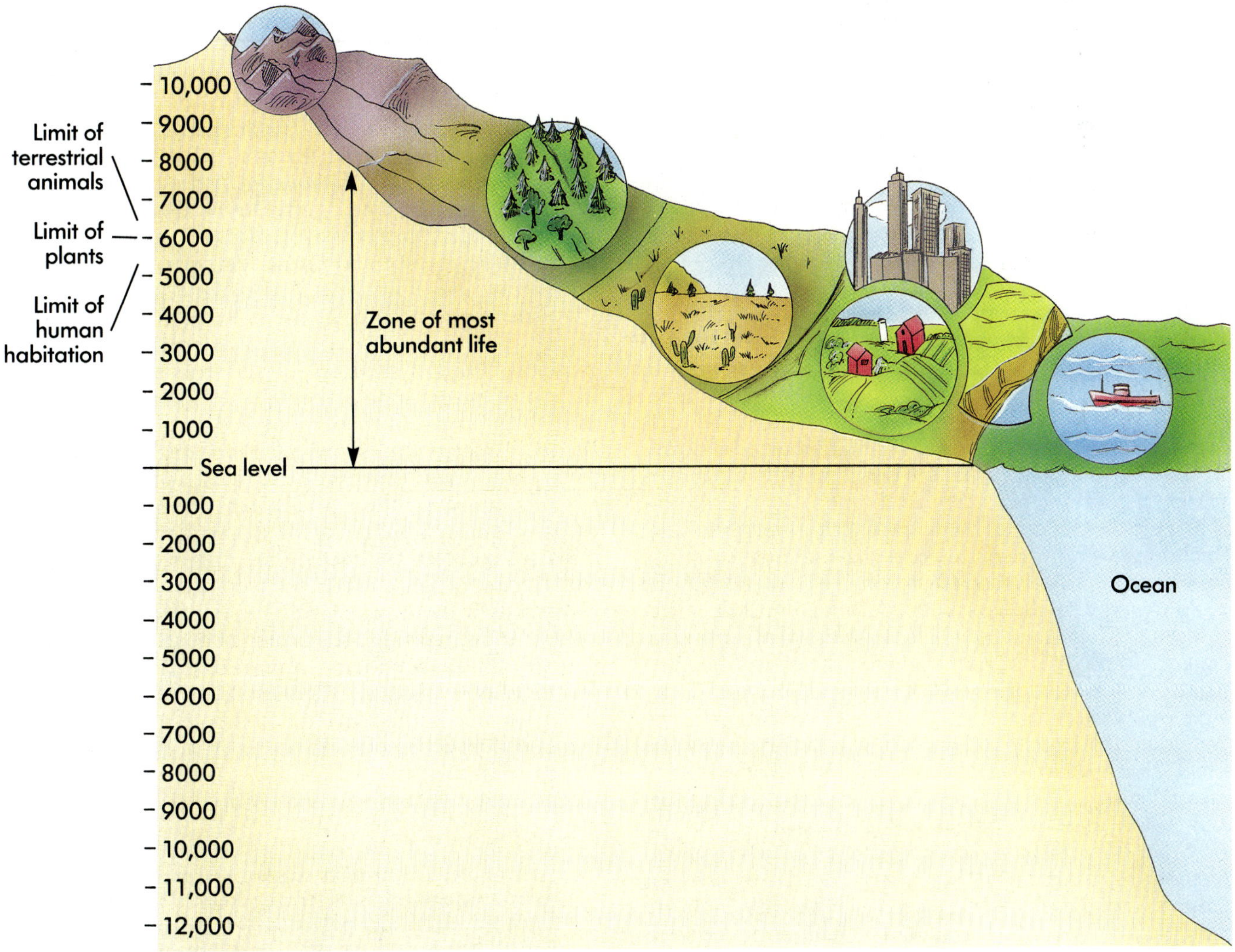

FIGURE 40-1 The biosphere.
The biosphere is that part of the Earth where life can be found, from the highest mountain peaks to the depths of the ocean. Most life, however, is found in the shaded areas.

The land

Humans interact with the land and its inhabitants in a myriad of ways. Its natural resources are used. Wastes are produced that are filling and in some cases polluting the land. Forests are cut down, stripping many species of their habitats and therefore their life. And people continue to procreate at a rate that will result in the doubling of the world's population by the year 2030. What can people do to stop using the land in destructive ways? What can people do to help heal the Earth?

Diminishing natural resources

The land gives humans many things: fossil fuels, timber, food, and minerals. Water, too, is an important natural resource. Some of these resources—fossil fuels and minerals—are finite; that is, they cannot be replaced. Forests *can* be used and replaced *if managed wisely*. However, when whole sections of forest are cut and burned and their soil depleted, they cannot be replaced . . . nor can the wildlife that once lived there.

Fossil fuels

Many of the fuels used to heat homes and run cars are **fossil fuels.** These substances—coal, oil, and natural gas—are formed over time (acted on by heat and pressure) from the undecomposed carbon compounds of organisms that died millions of years ago. Environmental scientists suggest that instead of depending on these fuels for energy—fuels that are finite—people should work toward a **sustainable society** that uses nonfinite and renewable sources of fuel. In a sustainable society, the needs of the society are satisfied without compromising the ability of future generations to survive and without diminishing the available natural resources.

The primary renewable and nonfinite sources of power that are available are the sun, the wind, moving water, geothermal energy, and bioenergy. Technology is available now to harness these energies but are not yet in widespread use. **Nuclear power,** once considered a viable alternative energy source, will probably contribute only 6% to 8% of the world's energy by the year 2000. Although nuclear power is dependent on uranium, a finite but abundant natural resource, its problems lie primarily with the cost of building nuclear power plants, disposal of highly radioactive wastes, and the public fear regarding safety. The waste disposal problem would be virtually eliminated if nuclear fusion reactors can be developed to replace today's reactors, all of which use nuclear fission (splitting) reactions. Nuclear power could become a primary source for generating electricity.

Solar power is not a new technology; panels mounted on rooftops, for example, heat water that can be used for bathing, cooking, and space heating (Figure 40-2, *A*). In addition, homes and businesses can be built using the strategies of passive solar architecture (Figure 40-2, *B*). The technology is also available to use solar power to produce electricity by converting the sun's energy to electricity after it has been captured within a fluid such as oil or by the use of photovoltaic solar cells that can convert sunlight directly into electricity.

Water has been used as an energy source for decades (Figure 40-2, *C*) and is currently supplying 20% of the world's electricity. A new form of hydropower called **wave power** is now being tested. Wave power uses the vertical motion of sea waves to produce electricity. In addition, **tidal power,** the use of the movement of the tides of the oceans to generate electricity, is currently under study in France and England (Figure 40-2, *D*). The wind, too, has been used not for decades but for centuries as a source of energy. Today, windmills are being used in developing countries (those not yet industrialized) for pumping water to livestock and to irrigate the land. In developed countries, however, windmills are being used in an entirely new way—to generate electricity. This "new breed" of windmill has rigid blades fashioned from lightweight materials (Figure 40-2, *E*), and looks unlike the picturesque windmills of Holland and Cape Cod, Massachusetts. Currently, the United States and China are the two countries that lead in the use of **wind power** to generate electricity. In fact, some scientists predict that the United States will be generating 10% to 20% of its electricity from wind power by the year 2030.

Geothermal energy refers to the use of heat deep within the Earth. In some places, reservoirs of hot water exist that can be extracted from the Earth using drilling procedures much like those used to tap into the Earth's oil and natural gas reserves. Alternatively, dry, hot rock can be drilled and water flushed through it. After the water is heated within the Earth, it can be used directly for heating purposes or as part of a process to produce electricity. Currently, this technology is being used and further developed in the United States, England, Italy, New Zealand, and Japan.

Bioenergy refers to the use of living plants to produce energy. The most obvious type of bioenergy, the burning of wood, was first used by our ancestors approximately 1 million years ago. In fact, until the Industrial Revolution of the 1800s, wood, not coal, supplied most of the world's energy. Today, wood supplies 12% of the world's energy, primarily in Latin America, Asia, India, and Africa. Unfortunately, however, the world is experiencing a fuel wood crisis—demand is exceeding the supply. The reasons for this crisis are complex but include the high world population and therefore high demand for wood, the degradation of woodlands without proper reforestation (replanting) techniques, and the cutting and burning of huge areas of tropical rain forests. In developing countries, the use of **bio-gas machines** is helping ease the shortage of fuel wood. These stoves use microorganisms to decompose animal or human excrement or other organic wastes in a closed container. This process yields a methane-rich gas that can be used as to fuel stoves, light lamps, and produce electricity.

Environmental scientists suggest that humans use renewable and nonfinite energy sources such as solar, wind, and water power, as well as tap the Earth's geothermal energy. The use of such fuels rather than the extensive use of fossil fuels will help satisfy people's needs without diminishing the available natural resources.

A

B

C

D

E

FIGURE 40-2 Alternative energy sources.
To decrease reliance on fossil fuels, many individuals and nations are experimenting with alternative energy sources.
A Solar power. These panels collect the energy from sunlight and use this energy to heat homes and water.
B A passive solar home, with solar panels mounted on the roof.
C Hoover Dam in Arizona. Water power has been used for many centuries.
D Tidal power plant in the Bay of Fundy. Using the ocean's tides to generate power is still in the experimental stages.
E Modern windmills in southern California.

A newer form of bioenergy is the use of plants such as corn and sugar cane to produce carbohydrates that are fermented, producing liquid fuels like ethanol. This technology, however, produces a product that appears to be an expensive alternative to fossil fuels. In addition, many cars do not function well on gasoline with ethanol additives.

Mineral resources

Minerals are inorganic substances that occur naturally within the Earth's crust. Zinc, lead, copper, aluminum, and iron are the minerals that humans use in the greatest quantities. However, other minerals are also mined, such as gold, silver, copper, and mercury. These minerals are present in the Earth in fixed amounts; once they are used, they are gone. Figure 40-3 shows the number of years the world's reserves of various minerals will last based on the rate of current consumption.

What can be done to avert a mineral shortage in years to come? First of all, people can cut down on their consumption. An American uses, on the average, 6 times more zinc, lead, and iron than a person living outside of the United States. In addition, an American uses 11 times more copper and 14 times more aluminum! Researchers agree that increased use of plastics and technology such as microelectronics will lower the demand for certain minerals. Plastics, however, are derived from petroleum products, so their manufacture increases consumption of fossil fuels.

Recycling is another way to help avert a mineral shortage in the coming years. Aluminum soda cans are produced from the clay-like mineral bauxite. Producing cans from recycled aluminum uses 95% less energy than producing them from bauxite, saves mineral resources, and saves waste disposal costs and problems. Plastics are not easily recyclable and are not "truly" recyclable. That is, plastics cannot be

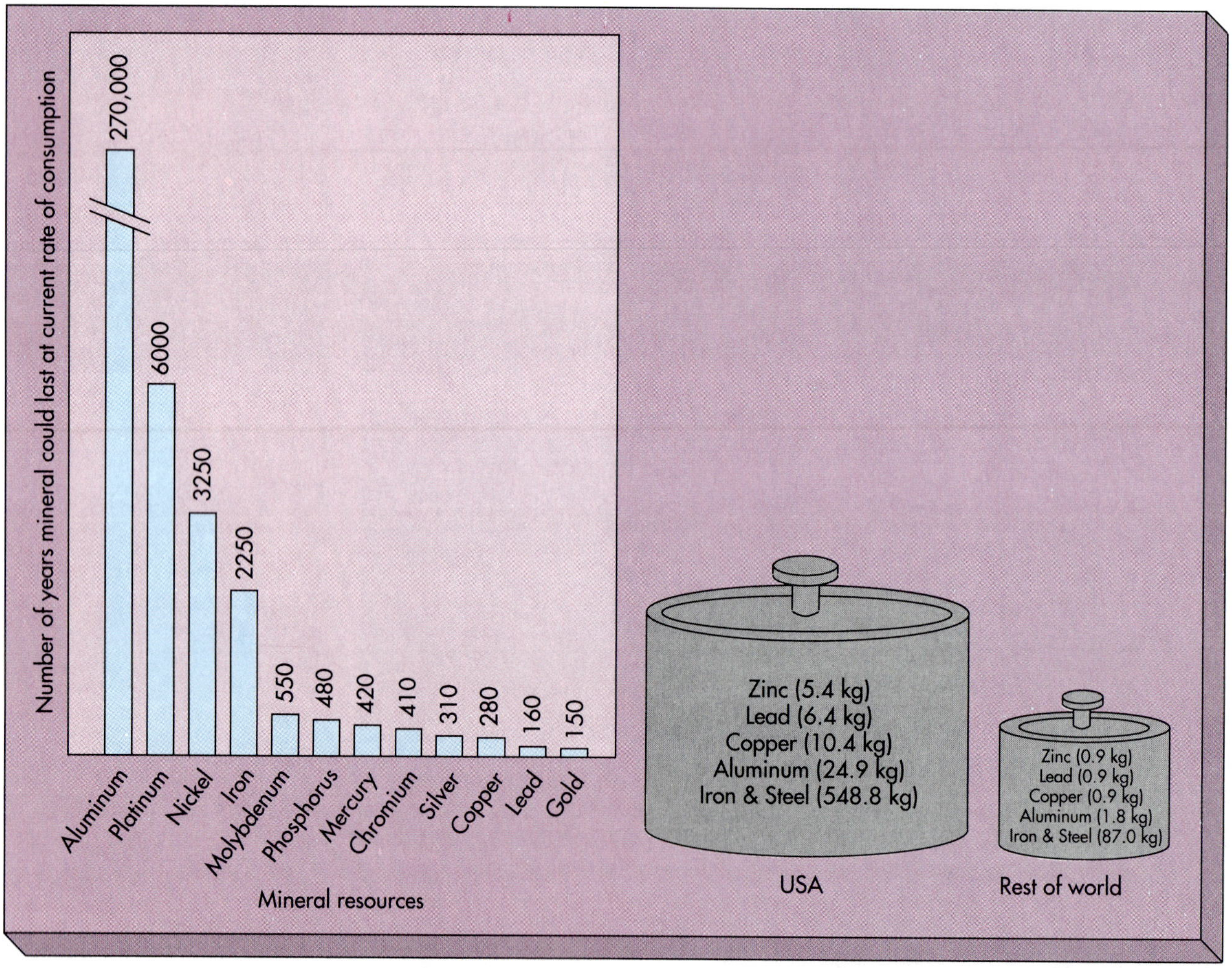

FIGURE 40-3 Consumption of mineral resources.
According to *The Earth Report,* modern industrialized societies require large amounts of minerals to sustain their high standard of living. The left shows a number of major minerals and the years they could last at the current rates of consumption. The right compares the consumption of major minerals by the United States with the rest of the world.

refashioned into the product from which they were claimed; plastic milk jugs cannot be made into plastic milk jugs but must be made into park benches or parking lot curbs. Grocery store plastic bags are the closest truly recycled plastic products, being made from used plastic bags and other additives. Therefore purchasing goods in recyclable containers such as paper, glass, and aluminum *and recycling those containers* is a wise choice. Scrap metal and old automobiles are also recyclable, as are automobile tires and batteries.

Recycling paper, glass, and various metals will help preserve mineral resources as well as other natural resources.

Deforestation

Paper makes up 40% of the garbage in the typical American household. Recycling just your daily newspaper will save four trees every year. Buying products made from recycled paper will save even more. But recycling is not enough. The forests of the world are in severe crisis and will be lost if steps are not taken to stop their destruction.

The most severe crisis is that of **tropical rain forest deforestation.** Tropical rain forests (see Chapter 39) are located in Central and South America, tropical Asia, and central Africa, forming a belt around the equatorial "waist" of the Earth. Although this belt of forest covers only 2% of the Earth's surface, it is home to over *half* the world's species of plants, animals and insects. These organisms contribute 25% of medicines, along with fuel wood, charcoal, oils, and nuts. In addition, the tropical rain forests play an important role in the world climate.

Population and poverty are both high in rain forest countries. People with few resources move from towns and cities to the rain forest and cut the trees for sale as lumber or burn them to clear a patch of land to grow crops and raise cattle to sustain themselves and their families (Figure 40-4). Commercial ranchers also cut and burn the forests to make way for pastureland to feed beef-producing cattle. Unfortunately, the soil of the rain forests is poor, with few nutrients, and does not support crops. Before it is cut, the forest sustains itself because of symbiotic relationships between the trees and microorganisms that quickly decompose dead and dying material on the forest floor. These "processed" nutrients are quickly reabsorbed by the tree roots. Few nutrients stay in the soil; most of the nutrients are in the vegetation. Cutting these trees down and burning them releases the nutrients from the trees. Crops grow poorly on this land after it is stripped. After a year or two, crops will not grow at all. The people move on, cutting yet another portion of the forest.

Commercial logging also takes its toll on the tropical rain forests. Many of the trees are cut to supply fuel wood, paper, wood panels such as plywood, and charcoal and to supply furniture manufacturers with mahogany and other woods demanded by consumers around the world. At this time, the tropical forest is being slashed and burned at a devastating rate. By 1950, two thirds of the forests of Central America had been cleared. Madagascar, an island off the southwestern coast of Africa, had lost about half its rain forest by 1950, but as Figure 40-5 shows, has lost half again since then! The United Nations Food and Agriculture Organization estimates that 100,000 square kilometers (62,000 square miles) of rain forest are now being lost each year worldwide—about the size of New England. At the present rate, scientists estimate, nearly all the tropical rain forests will be gone—including their rich diversity of animal life—within the next 50 years.

In addition to losing a rich natural resource, many scientists agree the the burning of the tropical forests is adding tremendous quantities of carbon dioxide (CO_2) to the air.

FIGURE 40-4 Deforestation: a desperate effort to stave off poverty results in an environmental disaster.
These farmers live near the Andasibe reserve in Madagascar, an island where the per capita income is less than $250 a year. Clearing the rain forest allows these impoverished farmers to plant rice and graze cattle. But the environmental price may be too steep. In Madagascar alone, 80% of the rain forest has been destroyed.

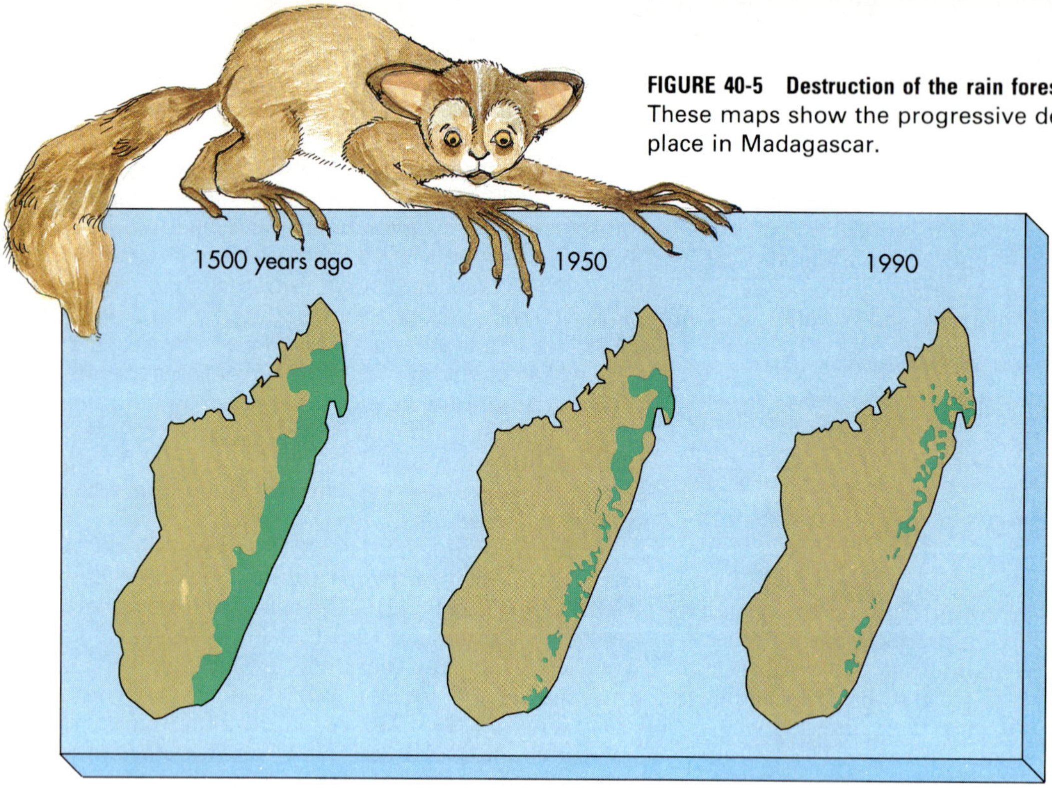

FIGURE 40-5 Destruction of the rain forests in eastern Madagascar. These maps show the progressive deforestation that has taken place in Madagascar.

CO_2 acts like the glass in a greenhouse (or the windows in your car), allowing heat to enter the Earth's atmosphere but preventing it from leaving. This selective energy absorption by CO_2 in the atmosphere is called the **greenhouse effect.** Rising levels of atmospheric CO_2 may therefore result in a worldwide temperature increase, a situation called **global warming.** Global warming could melt the arctic and antarctic ice packs, causing a rise in the sea level and the flooding of one fifth of the world's land area. In addition, a rise in global temperatures could affect rain patterns and agricultural lands.

FIGURE 40-6 A rain forest reserve.
In 1979, the World Wildlife Fund established a research program—the Biological Dynamics of Forest Fragments Project. The program isolates "islands" of rain forests out of the larger forest during conversion to pastureland. Scientists can study these islands to determine what their optimum size and shape should be if they are to provide a suitable habitat for various species.

To help stop rain forest destruction, individuals can join and support organizations involved in rain forest conservation, as well as avoid buying furniture constructed from tropical hardwoods. Research programs are currently being conducted that are exploring the concept of conserving forest "fragments" as reserves that will provide suitable habitats for species (Figure 40-6).

The world's forests are being destroyed at an alarming rate, resulting in the loss not only of this precious resource but of the habitats of thousands of species of organisms. In addition, the burning of the forests adds tremendous quantities of carbon dioxide to the air, a situation that may result in a worldwide temperature increase, or global warming.

Species extinction

Although scientists have classified approximately 1,700,000 species of organisms, they estimate that approximately 40,000,000 exist. But many of these species are becoming extinct—dying out—and will never be seen again. An estimated 1000 extinctions are taking place each year, a number that translates into more than two species per day. Although such mass extinctions have occurred in the past, as with the dying out of the dinosaurs, these former extinctions were caused by climatic and geophysical factors. The extinctions today are caused by human activity.

One way in which humans destroy species is by destroying their habitats. Widespread habitat destruction is taking place as the tropical rain forests are cut down. This one factor alone will cause the extinction of one third of the world's species.

FIGURE 40-7 A germ plasm bank at the United States Department of Agriculture.
Researchers are in a cold storage room that contains seeds from around the world.

Humans are also destroying coral reefs at an alarming rate. Because the base of the coral reef food chain is algae, the reef ecosystem depends on sunlight for its existence. In areas where soil erosion muddies the water, the reef dies from insufficient sunlight. In some cases the water may become polluted with fertilizer runoff or with sewage, which causes the algae to overgrow, smothering the corals. In some cases, humans use dynamite to kill and harvest fish, a practice that obviously destroys the coral reef. In addition, coral is often harvested to sell to tourists.

Although habitat destruction and pollution is the most serious threat to the existence of certain species, exploitation of commercially valuable species threatens them as well. The Convention on International Trade in Endangered Species of Wild Fauna and Flora (CITES) regulates trade in live wildlife and products. However, illegal trade still takes place because of smuggling and the inability of law enforcement inspectors to check all shipments of goods. Certain types of alligator, crocodile, sea turtle, snake, and lizard trade are illegal and endanger the existence of various species. Buying other products such as coral and ivory endangers the existence of coral reefs and African elephants. Even certain plants such as cacti and succulents should be purchased only if cultivated in greenhouses; they should not be taken from the wild. Organizations such as the World Wildlife Fund are invaluable sources of information regarding which products should be avoided by consumers so that they can help stop trade that threatens certain species.

Humans are destroying species by destroying their habitats, polluting their habitats, and illegally trading endangered species.

The extinction of many species leads to a reduction in **biological diversity,** a loss of richness of species. This loss is certainly regrettable; each species is not only the result of millions of years of evolution and can *never* be reproduced once it is gone but also is an intricate part of the interwoven relationships among organisms within the ecosystems of the world. The extinction of just one species can affect ecosystems in many unforeseen ways. In addition, many people feel that humans have no right to influence the world in such a devastating way. And looking at the issue from a utilitarian standpoint, humans rely on the other species of the world as sources of food, medicine, and other substances.

Recently, for example, the genetic diversity in crop plants has declined as scientists have used selective breeding techniques to produce plants with specific characteristics, such as resistance to particular diseases or pests, hardiness, and other characteristics considered important. However, when only a few species or varieties of a species are cultivated or survive, the genetic diversity of the organism declines. Populations of species that have little diversity are more vulnerable to being wiped out by new diseases or climatic changes.

In the 1960s, the United Nations Food and Agricultural Organization (FAO) made recommendations that have led to the establishment of "banks" that store plant seeds and genetic material, or germ plasm (Figure 40-7). Zoos around the world are making similar efforts to preserve genetic diversity among animals by establishing and carrying out sophisticated breeding programs to increase genetic diversity among endangered species (Figure 40-8).

FIGURE 40-8 Zoos are now responsible for preserving genetic diversity in endangered species.
These workers at the San Diego Zoo are helping this baby elephant to nurse. The efforts of zoos around the world will be instrumental in saving some endangered species from extinction.

BIOLOGY TECHNOLOGY & society

Ten things you can do to change the world

1. Make up your mind that the conditions on the planet are serious enough to merit your attention and commitment. Human beings have had a tremendously negative effect on the environment but can have a positive impact. Knowledge and awareness precede action.
2. Water: Install low-flow systems on all faucets. Use less. Take short showers. Clean water is a finite resource. Landscape for low water use.
3. Trash: How many things do you really need? Buy products in minimal packaging. Reuse what you can. Recycle everything else you can.
4. Life-style: Choose carefully where you live. Can you walk, bike, or take public transportation to work, to play, and to buy necessities? How much space do you need to live?
5. Energy: Use less. Turn out lights. Install compact fluorescents where possible. Encourage energy conservation. Learn about how conservation has an impact.
6. Transportation: If you need a car, buy for high mileage. Support mass-transit legislation. Walk, ride a bike, or take the bus.
7. Population: Learn about population as a problem and what has worked to control population growth in other parts of the world.
8. Work: Find a job that supports positive environmental change whether it is actually working directly for change, or teaching, or bringing positive changes to your workplace.
9. Attitude: Learn about what sustainable living means. How you manage resources today will affect how the next generations can live.
10. Political action: Be informed about legislation in process at state and national levels. Write and call your representatives. Find out how candidates have voted on environmental issues. Vote for the ones who are concerned.

Solid waste

The average American produces about 19 pounds of garbage and trash—referred to as **solid waste**—per week. If that does not sound like very much, multiply that number by 52, and you will realize that each person produces approximately a *half ton* of solid waste per year. In 1992, the population of the United States was over 250 million—a population that produced approximately 125 million tons of waste per year, not including the waste produced by schools, stores, and manufacturing. Americans, often described as having a "disposable society," are beginning to realize much of the wastes produced should be viewed as resources that must be reclaimed rather than as substances to throw away and clog the landscapes.

What happens to your trash when you put it out for collection? The burial site for your throwaways is the **sanitary landfill,** an enormous depression in the ground where trash and garbage is dumped, compacted, and then covered with dirt (Figure 40-9). In 1983, the Federal Resource Conservation and Recovery Act forced the closing of all **open dumps** or required them to be converted to landfill sites. At an open dump, solid waste is heaped on the ground, periodically burned, and left uncovered. Landfills are considered superior to dumps because landfill wastes are covered, reducing the number of flying insects and rodents that are attracted to the site and reducing the odor produced by open, rotting organic material. In addition, wastes are not burned at landfills, decreasing the problem of air pollution. Additionally, when the capacity of a landfill site is reached, it may be used as a building site or recreational area. Examples of landfill reuse are Mount Trashmore recreational complex in Evanston, Illinois, and Mile High Stadium in Denver, Colorado.

FIGURE 40-9 Sanitary landfill, New York City.
Sanitary landfills around the country are filling up rapidly. Communities must quickly find new ways to dispose of their trash. One of the most popular and efficient methods is recycling.

Problems do exist with landfills, however. First of all, space is running out! In just one year, the population of New York City alone will produce enough trash to cover over 700 acres of land 10 feet deep. Another problem with landfills is that liquid wastes can trickle down through a landfill, reaching and contaminating ground water below. Liquids leaching from landfills can also pollute nearby streams, lakes, and wells (therefore *never* put batteries, paint solvents, drain cleaners, and pesticides in with your trash). In addition, as the organic material compacted in landfills is decomposed in the absence of oxygen, methane gas is produced. This highly explosive gas rises from landfills and can seep into buildings constructed on or near reclaimed sites.

To reduce solid waste and landfill problems, everyone should recycle paper, glass, aluminum cans, and even clothing. An important part of recycling is buying recycled goods so that a market is maintained for them. In addition, people should purchase products in recyclable containers, avoid purchasing "overpackaged" products, and compost yard waste. Composting involves piling (and periodically turning to aerate) grass clippings, wood shavings, and similar yard wastes. Bacteria will degrade these substances, and they can be used to fertilize flower beds and vegetable gardens. By composting, you will be eliminating the second largest waste by volume in landfills.

In the future, careful planning will be necessary to ensure that landfill sites are located away from streams, lakes, and wells and that they have proper drainage and venting systems. Most likely, landfills will be only a portion of a solid waste disposal system that incorporates recycling, safe incineration, and waste-to-energy reclamation.

Solid waste is disposed of in sanitary landfills, depressions in the ground where trash and garbage is dumped, compacted, and covered with dirt. Solid waste management could be improved with increased and better-organized recycling efforts, safe incineration, better-engineered landfill sites, and waste-to-energy reclamation.

FIGURE 40-10 A toxic chemical dump in northern New Jersey. Toxic chemical dumps are serious threats to ground and surface water. Pollution occurs when the drums rust through and release their contents, which then enter the surface water and may eventually percolate down to the ground water.

The water

The water, or hydrosphere, of this planet lies mainly in the oceans but is also found in fresh water lakes and ponds, in the atmosphere as water vapor, and as subsurface reservoirs called **ground water.** Heated by the sun, water continually cycles from the land to the air, condenses, and falls back to the Earth (see Chapter 38). As this cycling of water repeats continually, contaminants may mix with the water, **polluting** it—causing physical or chemical changes in the water that harm living and nonliving things.

Surface water pollution

Surface water can be polluted by factories, power plants, and sewage treatment plants that dump waste chemicals, heated water, or human sewage into a lake, stream, or river. These sources of pollutants are called **point sources** because they enter the water at one or a few distinct places. Other types of pollutants may enter surface water at a variety of places and are called **nonpoint sources.** Examples of nonpoint sources of pollution are (1) sediments in land runoff caused by erosion from poor agricultural practices (a major type of water pollution), (2) metals and acids draining from mines, (3) poisons leaching from hazardous waste dumps (Figure 40-10), and (4) pesticides, herbicides, and fertilizers washing into surface waters after a rain. These pollutants affect aquatic organisms and also affect terrestrial organisms that drink the water. The manner in which a pollutant affects living things depends on its type: nutrient, infectious agent, toxin (poison), sediment, or thermal pollutant.

Organic nutrients are sometimes discharged into rivers or streams by sewage treatment plants, paper mills, and meat-packing plants. These "organics" are food for bacteria. If high amounts of organic nutrients are available to bacteria, their populations will grow exponentially. As they grow and reproduce, they use oxygen—oxygen that fish need. Therefore as the bacterial populations rise, the only organisms that survive are those that can live on little oxygen. So-called "trash fish" such as carp can out survive other species such as trout and bass, but if oxygen levels become extremely low, all the fish die, survived only by various worms and insects.

The accumulation of **inorganic nutrients** in a lake is called **eutrophication,** meaning "good feeding." Certain inorganics such as nitrogen and phosphorus, which come from croplands or laundry detergents, stimulate plant growth. Although heavy plant growth makes swimming, fishing, or boating difficult, it does not cause most of its problems until the autumn (in most regions of the United States), when the plants die. At that time, bacteria decompose the dead plant material, and problems similar to those of organic nutrient pollution arise. In addition, the decomposed materials begin to fill the bottom of the lake. Eventually, the lake may become transformed into a marsh and then into a terrestrial community by an accelerated process of succession (see Chapter 37). **Sediment** that flows into lakes from erosion of the land caused by certain agricultural practices, mining, and road construction also fills in lakes and hastens natural succession. An example of eutrophication that began leading to the successional "death" of a lake occurred in Lake Washington near Seattle. The problem stemmed from local communities dumping their wastes into the lake. In 1968 the dumping was halted, and Lake Washington is now fully recovered.

Surface waters are rarely polluted with **infectious agents,** or disease-causing microbes, in the United States. However, in areas such as Africa, Asia, and Latin America, waterborne diseases are common. Surface waters become polluted from untreated human wastes and from animal wastes, causing diseases such as hepatitis, polio, amebic dysentery, and cholera.

An array of **toxic substances** pollutes surface waters worldwide. Toxic substances include both organic compounds such as PCBs (polychlorinated biphenyls) and phenols and inorganic substances such as metals, acids, and salts. These toxic, or poisonous, substances come from a diverse array of sources such as industrial discharge, mining, air pollution, soil erosion, old lead pipes, and many natural sources. The effects on humans from drinking these substances in water range from numbness, deafness, vision problems, digestive problems, and the development of cancers.

Unfortunately, most toxic pollutants do not **degrade,** or break down, and are therefore present in bottom sediments of surface waters for decades or more. In fact, some organisms accumulate certain chemicals (often deadly ones) within their bodies, a process called **biological concentration.** Oysters, for example, accumulate heavy metals such as mercury, so these organisms might be highly toxic when living in waters with relatively low concentrations of this metal. Also, as organisms higher on the food chain eat organisms lower on the food chain, toxins accumulate in the predators in a high concentration, a concept called **biological magnification.** As discussed in Chapter 38, the relationships between trophic levels of food chains are pyramidal, so a smaller number of organisms eat a larger number of organisms as you progress "up" the chain. (Figure 40-11 illustrates this concept). Scientists estimate that the concentration of a toxin in polluted water may be magnified from 75,000 to 150,000 times in humans consuming tainted fish.

The electric power industry and various other industries such as steel mills, refineries, and paper mills use river water for cooling purposes, discharging the subsequently heated water back into the river (Figure 40-12). Small levels of **thermal pollution** do not cause serious problems in aquatic

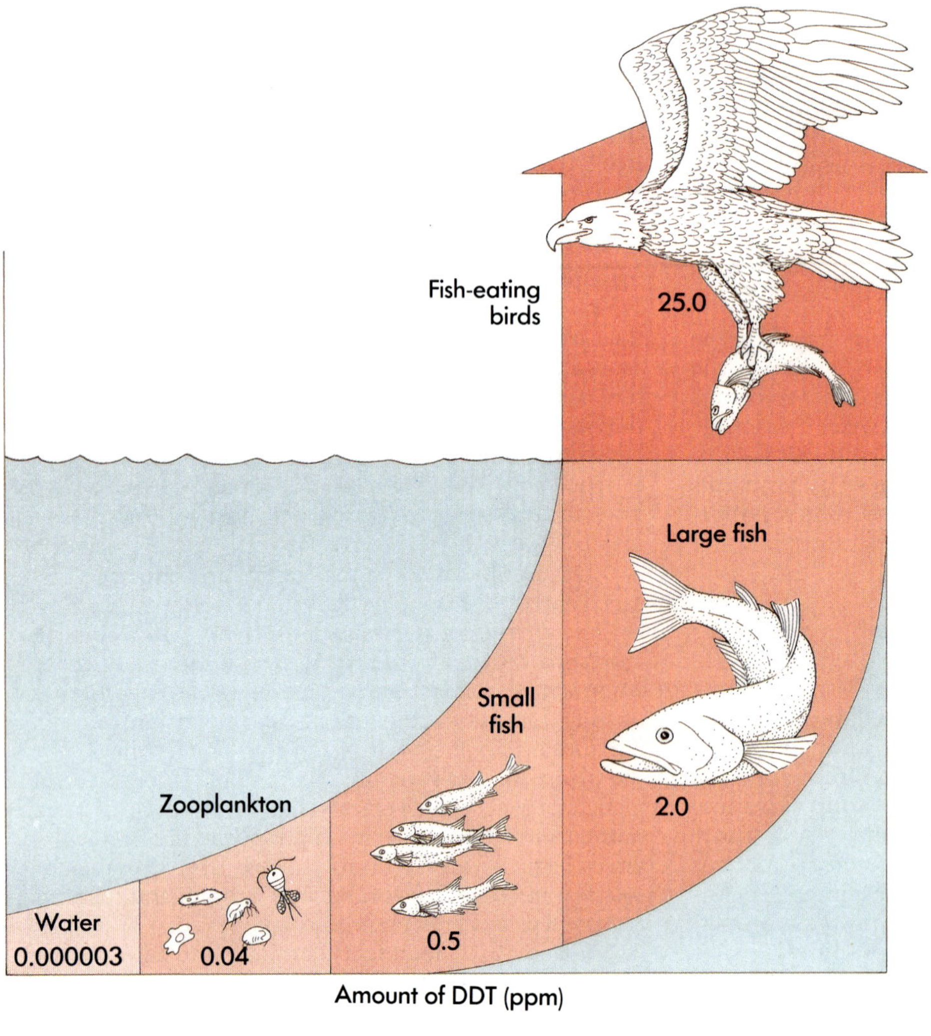

FIGURE 40-11 Biological magnification. The amount of DDT in an organism increases as you go up the food chain.

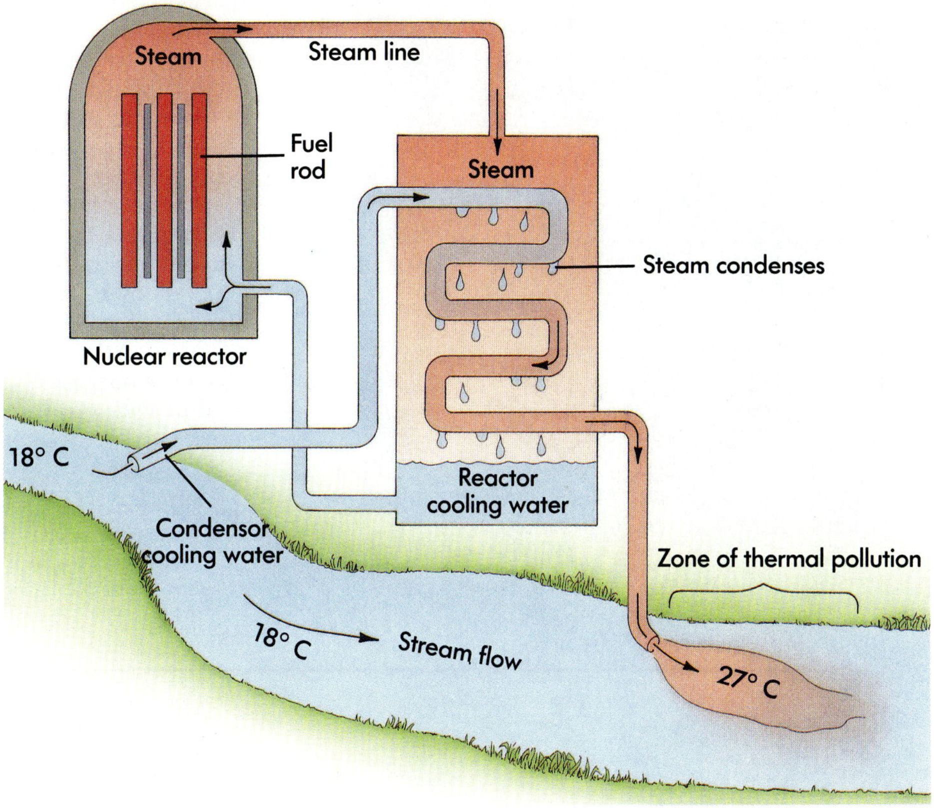

FIGURE 40-12 Thermal pollution. Water from streams or rivers is used to cool nuclear power plants and is used to condense steam to water. As it cools the reactor, the water is heated forming steam, which is then cooled again in the condenser. The cool stream water used in the condenser is discharged into the stream after it has absorbed heat from the steam.

ecosystems, but sudden, large temperature changes kill heat-intolerant plants and animals. Interestingly, ecosystems adjust to artificially heated waters and are damaged if the heat source is shut down, as when a power plant closes.

Surface water can be polluted by waste chemicals, heated water, or human sewage put into the water by factories, power plants, and sewage treatment plants or by silt and chemicals that can leach into the water from surrounding sites. These substances all affect the aquatic ecosystem in different ways.

Ground water pollution

The surface water and rainwater that trickles through the soil to underground reservoirs is ground water. This water can be contaminated with some of the same substances as surface water. The primary ground water contaminants are toxic chemicals that seep into the ground from hazardous waste dump sites and chemicals such as pesticides used in agriculture. Although ground water does get filtered and "cleansed" of some substances as it trickles through the soil, toxic chemicals consist of molecules too small to be filtered in this manner. At this time, prevention of contamination is the cheapest and most feasible way to end ground water contamination. Pumping contaminated water from underground sources to the surface, purifying it, and returning it to the ground is extremely costly.

Acid rain

In recent years, the water in the atmosphere has become polluted with sulfur dioxide and nitrogen dioxide, two chemical compounds that form acids when combined with water. As this water vapor condenses and falls to the ground, it is commonly referred to as **acid rain,** although acid precipitation also falls as snow or as dry "micro" particles, mixing with water when it reaches surfaces on the ground. "Normal" rain has a pH of approximately 5.7, primarily because of dissolved carbon dioxide (see Chapter 2 for a discussion of pH). The pH of acid rain is lower than 5.7 and usually falls between 3.5 and 5.5. However, rainfall samples taken in the eastern United States have measured as low as 1.5—a pH lower than that of lemon juice and approaching that of battery acid!

Acid rain results in many devastating effects on the environment. As it mixes with surface water, it acidifies lakes and streams, killing fish and other aquatic life. It seeps into ground water, causing heavy metals to leach out of the soil. The result is that these heavy metals enter the ground and surface water, posing health problems for humans as well as fish. Acid rain also eats away stone buildings and monuments as well as metal and painted surfaces (Figure 40-13).

The effect of acid rain falling on plant life has been hypothesized but undocumented until recently. Botanists realized that acid rain not only leached many of the minerals essential to plant growth from the soil but also liberated toxic minerals such as aluminum. Recent research suggests that acid rain also kills or damages microorganisms that live in symbiotic associations with forest trees, helping them extract water and needed minerals from the soil. Without these organisms, the trees die (Figure 40-14).

Although sulfur and nitrous oxides are produced naturally during volcanic eruptions and forest fires, humans produce more than half these chemicals from the burning of coal by electricity generating plants, industrial boilers, and large smelters that obtain metals from ores. In addition, nitrogen oxides are emitted by cars and trucks. The situation also becomes complicated because countries pollute their own air as well as the air of other countries. Emissions produced in the Midwest and eastern portions of the United States not only affect those areas, for example, but are carried by the wind into Canada. Emissions produced in England move into the Scandinavian countries.

FIGURE 40-13 Acid rain damage.
This statue has been eroded by acid precipitation.

Sulfur dioxide and nitrogen dioxide, gases emitted primarily by coal-fired power plants, combine with water in the atmosphere to produce acid rain, a precipitation that harms both living and nonliving things.

Solving the problem of acid rain and curtailing these emissions is not easy or inexpensive. One technique used to reduce sulfur dioxide is to put **scrubbers** on coal-burning power plants. This technology can remove up to 95% of the sulfur dioxide emissions produced. Unfortunately, the United States lags behind most sulfur-emitting countries of the world in implementing this technology. A solution to this problem depends on international agreements to reduce emissions, energy conservation measures, and the increased use of public transportation.

FIGURE 40-14 Effects of acid rain.
These balsam fir trees in North Carolina have been killed by acid rain. The chemicals in the acid rain, which are carried on prevailing winds, come from as far away as the Midwest.

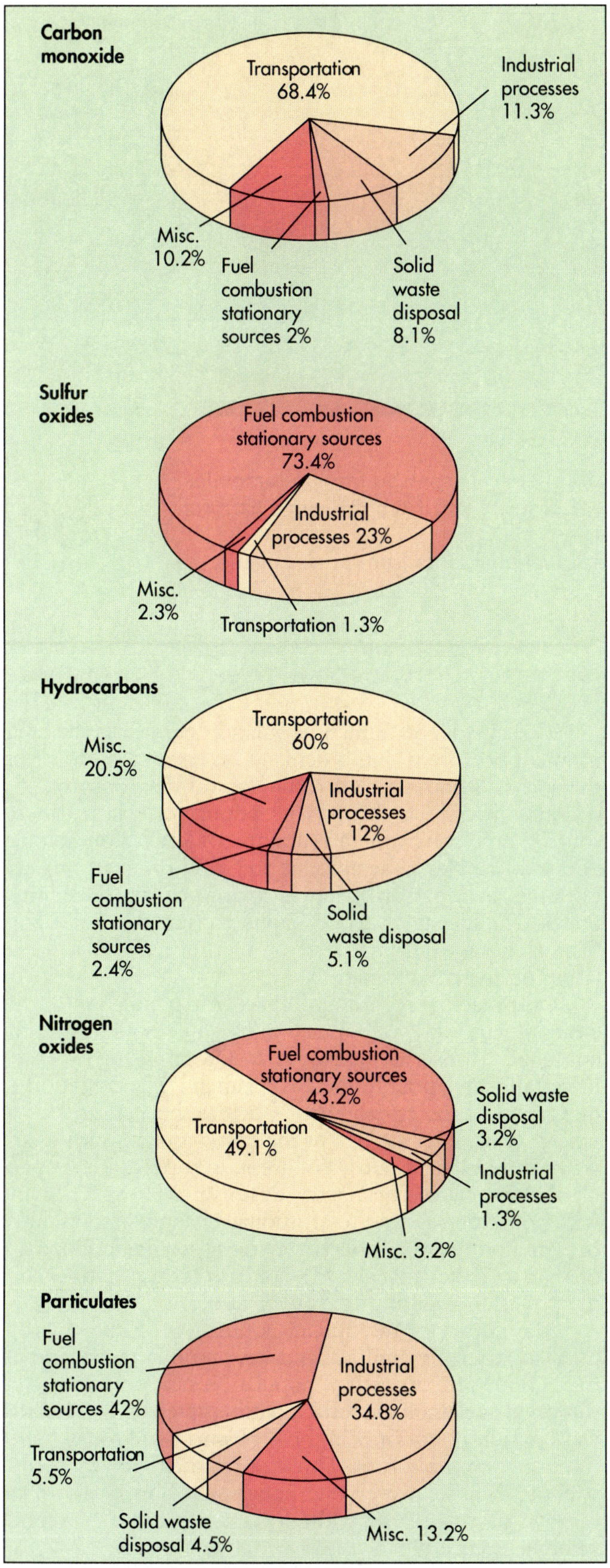

FIGURE 40-15 Major air pollutants and their sources. Transportation and fuel combustion at stationary sources are the main contributors to pollution.

The atmosphere

The Earth's atmosphere actually extends much higher than the portion within the biosphere, a part of the atmosphere more technically called the **troposphere.** The troposphere extends approximately 11 kilometers (36,000 feet) into the atmosphere but slopes downward toward the poles and upward toward the equator. The word *troposphere* literally means "turning over," a name extremely descriptive of the atmosphere. As the Earth is heated by the sun, air rises from its surface, cooling as it ascends. Cooler air then falls, resulting in a constant turnover of the air, aided by the prevailing winds.

Air pollution

About 99% of the clouds, dust, and other substances in the atmosphere are located in the troposphere. Some of these "other substances" are nitrogen and sulfur oxide pollutants. Other major air pollutants are carbon monoxide, hydrocarbons, and tiny particles, or particulates. As Figure 40-15 shows, sulfur oxides and particulates are produced primarily by the burning of coal in electricity-generating plants. Carbon monoxide and hydrocarbons are emitted primarily by cars, buses, and trucks; all these sources spew out nitrogen oxides.

The type of air pollution in a city depends not only on which of the pollutants are in the air, but on the climate of the city. **Gray-air cities,** such as New York, Philadelphia, and Pittsburgh, are located in relatively cold but moist climates and have an abundance of sulfur oxides and particulates in the air. The haze, or **smog,** that can be seen in the air is the result of the the burning of fossil fuels in power plants, other industries, and homes. **Brown-air cities,** such as Los Angeles, Denver, and Albuquerque, have an abundance of hydrocarbons and nitrogen oxides in the air. In these sun-drenched cities the hydrocarbons and nitrogen oxides undergo photochemical reactions that produce "new" pollutants called **secondary pollutants.** The principal secondary pollutant formed is **ozone** (O_3), a chemical that is extremely irritating to the eyes and the upper respiratory tract. Smog caused by pollutants reacting in the presence of sunlight is called **photochemical smog** (Figure 40-16).

The solutions to cutting down on smog are the same solutions to problems of acid rain formation and fossil fuel depletion: the use of coal-fired power plant scrubbers, energy conservation, and the recycling of resources leading to a reduction in energy use for manufacturing.

The primary type of air pollution in the first 11 kilometers of the atmosphere consists of smoke and fog, or smog. Smog is caused by sulfur oxides, nitrogen oxides, hydrocarbons, and particulates in the air as a result of the combustion of fossil fuels.

A  B

FIGURE 40-16 Air pollution.
A "Brown air" in Los Angeles. Sunlight heats the chemicals spewed from automobiles to create ozone, which in the upper atmosphere protects you from the sun's harmful rays but in the lower atmosphere is poisonous.
B "Gray air" in New York City is caused by fossil fuel pollutants from power plants.

Ozone depletion

Ironically, humans are producing ozone in the troposphere that is polluting the environment but destroying it in the **stratosphere** where it is needed. The stratosphere is the "layer" of the atmosphere directly above the biosphere. It contains a layer of ozone that is formed when sunlight reacts with oxygen. Although ozone is harmful in air that is breathed, it is helpful in the stratosphere, acting as a shield against the sun's powerful ultraviolet (UV) rays. Excess exposure to UV rays can cause serious burns and skin cancers and can harm or kill bacteria and plants.

Scientists have measured "holes" in the ozone layer where it is thinnest, over the polar ice caps. The chemical chlorofluorocarbon, or CFC, is the main culprit. The spray-can propellant and refrigerant used in air conditioners, freon, is a CFC. In addition, CFCs are used in the manufacture of styrofoam and foam insulation. To protect the ozone layer, people should have their air conditioners serviced by persons who use equipment that does not allow the escape of freon. In addition, they should not use aerosol products containing CFC and should avoid the use of styrofoam products.

Overpopulation and environmental problems

Some scientists hold that the enormous world population is the key to the problems of pollution and diminishing natural resources. Others state that **technology,** the application of science to industrial use, is the culprit. Most, however, would agree that neither factor alone is the cause of these complex problems. Many factors interact to affect population size and the "health" of the environment.

The size of the human population influences the environment because it puts a demand on resources. However, the social, economic, and technological development of a country affects the demand its population places on resources. A person living in a rural area of Kenya, for example, does not use the same amount and kind of environmental resources as a person living in a large city in the United States. The life-style and per capita consumption of a population make a big difference on the impact of that population on the environment.

Other factors regarding impact of a population on the environment is how a population uses its resources and the effects of that use. Using resources and disposing of wastes in certain ways pollute the environment. The pollution of one aspect of the environment (such as the air) can lead to pollution of other aspects of the environment (such as the water through acid rain). Pollution, in turn, can limit population size by increasing the death rate. However, living with pollution can lead to attitudinal changes among members of a population, which may result in the development of laws to better manage the use of resources and to curb the pollution resulting from their use.

The impact of the human population on the environment is extremely complex and has multiple causes and effects. Many factors, such as population size, per capita consumption, technology, and politics, interact in complex ways, resulting not only in the problems faced today but in solutions for the future. Through research, environmental scientists will help everyone understand how each person can live on this planet without harming it and everyone's future existence.

Summary

1. All life on Earth exists in the biosphere. The biosphere is the interface of the land, air, and water, extending from approximately 9 kilometers (30,000 feet) above sea level to 11 kilometers (36,000 feet) below.
2. Many of the natural resources of the Earth are finite or nonrenewable; that is, they cannot be replaced. Coal, oil, and natural gas are fossil fuels, nonrenewable resources formed over time from the remains of organisms that lived long ago. To curb the use of these resources, scientists are testing and refining alternative renewable energy sources, such as solar, wind, water, and geothermal power.
3. The most severe crisis of natural resource destruction is occurring today in the tropical rain forests. Approximately 100,000 square kilometers (62,000 square miles) of rain forest are being lost each year because of slashing and burning for agriculture and logging and to create pastures for cattle. Scientists estimate that at the present rate, nearly all the tropical rain forest will be gone within the next 50 years.
4. The atmospheric rise in carbon dioxide levels, which is caused by the burning of the tropical forests and the combustion of fossil fuels, is acting as a barrier against the escape of heat from the surface of the Earth. This greenhouse effect could affect global temperatures, rainfall patterns, and agricultural lands.
5. Species are dying out at a rate of 1000 extinctions per year. A major cause of species extinction is habitat loss, which occurs when the tropical rain forests are cut. The extinction of many species leads to a reduction in biological diversity, a loss of species richness. In addition, the loss of species affects ecosystems and diminishes future sources of food, medicine, and other substances.
6. The populations of the world are literally drowning in their trash and garbage, running out of room in which to put it. Solutions to this problem lie in recycling, lowering consumption, composting yard waste, and implementing landfill technology that incorporates recycling, safe incineration, and energy reclamation.
7. Surface water is contaminated by factories, power plants, and sewage treatment plants that dump waste chemicals, heated water, and human sewage into lakes, streams, and rivers. Other sources of surface water pollution are sediment runoff during erosion and the leaching of chemicals from mines, hazardous waste dumps, and croplands. These pollutants affect aquatic life in different ways and can become concentrated in their bodies.
8. Underground reservoirs of water can be contaminated by the same sources as surface water when the contaminants trickle through the earth; however, the primary ground water contaminants are toxic chemicals. At this time, prevention of contamination is the cheapest and most feasible way to end ground water contamination.
9. The atmosphere has become polluted with sulfur and nitrogen oxides, carbon monoxide, hydrocarbons, and particulates. The primary source of these pollutants is the combustion of fossil fuels in automobiles and electricity-generating plants. As the sulfur and nitrogen oxides mix with water vapor in the atmosphere, they form acid rain, which harms both living and nonliving things. As the nitrogen oxides and hydrocarbons react with sunlight, they form photochemical smog, an upper respiratory irritant that is a health hazard.
10. The environmental problems of this and the next century are the result of many interacting factors. Through research, environmental scientists will help people understand how they can live on this planet without harming it and their future existence.

REVIEW QUESTIONS

1. What is the biosphere?
2. Many scientists feel people should work toward creating a "sustainable society." What does this mean?
3. Discuss several energy sources that could be used to help create a sustainable society.
4. What can be done to help prevent a mineral shortage in the future? Why are preventative measures important?
5. Why is the destruction of tropical rain forests a serious problem for everyone—not just the people living in tropical countries?
6. What is meant by global warming? Why is it dangerous?
7. Define *biological diversity*. How are humans affecting the biological diversity of the world's species? Why is this serious?
8. How can the amount of solid waste thrown away be reduced and waste management improved?
9. Distinguish between point and nonpoint sources of surface water pollution. Give an example of each.
10. Describe the different effects of pollution by organic nutrients, inorganic nutrients, infectious agents, and toxic substances in surface water.
11. How is acid rain produced? What are its effects?
12. Explain the terms *gray-air cities* and *brown-air cities*. What pollutants are involved?
13. Ozone is a dangerous pollutant in air that is breathed. Why should people be concerned that the ozone layer in the stratosphere is being depleted?
14. Suppose the President of the United States asked you to put together a plan of action for addressing environmental problems. What steps would you recommend?

KEY TERMS

acid rain 705
bioenergy 696
bio-gas machines 696
biological concentration 704
biological diversity 701
biological magnification 704
biosphere 695
brown-air cities 707
degrade 704
environment 695
eutrophication 703
fossil fuels 696
geothermal energy 696
gray-air cities 707
greenhouse effect 700
ground water 703
infectious agent 704
inorganic nutrient 703
nonpoint source 703
nuclear power 696
open dump 702
organic nutrient 703
ozone 707
photochemical smog 707
point source 703
sanitary landfill 702
scrubber 706
secondary pollutant 707
sediment 703
smog 707
solar power 696
solid waste 702
stratosphere 708
sustainable society 696
thermal pollution 704
tidal power 696
toxic substance 704
tropical rain forest deforestation 699
troposphere 707
wave power 696
wind power 696

THOUGHT QUESTIONS

1. The most tragic of environmental injuries are those that needlessly consume or destroy irreplaceable resources. In the United States, two of the most important resources that are being seriously compromised today are ground water and topsoil. List potential sources of damage to these resources. What might be done to limit the damage?
2. Many environmentalists believe that the economic system of the United States subsidizes pollution of the environment. Consumption is kept higher than it should be, they claim, because the price a consumer pays for a manufactured item does not include the entire cost—the costs of injury to the environment are instead passed on to a third party (the future). As a result, consumers buy more of the item than they would if it were priced at its true cost. What might realistically be done to address this problem?
3. Imagine for a moment that no serious effective program is mounted in the next decades to address the Earth's environmental woes. Outline a scenario of the likely consequences—what would such a world be like in 20 years?

FOR FURTHER READING

Brown, L., editor. (1990). *State of the world, 1990,* New York: W. W. Norton & Co.
This is a highly recommended, easily read summary of the ecological problems faced by an overcrowded and hungry world, with an excellent chapter on suggested solutions. A new edition appears every year.

Matthews, S., & Sugar, J. (1990, October). Is our world warming? *National Geographic,* pp. 66-99.
This is a vivid, well-illustrated discussion of the greenhouse effect, its causes, and the possible consequences of global warming.

National Wildlife. (1990). What on Earth are we doing? *28*(2).
An excellent issue of this outstanding magazine is entirely devoted to the state of the environment and presents many actions that you can take to help solve the problems discussed in this chapter.

National Wildlife. (1992, April-May). Endangered species: preserving pieces of the puzzle. *30*(5).
This entire issue is devoted to the fight to protect plants and animals from extinction. Indepth features include the endangered species in the tropical rain forest, the effects of poaching on certain species, and the success stories of bald eagles, gray whales, and alligators.

APPENDIX

A

Answers to review questions

CHAPTER 1

1. The study of biology includes a wide variety of topics, for example, studying hormonal changes in one organism, researching changes in populations over time, and learning how your body protects itself against disease.
2. Inductive reasoning develops generalizations from specific instances (for instance, the large number of dying fish in a particular river could be due to rising pollution from factories in the area). Deductive reasoning begins with a general statement and proceeds to a specific statement (pollution from factories could kill wildlife; we will measure the effects of pollution on fish in a specific river).
3. Your hypothesis could be the following: If the students walk 1 hour at a time, 4 days a week, then they will each lose at least 5 pounds during the semester, assuming that this exercise is in addition to any exercise they're currently getting. The independent variable is the walking that the students will do; the dependent variable refers to the amount of weight they may or may not lose.
4. The control is the standard against which the subjects who go through the experiment's treatment would be compared. In this case, it would be a group of students who do not take the hour-long walks that the other students take.
5. The scientific method is a process that scientists use to answer the questions they ask. Its steps are (1) posing a question, (2) forming a hypothesis, (3) testing the hypothesis with a controlled experiment, (4) recording and analyzing the data, (5) drawing a conclusion, and (6) repeating the experiment.
6. Scientists cannot actually prove hypotheses; they can only disprove them or support them with evidence. However, there is no guarantee that further evidence might not disprove a hypothesis.
7. A theory is a synthesis of hypotheses that are supported by so much evidence they are commonly accepted as "true." However, scientists cannot state with certainty that they are true, since future evidence could disprove any theory.
8. (1) Living things display both diversity and unity. (2) Living things are composed of cells and are hierarchically organized. (3) Living things interact with each other and with their environments. (4) Living things transform energy and maintain a steady internal environment. (5) Living things exhibit forms that fit their functions. (6) Living things reproduce and pass on biological information to their offspring. (7) Living things change over time, or evolve.
9. The levels are (1) the cell (microscopic mass of protoplasm), (2) tissue (group of similar cells that work together to perform a function), (3) organ (group of tissues forming a structural and functional unit), and (4) organ system (group of organs that function together to carry out the principle activities of the organism).
10. The four levels are (1) population (individuals of a given species that occur together at one place and time), (2) community (populations of different species that interact with each other), (3) ecosystem (community of plants, animals, and microorganisms that interact with each other and with their environment), and (4) biosphere (the part of the Earth where biological activity takes place).
11. Unlike all other organisms, people use many nonrenewable resources and fill the biosphere with wastes. In addition, people are overpopulating the land and destroying habitats, killing off various species, and using up resources that cannot be renewed.
12. Ultimately, all energy comes from the sun. Solar energy is captured by producers, which use photosynthesis to convert it into chemical energy. Other organisms (consumers) feed on the producers and on each other, passing the energy along. When you eat plants or animals, you are harvesting their stored energy.
13. DNA is the code of life that contains the instructions translated into a working organism. Using only four different nucleotides, DNA codes for all of the structural and functional components of an organism.
14. The theory of evolution states that organisms alive today are descendants of organisms that lived long ago and that organisms have changed and diverged from one another over time. Fossils are remains of earlier cells and organisms that have been preserved in rocks, they provide a record of how living things have changed over billions of years.
15. (1) Monera (bacteria), (2) Protista (algae), (3) Plantae (rosebushes), (4) Fungi (mushrooms), (5) Animalia (you).
16. Kingdom, phylum, class, order, family, genus, species.
17. a. True.
    b. False. (they might be closely enough related to belong to the same genus, but this cannot be assumed).
    c. False. (Binomial nomenclature only gives the genus and species of an organism. These organisms may not belong to the same genus.)

CHAPTER 2

1. a. See Figure 2-1. Because it has an atomic number of 3, your atom would have three protons and three electrons, although not necessarily three neutrons.
   b. The atom would still have one proton and one electron, but it would have a different number of neutrons.
2. Protons; electrons.
3. Atoms are the tiny parts that make up all matter. Molecules are combinations of tightly bonded atoms of the same element. Compounds are molecules that consist of atoms from two or more elements. For example, one atom is oxygen (O), a molecule of two oxygen atoms is atmospheric oxygen (O_2), and a compound that includes oxygen is water (H_2O).
4. The three chemical forces that influence how an atom interacts with other

atoms are (1) the tendency of electrons to occur in pairs, (2) the tendency of atoms to balance positive and negative charges, and (3) the tendency of the outer shell (energy level) of electrons to be full. The third point is known as the *octet rule,* since an atom with an unfilled outer shell tends to interact with other atoms to gain a complete set of eight electrons in this outer shell.

5. An ionic bond forms when atoms are attracted to each other by opposite electrical charges. A common example is table salt (NaCl). A covalent bond forms when two atoms share one or more pairs of electrons; an example is hydrogen gas (H—H).
6. Water is unusual because it is the only common molecule on Earth that exists as a liquid in the natural environment. This liquid enables other molecules to move and interact; life evolved as a result. A second important characteristic is its ability to form hydrogen bonds because of its polar molecule. Water molecules are strongly attracted to ions and other polar molecules, giving water another important trait: it is an excellent solvent. Chemical interactions readily take place in water because so many molecules are water soluble.
7. Oil is a nonpolar molecule, which means that is cannot form hydrogen bonds with water and therefore does not dissolve.
8. The pH value refers to the concentration of H^+ ions in a solution. The values on the pH scale are expressed as positive numbers, which are determined by taking the negative value of the exponent of the hydrogen ion concentration. Pure water has a pH of 7.0. A substance with a lower pH is an acid: one with a higher pH is a base.
9. Organic molecules tend to be large, contain carbon, and interact with each other via covalent bonding. Inorganic molecules tend to be small, do not usually contain carbon, and interact by means of ionic bonding. If you wanted to study living things—such as human beings—you would learn about organic molecules.
10. Polymers and composite molecules are types of macromolecules, large organic molecules with many functional groups. They differ in their structures. Polymers are built by forming covalent bonds between a long chain of similar components (such as complex carbodyrates, proteins, and nucleic acids). Composite molecules (such as lipids) have several different components.
11. Monosaccharides, disaccharides, and polysaccharides are all carbohydrates. Most organisms use carbohydrates as an important fuel. Monosaccharides are among the least complex carbohydrates. Many organisms link monosaccharides to form disaccharides that are less readily broken down as they are transported within the organism. To store the energy from carbohydrates, organisms convert monosaccharides and disaccharides into polysaccharides, long polymers of soluble sugars.
12. a. The formula for the monosaccharides glucose, fructose, and galactose. Glucose is the primary energy-storage molecule used by living things.
    b. A generalized formula for an amino acid. Proteins perform many important functions; proteins called *enzymes,* for example, speed up chemical reactions.
13. The three major types of lipids are (1) oils, fats, and waxes; (2) phospholipids; and (3) steroids. Lipids have various functions. Fats are important energy-storing molecules, and land plants and some animals use waxes as a waterproofing material. Phospholipids and steroids are important components of many cell membranes; male and female sex hormones are steroids.
14. Saturated fats are fatty acids that carry as many hydrogen atoms as possible because they have only single bonds between their component carbon atoms. Polyunsaturated fats are fatty acids that carry fewer hydrogen atoms because they have more than one double bond between their carbon atoms. This product contains lipids (fats.) (It is olive oil, to be exact.)
15. Proteins are the third major group of macromolecules that make up the bodies of organisms. They play diverse roles in living things. Some speed up chemical reactions; others are chemical messengers throughout the body. Still others are structural proteins, an important component of bones, cartilage, and tendons.
16. The primary structure of a protein is the sequence of amino acids that make up the polypeptide chain. The amino acids interact by forming hydrogen bonds that cause the chains to fold into sheets or wrap into coils (secondary structure). The secondary structure of a protein largely determines its tertiary structure (three-dimensional shape). Proteins made up mostly of sheets often form fibers, whereas those with regions forming coils frequently fold into globular shapes. (When two protein chains associate to form a subunit (functional unit), these subunits assemble to form the quaternary structure.
17. The two forms of nucleic acids are DNA (deoxyribonucleic acid) and RNA (ribonucleic acid.) DNA stores the information for making proteins; RNA directs the production of proteins.

CHAPTER 3

1. Cells are the smallest unit of life that can exist independently. They can take in food, break it down to release energy, and expel wastes. They can also reproduce, react to stimuli, and maintain an internal environment that is different from their surroundings.
2. The cell theory includes three basic principles: (1) all living things are made up of one or more cells, (2) the smallest living unit of structure and function of all organisms is the cell, and (3) all cells arise from preexisting cells.
3. Cells must maintain a large surface area-to-volume ratio because they are constantly working. A relatively large surface area allows them to move substances in and out fast enough to survive.
4. Large, complex cells that are active often have many nuclei. A single nucleus could not control the activities of such a large cell.
5. Almost all cells share these four traits: (1) a surrounding membrane, (2) protoplasm that is enlosed by this membrane, (3) organelles (in eukaryotes) that carry out certain metabolic functions, and (4) a control center (nucleus or nucleoid) that contains DNA.
6. a. Eukaryotes.
   b. Eukaryotes.
   c. Prokaryotes.
   d. Eukaryotes.
7. See Figures 3-5 and 3-6.
8. Plasma membrane.
9. The cytoplasm, a viscous fluid inside the cell's plasma membrane, contains a gel-like fluid, storage substances, a network of interconnected filaments and fibers, and organelles. The cytoskeleton is a network of filaments and fibers that helps support and shape the cell, anchor the organelles, and move substances within the cell.
10. See Table 3-1.
11. Rough ER (endoplasmic reticulum) has surfaces studded with ribosomes; the ribosomes on its surface are the sites of protein synthesis. It transports proteins destined to leave the cell. Smooth ER, which lacks ribosomes, helps build carbohydrates and lipids.
12. If the lysosomes stopped working, your cells would soon fill up with old cell parts and foreign substances such as bacteria. In short, they would die.
13. DNA and RNA are both nucleic acids found in a eukaryotic cell's nucleus. DNA is the hereditary material that

directs the synthesis of RNA. RNA directs the synthesis of proteins.

14. Chromosomes. (This takes place inside the nucleus of the cell.)
15. Both mitochrondria and chloroplasts may have originated as endosymbiotic bacteria. They differ in that mitochondria use oxygen to break down cell fuel and release energy; chloroplasts, found in plants and algae, use carbon dioxide, water, and light energy to produce the cell's fuel (the process of photosynthesis.)
16. Flagella.
17. (1) Prokaryotic cells are smaller. (2) They lack membrane-bound compartments. (3) They have a nucleoid instead of a nucleus.
18. See Table 3-2 for a summary of the differences. For example, the cell with chloroplasts would be the plant, the cell with mitochondria but no chloroplasts would be the animal, and the one with neither mitochondria nor chloroplasts would be the bacterium.

CHAPTER 4

1. A lipid bilayer is the basic foundation of biological membranes. It consists of phospholipid molecules, with the nonpolar tails pointed toward the interior of the cell and the polar heads pointed outward. It forms a fluid, flexible cell covering and separates the cell's watery contents from its watery environment.
2. Molecules can move into and out of cells because of proteins floating in the cell's membrane. These proteins form specialized channels that regulate the types and amounts of substances that enter and leave the cell. These varied proteins allow the cell to interact with its environment; without them, the cell would soon die.
3. The carbohydrates attached to the cell membrane serve as cell name tags that help cells to recognize one another.
4. A gradient is the difference in concentration, pressure, or electrical charge between two areas. Molecules undergo net movement in response to this difference.
5. Passive transport is molecular movement down a gradient and across a cell membrane from an area of higher concentration to one of lower concentration. It does not require energy. Active transport involves molecular movement against the concentration gradient; it requires the cell to expend energy.
6. a. Facilitated diffusion (passive).
   b. Osmosis (passive).
   c. Diffusion (passive).
   d. Active transport (active).
7. The solute is the instant coffee crystals, the solvent is water, and the solution is the liquid coffee.
8. Hypotonic; hypertonic; isotonic.
9. Diffusion is the net movement of molecules down the concentration gradient as a result of random molecular motion, until the molecules are uniformly distributed. Facilitated diffusion is the movement of selected molecules across the cell membrane and down the concentration gradient; it requires specific transport proteins.
10. The sodium-potassium pump, a type of active transport, uses enegy and a transport protein to move sodium ions out of the cell and potassium ions into the cell. Maintaining appropriate concentrations of each ion, both inside and outside the cell, is important to the functioning of muscle and nerve cells.
11. See Figure 4-10. White blood cells use endocytosis to engulf bacteria and other foreign substances.
12. See Figure 4-11. In plants, exocytosis is the main way that cells move the materials out of the cytoplasm that are needed to build the cell wall.

CHAPTER 5

1. Chromatin is a complex of DNA and proteins found in the cell nucleus. DNA is the hereditary material that contains the code of life.
2. A karyotype is the particular array of chromosomes that belongs to an individual. The sex chromosomes are those that determine the gender of an individual. In addition to the pair of sex chromosomes, humans have 22 pairs of chromosomes, called *autosomes*. Chromosomes carry the genes (pieces of hereditary information) that determine the characteristics of the individual.
3. Histones; supercoils; sister chromatids; centromere.
4. Eukaryotic cells divide for three reasons: growth, repair, and reproduction. Generally, they undergo mitosis for the first two reasons, and reproduce by meiosis. Some simple animals and many plants, however, reproduce by mitosis as well.
5. Mitosis produces two daughter cells from one parent cell; the daughter cells are identical with each other and with the parent. Meiosis produces four sex cells, each of which contains half the amount of the parent cell's hereditary material.
6. Interphase is the portion of the cell cycle during which the cell grows and produces an exact copy of its DNA as it prepares for cell division.
7. The correct sequence is c (prophase), a (metaphase), d (anaphase), b (telophase). this is the process of mitosis.
8. The correct sequence is c (prophase I), d (metaphase I), a (anaphase I), b (telophase I), e (interphase). This is the process of meiosis I.
9. Cytokinesis is the physical division of the cytoplasm of a eukaryotic cell into two daughter cells. In animal cells and in eukaryotes that lack cell walls, cytokinesis involves "pinching" the cell in two by forming a cleavage furrow around its circumference. Plant cells divide by forming a cell plate (partition) inside which cellulose forms, producing a new cell wall and separating the two cells.
10. Diploid; haploid.
11. Homologous; synapsis.
12. Meiosis II produces four haploid cells that may function as gametes (animals) or spores (plants).
13. Genetic recombination during meiosis allows the hereditary material to be recombined during sexual reproduction. It is the principal factor that has made the evolution of eukaryotic organisms possible.

CHAPTER 6

1. Free energy of activation; substrates; products.
2. Exergonic reactions break apart substrates to produce chemically smaller products; they release energy. Endergonic reactions bond substrates together to form chemically larger products; they store enegy. In coupled reactions, exergonic reactions supply the energy needed for endergonic reactions.
3. Kinetic energy; potential energy.
4. The first law of thermodynamics states that energy cannot be created or destroyed, it can only be changed from one form to another. The second law states that disorder in the universe constantly increases, energy spontaneously converts to less organized forms.
5. Entrophy is the energy lost to disorder. Your friend is restating the second law of thermodynamics, which states that disorder in the universe constantly increases.
6. The Earth constantly receives energy from the sun. Photosynthetic organisms change this energy to other forms of energy that drive life processes.
7. Metabolic pathways are the chains of reactions within your body that move, store, and free energy. By means of metabolic pathways, people obtain energy from food, repair damaged tissues, and in general avoid increasing entropy.
8. Enzymes are proteins that act as biological catalysts. They speed up chemical reactions within living organisms. This reduces the amount of activation

energy needed to perform vital chemical reactions; without enzymes, these reactions would occur too slowly or require excessive amounts of energy to support life.

9. Enzymes catalyze reactions by bringing substrates together so that they react more easily and by placing stress on bonds that must break before a reaction can proceed.
10. Most human enzymes work best within a fairly narrow temperature range. If your body temperature became too high, this could affect the bonds between the amino acids that are responsible for the enzymes' three-dimensional shapes. Eventually this could stop chemical reactions from occurring.
11. Cofactors are special nonprotein molecules that help enzymes catalyze chemical reactions; one example is a zinc ion that helps digestive enzymes break down proteins in food. Coenzymes are cofactors that are nonprotein organic molecules. Many vitamins are synthesized into coenzymes in your body.
12. Your body can store energy for later use by converting it to fat, glycogen, or ATP (adenosine triphosphate). This is important because otherwise you would expend your available energy too quickly.
13. ATP is a high-energy compound that can be broken down easily to release energy for use by cells. It is called "energy currency" because cells can save energy released in exergonic reactions by storing it as ATP; cells can also spend ATP to provide necessary energy for endergonic reactions.
14. ATP, APD, and phosphate are continually being recycled within living cells. ATP is split into ADP plus phosphate to drive endergonic reactions, and ATP is constantly being re-formed from ADP, phosphate, and energy from exergonic reactions. These interrelating reactions allow cells to store and release energy easily.
15. Most enzymes function best within a narrow range of pH. If an environment becomes too acidic, enzymes could be affected, causing chemical reactions to stop and ultimately affecting the metabolic pathways of organisms.

CHAPTER 7

1. ATP, created by chemical energy, is the energy currency of cells. The splitting of ATP is coupled to endergonic reactions that provide the energy necessary to keep cells alive and functioning.
2. Cellular respiration is the process by which living things make the ATP that powers cellular operations. During respiration, fuel molecules are broken down in the presence of oxygen to release energy. Fermentation is the process by which organisms that do not breathe oxygen make ATP.
3. This simplified formula summarizes the process of cellular respiration. It shows that the net effect is to break down one molecule of glucose in the presence of six oxygen molecules to yield six molecules of carbon dioxide and six molecules of water.
4. Glycolysis; Krebs cycle; electron transport chain.
5. Oxidized; increases; reduced; decreases.
6. This is an oxidation-reduction reaction, an essential part of the flow of energy in cellular respiration.
7. Your body metabolizes foods by breaking down complex molecules into simple ones. These simple molecules are then broken apart through metabolic pathways (the primary pathways are glycolysis, the Krebs cycle, and the electron transport chain) to release their stored energy and capture it as ATP.
8. Three changes occur during glycolysis: (1) glucose is converted to pyruvate, (2) ADP + P_i is converted to ATP, and (3) NAD^+ is converted to NADH.
9. Glycolysis is inefficient because it captures only about 2% of the available chemical energy of glucose. However, cells continue to do it, probably because it was one of the first biochemical processes to evolve. As cells evolved, they retained glycolysis as the first step in catabolic metabolism but added additional steps that were more efficient.
10. The Krebs cycle consists of nine reactions. The first three reactions are preparation reactions in which acetyl-CoA joins a four-carbon molecule from the previous cycle to form citric acid, and the chemical groups are rearranged. The last six reactions are energy extraction reactions. At every turn of the cycle, acetyl-CoA enters and is oxidized to CO_2 and H_2O, and the hydrogen ions and electrons are donated to electron carriers.
11. See Table 7-1.
12. The electron transport chain is a series of carrier proteins that pass along electrons harvested from glucose. These electrons drive protons out across the inner membrane; the return of the protons by diffusion generates ATP.
13. See below.

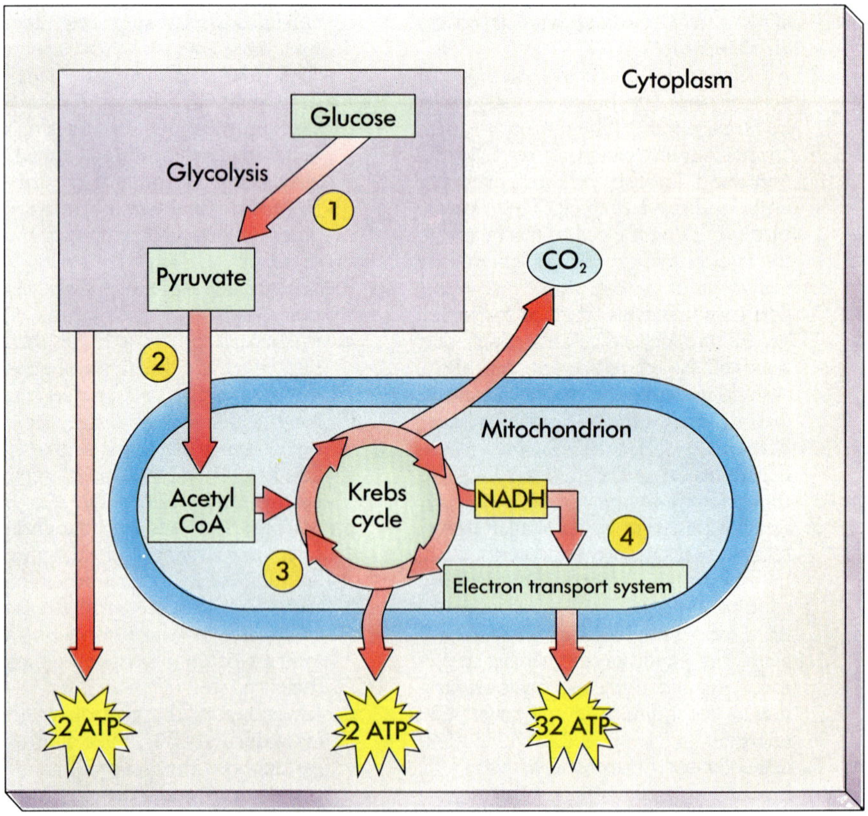

CHAPTER 8

1. Autotrophs are organisms that produce their own food either by photosynthesis or by chemical energy. Heterotrophs are organisms that cannot produce their own food; they depend on the autotrophs and the sun for survival. Human beings are heterotrophs.
2. The visible light in sunlight consists of many wavelengths. The prism separated the various wavelengths so that each color became visible as the light shone on the floor.
3. A pigment is a molecule that absorbs some wavelengths of light and reflects others. An object appears a particular color because it contains (or is covered with) a pigment that absorbs all the wavelenghts except that color. Those wavelengths bounce off the object and are reflected back to your eyes.
4. Carotenoids; chlorophylls.
5. This equation describes the process of photosynthesis: carbon dioxide and water, in the presence of light, produce glucose and oxygen. By means of photosynthesis, energy from the sun is captured and stored in molecules of food. This food ultimately builds and fuels the bodies of virtually all living things.
6. Photosynthesis involves three processes: (1) synthesis of ATP and NADPH, (2) the use of ATP and NADPH to produce glucose, and (3) the recycling of the photosynthetic pigment. The reactions that produce ATP and NADPH are called "light-dependent reactions" because they occur only in the presence of light. During the light-independent reactions (which do not need light to take place), ATP and NADPH are used to produce glucose from carbon dioxide.
7. A photocenter is an array of pigment molecules within the thylakoid membranes of a chloroplast. It channels photon energy to chlorophyll *a*, a special molecule of chlorophyll that will participate in photosynthesis.
8. In Photosystem I, the first to evolve in photosynthetic organisms, energy is transferred to a molecule of chlorophyll *a* called P700. In Photosystem II a different chlorophyll *a* molecule, P680, acts as a reaction center. Most bacteria use Photosystem I to produce ATP, whereas plants and algae use a two-stage photosystem. (Organisms that use a two-stage photosystem, however, can use Photosystem I to produce ATP without producing NADPH.)
9. During noncyclic electron flow, energized electrons ejected from Photosystem II are passed down an electron transport chain, triggering ATP production. Their places in Photosystem II are taken by electrons from the breakdown of water, a process that releases oxygen as a by-product. Energized electrons from Photosystem I, along with hydrogen ions, reduce $NADP^+$ to NADPH.
10. During the light-independent reactions (1) carbon is fixed with the attachment of CO_2 to an organic molecule, (2) glucose is produced from molecules of CO_2, and (3) the organic molecule that hooks to CO_2 re-forms to begin the cyclic again.
11. The metabolic pathways of photosynthesis and cellular respiration form a continuous cycle. The by-products of cellular respiration are carbon dioxide and water, the substrates of photosynthesis; the by-product of photosynthesis, oxygen gas, is essential to cellular respiration. Reducing one component of the cycle could harm animal life in the area; it could also affect the soil and the atmosphere.
12. Chloroplasts.
13. Such a dust cloud would interfere with photosynthesis, on which life on Earth ultimately depends. After a certain point, all living organisms would die.
14. Electromagnetic energy; radiation; waves; photons.

CHAPTER 9

1. A cell is the smallest structure in an organism that is capable of performing all the functions necessary for life. An example is a cardiac muscle cell. The cells of the body are organized into tissues (groups of similar cells that work together to perform a function, such as cardiac muscle tissue). Several different tissues group together to form a structural and functional unit—an organ (the heart.) An organ system is a group of organs that works together to carry out the body's principal activities (such as the circulatory system).
2. The four types of tissue in the body are (1) epithelial (skin cells), (2) connective (chondrocytes), (3) muscle (skeletal muscles such as leg muscles), and (4) nervous (neurons in your brain.)
3. Epithelial cells cover and line the surfaces of the body, both internal and external. They have six major functions: (1) protect tissues, (2) help or hinder absorption, (3) provide a sensory surface, (4) secrete substances, (5) excrete waste products, and (6) transport substances.
4. See Table 9-2. Squamous cells are found in the air sacs of the lungs, cuboidal cells line ducts in glands, and columnar cells line much of the digestive tract.
5. Connective tissue and its cells provide a framework for the body, join its tissues, help defend it from foreign invaders, and store specific substances.
6. Lymphocytes, erythrocytes, and fibroblasts are all cells found in connective tissue, but tney have different functions. Lymphocytes defend the body against infection, erythrocytes pick up and deliver gases in the blood, and fibroblasts produce fibers that are found in various types of connective tissue.
7. This is part of the body's inflammation response to injury. Mast cells produce histamine, which dilates blood vessels and increases blood flow to the area. This makes the area turn red and feel warm.
8. Fibroblasts; chondrocytes; osteocytes.
9. Chondrocytes; osteocytes; spongy; compact.
10. Platelets (e)
    Plasma (d)
    Erythrocytes (a)
    Leukocytes (b)
    Hemoglobin (c)
11. Yes, fat cells have their uses. In addition to storing fuel, fat tissue helps shape and pad the body and insulate it against heat loss.
12. The three kinds of muscle cells are (1) smooth muscle (intestinal walls), (2) skeletal muscle (arm muscles), and (3) cardiac muscle (the heart).
13. Neurons, which transmit nerve impulses; supporting cells, which nourish and protect the neurons.
14. See Table 9-1.
15. Homeostasis is the maintenance of a stable environment inside your body despite conditions in the external environment. The molecules, cells, tissues, organs, and organ systems in your body must all work together to maintain this internal equilibrium.

CHAPTER 10

1. The six classes of nutrients are carbohydrates, fats, proteins, vitamins, minerals, and water. The first three are organic compounds that your body uses as a source of energy; they are also used as building blocks for growth and repair and to produce other important substances. Vitamins, minerals, and water help body processes take place. Some minerals are also part of body structures.
2. Three types of enzymes help to digest the energy nutrients. Proteases break down proteins into peptides and amino acids. Amylases break down carbohydrates (starches and glycogen) to sugars. Lipases break down the tri-

glycerides in lipids to fatty acids and glycerol.
3. Saliva is the solution secreted by the salivary glands. It moistens food and begins the digestive process of breaking down food into component molecules.
4. Herbivore; carnivore; omnivore. Human beings are omnivores.
5. You would bite off the mouthful with your incisors, your canines would help you tear the food away, and your premolars and molars allow you to crush and grind the food thoroughly.
6. Pharynx; swallowing reflex; esophagus; peristalsis.
7. See Table 10-2.
8. Small intestine.
9. Two important accessory organs are the pancreas and liver. The pancreas secretes digestive enzymes. The liver, among other functions, produces bile, which helps digest lipids.
10. The duodenum is the first part of the small intestine; digestion is completed there. Starch and glycogen are broken down to disaccharides and then to monosaccharides. Proteins and polypeptides are broken down to shorter peptides and then to amino acids. Triglycerides are digested to fatty acids and glycerol.
11. These are hormones that help regulate digestion. Gastrin controls the release of hydrochloric acid in the stomach. Secretin stimulates the release of sodium bicarbonate to neutralize acid in the chyme and increases bile secretion in the liver. Cholecystokinin stimulates the gallbladder to release bile into the small intestine and stimulates the pancreas to release digestive enzymes.
12. Sodium bicarbonate neutralizes the acid in chyme. Thus it can help reduce the uncomfortable symptoms of acid indigestion.
13. Increasing the internal surface area of the small intestine helps absorb more nutrients from the same amount of food. Three features accomplish this: inner folds, projections (villi) on the folds, and additional projections on the projections (microvilli).
14. (1) Stomach (collects and partially digests food), (2) small intestine (finishes the digestive process), (3) large intestine (reabsorbs water and sodium and eliminates the rest as digestive wastes), and (4) anus (the opening through which feces are excreted.)
15. Glucose is broken down by glycolysis, the Krebs cycle, and the electron transport chain. Amino acids and triglycerides are converted to substances that can be metabolized by two of these pathways. Fatty acids are metabolized by another pathway.
16. The essential amino acids are eight amino acids that must be obtained in the diet. Humans cannot manufacture them.
17. See Figure 10-16. Answers will vary.

CHAPTER 11

1. Respiration is the uptake of oxygen and the release of carbon dioxide by the body. Cellular respiration is the process at the cellular level that uses the oxygen you breathe in and produces the carbon dioxide you breathe out. Internal respiration is the exchange of oxygen and carbon dioxide between the blood and tissue fluid; the exchange of the two gases between the blood and alveoli is external respiration.
2. Inspiration; expiration.
3. Your vocal cords produced the sound for speech as air rushed by and made them vibrate. Your lungs served as a power supply and volume control for you voice, and your lips and tongue formed the sounds into words.
4. Trachea; bronchi; respiratory bronchioles; alveoli.
5. Increases; decreases.
6. As the volume of the thoracic cavity increases, the air pressure within it decreases. Outside air is pulled in, equalizing the pressures inside and outside the thoracic cavity. As the volume of the thoracic cavity decreases, the air pressure inside it increases and forces air out, equalizing the pressures.
7. Tidal volume; residual air.
8. Deoxygenated blood has collected carbon dioxide during its passage through the body, so it contains more carbon dioxide than oxygen. Oxygenated blood contains more oxygen than carbon dioxide.
9. The circulatory system provides a transport system that distributes gases throughout the body. Without it you would not survive because it would take too long for oxygen to diffuse from the lungs to the rest of the body.
10. At the alveoli, carbon dioxide diffuses out of the blood into the alveoli because it moves down the pressure gradient (its pressure in the blood is greater than its pressure within the alveoli.) Meanwhile, oxygen diffuses from the alveoli into the blood because its pressure within the lungs is greater than in the blood.
11. At high altitudes the pressure of the oxygen molecules in the air is lower than at sea level. This means that the pressure gradient is lower at the alveoli and less oxygen diffuses into the blood. This can cause you to feel dizzy and short of breath.
12. Within the capillaries, blood pH decreases as carbon dioxide (in the form of bicarbonate ions) diffuses in and hydrogen ions accumulate. This more acid environment helps split oxygen from hemoglobin and enhances the diffusion of oxygen into the tissue fluid.
13. At the capillaries, oxygen moves down a pressure gradient to diffuse from the blood into the tissue fluid. Meanwhile, carbon dioxide moves down its own pressure gradient and diffuses from the tissue fluid into the blood.
14. Cigarette smoking paralyzes ciliated epithelial cells lining respiratory passageways, causing mucus to build up and plug passageways and the passageways to narrow. Smoking also impairs the action of the macrophages at the alveoli. The result is that smokers increase their risk of infection and their lungs and passageways become less elastic and efficient. The accumulated mucus, bacteria, viruses, and particles apparently contribute to the development of cancer.

CHAPTER 12

1. The circulatory system has four main functions: (1) nutrient and waste transport, (2) oxygen and carbon dioxide transport, (3) temperature maintenance, and (4) hormone circulation.
2. The blood carriers sugars, amino acids, and fatty acids to the liver, where some of the molecules are converted to glucose and released into the bloodstream. Excess energy molecules are stored in the liver for later use. Essential amino acids and vitamins pass through the liver into the bloodstream. The cells releases their metabolic waste products into the blood, which carries them to the kidneys.
3. A regulatory center in your brain constantly monitors your body temperature. If your temperature gets too high, signals from this center dilate your surface blood vessels and increase blood flow to the surface of your skin. This increases heat loss and lowers your body temperature.
4. The circulatory system has three components: the heart, the blood vessels, and the blood. The term *cardiovascular system* refers to the plumbing (heart and blood vessels) of the circulatory system.
5. Arteries; arterioles; capillaries; venules; veins.
6. When people are scared (or cold), the walls of the arterioles can contract and the blood flow decreases, routing more blood to other areas of the body. This can cause light-skinned people to "turn pale."

7. The entry to a capillary is guarded by a ring of muscle, which can constrict to block blood flow into that vessel. This is an important means of limiting heat loss. It also permits adjustments in blood flow, allowing more blood to go to areas where it is needed, such as to the intestines after a meal.
8. The walls of arteries have more elastic tissue and smooth muscle than those of veins; this helps arteries accommodate the pulses and high pressures of blood pumped from the heart. The lumen of veins is larger because blood flows through them at lower pressures; a larger diameter offers less resistance. The walls of capillaries are only one cell thick, which permits the exchange of substances between the cells and the blood.
9. See Figures 12-10 and 12-11.
10. Pulmonary circulation; systemic circulation.
11. Lower body.
12. The SA node is the pacemaker of the heart; this cluster of cells initiates the excitatory impulse that causes the atria to contract and the impulse to be passed along to other cardiac cells. The AV node conducts the impulse from the SA node to other cardiac cells, initiating the contraction of the ventricles.
13. Plasma is the liquid portion of blood in which the formed elements (cells and cell parts) are suspended. Plasma contains nutrients, hormones, respiratory gases, wastes, ions, salts, and blood proteins.
14. a. Erythrocytes are red blood cells; they carry oxygen.
    b. Leukocytes are white blood cells; they defend the body against microorganisms and other foreign substances.
    c. Platelets play an important role in blood clotting. These are all types of formed elements in blood.
15. Neutrophils (c)
    Basophils (a)
    Macrophages (d)
    Lymphocytes (b)
    All four cells are types of leukocytes.
16. The lymphatic system is a system of blind-ended vessels that collects the tissue fluid diffusing from blood plasma and returns it to the bloodstream. This counteracts the effects of the net movement of fluid out of the blood. The structures and organs of the lymphatic system also play a role in the immune system.
17. A high-fat diet can encourage the development of fatty deposits (plaque) on the inner walls of arteries, making the vessels narrower and less elastic. This can lead to atherosclerosis, heart attacks, and strokes.

CHAPTER 13

1. Nonspecific defenses act against any foreign invader, whereas specific defenses work against particular types of microbes. Nonspecific defenses are the first line of defense because they are the first barriers that a microorganism must bypass to enter your body.
2. The nonspecific defenses include (1) the skin (which if unbroken provides a barrier against invasion, (2) mucous membranes (mucus can trap particles and move then away from delicate areas), (3) chemicals that kill bacteria (many membranes are washed by fluids deadly to bacteria or have an acidic environment), (4) the inflammatory process (a series of events that remove the irritation, repair the damage, and protect the body from further damage).
3. The immune system is the body's specific defense; it protects you from particular microorganisms and other invaders.
4. The immune system differs from other body systems because it is not a system or organs and it lacks a single "controlling" organ. Instead, it consists of cells scattered throughout the body that have a common function: reacting to specific foreign molecules.
5. Antigens.
6. Phagocytes; T cells; B cells; NK (natural killer) cells; leukocytes.
7. Monocytes are precursors of macrophages; they develop into macrophages in response to infection. Macrophages are the phagocytes that play a key role in the immune response; they phagocytize foreign particles and cells and secrete proteins that activate the immune response. NK cells recognize and destroy body cells that have become cancerous or infected.
8. The four principal types of T cells are (1) helper T cells (which initiate the immune response), (2) cytotoxic T cells (which break apart infected and foreign cells), (3) inducer T cells (which oversee the development of T cells in the thymus), and (4) suppressor T cells (which limit the immune response).
9. During the cell-mediated immune response, cytotoxic T cells recognize and destroy infected body cells in addition to destroying transplanted and cancer cells. Helper T cells initiate the response, activating cytotoxic T cells, macrophages, inducer T cells, and finally, suppressor T cells.
10. When you had measles, B cell clones that did not become plasma cells became circulating lymphocytes—memory B cells. These cells give you an accelerated response during subsequent exposures to measles, promptly defending you against infection.
11. During the humoral response, B cells recognize foreign antigens and, if activated by helper T cells, produce antibodies. The antibodies bind to the antigens and mark them for destruction.
12. An antibody molecule recognizes a specific antigen because it has binding sites into which the antigen can fit. This allows your body to defend against specific microorgansisms.
13. Injection with antigens causes an antibody (B cell) response, with the production of memory cells. A booster shot induces these memory cells to differentiate into antibody-producing cells and form still more memory cells.
14. Active.
15. The AIDS virus is dangerous because it destroys helper T cells, thus destroying the immune system's ability to mount a defense against any infection.
16. An allergic reaction occurs when the immune system mounts a defense against a harmless antigen. When class E antibodies are involved and bind to the antigens, they cause a strong inflammatory response that can dilate blood vessels and lead to symptoms ranging from uncomfortable to life threatening.

CHAPTER 14

1. Excretion is the removal of metabolic wastes and excess water from the body. Without excretion, your body would soon become poisoned by the buildup of metabolic waste products. The process also helps maintain the balance of water and ions that is necessary for life.
2. The main metabolic waste products are carbon dioxide, water, salts, and nitrogen-containing molecules.
3. Urea, urine, and uric acid are all forms of metabolic waste products (nitrogenous wastes). Urea consists of ammonia and carbon dioxide. Uric acid forms from the breakdown of nucleic acids you eat and from the metabolic replacement of your nucleic acids and ATP. Urine is the excretion product consisting of water, urea, creatinine, uric acid, and other substances.
4. Lungs; kidneys.
5. Urea (d).
6. The kidneys maintain water balance, retain substances the body needs, and eliminate metabolic wastes by filtering most substances out of the blood and then reabsorbing what is needed.
7. See Figures 14-4 and 14-5.
8. A nephron is a filtering system that is the workhorse of the kidney. It has three major functions: filtration, selec-

tive reabsorption, and tubular secretion.

9. Kidneys help conserve water by reabsorbing it from the filtrate. By changing the concentrations of salts and urea in the filtrate, the kidneys create an osmotic gradient that results in water moving from the filtrate into the surrounding tissue fluid.
10. Urine can reveal the presence of certain substances in the body because the kidney detoxifies the blood. When these substances, such as drugs, are removed from the blood, they become part of the filtrate.
11. There is no right or wrong answer. Be sure to back up your opinion with reasons, however.
12. Urine is formed when the blood that flows through the kidneys is filtered, removing most of its water and all but its largest molecules and cells. The kidneys then selectively reabsorb certain substances.
13. The urinary system is a set of interconnected organs that removes excess wastes, water, and ions from the blood and stores them as urine until they are excreted. Its major components are the kidneys, ureters, urinary bladder, and urethra.
14. Homeostasis is the maintenance of constant physiological conditions in the body. ADH and aldosterone are two hormones that help regulate the water and ion balance necessary for homeostasis. ADH regulates that amount of water reabsorbed at the collecting ducts; aldosterone promotes the retention of sodium (and therefore water), while promoting the excretion of potassium.
15. Two problems with kidney function are kidney stones and renal failure. Kidney stones are crystals of certain salts that can block urine flow. Renal failure is a reduction in the rate of filtration of blood in the glomerulus.

CHAPTER 15

1. Neurons and hormones are both forms of communication that integrate and coordinate body functions. They differ in that neurons transmit rapid signals that report information or initiate quick responses in specific tissues. Hormones are chemical messengers that trigger widespread prolonged responses, often in a variety of tissues.
2. Brain; neurons; nerve impulses.
3. Dendrites and axons are cellular projections that extend from the cell body of a neuron. Dendrites bring messages in to the cell body; axons carry messages away from the cell body.
4. Sodium; potassium.
5. d, c, a, e, b; this describes the development and conduction of a nerve impulse.
6. *Resting potential* refers to the difference in electrical charge along the membrane of the resting neuron. The action of sodium-potassium transmembrane pumps and ion-specific membrane channels separates positive and negative ions along the inside and outside of the membrane, which creates the resting potential. The difference in electrical charges is the basis for the transmission of nerve impulses.
7. Stimulus; receptors; depolarization.
8. A nerve impulse travels because nearby transmembrane proteins respond to the electrical changes that accompany depolarization of the nerve cell membrane. The adjacent section of membrane depolarizes, followed by another section, leading to a wave of depolarization.
9. Saltatory conduction occurs along myelinated neurons when impulses jump to unmyelinated areas and skip over myelinated portions. Impulses travel much more quickly by this method than by continuous waves of depolarization that occur along continuous portions of the membrane.
10. Mylinated; unmylinated.
11. A synapse is a narrow gap between an axon tip of a neuron and the membrane of a muscle cell or another neuron. At a chemical synapse the impulse crosses this gap by stimulating the release of neurotransmitters that diffuse to the other side of the gap, combining with receptors in the membrane of the postsynaptic cell.
12. Synapses between neurons and skeletal muscle cells are neuromuscular junctions. At a neuromuscular junction, acetylcholine released from an axon tip depolarizes the muscle cell membrane, releasing calcium ions that trigger muscle contraction.
13. The enzyme acetylcholinesterase breaks down acetylcholine. If the neurotransmitter were not broken down, it would continue to signal the muscles to contract.
14. The postsynaptic neuron integrates the information it receives from presynaptic neurons. The summed effect of excitatory and inhibitory signals either facilitates or inhibits depolarization.

CHAPTER 16

1. The nervous system is a complex network of neurons that gathers information about the internal and external environment and processes and responds to this information. The central nervous system (brain and spinal cord) is the site of information processing. The peripheral nervous system (nerves) brings messages to and from the central nervous system.
2. Sensory neurons; afferent neurons; motor neurons; efferent neurons.
3. The somatic nervous system consists of motor neurons that control voluntary responses (for example, moving your leg to take a step). The autonomic nervous system consists of motor neurons that control involuntary activities (for example, digesting a meal.)
4. See Figure 16-3.
5. Cerebral cortex; gray matter; association areas; motor area; sensory area.
6. The thalamus and hypothalamus are masses of gray matter at the base of the cerebrum. The thalamus receives sensory stimuli, interprets some of them, and sends the rest to appropriate locations in the cerebrum. The hypothalamus controls various body organs and the secretions of some hormones.
7. a. Bundles of nerve fibers in the peripheral nervous system; carry information to and from the central nervous system.
   b. Network of neurons in the hypothalamus; responsible for many drives and emotions.
   c. The part of the brain that coordinates unconscious movements of skeletal muscles.
   d. Fluid in the spinal cord and surrounding the brain; acts as shock absorber and carries dissolved gases.
8. The brainstem (midbrain, pons, medulla) contains tracts of nerve fibers that carry messages to and from the spinal cord. Nuclei located there control important body reflexes.
9. The spinal cord is the part of the central nervous system that runs down the neck and back. It receives information from sense organs, carries it to the brain, and sends information from the brain to the rest of the body. Without it the brain could not control the rest of the body.
10. Sensory receptors change stimuli into nerve impulses. The nerve fibers of sensory neurons carry this information to the central nervous system, where interneurons interpret them and direct a response (in this case, turning your head.) The axons of motor neurons conduct these impulses to the appropriate muscles.
11. How you perceive stimuli depends on which receptors are stimulated. You registered this stimulus as a sound because it stimulated your auditory nerve.
12. A reflex is an automatic response to

nerve stimulation. A familiar example is the knee jerk reaction when tapped just below the kneecap.

13. These systems act in opposition to each other to maintain homeostasis and help you respond to environmental changes. The sympathetic nervous system generally mobilizes the body for greater activity (faster heart rate, increased respiration), whereas the parasympathetic nervous system stimulates normal body functions such as digestion.

CHAPTER 17

1. Sensory receptors are cells that can change environmental stimuli into nerve impulses. They provide information about the body's internal environment, its position in space, and its external environment.
2. For you to sense a stimulus, it must be of sufficient magnitude to open ion channels within the membrane of the receptor cell. This depolarizes the membrane, creating a generator potential that leads to an action potential (nerve impulse) in the sensory neurons with which the receptor synapses. Nerve fibers conduct the impulse to the central nervous system.
3. Receptors that sense the body's internal environment monitor body temperature, levels of carbon dioxide, blood pressure, and other bodily functions. This continual monitoring allows your body to respond and maintain homeostasis.
4. Proprioceptors are receptors that sense the body's position in space. These receptors help you to keep your balance, control your muscular movements, and protect your muscles, tendons, and joints from excessive tension and pulling.
5. The general senses are touch, pressure, pain, and temperature; their receptors are relatively simple and are located within the skin. Receptors for the special senses—smell, taste, sight, hearing, and balance—are more specialized and complex. Their receptors are located within sense organs.
6. Olfactory receptors detect different smells because of specific binding of airborne gases with the receptor chemicals located in the cilia of the nasal epithelium.
7. On its way to the olfactory area of the cerebral cortex, the nerve impulse travels through the limbic system, the area of the brain that is responsible for many of your drives and emotions. Thus certain odors become linked in your memory with emotions and events.
8. Taste buds are taste receptors concentrated on the tongue. They detect chemicals in food and register an overall taste that consists of different combinations of sweet, salty, sour, and bitter. The sense of taste interacts with the sense of smell to produce a taste sensation.
9. Rods and cones are sensory receptors located in the eye that detect certain wavelengths of electromagnetic energy. The pigment inside them absorbs photons of light, which causes a series of events leading to hyperpolarization of the receptor cells and firing of adjacent neurons.
10. Color vision is caused by cone cells. There are three types of cones, each of which absorbs a specific wavelength of light. The color you perceive depends on how strongly each group of cones is stimulated by a light source.
11. You see objects because light enters the eye through the pupil and the lens focuses it on the fovea, which is rich in cone cells. These cells, along with surrounding rods and cones, initiate nervous impulses that travel to the cerebrum via the optic nerve.
12. See Figure 18-8.
13. You have depth perception because of the positioning of your two eyes on either side of the head, which sends slightly different images to the brain. Covering one eye makes it more difficult to judge distance.
14. See Figure 18-12.
15. Your outer ear funnels sound waves toward the eardrum, which changes these waves into mechanical energy. This energy is then transmitted to the bones of the middle ear, which increase its force. Receptor cells in the inner ear change the mechanical energy into nerve impulses.
16. You can detect movements of your head because of the otoliths embedded in the jelly-like layer that covers the cilia in your inner ear. When you move your head, the otoliths slide and bend the underlying cilia. This generates signals to the brain, which interprets the type and degree of movement.

CHAPTER 18

1. Your movement resulted from the contraction of your skeletal muscles, which are anchored to bones. The muscles use bones like levers to direct force against an object. When you raised your hand, your skin stretched to accommodate the change in position.
2. Skin.
3. Skin has several important functions: (1) it serves as a protective barrier, (2) it provides a sensory surface, (3) it compensates for body movement, and (4) it helps control your internal temperature.
4. Bone is a type of connective tissue consisting of living cells that secrete collagen fibers into the surrounding matrix. The bones of the skeletal system support the body and permit movement by serving as points of attachment and acting as levers against which muscles can pull. They also protect delicate internal structures, store important minerals, and produce red and white blood cells.
5. Compact bone runs the length of long bones and has no spaces within its structure visible to the naked eye. Spongy bone is found in the ends of long bones and within short, flat, and irregularly shaped bones. It is an open latticwork of thin plates of bone, and its spaces are filled with red bone marrow. Compact bone gives bones strength, whereas spongy bone provides some support and stores red bone marrow as well as lightens the bones.
6. The axial skeleton includes the central axis (skull, vertebral column, and rib cage). The appendicular skeleton consists of the appendages (arms and legs) and the bones that help attach the appendages to the axial skeleton.
7. A "slipped disk" refers to one of the intervertebral disks of fibrocartilage that separate the vertebrae from each other (except in the sacrum and coccyx). These disks act as shock absorbers, provide the means of attachment between vertebra, and allow the vertebral column to move.
8. Pectoral girdle; pelvic girdle; appendicular.
9. An articulation is a joint—a place where bones, or bones and cartilage, come together. Joints are classified according to the degree to which they permit movement: immovable joints (the sutures in your skull), slightly movable joints (joints between your vertebra), and synovial joints (wrists).
10. These are all types of synovial (freely movable) joints. They differ in the type of movement they allow. A hinge joint allows movement in one plane only, a ball-and-socket joint allows rotation, and a pivot joint permits side-to-side movement.
11. A muscle consists of specialized cells packed with intracellular fibers capable of shortening. The three types of muscle tissue are (1) smooth muscle tissue (interior of arteries), (2) cardiac tissue (heart), and (3) skeletal muscle tissue (calf muscles).
12. Tendons; origin; insertion; origin; insertion.
13. Actin is a protein that makes up the

thin myofilaments of muscle; the protein myosin makes up the thick myofilaments. Myofilaments are the microfilaments of muscle cells. Muscle cells contract when the actin and myosin filaments slide past each other. Changes in the shape of the ends of the myosin molecules (located between adjacent actin filaments) cause the myosin molecule to move along the actin, causing the myofilament to contract.

14. When depolarization from a nerve impulse reaches a neuromuscular junction, it trigges the release of the neurotransmitter acetylcholine. The acetylcholine crosses over to the muscle fiber membrane and opens the ion channels of that membrane, depolarizing it. This sets off a series of reactions that release calcium ions; the calcium ions initiate the chemical reactions of contraction.

CHAPTER 19

1. Hormones are chemical messages secreted by cells that affect other cells. They are an important method the body uses to integrate the functioning of various tissues, organs, and organ systems.
2. The endocrine system consists of 10 glands that together produce over 30 different hormones. The messages carried by the endocrine hormones fall into four categories: (1) regulating the body's internal environment, (2) helping it respond to changes, (3) regulating reproduction, and (4) aiding normal growth and development.
3. The nervous system sends messages to glands and muscles, regulating glandular secretion and muscular contraction. Endocrine hormones carry messages to virtually any type of cell in the body.
4. The two main classes of endocrine hormones are peptide and steroid hormones. Peptide hormones bind to receptors on the cell membrane of target cells, ultimately triggering enzymes that alter cell functioning. Steroid hormones bind to receptors within the cytoplasm of target cells and ultimately cause the cell's hereditary material to produce specific proteins.
5. A feedback loop controls hormone production by initially stimulating a gland to produce the hormone. After the hormone has exerted its effect on the target cell, the body feeds back information to the endocrine gland. In a positive feedback loop, the feedback causes the gland to produce more hormone; in a negative feedback loop, it causes the gland to slow down or stop hormone production.
6. The hypothalamus stimulates or inhibits the secretion of the pituitary gland's hormones by means of chemicals called *regulating factors*. In addition, the hypothalamus produces two hormones that it stores in the pituitary.
7. Tropic hormones stimulate other endocrine glands. The four are the following: (1) Follicle-stimulating hormone (FSH): In women, FSH triggers the maturation of eggs in the ovaries and stimulates the secretion of estrogens. In men, it triggers the production of sperm. (2) Luteinizing hormone (LH): In women, LH stimulates the release of an egg from the ovary and fosters the development of progesterone. In men, it stimulates the production of testosterone. (3) Adrenocorticotropic hormone (ACTH): ACTH stimulates the adrenal cortex to produce steroid hormones. (4) Thyroid-stimulating hormone (TSH): TSH triggers the thyroid gland to produce the thyroid hormones.
8. a. Produced by the anterior pituitary; helps control normal growth.
   b. Produced by the posterior pituitary; enhances uterine contractions.
   c. Produced by the adrenal cortex; stimulates the kidneys to reabsorb sodium ions and water.
   d. Local hormones produced by cells throughout the body; stimulate smooth muscle contraction and the dilation and constriction of blood vessels.
9. No. Alcohol suppresses the release of ADH (antidiuretic hormone), which means that it encourages more water to leave your body in your urine. If you are already hot and thirsty, this will only dehydrate you further.
10. The thyroid gland produces thyroxine (T_4), triiodothyronine (T_3), and calcitonin (CT). Thyroid hormones regulate the body's metabolism.
11. Parathyroid hormone (PTH) and calcitonin (CT) work antagonistically to maintain appropriate calcium levels. If the level becomes too low, PTH stimulates osteoclasts to liberate calcium from the bones and stimulates the kidneys and intestines to reabsorb more calcium. When levels grow too high, more CT is secreted, which inhibits the release of calcium from bones and speeds up its absorption.
12. Both are produced by the adrenal cortex. The mineralocorticoids regulate the levels of specific ions in body fluids. The glucocorticoids affect glucose metabolism.
13. Over a prolonged period of stress, the body reacts in three stages: (1) alarm reaction (quickened metabolism triggered by adrenaline and noradrenaline), (2) resistance (glucose production and blood pressure rise; hormones involved are ACTH, GH, TSH, mineralocorticoids and glucocorticoids, and (3) exhaustion (loss of potassium and glucose; organs become weak and may stop functioning).
14. Chronic stress leads to long-term stimulation by the autonomic nervous system, which heightens metabolism, raises blood pressure, and speeds up internal chemical reactions. Over time this can physically stress the body and lead to health problems.
15. The pancreatic islets of Langerhans secrete two hormones that act antagonistically to one another to regulate glucose levels. Glucagon raises the glucose level by stimulating the liver to convert glycogen and other nutrients into glucose, whereas insulin decreases glucose levels in the blood by helping cells transport it across their membranes.
16. a. Tiny gland embedded in the brain; possible site of the biological clock that regulates daily rhythms.
    b. Small gland located in the neck; produces hormones that trigger maturation of T lymphocytes.
    c. In women, glands that produce female sex cells and the hormones estrogen and progesterone.
    d. In men, produce male sex cells and testosterone.

CHAPTER 20

1. Sexual reproduction is the process in which a male and a female sex cell combine to form the first cell of a new individual.
2. Spermatozoa; testes; scrotum; spermatogonia; spermatids.
3. Follicle-stimulating hormone (FSH) triggers sperm production. Luteinizing hormone (LH) regulates the testes' secretion of testosterone, a hormone responsible for the development of male secondary sexual characteristics.
4. A spermatozoon has a head that contains the hereditary material. Located at its leading tip is an acrosome that contains enzymes helping the sperm penetrate the egg's membrane. The sperm also has a flagellum that propels it and mitochondria that produce the ATP from which sperm derive the energy to power its flagellum.
5. These are accessory glands that add fluid to the sperm to produce semen. The seminal vesicles supply a fluid containing fructose, which serves as a source of energy for the sperm. The

prostate gland adds an alkaline fluid that neutralizes the acidity of any urine in the urethra and the acidity of the female vagina. The bulbourethral glands also contribute an alkaline fluid.

6. At birth, females have all of the oocytes that they will ever produce. As the woman ages, so do her oocytes, and the odds of a harmful mutation increase appreciably after age 35.
7. Oocyte; uterine tube; uterus; sperm; zygote.
8. The reproductive cycle of females occurs roughly every 28 days. The primary oocyte matures, is released from the ovary during ovulation, and journeys through the uterine tube to the uterus. The endometrial lining of the uterus has thickened to prepare for implantation; if fertilization does not occur, it sloughs off during menstruation. The hormones FSH, LH, estrogen, and progesterone orchestrate these events.
9. Fertilization occurs in the uterine tube. After the egg is fertilized, it completes a second meiotic division, joins its hereditary material with that of the sperm, begins dividing by mitosis, and implants itself on the thickened uterine wall. The placenta begins to form and human chorionic gonadotropin (HCG) is secreted.
10. The sexual response includes four phases: (1) excitement (sexual activity that precedes intercourse), (2) plateau (intercourse), (3) orgasm (reflexive muscular contractions; ejaculation), and (4) resolution (return to the normal physiological state).
11. The only methods of birth control that may prevent the transmission of sexually transmitted diseases are condoms and abstinence.
12. Birth control pills contain estrogen and progesterone, which shut down the production of FSH and LH. By maintaining high levels of estrogen and progesterone, the pills cause the body to act as if ovulation has already occurred; the ovarian follicles do not mature and ovulation does not occur.
13. The most effective birth control methods are vasectomies and tubal ligation. Birth control pills are very effective. Condoms and diaphragms are effective when used correctly, but mistakes are common. Least reliable are the rhythm method and withdrawal, which have high failure rates.

CHAPTER 21

1. Prenatal development refers to the growth and development that occur between conception and birth (about 8½ months).
2. After fertilization, the oocyte's surface changes so that no other sperm can penetrate, and oocyte meiosis is completed. The sperm sheds its tail and the sperm and egg nuclei fuse.
3. Zygote; cleavage; uterine tube; uterus; morula.
4. During implantation the blastocyst attaches to the inner wall of the uterus. This anchors it during development, allows it to obatin nutrients from the mother, and keeps it from being swept out of the uterus.
5. The three primary germ layers are the (1) ectoderm (outer layer of skin, the nervous system, and portions of sense organs), (2) endoderm (lining of the digestive tract and digestive organs, respiratory tract and lungs, and urinary bladder and urethra), and (3) mesoderm (skeleton, muscles, blood, reproductive organs, connective tissues, and innermost layer of skin).
6. The extraembryonic membranes play a role in the life support of the pre-embryo/embryo/fetus. They are the (1) amnion (cushions the embryo in amniotic fluid and keeps the temperature constant), (2) chorion (facilitates the exchange of nutrients, gases, and wastes between the embryo and mother), (3) yolk sac (produces blood for the embryo before its liver is functional and becomes the lining of the digestive tract), and (4) allantois (responsible for formation of the embryo's blood cells and vessels).
7. During gastrulation, groups of inner mass cells differentiate into the three primary germ layers from which all the organs and tissues will develop.
8. During neurulation, cells lying over the notochord curl upward to form a tube that will develop into the central nervous system.
9. The notochord is a structure that forms the midline axis of an embryo. In humans and in most other vertebrates, it develops into the vertebral column.
10. The correct sequence is b. morula (a mass of cells formed about three days after fertilization), d. blastula (a hollow ball of cells formed by cell migration; begins about 4 days after fertilization), a. gastrula (inner mass cells migrate and differentiate into the three primary germ layers; begins at the end of the second week and ends midway through the third week), c. neurula (development of a hollow nerve cord that later becomes the central nervous system and related structures; begins at the third week and ends at about 4½ weeks). These refer to the four stages of development of the pre-embryo and embryo.
11. Between the fourth and eighth weeks, the embryo grows much longer, establishes a primitive circulation, and begins to exchange gases, nutrients, and wastes with the mother. The central nervous system and body form begin to develop.
12. The first trimester is primarily a time of development; by the ninth week, most body systems are functional. During the second trimester the fetus grows, ossification is well underway, and it has a heartbeat. The third trimester is primarily a period of growth; by birth the fetus is able to exist on its own.
13. During pregnancy the fetus obtains all of its nutrients from the mother. Poor nutrition in the mother can damage the child, possibly resulting in retardation and stunted growth.
14. At birth the circulation of the newborn changes as the lungs, rather than the placenta, become the organ of gas exchange.

CHAPTER 22

1. Offspring of sexual reproduction inherit characteristics from both parents; those of asexual reproduction are exact duplicates of one parent. Sexual reproduction introduces variation within a species; the asexual variety does not.
2. Self-fertilization occurs when the gametes of one plant fertilize each other so that the offspring derive from one organism. Cross-fertilization occurs when the male gametes of one plant fertilize the female gametes of another.
3. A gene is the unit of transmission of hereditary characteristics in an organism. An allele is each member of a "factor pair"; each human receives one allele for each gene from the mother's egg and the allele for that gene from the father's sperm.
4. Mendel's Law of Segregation states that each gamete receives only one of an organism's pair of alleles. Random segregation occurs during meiosis.
5. Homozygous; heterozygous.
6. Genotype refers to the genetic makeup of an organism. Its phenotype is the expression of those genes in its outward appearance.
7. All of the F_1 offspring would have an Ll genotype and the phenotype of long leaves.

	L	L
l	Ll	Ll
l	Ll	Ll

8. The F_2 genotype would be 1:2:1

(LL:Ll:ll); the phenotype would be 3:1 (long-leaved:short-leaved).

	L	l
L	LL	Ll
l	Ll	ll

9. Mendel used a testcross to determine whether a phenotypically dominant plant was homozygous or heterozygous for the dominant trait. He crossed the plant with a homozygous recessive plant. When the test plant was homozygous, the offspring were hybrids and phenotypically dominant. When it was heterozygous, half the progeny were heterozygous and looked like the test plant, and half were homozygous recessive and resembled the recessive parent.
10. Monohybrids; dihybrids.
11. Mendel's Law of Independent Assortment states that the distribution of alleles for one trait into the gametes does not affect the distribution of alleles for other traits, unless they are on the same chromosome.
12. The probability is $\frac{1}{4}$.
13. a. $\frac{1}{16}$.
    b. $\frac{9}{16}$.
14. Sex-linked.

GENETICS PROBLEMS

1. The probability of getting two genes on the same chromosome is $(\frac{1}{2})^{23}$.
2. Somewhere in your herd you have cows and bulls that are not homozygous for the dominant gene "polled." Because you have many cows and probably only one or some small number of bulls, it would make sense to concentrate on the bulls. If you have only homozygous "polled" bulls, you could never produce a horned offspring regardless of the genotype of the mother. The most efficient thing to do would be to keep track of the matings and the phenotype of the offspring resulting from these matings and prevent any bull found to produce horned offspring from mating again.
3. Albinism, *a*, is a recessive gene. If heterozygotes mated, you would have the following:

	A	*a*
A	*AA*	*Aa*
a	*Aa*	*aa*

One-fourth would be expected to be albinos.

4. The best thing to do would be to mate Dingleberry to several dames homozygous for the recessive gene that causes the brittle bones. Half of the offspring would be expected to have brittle bones if Dingleberry were a heterozygous carrier of the disease gene. Although you could never be 100% certain Dingleberry was not a carrier, you could reduce the probability to a reasonable level.
5. Your mating of *DDWw* and Ddww individuals would look like the following:

	Dw	*Dw*	*dw*	*dw*
DW	*DDWw*	*DDWw*	*DdWw*	*DdWw*
Dw	*DDww*	*DDww*	*Ddww*	*Ddww*
DW	*DDWw*	*DDWw*	*DdWw*	*DdWw*
Dw	*DDww*	*DDww*	*Ddww*	*Ddww*

Long-wing, red-eyed individuals would result from 8 of the possible 16 combinations, and dumpy, white-eyed individuals would never be produced.

6. Breed Oscar to Heidi. If half of the offspring are white eyed, Oscar is a heterozygote.
7. Both parents carry at least one of the recessive genes. Because it is recessive, the trait is not manifested until they produce an offspring who is homozygous.
8. To solve this problem, let's first look at the second cross, where the individuals were crossed with the homozygous-recessive sepia flies *se/se*. In one case, all the flies were red eyed:

		Unknown genotype *Se*	*Se*
Sepia	*Se*	*Se/Se*	*Se/Se*
	se	*Se/se*	*Se/se*

The only way to have all red-eyed flies when bred to homozygous sepia flies is to mate the sepia fly with a homozygous red-eyed fly. In the other case, half of the offspring were black eyed and the other half red eyed.

		Unknown genotype *Se*	*se*
Sepia	*se*	*Se/se*	*se/se*
	se	*Se/se*	*se/se*

The unknown genotype in this case must have been *Se/se,* since this is the only mating that will produce the proper ratio of sepia-eyed flies to red-eyed flies. Since the ratio of this unknown genotype and the one previously determined was 1:1, you must deduce the genotype of the original flies, which when mated, will produce a 1:1 ratio of *Se/se* to *Se/Se* flies.

		Unknown original 1 *Se*	*Se*
Original 2	*Se*	*Se/Se*	*Se/Se*
	se	*Se/se*	*Se/se*

You can see from this diagram that if one of the original flies was homozygous for red eyes and the other was a heterozygous individual, the proper ratio of heterozygous and a homozygous offspring would be obtained.

CHAPTER 23

1. Sex chromosomes; 22; autosomes.
2. Nondisjunction occurs if homologous chromosomes fail to separate after synapsis. This results in gametes with abnormal numbers of chromosomes.
3. Down syndrome (trisomy 21) is a genetic disorder caused by having three 21 chromosomes. Individuals who inherit this condition show delayed maturation of the skeletal system and are often mentally retarded.
4. People who inherit abnormal numbers of sex chromosomes often have abnormal features and may be mentally retarded. Examples are (1) triple X females (XXX zygote), underdeveloped females who may have lower-than-average intelligence; (2) Klinefelter syndrome (XXY zygote), sterile males with some female characteristics; (3) Turner syndrome (XO zygote), sterile females with immature sex organs; and (4) XYY males, fertile males of normal appearance.
5. X.
6. A chromosomal duplication involves the replication of a portion of the chromosome. A translocation occurs when a section of a chromosome breaks off and reattaches to another chromosome. An inversion occurs if the broken piece reattaches to the same chromosome but is reversed.
7. The three major sources of damage to chromosomes are high-energy radiation (such as x-rays), low-energy radiation (such as UV light), and chemicals (such as certain legal and illegal drugs).
8. Heavy use of drugs could affect ova and sperm by damaging the DNA. In general, chemicals add or delete molecules from the structure of DNA. This can result in chromosomal breaks or changes as the cell works to repair the damage.
9. A point (gene) mutation is a change in the genetic message of a chromosome due to alterations of molecules within the chromosomal DNA. This produces a new allele of a gene.
10. a. Sex-linked.
    b. Dominant.
    c. Recessive.
    d. Autosomal.
11. Most human genetic disorders are recessive because those genes are able to persist in the population among carriers; people carrying lethal dominant disorders are more likely to die before

reproducing. Recessive disorders include cystic fibrosis, sickle-cell anemia, and Tay-Sachs disease.

12. In incomplete dominance, alternative forms of an allele are neither dominant nor recessive; heterozygotes are phenotypic intermediates. In codominance, alternative forms of an allele are both dominant, so heterozygotes exhibit both phenotypes.
13. Their children's genotypes would be 1:2:1 (AA [type A], AB [type AB], BB [type B]).

	A	B
A	AA	AB
B	AB	BB

14. Couples who suspect they may be at risk for genetic disorders can undergo genetic couseling to determine the probability of this risk. When a pregnancy is diagnosed as high risk, a woman can undergo amniocentesis (analysis of a sample of amniotic fluid) to test for many common genetic disorders.

GENETICS PROBLEMS

1.

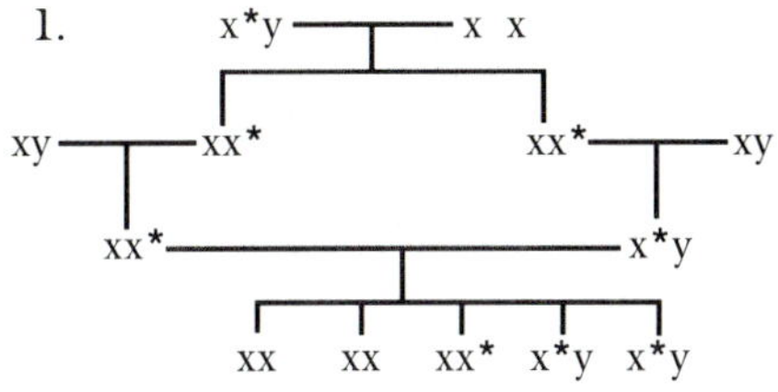

2. $I^AI^O \times I^BI^O \rightarrow I^OI^O$
3. 45 (44 autosomes + one X)
4. AO (blood type A)
5.

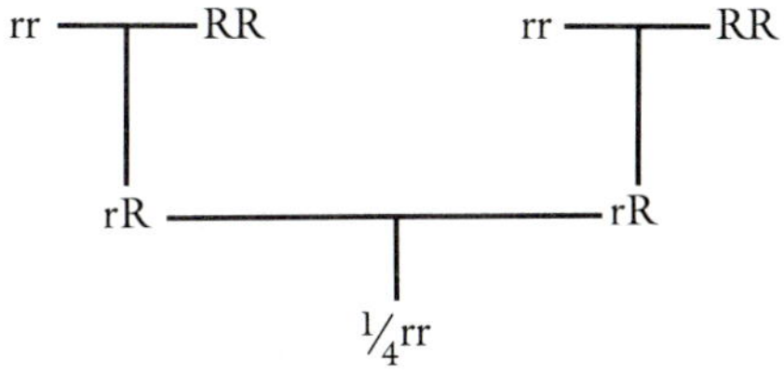

6. c—sex-linked recessive

CHAPTER 24

1. Mendel's work suggested that traits are inherited as discrete packets of information. Sutton theorized that these packets were located on chromosomes, and Morgan's experiments provided evidence that they were. Hammerling's experiments showed that this hereditary information is located in the nucleus of the cell.
2. The two types of nucleic acids are DNA (deoxyribonucleic acid) and RNA (ribonucleic acid). Both consist of nucleotides that are made up of three molecular parts: a sugar, a phosphate group, and a base. The sugar in RNA is ribose, whereas the sugar in DNA is deoxyribose. DNA contains the bases thymine and cytosine; RNA contains uracil instead of thymine.
3. Chargaff's experiments showed that the proportion of bases varied in the DNA of different types of organisms but that the amount of adenine always equalled the amount of thymine and guanine always equalled cytosine. This suggested that DNA was the hereditary material that encoded the information used by cells for growth, development, and repair.
4. *Double helix* refers to the DNA molecule, which is shaped like a double-stranded helical ladder. The bases of each strand together form rungs of uniform length, and alternating sugar-phosphate units form the ladder uprights.
5. DNA begins replicating by unwinding at intervals along its helix. Free nucleotides bind to the exposed bases, producing two new double-stranded DNA molecules that are identical to each other and to the original double strand.
6. Gene; nucleotides; amino acid.
7. The type of RNA found in ribosomes, along with the proteins also found there, is ribosomal RNA (rRNA). Transfer RNA (tRNA) is in the cytoplasm; during polypeptide synthesis, tRNA molecules transport amino acids to the ribosomes and position each amino acid at the correct place on the elongating polypeptide chain. Messenger RNA (mRNA) brings information from the DNA in the nucleus to the ribosomes in the cytoplasm to direct which polypeptide is assembled.
8. During transcription, the DNA message is copied onto a strand of mRNA. The mRNA is constructed from free RNA nucleotides by using a single strand of DNA as a template. The mRNA leaves the nucleus and travels to the ribosomes where polypeptide synthesis occurs.
9. c, a, d, b. This describes the process of translation (polypeptide synthesis).
10. Exons are the nucleotide sequences on a gene that encode the amino-acid sequence of the polypeptide. Introns are sequences of nucleotides that intervene between exons (introns may have evolutionary significance by allowing easier rearrangement of exons). Codons are the triplet-nucleotide code words on mRNA that reflect the sequence of the DNA code.
11. Regulatory genes are special sequences of nucleotides that control the transcription of structural genes (the genes that code for polypeptide synthesis). Regulatory genes are responsible for deciding which polypeptides are produced and when.
12. Repression; repressor protein; activation; activator protein.
13. In gene therapy, scientists attempt to cure genetic disorders by inserting genes with the correct genetic message into patients with incorrect genes.
14. Answers will vary. Be sure to back up your opinion with reasons.

CHAPTER 25

1. Darwin's observations led him to conclude that (1) organisms of the past and present are related, (2) factors other than or in addition to climate are involved in the development of plant and animal diversity, (3) members of the same species often change slightly in appearance after becoming geographically isolated from each other, and (4) organisms living on oceanic islands often resembled those living on a nearby mainland.
2. Three other factors that influenced Darwin were (1) evidence of geological layers and fossils that suggested the Earth was much older than traditionally thought, (2) the observed changes in species that occur with artificial selection, and (3) Malthus' writings on geometric population growth.
3. The two dogs look so different today because of artificial selection; people who bred them selected for desired characteristics, so over time the breeds changed. Artificial selection is based on the principle of natural variation exhibited by all organisms.
4. A geometric progression increases more rapidly because it involves multiplying a number by a constant factor; an arithmetic progression is one in which the elements increase by a constant difference. Malthus theorized that populations grow geometrically, but certain factors limit this growth. Darwin realized that nature acts to limit population numbers.
5. Adaptations are naturally occurring inheritable traits found within populations that confer a reproductive advantage on organisms that possess them. As adaptive (advantageous) traits are passed on from surviving individuals to their offspring, the individuals carrying those traits will increase, and the nature of the general population will gradually change.
6. *Survival of the fittest* refers to the fact that natural selection tends to favor those organisms that are most fit to survive to reproductive age in a particular environment and produce offspring.
7. Polar bears are white because they live

in the Arctic where there is snow year round. Thus their white color is an adaptation that camouflages them and has made them more fit to survive in this environment.

8. In adaptive radiation, the population of a species changes as it disperses into different habitats. Eventually some of these populations change so sufficiently that interbreeding is no longer possible. This is speciation, the formation of new species through the process of evolution.
9. Fossil; sedimentary; radioactive isotopes.
10. The carbon-14 method can help establish a date by estimating the relative amounts of the different isotopes of carbon present in a fossil. The half-life of carbon-14 is 5730 years (the amount of time for half of the ^{14}C to decay into nitrogen). Scientists can estimate the length of time that the carbon-14 has been decaying, which means the time that has elapsed since the organism died.
11. According to radioctive dating, the Earth is 4.6 billion years old. This is significant because it allows sufficient time for evolution to occur.
12. Vestigial organs are structures that are present in an organism of today but are no longer useful. A human example is the appendix.
13. Homologous structures share the same evolutionary origin but now differ in structure and function. Analogous structures have a similar form and function but different evolutionary origins.
14. Comparative studies of anatomy show that many organisms have groups of bones, nerves, muscles, and organs with the same anatomical plan but with different functions. These homologous structures imply evolutionary relatedness. Studies in comparative embryology show that many organisms have early developmental stages that are similar. This again implies evolutionary relatedness.
15. A phylogenetic tree diagrams the degree of relatedness among groups of organisms. Phylogenetic trees support evolution by showing the same evolutionary relationships revealed by anatomical studies.

CHAPTER 26

1. The primordial soup theory attempts to explain how life began on Earth. It states that life arose in the sea as elements and simple compounds in the atmosphere interacted to form simple organic molecules. Another theory holds that life might have arisen in hydrothermal vents in the oceans.
2. Dehydration synthesis is a type of chemical reaction that links molecules by the removal of water and the input of energy. These reactions build polymers (complex molecules). The synthesis of complex organic molecules would have been the first step toward the evolution of living organisms.
3. The formation of some sort of membrane would have allowed chemical reactions to occur in a closed environment. Enzymes could be organized to carry on life functions, chemicals could be selectively absorbed, wastes could be eliminated, and hereditary material could be passed on to future cells.
4. See Table 26-1.
5. Cyanobacteria are single-celled organisms that are similar to the first oxygen-producing bacteria. These evolved during the Archean Era and played an important evolutionary role by gradually oxygenating the atmosphere and oceans.
6. *Proterozoic* is actually inaccurate, since it comes from the Greek for "prior to life." During the Proterozoic Era the first eukaryotic organisms appeared.
7. The Cambrian Period was the oldest period in the Paleozoic Era (roughly 590 to 500 million years ago). All of the main phyla and divisions of organisms that exist today (except chordates and land plants) evolved by the end of the Cambrian Period.
8. The Devonian Period has been nicknamed "the age of fishes" because during this time fishes became abundant and diverse. Amphibians appeared by the end of this period.
9. During the Carboniferous Period, reptiles developed from amphibians and arthropods moved from the sea onto land. Insects, one arthropod group, evolved water-conserving characteristics such as a cuticle. Fungi evolved during the late Carboniferous Period. Much of the land was low and swampy with extensive forests; the worldwide climate was warm and moist.
10. To date, five major extinction events have occurred: (1) at the end of the Cambrian Period, roughly 500 million years ago; (2) about 440 million years ago; (3) about 360 million years ago; (4) during the Permian Period; and (5) at the end of the Mesozoic Era (65 million years ago).
11. Reptiles; mammals; birds; dinosaurs.
12. During this extinction event, many of the larger plankton disappeared, as did the ammonites and the dinosaurs. One theory suggests that a meteorite struck the Earth, creating a huge dust cloud that blocked photosynthesis, killing many organisms. Other theories are that the climate cooled, the sea level changed, or that the vegetation cover of the Earth changed.
13. The plate tectonics theory states that the Earth's outer shell consists of plates that drift over underlying rocks. This constant movement eventually moved land masses apart or closer together, changing habitats. Populations underwent adaptive radiation, speciation, and/or extinction in response to the changing pressures of natural selection.
14. The Cenozoic Era, which began 65 million years ago, saw the rapid evolution and growth of mammals, including primates.
15. Anthropoid primates; mammals.
16. Alas for Hollywood, no. Dinosaurs died out at the end of the Cretaceous Period (about 65 million years ago), long before the first primates appeared.

CHAPTER 27

1. Chordates are distinguished by three features: (1) a single, hollow nerve cord along the back, (2) a rod-shaped notochord located between the nerve cord and the developing gut, and (3) pharyngeal arches.
2. The vertebrates evolved from the chordates. Vertebrates differ from other chordates in that adult vertebrates have a vertebral column replacing the embryologic notochord. Most vertebrates also have a distinct head and bony skeleton.
3. Amphibians; reptiles.
4. c.
5. Warm; hair; milk; mammary glands.
6. Warm bloodedness is the ability to maintain a constant internal body temperature. This ability helps warm-blooded animals stay alive in environments that have extreme temperatures or widely varying temperatures.
7. These are all subclasses of mammals that are living today. Monotremes lay eggs with leathery shells and incubate them in a nest. Marsupials bear immature young and nurse them in a pouch until they are old enough to be on their own. In placental mammals, the young develop to maturity inside the mother.
8. Primates are the order of mammals with characteristics reflecting an arboreal life-style. They have developed two especially helpful characteristics: depth perception resulting from stereoscopic vision and flexible, grasping hands with opposable thumbs.
9. These are both suborders of the primates. Prosimians (lower primates such as lemurs) are small animals, usually nocturnal, with a well-developed

sense of smell. Anthropoids (higher primates such as apes and humans) have larger brains, flatter faces, eyes that are closer together, and relatively long front and hind limbs. Anthropoids are also diurnal and possess color vision.

10. (1) d, (2) c, (3) a, (4) b.
11. Statement (a) is correct; statement (b) is false. All hominids are a family within the hominoid superfamily.
12. *Australopithecus afarensis* is the oldest hominid fossil skeleton that has yet been found; it is not considered human. The first human fossils belong to the extinct species *Homo habilis,* which was more intelligent than its ancestors and able to make tools and clothing. The extinct species *Homo erectus* was close to modern human size, walked upright, and had a larger brain than its ancestors. Of the three, it is most like modern humans.
13. Using "neanderthal" in this way actually contradicts what scientists have learned about the Neanderthal subspecies of *Homo sapiens.* Neanderthals had larger brains than modern humans and made diverse tools. Neanderthal graves indicate that they cared for their injured and sick, believed in a life after death, and were capable of symbolic thought.
14. Early members of the species had smaller heads, brow ridges, teeth, jaws, and faces than earlier species did. They also fashioned sophisticated tools from stone and from other substances such as bone and ivory. Over time, groups of *Homo sapiens sapiens* moved away from the hunter-gatherer life-style to develop agriculture, complex social structures, and civilization.

CHAPTER 28

1. Viruses are infectious agents that lack a cellular structure, so they are not living. Instead, they are nonliving obligate parasites that must exist in association with and at the expense of other organisms.
2. See Figure 28-3.
3. Both cycles are patterns of viral replication. In the lytic cycle, viruses enter a cell, replicate, and cause the cell to burst and release new viruses. In the lysogenic cycle, viruses enter into a long-term relationship with the host cells, their nucleic acid replicating as the cells multiply.
4. A cold sore results from the lysogenic cycle of cell damage caused by the herpes simplex virus. The virus remains latent in nervous tissues until something triggers it, such as a cold.
5. Tumors are large masses of cells resulting from viral infections that caused these cells to grow faster than normal cells. A benign tumor is walled off from the rest of the body so that these cells cannot spread. A malignant (cancerous) tumor—the more dangerous of the two—consists of cells that invade and destroy other body tissues.
6. Bacteria are classified into their own kingdom because their cell structure is different from that of all other organisms. (They do not contain membrane-bounded organelles, they lack a nucleus, and the bacterial cell is bounded by a membrane encased within a cell wall.)
7. Bacteria perform many vital functions, such as decomposing organic materials and recycling inorganic compounds. They were also largely responsible for creating the properties of the atmosphere and soil that are present today.
8. Binary fission; exponential growth.
9. a. Chemoautotrophs (nitrifying bacteria).
   b. Heterotrophs (decomposers).
   c. Photoautotrophs.
   d. Archaebacteria.
10. These terms refer to the purposeful manipulation of genes within organisms.
11. These are all mechanisms by which genes move from one prokaryote to another. In transformation, free pieces of DNA move from a donor to a recipient cell. In transduction, DNA from a donor cell is carried to a recipient cell by a virus. During conjugation, DNA from the donor cell is injected into the recipient cell when the two cells make contact.
12. Researchers use restriction enzymes to cut desired pieces of human DNA from nuclei, open bacterial plasmid DNA, and insert the human fragment. These hybrid plasmids can be taken up by bacteria, which then express the human gene by producing a particular human protein. This technique can provide a relatively inexpensive and plentiful source of clinically useful substances; because the source is human, patients are less likely to suffer allergic reactions.
13. Gene therapy is the treatment of a genetic disorder by inserting normal genes into the patient's cells.
14. Genetic engineering offers many potential benefits, such as improving crops and forest trees, producing medically important substances, and treating some genetic disorders. It is important to establish appropriate safeguards for such experiments.

CHAPTER 29

1. Protists are single-celled eukaryotic organisms; this kingdom contains organisms that are animal-like (protozoa), plant-like (algae and diatoms), and fungus-like (slime molds).
2. An ameba is a protozoan that takes on changing shapes. It moves and obtains food by means of pseudopods (cytoplasmic extensions). Its pseudopods engulf organic matter, which the ameba then digests.
3. Both flagellates and ciliates are protozoans, and both have hair-like cellular processes that help them move and obtain food. However, flagellates have flagella (long hair-like processes), whereas ciliates have cilia (short hair-like extensions). Ciliates have a more complex internal organization then flagellates do.
4. Sporozoans are nonmotile protozoans that parasitize animals. They undergo complex life cycles in which they are passed from host to host by means of a vector.
5. Euglenoids, dinoflagellates, and golden algae are all phyla of plant-like protists. Euglenoids are flagellated protists with chloroplasts. Dinoflagellates have stiff outer coverings and their flagella beat in two grooves. The golden algae have gold-green photosynthetic pigments and store food as oil.
6. All three organisms belong to the phylum of golden algae.
7. Slime molds; fungus; ameba.
8. Saprophytic; parasite.
9. All fungi could be described as helpful, since as decomposers they are crucial to the cycling of materials in ecosystems. More specifically, helpful fungi include those that produce antibiotic drugs. Yeasts are fungi that are crucial in the production of bread, cheese, and beer. Other fungi, however, cause diseases in plants and animals, such as Dutch elm disease, wheat rust, and yeast infections.
10. Club fungi (c), zygote-forming fungi (d), sac fungi (a), water molds (e), imperfect fungi (b).
11. Lichens are symbiotic associations between fungi and photosynthetic partners. Lichens are able to tolerate harsh living conditions, so they are found in a wide range of habitats (such as deserts, extreme northern and southern latitudes, and mountaintops).
12. Radiolarians; foraminifera.

CHAPTER 30

1. Vascular; nonvascular.
2. Sporophytes; gametophytes; gametophytes; sporophytes.
3. The three classes are the mosses, liv-

erworts, and hornworts. All three share a fairly uniform life cycle involving two phases. The sporophyte generation grows out of or is embedded in the tissues of the gametophyte generation and depends on it for nutrition. Neither generation tends to dominate the life cycle.

4. Seeds are structures that protect the embryonic sporophyte plant from drying out or being eaten and contain stored food.
5. In seedless vascular plants, the sporophyte generation is dominant and lives separately from the gametophyte. The gametophytes produce motile sperm that swim to the eggs. The fertilized eggs produce sporophytes that at first grow within gametophyte tissues but eventually become free living.
6. Gymnosperms (c), angiosperms (d), bryophytes (b), ferns (a).
7. See Figure 30-7.
8. Sexual reproduction in gymnosperms and angiosperms produces seeds by the union of male and female gametes. Germinating pollen tubes convey the sperm to the eggs. The seeds protect and nourish the young sporophytes as they develop into new plants.
9. The relationship between flowering plants and their pollinators has influenced the evolution of both groups. Flowers have evolved colors and forms that attract certain pollinators, whereas insects, for example, have evolved special traits that are useful in obtaining food more efficiently from the flowers.
10. Fruits have evolved a range of shapes, textures, and tastes that help disperse their seeds more effectively. Fruits with brightly colored flesh are often eaten by animals and thus dispersed to different habitats. Other fruits have evolved hooked spines that stick to animals' fur; others have wings that are blown by the wind.
11. b, d, e, a, c.
12. Monocots and dicots are both angiosperms. They differ in the placement of the stored food in their seeds. In dicots, most of the stored food is in the cotyledons (seed leaves), whereas in monocots, most of it is stored in endosperm. In addition, monocots usually have parallel veins in their leaves, and dicots have net-like veins. Monocots also have flower parts in threes, whereas dicots have flower parts in fours and fives.
13. Vegetative propagation is an asexual reproductive process in which a new plant develops from a portion of a parent plant. An example is the white potato, a tuber which can grow new plants from its "eyes."
14. Hormones are chemical substances produced in one part of an organism and then transported to another part of the organism, where they bring about physiological responses. The five kinds of plant hormones are auxins, gibberellins, cytokinins (promote and regulate growth), abscisic acid (inhibits growth), and ethylene (affects the ripening of fruit).

CHAPTER 31

1. Vascular plants are plants having a system of vessels that transports water and nutrients.
2. Root; shoot; leaves; stem.
3. (1) c, (2) b, (3) a, (4) d.
4. Xylem and phloem are both types of vascular tissue, but they have different functions and contain different types of conducting cells. Xylem conducts water and dissolved minerals, whereas phloem conducts carbohydrates and other substances.
5. a. Cells in ground tissue that function in photosynthesis and storage.
   b. The most abundant type of cell in dermal tissue, often covered with a thick waxy layer that protects the plant and retards water loss.
   c. Openings that connect air spaces inside dicot leaves with the exterior; they are opened and closed by the change in shape of surrounding guard cells. Water vapor and gases move into and out of the leaves through these openings.
   d. Thimble-like mass of unorganized cells that covers and protects a root's apical meristem as it grows through soil.
6. Primary growth occurs mainly at the tips of roots and shoots, making plants taller; secondary growth occurs in the lateral meristem and makes plants larger in diameter.
7. Vegetables are a good source of minerals. Plants take in minerals from the soil to form proteins, fats, and vitamins.
8. See Figures 31-5 and 31-6.
9. A dandelion has a taproot, a single major root that can grow deep as a firm anchor. Some taproots also store food for the plant. An ivy plant has adventitious roots that develop from the lower part of the stem and help anchor the plant. Grass has fibrous roots, a root system without one major root; this type of system anchors the plant and absorbs nutrients over a large surface area.
10. Annual rings provide a clue to a tree's age by illustrating the annual pattern of rapid and slow growth in the cambium (lateral meristem tissue). This results in secondary growth.
11. Water rises beyond the point where it would be supported by air pressure because evaporation from the plant's leaves produces a force that pulls upward on the entire column of water. The forces invovled include air pressure, adhesion, cohesion, root pressure, and transpiration.
12. See Figure 31-19.
13. In general, nonvascular plants are shorter than vascular plants. This is because they lack the sophisticated transport systems of vascular plants.

CHAPTER 32

1. Both you and jellyfish are animals. An important difference is that jellyfish are invertebrates (they lack a backbone), whereas you are a vertebrate (with a backbone and dorsal nerve cord).
2. Eukaryotic organisms have a cellular structure different from that of bacteria. Animals are also multicellular (having more than one cell), and as heterotrophs, they are unable to make their own food. Therefore they must eat other organic matter for food. Beyond these basic similarities, however, animals are a diverse group.
3. Radial symmetry; bilateral symmetry.
4. Coelomates have lined, fluid-filled body cavities in which organs are suspended by thin sheets of connective tissue (such as in humans). Pseudocoelomates have body cavities, but the cavities are not lined and their organs are not suspended by connective tissue (such as in roundworms). Acoelomates have no body cavities (such as in flatworms).
5. Sponges are called the simplest animals because they lack tissues and organs.
6. Cnidarians exist as either polyps (cylindrically shaped animals that anchor to rocks) or as medusae (free-floating, umbrella-shaped animals).
7. The acoelomates are the most primitive bilaterally symmetrical animals. They have organs and some organ systems, including a primitive brain. Their organs are embedded within body tissues.
8. Roundworms can be dangerous if they parasitize humans, absorbing digested food from their hosts. Roundworms are cylindrical pseudocoelomates encased in a flexible outer cuticle; they have primitive excretory and nervous systems. They move by using muscles attached to the cuticle that push against the pseudocoelom.
9. In a closed circulatory system, blood is enclosed within vessels as it travels through the body. In an open system, blood flows through irregular chan-

nels (blood sinuses) in many parts of the body. Since you have blood vessels, you have a closed system.

10. Mollusks are coelomates with a muscular foot and soft body, usually covered by a shell. They use gills for respiration in water, whereas terrestrial species have adaptations to breathing on land. They show both open and closed circulatory systems and well-developed excretory systems.
11. Like many arthropods, flies have compound eyes that contain many independent visual units, each with its own lens. This feature helps them detect motion quickly.
12. Arthropods are a diverse phylum of organisms having jointed appendages and rigid exoskeletons.
13. Insects belong to a group of arthropods that have jaws (mandibles). Spiders belong to a different group of arthropods that lack mandibles.
14. Echinoderms are deuterostomes, whereas the other invertebrates are protostomes. These embryological differences make echinoderms more similar to the chordates, which implies that echinoderms and chordates evolved from a common ancestor.

CHAPTER 33

1. Nerve cord; notochord; pharyngeal (gill) slits.
2. The nerve cord develops into a brain and spinal cord; the notochord becomes a vertebral column enclosing the nerve cord; and the pharyngeal arches develop into gills (in fishes) or into ear, jaw, and throat structures (terrestrial vertebrates).
3. (1) b, (2) c, (3) a.
4. Vertebrates have (1) a vertebral column, (2) a distinct head with a skull enclosing their brain, (3) a closed circulatory system, and (4) a heart that pumps their blood.
5. The *Jaws* films starred sharks (class Chondrichthyes) and humans (class Mammalia).
6. A lateral line system (found in all fishes and amphibians) is a system of mechanoreceptors that detects sound, pressure, and movement.
7. Osmoregulation is the control of water movement into and out of organisms' bodies. Maintaining the proper balance of water and solutes is vital to life.
8. Ovoviviparous organisms (some fishes, reptiles; many insects) retain fertilized eggs within their oviducts until the young hatch; the young receive nourishment from the egg. Oviparous animals (birds) lay eggs and their young hatch outside the mother. Viviparous organisms (humans) bear their young alive, and the young are nourished by the mother, not the egg.
9. *Amphibian* means "two lives." This refers to the fact that amphibians live both aquatic and terrestrial existences.
10. Ectothermic animals (fish, amphibians, reptiles) regulate body temperature by taking in heat from the environment. Endothermic animals (birds, mammals) regulate their body temperatures internally.
11. Reptiles have dry skins covered with scales, retarding water loss. They also lay amniotic eggs, which contain nutrients and water for the embryos and protect the embryos from drying out.
12. Feathers are flexible, light, waterproof epidermal structures; birds are the only animals that have them. Feathers are important to flight; they also give birds waterproof coverings and insulate them against temperature changes.
13. Mammals are endothermic vertebrates that have hair and whose females secrete milk from mammary glands to feed their young.

CHAPTER 34

1. Behaviors include the patterns of movement, sounds, and body positions exhibited by animals. They also include any type of change in an animal that can trigger behaviors in other animals.
2. This statement expresses a fundamental idea in the study of animal behavior, that animal behaviors are evolutionary adaptations that make an organism more fit to survive in and adjust to its environment.
3. Innate (inborn) behaviors; learned behaviors.
4. Kineses; taxes; reflexes; fixed action patterns.
5. A kinesis is a change in the speed of an animals's random movements in response to environmental stimuli (for example, insects moving around underneath a fallen log). A taxis is a directed movement toward or away from a stimulus (for example, moths being attracted to a light at night). A reflex is an automatic response to nerve stimulation (for example, knee jerk reflex in humans). A fixed action pattern is a sequence of innate behaviors in which the actions follow an unchanging order of muscular movements (for example, a mother bird stuffing an insect into the gaping beak of her offspring).
6. Innate behaviors protect an animal from environmental hazards without it having learned to do so; many social behaviors also depend on innate behaviors. Organisms with simple nervous systems rely primarily on innate behaviors for survival. Learned behaviors, however, allow an animal to adjust its behavior based on its experiences, helping it to adapt better to its environment or to environmental changes.
7. Imprinting; habituation; trial-and-error learning; classical conditioning; insight.
8. a. Taxis (phototaxis).
   b. Habituation.
   c. Trial-and-error conditioning (operant conditioning).
9. Operant conditioning is a form of learning in which an animal associates something that it does with a reward or punishment. In classical conditioning, an animal learns to associate a new stimulus with a natural stimulus that normally evokes a particular response. Eventually the animal will respond to the new stimulus alone.
10. Insight (reasoning) is the most complex form of learning in which an animal recognizes a problem and solves it mentally before trying out the solution. Insight allows an animal to perform a correct behavior the first time it tries; it helps the animal adjust and respond to new situations and perhaps determine better ways to deal with its environment.
11. a. Locality imprinting (migration).
    b. Reflex.
    c. Fixed action pattern.

CHAPTER 35

1. Behaviors are the patterns of movement, vocalizations, and body positions exhibited by animals. Social behaviors are behaviors that help members of the same species communicate and interact.
2. Social behaviors help sexually reproducing species to reproduce, care for offspring, and defend territories. Other behaviors may help animals to hunt for and share food and to warn and defend against predators.
3. Competitive behavior results when two or more individuals are competing for the same resource (such as food, territories, or mates) and there is not enough of the resource to be shared. This competition encourages the population to disperse among the resources, conserving time and energy, rather than killing each other.
4. a. Threat display.
   b. Submissive behavior.
5. An animal exhibiting territorial behavior marks off an area as its own and defends it against same-sex members of its species. Territorial behavior limits competition and aggression within

a species and encourages the founding of new populations. It also rations resources among individuals, ensures that at least some individuals will have enough resources to survive and reproduce, and places the peak time of competition/aggression before the time of reproduction and parenting.

6. Sexual reproduction involves the union of male and female sex cells to form the first cells of new individuals. It provides genetic variability to a species, giving it the raw materials needed to adapt to environmental change.
7. Mating includes male-female behaviors that result in fertilization; courtship includes the behavior patterns that lead to mating.
8. a. Parenting behavior.
   b. Courtship ritual (reproductive behavior).
   c. Territorial behavior.
9. Animals form social groups to work together toward common goals: to protect against predators, hunt for food, and create a division of labor.
10. A honeybee colony consists of three castes: (1) the queen, which lays the eggs; (2) the drones, male bees that fertilize the queen's eggs; and (3) the workers, female bees that develop from fertilized eggs and do everything around the hive except lay eggs and fertilize them. (This includes feeding larvae, guarding the hive, and foraging for food).
11. A rank order is the social hierarchy that appears in groups of fishes, reptiles, birds, and mammals. This hierarchy helps reduce aggression and fighting within social groups, focusing these behaviors to a short time when the rank order develops.
12. Answers will vary; human behavior appears to result from a combination of hereditary and environmental factors.

CHAPTER 36

1. Ecology is the study of interactions between organisms and the environment.
2. Population change; this equation shows that the size of a population results from the number of births and of organisms moving into the population (immigrants), balanced against the number of deaths and organisms moving out of the population.
3. Exponential growth; carrying capacity.
4. A large population is more likely to survive because it is less vulnerable to random events and natural disasters and because its size promotes genetic diversity.
5. Density refers to the number of organisms per unit of area; dispersion refers to the way in which the individuals of a population are arranged. If individuals are too far apart, they may not be able to reproduce. If population becomes too dense, however, factors can arise that limit the population's size (disease, predation, starvation).
6. The three patterns are (1) uniform (evenly spaced), (2) random, and (3) clumped (organisms are grouped in areas of the environment that have the necessary resources). Human populations are distributed in the clumped pattern.
7. Density-dependent limiting factors (such as competition) come into play when a population size is large, whereas density-independent factors (such as weather) operate regardless of population size.
8. Both predation and parasitism are density-dependent limiting factors. Predators are organisms that kill and eat organisms of another species; parasites live on or in other species and derive nourishment from them but do not necessarily kill them.
9. See Figure 36-7. Type I is typical of organisms that tend to survive past their reproductive years (such as humans). Type II characterizes organisms that reproduce asexually and are equally likely to die at any age (such as the hydra). Type III is characteristic of organisms that produce offspring that must survive on their own and therefore die in large numbers when young (oysters).
10. Developing countries tend to have a population pyramid with a broad base, reflecting a rapidly growing population with large numbers of individuals entering their reproductive years. Industrialized countries tend to have populations with similar proportions of their populations in each age group and are growing slowly or not at all.
11. 1.8%; 95 million; double.
12. People are already putting pressure on our natural resources (land, water, forests, atmosphere). Increasing industrialization leads to rising levels of pollution and solid waste. Even though it is estimated that there is enough food produced to feed the world's population, many people live—and die—in poverty and hunger.
13. Answers will vary.

CHAPTER 37

1. A population consists of individuals of the same species. Populations that interact with each other form a community. An ecosystem is a community of organisms along with the abiotic factors with which the community interacts.
2. Habitat; niche.
3. Interactions can be grouped into (1) competition (organisms of different species living near each other strive to obtain the same limited resources), (2) predation (organisms of one species kill and eat organisms of another species), and (3) symbiosis (organisms of different species live together in close association over a period of time).
4. False. Competition is often greatest between organisms that obtain their food in similar ways.
5. The principle of competitive exclusion states that if two species are competing for the same limited resource in a location, the species that can use the resource most efficiently will eventually eliminate the other species in that area.
6. An organism's realized niche is the role it actually plays in the ecosystem; its fundamental niche is the role it might play if competitors were not present.
7. Predator; prey; parasite.
8. Gause's experiments showed that predator-prey relationships follow a cyclical pattern: as the number of prey increases, the number of predators increases. When the prey decrease because of predation, the predator population also declines until an upsurge in the prey population starts the cycle again.
9. The balance between predator and prey populations is complex because interactions depend on many factors: interactions with other organisms, movement of new organisms into or out of the community, and abiotic factors in the ecosystem.
10. Coevolution is the long-term, mutual evolutionary adjustment of characteristics of members of biological communities is relation to one another. Cattle are one example: their digestive systems have adapted (grinding teeth, strong jaws, a rumen) to allow them to digest silica in certain grasses.
11. a. Warning coloration.
    b. Camouflage.
12. These are all types of symbiotic relationships. In commensalism, one species benefits and the other is neither hurt nor helped (epiphytes). In mutualism, both participating species benefit (ants and aphids). In parasitism, one species benefits and the other is harmed (viruses).
13. Primary succession takes place in areas not previously changed by organisms. It could begin, for instance, with lichens growing on bare rock, producing acids that break down the rock to form soil. This gives rise to a pioneer community, which in turn leads to

more vegetation, eventually making the area able to support other forms of life. At some point the mix of plants and animals becomes somewhat stable, forming a climax community.

14. Secondary succession.

CHAPTER 38

1. An ecosystem is a community of plants, animals, and microorganisms that interact with one another and their environment and that are interdependent. The living (biotic) groups it contains include producers (organisms that can make their own food), consumers (organisms that eat other organisms for food), and decomposers (consumers that eat dead matter).
2. Human beings are consumers.
3. See Figure 38-1.
4. Primary consumers; secondary consumers; decomposers; detritus.
5. A trophic level is a feeding level within a food chain. A food chain is a linear relationship among organisms that feed one on another. Food webs are food chains that interweave with one another.
6. A green plant captures some of the sun's energy. The successive members of a food chain, in turn, incorporate into their own bodies about 10% of the energy in the organisms on which they feed. The rest is released as heat or is used to power metabolic activities.
7. Chemicals are stored in the atmosphere and in rocks (reservoirs). These substances enter the soil, are incorporated into the bodies of living organisms, and are passed along the food chain. Ultimately, the chemicals, with help from decomposers, return to the nonliving reservoirs.
8. Water in the atmosphere falls to the Earth as precipitation. Plants take up water from the soil, and animals obtain water from drinking it or from eating plants or other animals. Water returns to the atmosphere through the evaporation of surface water and transpiration by plants.
9. Ground water includes subsurface bodies of fresh water. It can be polluted by pesticides that are washed from plants and topsoil by the rain and by chemical wastes that are dumped in surface water. It is serious because at the present time there is no technology to remove these pollutants from ground water; already about 2% of ground water in the United States is polluted, and the situation is getting worse.
10. Carbon is held in the atmosphere as carbon dioxide. Photosynthetic organisms take in this gas and use it to build carbon compounds. Both producers and consumers use carbon compounds as an energy source, metabolizing them and releasing carbon dioxide to the atmosphere again.
11. Crop rotation can improve soil by alternating a nonleguminous crop (which consumes nitrogen from the soil) with a legumiunous crop (which can enrich the nitrate level of depleted soil by forming mutualistic associations with nitrogen-fixing bacteria). This relates to the nitrogen cycle.
12. Adding more phosphates than crops need causes the extra phosphates to wash off or erode into water. Bodies of water can become eutrophic; algae and other aquatic plants overgrow and decomposers feed on them, using much of the water's oxygen and choking out other organisms.

CHAPTER 39

1. A biome is an ecosystem of plants and animals that occurs over wide areas within specific climatic regions.
2. Australia is in the southern hemisphere, whereas the United States is in the northern hemisphere. Thus when the Earth tilts so that the United States is further from the sun, people in the United States experience winter but it is summer in Australia. When the Earth tilts so that the North Pole is closer to the sun than the South Pole, it is summer in America but winter in Australia.
3. Near the equator, warm air rises and flows toward the poles; the Earth's rotation breaks the air into six coils of rising and falling air that surround the Earth. The air in each coil rises at the region of its lowest latitude, moves toward the poles, sinks at the region of its highest latitude, and flows back to the equator. These air masses affect wind patterns and precipitation worldwide.
4. The climate of an area is determined by its latitude, wind patterns, and surface features of the Earth.
5. a. Tropical rainforests.
   b. Temperate deciduous forests.
   c. Arctic tundra.
   d. Savannahs.
   e. Taiga.
   f. Temperate grasslands.
   g. Desert.
6. Tropical rainforests make poor farmland, since most nutrients are found in the vegetation, not the soil. The soil itself is thin, often only a few centimeters deep. Cutting down and burning the trees removes the nutrients and breaks down the organic matter to carbon dioxide. In 2 to 3 years the land becomes barren.
7. Desert rainfall is limited by latitude because the air, which has released its moisture over the rain forests, is dry over desert latitudes. Deserts can also occur in the interior of continents, far from the ocean, and on the leeward side of mountains that block rainfall.
8. The taiga is the northern coniferous forest with long, cold winters and little precipitation. The tundra lies further north, and is open, often boggy, with desert-like levels of precipitation.
9. The three fresh water life zones are (1) shore zone (shallow water in which plants with roots may grow; some consumers live here), (2) open-water zone (the main body of water through which light penetrates; floating plants, microscopic floating animals, and fish live here), (3) deep zone (light does not penetrate, so there are no producers; inhabited mainly by decomposers or organisms that feed on organic material that filters down).
10. An estuary is an environment where fresh water and salt water meet. All organisms living here are adapted to living in an area of moving water and changing salt concentrations. Some are bottom-dwellers that attach to bottom material or burrow into the mud; others spawn in less salty water.
11. Estuaries are ecosystems where fresh water rivers and streams meet the ocean. They are vulnerable to pollutants that flow from upstream as well as to pesticides that leach from the land and soil.
12. The three major life zones of the ocean are (1) intertidal (area between high and low tides), (2) neritic (shallow waters along the coasts of the continents, and (3) open-sea (the rest of the ocean).
13. (1) Photic zone (well-lit layer of the ocean, which contains phytoplankton, zooplankton, fish, and air-breathing aquatic mammals (nekton), (2) mesopelagic zone (organisms exhibiting bioluminescence, some fish [swordfish, certain sharks and whales], and cephalopods), (3) abyssal zone (benthos or bottom-dwellers that eat detritus).

CHAPTER 40

1. The biosphere is the global ecosystem in which all other ecosystems on the Earth exist.
2. A sustainable society uses nonfinite, renewable sources of energy. Society's needs are satisfied without compromising future generations and natural resources.
3. People could use solar, wind, and water power, as well as the Earth's geothermal energy.

4. People can recycle paper, glass, and metal products to help preserve mineral resources. This is important because mineral supplies are finite; once used up, they are gone forever.
5. Tropical rainforests are home to over half the world's species of plants, animals, and insects from which people get many medicines as well as wood, charcoal, oils, and nuts. Burning the forests adds carbon dioxide to the air, which may raise temperatures worldwide. Destroying the forests also destroys the habitat for many species and leads to extinctions.
6. Global warming refers to a worldwide increase in temperature. Even a small increase in temperature could melt ice at the poles, raising the sea level, and flooding one fifth of the world's land area. Higher temperatures also affect rain patterns and agriculture.
7. Biological diversity refers to the richness of species. Humans are reducing the world's biological diversity through extinctions and selective breeding. Reducing diversity makes species more vulnerable to being wiped out by disease or environmental changes.
8. Humans can improve solid waste management by increasing and improving recycling, using safe incineration, engineering better landfill sites, and practicing waste-to-energy reclamation.
9. Point sources are sources of pollutants that enter the water at one or a few distinct places (industrial waste dumped by a specific factory). Nonpoint sources refer to pollutants that enter water at a variety of places (metals and acids draining from mines).
10. Organic nutrients are food for bacteria; as the bacteria multiply, they use up the oxygen in the water, killing the fish. Inorganic nutrients can stimulate plant growth in the water. Bacteria decompose the plants after they die; again, when the bacteria multiply, they consume the water's oxygen. The decomposed materials also begin to fill in the body of water, eventually turning it into a terrestrial community. Pollution with infectious agents can cause serious diseases. Toxic substances in water can poison the organisms that take in this water; sometimes organisms accumulate these substances in their bodies.
11. Acid rain results from gases such as sulfur dioxide and nitrogen dioxide that combine with water in the atmosphere. When this acidic rain falls, it kills aquatic life, pollutes groundwater, damages plants, and eats away at stone, metal, and painted surfaces.
12. Gray-air cities are those with relatively cold, moist climates. They develop a layer of smog because of the burning of fossil fuels and the abundance of sulfur oxides and particulates in the air. The air over brown-air cities contains hydrocarbons and nitrogen oxides that react with the sunny climate to produce secondary pollutants such as ozone.
13. Ozone in the stratosphere helps shield you from the sun's ultraviolet rays. Depletion of this layer can damage bacteria and plants and cause burns and skin cancers.
14. Answers will vary.

APPENDIX

B Classification of organisms

The purpose of the classification scheme presented here is for student reference. It was compiled from a variety of sources in microbiology, botany, and zoology, taking into account the most widely accepted schemes at this time. It is not intended to include all of the phyla of organisms currently identified by taxonomists, but it includes all of the phyla and divisions described and referred to in *Biology Today*. Because this textbook describes humans in its extensive physiology section and includes a chapter on chordate and human evolution, the classification of the chordates is listed in more detail than other phyla.

Kingdom Monera

Prokaryotic, single-celled organisms that have neither membrane-bounded nuclei nor membrane-bounded organelles.

Subkingdom Arcaebacteria: Methanogens (methane-producing bacteria), halophiles (bacteria that live in salt marshes), thermaocidophiles (bacteria that live in hot springs and deep-sea vents). A separate kingdom has been proposed for this subkingdom.

Subkingdom Eubacteria: All other bacteria, including nitrogen-fixing bacteria and cyanobacteria.

Kingdom Protista

A varied group of eukaryotic organisms. Many are single celled, although some phyla include multicellular or colonial forms. Some forms are photosynthetic.

Phylum Rhizopoda: Amebas
Phylum Zoomastigina: Flagellates
Phylum Ciliophora: Ciliates
Phylum Sporozoa: Sporozoans
Phylum Dinoflagellata: Dinoflagellates
Phylum Chrysophyta: Diatoms, yellow-green algae, golden-brown algae
Phylum Phaeophyta: Brown algae
Phylum Chlorophyta: Green algae
Phylum Rhodophyta: Red algae
Phylum Acrasiomycota: Cellular slime molds
Phylum Myxomycota: Plasmodial slime molds
Phylum Oomycota: Water molds, white rusts, downy mildews

Kingdom Fungi

Mostly multicellular eukaryotic organisms that are saprophytes, feeding on dead or decaying organic material. Some fungi are parasitic, feeding on living organisms.

Division Zygomycota: Zygote-forming fungi (black bread mold)
Division Ascomycota: Sac fungi (yeasts, cup fungi)
Division Basidiomycota: Club fungi (mushrooms, puffballs, shelf fungi)
Division Fungi Imperfect: Imperfect fungi (fungi that have no sexual stages of reproduction, such as *Penicillium* and *Aspergillus*)
Lichens: Associations between fungi and green algae and/or cyanobacteria

Kingdom Plantae

Multicellular, eukaryotic organisms that evolved on land and that perform photosynthesis, producing their own food by using energy from the sun and carbon dioxide from the atmosphere.

- Division Bryophyta: Mosses, hornworts, liverworts
- Division Psilophyta: Whisk ferns
- Division Lycophyta: Club mosses
- Division Sphenophyta: Horsetails
- Division Pterophyta: Ferns
- Division Coniferophyta: Conifers
- Division Cycadophyta: Cycads
- Division Ginkgophyta: Ginkgos
- Division Gnetophyta: Gnetae
- Division Anthophyta: Flowering plants (angiosperms)
    - Class Monocotyledons: Grasses, irises
    - Class Dicotyledons: Flowering trees, shrubs, roses

Kindgom Animalia

Multicellular, eukaryotic organisms that are heterotrophic, eating other organisms for food.

- Phylum Porifera: Sponges
- Phylum Cnidaria
    - Class Hydrozoa: Hydra
    - Class Scyphozoa: Jellyfish
    - Class Anthozoa: Corals, sea anemones
- Phylum Ctenophora: Comb jellies, sea walnuts
- Phylum Platyhelminthes
    - Class Turbellaria: Free-living flatworms
    - Class Trematoda: Flukes
    - Class Cestoda: Tapeworms
- Phylum Nematoda: Roundworms
- Phylum Mollusca
    - Class Polyplacophora: Chitons
    - Class Gastropoda: Snails and slugs
    - Class Bivalvia: Bivalves
    - Class Cephalopoda: Octopuses, squids, nautilus
- Phylum Annelida
    - Class Polychaeta: Marine worms
    - Class Oligochaeta: Earthworms and fresh water worms
    - Class Hirundinea: Leeches
- Phylum Arthropoda
    - Subphylum Chelicerata
        - Class Arachnida: Spiders, mites, ticks
        - Class Merostomata: Horseshoe crabs
        - Class Pycnogonida: Sea spiders
    - Subphylum Crustacea
        - Class Crustacea: Lobsters, Crayfish, shrimps, crabs
    - Subphylum Uniramia
        - Class Chilopoda: Centipedes
        - Class Diplopoda: Millipedes
        - Class Insecta: Insects
- Phylum Echinodermata
    - Class Crinoidea: Sea lilies
    - Class Asteroidea: Sea stars (starfish)
    - Class Ophiuroidea: Brittle stars
    - Class Echinoidea: Sea urchins and sand dollars
    - Class Holothuroidea: Sea cucumbers

- Phylum Chordata: Chordates
  - Subphylum Urochordata: Tunicates (sea squirts)
  - Subphylum Cephalochordata: Lancelets
  - Subphylum Vertebrata: Vertebrates
    - Class Agnatha: Jawless fishes (lamprey eels, hagfishes)
    - Class Chondrichthyes: Cartilaginous fishes (sharks, skates, rays)
    - Class Osteichthyes: Bony fishes (perch, cod, and trout)
    - Class Amphibia: Salamanders, frogs, toads
    - Class Reptilia: Reptiles (lizards, snakes, turtles, crocodiles)
    - Class Aves: birds
    - Class Mammalia: Mammals
      - Subclass Prototheria: Egg-laying mammals (duck-billed platypus, spiny anteater)
      - Subclass Metatheria: Pouched mammals or marsupials (oppossums, kangaroos, wombats)
      - Subclass Eutheria: Placental mammals
        - Order Edentata: Anteaters, armadillos, sloths
        - Order Lagomorpha: Rabbits, hares, pikas
        - Order Rodentia: Squirrels, rats, woodchucks
        - Order Insectivora: Hedgehogs, tenrecs, moles, shrews
        - Order Carnivora: Dogs, wolves, cats, bears, weasels
        - Order Pinnipeda: Sea lions, seals, walruses
        - Order Cetacea: Whales, dolphins, porpoises
        - Order Proboscidea: Elephants
        - Order Perissodactyla: Horses, asses, zebras, tapirs, rhinoceroses
        - Order Artidactyla: Swine, camels, deer, hippopotamuses, antelopes, cattle, sheep, goats
        - Order Tubulidentia: Aardvarks
        - Order Scandentia: Tree shrews
        - Order Chiroptera: Bats
        - Order Primates: Lemurs, monkeys, humans
          - Suborder Strepsirhini: Prosimians (lemurs, aye-ayes)
          - Suborder Haplorhini: Anthropoids
            - Superfamily Tarsioidea: Tarsiers
            - Superfamily Ceboidea: New World monkeys
            - Superfamily Cercopithecoidea: Old World monkeys
            - Superfamily Hominoidea
              - Family Hylobatidae: Gibbons, orangutans, chimpanzees
              - Family Pongidae: Gorillas
              - Family Hominidae: Humans

Glossary

abscisic acid
(ab sis′ik as′əd) (L. ab, away, off + scisso, dividing) A hormonal growth inhibitor that induces and maintains dormancy in plants. p. 541

acetyl CoA
(ə sēt′əl kō′ā) A compound combining the two-carbon acetyl fragment removed during the oxidation of pyruvate with the carrier molecule coenzyme A (CoA). p. 123

acetylcholine
(ə sēt′əl kō′lēn′) The neurotransmitter found at neuromuscular junctions that depolarizes the muscle cell membrane, releasing calcium ions that trigger muscle contraction. p. 257

acetylcholinesterase
(ə sēt′əl kō′lə nes′tə rās′) The enzyme that stops the action of acetylcholine; it is one of the fastest-acting enzymes in the blood. p. 257

acid
(as′əd) Any substance that dissociates to form H^+ ions when it is dissolves in water. p. 26

acid rain
(as′əd rān) An acid precipitation that falls to the ground as rain, caused primarily by coal-fired plants' emission of sulfur dioxide and nitrogen dioxide, which then combines with water in the atmosphere; results in many devastating environmental effects. p. 705

acrosome
(ak′rə sōm) (Gr. akron, extremity + soma, body) A vesicle located at the tip of a sperm cell. p. 336

ACTH
(ā′sē′tē′āch′) *See* adrenocorticotropic hormone.

actin
(ak′tən) (Gr. actis, ray) A protein that makes up the thin myofilaments in a muscle fiber; provides support and helps determine cell shape and movement. p. 307

action potential
(ak′shən pəten′chəl) The rapid change in a membrane's electrical potential caused by the depolarization of a neuron to a certain threshold. p. 254

active immunity
(ak′tiv i myo͞on′ə tē) A type of immunity conferred by vaccination, which causes your body to build up antibodies against a particular disease without getting the disease. p. 227

active site
(ak′tiv sīt) The grooved or furrowed location on the surface of an enzyme where catalysis occurs. p. 105

active transport
(ak′tiv trans′pôrt) The movement of a solute across a membrane against the concentration gradient with the expenditure of chemical energy. This process requires the use of a transport protein specific to the molecule(s) being transported. p. 75

activator
(ak′tə vāt′ər) A chemical that binds to an enzyme and changes its shape so that catalysis can occur. p. 108

adaptation
(ad ap tā′shən) (L. adaptare, to fit) A naturally occurring inherited trait found within a population that makes an individual better suited to a particular environment and produce more offspring than an individual lacking this trait. p. 432

adaptive radiation
(ə dap′tiv rād ē ā′shən) The phenomenon by which a population of a species changes as it is dispersed within a series of different habitats within a region. p. 433

adenine
(ad′ən ēn′) An organic compound that is one of the two purine bases of RNA and DNA. p. 39

adenosine diphosphate (ADP)
(ə den′ə sēn′ dī fos′fāt) The molecule remaining from the breaking off of a phosphate group when adenosine triphosphate is used to drive an endergonic reaction. p. 111

adenosine triphosphate (ATP)
(ə den′ə sēn′ trī fos′fāt) A molecule composed of three subunits: ribose, adenine, and a triphosphate group; ATP captures energy in its high-energy bonds and later releases this energy. ATP is used to fuel a variety of cell processes and is the universal energy currency of all cells. p. 110

ADP
(ā′dē′pē′) *See* adenosine diphosphate.

adrenal cortex
(ə drē′nəl kôr′teks) (L. near, + ren, kidney; L. rind) The outer, yellowish portion of each adrenal gland that secretes a group of hormones known as corticosteroids in response to ACTH. p. 325

adrenal gland
(ə drē′nəl gland) (L. near, + ren, kidney) Either of two triangular glands, named for their position in the body, having two parts with two different functions, the adrenal cortex and the adrenal medulla. p. 325

adrenal medulla
(ə drē′nəl mə dul′ə) (L. near, + ren, kidney; L. marrow) The inner, reddish portion of each adrenal gland surrounded by the cortex that secretes the hormone adrenaline and noradrenaline, p. 325

adrenaline
(ə dren′əl ən) *See* epinephrine. p. 326

adrenocorticotropic hormone (ACTH)
(ə drē′nō kôrt′ə kō träp′ik hôr′mōn) (L. near, + ren, kidney + cortex, bark + Gr. tropikos, turning) A tropic hormone secreted by the anterior pituitary that triggers the adrenal cortex to produce certain steroid hormones. p. 320

aerobic
(er rō′bik) (Gr. aer, air + bios. life) Oxygen dependent. p. 116

aerobic respiration
(er rō′bik res′pə rā′shən) The process in which the original glucose molecule has been consumed entirely by the combined actions of glycolysis and the Krebs cycle.

afferent neuron
(af′ər ənt no͞or′on) *See* sensory neuron. p. 264

agranulocyte
(ā gran′yə lō sīt′) (Gr. a-, not + L. granulum, granule + Gr. kytos, cell) One of the two major groups of leukocytes; they have neither a cytoplasmic granule nor a lobed nuclei. p. 210

albumin
(al byo͞o′mən) (L. albumen, white of egg) A protein found in blood that serves to elevate the solute concentration of the blood to match that of the tissues so that the movement of water molecules is regulated. p. 74

aldosterone
(al dôs'tə rōn) A mineralocorticoid hormone produced by the adrenal cortex that regulates the level of sodium and potassium ions in the blood, thereby promoting the conservation of sodium and water and the excretion of potassium. p. 244

alga, pl. algae
(al'gə, al'jē) (L. seaweed) A plant-like, photosynthetic eukaryotic organism that contains chlorophyll. p. 503

allantois
(ə lan'tō əs) (Gr. allas, sausage + eidos, form) An extraembryonic membrane that gives rise to the umbilical arteries and vein as the umbilical cord develops. p. 360

allele
(ə lēl') (Gr. allelon, of one another) Each member of a factor pair containing information for an alternative form of a trait that occupy corresponding positions on paired chromosomes. p. 378

allergen
(al'ər jən) (Gr. allos, other + ergon, work) Any substance that causes manifestations of allergy. p. 230

all-or-nothing
(ôl' ôr' nuth'ing) Refers to the nerve impulse response; the amount of stimulation applied to the receptor or neuron must always be sufficient to open enough sodium channels to generate an action potential; otherwise, the cell membrane will simply return to the resting potential. p. 255

alternation of generations
(ôl tər nā'shen uv jen ə rā'shənz) A type of life cycle that has both a multicellular haploid phase and a multicellular diploid phase. p. 529

alveolus, pl. alveoli
(al vē'ə ləs, al vē'ə lī) (L. a small cavity) Microscopic air sacs in the lungs where oxygen enters the blood and carbon dioxide leaves. p. 185

amino acid
(ə mē'nō as'əd) (Gr. Ammon, referring to the Egyptian sun god, near whose temple ammonium salts were first prepared from camel dung) A molecule containing an amino acid group (-NH_2), a carboxyl group (-COOH), a hydrogen atom, a carbon atom, and a functional group that differs among amino acids; an extremely diverse array of proteins is made from the 20 common amino acids. p. 34

amniocentesis
(am'nē ō sen tē'səs) (Gr. amnion, membrane around the fetus + centes, puncture) A prenatal diagnostic procedure in which a sampling of amniotic fluid is obtained by insertion of a needle into the amniotic cavity and withdrawn into a syringe; the removed fetal cells are then grown in tissue culture and tests are performed to determine if genetic abnormalities are present. p. 400

amnion
(am'nē ən) (Gr. membrane around the fetus) A thin, protective membrane that grows around the embryo during the third and fourth weeks, fully enclosing the embryo in a membranous sac. p. 360

amniotic egg
(am'nē ot'ik eg) (Gr. membrane around the fetus) An egg, characteristic of reptiles, birds, and monotremes, which protects the embryo from drying out, nourishes it, and enables it to develop outside of water. p. 468

amniotic fluid
(am'nē ot'ik flo͞o'əd) (Gr. membrane around the fetus) A fluid in which the fetus floats and moves; it also helps keep the temperature constant for fetal development. p. 360

amphibian
(am fib'ē ən) (Gr. amphibios, double life) An animal capable of living on land and in the water. p. 466

amylase
(am'ə lās) (Gr. amylon, starch + asis, colloid enzyme) An enzyme that breaks down starches and glycogen to sugars. p. 168

anaerobic
(an'er ō'bik) (Gr. an, without + aer, air + bios, life) Literally, without oxygen; any process that can occur without oxygen; includes glycolysis and fermentation. p. 116

analagous
(ə nal'ə gəs) (Gr. proportionate) Describes a part or organ of an organism that has a similar form and function but possesses different evolutionary origins. p. 439

anaphase
(an'ə fāz) (Gr. ana, up + phasis, form) The stage of mitosis characterized by the physical separation of sister chromatids and their movement to opposite poles of the cell. p. 90

angiosperm
(an'jē ə sperm') (Gr. angeion, vessel + sperma, seed) A vascular, flowering plant with protected seeds. p. 534

annual ring
(an'yo͞o əl ring) A growth ring in a tree, the result of cambium cells dividing; one ring equals one year's growth and can thus be used to calculate the age of a tree. p. 552

antagonists
(an tag'ə nəsts) (Gr. antagonizesthai, to struggle against) An opposing pair of skeletal muscles. p. 305

anther
(anth'ər) (Gr. anthos, flower) The compartmentalized structure in the male stamen of a flower where haploid pollen grains are produced. p. 534

antheridium, pl. antheridia
(an'thə rid'ē əm, an'thə rid'ē ə) (Gr. anthos, flower) The structure where sperm is produced in certain algae, the bryophytes, and several divisions of vascular plants. p. 529

anthropoid
(an'thrə poid) (Gr. anthropos, man + eidos, form) A suborder of mammals that includes monkeys, apes, gorillas, chimpanzees, and humans; they differ from prosimians in the structure of the teeth, brain, skull, and limbs; they are diurnal, live in groups with complex social interactions, and care for their young for prolonged periods. p. 460

antibody
(ant'i bäd'ē) (Gr. anti, against) A protein produced in the blood by a B lymphocyte that specifically binds circulating antigen and marks cells or viruses bearing antigens for destruction. p. 219

anticodon
(ant'i kō'don) (Gr. anti, against + L. code) A portion of a tRNA molecule with a sequence of three base pairs complementary to a specific mRNA codon. p. 416

antidiuretic hormone (ADH)
(ant'i dī yo͞o ret'ik hôr'mōn) (Gr. anti, against + dia, intensive + ouresis, urination) A hormone produced by the hypothalamus but stored and released by the posterior lobe of the pituitary that helps control the volume of the blood by regulating the amount of water reabsorbed by the kidneys. p. 244

antigen
(ant'i jən) (Gr. anti, against + genos, origin) A foreign substance inducing the formation of antibodies that specifically bind to the foreign substance, marking it for destruction. p. 219

anus
(ā'nəs) The opening of the rectum for the elimination of feces. p. 180

aorta
(ā ôrt'ə) (Gr. aeirein, to lift) The largest artery in the body, it carries the blood from the left side of the heart throughout the body except the lungs. p. 205

aortic semilunar valve
(ā ôrt'ik sem'i lo͞o'nər valv') A one-way valve that permits blood flow from the left side of the heart and then snaps shut, preventing a backflow from the aorta to the heart. p. 205

apical meristem
(ap′ə kəl *or* ā′pə kəl mer′ə stem) (L. apex, top + Gr. meristos, divided) A growth tissue at the tips of roots and the tips of shoots in a plant that allows a plant to grow taller and the roots to grow deeper into the ground. p. 548

appendicular skeleton
(ap′ən dik′yə lər skel′ə tən) (L. appendicula, a small appendage) The portion of the human skeleton, containing 126 bones, that consists of the bones of the appendages (arms and legs) and the bones that help attach the appendages to the axial skeleton. p. 299

appendix
(ə pen′diks) (L. appendere, to hang) A pouch that hangs from the beginning of the large intestine and serves no essential purpose in humans. p. 179

aqueous humor
(äk′wē əs hyo͞o′mər) (L. aqua, water + humor, fluid) A watery fluid filling a chamber behind the cornea that nourishes the cornea and the lens and, together with vitreous, creates a pressure within the eyeball maintaining the eyeball's shape. p. 287

archebacteria
(är′kē bak tir′ē ə) (Gr. arche, beginning + Gr. bakterion, rod) A taxonomic section consisting of bacteria that are chemically different in certain structures and metabolic processes from all other bacteria p. 450

archegonium, pl. archegonia
(är′kə gō′nē əm, är′kə gō′nē ə) (Gr. archegonos, first of a race) A structure in which eggs are formed in certain algae, the bryophytes, and several divisions of vascular plants. p. 529

arteriole
(är tir′ē ōl) A smaller artery that leads from an artery to a capillary. p. 201

artery, pl. arteries
(ärt′ə rē, ärt′ə rēz) (Gr. arteria, artery) A blood vessel that carries blood away from the heart and through the body. p. 201

articulation
(är tik′yə lā′shən) *See* joint.

artificial selection
(ärt′ə fish′əl sə lek′shən) The process of selecting organisms for a desirable characteristic and then breeding that organism with another organism exhibiting the same trait. p. 429

asexual reproduction
(ā′seksh′ə wəl *or* ā′sek′shəl rē′prə duk′shən) A type of reproduction in which a parent organism divides by mitosis and produces two identical organisms; does not introduce variation. p. 85

association area
(ə sō′sē ā′shən er′ē ə) The area in the brain that connects all parts of the cerebral cortex and appears to be the site of higher cognitive activities such as memory, reasoning, intelligence, and personality. p. 266

aster
(as′tər) In animal mitosis, an array of microtubules that radiates outward from the centrioles when the centrioles reach the poles of the cell. p. 89

atherosclerosis
(ath′ə rō sklə rō səs) (Gr. athere, porridge + sklerosis, hardness) A disease in which the inner walls of the arteries accumulate fat deposits, narrowing the passageways and leading to elevated systolic blood pressure. p. 207

atom
(at′əm) (Gr. atomos, indivisible) A core (nucleus) of protons and neutrons surrounded by orbiting electrons. The electrons largely determine the chemical properties of an atom. p. 17

atomic mass
(ə täm′ik mas) The combined mass of all the protons and neutrons of an atom without regard to its electrons. p. 17

atomic number
(ə täm′ik num′bər) The number of protons in an atom; the atomic number is also the same as the number of electrons in that atom. p.17

ATP
(ā′tē′pē′) *See* adenosine triphosphate.

atrioventricular (AV) node
(ā′trē ō ven trik′yə lər nōd) A group of specialized cardiac muscle cells that receives the impulse initiated by the sinoatrial node and conducts them by way of the bundle of His. p. 207

atrium, pl. atria
(ā′trē əm, ā′trē ə) (L. main room) The upper chamber of each half of the heart; the right atrium receives deoxygenated blood from the body (except the lungs), and the left atrium receives oxygenated blood from the lungs through the pulmonary veins. p. 204

auditory canal
(ôd′ə tōr′ē kə nal′) (L. audire, to hear) A 1-inch long canal that receives sound waves and leads directly to the eardrum. p. 289

Australopithecus afarensis
(ôs′trə lō pith′ə kəs af′ə ren′səs) (L. australis, southern + Gr. pithekos, ape; L. Afar region of Ethiopia) A species of australopithecines considered to be the first hominids, although not human (members of the genus *Homo*); its oldest fossil skeleton is 3.5 million years old. p. 477

autonomic nervous system
(ô′tə nom′ik nurv′əs sis′təm) (Gr. autos, self + nomos, law) A branch of the peripheral nervous system consisting of motor neurons that control the involuntary and automatic responses of the glands and the nonskeletal muscles of the body. p. 264

autosome
(ô′tə sōm′) (Gr. autos, self + soma, body) Any of the 22 pairs of human chromosomes that carry the majority of the genetic information but have no genes that determine gender. p. 81

autotroph
(ô′tə trōf) (Gr. autos, self + trophos, feeder) Self-feeder; an organism that produces its own food by photosynthesis; all organisms live on food produced by autotrophs, including the autotrophs themselves. p. 131

auxin
(ôk′sən) (Gr. auxein, to increase) A hormone that promotes growth through cell elongation and controls phototropism and geotropism in plants. p. 541

AV node
(ā′vē′ nōd′) *See* atrioventricular (AV) node.

axial skeleton
(ak′sē əl skel′ə tən) The central axis of the human skeleton, consisting of 80 bones, that includes the skull, vertebral column, and rib cage. p. 299

axon
(ak′son) (Gr. axle) A single projection extending from a neuron that conducts impulses away from the cell body. p. 250

B lymphocyte (B cell)
(bē′ lim′fə sīt′) A type of lymphocyte that matures in human bone marrow and secretes antibodies during the humoral immune response; named after a digestive organ in birds (the bursa of Fabricius) in which these lymphocytes were discovered. p. 223

bacteriophage (phage)
(bak tir′ē ə fāj) (Gr. bakterion, little rod + phagein, to eat) A virus that infects a bacterial cell. p. 489

bacterium, pl. bacteria
(bak tir′ē əm, bak tir′ē ə) (Gr. bakterion, a little rod) The oldest, simplest, and most abundant organism; it is the only organism with a prokaryotic cellular organization. p. 45

bark
(bärk) The outer protective covering of a woody plant or tree that consists of the

cork, the cork cambium, and the parenchyma cells. p. 552

basal body
(bā'səl bäd'ē) A structure composed of microtubules that serves to anchor a cilium or flagellum to the cell. p. 61

basal ganglion, pl. ganglia or ganglions
(bā'səl gang'glē ən, gang'glē ə *or* gang'glē ənz) Groups of nerve cell bodies located at the base of the cerebrum that control large, subconscious movements of the skeletal muscles, p. 267

base
(bās) Any substance that combines with H^+ ions; a substance that has a pH value higher than water's neutral value of 7. p. 26

basidium, pl. basidia
(bə sid'ē əm, bə sid'ē ə) (Gr. basis, base) In a club fungus, a club-shaped structure from which its sexual spores hang. p. 521

basophil
(bā'sə fil) (Gr. basis, base + philein, to love) A type of granulocyte containing granules that rupture and release chemicals enhancing the body's response to injury or infection; they also play a role in causing allergic responses. p. 210

behavior
(bə hā' vyər) The range of the pattern of movement, sound (vocalization), and body position (posture) exhibited by an animal; it also embraces competitive, social, submissive, territorial, and reproductive behavior patterns. p. 599

bilateral symmetry
(bī lat'ə rəl sim'ə trē) (L. bi, two + lateris, side; Gr. symmetria, symmetry) Describing an object or an organism whose right side is a mirror image of its left side. p. 562

bile
(bīl) (L. bilis) A collection of molecules secreted by the liver than helps in the digestion of lipids. p. 176

binary fission
(bī'nə rē fish'ən *or* fizh'ən) (L. binarius, consisting of two things or parts + fissus, split) Asexual reproduction in which one cell divides into two with no exchange of genetic material among cells; bacteria use this process. p. 491

binomial nomenclature
(bī nō'mē əl nō'mən klā'chər or nō men'klə chər) (L. bi, twice, + Gr. nomos, law; L. nomen, name + calare, to call) The system of using the last two categories in the hierarchy of classification, genus and species, to provide the scientific name of an organism, usually written in Latin. p. 14

biome
(bī'ōm) (Gr. bios, life + -oma, group) A distinctive, broad, terrestrial ecosystem of plants and animals that occurs over wide areas within specific climatic regions. p. 671

biosphere
(bī'ə sfir) (Gr. bios, life + sphaira, ball) The global ecosystem of life on Earth that extends from the tops of the tallest mountains to the depths of the deepest seas. p. 695

blade
(blād) The flattened portion of a leaf. p. 511

blastocyst
(blas'tə sist) (Gr. blastos, germ + kytos, cell) The stage of development following the morula in which the embryo is a hollow ball of cells. p. 357

blastula
(blas'chə lə) (Gr. a little sprout) The general term used to describe the sac-like blastocyst of mammals and, in other animals, the stage that develops a similar fluid-filled cavity. p. 357

blood pressure
(blud presh'ər) The pressure, determined indirectly, existing in the large arteries at the height of the pulse wave; the systolic intraarterial pressure; normal blood pressures values are 70 to 90 (mm of Hg) diastolic and 110 to 130 systolic. p. 207

bone
(bōn) (A.S. ban, bone) A hard material that forms the vertebrate skeleton; composed of collagen fibers that contribute flexibility and needle-shaped crystals of calcium that impart rigidity. p. 296

bony fish
(bōn'ē fish) A kind of fish that lives in both salt and fresh water and derives its name from its bony internal skeleton; by means of gas exchange between its swim bladder and its blood, a body fish regulates its buoyancy. p. 586

Bowman's capsule
(bō'mənz kap'səl *or* kap'sōōl) (after Sir William Bowman, British physician) An apparatus at the front end of each nephron tube in a kidney that functions as a filter in the formation of urine. p. 235

brain
(brān) That part of the central nervous system comprised of four main parts: the cerebrum, the cerebellum, the diencephalon, and the brainstem. p. 250

brainstem
(brān'stem) The part of the brain consisting of the midbrain, pons, and medulla that brings messages to and from the spinal cord and controls important body reflexes such as the rhythm of the heartbeat and rate of breathing. p. 265

bronchiole
(brong'kē ōl) (L. bronchiolus, air passage) The last division of bronchi, having thousands of tiny air passageways, whose walls have clusters of tiny pouches, or alveoli. p. 188

bronchus, pl. bronchi
(brong'kəs, brong'kī) (Gr. bronchos, windpipe) One of a pair of airway structures that branches from the lower end of the trachea into each lung, p. 188

brown algae
(broun al'jē) Large, multicellular algae found predominantly on northern, rocky shores, some growing to enormous sizes; kelp is a brown algae. p. 509

bryophyte
(brī'ə fit) (Gr. bryon, moss + phyton, plant) A single division of small, low-growing plants that are commonly found in moist places, including the mosses, liverworts, and hornworts. p. 529

bulbourethral glands
(bul'bō yōō rē'thrəl glandz) (L. bulbus, bulbous root + Gr. ourethra, urethra) A set of tiny accessory glands lying beneath the prostate that secretes an alkaline fluid into the semen. p. 338

bundle of His
(bun'dəl uv his) (after Wilhelm His, Jr, German physician) A strand of impulse-conducing muscle that conducts heartbeat impulses from the right atrium to the ventricles of the heart. p. 207

C_3 photosynthesis
(sē'thrē' fō'tō sin'thə səs) A process that uses the normal Calvin-Benson process to fix carbon. p. 141

C_4 photosynthesis
(sē'fôr' fō'tō sin'thə səs) In hot climates, a process by which plants deal with the problem of CO_2 being released by photorespiration by concentrating CO_2 within the cells where carbon fixation takes place. p. 142

calcitonin (CT)
(kal'sə tō'nən) (L. calcem, lime) A hormone secreted by the thyroid that works with parathyroid hormone to regulate the concentration of calcium in the bloodstream. p. 322

calorie
(kal'ə rē) (L. calor, heat) The measurement of the unit of energy in food. p. 166

Calvin-Benson cycle
(kal'vən ben'sən sī'kəl) (after Calvin and

Benson, American chemists) *See* light-independent reactions.

cambium
(kam′bē əm) (L. combiare, to exchange) Lateral meristem tissue in a dicot with a woody stem or trunk that consists of vascular cambium and cork cambium. p. 552

capillary
(kap′ə ler′ē) (L. capillaris, hair-like) A microscopic blood vessel with a wall only one cell thick that connects the end of an arteriole with the beginning of a venule; the site where gases are exchanged between the blood and tissues, nutrients delivered, and wastes picked up. p. 201

capsid
(kap′sid) (L. capsa, box) A protein covering that covers the nucleic acid core of a virus. p. 487

carbohydrate
(kär′bō hī′drāt) (L. carbo, charcoal + hydro, water) A molecule that contains carbon, hydrogen, and oxygen, with the concentration of hydrogen and oxygen atoms in a 2:1 ratio. p. 30

carbon fixation
(kär′bən fik sā′shən) A process by which organisms use the ATP and NADPH produced by photosynthetic reactions to build organic molecules from atmospheric carbon dioxide. p. 141

carcinogen
(kär sin′ə jən) (Gr. karkinos, cancer + -gen) Any cancer-causing agent. p. 490

cardiac muscle
(kärd′ē ak′ mus′əl) A type of striated muscle fiber arranged in a special way critical for cardiac function; cardiac muscle makes up the heart, acting as a pump for the circulatory system. p. 158

carnivore
(kär′nə vôr) (L. carnivorous, flesh eating) An animal that eats meat. p. 170

carpal
(kär′pəl) (Gr. karpos, wrist) Any of the eight short bones, lined up in two rows of four, that make up the wrist. p. 300

carrying capacity
(ker′ē′ing kə pas′ət ē) The number of individuals within a population that can be supported within a particular environment for an indefinite period. p. 627

cartilage
(kärt′əl ij *or* kärt′lij) (L. cartilago, gristle) A specialized connective tissue that is hard and strong; composed of chondrocytes that secret a matrix consisting of a semisolid gel and fibers. Laid down in long parallel arrays, the result is a firm and flexible tissue that does not stretch. p. 153

cartilaginous fish
(kär′tə laj′ə nəs fish) (L. cartilago, gristle) A marine fish such as a shark, skate, or ray having denticles that cover their skin; many sharks are predators, whereas skates and rays feed on invertebrates that dwell on the ocean floor. p. 584

catabolic
(kat′ə bäl′ik) (Gr. katabole, throwing down) Referring to a process in which complex molecules are broken down into simpler ones. p. 121

catalyst
(kat′əl əst) (Gr. kata, down + lysis, a loosening) A substance that increases the rate of a chemical reaction but is not chemically changed by the reaction; an enzyme is a catalyst. p. 105

cell
(sel′) (L. cella, a chamber or small room) A membrane-bounded unit containing hereditary material, cytoplasm, and organelles; a cell can release energy from fuel and use that energy to grow and reproduce. p. 47

cell body
(sel bäd′ē) The body of a nerve cell or neuron, which contains a nucleus and other cell organelles and two types of cellular projections, axons and dendrites. p. 252

cell cycle
(sel sī′kəl) The life of a cell; the major portion concerned with the activitites of interphase and the minority portion with mitosis. p. 86

cell-mediated immune response
(sel′ mēd′ē āt′əd ri spons′) A chain of events unleashed by T cells during which cytotoxic T cells attack infected body cells. p. 222

cell membrane
(sel′ mem′brān′) A bilayer of phospholipid molecules studded with proteins to which carbohydrates are attached. The membrane proteins allow the cell to interact with the environment and perform various cellular functions. p. 67

cell plate
(sel′ plāt′) During cytokinesis in plants and some algae, the formation of a partition, which begins by the plants manufacturing new membrane sections that then accumulate at the metaphase plate and fuse. p. 92

cell theory
(sel′ thē′ə rē *or* thir′ē) A statement first formulated in the mid-1800s, which states that all living things are made up of cells, that cells are the smallest living units of life, and that all cells arise from preexisting cells. p. 45

cell wall
(sel′ wôl) The structure that surrounds the plasma membrane in cells. In many plants, algae, and fungi the wall is rigid and composed of cellulose and imparts a stiffness to the tissues; in single-celled organisms, cell walls give shape to the organisms and help protect them. p. 61

cellular respiration
(sel′yə lər res′pə rā′shən) The complex series of chemical reactions—glycolysis, the Krebs cycle, and the electron transport chain—by which cells break down fuel molecules to release energy, using oxygen and producing carbon dioxide. p. 116

cellulose
(sel′yə lōs) (L. cellula, little cell) A polysaccharide that is the chief component of plant cell walls; because cellulose cannot readily be broken down, it works well as a biological structural material and occurs widely in plants. p. 32

central nervous system
(sen′trəl nur′vəs sis′təm) The site of information processing within the nervous system, comprising the brain and spinal cord. p. 264

centriole
(sen′trē ōl) (Gr. kentron, center of a circle + L. olus, little one) An organelle surrounded by microtubule-organizing centers that play a key role in the mitotic process; most plants and fungi lack centrioles. p. 86

centromere
(sen′trə mir) (Gr. kentron, center + meros, a part) A constricted region of the chromosome where two identical structures (sister chromatids) are joined. p. 81

cerebellum
(ser′ə bel′əm) (L. little brain) The part of the brain located below the occipital lobes of the cerebrum that coordinates subconscious movements of the skeletal muscles. p. 265

cerebral cortex
(sə rē′brəl kôr′teks) The thin layer of tissue, gray matter, that forms the outer layer of the cerebrum; the major site of higher cognitive processes such as sense perception, thinking, learning, and memory. p. 266

cerebrospinal fluid
(sə rē′brō spī′nəl flo͞o′əd) A liquid that cushions the brain and the spinal cord, acting like a shock absorber. p. 272

cerebrum
(sə rē′brəm) (L. brain) The largest and most dominant part of the human brain, divided into two hemispheres connected by the corpus callosum. p. 265

cervix
(sur′viks) (L. neck) The narrower part of the uterus that opens into the vagina. p. 343

chemoautotroph
(kē′mō ô′tə trōf) (Gr. chemeia, chemistry + auto, self + trophos feeder) A eubacterium that makes its own food by deriving energy from inorganic molecules. p. 492

chiasma, pl. chiasmata
(kī az′mə, kī az′mə tə) (Gr. a cross) In meiosis, the point of crossing-over where parts of chromosomes have been exchanged during synapsis; under a light microscope, a chiasma appears as an X-shaped structure. p. 93

chitin
(kīt′ən) (Gr. chiton, tunic) A modified form of cellulose, relatively indigestible, that is the structural material in insects and many fungi. p. 29

chlorophyll
(klôr′ə fil) (Gr. chloros, green + phyllon, leaf) The pigment that absorb photons of green, blue, and violet wavelengths; chlorophyll is the primary light gatherer in all plants and algae and in almost all photosynthetic bacteria. p. 133

chloroplast
(klôr′ə plast) (Gr. chloros, green + plastos, molded) An energy-producing organelle found in the cells of plants and algae; the site of photosynthesis in plants. p. 59

cholecystokinin (CCK)
(kol′ə sis′tə kī′nən) (Gr. chole, bile + kystis, bladder + kinein, to move) A hormone that helps control digestion in the small intestine. p. 177

chondrocyte
(kon′drə sīt) (Gr. chondros, cartilage + kytos, cell) A cell that produces cartilage, a specialized connective tissue that is hard and strong. p. 153

chordate
(kôr′dāt) (Gr. chorde, cord) An organism distinguished by three principal features: a nerve cord, a notochord, and pharyngeal slits; includes fishes, amphibians, reptiles, birds, mammals, and humans. p. 454

chorion
(kôr′ə on) An extraembryonic membrane that facilitates the transfer of nutrients, gases, and wastes between the embryo and the mother's body. p. 360

choroid
(kôr′oid′) (Gr. chorioeides, skin-like) A thin, dark-brown membrane that lines the sclera, containing blood-carrying vessels that nourish the retina and a dark pigment that absorbs light rays so that they will not be reflected within the eyeball. p. 288

chromatin
(krō′mə tən) (Gr. chroma, color) The complex of proteins and the hereditary material, DNA, which make up the chromosomes of eukaryotes. p. 81

chromosome
(krō′mə sōm′) (Gr. chroma, color + soma, body) A discrete, thread-like body that forms as the nuclear material condenses during meiosis and that carries genetic information. p. 57

chyme
(kīm) (Gr. chymos, juice) The mixture of partly digested food, gastric juice, and mucus, having the consistency of pea soup, found in the stomach during digestion. p. 175

ciliary muscle
(sil′ē er′ē mus′əl) (L. ciliaris, pert. to eyelid) A tiny circular muscle that slightly changes the shape of the lens by contracting or relaxing. p. 287

ciliate
(sil′ē ət *or* sil′ē āt′) (L. cilium, eyelash) A protozoan characterized by fine, short, hairline cellular extensions called cilia. p. 507

cilium, pl. cilia
(sil′ē əm, sil′ē ə) (L. eyelash) A whip-like organelle of motility that protrudes from some eukaryotic cells. p. 60

citric acid cycle
(sit′rək as′əd sī′kəl) *See* Krebs cycle.

cleavage
(klē′vij) The process of cell division that occurs without cell growth. p. 356

cleavage furrow
(klē′vij fur′ō) During cytokinesis in animal cells, the area where a belt of microfilaments pinches the cell at the metaphase plate. p. 91

climax community
(klī′maks′ kə myo͞o′nət ē) A community in which the mix of plants and animals becomes stable; the last stage of succession. p. 652

clitellum, pl. clitella
(klī tel′əm, klī tel′ə) (L. clitellae, a packsaddle) A thickened band on an earthworm's body that secretes a mucus holding the worms together as they exchange sperm. p. 574

clitoris
(klit′ə rəs) (Gr. kleitoris) A small mass of erectile and nervous tissue in the female genitalia that responds to sexual stimulation; it is homologous to the tip of the penis in males. p. 346

cloaca
(klō ā′kə) (L. sewer) In an aquatic animal, the terminal part of the gut into which ducts from the kidney and reproductive systems open. p. 586

clone
(klōn) (Gr. klon, a cutting used for propagation) A group of identical cells that arises by repeated mitotic divisions from one original cell. p. 224

cnidarian
(nī der′ē ən) (Gr. knide, sea nettle) A member of the phylum Cnidaria, which includes jellyfish, hydra, sea anemones, and corals; they have radially symmetrical bodies. p. 566

cochlea
(kok′lē ə) (Gr. kokhlos, land snail) A winding, cone-shaped tube forming a portion of the inner ear that contains the organ of Corti, the organ of hearing. p. 289

cochlear duct
(kok′lē ər dukt) The inner tube of the cochlea containing specialized cells that are the receptors of hearing. p. 289

codominant
(kō′däm′ə nənt) Refers to traits in which the alternative forms of an allele are both dominant. p. 400

codon
(kō′dän) (L. code) A sequence of three nucleotide bases in transcribed mRNA that code for an amino acid, which is the building block of a polypeptide. p. 416

coelom
(sē′ləm) (Gr. koilos, a hollow) A body cavity that is a fluid-filled enclosure within a bilaterally symmetrical organism. p. 153

coenzyme
(kō′en′zīm) (L. co-, together + Gr. en, in + zyme, leaven) A cofactor that is a nonprotein organic (carbon-containing) molecule helping an enzyme catalyze a chemical reaction. p. 110

coevolution
(kō′ev′ə lo͞o′shən) (L. co-, together + e-, out + volvere, to fill) Interactions that involve the long-term, mutual evolutionary adjustment of the characteristics of the members of biological communities in relation to one another. p. 457

cofactor
(kō′fak′tər) A special nonprotein molecule that helps an enzyme catalyze a chemical reaction. p. 110

collagen
(käl′ə jən) (Gr. kolla, glue + gennan, to produce) The most abundant protein in the human body, whose fibers are strong and wavy; found in the connective tissue of skin, bone, and cartilage. p. 34

collar cell
(käl′ər sel) A specialized, flagellated cell that lines the inside cavity of the body of a sponge. p. 565

collecting duct
(kə lek'ting dukt) A small duct that receives urine from the kidney nephrons. p. 235

colon
(kō'lən) (Gr. kolon) The large intestine from the small intestine to the rectum; its function is to absorb sodium and water, to eliminate wastes, and to provide a home for friendly bacteria. p. 179

commensalism
(kə men'sə liz'əm) (L. cum, together with + mensa, table) A type of symbiotic relationship in which one species benefits and the other species neither benefits nor is harmed. p. 649

community
(kə myo͞o'nət ē) (L. communitas, community, fellowship) The various populations of organisms that live together in a particular place. p. 639

compact bone
(käm'pakt' bōn) A type of bone in the human skeleton that runs the length of long bones and has no spaces within its structure visible to the naked eye; it is hard and dense, which gives the bone the strength to withstand mechanical stress. p. 156

companion cell
(kəm pan'yən sel') A specialized cell in a vascular plant that actively transports fluid to and from a phloem cell. p. 546

competition
(käm'pə tish'ən) The striving by organisms of different species that live near one another to obtain the same limited resources. p. 640

complement
(käm'plə mənt) A group of proteins that kill foreign cells by creating a hole in their membranes. p. 224

compound
(käm'pound') A molecule made up of the atoms of two or more elements. p. 18

compound eye
(käm'pound' ī) A structure that is composed of many independent visual units, each containing a lens; an important structure of many arthropods such as bees, flies, moths, and grasshoppers. p. 576

cone
(kōn) A light receptor located within the retina at the back of the eye that functions in bright light and detects color. p. 286

conjugation
(kän'jə gā'shən) (L. conjugare, to yoke together) A method of genetic recombination in bacteria during which a donor and a recipient cell make contact and the DNA from the donor is injected into the recipient cell. p. 494

connective tissue
(kə nek'tiv tish'o͞o) A collection of tissues and its cells that provides a framework for the body, joins its tissues, helps defend it from foreign invaders, and acts as storage sites for specific substances. p. 147

consumer
(kən so͞om'ər) An organism in an ecosystem that feeds on producers and other consumers, passing energy along that was once captured from the sun. p. 657

control
(kən trōl') In a scientific experiment, a standard against which observations or conclusions may be checked to establish their validity. p. 5

conus arteriosus
(kō'nəs är tir'ē ō'səs) A chamber in the heart of cartilaginous and bony fishes that pumps blood to the gills. p. 586

convergent evolution
(kən vur'jənt ev'ə lo͞o'shən) The changes over time among different species of organisms having different ancestors that result in similar structures and adaptations. p. 439

coral
(kôr'əl) A general term that refers to a variety of invertebrate animals of the phylum Cnidaria. p. 561

coral reef
(kôr'əl rēf) A reef built by the marine animals corals (phylum Cnidaria) that secrete carbonate, a hard, shell-like substance; coral reefs are complex ecosystems that provide habitat for a variety of invertebrates and fish. p. 688

cork cambium
(kôrk' kam'bē əm) A lateral meristem tissue whose outermost cells produce cork cells that are waterproof and decay resistant; on maturity, the cork cells lose their cytoplasms and become hardened. p. 552

cornea
(kôr'nē ə) (L. corneus, horny) The transparent portion of the eye's outer layer that permits light to enter the eye. p. 287

corpus callosum
(kôr'pəs kə lō'səm) (N.L. callous body) The single, thick bundle of nerve fibers that connects the two hemispheres of the cerebrum in humans and primates. p. 266

corpus luteum
(kôr'pəs lo͞o'tē əm) (N.L. yellow body) A structure that emerges from a ruptured follicle in the ovary after ovulation; it secretes increased estrogen and progesterone, preparing the endometrium for the implantation of the fertilized egg. p. 341

cortex
(kôr'teks) (L. rind, bark) *(1)* The outer layer of an organ as distinguished from the inner, as in the cerebral cortex. *(2)* The outer superficial portion of the root of a vascular plant. p. 235

cotyledon
(kät'əl ēd'ən) (Gr. kotyledon, a cup-shaped hollow) A seed leaf; in a dicot, the place where most of the food is stored for the developing embryo. p. 538

coupled channel
(kup'əld chan'əl) A type of channel that has binding sites on one membrane transport protein for both sodium (Na^+) and potassium (K^+) ions, which are transported across the cell membrane in opposite directions. p. 75

coupled reaction
(kup'əld rē ak'shən) A reaction in which the energy released in an exergonic reaction is used to drive an endergonic reaction. p. 102

covalent bond
(kō vā'lənt bänd) (L. co-, together + valare, to be strong) A chemical bond created by atoms sharing one or more pairs of electrons. p. 19

cranial nerves
(krā'nē əl nurvz) (Gr. kranion, skull) Any of the twelve pairs of nerves that enter the brain through holes in the skull. p. 275

creatinine
(krē at'ə nēn) (Gr. kreas, flesh) A nitrogenous waste found in urine, derived primarily from creatinine found in muscle cells. p. 239

cri du chat syndrome
(krē do͞o shä sin'drōm) A congenital disorder so named because the infant's cry resembles that of a cat; other symptoms include severe mental retardation and a moon face; caused by a deletion located on one of the 5 chromosomes. p. 393

crista, pl. cristae
(kris'tə, kris'tē) (L. crest) The enfoldings of the inner membrane of a mitochondrion. p. 58

Cro-Magnon
(krō mag'nən) (after a cave in southwestern France) An early member of *H. sapiens sapiens* whose anatomical features were more similar to modern humans; they used sophisticated tools, hunted, and made elaborate cave paintings of animals and hunt scenes. p. 481

crossing-over
(krôs'ing ō'vər) An essential element of meiosis, which produces sister chromatids that are not identical with

each other, resulting in new combinations of genes. p. 93

cuboidal
(kyo͞o boid'əl) (Gr. kubos, cube, + eidos, form) One of the main shapes of epithelial cells. Cuboidal cells have complex shapes but look like cubes when the tissue is cut at right angles to the surface; found lining tubules in the kidney and the ducts of glands. p. 150

cyanobacteria
(sī'ə nō bak tir'ē ə) (Gr. kyanos, dark-blue + bakterion, dim. of baktron, a staff) Formerly blue-green algae; a group of photosynthetic bacteria that played a key role in the evolution of life by gradually oxygenating the atmosphere and the oceans around 2 billion years ago. p. 450

cyclic adenosine monophosphate
(sī'klik *or* sik'lik ə den'ə sēn' mon'ə fos'fāt) A cousin of ATP that triggers enzymes causing a cell to alter its functioning in response to a hormone. p. 59

cyclic electron flow
(sī'klik *or* sik'lik i lek'tron flō) A system that harvests energy in a process involving Photosystem I working independently of Photosystem II. This method of generating energy produces ATP but does not directly produce NADPH as noncyclic electron flow does. Bacteria were the first organisms to use this method. p. 140

cystic fibrosis
(sis'tik fī brō'səs) (Gr. kystis, bladder + L. fibra, fiber + osis, condition) The most common fatal genetic disease of Caucasians in which affected individuals secrete a thick mucus that clogs the airways of the lungs and the passages of the pancreas and liver. p. 399

cytokinesis
(sī'tō kə nē'səs) (Gr. kytos, hollow vessel + kinesis, movement) The physical division of the cytoplasm of a eukaryotic cell into two daughter cells. p. 90

cytoplasm
(sī'tə plaz'əm) (Gr. kytos, hollow vessel + plasma, anything molded) The viscous or gel-like fluid within a cell that contains storage substances, a network of interconnected filaments and fibers, and cell organelles. p. 50

cytoskeleton
(sī'tō skel'ə tən) (Gr. kytos, hollow vessel + skeleton, a dried body) A network of filaments and fibers within the cytoplasm that helps maintain the shape of the cell, move substances within cells, and anchor various structures in place. p. 48

cytotoxic T cell
(sī'tə tôk'sik tē sel) (Gr. kytos, cell + toxikon, poison) A type of T cell that breaks apart cells infected by viruses and foreign cells such as incompatible organ transplants. p. 222

deamination
(dē am'ə nā'shən) The process in which an amino compound loses a (-NH_2) group; takes place in the human liver. p. 234

deciduous
(də sij'yo͞o əs) (L. decidere, to fall off) A tree that drops its leaves and remains dormant throughout the winter. p. 682

decomposer
(dē'kəm pō'zər) An organism in an ecosystem that breaks down organic molecules of dead organisms and contributes to the recycling of nutrients to the environment. p. 657

dehydration synthesis
(dē'hī drā'shən sin'thə səs) The process by which monomers are put together to form polymers. p. 28

deletion
(dē lē'shən) Refers to an abnormally short chromosome that has lost a chromosomal section. p. 393

dendrite
(den'drīt) (Gr. dendron, tree) A projection extending from a neuron that acts as an antenna for the reception of nerve impulses and conducts these impulses toward the cell body. p. 250

density
(den'sə tē) The number of organisms or individuals in a population per unit of area. p. 628

denticle
(den'tə kəl) (L. denticulus, dim. of dentem, tooth) A small, pointed, tooth-like scale that covers the skin of cartilaginous fishes and gives the skin a sandpaper-like texture. p. 585

deoxyribonucleic acid (DNA)
(de ok'sə rī'bō no͞o klē'ik as'əd) A nucleic acid present in the chromosomes that is the chemical basis of heredity and the carrier of genetic information; arranged as two long chains that twist around each other to form a double helix. p. 56

depolarization
(dē pō'lə rə zā'shən) The change in electrical potential of a receptor cell or nerve cell membrane. p. 254

dermal tissue
(dur'məl tish'o͞o) The outer protective covering of all plants with the exception of woody shrubs and trees, which have bark. p. 546

detritus
(di trīt'əs) (L. worn down) The refuse or waste material of an ecosystem. p. 659

deuterostome
(do͞ot'ə rə stōm') (Gr. deutero, second + stoma, mouth) A member of a branch of coelomate animals that during the gastrula stage of development, gives rise to a two-layered embryo; the first indentation becomes the anus and the second indentation becomes the mouth. p. 564

diabetes
(dī'ə bēt'ēz *or* dī'ə bēt'əs) (Gr. passing through) A set of disorders characterized by a high level of glucose in the blood, the underlying cause being the lack or partial lack of insulin; in type I or juvenile onset diabetes, a person has no effective insulin; in type II or maturity onset diabetes, a person has some effective insulin but not enough to meet body needs. p. 328

dialysis
(dī al'ə səs) (Gr. dia, through + lysis, dissolution) The filtering of blood through a selectively permeable membrane to remove toxic wastes, regulate blood pH, and regulate ion concentration; used in renal failure. p. 247

diastolic period
(dī'ə stol'ik pir'ē əd) (Gr. diastole, expansion) The period during the first part of a heartbeat when the atria are filling; at this time the pressure in the arteries leading from the left side of the heart to the tissues of the body decreases slightly as the blood moves out of the arteries, through the vascular system, and into the atria. p. 207

diatom
(dī'ə tom) (Gr. diatemnein, to cut through) A type of microscopic golden algae that possesses a siliceous or calcium-containing cell wall; fossil diatoms are sometimes mined commercially as diatomaceous earth. p. 503

diatomic molecule
(dī'ə tom'ik môl'ə kyo͞ol') A molecule having two atoms. p. 21

dicot
(dī'kot) An angiosperm in which the food reserves of its seeds is stored in the cotyledons, or seed leaves; a dicot usually has net-like veins in its leaves and its flower parts are in fours or fives. p. 538

diencephalon
(dī'ən sef'ə lôn) (Gr. dia, between + enkephalon, within the head) The part of the brain consisting of the thalamus and hypothalamus. p. 265

diffusion
(di fyo͞o'zhən) (L. diffundere, to pour

out) The net movement of molecules from a region of higher concentration to a region of lower concentration, eventually resulting in a uniform distribution of the molecules. This movement is the result of random, spontaneous molecular motions. p. 72

dihybrid
(dī hī'brəd) (Gr. dis, twice + L. hybrida, mongrel) The product of two plants that differ from one another in two traits. p. 380

dinoflagellate
(dī'nō flaj'ə lət *or* dī'nō flaj'ə lāt) (Gr. dinos, rotation + N.L. flagellum, whip) A type of flagellated unicellular red algae, characterized by stiff outer coverings; its flagella beat in two grooves, one encircling the cell like a belt and the other perpendicular to it. p. 509

diploid
(dip'loid) (Gr. diploos, double + eidos, form) A cell that contains double the haploid amount; a double set of the genetic information. p. 83

disaccharide
(dī sak'ə rīd) (Gr. dis, twice + sakcharon, sugar) Two monosaccharides linked together; sucrose (table sugar) is a disaccharide formed by linking a molecule of glucose to a molecule of fructose. p. 30

distal convoluted tubule
(dis'təl kôn'və lo͞o'təd to͞o'byo͞ol) The portion of the kidney tubule whose walls are permeable to water. p. 239

DNA
(dē' en' ā) *See* deoxyribonucleic acid.

dominant
(dôm'ə nənt) The form of a trait that will be expressed in a hybrid offspring. p. 375

Down syndrome
(doun sin'drōm) (after J. Langdon Down, British physician) A genetic disorder produced when an individual receives three (instead of two) 21 chromosomes; also called trisomy 21. p. 389

duodenum
(do͞o'ə dē'nəm *or* do͞o äd'ən əm) (L. duodeni, twelve each; with reference to its length, about twelve finger breadths) The initial, short segment of the small intestine that is actively involved in digestion. p. 175

duplication
(do͞o'pli kā'shən) A chromosomal abnormality in which a section of a chromosome has been duplicated; in a karyotype, the duplicated chromosome appears longer than its homologue. p. 393

ecology
(i kôl'ə jē *or* ē kôl'ə jē) (Gr. oikos, house + logos, word) The special field of biology that studies the interactions among organisms and between organisms and their environments in the laboratory and in nature. p. 690

ecosystem
(ē'kō sis'təm) (Gr. oikos, house + systema, that which is put together) A community of plants, animals, and microorganisms that interacts with one another and with their environments and is interdependent on one another for survival. p. 639

ectoderm
(ek'tə durm') (Gr. ectos, outside + derma, skin) The outer layer of cells formed during the development of embryos; forms the outer layer of skin, the nervous system, and portions of the sense organs. p. 359

ectothermic
(ek'tə thur'mik) (Gr. ectos, outside + therme, heat) Referring to animals like reptiles, amphibians, and fishes that regulate body temperature by taking in heat from the environment. p. 589

effector
(i fek'tər) A muscle or gland that effects (or causes) responses when stimulated by nerves. p. 257

efferent neuron
(ef'ər ənt no͞or'on) *See* motor neuron. p. 264

electrocardiogram (ECG)
(i lek'trō kärd'ē ə gram') (Gr. elektron, amber + kardia, heart + -gram, written) A recording of the electrical impulses initiated at the SA node as they pass throughout the heart as an electric current and then as a wave throughout the body. p. 208

electron
(i lek'tron) A subatomic particle, having very little mass and carrying a negative charge, that orbits the nucleus of an atom. p. 17

electron-transport chain
(i lek'tron trans'pōrt chān) A term that describes the membrane-associated electron carriers produced by the citric acid cycle; the electrons gleaned from the oxidation of glucose then work as proton pumps. p. 117

element
(el'ə mənt) A pure substance that is made up of a single kind of atom and that cannot be separated into different substances by ordinary chemical methods. p. 18

elimination
(i lim'ə nā'shən) A process that takes place as digestive wastes leave the body during defecation. p. 233

embryo
(em'brē ō') (Gr. embryon, something that swells in the body) The early stage of development in humans between the second and eighth weeks. p. 361

emigration
(em'ə grā'shən) (L. emigrare, to migrate) The movement of organisms out of a population. p. 625

endergonic
(en'dər gôn'ik) (Gr. endon, within + ergos, work) Used to describe a reaction in which the products of the reaction contain more energy than the reactants, so the extra energy must be supplied for the reaction to proceed. p. 101

endocrine gland
(en'də krən gland) (Gr. endon, within + krinein, to separate) A ductless gland that secretes hormones and spills these chemicals directly into the bloodstream. p. 315

endocrine system
(en'də krən sis'təm) (Gr. endon, within + krinein, to separate) The collective term for the 10 different endocrine glands that secrete 30 different hormones. p. 315

endocytosis
(en'dō sī tō'səs) (Gr. endon, within + kytos, cell) A process in which cells engulf large molecules or particles and bring these substances into the cell packaged within vesicles. p. 77

endoderm
(en'dō durm') (Gr. endon, within + derma, skin) The inner layer of cells formed during the early development of embryos; it gives rise to the digestive tract lining, the digestive organs, the respiratory tract, the lungs, the urinary bladder, and the urethra. p. 359

endodermis
(en'dō durm'əs) (Gr. endon, within + derma, skin) In the root of a vascular plant, the innermost layer of the cortex, which consists of specialized cells that regulate the flow of water between the vascular tissues and the outer portion of the root. p. 549

endometrium
(en'dō mē'trē əm) (Gr. endon, within + metrios, of the womb) The inner lining of the uterus, which has two layers. p. 341

endoplasmic reticulum
(en'dō plaz'mik ri tik'yə ləm) (Gr. endon, within + plasma, from cytoplasm; L. reticulum, network) An extensive system of membranes that divides the interior of eukaryotic cells into compartments and channels. p. 53

endosperm
(en'dō spurm) (Gr. endon, within + sperma, seed) An extraembryonic tissue in a monocot where most of the food is

stored for the developing embryo. p. 538

endosymbiont
(en'dō sim'bī änt) (Gr. endon, within + bios, life) An organism that is symbiotic within another; the major endosymbionts that occur in eukaryotic cells are mitochondria and chloroplasts. p. 48

endosymbiotic theory
(en'dō sim'bē ôt'ik thir'ē) (Gr. endon, within + bios, life) The idea that mitochondria and chloroplasts originated symbiotically, mitochondria from aerobic bacteria and chloroplasts from anaerobic bacteria. p. 452

endothermic
(en'dō thur'mik) (Gr. endon, within + therme, heat) Referring to organisms such as birds and mammals that regulate body temperature internally. p. 592

end-product inhibition
(end prôd'əkt in'ə bish'ən) The process in which the enzyme catalyzing the first step in a series of chemical reactions has an inhibitor binding site to which the end product of the pathway binds. As the concentration of the end product builds up in the cell, it begins to bind to the first enzyme in the metabolic pathway, shutting off that enzyme. In this way, the end product is feeding information back to the first enzyme in the pathway, shutting the pathway down when additional end product is not needed. p. 108

energy level
(en'ər jē lev'əl) An electron shell that surrounds the nucleus of an atom. p. 18

entropy
(en'trə pē) (Gr. en, in + tropos, change in manner) The energy lost to disorder; it is a measure of the disorder of a system. p. 105

environment
(in vī'rən mənt) (M.E. envirounen, encircle) A general term for the biosphere; the land, air, water, and every living thing on the Earth. p. 695

enzyme
(en'zīm) (Gr. enzymos, leavened, from en, in + zyme, leaven) A protein that lowers the free energy of activation, thus allowing chemical reactions to take place. p. 105

eosinophil
(ē'ə sin'ə fil) (Gr. eos, dawn (rose-colored) + philein, to love) A kind of granulocyte believed to be involved in allergic reactions; they also act against certain parasitic worms. p. 210

epidermis
(ep'ə dur'məs) (Gr. epi, upon + derma, skin) In the root of a vascular plant, the outer layer of the cortex, which absorbs water and minerals. p. 549

epididymis
(ep'ə did'ə məs) (Gr. epi, upon + didymos, testis) A long, coiled tube that sits on the back side of the testes where sperm undergo further development after their formation within the testes. p. 338

epiglottis
(ep'ə glot'əs) (Gr. epi, upon + glotta, tongue) A flap of tissue that folds back over the opening to the larynx, thus preventing food or liquids from entering the airway. p. 172

epinephrine
(ep'ə nef'rən) (Gr. epi, on, over + nephros, kidney) A hormone produced by the adrenal medulla that readies the body to react to stress p. 326

epiphyte
(ep'ə fīt) (Gr. epi, on, over + phyton, plant) A plant of the tropical rain forest that grows on trees or other plants for support but draws its nourishment from rain water. p. 676

epithelial tissue
(ep'ə thē'lē əl tish'o͞o) A collection of tissues that cover and line internal and external surfaces of the body and compose the glands. p. 147

epithelium
(ep'ə thē'lē əm) (Gr. epi, on + thele, nipple) The collective term for all epithelial cells, which have six different functions in the body: protection, absorption, sensation, secretion, excretion, and surface transport. p. 149

epoch
(ep'ək *or* ē'pôk) (Gr. epechein, to hold on, check) A subdivision of a period of the Cenozoic Era. p. 448

era
(ir'ə *or* er'ə) One of the five major periods of geological time since the formation of the Earth until the present day. p. 448

erythrocyte
(i rith'rō sīt) (Gr. erythros, red + kytos, hollow vessel) A red blood cell, packed with hemoglobin, which acts as a mobile transport unit, picking up and delivering gases. p. 157

esophagus
(i sof'ə gəs) (Gr. oiso, carry + phagein, to eat) The food tube that connects the pharynx to the stomach. p. 172

essential amino acids
(i sen'chəl ə mēn'ō as'ədz) The eight amino acids that humans cannot manufacture and therefore must be obtained from proteins in the food they eat. p. 181

estrogen
(es'trə jən) (Gr. oistros, frenzy + genos, origin) Any of various hormones that develop and maintain the female reproductive structures such as the ovarian follicles, the lining of the uterus, and the breasts. p. 341

estuary
(es'cho͞o er'ē) (L. aestus, tide) A place where the fresh water of rivers and streams meets the salt water of oceans. p. 685

ethology
(ē thôl'ə jē) (Gr. ethos, habit or custom + logos, discourse) The study of animal behavior in the natural environment, which examines the biological basis of the patterns of movement, sounds, and body positions of animals. p. 599

eubacteria
(yo͞o'bak tir'ē ə) (Gr. eus, good + bakterion, rod) The so-called true bacteria, which include all bacteria except the archaebacteria. p. 492

euglenoid
(yo͞o glē'noid) (Gr. eu, good + glene, eyeball) A flagellate member of the genus *Euglena,* having chloroplasts and making its own food by photosynthesis. p. 506

eukaryotic cell
(yo͞o kar'ē ōt'ik sel') (Gr. eu, good + karyon, kernel) A type of cell more complex than a prokaryote that makes up the bodies of plants, animals, protists, and fungi; a plasma membrane encloses the cytoplasm, which contains organelles and the nucleoid. p. 47

eustachian tube
(yo͞o stā'kē ən *or* yo͞o stā'shən to͞ob) (after Bartolomeo Eustachio, Italian anatomist) A structure that connects the middle ear with the nasopharynx; it equalizes air pressure on both sides of the eardrum when the outside air pressure is not the same. p. 289

eutrophication
(yo͞o trof'ə kā'shən) (Gr. eutrophic, thriving) The accumulation of inorganic nutrients in a lake. p. 668

evolution
(ev'ə lo͞o'shən) (L. evolvere, to unfold) Genetic change in a population of organisms over generations; Darwin proposed that natural selection was the mechanism behind evolutionary change. p. 11

excitatory synapse
(ik sīt'ə tōr'ē sin'aps) A type of synapse in which a neurotransmitter depolarizes the postsynaptic membrane, resulting in the continuation of the nerve impulse. p. 259

excretion
(ik skrē'shən) A process whereby metabolic wastes, excess water, and excess salts are removed from the blood and passed out of the body. p. 233

exergonic
(ek'sər gon'ik) (L. ex, out + Gr. ergon, work) Used to describe a reaction in which the products contain less energy than the reactants and the excess energy is released; exergonic reactions take place spontaneously. p. 101

exocrine gland
(ek'sə krən gland) (Gr. exo, outside + krinein, to separate) A term applied to a gland whose secretion reaches its destination by means of ducts. p. 315

exocytosis
(ek'sō sī tō'səs) (Gr. ex, out of + kytos, cell) The reverse of endocytosis; the discharge of material by a cell by packaging it in a vesicle and moving the vesicle to the cell surface. p. 77

exon
(ek'son) (Gr. exo, outside) A nucleotide sequence that encodes the amino-acid sequence of a polypeptide. p. 419

expiration
(ek'spə rā'shən) (Gr. ex, out + L. spirare, to breathe) The expelling of the air from the lungs in breathing; it occurs when the volume of the thoracic cavity is decreased and the resulting positive pressure forces air out of the lungs. p. 192

exponential growth
(ek'spō nen'shəl grōth) A growth rate in which although the rate of increase in population size may stay the same, the actual increase in the number of individuals grows. p. 492

external fertilization
(ek sturn'əl furt'əl ə zā'shən) Fertilization that takes place outside the body of the female. p. 584

external respiration
(ek sturn'əl res'pə rā'shən) The exchange of carbon dioxide and oxygen gases by the red blood cells and alveoli in the lungs. p. 185

extinction
(ik stingk'shən) (L. exstinguere, to quench) Drying out; refers to the serious and ongoing problem of species extinction of organisms taking place on Earth. p. 455

extraembryonic membrane
(ek'strə em'brē on'ik mem'brān) A structure that forms from the trophoblast and provides nourishment and protection; so named because it is not a part of the embryo. p. 360

F plasmid
(ef' plāz'məd) A certain plasmid with a fertility factor, which are several special genes that promote the transfer of the plasmid to other cells. p. 494

facilitated diffusion
(fə sil'ə tā'təd di fyo͞o'shən) The movement of selected molecules across the cell membrane by specific transport proteins along the concentration gradient and without an expenditure of energy. p. 74

fatty acid
(fat'ē as'əd) A long hydrocarbon chain ending in a carboxyl (-COOH) group; a fatty acid can be saturated, unsaturated, or polyunsaturated. p. 33

feces
(fē'sēz) (L. faeces) Body waste discharged by way of the anus. p. 178

feedback loop
(fēd'bak' lo͞op) A mechanism by which information regarding the status of a physiological situation or system is fed back to the system so that appropriate adjustments can be made. p. 162

femur
(fē'mur) The thigh bone in the appendicular skeleton of a human. p. 300

fermentation
(fur'mən tā'shən) (L. fermentum, ferment) An anaerobic process by which certain organisms make ATP. p. 116

fertilization
(furt'ə lə zā'shən) (L. ferre, to bear) The union of a male gamete (sperm) and a female gamete (egg). p. 355

fibroblast
(fī'brə blast') (L. fibra, fiber + Gr. blastos, sprout) The most numerous of the connective tissue cells, they are flat, irregular, branching cells that secrete fibers into the matrix between them. p. 153

fibrocartilage
(fī'brə kärt'əl ij *or* fī'brə kärt'lij) (L. fibra, fiber + cartilago, gristle) A type of cartilage that has collagen fibers embedded in its matrix; used by the body as a "shock absorber" in the knee joint and as disks between the vertebrae. p. 153

filtrate
(fil'trāt) The water and dissolved substances that are first filtered out of the blood during the formation of urine. p. 234

filtration
(fil'trā'shən) The process by which the blood is passed through nephron membranes that separate blood cells and proteins from the water and small molecules of the blood. p. 235

flagellum, pl. flagella
(flə jel'əm, flə jel'ə) (L. flagellum, whip) A long, whip-like organelle of motility that protrudes from some cells. p. 60

fluid mosaic model
(flo͞o'əd mō zā'ik mod'əl) The model of the cell membrane that describes the fluid nature of a lipid bilayer studded with a mosaic of proteins. p. 69

follicle
(fol'e kəl) (L. folliculus, little bag) A term for the follicular cells and the oocyte in the ovary. p. 339

follicle-stimulating hormone (FSH)
(fol'ə kəl-stim'yə lāt'ing hôr'mōn) A gonadotropic hormone secreted by the anterior pituitary that triggers the maturation of one egg each month in females; it triggers sperm production in males. p. 320

food chain
(fo͞od chān) A series of organisms, from each trophic level, that feed on one another. p. 659

food web
(fo͞od web) The interwoven food chain of an ecosystem; a diagram of who eats whom. p. 661

foramen, pl. foramina
(fə rā'mən, fə ram'ən ə) The inner space or tunnel in the vertebral column in which the spinal cord runs down the neck and back; this body casing protects the spinal cord from injury. p. 271

foraminifer, pl. foraminifera
(fôr'ə min'ə fər, fə ram'ə nif'ər ə) (L. forare, to bore) A type of ameba that secretes beautifully sculpted shells made out of calcium carbonate or limestone; also called a foram. p. 505

formed elements
(fôrmd el'ə məntz) The solid portion of blood plasma composed principally of erythrocytes, leukocytes, and platelets. p. 208

fossil
(fos'əl) (L. fodere, to dig) Any record of a dead organism; any trace or impression of an animal or plant that has been preserved in the Earth's crust. p. 434

fovea
(fō'vē ə) (L. a pit) The area of sharpest vision within the retina due to its high concentration of cones. p. 288

free energy of activation
(frē en'ər jē uv ak'tə vā'shən) The energy needed to initiate a chemical reaction. p. 101

free radical
(frē rad'ə kəl) A charged molecule fragment with an unpaired electron that is highly reactive. p. 393

frond
(frond) (L. foliage) The leaf of a fern. p. 531

frontal lobe
(frunt'əl lōb) A section on both hemispheres of the cerebral cortex

dealing with the motor activity movement of the body. p. 266

fruit
(fro͞ot) A vessel containing seeds produced by an angiosperm. p. 534

functional group
(fungk'shən əl gro͞op) Special groups of atoms attached to an organic molecule; important because most chemical reactions that occur within organisms involve the transfer of a functional group from one molecule to another. p. 28

fungus, pl. fungi
(fung'gəs, fun'jī) (L. mushroom) A multicellular, eukaryotic organisms that feeds on dead or decaying organic material. p. 516

gall bladder
(gôl blad'ər) In human beings, a sac attached to the underside of the liver, where excess bile is stored and concentrated. p. 175

gamete
(gam'ēt *or* gə mēt') (Gr. wife) A sex cell; the female gamete is the egg, and the male gamete is the sperm. p. 83

gametophyte
(gə mē'tō fīt) (Gr. gamete, wife + phyton, plant) The haploid phase of a plant life cycle that alternates with the diploid phase. p. 529

ganglion, pl. ganglia
(gang'glē ən, gang'glē ə) A nerve cell body located within the peripheral nervous system. p. 267

gastric glands
(gas'trik glandz) Glands dotting the inner surface of the stomach that secrete a gastric juice of hydrochloric acid and pepsinogen. p. 174

gastrin
(gas'trən) A digestive hormone of the stomach that controls the production of acid. p. 174

gastrulation
(gas'trə lā'shən) (L. little belly) The process by which groups of inner cells mass, migrate, divide, and differentiate into three primary germ layers from which all the organs and tissues of the body develop. p. 361

gene
(jēn) (Gr. genos, birth, race) A piece of hereditary information carried on the X and Y sex chromosomes. p. 81

gene families
(jēn fam'lēz *or* fam'ə lēz) Multiple copies of genes in eukaryotes, which are derived from a common ancestral gene and are a reflection of the evolutionary process. p. 422

gene mutation
(jēn myo͞o tā'shən) A change in the genetic message of a chromosome due to alterations of molecules within the structure of the chromosomal DNA. p. 394

gene therapy
(jēn ther'ə pē) The technique in which scientists try to cure inherited genetic disorders by inserting genes with the proper genetic message into patients having defective genes. p. 405

general adaptation syndrome
(jen'ə rəl ad ap tā shən sin'drōm) the three stages of reaction by the body to stress: the alarm reaction, resistance, and exhaustion. p. 326

genetic counseling
(jə net'ik coun'səl ing) The process in which geneticists identify couples at risk of having children with genetic defects and help them have healthy children.

genetic engineering (recombinant DNA technology) (jə net'ik en'jə nir'ing) New techniques of molecular biology that involve the purposeful manipulation of genes within organisms. p. 494

genetics
(jə net'iks) (Gr. genos, birth, race) The branch of biology dealing with the principles of heredity and variation in organisms. p. 377

genotype
(jēn'ə tīp) (Gr. genos, offspring + typos, form) The total set of genes that constitutes an organism's genetic makeup. p. 378

geographic isolation
(jē'ə graf'ik ī'sə lā'shən) A term describing organisms that live in sharply discontinuous habitats from one another. p. 432

germination
(jur'mə nā'shən) (L. germinare, to sprout) The sprouting of a seed that begins when it receives water and appropriate environmental cues. p. 538

gestation
(jes tā'shən) (L. gestare, to bear) The time of a developing human from conception until birth, approximately 8½ months. p. 355

gibberellin
(jib'ə rel'ən) (*Gibberella,* a genus of fungi) A hormone that together with auxin, promotes growth in plants through cell elongation. p. 541

gill
(gil) In the body of a mollusk, a filamentous projection of the mantle tissue that is rich in blood vessels and functions as an organ of respiration. p. 571

gizzard
(giz'ərd) The digestive organ of a bird, often filled with grit, that grinds food. p. 591

gland
(gland) (L. glans, acorn) An epithelial cell specialized to produce and discharge substances. p. 149

glial cell
(glē'əl *or* glī'əl sel) (Gr. glia, glue) A supporting nerve cell of the brain and spinal cord. p. 160

glomerulus
(glə mer'ə ləs) (L. a little ball) A tuft of capillaries surrounded by Bowman's capsule that acts as filtration devices in the formation of urine. p. 237

glucagon
(glo͞o'kə gon) A hormone produced in the islets of Langerhans that raises blood glucose level by converting glycogen into glucose. p. 327

glycogen
(glī'kə jən) (Gr. glykys, sweet + gen, of a kind) Animal starch; a storage form of glucose within animals. p. 30

glycolysis
(glī kol'ə səs) (Gr. glykys, sweet + lyein, to loosen) The first of three series of chemical reactions in cellular respiration, which results in the formation of two ATP molecules and two molecules of pyruvate. p. 117

Golgi complex
(gol'jē käm'pleks) (after Camillo Golgi, Italian physician) The delivery system of the eukaryotic cell; it collects, modifies, packages, and distributes molecules that are made at one location within the cell and used at another. p. 54

gonad
(gō'nad) (Gr. gone, seed) A male or female reproductive sex organ that produces sex cells, or gametes. p. 335

gradient
(grā'dē ənt) Describing the differences in concentration, pressure, or electrical charge in the random motion of molecules, which often results in a net movement of molecules in a particular direction. p. 72

grana, sing. granum
(grā'nə, grā'nəm) (L. grain or seed) Within chloroplasts, stacks of membrane thylakoid sacs. p. 135

granulocyte
(gran'yə lō sīt) (L. granulum, little grain + Gr. kytos, cell) A circulating leukocyte that gets its name from the tiny granules in its cytoplasm; granulocytes are classified into three groups by their staining properties. p. 210

greenhouse effect
(grēn'hous' ə fekt') The selective energy

absorption of carbon dioxide (CO_2), which allows heat to enter the Earth's atmosphere but prevents it from leaving; may be a contributing factor to global warming. p. 700

growth hormone (GH)
(grōth hôr′mōn) A hormone secreted by the anterior pituitary that works with the thyroid hormones to control normal growth. p. 320

guard cells
(gärd selz) A pair of cells that brackets a stoma and regulates its opening and closing. p. 554

gymnosperm
(jim′nə spurm′) (Gr. gymnos, naked + sperma, seed) A vascular plant whose seeds are not completely enclosed by the tissues of the parent at the time it is pollinated; the most familiar gymnosperms are the conifers. p. 533

habitat
(hab′ə tat′) (L. habitare, to inhabit) An area or space within an ecosystem, including the factors within it, in which each living thing resides. p. 639

half-life
(haf′ līf′) The length of time it takes for half of the ^{14}C present in a sample to be converted to ^{12}C, which is 5730 years; a useful tool for dating fossils. p. 436

haploid
(hap′loid) (Gr. haploos, single + eidos, form) A sex cell that contains half the amount of hereditary material of the original parent cell. p. 83

haversian canal
(hə vur′shən kə nal′) (after Clopton Havers, British physician and anatomist) A narrow channel that runs parallel to the length of the bone that contains blood vessels and nerve cells. p. 296

heart
(härt) (A.S. heorte) The muscular pump that is the center of the cardiovascular or circulatory system. p. 199

helper T cell
(help′ər tē′ sel) A kind of T lymphocyte that initiates the immune response by identifying foreign invaders and stimulating the production of other cells to fight an infection. p. 222

hemoglobin
(hē′mə glō′bən) (Gr. haima, blood + L. globus, a ball) The iron-containing pigment that imparts the color to red blood cells; hemoglobin is produced within the red bone marrow and carries oxygen in the blood. p. 193

hemophilia
(hē′mə fil′ē ə) (Gr. haima, blood + philein, to love) A hereditary condition in which the blood is slow to clot or does not clot at all. p. 389

herbivore
(hur′bə vôr) (L. herba, grass + vorare, to eat) An animal that feeds on plants. p. 170

hermaphrodite
(hər maf′rə dīt) (Gr. Hermaphroditos, son of Hermes and Aphrodite, who was man and woman combined) An animal or plant having both male and female reproductive organs; a sponge is a hermaphrodite. p. 566

heterotroph
(het′ər ə trōf′) (Gr. heteros, other + trophos, feeder) An organism that cannot produce its own food; a consumer in an ecosystem or food chain. p. 131

hetereozygous
(het′ər ə zī′gəs) (Gr. heteros, other + zygotos, a pair) Refers to a individual having two different alleles for a trait. p. 378

homeostasis
(hō′mē ə stā′səs) (Gr. homeos, similar + stasis, standing) The maintenance of a stable internal environment in spite of a possibly very different external environment. p. 162

hominid
(hom′ə nid) (L. homo, man) The family of hominoids consisting of human beings; the only living hominid is *Homo sapiens sapiens*. p. 475

hominoid
(hom′ə noid) (L. homo, man) A superfamily of the anthropoid suborder that includes apes and humans. p. 474

Homo erectus
(hō′mō i rek′təs) (L. homo, man + erectus, upright) An extinct species of hominids whose fossil record dates back to 1.6 million years ago; *H. erectus* was fully adapted to upright walking, made sophisticated tools, built shelters, used fire, and probably communicated with language. p. 480

Homo habilis
(hō′mō hab′ə ləs) (L. homo, man + habilis, skillful) An extinct species of hominids whose fossil record dates back about 2 million years; considered human because they exhibited a far greater intelligence than their ancestors by making tools and clothing. p. 480

Homo sapiens
(hō′mō sā′pē ənz *or* sap′ē ənz) (L. homo, man + sapiens, wise) A hominid whose fossil record dates back about 200,000 years and most likely evolved from *H. erectus;* their anatomy featured larger brains, flatter heads, more sloping foreheads, and more protruding brow ridges than modern humans. p. 475

Homo sapiens sapiens
(hō′mō sā′pē ənz sā′pē ənz) (L. homo, man + sapiens, wise) Modern man; the subspecies of hominids who made their appearance 10,000 years before the Neanderthal subspecies died out and whose early members are called Cro-Magnons. p. 481

homologous
(hō mol′ə gəs) (Gr. homologia, agreement) Said of the bones of different vertebrates that now differ in structure and function, although having the same evolutionary origin. p. 438

homozygous
(hō′mə zī′gəs) (Gr. homos, same or similar + zygotos, a pair) Refers to an individual having two identical alleles for a trait. p. 378

hormone
(hôr′mōn) (Gr. hormaien, to excite) A chemical messenger secreted and sent by a gland to other cells of the body. p. 315

human chorionic gonadotropic (HCG)
(hyo͞o′mən kōr′ēän′ik gō′nad ə trō′pik) (Gr. chorion, chorion + gonos, genitals + trope, turning) A hormone secreted by embryonic placental tissue that maintains the corpus luteum so that it will continue to secrete progesterone and estrogen; the detection of this hormone in the urine is the basis for home pregnancy tests. p. 341

human genome project
(hyo͞o′mən jē′nōm proj′ekt) A monumental, worldwide scientific project, the goal of which is to decipher the DNA code of all 46 human chromosomes. p. 410

human immunodeficiency virus (HIV)
(hyo͞o′mən im yo͞o′nō dē fish′ən sē vī′rəs) The virus that causes AIDS, especially deadly because it destroys the ability of the immune system to mount a defense against any infection because it attacks and destroys helper T cells. p. 227

humerus
(hyo͞o′mər əs) (L. upper arm) The upper arm bone in the appendicular skeleton of a human. p. 300

humoral immune response
(hyo͞o′mər əl i myo͞on ri spons′) A second, longer-range defense than the cell-mediated immune reseponse, in which B cells are converted to plasma cells that secrete the proteins antibodies. p. 222

Huntington's disease
(hunt′ing tənz diz ēz′) (after G. Huntington, U.S. physician) A fatal genetic disorder caused by a mutant dominant allele that causes progressive deterioration of brain cells. p. 398

hyaline cartilage
(hī′ə lən *or* hī′ə līn kärt′əl ij *or* kärt′lij) (Gr. hyalos, glass + L. cartilago, gristle)

A type of cartilage that has very fine collagen fibers in its matrix; it is found on the ends of long bones, ringing the windpipe, and in the ribs and nose. p. 153

hybrid
(hī'brəd) (L. hybrida, mongrel) The offspring of the cross between two different varieties of plants of the same species; also the cross between two different species of organisms. p. 374

hydrogen bond
(hī'drə jən bänd) A molecule formed when the partial negative charge at one end of a polar molecule is attracted to the partial positive charge of another polar molecule. p. 23

hydrolysis
(hī drōl'ə səs) (Gr. hydro, water + lyse, to break) The process by which a polymer is disassembled by adding a molecule of water. p. 30

hydrolyzing enzyme
(hī'drə līz'ing en'zīm) (Gr. hydro, water + lyse, to break) An enzyme that breaks down substances by hydrolysis. p. 170

hydrophilic
(hī drə fil'ik) (Gr. hydor, water + philos, loving) Referring to the phosphate functional groups in polar molecules, which form hydrogen bonds with water and, consequently, are soluble or dissolve in water. p. 69

hydrophobic
(hī'drə fō'bik) (Gr. hydor, water + phobos, hating) Refers to nonpolar molecules like oil that cannot form hydrogen bonds with water; this is why oil and water do not mix. p. 69

hyperpolarization
(hī'pər pō'lər ə zā'shən) (Gr. hyper, above + polaris, pole) The resulting state in the interior of a rod cell when it becomes even more negatively charged than before because sodium channels have closed because of stimulation. p. 286

hypertonic
(hī'pər ton'ik) (Gr. hyper, above + tonos, tension) Refers to a solution with a solute concentration higher than that of another fluid. p. 73

hypha, pl. hyphae
(hī'fə, hī'fē) (Gr. hyphe, web) A slender filament of a fungus barely visible to the naked eye. p. 516

hypothalamus
(hī'pō thal'ə məs) (Gr. hypo. under + thalamos, inner room) A mass of gray matter lying at the base of the cerebrum that produces two hormones and controls the secretion of hormones by the pituitary gland. p. 269, 318

hypothesis
(hī poth'ə səs) (Gr. hypo, under + tithenai, to put) A plausible answer to a scientific question, based on available knowledge and generalizations made from observations. p. 3

hypotonic
(hī'pə ton'ik) (Gr. hypo, under + tonos, tension) Refers to a solution with a solute concentration lower than that of another fluid. p. 73

ileum
(il'ē əm) (L. groin, flank) The third and final part of the small intestine, following the jejunum. p. 175

immigration
(im'ə grā'shən) (L. immigrare, to migrate) The movement of organisms into a population. p. 625

immune system
(i myo͞on' sis'təm) The collective term for populations of white blood cells that resist disease. p. 221

immunity
(i myo͞on'ə tē) (L. immunitas, safe) The specific defense of a human body that consists of cellular and molecular responses to particular foreign invaders. p. 217

implantation
(im'plan tā'shən) (L. in, into + plantare, to plant) The embedding of the developing blastocyst into the posterior wall of the uterus approximately 1 week after fertilization. p. 357

imprinting
(im'print ing) A rapid and irreversible type of learning that takes place during an early developmental stage of some animals. p. 603

incomplete dominance
(in'kəm plēt' däm'ə nəns) A situation in which neither member of a pair of alleles exhibits dominance over the other. p. 399

induction
(in duk'shən) (L. inductio, leading in) The process by which some cells turn "off and on" switches for the genes of neighboring cells. p. 364

inflammation
(in'flə mā'shən) (L. inflammare, to flame) A reaction when cells are damaged by microbes, chemicals, or physical substances. p. 217

inhibitory synapse
(in hib'ə tôr'ē sin'aps) A type of neurotransmitter that reduces the ability of the postsynaptic membrane to depolarize. p. 259

inner ear
(in'ər ir) A complex of fluid-filled canals in the ear where hearing actually takes place; receptor cells of this organ change the mechanical "sound" energy into nerve impulses. p. 289

insertion
(in sur'shən) The attachment of one end of a skeletal muscle to a bone that will move. p. 305

inspiration
(in'spə rā'shən) (L. in, in + spirare, to breathe) Drawing air into the lungs; it occurs when the volume of the thoracic cavity is increased and the resulting negative pressure causes air to be sucked into the lungs; opposite of expiration. p. 190

insulin
(in'sə lən) (L. insula, island) A hormone produced in the islets of Langerhans of the pancreas that promotes the regulation of glucose metabolism by the cells. p. 327

integration
(int'ə grā'shən) A function of the central nervous system in which the brain and spinal cord make sense of incoming sensory information and then produce outgoing motor impulses. p. 265

integument
(in teg'yə mənt) (L. integumentum, covering) The skin, hair, and nails of the human body. p. 296

interneuron
(int'ər no͞or'on) (L. inter, between + Gr. neuron, nerve) A nerve cell found in the spinal cord and brain situated between other neurons that receives incoming messages and sends outgoing messages in response. p. 251

interphase
(int'ər fāz') The portion of the cell cycle preceding mitosis in which the cell grows and carries out normal life functions. During this time the cell also produces an exact copy of the hereditary material, DNA, as it prepares for cell division. p. 86

intron
(in'tron) (L. intra, within) A segment of DNA transcribed into mRNA but removed before translation. p. 419

inversion
(in vur'zhən) Refers to the process in which a broken piece of chromosome reattaches to the same chromosome but in a reversed direction. p. 393

ion
(ī'ən *or* ī'on) (Gr. going) An electrically charged particle that results from an exchange of electrons in an atom. p. 20

ionic bond
(ī on'ik bänd) An attraction between ions of opposite charge. p. 19

iris
(ī'rəs) (L. rainbow) A diaphragm lying between the cornea and the lens that controls the amount of light entering the eye. p. 288

islets of Langerhans
(ī'ləts uv läng'ər hänz') (after Paul Langerhans, German anatomist) The separate types of cells within the exocrine cells of the pancreas that produce the hormones insulin and glucagon. p. 327

isomer
(ī'sə mər) (Gr. isos, equal + meros, part) One of two or more molecules that have the same molecular formula but are arranged slightly differently; fructose is an isomer of glucose. p. 447

isotonic
(ī'sə ton'ik) (Gr. isos, equal + tonos, tension) Refers to solutions having equal solute concentrations to one another. p. 73

isotope
(ī'sə tōp) (Gr. isos, equal, + topos, place) An atom of an element that has the same number of protons but different numbers of neutrons in its nuclei. p. 17

jawless fish
(jô'ləs fish) A tubular, scaleless, finless, jawless organism that lives in the sea or in brackish water; a lamprey is a jawless fish that parasitizes bony fish and poses a serious threat to some commercial fisheries. p. 584

jejunum
(ji jo͞o' nəm) (L. empty) The second portion of the small intestine extending from the duodenum to the ileum. p. 175.

joint
(joint) An articulation; a place where bones or bones and cartilage come together. p. 302

karyotype
(kar'ē ə tīp') (Gr. karyon, kernel + typos, stamp or print) The particular array of chromosomes that belongs to an individual. p. 81

kidney
(kid'nē) The organ in all vertebrates that carries out the processes of filtration, reabsorption, and excretion. p. 234

kidney stone
(kid'nē stōn) Crystals of certain salts that develop in the kidney and block urine flow. p. 234

kinetic energy
(kə net'ik en'ər jē) The energy of motion. p. 103

Klinefelter syndrome
(klīn'felt ər sin'drōm) (after Harry F. Klinefelter, Jr, U.S. physician) A genetic condition resulting from an XXY zygote that develops into a human male who is sterile, has many female body characteristics, and in some cases, has diminished mental capacity. p. 391

Krebs cycle
(krebz sī'kəl) (after Hans A. Krebs, German-born English biochemist) The series of nine reactions during which pyruvate, the end product of glycolysis, enters the cycle to form citric acid and is finally oxidized to carbon dioxide. Also called citric acid cycle. p. 117

lacteal
(lak'tē əl) (L. lacteus, of milk) A lymphatic vessel within a villus that helps pass nutrients into the lymph and blood. p. 178

lacuna, pl. lacunae
(lə kyo͞o'nə, lə kyo͞o'nē) (L. a pit) The tiny chambers where cartilage cells lie. p. 153

laguno
(lə go͞o'nə) A fine body hair that appears over the body of a fetus toward the end of the third month of pregnancy but that is lost before birth. p. 366

lancelet
(lans'lət) A scaleless, fish-like marine chordate. p. 583

larynx
(lar'ingks) (Gr.) The voice box located at the upper end of the human windpipe. p. 186

lateral line system
(lat'ər əl līn sis'təm) A complex system of mechanoreceptors possessed by all fishes and amphibians that detects mechanical stimuli such as sound, pressure, and movement. p. 585

lens
(lenz) (L. lentil) A body in the eye lying just behind the aqueous humor that plays a major role in focusing the light entering the eye. p. 287

leukocyte
(lo͞o'kə sīt) (Gr. leukos, white + kytos, hollow vessel) Any of several kinds of white blood cells including macrophages and lymphocytes, all functioning to defend the body against invading microorganisms and foreign substances. p. 157

ligament
(lig'ə mənt) (L. ligare, to bind) A bundle or strip of dense connective tissue that holds a bone to a bone. p. 302

light-dependent reactions
(līt di pen'dənt rē ak'shənz) The reactions of photosynthesis that produce ATP and NADPH; so called because they take place only in the presence of light. p. 135

light-independent reactions
(līt in'də pen'dənt rē ak'shənz) The reactions of photosynthesis that use ATP and NADPH to drive the formation of glucose from carbon dioxide; so called because they do not need light to occur as long as ATP is available; also called the Calvin-Benson cycle. p. 136

limb bud
(lim bud) The appearance of arms and legs that appear as microscopic flippers during the fourth week in an embryo. p. 366

limbic system
(lim'bik sis'təm) (L. limbus, border) A network of neurons, which together with the hypothalamus, forms a ring-like border around the top of the brainstem; responsible for many of the most deep-seated drives and emotions of vertebrates. p. 269

lipase
(lī'pās *or* lip'ās) (Gr. lipos, fat + -ase, enzyme) An enzyme that breaks down the triglycerides in lipids to fatty acids and glycerol. p. 168

lipid
(lip'id) (Gr. lipos, fat) Any of a wide variety of molecules, all of which are soluble in oil but insoluble in water; important categories of lipids are oils, fats, and waxes; phospholipids; and steroids. p. 33

lipid bilayer
(lip'id bī'lā'ər) The basic foundation of biological membranes; it forms a fluid, flexible covering for a cell and keeps the watery contents of the cell on one side of the membrane and the water environment on the other. p. 69

liver
(liv'ər) A large, complex organ weighing over 3 pounds, lying just under the diaphragm, that performs over 500 functions in the body, including aiding in the digestion of lipids. p. 175

loop of Henle
(lo͞op uv hen'lē) (after F.G.J. Henle, German anatomist) The descending and ascending loops of the renal tubule. p. 237

lumen, pl. lumina
(lo͞o'mən, lo͞o'mə nə) (L. light) The hollow core in the three layers of tissue that makes up the walls of the arteries through which blood flows. p. 201

luteinizing hormone (LH)
(lo͞o'tē ən īz'ing hôr'mōn) A gonadotropic hormone secreted by the anterior pituitary that stimulates the release of an egg in females and

stimulating the production of testosterone in males. p. 320

lymph
(limf) (L. lympha, clear water) A tissue fluid diffused out of the blood through the capillaries. p. 211

lymph node
(limf nōd) A small, ovoid "spongy" structure located in various places of the body along the route of the lymphatic vessels filtering the lymph as it passes through. p. 211

lymphatic system
(lim fat′ik sis′təm) The system of one-way, blind-ended vessels that collects and returns to the blood the approximately 10% of the fluid that does not return to the blood directly. p. 211

lysogenic cycle
(lī′sə jen′ik sī′kəl) (Gr. lysis, dissolution + gennan, to produce) A pattern of viral replication in which a virus integrates its genetic material with that of a host and is replicated each time the host cell replicates. p. 489

lysosome
(lī′sə sōm) (Gr. lysis, a loosening + soma, a body) A membrane-bounded vesicle containing digestive enzymes that break down old cell parts or materials brought into the cell from the environment and are extremely important to the health of a cell. p. 56

lytic cycle
(lit′ik sī′kəl) (Gr. lysis, dissolution) A pattern of viral replication in which a virus enters a cell, replicates, and then causes the cell to burst, releasing new viruses. p. 489

macromolecule
(mak′rə mol′ə kyo͞ol) (Gr. makros, large + L. moliculus, a little mass) A large organic molecule having many functional groups. p. 28

macrophage
(mak′rō fāj) (Gr. makros, long + -phage, eat) A phagocytic cell in the bloodstream that engulfs foreign bacteria and antibody-coated cells or particles in the process of phagocytosis; macrophages act as the body's scavengers. p. 152

malpighian tubule
(mal pig′ē ən to͞o′byo͞ol) (after Marcello Malpighi, Italian anatomist) A slender projection from the digestive tract in the body of a terrestrial arthropod that serves as an excretory system. p. 576

mammal
(mam′əl) (L. mamma, breast) A warm-blooded vertebrate that has hair in which the female secretes milk from mammary glands to feed her young. p. 468

mantle
(mant′əl) In the body of a mollusk, a soft epithelium that arises from the dorsal body wall and encloses a cavity between itself and the visceral mass. p. 571

marsupial
(mär so͞o′pē əl) (L. marsupium, pouch) A subclass of mammals that give birth to immature young that are carried in a pouch; a kangaroo is a marsupial. p. 471

medulla
(mə dul′ə) (L. marrow) The lowest portion of the brainstem, continuous with the spinal cord below; the site of neuron tracts, which cross over one another delivering sensory information from the right side of the body to the left side of the brain and vice versa. p. 235

medusa, pl. medusae
(mə do͞o′sə *or* mə do͞o′zə, mə do͞o′sē *or* mə do͞o′zē) (Gr. Medousa, a Gorgon slain by Perseus, who gave her head to Athena) A cnidarian that is a free-floating, umbrella-shaped animal. p. 567

megakaryocyte
(meg′ə kar′ē ə sīt) (Gr. megas, large + karyon, nucleus + kytos, cell) A large bone marrow cell that pinches off bits of its cytoplasm, resulting in a cell fragment, or platelet.

meiosis
(mī ō′ səs) (Gr. meioun, to make smaller) The two-staged process of nuclear division in which the number of chromosomes in cells is halved during gamete formation. p. 83

meninges, sing. meninx
(mə nin′jēz, mē′ningks) (Gr. membrane) Any of three layers of membranes covering both the brain and spinal cord. p. 272

menopause
(men′ə pôz) (L. mens, month + pausis, cessation) The period that marks the permanent cessation of menstrual activity in a woman, usually between the ages of 50 and 55; the end of the menses. p. 343

menstruation
(men′stro͞o ā′shən) (L. mens, month) The monthly sloughing off of the blood-enriched lining of the uterus when pregnancy does not occur; the lining degenerates and causes a flow of blood, tissue, and mucus from the uterus out the vagina. p. 342

meristematic tissue
(mer′ə stə mat′ik tish′o͞o) (Gr. merizein, to divide) An undifferentiated type of tissue in a vascular plant that produces new plant cells during growth; also called meristem. p. 546

mesoderm
(mez′ə durm′) (Gr. mesos, middle + derma, skin) The layer of cells in the developing embryo that differentiates into the skeleton, muscles, blood, reproductive organs, connective tissue, and the innermost layer of the skin. p. 568

mesophyll
(mez′ə fil′) (Gr. mesos, middle + phyllon, leaf) The parenchymal layer of a leaf, bounded by epidermis, where photosynthesis takes place. p. 554

messenger RNA (mRNA)
(mes′ən jər är′en′ā′) A type of RNA that brings information from the DNA within the nucleus to the ribosomes in the cytoplasm to direct which polypeptide is assembled. p. 413

metabolism
(mə tab′ə liz′əm) (Gr. metabole, change) All the chemical reactions that take place within a living organism. p. 121

metaphase
(met′ə fāz) (Gr. meta, middle + phasis, form) The stage of mitosis characterized by the alignment of the chromosomes in a ring, equidistant from the two poles of the cell. p. 89

microfilament
(mī′krō fil′ə mənt) (Gr. mikros, small + L. filum, a thread) A thin, twisted double-chain fiber of protein within the cytoskeleton that helps support and shape eukaryotic cells. p. 50

microtubule
(mī′krō to͞o′byo͞ol) (Gr. mikros, small + tubulus, little pipe) A spiral array composed of protein subunits within the cytoskeleton that provide intracellular support in the nondividing cell. p. 50

microvillus, pl. microvilli
(mī′krō vil′əs, mī krō vil′ī) (Gr. mikros, small + L. villus, tuft of hair) A microscopic, cytoplasmic projection that covers the epithelial cells of the villi on their exposed surfaces. p. 178

midbrain
(mid′brān′) The top part of the brainstem; it contains nerve tracts connecting the upper and lower parts of the brain and nuclei that acts as reflex centers for movement. p. 271

middle ear
(mid′əl ir) The middle portion of the ear containing three bones that act together like an amplifier to increase the force of sound vibrations. p. 289

migration
(mī grā′shən) Long-range, two-way movements by animals, often occurring yearly with the change of seasons. p. 603

mineral
(min′ə rəl) (L. minerale) An inorganic substance transported around as ions dissolved in blood and other body fluids; a variety of minerals perform a variety of functions in the human body. p. 166

mitochondrion, pl. mitochondria
(mīt′ə kon′drē ən, mīt′ə kon′drē ə) (Gr. mitos, thread + chondrion, small grain) An oval, sausage-shaped, or thread-like organelle about the size of a bacterium, bounded by a double membrane whose function is to break down fuel molecules, thus releasing energy for cell work. p. 58

mitosis
(mī tō′səs *or* mi tō′səs) (Gr. mitos, thread) A process of cell division that produces two identical cells from an original parent cell.

molecule
(môl′ə kyo͞ol′) (L. molecula, little mass) A combination of tightly bound atoms. p. 83

Monera
(mə ner′ə) (Gr. moneres, individual) The kingdom that consists of the bacteria. p. 491

monocot
(mon′ə kot) An angiosperm in which the food reserves within its seeds is stored in extraembryonic tissue called endosperm; a monocot usually has parallel veins in its leaves and its flowers parts are often in threes. p. 538

monohybrid
(mon′ə hī′brəd) (Gr. monos, single + L. hybrida, mongrel) The progeny or product of two plants that differ from one another in a single trait. p. 374

monosaccharide
(mon′ə sak′ə rīd) (Gr. monos, one + sakcharon, sugar) A simple sugar. p. 30

monotreme
(mon′ə trēm′) (Gr. mono. single + treme, hole) A mammal that lays eggs having leathery shells similar to those of a reptile; the only extant monotremes are the duckbilled platypus and two genera of spiny anteaters. p. 471

morphogenesis
(môr′fə jen′ə səs) (Gr. morphe, form + genesis, origin) The early stage of development in a vertebrate when cells begin to move, or migrate, thus shaping the new individual. p. 357

motor area
(mōt′ər er′ē ə) The part of the brain straddling the rearmost portion of the frontal lobe that sends messages to move the skeletal muscles. p. 266

motor neuron
(mōt′ər no͞or′on) A neuron of the peripheral nervous system that transmits commands away from the central nervous system. p. 251

multiple alleles
(mul′tə pəl a lēlz′) A system of alleles in a gene that exhibits either complete dominance or codominance. p. 400

muscle fiber
(mus′əl fī′bər) A long, multinucleated cell packed with organized arrangements of microfilaments capable of contraction. p. 305

muscle tissue
(mus′əl tish′o͞o) Any of three different kinds of muscle cells—smooth, skeletal, or cardiac—that are the workhorses of the body; characterized by an abundance of special thick and thin microfilaments. p. 147

mutation
(myo͞o tā′shən) (L. mutare, to change) A permanent change in the genetic material. p. 649

mutualism
(myo͞och′ə wə liz′əm) (L. mutuus, lent, borrowed) A type of symbiosis in which both participating species benefit.

mycelium, pl. mycelia
(mī sē′lē əm, mī sē′lē ə) (Gr. mykes, fungus) In fungi, a mass of hyphae. p. 516

myelin sheath
(mī′ə lən shēth) (Gr. myelinos, full of marrow) The fatty wrapping created by mutliple layers of Schwann cell membranes; the myelin sheath insulates the axon. p. 252

myofibril
(mī′ə fī′brəl) (Gr. myos, muscle + L. fibrilla, little fiber) A cylindrical, organized arrangement of special thick and thin microfilaments capable of shortening a muscle fiber. p. 305

myosin
(mī′ə sin) (Gr. mys, muscle + in, belonging to) One of the two protein components of myofilaments in a muscle fiber; actin is the other. p. 307

natural selection
(nach′ə rəl sə lek′shən) The process in which organisms having adaptive traits survive in greater numbers than organisms without such traits. p. 432

Neanderthal
(nē an′dər thôl′ *or* nē an′dər täl′) (Neander, valley in western Germany) A subspecies of *H. sapiens* that lived from about 125,000 to 35,000 years ago in Europe and the Middle East; Neanderthals were short and powerfully built, with large brains; they made diverse tools, took care of the sick and injured, and buried their dead. p. 480

negative feedback
(neg′ət iv fēd′bak′) The process by which enzyme activity is regulated by inhibitors. p. 108

nephridium, pl. nephridia
(ni frid′ē əm, ni frid′ē ə) (Gr. nephros, kidney) In the body of a mollusk, a tubular structure through which wastes are removed. p. 573

nephron
(nef′ron) (Gr. nephros, kidney) Any of the millions of microscopic tubular units of the kidney where urine is formed. p. 234

nerve
(nurv) A cluster of axons and dendrites surrounded by numerous supporting cells. p. 269

nerve cord
(nurv kôrd) A single, hollow cord along the back that is a principal feature of chordates; in vertebrates, the nerve cord differentiates into a brain and spinal cord. p. 464

nerve impulse
(nurv im′pəls) A rapid electrical signal of a neuron that reports information or initiates a quick response in specific tissues. p. 250

neuromuscular junction
(nyo͞or′ō mus′kyə lər jungk′shən) A synapse between a neuron and a skeletal muscle cell. p. 259

neuron
(no͞or′on) (Gr. nerve) A nerve cell specialized to conduct an electric current. p. 159

neurotransmitter
(no͞or′ō trans′mit ər) (Gr. neuron, nerve + L. trans, across + mitere, to send) A chemical released when a nerve impulse reaches the axon tip of a nerve cell. p. 257

neurulation
(no͞or′ə lā′shən) (Gr. neuron, nerve) The development of a hollow nerve cord, which later develops into the central nervous system. p. 363

neutron
(no͞o′tron) (L. neuter, neither) A subatomic particle found at the nucleus of atom, similar to a proton in mass but neutral and carrying no charge. p. 17

neutrophil
(no͞o′trə fil) (L. neuter, neither + Gr. philein, to love) A type of granulocyte that migrates to the site of an injury and sticks to the interior walls of blood vessels, where it forms projections and phagocytizes microorganisms and other foreign particles. p. 210

niche
(nich) (L. nidus, nest) The special role each organism plays within an ecosystem;

the terms of space, food, temperature, mating conditions, moisture requirements, and behavior of an organism within an ecosystem. p. 639

nitrifying bacteria
(nī'trə fī'ing bak tir'ē ə) (Gr. nitron, salt) Bacteria that live in nodules in the roots of legumes and cycle nitrogen in ecosystems by converting ammonia to nitrates, a form of nitrogen used by plants. p. 493

node of Ranvier
(nōd uv räN vyā') (after L. A. Ranvier, French histologist) An uninsulated spot between two Schwann cells. p. 252

noncyclic electron flow
(non sik'lik i lek'trōn flō) A system that harvests energy in a process that involves photosystems I and II working together to produce NAPH and ATP; these two products power the generation of glucose from carbon dioxide in light-independent reactions. p. 137

nondisjunction
(non dis jungk'shən) The failure of homologous chromosomes to separate after synapsis, resulting in gametes with abnormal numbers of chromosomes. p. 389

nonspecific defense
(non spə sif'ik di fens') A set of defenses that the body uses to act against foreign invaders; they include the skin and mucous membranes, chemicals that kill bacteria, and the inflammatory process. p. 217

notochord
(nō'tə kôrd) (Gr. noto, back + L. chorda, cord) A structure that forms the midline axis along which the vertebral column (backbone) develops in all vertebrate animals. p. 361

nuclear envelope
(nōō'klē ər en'və lōp) The outer, double membrane surrounding the surface of the nucleus of a eukaryotic cell. p. 56

nucleic acid
(nōō klē'ik *or* nōō klā'ik as'əd) A long polymer of repeating subunits called nucleotides; the two types of nucleic acid within cells are deoxyribonucleic acid (DNA) and ribonucleic acid (RNA). p. 39

nucleolus, pl. nucleoli
(nōō klē'ə ləs, nōō klē'ə lī') (L. a small nucleus) The site within the nucleus of ribosomal RNA synthesis; consists of ribosomal RNA plus some ribosomal proteins. p. 57

nucleotide
(nōō'klē ə tīd) A single unit of nucleic acid consisting of a five-carbon sugar, a phosphate group, and an organic nitrogen-containing molecule, or base. p. 39

nucleus
(nōō'klē əs) (L. a kernal, dim. fr. nux, nut) (1) The central core of an atom containing protons and neutrons. (2) The double membrane vesicle of a eukaryotic cell that contains the hereditary material, or DNA. p. 267

nutrient
(nōō'trē ənt) (L. nutritio, nourish) A raw material of food; the six classes of nutrients are carbohydrates, fats, proteins, vitamins, minerals, and water. p. 166

occipital lobe
(ok sip'ə təl lōb) The section of the cerebral cortex in each hemisphere of the brain having to do with vision, with different sites corresponding to different positions on the retina. p. 266

omnivore
(om'nə vôr) (L. omnis, all + vorare, to eat) An organism that eats both plant and animal foods; human beings are omnivores p. 170

oogenesis
(ō'ə jen'ə səs)(Gr. oon, egg + genesis, generation, birth) The process of meiosis and development that produces mature female sex cells, or eggs. p. 334

operator
(op'ə rāt'ər) In bacteria and viruses, a gene that acts as an on or off switch for the transcription of the genes. p. 420

operculum
(ō pur'kyə ləm) (L. operire, to shut, cover) A flap in a bony fish that protects the gills and enhances water flow over the gills, thus bringing more oxygen in contact with their gas-exchanging surfaces and allowing a fish to breath while stationary. p. 586

operon
(op'ər on) (L. operis, work) In bacteria and viruses, a cluster of genes having related functions that are regulated as a unit. p. 420

optic nerve
(op'tik nurv) The nerve carrying impulses for the sense of sight. p. 288

organ
(ôr'gən) (L. organon, tool) Grouped tissues that form a structural and functional unit. p. 147

organ of Corti
(ôr'gən uv kôrt'ē) (after Alfonso Corti, Italian anatomist) The organ of hearing; the collective term for the hair cells, the supporting cells of the basilar membrane, and the overhanging tectorial membrane. p. 290

organ system
(ôr'gən sis'təm) A group of organs that function together to carry out the principal activities of the organism. p. 149

organelle
(ôr'gə nel') (Gr. organella, little tool) Any of a number of highly specialized, membrane-bound, intracellular structures within a cell that perform specific cellular functions; a feature common to most eukaryotes but lacking in bacteria. p. 12

origin
(ôr'ə jən) (L. oriri, to arise) The attachment of one end of a skeletal muscle to a stationary bone. p. 305

osmoregulation
(oz'mō reg'yə lā'shən *or* os'mō reg'yə lā'shən) The control of water movement in and out of an organism's body. p. 585

osmosis
(oz mō'səs *or* os mō'səs) (Gr. osmos, impulse + osis, condition) A special form of diffusion in which water molecules move from an area of higher concentration to an area of lower concentration across a differentially permeable membrane. p. 72

osmotic pressure
(oz mot'ik *or* os mot'ik presh'ər) The increase that water pressure exerts on a cell as water molecules continue to diffuse into a cell. p. 73

osteoblast
(os'tē ə blast) (Gr. osteon, bone + blastos, cell) A cell that forms new bone. p. 296

osteocyte
(os'tē ə sīt) (Gr. osteon, bone + kytos, hollow vessel) A cell that produces bone. p. 153

otolith
(ō'tə lith) (Gr. otos, ear + lithos, stone) Small pebbles of calcium carbonate embedded in a layer of jelly-like material that is spread over the surface of ciliated and nonciliated cells within the saccule. p. 291

outer ear
(out'ər ir) The part of the ear that funnels sound waves in toward the eardrum; includes the flaps of skin on the outside of the head called ears. p. 289

oval window
(ō'vəl win'dō) The entrance to the inner ear. p. 289

ovary
(ōv'ə rē) (L. ovum, egg) *(1)* A female gonad, located in the pelvic cavity of an animal, where egg production occurs. *(2)* In flowering plants, a chamber at the base of the female pistil that completely encloses and protects the ovules. p. 339

oviparous
(ō vip'ər əs) (L. ovum, egg + parere, to

bring forth) A method of reproduction in which the female lays her eggs and the young hatch outside of the mother; birds and most reptiles are oviparous. p. 586

ovoviviparous
(ō′vō vī vip′ ər əs) (L. ovum, egg + vivus, alive + parere, to bring forth) A method of reproduction in which the female retains the fertilized eggs within her oviducts until the young hatch; the young receives its nutrition from the egg yolk and not from the mother. p. 586

ovulation
(ov′yə lā′ shən) (L. ovulum, little egg) The monthly process by which an egg is produced and released by the ovary. p. 340

ovule
(ō′vyo͞ol) (L. ovulum, little egg) A protective structure in which egg cells grow in a naked seed plant. p. 533

ovum, pl. ova
(ō′ vəm, ō′və) (L. egg) A mature egg cell, p. 340, 355

oxidation
(ok′sə dā′shən) (Fr. oxider, to oxidize) The loss of an electron by an atom or a molecule. p. 119

oxidation-reduction
(ok′sə dā′shən ri duk′shən) In some chemical reactions, the passing of electrons from one atom or molecule to another; critically important to the flow of energy through living systems and essential to the flow of energy in cellular respiration. p. 119

oxygen debt
(ok′si jən det) A term used to describe the oxygen needed to break down the lactic acid in the liver, delivered by the bloodstream from the muscles during strenuous exercise. p. 128

oxytocin
(ok′si tō′sən) (Gr. oxys, sharp + tokos, birth) A hormone produced by the hypothalamus but stored and released in the posterior lobe of the pituitary that affects the contraction of the uterus during childbirth and stimulates the mammary glands, allowing a new mother to nurse her child. p. 321

ozone
(ō′zōn) (Gr. ozein, to smell) A principal chemical air pollutant, (O_3), formed by photochemical reactions on hydrocarbons and nitrogen oxides in the air, that is extremely irritating to the eyes and upper respiratory tract. p. 707

pancreas
(pang′krē əs *or* pan′krē əs) (Gr. pan, all + kreas, flesh) A long gland that lies beneath the stomach and is surrounded on one side by the curve of the duodenum; it secretes a number of digestive enzymes and the hormones insulin and glucagon. p. 195

parasite
(par′ə sīt′) (Gr. para, beside + sitos, food) The organism in a parasitic relationship that is the beneficiary; the host organism is harmed. p. 516

parasitism
(par′ə sə tiz′əm) (Gr. para, beside + sitos, food) A type of symbiosis in which one species benefits but the other (the host) is harmed. p. 649

parasympathetic system
(par′ə sim′pə thet′ik sis′təm) (Gr. para, beside + syn, with + pathos, feeling) A subdivision of the autonomic nervous system that generally stimulates the activities of normal internal body functions and inhibits alarm responses; opposite of sympathetic nervous system. p. 276

parathyroid gland
(par′ə thī′roid gland) (Gr. para, beside + thyreos, shield + eidos, form, shape) One of four small glands embedded in the posterior side of the thyroid that produces parathyroid hormone. p. 325

parathyroid hormone (PTH)
(par′ə thī′roid hôr′mōn) (Gr. para, beside + thyreos, shield + eidos, form, shape) A hormone secreted by the parathyroid glands that works antagonistically to calcitonin to help maintain the proper blood levels of various ions, primarily calcium. p. 322

parenchyma cell
(pə reng′kə mə sel) (Gr. para, beside + en, in + chein, to pour) A cell in the ground tissue of a vascular plant that functions in photosynthesis and storage; the most common of all plant cell types. p. 547

parietal lobe
(pə rī′ə təl lōb) The section of the cerebral cortex of each hemisphere of the brain containing sensory receptors from different parts of the body. p. 266

passive immunity
(pas′iv i myo͞on′ə tē) The type of immunity produced by injection of antibodies into the subject to be protected or acquired by the fetus through the placenta. p. 227

passive transport
(pas′iv trans′pôrt) Molecular movement down a gradient but across a cell membrane; the three types of passive transport are diffusion, osmosis, and facilitated diffusion. p. 72

pectoral girdle
(pek′tə rəl gurd′əl) (L. pectus, chest) The part of the appendicular skeleton made up of two pairs of bones: the clavicles, or collarbones, and the scapulae, or shoulder blades. p. 300

pelvic girdle
(pel′vik gurd′əl) (L. pelvis, basin) The part of the appendicular skeleton made up of the two bones called coxal bones, pelvic bones or hip bones. p. 300

penis
(pē′nəs) A cylindrical organ that transfers sperm from the male reproductive tract to the female reproductive tract; the male urinary organ. p. 338

pepsin
(pep′sən) (Gr. pepsis, digestion) An enzyme of the stomach that digests only proteins, breaking them down into short peptides. p. 174

peptide bond
(pep′tīd bänd) A covalent bond that links two amino acids formed during dehydration synthesis when the amino group at one end and the carboxyl group at the other end lose a molecule of water between them. p. 34

pericycle
(per′ə sī′ kəl) (Gr. peri, around + kykos, circle) In the roots of vascular plants, a cylinder of cells made up of parenchymal cells that are able to undergo cell division to produce either branch roots or new vascular tissue. p. 549

peripheral nervous system
(pə rif′ə rəl nurv′əs sis′təm) (Gr. peripherein, to carry around) The part of the nervous system made up of the nerves of the body that bring messages to and from the brain and spinal cord. p. 264

peristalsis
(per′ə stôl′səs) (Gr. peri, around + stellein, to wrap) The rhythmic wave of contractions by the muscles of the esophagus that moves food down toward the stomach. p. 172

permafrost
(pur′mə frōst′) (Perma[nent] + frost) A layer of permanently frozen subsoil, found about 1 meter down from the surface, that is common throughout most of the Arctic regions. p. 684

petal
(pet′əl) (Gr. petalon, leaf) A part of a flower, frequently prominent and usually colorful, that serves to attract pollinators, usually insects. p. 534

petiole
(pet′ē ōl) (L. petiolus, a little foot) The slender stalk of a leaf. p. 554

pH scale
(pē′āch′ skāl) A scale that indicates the relative concentration of H^+ ions in a solution. Low pH values indicate high concentrations of H^+ ions (acids), and

high pH values indicate low concentrations. p. 25

phagocytosis
(fag′ə sī tō′səs) (Gr. phagein, to eat + kytos, hollow vessel) A type of endocytosis in which a cell ingests an organism or some other fragment of organic matter; macrophages and neutrophils are phagocytes. p. 77

pharyngeal (gill) slit
(fə rin′jē əl *or* far′in jē′əl) (Gr. pharynx, gullet) A principal feature of chordates that develops into the gill structure of a fish and into the ear, jaw, and throat structures of a terrestrial vertebrate. p. 464

pharynx
(far′ingks) (Gr. gullet) The upper part of the throat that extends from behind the nasal cavities to the openings of the esophagus and layrnx. p. 172

phenotype
(fē′nə tīp) (Gr. phainein, to show + typos, print) The outward appearance or expression of an organism's genes. p. 378

phloem
(flō′em) (Gr. phloos, bark) A type of vascular tissue that conducts carbohydrates a plant uses as food as well as other needed substances. p. 546

phospholipid
(fos′fō lip′id) (Gr. phosphoros, light-bearer + lipos, fat) A molecule made up of a portion of a fat molecule with a phosphate functional group attached. Because these molecules have polar and nonpolar parts, they form a double layer of molecules (a bilayer) when in a watery environment. A phospholipid bilayer is the foundation of cell membranes. p. 68

photoautotroph
(fōt′ō ôt′ə trōf) (Gr. photos, light + auto, self + trophos, feeder) A eubacterium that makes its own food by photosynthesis using the energy of the sun. p. 492

photocenter
(fōt′ō sint′ər) An array of pigment molecules within the thylakoid membranes of a chloroplast. It acts like a light antenna, capturing and directing photon energy toward a single molecule of a chlorophyll *a* that will participate in photosynthesis. p. 136

photochemical smog
(fōt′ō kem′i kəl smôg) (Gr. photos, light + chemeia, chemistry) Smog that is caused by pollutants reacting in the presence of sunlight. p. 707

photon
(fō′ton) (Gr. photos, light) A discrete packet of energy from sunlight. p. 132

photosynthesis
(fōt′ō sin′thə səs) (Gr. photos, light -syn, together + tithenai, to place) The process whereby energy from the sun is captured by living organisms and used to produce molecules of food; it takes place in the chloroplasts of photosynthetic eukaryotic cells.

Photosystem I
(fōt′ō sis′təm wun) The first photosynthetic pathway to evolve; it is used by most bacteria to produce ATP using energy from the sun. p. 137

Photosystem II
(fōt′ō sis′təm to͞o) A photosynthetic pathway in which a molecule of chlorophyll *a* (P680) can absorb more photons of higher energy than in Photosystem I. p. 137

phylum, pl. phyla
(fī′ləm, fī′lə) (Gr. phylon, race, tribe) A major taxonomic group, ranking above a class. p. 13

pineal gland
(pin′ē əl gland) (L. pinus, pine tree) A tiny gland lying deep within the brain whose exact function remains a mystery; it is the possible site of an individual's biological clock. p. 328

pinna
(pin′ə) (L. feather) The projected part of the outer ear, the ear flap. p. 289

pinocytosis
(pi′nō sī tō′səs) (Gr. pinein, to drink + kytos, vessel) A type of endocytosis in which a cell ingests liquid material containing dissolved molecules. p. 77

pistil
(pist′əl) (L. pistillum, pestle) The female sex organ at the center of a flower. p. 534

pith
(pith) A parenchymal (storage) tissue located in the center of the roots of a monocot. p. 549

pituitary
(pə to͞o′ə ter′ē) (L. pituita, phlegm) A tiny gland hanging from the underside of the brain, under the control of the hypothalamus, that secretes nine different major hormones. p. 318

placenta
(plə sen′tə) (L. a flat cake) A flat disk of tissue that grows into the uterine wall, through which the mother supplies the offspring with food, water, and oxygen and through which she removes wastes. p. 360

placental mammal
(plə sent′əl mam′əl) (L. a flat cake) A mammal that nourishes its developing embryo with the body of the mother by means of a placenta until development is almost complete. p. 471

plankton
(plangk′tən) (Gr. planktos, wandering) Small organisms that drift in the water, especially at or near the surface. p. 662

plasma
(plaz′mə) (Gr. form) The fluid intercellular matrix within which blood cells float; contains practically every substance used and discarded by cells as well as nutrients, hormones, proteins, salts, ions, and albumin. p. 157

plasma cell
(plaz′mə sel) Any of several different kinds of cells and cell parts suspended within plasma, including erythrocytes, leukocytes, and platelets. p. 224

plasma membrane
(plaz′mə mem′brān) A thin, nonrigid structure that encloses the cell and regulates interactions between the cell and its environment. p. 48

plasmid
(plaz′mid) (Gr. plasma, form) A small fragment of DNA that replicates independently of the main chromosome. p. 63

plasmodium, pl. plasmodia
(plaz mō′dē əm, plaz mo′dē ə) (L. L. plasma, form + Gr. eidos, shape) A nonwalled, multinucleate mass of cytoplasm in a slime mold. p. 515

plate tectonics theory
(plāt tek ton′iks thir′ē *or* thē′ə rē) The theory that the outer shell of the Earth is composed of six large and several smaller plates that are rigid pieces of the Earth's crust; the theory provides insight into the movement of land masses over time. p. 458

platelet
(plāt′lət) (Gr. dim of plattus, flat) A cell fragment present in blood that plays an important role in the clotting of blood. p. 211

predator
(pred′ə tər) (L. praeda, prey) An organism of one species that kills and eats organisms of another. p. 642

prey
(prā) (L. prehendere, to grasp or seize) An organism killed and eaten by a predator. p. 642

primary consumer
(prī′mer ē kən so͞om′ər) A herbivore that feeds directly on green plants. p. 658

primate
(prī′māt or prī′mət) (L. primus, first) A mammal that has characteristics reflecting a tree-dwelling life-style, such as hands and feet able to grasp things, flexible limbs, and a flexible spine. p. 471

producer
(prə do͞os′ər) An organism in an

ecosystem capable of capturing energy from the sun and converting it to chemical energy usable to themselves and consumers. p. 10

prokaryotic cell
(prō kar′ē ot′ik sel) (Gr. pro, before + karyon, kernel) A cell smaller than a eukaryote, it has a simple interior organization with a single, circular strand of hereditary material that is not enclosed with a membrane; bacteria and the blue-green bacteria are prokaryotes. p. 47

prolactin
(prō lak′tən) (L. pro, before + lac, milk) A hormone secreted by the anterior pituitary that in association with estrogen, progesterone, and other hormones, stimulates the mammary glands in the breasts to secrete milk after a woman has given birth to a child. p. 321

prophase
(prō fāz′) (Gr. pro, before + phasis, form) The stage of mitosis characterized by the appearance of visible chromosomes. p. 86

proprioceptor
(prō′prē ə sep′tər) (L. proprius, one's own + ceptor, a receiver) A receptor located within the skeletal muscles, tendons, and inner ear that gives the body information about the position of its parts relative to each other and to the pull of gravity. p. 282

prosimian
(prō sim′ē ən) (Gr. pro, before + L. simia, an ape) A suborder of primates that includes lemurs, indris, aye-ayes, and lorises; they are small animals, mostly nocturnal, with large ears and eyes, elongated snouts, and rear limbs. p. 473

prostaglandin
(pros′tə glan′dən) (from prosta[te] gland + -in) A hormone secreted by cell membranes throughout the body that stimulates smooth muscle contraction and the dilation and constriction of blood vessels. p. 330

pleura, pl. pleurae
(plo͝or′ə, plo͝or′ē) (Gr. side) A thin, delicate, sheet-like membrane that lines the interior walls of the thoracic cavity and folds back on itself to cover each lung; covered by a thin film of liquid, the pleura reduces friction during respiratory movements of the lungs. p. 191

pollen grain
(pol′ən grān) (L. fine dust) A male gametophyte produced by an anther within the male stamen of a flower. p. 534

pollination
(pol′ə nā′shən) The fertilization of the stigmas of flowers by the transfer of pollen from the anthers, usually accomplished by the wind, insects, or birds. p. 534

pollution
(pə lo͞o′shən) (L. polluere, to pollute) Contamination of the water sources and air of the Earth, causing physical and chemical changes that harm living and nonliving things. p. 703

polymer
(pol′ə mər) (Gr. polus, many + meris, part) A giant molecule formed of long chains of similar molecules. p. 28

polyp
(pol′əp) (Gr. polypous, many feet) A cnidarian that exists as an aquatic, cylindrical animal with a mouth at one end ringed with tentacles; it anchors itself to rocks. p. 567

polypeptide
(pol′ē pep′tīd) (Gr. polys, many + peptein, to digest) A long chain of amino acids linked end to end by peptide bonds; proteins are long, complex polypeptides, p. 34

polysaccharide
(pol′ē sak′ə rīd) (Gr. polys, many + sakcharon, sugar) A long polymer composed of insoluble sugar. p. 30

polyunsaturated
(pol′ē un sach′ə rā′təd) Referring to a fat composed of fatty acids that has more than one double bond; polyunsaturated fats have low melting points and are therefore liquid fats, or oils. p. 33

pons
(ponz) (L. bridge) The part of the brainstem consisting of a band of nerve fibers that acts as "bridges" and connects various parts of the brain to one another; it also brings messages to and from the spinal cord. p. 271

population
(pop′yə lā′shən) (L. populus, the people) A group that consists of the individuals of a given species that occur together at one place and at one time. p. 639

positive feedback loop
(poz′ət ɪv fēd′ bak′ lo͞op) A feedback loop in which the response of the regulating mechanism is positive with respect to the outcome. p. 163

postsynaptic membrane
(pōst′sə nap′tik mem′brān) The membrane of the target cell of the synaptic cleft. p. 257

potential energy
(pə ten′chəl en′ər jē) Stored energy; energy not actively doing work but having the capacity to do so. p. 103

predation
(pri dā′shən) (L. praeda, prey) The killing and eating of an organism of one species by an organism of another species; an animal that kills and eats members of its own species is a cannibal. p. 629

prostate
(pros′tāt) (Gr. prostates, one standing in front) A gland surrounded by the male urethra that adds a milky alkaline fluid to semen, neutralizing the acidity of the female vagina. p. 338

protein
(prō′tēn *or* prō′tē ən) (Gr. proteios, primary) A linear polymer of amino acids. p. 34

protist
(prō′tist) (Gr. protos, first) A member of the kingdom Protista, which includes unicellular eukaryotes as well as some multicellular forms. p. 503

protocell
(prōt′ō sel′) (Gr. protos, first + L. cella, chamber) An early precursor of a cell. p. 448

proton
(prō′tôn) A subatomic particle in the nucleus of an atom that has mass and carries a positive charge. p. 17

protosome
(prō′tə stōm′) (Gr. proto, first + stoma, mouth) A member of a branch of coelomate animals that during the gastrula stage of development, gives rise to a two-layered embryo; this first indentation becomes the mouth of the organism. p. 564

protozoan
(prō′tə zō′ən) (Gr. protos, first + zoon, animal) An animal-like heterotrophic protist that eats organic matter for energy. p. 503

proximal convoluted tubule
(prôk′sə məl kän′və lo͞o′təd to͞o′byo͞ol) A coiled portion of the nephron closest to Bowman's capsule lying in the cortex of the kidney, the site where reabsorption begins as the filtrate passes through the proximal tubule. p. 237

pseudocoelomate
(so͞od′ōsē′lə māt′) (Gr. pseudos, false + koiloma, a hollow) An organism, such as a roundworm, that has a false body cavity. p. 562

pseudopod, pl. pseudopodia
(so͞o′də pod, so͞o′də pō′dē ə) (Gr. pseudos, false + pous, foot) A temporary, pushed out cell part of an ameba used for food procurement and locomotion. p. 504

pseudostratified
(so͞od′ō strat′ə fīd′) (Gr. pseudos, false + L. stratificare, to arrange in layers) Refers to a type of epithelial

tissue that gives the false appearance of being stratified. p. 151

pulmonary circulation
(pul'mə ner'ē sur'kyə lā'shən) The part of the human circulatory system that circulates blood to and from the lungs. p. 204

Punnett square
(pun'ət skwar) (after Reginald C. Punnett, English geneticist) A diagram that visualizes the genotypes of progeny in simple Mendelian crosses and illustrates their expected ratios. p. 378

pupil
(pyo͞o'pəl) (L. pupilla, little doll) The opening in the center of the iris through which light passes. p. 288

pyruvate
(pī ro͞o'vāt) The three-carbon molecule left when glycolysis is completed and the beginning material of the citric acid cycle. p. 117

radial symmetry
(rād'ē əl sim'ə trē) (L. radius, a spoke of a wheel; Gr. symmetros, symmetry) Describing the distribution of an animal's body parts that emerge, or radiate, from a central point, much like spokes on a wheel. p. 562

radicle
(rad'ə kəl) (L. radicula, root) The young root that is usually the first portion of the embryo to emerge from the germinating seed. p. 538

radiolarian
(rād'ē ō lar'ē ən) (L. radius, ray) A kind of ameba that secretes a shell made of silica that is glass-like and delicate. p. 505

receptor
(ri sep'tər) A specialized cell component that detects stimuli; it provides the body with information about the internal environment, the position in space, and the external environment. p. 254

recessive
(ri ses'iv) The form of a trait that recedes or disappears entirely in a hybrid offspring. p. 375

rectum
(rek'təm) (L. straight) The lower part of the large intestine, which terminates at the anus. p. 180

red blood cell
(red blud sel) *See* erythrocyte.

red bone marrow
(red bōn mar'ō) The soft tissue that fills the spaces within the bony latticework of spongy bone; it is the place where most of the body's blood cells are formed. p. 297

reduction
(ri duk'shən) (L. reduction, a bringing back: originally "bringing back" a metal from its oxide) The gain of an electron by an atom or a molecule. p. 119

reflex
(rē'fleks) (L. reflectare, to bend back) An automatic response to a nerve stimulation. p. 264

reflex arc
(rē'fleks ärk) The pathway of nervous activity in a reflex. p. 271

refractory period
(ri frak'tə rē pir'ē əd) The recovery period after membrane depolarization during which the membrane is unable to respond to additional stimulation. p. 255

releasing hormone
(ri lēs'ing hôr'mōn) A hormone produced by the hypothalamus that affects the secretion of specific hormones from the anterior pituitary. p. 318

replication fork
(rep'lə kā'shən fôrk) The split in a DNA molecule where the double-stranded DNA molecule separates during DNA replication. p. 410

respiration
(res'pə rā'shən) (L. respirare, to breathe) The uptake of oxygen and the release of carbon dioxide by the body. Cellular, internal, and external respiration are all part of the general process of respiration. p. 185

respiratory assembly
(res'pə rə tōr'ē ə sem'blē) A special channel that allows protons to pass back from the outer compartment of a mitochondrion into the inner compartment; each passage of a proton back into the inner compartment through a respiratory assembly is coupled to the synthesis of an ATP molecule. p. 125

response
(ri spons') (L. respondere, to reply) The reply or reaction of an individual organism to a stimulus; it can be an innate or inborn reflex or a learned response such as by operant or classical conditioning. p. 255

resting potential
(rest'ing pə ten'chəl) An electrical potential difference, or electrical charge, along the membrane of the resting neuron. p. 254

restriction enzyme
(ri strik'shən en'zīm) An enzyme that recognizes certain base sequences in a DNA strand and breaks the bonds at that point. p. 496

retina
(ret'ən ə) (L. a small net) The structure in the eye, composed of rod and cone cells, that is sensitive to light. p. 286

rhizoid
(rī'zoid) (Gr. rhiza, root) In a bryophyte, a slender, root-like projection that anchors the plant to its substrate; unlike roots, however, it consists of only a few cells and does not play a major role in the absorption of water or minerals. p. 557

rhizome
(rī'zōm) (Gr. rhizoma, mass of roots) An underground stem of a plant that can produce a new plant by the process of vegetative propagation. p. 540

rhodopsin
(rō dop'sən) (Gr. rhodon, rose + opsis, vision) A complex in the retina formed by the coupling of retinal and opsin. p. 286

ribonucleic acid (RNA)
(rī'bō no͞o klē'ik as'əd) One of two types of nucleic acid found in cells; differs from DNA in that its sugar is ribose and uracil is present rather than thymine. p. 406

ribosomal RNA (rRNA)
(rī'bə sō'məl är'en' ā') The type of RNA found in ribosomes that plays a role in the manufacture of polypeptides. p. 57

ribosome
(rī'bə sōm') A minute, round structure found in endoplasmic reticulum; ribosomes are the places where proteins are manufactured. p. 53

ribulose biphosphate (RuBP)
(rib'yə lōs' bī'fos'fāt) A molecule which joins with carbon dioxide in the first stage of the Calvin-Benson cycle and results in the process of carbon fixation. p. 142

RNA polymerase
(är'en'ā' pol'ə mə rās') The special enzyme that transcribes RNA from DNA. p. 413

rod
(räd) A light receptor located within the retina at the back of the eye that functions in dim light and detects white light only. p. 286

root
(ro͞ot *or* ro͝ot) The part of a plant usually found below ground; it absorbs water and minerals and anchors the plant. p. 545

root cap
(ro͞ot *or* ro͝ot kap) A thimble-like mass of cells that covers and protects the root's apical meristem as it grows through the soil. p. 550

root pressure
(ro͞ot *or* ro͝ot presh'ər) The process in which root cells, by means of special cellular pumps, maintain a higher concentration of dissolved minerals than the concentration of minerals in the water of the soil. p. 556

rough endoplasmic reticulum
(ruf en′dō plaz′mik ri tik′yə ləm) (Gr. endon, within + plasma, from cytoplasm; L. reticulum, network) The type of endoplasmic reticulum that makes and transports proteins destined to leave the cell; rough refers to the minute, round structures called ribosomes covering the surface where proteins are made. p. 53

round window
(round win′dō) A membrane-covered hole at the wider end of the cochlea. p. 289

SA node
(es′ā′ nōd) *See* sinoatrial (SA) node.

saccule
(sak′yo͞ol) (N.L. sacculus, small bag) A sac inside a bulge in the vestibule of the inner ear containing both ciliated and nonciliated cells. p. 291

saliva
(sə lī′və) (L. spittle) The secretion of the salivary glands; a solution consisting primarily of water, mucus, and the digestive enzyme salivary amylase. p. 170

salivary amylase
(sal′ə ver′ē am′ə lās) (L. salivarius, slimy + Gr. amylon, starch -asis, coloid enzyme) A digestive enzyme that breaks down starch into molecules of the disaccharide maltose. p. 170

salivary glands
(sal′ə ver′ē glandz) (L. salivarius, slimy + glans, acorn) The paired glands of the mouth that secrete saliva. p. 170

saltatory conduction
(sal′tə tōr′ē kən duk′shən) (L. saltatio, leaping) A very fast form of nerve impulse conduction in which impulses "jump" along myelinated neurons. p. 256

saprophytic
(sap′rō fit′ik) (Gr. sapros, putrid + phyton, plant) Feeding on dead or decaying organic material; fungi are saprophytic. p. 516

sarcomere
(sär′kə mir′) (Gr. sarx, flesh + meris, part of) The repeating bands of actin and myosin myofilaments that appear between two Z lines in a muscle fiber. p. 307

sarcoplasmic reticulum
(sär′kō plaz′mik ri tik′yə ləm) (Gr. sarx, flesh + plassein, to form, mold; L. reticulum, network) A tubular, branching latticework of endoplasmic reticulum that wraps around each myofibril like a sleeve. p. 310

Schwann cells
(shwän *or* shvän selz) (after Theodor Schwann, German anatomist) The supporting cells associated with nerve fibers of all the other cells that make up the peripheral nervous system. p. 252

scientific method
(sī′ən tif′ik meth′əd) A set of procedures used to answer questions that is common among the various scientific disciplines. p. 3

sclera
(sklir′ə) (Gr. skleros, hard) The tough outer layer of connective tissue that covers and protects the eye. p. 287

sclerenchyma cell
(sklə reng′kə mə sel) (Gr. skleros, hard + en, in + chymein, to pour) A dead, hollow cell with a strong wall found in the ground tissue of a vascular plant that helps support and strengthen the ground tissue. p. 547

scrotum
(skrō′təm) (L. a bag) A sac of skin, located outside the lower pelvic area of the male, which houses the testicles, or testes. p. 335

sea squirt
(sē skwurt) The common name given to most species of tunicates because of its habit of forcefully squirting water out of its excurrent siphons when disturbed. p. 581

secondary consumer
(sek′ən der′ ē kən so͞om′ər) A meat eater, or carnivore, that feeds on a herbivore. p. 658

seed
(sēd) A structure from which a new plant grows. p. 531

selective reabsorption
(sə lek′tiv rē′əb sôrp′shən) The process in which the kidneys select specific substances according to the body's needs from the filtrate. p. 235

semen
(sē′mən) (L. seed) Fluid produced by the accessory glands of the male reproductive system combined with sperm. p. 338

semicircular canal
(sem′i sur′kyə lər kə nal′) Any of three fluid-filled canals in the inner ear that detect the direction of an individual's movement. p. 291

seminal vesicle
(sem′ ən əl ves′ə kəl) One of two accessory glands that secrete a thick, clear fluid forming a part of the semen. p. 338

seminiferous tubule
(sem′ə nif′ər əs to͞o′byo͞ol) (L. semen, seed + ferre, to produce) A tightly coiled tube within a testis where sperm cells develop. p. 336

sensory neuron
(sens′ə rē no͞or′on) A neuron of the peripheral nervous system that transmits information to the central nervous system. p. 250, 264

sepal
(sē′pəl) (L. sepalum, a covering) A member of the outermost whorl, or ring, of a flowering plant. p. 534

septum, pl. septa
(sep′təm, sep′tə) (L. saeptum, a fence) (1) The tissue that separates the two sides of the heart. (2) In a fungus, a cross wall that divides a hypha into cells. (3) In an annelid, an internal partition that divides the segments. p. 516

seta, pl. setae
(sē′tə, sē′tē) (L. bristle) A bristle in the body of an earthworm that helps anchor the worm during locomotion or when it is in its burrow. p. 574

sex chromosome
(seks krō′mə sōm′) The X and Y chromosomes that determine the gender of an individual as well as certain other characteristics. p. 81

sexual reproduction
(seksh′ə wəl *or* sek′shəl rē prə duk′shən) The type of reproduction that involves the fusion of gametes to produce the first cell of a new individual. p. 83

sickle-cell anemia
(sik′əl sel′ ə nē′mē ə) (A.S. sicol, curved + L. cella, chamber + Gr. a, not + haima, blood) A recessive genetic disorder common to African blacks and their descendants in which affected individuals cannot transport oxygen to their tissues properly because the molecules within the red blood cells that carry oxygen, hemoglobin proteins, are defective. p. 398

simple eye
(sim′pəl ī) A structure composed of a single visual unit having one lens; found in some arthropods and some flying insects. p. 576

sinoatrial (SA) node
(sī′nō ā′trē əl nōd) A small cluster of specialized cardiac muscle cells embedded in the upper wall of the right atrium of the heart that automatically and rhythmically sends out impulses initiating each heartbeat. p. 206

sinus venosus
(sī′nəs və nō′səs) A chamber in the hearts of cartilaginous and bony fishes that collects blood from the organs. p. 586

sister chromatid
(sis′tər krō′mə tid) Either of two identical structures held together at the centromere, composed of chromatin material that coils and condenses just

before and during cell division to form chromosomes. p. 81

skeletal muscle
(skel′ət əl mus′əl) A type of muscle that is voluntary because of conscious control over its action; skeletal muscle is connected to bones and allows for body movement. p. 158

slime mold
(slīm mōld) A protist organism that is fungus-like in one phase of its life cycle and ameba-like in another phase of its life cycle. p. 504

small intestine
(smôl in tes′tən) The tube-like portion of the digestive tract that begins at the pyloric sphincter and ends at its T-shaped junction with the large intestine. p. 175

smooth endoplasmic reticulum
(smo͞oth en′dō plaz′mik ri tik′yə ləm) (Gr. endon, within + plasma, from cytoplasm; L. reticulum, network) The type of endoplasmic reticulum that helps build carbohydrates and lipids within the cytoplasm; it does not have ribosomes attached to its surface and does not manufacture proteins. p. 53

smooth muscle
(smo͞oth mus′əl) A type of muscle that contracts involuntarily and is located in the walls of certain internal structures such as blood vessels and the stomach. p. 158

sociobiology
(sō′sē ō bī ol′ə jē) The biology of social behavior that applies the knowledge of evolution to the study of animal behavior. p. 611

sodium-potassium pump
(sōd′ē əm pə tas′ē əm pump) The term given to the coupled channel that uses energy to move sodium (Na^+) and potassium (K^+) ions across the cell membrane. p. 75

soft palate
(sôft pal′ət) (L. palatum, palate) The tissue at the back of the roof of the mouth. p. 172

solute
(sol′yo͞ot) The other kinds of molecules dissolved in water. *See* solution, solvent. p. 73

solution
(sə lo͞o′shən) A mixture of molecules and ions dissolved in water. p. 73

solvent
(solv′ənt) The most common of the molecules in a solution, usually water. p. 73

somatic nervous system
(sō mat′ik nurv′əs sis′təm) (Gr. soma, body) The branch of the peripheral nervous system consisting of motor neurons that send messages to the skeletal muscles and control voluntary responses. p. 264

somite
(sō′mīt) (Gr. soma, body) A chunk of mesoderm that gives rise to most of the axial skeleton and most of the dermis of the body. p. 364

spawn
(späwn) (M.E. spawnen, to spread out) The depositing of sperm and eggs directly into water by males and females; a method of reproduction used by most fishes, amphibians, and shellfish. p. 584

speciation
(spē′shē ā′shən) The process by which new species are formed during the process of evolution. p. 433

species
(spē′shēz *or* spē′sēz) A group of related organisms that shares common characteristics and is able to interbreed and produce viable offspring. p. 13

specific defense
(spə sif′ik di fens′) The immune response of the body. p. 217

spermatid
(spur′mə tid) A haploid cell in the testes arising from a diploid cell called a spermatogonium. p. 336

spermatogenesis
(spur′mə tō jen′ə səs) (Gr. sperma, sperm, seed + gignesthai, to be born) The development of sperm cells within the coiled tubules of the testis triggered by follicle-stimulating hormones. p. 336

spinal cord
(spīn′əl kôrd) The part of the central nervous system that runs down the neck and back, receives information from the body, carries this information to the brain, and sends information from the brain to the body. p. 264

spinal nerve
(spīn′əl nurv) A nerve by which the spinal cord receives information from the body; includes sensory and motor nerves. p. 271

spindle fibers
(spin′dəl fī′bərz) Special microtubules that are assembled from a pair of related centrioles during the prophase stage of mitosis. p. 87

spiracle
(spir′ə kəl *or* spī′rə kəl) (L. spirare, to breathe) A specialized opening in the body of most terrestrial arthropods that allows air to pass into the trachea. p. 576

spleen
(splēn) (Gr. splen) An organ of the lymphatic system that stores an emergency blood supply and also contains white blood cells. p. 211

spongy bone
(spun′jē bōn) A type of bone in the human skeleton that is composed of an open latticework of thin plates of bone; spongy bone is found at the ends of long bones and within short, flat, and irregularly shaped bones. p. 156

spore
(spôr) (Gr. spora, seed) A reproductive body formed by meiosis in a diploid parent and by mitosis in a haploid parent; the spores are always haploid. p. 514

sporophyte
(spôr′ə fīt) (Gr. spora, seed + phyton, plant) In the life cycle of a plant, the diploid (spore-plant) generation that alternates with the haploid phase. p. 529

sporozoan
(spôr′ə zō′ən) (Gr. spora, seed + zoon, animal) A nonmotile protozoan that is a parasite of vertebrates, including humans; they are passed from host to host by various insect species. p. 509

squamous
(skwā′məs) (L. squama, scale) One of the main shapes of epithelial cells. Squamous cells are thin and flat and are found in the air sacs of the lungs, the lining of blood vessels, and the skin. p. 150

stamen
(stā′mən) (L. thread) The male sex organ of a flower. p. 534

starch
(stärch) (A.S. stercan) Stored energy in plants formed by using glucose to form polysaccharides. p. 30

stem
(stem) The framework of a plant that supports and positions the leaves where most photosynthesis occurs. p. 545

stereoscopic vision
(ster′ē ō skôp′ik vizh′ən) (Gr. stereos, solid + skopein, to view) Vision created by two eyes focusing on the same object. p. 472

stoma, pl. stomata
(stō′mə, stō′mə tə) (Gr. mouth) An opening in the epidermis of a leaf that allows the carbon dioxide needed for photosynthesis to enter and the oxygen produced by photosynthesis to escape; it also permits water vapor to escape. p. 554

stomach
(stum′ək) (Gr. stomachos, mouth) A muscular sac in which food is collected and partially digested by hydrochloric acid and proteases. p. 177

stratified
(strat′ə fīd′) (L. stratificare, to arrange in layers) Refers to epithelium that is made up of two or more layers. p. 151

stretch receptor
(strech ri sep′tər) A special nerve that is sensitive to any stretching in its tissue. p. 283

stroma, pl. stromata
(strō′mə, strō′mə tə) (Gr. anything spread out) The fluid that surrounds the thylakoids inside chloroplasts. p. 135

structural gene
(struk′chə rəl jēn) In bacteria and viruses, a gene that directs the synthesis of enzymes having related functions. p. 420

style
(stīl) (L. stilus, stylus) A narrow stalk arising from the top of the ovary that bears the stigma in a flowering plant. p. 534

succession
(sək sesh′ən) The dynamic process of change, during which a sequence of communities replaces one another in an orderly and predictable way. p. 651

suppressor T cell
(sə pres′ər tē sel) A kind of T cell that limits the immune response. p. 222

symbiosis
(sim′bē ō′səs *or* sim′bī ō′səs) (Gr. a living together) The living together of two or more organisms in a close association. p. 48

sympathetic nervous system
(sim′pə thet′ik nurv′əs sis′təm) A subdivision of the autonomic nervous system that generally mobilizes the body for greater activity; produces responses that are the opposite of the parasympathetic nervous system. p. 276

synapse
(sin′aps *or* sə naps′) (Gr. synapsis, a union) A junction between an axon tip and another cell, usually including a narrow gap separating the two cells. p. 257

synapsis
(sə nap′səs) (Gr. union) The lining up of homologous chromosomes during prophase I of meiosis, initiating the process of crossing-over. p. 93

synaptic cleft
(sə nap′tik kleft) The space or gap between two adjacent neurons. p. 257

synovial joint
(sənō′vē əl joint) (L. synovia, joint fluid) A freely movable joint in which a space exists between articulating bones. p. 302

systemic circulation
(sis tem′ik sur′kyə lā′shən) The pathway of blood vessels to the body regions and organs other than the lung. p. 205

systolic period
(sis tol′ik pir′ē əd) (Gr. systole, contraction) The pushing period of heart contraction, which ends with the closing of the aortic valve, during which a pulse of blood is forced into the systemic arterial system, immediately raising the blood pressure within these vessels. p. 207

T cell (T lymphocyte)
(tē sel) A type of lymphocyte that carries out the cell-mediated immune response. p. 222

T lymphocyte
(tē lim′fə sīt) *See* T cell.

taiga
(tī gä′) (Russ.) The northern coniferous forest extending across vast areas of Eurasia and North America; it is a marshy region south of the tundra that has long, cold winters. p. 682

tarsal
(tär′səl) (Gr. tarsos, a broad, flat surface) Any of the seven short bones that make up the ankle in the appendicular skeleton of a human. p. 302

taste bud
(tāst bud) A microscopic chemoreceptor embedded within the papillae of the tongue that works with the olfactory receptors to produce the taste sensation. p. 285

taxonomy
(tak sän′ə mē) (Gr. taxis, arrangement + nomos, law) The classification of the diverse array of species by categorizing organisms based on their common ancestry. p. 12

Tay-Sachs disease
(tā′saks′ diz ēz′) (after Warren Tay, British physician, and Bernard Sachs, U.S. neurologist) An incurable, fatal, recessive, hereditary disorder in which brain deterioration causes death by the age of 5 years; has a high incidence of occurrence among Jews of Eastern and Central Europe and among American Jews. p. 398

tectorial membrane
(tek tôr′ē əl mem′brān) (L. tectum, roof) The membrane that covers the hairs that stick up from the cochlear duct. p. 290

telophase
(tel′ə fāz) (Gr. telos, end + phasis, form) The stage of mitosis during which the mitotic apparatus assembled during prophase is disassembled, the nuclear envelope is reestablished, and the normal use of the genes present in the chromosomes is reinitiated. p. 90

template
(tem′plət) A pattern used as a guide to duplicate a shape or structure; refers to the ability of any polynucleotide, such as a RNA molecule, that can act as a guide for the synthesis of a second polynucleotide based on the complementarity of its bases. p. 447

temporal lobe
(tem′pə rəl lōb) (L. temporalis, the temples) The section of the cerebral cortex in each hemisphere of the brain dealing with hearing; different surface areas correspond to different tones and rhythms. p. 266

tendon
(ten′dən) (Gr. tenon, stretch) A tissue that connects muscles to bones. p. 305

teratogen
(tə rat′ə jən) (Gr. teratos, monster + gennan, to produce) An agent, such as alcohol, that can induce malformations in rapidly developing tissues and organs. p. 361

testcross
(test′krôs′) A cross between a phenotypically dominant test plant with a known homozygous recessive plant; devised by Mendel to further test his conclusions. p. 380

testis, pl. testes
(tes′təs, tes′tēz′) (L. witness) The male gonads where sperm production occurs. p. 335

testosterone
(tes′tos′tə rōn) (Gr. testis, testicle + steiras, barren) A sex hormone secreted by the testes responsible for the development and maintenance of male secondary sexual characteristics. p. 320

tetrad
(te′trad′) (Gr. tetras, four) During prophase I of meiosis I, the name given to the paired homologous chromosomes that together have four chromatids. p. 93

tetrapod
(tet′rə pod) (Gr. tetrapodos, four-footed) A land-living four-limbed vertebrate. p. 584

thalamus
(thal′ə məs) (Gr. thalamos, chamber) A mass of gray matter lying at the base of the cerebrum that receives sensory stimuli, interprets some of these stimuli, and sends the remaining sensory messages to appropriate locations in the cerebrum. p. 269

theory
(thē′ə rē or thir′ē) A synthesis of hypotheses that has withstood the test of time and is therefore a powerful concept that helps scientists made dependable predictions about the world. p. 7

thrombus
(throm′bəs) (Gr. clot) A blood clot that forms in a blood vessel and interferes with the flow of blood. p. 213

thylakoid
(thī'lə koid) (Gr. thylakos, sac + -oides, like) A flat, sac-like membrane in the chloroplast of a eukaryote; stacks of thylakoids are called grana. p. 60

thymus
(thī'məs) (Gr. thymos) A small gland located in the neck that plays an important role in the maturation of certain lymphocytes called T cells, which are an essential part of the immune system. p. 211

thyroid gland
(thi'roid gland) An important gland located in the neck near the voice box that produces hormones regulating the body's metabolism. p. 322

thyroid-stimulating hormone (TSH)
(thi'roid stim'yə lāt ing hôr'mōn) A tropic hormone produced by the anterior pituitary that triggers the thyroid gland to produce the three thyroid hormones. p. 320

thyroxine (T_4)
(thī rok'sēn *or* thī rok'sən) A hormone produced by the thyroid gland that helps regulate the body's metabolism. p. 322

tidal volume
(tīd'əl väl' yəm *or* väl'yum) The amount or volume of air inspired or expired with each breath. p. 192

tissue
(tish'o͞o) (L. texere, to weave) A member of a group of similar cells that works together to perform a function. p. 8

trachea
(trā'kē ə) (L. windpipe) The windpipe; the air passageway that runs down the neck in front of the esophagus and brings air to the lungs. p. 186

tracheid
(trākē əd) (Gr. tracheia, rough) A cell in the xylem that has holes along the length of its walls; it carries water and dissolved minerals through a plant and also provides support. p. 546

trait
(trāt) A distinguishing feature or characteristic of a plant or person due to the transmission of genes. p. 374

transcription
(trans krip'shən) (L. trans, across + scribere, to write) The first step in the process of polypeptide synthesis in which a gene is copied into a strand of messenger RNA. p. 413

transfer RNA (tRNA)
(trans'fur är'en'ā') A type of RNA that transports amino acids, used to build polypeptides, to the ribosomes; they also align each amino acid at the correct place on the elongating chain. p. 413

translation
(trans lā'shən) (L. trans, across + locare, to put or place) The second step of gene expression in which mRNA, using its copied DNA code, directs the synthesis of a polypeptide. p. 416

translocation
(trans'lō kā'shən) (L. trans, across + locare, to put or place) A situation in which a section of a chromosome breaks off and then reattaches to another chromosome, producing an abnormally long chromosome. p. 393

transpiration
(trans'pə rā'shən) (L. trans, across + spirare, to breathe) The process by which water leaves a leaf through a stoma in the form of water vapor. p. 555

trophic level
(trō'fik lev'əl) (Gr. trophos, feeder) A feeding level in an ecosystem. p. 659

tropomyosin
(trō'pə mī'ə sən *or* trop'ə mī'ə sən) (Gr. tropos, turn + myos, muscle) A muscle protein that is involved in the formation of cross-bridges during muscle contraction. p. 310

troponin
(trō'pə nən *or* trop'ə nən) (Gr. tropos, turn) A muscle protein that attaches to both actin and tropomyosin; it is concerned with calcium binding and inhibiting cross-bridge formation. p. 310

true breeding
(tro͞o brēd'ing) Said of plants that produce offspring consistently identical to the parent with respect to certain defined characteristics after generations of self-fertilization. p. 374

trypsin
(trip'sən) (Gr. tripsis, friction) An enzyme produced by the pancreas, which together with chymotrypsin and carboxypeptidase, completes the digestion of proteins. p. 177

tubular secretion
(to͞o'byo͞o lər sə krē'shən) The process by which the kidneys excrete a variety of potentially harmful substances from the blood. p. 235

tumor
(to͞o'mər) (L. swollen) An uncontrolled growth of a large mass of cells. p. 490

tundra
(tun'drə) (Russ.) A vast, uniform, virtually treeless region encircling the top of the world across North America and Eurasia that covers one-fifth of the Earth's land surface; it has long and cold winters, and the ground below 1 meter is frozen even in summer. p. 684

turgor pressure
(tur'gər presh'ər) (L. turgor, a swelling) Pressure inside a cell that results from water moving into the cell. p. 74

Turner syndrome
(tur' nər sin' drōm) (after H.H. Turner, U.S. physician) A genetic condition resulting from an XO zygote that develops into a human female who is sterile, of short stature, with a webbed neck, low-set ears, a broad chest, infantile sex organs, and low-normal mental abilities. p. 392

tympanic membrane
(tim pan'ik mem'brān) (Gr. tympanon, drum) The thin piece of fibrous connective tissue stretched over the opening to the middle ear; the eardrum. p. 289

ultrasound
(ul'trə sound') A noninvasive procedure that uses sound waves to produce an image of the fetus but that harms neither the mother nor the fetus; allows the fetus to be examined for major abnormalities. p. 401

umbilical artery and vein
(um bil'ə kəl ärt'ə rē and vān) (L. umbilicus, navel) The blood lines of the umbilical cord; the artery provides nourishment and sends wastes from the embryo to the placenta; the fetal blood then travels through the umbilical vein back to the embryo. p. 360

umbilical cord
(um bil'ə kəl körd) (L. umbilicus, navel) The attachment connecting the fetus with the placenta. p. 360

unsaturated
(un sach'ə rā'təd) Referring to a fat composed of fatty acids with double bonds that replace some of the hydrogen atoms. p. 33

urea
(yo͞o rē'ə *or* yo͞or'ē ə) (Gr. ouron, urine) The primary excretion product from the deamination of amino acids. p. 234

ureter
(yo͞o rē'tər *or* yo͞or'ə tər) (Gr. oureter) The tube that carries urine from the kidney to the bladder. p. 235

urethra
(yo͞o rē'thrə) (Gr. ourein, to urinate) A muscular tube that brings urine from the urinary bladder to the outside; in men, the urethra also carries semen to the outside of the body during ejaculation. p. 338

uric acid
(yo͞or'ik as'əd) A nitrogenous waste in the urine formed from the breakdown of nucleic acids (DNA and RNA) found in the cells of ingested food and from the metabolic turnover of bodily nucleic acids and ATP. p. 234

urinary bladder
(yo͞or'ə ner'ē blad'ər) A hollow muscular

organ that acts as a storage pouch for urine. p. 244

urine
(yo͞or'ən) (Gr. ouron, urine) The fluid produced by the kidneys made up of water and dissolved waste products. p. 234

urochordate
(yo͞or'ə kôrd'āt) (Gr. oura, tail + chorde, cord) A tunicate that has a larval notochord. p. 581

uterus
(yo͞o'tər əs) (L. womb) The organ in females in which a fertilized ovum can develop; the womb. p. 340

vaccination
(vak'sə nā'shən) (L. vaccinus, pert. to cows) An injection with disease-causing microbes or toxins that have been killed or changed in some way so as to be harmless; it causes the body to build up antibodies against a particular disease. p. 218

vacuole
(vak'yo͞o ōl') (L. vacuus, empty) A membrane-bounded storage sac within which such substances as water, food, and wastes can be found; most often found in plant cells, vacuoles play a major role in helping plant tissues stay rigid. p. 56

vagina
(və jīn'ə) (L. sheath) An organ in the body of a female whose muscular, tube-like passageway to the exterior has three functions; it accepts the penis during intercourse, it is the lower portion of the birth canal, and it provides an exit for the menstrual flow. p. 339

valve
(valv) (L. valva, leaf of a folding door) Any of the one-way valves found in the heart and blood vessels, similarly constructed, that prevent the backflow of blood. p. 206

vas deferens
(vas def'ə rənz) A long connecting tube that ascends from the epididymis into the pelvic cavity, looping over the side of the urinary bladder. p. 338

vascular bundle
(vas'kyə lər bund'əl) A cylinder of tissue made up of strands of xylem and phloem tissue positioned next to each other that constitute the transportation system of the stem of a plant. p. 551

vascular cambium
(vas'kyə lər kam'bē əm) A growth tissue in the woody stems of vascular plants that increases the diameter of a stem or root and gives rise to secondary xylem and secondary phloem. p. 552

vascular plant
(vas'kyə lər plant) A plant having a system of specialized vessels within itself that transports water and nutrients. p. 527

vascular tissue
(vas'kyə lər tish'o͞o) A differentiated tissue in vascular plants that conducts water, minerals, carbohydrates, and other substances throughout the plant; the principal types are xylem and phloem. p. 546

vector
(vek'tər) The infected genome of a virus having foreign DNA, which then carries it to a bacterial cell it is capable of infecting. p. 405

vein
(vān) (L. vena, a blood vessel) A small vein that collects blood from the capillary beds and brings it to larger veins. p. 201

ventricle
(ven'trə kəl) (L. ventriculus. belly) Either of two lower chambers of the heart; the right ventricle pumps blood into the pulmonary artery and then the lungs, whereas the left ventricle pumps blood through the aorta into the arteries. p. 204

venule
(ven'yo͞ol) (L. vena, vein) A small vein that starts at a capillary and connects it to a larger vein. p. 201

vertebral column
(ver'tə brəl kol'əm) The collection of 26 bones in the middle of the back, stacked one on top of the other, that acts like a strong, flexible rod and supports the head in a human skeleton; the spine or backbone. p. 300

vertebrate
(ver'tə brət *or* ver'tə brāt) (L. vertebra, vertebra) A member of a subphylum of chordates characterized by having a vertebral column surrounded by a dorsal nerve cord. p. 583

villus, pl. villi
(vil'əs, vil'ī) (L. a tuft of hair) Fine, finger-like projections that increase the surface absorption capability of the small intestine. p. 178

visceral mass
(vis'ə rəl mas) In a mollusk, a group of organs consisting of the digestive, excretory, and reproductive organs. p. 571

vitamin
(vīt'ə mən) (L. vita, life + amine, of chemical origin) An organic molecule that performs functions such as helping the body use the energy of carbohydrates, fats, and proteins; 13 different vitamins play a vital role in the human body. p. 166

viviparous
(vī vip'ər əs) (L. vivus, alive + parere, to bring forth) A method of reproduction in which the young is born alive and the developing embryo derives primary nourishment from the mother and not the egg. p. 586

vocal cord
(vō'kəl kôrd) Two pieces of elastic tissue covered with a mucous membrane stretched across the larynx that are involved in the production of sound. p. 186

vulva
(vul'və) (L. covering) The collective term for the external genitals of a female. p. 343

water mold
(wôt'ər mōld) A mold that thrives in moist places and aquatic environments, parasitizing plants and animals; during sexual reproduction, they produce large egg cells. p. 504

white blood cell
(hwīt *or* wīt blud sel) *See* leukocyte.

xylem
(zīləm) (Gr. xylon, wood) A principal type of vascular tissue that conducts water and dissolved minerals in a plant. p. 546

yellow marrow
(yel'ō mar'ō) A soft, fatty connective tissue that fills the hollow cylindrical core of long bones. p. 297

yolk
(yōk) (O.E. geolu, yellow) The nutrient material of an ovum that the developing individual can live on until nutrients can be derived from the mother. p. 355

yolk sac
(yōk sak) A membranous sac surrounding the food yolk in the embryo. p. 361

zona pellucida
(zō'nə pə lo͞o'sə də) (L. zona, girdle + pellucidus, transparent) A jelly-like membranous covering of the ovum. p. 355

zygote
(zī'gōt) (Gr. zygotos, paired together) A cell produced by the haploid nuclei of the sperm and egg; the fertilized ovum. p. 340

CREDITS

CHAPTER 1

Opener (photo) James Aronovsky-(Illu) Frank Crymble/1991 *Discover Magazine;* Laura J. Edwards; Chuck Dressner/St. Louis Zoo; Laura J. Edwards; Mark Moffett/Minden Pictures.
1-1, 1-4, 1-6 Nadine Sokol.
1-2, *A* Stephen Dalton/Natural History Photographic Agency.
1-2, *B* John Trager/Visuals Unlimited.
1-2, *C* Richard Walters/Visuals Unlimited.
1-2, *D* Cabisco/Visuals Unlimited.
1-2, *E* Balkwill-D. Maratea/Visuals Unlimited.
1-3, *A* Triarch/Visuals Unlimited.
1-3, *B* Biophoto Associates/Photo Researchers, Inc.
1-3, *C,* 1-3, *F* E. Rhone Rudder.
1-3, *D,* 1-7 Raychel Ciemma.
1-3, *E* Brian Milne/Animals Animals.
1-3, *G* Jeff Fott/Bruce Coleman Ltd.
1-3, *H* Robert Maier/Animals Animals.
1-3, *I* NASA.
1-5, *A* Patti Murray/Animals Animals.
1-5, *B* Wildtype Products/Bruce Coleman Ltd.
1-5, *C* Adrian Davies/Bruce Coleman Ltd.
1-A John H. Gerard

CHAPTER 2

Opener Marc Gottfried.
2-1, 2-2, 2-3, *A-B,* 2-7, 2-8, 2-9, 2-17, *B* Nadine Sokol.
2-3, *C* Michael Gadomski/Tom Stack & Associates.
2-4 The Bettmann Archives.
2-5, *A* Jurgen Schmitt/The Image Bank.
2-5, *B,* 2-5, *D* Frank T. Awbrey/Visuals Unlimited.
2-5, *C* Nicholas Foster/The Image Bank.
2-6 Michael & Patricia Fogden.
2-8, *A* Joseph Devenney/The Image Bank.
2-8, *B,* 2-24, *C* George Bernard/Animals Animals.
2-8, *C* Eastcott/Momatiuk/The Image Works.
2-10 Lilli Robbins.
2-15, *B* E.S. Ross.
2-16, *A* Manfred Kage/Peter Arnold, Inc.
2-17, *A* J.D. Litvay/Visuals Unlimited.
2-18 Scott Johnson/Animals Animals.
2-23 Nadine Sokol after Bill Ober.
2-24, *A* Manfred Kage/Peter Arnold, Inc.
2-24, *B* Michael Pasdizor/The Image Bank.
2-24, *D* Oxford Scientific Films/Animals Animals.
2-24, *E* Scott Blakeman/Tom Stack & Associates.
2-A Simon Frasher/Medical Phisics, RVI, Newcastle/Science Photo Library/Custom Medical Stock Photo, Inc.

CHAPTER 3

Opener Allen R.D.: The microtuble; Scientific American, Inc, 1987, p. 43.
3-1, *A* M. Abbey/Visuals Unlimited.
3-1, *B-C,* 3-4, *A,* 3-23, *B* David M. Phillips/Visuals Unlimited.
3-1, *D* John D. Cunningham/Visuals Unlimited.
3-1, *E* Bruce Iverson/Visuals Unlimited.
3-1, *F* Triarch/Visuals Unlimited.
3-1, *G* Carolina Biological Supply Co.
3-2 L.L. Sims/Visuals Unlimited.
3-4, *B,* 3-25, *B* Barbara Cousins.
3-5, 3-6, 3-9, 3-11, *B,* 3-15, *A,* 3-18, *B,* 3-19, *B* Bill Ober.
3-7, *A,* 3-20 Manfred Kage/Peter Arnold, Inc.
3-7, *B* J. David Robertson.
3-8, *A-C* Klaus Weber & Mary Osborn, for **Scientific American,** vol. 153, pp. 110-121, 1985 (unpublished).
3-8 (art) Barbara Cousins after Bill Ober.
3-10, *A* Dr. A. Brody/Science Photo Library/Photo Researchers, Inc.
3-10, *B* Kevein Somerville after Bill Ober.
3-11, *A* Richard Rodewald, University of Virginia.
3-12, *A,* 3-15, *C,* 3-18, *A* Charles J. Flickinger.
3-12, *B,* 3-13 Nadine Sokol.
3-14 K.G. Murti/Visuals Unlimited.
3-15, *B* Dr. Thomas W. Tillack.
3-16, 3-23, *C-D* Ed Reschke.
3-17 C.P. Morgan & R.A. Jersild, *Anatomical Record,* vol. 166, p. 575-586, 1970.
3-19, *A* Kenneth R. Miller.
3-21 Raychel Ciemma.
3-22 Cabisco/Visuals Unlimited.
3-23, *A* J.J. Candamse & B.C. Pugasheti/Tom Stack & Associates.
3-25, *A* L.M. Pope/BPS/Tom Stack & Associates.
3-26 T.D. Pugh and E.H. Newcomb, University of Wisconsin.
3-A Boehringer Ingelham International GmbH/Lennart Nilsson/Bonnier Fakta.

CHAPTER 4

Opener Trevor Wood/Tony Stone Worldwide.
4-1, 4-9 Nadine Sokol.
4-2, 4-3, 4-8 Nadine Sokol after Bill Ober.
4-4 Kevin Somerville after Bill Ober.
4-5, 4-6, 4-7, 4-11, *A* Barbara Cousins.
4-10, *A-B* Christy Krames.
4-10, *C-D* Jim Pearson/Taurus Association.
4-11, *B* Dr. Birgit H. Satir.
4-A William L. Mathews.

CHAPTER 5

Opener CNRI/SPL/Science Source.
5-1 U.K. Laemmli & J.R. Paulson.
5-2, 5-18 Raychel Ciemma.
5-3, *A* Ada L. Olins/Biological Photo Service.
5-3, *B,* 5-4, 5-8 (art), 5-9 (art), 5-10, *B,* 5-13, 5-17 (art) Barbara Cousins.
5-5 Carlyn Iverson.
5-6 John D. Cunningham/Visuals Unlimited.
5-7 Lilli Robins.
5-8 (photos) Carolina Biological Supply Company.
5-9 (photos) Lester V. Bergman, NY.
5-10, *A* Dr. A.S. Bajer.
5-11 David M. Phillips/Visuals Unlimited.
5-12, *A,* 5-14 Nadine Sokol.
5-12, *B* B.A. Palevitz & E.H. Newcomb/BPS/Tom Stack & Associates.
5-15 Kevin Somerville.
5-16 James Kezer, University of Oregon.
5-17 (photo) C.A. Hasenkampf, University of Toronto/Biological Photo Service.

CHAPTER 6

Opener Walter Iooss Jr./The Image Bank.
6-3, 6-5, 6-11, 6-14 Barbara Cousins.
6-6 Gunter Ziesler/Peter Arnold, Inc.
6-7 John Sohlden/Visuals Unlimited.
6-9 Bill Ober.
6-10 Nadine Sokol after Bill Ober.
6-A Tom Tracey/Photographic Resources.

CHAPTER 7

Opener Marc Gottfried.
7-1, 7-6, 7-11, 7-12, *B,* 7-13 Nadine Sokol.
7-2 Robert Barclay/Grant Heilman Photography.
7-3 Barbara Cousins.
7-4 George Klatt.
7-8 Grant Heilman/Grant Heilman Photography.
7-12, *A* Efriam Racker.

7-15, *A* Custom Medical Stock Photo, Inc.
7-15, *B* Michael Freeman/Bruce Coleman Ltd.

CHAPTER 8

Opener Dennis Di Cicco/Peter Arnold, Inc.
8-1 E.S. Ross.
8-3 Nadine Sokol.
8-5, 8-8, 8-14 Raychel Ciemma.
8-6 Manfred Kage/Peter Arnold, Inc.
8-7, 8-10, 8-12 Barbara Cousins.
8-9 Sherman Thomson/Visuals Unlimited.
8-11 Kevin Somerville after Bill Ober.

CHAPTER 9

Opener, 9-8, 9-11, *A* Lennart Nilsson, ***Behold man,*** Little, Brown and Co.
9-1, 9-11, *B*, 9-13 Nadine Sokol.
9-2 (art) Christine Oleksyk.
9-2 (photo) Tom Tracy/Photographic Resources.
9-3 Barbara Cousins.
9-4 Emma Shelton.
9-5 J.V. Small & F. Rinnerthaler.
9-6 *St. Louis Globe-Democrat.*
9-7, *A* David J. Hascaro and Associates.
9-7, *B-C* John Hagen.
9-9 David M. Phillips/Visuals Unlimited.
9-10 Bill Ober.
9-12 Cynthia Truner Alexander/Terry Cockerham, Snapse Media Production/Christine Oleksyk.
Table 9-2, Table 9-3 Ed Reschke.
Table 9-4, *A* Triarch/Visuals Unlimited.
Table 9-4, *B-C* John D. Cunningham/Visuals Unlimited.

CHAPTER 10

Opener, 10-13 Chet Hanchet/Photographic Resources.
10-1, 10-2, *A*, 10-5, 10-8, 10-10 Nadine Sokol.
10-2, *B* G. David Brown.
10-3, 10-9, *A* Barbara Cousins.
10-4 Kate Sweeney.
10-6 Bill Ober.
10-7 Christy Krames after Bill Ober.
10-9, *B* David M. Phillips/Visuals Unlimited.
10-12 Lilli Robbins.
10-A Benjamin, M.D./Custom Medical Stock Photos, Inc.

CHAPTER 11

Opener *A* American Cancer Society.
Opener *B* James Stevenson/SPL.
11-1, 11-11 Kate Sweeney.
11-2 Courtesy of AT&T Archives.
11-3 Ellen Dirkson/Visuals Unlimited.
11-4 Lennart Nilsson, ***Behold man,*** Little, Brown and Co.
11-5 Art Siegel, University of Pennsylvania.
11-6, 11-10 Nadine Sokol.
11-7, 11-8, 11-9 Barbara Cousins.
11-12 M. Moore/Visuals Unlimited.

CHAPTER 12

Opener William Strode/Humana, Inc./Black Star.
12-1, 12-13 Nadine Sokol.
12-2, 12-17 Kate Sweeney.
12-3 Christy Krames.
12-4, 12-7, 12-18 Ed Reschke.
12-5, 12-12 Bill Ober.
12-6 D.W. Fawcett-T. Kuwabara/Visuals Unlimited.
12-8 John D. Cunningham/Visuals Unlimited.
12-9, 12-10, 12-11 Barbara Cousins.
12-14 Makio Murayama/BPS/Tom Stack & Associates.
12-15 Raychel Ciemma.
12-16 Manfred Kage/Peter Arnold, Inc.
12-19 Harry Ransom Humanities Research Center.
12-A Chet Hanchet/Photographic Resources.

CHAPTER 13

Opener Lennart Nilsson.
13-1 Raychel Ciemma.
13-2 Science VU-NLM/Visuals Unlimited.
13-3, 13-12, *A* Bill Ober.
13-5 Kate Sweeney.
13-6, 13-8, 13-9, 13-15 Barbara Cousins.
13-7 Dr. A. Liepins/SPL/Photo Researchers, Inc.
13-10 Secchi, LeCaque, Roussel, Ucalf, CNRI/Science Source/Photo Researchers, Inc.
13-11 Manfred Kage/Peter Arnold, Inc.
13-12, *B* Nadine Sokol after Bill Ober.
13-16 Larry G. Arlain, Wright State University.
13-A M. English/Custom Medical Stock Photo, Inc.

CHAPTER 14

Opener M.P. Kahl/Photo Researchers, Inc.
14-1 Kate Sweeney.
14-2 Dan Gotshall.
14-3 Larry Brock/Tom Stack & Associates.
14-4, 14-5, 14-6, 14-14 Barbara Cousins.
14-13 Raychel Ciemma.
14-16 Organon Teknika/Visuals Unlimited.
14-A William B. Folsom/Uniphoto Picture Agency.

CHAPTER 15

Opener Tom McCarthy 1988 Discover Publications.
15-1 Peter Cohen—Custom Medical Stock Photo, Inc.
15-2, 15-3 (art), 15-7, 15-8, 15-9, 15-10, 15-11, 15-14 Nadine Sokol.
15-3 C.S. Raines/Visuals Unlimited.
15-4 E.S. Ross.
15-6, 15-12, *B*, 15-13 Barbara Cousins.
15-12, *A* Heimer L.: ***Human brain and spinal cord,*** Springer-Verlag.

CHAPTER 16

Opener James H. Karales/Peter Arnold, Inc.
16-1 Kate Sweeny.
16-3, 16-4, 16-6, 16-9 Nadine Sokol.
16-5 Bill Ober.
16-7, 16-12 Barbara Cousins.
16-8, 16-10 Michael P. Schenk.
16-11 John Daugherty.
16-13 Raychel Ciemma.
16-A Sobel/Klonsky/The Image Bank.

CHAPTER 17

Opener Rod Cardoza.
17-1, *A* Martha Swope, 1990.
17-1, *B-C* Martin/Custom Medical Stock Photo, Inc.
17-2, 17-8, 17-11 Nadine Sokol.
17-3, 17-6, *B*, 17-9 Barbara Cousins.
17-4, *A* Marsha J. Dohrman.
17-4, *B* Christine Oleksyk.
17-5, 17-10, 17-12, 17-13, 17-14 Raychel Ciemma.
17-6, *A* Scott Mittman and Maria T. Maglio.
17-7 Bill Ober.
17-A Fritz Prenzel/Animals Animals.

CHAPTER 18

Opener Vince Rodriguez.
18-1, 18-13, *B* Raychel Ciemma.
18-2 Bill Ober.
18-3, 18-9 Kate Sweeney.
18-4, 18-10 Nadine Sokol.
18-5, 18-6, 18-7, 18-11, 18-15 Barbara Cousins.
18-8 Scott Bodell.
18-12 Richard Rodewald, University of Virginia.
18-13, *A* Raychel Ciemma after Bill Ober.
18-14 John D. Cunningham/Visuals Unlimited.
18-A Jacques Cochin/The Image Bank.
Table 18-1 Nadine Sokol after Rusty Jones.

CHAPTER 19

Opener U.C. Davis Magazine, University of California, Davis.
19-1 Kate Sweeney.
19-2, 19-3, 19-9 Barbara Cousins.
19-4, 19-5, 19-13, 19-14 Nadine Sokol.
19-6 Bettina Cirone/Photo Researchers, Inc.
19-7 ***Akhenaten,*** The National Museum at Berlin, DDR, Egyptian Museum, no. 14512.
19-8 NMSB/Custom Medical Stock Photo, Inc.
19-10 Custom Medical Stock Photo, Inc.
19-11, 19-12 Raychel Ciemma.
19-15 Ed Reschke.
19-A Susan Lapides/Time Magazine, Inc.

CHAPTER 20

Opener, 20-1 Lennart Nilsson, ***Behold man,*** Little, Brown, and Co.
20-2, 20-6 Kate Sweeney.

20-3, 20-5 Nadine Sokol after Bill Ober.
20-4 Barbara Cousins.
20-7 Raychel Ciemma.
20-8 Ed Reschke.
20-9 Kevin Somerville.
20-10 David J. Hascuro and Associates.
20-12, 20-13 Joel Gordon.
20-14 Scott Bodell.
20-A Peter Read Miller/Sports Illustrated.

CHAPTER 21

Opener Lauros Giraudon/Art Resource, NY/ from Alfred Roll, The Nurse (detail) Lille, Musee des Beaux-Arts, Paris.
21-1, *A* David M. Phillips/Visuals Unlimited.
21-1, *B*, 21-10, 21-14 Barbara Cousins.
21-2, 21-3, 21-4, 21-5, 21-9, 21-12, 21-13, 21-15, 21-16, 21-17, 21-18 Lennart Nilsson, *A child is born*.
21-4, 21-6, 21-9 Raychel Ciemma.
21-7, 21-20, *A* Scott Bodell.
21-8 Kate Sweeney.
21-11 Kevin Somerville after Bill Ober.
21-19 Nadine Sokol.
21-20, *B* Martin/Custom Medical Stock Photo, Inc.

CHAPTER 22

Opener Bob Krist.
22-1 E. Rhone Rudder.
22-2, 22-3 Nadine Sokol after Bill Ober.
22-4, 22-5, 22-6, 22-8, 22-10 Nadine Sokol.
22-11 Carolina Biological Supply Company.
22-12 Raychel Ciemma.

CHAPTER 23

Opener Science VU-NLM/Visuals Unlimited.
23-1 CNRI/SPL/Science Source/Photo Researchers, Inc.
23-2, *A* The Children's Hospital, Denver, Cytogenetics Laboratory.
23-2, *B* R. Hutchings/Photo Researchers, Inc.
23-3, 23-8 Barbara Cousins.
23-5 Victor A. McKusick/Blackwell Scientific Publications.
23-6 Margaret A. Davee, M.S.; David D. Weaver M.D.; Department of Medical Genetics, Indiana University School of Medicine/Blackwell Scientific Publications.
23-10 Adam Hart-Davis/SPL/Custom Medical Stock Photo, Inc.
23-13 The Bettmann Archive.
23-14 M. Murayama/Biological Photo Service.
23-16 Kevin Somerville after Bill Ober.
23-17 Washington University School of Medicine.
23-A Dr. W. French Anderson/National Institutes of Health.

CHAPTER 24

Opener Richard Feldmann/National Institutes of Health.
24-1 Dennis D. Kunkel/Biological Photo Service.
24-2 Raychel Ciemma.
24-6, 24-7 Cold Springs Harbor Laboratory Archives.
24-8, 24-9 Bill Ober.
24-11 Barbara Cousins.
24-12, 24-17, *A*, 24-*B* Nadine Sokol.
24-13, 24-16 Molly Babich.
24-15, 24-18 Carlyn Iverson.
24-17, *B* 1982 C. Franke, J.E. Edstrom, A.W. McDowall, & O.L. Miller Jr.
24-20 Christian Vioujard/Gamma Liason.
24-A Herb Weitman/Washington University in St. Louis.

CHAPTER 25

Opener Dan Dreyfus & Associates Photography.
25-2 Ron Ervin.
25-3, *A* Cleveland P. Hickman, Jr.
25-3, *B* Peter Gregg/Imagery.
25-4, *A* Frank B. Gill/VIREO.
25-4, *B* John S. Dunning/VIREO.
25-5 Louise Van der Meid.
25-6 E. Rhone Rudder.
25-8, 25-14 Bill Ober.
25-9 The National Portrait Gallery, London.
25-10 Frank S. Balthis.
25-11 Molly Babich.
25-12 John D. Cunningham/Visuals Unlimited.
25-13 S.M. Awramik, University of California/Biological Photo Service.
25-15, 25-18 Nadine Sokol after Bill Ober.
25-16 Don & Pat Valenti/Tom Stack & Associates.
25-17, *A* Johnny Johnson/Animals Animals.
25-17, *B* Dave Watts/Tom Stack & Associates.
25-17, *C* Arthur Gloor/Earth Scenes.
25-17, *D* George H.H. Huey/Earth Scenes.
25-19 Raychel Ciemma.

CHAPTER 26

Opener NASA.
26-1, *A*, 26-6 Barbara Cousins.
26-1, *C* Kevin Walsh, University of California, San Diego.
26-2 Dudley Foster/Woods Hole Oceanographic Institution.
26-4, *A* M.R. Walter, Macquarie University, Australia.
26-4, *B* Nadine Sokol.
26-4, *C-D* Paul F. Hoffman.
26-5 Andrew H. Knoll, Botanical Museum of Harvard University.
26-6 Barbara Cousins.
26-6, *A*, *D* David Bruton.
26-6, *B*, *C*, *E* Simon Conway Morris, University of Cambridge.
26-8 William Boehm.
26-9, *A* William Boehm.
26-9, *B* Frans Lanting/Minden Pictures.
26-10, *A* F.M. Carpenter.
26-10, *B* John Gerlach/Animals Animals.
26-11 Michael W. Nickell.
26-12 Barbara Laing/Black Star.
26-A Mark Mamawal.

CHAPTER 27

Opener, 27-21, *B* John Reader.
27-1, *A* Smithsonian Institution photo no. USNM 57628: Walcott, 1911.
27-1, *B* Heather Angel/Biofotos.
27-2 Raychel Ciemma.
27-3, 27-4 Nadine Sokol after Bill Ober.
27-5 Steve Martin/Tom Stack & Associates.
27-6, *A*, 27-7, *C* Norbert Wu.
27-6, *B* Marty Snyderman/Visuals Unlimited.
27-6, *C*, 27-7, *B* Marty Snyderman.
27-7, *A* Alex Kerstitch.
27-7, D John D. Cunningham/Visuals Unlimited.
27-8 Gary Milburn/Tom Stack & Associates.
27-9, 27-11 Bill Ober.
27-10 Stouffer Productions Ltd./Animals Animals.
27-13 Ron Ervin.
27-14, 27-19 Alan E. Mann.
27-15, *A* Luis Castaneda/The Image Bank.
27-15, *B* Alan Nelson/Animals Animals.
27-16, *A* Doug Wechsler/Animals Animals.
27-16, *B* C.C. Lockwood/Animals Animals.
27-17 Russell A. Mittermeier.
27-20 Molly Babich.
27-21, *A* The Living World, St. Louis Zoo, St. Louis, MO.
27-22 Kate Sweeney.
27-23 Douglas Waugh/Peter Arnold, Inc.
27-24 SCODE/Overseas.
27-A Redrawn from Vigilant, L., Stoneking, M., and others: African populations and the evolution of human mitochondrial DNA, *Science*, vol. 253, September 27, 1991, pp. 1205-1207.

CHAPTER 28

Opener Science Source.
28-1, *A* K.G. Murti/Visuals Unlimited.
28-1, *B* K. Namba and D.L.D. Caspar.
28-3, *A* Science VU/Visuals Unlimited.
28-3, *B* Carlyn Iverson.
28-4, 28-5, 28-9, 28-11, 28-14, 28-15 Nadine Sokol.
28-6 Elizabeth Gentt/Visuals Unlimited.
28-7 Runk/Schoenberger/Grant Heilman Photography, Inc.
28-8 Jo Handelsman and Steven A. Vicen.
28-10 Dr. Huntington Potter and Dr. David Dressler/Harvard Medical School.
28-12 Eli Lilly & Co.
28-13 Barbara Cousins.
28-A Monsanto Company.

CHAPTER 29

Opener L.L.T. Rhodes/Earth Scenes.
29-1 Courtesy of Stanley Erlandson.
29-2, *A* Dr. E.W. Daniels, Argonne National Laboratory, Argonne, IL; and University of

Illinois College of Medicine, Dept. of Anatomy & Cell Biology, Chicago.
29-2, ***B-C,*** **29-18,** ***A*** Cabisco/Visuals Unlimited.
29-3, ***A*** M. Abbey/Visuals Unlimited.
29-3, ***B,*** **29-22, 29-23,** ***C,*** **29-24,** ***B,*** **29-27** Carlyn Iverson.
29-4, 29-5, ***B*** Manfred Kage/Peter Arnold, Inc.
29-5, ***A*** Richard H. Gross.
29-6 Courtesy of the British Tourist Authority.
29-7, 29-12, 29-A Bill Ober.
29-8, 29-14, ***A,*** **29-28,** ***A*** David M. Phillips/Visuals Unlimited.
29-9 K.G. Murti/Visuals Unlimited.
29-10 Nadine Sokol after Bill Ober.
29-11 A.M. Siegelman/Visuals Unlimited.
29-13 Sanford Berry/Visuals Unlimited.
29-14, ***B*** William Patterson/Tom Stack & Associates.
29-15, ***A*** Philip Sze/Visuals Unlimited.
29-15, ***B,*** **29-17,** ***A*** William C. Jorgensen/Visuals Unlimited.
29-15, ***C*** Bob Evans/Peter Arnold, Inc.
29-16, ***A*** BioPhoto Associates/Photo Researchers, Inc.
29-16, ***B-C*** John D. Cunningham/Visuals Unlimited.
29-16, ***D*** London Scientific Films/Oxford Scientific Films.
29-17, ***B*** Gary K. Robinson/Visuals Unlimited.
29-17, ***C*** Brian Parker/Tom Stack & Associates.
29-18, ***B-F*** Higuchi Bioscience Laboratory.
29-19, 29-30, ***A,*** **29-30,** ***C,*** E.S. Ross.
29-20, ***A-C*** John Shaw/Tom Stack & Associates.
29-21, ***A*** Dwight Kuhn.
29-21, ***B,*** **29-23,** ***B*** Jeremy Burgess/Science Photo Library/Photo Researchers, Inc.
29-23, ***A*** Sherman Thomson/Visuals Unlimited.
29-24, ***A*** Bob Evans/Peter Arnold, Inc.
29-25 Gordon Langsbury/Bruce Coleman Ltd.
29-26, ***A*** Walt Anderson/Tom Stack & Associates.
29-26, ***B*** Michael and Patricia Fogden.
29-26, ***C*** Jeff Lepore/Photo Researchers, Inc.
29-28, ***B*** Biophoto Associates/Photo Researchers, Inc.
29-28, ***C*** Jack M. Bostrack/Visuals Unlimited.
29-29 Ken Greer/Visuals Unlimited.
29-30, ***B*** James L. Caster.
29-31 Ed Reschke.

CHAPTER 30

Opener Sneads Ferry/Photo Researchers, Inc.
30-1 Barbara Cousins.
30-2, ***A*** Kirtley-Perkins/Visuals Unlimited.
30-2, ***B*** Ken Davis/Tom Stack & Associates.
30-3, ***A*** John Shaw/Tom Stack & Associates.
30-3, ***B,*** **30-17** John D. Cunningham/Visuals Unlimited.
30-4, 30-6 Nadine Sokol after Bill Ober.
30-5 Bill Ober.
30-7, 30-9, 30-11 Carlyn Iverson.
30-8, ***A,*** **30-12,** ***B, C, F*** Whit Bronaugh.
30-8, ***B*** E.S. Ross.
30-8, ***C*** Michael & Patricia Fogden.
30-10, ***A*** Tom J. Ulrich/Visuals Unlimited.
30-10, ***B*** Ron Spomer/Visuals Unlimited.
30-10, ***C-D*** James L. Castner.
30-12, ***A, D, E*** Richard H. Gross.
30-13 Peter C. Hoch.
30-14 Jack M. Bostrack/Visuals Unlimited.
30-15 E. Webber/Visuals Unlimited.
30-16 Sylvan H. Wittwer, Michigan State University.
30-A, ***1*** George Huey/Earth Scenes.
30-A, ***2*** Rodale Institute.
30-A, ***3*** T. Lawrence Mellichamp/Visuals Unlimited.

CHAPTER 31

Opener Ernst van Jaarsueld/Kirstenbosch National Botanic Garden/South Africa.
31-1, 31-4 (art), 31-5 (art), 31-6 (art), 31-10, 31-11 (art), 31-13, 31-17, 31-18 Carlyn Iverson.
31-2 (art), 31-15, 31-16, ***A-B*** Bill Ober.
31-2, ***B*** SUNY College of Environmental Science and Forestry Center for Ultrastructure Studies.
31-2, ***C*** Randy Moore/Visuals Unlimited.
31-3, ***A*** George J. Wilder/Visuals Unlimited.
31-3, ***B,*** **31-6** Richard H. Gross.
31-4, ***A*** Cabisco/Visuals Unlimited.
31-4, ***B,*** **31-16 (photo)** Ed Reschke/Peter Arnold, Inc.
31-4, ***C*** Fred E. Hossler/Visuals Unlimited.
31-5 Stan Elms/Visuals Unlimited.
31-7 Jack M. Bostrack/Visuals Unlimited.
31-8, ***A, C,*** **31-9,** ***B*** John D. Cunningham/Visuals Unlimited.
31-8, ***B*** Salvadore Giordano III.
31-9, ***A*** Ed Reschke.
31-11 Ray F. Evert.
31-12 USDA Forest Service, Forest Products Laboratory.
31-14, ***A-B*** Kjell B. Sandved.

CHAPTER 32

Opener, 32-23, ***A*** Alex Kerstitch.
32-1 Raychel Ciemma.
32-2, 32-6, ***B,*** **32-15,** ***E,*** **32-20, 32-22, 32-23,** ***D,*** **32-24** Bill Ober.
32-3 Barbara Cousins.
32-4 Scott Bodell after Bill Ober.
32-5, ***A,*** **32-8, 32-9, 32-11, 32-A** Nadine Sokol after Bill Ober.
32-5, ***B*** D.J. Wrobel, Monterey Bay Aquarium/Biological Photo Service.
32-5, ***C*** Mickey Gibson/Animals Animals.
32-6, ***A*** Marty Snyderman.
32-6, ***C,*** **32-16,** ***C*** Gwen Fidler/Tom Stack & Associates.
32-7, 32-23, ***E*** Jeff Rotman.
32-10, ***A*** Jim & Cathy Church.
32-10, ***B*** E.J. Cable/Tom Stack & Associates.
32-12 V.R. Ferris.
32-13 M.G. Schultz from H. Zaimen (ed.) pictorial presentation of parasites.
32-14, 32-17 Nadine Sokol.
32-15, ***A*** Milton Rand/Tom Stack & Associates.
32-15, ***B*** Fred Bavendam.
32-15, ***C*** A. E. Sprietzer/O.S.U. Museum of Zoology.
32-15, ***D,*** **32-16,** ***A,*** **32-18,** ***C,*** **32-19,** ***A, 32-21*** Kjell Sandved.
32-16, ***B*** David Dennis/Tom Stack & Associates.
32-18, ***A*** J. MacGregor/Peter Arnold, Inc.
32-18, ***B*** J. Alcock/Visuals Unlimited.
32-19, ***B*** Cleveland P. Hickman, Jr.
32-23, ***B*** Carl Roessler/Tom Stack & Associates.
32-23, ***C*** David L. Pawson/Harbor Branch Oceanographic Institution, Inc.

CHAPTER 33

Opener D. Griffiths/National Photographic Index of Australian Wildlife.
33-1, 33-2, ***C,*** **33-5,** ***B,*** **33-12, 33-13** Bill Ober.
33-2, ***A*** Rick Harbo.
33-2, ***B*** Jim & Cathy Church.
33-3 Richard Olson.
33-4 Nadine Sokol after Bill Ober.
33-5, ***A*** Heather Angel.
33-6, ***A-B,*** **33-18,** ***E*** Marty Snyderman.
33-6, ***C*** Carl Roessler/Tom Stack & Associates.
33-9, ***A,*** **33-14,** ***D*** John D. Cunningham/Visuals Unlimited.
33-9, ***B*** David Dennis/Tom Stack & Associates.
33-9, ***C*** Cleveland P. Hickman, Jr.
33-9, ***D-E*** John Shaw/Stack & Associates.
33-10 Michael Fogden.
33-11, ***A,*** **33-18,** ***A*** Brian Parker/Tom Stack & Associates.
33-11, ***B,*** **33-14,** ***E*** William J. Weber/Visuals Unlimited.
33-11, ***C*** John Cancalosi/Tom Stack & Associates.
33-11, ***D,*** **33-18,** ***B*** Rod Planck/Tom Stack & Associates.
33-14, ***A-B*** E.S. Ross.
33-14, ***C*** Alan Nelson/Tom Stack & Associates.
33-15 Raychel Ciemma.
33-16, ***A*** J.E. Wapstra/NPIAW.
33-16, ***B*** J. Alcock/Visuals Unlimited.
33-17, ***A*** Fritz Prenzel/Animals Animals.
33-17, ***B*** B.G. Murray/Animals Animals.
33-18, ***C*** Stephen Dalton/Animals Animals.

CHAPTER 34

Opener Paul Lemmons/Frank Lane Picture Agency.
34-1 Jeff Foott/Bruce Coleman, Inc.
34-2 Frans Lanting/Minden Pictures.
34-3 Don & Pat Valenti/Tom Stack & Associates.
34-4, 34-11 Raychel Ciemma.

34-5 Stoufer Productions Ltd./Animals Animals.
34-6 E. Rhone Rudder.
34-7 E.S. Ross.
34-8 Thomas McAvoy, Life Magazine, 1955, Time Warner, Inc.
34-9 James R. Risher.
34-10 FPG International Corp.
34-12 Omikron/Photo Researchers, Inc.
34-13 Lee Boltin.
34-14 Babkin Papers, Osler Library of the History of Medicine, McGill University, Montreal, Canada.
34-15 Bill Eades.
34-A Dale & Marian Zimmerman/Animals Animals.

CHAPTER 35

Opener, 35-1 Jane Burton/Bruce Coleman Ltd.
35-2 Bill Eades.
35-3, 35-9 Nadine Sokol.
35-4 Tui De Roy.
35-5 Dale Sarver/Animals Animals.
35-6 Mark D. Phillips/Photo Researchers, Inc.
35-7 Bob and Clara Calhoun/Bruce Coleman Ltd.
35-8 Oxford Scientific Films/Animals Animals.
35-8 (art) Bill Ober.
35-10 Herman Kacher.
35-11 Fritz Prenzel/Animals Animals.
35-12 Fred Bruemmer.
35-13, *A* John Stern/Animals Animals.
35-13, *B* J.C. Stevenson/Animals Animals.
35-14, 35-15 E.S. Ross.
35-16 E. Rhone Rudder.
35-17 Mark Moffett.
35-18 Tom Stack & Associates.
35-A Mike McKavett/Bruce Coleman Ltd.

CHAPTER 36

Opener Thomas D. Mangelsen/Images of Nature.
36-1 David Scharf/Peter Arnold, Inc.
36-2 Alan Fletcher Research Station, Queensland Department of Lands, Queensland, Australia.
36-4 Nadine Sokol.
36-7 E.S. Ross.
36-9, *A* Bonnie Freer/Peter Arnold, Inc.
36-9, *B* S.L. Nou/Photo Researchers, Inc.
36-11 Fred Griffing/UNICEF.
36-A Stephanie Maze/Woodfin Camp, Inc.

CHAPTER 37

Opener, 37-5 D.P. Wilson/Eric & David Hosking.
37-1 David Swanlung, Save-the-Redwoods League.
37-2, 37-12, 37-13, *D*, 37-15, 37-16, *A-B*, 37-21, 37-23 E.S. Ross.
37-3, *A* Fred Whitehead/Animals Animals-Earth Scenes.
37-3, *B* Stephen J. Krasemann/DRK Photo.
37-3, *C* Jeri Clark.
37-4 Roger del Moral, University of Washington.
37-6 Barbara Cousins.
37-7, *A* George H. Harrison.
37-7, *B* M.P. Kahl/National Geographic Society.
37-7, *C* J.A.L. Cooke/Oxford Scientific Films.
37-8 M. Abbey/Visuals Unlimited.
37-10 Joe McDonald/Animals Animals.
37-11 Calgene, Inc.
37-13, *A* James L. Castner.
37-13, *B* William J. Weber/Visuals Unlimited.
37-13, *C* Alex Kerstitch.
37-14, *A* Paul A. Opler.
37-14, *B* Patty Murray/Animals Animals.
37-14, *C-D*, 37-22 Dwight Kuhn.
37-16, *C-E* James L. Castner.
37-18, *A* Kennan Ward.
37-18, *B* Frank S. Balthis.
37-19 Bill Wood/Bruce Coleman, Inc.
37-20 Cabisco/Visuals Unlimited.
37-24 Gary Braasch.
37-A Scott Camazine/Photo Researchers, Inc.

CHAPTER 38

Opener A. J. Stevens/Bruce Coleman Ltd.
38-1, 38-9, 38-10, 38-11, 38-12, 38-14 Nadine Sokol.
38-2, 38-3 E.S. Ross.
38-4, 38-5 Raychel Ciemma.
38-6 Lilli Robins.
38-13, *A* Gerald A. Peters, Virginia Commonwealth University.
38-13, *B* Tom Stack & Associates.
38-13, *C* Kjell Sandved.
38-15 Norbert Wu.
38-A John D. Cunningham/Visuals Unlimited.

CHAPTER 39

Opener, 39-8, 39-10 James L. Castner.
39-1, 39-4, 39-7 Bill Ober.
39-8, *B*, 39-9, *A*, *C*, *D* Michael & Patricia Fogden.
39-9, *B* Doug Wechsler/VIREO.
39-9, *E* E.S. Ross.
39-11 W. Perry Conway/Tom Stack & Associates.
39-12 Patti Murray/Animals Animals.
39-13, *A* Richard Kolar/Earth Scenes.
39-13, *B* Breck P. Kent/Earth Scenes.
39-13, *C* Arthur Gloor/Earth Scenes.
39-14, *A* John Cancalosi/Tom Stack & Associates.
39-14, *B* William J. Hamilton III.
39-15 W.J. Weber/Visuals Unlimited.
39-16 Whit Bronaugh.
39-17 Lee Casebere.
39-18, 39-20, *D* Cleveland P. Hickman, Jr.
39-19 Bill McRae.
39-20, *A* John Gerlach/Tom Stack & Associates.
39-20, *B* Rich McIntyre/Tom Stack & Associates.
39-20, *C* M. & C. Ederegger/Peter Arnold, Inc.
39-21, 39-27 Raychel Ciemma.
39-22, *B-E* Chesapeake Bay Foundation.
39-23 Dan Gotshell.
39-24 Jeff Foott/Tom Stack & Associates.
39-25 Robert Leo Smith.
39-26, *A-B*, 39-28, *D* Norbert Wu.
39-26, *C-E* Dale Glantz.
39-28, *A* Scott D. Taylor, Duke University Marine Laboratory.
39-28, *B* Doc White/Images Unlimited.
39-28, *C* Jeff Rotman/Jeff Rotman Photography.
39-28, *E* Dudley Foster/Woods Hole Oceanographic Institution.
39-A NASA.

CHAPTER 40

Opener Chet Hanchet/Photographic Resources.
40-1 Nadine Sokol.
40-2, *A-B*, 40-14 Thomas A. Schneider.
40-2, *C* Grant Heilman/Grant Heilman Photography.
40-2, *D* Kevin Schafer/Peter Arnold, Inc.
40-2, *E* Stan W. Elms/Visuals Unlimited.
40-4 Frans Lanting/Minden Pictures.
40-6 J.M. Rankin/WorldWide Fund for Nature.
40-7 Agricultural Research Service, USDA.
40-8 Zoological Society of San Diego.
40-9, 40-10 Ray Pfortner/Peter Arnold, Inc.
40-11 Raychel Ciemma.
40-13 Gamma Liaison Network.
40-16, *A* Frank S. Balthis.
40-16, *B* Bruno J. Zehnder/Peter Arnold, Inc.

Index

A

Aberration, chromosomal, 394
Abiotic, 639, 657
Abortion, spontaneous, 391
Abscisic acid, 541
Absorption
  epithelial tissue and, 149
  in small intestine, 175
Absorption spectrum, 133, 286
Abstinence in birth control, 347-350
Abyssal zone, 690
Acacia, 650
Accessory gland, 338
Acetabularia, 46, 405
  in reciprocal graft experiment, 406
Acetaldehyde, 127
Acetyl-CoA, 123
Acetyl group, 123
Acetylcholine, 258, 259
Acetylcholinesterase, 259
Achondroplasia, 398
Acid, 26
  fatty, 33
  folic, 167
  linolenic, 34
  palmitic, 34
  pantothenic, 167
Acid environment, 218
Acid rain, 705-706
Acoelomate, 563
Acromegaly, 320
Acrosome, 336
ACTH; *see* Adrenocorticotropic hormone
Actin, 51, 307
Action potential, 253, 254-255
Activation energy, free, 101
Activator, 109, 422
Active immunity, 227
Active sites, enzymes and, 105, 106
Active transport, 75, 178
Adam's apple, 172
Adaptation, 432
Adaptation syndrome, general, 326
Adaptive radiation, 433, 455
Adenine, 39
  in adenosine triphosphate, 39, 110
Adenosine, 110
Adenosine diphosphate, 111-112
Adenosine triphosphate, 59, 116-127; *see also* Cellular respiration
  adenine in, 39, 110
  definition of, 110-112
  glycolysis and, 121
  production of, 116
ADH; *see* Antidiuretic hormone
Adhesion, 555
Adipose cell, 147
Adipose connective tissue, 155
ADP; *see* Adenosine diphosphate
Adrenal cortex, 325, 326, 329
Adrenal gland, 325-327
Adrenal medulla, 325, 326-327, 329
Adrenaline; *see* Epinephrine
Adrenocorticotropic hormone, 320
Adventitious root, 551
Aegyptopithecus zeuxis, 477
Aerobic metabolism, 121
Aerobic process, 116
Afferent arteriole, 237
Afferent neuron, 250-251, 267
African grasshopper, 647
African savanna, 679
Afterbirth, 368
Age distribution of population, 630
Agnatha, 465
Agranulocytes, 210
Agriculture in recombinant deoxyribonucleic acid technology, 499
Agrobacterium tumefaciens, 499
Ahnfeltia concinna, 513
Ahnfeltia plicata, 513
AIDS virus, 227-228, 229
Air pollution, 707, 708
Air pressure, 555
Albumin, 74
Aldosterone, 244, 245, 326
Algae, 509-513
  brown, 509, 511
  dinoflagellate, 509-510
  golden, 509, 510
  green, 509, 512
  red, 509, 513
All-or-nothing response, 255
Allantois, 360
Allele, 378
  codominant, 400
  multiple, 400
Allergen, 230
Allergic reaction, 229
Alternation of generation, 529
Alternative energy sources, 697
Aluminum, 27
Alveoli, 185
Ameba, 504-505
Amine, 322
Amino acid, 34, 35, 37, 38, 63
  essential, 181
Amino group, 34
Ammonia, 241
Amniocentesis, 400
Amnion, 360
Amniotic egg, 468, 589, 590
Amniotic fluid, 360
Amphibian, 587-588
  egg protection of, 588
  evolutionary history of, 466-468
Amylase, 168, 169
Amylopectin, 30
Amylose, 30
Anabaena azollae, 667
Anabolic reaction, 101
Anaerobic, defined, 116
Anaerobic metabolism, 121, 127
Anal pore, 507
Anaphase, 87, 88, 90, 91
Anemia, 167
Angina pectoris, 213
Angiosperm, 534, 548
  life cycle of, 535
Animal, 12, 13
  ancestral tree of, 562
Animal—cont'd
  bacterial, compared with plant cell, 48
  bodies of, 563
  cell of, 48
  characteristics of, 561-564
  invertebrate; *see* Invertebrate animal
  vertebrate; *see* Vertebrate
Animal cell, 48
Animalia, 12, 13
Annelid, 573
Annual ring, 552
Ant, bulldog, 575
Ant society, 619
Antagonistic muscle, 305
Anterior pituitary gland, 320-321, 329
Anther, 534
Antheridium, 529
Anthropoid, 473, 474-475
  evolution of, 476-477
Anthropoid primates, 460
Antibody, 219, 224
  heavy or light chain, 226
  molecular structure of, 226
Antidiuretic hormone, 244, 245, 321
Antigen, 219
Antitoxin, 227
Anus, 180
Anvil of ear, 289
Aorta, 204, 205
Aortic semilunar valve, 204, 205
Ape, 474, 475
Apical meristem, 548
Appendages, 300
Appendicitis, 179
Appendicular skeleton, 298, 299, 300-302
Appendix, 179
Aqueous humor, 287
Aquifer, 665
Archaebacteria, 450, 492
Archaeopteryx fossil, 435, 456
Archean Era, 448-452
Archegonium, 529
Arcyria, 515
Arm, 300, 301
Arrowhead spider, 575
Arteriole, efferent, 238, 239, 240, 241
Arteriosclerosis, 213
Artery, 201
  umbilical, 360
Arthropod, 574, 575
Articulation, 302
Artificial breeding, 429-431
Artificial kidney, 247
Artificial selection, 429, 430
Ascaris, anatomy of, 570
Ascending arm of Henle's loop, 237, 238
Asexual reproduction, 85
Aspen, clones of, 540
Aspergillus, 522
Assembly, respiratory, 125
Association area of brain, 266
Aster, 89
Asymmetry, 565-566
Atherosclerosis, 207, 213
Athlete's foot, 522
Atmosphere, 707-708
Atmospheric effect on climate, 672-675
Atom, 17-18
  functional groups of, 28, 29
Atomic energy level, 18
Atomic mass, 17
Atomic number, 17
ATP; *see* Adenosine triphosphate
Atrioventricular node, 207
Atrium
  in bony fish, 586
  left or right, 204, 207
Auditory canal, 289
Australopithecus, 477-479
Autonomic nervous system, 264, 276-277
Autosome, 81, 389
Autotroph, 131, 659
Auxin, 541
AV node; *see* Atrioventricular node
Aves, 591
Axial skeleton, 298, 299-300
Axon, 250
  in giraffe, 251

B

B cell, 209, 223
  memory, 224
B lymphocyte; *see* B cell
Bacteria, 45, 47, 217, 491-494
  binary fission in, 491
  as disease producer, 493-494
  diversity and classification of, 492-493
  eukaryotic cells compared with, 64, 65
  exponential growth in population of, 626
  fossils of, 451
  gene transfer in human-engineered, 496-498
  gene transfer in natural, 494, 495
  nitrogen fixing, 493, 667
  photosynthetic, 59
  reproduction of, 491-492
  size of, 488
Bacteria-like organelle, 58-60
Bacterial cell, 47
Bacteriophage, 489
Balance, 290-291
Balanus, 641, 642
Ball-and-socket joint, 304
Bark of plant, 552, 553
Barnacle, 650
Barrel sponge, 565
Basal body, 61
Basal ganglia, 267, 269
Base, 26
  organic nitrogen-containing, 39
Basidia, 521
Basidiomycota, 520
Basophil, 209, 210
Batesian mimicry, 647

H.M.S. *Beagle*, 427
Beak, 591
Bee colony, 619
Beetle, 681
Behavior
innate, 598-609
ethology in, 599
fixed action pattern in, 602-603
genetics and, 600
kinesis in, 600
learning and, 603-606, 607
reflex in, 600
taxis in, 600
social, 610-623
competitive, 611-613, 614
group, 618-621
human, 621
parenting, 618
reproductive, 615-616, 617
Benign tumor, 490
Benthos, 690
Beriberi, 167
Bicarbonate, pancreas and, 177
Biceps, 305
Bicuspid valve, 204
Bile, 176
Bile pigment, 233
Bile salts, 176
Binary fission, 491
Binomial nomenclature, 14
Bio-gas machine, 696
Bioenergy, 696
Biological concentration, 704
Biological diversity, 701
Biological magnification, 704
Biology, 2-15
classification in, 12-14
scientific process in, 3-7
unifying themes, 7-12
Biomass, 662
Biome, 671-693
climate of, 672-675
desert, 680-681
distribution of, 675
life zone and
estuary, 685-687
fresh water, 684-685
ocean, 687-690
savanna, 678, 679
taiga, 682-683
temperate
deciduous forest, 682
grassland, 681
tropical rain forest, 676, 677
tundra, 684
Biosphere in future, 10, 694-710
atmosphere of, 707-708
diminishing resources of, 696-700
land in, 696-703
solid waste and, 702-703
species extinction and, 700-701
water in, 703-706
acid rain and, 705-706
pollution of, 703-705
Biotic, defined, 639, 657
Biotin, 167
Bipedal, 475, 478
Bipolar cell, 288
Bird, 591-592
evolution of, 466-468
feather of, 11, 592
song development in, 603, 604
Birth, 368, 369
labor in, 368
Birth canal, 368
Birth control, 347-351
abstinence, 347-350
foam, 347
implant for, 350
oral, 347
pill, 350
prevention of egg maturation in, 350-351
rhythm method of, 347
sperm blockage in, 350
sperm destruction in, 350
surgical intervention in, 351
Bivalve, 571
Black bread mold, 518
Bladder, 233, 244
Blade
of leaf, 554
of seaweed, 511
Blastocyst, 357, 358
Blastula, 564
Blending inheritance, refutation of, 376
Blood, 155
heart pathway of, 204
Blood cell, 147, 157
Blood clot, 211
Blood glucose, 180
Blood plasma, 157
Blood sinuses, 571
Blue shark, 584
Blue tunicate, 582
Body cavity, 562-563
Body organization, 146-164
organs in, 160-162
interaction of, 162-163
systems of, 147-149
tissues in, 149-160
connective, 152-157
epithelial, 149-151
muscle, 158-159
nervous, 159-160
Bohr, N., 17
Bolus, 172
Bond, high-energy, 110
Bone, 147, 155, 296-304
compact, 156, 296
coxal, 300
organization of, 297
spongy, 156, 297
Bone cell, 45, 147, 153, 156
Bone marrow, 156, 220
Bone suture, 302
Bony fish, 467, 586
Boron, 27
Bossiella, 513
Bowman's capsule, 235, 237-241
Brachiation, 478
Brachii, 305
Brain, 149
association area of, 266
human, 265
lobes of, 266
in nervous system, 264, 265-271
neurons in, 250
underside view of, 270
weight of, 265
Brain grafting, 262
Brainstem, 265, 270
Branchial arch, 365
Breastbone, 300
Breathing, 185, 191-192
Bristleworm, shiny, 573
Brittle star, 577
Bronchi, 188
Bronchitis, chronic, 196
Brown-air city, 707, 708
Brown algae, 509, 511
Brownian movement, 72
Bryophyte, 529-530
Bulbourethral gland, 338
Bull elk, 683
Bulldog ant, 575
Bundle of His, 207
Bundle branches, 207
Burgess shale stratification, 453
Bursa of Fabricius, 223
Butterfly, 31

C

Cabbage butterfly, 648
Cactus, prickly pear, 627
Calcitonin, 322, 324
Calcium, 27, 168
Calories, 166
Calvin cycle, 141
Cambium, cork or vascular, 552
Cambrian Period, 452
fossils of, 453
Camouflage, 648, 649
Cancer, 490
Canine teeth, 171
Capillary, 201, 202-203
Capillary bed, 202
Capillus, 201
Capsid, viral, 487
Capsule, Bowman's, 235, 237-241
Carbohydrate, 30-32, 166
plasma membrane and, 71
Carbon, 27
covalent bonds in, 21
Carbon cycle, 665-666
oxygen cycling and, 665
Carbon fixation, 142
Carboniferous Period, 454
Carboxyl group, 34
Carboxypeptidase, 177
Cardiac muscle, 159-160
Cardiac opening, 172
Cardiac sphincter, 173
Cardio, as prefix, 199
Cardiovascular system, 199
of embryo, 365
Caribou migration, 604
Carnivore, 170
Carotene, 321
Carotenoid, 133
Carpal, defined, 300
Carrying capacity, 627
Cartilage
elastic, 153, 155
hyaline, 153, 154
Cartilage cell, 147, 153
Cartilaginous fish, 584-586
Caste, 619
Catabolic process, 121
Catabolic reaction, 101
Catalysis, 105
Catalyst, 105
Caudal, as term, 562
Cave painting, 481
Cavity
cranial, 149
pelvic, 149, 339
Cayuga Lake, 661
CCK; *see* Cholecystokinin
Cecum, 180
Cell
animal, 48
bacterial, 47
bipolar, 288
cone, 286
glial, 160
Golgi complex and, 47, 48, 54, 55
of hair, 289
in human body, 147-149
in immune system, 219
in living things, 7-8, 9
muscle, 307
plant, 48
rod, 286
Schwann, 160, 252
structure of; *see* Cell structure
target, 315
white blood, 157, 210
Cell body of neuron, 252
Cell culture technique, 540, 541
Cell cycle, 86
Cell division, 80-98
chromosomes in, 81, 82
cytokinesis in, 91-92
eukaryotic versus prokaryotic, 64, 65
in growth and reproduction, 83-85
meiosis in, 92-94, 95
meiotic recombination in, 96, 97
mitosis in, 86-90
Cell identity, 71
Cell-mediated immune response, 222, 223
Cell membrane, 67-79
bacteria and, 47
foundation of, 68-69
movement across, 72-77
active transport in, 75
diffusion in, 72
endocytosis in, 76, 77
exocytosis in, 77
facilitated diffusion in, 74-75
osmosis in, 72-74
proteins in, 70-72
Cell plate, 92
Cell structure, 44-66
bacteria in, 47, 61-63, 64
characteristics of, 47, 48
eukaryotic
cilia of, 60-61
cytoplasm of, 50-51, 52
cytoskeleton of, 50-51, 52
endoplasmic reticulum of, 53
flagella of, 60-61
Golgi complex in, 47, 48, 54, 55
lysosomes of, 56
nucleus of, 56-58
organelles of, 53-58, 58-60
overview of, 49-61
prokaryotic compared to, 64, 65
vacuoles of, 56
wall of, 48, 49, 61
plasma membrane of, 47, 48, 49-50, 60, 61
prokaryotic, 61-63, 64
eukaryotic compared to, 64, 65
interior organization of, 63, 64
wall of, 48, 61
size of, 46-47
Cell theory, 44-66
Cell wall, 48, 49, 61, 62, 63
Cellular respiration, 58, 114-129, 185
adenosine triphosphate production in, 116

Cellular respiration—cont'd
adenosine triphosphate release, 116-128
electron carriers in, 119
electron transport train in, 125-127
glycolysis in, 121-123
Krebs cycle in, 123, 124
oxidation-reduction in, 119
chemical energy of, 115
overview of, 116-117
Cellular slime mold, 514
Cellulose, 29, 32
Cementum, 171
Cenozoic Era, 458-460
Central nervous system, 264, 265-273
ganglia in, 267
nuclei in, 267
Centriole, 48
animal cells and, 86
Centromere, 81
Cephalic, as term, 562
Cephalochodate, 583
Cephalopod, 571
Cerebellum, 265, 270
Cerebral cortex, 265, 266
Cerebral ganglia, 574
Cerebrospinal fluid, 272
Cerebrum, 265, 266-269
Cervical cap, 350
Cervix, 343
Channel, 50
coupled, 75
thoroughfare, 202
Chaparral, 658
Chase, M., 414, 415
Chemical bond, 19
Chemical cycling in ecosystems, 664-668
Chemical energy to drive metabolism, 115
Chemical forces, 18-19
Chemical formula, 18
Chemical groups, 28, 29
Chemical reaction
regulating, 105-110
starting, 101-102, 103
Chemistry of life, 16-41
atoms in, 17-18
carbohydrates in, 30-32
chemical bonds in, 19
chemical building blocks in, 28-40
elements and molecules in, 18-22
fats and lipids in, 32-34
nucleic acids in, 39
proteins in, 34-38
water in, 22-26
Chemoautotroph, 492
Chemoreceptors, 170
Chesapeake Bay, 686
Chiasmata, 93, 94
Chimpanzee, 475
Chitin, 29, 32
Chiton, 571
Chlamydia, 116
Chlamydomonas, 512
Chloride, 20
Chlorine, 19-20, 168
Chlorophyll, 60, 133
Chlorophyll *a*, 136
Chloroplast, 48, 49, 59
adenosine triphosphate in, 139
photosynthesis and, 135
Choking, Heimlich maneuver in, 195
Cholecystokinin, 175, 177
Chondrocytes, 147, 153
Chordate, 454, 464, 465
embryo, 581
and vertebrate; *see* Vertebrate
Chorionic gonadotropin, 341, 358
Chorionic villus sampling, 364
Choroid, 288
Chromatid
nonsister, 93
sister, 81, 94
Chromatin, 57, 81, 82
Chromium, 27, 168
Chromosomal aberrations, 394
Chromosomal rearrangement, 394
Chromosomal theory of inheritance, 383
Chromosome, 57, 81, 82
crossing over, 93
deoxyribonucleic acid in, 81, 82
homologous, 93
human, 392-393, 394
changes in structure of, 392-393, 394
sex, 81, 383-384, 389
X, 384
Y, 384
Chronic obstructive pulmonary disease, 196
Chthamalus, 641, 642
Chyme, 175
Chymotrypsin, 177
Cilia, 49, 60
respiratory, 186
Ciliary muscle, 287
Ciliate, 507-509
Circulation, 198-215
arteries and arterioles in, 201-202
atmospheric, 672
blood, 208-211
capillaries in, 202-203
heart and blood vessel disease and, 213
hormone, 201
lymphatic system in, 211, 212
nutrient and waste transport in, 199
ocean, 674
oxygen and carbon dioxide transport in, 199
pathways of, 204-208
temperature maintenance in, 199, 200
veins and venules in, 203, 204
Circulatory system, 148, 199, 200
closed, 204, 571
open, 571
Cisternae, 55
Citric acid cycle, 117
Civilization, beginning of, 481
Clam, 570-573
Classical conditioning, 606
Classification
of living things, 12
of mammal, 472
of squirrel, 13
of virus, 490
Cleavage, 357
glycolysis and, 121
Cleft, synaptic, 257, 258
Climate, atmospheric effect on, 672-675
Climax in sexual intercourse, 346
Climax community, 652
Clitellum, 574
Clitoris, 346
Cloaca of cartilaginous fish, 586
Clones of aspen, 540
Closed circulatory system, 204, 571
Closely related, 12
Clot, blood, 211
Clownfish, 650
Club fungi, 517, 520-521
Cnidarian, 566, 567
Cobalt, 27, 168
Cochlea, 289
Cochlear duct, 289
Code of life; *see* Deoxyribonucleic acid
Codominance, incomplete, 399-400
Codominant allele, 400
Codon, 416
Coelom, 562
diversity in, 565-578
human, 149
Coelomate, 562
Coenzyme, 110, 166
Coevolution, 457644-645
Cofactor, 110
Cohesion, 555
Coitus, 346
Cold blooded, defined, 471
Collagen, 34, 38, 153
Collection duct, 236, 238, 239
Colon, 179-180
Colonial organism, 7
Colonial tunicate, 582
Colony of bees, 619
Color blindness, 395-396
Coloration
protective, 647-648, 649
warning, 645, 647
Columnar, defined, 147, 150
Comb jelly, 566
Commensalism, 649-650
Community, 10
climax, 652
redwood, 639
Community interaction, 638-655
competition in, 640-642
in ecosystem, 639
predation in, 642-648
succession in, 651-653
symbiosis in, 649-651
Compact bone, 156, 296
Comparative anatomy, 438-439
Comparative embryology, 440
Competent cell, 494
Competition in nature, 640-642
Competitive behavior, 611-613, 614
Competitive exclusion principle, 640-641
Complement, 224
Compound, 18
Compound eye, 576
Concentration stimulation of endocrine gland, 318
Condensation in cell cycle, 86
Conditioned behavior, 603-606, 607
Condom, 347, 350
Cone cell, 286
Coniferous forest, 682-683
Conjugation, 494, 495, 507
Connective tissue, 152-157, 201
adipose, 155
defensive, 152
dense fibrous, 153, 154
elastic, 153, 154
isolating, 152, 156-157
loose, 153, 154
Connective tissue—cont'd
matrix, 152
reticular, 153, 154
structural, 152
as tissue type, 147
Connell, J.H., 641
Consumer, 10, 657
primary or secondary, 658
Continental shelf, 688
Contraceptive; *see* Birth control
Contraction
of heart, 206-207
of muscle, 305-308, 309
Control, 5
positive and negative, 422
Controlled experiment, 5
Conus arteriosus in bony fish, 586
Convergent evolution, 439, 470, 471
Convolutions, 266
Cookeina tricholoma, 519
COPD; *see* Chronic obstructive pulmonary disease
Copper, 27, 168
Copulation, 346
of cartilaginous fish, 586
Coral, 561
Coral reef, 688, 689
destruction of, 701
Core, viral, 487
Cork cambium, 552
Cornea, 287
Corpus callosum, 265, 266
Corpus luteum, 341
Corpuscle
Meissner's, 283
pacinian, 284
Cortex, 236
adrenal, 325, 326, 329
of plant root, 549
Corti, organ of, 290
Corticosteroid, 326
Cotyledon, 538
Counseling, genetic, 400-401
Coupled channel, 75
Coupled reaction, 102, 103, 115
Courtship, 615, 616
hummingbird, 617
prairie chicken, 616
Covalent bond, 20-22
Cowpox, 218, 220
Coxal bone, 300
Cranial cavity, 149
Cranial nerve, 275
Crayfish, freshwater, 575
Creatinine, 234
Cretaceous Period, 456
Cri du chat syndrome, 393
Crick, F., 409
Cristae, 58
Cro-Magnon, 481
Cross bridge binding site, 310
Cross-fertilization, 374
Crossing-over, 93, 94
Crown, 171
Crown gall disease, 493
Crustacean, 574
CT; *see* Calcitonin
Cuboidal cell, 147, 150
Culture, 480
Cup fungi, 519
Cuticle, plant, 547, 552
Cyanobacteria, 450, 493
Cyanocobalamin, 167

Cycle
Calvin, 141
citric acid, 117
Krebs, 117, 123, 124
lysogenic, 489, 490
menstrual, 341
ovarian, 341
reproductive, 341-343
Cyclic electron flow, 140
Cylinder, vascular, 549
Cytokinesis, 90, 91-92
Cytokinin, 541
Cytoplasm, 48, 50
Cytosine, 39
Cytoskeleton, 49, 50, 51, 52
Cytosol, 49
Cytotoxic T cell, 222, 223

D

Dancer, interoception and, 281
Darwin
observations of, 427-429
theory developed by, 427-440, 441; *see also* Evolution
Data collection, 5
Dead air space, 192
Deamination, 234
Deciduous forest, 682
Decomposer, 657, 658
bacteria as, 493
Deductive reasoning, 5
Deep breathing, 191-192
Defecation, 180
Defense, 645
disease, 216-231
allergy in, 229-230
antigen receptors in, 226-227
cells of immune system in, 220, 221
discovery of, 218-219, 220
immunization in, 227
mechanisms of, 221-225
specific and nonspecific resistance in, 217, 218
genetic engineering and, 645
Defensive connective tissue, 152
Deforestation, 699-700
Dehydration, 28
Dehydration synthesis, 447
Deletion, 393
Delta, 685
Demography, 631-632
Demoiselle dragonfly, 613
Dendrite, 160, 250
Dense fibrous connective tissue, 153, 154
Density of population, 628
Density-dependent limiting factor, 628-630
Density-independent limiting factor, 628
Denticle, 585
Dentin, 171
Deoxyribonucleic acid, 11, 56-57, 63, 406-408
base pairing in, 409
in chromosomes, 81, 82
double helix and size of, 409
fragment production method for, 497
genes and, 412
hereditary information and, 39
as macromolecule, 29
Deoxyribonucleic acid—cont'd
nucleotide subunit of, 39, 408
replication and, 410, 411
structure of, 39, 408-411
supercoil of, 81, 82
Watson-Crick model of, 408
Dependent variable, 5
Depolarization, 253, 254
Depolarization stimulus, 288
Depolarizing, 286
Depth perception, 288
in evolution, 472, 473
Dermal tissue of plant, 546, 547-548
Descending arm of Henle's loop, 237, 238
Desert, 680-681
Detritus, 659
Deuterostome, 564
Development
prenatal, 355
of seed, 538
Devonian Period, 454
Diabetes mellitus, 328
Dialysis, kidney, 246, 247
Diamond sting ray, 466, 584
Diaphragm
in birth control, 347, 350
respiratory, 190
experiment for, 191
human, 149
Diastolic period, 207
Diatom, 503, 510
Diatomic molecule, 21
Dicot, 538, 539
leaf of, 554
root of, 549
Didnium, 642, 643, 644
Diencephalon, 265
Diet, 180-182
Dietary fiber, 180
Differentially permeable membrane, 72
Diffusion, 72
facilitated, 74-75
Digestion, 164-183; *see also* Digestive System
diet and, 180-182
hormones in, 175
large intestine in, 179-180
mouth in, 170-171
nutrition and, 166-168, 169, 180-182
organs supporting, 175-176
small intestine in, 175-179
stomach in, 174-175
journey to, 172-173
Digestive enzymes, 169
Digestive system; *see also* Digestion
function and components of, 148
human, 148, 173
organs of, 148, 173
upper, 174
Dihybrid, 380
and probability theory, 381-382
and Punnett square, 382-383
Diminishing natural resource, 696-700
Dinoflagellate, 509-510
Diploid, 83, 85, 529
Disaccharidase, 176
Disaccharide, 30
Dispersion of seed, 536
Distal convoluted tubule, 237, 238, 239, 240, 241
Distant relative, 12, 428
Distribution pattern in population, 628
Diurnal mammals, 474
Diversity
and classification of bacteria, 492-493
of living things, 7, 8
Division, 13
plant, 527
reduction, 92
DNA; *see* Deoxyribonucleic acid
Dodder plant, 651
Dominance, incomplete, 399-400
Dominant genetic disorder, 398, 399
Dominant trait, 375
Dopamine, 263
Dorsal side, 562
Double bond, 21
Double helix, 39
Down syndrome, 389-391
Dragonfly fossil, 455
Drosophila melanogaster, 383-384
Drug testing, 241
Duct, cochlear, 289
Ductus arteriosus, 369
Duodenal ulcer, 175
Duodenum, 175
Duplication, 393
Dura mater, 272
Dutch elm disease, 519

E

Ear, 289, 290
Earthworm, 573-574
ECG; *see* Electrocardiogram
Echinoderm, 577
Ecosystem, 10, 639, 656-670
aquatic, 662
carbon cycling in, 665-666
community and, 657-658
food chain and web in, 659-661
food pyramid in, 661-662
forest in, 10
nitrogen cycling in, 666-667
ocean, 687-690
phosphorus cycling in, 667-668
water cycling in, 664-665
Ectoderm, 359, 568
Ectopic pregnancy, 357
Ectotherm, 471
Ectothermic, defined, 589
Effectors, 257
in peripheral nervous system, 273
Efferent arteriole, 238, 239, 240, 241
Efferent neuron, 251, 264
Egg, 355
Elastic cartilage, 153, 155
Elastic connective tissue, 153, 154
Elastic fiber, 153
Elastin, 153
Electrical potential, 253-254
Electrocardiogram, 208
Electrolyte, 208
Electromagnetic energy, 132
Electromagnetic spectrum, 132
Electron, 17
Electron carrier, 117, 119
Electron flow
cyclic, 140
noncyclic, 137-139
Electron gain and reduction, 20
Electron shell, 18
Electron transport chain, 117, 125-127, 139
Elements, 18-19
periodic table of, 28
Elephantiasis, 570
Elimination, 233
Embolus, 213
Embryo, 361-366
development patterns of, 564
in evolutionary history, 440
Emigration, 625
Emperor penguin, 613
Emphysema, 196
Enamel, 171
End-product inhibition, 109
Endergonic reaction, 102, 115
Endocrine gland
growth and development of, 315
hormones of, 250, 315-328, 329; *see also* Hormone
in reproduction, 315
response in, 315
stimulation of, 318
Endocytosis, 77
Endoderm, 359, 568
Endodermis, root, 549
Endometrium, 341
Endoplasmic reticulum, 49, 52
rough or smooth, 48, 53
Endorphin, 330
Endosperm, 538
Endosymbiont, 49
Endosymbiotic theory, 452
Endothelial cell, 201
Endotherm, 471
Endothermic, as term, 592
Endplate, motor, 310
Energy
alternative sources of, 697
atomic, 18
electromagnetic, 132
free activation, 101
geothermal, 696
kinetic, 103
of motion, 104
potential, 103
in sunlight, 132
Energy extraction, 123
Energy flow, 100-113
change in living systems and, 103-104
chemical reactions and enzymes in, 105-110
start of, 101-102, 103
in ecosystem, 658
in Lake Cayuga, 661
storing and transferring in, 110-112
Energy level, 18
Energy pyramid, 10, 662
Energy transformation in living things, 10
Enkephalin, 330
Entropy, 104
Envelope, viral, 487
Environment, 695
Enzyme, 105-110
active sites and, 105, 106
catalysis and, 29
digestive, 169
function of, 105-107, 108
hydrolyzing, 170
induced fit and, 105, 106
substrate and, 108
Eocene Epoch, 460
Eosinophil, 209, 210
Epidermal cell, 547

Epidermis of root, 549
Epididymis, 335, 336, 338
Epidural space, 272
Epiglottis, 172
Epinephrine, 326
Epiphyte, 649, 676
Epithelial cell, 147, 149
Epithelial tissue, 147, 149-151
Epithelium, 149
pseudostratified, 151
simple, 151
of stomach, 174
stratified, 151
transitional, 151
Epoch, 448
Eocene, 460
Oligocene, 460
Paleocene, 460
Era, 448
Archean, 448-452
Cenozoic, 458-460
Mesozoic, 455-458
Paleozoic, 452-455
Proterozoic, 452
Erythrocyte, 157, 209, 210
oxygen transport and, 156
Esophageal hiatus, 172
Esophageal sphincter, 172
Esophagus, 172
Essential amino acids, 181
Estrogen, 320, 341
Estuary, 685-687
Ethology, 599
Ethylene, 541
Eubacteria, 492-493
Euglena, 61, 506
Euglenoid, 135, 506
Eukaryote, 13
chromosomes of, 57
flagellum of, 47, 49, 60, 61
Eukaryotic cell, 47, 49-61
bacterial cells versus, 64, 65
nucleus of, 49
structure of, 48, 52
Eustachian tube, 289
Eutrophic lake or pond, 668
Eutrophication, 703
Evolution, 11-12, 426-443
convergent, 439, 470, 471
Darwin's theory of, 427-433
artificial breeding in, 429-431
finches in, 432-433
geology in, 429
natural selection in, 432
observations in, 427-429
populations in, 431
publication of, 433-434
testing of, 434-440, 441
defined, 426-427
dinosaur extinction in, 455
drifting continents and, 458, 459, 460
early history of, 445-446, 447
embryos in, 440
evidence of, 427-433
fossils and, 434-436, 437
of horse, 437
human; *see* Human evolution
of kingdoms, 444-462
in Archean Era, 448-452
and cell origin, 447-448
in Cenozoic Era, 458-460
and mass extinction, 455
in Mesozoic Era, 455-458
Evolution—cont'd
of kingdoms—cont'd
and organic molecule origin, 445-446, 447
in Paleozoic Era, 452-455
in Proterozoic Era, 452
Excitatory synapse, 259
Excitement phase, 346
Exclusion, competitive, 640-641
Excretion, 232-248
epithelial tissue and, 149
kidney in, 234-243
anatomy of, 235, 236
filtration of, 235-237
homeostasis and, 244-245
nephron of, 235-243
problems of, 245-247
renal failure and, 246, 247
selective reabsorption of, 237-241
stones of, 245-246
tubular secretion of, 237-241
organs of, 233, 234
substances from body in, 233-234
urinary system in, 244
Exergonic reaction, 101, 102, 106
Exocrine gland, 315
Exocytosis, 77
Exoskeleton of arthropod, 574, 575
Experiment
Hershey-Chase, 414, 415
Miller-Urey, 445, 446
of Morgan in sex linkage, 383-384
Experimentation, 3, 4, 6
Expiration, 186, 191, 192
Exponential growth, 492, 625-626, 627
External fertilization, 584
External genitals, 338
in female reproductive system, 343-346
External intercostal muscle, 190
External respiration, 185
Extraembryonic membrane, development of, 360-361
Eye
compound, 576
human, 286, 287
simple, 576
Eyespot, 506

F

F plasmid, 494
Facilitated diffusion, 74-75
Fallopian tube, 340
False rib, 300
Family, 13
Family tree of primate, 476
Fat, 32
macromolecule of, 29
nutrition and, 166
polyunsaturated, 33
saturated, 33, 34
triglyceride, 33
unsaturated, 33, 34
Fat cell, 45
Fatty acids, 33, 68
Feather, 11, 592
Feces, 178
Feedback, 162-163
negative, 109, 162
positive, 163
Feedback loop, 162, 318
negative, 322, 323
Feedback loop—cont'd
positive, 322
Felis domesticus, 14
Female external genitals, 343-346
Female gamete, 335
Female reproductive system; *see* Reproductive system, female
Femur, 300
Fermentation, 116, 121, 123
Fern, 532
Fertilization, 83, 85, 355-356, 358
external, 584
Fetus, human, 366-368
Fever, 222
Fiber
cellulose, 32
protein, 38
skeletal muscle, 159, 305
spindle, 86
Fibrin, 38
Fibrin thread, 211
Fibroblast, 147, 153
Fibrocartilage, 153, 155
Fibrous protein, 38
Fibrous root, 551
Filial generation
first, 374
second, 375
Filtration, kidney, 234, 235
Finch
Darwin's theory of evolution and, 429, 432-433
ground, 429, 433
tree, 433
warbler, 433
First filial generation, 374
First Law
of thermodynamics, 104
First law
of Mendel, 378
First trimester of pregnancy, 366
Fish
bony, 466, 467, 586
cartilaginous, 584-586
evolution of, 465-466, 467
fleshy-finned, 466
jawless, 465, 584
lobe-finned, 466, 586
lung, 466, 467
ray-finned, 466, 586
Fission, transverse, 507
Five-carbon sugar, 39
Fixed action pattern, 602-603
Flagellum, 47, 49, 60, 61, 506
Flatworm, 568-569
Fleshy-finned fish, 466
Floating rib, 300
Fluid mosaic model, 69
Fluoride, 168
Fluorine, 27
Foam, birth control, 347
Focusing of human eye, 287
Folic acid, 167
Follicle, 339
Follicle-stimulating hormone, 320, 336
Food chain, 659-661
Food pyramid, 661-662
Food web, 659-661
Foramen, 271
Foramen ovale, 369
Foraminifera shell, 505
Forest
coniferous, 682-683
deciduous, 682
Forest—cont'd
ecosystem of, 10
Formation, reticular, 271
Formed elements, 208
Fossil, 12, 434
Archaeopteryx, 435, 456
Burgess shale, 453
dating of, 436-438
dragonfly, 455
evolution of horse based on, 437
formation of, 435
Mawsonites spriggi, 436
mold, 436
Pikaia gracilens, 464
process of formation of, 435
sediment, 434
trilobite, 454
Fossil fuel, 696-698
Fossil record, 434-436, 437
Fovea, 288
Free activation energy, 101
Free radical, 393
Fresh water life zone, 684-685
Freshwater crayfish, 575
Frog, warning coloration in, 647
Frond, 531
Frontal lobe, 266
Fruit in reproduction, 534, 536
FSH; *see* Follicle-stimulating hormone
Fuligo, 515
Fundamental niche, 642
Fungus, 516-523
club, 517, 520-521
cup, 519
hyphae of, 516
imperfect, 517, 522
as kingdom, 12, 13
lichen and, 523
life cycle of, 517
mycelium of, 516
sac, 519-520
spores of, 517
zygote-forming, 517, 518
Fungus-like protist, 513-514
Fusion of nuclei, 356

G

Galapagos Islands, 428
Galapagos tortoises, 428
Gallbladder, 176
Gamete, 83, 85
female and male, 335
Gametophyte, 84, 85, 529
Ganglia, 275
basal, 267, 269
in central nervous system, 267
cerebral, 574
Gas
hydrogen, 17, 21
transport and exchange of
hemoglobin and, 192
respiration and, 192-195
Gastric fluid, 174
Gastric gland, 174
Gastrin, 174, 175
Gastropod, 571
Gastrula, 564
Gastrulation, 361, 363
Gate-control theory of pain, 281
Gause, G.F., 642, 643, 644
Gene, 81
defined, 413
deoxyribonucleic acid and, 412

Gene—cont'd
hereditary information in, 412-413
structural, 420
Gene expression, 413-419
Gene family, 422
Gene mutation, 394-395
Gene therapy, 422-423
germ line in, 423
in recombinant deoxyribonucleic acid technology, 499
somatic cell, 423
General adaptation syndrome, 326
Generation
alternation of, 529
first filial, 374
parental, 374
second filial, 375
Generator potential, 275, 282
Genetic code, 416
Genetic counseling, 400-401
Genetic engineering, 494-498
defense by, 645
Genetics, 377
behavioral, 600
human; *see* Human genetics
Genitals, external, 338, 343-346
Genome, 405
Genotype, 378
Genus, 13
Geographical isolation, 432
Geothermal energy, 696
Germ layer, 359
Germ line gene therapy, 423
Germ plasm bank, 701
Germination, 538, 539
Gestation, 355
GH; *see* Growth hormone
Giantism, 320
Giardia lamblia, 503
Gibberellin in plant, 541
Gila monster, 647
Gill of mollusk, 571
Gizzard, 591
Gland, 149
accessory, 338
adrenal, 325-327, 329
bulbourethral, 338
endocrine; *see* Endocrine gland
exocrine, 315
mammary, 346
parathyroid, 324, 325, 329
pineal, 328
pituitary, 318-322
anterior, 320-321, 329
posterior, 321-322, 329
prostate, 244, 338
salivary, 169
thymus, 328
thyroid, 322, 323, 329
Glandular epithelial tissue, 149
Glial cell, 160
Global warming, 700
Globin gene, 441
Globular protein, 38
Glomerulus, 237, 238, 239, 240, 241
Glottis, 172, 186
Glucagon, 327, 328
Glucocorticoid, 326
Glucose, 30, 31
blood, 180
mobilization of, 121
oxidation of, 126
Glyceraldehyde phosphate, 141
Glycerol, 33
phospholipid, 68
Glycogen, 29, 30, 31
Glycolipid, 55
Glycolysis, 117, 121-123
Glycoprotein, 55
Goiter, 322
Gold tunicate, 582
Golden algae, 509, 510
Golden-brown algae, 510
Golgi, C., 54
Golgi body, 53, 54, 55
Golgi complex, 48, 49, 54, 55
Gonad, 335
Gonadotropin hormone, 320
Gondwana, 458
Goose, 617
Gorilla, 475
Gradient, 72
Grand Canyon, 435
Granulocyte, 210
Granum, 60, 135
Grasshopper
African, 647
anatomy of, 576
copulating, 575
Grassland, 658
temperate, 681
Gravitaxis, 601
Gray-air city, 707, 708
Gray matter, 266, 271
Gray whale, 650
Great ape, 474
Green algae, 509, 512
Green caterpillar, 648
Greenhouse effect, 700
Grinding teeth, 171
Ground finch, 429, 433
Ground tissue, 546, 547
Ground water, 664, 703
pollution of, 705
Group behavior, 618-621
Growth
endocrine gland and, 315
exponential, 625-626, 627
plant, 548
population, 625-627
Growth curve
of human population, 633
sigmoid, 627
Growth hormone, 320
Growth rate, 625
Guanine, 39
Guano coast, 668
Guard cell, 554, 555
Gymnosperm, 533

H

Habitat, 639
Habituation, 605
Hair, 38
Hair cell, 289
Hairy-cap moss, 530
Half-life of carbon-14 isotope, 436
Haploid, 83, 85, 529
Hardwood, 552
Haversian canal, 296
Hawkmoth, 649
Head chordate, 583
Hearing, 288-290
Heart, 199
contraction of, 206-207
Heart attack, 185, 212, 213
Heart disease, prevention of, 206
Heart murmur, 207
Heat regulation, 200
Heavy chain, 226
Heimlich maneuver, 195
Helium, 18
Helix, 39
Helper T cell, 222, 223, 224, 227
Hemoglobin, 210
and gas transport, 192
Hemophilia, 389, 397
Henle's loop, 237, 238-241, 243
ascending or descending arm of, 237, 238
Herbivore
food plants and, 644-645
teeth of, 170
Heredity of living things, 11; *see also* Human genetics
Hermaphrodite, 566
Hershey, A., 414
Hershey-Chase experiment, 414, 415
Heterotroph, 131, 659
bacteria as, 493
Heterozygous, defined, 378
Hiatal hernia, 172
High altitude sickness, 194
High-energy bond, 110
Hinge joint, 304
Hip, 300
His bundle, 207
HIV; *see* Human immunodeficiency virus
Holdfast, 511
Homeostasis, 162, 245, 276
Hominid, 475
evolution of, 477-481
Hominoid, 474
Homo erectus, 480
Homo habilis, 480
Homo sapiens, 14, 475, 480-481
Homologous chromosomes, 93
Homology, 438
Homozygous, defined, 378
Homunculus, 373
Hookworm, 651
Hormone, 162, 244, 314-331
adrenocorticotropic, 320
aldosterone, 244, 245
antidiuretic, 244, 245, 321
in communication, 250
digestion and, 175
endocrine gland, 250, 315-328
adrenal, 325-327
ovarian and testicular, 328
pancreatic, 327-328
parathyroid, 324, 325
pineal, 328
pituitary, 318-322
thymic, 328
thyroid, 322, 323
follicle-stimulating, 320, 336
gonadotropin as, 320
growth, 320
local, 315
luteinizing, 320, 336
melanocyte-stimulating, 321
nonendocrine, 328-330
parathyroid, 322, 324
peptide, 315, 317
of plant, 541
prolactin, 321
releasing, 318
steroid, 315, 317
Hormone—cont'd
thyroid-stimulating, 320
tropic, 320
in water and salt balance in kidney, 245
Hornwort, 529, 530
Horseshoe crab, 454
House cat, 14
Human
body organization of; *see* Body organization
brain of, 265
chromosome of, 392-393, 394
digestive system of, 148, 173
eye of, 286, 287
nervous system of, 148
tissue of, 149-160
Human-engineered gene transfer, 496-498
Human evolution, 463-483
amphibian and, 466-468
bird and, 466-468
in Cenozoic Era, 458-460
fish and, 465-466, 467
hominid in, 477-481
Australopithecus, 477-479
Homo erectus, 480
Homo habilis, 480
Homo sapiens, 480-481
mammal in, 468-471
primate in, 471-475, 476
anthropoid and, 473, 474-475, 476-477
prosimian and, 473
vertebrate in, 464-471
Human fetus, 366-368
Human genetics
counseling in, 400-401
dominant and recessive disorders in, 398, 399
incomplete dominance and codominance in, 399-400
karyotypes in, 389-395
autosomes and, 389-391
chromosome changes and, 392-393, 394
gene mutations and, 394-395
Klinefelter syndrome and, 391
triple X females and, 391
Turner syndrome and, 392
XYY males and, 392
multiple alleles in, 400
patterns of inheritance in, 389-397
pedigrees and, 395-397
Human immunodeficiency virus, 228
Human life cycle, 83
Human population, 632-636
Humerus, 300
Hummingbird, 617
Humoral immune response, 223-224, 225
Humulin, 497
Huntington's disease, 398
Hyaline cartilage, 153, 154
Hydra, 566-568
Hydration shell, 24
Hydrocarbon, 33
Hydrochloric acid, 26, 174
Hydrogen, 17, 20-21, 27
covalent bond of, 20-21
Hydrogen bond, 23
Hydrogen gas, 17, 21
Hydrolysis, 30, 170
Hydrolyzing enzyme, 170

Hydrophilic, defined, 69
Hydrophobic, defined, 69
Hydrophobic bond, 24-25
Hydrophobic compound, 24
Hydrothermal vent, 447
Hydroxyl group, 29
Hypercholesterolemia, 398
Hyperosmotic, defined, 587
Hyperpolarization, rod cell and, 286
Hyperthyroidism, 322, 323
Hypertonic solution, 73
Hyphae of fungi, 516
Hypoosmotic, defined, 587
Hypopituitary dwarfism, 320
Hypothalamus, 163, 265, 329
in central nervous system, 269, 270
pituitary gland and, 318
Hypotheses, 3, 6
Hypothyroid goiter, 322
Hypothyroidism, 322
Hypotonic solution, 73

I

Ice age, 480
Ice formation, 24
Ichthyostega, 468
IgM; *see* Immunoglobulin M
Immigration, 625
Immune response, 219, 221
Immune system, 148, 218-230; *see also* Defense, disease
Immunity, 217
active or passive, 227
Immunodeficiency virus, 228
Immunoglobulin, 224
Immunoglobulin M, 227
Imperfect fungi, 517, 522
Implant, contraceptive, 350
Implantation, 357, 358
Imprinting, 603-604
Inborn behavior, 600-602
Inchworm caterpillar, 649
Incisors, 171
Inclusions, 63
Incomplete dominance or codominance, 399-400
Incus of ear, 289
Independent assortment, 381
Independent variable, 3
Induced fit, 105, 106
Inducer T cell, 222
Induction in human development, 364
Inductive reasoning, 3
Infection, 217
Infectious agent, 704
Inferior vena cava, 204, 206
Inflammation, 217
Information
storage of hereditary, 405, 406
transmission of, 249-261
communication system of body and, 250
nerve impulse in, 252-257
nerve to target tissue, 257-259
neuron in, 250-252
Inheritance
historical view of, 373-374
karyotypes and, 389-395
Mendelian, 374-383
analyzing, 378-379
conclusions of, 377-378
crossing-over in, 374
dihybrid crosses in, 381-383
Inheritance—cont'd
Mendelian—cont'd
factors and chromosomes in, 383
gene interaction in, 377
questions of, 380-381
testing of, 379-380
molecular basis of, 404-424
deoxyribonucleic acid replication in, 411
gene in, 412-413
gene expression in, 413-419
gene therapy in, 422-423
location of hereditary material in, 405, 406
nucleic acid in, 406-411
prokaryotic and eukaryotic genes in, 419, 420-422
mysteries of, 385
pedigrees and, 395-397
sex linkage, 383-384
study of, 374
of traits, 373, 374
variations in species of, 373-374
Inhibition, end-product, 109
Inhibitor, 109
Inhibitory synapse, 259
Innate behavior; *see* Behavior, innate
Inner cell mass, 357
Inner ear, 289, 290
Inorganic nutrient, 703
in plant, 548
Insect, 574-576
society behavior of, 619-621
Insertion, 305
Insight, 606, 607
Inspiration, 186, 190-191
Instinctive behavior, 600-602
Insulin, 243, 327, 328
genetic engineering and, 498
Integration, 265, 275
Integrator, 259
Integument, 296
Integumentary system, 148, 296
Interaction of living things, 8-10
Intercostal muscle, 190
Intercourse, 346
Interleukin-1, 222
Intermediate fibers, 50, 51
Internal intercostal muscle, 190
Internal respiration, 186
Interneuron, 251, 275
Interphase, 86, 87, 88, 90
Intertidal zone, 687-688
Intervertebral disk, 300
Intestinal cell, 178
Intestine, 233
large, 179-180
small, 175-179
digestive enzymes and, 169
Intimidation display, 611
Inversion, 393
Invertebrate animal, 560-579
asymmetry in, 565-566
bilateral symmetry in, 568-578
clam, snail, and octopus, 570-573
earthworm and leech, 573-574
flatworm, 568-569
lobster, insect, and spider, 574-576
roundworm, 569-570
sea urchin and starfish, 577-578
characteristics of, 561-564
radial symmetry in, 566-568
Involuntary response, 264
Iodine, 27, 168
Ion, 19-20
in vertebrate, 585
Ionic bond, 19-20
Ionization, 25-26
Ionizing radiation, 392
Iris, 288
Iron, 27, 168
Islets of Langerhans, 327, 329
Isolation, geographical, 432
Isotonic solution, 73
Isotopes, 17
half-life of carbon-14, 436

J

Jackrabbit, 649
Jaw, evolution of, 465
Jawless fish, 584
Jellyfish, 566-568
Joint, 302-304
Jumping spider, 575
Junction, neuromuscular, 258, 259, 310
Jurassic Period, 456

K

Kalanchoe daigremontiana, 540
Kangaroo rat, 235
Karyotype, 81
and inheritance patterns, 389-395
Kelp, 511
Keratin, 38
Ketone, 328
Kidney
of fish, 235
human, 233, 234
artificial, 247
dialysis and, 246, 247
failure of, 246
filtration of, 234, 235-237
nephron in, 235-243
selective reabsorption in, 235, 237-241
structure of, 236
tubular secretion in, 235, 241
of kangaroo rat, 235
water balance and, 234
Kidney stones, 245-246, 325
Killer cell, 222
Kineses, 600
Kinetic energy, 103, 104
Kinetochore, 89
Kingdoms of life, 12
Klinefelter syndrome, 391
Knee-jerk reflex, 276
Koran angelfish, 467
Krebs cycle, 117, 123, 124

L

Labia majora or minora, 346
Labor in birth, 368
Lac operon, 420, 421
Lac repressor, 422
Lactase, 169
Lacteal, 178
Lactose, 30
Lacunae, 153
Lake Cayuga, 661
Lake zones, 685
Lamprey, 465
Lancelet, 583
Land biome, 675-684, 696-703
Langerhans islets, 327, 329
Lanugo, 366
Large intestine, 179-180
Larval tunicate, 583
Larynx, 172, 186, 187
Lateral bud, 551
Lateral line system, 585
Lateral meristem, 548
Laurasia, 458
Law
of independent assortment, 381
of segregation, 378
of thermodynamics, 103
Leaf, 545, 554-555
Leafcutter ant, 620-621
Learning, 603-606, 607
Leech, 573-574
Left atrium, 204
Left ventricle, 204, 207
Left ventricle assist device, 199
Leg, 300, 301
Lens of eye, 287
Lesser ape, 474
Leukocyte, 157, 210
Levels of human body organization; *see* Body organization
LH; *see* Luteinizing hormone
Lichen, 523
Life cycle
of angiosperm, 535
of fern, 532
of fungi, 517
of mushroom, 521
of *Obelia*, 567
of pine, 533
of plant, 527
sexual, 528
Ligament, 302
Ligase, 497
Light chain, 226
Light-dependent reaction, 135, 136-140
Light-independent reaction, 135, 141-142
Limb bud, 366
Limbic system, 269, 270
Linolenic acid, 34
Lipase, 168, 177
Lipid, 29, 32-34, 219
Lipid bilayer, 69, 219
Liver, 175, 233
Liverwort, 529, 530
Living things, 7-12
Lobe of brain, 266
Lobe-finned fish, 586
Lobster, 574-576
Local hormone, 315, 328
Loop of Henle, 239, 240, 241, 243
ascending or descending arm of, 237, 238
Loose connective tissue, 153, 154
Lorenz K., 603, 604
Lower esophageal sphincter, 172
Lucy, *Australopithecus*, 478
Lumen, 201
Lung, 205, 233
Lung volume, 192
Lungfish, 467
Luteinizing hormone, 320, 336
Lymph node, 212, 220
Lymphatic system, 211, 212, 221
Lymphocyte, 210, 221
B, 209, 223, 224

Lymphocyte—cont'd
immune system and, 147, 150
T, 209, 211, 222, 223, 328
Lymphokine, 222, 223, 224
Lyse, 74, 489
Lysogenic cycle, 489, 490
Lysosomes, 48, 49, 56
Lysozyme, 37
Lytic cycle, 489-490

M

Macromolecule, 28, 29
Macrophage, 188, 210
in body defense, 147, 152, 221, 223, 224
cell ingestion by, 221
immune system and, 147, 152
Magnesium, 27, 168
Malaria, 508, 509
Male gamete, 335
Male reproductive system, 335-339
Malleus, 289
Malpighian tubule, 576
Maltase, 169
Mammal, 593, 594
classification of, 472
and evolution, 468-471
placental, 471, 593, 594
Mammalia, 593
Mammary gland, 346
Mandible of arthropod, 574, 575
Manganese, 27, 168
Manipulated variable, 3
Manta ray, 466, 584
Mantle, 571
Marchantia, 529
Marfan's syndrome, 398
Marrow
bone, 156
red, 297
yellow, 297
Marsupial, 471, 593, 594
with young, 618
Mass extinction, 455
Mass flow, 557
Mast cell, 147, 152
Mathematical progressions, 431
Mating, 615
Matrix in connective tissue, 152
Mawsonites spriggi fossil, 436
Mechanoreceptor, 585
Medulla, 191, 236, 265, 271
adrenal, 325, 326-327
Medusa, 567
Megakaryocyte, 211
Meiosis, 83, 85, 92-94, 95
mitosis compared to, 96
Meiotic recombination, 97
Meissner's corpuscle, 283
Melanin, 295, 321
Melanocyte, 321
Melanocyte-stimulating hormone, 321
Melzack, R., 281
Membrane, 48
nuclear, 56, 57
photosynthetic, 135
pleura, 191
postsynaptic, 257
presynaptic, 257
tectorial, 289-290
Membranous organelle, 53
Memory B cell, 224
Memory T cell, 223
Mendel, G.J., 374
first law of, 378
second law of, 381
Mendelian genetics; *see* Inheritance, Mendelian
Mendelian trait, 374
Meninges, 272
Menstrual cycle, 341, 342
Menstruation, 342
Meristem, 546, 548
Meristematic tissue, 546, 548
Merkel's disk, 283
Mesentery, 174
Mesoderm, 359, 568
Mesopelagic, defined, 690
Mesophyll, 554
Mesosome, 63
Mesozoic Era, 455-458
Messenger, second, 315
Messenger ribonucleic acid, 413
Metabolic machine, 142
Metabolic pathway, 105, 115
Metabolism, 121, 127
Metamorphosis, 588
Metaphase, 87, 88, 89, 91
Metaphase plate, 89, 93
Microbe, 221
Microfibrils, 32
Microfilament, 50, 51, 52
Microtubule, 50, 52, 60
tubulin and, 51
Microtubule organizing center, 44, 86
Microvilli, 178
Midbrain, 265
Middle ear, 289
Migration, 603
caribou, 604
Milk letdown, 346
Miller-Urey experiment, 445, 446
Mimicry, 647-648
Mineral resource, 698-699
Mineralocorticoid, 326
Minerals, 166, 167, 168, 169
Mite, 229
Mitochondria, 48, 49, 52, 58-59
Mitosis, 83, 86-90
in animal cell, 87, 90, 91
meiosis compared to, 96
in plant cell, 88, 90, 91
Mixed nerve, 275
Molars, 171
Mold
black bread, 518
slime, 504, 514
structure of, 516
water, 504, 515
Mold fossil, 436
Molecular basis of inheritance; *see* Inheritance, molecular basis of
Molecular biology, 440, 441
Molecular motion, kinetic energy of, 104
Molecule, 18-22
diatomic, 21
of glucose, 30
organic, 26, 27, 28
polar, 23
stable, 21
Mollusca, 570
Mollusk, 570, 571
Molybdenum, 27
Monarch butterfly, 645
Monera, 12, 13, 491
Monkey, 474
Monocot, 538, 539
leaf, 554
root, 549
Monocyte, 152, 209, 221, 222
Monomer, 447
Monosaccharide, 30
Monosynaptic reflex arc, 276
Monotreme, 471, 593
Mons pubis, 343
Moray eel, 467
Morphogenesis, 357, 361
Mortality of population, 630-631
Morula, 357, 564
Mosaic model, fluid, 69
Moss, 530
Motion, energy of, 104
Motor area of brain, 266, 267
Motor endplate, 310
Motor neuron, 251, 264
Motor pathway, 275-277
Mouth in digestion, 170-171
Movement of water in plants, 555-557
MTOC; *see* Microtubule organizing center
Mucin, 355
Mucosa, 174
Mucous membrane, 218
Mucus, 174, 175
Mueller gibbon, 475
Müllerian mimicry, 648
Multicelled, defined, 84
Multiple allele, 400
Muscle, 158-159, 304, 305-311
ciliary, 287
intercostal, 190
microfilament in, 158, 159, 305, 309
spindle, 283
as tissue type, 147
Muscle cell, 307
Muscle fiber, 305
nerve impulses and, 305, 310, 311
Muscle tissue, 158-159
Muscular system, 304, 305-311
function and components of, 148
Mushroom, life cycle of, 521
Musk-oxen, 618
Mutation, 389
gene, 394-395
point, 394
Mutualism, 649, 650-651
Mycelium of fungi, 516
Myelin sheath, 252
Myofibril, 305
Myofilament, 305, 309
Myosin, 307

N

Naked seed, 533
Nasal cavity, 172, 185, 186, 187
Natural gene transfer, 494, 495
Natural history, 3
Natural killer cell, 221, 222
Natural resources, 696-700
Natural selection, 432
Natural variation, 429
Naturalist, 3
Neanderthal man, 480
Nectar, 535
Negative control, 422
Negative feedback, 109
Negative feedback loop, 322, 323
Nekton, 690
Nephridia, 573
Nephron, 235-243
Neritic zone, 688, 689
Nerve, 160
cranial, 275
mixed, 275
optic, 288
peripheral nervous system and, 269
spinal, 271
structure, 274
Nerve cell, 45
Nerve cord, 464
spinal, 581
Nerve impulse, 250, 252-257
muscle and, 310, 311
Nerve ring, 578
Nervous system, 148, 250
autonomic, 264, 276-277
central, 264, 265-273
brain in, 265-271
spinal cord in, 271-272
organization of, 148, 263-265
parasympathetic, 265, 276, 277
peripheral, 264, 273-277
somatic, 264, 275-276
sympathetic, 265, 276, 277
Nervous tissue, 8, 159-160
as tissue type, 147
Net movement, 72
Neural fold, 364
Neural groove, 364
Neuroglia, 252
Neuromuscular junction, 258, 259, 310
Neuron, 159-160, 250-252, 264
Neurotransmitter, 257, 258
Neurulation, 363, 364
Neutral atom, 17
Neutron, 17
Neutrophil, 209
New Jersey toxic chemical dump, 703
New World monkey, 474
New York City dump, 702
Newborn, 368, 369
Niacin, 167
Niche, 639
fundamental, 642
realized, 642
Nicholas II of Russia, Czar, 397
Night blindness, 167
Nitrifying bacteria, 493, 667
Nitrogen, 27
energy levels for, 18
Nitrogen cycle, 666-667
Nitrogen-fixing bacteria, 493, 667
Nitrogenous waste, 234
NK cell; *see* Natural killer cell
Noble gases, 19
Nocturnal, defined, 468
Nodes
of plant stem, 551
of Ranvier, 252
Noncyclic electron flow, 137-139
Nondisjunction, 389, 390, 391, 392
Nonendocrine hormone, 328
Nonpoint source, 703
Nonsister chromatid, 93
Nonspecific defense, 217, 218
Nonvascular plant, 527, 557
patterns of reproduction in, 529-530, 531
Noradrenaline; *see* Norepinephrine
Norepinephrine, 326

Notochord, 361, 454, 464
in chordates, 581
Nuclear envelope, 48, 56, 57
Nuclear membrane, 56, 57
Nuclear pores, 48, 56, 57
Nuclear power, 696
Nuclei, 47
in central nervous system, 267
fusion of, 356
Nucleic acid, 29, 39, 405, 406-411
Nucleoid, 47, 63
Nucleolus, 48, 49, 57, 58
Nucleosome, 82
Nucleotide, 11, 39
structure of, 406
Nucleus, 47, 48, 56-58
Number pyramid, 662
Nutrition, 180-182

O

Obelia, 567
Obligate parasite, 487
Obstructive pulmonary disease
chronic, 196
Occipital lobe, 266
Ocean
ecosystems of, 687-690
patterns of circulation in, 674
zones of, 687, 690
Octet rule, 19
Octopus, 570-573
Oil, 33
Old World monkey, 474
Oleic acid, 34
Olfactory receptor, 284
Oligocene Epoch, 460
Omnivore, 170
On the Origin of Species, 434
One gene–one enzyme theory, 412
Onymacris unguicularis beetle, 681
Oocyte, 339, 340
Oogenesis, 340
Oogonia, 340
Open circulatory system, 571
Open dump, 702
Open-sea zone, 684, 687, 690, 691
Operant conditioning, 605-606
Operator, 420
Operculum, 586
Operon, 420
Opsin, 286
Optic nerve, 288
Oral contraceptive, 347
Orangutan, 475
Orb-weaving spider, 575
Orbital, 17
Order, 13
Ordovician Period, 454
Organ, 8
of Corti, 290
defined, 147, 160
of vascular plant, 548-555
in vertebrate, 160
vestigial, 438
Organ system, 8
in human body organization, 148, 149, 160-162, 163
Organelle, 12, 49
bacteria-like, 58-60
membranous, 53
Organic molecule, 26, 27, 28
origin of, 445-446, 447
Organic nitrogen-containing base, 39
Organic nutrient, 703
Organism, 7
Organization
bone, 297
of living things, 7-8, 9
Organizing center in microtubule, 86
Organogenesis, 361
Orgasm, 346
Osmoregulation, 585
Osmosis, 72-74
Osmotic pressure, 73
Ossification, 367
Osteoblast, 296
Osteocyte, 147, 153, 156
Osteoporosis, 325
Ostracoderm, 454
Outer ear, 289
Oval window, 289
Ovarian cycle, 341
Ovarian follicle of cat, 341
Ovary, 328, 329, 339-340, 534
Overpopulation, 708
Oviduct, 340
Oviparous, defined, 586
Ovoviviparous, defined, 586
Ovulation, 340
Ovule, 533
Oxidation, 20, 119
of glucose, 126
glycolysis and, 121
Oxidation-reduction, 119
Oxidized, defined, 119
Oxygen, 20
Oxygen cycling and carbon cycle, 665
Oxygen debt, 128
Oxygen transport and erythrocyte, 156
Oxytocin, 321, 322
Ozone depletion, 708

P

Pacinian corpuscle, 284
Paleocene Epoch, 460
Paleozoic Era, 452-455
Palisade layer, 554
Palmitic acid, 34
Pancreas, 327-328, 329
digestion and, 169, 175
Pancreatic amylase, 176
Pancreatic lipase, 177
Pangea, 458
Pantothenic acid, 167
Papillae, 285
Paramecium, 45, 61, 507, 642, 643, 644
microscopic view of, 47
pinocytosis and, 76
Parasite, 516
Parasitism, 630, 649, 651
Parasympathetic nervous system, 265, 276
Parathyroid gland, 324, 325, 329
Parathyroid hormone, 322, 324
Parenchyma cell, 547
Parental generation, 374
Parenting behavior, 618
Parietal lobe, 266
Parotid gland, 170
Passive facilitated diffusion, 75
Passive immunity, 227
Passive transport, 72
Pathway
of air, 186-189
metabolic, 105, 115
Pathway—cont'd
motor, 275-277
sensory, 273-275
Pavlov, I., 606
Pea plant
anatomy of, 374
contrasting traits in, 375
Peacock, 615
Peat moss, 530
Pectoral girdle, 300, 301
Pedigree
and inheritance patterns in humans, 395-397
red-green color blindness, 395
Pelagic zone, 684, 687, 690, 691
Pelican hunting behavior, 619
Pellicle, 61
Pelvic cavity, 149, 339
Pelvic clasper of cartilaginous fish, 585
Pelvic girdle, 300, 301
Penicillium, 522
Penis, 244, 335, 338-339
Pepsin, 169, 174
Peptidases, 169, 177
Peptide, 34
Peptide bond, 34, 36
Peptide hormone, 315, 317
Peptidoglycan, 62, 63
Perception, depth, 288
Pericycle of plant root, 549
Periodic table of elements, 28
Peripheral nervous system, 264, 273-277
Peristalsis, 172, 173, 244
Permafrost, 684
Permeable membrane
differentially, 72
selectively, 73
Permian Period, 455
Pernicious anemia, 167
Petiole, 554
pH scale, 25, 26
Phage, 489
Phagocyte, 221
Phagocytosis, 76, 77, 152, 210, 221
Pharyngeal arch, 464, 581
Pharynx, 172, 186
Phenotype, 378
Pheromone, 620
Phloem, 546
Phosphate, 39, 68
glyceraldehyde, 141
Phospholipid, 29, 34, 68
Phosphorus, 27, 168
Phosphorus cycle, 667-668
Photic zone, 690
Photoautotroph, 492
Photocenter, 136, 137
Photochemical smog, 707
Photon, 132, 286
Photoreceptor, 506
Photosynthesis, 10, 59, 116, 130-144
Photosynthetic membrane, 65, 135
Photosystem, 136-140
Phototaxis, 601
Phototropism, 541
Phyla, 464
bilateral symmetry in, 568-578
Phylogenetic tree, 440, 476
Phylum, 13
Phytochrome, 541
Pigment, 133
Pikaia gracilens, 464
Pili, 47
Pill, birth control, 350
Pilus, 494
Pine
life cycle of, 533
ponderosa, 32
Pineal gland, 328
Pinna, 289
Pinocytosis, 76, 77
Pipevine swallowtail, 648
Pith, 549
Pituitary gland, 318-322
Pivot joint, 304
Placenta, 360
Placental-fetal circulation, 368
Placental mammal, 471, 593, 594
Plant, 12, 13
bark of, 552, 553
competition of, 641
cuticle of, 547, 552
dermal tissue of, 546, 547-548
divisions of, 527
dodder, 651
gibberellin in, 541
growth regulation in, 548
guard cell of, 554, 555
herbivore relationship to, 644-655
hormone in, 541
hornwort, 530
life cycle of, 85, 527
liverwort, 530
moss, 530
nonvascular, 527
reproduction of; *see* Plant reproduction and diversity
root of, 548-551
secondary growth of, 548, 553
seed, 538, 539
naked, 533
protected, 534-535
shoot of, 551-555
stem of, 551
structure of, 545
transpiration in, 548
vegetative propagation in, 540, 541
water movement in, 555-557
Plant cell, 45, 48
Plant defense, 645
Plant reproduction and diversity, 544-559
patterns of, 526-543, 557
vascular, 545-557
patterns of, 531-540, 541
reproduction in, 531-540, 541
Plantae, 12, 13
Plaque, 213
Plasma, blood, 157
Plasma cell, 224
immune system and, 157
Plasma membrane, 68, 76, 77
eukaryotic cell, 47, 48, 49-50, 60, 61
protein, in, 71
Plasmid, 63, 494, 495
Plasmodial slime mold, 515
Plasmodium, 508, 509, 515
Plasmolysis, 74
Plate tectonic theory, 458, 459
Plateau phase, 346
Platelet, 157, 209, 211
Platyhelminthes, 568
Pleura membrane, 191
PMS; *see* Premenstrual tension syndrome
Point mutation, 394
Polar body, 340
Polar molecule, 23

Polarized nerve impulse, 254
Pollen grains, 534
Pollination, 534, 535
  mechanisms of, 535
  modes of, 534
Pollinator, 535
Pollution
  air, 707
  water, 703-705
Polydactyly, 398
Polymer, 28, 447
Polyp, 567
Polypeptide, 34, 36
  synthesis of, 420
Polysaccharides, 30
Polyunsaturated fat, 33
Pond food chain, 659
Ponderosa pine, 32
Pons, 191, 265, 271
Population, 8, 431, 625-636
Population ecologist, 625
Population ecology, 624-637
Population mortality, 630-631
Population pyramid, 631
Population size, 627, 628-630
Pore, taste, 285
Porifera, 565
Positive control, 422
Positive feedback, 163
Positive feedback loop, 322
Positive phototaxis, 506
Postcopulatory behavior in goose, 617
Posterior pituitary gland, 321-322, 329
Postsynaptic membrane, 257
Potassium, 27, 168
Potassium pump, sodium, 253, 254
Potential energy, 103
Power
  nuclear, 696
  solar, 696, 697
  tidal, 696, 697
  water, 697
  wave, 696
  wind, 696, 697
Prairie chicken, courtship dance, 616
Pre-embryo, 356-359, 360
Precambrian Period, 453
Precapillary sphincter, 202
Predation, 629, 640, 642-648
Predator, 642
Preformationist, 373
Pregnancy, 355
  ectopic, 357
Premenstrual tension syndrome, 343
Premolars, 171
Prenatal development, 354-371
Preparation reaction, 123
Pressure
  air, 555
  root, 556
  turgor, 74
Pressure gradient, 192
Presynaptic membrane, 257
Prey, 642
Prickly pear cactus, 627
Primary amino acid, 35
Primary bronchi, 188
Primary consumer, 658
Primary germ layer, 359
Primary oocyte, 339
Primary plant growth, 548
Primary structure of protein, 35
Primary succession, 651
Primate
  evolution of, 471-475, 476
  family tree of, 476
Primitive streak, 361
Probability, 381
Probability theory and dihybrid cross, 381-382
Procambarus, 575
Producer, 10, 657
Product, changed substrate, 101
Prokaryote, 12, 63
  cell, 47, 61-63, 64
  and eukaryote gene, 419, 420
  gene expression, 420-422
Prolactin, 321
Promoter, 420
Promoter site, 413
Prophase, 86-89, 90
Proprioceptor, 282
Prosimian, 473
Prostaglandins, 330, 368
Prostate gland, 244, 338
Proteases, 168
Protection
  epithelial tissue and, 149
  support and movement and, 294-313
    bones in, 296-304
    muscles in, 304, 305-311
    skeletal system in, 298-304
    skin in, 295-296
Protective coloration, 647-648, 649
Protein, 50, 53, 55, 166
  chain, 71
  fibrous, 38
  globular, 38
  molecule of, 29
  of plasma membrane, 71
  structure of, 37, 38
  subunits of, 37, 38
  synthesis of, 418
Protein fiber, 38
Proterozoic Era, 452
Prothrombin activator, 211
Protist, 502-515
  fungus and; *see* Fungus
  as kingdom, 12, 13
  phagocytosis and, 76
Protoavis, 456, 457
Protocell, 448
Proton, 17
Protoplasm, 47
Protostome, 564
Protozoan, 503, 504-509
Proximal convoluted tube, 237, 238, 240, 241
Pseudocoelomate, 562
Pseudomonas, 62
Pseudopods, 504
Pseudostratified epithelium, 151
PTH; *see* Parathyroid hormone
Puberty, 336
Puffer, 467
Pulmonary artery, 204, 206
Pulmonary circulation, 204, 205, 206
Pulmonary disease
  chronic obstructive, 196
Pulmonary semilunar valve, 204, 206
Pulmonary vein, 204
Pump, sodium-potassium, 75
Punnett square, 378, 379, 382, 396
  and dihybrid cross, 382-383
Pupil, 288
Purine, 407
Purkinje fiber, 207
Pyloric sphincter, 174, 175
Pyramid of biomass, 662
Pyramid of energy, 662
Pyramid of number, 662
Pyramid of population, 631
Pyramids for aquatic ecosystem, 662
Pyridoxine, 167
Pyrimidine, 407
Pyruvate, 117, 123
Python, vestigial organ of, 438

Q

Qualitative or quantitative data, 5
Quaternary Period, 458
Quaternary structure of protein, 37, 38
Queen Victoria, 396, 397

R

R group, 35
Radial symmetry, 562, 566-568
Radiation, 132
  adaptive, 433, 455
  ionizing, 392
  ultraviolet, 394
Radicle in seed germination, 538
Radioactive isotope, 436
Radiolarian shell, 504, 505
Rain shadow effect, 674
Rank order in group behavior, 621
Ranvier, node of, 252
Rate of growth, 625
Ray in ocean, 466
Ray-finned fish, 586
Reaction
  allergic, 229
  anabolic, 101
  catabolic, 101
  chemical, 101
  coupled, 102, 103, 115
  endergonic, 102
  exergonic, 101, 102, 106
  hydrolysis, 30
  light-dependent, 135, 136-140
  light-independent, 135, 141-142
  preparation for, 123
  redox, 20, 119
Realized niche, 642
Rearrangement, chromosomal, 394
Reasoning, 3, 5
  insight and, 606, 607
Receptor, 50
  in human skin, 283
  olfactory, 284
  in peripheral nervous system, 273
  stretch, 191, 244, 283
Recessive genetic disorder, 398, 399
Recessive trait, 375
Recognition sequence, 496-497
Recombinant deoxyribonucleic acid technology, 498-499
Recombination, meiotic, 97
Rectum, 179, 180
Red algae, 509, 513
Red-and-black African grasshopper, 647
Red blood cell, 157, 202, 210
Red boring sponge, 565
Red-eyed *Drosophila melanogaster*, 383
Red-green color blindness, 395
Red marrow, 297
Red tide, 509, 510
Redox reaction, 20, 119
Reduced, defined, 20, 119
Reduction, 119
  electron gain and, 20
Reduction-division, 92
Redwood community, 639
Reflex, 264, 602
  knee-jerk, 276
  in somatic nervous system, 276
Reflex arc, 271, 276
Refractory period, 255
Regulation, endocrine hormone, 315
Regulatory site, 420
Relation, 12
Releasing hormone, 318
Renal artery, 235, 236
Renal failure, 246, 247
Renal pelvis, 235, 236, 244
Renal pyramid, 235
Renal vein, 236
Repeated trial, 7
Replication
  of deoxyribonucleic acid, 411
  viral, 489-490
Replication fork, 411
Reproduction
  asexual, 85
  of endocrine gland, 315
  sexual, 335
Reproductive behavior, 615-616, 617
Reproductive cycle in human, 341-343
Reproductive system, 148
  female, 339-346
  as major organ system, 148
  male, 335-339
Reptile, 588-591
  evolution and, 466-468
Reptillia, 589
Reservoir, 664
Respiration, 184-197
Respiratory assembly, 125
Respiratory chain, 125
Respiratory cilia, 218
Respiratory diaphragm, 190
Respiratory system, 148
Respiratory tract, 218
Responding variable, 3
Response
  all-or-nothing, 255
  endocrine gland and, 315
  involuntary or voluntary, 264
Resting potential, 253-254
Reticular connective tissue, 153, 154
Reticular formation, 271
Reticulin fiber, 153
Reticulum, 53
  sarcoplasmic, 310
Retina, 286, 288
Retinol, 167
Rhizoid, 557
Rhizome, 540
Rhodopsin, 286
Rhythm method of birth control, 347
Rib, 300
Riboflavin, 167
Ribonucleic acid, 56, 57, 406-408
  messenger, 413
  molecules of, 29
  ribosomal, 413
  structure of, 39
  transfer, 413, 417
Ribonucleic acid polymerase, 413
Ribose, 110

Ribosomal ribonucleic acid, 57, 58, 413
Ribosome, 47, 48, 49, 52, 53, 63, 413
Ribulose bisphosphate, 142
Rickets, 167
Right atrium, 204, 207
RNA; *see* Ribonucleic acid
Rockweed, 511
Rod cell, 286
Root, 545, 548-551
Root canal, 171
Root pressure, 556
Rotifer, 562
Rough endoplasmic reticulum, 48, 53
Round window, 289
Roundworm, 569-570
rRNA; *see* Ribosomal ribonucleic acid
RuBP; *see* Ribulose bisphosphate
Runner, 540
Runner's high, 330

S

SA node; *see* Sinoatrial node
Sac fungi, 519-520
Saccule, 291
Salinity, 685
Saliva, 170
Salivary gland, 169, 170
Salt
  blood plasma and, 208
  dissolution of, 25
  table, 20, 25
Salt marsh food web, 660
Saltatory conduction, 256-257
Sand dollar, 577
Sanitary landfill, 702
Sap, 557
Saprophytic, defined, 516
Sarcomere, 307
Sarcoplasmic reticulum, 310
Saturated fat, 33, 34
Savanna, 678, 679
Schleiden, M., 45
Schwann, T., 45
Schwann cell, 160, 256
  neuron and, 252
Scientific law, 434
Scientific method, 3
Sclera, 287
Sclerenchymal cell, 547
Scrotum, 335
Scrubber in coal-burning power plant, 706
Scurvy, 167
Sea anemone, 650
Sea cucumber, 577
Sea horse, 467
Sea lilies, 577
Sea otter, 599
Sea perch, 582
Sea squirt, 581
Sea star, 577
Sea urchin, 577-578
Seaweed, 511
Second filial generation, 375
Second law
  of Mendel, 378
  of thermodynamics, 104
Second messenger, 315
Second trimester of pregnancy, 367
Secondary amino acid, 37, 38
Secondary bronchi, 188
Secondary consumer, 658
Secondary oocyte, 340
Secondary phloem, 552
Secondary plant growth, 548, 553
Secondary pollutant, 707
Secondary protein structure, 37, 38
Secondary succession, 653
Secondary xylem, 552
Secretin, 175, 177
Secretion in epithelial tissue, 149
Secretory vesicle, 55
Sediment, 703
Sedimentary rock, 434-435
Seed
  dispersion of, 536
  formation of, 538, 539
  germination, 538, 539
  of vascular plants, 531
Seedless vascular plant, 527, 531-532
Segregation, law of, 378
Selective reabsorption in kidney, 235, 237-241
Selectively permeable membrane, 73
Selenium, 27
Self-fertilization, 374
Self-pollination, 535
Semen, 338
Semicircular canal, 291
Seminal vesicles, 338
Seminiferous tubule, 336
Semiterrestrial, defined, 512
Sensation in epithelial tissue, 149
Senses, 280-293
Sensory area of brain, 266, 267
Sensory neuron, 250-251, 264
Sensory pathway, 273-275
Septum, 207, 573
  of reptile, 588
Serosa, 174
Serum albumin, 208
Sessile, 562
Seta of annelid, 574
Sex chromosome, 81, 383-384, 389
Sex flush, 346
Sex linkage, 383-384
Sex and reproduction, 334-353
  contraception and birth control in; *see* Birth control
Sexual life cycle, 528
Sexual reproduction, 83, 335
Sexual response, 346-347
Shark, 466
Shell, 18
Shell hydration, 24
Shinbone, 300
Shiny bristleworm, 573
Shockwave lithotripsy, 246
Shoot of plant, 545, 551-555
Shore zone, 684, 687-688
Shoulder, 300
Shrew, 468
Sickle-cell anemia, 398
Sickle-shaped red blood cell, 398
Sight, 286-288
Sigmoid growth curve, 627
Silicon, 27
Silk, 38
Silurian Period, 454
Simple epithelium, 151
Simple eye, 576
Simple sugar, 30
Single bond, 21
Single-celled protist, 503
Sinoatrial node, 206
Sinus venosus in bony fish, 586
Sister chromatid, 81, 94
  anaphase and, 90
Skeletal muscle, 159
  bone attachment of, 305
  contraction of, 305-308, 309
  structure of, 306
  voluntary function and, 158
Skeletal system, 298-304
  function and components of, 148
Skeleton
  appendicular, 298, 299, 300-302
  axial, 299-300
  joint in, 302-304
Skin, 295-296, 296
Skinner, B.F., 605
Skinner box, 605
Skull, 265, 298, 299
Skunk, 647
Slime mold, 504
Small intestine, 175-179
  digestive enzymes and, 169
Smallpox, 218
  immunity to, 220
Smell, 284-285
Smelt, 661
Smog, 707
Smooth endoplasmic reticulum, 48, 53
Smooth muscle, 158-159
Snail, 570-573
Social behavior in animal; *see* Behavior, social
Society
  ant, 619
  insect, 619-621
Sodium, 19-20, 168
Sodium bicarbonate, 177
Sodium chloride, 19-20
  water and, 24, 25
Sodium hydroxide, 26
Sodium ion, 19-20
Sodium-potassium pump, 75, 253, 254
Soft palate, 172
Softwood, 552
Solar power, 696, 697
Solid waste, 702-703
Soluble water, 24
Solutes, 73
Solution, 73
Solvent, 73
  water as, 24, 25
Somatic cell gene therapy, 423
Somatic nervous system, 264, 275-276
Somite, 364, 581
Soybean root nodule, 493
Spadefoot toad, 681
Spanish moss, 530
Spawn, 584
Speciation, 433
Species, 13
  defined, 427
Species extinction, 700-701
Specific facilitated diffusion, 75
Spectrum
  absorption, 133
  electromagnetic, 132
Sperm, 45, 335, 336, 337
  blockage or destruction of, in birth control, 350
Spermatids, 336
Spermatogenesis, 336, 337
Spermatogonia, 336
Spermatozoon, 335
Spermicide, 347, 350
Sphagnum moss, 530
Sphincter
  lower esophageal, 172
  pyloric, 174, 175
Spider, 574-576
Spinal cord, 149, 264, 271-272
Spinal nerve, 271
Spindle fiber, 86
Spiracle, 576
Spirillum, 62
Spirogyra, 512
Spleen, 211, 220
Sponge, 565-566
  birth control and, 347
Spongy bone, 156, 297
Spongy layer, 554
Spontaneous abortion, 391
Spontaneous generation, 46
Spore, 514, 517
Sporophyte, 84, 85, 529
Sporozoa, 509
Spotted skunk, 647
Squamous cell, 147, 150
Squirrel, 13
Stable molecule, 21
Stable population, 631
Stapes of ear, 289
Staphylococcus, 62
Starch, 29, 30
Starfish, 577-578
Starvation, 635
Stem, 545, 551-552, 553
Stereoscopic vision, 472
Sternum, 300
Steroids, 315, 317
  lipid membrane and, 34
  molecules of, 29
Stickleback courtship, 616
Stigma, 534
Stipe, 511
Stirrup of ear, 289
Stolon, 540
Stoma, 554
Stomach
  digestion in, 174-175
  digestive enzymes and, 169
  epithelium of, 174
Stone Age, 480
Stratification
  of epithelium, 151
  of shale, 453
Stratosphere, 708
Streptococcus, 62
Stress, 326, 327
Stretch receptor, 191, 244, 283
Striated muscle, 159
Stroke, 213
Stromatolite, 450, 451
Style, 534
Subarachnoid space, 272
Submissive behavior, 611
Submucosa, 174
Substrate, 101
  enzymes and, 108
Subunits, 39
Succession, 651-653
Sucrose, 30, 31
Sugar
  five-carbon, 39
  simple, 30
Sugar chain, 30
Sulfur, 27, 168
Sunlight, 132
  climate and, 672, 673
Supercoil, 81, 82

Superior vena cava, 204
Suppressor T cell, 222, 223
Surface tension, 24
Surface transport, 149-150
Surface-to-volume ratio, cell, 46
Surface water pollution, 703-705
Survivorship, population, 630-631
Sustainable society, 696
Suture, bone, 302
Swallowing, 172
Swim bladder, 585
Symbiosis, 49, 640, 649-651
Symmetry, 562, 568-578
Sympathetic nervous system, 265, 276
Synapse, 257, 259
Synapsis, 93
Synaptic cleft, 257, 258
Synaptic transmission, 257, 258
Synovial fluid, 302
Synovial joint, 302
Systemic capillary, 205
Systemic circulation, 205
Systolic period, 207

T

T_3; *see* Triiodothyronine
T_4; *see* Thyroxine
T cell, 209, 211, 328
  cytotoxic or suppressor, 222, 223
T tubule, 310
Table salt, 20
Taiga, 682-683
Taproot, 550
Target cell, 315
Tarsal, defined, 302
Taste, 285
Taste bud, 285
Taste pore, 285
Taxis, 600-601, 602
Taxonomy, 12
Tay-Sachs disease, 398
Technicolor fish, 612
Technology as environmental problem, 708
Tectonic plate, 458, 459
Tectorial membrane, 289-290
Teeth, 170-171
Telophase, 87, 88, 90, 91
Temperate deciduous forest, 682
Temperate grassland, 681
Template, 447
Temporal lobe, 266
Tendon, 305
Teratogen, 361
Termite mound, 679
Territorial behavior, 612-613
Tertiary amino acid, 37, 38
Tertiary Period, 458
Testcross, 380
Testis, 328, 329, 335-337
Tetrapod, 466, 584
Thalamus, 265, 269, 270
Theory, 7
  endosymbiotic, 452
  of evolution, 427-434
  plate tectonic, 458, 459
Thermal pollution, 704, 705
Thermodynamics, 103, 104
Third trimester of pregnancy, 368
Thoracic cavity, 149, 190
Thoroughfare channel, 202
Threat display, 611
Threshold potential, 254
Thromboplastin, 211
Thrombus, 213
Thumb, opposable, 473
Thylakoid, 60, 135
Thymine, 39
Thymus gland, 211, 220, 328
Thyroid gland, 322, 323, 329
Thyroid hormone, 322
Thyroid-stimulating hormone, 320
Thyroxine, 322
Tidal power, 696, 697
Tidal volume, 192
Tin, 27
Tissue, 8, 546, 547-548
  in human body, 149-160
Tissue differentiation, 364
Toad, 681
Tobacco mosaic virus, 487
Tocopherol, 167
Tongue, 172
Toxic substance, 704
Toxin, 494
Trachea, 172, 186, 187
  of insect, 576
Tracheid, 546
Tracheole, 576
Trait, 373-375
Transcription, 413, 415
Transduction, 494
Transfer ribonucleic acid, 413, 417
Transformation, 494
Transitional epithelium, 151
Transitional form, 468
Translation, 416, 417
Translocation, 393
Transmission, 250, 255-259
Transpiration in plants, 548, 555, 556
Transport, 72, 75
  surface, 149-150
Transport chain, electron, 117, 125-127, 139
Transverse fission, 507
Transverse tubule, 310
Tree finch, 433
Trial-and-error learning, 605-606
Triassic Period, 454
Tricuspid valve, 204, 206
Triglyceride, 33
Triiodothyronine, 322
Trilobite fossil, 454
Triphosphate group, 110
Triple bond, 21
Triple female, 391
Trophic level, 659
Trophoblast, 357
Tropic hormone, 320
Tropical rain forest, 676, 677
  deforestation of, 699, 700
Tropomyosin, 310
Troponin, 310
True-breeding, 374, 375, 377
True rib, 300
Trypsin, 169, 177
TSH; *see* Thyroid-stimulating hormone
Tubal ligation, 351
Tubifera, 515
Tubular secretion, 235, 241
Tubule
  convoluted, 237-241
  malpighian, 576
  seminiferous, 336
  transverse, 310
Tubulin, 51
Tundra, 684
Tunicate, 581-582
Turbellaria, 568
Turgor pressure, 74
Turner syndrome, 392
Tympanic canal, 289

U

Ulcer, duodenal, 175
Ultrasound in genetic counseling, 401
Ultraviolet radiation, 394
Umbilical artery, 360
Umbilical cord, 360
Unicorn caterpillar, 649
Unifying themes of biology today, 7-12
Unity of living things, 7, 8
Unsaturated fat, 33, 34
Upper digestive system, 174
Urea, 234
  in cartilaginous fish, 585
Ureter, 235, 236, 244
Urethra, 244, 338
Urey, H.C., 445, 446
Uric acid, 234
Urinary bladder, 244
Urinary system, 148, 244
Urine, 234, 235, 238-241, 243
Uterine tube, 340-341
Uterus, 339, 341, 357
Utricle, 291

V

Vaccinia virus, 218
Vacuole, 48, 56
Vagina, 339, 343
Vanadium, 27
Variable
  dependent, 5
  independent, 3
  manipulated, 3
  responding, 5
Variation in gene interaction, 373-374
Varieties, 374
Vas deferens, 335, 336, 338
Vascular, defined, 199
Vascular cambium, 552
Vascular cylinder, 549
Vascular plant, 545
  life cycle of, 531
  naked seeds in, 533
  organs of, 548-555
  patterns of reproduction in, 531-540
  protected seed in, 534-535
  seedless, 527, 531-532
  tissue of, 545-548
  transpiration in, 548
Vascular tissue, 546-547
Vasectomy, 351
Vasoconstriction, 200
Vegetative propagation in plant, 540, 541
Vein, 201
Ventral side, 562
Ventricle
  in bony fish, 586
  of brain, 272
Venule, 201, 202, 203
Vertebrae, 300
Vertebral column, 149, 300, 465
Vertebrate, 580-596
  digestive tract of, 174
  evolution of, 464-471
  neurons of, 159
Vertebrate—cont'd
  society behavior of, 621
Vesicle, 55
Vestibular canal, 289
Vestibule in inner ear, 290
Vestigial organ, 438
Vibrio cholerae, 64
Viceroy butterfly, 648
Villi
  of chorion, 364
  of small intestine, 178
Vimentin, 51
Viral core, 487
Viral envelope, 487
Virchow, R., 45
Virus, 486-490
  tobacco mosaic, 487
Visceral mass, 571
Vitamins, 166-169, 180
Vitreous humor, 287
Viviparous, defined, 586
Vocal cord, 186
Voluntary muscle, 159
Voluntary response, 264
Volvox, 512
Vulva, 343

W

Waggle dance, 620
Wall, P., 281
Wallace, A. R., 434
Warbler finch, 433
Warm blooded, 471
Warning coloration, 645, 647
Waste, nitrogenous, 234
Water, 22-26, 166
Water balance and kidney, 234
Water cycle, 664-665
Water mold, 504, 515
Water power, 697
Watson-Crick model of deoxyribonucleic acid, 409
Wave power, 696
Whale, 650
White blood cell, 157, 210
White-eyed *Drosophila melanogaster*, 383
White matter, 271
Wind power, 696, 697
Window, oval or round, 289
Wood, 32
World population distribution, 635

X

X Chromosome, 384
Xylem, 546
XYY males, 392

Y

Y Chromosome, 384
Yellow-green algae, 510
Yellow marrow, 297
Yolk sac, 361

Z

Zebra, 645
Zinc, 27, 169
Zona pellucida, 355
Zygospore, 518
Zygote, 340, 356
Zygote-forming fungi, 517, 518